Expanded Study and Solutions Guide for

CALCULUS

SIXTH EDITION

Larson / Hostetler / Edwards

Bruce H. Edwards
University of Florida

David E. Heyd
The Pennsylvania State University
The Behrend College

Houghton Mifflin Company Boston New York

Editor in Chief, Mathematics: Charles Hartford
Managing Editor: Cathy Cantin
Senior Associate Editor: Maureen Brooks
Associate Editor: Michael Richards
Assistant Editor: Carolyn Johnson
Supervising Editor: Karen Carter
Art Supervisor: Gary Crespo
Marketing Manager: Sara Whittern
Associate Marketing Manager: Ros Kane
Marketing Assistant: Carrie Lipscomb
Design: Henry Rachlin
Composition and Art: Meridian Creative Group

Printed in the U.S.A.

ISBN: 0-395-93896-1

456789-B-01 00

Preface

This *Expanded Study and Solutions Guide* is designed as a supplement to *Calculus,* Sixth Edition, by Roland E. Larson, Robert P. Hostetler, and Bruce H. Edwards. All references to chapters, theorems, and exercises relate to the main text. Although this supplement is not a substitute for good study habits, it can be valuable when incorporated into a well-planned course of study. The following suggestions may assist you in the use of the text, your lecture notes, and this *Guide*.

- *Read the section in the text for general content before class.* You will be surprised at how much more you will absorb from the lecture if you are aware of the objectives of the section and the types of problems that will be solved. If you are familiar with the topic, you will understand more of the lecture, and you will be able to take fewer (and better) notes.
- *As soon after class as possible, work problems from the exercise set.* The exercise sets in the text are divided into groups of similar problems and are presented in approximately the same order as the section topics. Try to get an overall picture of the various types of problems in the set. As you work your way through the exercise set, reread your class notes and the portion of the section that covers each type of problem. Pay particular attention to the solved examples.
- *Learning calculus takes practice.* You cannot learn calculus merely by reading, any more than you can learn to play the piano or to bowl merely by reading. Only after you have practiced the techniques of a section and have discovered your weak points can you make good use of the supplementary solutions in this *Guide*.
- *Technology.* Graphing utilities and symbolic algebra systems are now readily available. The computer and calculator are merely tools. Your ability to use these tools effectively requires that you continually sharpen your problem-solving skills and your understanding of fundamental mathematical principles.

Good study habits are essential for success in mathematics. Many students have found the following additional suggestions to be helpful in making the best use of their time.

- *Write neatly in pencil.* A notebook filled with unorganized scribbling has little value.
- *Work at a deliberate and methodical pace, without skipping steps.* When you hurry through a problem you are more apt to make careless arithmetic or algebraic errors that, in the long run, waste time.
- *Keep up with the work.* This suggestion is crucial because calculus is a very structured topic. If you cannot do the problems in one section, you are not likely to be able to do the problems in the next. The night before a quiz or test is not the time to start working problems. In some instances cramming may help you pass an examination, but it is an inferior way to learn and retain essential concepts.
- After working some of the assigned exercises with access to the examples and answers, *try at least one of each type of exercise with the book closed.* This will increase your confidence on quizzes and tests.
- Do not be overly concerned with finding the most efficient way to solve a problem. *Your first goal is to find one way that works.* Short cuts and clever methods come later.
- If you have trouble with the algebra of calculus, *refer to the algebra review at the beginning of this guide.*

If you have any corrections or suggestions for improving this *Study Guide*, we would appreciate hearing from you. Good Luck with your study of calculus.

Bruce H. Edwards
358 Little Hall
University of Florida
Gainesville, Florida 32611
(be@math.ufl.edu)

David E. Heyd
The Pennsylvania State University
The Behrend College

Contents

CHAPTER P
Preparation for Calculus

CHAPTER P
Preparation for Calculus

Section P.1 Graphs and Models

Solutions to Odd-Numbered Exercises

1. $y = -\frac{1}{2}x + 2$

x-intercept: $(4, 0)$

y-intercept: $(0, 2)$

Matches graph (b)

3. $y = 4 - x^2$

x-intercepts: $(2, 0), (-2, 0)$

y-intercept: $(0, 4)$

Matches graph (a)

5. $y = x^2 + x - 2$

y-intercept: $y = 0^2 + 0 - 2$

$y = -2; (0, -2)$

x-intercepts: $0 = x^2 + x - 2$

$0 = (x + 2)(x - 1)$

$x = -2, 1; (-2, 0), (1, 0)$

7. $y = x^2\sqrt{9 - x^2}$

y-intercept: $y = 0^2\sqrt{9 - 0^2}$

$y = 0; (0, 0)$

x-intercepts: $0 = x^2\sqrt{9 - x^2}$

$0 = x^2\sqrt{(3 - x)(3 + x)}$

$x = 0, \pm 3; (0, 0), (\pm 3, 0)$

9. $x^2y - x^2 + 4y = 0$

y-intercept:

$0^2(y) - 0^2 + 4y = 0$

$y = 0; (0, 0)$

x-intercept:

$x^2(0) - x^2 + 4(0) = 0$

$x = 0; (0, 0)$

11. Symmetric with respect to the y-axis since

$$y = (-x)^2 - 2 = x^2 - 2.$$

13. Symmetric with respect to the x-axis since

$$(-y)^2 = y^2 = x^3 - 4x.$$

15. Symmetric with respect to the origin since

$$(-y) = (-x)^3 - (-x)$$

$$-y = -x^3 - x$$

$$y = x^3 + x.$$

17. Symmetric with respect to the origin since

$$-y = \frac{-x}{(-x)^2 + 1}$$

$$y = \frac{x}{x^2 + 1}.$$

19. $y = -3x + 2$

Intercepts:

$\left(\frac{2}{3}, 0\right), (0, 2)$

Symmetry: none

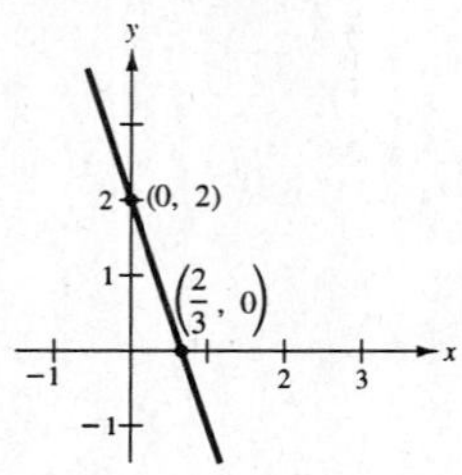

21. $y = \dfrac{x}{2} - 4$

Intercepts:

$(8, 0), (0, -4)$

Symmetry: none

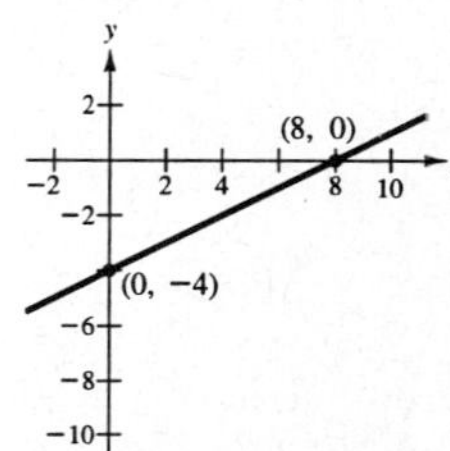

23. $y = 1 - x^2$

Intercepts:

$(1, 0), (-1, 0), (0, 1)$

Symmetry: y-axis

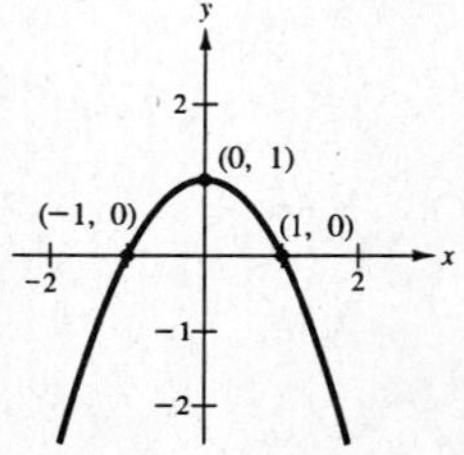

25. $y = x^3 + 2$

Intercepts:

$\left(-\sqrt[3]{2}, 0\right), (0, 2)$

Symmetry: none

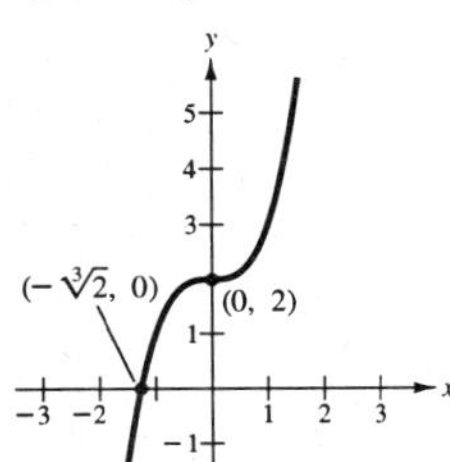

27. $y = (x + 2)^2$

Intercepts:

$(-2, 0), (0, 4)$

Symmetry: none

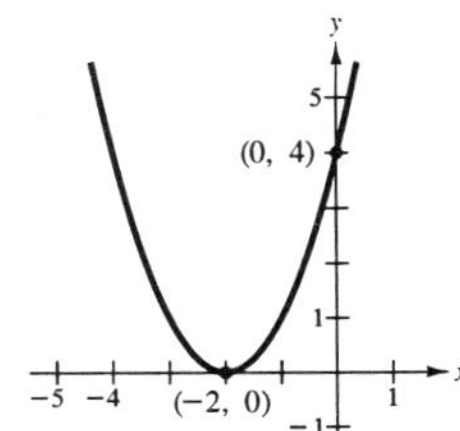

29. $y = \dfrac{1}{x}$

Intercepts: none

Symmetry: origin

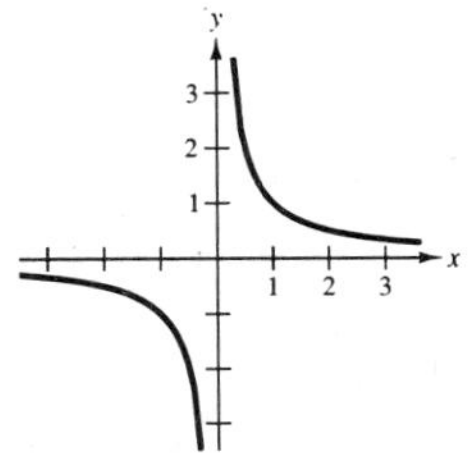

31.

Xmin = -3
Xmax = 5
Xscl = 1
Ymin = -3
Ymax = 5
Yscl = 1

33. $y = -2x^2 + x + 1$

$= (2x + 1)(-x + 1)$

Intercepts:

$\left(-\frac{1}{2}, 0\right), (1, 0), (0, 1)$

Symmetry: none

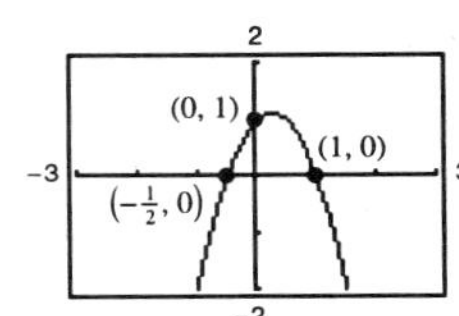

35. $y = \dfrac{5}{x^2 + 1} - 1$

Intercepts: $(0, 4), (-2, 0), (2, 0)$

Symmetry: y-axis

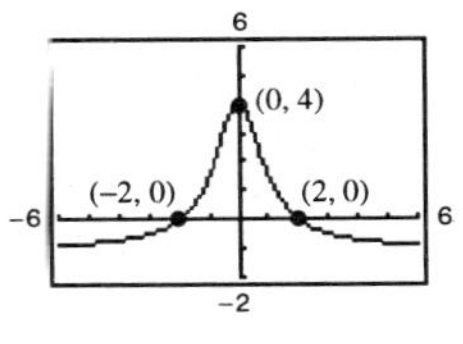

37.

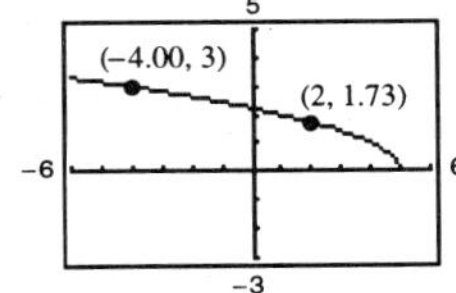

(a) $(2, y) = (2, 1.73)$ $\left(y = \sqrt{5 - 2} = \sqrt{3} \approx 1.73\right)$

(b) $(x, 3) = (-4, 3)$ $\left(3 = \sqrt{5 - (-4)}\right)$

39. $y = (x + 2)(x - 4)(x - 6)$ (other answers possible)

41. Some possible equations:

$y = x$

$y = x^3$

$y = 3x^3 - x$

$y = \sqrt[3]{x}$

43. $x + y = 2 \Rightarrow y = 2 - x$

$2x - y = 1 \Rightarrow y = 2x - 1$

$2 - x = 2x - 1$

$3 = 3x$

$1 = x$

The corresponding y-value is $y = 1$.

Point of intersection: $(1, 1)$

45. $x + y = 7 \Rightarrow y = 7 - x$

$3x - 2y = 11 \Rightarrow y = \dfrac{3x - 11}{2}$

$7 - x = \dfrac{3x - 11}{2}$

$14 - 2x = 3x - 11$

$-5x = -25$

$x = 5$

The corresponding y-value is $y = 2$.

Point of intersection: $(5, 2)$

47. $x^2 + y^2 = 5 \Rightarrow y^2 = 5 - x^2$

$$x - y = 1 \Rightarrow y = x - 1$$
$$5 - x^2 = (x - 1)^2$$
$$5 - x^2 = x^2 - 2x + 1$$
$$0 = 2x^2 - 2x - 4 = 2(x + 1)(x - 2)$$
$$x = -1 \text{ or } x = 2$$

The corresponding y-values are $y = -2$ and $y = 1$.
Points of intersection: $(-1, -2)$, $(2, 1)$

49.
$$y = x^3$$
$$y = x$$
$$x^3 = x$$
$$x^3 - x = 0$$
$$x(x + 1)(x - 1) = 0$$
$$x = 0, x = -1, \text{ or } x = 1$$

The corresponding y-values are $y = 0$, $y = -1$, and $y = 1$.
Points of intersection: $(0, 0)$, $(-1, -1)$, $(1, 1)$

51.
$$y = x^3 - 2x^2 + x - 1$$
$$y = -x^2 + 3x - 1$$
$$x^3 - 2x^2 + x - 1 = -x^2 + 3x - 1$$
$$x^3 - x^2 - 2x = 0$$
$$x(x - 2)(x + 1) = 0$$
$$x = -1, 0, 2$$

$(-1, -5)$, $(0, -1)$, $(2, 1)$

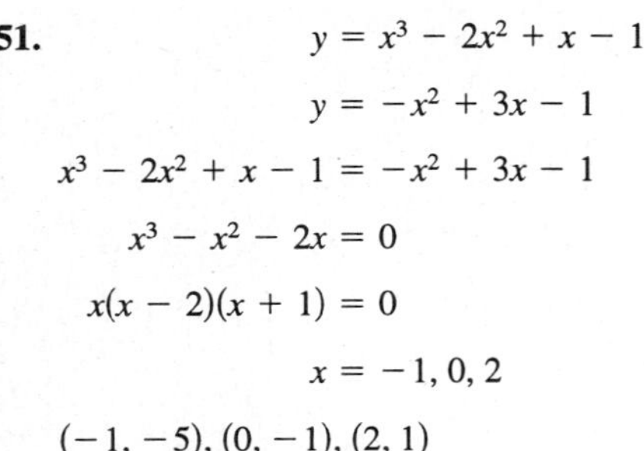

53. $5.5\sqrt{x} + 10{,}000 = 3.29x$

$$\left(5.5\sqrt{x}\right)^2 = (3.29x - 10{,}000)^2$$
$$30.25x = 10.8241x^2 - 65{,}800x + 100{,}000{,}000$$
$$0 = 10.8241x^2 - 65{,}830.25x + 100{,}000{,}000 \quad \text{Use the Quadratic Formula.}$$
$$x \approx 3133 \text{ units}$$

The other root, $x \approx 2949$, does not satisfy the equation $R = C$. This problem can also be solved by using a graphing utility and finding the intersection of the graphs of C and R.

55. (a) Using a graphing utility, you obtain

$$y = 0.0127t^2 + 4.4181t + 36.2896.$$

(b)

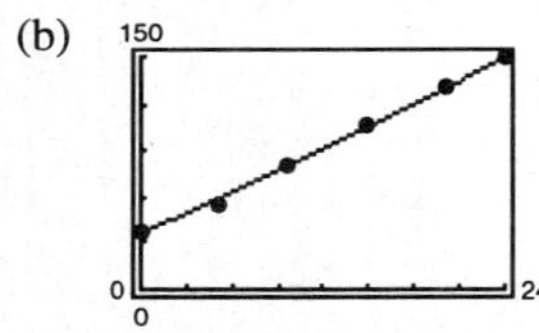

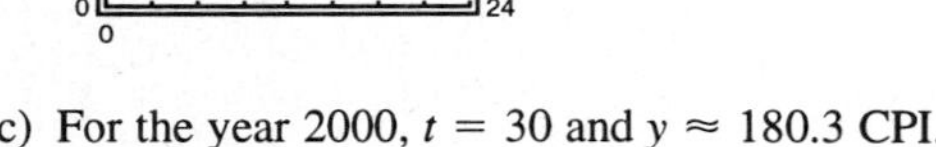

(c) For the year 2000, $t = 30$ and $y \approx 180.3$ CPI.

57.

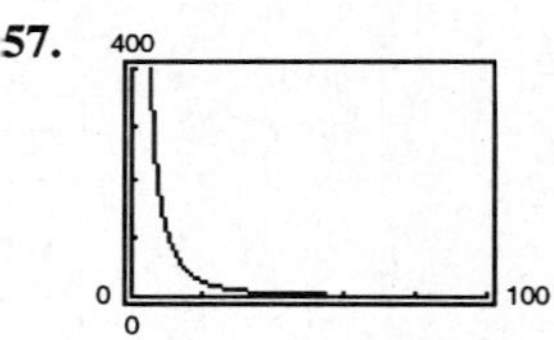

If the diameter is doubled, the resistance is changed by approximately a factor of $(1/4)$. For instance, $y(20) \approx 26.555$ and $y(40) \approx 6.36125$.

59. False; x-axis symmetry means that if $(1, -2)$ is on the graph, then $(1, 2)$ is also on the graph.

61. True; the x-intercepts are

$$\left(\frac{-b \pm \sqrt{b^2 - 4ac}}{2a}, 0\right).$$

63. Distance to the origin $= K \times$ Distance to $(2, 0)$

$$\sqrt{x^2 + y^2} = K\sqrt{(x - 2)^2 + y^2},\ K \neq 1$$
$$x^2 + y^2 = K^2(x^2 - 4x + 4 + y^2)$$
$$(1 - K^2)x^2 + (1 - K^2)y^2 + 4K^2x - 4K^2 = 0$$

Note: This is the equation of a circle!

Section P.2 Linear Models and Rates of Change

1. $m = 1$ **3.** $m = 0$ **5.** $m = -12$

7.

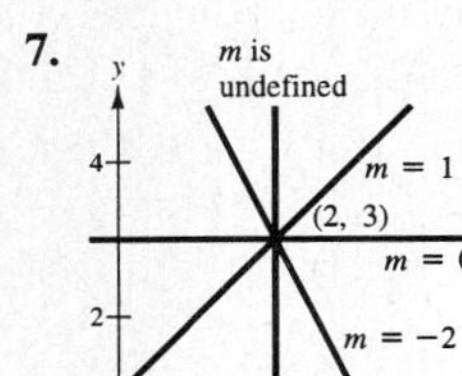

9. $m = \dfrac{2 - (-4)}{5 - 3}$

$= \dfrac{6}{2} = 3$

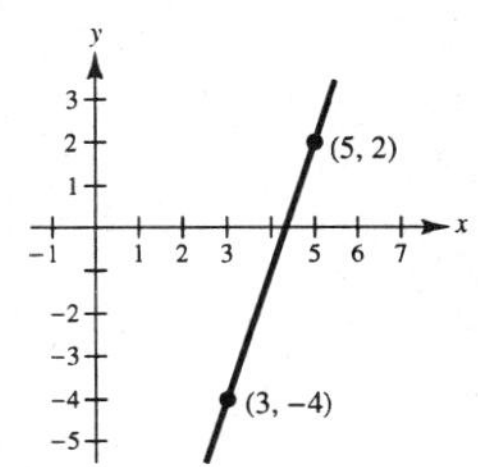

11. $m = \dfrac{4 - 2}{-2 - 1}$

$= -\dfrac{2}{3}$

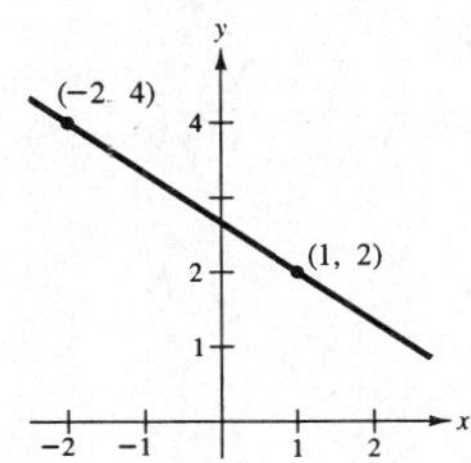

13. Since the slope is 0, the line is horizontal and its equation is $y = 1$. Therefore, three additional points are (0, 1), (1, 1), and (3, 1).

15. The equation of this line is

$$y - 7 = -3(x - 1)$$
$$y = -3x + 10.$$

Therefore, three additional points are (0, 10), (2, 4), and (3, 1).

17. The slopes of the line segments are

$$\frac{1.63 - 1.25}{1} = 0.38$$
$$\frac{2.53 - 1.63}{1} = 0.9$$
$$\frac{2.32 - 2.53}{1} = -0.21$$
$$\frac{2.87 - 2.32}{1} = 0.55$$
$$\frac{2.99 - 2.87}{1} = 0.12$$
$$\frac{3.10 - 2.99}{1} = 0.11$$
$$\frac{2.95 - 3.10}{1} = -0.15.$$

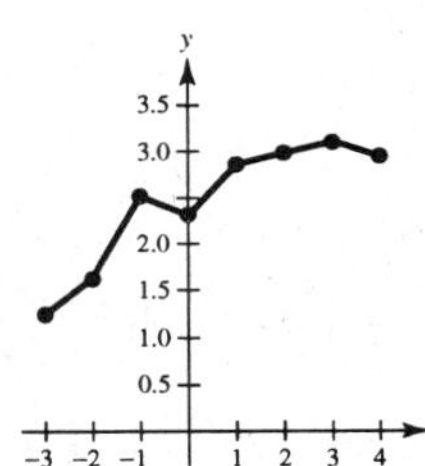

Earnings decreased most rapidly from 1989 to 1990 ($m = -0.21$). Earnings increased most rapidly from 1988 to 1989 ($m = 0.9$).

19. Given a line L, you can use any two distinct points to calculate its slope. Since a line is straight, the ratio of the change in y-values to the change in x-values will always be the same. See Section P.2 Exercise 85 for a proof.

21. $x + 5y = 20$

$$y = -\tfrac{1}{5}x + 4$$

Therefore, the slope is $m = -\frac{1}{5}$ and the y-intercept is (0, 4).

23. $x = 4$

The line is vertical. Therefore, the slope is undefined and there is no y-intercept.

25. $m = \dfrac{1-(-3)}{2-0} = 2$

$y - 1 = 2(x - 2)$

$y - 1 = 2x - 4$

$0 = 2x - y - 3$

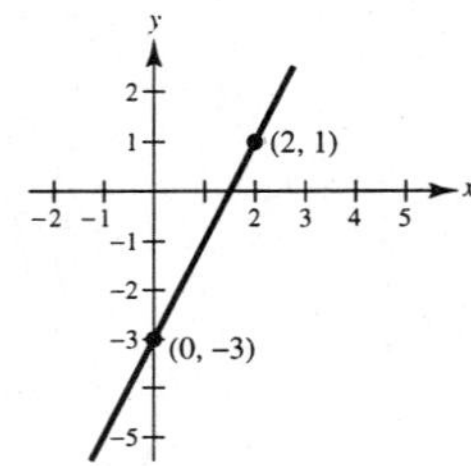

27. $m = \dfrac{3-0}{-1-0} = -3$

$y - 0 = -3(x - 0)$

$y = -3x$

$3x + y = 0$

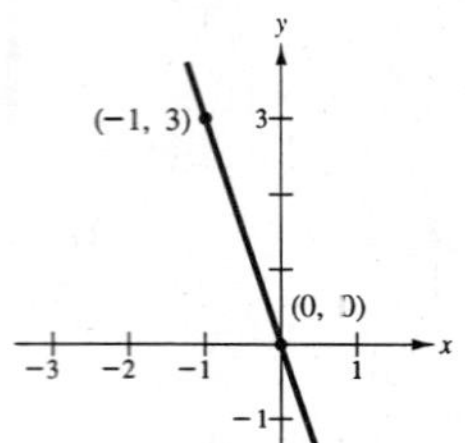

29. $m = 0$

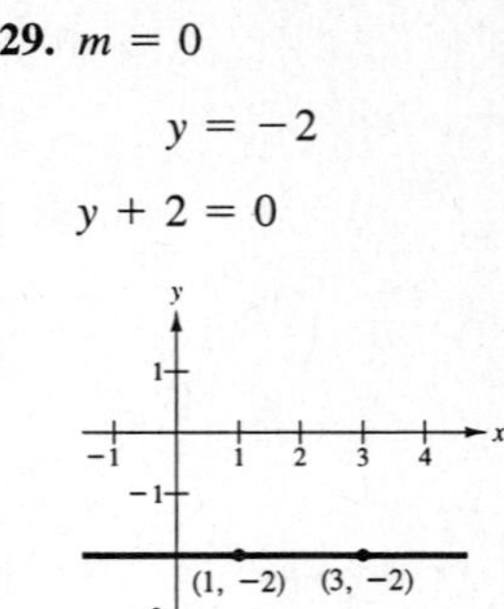

$y = -2$

$y + 2 = 0$

31. $y = \frac{3}{4}x + 3$

$4y = 3x + 12$

$0 = 3x - 4y + 12$

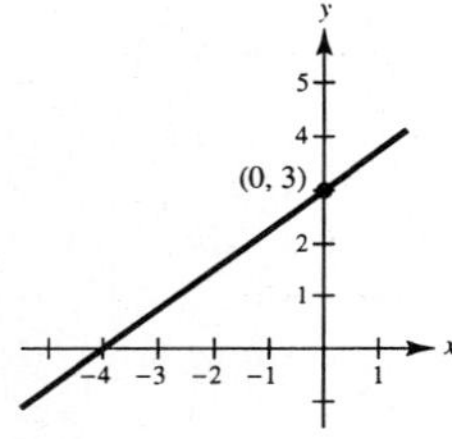

33.

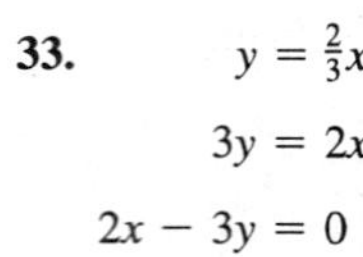

$y = \frac{2}{3}x$

$3y = 2x$

$2x - 3y = 0$

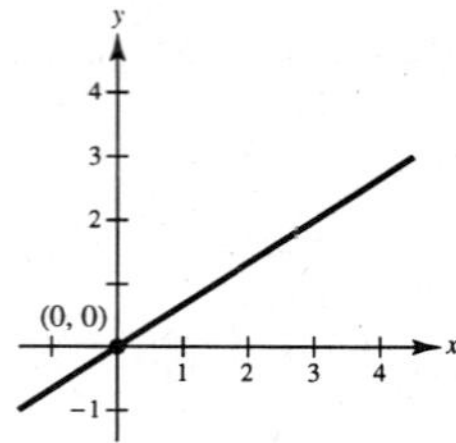

35. $y - 2 = 4(x - 0)$

$y = 4x + 2$

$0 = 4x - y + 2$

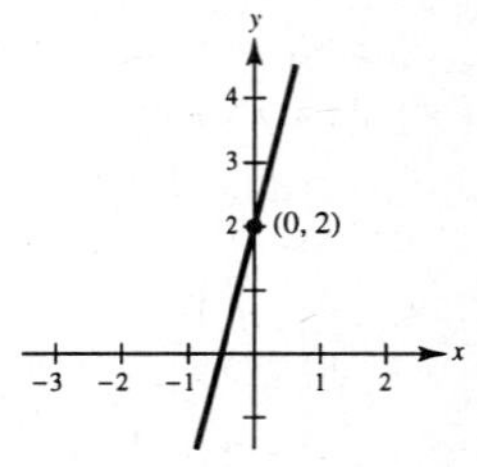

37. $x = 3$

$x - 3 = 0$

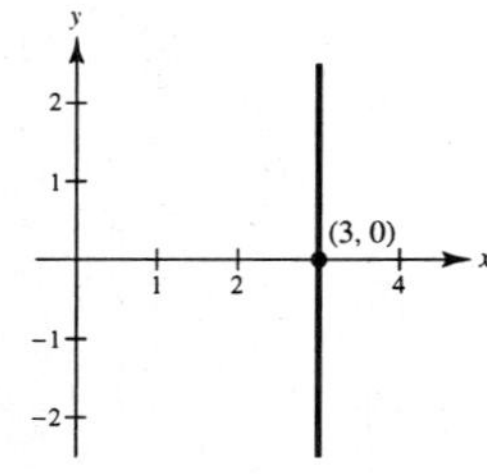

39. $\dfrac{x}{2} + \dfrac{y}{3} = 1$

$3x + 2y - 6 = 0$

41. $\dfrac{x}{a} + \dfrac{y}{a} = 1$

$\dfrac{1}{a} + \dfrac{2}{a} = 1$

$\dfrac{3}{a} = 1$

$a = 3 \Longrightarrow x + y = 3$

$x + y - 3 = 0$

43. $4x - 2y = 3$

$y = 2x - \frac{3}{2}$

$m = 2$

(a) $y - 1 = 2(x - 2)$

$y - 1 = 2x - 4$

$2x - y - 3 = 0$

(b) $y - 1 = -\frac{1}{2}(x - 2)$

$2y - 2 = -x + 2$

$x + 2y - 4 = 0$

45. $5x + 3y = 0$

$y = -\frac{5}{3}x$

$m = -\frac{5}{3}$

(a) $y - \frac{3}{4} = -\frac{5}{3}\left(x - \frac{7}{8}\right)$

$24y - 18 = -40x + 35$

$40x + 24y - 53 = 0$

(b) $y - \frac{3}{4} = \frac{3}{5}\left(x - \frac{7}{8}\right)$

$40y - 30 = 24x - 21$

$24x - 40y + 9 = 0$

47. (a) $x = 2 \Rightarrow x - 2 = 0$

(b) $y = 5 \Rightarrow y - 5 = 0$

49. $y = -3$

$y + 3 = 0$

51. $2x - y - 3 = 0$

$y = 2x - 3$

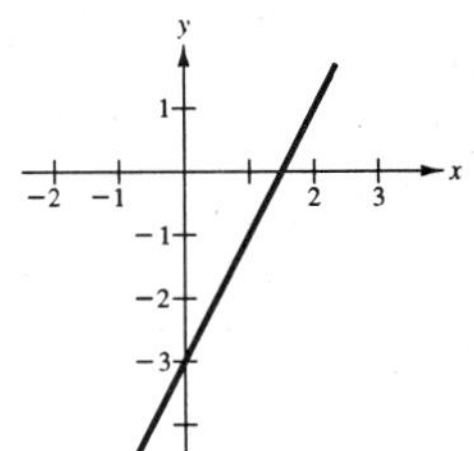

53. $y = -2x + 1$

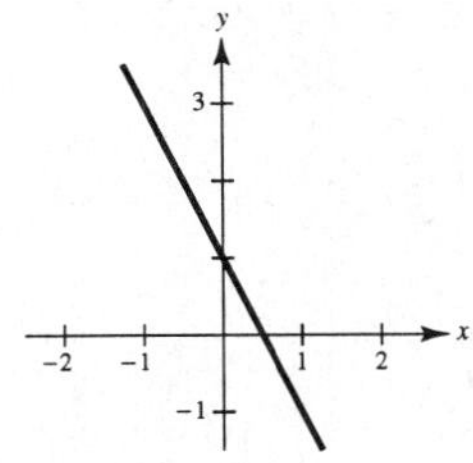

55. $y = 0.5x - 3$. The second viewing rectangle shows the important features of the line (intercepts).

57. The slope is 125. Hence

$$V = 125(t - 8) + 2540$$
$$= 125t + 1540.$$

59. The slope is -2000. Hence,

$$V = -2000(t - 8) + 20{,}400$$
$$= -2000t + 36{,}400.$$

61. You can use the graphing utility to determine that the points of intersection are (0, 0) and (2, 4). Analytically,

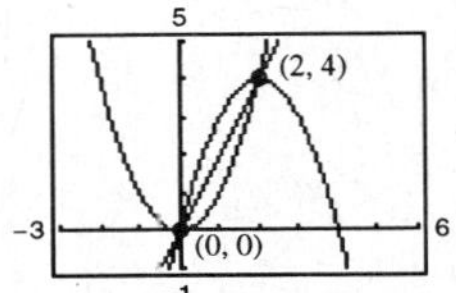

$$x^2 = 4x - x^2$$
$$2x^2 - 4x = 0$$
$$2x(x - 2) = 0$$

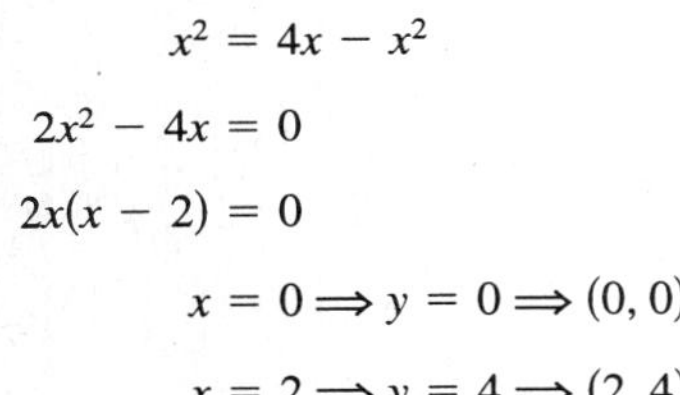

$$x = 0 \Rightarrow y = 0 \Rightarrow (0, 0)$$
$$x = 2 \Rightarrow y = 4 \Rightarrow (2, 4).$$

The slope of the line joining (0, 0) and (2, 4) is $m = (4 - 0)/(2 - 0) = 2$. Hence, an equation of the line is

$$y - 0 = 2(x - 0)$$
$$y = 2x.$$

63. $m_1 = \dfrac{1 - 0}{-2 - (-1)} = -1$

$m_2 = \dfrac{-2 - 0}{2 - (-1)} = -\dfrac{2}{3}$

$m_1 \neq m_2$

The points are not collinear.

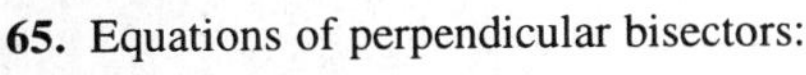

65. Equations of perpendicular bisectors:

$$y - \frac{c}{2} = \frac{a - b}{c}\left(x - \frac{a + b}{2}\right)$$

$$y - \frac{c}{2} = \frac{a + b}{-c}\left(x - \frac{b - a}{2}\right)$$

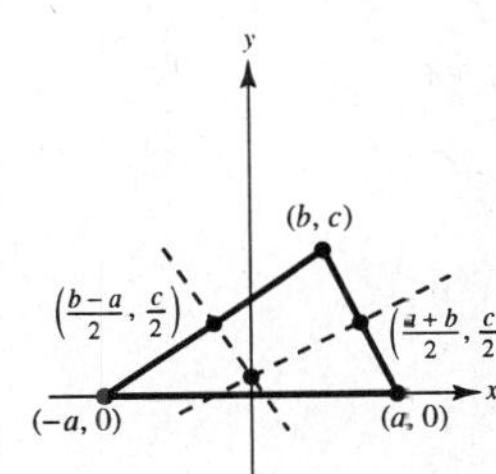

Letting $x = 0$ in either equation gives the point of intersection:

$$\left(0, \frac{-a^2 + b^2 + c^2}{2c}\right)$$

67. Equations of altitudes:

$$y = \frac{a-b}{c}(x+a)$$

$$x = b$$

$$y = -\frac{a+b}{c}(x-a)$$

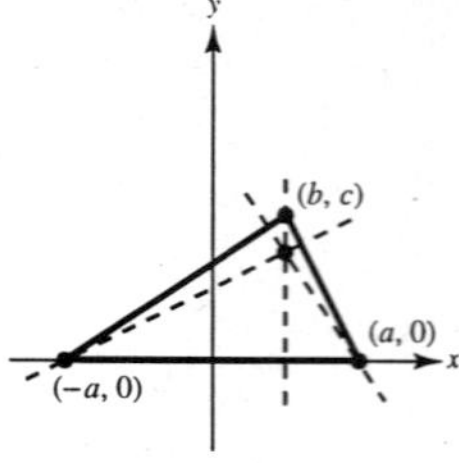

Solving simultaneously, the point of intersection is

$$\left(b, \frac{a^2-b^2}{c}\right).$$

69. Find the equation of the line through the points (0, 32) and (100, 212).

$$m = \tfrac{180}{100} = \tfrac{9}{5}$$

$$F - 32 = \tfrac{9}{5}(C - 0)$$

$$F = \tfrac{9}{5}C + 32$$

$$5F - 9C - 160 = 0$$

For $F = 72°$, $C \approx 22.2°$.

71. (a) $W_1 = 0.75x + 12.50$

$W_2 = 1.30x + 9.20$

(b)

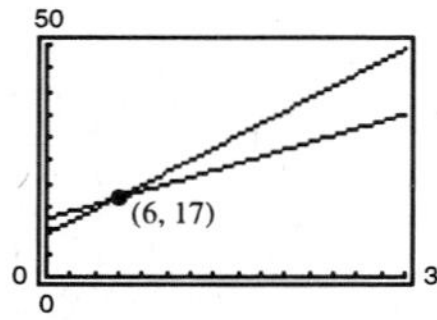

Using a graphing utility, the point of intersection is approximately (6, 17). Analytically,

$$0.75x + 12.50 = 1.30x + 9.20$$

$$3.3 = 0.55x \Rightarrow x = 6$$

$$y = 0.75(6) + 12.50 = 17.$$

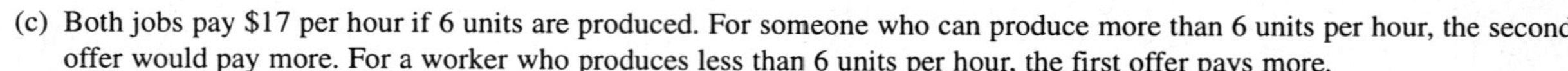
(c) Both jobs pay $17 per hour if 6 units are produced. For someone who can produce more than 6 units per hour, the second offer would pay more. For a worker who produces less than 6 units per hour, the first offer pays more.

73. (a) Two points are (50, 580) and (47, 625). The slope is

$$m = \frac{625 - 580}{47 - 50} = -15.$$

$$p - 580 = -15(x - 50)$$

$$p = -15x + 750 + 580 = -15x + 1330$$

$$\text{or } x = \tfrac{1}{15}(1330 - p)$$

(b)

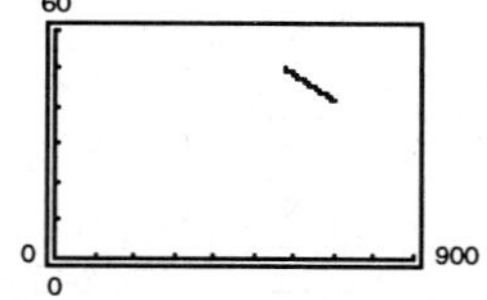

If $p = 655$, $x = \frac{1}{15}(1330 - 655) = 45$ units.

(c) If $p = 595$, $x = \frac{1}{15}(1330 - 595) = 49$ units.

75. $4x + 3y - 10 = 0 \Rightarrow d = \dfrac{|4(0) + 3(0) - 10|}{\sqrt{4^2 + 3^2}} = \dfrac{10}{5} = 2$

77. $x - y - 2 = 0 \Rightarrow d = \dfrac{|1(-2) + (-1)(1) - 2|}{\sqrt{1^2 + 1^2}} = \dfrac{5}{\sqrt{2}} = \dfrac{5\sqrt{2}}{2}$

79. A point on the line $x + y = 1$ is (0, 1). The distance from the point (0, 1) to $x + y - 5 = 0$ is

$$d = \frac{|1(0) + 1(1) - 5|}{\sqrt{1^2 + 1^2}} = \frac{|1 - 5|}{\sqrt{2}} = \frac{4}{\sqrt{2}} = 2\sqrt{2}.$$

81. If $A = 0$, then $By + C = 0$ is the horizontal line $y = -C/B$. The distance to (x_1, y_1) is

$$d = \left|y_1 - \left(\frac{-C}{B}\right)\right| = \frac{|By_1 + C|}{|B|} = \frac{|Ax_1 + By_1 + C|}{\sqrt{A^2 + B^2}}.$$

If $B = 0$, then $Ax + C = 0$ is the vertical line $x = -C/A$. The distance to (x_1, y_1) is

$$d = \left|x_1 - \left(\frac{-C}{A}\right)\right| = \frac{|Ax_1 + C|}{|A|} = \frac{|Ax_1 + By_1 + C|}{\sqrt{A^2 + B^2}}.$$

(Note that A and B cannot both be zero.)

The slope of the line $Ax + By + C = 0$ is $-A/B$. The equation of the line through (x_1, y_1) perpendicular to $Ax + By + C = 0$ is:

$$y - y_1 = \frac{B}{A}(x - x_1)$$

$$Ay - Ay_1 = Bx - Bx_1$$

$$Bx_1 - Ay_1 = Bx - Ay$$

The point of intersection of these two lines is:

$$Ax + By = -C \implies A^2x + ABy = -AC \quad (1)$$

$$Bx - Ay = Bx_1 - Ay_1 \implies B^2x - ABy = B^2x_1 - ABy_1 \quad (2)$$

$$(A^2 + B^2)x = -AC + B^2x_1 - ABy_1 \quad \text{(By adding equations (1) and (2))}$$

$$x = \frac{-AC + B^2x_1 - ABy_1}{A^2 + B^2}$$

$$Ax + By = -C \implies ABx + B^2y = -BC \quad (3)$$

$$Bx - Ay = Bx_1 - Ay_1 \implies -ABx + A^2y = -ABx_1 + A^2y_1 \quad (4)$$

$$(A^2 + B^2)y = -BC - ABx_1 + A^2y_1 \quad \text{(By adding equations (3) and (4))}$$

$$y = \frac{-BC - ABx_1 + A^2y_1}{A^2 + B^2}$$

$$\left(\frac{-AC + B^2x_1 - ABy_1}{A^2 + B^2}, \frac{-BC - ABx_1 + A^2y_1}{A^2 + B^2}\right) \text{ Point of intersection}$$

The distance between (x_1, y_1) and this point gives us the distance between (x_1, y_1) and the line $Ax + By + C = 0$.

$$d = \sqrt{\left[\frac{-AC + B^2x_1 - ABy_1}{A^2 + B^2} - x_1\right]^2 + \left[\frac{-BC - ABx_1 + A^2y_1}{A^2 + B^2} - y_1\right]^2}$$

$$= \sqrt{\left[\frac{-AC - ABy_1 - A^2x_1}{A^2 + B^2}\right]^2 + \left[\frac{-BC - ABx_1 - B^2y_1}{A^2 + B^2}\right]^2}$$

$$= \sqrt{\left[\frac{-A(C + By_1 + Ax_1)}{A^2 + B^2}\right]^2 + \left[\frac{-B(C + Ax_1 + By_1)}{A^2 + B^2}\right]^2}$$

$$= \sqrt{\frac{(A^2 + B^2)(C + Ax_1 + By_1)^2}{(A^2 + B^2)^2}} = \frac{|Ax_1 + By_1 + C|}{\sqrt{A^2 + B^2}}$$

83. For simplicity, let the vertices of the rhombus be $(0, 0)$, $(a, 0)$, (b, c), and $(a + b, c)$, as shown in the figure. The slopes of the diagonals are then

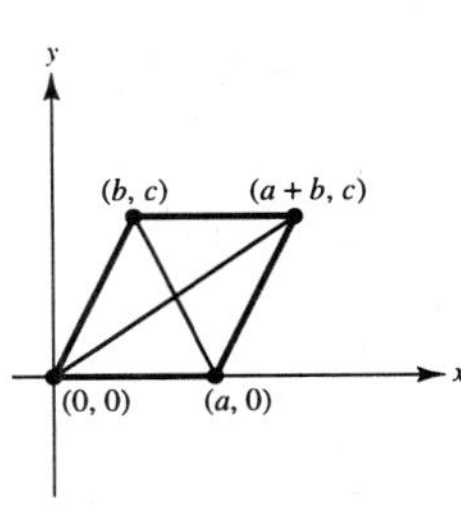

$$m_1 = \frac{c}{a + b} \text{ and } m_2 = \frac{c}{b - a}.$$

Since the sides of the Rhombus are equal, $a^2 = b^2 + c^2$, and we have

$$m_1 m_2 = \frac{c}{a + b} \cdot \frac{c}{b - a} = \frac{c^2}{b^2 - a^2} = \frac{c^2}{-c^2} = -1.$$

Therefore, the diagonals are perpendicular.

85. Consider the figure below in which the four points are collinear. Since the triangles are similar, the result immediately follows.

$$\frac{y_2^* - y_1^*}{x_2^* - x_1^*} = \frac{y_2 - y_1}{x_2 - x_1}$$

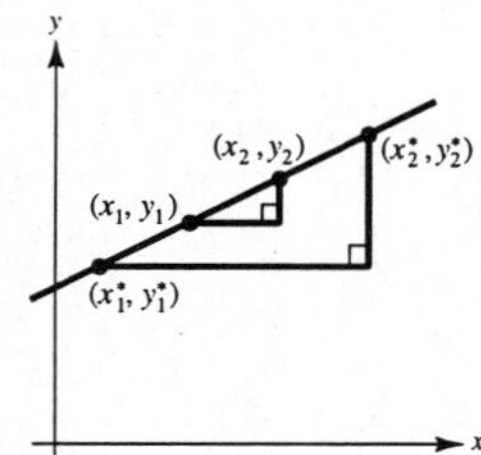

87. True.

$$ax + by = c_1 \Rightarrow y = -\frac{a}{b}x + \frac{c_1}{b} \Rightarrow m_1 = -\frac{a}{b}$$

$$bx - ay = c_2 \Rightarrow y = \frac{b}{a}x - \frac{c_2}{a} \Rightarrow m_2 = \frac{b}{a}$$

$$m_2 = -\frac{1}{m_1}$$

Section P.3 Functions and Their Graphs

1. (a) $f(0) = 2(0) - 3 = -3$

(b) $f(-3) = 2(-3) - 3 = -9$

(c) $f(b) = 2b - 3$

(d) $f(x - 1) = 2(x - 1) - 3 = 2x - 5$

3. (a) $f(-1) = 2(-1) + 1 = -1$

(b) $f(0) = 2(0) + 2 = 2$

(c) $f(2) = 2(2) + 2 = 6$

(d) $f(t^2 + 1) = 2(t^2 + 1) + 2 = 2t^2 + 4$

(**Note:** $t^2 + 1 \geq 0$ for all real t.)

5. (a) $f(0) = \cos(2(0)) = \cos 0 = 1$

(b) $f\left(-\frac{\pi}{4}\right) = \cos\left(2\left(-\frac{\pi}{4}\right)\right) = \cos\left(-\frac{\pi}{2}\right) = 0$

(c) $f\left(\frac{\pi}{3}\right) = \cos\left(2\left(\frac{\pi}{3}\right)\right) = \cos\frac{2\pi}{3} = -\frac{1}{2}$

7. $$\frac{f(x + \Delta x) - f(x)}{\Delta x} = \frac{(x + \Delta x)^3 - x^3}{\Delta x}$$
$$= \frac{x^3 + 3x^2\Delta x + 3x(\Delta x)^2 + (\Delta x)^3 - x^3}{\Delta x}$$
$$= 3x^2 + 3x\Delta x + (\Delta x)^2, \Delta x \neq 0$$

9. $$\frac{f(x) - f(2)}{x - 2} = \frac{(1/\sqrt{x - 1} - 1)}{x - 2}$$
$$= \frac{1 - \sqrt{x - 1}}{(x - 2)\sqrt{x - 1}} \cdot \frac{1 + \sqrt{x - 1}}{1 + \sqrt{x - 1}} = \frac{2 - x}{(x - 2)\sqrt{x - 1}(1 + \sqrt{x - 1})} = \frac{-1}{\sqrt{x - 1}(1 + \sqrt{x - 1})}, x \neq 2$$

11. $f(x) = 4 - x$

Domain: $(-\infty, \infty)$
Range: $(-\infty, \infty)$

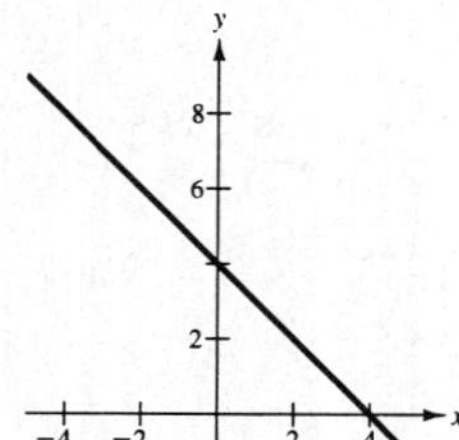

13. $h(x) = \sqrt{x - 1}$

Domain: $[1, \infty)$
Range: $[0, \infty)$

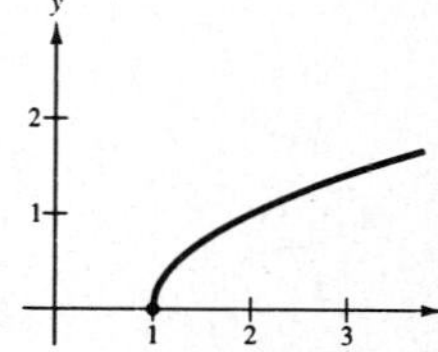

15. $f(x) = \sqrt{9 - x^2}$

Domain: $[-3, 3]$
Range: $[0, 3]$

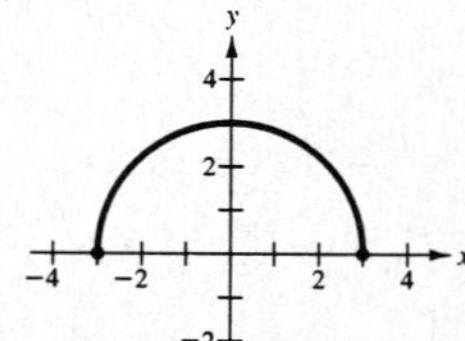

17. $g(t) = 2\sin \pi t$

Domain: $(-\infty, \infty)$

Range: $[-2, 2]$

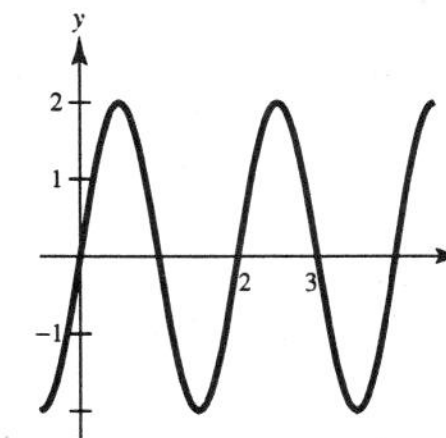

19. $x - y^2 = 0 \Longrightarrow y = \pm\sqrt{x}$

y is not a function of x.

21. $f(x) = |x| + |x - 2|$

If $x < 0$, then $f(x) = -x - (x - 2) = -2x + 2 = 2(1 - x)$.
If $0 \le x < 2$, then $f(x) = x - (x - 2) = 2$.
If $x \ge 2$, then $f(x) = x + (x - 2) = 2x - 2 = 2(x - 1)$.

Thus,

$$f(x) = \begin{cases} 2(1 - x), & x < 0 \\ 2, & 0 \le x < 2 \\ 2(x - 1), & x \ge 2. \end{cases}$$

23. $x^2 + y^2 = 4 \Longrightarrow y = \pm\sqrt{4 - x^2}$

y is not a function of x since there are two values of y for some x.

25. $y^2 = x^2 - 1 \Longrightarrow y = \pm\sqrt{x^2 - 1}$

y is not a function of x since there are two values of y for some x.

27. No, the relation is not a function. The element "National League" in the domain corresponds to more than one team name.

29. The function is $g(x) = cx^2$. Since $(1, -2)$ satisfies the equation, $c = -2$. Thus, $g(x) = -2x^2$.

31. The function is $r(x) = c/x$, since it must be undefined at $x = 0$. Since $(1, 32)$ satisfies the equation, $c = 32$. Thus, $r(x) = 32/x$.

33. (a) For each time t, there corresponds a depth d.

(b) Domain: $0 \le t \le 5$

Range: $0 \le d \le 30$

(c)

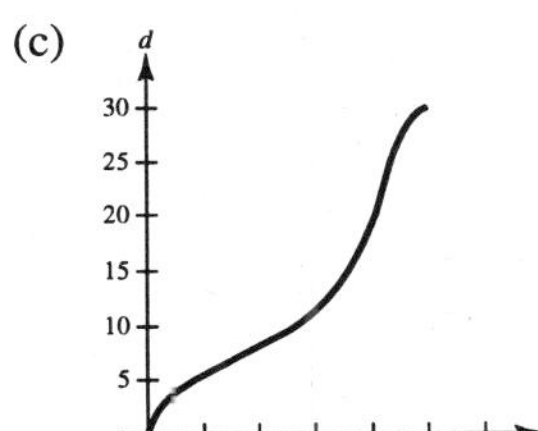

35. By trying each viewing rectangle, you see that the third viewing rectangle is best.

37. (a) $y = \sqrt{x} + 2$

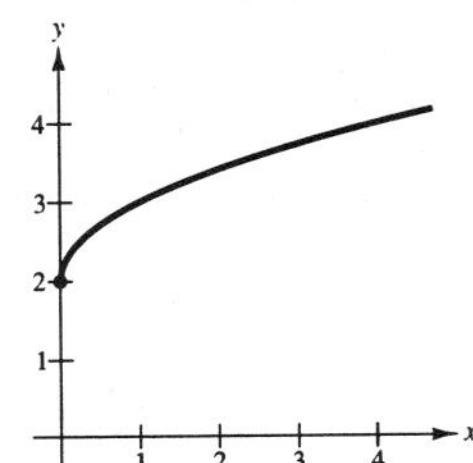

Vertical shift 2 units upward

(b) $y = -\sqrt{x}$

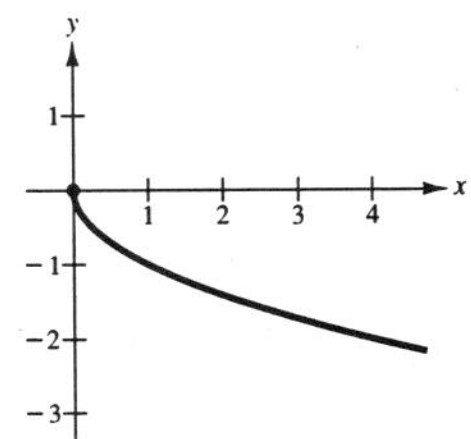

Reflection about the x-axis

(c) $y = \sqrt{x - 2}$

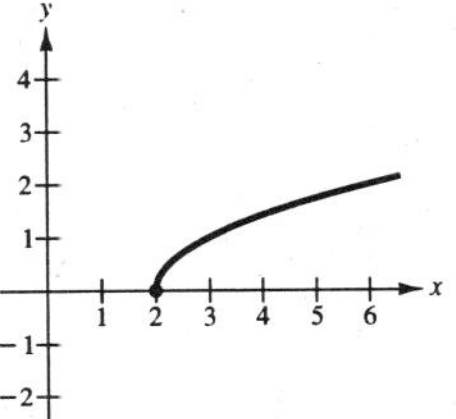

Horizontal shift 2 units to the right

39. (a) $T(4) = 16°$, $T(15) \approx 23°$

(b) If $H(t) = T(t - 1)$, then the program would turn on (and off) one hour later.

(c) If $H(t) = T(t) - 1$, then the overall temperature would be reduced 1 degree.

41. $f(x) = x^2,\ g(x) = \sqrt{x}$

$(f \circ g)(x) = f(g(x)) = f(\sqrt{x}) = (\sqrt{x})^2 = x, \quad x \geq 0$

Domain: $[0, \infty)$

$(g \circ f)(x) = g(f(x)) = g(x^2) = \sqrt{x^2} = |x|$

Domain: $(-\infty, \infty)$

$(f \circ g) = (g \circ f)$ for $x \geq 0$.

43. $(f \circ g)(x) = f(x^2 + 1) = \dfrac{1}{x^2 + 1}$

Domain: $(-\infty, \infty)$

$(g \circ f)(x) = g\left(\dfrac{1}{x}\right) = \left(\dfrac{1}{x}\right)^2 + 1$

Domain: $(-\infty, 0), (0, \infty)$

The composite functions are not equal.

45. $(A \circ r)(t) = A(r(t)) = A(0.6t) = \pi(0.6t)^2 = 0.36\pi t^2$

$(A \circ r)(t)$ represents the area of the circle at time t.

47. $f(-x) = 4 - (-x)^2 = 4 - x^2 = f(x)$

Even

49. $f(-x) = (-x)\cos(-x) = -x\cos x = -f(x)$

Odd

51. (a) If f is even, then $\left(\frac{3}{2}, 4\right)$ is on the graph

(b) If f is odd, then $\left(\frac{3}{2}, -4\right)$ is on the graph.

53. $f(-x) = a_{2n+1}(-x)^{2n+1} + \cdots + a_3(-x)^3 + a_1(-x)$

$= -[a_{2n+1}x^{2n+1} + \cdots + a_3x^3 + a_1x]$

$= -f(x)$

Odd

55. Let $F(x) = f(x)g(x)$ where f and g are even. Then

$F(-x) = f(-x)g(-x) = f(x)g(x) = F(x).$

Thus, $F(x)$ is even. Let $F(x) = f(x)g(x)$ where f and g are odd. Then

$F(-x) = f(-x)g(-x)$

$= [-f(x)][-g(x)] = f(x)g(x) = F(x).$

Thus, $F(x)$ is even.

57. $f(x) = x^2 + 1$ and $g(x) = x^4$ are even.

$f(x)g(x) = (x^2 + 1)(x^4) = x^6 + x^4$ is even.

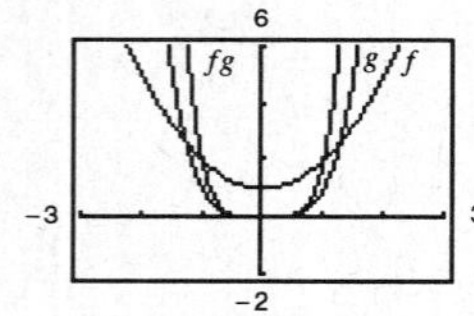

$f(x) = x^3 - x$ is odd and $g(x) = x^2$ is even.

$f(x)g(x) = (x^3 - x)(x^2) = x^5 - x^3$ is odd.

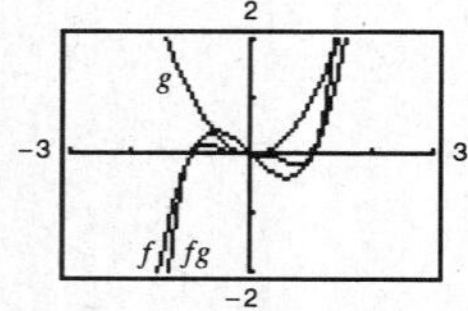

59. (a)

x

50 − x

$100 =$ perimeter

(b) $A = x(50 - x) = 50x - x^2$

(c)

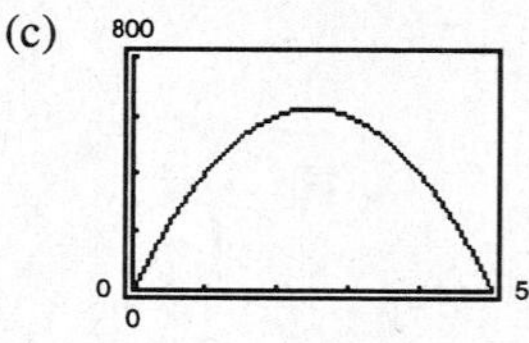

Domain: $0 < x < 50$

(d) The area is maximum when $x = 25$ (a square!).

61. (a)

x	length and width	volume V
1	$24 - 2(1)$	484
2	$24 - 2(2)$	800
3	$24 - 2(3)$	972
4	$24 - 2(4)$	1024
5	$24 - 2(5)$	980
6	$24 - 2(6)$	864

The maximum volume appears to be 1024 cm^3.

(b) 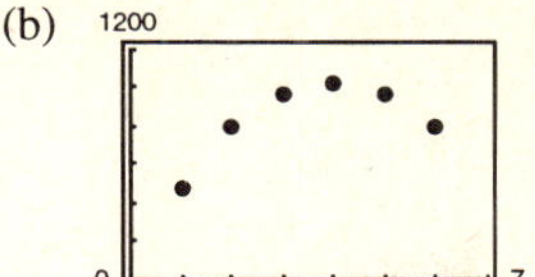

Yes, V is a function of x.

(c) $V = x(24 - 2x)$

Domain: $0 < x < 12$

(d) 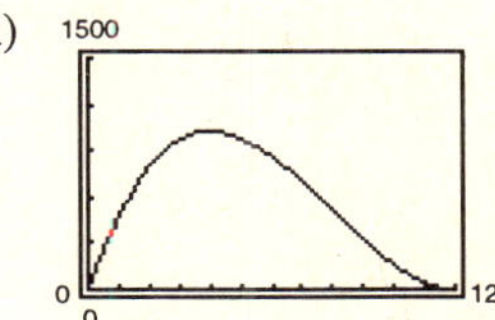

Maximum volume is $V = 1024 \text{ cm}^3$ for box having dimensions $4 \times 16 \times 16$.

63. False; let $f(x) = x^2$.
Then $f(-3) = f(3) = 9$, but $-3 \neq 3$.

65. True, the function is even.

Section P.4 Fitting Models to Data

1. Quadratic function

3. Linear function

5. (a) 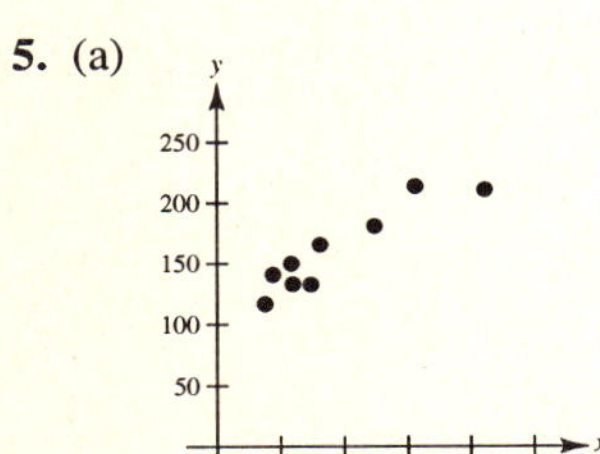

Yes. The cancer mortality increases linearly with increased exposure to the carcinogenic substance.

(b) 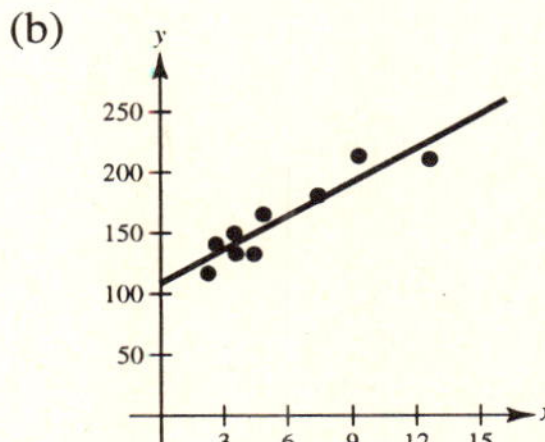

(c) If $x = 3$, then $y \approx 136$.

7. (a) $d = 0.066F$ or $F = 15.1d + 0.1$

(b) 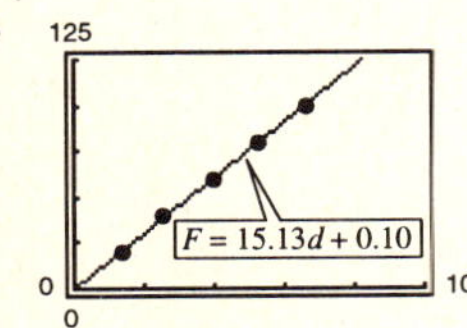

The model fits well.

(c) If $F = 55$, then
$d \approx 0.066(55) = 3.63$ cm.

9. (a) $y = 2.62x + 2.66$

(b) 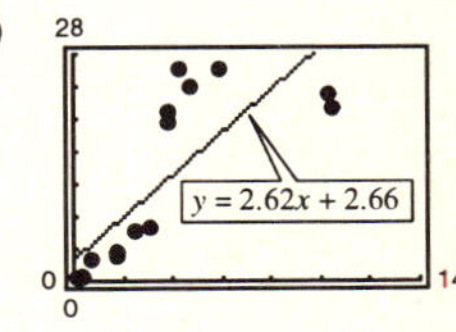

(c) Denmark, France, and Italy differ substantially.

11. (a) $y_1 = 0.16t^2 - 2.43t + 13.96$

$y_2 = 0.17t + 0.38$

$y_3 = 0.04t + 0.44$

(b) 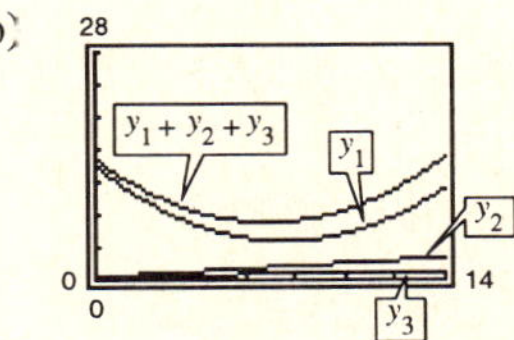

(c) For 1998, $t = 18$ and
$y_1 + y_2 + y_3 \approx 26.66$ cents/mile.

13. (a) $y_1 = 3.09t + 6.67$

$y_2 = 0.06t^3 - 1.65t^2 + 17.80t - 34.09$

(b) 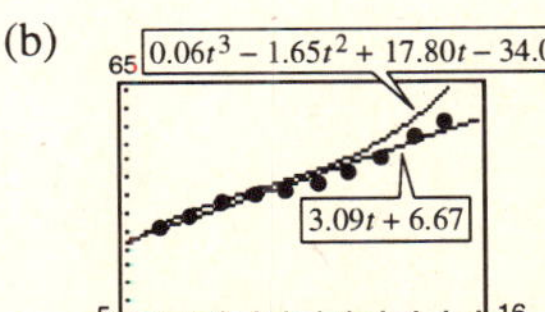

—CONTINUED—

13. —CONTINUED—

(c) The cubic model is better.

(d) $y_3 = 0.17t^2 - 0.44t + 23.82$

(e) The slope represents the average increase per year in the number of people in HMO's.

(f) For $t = 10$, $y_1 = 68.5$ (linear) and $y_2 = 141.9$ (cubic).

15. (a) $y = -1.81x^3 + 14.58x^2 + 16.39x + 10$

(b)

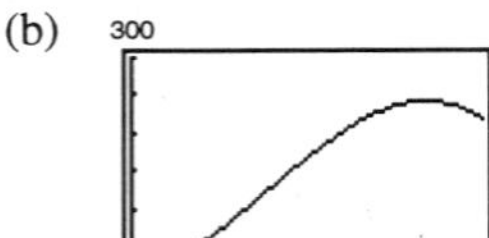

(c) If $x = 4.5$, $y \approx 214$ horsepower.

17. (a) Yes, y is a function of t. At each time t, there is one and only one displacement y.

(b) The amplitude is approximately

$$(2.35 - 1.65)/2 = 0.35.$$

The period is approximately

$$2(0.375 - 0.125) = 0.5.$$

(c) One model is $y = 0.35 \sin(4\pi t) + 2$.

(d)

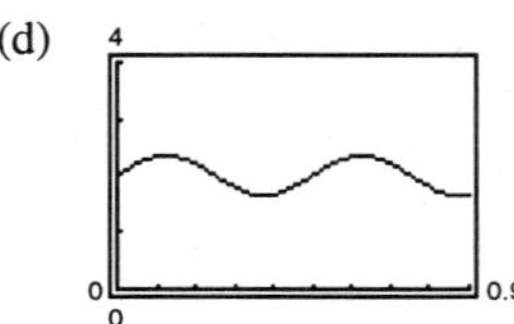

Review Exercises for Chapter P

1. $y = 2x - 3$

$x = 0 \Rightarrow y = 2(0) - 3 = -3 \Rightarrow (0, -3)$ y-intercept

$y = 0 \Rightarrow 0 = 2x - 3 \Rightarrow x = \frac{3}{2} \Rightarrow \left(\frac{3}{2}, 0\right)$ x-intercept

3. $y = \dfrac{x - 1}{x - 2}$

$x = 0 \Rightarrow y = \dfrac{0 - 1}{0 - 2} = \dfrac{1}{2} \Rightarrow \left(0, \dfrac{1}{2}\right)$ y-intercept

$y = 0 \Rightarrow 0 = \dfrac{x - 1}{x - 2} \Rightarrow x = 1 \Rightarrow (1, 0)$ x-intercept

5. Symmetric with respect to y-axis since

$$(-x)^2y - (-x)^2 + 4y = 0$$
$$x^2y - x^2 + 4y = 0.$$

7. $y = -\frac{1}{2}x + \frac{3}{2}$

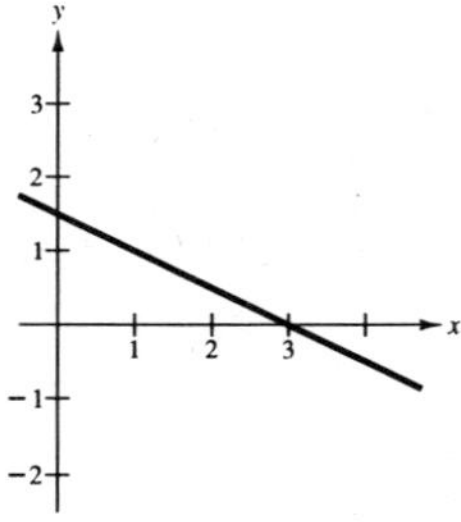

9. $-\frac{1}{3}x + \frac{5}{6}y = 1$

$-\frac{2}{5}x + y = \frac{6}{5}$

$y = \frac{2}{5}x + \frac{6}{5}$

Slope: $\frac{2}{5}$

y-intercept: $\frac{6}{5}$

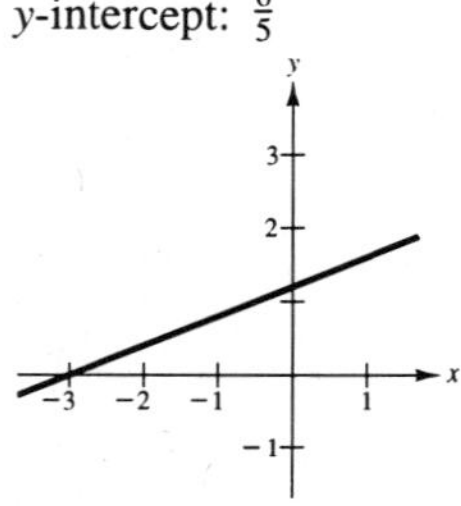

11. $y = 7 - 6x - x^2$

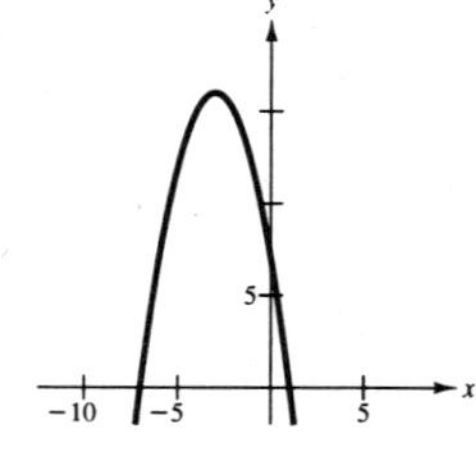

13. $y = \sqrt{5 - x}$

Domain: $(-\infty, 5]$

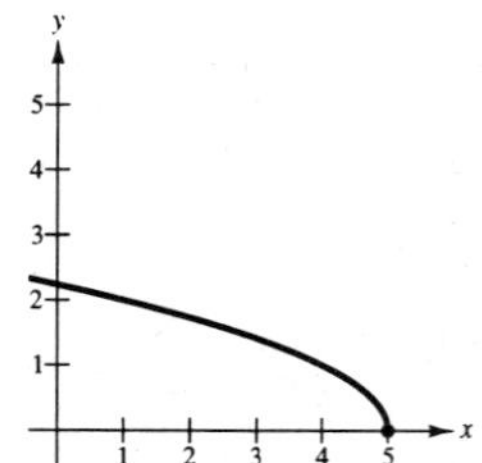

15. $y = 4x^2 - 25$

Xmin = -5
Xmax = 5
Xscl = 1
Ymin = -30
Ymax = 10
Yscl = 5

17.
$$\begin{aligned} 3x - 4y &= 8 \\ 4x + 4y &= 20 \\ \hline 7x \quad &= 28 \\ x &= 4 \\ y &= 1 \end{aligned}$$

Point: $(4, 1)$

19. You need factors $(x + 2)$ and $(x - 2)$. Multiply by x to obtain origin symmetry

$$\begin{aligned} y &= x(x + 2)(x - 2). \\ &= x^3 - 4x. \end{aligned}$$

25. (a)
$$\begin{aligned} y - 4 &= \frac{7}{16}(x + 2) \\ 16y - 64 &= 7x + 14 \\ 0 &= 7x - 16y + 78 \end{aligned}$$

(b) Slope of line is $\frac{5}{3}$.

$$\begin{aligned} y - 4 &= \frac{5}{3}(x + 2) \\ 3y - 12 &= 5x + 10 \\ 0 &= 5x - 3y + 22 \end{aligned}$$

(c)
$$\begin{aligned} m &= \frac{4 - 0}{-2 - 0} = -2 \\ y &= -2x \\ 2x + y &= 0 \end{aligned}$$

(d)
$$\begin{aligned} x &= -2 \\ x + 2 &= 0 \end{aligned}$$

21. Slope $= \dfrac{(5/2) - 1}{5 - (3/2)} = \dfrac{3/2}{7/2} = \dfrac{3}{7}$

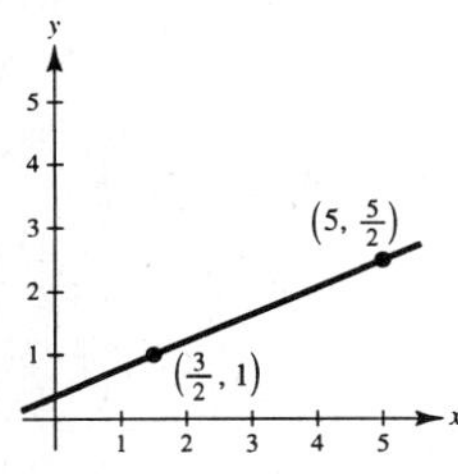

23.
$$\begin{aligned} \frac{1 - t}{1 - 0} &= \frac{1 - 5}{1 - (-2)} \\ 1 - t &= -\frac{4}{3} \\ t &= \frac{7}{3} \end{aligned}$$

27. The slope is -850. $V = -850t + 12{,}500$.

$V(3) = -850(3) + 12{,}500 = \$9950$

29. $x - y^2 = 0$

$y = \pm\sqrt{x}$

Not a function of x since there are two values of y for some x.

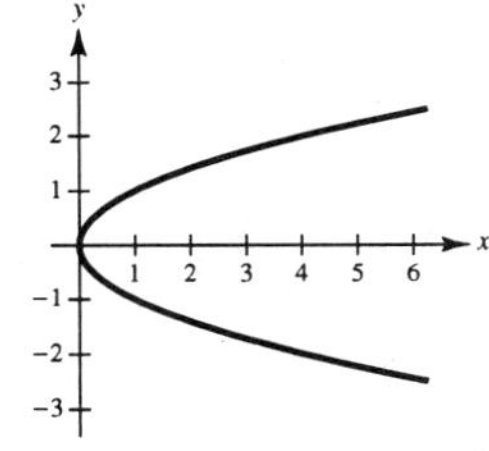

31. $y = x^2 - 2x$

Function of x since there is one value of y for each x.

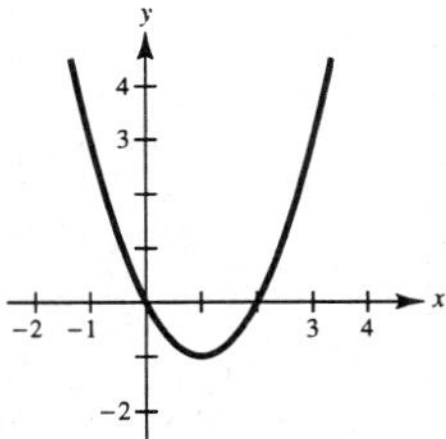

33. $f(x) = \dfrac{1}{x}$

(a) $f(0)$ does not exist.

(b)
$$\begin{aligned} \frac{f(1 + \Delta x) - f(1)}{\Delta x} &= \frac{\dfrac{1}{1 + \Delta x} - \dfrac{1}{1}}{\Delta x} = \frac{1 - 1 - \Delta x}{(1 + \Delta x)\Delta x} \\ &= \frac{-1}{1 + \Delta x}, \ \Delta x \neq -1, 0 \end{aligned}$$

35. (a) $f(x) = x^3 + c$

$c = -2, 0, 2$

(b) $f(x) = (x - c)^3$

$c = -2, 0, 2$

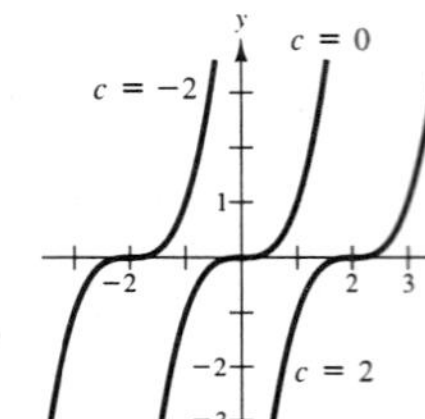

(c) $f(x) = (x - 2)^3 + c$

$c = -2, 0, 2$

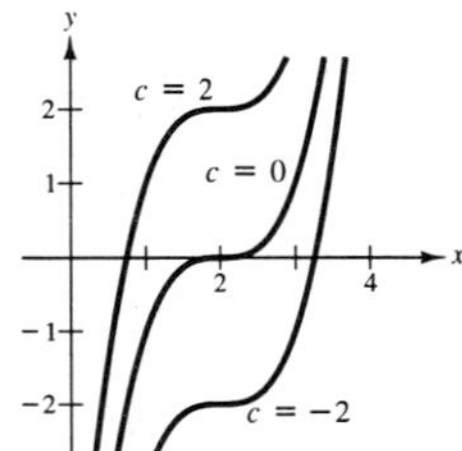

(d) $f(x) = cx^3$

$c = -2, 0, 2$

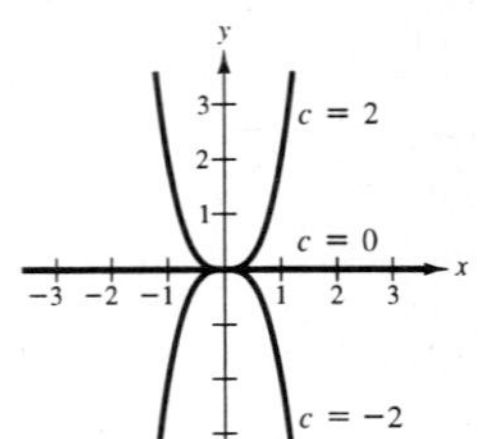

37. (a) Odd powers: $f(x) = x$, $g(x) = x^3$, $h(x) = x^5$

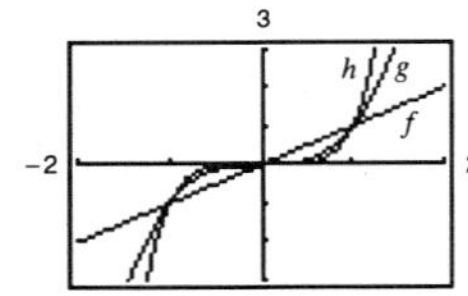

The graphs of f, g, and h all rise to the right and fall to the left. As the degree increases, the graph rises and falls more steeply. All three graphs pass through the points $(0, 0)$, $(1, 1)$, and $(-1, -1)$.

Even powers: $f(x) = x^2$, $g(x) = x^4$, $h(x) = x^6$

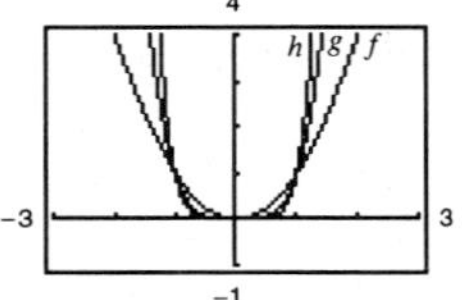

The graphs of f, g, and h all rise to the left and to the right. As the degree increases, the graph rises more steeply. All three graphs pass through the points $(0, 0)$, $(1, 1)$, and $(-1, 1)$.

(b) $y = x^7$ will look like $h(x) = x^5$, but rise and fall even more steeply.

$y = x^8$ will look like $h(x) = x^6$, but rise even more steeply.

39. (a)

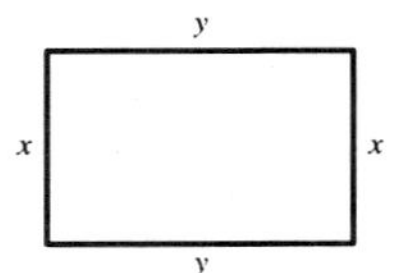

$$2x + 2y = 24$$

$$y = 12 - x$$

$$A = xy = x(12 - x) = 12x - x^2$$

(b) Domain: $0 < x < 12$

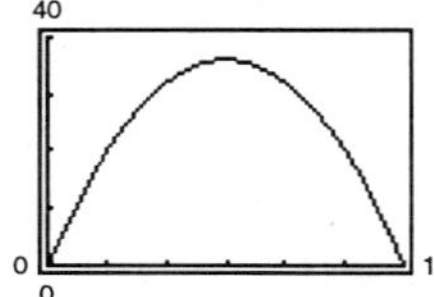

(c) Maximum area is $A = 36$. In general, the maximum area is attained when the rectangle is a square. In this case, $x = 6$.

41. (a) 3 (cubic), negative leading coefficient

(b) 4 (quartic), positive leading coefficient

(c) 2 (quadratic), negative leading coefficient

(d) 5, positive leading coefficient

43. (a) Yes, y is a function of t. At each time t, there is one and only one displacement y.

(b) The amplitude is approximately

$$\frac{0.25 - (-0.25)}{2} = 0.25.$$

The period is approximately 1.1.

(c) One model is $y = \frac{1}{4}\cos\left(\frac{2\pi}{1.1}t\right) \approx \frac{1}{4}\cos(5.7t)$

(d)

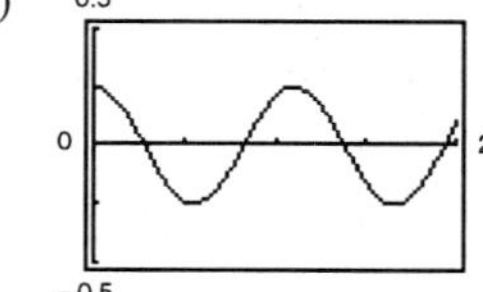

CHAPTER 1
Limits and Their Properties

CHAPTER 1
Limits and Their Properties

Section 1.1 A Preview of Calculus

Solutions to Odd-Numbered Exercises

1. Precalculus: (20 ft/sec)(15 seconds) = 300 feet

3. Calculus required: slope of tangent line at $x = 2$ is rate of change, and equals about 0.15.

5. Precalculus: Area $= \frac{1}{2}bh = \frac{1}{2}(5)(3) = \frac{15}{2}$

7. Precalculus: Volume = (2)(4)(3) = 24

9. (a)

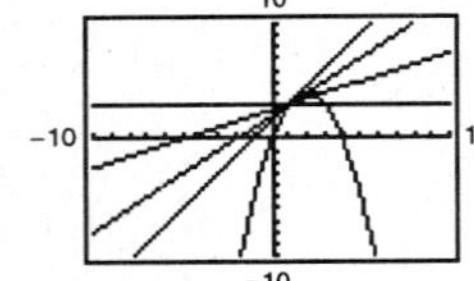

(b) The graphs of y_2 are approximations to the tangent line to y_1 at $x = 1$.

(c) The slope is approximately 2. For a better approximation make the list numbers smaller:

$\{0.2, 0.1, 0.01, 0.001\}$

11. (a) $D_1 = \sqrt{(5-1)^2 + (1-5)^2} = \sqrt{16+16} \approx 5.66$

(b) $D_2 = \sqrt{1 + \left(\frac{5}{2}\right)^2} + \sqrt{1 + \left(\frac{5}{2} - \frac{5}{3}\right)^2} + \sqrt{1 + \left(\frac{5}{3} - \frac{5}{4}\right)^2} + \sqrt{1 + \left(\frac{5}{4} - 1\right)^2}$

$\approx 2.693 + 1.302 + 1.083 + 1.031 \approx 6.11$

(c) Increase the number of line segments.

Section 1.2 Finding Limits Graphically and Numerically

1. $C(t) = 0.75 - 0.50[\![-(t-1)]\!]$

(a)

(b)

t	3	3.3	3.4	3.5	3.6	3.7	4
C	1.75	2.25	2.25	2.25	2.25	2.25	2.25

$\lim_{t \to 3.5} C(t) = 2.25$

(c)

t	2	2.5	2.9	3	3.1	3.5	4
C	1.25	1.75	1.75	1.75	2.25	2.25	2.25

$\lim_{t \to 3} C(t)$ does not exist. The values of C jump from 1.75 to 2.25 at $t = 3$.

3.

x	1.9	1.99	1.999	2.001	2.01	2.1
$f(x)$	0.3448	0.3344	0.3334	0.3332	0.3322	0.3226

$\lim_{x \to 2} \frac{x-2}{x^2 - x - 2} \approx 0.3333$ (Actual limit is $\frac{1}{3}$.)

5.

x	−0.1	−0.01	−0.001	0.001	0.01	0.1
$f(x)$	0.2911	0.2889	0.2887	0.2887	0.2884	0.2863

$\lim_{x \to 0} \frac{\sqrt{x+3} - \sqrt{3}}{x} \approx 0.2887$ (Actual limit is $1/(2\sqrt{3})$.)

7.

x	2.9	2.99	2.999	3.001	3.01	3.1
$f(x)$	-0.0641	-0.0627	-0.0625	-0.0625	-0.0623	-0.0610

$$\lim_{x\to 3} \frac{[1/(x+1)] - (1/4)}{x-3} \approx -0.0625 \quad \left(\text{Actual limit is } -\tfrac{1}{16}.\right)$$

9.

x	-0.1	-0.01	-0.001	0.001	0.01	0.1
$f(x)$	0.9983	0.99998	1.0000	1.0000	0.99998	0.9983

$$\lim_{x\to 0} \frac{\sin x}{x} \approx 1.0000 \quad \text{(Actual limit is 1.) (Make sure you use radian mode.)}$$

11. $\lim_{x\to 3} (4 - x) = 1$

13. $\lim_{x\to 2} f(x) = \lim_{x\to 2} (4 - x) = 2$

15. $\lim_{x\to 5} \frac{|x-5|}{x-5}$ does not exist. For values of x to the left of 5, $\frac{|x-5|}{x-5}$ equals -1, whereas for values of x to the right of 5, $\frac{|x-5|}{x-5}$ equals 1.

17. $\lim_{x\to \pi/2} \tan x$ does not exist since the function increases and decreases without bound as x approaches $\pi/2$.

19. $\lim_{x\to 0} \cos(1/x)$ does not exist since the function oscillates between -1 and 1 as x approaches 0.

21. $\lim_{x\to 2} (3x + 2) = 8 = L$

$$|(3x+2) - 8| < 0.01$$
$$|3x - 6| < 0.01$$
$$3|x - 2| < 0.01$$
$$0 < |x - 2| < \frac{0.01}{3} \approx 0.0033 = \delta$$

23. $\lim_{x\to 2} (x^2 - 3) = 1 = L$

$$|(x^2 - 3) - 1| < 0.01$$
$$|x^2 - 4| < 0.01$$
$$|(x+2)(x-2)| < 0.01$$
$$|x+2|\,|x-2| < 0.01$$
$$|x - 2| < \frac{0.01}{|x+2|}$$

If we assume $1 < x < 3$, then $\delta = 0.01/5 = 0.002$.

25. $\lim_{x\to 2} (x + 3) = 5$

Given $\epsilon > 0$:

$$|(x+3) - 5| < \epsilon$$
$$|x - 2| < \epsilon = \delta$$

Hence, let $\delta = \epsilon$.

27. $\lim_{x\to -4} \left(\frac{1}{2}x - 1\right) = \frac{1}{2}(-4) - 1 = -3$

Given $\epsilon > 0$:

$$\left|\left(\tfrac{1}{2}x - 1\right) - (-3)\right| < \epsilon$$
$$\left|\tfrac{1}{2}x + 2\right| < \epsilon$$
$$\tfrac{1}{2}|x - (-4)| < \epsilon$$
$$|x - (-4)| < 2\epsilon$$

Hence, let $\delta = 2\epsilon$.

29. $\lim_{x\to 6} 3 = 3$

Given $\epsilon > 0$:

$$|3 - 3| < \epsilon$$
$$0 < \epsilon$$

Hence, any $\delta > 0$ will work.

31. $\lim_{x\to 0} \sqrt[3]{x} = 0$

Given $\epsilon > 0$: $\left|\sqrt[3]{x} - 0\right| < \epsilon$

$$\left|\sqrt[3]{x}\right| < \epsilon$$
$$|x| < \epsilon^3 = \delta$$

Hence, let $\delta = \epsilon^3$.

33. $\lim_{x\to -2} |x - 2| = |(-2) - 2| = 4$

Given $\epsilon > 0$:

$$\big||x - 2| - 4\big| < \epsilon$$
$$|-(x - 2) - 4| < \epsilon \quad (x - 2 < 0)$$
$$|-x - 2| = |x + 2| = |x - (-2)| < \epsilon$$

Hence, $\delta = \epsilon$.

35. $\lim_{x\to 1} (x^2 + 1) = 2$

Given $\epsilon > 0$:

$$|(x^2 + 1) - 2| < \epsilon$$

$$|x^2 - 1| < \epsilon$$

$$|(x + 1)(x - 1)| < \epsilon$$

$$|x - 1| < \frac{\epsilon}{|x + 1|}$$

If we assume $0 < x < 2$, then $\delta = \epsilon/3$.

37. $f(x) = \dfrac{\sqrt{x+5} - 3}{x - 4}$

$$\lim_{x\to 4} f(x) = \frac{1}{6}$$

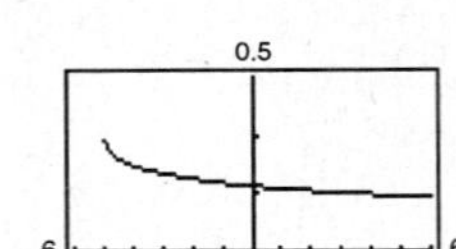

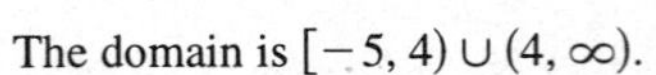

The domain is $[-5, 4) \cup (4, \infty)$.
The graphing utility does not show the hole at $\left(4, \frac{1}{6}\right)$.

39. $f(x) = \dfrac{x - 9}{\sqrt{x} - 3}$

$$\lim_{x\to 9} f(x) = 6$$

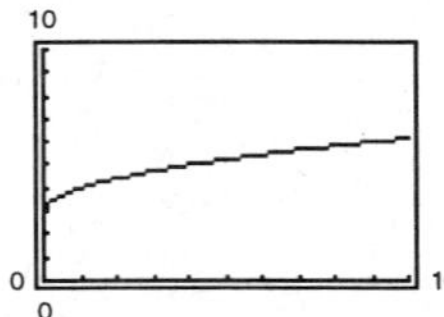

The domain is all $x \geq 0$ except $x = 9$. The graphing utility does not show the hole at $(9, 6)$.

41.

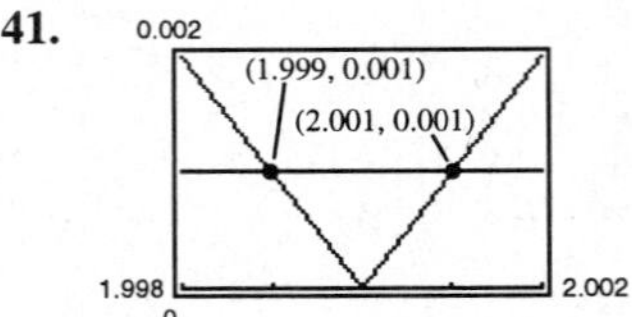

Using the zoom and trace feature, $\delta = 0.001$. That is, for

$$0 < |x - 2| < 0.001, \left|\frac{x^2 - 4}{x - 2} - 4\right| < 0.001.$$

43. False; $f(x) = (\sin x)/x$ is undefined when $x = 0$. From Exercise 9, we have

$$\lim_{x\to 0} \frac{\sin x}{x} = 1.$$

45. False; let

$$f(x) = \begin{cases} x^2 - 4x, & x \neq 4 \\ 10, & x = 4 \end{cases}.$$

$f(4) = 10$

$\lim_{x\to 4} f(x) = \lim_{x\to 4} (x^2 - 4x) = 0 \neq 10$

47. Answers will vary.

49. If $\lim_{x\to c} f(x) = L_1$ and $\lim_{x\to c} f(x) = L_2$, then for every $\epsilon > 0$, there exists $\delta_1 > 0$ and $\delta_2 > 0$ such that $|x - c| < \delta_1 \Rightarrow |f(x) - L_1| < \epsilon$ and $|x - c| < \delta_2 \Rightarrow |f(x) - L_2| < \epsilon$. Let δ equal the smaller of δ_1 and δ_2. Then for $|x - c| < \delta$, we have

$$|L_1 - L_2| = |L_1 - f(x) + f(x) - L_2| \leq |L_1 - f(x)| + |f(x) - L_2| < \epsilon + \epsilon.$$

Therefore, $|L_1 - L_2| < 2\epsilon$. Since $\epsilon > 0$ is arbitrary, it follows that $L_1 = L_2$.

51. $\lim_{x\to c} [f(x) - L] = 0$ means that for every $\epsilon > 0$ there exists $\delta > 0$ such that if

$$0 < |x - c| < \delta,$$

then

$$|(f(x) - L) - 0] < \epsilon.$$

This means the same as $|(f(x) - L| < \epsilon$ when

$$0 < |x - c| < \delta.$$

Thus, $\lim_{x\to c} f(x) = L$.

Section 1.3 Evaluating Limits Analytically

1. 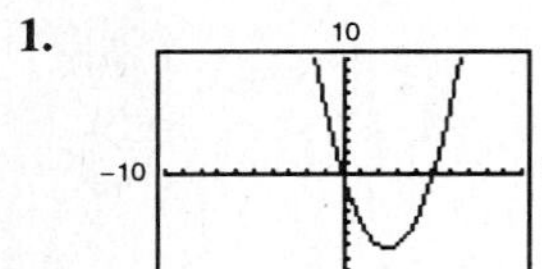

$h(x) = x^2 - 5x$

(a) $\lim_{x \to 5} h(x) = 0$

(b) $\lim_{x \to -1} h(x) = 6$

3. 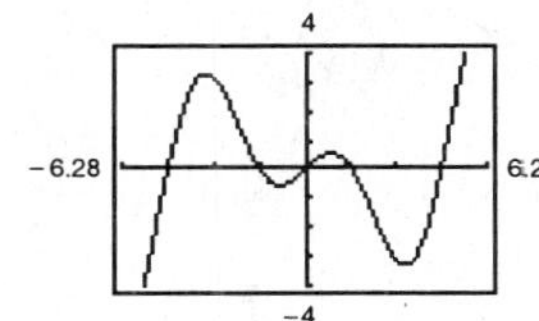

$f(x) = x \cos x$

(a) $\lim_{x \to 0} f(x) = 0$

(b) $\lim_{x \to \pi/3} f(x) \approx 0.524$ $\left(= \frac{\pi}{6}\right)$

5. $\lim_{x \to 4} x^2 = 4^2 = 16$

7. $\lim_{x \to 0} (2x - 1) = 2(0) - 1 = -1$

9. $\lim_{x \to 2} (-x^2 + x - 2) = -(2)^2 + (2) - 2 = -4$

11. $\lim_{x \to 3} \sqrt{x + 1} = \sqrt{3 + 1} = 2$

13. $\lim_{x \to -4} (x + 3)^2 = (-4 + 3)^2 = 1$

15. $\lim_{x \to 2} \frac{1}{x} = \frac{1}{2}$

17. $\lim_{x \to -1} \frac{x^2 + 1}{x} = \frac{(-1)^2 + 1}{-1} = -2$

19. $\lim_{x \to \pi/2} \sin x = \sin \frac{\pi}{2} = 1$

21. $\lim_{x \to 1} \cos \pi x = \cos \pi = -1$

23. $\lim_{x \to 0} \sec 2x = \sec 0 = 1$

25. $\lim_{x \to 5\pi/6} \sin x = \sin \frac{5\pi}{6} = \frac{1}{2}$

27. $\lim_{x \to 3} \tan\left(\frac{\pi x}{4}\right) = \tan \frac{3\pi}{4} = -1$

29. (a) $\lim_{x \to c} [5g(x)] = 5 \lim_{x \to c} g(x) = 5(3) = 15$

(b) $\lim_{x \to c} [f(x) + g(x)] = \lim_{x \to c} f(x) + \lim_{x \to c} g(x) = 2 + 3 = 5$

(c) $\lim_{x \to c} [f(x)g(x)] = \left[\lim_{x \to c} f(x)\right]\left[\lim_{x \to c} g(x)\right] = (2)(3) = 6$

(d) $\lim_{x \to c} \frac{f(x)}{g(x)} = \frac{\lim_{x \to c} f(x)}{\lim_{x \to c} g(x)} = \frac{2}{3}$

31. (a) $\lim_{x \to c} [f(x)]^3 = \left[\lim_{x \to c} f(x)\right]^3 = (4)^3 = 64$

(b) $\lim_{x \to c} \sqrt{f(x)} = \sqrt{\lim_{x \to c} f(x)} = \sqrt{4} = 2$

(c) $\lim_{x \to c} [3f(x)] = 3 \lim_{x \to c} f(x) = 3(4) = 12$

(d) $\lim_{x \to c} [f(x)]^{3/2} = \left[\lim_{x \to c} f(x)\right]^{3/2} = (4)^{3/2} = 8$

33. $f(x) = -2x + 1$ and $g(x) = \frac{-2x^2 + x}{x}$ agree except at $x = 0$.

(a) $\lim_{x \to 0} g(x) = \lim_{x \to 0} f(x) = 1$

(b) $\lim_{x \to -1} g(x) = \lim_{x \to -1} f(x) = 3$

35. $f(x) = x(x + 1)$ and $g(x) = \frac{x^3 - x}{x - 1}$ agree except at $x = 1$.

(a) $\lim_{x \to 1} g(x) = \lim_{x \to 1} f(x) = 2$

(b) $\lim_{x \to -1} g(x) = \lim_{x \to -1} f(x) = 0$

37. $f(x) = \frac{x^2 - 1}{x + 1}$ and $g(x) = x - 1$ agree except at $x = -1$.

$\lim_{x \to -1} f(x) = \lim_{x \to -1} g(x) = -2$

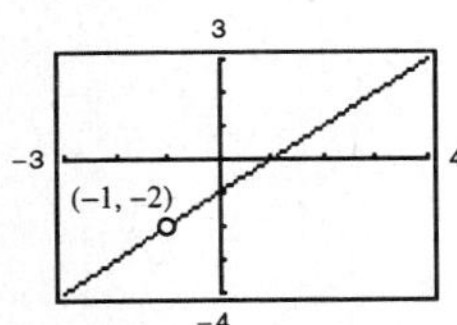

39. $f(x) = \frac{x^3 + 8}{x + 2}$ and $g(x) = x^2 - 2x + 4$ agree except at $x = -2$.

$\lim_{x \to -2} f(x) = \lim_{x \to -2} g(x) = 12$

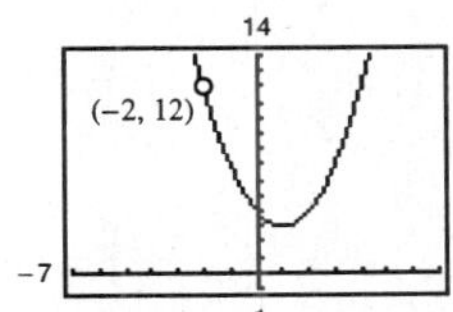

41. $\lim_{x\to 5}\dfrac{x-5}{x^2-25}=\lim_{x\to 5}\dfrac{x-5}{(x+5)(x-5)}$

$=\lim_{x\to 5}\dfrac{1}{x+5}=\dfrac{1}{10}$

43. $\lim_{x\to 1}\dfrac{x^2+x-2}{x^2-1}=\lim_{x\to 1}\dfrac{(x+2)(x-1)}{(x+1)(x-1)}=\lim_{x\to 1}\dfrac{x+2}{x+1}=\dfrac{3}{2}$

45. $\lim_{x\to 0}\dfrac{\sqrt{3+x}-\sqrt{3}}{x}=\lim_{x\to 0}\dfrac{\sqrt{3+x}-\sqrt{3}}{x}\cdot\dfrac{\sqrt{3+x}+\sqrt{3}}{\sqrt{3+x}+\sqrt{3}}$

$=\lim_{x\to 0}\dfrac{3+x-3}{\left(\sqrt{3+x}+\sqrt{3}\right)x}=\lim_{x\to 0}\dfrac{1}{\left(\sqrt{3+x}+\sqrt{3}\right)}=\dfrac{1}{2\sqrt{3}}=\dfrac{\sqrt{3}}{6}$

47. $\lim_{x\to 0}\dfrac{\dfrac{1}{2+x}-\dfrac{1}{2}}{x}=\lim_{x\to 0}\dfrac{\dfrac{2-(2+x)}{2(2+x)}}{x}=\lim_{x\to 0}\dfrac{-1}{2(2+x)}=-\dfrac{1}{4}$

49. $\lim_{\Delta x\to 0}\dfrac{2(x+\Delta x)-2x}{\Delta x}=\lim_{\Delta x\to 0}\dfrac{2x+2\Delta x-2x}{\Delta x}$

$=\lim_{\Delta x\to 0}2=2$

51. $\lim_{\Delta x\to 0}\dfrac{(x+\Delta x)^2-2(x+\Delta x)+1-(x^2-2x+1)}{\Delta x}=\lim_{\Delta x\to 0}\dfrac{x^2+2x\Delta x+(\Delta x)^2-2x-2\Delta x+1-x^2+2x-1}{\Delta x}$

$=\lim_{\Delta x\to 0}(2x+\Delta x-2)=2x-2$

53. $\lim_{x\to 0}\dfrac{\sqrt{x+2}-\sqrt{2}}{x}\approx 0.354$

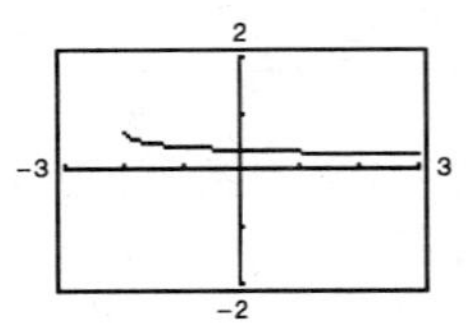

x	-0.1	-0.01	-0.001	0	0.001	0.01	0.1
$f(x)$	0.358	0.354	0.345	?	0.354	0.353	0.349

Analytically, $\lim_{x\to 0}\dfrac{\sqrt{x+2}-\sqrt{2}}{x}=\lim_{x\to 0}\dfrac{\sqrt{x+2}-\sqrt{2}}{x}\cdot\dfrac{\sqrt{x+2}+\sqrt{2}}{\sqrt{x+2}+\sqrt{2}}$

$=\lim_{x\to 0}\dfrac{x+2-2}{x\left(\sqrt{x+2}+\sqrt{2}\right)}=\lim_{x\to 0}\dfrac{1}{\sqrt{x+2}+\sqrt{2}}=\dfrac{1}{2\sqrt{2}}=\dfrac{\sqrt{2}}{4}\approx 0.354.$

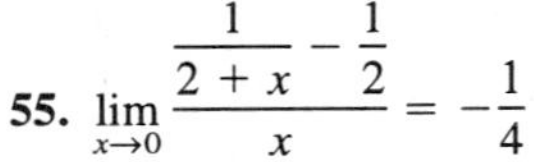

55. $\lim_{x\to 0}\dfrac{\dfrac{1}{2+x}-\dfrac{1}{2}}{x}=-\dfrac{1}{4}$

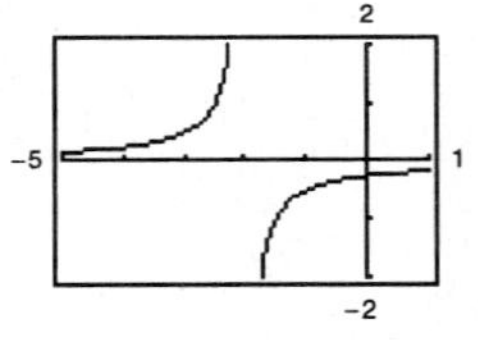

x	-0.1	-0.01	-0.001	0	0.001	0.01	0.1
$f(x)$	-0.263	-0.251	-0.250	?	-0.250	-0.249	-0.238

Analytically, $\lim_{x\to 0}\dfrac{\dfrac{1}{2+x}-\dfrac{1}{2}}{x}=\lim_{x\to 0}\dfrac{2-(2+x)}{2(2+x)}\cdot\dfrac{1}{x}=\lim_{x\to 0}\dfrac{-x}{2(2+x)}\cdot\dfrac{1}{x}=\lim_{x\to 0}\dfrac{-1}{2(2+x)}=-\dfrac{1}{4}.$

57. $\lim_{x\to 0}\dfrac{\sin x}{5x}=\lim_{x\to 0}\left[\left(\dfrac{\sin x}{x}\right)\left(\dfrac{1}{5}\right)\right]=(1)\left(\dfrac{1}{5}\right)=\dfrac{1}{5}$

59. $\lim_{\theta\to 0}\dfrac{\sec\theta-1}{\theta\sec\theta}=\lim_{\theta\to 0}\dfrac{1-\cos\theta}{\theta}=0$

61. $\lim_{x\to 0}\dfrac{\sin^2 x}{x}=\lim_{\theta\to 0}\left[\dfrac{\sin x}{x}\sin x\right]=(1)\sin 0=0$

63. $\lim_{h\to 0}\dfrac{(1-\cos h)^2}{h}=\lim_{h\to 0}\left[\dfrac{1-\cos h}{h}(1-\cos h)\right]$

$=(0)(0)=0$

65. $\lim_{x\to\pi/2} \dfrac{\cos x}{\cot x} = \lim_{x\to\pi/2} \sin x = 1$

67. $\lim_{t\to 0} \dfrac{\sin^2 t}{t^2} = \lim_{t\to 0}\left(\dfrac{\sin t}{t}\right)^2 = (1)^2 = 1$

69. $f(t) = \dfrac{\sin 3t}{t}$

t	-0.1	-0.01	-0.001	0	0.001	0.01	0.1
$f(t)$	2.96	2.9996	3	?	3	2.9996	2.96

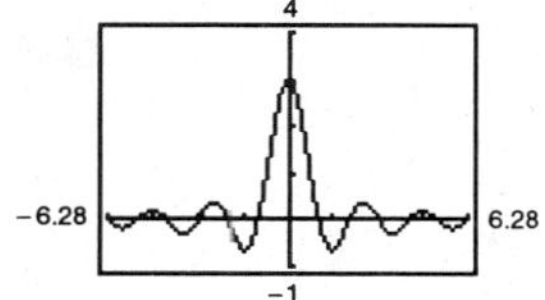

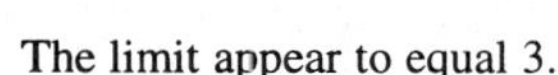
The limit appear to equal 3.

Analytically, $\lim_{t\to 0} \dfrac{\sin 3t}{t} = \lim_{t\to 0} 3\left(\dfrac{\sin 3t}{3t}\right) = 3(1) = 3.$

71. $f(x) = \dfrac{\sin x^2}{x}$

x	-0.1	-0.01	-0.001	0	0.001	0.01	0.1
$f(x)$	-0.099998	-0.01	-0.001	?	0.001	0.01	0.099998

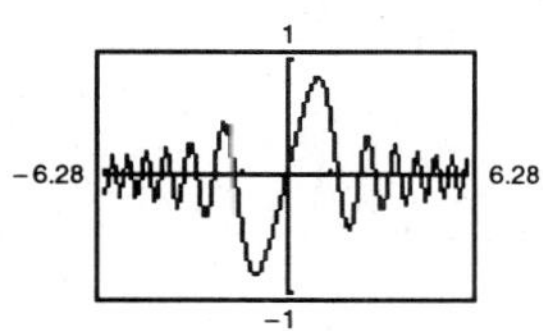

The limit appear to equal 0.

Analytically, $\lim_{x\to 0} \dfrac{\sin x^2}{x} = \lim_{x\to 0} x\left(\dfrac{\sin x^2}{x^2}\right) = 0(1) = 0.$

73. $\lim_{h\to 0} \dfrac{f(x+h) - f(x)}{h} = \lim_{h\to 0} \dfrac{2(x+h) + 3 - (2x+3)}{h}$

$= \lim_{h\to 0} \dfrac{2x + 2h + 3 - 2x - 3}{h}$

$= \lim_{h\to 0} \dfrac{2h}{h} = 2$

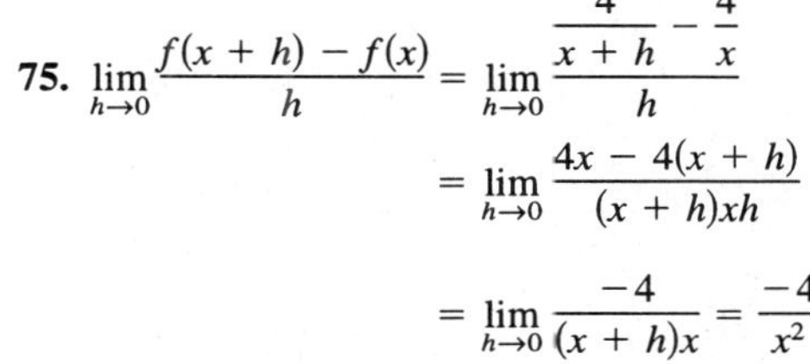

75. $\lim_{h\to 0} \dfrac{f(x+h) - f(x)}{h} = \lim_{h\to 0} \dfrac{\dfrac{4}{x+h} - \dfrac{4}{x}}{h}$

$= \lim_{h\to 0} \dfrac{4x - 4(x+h)}{(x+h)xh}$

$= \lim_{h\to 0} \dfrac{-4}{(x+h)x} = \dfrac{-4}{x^2}$

77. $\lim_{x\to 0} (4 - x^2) \le \lim_{x\to 0} f(x) \le \lim_{x\to 0} (4 + x^2)$

$4 \le \lim_{x\to 0} f(x) \le 4$

Therefore, $\lim_{x\to 0} f(x) = 4.$

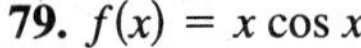

79. $f(x) = x \cos x$

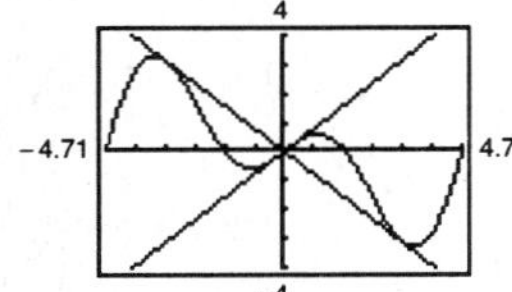

$\lim_{x\to 0}(x \cos x) = 0$

81. $f(x) = |x| \sin x$

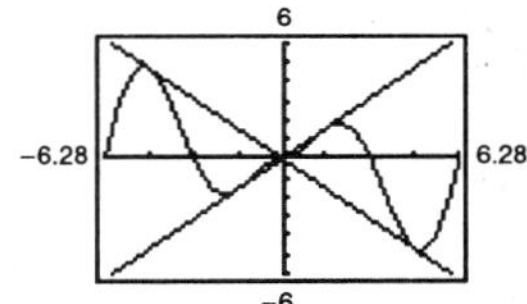

$\lim_{x\to 0} |x| \sin x = 0$

83. $f(x) = x \sin \dfrac{1}{x}$

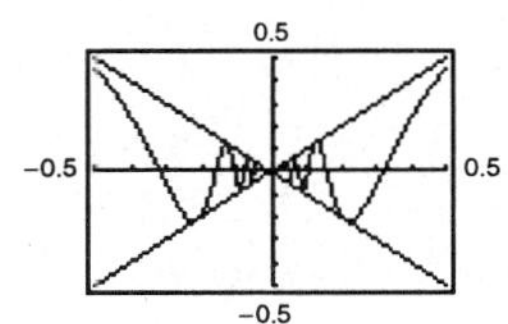

$\lim_{x\to 0}\left(x \sin \dfrac{1}{x}\right) = 0$

85. $f(x) = x$, $g(x) = \sin x$, $h(x) = \dfrac{\sin x}{x}$

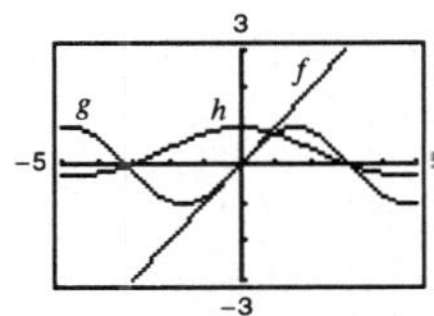

When you are "close to" 0 the magnitude of f is approximately equal to the magnitude of g. Thus, $|g|/|f| \approx 1$ when x is "close to" 0.

87. $s(t) = -16t^2 + 1000$

$$\lim_{t\to 5} \frac{s(5) - s(t)}{5 - t} = \lim_{t\to 5} \frac{600 - (-16t^2 + 1000)}{5 - t}$$

$$= \lim_{t\to 5} \frac{16(t + 5)(t - 5)}{-(t - 5)} = \lim_{t\to 5} -16(t + 5) = -160 \text{ ft/sec}$$

89. $s(t) = -4.9t^2 + 150$

$$\lim_{t\to 3} \frac{s(3) - s(t)}{3 - t} = \lim_{t\to 3} \frac{-4.9(3^2) + 150 - (-4.9t^2 + 150)}{3 - t} = \lim_{t\to 3} \frac{-4.9(9 - t^2)}{3 - t}$$

$$= \lim_{x\to 3} \frac{-4.9(3 - t)(3 + t)}{3 - t} = \lim_{x\to 3} -4.9(3 + t) = -29.4 \text{ m/sec}$$

91. (a)

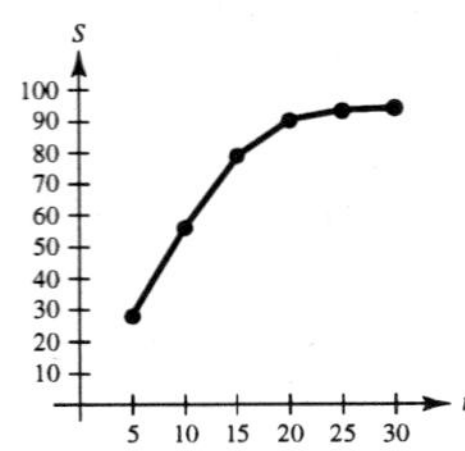

(b) Yes. As time increases, the typing speed seems to approach 95 or 100. The rate of increase is decreasing.

93. Suppose, on the contrary, that $\lim_{x\to c} g(x)$ exists. Then, since $\lim_{x\to c} f(x)$ exists, so would $\lim_{x\to c} [f(x) + g(x)]$, which is a contradiction. Hence, $\lim_{x\to c} g(x)$ does not exist.

95. Given $f(x) = x^n$, n is a positive integer, then

$$\lim_{x\to c} x^n = \lim_{x\to c} (xx^{n-1}) = \left[\lim_{x\to c} x\right]\left[\lim_{x\to c} x^{n-1}\right]$$

$$= c\left[\lim_{x\to c} (xx^{n-2})\right] = c\left[\lim_{x\to c} x\right]\left[\lim_{x\to c} x^{n-2}\right]$$

$$= c(c)\lim_{x\to c} (xx^{n-3}) = \cdots = c^n.$$

97. False. Let $f(x) = \frac{1}{2}x^2$ and $g(x) = x^2$. Then $f(x) < g(x)$ for all $x \neq 0$. But $\lim_{x\to 0} f(x) = \lim_{x\to 0} g(x) = 0$.

99.

$$-M|f(x)| \le f(x)g(x) \le M|f(x)|$$

$$\lim_{x\to c} (-M|f(x)|) \le \lim_{x\to c} f(x)g(x) \le \lim_{x\to c} (M|f(x)|)$$

$$-M(0) \le \lim_{x\to c} f(x)g(x) \le M(0)$$

$$0 \le \lim_{x\to c} f(x)g(x) \le 0$$

Therefore, $\lim_{x\to c} f(x)g(x) = 0$.

101. Given $\lim_{x\to c} f(x) = L$:

For every $\epsilon > 0$, there exists $\delta > 0$ such that $|f(x) - L| < \epsilon$ whenever $0 < |x - c| < \delta$. Since $||f(x)| - |L|| \le |f(x) - L| < \epsilon$ for $|x - c| < \delta$, then $\lim_{x\to c} |f(x)| = |L|$.

103. $$\lim_{x\to 0} \frac{1 - \cos x}{x} = \lim_{x\to 0} \frac{1 - \cos x}{x} \cdot \frac{1 + \cos x}{1 + \cos x}$$

$$= \lim_{x\to 0} \frac{1 - \cos^2 x}{x(1 + \cos x)} = \lim_{x\to 0} \frac{\sin^2 x}{x(1 + \cos x)} = \lim_{x\to 0} \frac{\sin x}{x} \cdot \frac{\sin x}{1 + \cos x}$$

$$= \left[\lim_{x\to 0} \frac{\sin x}{x}\right]\left[\lim_{x\to 0} \frac{\sin x}{1 + \cos x}\right] = (1)(0) = 0$$

105. $f(x) = \dfrac{\sec x - 1}{x^2}$

(a) The domain of f is all $x \neq 0, \pi/2 + n\pi$.

(b)

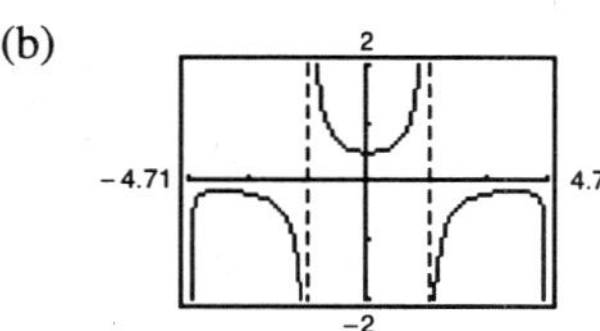

The domain is not obvious. The hole at $x = 0$ is not apparent.

(c) $\lim\limits_{x\to 0} f(x) = \dfrac{1}{2}$

(d) $\dfrac{\sec x - 1}{x^2} = \dfrac{\sec x - 1}{x^2} \cdot \dfrac{\sec x + 1}{\sec x + 1} = \dfrac{\sec^2 x - 1}{x^2(\sec x + 1)}$

$= \dfrac{\tan^2 x}{x^2(\sec x + 1)} = \dfrac{1}{\cos^2 x}\left(\dfrac{\sin^2 x}{x^2}\right)\dfrac{1}{\sec x + 1}$

Hence, $\lim\limits_{x\to 0} \dfrac{\sec x - 1}{x^2} = \lim\limits_{x\to 0} \dfrac{1}{\cos^2 x}\left(\dfrac{\sin^2 x}{x^2}\right)\dfrac{1}{\sec x + 1}$

$= 1(1)\left(\dfrac{1}{2}\right) = \dfrac{1}{2}.$

107. Two functions agree at all but one point means that the functions are identical for all x in their domain, except possibly at one value of x.

Section 1.4 Continuity and One-Sided Limits

1. (a) The limit does not exist at $x = c$.

(b) The function is not defined at $x = c$.

(c) The limit exists at $x = c$, but it is not equal to the value of the function at $x = c$.

(d) The limit does not exist at $x = c$.

3.

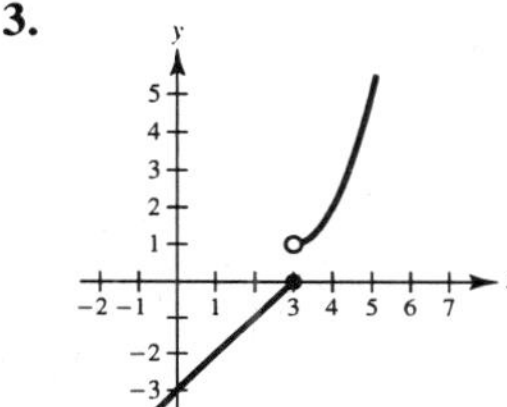

The function is not continuous at $x = 3$ because $\lim\limits_{x\to 3^+} f(x) = 1 \neq 0 = \lim\limits_{x\to 3^-} f(x)$.

5. (a) $\lim\limits_{x\to 3^+} f(x) = 1$

(b) $\lim\limits_{x\to 3^-} f(x) = 1$

(c) $\lim\limits_{x\to 3} f(x) = 1$

7. (a) $\lim\limits_{x\to 3^+} f(x) = 0$

(b) $\lim\limits_{x\to 3^-} f(x) = 0$

(c) $\lim\limits_{x\to 3} f(x) = 0$

9. (a) $\lim\limits_{x\to 3^+} f(x) = 3$

(b) $\lim\limits_{x\to 3^-} f(x) = -3$

(c) $\lim\limits_{x\to 3} f(x)$ does not exist.

11. $\lim\limits_{x\to 5^+} \dfrac{x - 5}{x^2 - 25} = \lim\limits_{x\to 5^+} \dfrac{1}{x + 5} = \dfrac{1}{10}$

13. $\lim\limits_{x\to 2^+} \dfrac{x}{\sqrt{x^2 - 4}}$ does not exist since $\dfrac{x}{\sqrt{x^2 - 4}}$ grows without bound as $x \longrightarrow 2^+$.

15. $\lim\limits_{x\to 0} \dfrac{|x|}{x}$ does not exist since $\lim\limits_{x\to 0^+} \dfrac{|x|}{x} = 1$ and

$\lim\limits_{x\to 0^-} \dfrac{|x|}{x} = -1.$

17. $\lim\limits_{\Delta x\to 0^-} \dfrac{\dfrac{1}{x + \Delta x} - \dfrac{1}{x}}{\Delta x} = \lim\limits_{\Delta x\to 0^-} \dfrac{x - (x + \Delta x)}{x(x + \Delta x)} \cdot \dfrac{1}{\Delta x} = \lim\limits_{\Delta x\to 0^-} \dfrac{-\Delta x}{x(x + \Delta x)} \cdot \dfrac{1}{\Delta x}$

$= \lim\limits_{\Delta x\to 0^-} \dfrac{-1}{x(x + \Delta x)}$

$= \dfrac{-1}{x(x + 0)} = -\dfrac{1}{x^2}$

19. $\lim_{x\to 3^+} f(x) = \lim_{x\to 3^+} \frac{12 - 2x}{3} = 2$

$\lim_{x\to 3^-} f(x) = \lim_{x\to 3^-} \frac{x + 2}{2} = \frac{5}{2}$

$\lim_{x\to 3} f(x)$ does not exist.

21. $\lim_{x\to 1^+} f(x) = \lim_{x\to 1^+} (x + 1) = 2$

$\lim_{x\to 1^-} f(x) = \lim_{x\to 1^-} (x^3 + 1) = 2$

$\lim_{x\to 1} f(x) = 2$

23. $\lim_{x\to \pi} \cot x$ does not exist since

$\lim_{x\to \pi^+} \cot x$ and $\lim_{x\to \pi^-} \cot x$ do not exist.

25. $\lim_{x\to 3^-} (2[\![x]\!] - 1) = 2(2) - 1 = 3$

$([\![x]\!] = 2$ for $2 < x < 3)$

27. $f(x) = \frac{1}{x^2 - 4}$

has discontinuities at $x = -2$ and $x = 2$ since $f(-2)$ and $f(2)$ are not defined.

29. $f(x) = \frac{[\![x]\!]}{2} + x$

has discontinuities at each integer k since $\lim_{x\to k^-} f(x) \neq \lim_{x\to k^+} f(x)$.

31. $f(x) = x^2 - 2x + 1$ is continuous for all real x.

33. $f(x) = x + \sin x$ is continuous for all real x.

35. $f(x) = \frac{1}{x - 1}$

has a nonremovable discontinuity at $x = 1$ since $\lim_{x\to 1} f(x)$ does not exist.

37. $f(x) = \frac{x}{x^2 + 1}$ is continuous for all real x.

39. $f(x) = \frac{x + 2}{(x + 2)(x - 5)}$

has a nonremovable discontinuity at $x = 5$ since $\lim_{x\to 5} f(x)$ does not exist, and has a removable discontinuity at $x = -2$ since

$$\lim_{x\to -2} f(x) = \lim_{x\to -2} \frac{1}{x - 5} = -\frac{1}{7}.$$

41. $f(x) = \frac{|x + 2|}{x + 2}$

has a nonremovable discontinuity at $x = -2$ since $\lim_{x\to -2} f(x)$ does not exist.

43. $f(x) = \begin{cases} x, & x \le 1 \\ x^2, & x > 1 \end{cases}$

has a **possible** discontinuity at $x = 1$.

1. $f(1) = 1$

2. $\left.\begin{array}{l} \lim_{x\to 1^-} f(x) = \lim_{x\to 1^-} x = 1 \\ \lim_{x\to 1^+} f(x) = \lim_{x\to 1^+} x^2 = 1 \end{array}\right\} \lim_{x\to 1} f(x) = 1$

3. $f(1) = \lim_{x\to 1} f(x)$

f is continuous at $x = 1$, therefore, f is continuous for all real x.

45. $f(x) = \begin{cases} \frac{x}{2} + 1, & x \le 2 \\ 3 - x, & x > 2 \end{cases}$ has a **possible** discontinuity at $x = 2$.

1. $f(2) = \frac{2}{2} + 1 = 2$

2. $\left.\begin{array}{l} \lim_{x\to 2^-} f(x) = \lim_{x\to 2^-} \left(\frac{x}{2} + 1\right) = 2 \\ \lim_{x\to 2^+} f(x) = \lim_{x\to 2^+} (3 - x) = 1 \end{array}\right\} \lim_{x\to 2} f(x)$ does not exist.

Therefore, f has a nonremovable discontinuity at $x = 2$.

47. $f(x) = \begin{cases} \csc \frac{\pi x}{6}, & |x - 3| \le 2 \\ 2, & |x - 3| > 2 \end{cases} = \begin{cases} \csc \frac{\pi x}{6}, & 1 \le x \le 5 \\ 2, & x < 1 \text{ or } x > 5 \end{cases}$ has **possible** discontinuities at $x = 1, x = 5$.

1. $f(1) = \csc \frac{\pi}{6} = 2$ $\qquad f(5) = \csc \frac{5\pi}{6} = 2$

2. $\lim_{x\to 1} f(x) = 2$ $\qquad \lim_{x\to 5} f(x) = 2$

3. $f(1) = \lim_{x\to 1} f(x)$ $\qquad f(5) = \lim_{x\to 5} f(x)$

f is continuous at $x = 1$ and $x = 5$, therefore, f is continuous for all real x.

49. $f(x) = \csc 2x$ has nonremovable discontinuities at integer multiples of $\pi/2$.

51. $f(x) = [\![x - 1]\!]$ has nonremovable discontinuities at each integer k.

53. $\lim_{x \to 0^+} f(x) = 0$

$\lim_{x \to 0^-} f(x) = 0$

f is not continuous at $x = -2$.

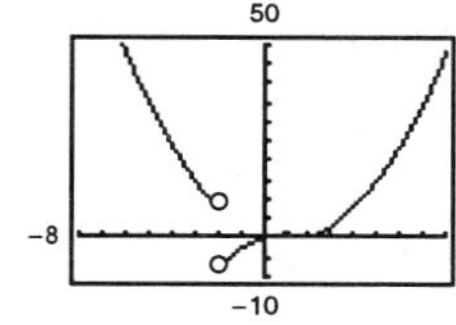

55. $f(2) = 8$

Find a so that $\lim_{x \to 2^+} ax^2 = 8 \Rightarrow a = \frac{8}{2^2} = 2$.

57. Find a and b such that $\lim_{x \to -1^+} (ax + b) = -a + b = 2$ and $\lim_{x \to 3^-} (ax + b) = 3a + b = -2$.

$$a - b = -2$$
$$(+)\ 3a + b = -2$$
$$4a = -4$$
$$a = -1$$
$$b = 2 + (-1) = 1$$

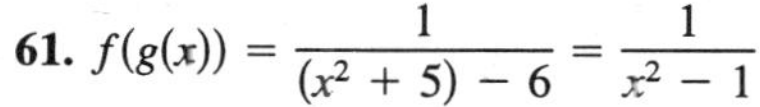

$$f(x) = \begin{cases} 2, & x \le -1 \\ -x + 1, & -1 < z < 3 \\ -2, & x \ge 3 \end{cases}$$

59. $f(g(x)) = (x - 1)^2$

Continuous for all real x.

61. $f(g(x)) = \frac{1}{(x^2 + 5) - 6} = \frac{1}{x^2 - 1}$

Nonremovable discontinuities at $x = \pm 1$

63. $y = [\![x]\!] - x$

Nonremovable discontinuity at each integer

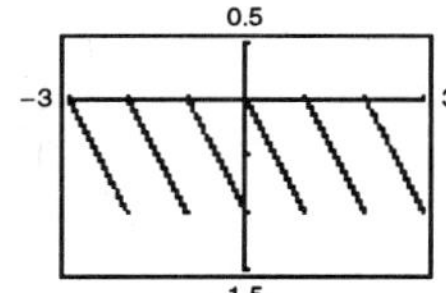

65. $f(x) = \begin{cases} 2x - 4, & x \le 3 \\ x^2 - 2x, & x > 3 \end{cases}$

Nonremovable discontinuity at $x = 3$

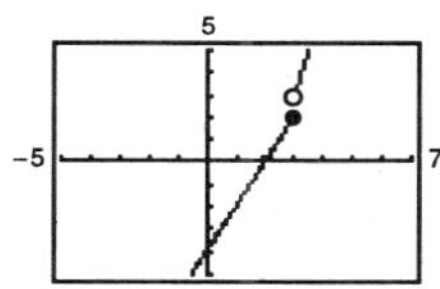

67. $f(x) = \frac{x}{x^2 + 1}$

Continuous on $(-\infty, \infty)$

69. 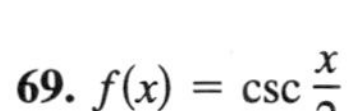 $f(x) = \csc \frac{x}{2}$

Continuous on: $\ldots, (-2\pi, 0), (0, 2\pi), (2\pi, 4\pi), \ldots$

71. $f(x) = \frac{\sin x}{x}$

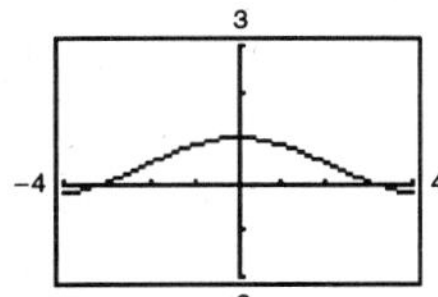

The graph **appears** to be continuous on the interval $[-4, 4]$. Since $f(0)$ is not defined, we know that f has a discontinuity at $x = 0$. This discontinuity is removable so it does not show up on the graph.

73. $f(x) = x^2 - 4x + 3$

$f(x)$ is continuous on $[2, 4]$.

$f(2) = -1$ and $f(4) = 3$

By the Intermediate Value Theorem, $f(c) = 0$ for at least one value of c between 2 and 4.

75. $f(x) = x^3 + x - 1$

$f(x)$ is continuous on $[0, 1]$.

$f(0) = -1$ and $f(1) = 1$

By the Intermediate Value Theorem, $f(x) = 0$ for at least one value of c between 0 and 1. Using a graphing utility, we find that $x \approx 0.6823$.

77. $g(t) = 2\cos t - 3t$.

g is continuous on $[0, 1]$.

$g(0) = 2 > 0$ and $g(1) \approx -1.9 < 0$.

By the Intermediate Value Theorem, $g(t) = 0$ for at least one value c between 0 and 1. Using a graphing utility, we find that $t \approx 0.5636$.

79. $f(x) = x^2 + x - 1$

f is continuous on $[0, 5]$.

$f(0) = -1$ and $f(5) = 29$

$-1 < 11 < 29$

The Intermediate Value Theorem applies.

$$x^2 + x - 1 = 11$$
$$x^2 + x - 12 = 0$$
$$(x + 4)(x - 3) = 0$$
$$x = -4 \text{ or } x = 3$$
$$c = 3 \ (x = -4 \text{ is not in the interval.})$$

Thus, $f(3) = 11$.

81. $f(x) = x^3 - x^2 + x - 2$

f is continuous on $[0, 3]$.

$f(0) = -2$ and $f(3) = 19$

$-2 < 4 < 19$

The Intermediate Value Theorem applies.

$$x^3 - x^2 + x - 2 = 4$$
$$x^3 - x^2 + x - 6 = 0$$
$$(x - 2)(x^2 + x + 3) = 0$$
$$x = 2$$
$$(x^2 + x + 3 \text{ has no real solution.})$$
$$c = 2$$

Thus, $f(2) = 4$.

83. Let $V = \frac{4}{3}\pi r^3$ be the volume of a sphere of radius r.

$$V(1) = \frac{4}{3}\pi \approx 4.19$$

$$V(5) = \frac{4}{3}\pi(5^3) \approx 523.6$$

Since $4.19 < 275 < 523.6$, the Intermediate Value Theorem implies that there is at least one value r between 1 and 5 such that $V(r) = 275$. (In fact, $r \approx 4.0341$.)

85. $N(t) = 25\left(2\left[\!\left[\frac{t+2}{2}\right]\!\right] - t\right)$

t	0	1	1.8	2	3	3.8
$N(t)$	50	25	5	50	25	5

Discontinuous at every positive even integer. The company replenishes its inventory every two months.

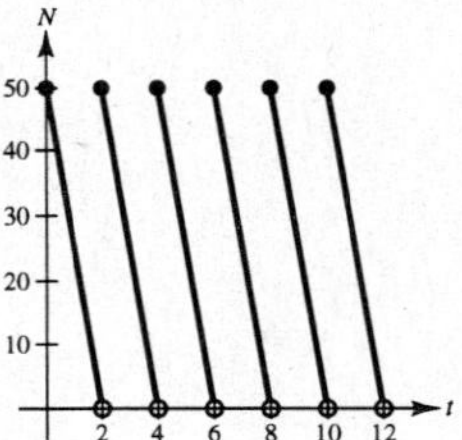

87. Suppose there exists x_1 in $[a, b]$ such that $f(x_1) > 0$ and there exists x_2 in $[a, b]$ such that $f(x_2) < 0$. Then by the Intermediate Value Theorem, $f(x)$ must equal zero for some value of x in $[x_1, x_2]$ (or $[x_2, x_1]$ if $x_2 < x_1$). Thus, f would have a zero in $[a, b]$, which is a contradiction. Therefore, $f(x) > 0$ for all x in $[a, b]$ or $f(x) < 0$ for all x in $[a, b]$.

89. If $x = 0$, then $f(0) = 0$ and $\lim_{x\to 0} f(x) = 0$. Hence, f is continuous at $x = 0$.

If $x \neq 0$, then $\lim_{t\to x} f(t) = 0$ for x rational, whereas $\lim_{t\to x} f(t) = \lim_{t\to x} kt = kx \neq 0$ for x irrational. Hence, f is not continuous for all $x \neq 0$.

91. True

1. $f(c) = L$ is defined.
2. $\lim_{x\to c} f(x) = L$ exists.
3. $f(c) = \lim_{x\to c} f(x)$

All of the conditions for continuity are met.

93. False; a rational function can be written as $P(x)/Q(x)$ where P and Q are polynomials of degree m and n, respectively. It can have, at most, n discontinuities.

95. (a)

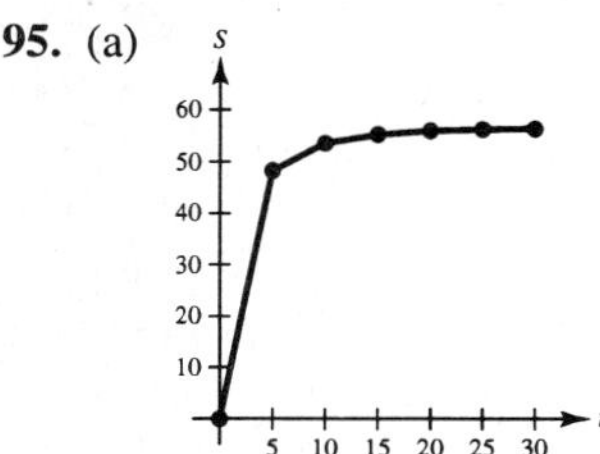

(b) There appears to be a limiting speed and a possible cause is air resistance.

97. Let y be a real number. If $y = 0$, then $x = 0$. If $y > 0$, then let $0 < x_0 < \pi/2$ such that

$$M = \tan x_0 > y$$

(this is possible since the tangent function increases without bound on $[0, \pi/2)$).

By the Intermediate Value Theorem, $f(x) = \tan x$ is continuous on $[0, x_0]$ and $0 < y < M$, which implies that there exists x between 0 and x_0 such that $\tan x = y$. The argument is similar if $y < 0$.

99. $f(x) = \dfrac{\sqrt{x + c^2} - c}{x}, \; c > 0$

Domain: $x + c^2 \geq 0 \Rightarrow x \geq -c^2$ and $x \neq 0$, $[-c^2, 0) \cup (0, \infty)$

$$\lim_{x\to 0} \frac{\sqrt{x+c^2} - c}{x} = \lim_{x\to 0} \frac{\sqrt{x+c^2} - c}{x} \cdot \frac{\sqrt{x+c^2} + c}{\sqrt{x+c^2} + c}$$

$$= \lim_{x\to 0} \frac{(x + c^2) - c^2}{x\left[\sqrt{x+c^2} + c\right]} = \lim_{x\to 0} \frac{1}{\sqrt{x+c^2} + c} = \frac{1}{2c}$$

Define $f(0) = 1/(2c)$ to make f continuous at $x = 0$.

Section 1.5 Infinite Limits

1. $\displaystyle\lim_{x\to -2^+} \frac{1}{(x+2)^2} = \infty$

$\displaystyle\lim_{x\to -2^-} \frac{1}{(x+2)^2} = \infty$

3. $\displaystyle\lim_{x\to -2^+} \tan \frac{\pi x}{4} = -\infty$

$\displaystyle\lim_{x\to -2^-} \tan \frac{\pi x}{4} = \infty$

5. $f(x) = \dfrac{1}{x^2 - 9}$

x	-3.5	-3.1	-3.01	-3.001	-2.999	-2.99	-2.9	-2.5
$f(x)$	0.308	1.639	16.64	166.6	-166.7	-16.69	-1.695	-0.364

$\displaystyle\lim_{x\to -3^-} f(x) = \infty$

$\displaystyle\lim_{x\to -3^+} f(x) = -\infty$

7. $f(x) = \dfrac{x^2}{x^2 - 9}$

x	-3.5	-3.1	-3.01	-3.001	-2.999	-2.99	-2.9	-2.5
$f(x)$	3.769	15.75	150.8	1501	-1499	-149.3	-14.25	-2.273

$\displaystyle\lim_{x\to -3^-} f(x) = \infty$

$\displaystyle\lim_{x\to -3^+} f(x) = -\infty$

9. $\lim_{x\to 0^+} \frac{1}{x^2} = \infty = \lim_{x\to 0^-} \frac{1}{x^2}$

Therefore, $x = 0$ is a vertical asymptote.

11. $\lim_{x\to 2^+} \frac{x^2 - 2}{(x-2)(x+1)} = \infty$

$\lim_{x\to 2^-} \frac{x^2 - 2}{(x-2)(x+1)} = -\infty$

Therefore, $x = 2$ is a vertical asymptote.

$\lim_{x\to -1^+} \frac{x^2 - 2}{(x-2)(x+1)} = \infty$

$\lim_{x\to -1^-} \frac{x^2 - 2}{(x-2)(x+1)} = -\infty$

Therefore, $x = -1$ is a vertical asymptote.

13. $\lim_{x\to -1^+} \frac{x^3}{x^2 - 1} = \infty$

$\lim_{x\to -1^-} \frac{x^3}{x^2 - 1} = -\infty$

Therefore, $x = -1$ is a vertical asymptote.

$\lim_{x\to 1^+} \frac{x^3}{x^2 - 1} = \infty$

$\lim_{x\to 1^-} \frac{x^3}{x^2 - 1} = -\infty$

Therefore, $x = 1$ is a vertical asymptote.

15. $f(x) = \tan 2x = \frac{\sin 2x}{\cos 2x}$ has vertical asymptotes at

$$x = \frac{(2n+1)\pi}{4} = \frac{\pi}{4} + \frac{n\pi}{2},\ n \text{ any integer.}$$

17. $\lim_{t\to 0^+} \left(1 - \frac{4}{t^2}\right) = -\infty = \lim_{t\to 0^-} \left(1 - \frac{4}{t^2}\right)$

Therefore, $t = 0$ is a vertical asymptote.

19. $\lim_{x\to -2^+} \frac{x}{(x+2)(x-1)} = \infty$

$\lim_{x\to -2^-} \frac{x}{(x+2)(x-1)} = -\infty$

Therefore, $x = -2$ is a vertical asymptote.

$\lim_{x\to 1^+} \frac{x}{(x+2)(x-1)} = \infty$

$\lim_{x\to 1^-} \frac{x}{(x+2)(x-1)} = -\infty$

Therefore, $x = 1$ is a vertical asymptote.

21. $f(x) = \frac{x^3 + 1}{x + 1} = \frac{(x+1)(x^2 - x + 1)}{x + 1}$

has no vertical asymptote since

$$\lim_{x\to -1} f(x) = \lim_{x\to -1} (x^2 - x + 1) = 3,$$

not infinity.

23. $s(t) = \frac{t}{\sin t}$ has vertical asymptotes at $t = n\pi$, n a nonzero integer. There is no vertical asymptote at $t = 0$ since

$$\lim_{t\to 0} \frac{t}{\sin t} = 1.$$

25. $\lim_{x\to -1} \frac{x^2 - 1}{x + 1} = \lim_{x\to -1} (x - 1) = -2$

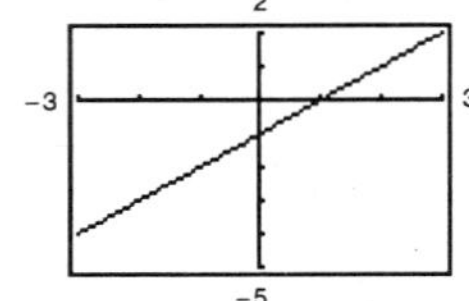

Removable discontinuity at $x = -1$

27. $\lim_{x\to -1^+} \frac{x^2 + 1}{x + 1} = \infty$

$\lim_{x\to -1^-} \frac{x^2 + 1}{x + 1} = -\infty$

Vertical asymptote at $x = -1$

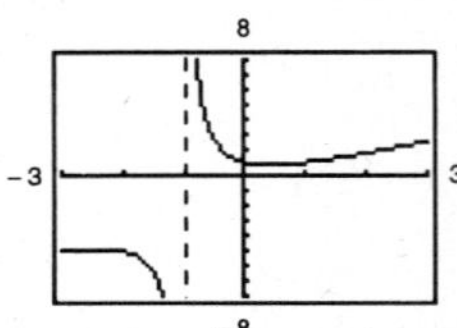

29. $\lim_{x\to 2^+} \frac{x-3}{x-2} = -\infty$

31. $\lim_{x\to 4^+} \frac{x^2}{x^2-16} = \infty$

33. $\lim_{x\to -3^-} \frac{x^2+2x-3}{x^2+x-6} = \lim_{x\to -3^-} \frac{x-1}{x-2} = \frac{4}{5}$

35. $\lim_{x\to 0^-} \left(1 + \frac{1}{x}\right) = -\infty$

37. $\lim_{x\to 0^+} \frac{2}{\sin x} = \infty$

39. $\lim_{x\to 1} \frac{x^2-x}{(x^2+1)(x-1)} = \lim_{x\to 1} \frac{x}{x^2+1} = \frac{1}{2}$

41. $f(x) = \frac{x^2+x+1}{x^3-1}$

$\lim_{x\to 1^+} f(x) = \lim_{x\to 1^+} \frac{1}{x-1} = \infty$

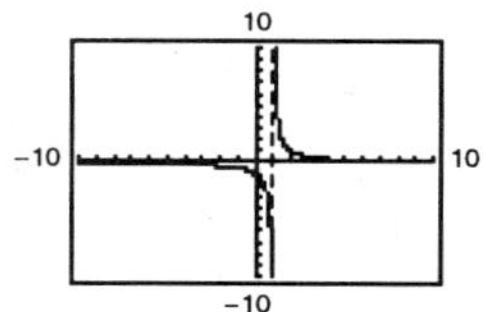

43. $f(x) = \frac{1}{x^2-25}$

$\lim_{x\to 5^-} f(x) = -\infty$

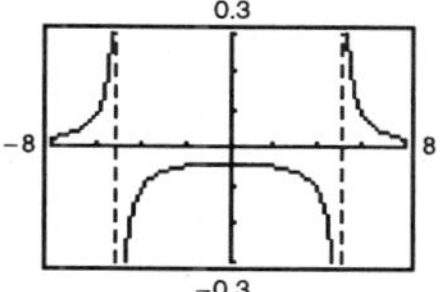

45. $S = \frac{k}{1-r}$, $0 < |r| < 1$. Assume $k \neq 0$.

$\lim_{r\to 1^-} S = \lim_{r\to 1^-} \frac{k}{1-r} = \infty$ (or $-\infty$ if $k < 0$)

47. (a) $r = \frac{2(7)}{\sqrt{625-49}} = \frac{7}{12}$ ft/sec

(b) $r = \frac{2(15)}{\sqrt{625-225}} = \frac{3}{2}$ ft/sec

(c) $\lim_{x\to 25^-} \frac{2x}{\sqrt{625-x^2}} = \infty$

49. $C = \frac{528x}{100-x}$, $0 \le x < 100$

(a) $C(25) = \$176$ million

(b) $C(50) = \$528$ million

(c) $C(75) = \$1584$ million

(d) $\lim_{x\to 100^-} \frac{528}{100-x} = \infty$ Thus, it is not possible.

51. $m = \frac{m_0}{\sqrt{1-(v^2/c^2)}}$

$\lim_{v\to c^-} m = \lim_{v\to c^-} \frac{m_0}{\sqrt{1-(v^2/c^2)}} = \infty$

53. (a) Because the circumference of the motor is half that of the saw arbor, the saw makes $1700/2 = 850$ revolutions per minute.

(c) $2(20 \cot \phi) + 2(10 \cot \phi)$: straight sections. The angle subtended in each circle is

$$2\pi - \left(2\left(\frac{\pi}{2} - \phi\right)\right) = \pi + 2\phi.$$

Thus, the length of the belt around the pulleys is

$$20(\pi + 2\phi) + 10(\pi + 2\phi) = 30(\pi + 2\phi).$$

Total length $= 60 \cot \phi + 30(\pi + 2\phi)$

Domain: $\left(0, \frac{\pi}{2}\right)$

(b) The direction of rotation is reversed.

(d)

ϕ	0.3	0.6	0.9	1.2	1.5
L	306.2	217.9	195.9	189.6	188.5

(e)

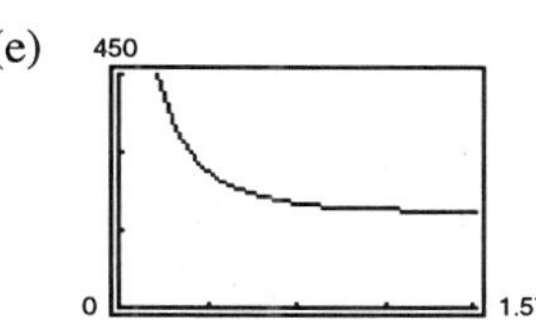

(f) $\lim_{\phi\to(\pi/2)^-} L = 60\pi \approx 188.5$

(All the belts are around pulleys.)

(g) $\lim_{\phi\to 0^+} L = \infty$

55. False; for instance, let

$$f(x) = \frac{x^2 - 1}{x - 1}.$$

The graph of f has a hole at $(1, 2)$, not a vertical asymptote.

57. True

59. Let $f(x) = \frac{1}{x^2}$ and $g(x) = \frac{1}{x^4}$, and $c = 0$.

$$\lim_{x \to 0} \frac{1}{x^2} = \infty \text{ and } \lim_{x \to 0} \frac{1}{x^4} = \infty, \text{ but}$$

$$\lim_{x \to 0}\left(\frac{1}{x^2} - \frac{1}{x^4}\right) = \lim_{x \to 0}\left(\frac{x^2 - 1}{x^4}\right) = -\infty \neq 0.$$

61. Given $\lim_{x \to c} f(x) = \infty$, let $g(x) = 1$. Then

$$\lim_{x \to c} \frac{g(x)}{f(x)} = \lim_{x \to c} \frac{1}{f(x)} = 0$$

by Theorem 1.15.

63. $f(x) = \frac{x + 2}{(x - 1)^2}$

(a) $\delta \approx 0.178$ for $M = 100$.

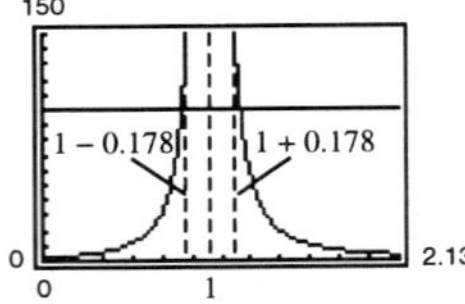

(b) $\delta \approx 0.055$ for $M = 1000$.

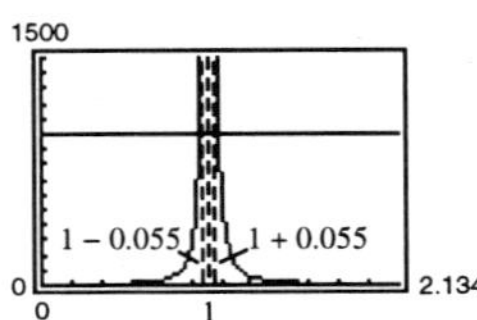

(c) As M increases, δ decreases.

Review Exercises for Chapter 1

1. Calculus required. Using a graphing utility, you can estimate the length to be 8.3. Or, the length is slightly longer than the distance between the two points, 8.25.

3.

x	-0.1	-0.01	-0.001	0.001	0.01	0.1
$f(x)$	-0.26	-0.25	-0.250	-0.2499	-0.249	-0.24

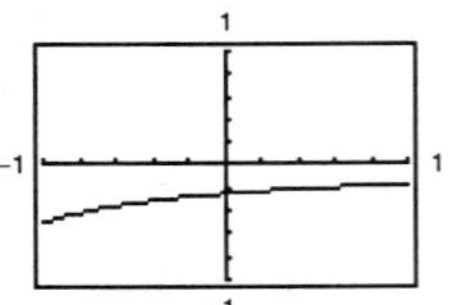

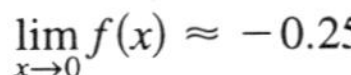

$\lim_{x \to 0} f(x) \approx -0.25$

5. $h(x) = \frac{x^2 - 2x}{x}$

(a) $\lim_{x \to 0} h(x) = -2$

(b) $\lim_{x \to -1} h(x) = -3$

7. $\lim_{x \to 2}(5x - 3) = 5(2) - 3 = 7$

9. $\lim_{x \to 2}(5x - 3)(3x + 5) = [5(2) - 3][3(2) + 5]$

$= 7 \cdot 11 = 77$

11. $\lim_{t \to 3} \frac{t^2 + 1}{t} = \frac{3^2 + 1}{3} = \frac{10}{3}$

13. $\lim_{t \to -2} \frac{t + 2}{t^2 - 4} = \lim_{t \to -2} \frac{1}{t - 2} = -\frac{1}{4}$

15. $\lim_{x \to 0} \frac{[1/(x + 1)] - 1}{x} = \lim_{x \to 0} \frac{1 - (x + 1)}{x(x + 1)}$

$= \lim_{x \to 0} \frac{-1}{x + 1} = -1$

17. $\lim_{x \to -5} \frac{x^3 + 125}{x + 5} = \lim_{x \to -5} \frac{(x + 5)(x^2 - 5x + 25)}{x + 5}$

$= \lim_{x \to -5} (x^2 - 5x + 25) = 75$

19. $\lim_{x \to 0^+}\left(x - \frac{1}{x^3}\right) = -\infty$

21. $\lim_{\Delta x \to 0} \dfrac{\sin[(\pi/6) + \Delta x] - (1/2)}{\Delta x} = \lim_{\Delta x \to 0} \dfrac{\sin(\pi/6)\cos \Delta x + \cos(\pi/6)\sin \Delta x - (1/2)}{\Delta x}$

$$= \lim_{\Delta x \to 0} \frac{1}{2} \cdot \frac{(\cos \Delta x - 1)}{\Delta x} + \lim_{\Delta x \to 0} \frac{\sqrt{3}}{2} \cdot \frac{\sin \Delta x}{\Delta x}$$

$$= 0 + \frac{\sqrt{3}}{2}(1) = \frac{\sqrt{3}}{2}$$

23. $\lim_{x \to -2^-} \dfrac{2x^2 + x + 1}{x + 2} = -\infty$

25. $\lim_{x \to -1^+} \dfrac{x + 1}{x^3 + 1} = \lim_{x \to -1^+} \dfrac{1}{x^2 - x + 1} = \dfrac{1}{3}$

27. $\lim_{x \to 1^-} \dfrac{x^2 + 2x + 1}{x - 1} = -\infty$

29. $\lim_{x \to 0^+} \dfrac{\sin 4x}{5x} = \lim_{x \to 0^+} \left[\dfrac{4}{5}\left(\dfrac{\sin 4x}{4x}\right)\right] = \dfrac{4}{5}$

31. $\lim_{x \to 0^+} \dfrac{\csc 2x}{x} = \lim_{x \to 0^+} \dfrac{1}{x \sin 2x} = \infty$

33. $f(x) = \dfrac{\sqrt{2x + 1} - \sqrt{3}}{x - 1}$

(a)

x	1.1	1.01	1.001	1.0001
$f(x)$	0.5680	0.5764	0.5773	0.5773

$\lim_{x \to 1^+} \dfrac{\sqrt{2x + 1} - \sqrt{3}}{x - 1} \approx 0.577$ (Actual limit is $\sqrt{3}/3$.)

(b)

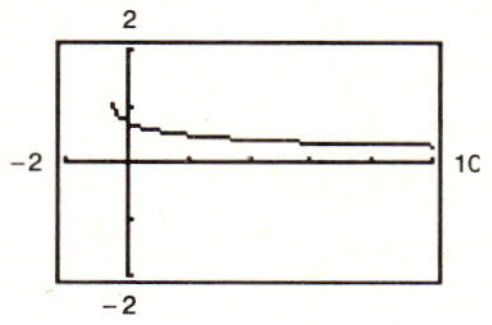

-2 2 10 -2

(c) $\lim_{x \to 1^+} \dfrac{\sqrt{2x + 1} - \sqrt{3}}{x - 1} = \lim_{x \to 1^+} \dfrac{\sqrt{2x + 1} - \sqrt{3}}{x - 1} \cdot \dfrac{\sqrt{2x + 1} + \sqrt{3}}{\sqrt{2x + 1} + \sqrt{3}}$

$$= \lim_{x \to 1^+} \frac{(2x + 1) - 3}{(x - 1)\left(\sqrt{2x + 1} + \sqrt{3}\right)}$$

$$= \lim_{x \to 1^+} \frac{2}{\sqrt{2x + 1} + \sqrt{3}}$$

$$= \frac{2}{2\sqrt{3}} = \frac{1}{\sqrt{3}} = \frac{\sqrt{3}}{3}$$

35. $f(x) = [\![x + 3]\!]$

$\lim_{x \to k^+} [\![x + 3]\!] = k + 3$ where k is an integer.

$\lim_{x \to k^-} [\![x + 3]\!] = k + 2$ where k is an integer.

Nonremovable discontinuity at each integer k

Continuous on $(k, k + 1)$ for all integers k

37. $f(x) = \begin{cases} \dfrac{3x^2 - x - 2}{x - 1}, & x \neq 1 \\ 0, & x = 1 \end{cases}$

$\lim_{x \to 1} f(x) = \lim_{x \to 1} \dfrac{3x^2 - x - 2}{x - 1}$

$= \lim_{x \to 1} (3x + 2) = 5 \neq 0$

Removable discontinuity at $x = 1$

Continuous on $(-\infty, 1) \cup (1, \infty)$

39. $f(x) = \dfrac{1}{(x - 2)^2}$

$\lim_{x \to 2} \dfrac{1}{(x - 2)^2} = \infty$

Nonremovable discontinuity at $x = 2$

Continuous on $(-\infty, 2) \cup (2, \infty)$

41. $f(x) = \dfrac{3}{x + 1}$

$\lim_{x \to 1^-} f(x) = -\infty$

$\lim_{x \to 1^+} f(x) = \infty$

Nonremovable discontinuity at $x = -1$

Continuous on $(-\infty, -1) \cup (-1, \infty)$

43. $f(x) = \csc \dfrac{\pi x}{2}$

Nonremovable discontinuities at each even integer. Continuous on

$(2k, 2k + 2)$

for all integers k.

45. $f(2) = 5$

Find c so that $\lim\limits_{x \to 2^+} (cx + 6) = 5$.

$$c(2) + 6 = 5$$

$$2c = -1$$

$$c = -\frac{1}{2}$$

47. $A = 5000(1.06)^{[\![2t]\!]}$

Nonremovable discontinuity every 6 months

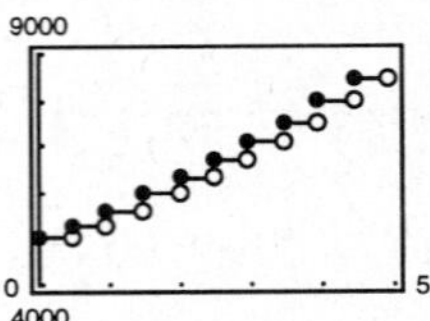

49. $g(x) = 1 + \dfrac{2}{x}$

Vertical asymptote at $x = 0$

51. $f(x) = \dfrac{8}{(x - 10)^2}$

Vertical asymptote at $x = 10$

53. $C = \dfrac{80{,}000p}{100 - p}, \quad 0 \le 0 < 100$

(a) $C(15) \approx \$14{,}117.65$ (b) $C(50) = \$80.000$

(c) $C(90) = \$720{,}000$ (d) $\lim\limits_{p \to 100^-} \dfrac{80{,}000p}{100 - p} = \infty$

55. $\lim\limits_{t \to a} \dfrac{s(a) - s(t)}{a - t} = \lim\limits_{t \to 4} \dfrac{(-4.9(4)^2 + 200) - (-4.9t^2 + 200)}{4 - t}$

$$= \lim_{t \to 4} \frac{4.9(t - 4)(t + 4)}{4 - t}$$

$$= \lim_{t \to 4} -4.9(t + 4) = -39.2 \text{ m/sec}$$

57. $\lim\limits_{x \to 0} \dfrac{|x|}{x} = 1$ is false.

$\lim\limits_{x \to 0} \dfrac{|x|}{x}$ does not exist, since $\lim\limits_{x \to 0^-} \dfrac{|x|}{x} = -1$ and $\lim\limits_{x \to 0^+} \dfrac{|x|}{x} = 1$.

59. True; see Theorem 1.7.

61. True

63. $\lim\limits_{x \to 3} f(x) = 1$ is true since

$\lim\limits_{x \to 3^-} (x - 2) = 1$ and $\lim\limits_{x \to 3^+} (-x^2 + 8x - 14) = 1$.

65. $f(x) = \sqrt{(x - 1)x}$

Domain: $(-\infty, 0] \cup [1, \infty)$

$$\lim_{x \to 0^-} f(x) = 0$$

$$\lim_{x \to 1^+} f(x) = 0$$

CHAPTER 2
Differentiation

CHAPTER 2
Differentiation

Section 2.1 The Derivative and the Tangent Line Problem

Solutions to Odd-Numbered Exercises

1. (a) $m = 0$

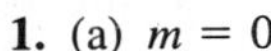

(b) $m = -3$

3. (a), (b)

$y = \frac{f(4) - f(1)}{4 - 1}(x - 1) + f(1) = x + 1$

$f(4) = 5$, $(4, 5)$, $f(4) - f(1) = 3$, $f(1) = 2$, $(1, 2)$

(c) $$y = \frac{f(4) - f(1)}{4 - 1}(x - 1) + f(1)$$
$$= \frac{3}{3}(x - 1) + 2$$
$$= 1(x - 1) + 2$$
$$= x + 1$$

5. $f(x) = 3$

$$f'(x) = \lim_{\Delta x \to 0} \frac{f(x + \Delta x) - f(x)}{\Delta x}$$
$$= \lim_{\Delta x \to 0} \frac{3 - 3}{\Delta x} = \lim_{\Delta x \to 0} 0 = 0$$

7. $f(x) = -5x$

$$f'(x) = \lim_{\Delta x \to 0} \frac{f(x + \Delta x) - f(x)}{\Delta x}$$
$$= \lim_{\Delta x \to 0} \frac{-5(x + \Delta x) - (-5x)}{\Delta x} = \lim_{\Delta x \to 0} -5 = -5$$

9. $f(x) = 2x^2 + x - 1$

$$f'(x) = \lim_{\Delta x \to 0} \frac{f(x + \Delta x) - f(x)}{\Delta x}$$
$$= \lim_{\Delta x \to 0} \frac{[2(x + \Delta x)^2 + (x + \Delta x) - 1] - [2x^2 + x - 1]}{\Delta x}$$
$$= \lim_{\Delta x \to 0} \frac{(2x^2 + 4x\Delta x + 2(\Delta x)^2 + x + \Delta x - 1) - (2x^2 + x - 1)}{\Delta x}$$
$$= \lim_{\Delta x \to 0} \frac{4x\Delta x + 2(\Delta x)^2 + \Delta x}{\Delta x} = \lim_{\Delta x \to 0} (4x + 2\Delta x + 1) = 4x + 1$$

11. $f(x) = x^3 - 12x$

$$f'(x) = \lim_{\Delta x \to 0} \frac{f(x + \Delta x) - f(x)}{\Delta x}$$
$$= \lim_{\Delta x \to 0} \frac{[(x + \Delta x)^3 - 12(x + \Delta x)] - [x^3 - 12x]}{\Delta x}$$
$$= \lim_{\Delta x \to 0} \frac{x^3 + 3x^2\Delta x + 3x(\Delta x)^2 + (\Delta x)^3 - 12x - 12\Delta x - x^3 + 12x}{\Delta x}$$
$$= \lim_{\Delta x \to 0} \frac{3x^2\Delta x + 3x(\Delta x)^2 + (\Delta x)^3 - 12\Delta x}{\Delta x}$$
$$= \lim_{\Delta x \to 0} (3x^2 + 3x\Delta x + (\Delta x)^2 - 12) = 3x^3 - 12$$

13. $f(x) = \dfrac{1}{x-1}$

$$f'(x) = \lim_{\Delta x \to 0} \frac{f(x+\Delta x) - f(x)}{\Delta x}$$

$$= \lim_{\Delta x \to 0} \frac{\dfrac{1}{x+\Delta x-1} - \dfrac{1}{x-1}}{\Delta x}$$

$$= \lim_{\Delta x \to 0} \frac{(x-1)-(x+\Delta x-1)}{\Delta x(x+\Delta x-1)(x-1)}$$

$$= \lim_{\Delta x \to 0} \frac{-\Delta x}{\Delta x(x+\Delta x-1)(x-1)}$$

$$= \lim_{\Delta x \to 0} \frac{-1}{(x+\Delta x-1)(x-1)} = -\frac{1}{(x-1)^2}$$

15. $f(x) = \sqrt{x-4}$

$$f'(x) = \lim_{\Delta x \to 0} \frac{f(x+\Delta x) - f(x)}{\Delta x}$$

$$= \lim_{\Delta x \to 0} \frac{\sqrt{x+\Delta x-4} - \sqrt{x-4}}{\Delta x} \cdot \frac{\sqrt{x+\Delta x-4} + \sqrt{x-4}}{\sqrt{x+\Delta x-4} + \sqrt{x-4}}$$

$$= \lim_{\Delta x \to 0} \frac{(x+\Delta x-4)-(x-4)}{\Delta x[\sqrt{x+\Delta x-4} + \sqrt{x-4}]}$$

$$= \lim_{\Delta x \to 0} \frac{1}{\sqrt{x+\Delta x-4} + \sqrt{x-4}} = \frac{1}{2\sqrt{x-4}}$$

17. (a) $f(x) = x^2 + 1$

$$f'(x) = \lim_{\Delta x \to 0} \frac{f(x+\Delta x) - f(x)}{\Delta x}$$

$$= \lim_{\Delta x \to 0} \frac{[(x+\Delta x)^2 + 1] - [x^2 + 1]}{\Delta x}$$

$$= \lim_{\Delta x \to 0} \frac{2x\Delta x + (\Delta x)^2}{\Delta x}$$

$$= \lim_{\Delta x \to 0} (2x + \Delta x) = 2x$$

At (2, 5), the slope of the tangent line is $m = 2(2) = 4$. The equation of the tangent line is

$$y - 5 = 4(x - 2)$$

$$y - 5 = 4x - 8$$

$$y = 4x - 3.$$

(b)

19. (a) $f(x) = x^3$

$$f'(x) = \lim_{\Delta x \to 0} \frac{f(x+\Delta x) - f(x)}{\Delta x}$$

$$= \lim_{\Delta x \to 0} \frac{(x+\Delta x)^3 - x^3}{\Delta x}$$

$$= \lim_{\Delta x \to 0} \frac{3x^2\Delta x + 3x(\Delta x)^2 + (\Delta x)^3}{\Delta x}$$

$$= \lim_{\Delta x \to 0} (3x^2 + 3x\Delta x + (\Delta x)^2) = 3x^2$$

At (2, 8), the slope of the tangent is $m = 3(2)^2 = 12$. The equation of the tangent line is

$$y - 8 = 12(x - 2)$$

$$y = 12x - 16.$$

(b)

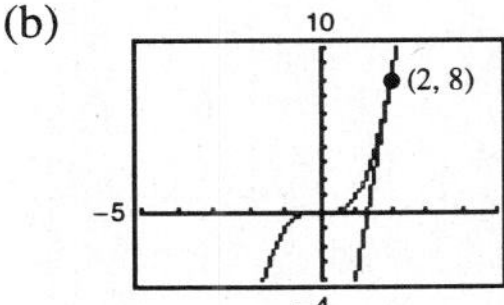

21. (a) $f(x) = x + \frac{1}{x}$

$$f'(x) = \lim_{\Delta x \to 0} \frac{f(x + \Delta x) - f(x)}{\Delta x}$$

$$= \lim_{\Delta x \to 0} \frac{\left[(x + \Delta x) + \frac{1}{x + \Delta x}\right] - \left[x + \frac{1}{x}\right]}{\Delta x}$$

$$= \lim_{\Delta x \to 0} \frac{\Delta x + \frac{x - (x + \Delta x)}{(x + \Delta x)x}}{\Delta x}$$

$$= \lim_{\Delta x \to 0} \left[1 + \frac{-\Delta x}{\Delta x(x + \Delta x)x}\right]$$

$$= \lim_{\Delta x \to 0} \left[1 - \frac{1}{(x + \Delta x)x}\right]$$

$$= 1 - \frac{1}{x^2}$$

At $(1, 2)$, the slope of the tangent line is

$$m = 1 - \frac{1}{1^2} = 0.$$

The equation of the tangent line is

$$y - 2 = 0(x - 1)$$
$$y = 2.$$

(b)

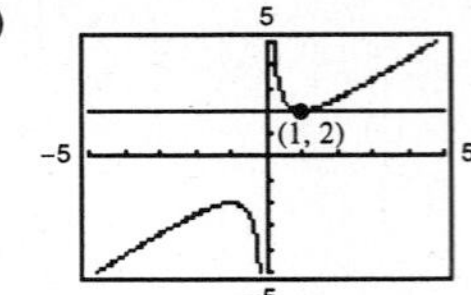

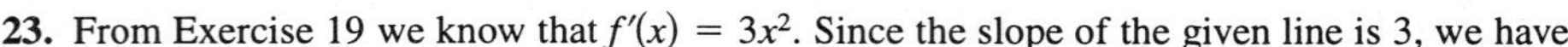

23. From Exercise 19 we know that $f'(x) = 3x^2$. Since the slope of the given line is 3, we have

$$3x^2 = 3$$
$$x = \pm 1.$$

Therefore, at the points $(1, 1)$ and $(-1, -1)$ the tangent lines are parallel to $3x - y + 1 = 0$. These lines have equations

$$y - 1 = 3(x - 1) \quad \text{and} \quad y + 1 = 3(x + 1)$$
$$y = 3x - 2 \qquad\qquad y = 3x + 2.$$

25. Let (x_0, y_0) be a point of tangency on the graph of f. By the limit definition for the derivative, $f'(x) = 4 - 2x$. The slope of the line through $(2, 5)$ and (x_0, y_0) equals the derivative of f at x_0:

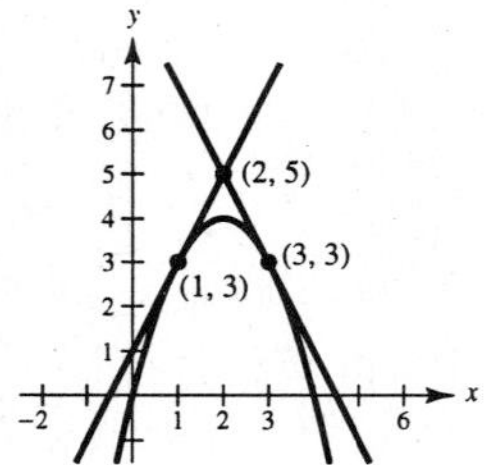

$$\frac{5 - y_0}{2 - x_0} = 4 - 2x_0$$

$$5 - y_0 = (2 - x_0)(4 - 2x_0)$$

$$5 - (4x_0 - x_0^2) = 8 - 8x_0 + 2x_0^2$$

$$0 = x_0^2 - 4x_0 + 3$$

$$0 = (x_0 - 1)(x_0 - 3) \Rightarrow x_0 = 1, 3$$

Therefore, the points of tangency are $(1, 3)$ and $(3, 3)$, and the corresponding slopes are 2 and -2. The equations of the tangent lines are

$$y - 5 = 2(x - 2) \qquad y - 5 = -2(x - 2)$$
$$y = 2x + 1 \qquad\qquad y = -2x + 9$$

27. $f(x) = x$

$f'(x) = 1$

Matches graph (b).

29. $f(x) = \sqrt{x}$

$f'(x) = \dfrac{1}{2\sqrt{x}}$

Matches graph (c).

31. $f(x) = |x| = \begin{cases} x, & \text{if } x \geq 0 \\ -x, & \text{if } x < 0 \end{cases}$

$f'(x) = \begin{cases} 1, & \text{if } x > 0 \\ -1, & \text{if } x < 0 \end{cases}$

Matches graph (f).

33. (a) $g'(0) = -3$

(b) $g'(3) = 0$

(c) Because $g'(1) = -\frac{8}{3}$, g is decreasing (falling) at $x = 1$.

(d) Because $g'(-4) = \frac{7}{3}$, g is increasing (rising) at $x = -4$.

(e) Because $g'(4)$ and $g'(6)$ are both positive, $g(6)$ is greater than $g(4)$, and $g(6) - g(4) > 0$.

(f) No, it is not possible. All you can say is that g is decreasing (falling) at $x = 2$.

35.

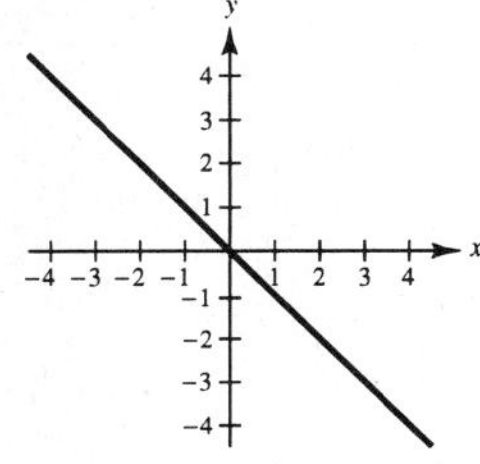

$y = -x$ is one possible answer. (slope is -1)

37. $f(x) = \frac{1}{4}x^3$

By the limit definition of the derivative we have $f'(x) = \frac{3}{4}x^2$.

x	-2	-1.5	-1	-0.5	0	0.5	1	1.5	2
$f(x)$	-2	$-\frac{27}{32}$	$-\frac{1}{4}$	$-\frac{1}{32}$	0	$\frac{1}{32}$	$\frac{1}{4}$	$\frac{27}{32}$	2
$f'(x)$	3	$\frac{27}{16}$	$\frac{3}{4}$	$\frac{3}{16}$	0	$\frac{3}{16}$	$\frac{3}{4}$	$\frac{27}{16}$	3

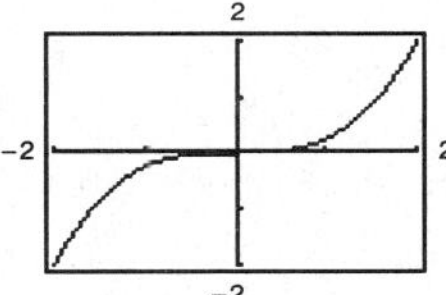

39. $g(x) = \dfrac{f(x + 0.01) - f(x)}{0.01}$

$= (2(x + 0.01) - (x + 0.01)^2 - 2x + x^2) \cdot 100$

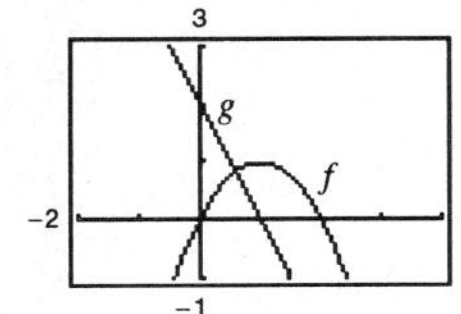

The graph of $g(x)$ is approximately the graph of $f'(x)$.

41. $f(x) = \dfrac{1}{\sqrt{x}}$ and $f'(x) = \dfrac{-1}{2x^{3/2}}$.

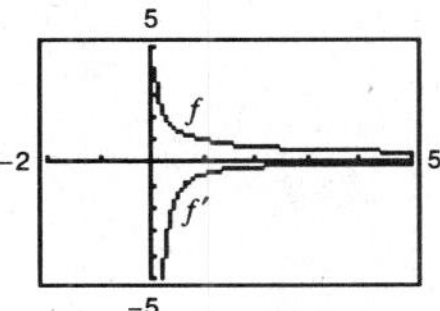

43. $f(x) = 4 - (x - 3)^2$

$S_{\Delta x}(x) = \dfrac{f(2 + \Delta x) - f(2)}{\Delta x}(x - 2) + f(2)$

$= \dfrac{4 - (2 + \Delta x - 3)^2 - 3}{\Delta x}(x - 2) + 3 = \dfrac{1 - (\Delta x - 1)^2}{\Delta x}(x - 2) + 3 = (-\Delta x + 2)(x - 2) + 3$

—CONTINUED—

43. —CONTINUED—

(a) $\Delta x = 1$: $S_{\Delta x} = (x - 2) + 3 = x + 1$

$\Delta x = 0.5$: $S_{\Delta x} = \left(\frac{3}{2}\right)(x - 2) + 3 = \frac{3}{2}x$

$\Delta x = 0.1$: $S_{\Delta x} = \left(\frac{19}{10}\right)(x - 2) + 3 = \frac{19}{10}x - \frac{4}{5}$

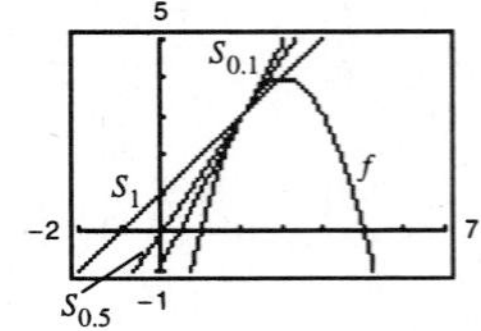

(b) As $\Delta x \to 0$, the line approaches the tangent line to f at (2, 3).

45. $f(x) = x^2 - 1, c = 2$

$$f'(2) = \lim_{x\to2} \frac{f(x) - f(2)}{x - 2} = \lim_{x\to2} \frac{(x^2 - 1) - 3}{x - 2} = \lim_{x\to2} \frac{(x - 2)(x + 2)}{x - 2} = \lim_{x\to2} (x + 2) = 4$$

47. $f(x) = x^3 + 2x^2 + 1, c = -2$

$$f'(-2) = \lim_{x\to-2} \frac{f(x) - f(-2)}{x + 2} = \lim_{x\to-2} \frac{x^2(x + 2)}{x + 2} = \lim_{x\to-2} \frac{(x^3 + 2x^2 + 1) - 1}{x + 2} = \lim_{x\to-2} x^2 = 4$$

49. $f(x) = (x - 1)^{2/3}, c = 1$

$$f'(1) = \lim_{x\to1} \frac{f(x) - f(1)}{x - 1} = \lim_{x\to1} \frac{(x - 1)^{2/3} - 0}{x - 1} = \lim_{x\to1} \frac{1}{(x - 1)^{1/3}}$$

The limit does not exist. Thus, f is not differentiable at $x = 1$.

51. $f(x)$ is differentiable everywhere except at $x = -3$. (Sharp turn in the graph.)

53. $f(x)$ is differentiable everywhere except at $x = -1$. (Discontinuity)

55. $f(x)$ is differentiable everywhere except at $x = 3$. (Sharp turn in the graph)

57. $f(x)$ is differentiable on the interval $(1, \infty)$. (At $x = 1$ the tangent line is vertical.)

59. $f(x)$ is differentiable everywhere except at $x = 0$. (Discontinuity)

61. $f(x) = |x - 1|$

The derivative from the left is $\lim_{x\to1^-} \frac{f(x) - f(1)}{x - 1} = \lim_{x\to1^-} \frac{|x - 1| - 0}{x - 1} = -1$.

The derivative from the right is $\lim_{x\to1^+} \frac{f(x) - f(1)}{x - 1} = \lim_{x\to1^+} \frac{|x - 1| - 0}{x - 1} = 1$.

The one-sided limits are not equal. Therefore, f is not differentiable at $x = 1$.

63. $f(x) = \begin{cases} (x - 1)^3, & x \le 1 \\ (x - 1)^2, & x > 1 \end{cases}$

The derivative from the left is

$$\lim_{x\to1^-} \frac{f(x) - f(1)}{x - 1} = \lim_{x\to1^-} \frac{(x - 1)^3 - 0}{x - 1} = \lim_{x\to1^-} (x - 1)^2 = 0.$$

The derivative from the right is

$$\lim_{x\to1^+} = \frac{f(x) - f(1)}{x - 1} = \lim_{x\to1^+} \frac{(x - 1)^2 - 0}{x - 1} = \lim_{x\to1^+} (x - 1) = 0.$$

These one-sided limits are equal. Therefore, f is differentiable at $x = 1$. $(f'(1) = 0)$

65. Note that f is continuous at $x = 2$. $f(x) = \begin{cases} x^2 + 1, & x \le 2 \\ 4x - 3, & x > 2 \end{cases}$

The derivative from the left is

$$\lim_{x \to 2^-} \frac{f(x) - f(2)}{x - 2} = \lim_{x \to 2^-} \frac{(x^2 + 1) - 5}{x - 2} = \lim_{x \to 2^-} (x + 2) = 4.$$

The derivative from the right is

$$\lim_{x \to 2^+} \frac{f(x) - f(2)}{x - 2} = \lim_{x \to 2^+} \frac{(4x - 3) - 5}{x - 2} = \lim_{x \to 2^+} 4 = 4.$$

The one-sided limits are equal. Therefore, f is differentiable at $x = 2$. $(f'(2) = 4)$

67. (a) The distance from $(3, 1)$ to the line $mx - y + 4 = 0$ is

$$d = \frac{|Ax_1 + By_1 + C|}{\sqrt{A^2 + B^2}}$$

$$= \frac{|m(3) - 1(1) + 4|}{\sqrt{m^2 + 1}} = \frac{|3m + 3|}{\sqrt{m^2 + 1}}.$$

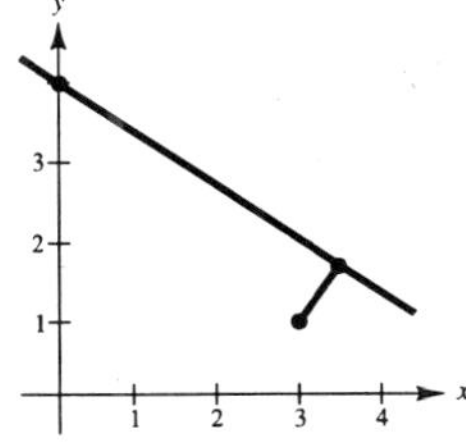

(b)

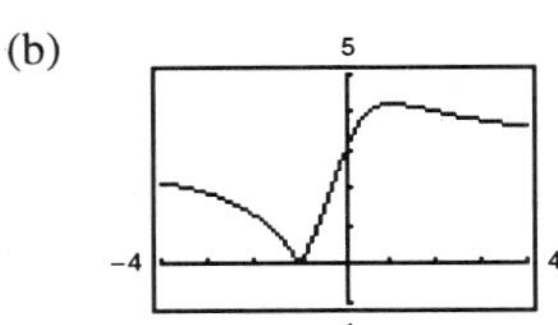

The function d is not differentiable at $m = -1$. This corresponds to the line $y = -x + 4$, which passes through the point $(3, 1)$.

69. False. $y = |x - 2|$ is continuous at $x = 2$, but is not differentiable at $x = 2$. (Sharp turn in the graph)

71. True—see Theorem 2.1

73. $f(x) = \begin{cases} x \sin(1/x), & x \ne 0 \\ 0, & x = 0 \end{cases}$

Using the Squeeze Theorem, we have $-|x| \le x \sin(1/x) \le |x|$, $x \ne 0$. Thus, $\lim_{x \to 0} x \sin(1/x) = 0 = f(0)$ and f is continuous at $x = 0$. Using the alternative form of the derivative we have

$$\lim_{x \to 0} \frac{f(x) - f(0)}{x - 0} = \lim_{x \to 0} \frac{x \sin(1/x) - 0}{x - 0} = \lim_{x \to 0} \left(\sin\frac{1}{x}\right).$$

Since this limit does not exist (it oscillates between -1 and 1), the function is not differentiable at $x = 0$.

$$g(x) = \begin{cases} x^2 \sin(1/x), & x \ne 0 \\ 0, & x = 0 \end{cases}$$

Using the Squeeze Theorem again we have $-x^2 \le x^2 \sin(1/x) \le x^2$, $x \ne 0$. Thus, $\lim_{x \to 0} x^2 \sin(1/x) = 0 = f(0)$ and f is continuous at $x = 0$. Using the alternative form of the derivative again we have

$$\lim_{x \to 0} \frac{f(x) - f(0)}{x - 0} = \lim_{x \to 0} \frac{x^2 \sin(1/x) - 0}{x - 0} = \lim_{x \to 0} x \sin\frac{1}{x} = 0.$$

Therefore, g is differentiable at $x = 0$, $g'(0) = 0$.

Section 2.2 Basic Differentiation Rules and Rates of Change

1. (a) $y = x^{1/2}$, $y' = \frac{1}{2}x^{-1/2}$, $y'(1) = \frac{1}{2}$

(b) $y = x^{3/2}$, $y' = \frac{3}{2}x^{1/2}$, $y'(1) = \frac{3}{2}$

(c) $y = x^2$, $y' = 2x$, $y'(1) = 2$

(d) $y = x^3$, $y' = 3x^2$, $y'(1) = 3$

3. $y = 3$
$y' = 0$

5. $f(x) = x + 1$
$f'(x) = 1$

7. $g(x) = x^2 + 4$
$g'(x) = 2x$

9. $f(t) = -2t^2 + 3t - 6$
$f'(x) = -4t + 3$

11. $s(t) = t^3 - 2t + 4$
$s'(t) = 3t^2 - 2$

13. $y = x^2 - \dfrac{1}{2}\cos x$
$y' = 2x + \dfrac{1}{2}\sin x$

15. $y = \dfrac{1}{x} - 3\sin x$
$y' = -\dfrac{1}{x^2} - 3\cos x$

	Function	*Rewrite*	*Derivative*	*Simplify*
17.	$y = \dfrac{1}{3x^3}$	$y = \dfrac{1}{3}x^{-3}$	$y' = -x^{-4}$	$y' = -\dfrac{1}{x^4}$
19.	$y = \dfrac{1}{(3x)^3}$	$y = \dfrac{1}{27}x^{-3}$	$y' = -\dfrac{1}{9}x^{-4}$	$y' = -\dfrac{1}{9x^4}$
21.	$y = \dfrac{\sqrt{x}}{x}$	$y = x^{-1/2}$	$y' = -\dfrac{1}{2}x^{-3/2}$	$y' = -\dfrac{1}{2x^{3/2}}$

23. $f(x) = \dfrac{1}{x}$, $(1, 1)$
$f'(x) = -\dfrac{1}{x^2}$
$f'(1) = -1$

25. $f(x) = -\dfrac{1}{2} + \dfrac{7}{5}x^3$, $\left(0, -\dfrac{1}{2}\right)$
$f'(x) = \dfrac{21}{5}x^2$
$f'(0) = 0$

27. $y = (2x + 1)^2$, $(0, 1)$
$= 4x^2 + 4x + 1$
$y' = 8x + 4$
$y'(0) = 4$

29. $f(\theta) = 4\sin\theta - \theta$, $(0, 0)$
$f'(\theta) = 4\cos\theta - 1$
$f'(0) = 4(1) - 1 = 3$

31. $f(x) = x^3 - 3x - 2x^{-4}$
$f'(x) = 3x^2 - 3 + 8x^{-5}$
$= 3x^2 - 3 + \dfrac{8}{x^5}$

33. $g(t) = t^2 - 4t^{-1}$
$g'(t) = 2t + 4t^{-2}$
$= 2t + \dfrac{4}{t^2}$

35. $f(x) = \dfrac{x^3 - 3x^2 + 4}{x^2} = x - 3 + 4x^{-2}$
$f'(x) = 1 - \dfrac{8}{x^3} = \dfrac{x^3 - 8}{x^3}$

37. $y = x(x^2 + 1) = x^3 + x$
$y' = 3x^2 + 1$

39. $h(s) = s^{4/5}$
$h'(s) = \dfrac{4}{5}s^{-1/5} = \dfrac{4}{5s^{1/5}}$

41. $f(x) = 4\sqrt{x} + 3\cos x = 4x^{1/2} + 3\cos x$
$f'(x) = 2x^{-1/2} - 3\sin x$
$= \dfrac{2}{\sqrt{x}} - 3\sin x$

43. (a) $y = x^4 - 3x^2 + 2$

$y' = 4x^3 - 6x$

At $(1, 0)$: $y' = 4(1)^3 - 6(1) = -2$.

Tangent line: $y - 0 = -2(x - 1)$

$2x + y - 2 = 0$

(b) 

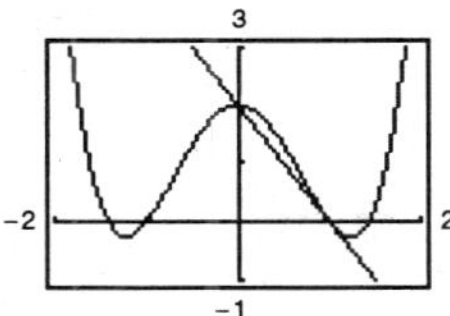

45. (a) $f(x) = \dfrac{1}{\sqrt[3]{x^2}} = x^{-2/3}$

$f'(x) = -\dfrac{2}{3}x^{-5/3} = -\dfrac{2}{3\sqrt[3]{x^5}}$

At $\left(8, \dfrac{1}{4}\right)$: $y' = -\dfrac{2}{3(\sqrt[3]{8})^5} = -\dfrac{1}{48}$.

Tangent line: $y - \dfrac{1}{4} = -\dfrac{1}{48}(x - 8)$

$-48y + 12 = x - 8$

$0 = x + 48y - 20$

(b)

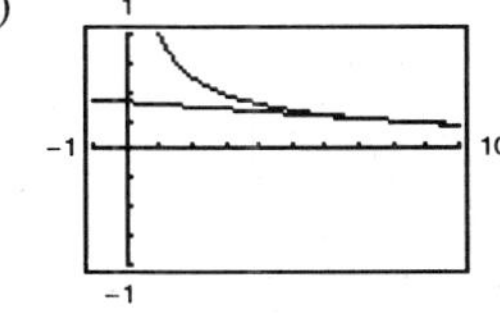

47. $y = x^4 - 8x^2 + 2$

$y' = 4x^3 - 16x$

$= 4x(x^2 - 4)$

$= 4x(x - 2)(x + 2)$

$y' = 0 \implies x = 0, \pm 2$

Horizontal tangents: $(0, 2)$, $(2, -14)$, $(-2, -14)$

49. $y = \dfrac{1}{x^2} = x^{-2}$

$y' = -2x^{-3} = \dfrac{-2}{x^3}$ cannot equal zero.

Therefore, there are no horizontal tangents.

51. $y = x + \sin x,\ 0 \le x < 2\pi$

$y' = 1 + \cos x = 0$

$\cos x = -1 \implies x = \pi$

At $x = \pi$, $y = \pi$.

Horizontal tangent: (π, π)

53. If f is linear then its derivative is a constant function.

$f(x) = ax + b$

$f'(x) = a$

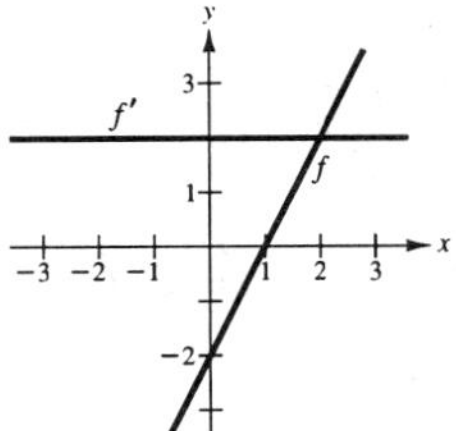

55. Let (x_1, y_1) and (x_2, y_2) be the points of tangency on $y = x^2$ and $y = -x^2 + 6x - 5$, respectively. The derivatives of these functions are

$y' = 2x \implies m = 2x_1$ and $y' = -2x + 6 \implies m = -2x_2 + 6$.

$m = 2x_1 = -2x_2 + 6$

$x_1 = -x_2 + 3$

—CONTINUED—

55. —CONTINUED—

Since $y_1 = x_1^2$ and $y_2 = -x_2^2 + 6x_2 - 5$,

$$m = \frac{y_2 - y_1}{x_2 - x_1} = \frac{(-x_2^2 + 6x_2 - 5) - (-x_1^2)}{x_2 - x_1} = -2x_2 + 6.$$

$$\frac{(-x_2^2 + 6x_2 - 5) - (-x_2 + 3)^2}{x_2 - (-x_2 + 3)} = -2x_2 + 6$$

$$(-x_2^2 + 6x_2 - 5) - (x_2^2 - 6x_2 + 9) = (-2x_2 + 6)(2x_2 - 3)$$

$$-2x_2^2 + 12x_2 - 14 = -4x_2^2 + 18x_2 - 18$$

$$2x_2^2 - 6x_2 + 4 = 0$$

$$2(x_2 - 2)(x_2 - 1) = 0$$

$$x_2 = 1 \text{ or } 2$$

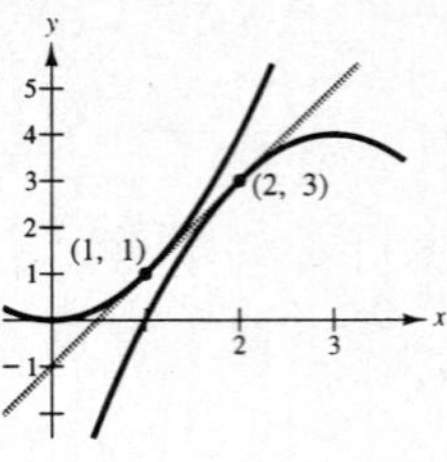

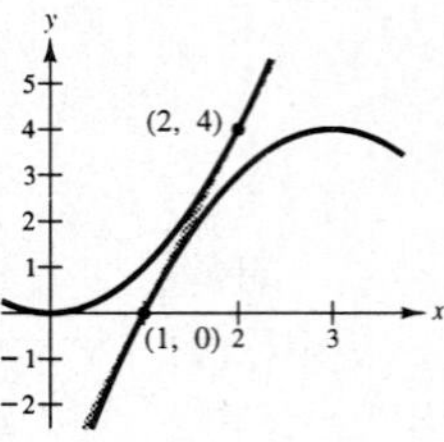

$x_2 = 1 \Rightarrow y_2 = 0, x_1 = 2$ and $y_1 = 4$

Thus, the tangent line through (1, 0) and (2, 4) is

$$y - 0 = \left(\frac{4 - 0}{2 - 1}\right)(x - 1) \Rightarrow y = 4x - 4.$$

$x_2 = 2 \Rightarrow y_2 = 3, x_1 = 1$ and $y_1 = 1$

Thus, the tangent line through (2, 3) and (1, 1) is

$$y - 1 = \left(\frac{3 - 1}{2 - 1}\right)(x - 1) \Rightarrow y = 2x - 1.$$

57. $f(x) = \sqrt{x}, \ (-4, 0)$

$$f'(x) = \frac{1}{2}x^{-1/2} = \frac{1}{2\sqrt{x}}$$

$$\frac{1}{2\sqrt{x}} = \frac{0 - y}{-4 - x}$$

$$4 + x = 2\sqrt{x}y$$

$$4 + x = 2\sqrt{x}\sqrt{x}$$

$$4 + x = 2x$$

$$x = 4, y = 2$$

The point (4, 2) is on the graph of f.

Tangent line: $y - 2 = \dfrac{0 - 2}{-4 - 4}(x - 4)$

$$4y - 8 = x - 4$$

$$0 = x - 4y + 4$$

59. (a) One possible secant is between (3.9, 7.7019) and (4, 8):

$$y - 8 = \frac{8 - 7.7019}{4 - 3.9}(x - 4)$$

$$y - 8 = 2.981(x - 4)$$

$$y = S(x) = 2.981x - 3.924$$

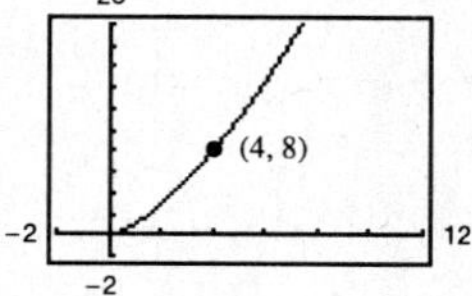

(b) $f'(x) = \frac{3}{2}x^{1/2} \Rightarrow f'(4) = \frac{3}{2}(2) = 3$

$$T(x) = 3(x - 4) + 8 = 3x - 4$$

$S(x)$ is an approximation of the tangent line $T(x)$.

—CONTINUED—

59. —CONTINUED—

(c) As you move further away from (4, 8), the accuracy of the approximation T gets worse.

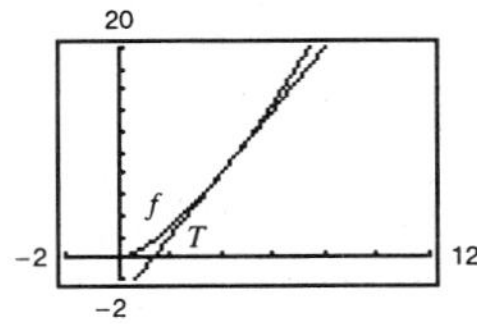

(d)

Δx	-3	-2	-1	-0.5	-0.1	0	0.1	0.5	1	2	3
$f(x)$	1	2.828	5.196	6.548	7.702	8	8.302	9.546	11.180	14.697	18.520
$T(x)$	-1	2	5	6.5	7.7	8	8.3	9.5	11	14	17

61. False. Let $f(x) = x^2$ and $g(x) = x^2 + 4$. Then $f'(x) = g'(x) = 2x$, but $f(x) \neq g(x)$.

63. False. If $y = \pi^2$, then $dy/dx = 0$. (π^2 is a constant.)

65. $f(t) = 2t + 7, [1, 2]$

$f'(t) = 2$

Instantaneous rate of change is the constant 2.
Average rate of change:

$$\frac{f(2) - f(1)}{2 - 1} = \frac{[2(2) + 7] - [2(1) + 7]}{1} = 2$$

(These are the same because f is a line of slope 2.)

67. $f(x) = -\frac{1}{x}, [1, 2]$

$f'(x) = \frac{1}{x^2}$

Instantaneous rate of change:

$(1, -1) \Rightarrow f'(1) = 1$

$\left(2, -\frac{1}{2}\right) \Rightarrow f'(2) = \frac{1}{4}$

Average rate of change:

$$\frac{f(2) - f(1)}{2 - 1} = \frac{(-1/2) - (-1)}{2 - 1} = \frac{1}{2}$$

69.

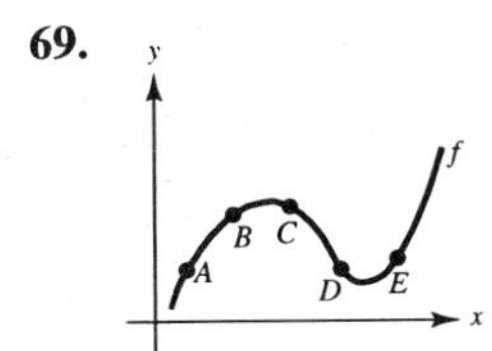

(a) The slope appears to be steepest between A and B.

(b) The average rate of change between A and B is **greater** than the instantaneous rate of change at B.

(c)

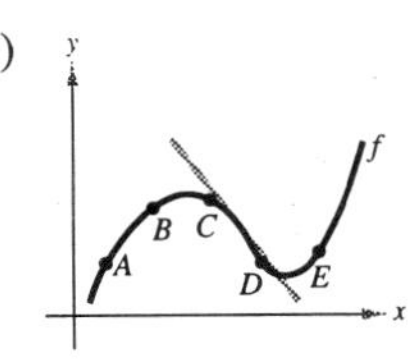

(d) The average rates of change are approximately equal between B and C, and between D and E.

71. (a) $s(t) = -16t^2 + 1362$

$v(t) = -32t$

(b) $\frac{s(2) - s(1)}{2 - 1} = 1298 - 1346 = -48$ ft/sec

(c) $v(t) = s'(t) = -32t$

When $t = 1$: $v(1) = -32$ ft/sec.

When $t = 2$: $v(2) = -64$ ft/sec.

(d) $-16t^2 + 1362 = 0$

$t^2 = \frac{1362}{16} \Rightarrow t = \frac{\sqrt{1362}}{4} \approx 9.226$ sec

(e) $v\left(\frac{\sqrt{1362}}{4}\right) = -32\left(\frac{\sqrt{1362}}{4}\right)$

$= -8\sqrt{1362} \approx -295.242$ ft/sec

73. $s(t) = -4.9t^2 + v_0t + s_0$

$= -4.9t^2 + 120t$

$v(t) = -9.8t + 120$

$v(5) = -9.8(5) + 120 = 71 \text{ m/sec}$

$v(10) = -9.8(10) + 120 = 22 \text{ m/sec}$

75. (The velocity has been converted to miles per hour)

77. $v = 40 \text{ mph} = \frac{2}{3} \text{ mi/min}$

$\left(\frac{2}{3} \text{ mi/min}\right)(6 \text{ min}) = 4 \text{ mi}$

$v = 0 \text{ mph} = 0 \text{ mi/min}$

$(0 \text{ mi/min})(2 \text{ min}) = 0 \text{ mi}$

$v = 60 \text{ mph} = 1 \text{mi/min}$

$(1 \text{ mi/min})(2 \text{ min}) = 2 \text{ mi}$

79. (a) Using a graphing utility, you obtain

$R = 0.167v - 0.02.$

(b) Using a graphing utility, you obtain

$B = 0.00586v^2 - 0.0239v + 0.46.$

(c) $T = R + B = 0.00586v^2 + 0.1431v + 0.44$

(d)

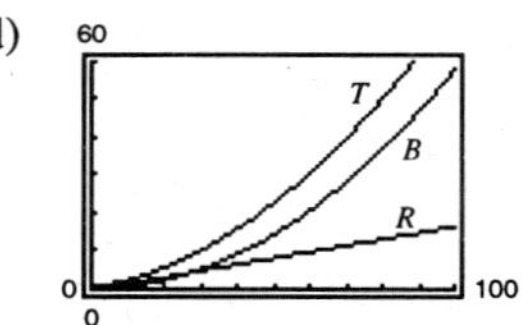

(e) $\dfrac{dT}{dv} = 0.01172v + 0.1431$

For $v = 40$, $T'(40) \approx 0.612$.

For $v = 80$, $T'(80) \approx 1.081$.

For $v = 100$, $T'(100) \approx 1.315$.

(f) For increasing speeds, the total stopping distance increases.

81. $A = s^2, \dfrac{dA}{ds} = 2s$

When $s = 4$ m, $dA/ds = 8 \text{ m}^2$.

83. $C = \dfrac{1{,}008{,}000}{Q} + 6.3Q$

$\dfrac{dC}{dQ} = -\dfrac{1{,}008{,}000}{Q^2} + 6.3$

$C(351) - C(350) \approx 5083.095 - 5085 \approx -\$1.91$

When $Q = 350$, $dC/dQ \approx -\$1.93$.

85. The runner was safe. Explanations will vary.

87. $y = ax^2 + bx + c$

Since the parabola passes through $(0, 1)$ and $(1, 0)$, we have

$(0, 1)$: $1 = a(0)^2 + b(0) + c \Rightarrow c = 1$

$(1, 0)$: $0 = a(1)^2 + b(1) + 1 \Rightarrow b = -a - 1.$

Thus, $y = ax^2 + (-a - 1)x + 1$. From the tangent line $y = x - 1$, we know that the derivative is 1 at the point $(1, 0)$.

$y' = 2ax + (-a - 1)$

$1 = 2a(1) + (-a - 1)$

$1 = a - 1$

$a = 2$

$b = -a - 1 = -3$

Therefore, $y = 2x^2 - 3x + 1$.

89. $y = x^3 - 9x$

$y' = 3x^2 - 9$

Tangent lines through $(1, -9)$:

$$y + 9 = (3x^2 - 9)(x - 1)$$
$$(x^3 - 9x) + 9 = 3x^3 - 3x^2 - 9x + 9$$
$$0 = 2x^3 - 3x^2 = x^2(2x - 3)$$
$$x = 0 \text{ or } x = \tfrac{3}{2}$$

The points of tangency are $(0, 0)$ and $\left(\frac{3}{2}, -\frac{81}{8}\right)$. At $(0, 0)$ the slope is $y'(0) = -9$. At $\left(\frac{3}{2}, -\frac{81}{8}\right)$ the slope is $y'\left(\frac{3}{2}\right) = -\frac{9}{4}$.

Tangent lines:

$$y - 0 = -9(x - 0) \quad \text{and} \quad y + \tfrac{81}{8} = -\tfrac{9}{4}\left(x - \tfrac{3}{2}\right)$$
$$y = -9x \qquad\qquad y = -\tfrac{9}{4}x - \tfrac{27}{4}$$
$$9x + y = 0 \qquad\qquad 9x + 4y + 27 = 0$$

91. $f(x) = \begin{cases} ax^3, & x \le 2 \\ x^2 + b, & x > 2 \end{cases}$

f must be continuous at $x = 2$ to be differentiable at $x = 2$.

$$\left.\begin{aligned} \lim_{x\to 2^-} f(x) &= \lim_{x\to 2^-} ax^3 = 8a \\ \lim_{x\to 2^+} f(x) &= \lim_{x\to 2^+} (x^2 + b) = 4 + b \end{aligned}\right\} \quad \begin{aligned} 8a &= 4 + b \\ 8a - 4 &= b \end{aligned}$$

$$f'(x) = \begin{cases} 3ax^2, & x < 2 \\ 2x, & x > 2 \end{cases}$$

For f to be differentiable at $x = 2$, the left derivative must equal the right derivative.

$$3a(2)^2 = 2(2)$$
$$12a = 4$$
$$a = \tfrac{1}{3}$$
$$b = 8a - 4 = -\tfrac{4}{3}$$

93. Let $f(x) = \cos x$.

$$f'(x) = \lim_{\Delta x\to 0} \frac{f(x + \Delta x) - f(x)}{\Delta x}$$
$$= \lim_{\Delta x\to 0} \frac{\cos x \cos \Delta x - \sin x \sin \Delta x - \cos x}{\Delta x}$$
$$= \lim_{\Delta x\to 0} \frac{\cos x(\cos \Delta x - 1)}{\Delta x} - \lim_{\Delta x\to 0} \sin x\left(\frac{\sin \Delta x}{\Delta x}\right)$$
$$= 0 - \sin x(1) = -\sin x$$

Section 2.3 The Product and Quotient Rules and Higher-Order Derivatives

1. $f(x) = \frac{1}{3}(2x^3 - 4)$

$f'(x) = \frac{1}{3}(6x^2) = 2x^2$

$f'(0) = 0$

3. $f(x) = (x^3 - 3x)(2x^2 + 3x + 5)$

$$f'(x) = (x^3 - 3x)(4x + 3) + (2x^2 + 3x + 5)(3x^2 - 3)$$
$$= 10x^4 + 12x^3 - 3x^2 - 18x - 15$$

$f'(0) = -15$

5. $f(x) = x\cos x$

$$f'(x) = (x)(-\sin x) + (\cos x)(1)$$
$$= \cos x - x\sin x$$
$$f'\left(\frac{\pi}{4}\right) = \frac{\sqrt{2}}{2} - \frac{\pi}{4}\left(\frac{\sqrt{2}}{2}\right)$$
$$= \frac{\sqrt{2}}{8}(4 - \pi)$$

	Function	Rewrite	Derivative	Simplify
7.	$y = \dfrac{x^2 + 2x}{x}$	$y = x + 2,\ x \neq 0$	$y' = 1$	$y' = 1$
9.	$y = \dfrac{7}{3x^3}$	$y = \dfrac{7}{3}x^{-3}$	$y' = -7x^{-4}$	$y' = -\dfrac{7}{x^4}$
11.	$y = \dfrac{3x^2 - 5}{7}$	$y = \dfrac{1}{7}(3x^2 - 5)$	$y' = \dfrac{1}{7}(6x)$	$y' = \dfrac{6x}{7}$

13. $f(x) = \dfrac{3x - 2}{2x - 3}$

$$f'(x) = \frac{(2x - 3)(3) - (3x - 2)(2)}{(2x - 3)^2}$$
$$= \frac{-5}{(2x - 3)^2}$$

15. $f(x) = \dfrac{3 - 2x - x^2}{x^2 - 1}$

$$f'(x) = \frac{(x^2 - 1)(-2 - 2x) - (3 - 2x - x^2)(2x)}{(x^2 - 1)^2}$$
$$= \frac{2x^2 - 4x + 2}{(x^2 - 1)^2} = \frac{2(x - 1)^2}{(x^2 - 1)^2}$$
$$= \frac{2}{(x + 1)^2},\ x \neq 1$$

17. $f(x) = \dfrac{x + 1}{\sqrt{x}}$

$$f'(x) = \frac{\sqrt{x}(1) - (x + 1)\left[1/(2\sqrt{x})\right]}{x}$$
$$= \frac{x - 1}{2x^{3/2}}$$

Alternate solution:

$$f(x) = \frac{x + 1}{\sqrt{x}} = x^{1/2} + x^{-1/2}$$
$$f'(x) = \frac{1}{2}x^{-1/2} - \frac{1}{2}x^{-3/2}$$
$$= \frac{1}{2x^{1/2}} - \frac{1}{2x^{3/2}}$$
$$= \frac{x - 1}{2x^{3/2}}$$

19. $h(s) = (s^3 - 2)^2 = s^6 - 4s^3 + 4$

$$h'(s) = 6s^5 - 12s^2 = 6s^2(s^3 - 2)$$

21. $h(t) = \dfrac{t + 1}{t^2 + 2t + 2}$

$$h'(t) = \frac{(t^2 + 2t + 2)(1) - (t + 1)(2t + 2)}{(t^2 + 2t + 2)^2}$$
$$= \frac{-t^2 - 2t}{(t^2 + 2t + 2)^2}$$

23. $f(x) = (3x^3 + 4x)(x - 5)(x + 1)$

$$f'(x) = (9x^2 + 4)(x - 5)(x + 1) + (3x^3 + 4x)(1)(x + 1) + (3x^3 + 4x)(x - 5)(1)$$
$$= (9x^2 + 4)(x^2 - 4x - 5) + 3x^4 + 3x^3 + 4x^2 + 4x + 3x^4 - 15x^3 + 4x^2 - 20x$$
$$= 9x^4 - 36x^3 - 41x^2 - 16x - 20 + 6x^4 - 12x^3 + 8x^2 - 16x$$
$$= 15x^4 - 48x^3 - 33x^2 - 32x - 20$$

25. $f(x) = \dfrac{x^2 + c^2}{x^2 - c^2}$

$$f'(x) = \frac{(x^2 - c^2)(2x) - (x^2 + c^2)(2x)}{(x^2 - c^2)^2}$$
$$= \frac{-4xc^2}{(x^2 - c^2)^2}$$

27. $f(x) = t^2 \sin t$

$$f'(t) = t^2 \cos t + 2t \sin t$$
$$= t(t \cos t + 2 \sin t)$$

29. $f(t) = \dfrac{\cos t}{t}$

$$f'(t) = \frac{-t \sin t - \cos t}{t^2}$$
$$= -\frac{t \sin t + \cos t}{t^2}$$

31. $f(x) = -x + \tan x$

$$f'(x) = -1 + \sec^2 x = \tan^2 x$$

33. $g(t) = \sqrt{t} + 4 \sec t$

$$g'(t) = \frac{1}{2}t^{-1/2} + 4 \sec t \tan t$$
$$= \frac{1}{2\sqrt{t}} + 4 \sec t \tan t$$

35. $y = 5x \csc x$

$$y' = -5x \csc x \cot x + 5 \csc x$$
$$= 5 \csc x(-x \cot x + 1)$$
$$= 5 \csc x(1 - x \cot x)$$

37. $y = -\csc x - \sin x$

$$y' = \csc x \cot x - \cos x$$
$$= \frac{\cos x}{\sin^2 x} - \cos x$$
$$= \cos x(\csc^2 x - 1)$$
$$= \cos x \cot^2 x$$

39. $y = x^2 \sin x + 2x \cos x$

$$y' = x^2 \cos x + 2x \sin x - 2x \sin x + 2 \cos x$$
$$= x^2 \cos x + 2 \cos x$$

41. $f(x) = x^2 \tan x$

$$f'(x) = x^2 \sec^2 x + 2x \tan x$$
$$= x(x \sec^2 x + 2 \tan x)$$

43. $g(x) = \left(\dfrac{x + 1}{x + 2}\right)(2x - 5)$

$g'(x) = \dfrac{2x^2 + 8x - 1}{(x + 2)^2}$ (form of answer may vary)

45. $g(\theta) = \dfrac{\theta}{1 - \sin \theta}$

$g'(\theta) = \dfrac{1 - \sin \theta + \theta \cos \theta}{(\sin \theta - 1)^2}$ (form of answer may vary)

47. $y = \dfrac{1 + \csc x}{1 - \csc x}$

$$y' = \frac{(1 - \csc x)(-\csc x \cot x) - (1 + \csc x)(\csc x \cot x)}{(1 - \csc x)^2} = \frac{-2 \csc x \cot x}{(1 - \csc x)^2}$$

$$y'\left(\frac{\pi}{6}\right) = \frac{-2(2)(\sqrt{3})}{(1 - 2)^2} = -4\sqrt{3}$$

49. $h(t) = \dfrac{\sec t}{t}$

$$h'(t) = \frac{t(\sec t \tan t) - (\sec t)(1)}{t^2} = \frac{\sec t(t \tan t - 1)}{t^2}$$

$$h'(\pi) = \frac{\sec \pi(\pi \tan \pi - 1)}{\pi^2} = \frac{1}{\pi^2}$$

51. (a) $f(x) = \dfrac{x}{x-1}$, $(2, 2)$

$$f'(x) = \frac{(x-1)(1) - x(1)}{(x-1)^2} = \frac{-1}{(x-1)^2}$$

$$f'(2) = \frac{-1}{(2-1)^2} = -1 = \text{slope at } (2, 2).$$

Tangent line: $y - 2 = -1(x - 2) \Rightarrow y = -x + 4$

(b)

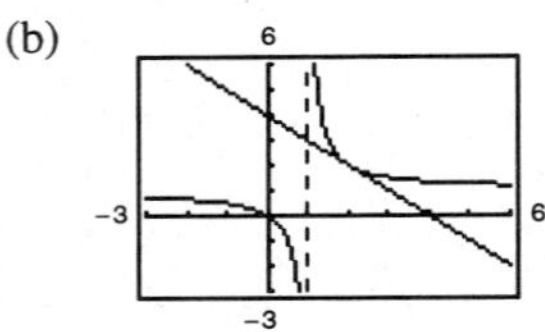

53. (a)

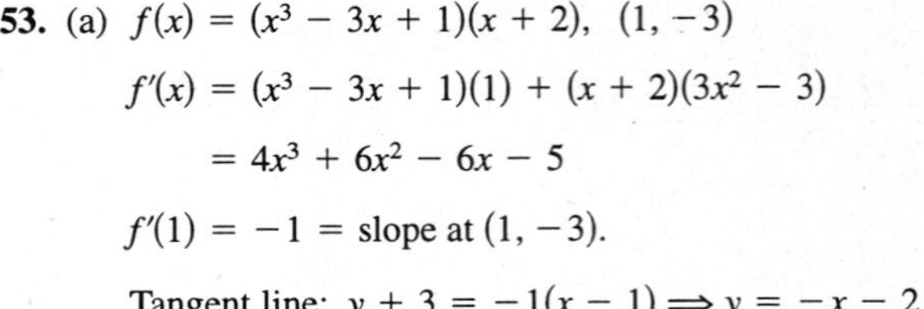

$f(x) = (x^3 - 3x + 1)(x + 2)$, $(1, -3)$

$$f'(x) = (x^3 - 3x + 1)(1) + (x + 2)(3x^2 - 3)$$
$$= 4x^3 + 6x^2 - 6x - 5$$

$f'(1) = -1 = \text{slope at } (1, -3).$

Tangent line: $y + 3 = -1(x - 1) \Rightarrow y = -x - 2$

(b)

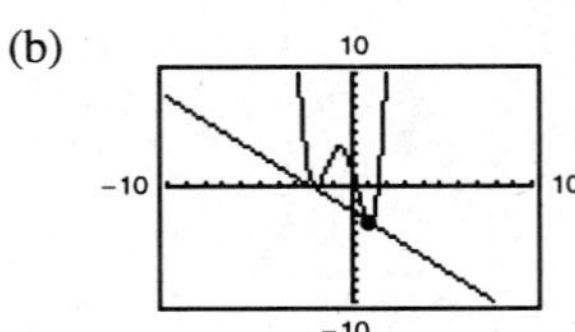

55. (a)

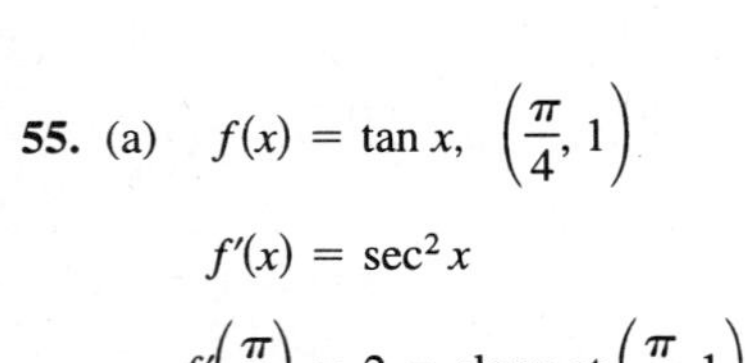

$f(x) = \tan x$, $\left(\dfrac{\pi}{4}, 1\right)$

$f'(x) = \sec^2 x$

$f'\left(\dfrac{\pi}{4}\right) = 2 = \text{slope at } \left(\dfrac{\pi}{4}, 1\right).$

Tangent line:

$$y - 1 = 2\left(x - \frac{\pi}{4}\right)$$

$$y - 1 = 2x - \frac{\pi}{2}$$

$4x - 2y - \pi + 2 = 0$

(b)

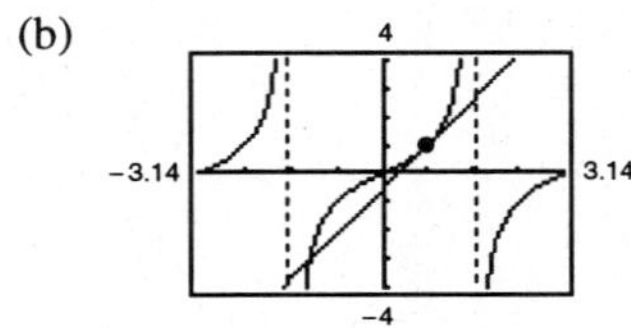

57. $f(x) = \dfrac{x^2}{x-1}$

$$f'(x) = \frac{(x-1)(2x) - x^2(1)}{(x-1)^2} = \frac{x^2 - 2x}{(x-1)^2} = \frac{x(x-2)}{(x-1)^2}$$

$f'(x) = 0$ when $x = 0$ or $x = 2$.

Horizontal tangents are at $(0, 0)$ and $(2, 4)$.

59.

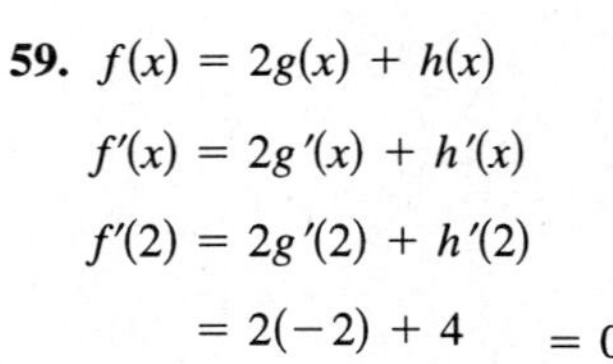

$f(x) = 2g(x) + h(x)$

$$f'(x) = 2g'(x) + h'(x)$$
$$f'(2) = 2g'(2) + h'(2)$$
$$= 2(-2) + 4 = 0$$

61. $f(x) = \dfrac{g(x)}{h(x)}$

$$f'(x) = \frac{h(x)g'(x) - g(x)h'(x)}{[h(x)]^2}$$

$$f'(2) = \frac{h(2)g'(2) - g(2)h'(2)}{[h(2)]^2} = \frac{(-1)(-2) - (3)(4)}{(-1)^2} = -10$$

63. $f(x) = x^n \sin x$

$f'(x) = x^n \cos x + nx^{n-1} \sin x$

$= x^{n-1}(x \cos x + n \sin x)$

When $n = 1$: $f'(x) = x \cos x + \sin x$.

When $n = 2$: $f'(x) = x(x \cos x + 2 \sin x)$.

When $n = 3$: $f'(x) = x^2(x \cos x + 3 \sin x)$.

When $n = 4$: $f'(x) = x^3(x \cos x + 4 \sin x)$.

For general n, $f'(x) = x^{n-1}(x \cos x + n \sin x)$.

65. $C = 100\left(\dfrac{200}{x^2} + \dfrac{x}{x + 30}\right), \quad 1 \le x$

$$\frac{dC}{dx} = 100\left(-\frac{400}{x^3} + \frac{30}{(x + 30)^2}\right)$$

(a) When $x = 10$: $\dfrac{dC}{dx} = -\$38.13$.

(b) When $x = 15$: $\dfrac{dC}{dx} = -\$10.37$.

(c) When $x = 20$: $\dfrac{dC}{dx} = -\$3.80$.

As the order size increases, the cost per item decreases.

67. $P(t) = 500\left[1 + \dfrac{4t}{50 + t^2}\right]$

$$P'(t) = 500\left[\frac{(50 + t^2)(4) - (4t)(2t)}{(50 + t^2)^2}\right] = 500\left[\frac{200 - 4t^2}{(50 + t^2)^2}\right] = 2000\left[\frac{50 - t^2}{(50 + t^2)^2}\right]$$

$P'(2) \approx 31.55$ per hour

69. (a) $\sec x = \dfrac{1}{\cos x}$

$$\frac{d}{dx}[\sec x] = \frac{d}{dx}\left[\frac{1}{\cos x}\right] = \frac{(\cos x)(0) - (1)(-\sin x)}{(\cos x)^2} = \frac{\sin x}{\cos x \cos x} = \frac{1}{\cos x} \cdot \frac{\sin x}{\cos x} = \sec x \tan x$$

(b) $\csc x = \dfrac{1}{\sin x}$

$$\frac{d}{dx}[\csc x] = \frac{d}{dx}\left[\frac{1}{\sin x}\right] = \frac{(\sin x)(0) - (1)(\cos x)}{(\sin x)^2} = -\frac{\cos x}{\sin x \sin x} = -\frac{1}{\sin x} \cdot \frac{\cos x}{\sin x} = -\csc x \cot x$$

(c) $\cot x = \dfrac{\cos x}{\sin x}$

$$\frac{d}{dx}[\cot x] = \frac{d}{dx}\left[\frac{\cos x}{\sin x}\right] = \frac{\sin x(-\sin x) - (\cos x)(\cos x)}{(\sin x)^2} = -\frac{\sin^2 x + \cos^2 x}{\sin^2 x} = -\frac{1}{\sin^2 x} = -\csc^2 x$$

71.

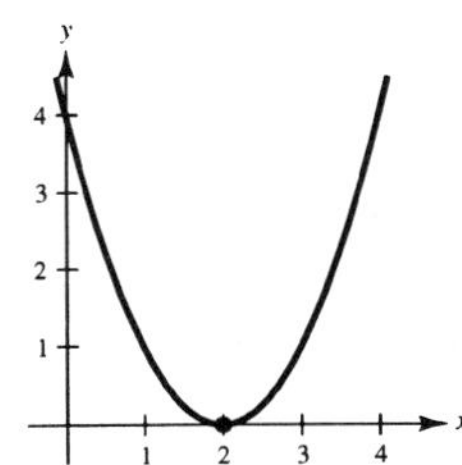

$f(2) = 0$

One such function is

$f(x) = (x - 2)^2$.

73. $f(x) = 4x^{3/2}$

$f'(x) = 6x^{1/2}$

$f''(x) = 3x^{-1/2} = \dfrac{3}{\sqrt{x}}$

75. $f(x) = \dfrac{x}{x - 1}$

$f'(x) = \dfrac{(x - 1)(1) - x(1)}{(x - 1)^2}$

$= \dfrac{-1}{(x - 1)^2}$

$f''(x) = \dfrac{2}{(x - 1)^3}$

77. $f(x) = 3 \sin x$

$f'(x) = 3 \cos x$

$f''(x) = -3 \sin x$

79. $f'(x) = x^2$

$f''(x) = 2x$

81. $f'''(x) = 2\sqrt{x}$

$f^{(4)}(x) = \dfrac{1}{2}(2)x^{-1/2}$

$= \dfrac{1}{\sqrt{x}}$

83. It appears that f is cubic; so f' would be quadratic and f'' would be linear.

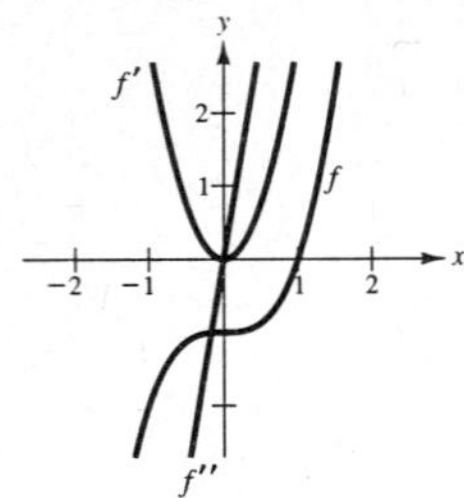

85. $f(x) = g(x)h(x)$

(a) $f'(x) = g(x)h'(x) + h(x)g'(x)$

$$f''(x) = g(x)h''(x) + g'(x)h'(x) + h(x)g''(x) + h'(x)g'(x)$$

$$= g(x)h''(x) + 2g'(x)h'(x) + h(x)g''(x)$$

$$f'''(x) = g(x)h'''(x) + g'(x)h''(x) + 2g'(x)h''(x) + 2g''(x)h'(x) + h(x)g'''(x) + h'(x)g''(x)$$

$$= g(x)h'''(x) + 3g'(x)h''(x) + 3g''(x)h'(x) + g'''(x)h(x)$$

$$f^{(4)}(x) = g(x)h^{(4)}(x) + g'(x)h'''(x) + 3g'(x)h'''(x) + 3g''(x)h''(x) + 3g''(x)h''(x) + 3g'''(x)h'(x) + g'''(x)h'(x) + g^{(4)}(x)h(x)$$

$$= g(x)h^{(4)}(x) + 4g'(x)h'''(x) + 6g''(x)h''(x) + 4g'''(x)h'(x) + g^{(4)}(x)h(x)$$

(b) $$f^{(n)}(x) = g(x)h^{(n)}(x) + \frac{n(n-1)(n-2)\cdots(2)(1)}{1[(n-1)(n-2)\cdots(2)(1)]}g'(x)h^{(n-1)}(x) + \frac{n(n-1)(n-2)\cdots(2)(1)}{(2)(1)[(n-2)(n-3)\cdots(2)(1)]}g''(x)h^{(n-2)}(x)$$

$$+ \frac{n(n-1)(n-2)\cdots(2)(1)}{(3)(2)(1)[(n-3)(n-4)\cdots(2)(1)]}g'''(x)h^{(n-3)}(x) + \cdots$$

$$+ \frac{n(n-1)(n-2)\cdots(2)(1)}{[(n-1)(n-2)\cdots(2)(1)](1)}g^{(n-1)}(x)h'(x) + g^{(n)}(x)h(x)$$

$$= g(x)h^{(n)}(x) + \frac{n!}{1!(n-1)!}g'(x)h^{(n-1)}(x) + \frac{n!}{2!(n-2)!}g''(x)h^{(n-2)}(x) + \cdots$$

$$+ \frac{n!}{(n-1)!1!}g^{(n-1)}(x)h'(x) + g^{(n)}(x)h(x)$$

Note: $n! = n(n-1)\ldots 3 \cdot 2 \cdot 1$ (read "n factorial.")

87. $v(t) = 36 - t^2,\ 0 \le t \le 6$

$a(t) = -2t$

$v(3) = 27$ m/sec

$a(3) = -6$ m/sec

The speed of the object is decreasing, but the rate of the decrease is increasing.

89. $s(t) = -8.25t^2 + 66t$

$v(t) = -16.50t + 66$

$a(t) = -16.50$

t(sec)	0	1	2	3	4
$s(t)$ (ft)	0	57.75	99	123.75	132
$v(t) = s'(t)$ (ft/sec)	66	49.5	33	16.5	0
$a(t) = v'(t)$ (ft/sec²)	−16.5	−16.5	−16.5	−16.5	−16.5

Average velocity on:

$[0, 1]$ is $\dfrac{57.75 - 0}{1 - 0} = 57.75.$

$[1, 2]$ is $\dfrac{99 - 57.75}{2 - 1} = 41.25.$

$[2, 3]$ is $\dfrac{123.75 - 99}{3 - 2} = 24.75.$

$[3, 4]$ is $\dfrac{132 - 123.75}{4 - 3} = 8.25.$

91. $f(x) = \cos x \qquad f\left(\frac{\pi}{3}\right) = \cos\frac{\pi}{3} = \frac{1}{2}$

$f'(x) = -\sin x \qquad f'\left(\frac{\pi}{3}\right) = -\sin\frac{\pi}{3} = -\frac{\sqrt{3}}{2}$

$f''(x) = -\cos x \qquad f''\left(\frac{\pi}{3}\right) = -\cos\frac{\pi}{3} = -\frac{1}{2}$

(a) $P_1(x) = f'(a)(x - a) + f(a) = -\frac{\sqrt{3}}{2}\left(x - \frac{\pi}{3}\right) + \frac{1}{2}$

$$P_2(x) = \frac{1}{2}f''(a)(x - a)^2 + f'(a)(x - a) + f(a)$$

$$= -\frac{1}{4}\left(x - \frac{\pi}{3}\right)^2 - \frac{\sqrt{3}}{2}\left(x - \frac{\pi}{3}\right) + \frac{1}{2}$$

(b)

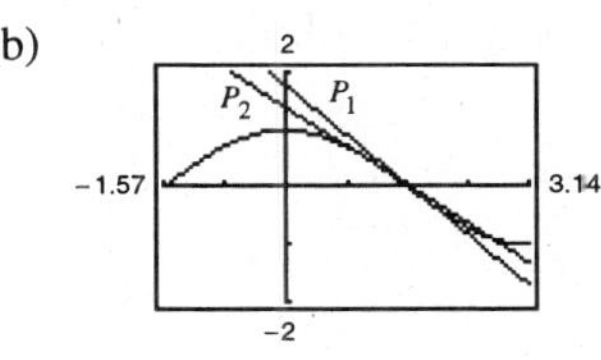

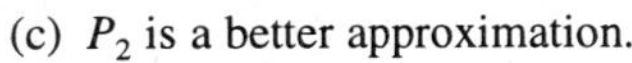

(c) P_2 is a better approximation.

(d) The accuracy worsens as you move farther away from $x = a = (\pi/3)$.

93. False. If $y = f(x)g(x)$, then

$$\frac{dy}{dx} = f(x)g'(x) + g(x)f'(x).$$

95. True

$$h'(c) = f(c)g'(c) + g(c)f'(c)$$
$$= f(c)(0) + g(c)(0) = 0$$

97. True

99. $f(x) = x|x| = \begin{cases} x^2, & \text{if } x \ge 0 \\ -x^2, & \text{if } x < 0 \end{cases}$

$$f'(x) = \begin{Bmatrix} 2x, & \text{if } x \ge 0 \\ -2x, & \text{if } x < 0 \end{Bmatrix} = 2|x|$$

$$f''(x) = \begin{cases} 2, & \text{if } x > 0 \\ -2, & \text{if } x < 0 \end{cases}$$

$f''(0)$ does not exist since the left and right derivatives are not equal.

Section 2.4 The Chain Rule

$y = f(g(x))$	$u = g(x)$	$y = f(u)$
1. $y = (6x - 5)^4$	$u = 6x - 5$	$y = u^4$
3. $y = \sqrt{x^2 - 1}$	$u = x^2 - 1$	$y = \sqrt{u}$
5. $y = \csc^3 x$	$u = \csc x$	$y = u^3$

7. $y = (2x - 7)^3$

$y' = 3(2x - 7)^2(2) = 6(2x - 7)^2$

9. $g(x) = 3(4 - 9x)^4$

$g'(x) = 12(4 - 9x)^3(-9) = -108(4 - 9x)^3$

11. $f(x) = (9 - x^2)^{2/3}$

$f'(x) = \frac{2}{3}(9 - x^2)^{-1/3}(-2x) = -\frac{4x}{3(9 - x^2)^{1/3}}$

13. $f(t) = (1 - t)^{1/2}$

$f'(t) = \frac{1}{2}(1 - t)^{-1/2}(-1) = -\frac{1}{2\sqrt{1 - t}}$

15. $y = (9x^2 + 4)^{1/3}$

$y' = \frac{1}{3}(9x^2 + 4)^{-2/3}(18x) = \frac{6x}{(9x^2 + 4)^{2/3}}$

17. $y = 2(4 - x^2)^{1/2}$

$y' = (4 - x^2)^{-1/2}(-2x) = -\frac{2x}{\sqrt{4 - x^2}}$

19. $y = (x - 2)^{-1}$

$$y' = -1(2 - x)^{-2}(1) = \frac{-1}{(x - 2)^2}$$

21. $f(t) = (t - 3)^{-2}$

$$f'(t) = -2(t - 3)^{-3} = \frac{-2}{(t - 3)^3}$$

23. $y = (x + 2)^{-1/2}$

$$\frac{dy}{dx} = -\frac{1}{2}(x + 2)^{-3/2} = -\frac{1}{2(x + 2)^{3/2}}$$

25. $f(x) = x^2(x - 2)^4$

$$\begin{aligned} f'(x) &= x^2[4(x - 2)^3(1)] + (x - 2)^4(2x) \\ &= 2x(x - 2)^3[2x + (x - 2)] \\ &= 2x(x - 2)^3(3x - 2) \end{aligned}$$

27. $y = x\sqrt{1 - x^2} = x(1 - x^2)^{1/2}$

$$\begin{aligned} y' &= x\left[\frac{1}{2}(1 - x^2)^{-1/2}(-2x)\right] + (1 - x^2)^{1/2}(1) \\ &= -x^2(1 - x^2)^{-1/2} + (1 - x^2)^{1/2} \\ &= (1 - x^2)^{-1/2}[-x^2 + (1 - x^2)] \\ &= \frac{1 - 2x^2}{\sqrt{1 - x^2}} \end{aligned}$$

29. $y = \dfrac{x}{\sqrt{x^2 + 1}} = x(x^2 + 1)^{-1/2}$

$$\begin{aligned} y' &= x\left[-\frac{1}{2}(x^2 + 1)^{-3/2}(2x)\right] + (x^2 + 1)^{-1/2}(1) \\ &= -x^2(x^2 + 1)^{-3/2} + (x^2 + 1)^{-1/2} \\ &= (x^2 + 1)^{-3/2}[-x^2 + (x^2 + 1)] \\ &= \frac{1}{(x^2 + 1)^{3/2}} \end{aligned}$$

31. $y = \dfrac{\sqrt{x} + 1}{x^2 + 1}$

$$y' = \frac{1 - 3x^2 - 4x^{3/2}}{2\sqrt{x}(x^2 + 1)^2}$$

The zero of y' corresponds to the point on the graph of y where the tangent line is horizontal.

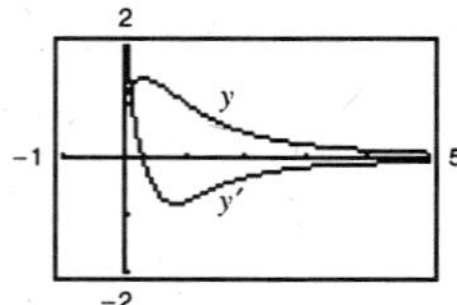

33. $g(t) = \dfrac{3t^2}{\sqrt{t^2 + 2t - 1}}$

$$g'(t) = \frac{3t(t^2 + 3t - 2)}{(t^2 + 2t - 1)^{3/2}}$$

The zeros of g' correspond to the points on the graph of g where the tangent lines are horizontal.

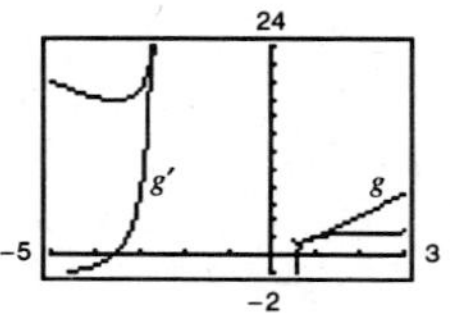

35. $y = \sqrt{\dfrac{x + 1}{x}}$

$$y' = -\frac{\sqrt{(x + 1)/x}}{2x(x + 1)}$$

y' has no zeros.

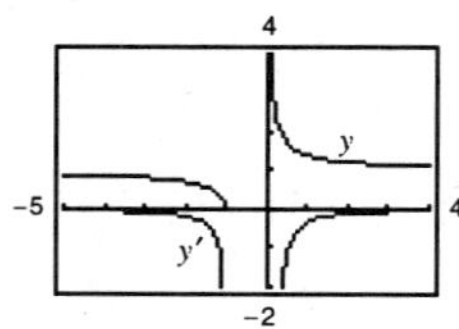

37. $s(t) = \dfrac{-2(2 - t)\sqrt{1 + t}}{3}$

$$s'(t) = \frac{t}{\sqrt{1 + t}}$$

The zero of $s'(t)$ corresponds to the point on the graph of $s(t)$ where the tangent line is horizontal.

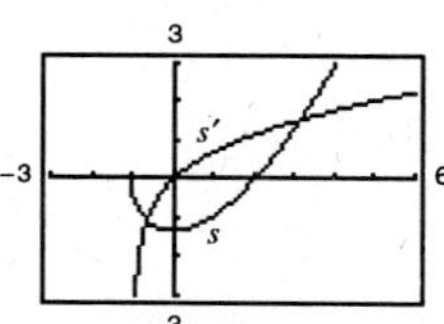

39. $y = \dfrac{\cos \pi x + 1}{x}$

$$\frac{dy}{dx} = \frac{-\pi x \sin \pi x - \cos \pi x - 1}{x^2}$$

$$= -\frac{\pi x \sin \pi x + \cos \pi x + 1}{x^2}$$

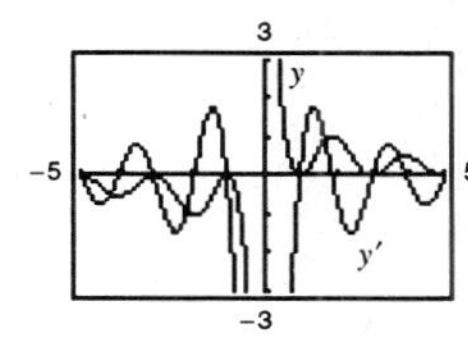

The zeros of y' correspond to the points on the graph of y where the tangent lines are horizontal.

41. (a) $y = \sin x$

$y' = \cos x$

$y'(0) = 1$

1 cycle in $[0, 2\pi]$

(b) $y = \sin 2x$

$y' = 2 \cos 2x$

$y'(0) = 2$

2 cycles in $[0, 2\pi]$

43. $y = \cos 3x$

$$\frac{dy}{dx} = -3 \sin 3x$$

45. $g(x) = 3 \tan 4x$

$g'(x) = 12 \sec^2 4x$

47. $f(\theta) = \frac{1}{4} \sin^2 2\theta = \frac{1}{4}(\sin 2\theta)^2$

$f'(\theta) = 2(\frac{1}{4})(\sin 2\theta)(\cos 2\theta)(2)$

$= \sin 2\theta \cos 2\theta = \frac{1}{2} \sin 4\theta$

49. $y = \sqrt{x} + \dfrac{1}{4} \sin(2x)^2 = \sqrt{x} + \dfrac{1}{4} \sin(4x^2)$

$$\frac{dy}{dx} = \frac{1}{2}x^{-1/2} + \frac{1}{4}\cos(4x^2)(8x) = \frac{1}{2\sqrt{x}} + 2x\cos(2x)^2$$

51. $y = \sin(\cos x)$

$$\frac{dy}{dx} = \cos(\cos x) \cdot (-\sin x) = -\sin x \cos(\cos x)$$

53. $s(t) = (t^2 + 2t + 8)^{1/2}, \quad (2, 4)$

$$s'(t) = \frac{1}{2}(t^2 + 2t + 8)^{-1/2}(2t + 2)$$

$$= \frac{t + 1}{\sqrt{t^2 + 2t + 8}}$$

$$s'(2) = \frac{3}{4}$$

55. $f(x) = \dfrac{3}{x^3 - 4} = 3(x^3 - 4)^{-1}, \quad \left(-1, -\dfrac{3}{5}\right)$

$$f'(x) = -3(x^3 - 4)^{-2}(3x^2) = -\frac{9x^2}{(x^3 - 4)^2}$$

$$f'(-1) = -\frac{9}{25}$$

57. $f(t) = \dfrac{3t + 2}{t - 1}, \quad (0, -2)$

$$f'(t) = \frac{(t - 1)(3) - (3t + 2)(1)}{(t - 1)^2} = \frac{-5}{(t - 1)^2}$$

$$f'(0) = -5$$

59. $y = 37 - \sec^3(2x), \quad (0, 36)$

$$y' = -3\sec^2(2x)[2\sec(2x)\tan(2x)]$$

$$= -6\sec^3(2x)\tan(2x)$$

$$y'(0) = 0$$

61. (a) $f(x) = \sqrt{3x^2 - 2}, \quad (3, 5)$

$$f'(x) = \frac{1}{2}(3x^2 - 2)^{-1/2}(6x)$$

$$= \frac{3x}{\sqrt{3x^2 - 2}}$$

$$f'(3) = \frac{9}{5}$$

Tangent line:

$$y - 5 = \frac{9}{5}(x - 3) \Longrightarrow 9x - 5y - 2 = 0$$

(b)

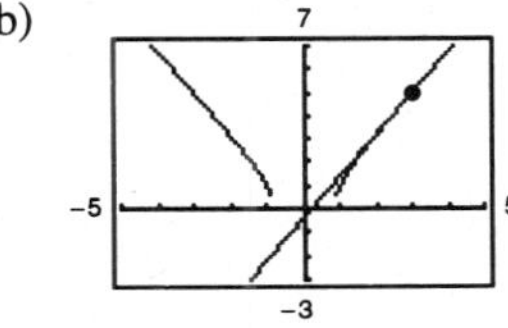

63. (a) $f(x) = \sin 2x, \quad (\pi, 0)$

$f'(x) = 2\cos 2x$

$f'(\pi) = 2$

Tangent line:

$y = 2(x - \pi) \Rightarrow 2x - y - 2\pi = 0$

(b)

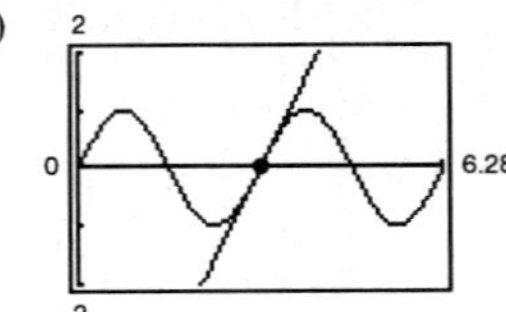

65.

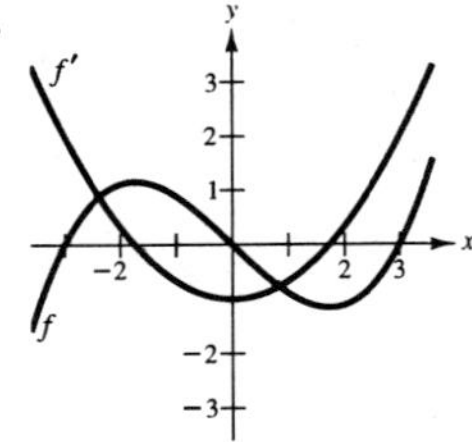

The zeros of f' correspond to the points where the graph of f has horizontal tangents.

67.

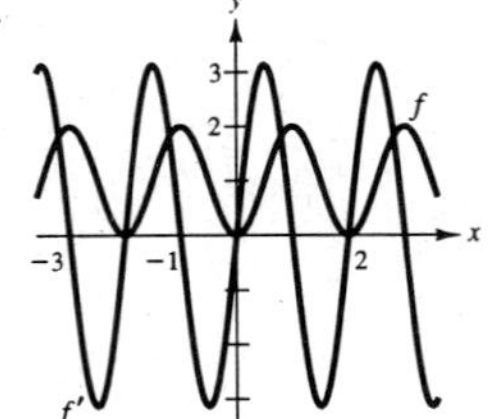

The zeros of f' correspond to the points where the graph of f has horizontal tangents.

69. $f(x) = 2(x^2 - 1)^3$

$f'(x) = 6(x^2 - 1)^2(2x)$

$= 12x(x^4 - 2x^2 + 1)$

$= 12x^5 - 24x^3 + 12x$

$f''(x) = 60x^4 - 72x^2 + 12$

$= 12(5x^2 - 1)(x^2 - 1)$

71. $f(x) = \sin x^2$

$f'(x) = 2x \cos x^2$

$f''(x) = 2x[2x(-\sin x^2)] + 2\cos x^2$

$= 2[\cos x^2 - 2x^2 \sin x^2]$

73. (a) $f(x) = g(x)h(x)$

$f'(x) = g(x)h'(x) + g'(x)h(x)$

$f'(5) = (-3)(-2) + (6)(3) = 24$

(b) $f(x) = g(h(x))$

$f'(x) = g'(h(x))h'(x)$

$f'(5) = g'(3)(-2) = -2g'(3)$

Need $g'(3)$ to find $f'(5)$.

(c) $f(x) = \dfrac{g(x)}{h(x)}$

$f'(x) = \dfrac{h(x)g'(x) - g(x)h'(x)}{[h(x)]^2}$

$f'(5) = \dfrac{(3)(6) - (-3)(-2)}{(3)^2} = \dfrac{12}{9} = \dfrac{4}{3}$

(d) $f(x) = [g(x)]^3$

$f'(x) = 3[g(x)]^2 g'(x)$

$f'(5) = 3(-3)^2(6) = 162$

75. (a) $f = 132{,}400(331 - v)^{-1}$

$f' = (-1)(132{,}400)(331 - v)^{-2}(-1)$

$= \dfrac{132{,}400}{(331 - v)^2}$

When $v = 30, f' \approx 1.461$.

(b) $f = 132{,}400(331 + v)^{-1}$

$f' = (-1)(132{,}400)(331 + v)^{-2}(1)$

$= \dfrac{-132{,}400}{(331 + v)^2}$

When $v = 30, f' \approx -1.016$.

77. $\theta = 0.2\cos 8t$

The maximum angular displacement is $\theta = 0.2$ (since $-1 \le \cos 8t \le 1$).

$\dfrac{d\theta}{dt} = 0.2[-8\sin 8\text{t}] = -1.6\sin 8t$

When $t = 3$, $d\theta/dt = -1.6\sin 24 \approx 1.4489$ radians per second.

79. $S = C(R^2 - r^2)$

$$\frac{dS}{dt} = C\left(2R\frac{dR}{dt} - 2r\frac{dr}{dt}\right)$$

Since r is constant, we have $dr/dt = 0$ and $\frac{dS}{dt} = (1.76 \times 10^5)(2)(1.2 \times 10^{-2})(10^{-5}) = 4.224 \times 10^{-2} = 0.04224.$

81. (a) $g(x) = f(x) - 2 \Rightarrow g'(x) = f'(x)$

(b) $h(x) = 2f(x) \Rightarrow h'(x) = 2f'(x)$

(c) $r(x) = f(-3x) \Rightarrow r'(x) = f'(-3x)(-3) = -3f'(-3x)$

Hence, you need to know $f'(-3x)$.

$r'(0) = -3f'(0) = (-3)\left(-\frac{1}{3}\right) = 1$

$r'(-1) = -3f'(3) = (-3)(-4) = 12$

(d) $s(x) = f(x + 2) \Rightarrow s'(x) = f'(x + 2)$

Hence, you need to know $f'(x + 2)$.

$s'(-2) = f'(0) = -\frac{1}{3}$, etc.

x	-2	-1	0	1	2	3
$f'(x)$	4	$\frac{2}{3}$	$-\frac{1}{3}$	-1	-2	-4
$g'(x)$	4	$\frac{2}{3}$	$-\frac{1}{3}$	-1	-2	-4
$h'(x)$	8	$\frac{4}{3}$	$-\frac{2}{3}$	-2	-4	-8
$r'(x)$		12	1			
$s'(x)$	$-\frac{1}{3}$	-1	-2	-4		

83. $f(x + p) = f(x)$ for all x.

(a) Yes,

$f'(x + p) = f'(x)$,

which shows that f' is periodic as well.

(b) Yes, let

$g(x) = f(2x)$, so $g'(x) = 2f'(2x)$.

Since f' is periodic, so is g'.

85.
$$\begin{aligned} g &= \sqrt{x(x + n)} \\ &= \sqrt{x^2 + nx} \\ \frac{dg}{dx} &= \frac{1}{2}(x^2 + nx)^{-1/2}(2x + n) \\ &= \frac{2x + n}{2\sqrt{x^2 + nx}} \\ &= \frac{(2x + n)/2}{\sqrt{x(x + n)}} \\ &= \frac{[x + (x + n)]/2}{\sqrt{x(x + n)}} = \frac{a}{g} \end{aligned}$$

87. $g(x) = |2x - 3|$

$$g'(x) = 2\left(\frac{2x - 3}{|2x - 3|}\right), \quad x \neq \frac{3}{2}$$

89. $h(x) = |x| \cos x$

$$h'(x) = -|x| \sin x + \frac{x}{|x|}\cos x, \quad x \neq 0$$

91. (a) $f(x) = \sin\frac{x}{2}$ $\qquad f(\pi) = \sin\frac{\pi}{2} = 1$

$f'(x) = \frac{1}{2}\cos\frac{x}{2}$ $\qquad f'(\pi) = \frac{1}{2}\cos\frac{\pi}{2} = 0$

$f''(x) = -\frac{1}{4}\sin\frac{x}{2}$ $\qquad f''(\pi) = -\frac{1}{4}\sin\frac{\pi}{2} = -\frac{1}{4}$

$P_1(x) = f'(\pi)(x - \pi) + f(\pi) = f(\pi) = 1$

$$\begin{aligned} P_2(x) &= \frac{1}{2}f''(\pi)(x - \pi)^2 + f'(\pi)(x - \pi) + f(\pi) \\ &= \frac{1}{2}\left(-\frac{1}{4}\right)(x - \pi)^2 + 1 = 1 - \frac{1}{8}(x - \pi)^2 \end{aligned}$$

(b)

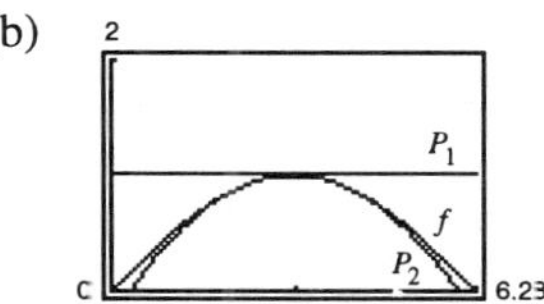

(c) P_2 is a better approximation than P_1.

(d) The accuracy worsens as you move away from $x = \pi$.

93. False. If $y = (1 - x)^{1/2}$, then $y' = \frac{1}{2}(1 - x)^{-1/2}(-1)$.

95. True

Section 2.5 Implicit Differentiation

1. $x^2 + y^2 = 16$

$2x + 2yy' = 0$

$y' = \frac{-x}{y}$

3. $x^{1/2} + y^{1/2} = 9$

$\frac{1}{2}x^{-1/2} + \frac{1}{2}y^{-1/2}y' = 0$

$y' = -\frac{x^{-1/2}}{y^{-1/2}} = -\sqrt{\frac{y}{x}}$

5. $x^3 - xy + y^2 = 4$

$3x^2 - xy' - y + 2yy' = 0$

$(2y - x)y' = y - 3x^2$

$y' = \frac{y - 3x^2}{2y - x}$

7. $x^3y^3 - y - x = 0$

$3x^3y^2y' + 3x^2y^3 - y' - 1 = 0$

$(3x^3y^2 - 1)y' = 1 - 3x^2y^3$

$y' = \frac{1 - 3x^2y^3}{3x^3y^2 - 1}$

9. $x^3 - 2x^2y + 3xy^2 = 38$

$3x^2 - 2x^2y' - 4xy + 6xyy' + 3y^2 = 0$

$2x(3y - x)y' = 4xy - 3x^2 - 3y^2$

$y' = \frac{4xy - 3x^2 - 3y^2}{2x(3y - x)}$

11. $\sin x + 2\cos 2y = 1$

$\cos x - 4(\sin 2y)y' = 0$

$y' = \frac{\cos x}{4\sin 2y}$

13. $\sin x = x(1 + \tan y)$

$\cos x = x(\sec^2 y)y' + (1 + \tan y)(1)$

$y' = \frac{\cos x - \tan y - 1}{x\sec^2 y}$

15. $y = \sin(xy)$

$y' = [xy' + y]\cos(xy)$

$y' - x\cos(xy)y' = y\cos(xy)$

$y' = \frac{y\cos(xy)}{1 - x\cos(xy)}$

17. $xy = 4$

$xy' + y(1) = 0$

$xy' = -y$

$y' = \frac{-y}{x}$

At $(-4, -1)$: $y' = -\frac{1}{4}$

19. $y^2 = \frac{x^2 - 9}{x^2 + 9}$

$2yy' = \frac{(x^2 + 9)(2x) - (x^2 - 9)2x}{(x^2 + 9)^2}$

$y' = \frac{18x}{(x^2 + 9)^2 y}$

At $(3, 0)$: y' is undefined.

21. $x^{2/3} + y^{2/3} = 5$

$\frac{2}{3}x^{-1/3} + \frac{2}{3}y^{-1/3}y' = 0$

$y' = \frac{-x^{-1/3}}{y^{-1/3}} = -\sqrt[3]{\frac{y}{x}}$

At $(8, 1)$: $y' = -\frac{1}{2}$.

23. $\tan(x + y) = x$

$(1 + y')\sec^2(x + y) = 1$

$y' = \frac{1 - \sec^2(x + y)}{\sec^2(x + y)}$

$= \frac{-\tan^2(x + y)}{\tan^2(x + y) + 1}$

$= -\frac{x^2}{x^2 + 1}$

At $(0, 0)$: $y' = 0$.

25. $\sqrt{x} + \sqrt{y} = 3$

$\frac{1}{2}x^{-1/2} + \frac{1}{2}y^{-1/2}y' = 0$

$y' = -\frac{(1/2)x^{-1/2}}{(1/2)y^{-1/2}} = -\frac{\sqrt{y}}{\sqrt{x}}$

At $(4, 1)$: $y' = -\frac{1}{2}$.

Tangent line:

$y - 1 = -\frac{1}{2}(x - 4)$

$y = -\frac{1}{2}x + 3$

$x + 2y - 6 = 0$

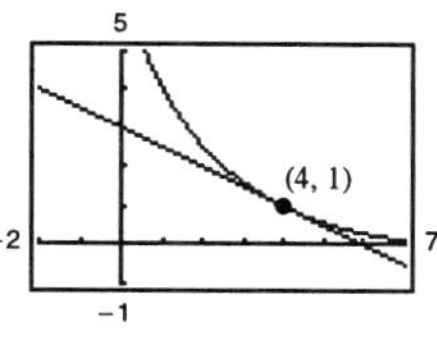

27. $(x^2 + 4)y = 8$

$(x^2 + 4)y' + y(2x) = 0$

$y' = \frac{-2xy}{x^2 + 4}$

$= \frac{-2x[8/(x^2 + 4)]}{x^2 + 4}$

$= \frac{-16x}{(x^2 + 4)^2}$

At $(2, 1)$: $y' = \frac{-32}{64} = -\frac{1}{2}$

$\left(\text{Or, you could just solve for } y\text{: } y = \frac{8}{x^2 + 4}\right)$

29. $(x^2 + y^2)^2 = 4x^2y$

$2(x^2 + y^2)(2x + 2yy') = 4x^2y' + y(8x)$

$4x^3 + 4x^2yy' + 4xy^2 + 4y^3y' = 4x^2y' + 8xy$

$4x^2yy' + 4y^3y' - 4x^2y' = 8xy - 4x^3 - 4xy^2$

$4y'(x^2y + y^3 - x^2) = 4(2xy - x^3 - xy^2)$

$y' = \frac{2xy - x^3 - xy^2}{x^2y + y^3 - x^2}$

At $(1, 1)$: $y' = 0$.

31. (a) $x^2 + y^2 = 16$

$y^2 = 16 - x^2$

$y = \pm\sqrt{16 - x^2}$

(b)

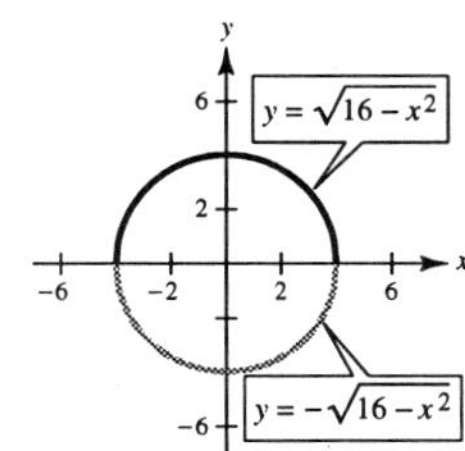

(c) Explicitly:

$\frac{dy}{dx} = \pm\frac{1}{2}(16 - x^2)^{-1/2}(-2x)$

$= \frac{\mp x}{\sqrt{16 - x^2}} = \frac{-x}{\pm\sqrt{16 - x^2}} = \frac{-x}{y}$

(d) Implicitly:

$2x + 2yy' = 0$

$y' = -\frac{x}{y}$

33. (a) $16y^2 = 144 - 9x^2$

$y^2 = \frac{1}{16}(144 - 9x^2) = \frac{9}{16}(16 - x^2)$

$y = \pm\frac{3}{4}\sqrt{16 - x^2}$

(b)

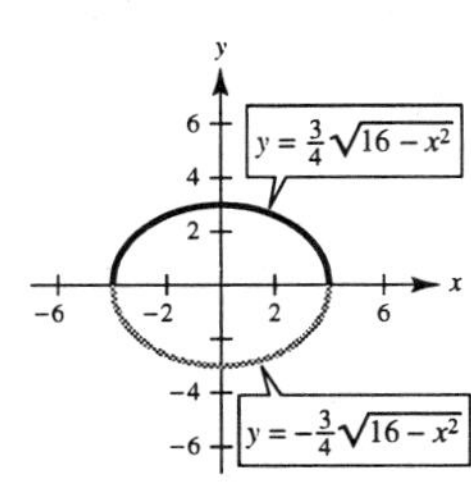

(c) Explicitly:

$\frac{dy}{dx} = \pm\frac{3}{8}(16 - x^2)^{-1/2}(-2x)$

$= \mp\frac{3x}{4\sqrt{16 - x^2}} = \frac{-3x}{4(4/3)y} = \frac{-9x}{16y}$

(d) Implicitly:

$18x + 32yy' = 0$

$y' = \frac{-9x}{16y}$

35.
$$x^2 + xy = 5$$
$$2x + xy' + y = 0$$
$$y' = \frac{-(2x + y)}{x}$$
$$2 + xy'' + y' + y' = 0$$
$$xy'' = -2(1 + y')$$
$$y'' = \frac{-2[1 - (2x + y)/x]}{x} = \frac{2(x + y)}{x^2}$$
$$y'' = \frac{2(x + y)}{x^2} \cdot \frac{x}{x} = \frac{10}{x^3}$$

(**Note:** You could write $y = (5 - x^2)/x$ and calculate y' and y'' directly.)

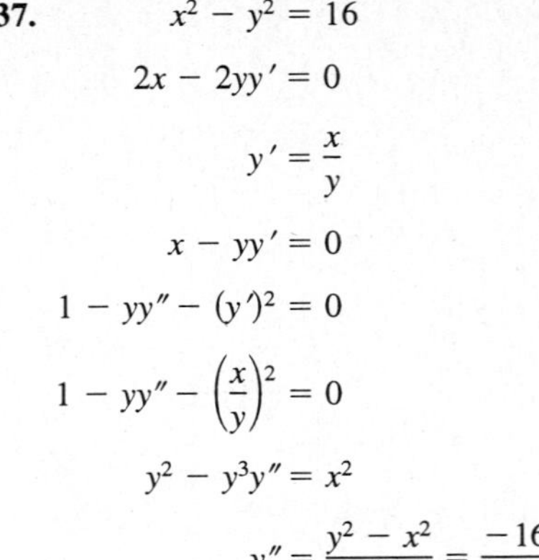

37.
$$x^2 - y^2 = 16$$
$$2x - 2yy' = 0$$
$$y' = \frac{x}{y}$$
$$x - yy' = 0$$
$$1 - yy'' - (y')^2 = 0$$
$$1 - yy'' - \left(\frac{x}{y}\right)^2 = 0$$
$$y^2 - y^3y'' = x^2$$
$$y'' = \frac{y^2 - x^2}{y^3} = \frac{-16}{y^3}$$

39. $y^2 = x^3$

$$2yy' = 3x^2$$
$$y' = \frac{3x^2}{2y} = \frac{3x^2}{2y} \cdot \frac{xy}{xy} = \frac{3y}{2x} \cdot \frac{x^3}{y^2} = \frac{3y}{2x}$$
$$y'' = \frac{2x(3y') - 3y(2)}{4x^2} = \frac{2x[3 \cdot (3y/2x)] - 6y}{4x^2} = \frac{3y}{4x^2} = \frac{3x}{4y}$$

41. $x^2 + y^2 = 25$

$$y' = \frac{-x}{y}$$

At (4, 3):

Tangent line: $y - 3 = \frac{-4}{3}(x - 4) \Rightarrow 4x + 3y - 25 = 0$

Normal line: $y - 3 = \frac{3}{4}(x - 4) \Rightarrow 3x - 4y = 0.$

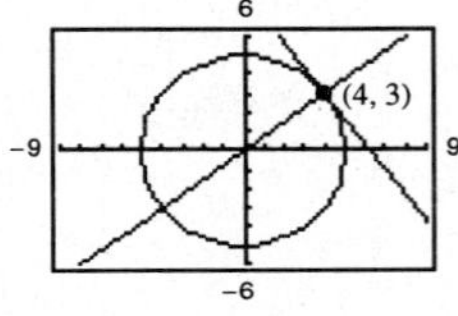

At (−3, 4):

Tangent line: $y - 4 = \frac{3}{4}(x + 3) \Rightarrow 3x - 4y + 25 = 0$

Normal line: $y - 4 = \frac{-4}{3}(x + 3) \Rightarrow 4x + 3y = 0.$

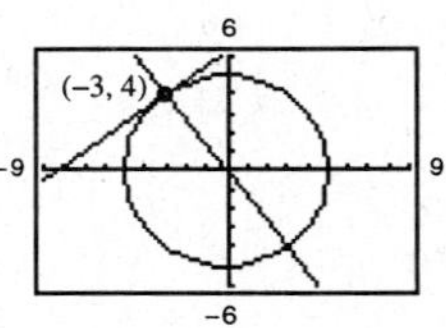

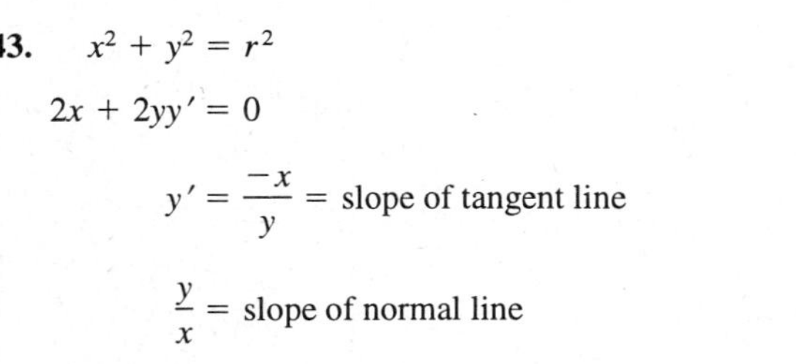

43. $x^2 + y^2 = r^2$

$$2x + 2yy' = 0$$
$$y' = \frac{-x}{y} = \text{slope of tangent line}$$
$$\frac{y}{x} = \text{slope of normal line}$$

Let (x_0, y_0) be a point on the circle. If $x_0 = 0$, then the tangent line is horizontal, the normal line is vertical and, hence, passes through the origin. If $x_0 \neq 0$, then the equation of the normal line is

$$y - y_0 = \frac{y_0}{x_0}(x - x_0)$$
$$y = \frac{y_0}{x_0}x$$

which passes through the origin.

45. $25x^2 + 16y^2 + 200x - 160y + 400 = 0$

$$50x + 32yy' + 200 - 160y' = 0$$

$$y' = \frac{200 + 50x}{160 - 32y}$$

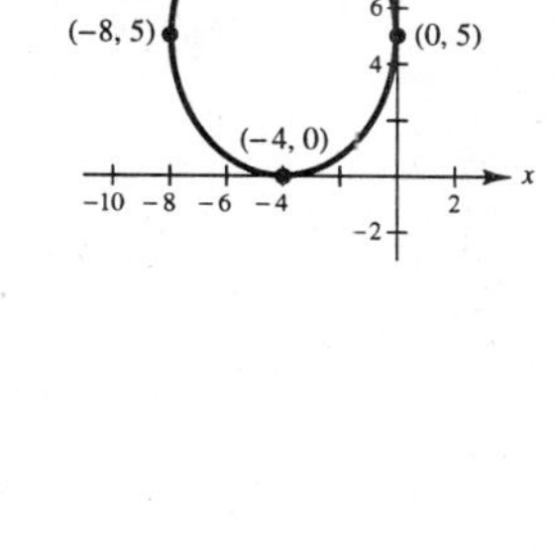

Horizontal tangents occur when $x = -4$:

$$25(16) + 16y^2 + 200(-4) - 160y + 400 = 0$$

$$y(y - 10) = 0 \Rightarrow y = 0, 10$$

Horizontal tangents: $(-4, 0), (-4, 10)$.

Vertical tangents occur when $y = 5$:

$$25x^2 + 400 + 200x - 800 + 400 = 0$$

$$25x(x + 8) = 0 \Rightarrow x = 0, -8$$

Vertical tangents: $(0, 5), (-8, 5)$.

47. Find the points of intersection by letting $y^2 = 4x$ in the equation $2x^2 + y^2 = 6$.

$2x^2 + 4x = 6$ and $(x + 3)(x - 1) = 0$

The curves intersect at $(1, \pm 2)$.

$y^2 = 4x$

$2x^2 + y^2 = 6$

Ellipse:	*Parabola:*
$4x + 2yy' = 0$	$2yy' = 4$
$y' = -\frac{2x}{y}$	$y' = \frac{2}{y}$

At $(1, 2)$, the slopes are:

$y' = -1$	$y' = 1.$

At $(1, -2)$, the slopes are:

$y' = 1$	$y' = -1.$

Tangents are perpendicular.

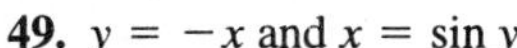

49. $y = -x$ and $x = \sin y$

Point of intersection: $(0, 0)$

$y = -x$:	$x = \sin y$:
$y' = -1$	$1 = y' \cos y$
	$y' = \sec y$

At $(0, 0)$, the slopes are:

$y' = -1$	$y' = 1.$

Tangents are perpendicular.

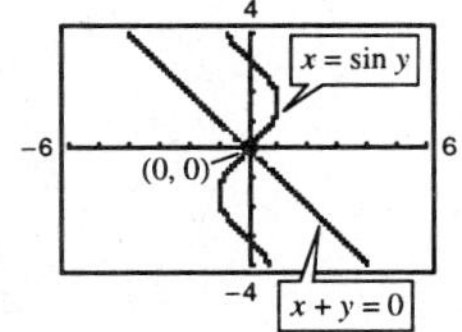

51.

$xy = C$	$x^2 - y^2 = K$
$xy' + y = 0$	$2x - 2yy' = 0$
$y' = -\frac{y}{x}$	$y' = \frac{x}{y}$

At any point of intersection (x, y) the product of the slopes is $(-y/x)(x/y) = -1$. The curves are orthogonal.

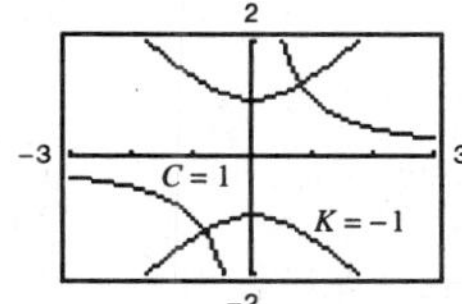

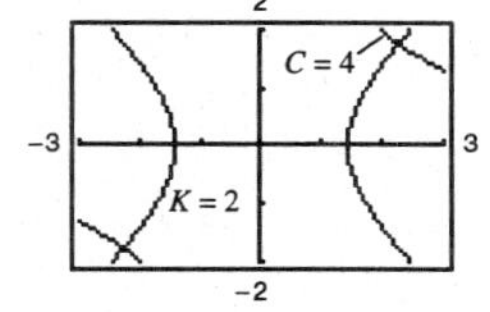

53. $2y^2 - 3x^4 = 0$

(a) $4yy' - 12x^3 = 0$

$$4yy' = 12x^3$$

$$y' = \frac{12x^3}{4y} = \frac{3x^3}{y}$$

(b) $4y\frac{dy}{dt} - 12x^3\frac{dx}{dt} = 0$

$$y\frac{dy}{dt} = 3x^3\frac{dx}{dt}$$

55. $\cos \pi y - 3 \sin \pi x = 1$

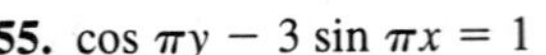

(a) $-\pi \sin(\pi y)y' - 3\pi \cos \pi x = 0$

$$y' = \frac{-3 \cos \pi x}{\sin \pi y}$$

(b) $-\pi \sin(\pi y)\frac{dy}{dt} - 3\pi \cos(\pi x)\frac{dx}{dt} = 0$

$$-\sin(\pi y)\frac{dy}{dt} = 3 \cos(\pi x)\frac{dx}{dt}$$

57. (a) $x^4 = 4(4x^2 - y^2)$

$$4y^2 = 16x^2 - x^4$$

$$y^2 = 4x^2 - \frac{1}{4}x^4$$

$$y = \pm\sqrt{4x^2 - \frac{1}{4}x^4}$$

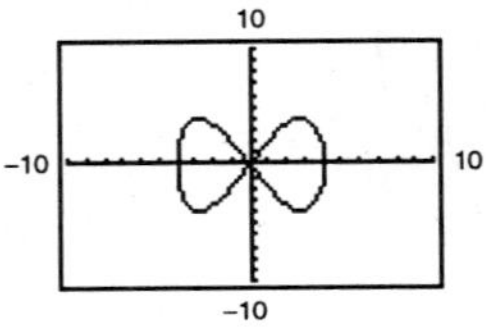

(b) $y = 3 \Longrightarrow 9 = 4x^2 - \frac{1}{4}x^4$

$$36 = 16x^2 - x^4$$

$$x^4 - 16x^2 + 36 = 0$$

$$x^2 = \frac{16 \pm \sqrt{256 - 144}}{2} = 8 \pm \sqrt{28}$$

Note that $x^2 = 8 \pm \sqrt{28} = 8 \pm 2\sqrt{7} = (1 \pm \sqrt{7})^2$.

Hence, there are four values of x:

$-1-\sqrt{7}, 1-\sqrt{7}, -1+\sqrt{7}, 1+\sqrt{7}$

To find the slope, $2yy' = 8x - x^3 \Longrightarrow y' = \frac{x(8 - x^2)}{2(3)}$.

For $x = -1 - \sqrt{7}$, $y' = \frac{1}{3}(\sqrt{7} + 7)$, and the line is

$$y_1 = \tfrac{1}{3}(\sqrt{7} + 7)(x + 1 + \sqrt{7}) + 3 = \tfrac{1}{3}[(\sqrt{7} + 7)x + 8\sqrt{7} + 23)].$$

For $x = 1 - \sqrt{7}$, $y' = \frac{1}{3}(\sqrt{7} - 7)$, and the line is

$$y_2 = \tfrac{1}{3}(\sqrt{7} - 7)(x - 1 + \sqrt{7}) + 3 = \tfrac{1}{3}[(\sqrt{7} - 7)x + 23 - 8\sqrt{7}].$$

For $x = -1 + \sqrt{7}$, $y' = -\frac{1}{3}(\sqrt{7} - 7)$, and the line is

$$y_3 = -\tfrac{1}{3}(\sqrt{7} - 7)(x + 1 - \sqrt{7}) + 3 = -\tfrac{1}{3}[(\sqrt{7} - 7)x - (23 - 8\sqrt{7})].$$

For $x = 1 + \sqrt{7}$, $y' = -\frac{1}{3}(\sqrt{7} + 7)$, and the line is

$$y_4 = -\tfrac{1}{3}(\sqrt{7} + 7)(x - 1 - \sqrt{7}) + 3 = -\tfrac{1}{3}[(\sqrt{7} + 7)x - (8\sqrt{7} + 23)].$$

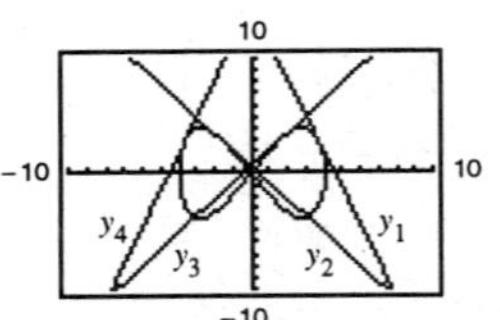

(c) Equating y_3 and y_4,

$$-\frac{1}{3}(\sqrt{7} - 7)(x + 1 - \sqrt{7}) + 3 = -\frac{1}{3}(\sqrt{7} + 7)(x - 1 - \sqrt{7}) + 3$$

$$(\sqrt{7} - 7)(x + 1 - \sqrt{7}) = (\sqrt{7} + 7)(x - 1 - \sqrt{7})$$

$$\sqrt{7}x + \sqrt{7} - 7 - 7x - 7 + 7\sqrt{7} = \sqrt{7}x - \sqrt{7} - 7 + 7x - 7 - 7\sqrt{7}$$

$$16\sqrt{7} = 14x$$

$$x = \frac{8\sqrt{7}}{7}$$

If $x = \frac{8\sqrt{7}}{7}$, then $y = 5$ and the lines intersect at $\left(\frac{8\sqrt{7}}{7}, 5\right)$.

59. Let $f(x) = x^n = x^{p/q}$, where p and q are nonzero integers and $q > 0$. First consider the case where $p = 1$. The derivative of $f(x) = x^{1/q}$ is given by

$$\frac{d}{dx}[x^{1/q}] = \lim_{\Delta x \to 0} \frac{f(x + \Delta x) - f(x)}{\Delta x} = \lim_{t \to x} \frac{f(t) - f(x)}{t - x}$$

where $t = x + \Delta x$. Observe that

$$\frac{f(t) - f(x)}{t - x} = \frac{t^{1/q} - x^{1/q}}{t - x} = \frac{t^{1/q} - x^{1/q}}{(t^{1/q})^q - (x^{1/q})^q}$$

$$= \frac{t^{1/q} - x^{1/q}}{(t^{1/q} - x^{1/q})(t^{1-(1/q)} + t^{1-(2/q)}x^{1/q} + \cdots + t^{1/q}x^{1-(2/q)} + x^{1-(1/q)})}$$

$$= \frac{1}{t^{1-(1/q)} + t^{1-(2/q)}x^{1/q} + \cdots + t^{1/q}x^{1-(2/q)} + x^{1-(1/q)}}.$$

As $t \to x$, the denominator approaches $qx^{1-(1/q)}$. That is,

$$\frac{d}{dx}[x^{1/q}] = \frac{1}{qx^{1-(1/q)}} = \frac{1}{q}x^{(1/q)-1}.$$

Now consider $f(x) = x^{p/q} = (x^p)^{1/q}$. From the Chain Rule,

$$f'(x) = \frac{1}{q}(x^p)^{(1/q)-1}\frac{d}{dx}[x^p]$$

$$= \frac{1}{q}(x^p)^{(1/q)-1}px^{p-1} = \frac{p}{q}x^{[(p/q)-p]+(p-1)} = \frac{p}{q}x^{(p/q)-1} = nx^{n-1} \left(n = \frac{p}{q}\right).$$

Section 2.6 Related Rates

1. $y = \sqrt{x}$

$$\frac{dy}{dt} = \left(\frac{1}{2\sqrt{x}}\right)\frac{dx}{dt}$$

$$\frac{dx}{dt} = 2\sqrt{x}\frac{dy}{dt}$$

(a) When $x = 4$ and $dx/dt = 3$,

$$\frac{dy}{dt} = \frac{1}{2\sqrt{4}}(3) = \frac{3}{4}.$$

(b) When $x = 25$ and $dy/dt = 2$,

$$\frac{dx}{dt} = 2\sqrt{25}\,(2) = 20.$$

3. $xy = 4$

$$x\frac{dy}{dt} + y\frac{dx}{dt} = 0$$

$$\frac{dy}{dt} = \left(-\frac{y}{x}\right)\frac{dx}{dt}$$

$$\frac{dx}{dt} = \left(-\frac{x}{y}\right)\frac{dy}{dt}$$

(a) When $x = 8$, $y = 1/2$, and $dx/dt = 10$,

$$\frac{dy}{dt} = -\frac{1/2}{8}(10) = -\frac{5}{8}.$$

(b) When $x = 1$, $y = 4$, and $dy/dt = -6$,

$$\frac{dx}{dt} = -\frac{1}{4}(-6) = \frac{3}{2}.$$

5. $y = x^2 + 1$

$$\frac{dx}{dt} = 2$$

$$\frac{dy}{dt} = 2x\frac{dx}{dt}$$

(a) When $x = -1$,

$$\frac{dy}{dt} = 2(-1)(2) = -4 \text{ cm/sec.}$$

(c) When $x = 1$,

$$\frac{dy}{dt} = 2(1)(2) = 4 \text{ cm/sec.}$$

(b) When $x = 0$,

$$\frac{dy}{dt} = 2(0)(2) = 0 \text{ cm/sec.}$$

(d) When $x = 3$,

$$\frac{dy}{dt} = 2(3)(2) = 12 \text{ cm/sec.}$$

7. $y = \tan x$

$$\frac{dx}{dt} = 2$$

$$\frac{dy}{dt} = \sec^2 x\frac{dx}{dt}$$

(a) When $x = -\pi/3$,

$$\frac{dy}{dt} = (2)^2(2) = 8 \text{ cm/sec.}$$

(b) When $x = -\pi/4$,

$$\frac{dy}{dt} = (\sqrt{2})^2(2) = 4 \text{ cm/sec.}$$

(c) When $x = 0$,

$$\frac{dy}{dt} = (1)^2(2) = 2 \text{ cm/sec.}$$

(d) When $x = 1$,

$$\frac{dy}{dt} = (\sec 1)^2 2 \approx 6.8510 \text{ cm/sec.}$$

9. (a) For increasing x, dy/dt decreases.

(b) For increasing y, dx/dt increases.

11. $D = \sqrt{x^2 + y^2} = \sqrt{x^2 + (x^2 + 1)^2} = \sqrt{x^4 + 3x^2 + 1}$

$$\frac{dx}{dt} = 2$$

$$\frac{dD}{dt} = \frac{1}{2}(x^4 + 3x^2 + 1)^{-1/2}(4x^3 + 6x)\frac{dx}{dt} = \frac{2x^3 + 3x}{\sqrt{x^4 + 3x^2 + 1}}\frac{dx}{dt} = \frac{4x^3 + 6x}{\sqrt{x^4 + 3x^2 + 1}}$$

13. $A = \pi r^2$

$$\frac{dr}{dt} = 2$$

$$\frac{dA}{dt} = 2\pi r\frac{dr}{dt}$$

(a) When $r = 6$,

$$\frac{dA}{dt} = 2\pi(6)(2) = 24\pi \text{ in}^2/\text{min.}$$

(b) When $r = 24$,

$$\frac{dA}{dt} = 2\pi(24)(2) = 96\pi \text{ in}^2/\text{min.}$$

15. (a) $\sin\frac{\theta}{2} = \frac{(1/2)b}{s} \Rightarrow b = 2s\sin\frac{\theta}{2}$

$$\cos\frac{\theta}{2} = \frac{h}{s} \Rightarrow h = s\cos\frac{\theta}{2}$$

$$A = \frac{1}{2}bh = \frac{1}{2}\left(2s\sin\frac{\theta}{2}\right)\left(s\cos\frac{\theta}{2}\right)$$

$$= \frac{s^2}{2}\left(2\sin\frac{\theta}{2}\cos\frac{\theta}{2}\right) = \frac{s^2}{2}\sin\theta$$

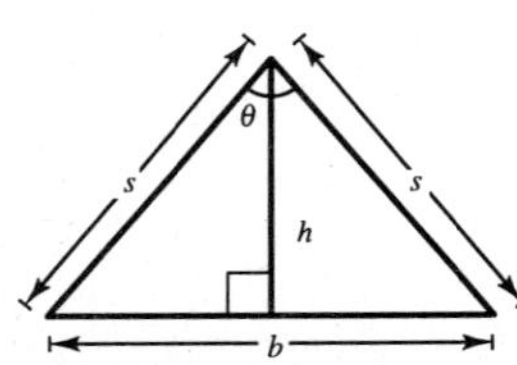

(b) $\frac{dA}{dt} = \frac{s^2}{2}\cos\theta\frac{d\theta}{dt}$ where $\frac{d\theta}{dt} = \frac{1}{2}$ rad/min.

When $\theta = \frac{\pi}{6}$, $\frac{dA}{dt} = \frac{s^2}{2}\left(\frac{\sqrt{3}}{2}\right)\left(\frac{1}{2}\right) = \frac{\sqrt{3}s^2}{8}$ rad/min.

When $\theta = \frac{\pi}{3}$, $\frac{dA}{dt} = \frac{s^2}{2}\left(\frac{1}{2}\right)\left(\frac{1}{2}\right) = \frac{s^2}{8}$ rad/min.

(c) If $d\theta/dt$ is constant, dA/dt is proportional to $\cos\theta$.

17. $V = \frac{4}{3}\pi r^3, \frac{dV}{dt} = 500$

$\frac{dV}{dt} = 4\pi r^2 \frac{dr}{dt}$

$\frac{dr}{dt} = \frac{1}{4\pi r^2}\left(\frac{dV}{dt}\right) = \frac{1}{4\pi r^2}(500)$

(a) When $r = 30$, $\frac{dr}{dt} = \frac{1}{4\pi(30)^2}(500) = \frac{5}{36\pi}$ cm/min.

(b) When $r = 60$, $\frac{dr}{dt} = \frac{1}{4\pi(60)^2}(500) = \frac{5}{144\pi}$ cm/min.

19. $s = 6x^2$

$\frac{dx}{dt} = 3$

$\frac{ds}{dt} = 12x\frac{dx}{dt}$

(a) When $x = 1$,

$\frac{ds}{dt} = 12(1)(3) = 36$ cm²/sec.

(b) When $x = 10$,

$\frac{ds}{dt} = 12(10)(3) = 360$ cm²/sec.

21. $V = \frac{1}{3}\pi r^2 h = \frac{1}{3}\pi\left(\frac{9}{4}h^2\right)h$ [since $2r = 3h$]

$= \frac{3\pi}{4}h^3$

$\frac{dV}{dt} = 10$

$\frac{dV}{dt} = \frac{9\pi}{4}h^2\frac{dh}{dt} \Rightarrow \frac{dh}{dt} = \frac{4(dV/dt)}{9\pi h^2}$

When $h = 15$, $\frac{dh}{dt} = \frac{4(10)}{9\pi(15)^2} = \frac{8}{405\pi}$ ft/min.

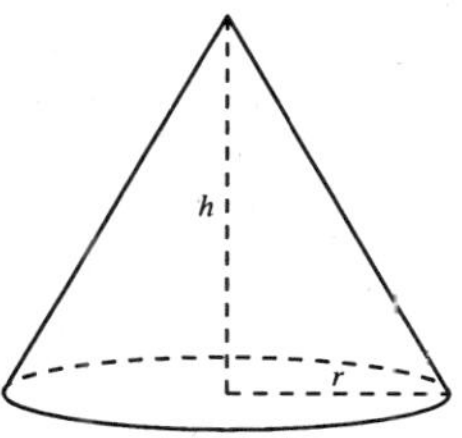

23.

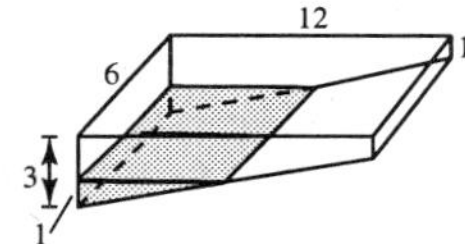

(a) Total volume of pool $= \frac{1}{2}(2)(12)(6) + (1)(6)(12) = 144$ m³

Volume of 1m. of water $= \frac{1}{2}(1)(6)(6) = 18$ m³

(see similar triangle diagram)

% pool filled $= \frac{18}{144}(100\%) = 12.5\%$

(b) Since for $0 \le h \le 2$, $b = 6h$, you have

$V = \frac{1}{2}bh(6) = 3bh = 3(6h)h = 18h^2$

$\frac{dV}{dt} = 36h\frac{dh}{dt} = \frac{1}{4} \Rightarrow \frac{dh}{dt} = \frac{1}{144h} = \frac{1}{144(1)} = \frac{1}{144}$ m/min.

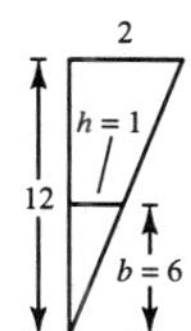

25. $x^2 + y^2 = 25^2$

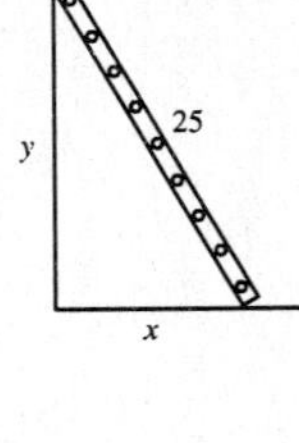

$$2x\frac{dx}{dt} + 2y\frac{dy}{dt} = 0$$

$$\frac{dy}{dt} = \frac{-x}{y} \cdot \frac{dx}{dt} = \frac{-2x}{y} \text{ since } \frac{dx}{dt} = 2.$$

(a) When $x = 7, y = \sqrt{576} = 24, \dfrac{dy}{dt} = \dfrac{-2(7)}{24} = \dfrac{-7}{12}$ ft/sec.

When $x = 15, y = \sqrt{400} = 20, \dfrac{dy}{dt} = \dfrac{-2(15)}{20} = \dfrac{-3}{2}$ ft/sec.

When $x = 24, y = 7, \dfrac{dy}{dt} = \dfrac{-2(24)}{7} = \dfrac{-48}{7}$ ft/sec.

(b) $A = \dfrac{1}{2}xy$

$$\frac{dA}{dt} = \frac{1}{2}\left(x\frac{dy}{dt} + y\frac{dx}{dt}\right)$$

From part (a) we have $x = 7, y = 24, \dfrac{dx}{dt} = 2$, and $\dfrac{dy}{dt} = -\dfrac{7}{12}$.

Thus, $\dfrac{dA}{dt} = \dfrac{1}{2}\left[7\left(-\dfrac{7}{12}\right) + 24(2)\right] = \dfrac{527}{24} \approx 21.96$ ft²/sec.

(c) $\tan\theta = \dfrac{x}{y}$

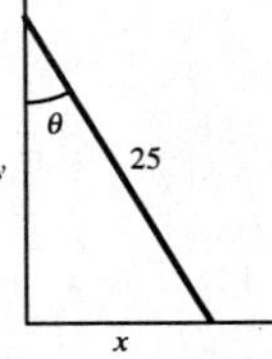

$$\sec^2\theta\frac{d\theta}{dt} = \frac{1}{y} \cdot \frac{dx}{dt} - \frac{x}{y^2} \cdot \frac{dy}{dt}$$

$$\frac{d\theta}{dt} = \cos^2\theta\left[\frac{1}{y} \cdot \frac{dx}{dt} - \frac{x}{y^2} \cdot \frac{dy}{dt}\right]$$

Using $x = 7, y = 24, \dfrac{dx}{dt} = 2, \dfrac{dy}{dt} = -\dfrac{7}{12}$ and $\cos\theta = \dfrac{24}{25}$, we have

$$\frac{d\theta}{dt} = \left(\frac{24}{25}\right)^2\left[\frac{1}{24}(2) - \frac{7}{(24)^2}\left(-\frac{7}{12}\right)\right] = \frac{1}{12} \text{ rad/sec.}$$

27. When $y = 6, x = \sqrt{12^2 - 6^2} = 6\sqrt{3}$, and

$s = \sqrt{x^2 + (12 - y)^2} = \sqrt{108 + 36} = 12.$

$$x^2 + (12 - y)^2 = s^2$$

$$2x\frac{dx}{dt} + 2(12 - y)(-1)\frac{dy}{dt} = 2s\frac{ds}{dt}$$

$$x\frac{dx}{dt} + (y - 12)\frac{dy}{dt} = s\frac{ds}{dt}$$

Also, $x^2 + y^2 = 12^2$

$$2x\frac{dx}{dt} + 2y\frac{dy}{dt} = 0 \Rightarrow \frac{dy}{dt} = \frac{-x}{y}\frac{dx}{dt}.$$

Thus, $x\dfrac{dx}{dt} + (y - 12)\left(\dfrac{-x}{y}\dfrac{dx}{dt}\right) = s\dfrac{ds}{dt}$

$$\frac{dx}{dt}\left[x - x + \frac{12x}{y}\right] = s\frac{ds}{dt} \Rightarrow \frac{dx}{dt} = \frac{sy}{12x} \cdot \frac{ds}{dt} = \frac{(12)(6)}{(12)(6\sqrt{3})}(-0.2) = \frac{-1}{5\sqrt{3}} = \frac{-\sqrt{3}}{15} \text{ m/sec (horizontal)}$$

$$\frac{dy}{dt} = \frac{-x}{y}\frac{dx}{dt} = \frac{-6\sqrt{3}}{6} \cdot \frac{(-\sqrt{3})}{15} = \frac{1}{5} \text{ m/sec (vertical).}$$

29. (a) $s^2 = x^2 + y^2$

$$\frac{dx}{dt} = -450$$

$$\frac{dy}{dt} = -600$$

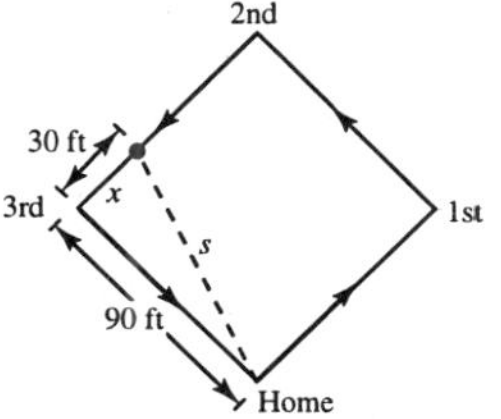

$$2s\frac{ds}{dt} = 2x\frac{dx}{dt} + 2y\frac{dy}{dt}$$

$$\frac{ds}{dt} = \frac{x(dx/dt) + y(dy/dt)}{s}$$

When $x = 150$ and $y = 200$, $s = 250$ and

$$\frac{ds}{dt} = \frac{150(-450) + 200(-600)}{250} = -750 \text{ mph.}$$

(b) $t = \frac{250}{750} = \frac{1}{3}$ hr $= 20$ min

31. $s^2 = 90^2 + x^2$

$$x = 30$$

$$\frac{dx}{dt} = -28$$

$$2s\frac{ds}{dt} = 2x\frac{dx}{dt} \Longrightarrow \frac{ds}{dt} = \frac{x}{s}\cdot\frac{dx}{dt}$$

When $x = 30$,

$$s = \sqrt{90^2 + 30^2} = 30\sqrt{10}$$

$$\frac{ds}{dt} = \frac{30}{30\sqrt{10}}(-28) = \frac{-28}{\sqrt{10}} \approx -8.85 \text{ ft/sec.}$$

33. (a) $\frac{15}{6} = \frac{y}{y - x} \Longrightarrow 15y - 15x = 6y$

$$y = \frac{5}{3}x$$

$$\frac{dx}{dt} = 5$$

$$\frac{dy}{dt} = \frac{5}{3}\cdot\frac{dx}{dt} = \frac{5}{3}(5) = \frac{25}{3} \text{ ft/sec}$$

(b) $\frac{d(y - x)}{dt} = \frac{dy}{dt} - \frac{dx}{dt} = \frac{25}{3} - 5 = \frac{10}{3}$ ft/sec

35. $x(t) = \frac{1}{2}\sin\frac{\pi t}{6}, x^2 + y^2 = 1$

(a) Period: $\frac{2\pi}{\pi/6} = 12$ seconds

(b) When $x = \frac{1}{2}, y = \sqrt{1^2 - \left(\frac{1}{2}\right)^2} = \frac{\sqrt{3}}{2}$ m.

Lowest point: $\left(0, \frac{\sqrt{3}}{2}\right)$

(c) When $x = \frac{1}{4}, y = \sqrt{1 - \left(\frac{1}{4}\right)^2} = \frac{\sqrt{15}}{4}$ and $t = 1$

$$\frac{dx}{dt} = \frac{1}{2}\left(\frac{\pi}{6}\right)\cos\frac{\pi t}{6} = \frac{\pi}{12}\cos\frac{\pi t}{6}$$

$$x^2 + y^2 = 1$$

$$2x\frac{dx}{dt} + 2y\frac{dy}{dt} = 0 \Longrightarrow \frac{dy}{dt} = \frac{-x}{y}\frac{dx}{dt}.$$

Thus,

$$\frac{dy}{dt} = -\frac{1/4}{\sqrt{15/4}}\cdot\frac{\pi}{12}\cos\left(\frac{\pi}{6}\right)$$

$$= \frac{-\pi}{\sqrt{15}}\left(\frac{1}{12}\right)\frac{\sqrt{3}}{2} = \frac{-\pi}{24}\frac{1}{\sqrt{5}} = \frac{-\sqrt{5}\pi}{120}.$$

$$\text{Speed} = \left|\frac{-\sqrt{5}\pi}{120}\right| = \frac{\sqrt{5}\pi}{120} \text{ m/sec}$$

37. Since the evaporation rate is proportional to the surface area, $dV/dt = k(4\pi r^2)$. However, since $V = (4/3)\pi r^3$, we have

$$\frac{dV}{dt} = 4\pi r^2\frac{dr}{dt}.$$

Therefore,

$$k(4\pi r^2) = 4\pi r^2\frac{dr}{dt} \Longrightarrow k = \frac{dr}{dt}.$$

39.
$$pv^{1.3} = k$$
$$1.3\,pv^{0.3}\frac{dv}{dt} + v^{1.3}\frac{dp}{dt} = 0$$
$$v^{0.3}\left(1.3p\frac{dv}{dt} + v\frac{dp}{dt}\right) = 0$$
$$1.3p\frac{dv}{dt} = -v\frac{dp}{dt}$$

41.
$$\tan\theta = \frac{y}{30}$$
$$\frac{dy}{dt} = 3 \text{ m/sec.}$$
$$\sec^2\theta \cdot \frac{d\theta}{dt} = \frac{1}{30}\frac{dy}{dt}$$
$$\frac{d\theta}{dt} = \frac{1}{30}\cos^2\theta \cdot \frac{dy}{dt}$$

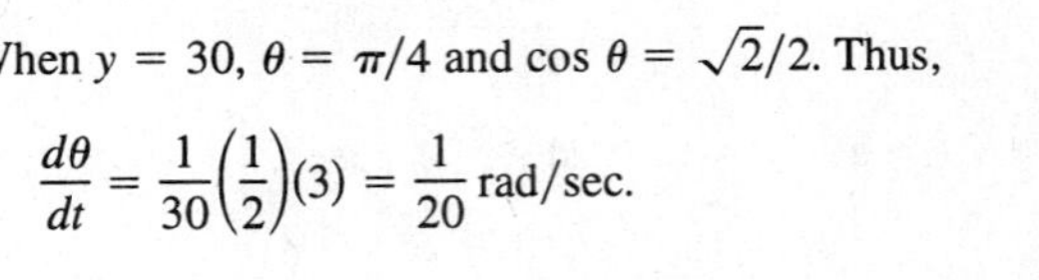
When $y = 30$, $\theta = \pi/4$ and $\cos\theta = \sqrt{2}/2$. Thus,
$$\frac{d\theta}{dt} = \frac{1}{30}\left(\frac{1}{2}\right)(3) = \frac{1}{20} \text{ rad/sec.}$$

43.
$$\tan\theta = \frac{y}{x},\ y = 5$$
$$\frac{dx}{dt} = -600 \text{ mi/hr}$$
$$(\sec^2\theta)\frac{d\theta}{dt} = -\frac{5}{x^2}\cdot\frac{dx}{dt}$$
$$\frac{d\theta}{dt} = \cos^2\theta\left(-\frac{5}{x^2}\right)\frac{dx}{dt} = \frac{x^2}{L^2}\left(-\frac{5}{x^2}\right)\frac{dx}{dt}$$
$$= \left(-\frac{5^2}{L^2}\right)\left(\frac{1}{5}\right)\frac{dx}{dt} = (-\sin^2\theta)\left(\frac{1}{5}\right)(-600) = 120\sin^2\theta$$

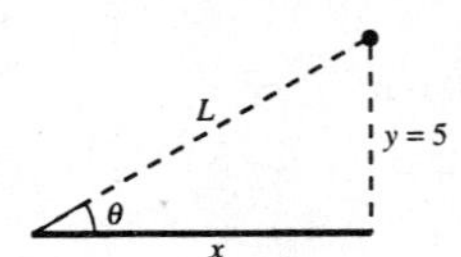

(a) When $\theta = 30°$, $\dfrac{d\theta}{dt} = \dfrac{120}{4} = 30 \text{ rad/hr} = \dfrac{1}{2} \text{ rad/min.}$

(b) When $\theta = 60°$, $\dfrac{d\theta}{dt} = 120\left(\dfrac{3}{4}\right) = 90 \text{ rad/hr} = \dfrac{3}{2} \text{ rad/min.}$

(c) When $\theta = 75°$, $\dfrac{d\theta}{dt} = 120\sin^2 75° \approx 111.96 \text{ rad/hr} \approx 1.87 \text{ rad/min.}$

45. $\dfrac{d\theta}{dt} = (10 \text{ rev/sec})(2\pi \text{ rad/rev}) = 20\pi \text{ rad/sec}$

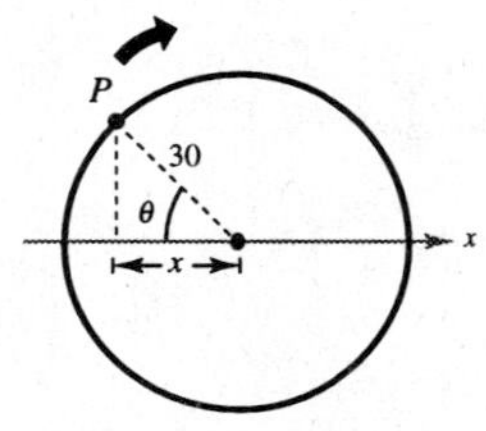

(a) $\cos\theta = \dfrac{x}{30}$
$$-\sin\theta\frac{d\theta}{dt} = \frac{1}{30}\frac{dx}{dt}$$
$$\frac{dx}{dt} = -30\sin\theta\frac{d\theta}{dt} = -30\sin\theta(20\pi) = -600\pi\sin\theta$$

(b)

2000

0 12.57

−2000

(c) $|dx/dt| = |-600\pi\sin\theta|$ is greatest when $\sin\theta = 1 \Rightarrow \theta = (\pi/2) + n\pi$ (or $90° + n \cdot 180°$)

$|dx/dt|$ is least when $\theta = n\pi$ (or $n \cdot 180°$).

(d) For $\theta = 30°$, $\dfrac{dx}{dt} = -600\pi\sin(30°) = -600\pi\dfrac{1}{2} = -300\pi \text{ cm/sec.}$

For $\theta = 60°$, $\dfrac{dx}{dt} = -600\pi\sin(60°) = -600\pi\dfrac{\sqrt{3}}{2} = -300\sqrt{3}\,\pi \text{ cm/sec}$

47. $\tan\theta = \dfrac{x}{50} \Rightarrow x = 50 \tan \theta$

$$\frac{dx}{dt} = 50 \sec^2 \theta \frac{d\theta}{dt}$$

$$2 = 50 \sec^2 \theta \frac{d\theta}{dt}$$

$$\frac{d\theta}{dt} = \frac{1}{25}\cos^2 \theta, \; -\frac{\pi}{4} \le \theta \le \frac{\pi}{4}$$

49. $x^2 + y^2 = 25$; acceleration of the top of the ladder $= \dfrac{d^2y}{dt^2}$

First derivative: $2x\dfrac{dx}{dt} + 2y\dfrac{dy}{dt} = 0$

$$x\frac{dx}{dt} + y\frac{dy}{dt} = 0$$

Second derivative: $x\dfrac{d^2x}{dt^2} + \dfrac{dx}{dt} \cdot \dfrac{dx}{dt} + y\dfrac{d^2y}{dt^2} + \dfrac{dy}{dt} \cdot \dfrac{dy}{dt} = 0$

$$\frac{d^2y}{dt^2} = \left(\frac{1}{y}\right)\left[-x\frac{d^2x}{dt^2} - \left(\frac{dx}{dt}\right)^2 - \left(\frac{dy}{dt}\right)^2\right]$$

When $x = 7, y = 24, \dfrac{dy}{dt} = -\dfrac{7}{12}$, and $\dfrac{dx}{dt} = 2$ (see Exercise 25). Since $\dfrac{dx}{dt}$ is constant, $\dfrac{d^2x}{dt^2} = 0$.

$$\frac{d^2y}{dt^2} = \frac{1}{24}\left[-7(0) - (2)^2 - \left(-\frac{7}{12}\right)^2\right] = \frac{1}{24}\left[-4 - \frac{49}{144}\right] = \frac{1}{24}\left[-\frac{625}{144}\right] \approx -0.1808 \text{ ft/sec}^2$$

51. (a) Using a graphing utility, you obtain

$$m(s) = -1.014s^2 + 31.685s - 214.437.$$

(b) $\dfrac{dm}{dt} = (-2.028s + 31.685)\dfrac{ds}{dt}$

(c) If $\dfrac{ds}{dt} = 1.2$, then $s = 16.5$ in 1995 $(t = 5)$.

Then $\dfrac{dm}{dt} = (-2.028(16.5) + 31.685)(1.2) \approx -2.1$.

Review Exercises for Chapter 2

1. $f(x) = x^2 - 2x + 3$

$$f'(x) = \lim_{\Delta x \to 0} \frac{f(x + \Delta x) - f(x)}{\Delta x}$$

$$= \lim_{\Delta x \to 0} \frac{[(x + \Delta x)^2 - 2(x + \Delta x) + 3] - [x^2 - 2x + 3]}{\Delta x}$$

$$= \lim_{\Delta x \to 0} \frac{(x^2 + 2x(\Delta x) + (\Delta x)^2 - 2x - 2(\Delta x) + 3) - (x^2 - 2x + 3)}{\Delta x}$$

$$= \lim_{\Delta x \to 0} \frac{2x(\Delta x) + (\Delta x)^2 - 2(\Delta x)}{\Delta x} = \lim_{\Delta x \to 0} (2x + \Delta x - 2) = 2x - 2$$

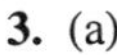

3. (a)

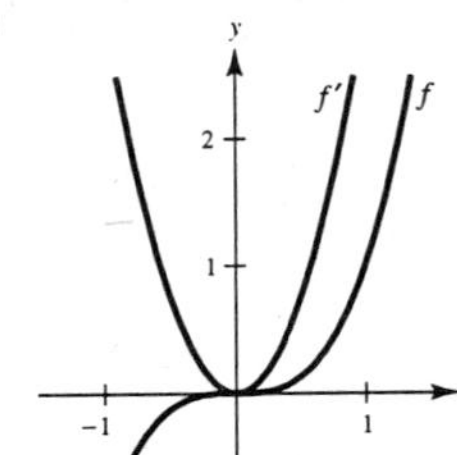

(b)

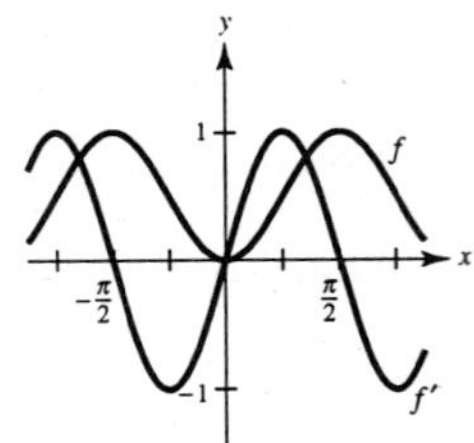

5. f is differentiable for all $x \neq 1$.

7. $f(x) = x^3 - 3x^2$

$f'(x) = 3x^2 - 6x = 3x(x - 2)$

9. $f(x) = 2x - x^{-2}$

$f'(x) = 2 + 2x^{-3} = 2\left(1 + \dfrac{1}{x^3}\right)$

$= \dfrac{2(x^3 + 1)}{x^3}$

11. $g(t) = \dfrac{2}{3}t^{-2}$

$g'(x) = \dfrac{-4}{3}t^{-3} = \dfrac{-4}{3t^3}$

13. $f(x) = (1 - x^3)^{1/2}$

$f'(x) = \dfrac{1}{2}(1 - x^3)^{-1/2}(-3x^2) = -\dfrac{3x^2}{2\sqrt{1 - x^3}}$

15. $f(x) = (3x^2 + 7)(x^2 - 2x + 3)$

$f'(x) = (3x^2 + 7)(2x - 2) + (x^2 - 2x + 3)(6x)$

$= 2(6x^3 - 9x^2 + 16x - 7)$

17. $f(x) = \left(x^2 + \dfrac{1}{x}\right)^5$

$f'(x) = 5\left(x^2 + \dfrac{1}{x}\right)^4\left(2x - \dfrac{1}{x^2}\right)$

19. $f(x) = \dfrac{x^2 + x - 1}{x^2 - 1}$

$f'(x) = \dfrac{(x^2 - 1)(2x + 1) - (x^2 + x - 1)(2x)}{(x^2 - 1)^2}$

$= \dfrac{-(x^2 + 1)}{(x^2 - 1)^2}$

21. $f(x) = (4 - 3x^2)^{-1}$

$f'(x) = -(4 - 3x^2)^{-2}(-6x) = \dfrac{6x}{(4 - 3x^2)^2}$

23. $y = 3\cos(3x + 1)$

$y' = -9\sin(3x + 1)$

25. $y = \dfrac{1}{2}\csc 2x$

$y' = \dfrac{1}{2}(-\csc 2x \cot 2x)(2)$

$= -\csc 2x \cot 2x$

27. $y = \dfrac{x}{2} - \dfrac{\sin 2x}{4}$

$y' = \dfrac{1}{2} - \dfrac{1}{4}\cos 2x(2)$

$= \dfrac{1}{2}(1 - \cos 2x) = \sin^2 x$

29. $y = \frac{2}{3}\sin^{3/2} x - \frac{2}{7}\sin^{7/2} x$

$y' = \sin^{1/2} x \cos x - \sin^{5/2} x \cos x$

$= (\cos x)\sqrt{\sin x}(1 - \sin^2 x) = (\cos^3 x)\sqrt{\sin x}$

31. $y = -x\tan x$

$y' = -x\sec^2 x - \tan x$

33. $y = \dfrac{\sin x}{x^2}$

$y' = \dfrac{(x^2)\cos x - (\sin x)(2x)}{x^4} = \dfrac{x\cos x - 2\sin x}{x^3}$

35. $f(t) = t^2(t - 1)^5$

$f'(t) = t(t - 1)^4(7t - 2)$

The zeros of f' correspond to the points on the graph of f where the tangent line is horizontal.

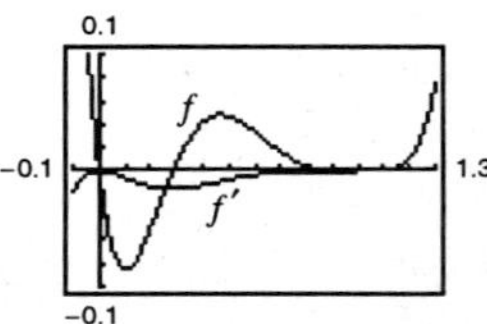

37. $g(x) = 2x(x+1)^{-1/2}$

$$g'(x) = \frac{x+2}{(x+1)^{3/2}}$$

g' does not equal zero for any value of x in the domain. The graph of g has no horizontal tangent lines.

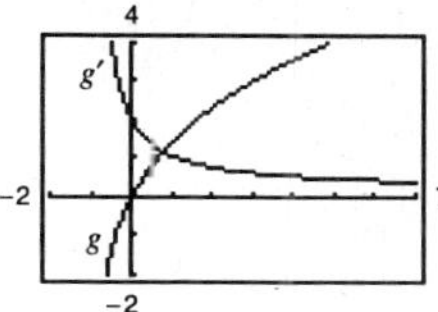

39. $f(t) = (t+1)^{1/2}(t+1)^{1/3} = (t+1)^{5/6}$

$$f'(t) = \frac{5}{6(t+1)^{1/6}}$$

f' does not equal zero for any x in the domain. The graph of f has no horizontal tangent lines.

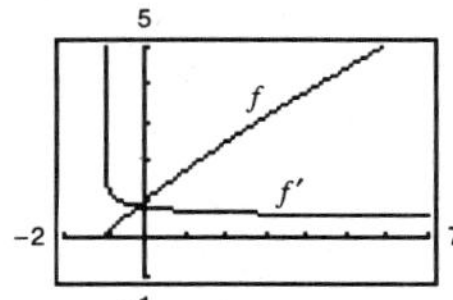

41. $y = \tan\sqrt{1-x}$

$$y' = -\frac{\sec^2\sqrt{1-x}}{2\sqrt{1-x}}$$

y' does not equal zero for any x in the domain. The graph has no horizontal tangent lines.

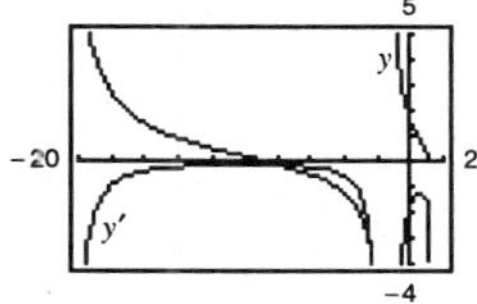

43. $f(x) = \cos\dfrac{\pi x}{2}$

$$f'(x) = -\frac{\pi}{2}\sin\frac{\pi x}{2}$$

$$f'\left(\frac{2}{3}\right) = -\frac{\pi}{2}\cdot\frac{\sqrt{3}}{2} = -\frac{\pi\sqrt{3}}{4}$$

45. $y = 2x^2 + \sin 2x$

$$y' = 4x + 2\cos 2x$$

$$y'' = 4 - 4\sin 2x$$

47. $f(x) = \cot x$

$$f'(x) = -\csc^2 x$$

$$f'' = -2\csc x(-\csc x \cdot \cot x)$$

$$= 2\csc^2 x \cot x$$

49. $f(t) = \dfrac{t}{(1-t)^2}$

$$f'(t) = \frac{t+1}{(1-t)^3}$$

$$f''(t) = \frac{2(t+2)}{(1-t)^4}$$

51. $g(x) = x\tan x$

$$g'(x) = x\sec^2 x + \tan x$$

$$g''(x) = 2\sec^2 x(x\tan x + 1)$$

53. $x^2 + 3xy + y^3 = 10$

$$2x + 3xy' + 3y + 3y^2y' = 0$$

$$3(x + y^2)y' = -(2x + 3y)$$

$$y' = \frac{-(2x+3y)}{3(x+y^2)}$$

55. $y\sqrt{x} - x\sqrt{y} = 16$

$$y\left(\frac{1}{2}x^{-1/2}\right) + x^{1/2}y' - x\left(\frac{1}{2}y^{-1/2}y'\right) - y^{1/2} = 0$$

$$\left(\sqrt{x} - \frac{x}{2\sqrt{y}}\right)y' = \sqrt{y} - \frac{y}{2\sqrt{x}}$$

$$\frac{2\sqrt{xy} - x}{2\sqrt{y}}y' = \frac{2\sqrt{xy} - y}{2\sqrt{x}}$$

$$y' = \frac{2\sqrt{xy} - y}{2\sqrt{x}} \cdot \frac{2\sqrt{y}}{2\sqrt{xy} - x} = \frac{2y\sqrt{x} - y\sqrt{y}}{2x\sqrt{y} - x\sqrt{x}}$$

57.
$$x \sin y = y \cos x$$
$$(x \cos y)y' + \sin y = -y \sin x + y' \cos x$$
$$y'(x \cos y - \cos x) = -y \sin x - \sin y$$
$$y' = \frac{y \sin x + \sin y}{\cos x - x \cos y}$$

59. $y = (x + 3)^3$

$y' = 3(x + 3)^2$

At $(-2, 1)$: $y' = 3$

Tangent line: $y - 1 = 3(x + 2)$

$3x - y + 7 = 0$

Normal line: $y - 1 = -\frac{1}{3}(x + 2)$

$x + 3y - 1 = 0$

61. $x^2 + y^2 = 20$

$2x + 2yy' = 0$

$y' = -\dfrac{x}{y}$

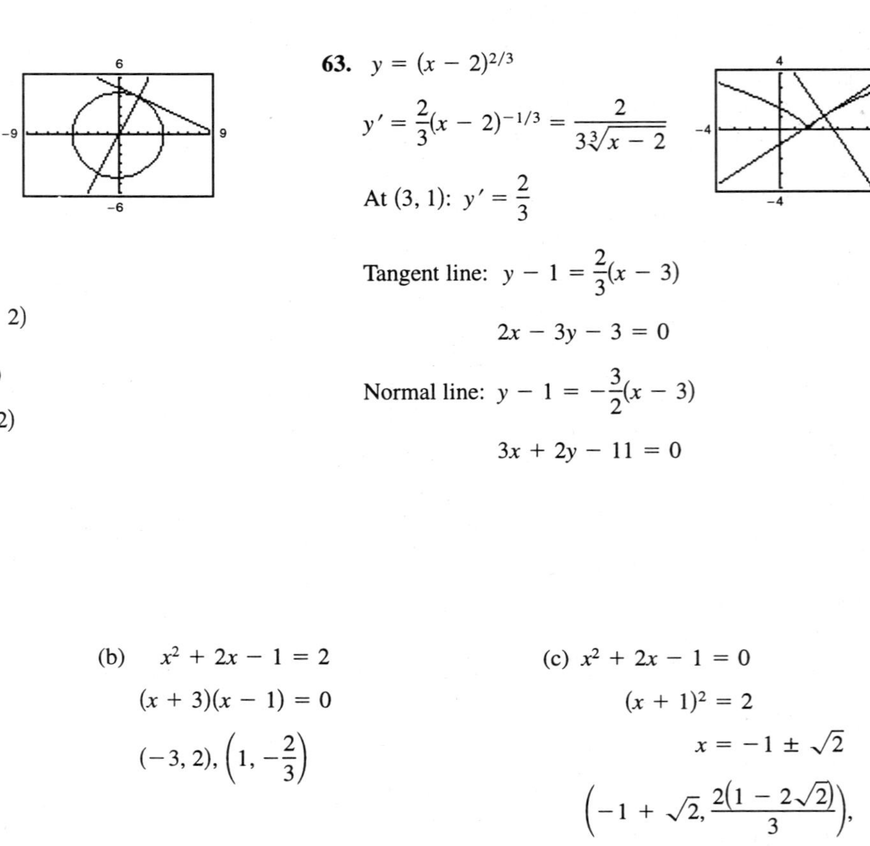

At $(2, 4)$: $y' = -\dfrac{1}{2}$

Tangent line: $y - 4 = -\dfrac{1}{2}(x - 2)$

$x + 2y - 10 = 0$

Normal line: $y - 4 = 2(x - 2)$

$2x - y = 0$

63. $y = (x - 2)^{2/3}$

$y' = \dfrac{2}{3}(x - 2)^{-1/3} = \dfrac{2}{3\sqrt[3]{x - 2}}$

At $(3, 1)$: $y' = \dfrac{2}{3}$

Tangent line: $y - 1 = \dfrac{2}{3}(x - 3)$

$2x - 3y - 3 = 0$

Normal line: $y - 1 = -\dfrac{3}{2}(x - 3)$

$3x + 2y - 11 = 0$

65. $f(x) = \dfrac{1}{3}x^3 + x^2 - x - 1$

$f'(x) = x^2 + 2x - 1$

(a) $x^2 + 2x - 1 = -1$

$x(x + 2) = 0$

$(0, -1), \left(-2, \dfrac{7}{3}\right)$

(b) $x^2 + 2x - 1 = 2$

$(x + 3)(x - 1) = 0$

$(-3, 2), \left(1, -\dfrac{2}{3}\right)$

(c) $x^2 + 2x - 1 = 0$

$(x + 1)^2 = 2$

$x = -1 \pm \sqrt{2}$

$\left(-1 + \sqrt{2}, \dfrac{2(1 - 2\sqrt{2})}{3}\right)$,

$\left(-1 - \sqrt{2}, \dfrac{2(1 + 2\sqrt{2})}{3}\right)$

67. $f(x) = 4 - |x - 2|$

(a) Continuous at $x = 2$.

(b) Not differentiable at $x = 2$ because of the sharp turn in the graph.

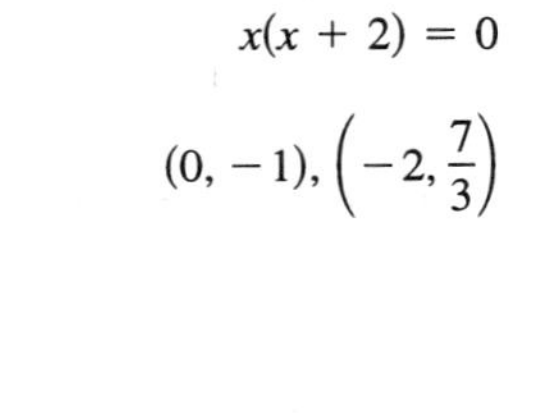

69.
$$y = 2 \sin x + 3 \cos x$$
$$y' = 2 \cos x - 3 \sin x$$
$$y'' = -2 \sin x - 3 \cos x$$
$$y'' + y = -(2 \sin x + 3 \cos x) + (2 \sin x + 3 \cos x)$$
$$= 0$$

71. $T = 700(t^2 + 4t + 10)^{-1}$

$$T' = \frac{-1400(t + 2)}{(t^2 + 4t + 10)^2}$$

(a) When $t = 1$,

$$T' = \frac{-1400(1 + 2)}{(1 + 4 + 10)^2} \approx -18.667 \text{ deg/hr.}$$

(b) When $t = 3$,

$$T' = \frac{-1400(3 + 2)}{(9 + 12 + 10)^2} \approx -7.284 \text{ deg/hr.}$$

(c) When $t = 5$,

$$T' = \frac{-1400(5 + 2)}{(25 + 30 + 10)^2} \approx -3.240 \text{ deg/hr.}$$

(d) When $t = 10$,

$$T' = \frac{-1400(10 + 2)}{(100 + 40 + 10)^2} \approx -0.747 \text{ deg/hr.}$$

73. $F = 200\sqrt{T}$

$$F'(t) = \frac{100}{\sqrt{T}}$$

(a) When $T = 4$, $F'(4) = 50$ vibrations/sec/lb.

(b) When $T = 9$, $F'(9) = 33\frac{1}{3}$ vibrations/sec/lb.

75. Assume that the stone is thrown from an initial height of $s_0 = 0$. Thus, the position equation is $s = -16t^2 + v_0t$. The maximum value of s occurs when $ds/dt = 0$ and thus

$$\frac{ds}{dt} = -32t + v_0 = 0$$

$$-32t = -v_0$$

$$t = \frac{v_0}{32}.$$

This means that the maximum height is

$$s = -16\left(\frac{v_0}{32}\right)^2 + v_0\left(\frac{v_0}{32}\right) = \frac{{v_0}^2}{64}.$$

If s is to attain a value of 49, then

$$\frac{{v_0}^2}{64} = 49$$

$${v_0}^2 = 3136$$

$$v_0 = 56 \text{ ft/sec.}$$

77. (a)

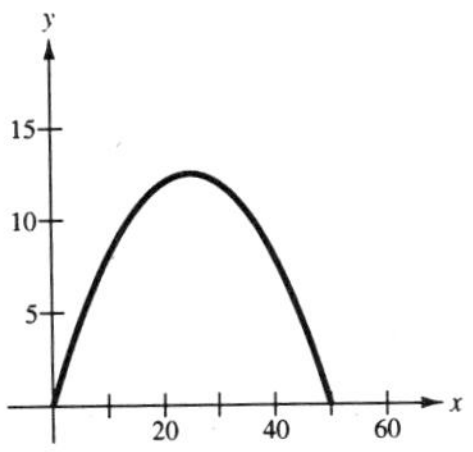

Total horizontal distance: 50

(b) $0 = x - 0.02x^2$

$$0 = x\left(x - \frac{x}{50}\right) \text{ implies } x = 50.$$

(c) Ball reaches maximum height when $x = 25$.

(d) $y = x - 0.02x^2$

$$y' = 1 - 0.04x$$

$$y'(0) = 1$$

$$y'(10) = 0.6$$

$$y'(25) = 0$$

$$y'(30) = -0.2$$

$$y'(50) = -1$$

(e) $y'(25) = 0$

79. $y = \sqrt{x}$

$L^2 = x^2 + y^2$

$\dfrac{dy}{dt} = 2 \text{ units/sec}$

$L^2 = y^4 + y^2$

$2L\dfrac{dL}{dt} = (4y^3 + 2y)\dfrac{dy}{dt}$

$\dfrac{dL}{dt} = \dfrac{4y^3 + 2y}{2L}\dfrac{dy}{dt} = \dfrac{4y^3 + 2y}{L} = \dfrac{(4x + 2)\sqrt{x}}{L}$

(a) When $x = \dfrac{1}{2}$, $L = \sqrt{\left(\dfrac{1}{2}\right)^2 + \left(\dfrac{1}{\sqrt{2}}\right)^2} = \dfrac{\sqrt{3}}{2}$ and $\dfrac{dL}{dt} = \dfrac{(2 + 2)(1/\sqrt{2})}{\sqrt{3}/2} = \dfrac{8}{\sqrt{6}} = \dfrac{4\sqrt{6}}{3}$ units/sec.

(b) When $x = 1$, $L = \sqrt{(1)^2 + (1)^2} = \sqrt{2}$ and $\dfrac{dL}{dt} = \dfrac{(4 + 2)(1)}{\sqrt{2}} = 3\sqrt{2}$ units/sec.

(c) When $x = 4$, $L = \sqrt{(4)^2 + (2)^2} = 2\sqrt{5}$ and $\dfrac{dL}{dt} = \dfrac{(16 + 2)(2)}{2\sqrt{5}} = \dfrac{18}{\sqrt{5}} = \dfrac{18\sqrt{5}}{5}$ units/sec.

81. $\dfrac{s}{h} = \dfrac{1/2}{2}$

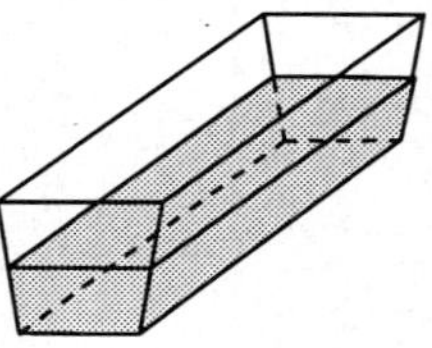

$s = \dfrac{1}{4}h$

$\dfrac{dV}{dt} = 1$

Width of water at depth h: $w = 2 + 2s = 2 + 2\left(\dfrac{1}{4}h\right) = \dfrac{4 + h}{2}$

$V = \dfrac{5}{2}\left(2 + \dfrac{4 + h}{2}\right)h = \dfrac{5}{4}(8 + h)h$

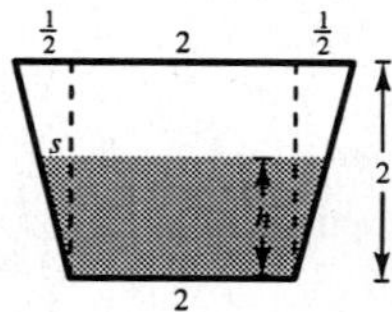

$\dfrac{dV}{dt} = \dfrac{5}{2}(4 + h)\dfrac{dh}{dt}$

$\dfrac{dh}{dt} = \dfrac{2(dV/dt)}{5(4 + h)}$

When $h = 1$, $\dfrac{dh}{dt} = \dfrac{2}{25}$ m/min.

83. $s(t) = 60 - 4.9t^2$

$s'(t) = -9.8t$

$s = 35 = 60 - 4.9t^2$

$4.9t^2 = 25$

$t = \dfrac{5}{\sqrt{4.9}}$

$\tan 30° = \dfrac{1}{\sqrt{3}} = \dfrac{s(t)}{x(t)}$

$x(t) = \sqrt{3}s(t)$

$\dfrac{dx}{dt} = \sqrt{3}\dfrac{ds}{dt} = \sqrt{3}(-9.8)\dfrac{5}{\sqrt{4.9}}$

≈ -38.34 m/sec

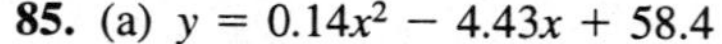

85. (a) $y = 0.14x^2 - 4.43x + 58.4$

(b)

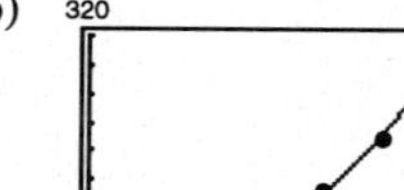

(c)

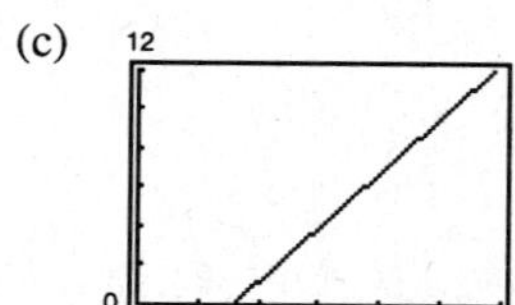

(d) If $x = 65$, $y \approx 362$ feet.

(e) As the speed increases, the stopping distance increases at an increasing rate.

CHAPTER 3
Applications of Differentiation

CHAPTER 3
Applications of Differentiation

Section 3.1 Extrema on an Interval

Solutions to Odd-Numbered Exercises

1. $f(x) = \dfrac{x^2}{x^2 + 4}$

$f'(x) = \dfrac{(x^2 + 4)(2x) - (x^2)(2x)}{(x^2 + 4)^2} = \dfrac{8x}{(x^2 + 4)^2}$

$f'(0) = 0$

3. $f(x) = x + \dfrac{32}{x^2}$

$f'(x) = 1 - \dfrac{64}{x^3}$

$f'(4) = 0$

5. $f(x) = (x + 2)^{2/3}$

$f'(x) = \frac{2}{3}(x + 2)^{-1/3}$

$f'(-2)$ is undefined.

7. $f(x) = x^2(x - 3) = x^3 - 3x^2$

$f'(x) = 3x^2 - 6x = 3x(x - 2)$

Critical numbers: $x = 0, x = 2$

9. $g(t) = t\sqrt{4 - t}$

$g'(t) = t\left[\dfrac{1}{2}(4 - t)^{-1/2}(-1)\right] + (4 - t)^{1/2}$

$= \dfrac{1}{2}(4 - t)^{-1/2}[-t + 2(4 - t)]$

$= \dfrac{8 - 3t}{2\sqrt{4 - t}}$

Critical numbers: $t = 4, t = \dfrac{8}{3}$

11. $h(x) = \sin^2 x + \cos x, 0 \le x < 2\pi$

$h'(x) = 2 \sin x \cos x - \sin x = \sin x(2 \cos x - 1)$

Critical numbers: $x = 0, x = \dfrac{\pi}{3}, x = \pi, x = \dfrac{5\pi}{3}$

13. $f(x) = 2(3 - x), [-1, 2]$

$f'(x) = -2 \Rightarrow$ No critical numbers

Left endpoint: $(-1, 8)$ Maximum
Right endpoint: $(2, 2)$ Minimum

15. $f(x) = -x^2 + 3x, [0, 3]$

$f'(x) = -2x + 3$

Left endpoint: $(0, 0)$ Minimum
Critical number: $\left(\frac{3}{2}, \frac{9}{4}\right)$ Maximum
Right endpoint: $(3, 0)$ Minimum

17. $f(x) = x^3 - 3x^2, [-1, 3]$

$f'(x) = 3x^2 - 6x = 3x(x - 2)$

Left endpoint: $(-1, -4)$ Minimum
Critical number: $(0, 0)$ Maximum
Critical number: $(2, -4)$ Minimum
Right endpoint: $(3, 0)$ Maximum

19. $f(x) = 3x^{2/3} - 2x, [-1, 1]$

$f'(x) = 2x^{-1/3} - 2 = \dfrac{2(1 - \sqrt[3]{x})}{\sqrt[3]{x}}$

Left endpoint: $(-1, 5)$ Maximum
Critical number: $(0, 0)$ Minimum
Right endpoint: $(1, 1)$

21. $h(t) = 4 - |t - 4|, [1, 6]$

From the graph of the function on the interval $[1, 6]$ you can determine the following.

Left endpoint: $(1, 1)$ Minimum

Critical number: $(4, 4)$ Maximum

Right endpoint: $(6, 2)$

23. $h(s) = \dfrac{1}{s - 2}, [0, 1]$

$$h'(s) = \frac{-1}{(s-2)^2}$$

Left endpoint: $\left(0, -\dfrac{1}{2}\right)$ Maximum

Right endpoint: $(1, -1)$ Minimum

25. $f(x) = \cos \pi x, \left[0, \dfrac{1}{6}\right]$

$f'(x) = -\pi \sin \pi x$

Left endpoint: $(0, 1)$ Maximum

Right endpoint: $\left(\dfrac{1}{6}, \dfrac{\sqrt{3}}{2}\right)$ Minimum

27. $f(x) = \tan x$

f is continuous on $[0, \pi/4]$ but not on $[0, \pi]$.

$\lim_{x \to \pi/2^-} \tan x = \infty$.

29. (a) Yes

(b) No

31. (a) No

(b) Yes

33. (a) Minimum: $(0, -3)$

Maximum: $(2, 1)$

(b) Minimum: $(0, -3)$

(c) Maximum: $(2, 1)$

(d) No extrem

35. $f(x) = x^2 - 2x$

(a) Minimum: $(1, -1)$

Maximum: $(-1, 3)$

(b) Maximum: $(3, 3)$

(c) Minimum: $(1, -1)$

(d) Minimum: $(1, -1)$

37. $f(x) = \begin{cases} 2x + 2, & 0 \le x \le 1 \\ 4x^2, & 1 < x \le 3 \end{cases}$

Left endpoint: $(0, 2)$ Minimum

Right endpoint: $(3, 36)$ Maximum

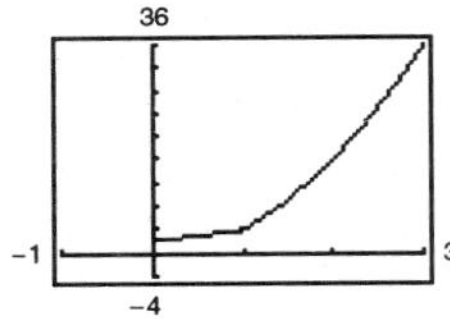

39. $f(x) = \dfrac{3}{x - 1}, (1, 4]$

Right endpoint: $(4, 1)$ Minimum

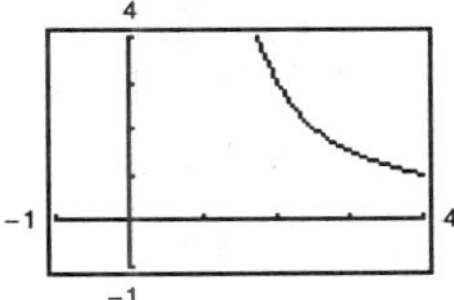

41. (a)

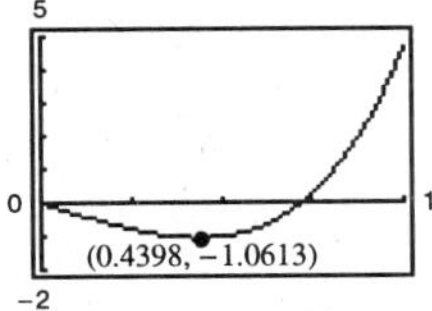

Maximum: $(1, 4.7)$

Minimum: $(0.4398, -1.0613)$

(b)

$$f(x) = 3.2x^5 + 5x^3 - 3.5x, [0, 1]$$

$$f'(x) = 16x^4 + 15x^2 - 3.5$$

$$16x^4 + 15x^2 - 3.5 = 0$$

$$x^2 = \frac{-15 \pm \sqrt{(15)^2 - 4(16)(-3.5)}}{2(16)} = \frac{-15 \pm \sqrt{449}}{32}$$

$$x = \sqrt{\frac{-15 + \sqrt{449}}{32}} \approx 0.4398$$

$$f(0) = 0$$

$$f(1) = 4.7 \text{ Maximum}$$

$$f\left(\sqrt{\frac{-15 + \sqrt{449}}{32}}\right) \approx -1.0613$$

Minimum: $(0.4398, -1.0613)$

43. $f(x) = (1 + x^3)^{1/2}, [0, 2]$

$f'(x) = \frac{3}{2}x^2(1 + x^3)^{-1/2}$

$f''(x) = \frac{3}{4}(x^4 + 4x)(1 + x^3)^{-3/2}$

$f'''(x) = -\frac{3}{8}(x^6 + 20x^3 - 8)(1 + x^3)^{-5/2}$

Setting $f''' = 0$, we have $x^6 + 20x^3 - 8 = 0$.

$x^3 = \frac{-20 \pm \sqrt{400 - 4(1)(-8)}}{2}$

$x = \sqrt[3]{-10 \pm \sqrt{108}} = \sqrt{3} - 1$

In the interval $[0, 2]$, choose

$x = \sqrt[3]{-10 \pm \sqrt{108}} = \sqrt{3} - 1 \approx 0.732$.

$\left|f''\left(\sqrt[3]{-10 + \sqrt{108}}\right)\right| \approx 1.47$ is the maximum value.

45. $f(x) = (x + 1)^{2/3}, [0, 2]$

$f'(x) = \frac{2}{3}(x + 1)^{-1/3}$

$f''(x) = -\frac{2}{9}(x + 1)^{-4/3}$

$f'''(x) = \frac{8}{27}(x + 1)^{-7/3}$

$f^{(4)}(x) = -\frac{56}{81}(x + 1)^{-10/3}$

$f^{(5)}(x) = \frac{560}{243}(x + 1)^{-13/3}$

$|f^{(4)}(0)| = \frac{56}{81}$ is the maximum value.

47. $P = VI - RI^2 = 12I - 0.5I^2, 0 \le I \le 15$

$P = 0$ when $I = 0$.

$P = 67.5$ when $I = 15$.

$P' = 12 - I = 0$

Critical number: $I = 12$ amps

When $I = 12$ amps, $P = 72$, the maximum output.

No, a 20-amp fuse would not increase the power output. P is decreasing for $I > 12$.

49. $x = \frac{v^2 \sin 2\theta}{32}, \frac{\pi}{4} \le \theta \le \frac{3\pi}{4}$

$\frac{d\theta}{dt}$ is constant.

$\frac{dx}{dt} = \frac{dx}{d\theta}\frac{d\theta}{dt}$ (by the Chain Rule)

$= \frac{v^2 \cos 2\theta}{16}\frac{d\theta}{dt}$

In the interval $[\pi/4, 3\pi/4]$, $\theta = \pi/4, 3\pi/4$ indicate minimums for dx/dt and $\theta = \pi/2$ indicates a maximum for dx/dt. This implies that the sprinkler waters longest when $\theta = \pi/4$ and $3\pi/4$. Thus, the lawn farthest from the sprinkler gets the most water.

51. $S = 6hs + \frac{3s^2}{2}\left(\frac{\sqrt{3} - \cos\theta}{\sin\theta}\right), \frac{\pi}{6} \le \theta \le \frac{\pi}{2}$

$\frac{dS}{d\theta} = \frac{3s^2}{2}\left(-\sqrt{3}\csc\theta\cot\theta + \csc^2\theta\right) = \frac{3s^2}{2}\csc\theta\left(-\sqrt{3}\cot\theta + \csc\theta\right) = 0$

$\csc\theta = \sqrt{3}\cot\theta$

$\sec\theta = \sqrt{3}$

$\theta = \operatorname{arcsec}\sqrt{3} \approx 0.9553$ radians

$S\left(\frac{\pi}{6}\right) = 6hs + \frac{3s^2}{2}\left(\sqrt{3}\right)$

$S\left(\frac{\pi}{2}\right) = 6hs + \frac{3s^2}{2}\left(\sqrt{3}\right)$

$S\left(\operatorname{arcsec}\sqrt{3}\right) = 6hs + \frac{3s^2}{2}\left(\sqrt{2}\right)$

S is minimum when $\theta = \operatorname{arcsec}\sqrt{3} \approx 0.9553$ radians.

53. True. See Exercise 17. **55.** True.

57. $f(x) = [\![x]\!]$

The derivative of f is undefined at every integer and is zero at any noninteger real number. All real numbers are critical numbers.

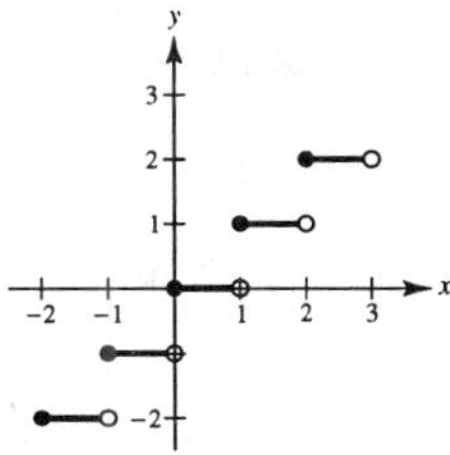

Section 3.2 Rolle's Theorem and the Mean Value Theorem

1. Rolle's Theorem does not apply to $f(x) = 1 - |x - 1|$ over $[0, 2]$ since f is not differentiable at $x = 1$.

3. $f(x) = x^2 - 2x, [0, 2]$

$f(0) = f(2) = 0$

f is continuous on $[0, 2]$. f is differentiable on $(0, 2)$. Rolle's Theorem applies.

$$f'(x) = 2x - 2$$

$$2x - 2 = 0 \Rightarrow x = 1$$

c value: 1

5. $f(x) = (x - 1)(x - 2)(x - 3), [1, 3]$

$f(1) = f(3) = 0$

f is continuous on $[1, 3]$. f is differentiable on $(1, 3)$. Rolle's Theorem applies.

$$f(x) = x^3 - 6x^2 + 11x - 6$$

$$f'(x) = 3x^2 - 12x + 11$$

$$3x^2 - 12x + 11 = 0 \Rightarrow x = \frac{6 \pm \sqrt{3}}{3}$$

$$c = \frac{6 - \sqrt{3}}{3}, c = \frac{6 + \sqrt{3}}{3}$$

7. $f(x) = x^{2/3} - 1, [-8, 8]$

$f(-8) = f(8) = 3$

f is continuous on $[-8, 8]$. f is not differentiable on $(-8, 8)$ since $f'(0)$ does not exist. Rolle's Theorem does not apply.

9. $f(x) = \dfrac{x^2 - 2x - 3}{x + 2}, [-1, 3]$

$f(-1) = f(3) = 0$

f is continuous on $[-1, 3]$. (**Note:** The discontinuity, $x = -2$, is not in the interval.) f is differentiable on $(-1, 3)$. Rolle's Theorem applies.

$$f'(x) = \frac{(x + 2)(2x - 2) - (x^2 - 2x - 3)(1)}{(x + 2)^2} = 0$$

$$\frac{x^2 + 4x - 1}{(x + 2)^2} = 0$$

$$x = \frac{-4 \pm 2\sqrt{5}}{2} = -2 \pm \sqrt{5}$$

c value: $-2 + \sqrt{5}$

11. $f(x) = \sin x,\ [0, 2\pi]$

$f(0) = f(2\pi) = 0$

f is continuous on $[0, 2\pi]$. f is differentiable on $(0, 2\pi)$. Rolle's Theorem applies.

$$f'(x) = \cos x$$

c values: $\frac{\pi}{2}, \frac{3\pi}{2}$

13. $f(x) = \sin 2x,\ \left[\frac{\pi}{6}, \frac{\pi}{3}\right]$

$$f\left(\frac{\pi}{6}\right) = f\left(\frac{\pi}{3}\right) = \frac{\sqrt{3}}{2}$$

f is continuous on $[\pi/6, \pi/3]$. f is differentiable on $(\pi/6, \pi/3)$. Rolle's Theorem applies.

$$f'(x) = 2\cos 2x$$

$$2\cos 2x = 0$$

$$x = \frac{\pi}{4}$$

c value: $\frac{\pi}{4}$

15. $f(x) = \tan x,\ [0, \pi]$

$f(0) = f(\pi) = 0$

f is not continuous on $[0, \pi]$ since $f(\pi/2)$ does not exist. Rolle's Theorem does not apply.

17. $f(x) = |x| - 1,\ [-1, 1]$

$f(-1) = f(1) = 0$

f is continuous on $[-1, 1]$. f is not differentiable on $(-1, 1)$ since $f'(0)$ does not exist. Rolle's Theorem does not apply.

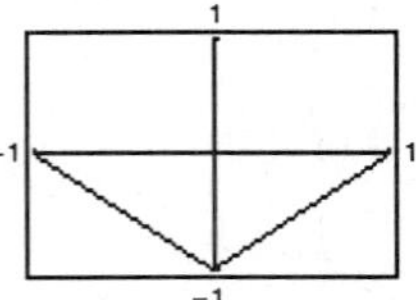

19. $f(x) = 4x - \tan \pi x,\ \left[-\frac{1}{4}, \frac{1}{4}\right]$

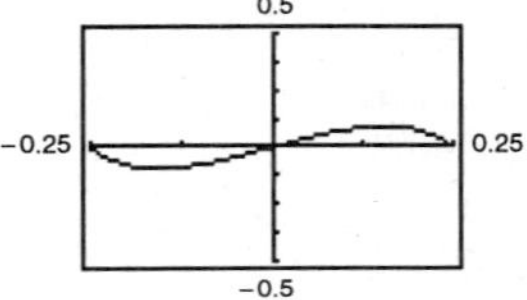

$$f\left(-\frac{1}{4}\right) = f\left(\frac{1}{4}\right) = 0$$

f is continuous on $[-1/4, 1/4]$. f is differentiable on $(-1/4, 1/4)$. Rolle's Theorem applies.

$$f'(x) = 4 - \pi \sec^2 \pi x = 0$$

$$\sec^2 \pi x = \frac{4}{\pi}$$

$$\sec \pi x = \pm\frac{2}{\sqrt{\pi}}$$

$$x = \pm\frac{1}{\pi}\operatorname{arcsec}\frac{2}{\sqrt{\pi}} = \pm\frac{1}{\pi}\arccos\frac{\sqrt{\pi}}{2}$$

$$\approx \pm 0.1533 \text{ radian}$$

c values: ± 0.1533 radian

21. $f(t) = -16t^2 + 48t + 32$

(a) $f(1) = f(2) = 64$

(b) $v = f'(t)$ must be 0 at some time in $[1, 2]$.

$$f'(t) = -32t + 48 = 0$$

$$t = \tfrac{3}{2} \text{ seconds}$$

23. No. Let $f(x) = x^2$ on $[-1, 2]$.

$$f'(x) = 2x$$

$f'(0) = 0$ and zero is in the interval $(-1, 2)$ but $f(-1) \neq f(2)$.

25. (a) f is continuous on $[-10, 4]$ and changes sign, $(f(-8) > 0,\ f(3) < 0)$. By the Intermediate Value Theorem, there exists at least one value of x in $[-10, 4]$ satisfying $f(x) = 0$.

—CONTINUED—

(b) There exist real numbers a and b such that $-10 < a < b < 4$ and $f(a) = f(b) = 2$. Therefore, by Rolle's Theorem there exists at least one number c in $(-10, 4)$ such that $f'(c) = 0$. This is called a critical number.

25. —CONTINUED—

(c)

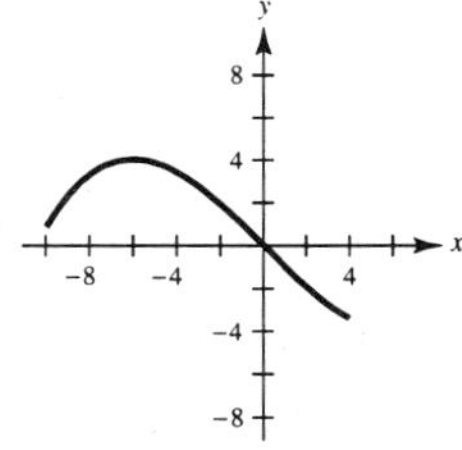

(d)

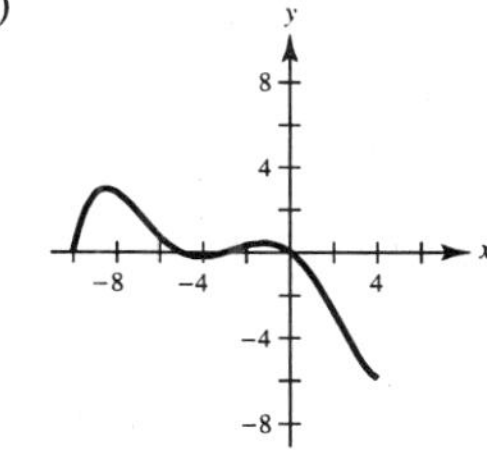

(e) No, f' did not have to be continuous on $[-10, 4]$.

27. $f(x) = x^2$ is continuous on $[-2, 1]$ and differentiable on $(-2, 1)$.

$$\frac{f(1) - f(-2)}{1 - (-2)} = \frac{1 - 4}{3} = -1$$

$f'(x) = 2x = -1$ when $x = -\frac{1}{2}$. Therefore, $c = -\frac{1}{2}$.

29. $f(x) = x^{2/3}$ is continuous on $[0, 1]$ and differentiable on $(0, 1)$.

$$\frac{f(1) - f(0)}{1 - 0} = 1$$

$$f'(x) = \frac{2}{3}x^{-1/3} = 1$$

$$x = \left(\frac{2}{3}\right)^3 = \frac{8}{27}$$

$$c = \frac{8}{27}$$

31. $f(x) = \sqrt{x - 2}$ is continuous on $[2, 6]$ and differentiable on $(2, 6)$.

$$\frac{f(6) - f(2)}{6 - 2} = \frac{2 - 0}{4} = \frac{1}{2}$$

$$f'(x) = \frac{1}{2\sqrt{x - 2}} = \frac{1}{2}$$

$$\sqrt{x - 2} = 1$$

$$c = 3$$

33. $f(x) = \sin x$ is continuous on $[0, \pi]$ and differentiable on $(0, \pi)$.

$$\frac{f(\pi) - f(0)}{\pi - 0} = \frac{0 - 0}{\pi} = 0$$

$$f'(x) = \cos x = 0$$

$$c = \frac{\pi}{2}$$

35. $f(x) = \dfrac{x}{x + 1}$ on $\left[-\dfrac{1}{2}, 2\right]$.

(a)

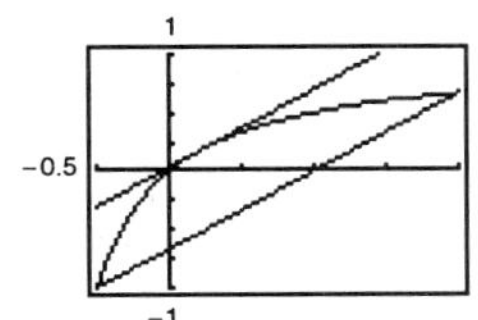

(b) Secant line:

$$\text{slope} = \frac{f(2) - f(-1/2)}{2 - (-1/2)} = \frac{2/3 - (-1)}{5/2} = \frac{2}{3}$$

$$y - \frac{2}{3} = \frac{2}{3}(x - 2)$$

$$3y - 2 = 2x - 4$$

$$3y - 2x + 2 = 0$$

(c) $f'(x) = \dfrac{1}{(x + 1)^2} = \dfrac{2}{3}$

$$(x + 1)^2 = \frac{3}{2}$$

$$x = -1 \pm \sqrt{\frac{3}{2}} = -1 \pm \frac{\sqrt{6}}{2}$$

In the interval $[-1/2, 2]$, $c = -1 + (\sqrt{6}/2)$.

$$f(c) = \frac{-1 + (\sqrt{6}/2)}{[-1 + (\sqrt{6}/2)] + 1} = \frac{-2 + \sqrt{6}}{\sqrt{6}} = \frac{-2}{\sqrt{6}} + 1$$

Tangent line: $y - 1 + \dfrac{2}{\sqrt{6}} = \dfrac{2}{3}\left(x - \dfrac{\sqrt{6}}{2} + 1\right)$

$$y - 1 + \frac{\sqrt{6}}{3} = \frac{2}{3}x - \frac{\sqrt{6}}{3} + \frac{2}{3}$$

$$3y - 2x - 5 + 2\sqrt{6} = 0$$

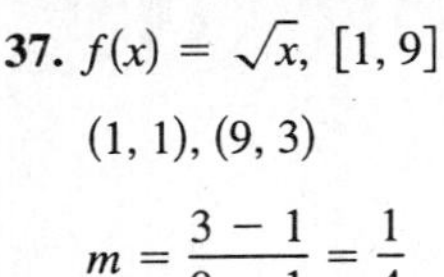

37. $f(x) = \sqrt{x},\ [1, 9]$

$(1, 1), (9, 3)$

$$m = \frac{3 - 1}{9 - 1} = \frac{1}{4}$$

(a)

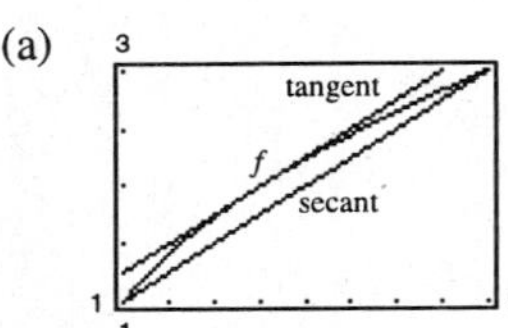

(b) Secant line: $y - 1 = \frac{1}{4}(x - 1)$

$$y = \frac{1}{4}x + \frac{3}{4}$$

$$0 = x - 4y + 3$$

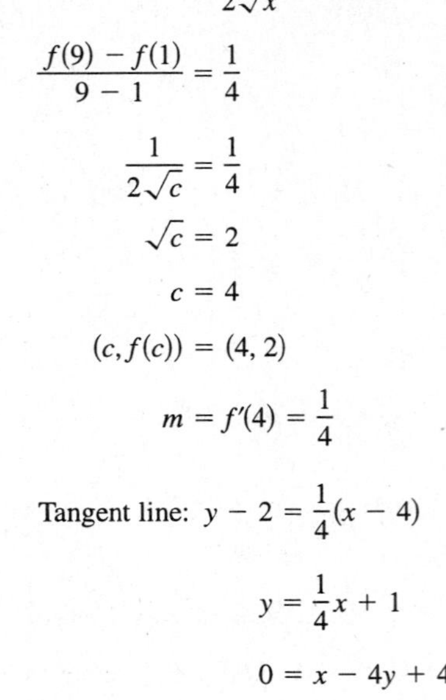

(c) $$f'(x) = \frac{1}{2\sqrt{x}}$$

$$\frac{f(9) - f(1)}{9 - 1} = \frac{1}{4}$$

$$\frac{1}{2\sqrt{c}} = \frac{1}{4}$$

$$\sqrt{c} = 2$$

$$c = 4$$

$$(c, f(c)) = (4, 2)$$

$$m = f'(4) = \frac{1}{4}$$

Tangent line: $y - 2 = \frac{1}{4}(x - 4)$

$$y = \frac{1}{4}x + 1$$

$$0 = x - 4y + 4$$

39. $f(x) = \dfrac{1}{x - 3},\ [0, 6]$

f has a discontinuity at $x = 3$.

41. f is continuous on $[-5, 5]$ and does not satisfy the conditions of the Mean Value Theorem.$\Rightarrow f$ is not differentiable on $(-5, 5)$.

Example: $f(x) = |x|$

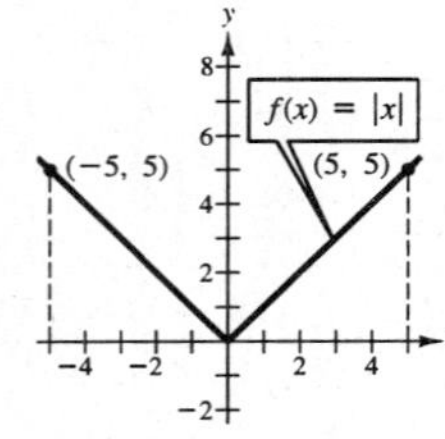

43. $s(t) = -4.9t^2 + 500$

(a) $$V_{\text{avg}} = \frac{s(3) - s(0)}{3 - 0} = \frac{455.9 - 500}{3} = -14.7 \text{ m/sec}$$

(b) $s(t)$ is continuous on $[0, 3]$ and differentiable on $(0, 3)$. Therefore, the Mean Value Theorem applies.

$$v(t) = s'(t) = -9.8t = -14.7$$

$$t = \frac{-14.7}{-9.8} = 1.5 \text{ seconds}$$

45. Let $S(t)$ be the postition function of the plane. If $t = 0$ corresponds to 2 P.M., $S(0) = 0$, $S(5.5) = 2500$ and the Mean Value Theorem says that there exists a time t_0, $0 < t_0 < 5.5$, such that

$$S'(t_0) = v(t_0) = \frac{2500 - 0}{5.5 - 0} \approx 454.54.$$

Applying the Intermediate Value Theorem to the velocity function on the intervals $[0, t_0]$ and $[t_0, 5.5]$, you see that there are at least two times during the flight when the speed was 400 miles per hour. $(0 < 400 < 454.54)$

47. False. $f(x) = 1/x$ has a discontinuity at $x = 0$.

49. True. A polynomial is continuous and differentiable everywhere.

51. Suppose that $p(x) = x^{2n+1} + ax + b$ has two real roots x_1 and x_2. Then by Rolle's Theorem, since $p(x_1) = p(x_2) = 0$, there exists c in (x_1, x_2) such that $p'(c) = 0$. But $p'(x) = (2n + 1)x^{2n} + a \neq 0$, since $n > 0, a > 0$. Therefore, $p(x)$ cannot have two real roots.

53. If $p(x) = Ax^2 + Bx + C$, then

$$p'(x) = 2Ax + B = \frac{f(b) - f(a)}{b - a} = \frac{(Ab^2 + Bb + C) - (Aa^2 + Ba + C)}{b - a}$$

$$= \frac{A(b^2 - a^2) + B(b - a)}{b - a}$$

$$= \frac{(b - a)[A(b + a) + B]}{b - a}$$

$$= A(b + a) + B.$$

Thus, $2Ax = A(b + a)$ and $x = (b + a)/2$ which is the midpoint of $[a, b]$.

55. $f(x) = \frac{1}{2}\cos x$ differentiable on $(-\infty, \infty)$.

$f'(x) = -\frac{1}{2}\sin x$

$-\frac{1}{2} \le f'(x) \le \frac{1}{2} \Rightarrow f'(x) < 1$ for all real numbers.

Thus, from Exercise 54, f has, at most, one fixed point. $(x \approx 0.4502)$

Section 3.3 Increasing and Decreasing Functions and the First Derivative Test

1. $f(x) = x^2 - 6x + 8$

Increasing on: $(3, \infty)$

Decreasing on: $(-\infty, 3)$

3. $y = \dfrac{x^3}{4} - 3x$

Increasing on: $(-\infty, -2), (2, \infty)$

Decreasing on: $(-2, 2)$

5. $f(x) = \dfrac{1}{x^2}$

Increasing on: $(-\infty, 0)$

Decreasing on: $(0, \infty)$

7. $f(x) = -2x^2 + 4x + 3$

$f'(x) = -4x + 4 = 0$

Critical number: $x = 1$

Test intervals:	$-\infty < x < 1$	$1 < x < \infty$
Sign of $f'(x)$:	$f' > 0$	$f' < 0$
Conclusion:	Increasing	Decreasing

Increasing on: $(-\infty, 1)$

Decreasing on: $(1, \infty)$

Relative maximum: $(1, 5)$

9. $f(x) = x^2 - 6x$

$f'(x) = 2x - 6 = 0$

Critical number: $x = 3$

Test intervals:	$-\infty < x < 3$	$3 < x < \infty$
Sign of $f'(x)$:	$f' < 0$	$f' > 0$
Conclusion:	Decreasing	Increasing

Increasing on: $(3, \infty)$

Decreasing on: $(-\infty, 3)$

Relative minimum: $(3, -9)$

11. $f(x) = 2x^3 + 3x^2 - 12x$

$f'(x) = 6x^2 + 6x - 12 = 6(x + 2)(x - 1) = 0$

Critical numbers: $x = -2, 1$

Test intervals:	$-\infty < x < -2$	$-2 < x < 1$	$1 < x < \infty$
Sign of $f'(x)$:	$f' > 0$	$f' < 0$	$f' > 0$
Conclusion:	Increasing	Decreasing	Increasing

Increasing on: $(-\infty, -2), (1, \infty)$

Decreasing on: $(-2, 1)$

Relative maximum: $(-2, 20)$

Relative minimum: $(1, -7)$

13. $f(x) = \dfrac{x^5 - 5x}{5}$

$f'(x) = x^4 - 1$

Critical numbers: $x = -1, 1$

Test intervals:	$-\infty < x < -1$	$-1 < x < 1$	$1 < x < \infty$
Sign of $f'(x)$:	$f' > 0$	$f' < 0$	$f' > 0$
Conclusion:	Increasing	Decreasing	Increasing

Increasing on: $(-\infty, -1), (1, \infty)$
Decreasing on: $(-1, 1)$
Relative maximum: $\left(-1, \frac{4}{5}\right)$
Relative minimum: $\left(1, -\frac{4}{5}\right)$

15. $f(x) = x^{1/3} + 1$

$f'(x) = \dfrac{1}{3}x^{-2/3} = \dfrac{1}{3x^{2/3}}$

Critical number: $x = 0$

Test intervals:	$-\infty < x < 0$	$0 < x < \infty$
Sign of $f'(x)$:	$f' > 0$	$f' > 0$
Conclusion:	Increasing	Increasing

Increasing on: $(-\infty, \infty)$
No relative extrema

17. $f(x) = 5 - |x - 5|$

$f'(x) = -\dfrac{x - 5}{|x - 5|} = \begin{cases} 1, & x < 5 \\ -1, & x > 5 \end{cases}$

Critical number: $x = 5$

Test intervals:	$-\infty < x < 5$	$5 < x < \infty$
Sign of $f'(x)$:	$f' > 0$	$f' < 0$
Conclusion:	Increasing	Decreasing

Increasing on: $(-\infty, 5)$
Decreasing on: $(5, \infty)$
Relative maximum: $(5, 5)$

19. $f(x) = \dfrac{x^2}{x^2 - 9}$

$f'(x) = \dfrac{(x^2 - 9)(2x) - (x^2)(2x)}{(x^2 - 9)^2} = \dfrac{-18x}{(x^2 - 9)^2}$

Critical number: $x = 0$
Discontinuities: $x = -3, 3$

Test intervals:	$-\infty < x < -3$	$-3 < x < 0$	$0 < x < 3$	$3 < x < \infty$
Sign of $f'(x)$:	$f' > 0$	$f' > 0$	$f' < 0$	$f' < 0$
Conclusion:	Increasing	Increasing	Decreasing	Decreasing

Increasing on: $(-\infty, -3), (-3, 0)$
Decreasing on: $(0, 3), (3, \infty)$
Relative maximum: $(0, 0)$

21. $f(x) = x^3 - 6x^2 + 15$

$f'(x) = 3x^2 - 12x = 3x(x - 4) = 0$

Crititcal numbers: $x = 0, 4$

Test intervals:	$-\infty < x < 0$	$0 < x < 4$	$4 < x < \infty$
Sign of $f'(x)$:	$f' > 0$	$f' < 0$	$f' > 0$
Conclusion:	Increasing	Decreasing	Increasing

Increasing on: $(-\infty, 0), (4, \infty)$
Decreasing on: $(0, 4)$
Relative maximum: $(0, 15)$
Relative minimum: $(4, -17)$

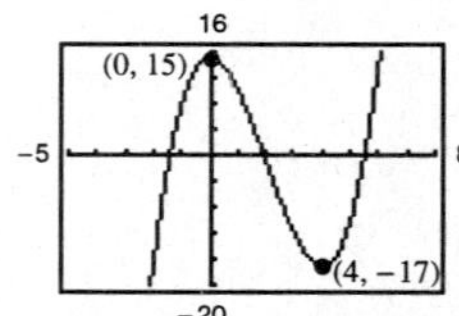

23. $f(x) = (x-1)^{2/3}$

$$f'(x) = \frac{2}{3(x-1)^{1/3}}$$

Critical number: $x = 1$

Test intervals:	$-\infty < x < 1$	$1 < x < \infty$
Sign of $f'(x)$:	$f' < 0$	$f' > 0$
Conclusion:	Decreasing	Increasing

Increasing on: $(1, \infty)$

Decreasing on: $(-\infty, 1)$

Relative minimum: $(1, 0)$

25. $f(x) = x + \dfrac{1}{x}$

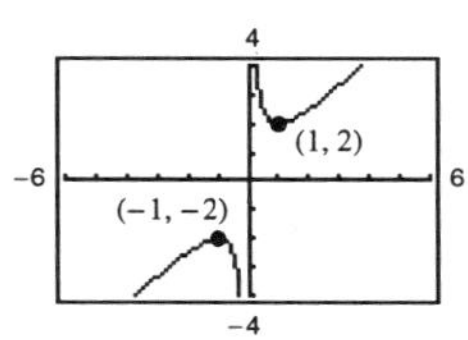

$$f'(x) = 1 - \frac{1}{x^2} = \frac{x^2-1}{x^2}$$

Critical numbers: $x = -1, 1$

Discontinuity: $x = 0$

Test intervals:	$-\infty < x < -1$	$-1 < x < 0$	$0 < x < 1$	$1 < x < \infty$
Sign of $f'(x)$:	$f' > 0$	$f' < 0$	$f' < 0$	$f' > 0$
Conclusion:	Increasing	Decreasing	Decreasing	Increasing

Increasing on: $(-\infty, -1), (1, \infty)$

Decreasing on: $(-1, 0), (0, 1)$

Relative maximum: $(-1, -2)$

Relative minimum: $(1, 2)$

27. $f(x) = \dfrac{x^2 - 2x + 1}{x+1}$

$$f'(x) = \frac{(x+1)(2x-2) - (x^2-2x+1)(1)}{(x+1)^2} = \frac{x^2+2x-3}{(x+1)^2} = \frac{(x+3)(x-1)}{(x+1)^2}$$

Critical numbers: $x = -3, 1$

Discontinuity: $x = -1$

Test intervals:	$-\infty < x < -3$	$-3 < x < -1$	$-1 < x < 1$	$1 < x < \infty$
Sign of $f'(x)$:	$f' > 0$	$f' < 0$	$f' < 0$	$f' > 0$
Conclusion:	Increasing	Decreasing	Decreasing	Increasing

Increasing on: $(-\infty, -3), (1, \infty)$

Decreasing on: $(-3, -1), (-1, 1)$

Relative maximum: $(-3, -8)$

Relative minimum: $(1, 0)$

29. $f(x) = \dfrac{x}{2} + \cos x,\ 0 < x < 2\pi$

$f'(x) = \dfrac{1}{2} - \sin x = 0$

Critical numbers: $x = \dfrac{\pi}{6}, \dfrac{5\pi}{6}$

Test intervals:	$0 < x < \dfrac{\pi}{6}$	$\dfrac{\pi}{6} < x < \dfrac{5\pi}{6}$	$\dfrac{5\pi}{6} < x < 2\pi$
Sign of $f'(x)$:	$f' > 0$	$f' < 0$	$f' > 0$
Conclusion:	Increasing	Decreasing	Increasing

Increasing on: $\left(0, \dfrac{\pi}{6}\right), \left(\dfrac{5\pi}{6}, 2\pi\right)$

Decreasing on: $\left(\dfrac{\pi}{6}, \dfrac{5\pi}{6}\right)$

Relative maximum: $\left(\dfrac{\pi}{6}, \dfrac{\pi + 6\sqrt{3}}{12}\right)$

Relative minimum: $\left(\dfrac{5\pi}{6}, \dfrac{5\pi - 6\sqrt{3}}{12}\right)$

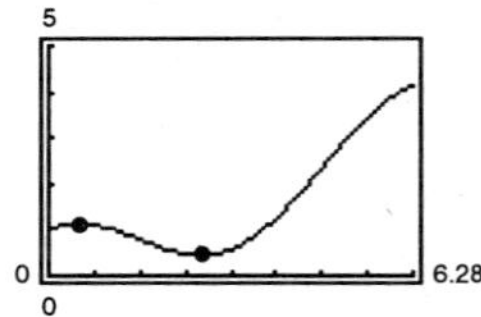

31. $f(x) = \sin^2 x + \sin x,\ 0 < x < 2\pi$

$f'(x) = 2\sin x \cos x + \cos x = \cos x(2\sin x + 1) = 0$

Critical numbers: $x = \dfrac{\pi}{2}, \dfrac{7\pi}{6}, \dfrac{3\pi}{2}, \dfrac{11\pi}{6}$

Test intervals:	$0 < x < \dfrac{\pi}{2}$	$\dfrac{\pi}{2} < x < \dfrac{7\pi}{6}$	$\dfrac{7\pi}{6} < x < \dfrac{3\pi}{2}$	$\dfrac{3\pi}{2} < x < \dfrac{11\pi}{6}$	$\dfrac{11\pi}{6} < x < 2\pi$
Sign of $f'(x)$:	$f' > 0$	$f' < 0$	$f' > 0$	$f' < 0$	$f' > 0$
Conclusion:	Increasing	Decreasing	Increasing	Decreasing	Increasing

Increasing on: $\left(0, \dfrac{\pi}{2}\right), \left(\dfrac{7\pi}{6}, \dfrac{3\pi}{2}\right), \left(\dfrac{11\pi}{6}, 2\pi\right)$

Decreasing on: $\left(\dfrac{\pi}{2}, \dfrac{7\pi}{6}\right), \left(\dfrac{3\pi}{2}, \dfrac{11\pi}{6}\right)$

Relative minima: $\left(\dfrac{7\pi}{6}, -\dfrac{1}{4}\right), \left(\dfrac{11\pi}{6}, -\dfrac{1}{4}\right)$

Relative maxima: $\left(\dfrac{\pi}{2}, 2\right), \left(\dfrac{3\pi}{2}, 0\right)$

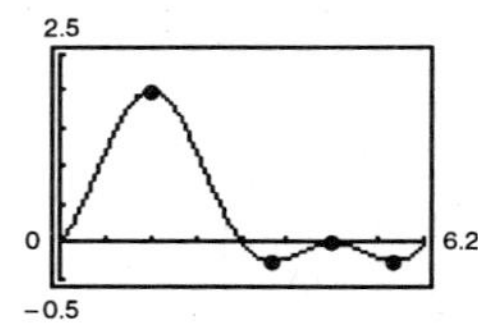

33. $f(x) = 2x\sqrt{9 - x^2},\ [-3, 3]$

(a) $f'(x) = \dfrac{2(9 - 2x^2)}{\sqrt{9 - x^2}}$

(b)

(c) $\dfrac{2(9 - 2x^2)}{\sqrt{9 - x^2}} = 0$

Critical numbers: $x = \pm\dfrac{3}{\sqrt{2}} = \pm\dfrac{3\sqrt{2}}{2}$

(d) Intervals:

$\left(-3, -\dfrac{3\sqrt{2}}{2}\right)$	$\left(-\dfrac{3\sqrt{2}}{2}, \dfrac{3\sqrt{2}}{2}\right)$	$\left(\dfrac{3\sqrt{2}}{2}, 3\right)$
$f'(x) < 0$	$f'(x) > 0$	$f'(x) < 0$
Decreasing	Increasing	Decreasing

f is increasing when f' is positive and decreasing when f' is negative.

35. $f(t) = t^2 \sin t, [0, 2\pi]$

(a) $f'(t) = t^2 \cos t + 2t \sin t$

$= t(t \cos t + 2 \sin t)$

(b)

(c) $t(t \cos t + 2 \sin t) = 0$

$t = 0$ or $t = -2 \tan t$

$t \cot t = -2$

$t \approx 2.2889, 5.0870$ (graphing utility)

Critical numbers: $t = 2.2889$, $t = 5.0870$

(d) Intervals:

$(0, 2.2889)$	$(2.2889, 5.0870)$	$(5.0870, 2\pi)$
$f'(t) > 0$	$f'(t) < 0$	$f'(t) > 0$
Increasing	Decreasing	Increasing

f is increasing when f' is positive and decreasing when f' is negative.

In Exercises 37–41, $f'(x) > 0$ on $(-\infty, -4)$, $f'(x) < 0$ on $(-4, 6)$ and $f'(x) > 0$ on $(6, \infty)$.

37. $g(x) = f(x) + 5$

$g'(x) = f'(x)$

$g'(0) = f'(0) < 0$

39. $g(x) = -f(x)$

$g'(x) = -f'(x)$

$g'(-6) = -f'(-6) < 0$

41. $g(x) = f(x - 10)$

$g'(x) = f'(x - 10)$

$g'(0) = f'(-10) > 0$

43. $f(x) = c$ is constant $\Rightarrow f'(x) = 0$.

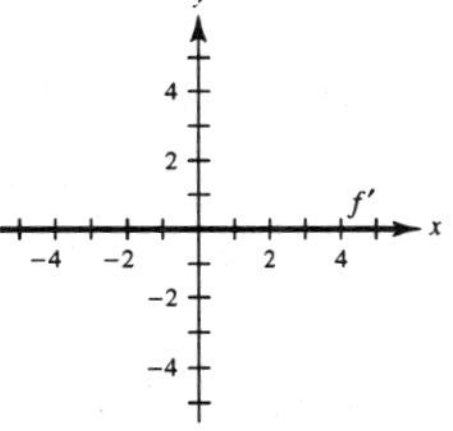

45. f is quadratic $\Rightarrow f'$ is a line.

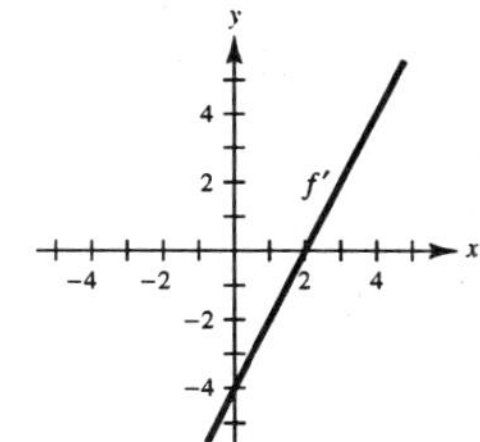

47. f has positive, but decreasing slope.

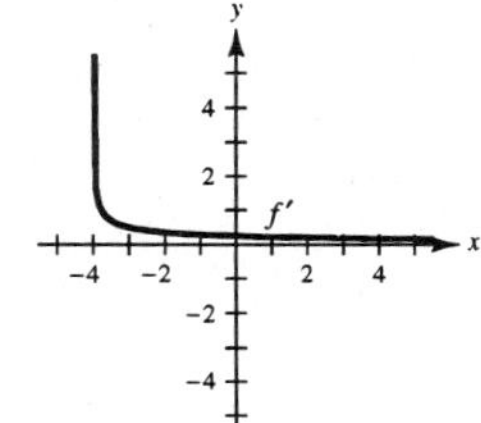

49. $f'(x) = \begin{cases} > 0, & x < 4 \Rightarrow f \text{ is increasing on } (-\infty, 4). \\ \text{undefined}, & x = 4 \\ < 0, & x > 4 \Rightarrow f \text{ is decreasing on } (4, \infty). \end{cases}$

Two possibilities for $f(x)$ are given below.

(a)

(b)

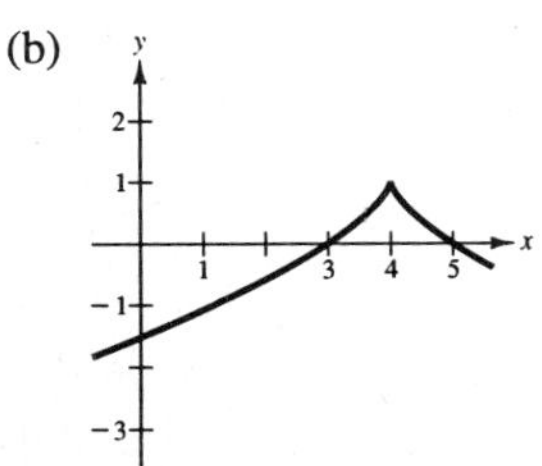

51. The critical numbers are in intervals $(-0.50, -0.25)$ and $(0.25, 0.50)$ since the sign of f' changes in these intervals. f is decreasing on approximately $(-1, -0.40)$, $(0.48, 1)$, and increasing on $(-0.40, 0.48)$.

Relative minimum when $x \approx -0.40$.

Relative maximum when $x \approx 0.48$.

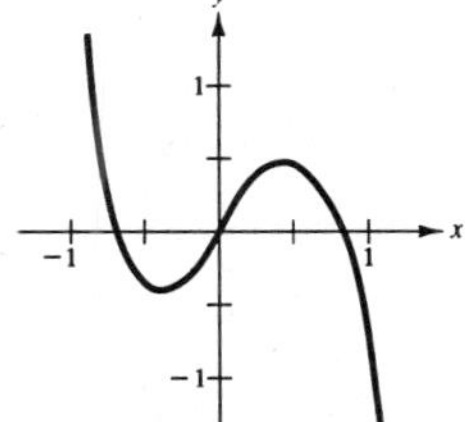

53. $f(x) = x$, $g(x) = \sin x$, $0 < x < \pi$

(a)

x	0.5	1	1.5	2	2.5	3
$f(x)$	0.5	1	1.5	2	2.5	3
$g(x)$	0.479	0.841	0.997	0.909	0.598	0.141

$f(x)$ seems greater than $g(x)$ on $(0, \pi)$.

(b)

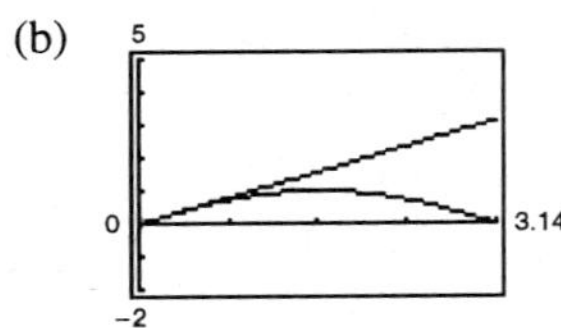
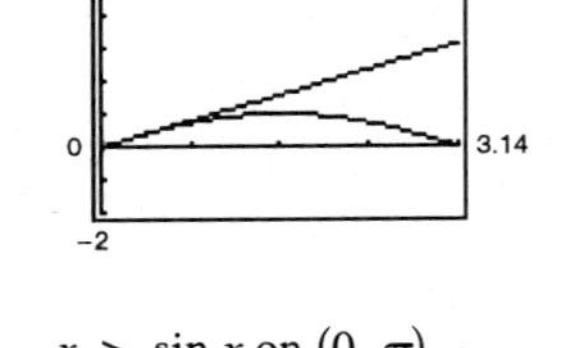

$x > \sin x$ on $(0, \pi)$

(c) Let $h(x) = f(x) - g(x) = x - \sin x$

$$h'(x) = 1 - \cos x > 0 \text{ on } (0, \pi).$$

Therefore, $h(x)$ is increasing on $(0, \pi)$. Since $h(0) = 0$, $h(x) > 0$ on $(0, \pi)$. Thus,

$$x - \sin x > 0$$

$$x > \sin x$$

$$f(x) > g(x) \text{ on } (0, \pi).$$

55. $v = k(R - r)r^2 = k(Rr^2 - r^3)$

$$v' = k(2Rr - 3r^2)$$

$$= kr(2R - 3r) = 0$$

$$r = 0 \text{ or } \frac{2}{3}R$$

Maximum when $r = \frac{2}{3}R$.

57. $P = \dfrac{vR_1R_2}{(R_1 + R_2)^2}$, v and R_1 are constant.

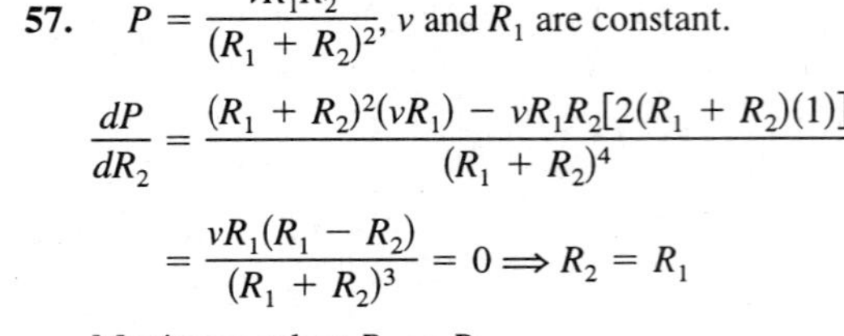

$$\frac{dP}{dR_2} = \frac{(R_1 + R_2)^2(vR_1) - vR_1R_2[2(R_1 + R_2)(1)]}{(R_1 + R_2)^4}$$

$$= \frac{vR_1(R_1 - R_2)}{(R_1 + R_2)^3} = 0 \Rightarrow R_2 = R_1$$

Maximum when $R_1 = R_2$.

59. (a) Using a graphing utility, you obtain

$$B = -0.1538t^4 + 3.5379t^3 - 19.7537t^2 + 42.3906t + 352.8234.$$

(b)

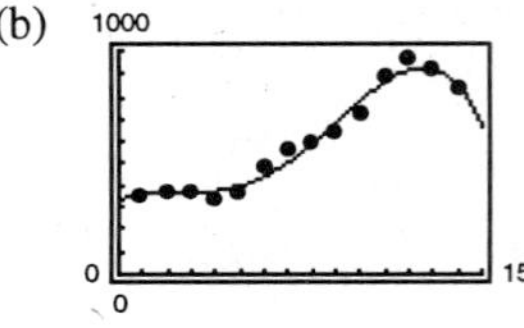

(c) The maximum of the model is $B = 951.5$ at $t = 12.58$.

61. (a) Use a cubic polynomial $f(x) = a_3x^3 + a_2x^2 + a_1x + a_0$.

(b) $f'(x) = 3a_3x^2 + 2a_2x + a_1$.

(0, 0):	$0 = a_0$	$(f(0) = 0)$
	$0 = a_1$	$(f'(0) = 0)$
(2, 2):	$2 = 8a_3 + 4a_2$	$(f(2) = 2)$
	$0 = 12a_3 + 4a_2$	$(f'(2) = 0)$

(c) The solution is $a_0 = a_1 = 0$, $a_2 = \frac{3}{2}$, $a_3 = -\frac{1}{2}$:

$$f(x) = -\frac{1}{2}x^3 + \frac{3}{2}x^2.$$

(d)

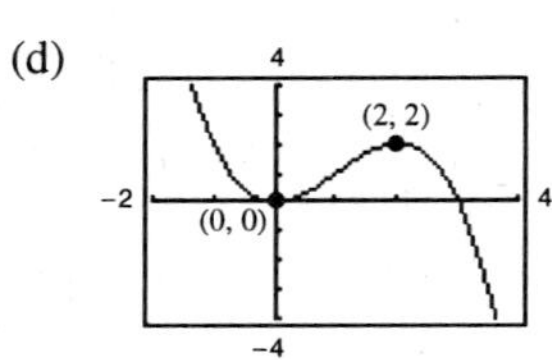

63. (a) Use a fourth degree polynomial $f(x) = a_4x^4 + a_3x^3 + a_2x^2 + a_1x + a_0$.

(b) $f'(x) = 4a_4x^3 + 3a_3x^2 + 2a_2x + a_1$

(0, 0):	$0 = a_0$	$(f(0) = 0)$
	$0 = a_1$	$(f'(0) = 0)$
(4, 0):	$0 = 256a_4 + 64a_3 + 16a_2$	$(f(4) = 0)$
	$0 = 256a_4 + 48a_3 + 8a_2$	$(f'(4) = 0)$
(2, 4):	$4 = 16a_4 + 8a_3 + 4a_2$	$(f(2) = 4)$
	$0 = 32a_4 + 12a_3 + 4a_2$	$(f'(2) = 0)$

(c) The solution is $a_0 = a_1 = 0$, $a_2 = 4$, $a_3 = -2$, $a_4 = \frac{1}{4}$.

$$f(x) = \frac{1}{4}x^4 - 2x^3 + 4x^2$$

(d)

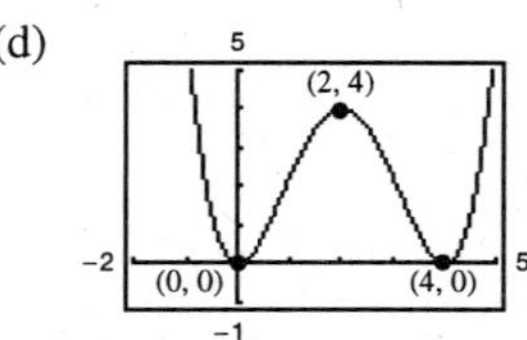

65. True

Let $h(x) = f(x) + g(x)$ where f and g are increasing. Then $h'(x) = f'(x) + g'(x) > 0$ since $f'(x) > 0$ and $g'(x) > 0$.

67. False

Let $f(x) = x^3$, then $f'(x) = 3x^2$ and f only has one critical number. Or, let $f(x) = x^3 + 3x + 1$, then $f'(x) = 3(x^2 + 1)$ has no critical numbers.

69. Assume that $f'(x) < 0$ for all x in the interval (a, b) and let $x_1 < x_2$ be any two points in the interval. By the Mean Value Theorem, we know there exists a number c such that $x_1 < c < x_2$, and

$$f'(c) = \frac{f(x_2) - f(x_1)}{x_2 - x_1}.$$

Since $f'(c) < 0$ and $x_2 - x_1 > 0$, then $f(x_2) - f(x_1) < 0$, which implies that $f(x_2) < f(x_1)$. Thus, f is decreasing on the interval.

71. Let $f(x) = (1 + x)^n - nx - 1$. Then

$$f'(x) = n(1 + x)^{n-1} - n$$

$$= n[(1 + x)^{n-1} - 1] > 0 \text{ since } x > 0 \text{ and } n > 1.$$

Thus, $f(x)$ is increasing on $(0, \infty)$.
Since $f(0) = 0 \Rightarrow f(x) > 0$ on $(0, \infty)$

$$(1 + x)^n - nx - 1 > 0 \Rightarrow (1 + x)^n > 1 + nx.$$

Section 3.4 Concavity and the Second Derivative Test

1. $y = x^2 - x - 2, y'' = 2$

Concave upward: $(-\infty, \infty)$

3. $f(x) = \dfrac{24}{x^2 + 12}, y'' = \dfrac{-144(4 - x^2)}{(x^2 + 12)^3}$

Concave upward: $(-\infty, -2), (2, \infty)$
Concave downward: $(-2, 2)$

5. $f(x) = \dfrac{x^2 + 1}{x^2 - 1}$

Concave upward: $(-\infty, -1), (1, \infty)$
Concave downward: $(-1, 1)$

7. $f(x) = 6x - x^2$

$f'(x) = 6 - 2x$

$f''(x) = -2$

Critical number: $x = 3$

$f''(3) < 0$

Therefore, $(3, 9)$ is relative maximum.

9. $f(x) = (x - 5)^2$

$f'(x) = 2(x - 5)$

$f''(x) = 2$

Critical number: $x = 5$

$f''(5) > 0$

Therefore, $(5, 0)$ is a relative minimum.

11. $f(x) = x^3 - 3x^2 + 3$

$f'(x) = 3x^2 - 6x = 3x(x - 2)$

$f''(x) = 6x - 6 = 6(x - 1)$

Critical numbers: $x = 0, x = 2$

$f''(0) = -6 < 0$

Therefore, $(0, 3)$ is a relative maximum.

$f''(2) = 6 > 0$

Therefore, $(2, -1)$ is a relative minimum.

13. $f(x) = x^4 - 4x^3 + 2$

$f'(x) = 4x^3 - 12x^2 = 4x^2(x - 3)$

$f''(x) = 12x^2 - 24x = 12x(x - 2)$

Critical numbers: $x = 0, x = 3$

However, $f''(0) = 0$, so we must use the First Derivative Test. $f'(x) < 0$ on the intervals $(-\infty, 0)$ and $(0, 3)$; hence, $(0, 2)$ is not an extremum. $f''(3) > 0$ so $(3, -25)$ is a relative minimum.

15. $f(x) = x^{2/3} - 3$

$$f'(x) = \frac{2}{3x^{1/3}}$$

$$f''(x) = \frac{-2}{9x^{4/3}}$$

Critical number: $x = 0$

However, $f''(0)$ is undefined, so we must use the First Derivative Test. Since $f'(x) < 0$ on $(-\infty, 0)$ and $f'(x) > 0$ on $(0, \infty)$, $(0, -3)$ is a relative minimum.

17. $f(x) = x + \frac{4}{x}$

$$f'(x) = 1 - \frac{4}{x^2} = \frac{x^2 - 4}{x^2}$$

$$f''(x) = \frac{8}{x^3}$$

Critical numbers: $x = \pm 2$

$f''(-2) < 0$

Therefore, $(-2, -4)$ is a relative maximum.

$f''(2) > 0$

Therefore, $(2, 4)$ is a relative minimum.

19. $f(x) = \cos x - x, \ 0 \le x \le 4\pi$

$f'(x) = -\sin x - 1 \le 0$

Therefore, f is non-increasing and there are no relative extrema.

21. $f(x) = x^3 - 12x$

$f'(x) = 3x^2 - 12 = 3(x + 2)(x - 2) = 0$ when $x = \pm 2$.

$f''(x) = 6x$

$f''(-2) = -12 < 0 \Rightarrow (-2, 16)$ is a relative maximum.

$f''(2) = 12 > 0 \Rightarrow (2, -16)$ is a relative minimum.

$f''(x) = 6x = 0$ when $x = 0$.

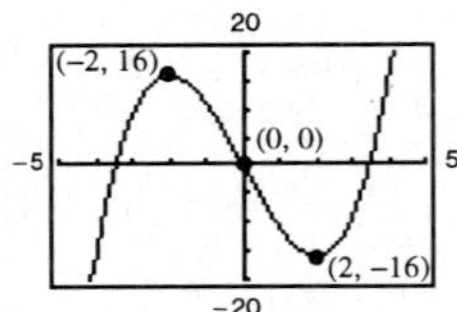

Test interval	$-\infty < x < 0$	$0 < x < \infty$
Sign of $f''(x)$	$f''(x) < 0$	$f''(x) > 0$
Conclusion	Concave downward	Concave upward

Point of inflection: $(0, 0)$

23. $f(x) = x^3 - 6x^2 + 12x$

$f'(x) = 3x^2 - 12x + 12$

$= 3(x - 2)^2 = 0$ when $x = 2$.

$f''(x) = 6(x - 2) = 0$ when $x = 2$.

Since $f'(x) > 0$ when $x \ne 2$ and the concavity changes at $x = 2$. $(2, 8)$ is a point of inflection.

25. $f(x) = \frac{1}{4}x^4 - 2x^2$

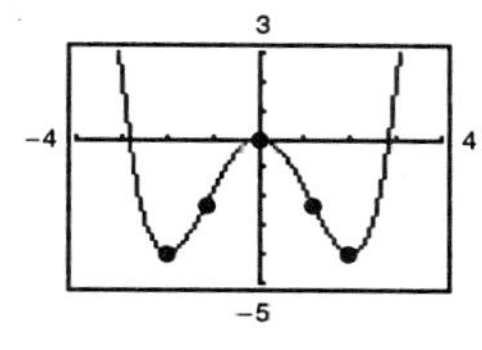

$f'(x) = x^3 - 4x = x(x+2)(x-2) = 0$ when $x = 0, \pm 2$.

$f''(x) = 3x^2 - 4$

$f''(-2) = 8 > 0 \Rightarrow (-2, -4)$ is a relative minimum.

$f''(0) = -4 < 0 \Rightarrow (0, 0)$ is a relative maximum.

$f''(2) = 8 > 0 \Rightarrow (2, -4)$ is a relative minimum.

$f''(x) = 3x^2 - 4 = 0$ when $x = \pm\frac{2}{\sqrt{3}}$.

Test interval	$-\infty < x < -\frac{2}{\sqrt{3}}$	$-\frac{2}{\sqrt{3}} < x < \frac{2}{\sqrt{3}}$	$\frac{2}{\sqrt{3}} < x < \infty$
Sign of $f''(x)$	$f''(x) > 0$	$f''(x) < 0$	$f''(x) > 0$
Conclusion	Concave upward	Concave downward	Concave upward

Points of inflection: $\left(\pm\frac{2}{\sqrt{3}}, -\frac{20}{9}\right)$

27. $f(x) = x(x-4)^3$

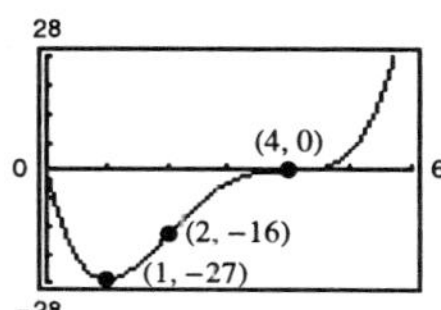

$f'(x) = x[3(x-4)^2] + (x-4)^3$

$= (x-4)^2(4x-4) = 4(x-1)(x-4)^2 = 0$ when $x = 1, 4$.

$f''(x) = 4(x-1)[2(x-4)] + 4(x-4)^2$

$= 4(x-4)[2(x-1) + (x-4)]$

$= 4(x-4)(3x-6) = 12(x-4)(x-2)$

$f''(1) = 36 > 0 \Rightarrow (1, -27)$ is a relative minimum.

$f''(4) = 0 \Rightarrow$ test fails

By the First Derivative Test we see that $x = 4$ does not yield a relative extrema.

$f''(x) = 12(x-4)(x-2) = 0$ when $x = 2, 4$.

Test interval	$-\infty < x < 2$	$2 < x < 4$	$4 < x < \infty$
Sign of $f''(x)$	$f''(x) > 0$	$f''(x) < 0$	$f''(x) > 0$
Conclusion	Concave upward	Concave downward	Concave upward

Points of inflection: $(2, -16), (4, 0)$

29. $f(x) = x\sqrt{x+3}$

Domain: $[-3, \infty)$

$$f'(x) = x\left(\frac{1}{2}\right)(x+3)^{-1/2} + \sqrt{x+3} = \frac{3(x+2)}{2\sqrt{x+3}} = 0 \text{ when } x = -2.$$

$$f''(x) = \frac{6\sqrt{x+3} - 3(x+2)(x+3)^{-1/2}}{4(x+3)} = \frac{3(x+4)}{4(x+3)^{3/2}}$$

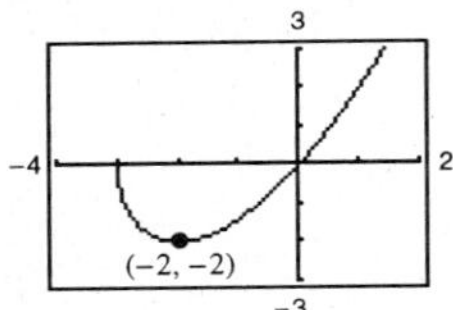

Since $f''(-2) > 0$, then $(-2, -2)$ is a relative minimum. $f''(x) > 0$ on the entire domain of f (except for $x = -3$, for which $f''(x)$ is undefined). There are no points of inflection.

31. $f(x) = \sin\left(\frac{x}{2}\right), 0 \le x \le 4\pi$

$$f'(x) = \frac{1}{2}\cos\left(\frac{x}{2}\right) = 0 \text{ when } x = \pi, 3\pi.$$

$$f''(x) = -\frac{1}{4}\sin\left(\frac{x}{2}\right)$$

$f''(\pi) < 0 \Rightarrow (\pi, 1)$ is a relative maximum.

$f''(3\pi) > 0 \Rightarrow (3\pi, -1)$ is a relative minimum.

$f''(x) = 0$ when $x = 0, 2\pi, 4\pi$.

Test interval	$0 < x < 2\pi$	$2\pi < x < 4\pi$
Sign of $f''(x)$	$f'' < 0$	$f'' > 0$
Conclusion	Concave downward	Concave upward

Point of inflection: $(2\pi, 0)$

33. $f(x) = \sec\left(x - \frac{\pi}{2}\right), 0 < x < 4\pi$

$$f'(x) = \sec\left(x - \frac{\pi}{2}\right)\tan\left(x - \frac{\pi}{2}\right) = 0 \text{ when } x = \frac{\pi}{2}, \frac{3\pi}{2}, \frac{5\pi}{2}, \frac{7\pi}{2}.$$

$$f''(x) = \sec^3\left(x - \frac{\pi}{2}\right) + \sec\left(x - \frac{\pi}{2}\right)\tan^2\left(x - \frac{\pi}{2}\right) \neq 0 \text{ for any } x \text{ in the domain of } f.$$

$f''\left(\frac{\pi}{2}\right) > 0 \Rightarrow \left(\frac{\pi}{2}, 1\right)$ is a relative minimum.

$f''\left(\frac{3\pi}{2}\right) < 0 \Rightarrow \left(\frac{3\pi}{2}, -1\right)$ is a relative maximum.

$f''\left(\frac{5\pi}{2}\right) > 0 \Rightarrow \left(\frac{5\pi}{2}, 1\right)$ is a relative minimum.

$f''\left(\frac{7\pi}{2}\right) < 0 \Rightarrow \left(\frac{7\pi}{2}, -1\right)$ is a relative maximum.

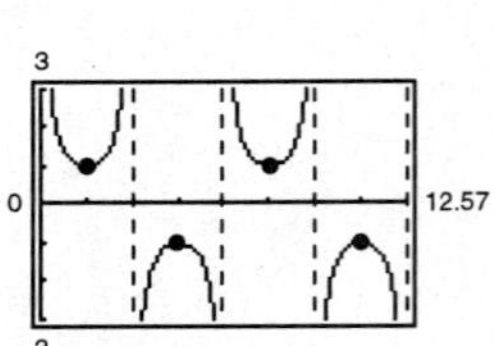

35. $f(x) = 2\sin x + \sin 2x, 0 \le x \le 2\pi$

$f'(x) = 2\cos x + 2\cos 2x = 0$

$2(2\cos^2 x + \cos x - 1) = 0$

$(\cos x + 1)(2\cos x - 1) = 0$ when $\cos x = \frac{1}{2}, -1$ or $x = \frac{\pi}{3}, \pi, \frac{5\pi}{3}.$

$f''(x) = -2\sin x - 4\sin 2x = -2\sin x(1 + 4\cos x)$

$f''\left(\frac{\pi}{3}\right) < 0 \Rightarrow \left(\frac{\pi}{3}, 2.598\right)$ is a relative maximum.

$f''\left(\frac{5\pi}{3}\right) > 0 \Rightarrow \left(\frac{5\pi}{3}, -2.598\right)$ is a relative minimum.

$f''(\pi) = 0$. By the first derivative test, $(\pi, 0)$ is not a relative extrema.

$f''(x) = 0$ when $x = 0, 1.823, \pi, 4.460.$

Test interval	$0 < x < 1.823$	$1.823 < x < \pi$	$\pi < x < 4.460$	$4.460 < x < 2\pi$
Sign of $f''(x)$	$f'' < 0$	$f'' > 0$	$f'' < 0$	$f'' > 0$
Conclusion	Concave downward	Concave upward	Concave downward	Concave upward

Points of inflection: $(1.823, 1.452), (\pi, 0), (4.46, -1.452)$

37. $f(x) = 0.2x^2(x-3)^3, [-1, 4]$

(a) $f'(x) = 0.2x(5x-6)(x-3)^2$

$f''(x) = (x-3)(4x^2 - 9.6x + 3.6)$

$= 0.4(x-3)(10x^2 - 24x + 9)$

(b) $f''(0) < 0 \Rightarrow (0, 0)$ is a relative maximum.

$f''\left(\frac{6}{5}\right) > 0 \Rightarrow (1.2, -1.6796)$ is a relative minimum.

Points of inflection:

$(3, 0), (0.4652, -0.7049), (1.9348, -0.9049)$

(c) f is increasing when $f' > 0$ and decreasing when $f' < 0$. f is concave upward when $f'' > 0$ and concave downward when $f'' < 0$.

39. $f(x) = \sin x - \frac{1}{3}\sin 3x + \frac{1}{5}\sin 5x, [0, \pi]$

(a) $f'(x) = \cos x - \cos 3x + \cos 5x$

$f'(x) = 0$ when $x = \frac{\pi}{6}, x = \frac{\pi}{2}, x = \frac{5\pi}{6}.$

$f''(x) = -\sin x + 3\sin 3x - 5\sin 5x$

$f''(x) = 0$ when $x = \frac{\pi}{6}, x = \frac{5\pi}{6}, x \approx 1.1731, x \approx 1.9685$

(b) $f''\left(\frac{\pi}{2}\right) < 0 \Rightarrow \left(\frac{\pi}{2}, 1.53333\right)$ is a relative maximum.

Points of inflection: $\left(\frac{\pi}{6}, 0.2667\right), (1.1731, 0.9638),$

$(1.9685, 0.9637), \left(\frac{5\pi}{6}, 0.2667\right)$

Note: $(0, 0)$ and $(\pi, 0)$ are not points of inflection since they are endpoints.

(c)

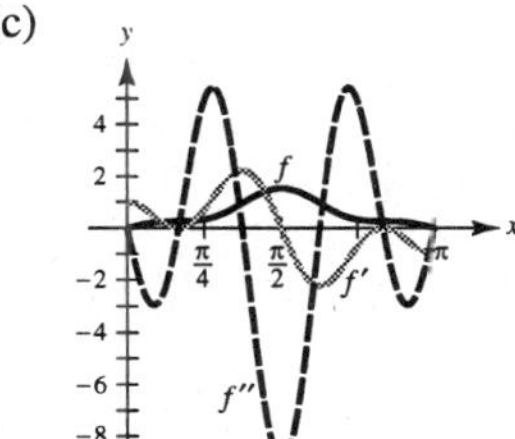

The graph of f is increasing when $f' > 0$ and decreasing when $f' < 0$. f is concave upward when $f'' > 0$ and concave downward when $f'' < 0$.

41.

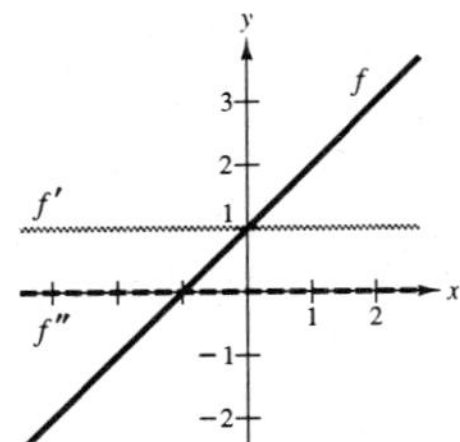

43.

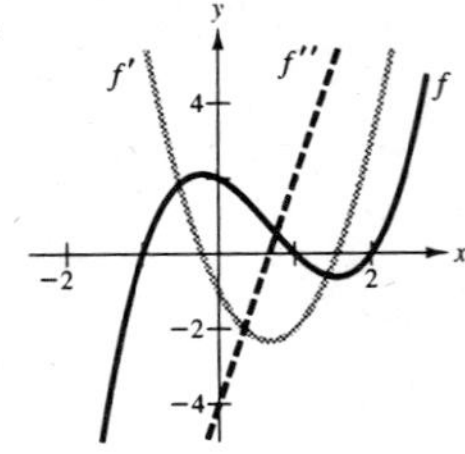

45. (a)

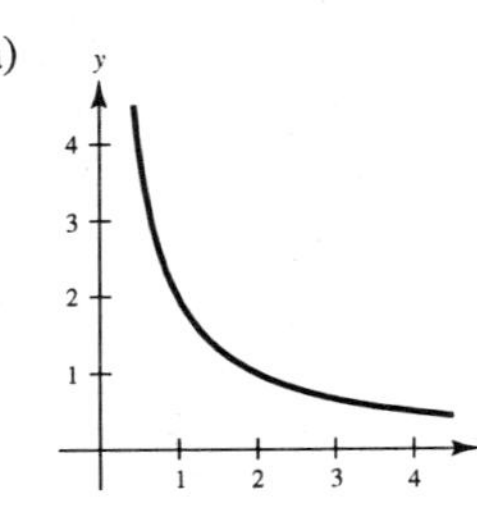

$f' < 0$ means f decreasing

f' increasing means concave upward

(b)

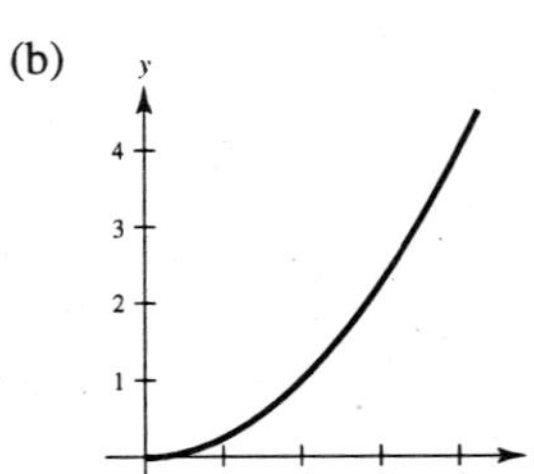

$f' > 0$ means f increasing

f' increasing means concave upward

47.

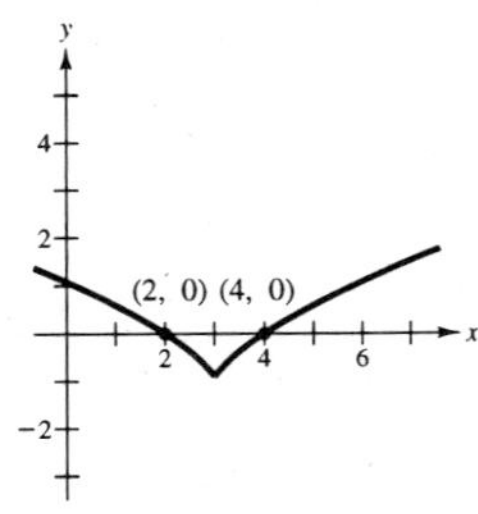

49.

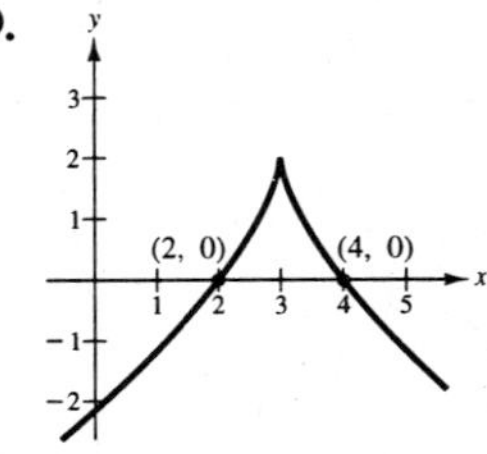

51. $f(x) = ax^3 + bx^2 + cx + d$

Relative maximum: $(3, 3)$

Relative minimum: $(5, 1)$

Point of inflection: $(4, 2)$

$f'(x) = 3ax^2 + 2bx + c, f''(x) = 6ax + 2b$

$$\left.\begin{array}{l} f(3) = 27a + 9b + 3c + d = 3 \\ f(5) = 125a + 25b + 5c + d = 1 \end{array}\right\} 98a + 16b + 2c = -2 \Rightarrow 49a + 8b + c = -1$$

$f'(3) = 27a + 6b + c = 0,\ f''(4) = 24a + 2b = 0$

$$\begin{array}{rlrl} 49a + 8b + c &= -1 & 24a + 2b &= 0 \\ 27a + 6b + c &= 0 & 22a + 2b &= -1 \\ \hline 22a + 2b &= -1 & 2a &= 1 \end{array}$$

$a = \frac{1}{2}, b = -6, c = \frac{45}{2}, d = -24$

$f(x) = \frac{1}{2}x^3 - 6x^2 + \frac{45}{2}x - 24$

53.

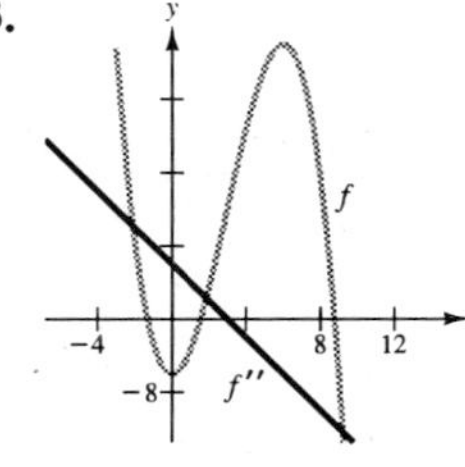

f'' is linear.

f' is quadratic.

f is cubic.

f concave upwards on $(-\infty, 3)$, downwards on $(3, \infty)$.

55. (a) $n = 1$:

$f(x) = x - 2$

$f'(x) = 1$

$f''(x) = 0$

No inflection points

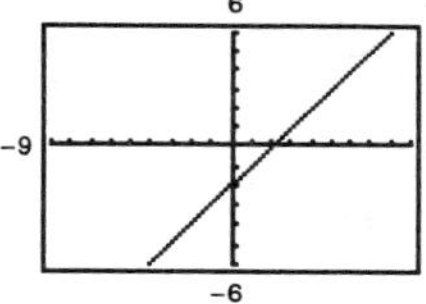

$n = 2$:

$f(x) = (x - 2)^2$

$f'(x) = 2(x - 2)$

$f''(x) = 2$

No inflection points

Relative minimum: $(2, 0)$

$n = 3$:

$f(x) = (x - 2)^3$

$f'(x) = 3(x - 2)^2$

$f''(x) = 6(x - 2)$

Inflection point: $(2, 0)$

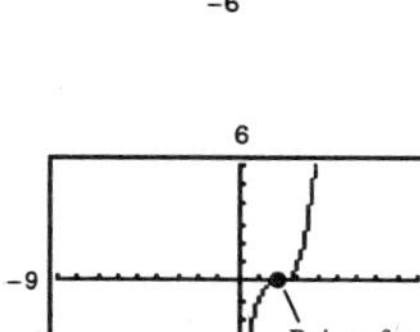

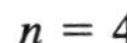

$n = 4$:

$f(x) = (x - 2)^4$

$f'(x) = 4(x - 2)^3$

$f''(x) = 12(x - 2)^2$

No inflection points:

Relative minimum: $(2, 0)$

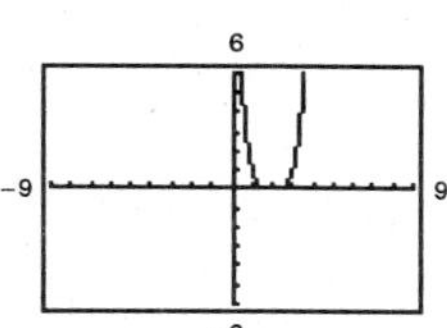

Conclusion: If $n \geq 3$ and n is odd, then $(2, 0)$ is an inflection point. If $n \geq 2$ and n is even, then $(2, 0)$ is a relative minimum.

(b) Let $f(x) = (x - 2)^n$, $f'(x) = n(x - 2)^{n-1}$, $f''(x) = n(n - 1)(x - 2)^{n-2}$.

For $n \geq 3$ and odd, $n - 2$ is also odd and the concavity changes at $x = 2$.

For $n \geq 4$ and even, $n - 2$ is also even and the concavity does not change at $x = 2$.

Thus, $x = 2$ is an inflection point if and only if $n \geq 3$ is odd.

57. (a) The rate of change of sales is increasing.
$S'' > 0$

(b) The rate of change of sales is decreasing.
$S' > 0, S'' < 0$

(c) The rate of change of sales is constant.
$S' = C, S'' = 0$

(d) Sales are steady.
$S = C, S' = 0, S'' = 0$

(e) Sales are declining, but at a lower rate.
$S' < 0, S'' > 0$

(f) Sales have bottomed out and have started to rise.
$S' > 0$

59. $f(x) = ax^3 + bx^2 + cx + d$

Maximum: $(-4, 1)$

Minimum: $(0, 0)$

(a) $f'(x) = 3ax^2 + 2bx + c, \quad f''(x) = 6ax + 2b$

$f(0) = 0 \Rightarrow d = 0$

$$f(-4) = 1 \Rightarrow -64a + 16b - 4c = 1$$

$$f'(-4) = 0 \Rightarrow 48a - 8b + c = 0$$

$$f'(0) = 0 \Rightarrow c = 0$$

Solving this system yields $a = \frac{1}{32}$ and $b = 6a = \frac{3}{16}$.

$f(x) = \frac{1}{32}x^3 + \frac{3}{16}x^2$

(b) The plane would be descending at the greatest rate at the point of inflection.

$$f''(x) = 6ax + 2b = \tfrac{3}{16}x + \tfrac{3}{8} = 0 \Rightarrow x = -2.$$

Two miles from touchdown.

61. $D = 2x^4 - 5Lx^3 + 3L^2x^2$

$D' = 8x^3 - 15Lx^2 + 6L^2x = x(8x^2 - 15Lx + 6L^2) = 0$

$$x = 0 \text{ or } x = \frac{15L \pm \sqrt{33}L}{16} = \left(\frac{15 \pm \sqrt{33}}{16}\right)L$$

By the Second Derivative Test, the deflection is maximum when

$$x = \left(\frac{15 - \sqrt{33}}{16}\right)L \approx 0.578L.$$

63. $C = 0.5x^2 + 15x + 5000$

$$\overline{C} = \frac{C}{x} = 0.5x + 15 + \frac{5000}{x}$$

$\overline{C}$ = average cost per unit

$$\frac{d\overline{C}}{dx} = 0.5 - \frac{5000}{x^2} = 0 \text{ when } x = 100$$

By the First Derivative Test, $\overline{C}$ is minimized when $x = 100$ units.

65. $v = -2400\pi \sin\theta$

$$\frac{d\theta}{dt} = (200)(2\pi) = 400\pi \text{ rad/min}$$

$$\frac{dv}{dt} = -2400\pi\cos\theta \cdot \frac{d\theta}{dt}$$

$$= -2400\pi(400\pi)\cos\theta$$

$$= 0 \text{ when } \theta = \frac{(2n+1)\pi}{2} = \frac{\pi}{2} + 2n\pi,\ n \text{ any integer}$$

By the First Derivative Test, v is maximum when $\theta = (3\pi/2) + 2n\pi$. The speed $|v|$ is maximum when $\theta = (\pi/2) + 2n\pi$ and $(3\pi/2) + 2n\pi$.

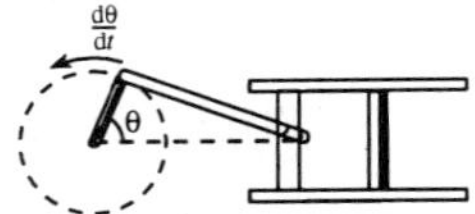

67. $f(x) = 2(\sin x + \cos x)$, $\quad f(0) = 2$

$f'(x) = 2(\cos x - \sin x)$, $\quad f'(0) = 2$

$f''(x) = 2(-\sin x - \cos x)$, $\quad f''(0) = -2$

$P_1(x) = 2 + 2(x - 0) = 2(1 + x)$

$P_1'(x) = 2$

$P_2(x) = 2 + 2(x - 0) + \frac{1}{2}(-2)(x - 0)^2 = 2 + 2x - x^2$

$P_2'(x) = 2 - 2x$

$P_2''(x) = -2$

The values of f, P_1, P_2, and their first derivatives are equal at $x = 0$. The values of the second derivatives of f and P_2 are equal at $x = 0$. The approximations worsen as you move away from $x = 0$.

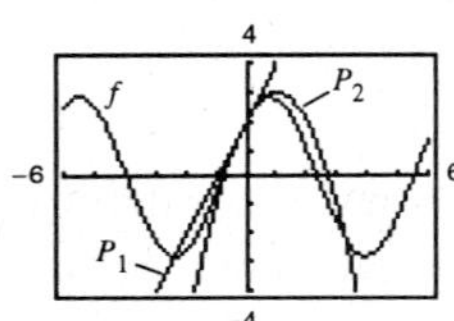

69. $f(x) = 2(\sin x + \cos x)$, $\quad f\left(\frac{\pi}{4}\right) = 2\sqrt{2}$

$f'(x) = 2(\cos x - \sin x)$, $\quad f'\left(\frac{\pi}{4}\right) = 0$

$f''(x) = 2(-\sin x - \cos x)$, $\quad f''\left(\frac{\pi}{4}\right) = -2\sqrt{2}$

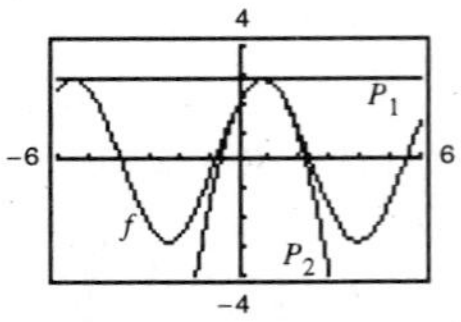

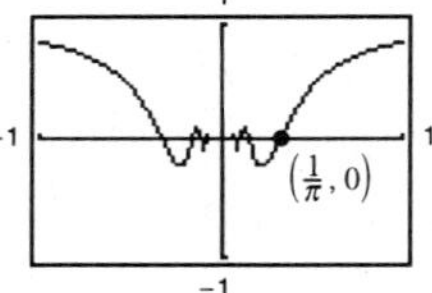

$$P_1(x) = 2\sqrt{2} + 0\left(x - \frac{\pi}{4}\right) = 2\sqrt{2}$$

$$P_1'(x) = 0$$

$$P_2(x) = 2\sqrt{2} + 0\left(x - \frac{\pi}{4}\right) + \frac{1}{2}\left(-2\sqrt{2}\right)\left(x - \frac{\pi}{4}\right)^2 = 2\sqrt{2} - \sqrt{2}\left(x - \frac{\pi}{4}\right)^2$$

$$P_2'(x) = -2\sqrt{2}\left(x - \frac{\pi}{4}\right)$$

$$P_2''(x) = -2\sqrt{2}$$

The values of f, P_1, P_2, and their first derivatives are equal at $x = \pi/4$. The values of the second derivatives of f and P_2 are equal at $x = \pi/4$. The approximations worsen as you move away from $x = \pi/4$.

71. $f(x) = x\sin\left(\frac{1}{x}\right)$

$$f'(x) = x\left[-\frac{1}{x^2}\cos\left(\frac{1}{x}\right)\right] + \sin\left(\frac{1}{x}\right) = -\frac{1}{x}\cos\left(\frac{1}{x}\right) + \sin\left(\frac{1}{x}\right)$$

$$f''(x) = -\frac{1}{x}\left[\frac{1}{x^2}\sin\left(\frac{1}{x}\right)\right] + \frac{1}{x^2}\cos\left(\frac{1}{x}\right) - \frac{1}{x^2}\cos\left(\frac{1}{x}\right) = -\frac{1}{x^3}\sin\left(\frac{1}{x}\right) = 0$$

$$x = \frac{1}{\pi}$$

Point of inflection: $\left(\frac{1}{\pi}, 0\right)$

When $x > 1/\pi$, $f'' < 0$, so the graph is concave downward.

73. Assume the zeros of f are all real. Then express the function as $f(x) = a(x - r_1)(x - r_2)(x - r_3)$ where r_1, r_2, and r_3 are the distinct zeros of f. From the Product Rule for a function involving three factors, we have

$$f'(x) = a[(x - r_1)(x - r_2) + (x - r_1)(x - r_3) + (x - r_2)(x - r_3)]$$

$$f''(x) = a[(x - r_1) + (x - r_2) + (x - r_1) + (x - r_3) + (x - r_2) + (x - r_3)]$$

$$= a[6x - 2(r_1 + r_2 + r_3)].$$

Consequently, $f''(x) = 0$ if

$$x = \frac{2(r_1 + r_2 + r_3)}{6} = \frac{r_1 + r_2 + r_3}{3} = \text{(Average of } r_1, r_2, \text{and } r_3).$$

75. Suppose $y_1 < d < y_2$ and let $g(x) = f(x) - d(x - a)$. g is continuous on $[a, b]$, and therefore, by the Extreme Value Theorem, attains a minimum value at some point c in $[a, b]$. This point c cannot be the endpoints a or b because

$$g'(a) = f'(a) - d = y_1 - d < 0$$

$$g'(b) = f'(b) - d = y_2 - d > 0.$$

Hence, g attains its minimum value at c, $a < c < b$. This implies that $g'(c) = 0$ and $f'(c) = d$. The proof for $y_2 < d < y_1$ is similar.

77. True.
Let $y = ax^3 + bx^2 + cx + d$, $a \neq 0$. Then $y'' = 6ax + 2b = 0$ when $x = -(b/3a)$, and the concavity changes at this point.

79. False.

$$f(x) = 3 \sin x + 2 \cos x$$

$$f'(x) = 3 \cos x - 2 \sin x$$

$$3 \cos x - 2 \sin x = 0$$

$$3 \cos x = 2 \sin x$$

$$\tfrac{3}{2} = \tan x$$

Critical number: $x = \tan^{-1}\left(\frac{3}{2}\right)$

$f\left(\tan^{-1} \frac{3}{2}\right) \approx 3.60555$ is the maximum value of y.

Section 3.5 Limits at Infinity

1. $f(x) = \dfrac{3x^2}{x^2 + 2}$

No vertical asymptotes

Horizontal asymptote: $y = 3$

Matches (f)

3. $f(x) = \dfrac{x}{x^2 + 2}$

No vertical asymptotes

Horizontal asymptote: $y = 0$

Matches (d)

5. $f(x) = \dfrac{4 \sin x}{x^2 + 1}$

No vertical asymptotes

Horizontal asymptotes: $y = 0$

Matches (b)

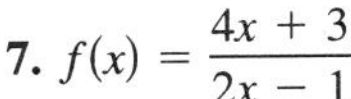

7. $f(x) = \dfrac{4x + 3}{2x - 1}$

x	10^0	10^1	10^2	10^3	10^4	10^5	10^6
$f(x)$	7	2.26	2.025	2.0025	2.0003	2	2

$\lim_{x \to \infty} f(x) = 2$

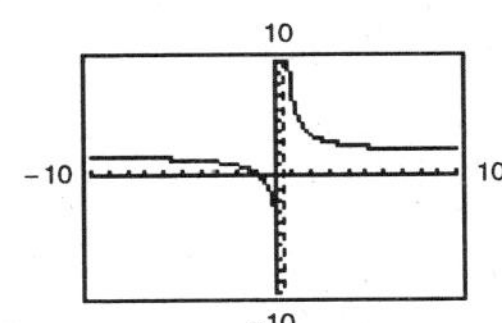

9. $f(x) = \dfrac{-6x}{\sqrt{4x^2 + 5}}$

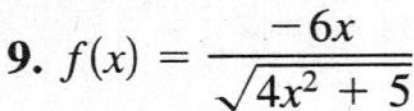

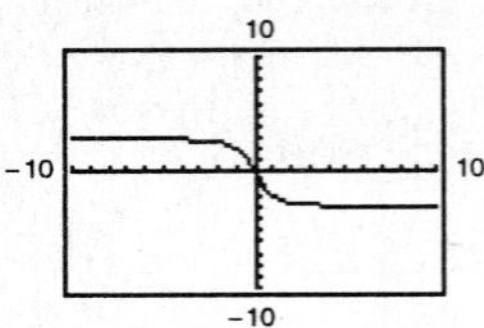

x	10^0	10^1	10^2	10^3	10^4	10^5	10^6
$f(x)$	-2	-2.98	-2.9998	-3	-3	-3	-3

$\lim_{x\to\infty} f(x) = -3$

11. $\displaystyle\lim_{x\to\infty} \frac{2x-1}{3x+2} = \lim_{x\to\infty} \frac{2-(1/x)}{3+(2/x)} = \frac{2-0}{3+0} = \frac{2}{3}$

13. $\displaystyle\lim_{x\to\infty} \frac{x}{x^2-1} = \lim_{x\to\infty} \frac{1/x}{1-(1/x^2)} = \frac{0}{1} = 0$

15. $\displaystyle\lim_{x\to-\infty} \frac{5x^2}{x+3} = \lim_{x\to-\infty} \frac{5x}{1+(3/x)} = -\infty$

Limit does not exist.

17. $\displaystyle\lim_{x\to-\infty} \frac{x}{\sqrt{x^2-x}} = \lim_{x\to-\infty} \frac{1}{\dfrac{\sqrt{x^2-x}}{-\sqrt{x^2}}}, \left(\text{for } x < 0 \text{ we have } x = -\sqrt{x^2}\right)$

$\displaystyle = \lim_{x\to-\infty} \frac{-1}{\sqrt{1-(1/x)}} = -1$

19. $\displaystyle\lim_{x\to\infty} \frac{2x+1}{\sqrt{x^2-x}} = \lim_{x\to\infty} \frac{2+(1/x)}{\sqrt{1-(1/x)}} = 2$

21. Since $(-1/x) \le (\sin(2x))/x \le (1/x)$ for all $x \ne 0$, we have by the Squeeze Theorem,

$$\lim_{x\to\infty} -\frac{1}{x} \le \lim_{x\to\infty} \frac{\sin(2x)}{x} \le \lim_{x\to\infty} \frac{1}{x}$$

$$0 \le \lim_{x\to\infty} \frac{\sin(2x)}{x} \le 0.$$

Therefore, $\displaystyle\lim_{x\to\infty} \frac{\sin(2x)}{x} = 0.$

23. $\displaystyle\lim_{x\to\infty} \frac{1}{2x+\sin x} = 0$

25. $\displaystyle\lim_{x\to\infty} x \sin \frac{1}{x} = \lim_{t\to0^+} \frac{\sin t}{t} = 1$

(Let $x = 1/t$.)

27. $\displaystyle\lim_{x\to-\infty} \left(x + \sqrt{x^2+3}\right) = \lim_{x\to-\infty} \left[\left(x + \sqrt{x^2+3}\right) \cdot \frac{x - \sqrt{x^2+3}}{x - \sqrt{x^2+3}}\right] = \lim_{x\to-\infty} \frac{-3}{x - \sqrt{x^2+3}} = 0$

29. $\displaystyle\lim_{x\to\infty} \left(x - \sqrt{x^2+x}\right) = \lim_{x\to\infty} \left[\left(x - \sqrt{x^2+x}\right) \cdot \frac{x + \sqrt{x^2+x}}{x + \sqrt{x^2+x}}\right]$

$\displaystyle = \lim_{x\to\infty} \frac{-x}{x + \sqrt{x^2+x}} = \lim_{x\to\infty} \frac{-1}{1 + \sqrt{1+(1/x)}} = -\frac{1}{2}$

31.

x	10^0	10^1	10^2	10^3	10^4	10^5	10^6
$f(x)$	1	0.513	0.501	0.500	0.500	0.500	0.500

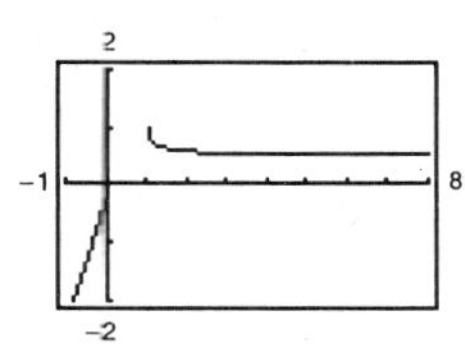

$$\lim_{x\to\infty}\left(x - \sqrt{x(x-1)}\right) = \lim_{x\to\infty}\frac{x - \sqrt{x^2 - x}}{1}\cdot\frac{x + \sqrt{x^2 - x}}{x + \sqrt{x^2 - x}}$$

$$= \lim_{x\to\infty}\frac{x}{x + \sqrt{x^2 - x}}$$

$$= \lim_{x\to\infty}\frac{1}{1 + \sqrt{1 - (1/x)}}$$

$$= \frac{1}{2}$$

33.

x	10^0	10^1	10^2	10^3	10^4	10^5	10^6
$f(x)$	0.479	0.500	0.500	0.500	0.500	0.500	0.500

Let $x = 1/t$.

$$\lim_{x\to\infty} x\sin\left(\frac{1}{2x}\right) = \lim_{t\to 0^+}\frac{\sin(t/2)}{t} = \lim_{t\to 0^+}\frac{1}{2}\,\frac{\sin(t/2)}{t/2} = \frac{1}{2}$$

35. $y = \dfrac{2 + x}{1 - x}$

Intercepts: $(-2, 0)$, $(0, 2)$

Symmetry: none

Horizontal asymptote: $y = -1$ since

$$\lim_{x\to-\infty}\frac{2 + x}{1 - x} = -1 = \lim_{x\to\infty}\frac{2 + x}{1 - x}.$$

Discontinuity: $x = 1$ (Vertical asymptote)

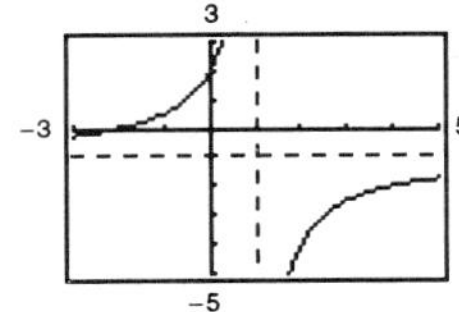

37. $y = \dfrac{x}{x^2 - 4}$

Intercept: $(0, 0)$

Symmetry: origin

Horizontal asymptote: $y = 0$

Vertical asymptote: $x = \pm 2$

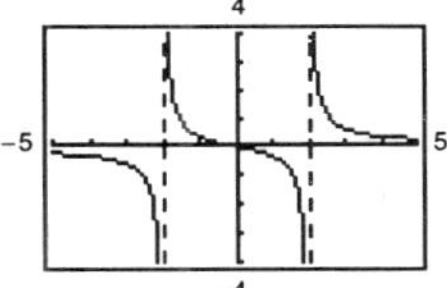

39. $y = \dfrac{x^2}{x^2 + 9}$

Intercept: $(0, 0)$

Symmetry: y-axis

Horizontal asymptote: $y = 1$ since

$$\lim_{x\to-\infty}\frac{x^2}{x^2 + 9} = 1 = \lim_{x\to\infty}\frac{x^2}{x^2 + 9}.$$

Relative minimum: $(0, 0)$

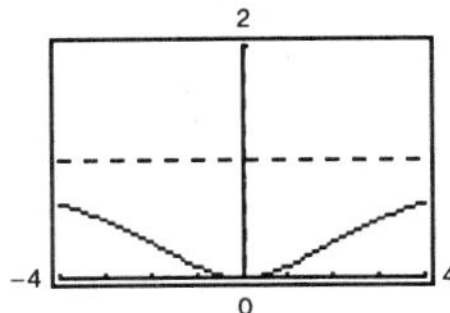

41. $y = \dfrac{2x^2}{x^2 - 4}$

Intercept: $(0, 0)$

Symmetry: y-axis

Horizontal asymptote: $y = 2$

Vertical asymptote: $x = \pm 2$

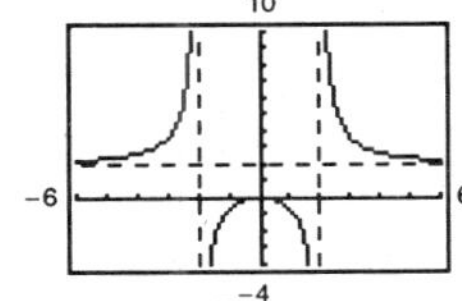

43. $xy^2 = 4$

Domain: $x > 0$

Intercepts: none

Symmetry: x-axis

Horizontal asymptote: $y = 0$ since

$$\lim_{x \to \infty} \frac{2}{\sqrt{x}} = 0 = \lim_{x \to \infty} -\frac{2}{\sqrt{x}}.$$

Discontinuity: $x = 0$ (Vertical asymptote)

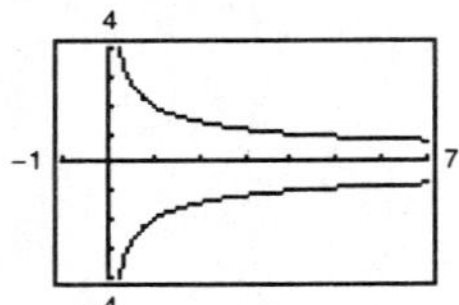

45. $y = \dfrac{2x}{1 - x}$

Intercept: $(0, 0)$

Symmetry: none

Horizontal asymptote: $y = -2$ since

$$\lim_{x \to -\infty} \frac{2x}{1 - x} = -2 = \lim_{x \to \infty} \frac{2x}{1 - x}.$$

Discontinuity: $x = 1$ (Vertical asymptote)

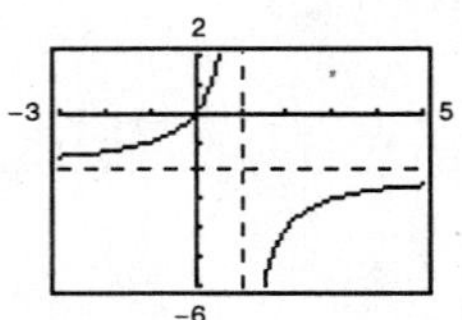

47. $y = 2 - \dfrac{3}{x^2}$

Intercepts: $\left(\pm\sqrt{3/2}, 0\right)$

Symmetry: y-axis

Horizontal asymptote: $y = 2$ since

$$\lim_{x \to -\infty} \left(2 - \frac{3}{x^2}\right) = 2 = \lim_{x \to \infty} \left(2 - \frac{3}{x^2}\right).$$

Discontinuity: $x = 0$ (Vertical asymptote)

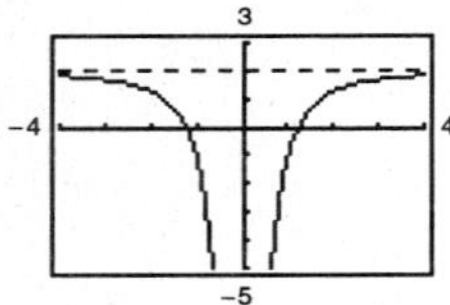

49. $y = 3 + \dfrac{2}{x}$

Intercept:

$$y = 0 = 3 + \frac{2}{x} \Rightarrow \frac{2}{x} = -3 \Rightarrow x = -\frac{2}{3} \quad \left(-\frac{2}{3}, 0\right)$$

Symmetry: none

Horizontal asymptote: $y = 3$

Vertical asymptote: $x = 0$

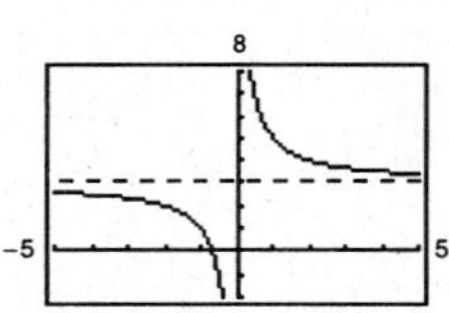

51. $y = \dfrac{x^3}{\sqrt{x^2 - 4}}$

Domain: $(-\infty, -2), (2, \infty)$

Intercepts: none

Symmetry: origin

Horizontal asymptote: none

Vertical asymptotes: $x = \pm 2$ (discontinuities)

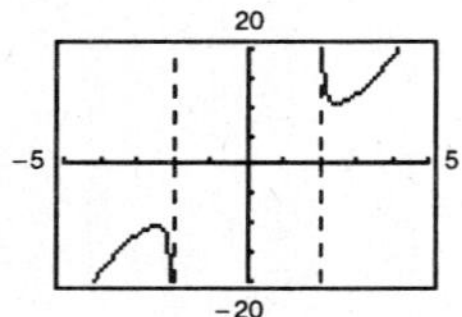

53. $f(x) = 5 - \dfrac{1}{x^2} = \dfrac{5x^2 - 1}{x^2}$

Domain: $(-\infty, 0), (0, \infty)$

$f'(x) = \dfrac{2}{x^3} \Rightarrow$ No relative extrema

$f''(x) = -\dfrac{6}{x^4} \Rightarrow$ No points of inflection

Vertical asymptote: $x = 0$

Horizontal asymptote: $y = 5$

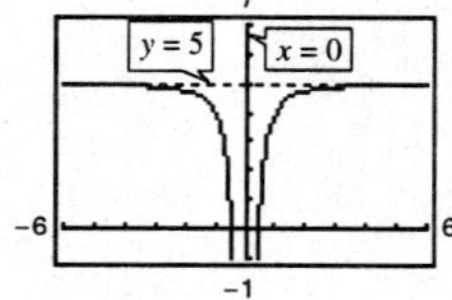

55. $f(x) = \dfrac{x}{x^2 - 4}$

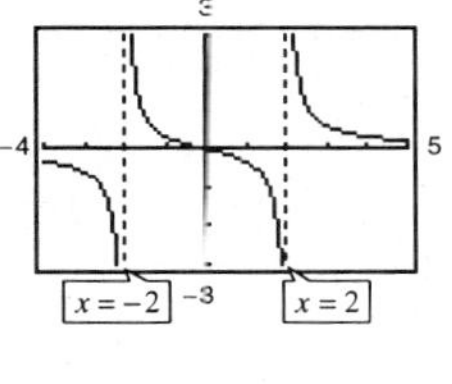

$$f'(x) = \frac{(x^2 - 4) - x(2x)}{(x^2 - 4)^2}$$

$$= \frac{-(x^2 + 4)}{(x^2 - 4)^2} \neq 0 \text{ for any } x \text{ in the domain of } f.$$

$$f''(x) = \frac{(x^2 - 4)^2(-2x) + (x^2 + 4)(2)(x^2 - 4)(2x)}{(x^2 - 4)^2}$$

$$= \frac{2x(x^2 + 12)}{(x^2 - 4)^3} = 0 \text{ when } x = 0.$$

Since $f''(x) > 0$ on $(-2, 0)$ and $f''(x) < 0$ on $(0, 2)$, then $(0, 0)$ is a point of inflection.

Vertical asymptotes: $x = \pm 2$

Horizontal asymptote: $y = 0$

57. $f(x) = \dfrac{x - 2}{x^2 - 4x + 3} = \dfrac{x - 2}{(x - 1)(x - 3)}$

$$f'(x) = \frac{(x^2 - 4x + 3) - (x - 2)(2x - 4)}{(x^2 - 4x + 3)^2} = \frac{-x^2 + 4x - 5}{(x^2 - 4x + 3)^2} \neq 0$$

$$f''(x) = \frac{(x^2 - 4x + 3)^2(-2x + 4) - (-x^2 + 4x - 5)(2)(x^2 - 4x + 3)(2x - 4)}{(x^2 - 4x + 3)^4}$$

$$= \frac{2(x^3 - 6x^2 + 15x - 14)}{(x^2 - 4x + 3)^3} = 0 \text{ when } x = 2.$$

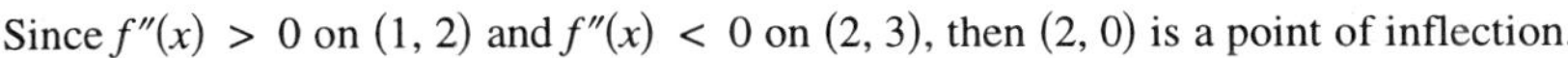

Since $f''(x) > 0$ on $(1, 2)$ and $f''(x) < 0$ on $(2, 3)$, then $(2, 0)$ is a point of inflection.

Vertical asymptote: $x = 1, x = 3$

Horizontal asymptote: $y = 0$

59. $f(x) = \dfrac{3x}{\sqrt{4x^2 + 1}}$

$$f'(x) = \frac{3}{(4x^2 + 1)^{3/2}} \Longrightarrow \text{No relative extrema}$$

$$f''(x) = \frac{-36x}{(4x^2 + 1)^{5/2}} = 0 \text{ when } x = 0.$$

Point of inflection: $(0, 0)$

Horizontal asymptotes: $y = \pm\dfrac{3}{2}$

No vertical asymptotes

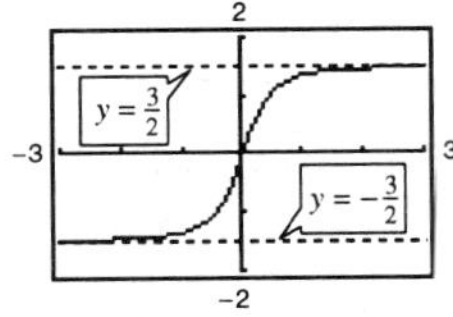

61. (a) $f(x) = \dfrac{|x|}{x + 1}$

$$\lim_{x \to \infty} \frac{|x|}{x + 1} = 1$$

$$\lim_{x \to -\infty} \frac{|x|}{x + 1} = -1$$

Therefore, $y = 1$ and $y = -1$ are both horizontal asymptotes.

(b) $f(x) = \dfrac{2x}{\sqrt{x^2 + 1}}$

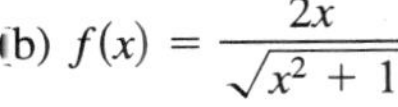

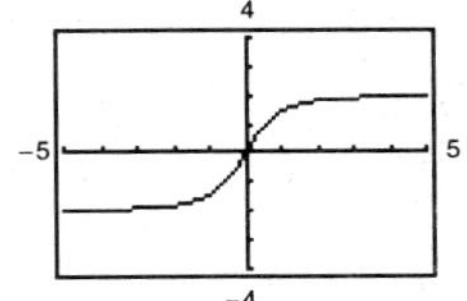

$$\lim_{x \to \infty} \frac{2x}{\sqrt{x^2 + 1}} = 2$$

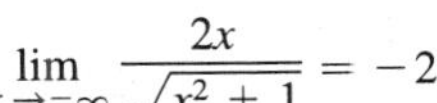

$$\lim_{x \to -\infty} \frac{2x}{\sqrt{x^2 + 1}} = -2$$

Therefore, $y = 2$ and $y = -2$ are both horizontal asymptotes.

63. $f(x) = \dfrac{x^3 - 3x^2 + 2}{x(x-3)}$, $g(x) = x + \dfrac{2}{x(x-3)}$

(a)

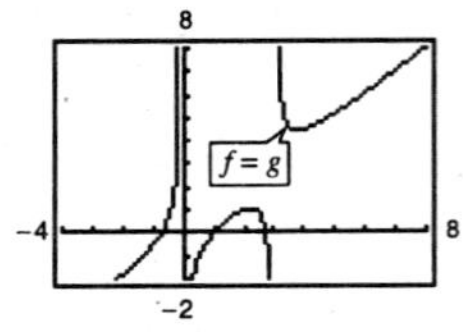

(b) $f(x) = \dfrac{x^3 - 3x^2 + 2}{x(x-3)}$

$$= \frac{x^2(x-3)}{x(x-3)} + \frac{2}{x(x-3)}$$

$$= x + \frac{2}{x(x-3)} = g(x)$$

(c)

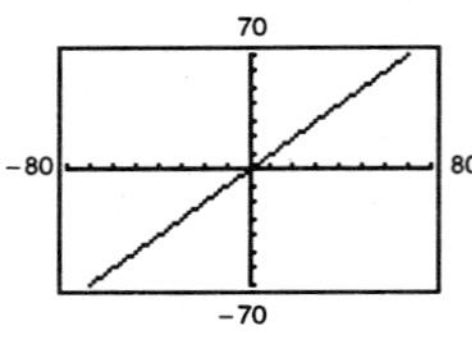

The graph appears as the slant asymptote $y = x$.

65. (a)

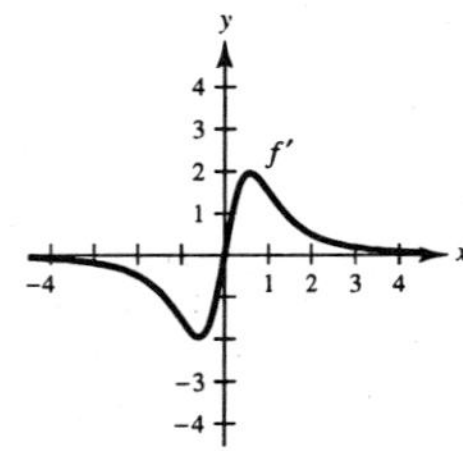

(b) $\lim_{x\to\infty} f(x) = 3$ $\qquad \lim_{x\to\infty} f'(x) = 0$

(c) Since $\lim_{x\to\infty} f(x) = 3$, the graph approaches that of a horizontal line, $\lim_{x\to\infty} f'(x) = 0$.

67. $C = 0.5x + 500$

$$\overline{C} = \frac{C}{x}$$

$$\overline{C} = 0.5 + \frac{500}{x}$$

$$\lim_{x\to\infty}\left(0.5 + \frac{500}{x}\right) = 0.5$$

69. (a) $T_1(t) = -0.003t^2 + 0.677t + 26.564$

(b)

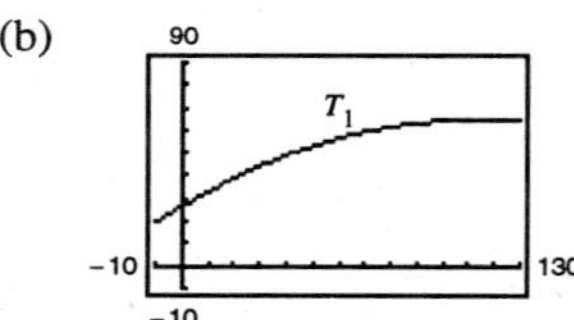

(c)

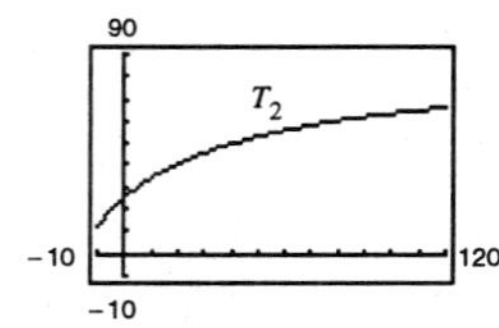

$$T_2 = \frac{1451 + 86t}{58 + t}$$

(d) $T_1(0) \approx 26.6$

$$T_2(0) = \frac{1451}{58} \approx 25.0$$

(e) $\lim_{t\to\infty} T_2 = \dfrac{86}{1} = 86$

(f) The limiting temperature is 86. T_1 has no horizontal asymptote.

71. False. Let $f(x) = \dfrac{2x}{\sqrt{x^2 + 2}}$. (See Exercise 2.)

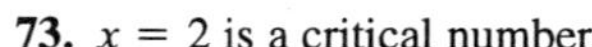

73. $x = 2$ is a critical number.

$f'(x) < 0$ for $x < 2$.

$f'(x) > 0$ for $x > 2$.

$$\lim_{x\to-\infty} f(x) = \lim_{x\to\infty} f(x) = 6$$

For example, let $f(x) = \dfrac{-6}{0.1(x-2)^2 + 1} + 6$.

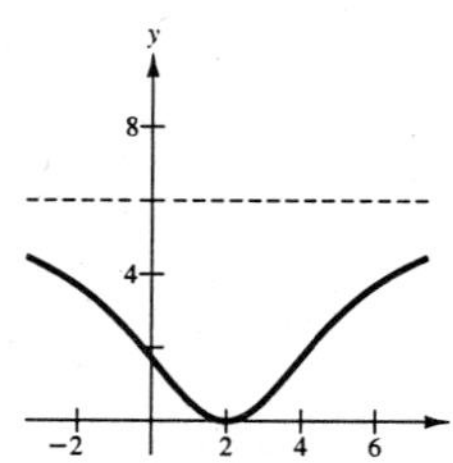

75. $\lim\limits_{x\to\infty} \dfrac{p(x)}{q(x)} = \lim\limits_{x\to\infty} \dfrac{a_n x^n + \cdots + a_1 x + a_0}{b_m x^m + \cdots + b_1 x + b_0}$

Divide $p(x)$ and $q(x)$ by x^m.

Case 1: If $n < m$: $\lim\limits_{x\to\infty} \dfrac{p(x)}{q(x)} = \lim\limits_{x\to\infty} \dfrac{\dfrac{a_n}{x^{m-n}} + \cdots + \dfrac{a_1}{x^{m-1}} + \dfrac{a_0}{x^m}}{b_m + \cdots + \dfrac{b_1}{x^{m-1}} + \dfrac{b_0}{x^m}} = \dfrac{0 - \cdots + 0 + 0}{b_m + \cdots + 0 + 0} = \dfrac{0}{b_m} = 0$

Case 2: If $m = n$: $\lim\limits_{x\to\infty} \dfrac{p(x)}{q(x)} = \lim\limits_{x\to\infty} \dfrac{a_n + \cdots + \dfrac{a_1}{x^{m-1}} + \dfrac{a_0}{x^m}}{b_m + \cdots + \dfrac{b_1}{x^{m-1}} + \dfrac{b_0}{x^m}} = \dfrac{a_n + \cdots + 0 + 0}{b_m + \cdots + 0 + 0} = \dfrac{a_n}{b_m}.$

Case 3: If $n > m$: $\lim\limits_{x\to\infty} \dfrac{p(x)}{q(x)} = \lim\limits_{x\to\infty} \dfrac{a_n x^{n-m} + \cdots + \dfrac{a_1}{x^{m-1}} + \dfrac{a_0}{x^m}}{b_m + \cdots + \dfrac{b_1}{x^{m-1}} + \dfrac{b_0}{x^m}} = \dfrac{\pm\infty + \cdots + 0}{b_m + \cdots + 0} = \pm\infty.$

Section 3.6 A Summary of Curve Sketching

1. (a) f has constant negative slope. Matches (D)

(b) The slope of f approaches ∞ as $x \to 0^-$, and approaches $-\infty$ as $x \to 0^+$. Matches (C)

(c) The slope is periodic, and zero at $x = 0$. Matches (A)

(d) The slope is positive up to approximately $x = 1.5$. Matches (B)

3. Since $f'(x) = \frac{2}{3}$ for all x, $f(x)$ is a line of slope $\frac{2}{3}$. The y-intercept is $(0, 1)$: $f(x) = \frac{2}{3}x + 1$. Hence, $f(6) = 5$.

5. (a) $f'(x) = 0$ for $x = -2$ and $x = 2$

f' is negative for $-2 < x < 2$ (decreasing function).

f' is positive for $x > 2$ and $x < -2$ (increasing function).

(c) f' is increasing on $(0, \infty)$. $(f'' > 0)$

(b) $f''(x) = 0$ at $x = 0$ (Inflection point).

f'' is positive for $x > 0$ (Concave upwards).

f'' is negative for $x < 0$ (Concave downward).

(d) $f'(x)$ is minimum at $x = 0$. The rate of change of f at $x = 0$ is less than the rate of change of f for all other values of x.

7. $y = x^3 - 3x^2 + 3$

$y' = 3x^2 - 6x = 3x(x - 2) = 0$ when $x = 0, x = 2$

$y'' = 6x - 6 = 6(x - 1) = 0$ when $x = 1$

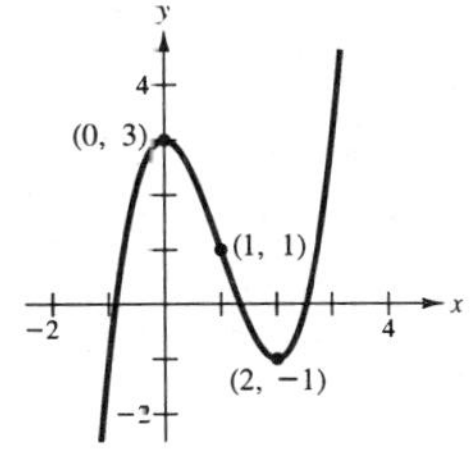

	y	y'	y''	Conclusion
$-\infty < x < 0$		+	−	Increasing, concave down
$x = 0$	3	0	−	Relative maximum
$0 < x < 1$		−	−	Decreasing, concave down
$x = 1$	1	−	0	Point of inflection
$1 < x < 2$		−	+	Decreasing, concave up
$x = 2$	−1	0	+	Relative minimum
$2 < x < \infty$		+	+	Increasing, concave up

9. $y = 2 - x - x^3$

$y' = -1 - 3x^2$

No critical numbers

$y'' = -6x = 0$ when $x = 0$.

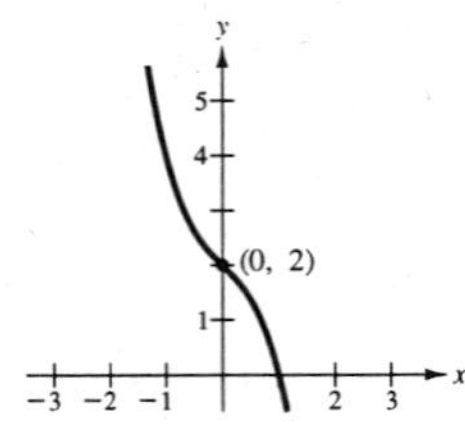

	y	y'	y''	Conclusion
$-\infty < x < 0$		$-$	$+$	Decreasing, concave up
$x = 0$	2	$-$	0	Point of inflection
$0 < x < \infty$		$-$	$-$	Decreasing, concave down

11. $f(x) = 3x^3 - 9x + 1$

$f'(x) = 9x^2 - 9 = 9(x^2 - 1) = 0$ when $x = \pm 1$.

$f''(x) = 18x = 0$ when $x = 0$.

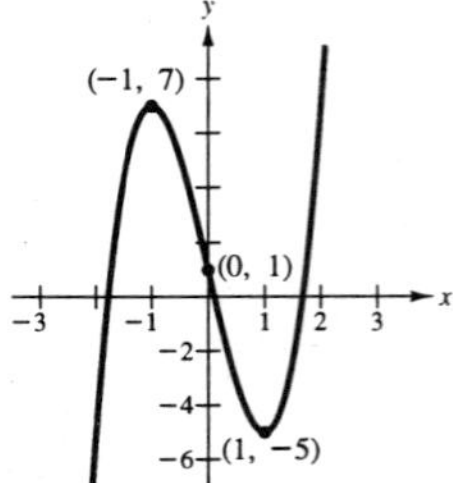

	$f(x)$	$f'(x)$	$f''(x)$	Conclusion
$-\infty < x < -1$		$+$	$-$	Increasing, concave down
$x = -1$	7	0	$-$	Relative maximum
$-1 < x < 0$		$-$	$-$	Decreasing, concave down
$x = 0$	1	$-$	0	Point of inflection
$0 < x < 1$		$-$	$+$	Decreasing, concave up
$x = 1$	-5	0	$+$	Relative minimum
$1 < x < \infty$		$+$	$+$	Increasing, concave up

13. $y = 3x^4 + 4x^3$

$y' = 12x^3 + 12x^2 = 12x^2(x + 1) = 0$ when $x = 0, x = -1$.

$y'' = 36x^2 + 24x = 12x(3x + 2) = 0$ when $x = 0, x = -\frac{2}{3}$.

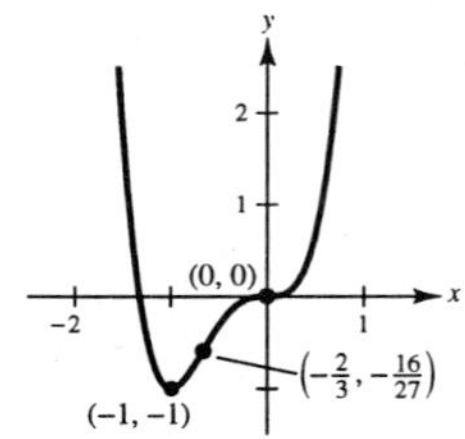

	y	y'	y''	Conclusion
$-\infty < x < -1$		$-$	$+$	Decreasing, concave up
$x = -1$	-1	0	$+$	Relative minimum
$-1 < x < -\frac{2}{3}$		$+$	$+$	Increasing, concave up
$x = -\frac{2}{3}$	$-\frac{16}{27}$	$+$	0	Point of inflection
$-\frac{2}{3} < x < 0$		$+$	$-$	Increasing, concave down
$x = 0$	0	0	0	Point of inflection
$0 < x < \infty$		$+$	$+$	Increasing, concave up

15. $f(x) = x^4 - 4x^3 + 16x$

$f'(x) = 4x^3 - 12x^2 + 16 = 4(x + 1)(x - 2)^2 = 0$ when $x = -1, x = 2$.

$f''(x) = 12x^2 - 24x = 12x(x - 2) = 0$ when $x = 0, x = 2$.

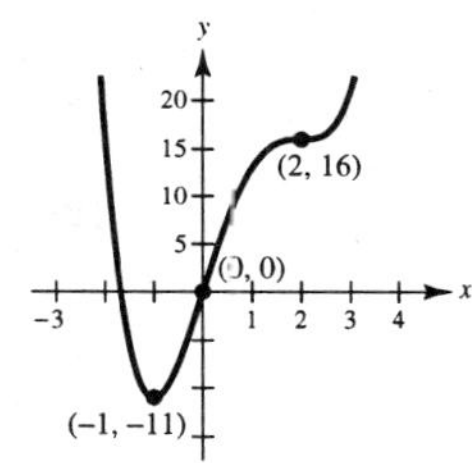

	$f(x)$	$f'(x)$	$f''(x)$	Conclusion
$-\infty < x < -1$		$-$	$+$	Decreasing, concave up
$x = -1$	-11	0	$+$	Relative minimum
$-1 < x < 0$		$+$	$+$	Increasing, concave up
$x = 0$	0	$+$	0	Point of inflection
$0 < x < 2$		$+$	$-$	Increasing, concave down
$x = 2$	16	0	0	Point of inflection
$2 < x < \infty$		$+$	$+$	Increasing, concave up

17. $y = x^5 - 5x$

$y' = 5x^4 - 5 = 5(x^4 - 1) = 0$ when $x = \pm 1$.

$y'' = 20x^3 = 0$ when $x = 0$.

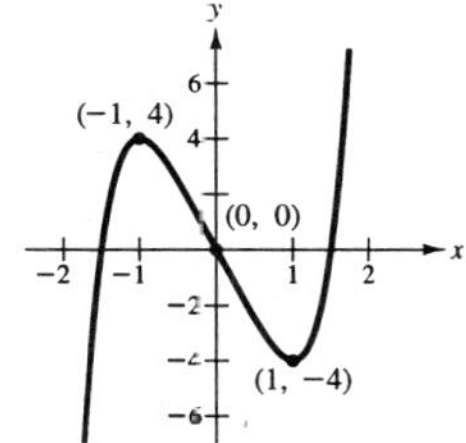

	y	y'	y''	Conclusion
$-\infty < x < -1$		$+$	$-$	Increasing, concave down
$x = -1$	4	0	$-$	Relative maximum
$-1 < x < 0$		$-$	$-$	Decreasing, concave down
$x = 0$	0	$-$	0	Point of inflection
$0 < x < 1$		$-$	$+$	Decreasing, concave up
$x = 1$	-4	0	$+$	Relative minimum
$1 < x < \infty$		$+$	$+$	Increasing, concave up

19. $y = |2x - 3|$

$y' = \dfrac{2(2x - 3)}{|2x - 3|}$ undefined at $x = \dfrac{3}{2}$.

$y'' = 0$

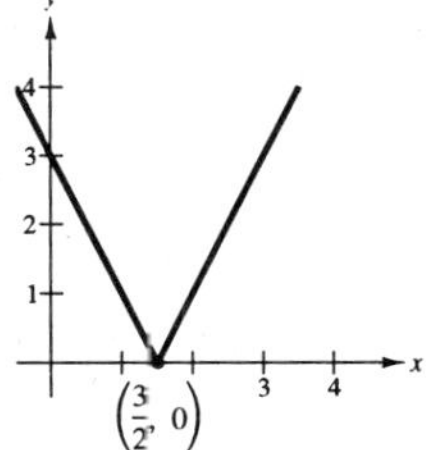

	y	y'	Conclusion
$-\infty < x < \frac{3}{2}$		$-$	Decreasing
$x = \frac{3}{2}$	0	Undefined	Relative minimum
$\frac{3}{2} < x < \infty$		$+$	Increasing

21. $y = x\sqrt{x - 4},$

Domain: $(-\infty, 4]$

$$y' = \frac{8 - 3x}{2\sqrt{4 - x}} = 0 \text{ when } x = \frac{8}{3} \text{ and undefined when } x = 4.$$

$$y'' = \frac{3x - 16}{4(4 - x)^{3/2}} = 0 \text{ when } x = \frac{16}{3} \text{ and undefined when } x = 4.$$

Note: $x = \frac{16}{3}$ is not in the domain.

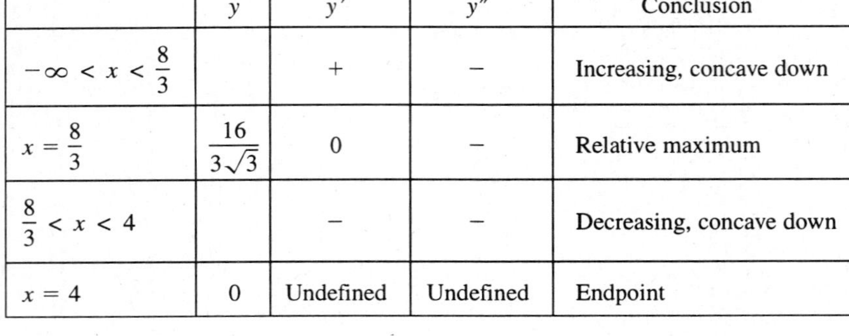

	y	y'	y''	Conclusion
$-\infty < x < \frac{8}{3}$		+	−	Increasing, concave down
$x = \frac{8}{3}$	$\frac{16}{3\sqrt{3}}$	0	−	Relative maximum
$\frac{8}{3} < x < 4$		−	−	Decreasing, concave down
$x = 4$	0	Undefined	Undefined	Endpoint

23. $y = 3x^{2/3} - 2x$

$$y' = 2x^{-1/3} - 2 = \frac{2(1 - x^{1/3})}{x^{1/3}}$$

$= 0$ when $x = 1$ and undefined when $x = 0$.

$$y'' = \frac{-2}{3x^{4/3}} < 0 \text{ when } x \neq 0.$$

	y	y'	y''	Conclusion
$-\infty < x < 0$		−	−	Decreasing, concave down
$x = 0$	0	Undefined	Undefined	Relative minimum
$0 < x < 1$		+	−	Increasing, concave down
$x = 1$	1	0	−	Relative maximum
$1 < x < \infty$		−	−	Decreasing, concave down

25. $y = \sin x - \frac{1}{18}\sin 3x, 0 \leq x \leq 2\pi$

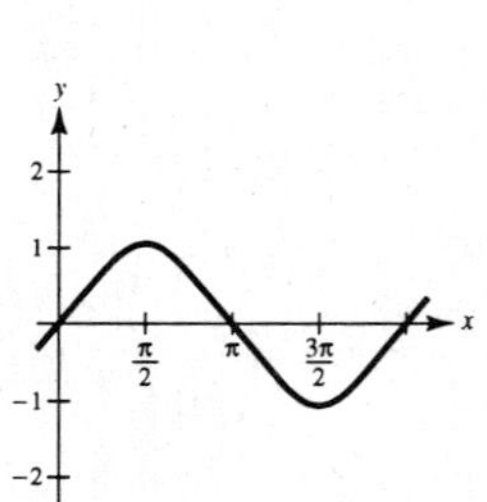

$$y' = \cos x - \frac{1}{6}\cos 3x = 0 \text{ when } x = \frac{\pi}{2}, \frac{3\pi}{2}.$$

$$y'' = -\sin x + \frac{1}{2}\sin 3x = 0 \text{ when } x = 0, \frac{\pi}{6}, \frac{5\pi}{6}, \pi, \frac{7\pi}{6}, \frac{11\pi}{6}.$$

Relative maximum: $\left(\frac{\pi}{2}, \frac{19}{18}\right)$

Relative minimum: $\left(\frac{3\pi}{2}, -\frac{19}{18}\right)$

Inflection points: $\left(\frac{\pi}{6}, \frac{4}{9}\right), \left(\frac{5\pi}{6}, \frac{4}{9}\right), (\pi, 0), \left(\frac{7\pi}{6}, -\frac{4}{9}\right), \left(\frac{11\pi}{6}, -\frac{4}{9}\right)$

27. $y = 2x - \tan x, \ -\frac{\pi}{2} < x < \frac{\pi}{2}$

$y' = 2 - \sec^2 x = 0$ when $x = \pm\frac{\pi}{4}$.

$y'' = -2\sec^2 x \tan x = 0$ when $x = 0$.

Relative maximum: $\left(\frac{\pi}{4}, \frac{\pi}{2} - 1\right)$

Relative minimum: $\left(-\frac{\pi}{4}, 1 - \frac{\pi}{2}\right)$

Inflection point: $(0, 0)$

Vertical asymptotes: $x = \pm\frac{\pi}{2}$

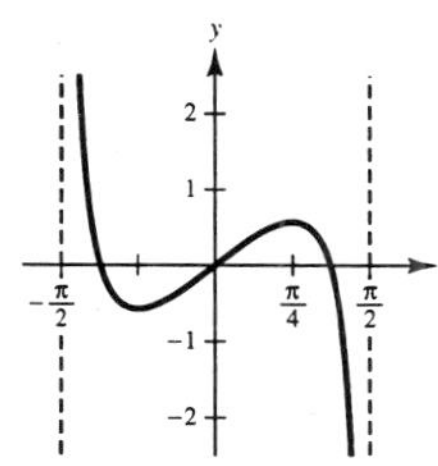

29. $y = \dfrac{x^2}{x^2 + 3}$

$y' = \dfrac{6x}{(x^2 + 3)^2} = 0$ when $x = 0$.

$y'' = \dfrac{18(1 - x^2)}{(x^2 + 3)^3} = 0$ when $x = \pm 1$.

Horizontal asymptote: $y = 1$

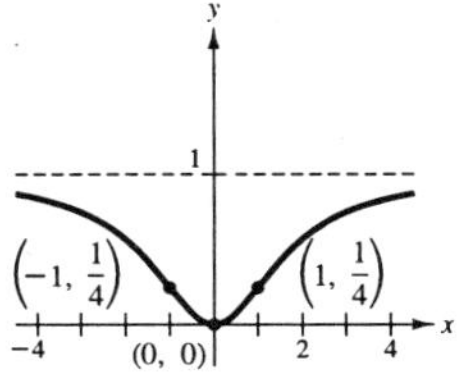

	y	y'	y''	Conclusion
$-\infty < x < -1$		$-$	$-$	Decreasing, concave down
$x = -1$	$\frac{1}{4}$	$-$	0	Point of inflection
$-1 < x < 0$		$-$	$+$	Decreasing, concave up
$x = 0$	0	0	$+$	Relative minimum
$0 < x < 1$		$+$	$+$	Increasing, concave up
$x = 1$	$\frac{1}{4}$	$+$	0	Point of inflection
$1 < x < \infty$		$+$	$-$	Increasing, concave down

31. $y = \dfrac{1}{x - 2} - 3$

$y' = -\dfrac{1}{(x - 2)^2} < 0$ when $x \neq 2$.

$y'' = \dfrac{2}{(x - 2)^3}$

No relative extrema, no points of inflection

Intercepts: $\left(\frac{7}{3}, 0\right), \left(0, -\frac{7}{2}\right)$

Vertical asymptote: $x = 2$

Horizontal asymptote: $y = -3$

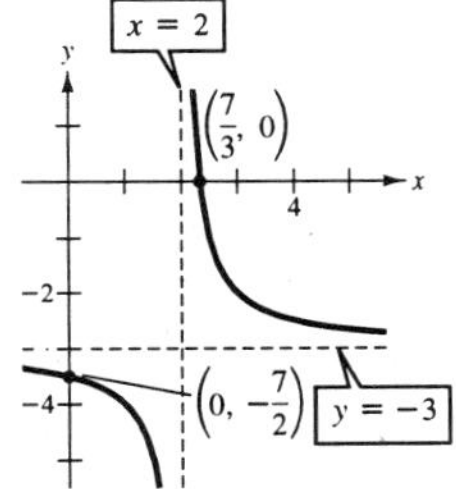

33. $y = \dfrac{2x}{x^2 - 1}$

$y' = \dfrac{-2(x^2 + 1)}{(x^2 - 1)^2} < 0$ if $x \neq \pm 1$.

$y'' = \dfrac{4x(x^2 + 3)}{(x^2 - 1)^3} = 0$ if $x = 0$.

Inflection point: $(0, 0)$

Intercept: $(0, 0)$

Vertical asymptote: $x = \pm 1$

Horizontal asymptote: $y = 0$

Symmetry with respect to the origin

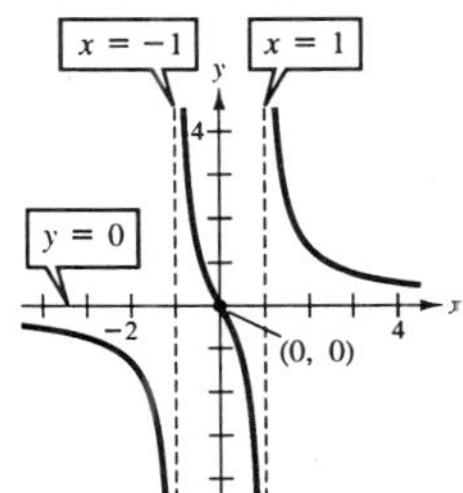

35. $g(x) = x + \dfrac{4}{x^2 + 1}$

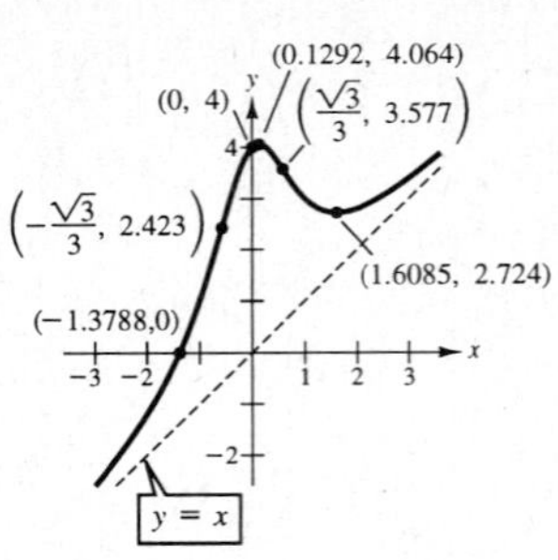

$g'(x) = 1 - \dfrac{8x}{(x^2 + 1)^2} = \dfrac{x^4 + 2x^2 - 8x + 1}{(x^2 + 1)^2} = 0$ when $x \approx 0.1292, 1.6085$

$g''(x) = \dfrac{8(3x^2 - 1)}{(x^2 + 1)^3} = 0$ when $x = \pm\dfrac{\sqrt{3}}{3}$

$g''(0.1292) < 0$, therefore, $(0.1292, 4.064)$ is relative maximum.

$g''(1.6085) > 0$, therefore, $(1.6085, 2.724)$ is a relative minimum.

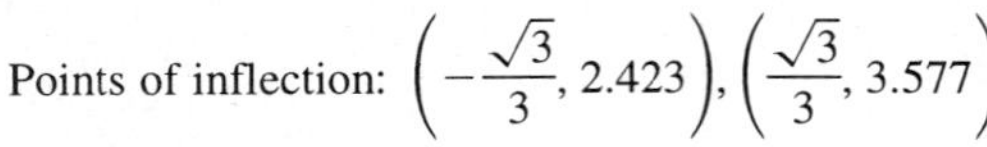

Points of inflection: $\left(-\dfrac{\sqrt{3}}{3}, 2.423\right), \left(\dfrac{\sqrt{3}}{3}, 3.577\right)$

Intercepts: $(0, 4), (-1.3788, 0)$

Slant asymptote: $y = x$

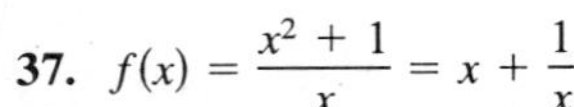

37. $f(x) = \dfrac{x^2 + 1}{x} = x + \dfrac{1}{x}$

$f'(x) = 1 - \dfrac{1}{x^2} = 0$ when $x = \pm 1$.

$f''(x) = \dfrac{2}{x^3} \neq 0$

Relative maximum: $(-1, -2)$

Relative minimum: $(1, 2)$

Vertical asymptote: $x = 0$

Slant asymptote: $y = x$

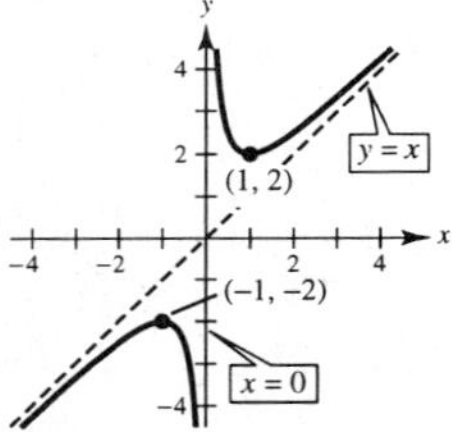

39. $y = \dfrac{x^2 - 6x + 12}{x - 4} = x - 2 + \dfrac{4}{x - 4}$

$y' = 1 - \dfrac{4}{(x - 4)^2}$

$= \dfrac{(x - 2)(x - 6)}{(x - 4)^2} = 0$ when $x = 2, 6$.

$y'' = \dfrac{8}{(x - 4)^3}$

$y'' < 0$ when $x = 2$.

Therefore, $(2, -2)$ is a relative maximum.

$y'' > 0$ when $x = 6$.

Therefore, $(6, 6)$ is a relative minimum.

Vertical asymptote: $x = 4$

Slant asymptote: $y = x - 2$

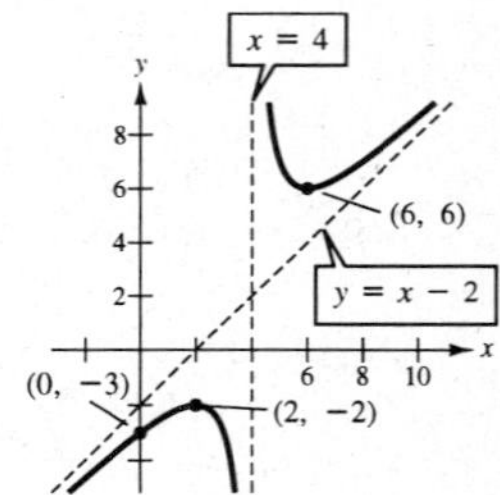

41. $f(x) = \dfrac{20x}{x^2 + 1} - \dfrac{1}{x} = \dfrac{19x^2 - 1}{x(x^2 + 1)}$

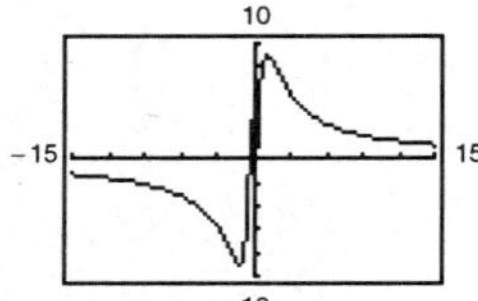

$x = 0$ vertical asymptote

$y = 0$ horizontal asymptote

43. $y = \dfrac{x}{\sqrt{x^2 + 7}}$

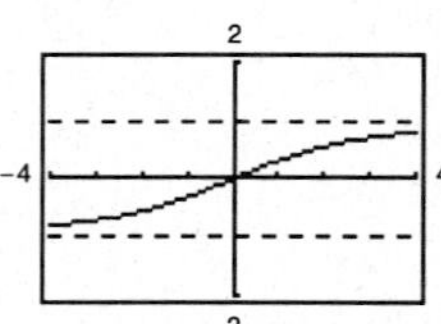

$(0, 0)$ point of inflection

$y = \pm 1$ horizontal asymptotes

45. $f(x) = \dfrac{4(x-1)^2}{x^2 - 4x + 5}$

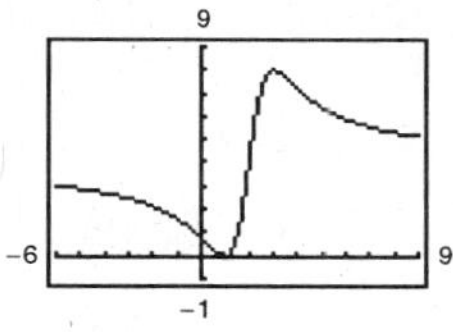

Vertical asymptote: none

Horizontal asymptote: $y = 4$

The graph crosses the horizontal asymptote $y = 4$. If a function has a vertical asymptote at $x = c$, the graph would not cross it since $f(c)$ is undefined.

47. $h(x) = \dfrac{6 - 2x}{3 - x}$

$$= \frac{2(3-x)}{3-x} = \begin{cases} 2, & \text{if } x \neq 3 \\ \text{Undefined}, & \text{if } x = 3 \end{cases}$$

The rational function is not reduced to lowest terms.

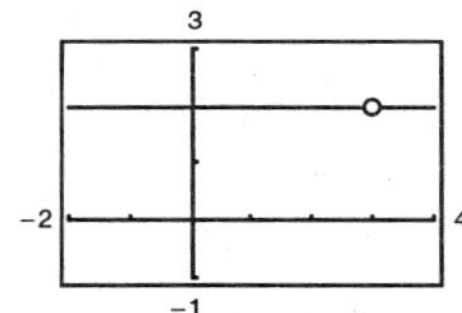

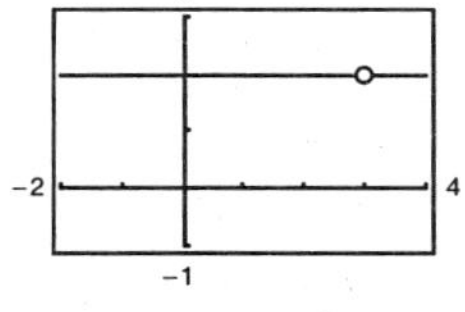

Hole at $(3, 2)$

49. $f(x) = -\dfrac{x^2 - 3x - 1}{x - 2} = -x + 1 + \dfrac{3}{x - 2}$

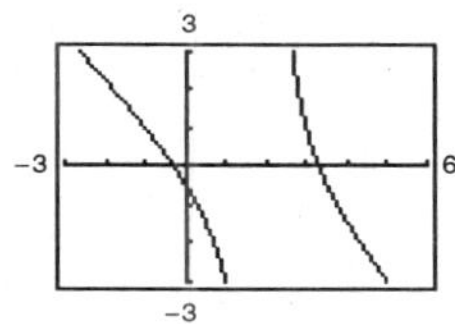

The graph appears to approach the slant asymptote $y = -x + 1$.

51. Vertical asymptote: $x = 5$

Horizontal asymptote: $y = 0$

$$y = \frac{1}{x - 5}$$

53. Vertical asymptote: $x = 5$

Slant asymptote: $y = 3x + 2$

$$y = 3x + 2 + \frac{1}{x - 5} = \frac{3x^2 - 13x - 9}{x - 5}$$

55. $f(x) = \dfrac{ax}{(x - b)^2}$

(a) The graph has a vertical asymptote at $x = b$. If $a > 0$, the graph approaches ∞ as $x \to b$. If $a < 0$, the graph approaches $-\infty$ as $x \to b$. The graph approaches its vertical asymptote faster as $|a| \to 0$.

(b) As b varies, the position of the vertical asymptote changes: $x = b$. Also, the coordinates of the minimum are changed.

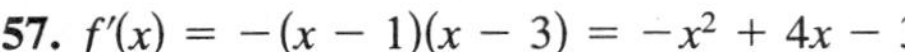

57. $f'(x) = -(x - 1)(x - 3) = -x^2 + 4x - 3$

Relative minimum when $x = 1$.

Relative maximum when $x = 3$.

$$f(x) = -\frac{x^3}{3} + 2x^2 - 3x + C, C \text{ is any constant}.$$

Let $C = 0$; $f(x) = -\dfrac{x^3}{3} + 2x^2 - 3x$

$$f(1) = -\frac{4}{3}$$

$$f(3) = 0.$$

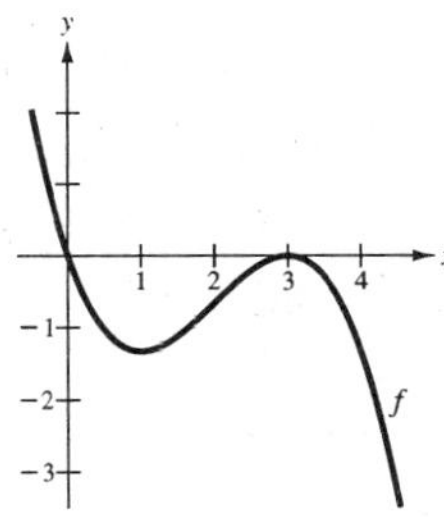

59. $f'(x) = 2$

$f(x) = 2x + C$, C is any constant.

If $C = 0, f(x) = 2x$.

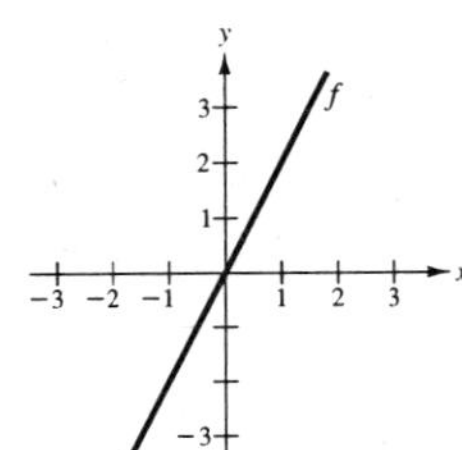

61. $f''(x) = 2 > 0 \Rightarrow f$ is concave up.

$f'(x) = 2x + C_1$

$f(x) = x^2 + C_1x + C_2$

Let $C_1 = C_2 = 0$, then $f(x) = x^2$.

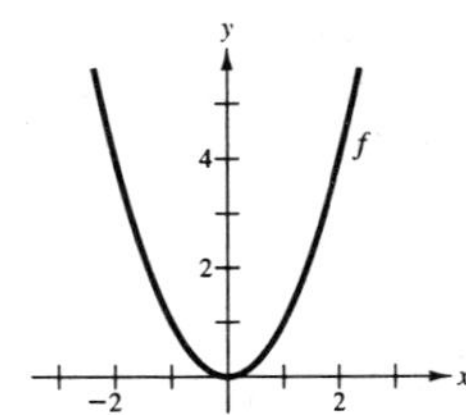

63. f is cubic.

f' is quadratic.

f'' is linear.

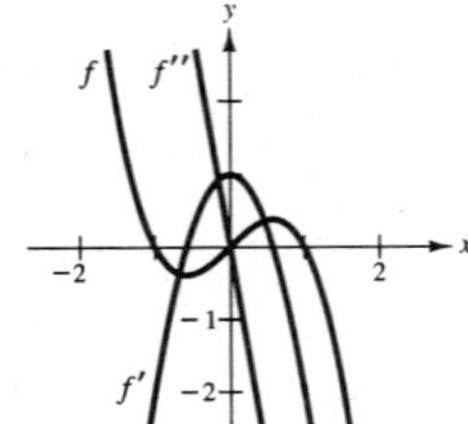

In Exercises 65, 67, and 69

$$f(x) = ax^3 + bx^2 + cx + d$$

$$f'(x) = 3ax^2 + 2bx + c$$

$$f''(x) = 6ax + 2b$$

$$f'(x) = 0 \text{ when } x = \frac{-2b \pm \sqrt{4b^2 - 12ac}}{6a} = \frac{-b \pm \sqrt{b^2 - 3ac}}{3a}.$$

Furthermore,

$$\lim_{x \to \infty} f(x) = \infty \text{ if and only if } a > 0.$$

$$\lim_{x \to \infty} f(x) = -\infty \text{ if and only if } a < 0.$$

65. Since $\lim_{x \to \infty} f(x) = -\infty, a < 0$.

Also, $f'(x) < 0$ for all x. Therefore, the discriminant is negative.

$b^2 - 3ac < 0 \Rightarrow b^2 < 3ac$

67. Since $\lim_{x \to \infty} f(x) = -\infty, a < 0$.

Also, since there is only one critical point, the discriminant is zero.

$b^2 - 3ac = 0 \Rightarrow b^2 = 3ac$

69. Since $\lim_{x \to \infty} f(x) = -\infty, a < 0$.

Also, since there are two critical points, the discriminant is positive.

$b^2 - 3ac > 0 \Rightarrow b^2 > 3ac$

71.

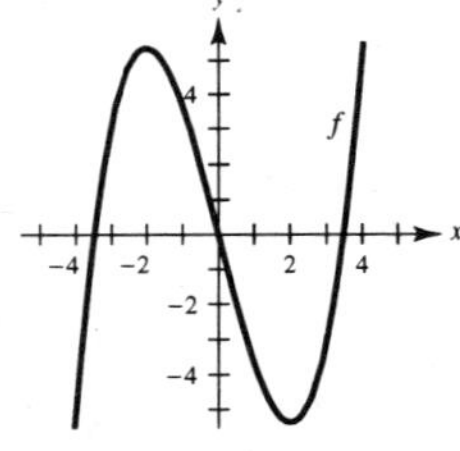

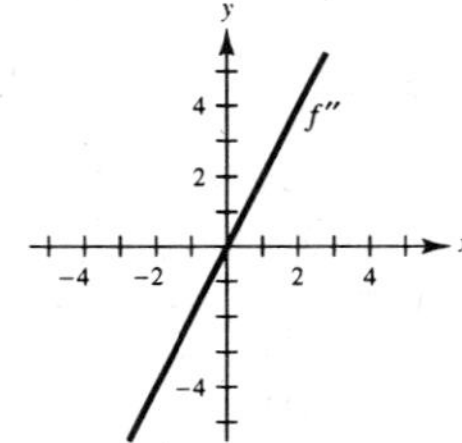

(any vertical translate of f will do)

73.

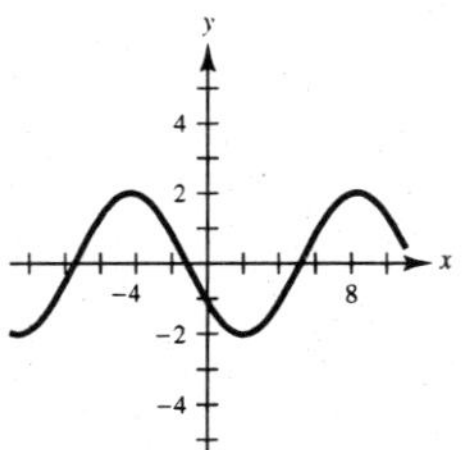

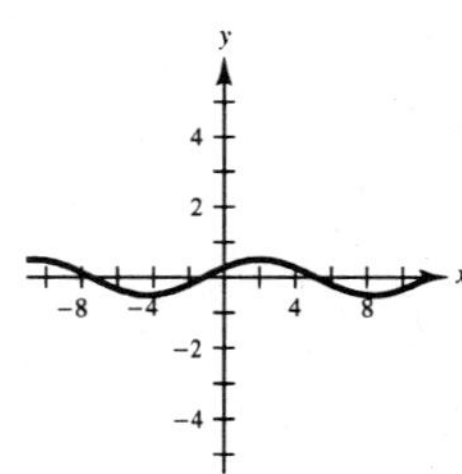

(any vertical translate of f will do)

Section 3.7 Optimization Problems

1. (a)

First Number, x	Second Number	Product, P
10	$110 - 10$	$10(110 - 10) = 1000$
20	$110 - 20$	$20(110 - 20) = 1800$
30	$110 - 30$	$30(110 - 30) = 2400$
40	$110 - 40$	$40(110 - 40) = 2800$
50	$110 - 50$	$50(110 - 50) = 3000$
60	$110 - 60$	$60(110 - 60) = 3000$

(b)

First Number, x	Second Number	Product, P
10	$110 - 10$	$10(110 - 10) = 1000$
20	$110 - 20$	$20(110 - 20) = 1800$
30	$110 - 30$	$30(110 - 30) = 2400$
40	$110 - 40$	$40(110 - 40) = 2800$
50	$110 - 50$	$50(110 - 50) = 3000$
60	$110 - 60$	$60(110 - 60) = 3000$
70	$110 - 70$	$70(110 - 70) = 2800$
80	$110 - 80$	$80(110 - 80) = 2400$
90	$110 - 90$	$90(110 - 90) = 1800$
100	$110 - 100$	$100(110 - 100) = 1000$

The maximum is attained near $x = 50$ and 60.

(c) $P = x(110 - x) = 110x - x^2$

(d)

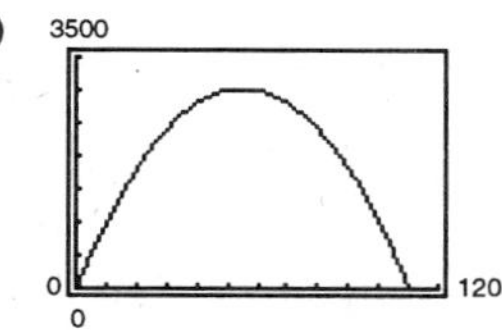

The solution appears to be $x = 55$.

(e) $\frac{dP}{dx} = 110 - 2x = 0$ when $x = 55$.

$$\frac{d^2P}{dx^2} = -2 < 0$$

P is a maximum when $x = 110 - x = 55$.

3. Let x and y be two positive numbers such that $xy = 192$.

$$S = x + y = x + \frac{192}{x}$$

$$\frac{dS}{dx} = 1 - \frac{192}{x^2} = 0 \text{ when } x = \sqrt{192}.$$

$$\frac{d^2S}{dx^2} = \frac{384}{x^3} > 0 \text{ when } x = \sqrt{192}.$$

S is a minimum when $x = y = \sqrt{192}$.

5. Let x be a positive number.

$$S = x + \frac{1}{x}$$

$$\frac{dS}{dx} = 1 - \frac{1}{x^2} = 0 \text{ when } x = 1.$$

$$\frac{d^2S}{dx^2} = \frac{2}{x^3} > 0 \text{ when } x = 1.$$

The sum is a minimum when $x = 1$ and $1/x = 1$.

7. Let x be the length and y the width of the rectangle.

$$2x + 2y = 100$$

$$y = 50 - x$$

$$A = xy = x(50 - x)$$

$$\frac{dA}{dx} = 50 - 2x = 0 \text{ when } x = 25.$$

$$\frac{d^2A}{dx^2} = -2 < 0 \text{ when } x = 25.$$

A is maximum when $x = y = 25$ meters.

9. Let x be the length and y the width of the rectangle.

$$xy = 64$$

$$y = \frac{64}{x}$$

$$P = 2x + 2y = 2x + 2\left(\frac{64}{x}\right) = 2x + \frac{128}{x}$$

$$\frac{dP}{dx} = 2 - \frac{128}{x^2} = 0 \text{ when } x = 8.$$

$$\frac{d^2P}{dx^2} = \frac{256}{x^3} > 0 \text{ when } x = 8.$$

P is minimum when $x = y = 8$ feet.

11. $d = \sqrt{(x-4)^2 + (\sqrt{x} - 0)^2}$

$= \sqrt{x^2 - 7x + 16}$

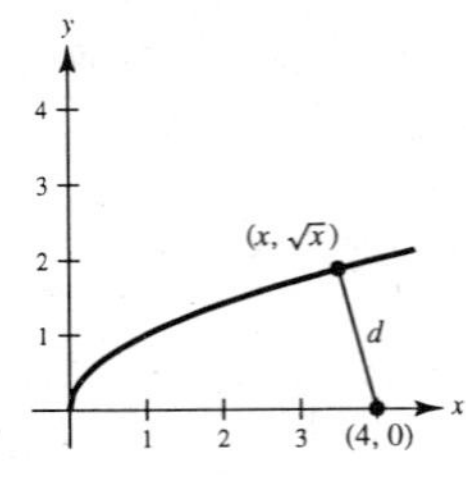

Since d is smallest when the expression inside the radical is smallest, you need only find the critical numbers of

$$f(x) = x^2 - 7x + 16.$$

$$f'(x) = 2x - 7 = 0$$

$$x = \tfrac{7}{2}$$

By the First Derivative Test, the point nearest to (4, 0) is $\left(7/2, \sqrt{7/2}\right)$.

13. $\dfrac{dQ}{dx} = kx(Q_0 - x) = kQ_0x - kx^2$

$$\frac{d^2Q}{dx^2} = kQ_0 - 2kx$$

$$= k(Q_0 - 2x) = 0 \text{ when } x = \frac{Q_0}{2}.$$

$$\frac{d^3Q}{dx^3} = -2k < 0 \text{ when } x = \frac{Q_0}{2}.$$

dQ/dx is maximum when $x = Q_0/2$.

15. $xy = 180{,}000$ (see figure)

$S = x + 2y = \left(x + \dfrac{360{,}000}{x}\right)$ where S is the length of fence needed.

$$\frac{dS}{dx} = 1 - \frac{360{,}000}{x^2} = 0 \text{ when } x = 600.$$

$$\frac{d^2S}{dx^2} = \frac{720{,}000}{x^3} > 0 \text{ when } x = 600.$$

S is a minimum when $x = 600$ meters and $y = 300$ meters.

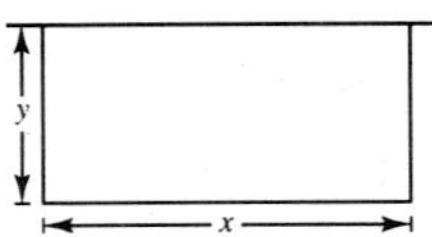

17. (a) $A = 4(\text{area of side}) + 2(\text{area of Top})$

(a) $A = 4(3)(11) + 2(3)(3) = 150$ square inches

(b) $A = 4(5)(5) + 2(5)(5) = 150$ square inches

(c) $A = 4(3.25)(6) + 2(6)(6) = 150$ square inches

(b) $V = (\text{length})(\text{width})(\text{height})$

(a) $V = (3)(3)(11) = 99$ cubic inches

(b) $V = (5)(5)(5) = 125$ cubic inches

(c) $V = (6)(6)(3.25) = 117$ cubic inches

(c) $S = 4xy + 2x^2 = 150 \Rightarrow y = \dfrac{150 - 2x^2}{4x}$

$$V = x^2y = x^2\left(\frac{150 - 2x^2}{4x}\right) = \frac{75}{2}x - \frac{1}{2}x^3$$

$$V' = \frac{75}{2} - \frac{3}{2}x^2 = 0 \Rightarrow x = \pm 5$$

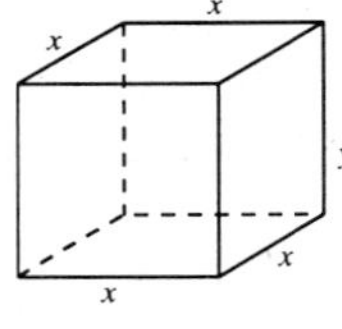

By the First Derivative Test, $x = 5$ yields the maximum volume. Dimensions: $5 \times 5 \times 5$. (A cube!)

19. (a) $V = x(s - 2x)^2, 0 < x < \dfrac{s}{2}$

$$\frac{dV}{dx} = 2x(s - 2x)(-2) + (s - 2x)^2$$

$$= (s - 2x)(s - 6x) = 0 \text{ when } x = \frac{s}{2}, \frac{s}{6}\ (s/2 \text{ is not in the domain}).$$

$$\frac{d^2V}{dx^2} = 24x - 8s$$

$$\frac{d^2V}{dx^2} < 0 \text{ when } x = \frac{s}{6}.$$

$$V = \frac{2s^3}{27} \text{ is maximum when } x = \frac{5}{6}.$$

(b) If the length is doubled, $V = \frac{2}{27}(2s)^3 = 8\left(\frac{2}{27}s^3\right)$. Volume is increased by a factor of 8.

21. $16 = 2y + x + \pi\left(\dfrac{x}{2}\right)$

$$32 = 4y + 2x + \pi x$$

$$y = \frac{32 - 2x - \pi x}{4}$$

$$A = xy + \frac{\pi}{2}\left(\frac{x}{2}\right)^2 = \left(\frac{32 - 2x - \pi x}{4}\right)x + \frac{\pi x^2}{8}$$

$$= 8x - \frac{1}{2}x^2 - \frac{\pi}{4}x^2 + \frac{\pi}{8}x^2$$

$$\frac{dA}{dx} = 8 - x - \frac{\pi}{2}x + \frac{\pi}{4}x = 8 - x\left(1 + \frac{\pi}{4}\right)$$

$$= 0 \text{ when } x = \frac{8}{1 + (\pi/4)} = \frac{32}{4 + \pi}.$$

$$\frac{d^2A}{dx^2} = -\left(1 + \frac{\pi}{4}\right) < 0 \text{ when } x = \frac{32}{4 + \pi}$$

$$y = \frac{32 - 2[32/(4 + \pi)] - \pi[32/(4 + \pi)]}{4} = \frac{16}{4 + \pi}$$

The area is maximum when $y = \dfrac{16}{4 + \pi}$ feet and $x = \dfrac{32}{4 + \pi}$ feet.

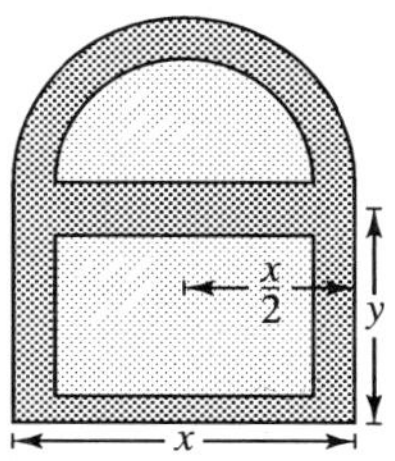

23. (a) $\dfrac{y - 2}{0 - 1} = \dfrac{0 - 2}{x - 1}$

$$y = 2 + \frac{2}{x - 1}$$

$$L = \sqrt{x^2 + y^2} = \sqrt{x^2 + \left(2 + \frac{2}{x - 1}\right)^2}$$

$$= \sqrt{x^2 + 4 + \frac{8}{x - 1} + \frac{4}{(x - 1)^2}}, \quad x > 1$$

(b)

(2.587, 4.162)

L is minimum when $x \approx 2.587$ and $L \approx 4.162$.

(c) Area $= \dfrac{1}{2}$(base)(height)

$$= \frac{1}{2}(x)(y)$$

$$= \frac{1}{2}(x)\left(2 + \frac{2}{x - 1}\right)$$

$$= x + \frac{x}{x - 1}$$

Use a graphing utility to approximate x:

$$y = x + \frac{x}{x - 1}.$$

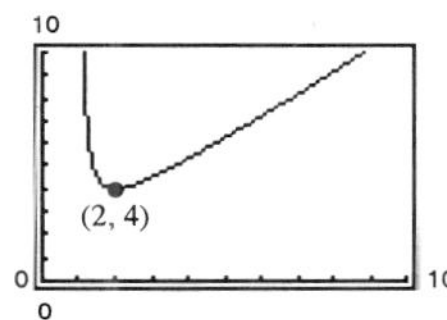

Area is minimum when $x = 2$ and $y = 4$.
Vertices: (0, 0), (2, 0), and (0, 4)

25. $A = 2xy = 2x\sqrt{25 - x^2}$ (see figure)

$$\frac{dA}{dx} = 2x\left(\frac{1}{2}\right)\left(\frac{-2x}{\sqrt{25 - x^2}}\right) + 2\sqrt{25 - x^2}$$

$$= 2\left(\frac{25 - 2x^2}{\sqrt{25 - x^2}}\right) = 0 \text{ when } x = y = \frac{5\sqrt{2}}{2} \approx 3.54.$$

By the First Derivative Test, the inscribed rectangle of maximum area has vertices

$$\left(\pm\frac{5\sqrt{2}}{2}, 0\right), \left(\pm\frac{5\sqrt{2}}{2}, \frac{5\sqrt{2}}{2}\right).$$

Width: $\dfrac{5\sqrt{2}}{2}$; Length: $5\sqrt{2}$

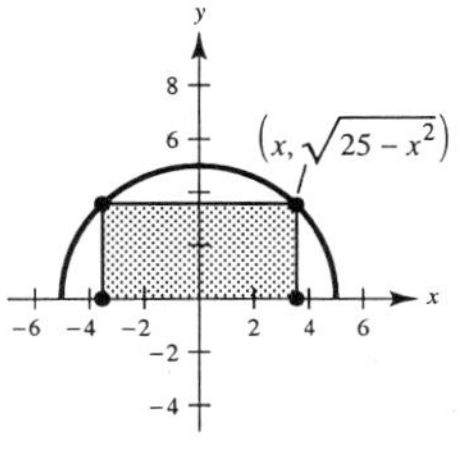

27. $A = \dfrac{1}{2}(2r + 2x)\sqrt{r^2 - x^2} = (r + x)\sqrt{r^2 - x^2}$ (see figure)

$$\frac{dA}{dx} = (r + x)\left(\frac{1}{2}\right)(r^2 - x^2)^{-1/2}(-2x) + \sqrt{r^2 - x^2}$$

$$= \frac{-x(r + x)}{\sqrt{r^2 - x^2}} + \sqrt{r^2 - x^2} = \frac{r^2 - rx - 2x^2}{\sqrt{r^2 - x^2}}$$

$$= \frac{(r - 2x)(r + x)}{\sqrt{r^2 - x^2}} = 0 \text{ when } x = \frac{r}{2}.$$

By the First Derivative Test, A will be a maximum when the trapezoid bases are r and $2r$, and the altitude is $(\sqrt{3}r)/2$.

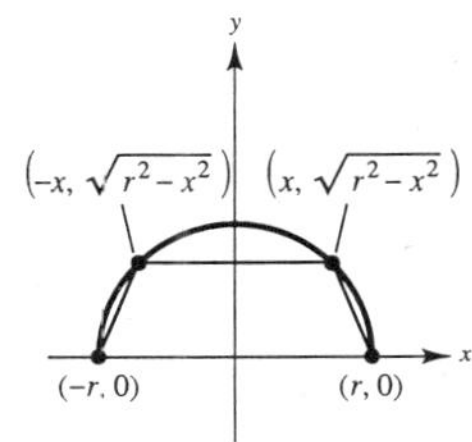

29. $V = \pi r^2 h = 22$ cubic inches or $h = \dfrac{22}{\pi r^2}$

(a)

Radius, r	Height	Surface Area
0.2	$\dfrac{22}{\pi(0.2)^2}$	$2\pi(0.2)\left[0.2 + \dfrac{22}{\pi(0.2)^2}\right] \approx 220.3$
0.4	$\dfrac{22}{\pi(0.4)^2}$	$2\pi(0.4)\left[0.4 + \dfrac{22}{\pi(0.4)^2}\right] \approx 111.0$
0.6	$\dfrac{22}{\pi(0.6)^2}$	$2\pi(0.6)\left[0.6 + \dfrac{22}{\pi(0.6)^2}\right] \approx 75.6$
0.8	$\dfrac{22}{\pi(0.8)^2}$	$2\pi(0.8)\left[0.8 + \dfrac{22}{\pi(0.8)^2}\right] \approx 59.0$

(b)

Radius, r	Height	Surface Area
0.2	$\dfrac{22}{\pi(0.2)^2}$	$2\pi(0.2)\left[0.2 + \dfrac{22}{\pi(0.2)^2}\right] \approx 220.3$
0.4	$\dfrac{22}{\pi(0.4)^2}$	$2\pi(0.4)\left[0.4 + \dfrac{22}{\pi(0.4)^2}\right] \approx 111.0$
0.6	$\dfrac{22}{\pi(0.6)^2}$	$2\pi(0.6)\left[0.6 + \dfrac{22}{\pi(0.6)^2}\right] \approx 75.6$
0.8	$\dfrac{22}{\pi(0.8)^2}$	$2\pi(0.8)\left[0.8 + \dfrac{22}{\pi(0.8)^2}\right] \approx 59.0$
1.0	$\dfrac{22}{\pi(1.0)^2}$	$2\pi(1.0)\left[1.0 + \dfrac{22}{\pi(1.0)^2}\right] \approx 50.3$
1.2	$\dfrac{22}{\pi(1.2)^2}$	$2\pi(1.2)\left[1.2 + \dfrac{22}{\pi(1.2)^2}\right] \approx 45.7$
1.4	$\dfrac{22}{\pi(1.4)^2}$	$2\pi(1.4)\left[1.4 + \dfrac{22}{\pi(1.4)^2}\right] \approx 43.7$
1.6	$\dfrac{22}{\pi(1.6)^2}$	$2\pi(1.6)\left[1.6 + \dfrac{22}{\pi(1.6)^2}\right] \approx 43.6$
1.8	$\dfrac{22}{\pi(1.8)^2}$	$2\pi(1.8)\left[1.8 + \dfrac{22}{\pi(1.8)^2}\right] \approx 44.8$
2.0	$\dfrac{22}{\pi(2.0)^2}$	$2\pi(2.0)\left[2.0 + \dfrac{22}{\pi(2.0)^2}\right] \approx 47.1$

The minimum seems to be about 43.6 for $r = 1.6$.

(c) $S = 2\pi r^2 + 2\pi rh$

$$= 2\pi r(r + h) = 2\pi r\left[r + \frac{22}{\pi r^2}\right] = 2\pi r^2 + \frac{44}{r}$$

(d)

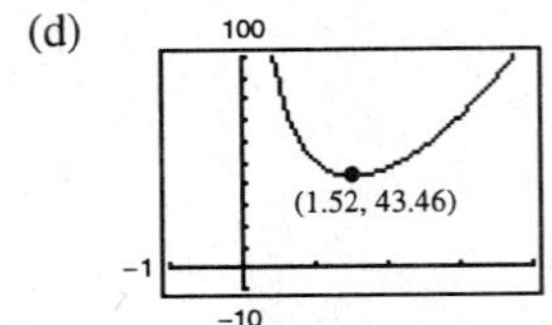

The minimum seems to be 43.46 for $r \approx 1.52$.

(e) $\dfrac{dS}{dr} = 4\pi r - \dfrac{44}{r^2} = 0$ when $r = \sqrt[3]{11/\pi} \approx 1.52$ in.

$$h = \frac{22}{\pi r^2} \approx 3.04 \text{ in.}$$

Note: Notice that

$$h = \frac{22}{\pi r^2} = \frac{22}{\pi(11/\pi)^{2/3}} = 2\left(\frac{11^{1/3}}{\pi^{1/3}}\right) = 2r.$$

31. Let x be the sides of the square ends and y the length of the package.

$$P = 4x + y = 108 \implies y = 108 - 4x$$

$$V = x^2y = x^2(108 - 4x) = 108x^2 - 4x^3$$

$$\frac{dV}{dx} = 216x - 12x^2 = 12x(18 - x) = 0 \text{ when } x = 18.$$

$$\frac{d^2V}{dx^2} = 216 - 24x = -216 < 0 \text{ when } x = 18.$$

The volume is maximum when $x = 18$ inches and $y = 108 - 4(18) = 36$ inches.

33. $V = \frac{1}{3}\pi x^2 h = \frac{1}{3}\pi x^2\left(r + \sqrt{r^2 - x^2}\right)$ (see figure)

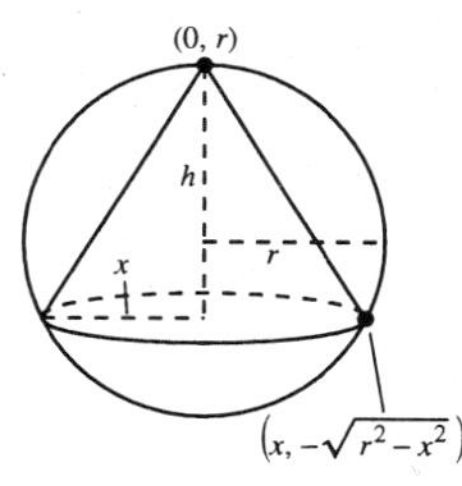

$$\frac{dV}{dx} = \frac{1}{3}\pi\left[\frac{-x^3}{\sqrt{r^2 - x^2}} + 2x\left(r + \sqrt{r^2 - x^2}\right)\right] = \frac{\pi x}{3\sqrt{r^2 - x^2}}\left(2r^2 + 2r\sqrt{r^2 - x^2} - 3x^2\right) = 0$$

$$2r^2 + 2r\sqrt{r^2 - x^2} - 3x^2 = 0$$

$$2r\sqrt{r^2 - x^2} = 3x^2 - 2r^2$$

$$4r^2(r^2 - x^2) = 9x^4 - 12x^2r^2 + 4r^4$$

$$0 = 9x^4 - 8x^2r^2 = x^2(9x^2 - 8r^2)$$

$$x = 0, \frac{2\sqrt{2}r}{3}$$

By the First Derivative Test, the volume is a maximum when $x = \dfrac{2\sqrt{2}r}{3}$ and $h = r + \sqrt{r^2 - x^2} = \dfrac{4r}{3}$.

Thus, the maximum volume is $V = \dfrac{1}{3}\pi\left(\dfrac{8r^2}{9}\right)\left(\dfrac{4r}{3}\right) = \dfrac{32\pi r^3}{81}$ cubic units.

35. $V = 12 = \frac{4}{3}\pi r^3 + \pi r^2 h$

$$h = \frac{12 - (4/3)\pi r^3}{\pi r^2} = \frac{12}{\pi r^2} - \frac{4}{3}r$$

$$S = 4\pi r^2 + 2\pi rh = 4\pi r^2 + 2\pi r\left(\frac{12}{\pi r^2} - \frac{4}{3}r\right)$$

$$= 4\pi r^2 + \frac{24}{r} - \frac{8}{3}\pi r^2 = \frac{4}{3}\pi r^2 + \frac{24}{r}$$

$$\frac{dS}{dr} = \frac{8}{3}\pi r - \frac{24}{r^2} = 0 \text{ when } r = \sqrt[3]{9/\pi} \approx 1.42 \text{ cm.}$$

$$\frac{d^2S}{dr^2} = \frac{8}{3}\pi + \frac{48}{r^3} > 0 \text{ when } r = \sqrt[3]{9/\pi} \text{ cm.}$$

The surface area is minimum when $r = \sqrt[3]{9/\pi}$ cm and $h = 0$. The resulting solid is a sphere of radius $r \approx 1.42$ cm.

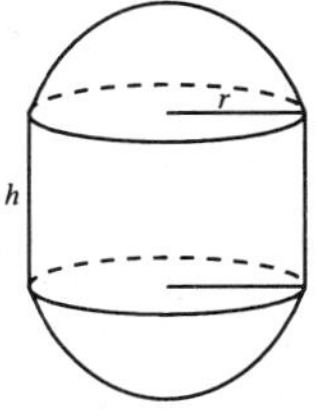

37. Let x be the length of a side of the square and y the length of a side of the triangle.

$$4x + 3y = 10$$

$$A = x^2 + \frac{1}{2}y\left(\frac{\sqrt{3}}{2}y\right)$$

$$= \frac{(10 - 3y)^2}{16} + \frac{\sqrt{3}}{4}y^2$$

$$\frac{dA}{dy} = \frac{1}{8}(10 - 3y)(-3) + \frac{\sqrt{3}}{2}y = 0$$

$$-30 + 9y + 4\sqrt{3}y = 0$$

$$y = \frac{30}{9 + 4\sqrt{3}}$$

$$\frac{d^2A}{dy^2} = \frac{9 + 4\sqrt{3}}{8} > 0$$

A is minimum when

$$y = \frac{30}{9 + 4\sqrt{3}} \text{ and } x = \frac{10\sqrt{3}}{9 + 4\sqrt{3}}.$$

39. Let S be the strength and k the constant of proportionality. Given $h^2 + w^2 = 24^2$, $h^2 = 24^2 - w^2$,

$$S = kwh^2$$

$$S = kw(576 - w^2) = k(576w - w^3)$$

$$\frac{dS}{dw} = k(576 - 3w^3) = 0 \text{ when } w = 8\sqrt{3},\ h = 8\sqrt{6}.$$

$$\frac{d^2S}{dw^2} = -6kw < 0 \text{ when } w = 8\sqrt{3}.$$

These values yield a maximum.

41. $R = \dfrac{v_0^2}{g}\sin 2\theta$

$$\frac{dR}{d\theta} = \frac{2v_0^2}{g}\cos 2\theta = 0 \text{ when } \theta = \frac{\pi}{4}, \frac{3\pi}{4}.$$

$$\frac{d^2R}{d\theta^2} = -\frac{4v_0^2}{g}\sin 2\theta < 0 \text{ when } \theta = \frac{\pi}{4}.$$

By the Second Derivative Test, R is maximum when $\theta = \pi/4$.

43. $f(x) = \frac{1}{2}x^2 \qquad g(x) = \frac{1}{16}x^4 - \frac{1}{2}x^2$ on $[0, 4]$

(a)

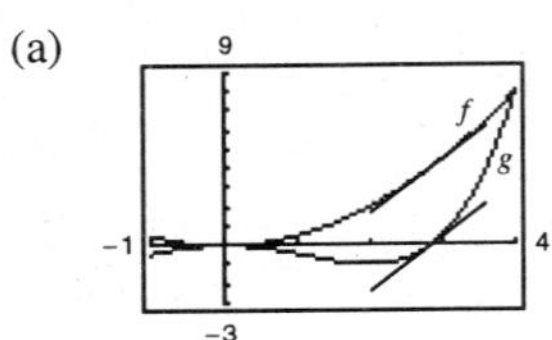

(b) $d(x) = f(x) - g(x) = \frac{1}{2}x^2 - \left(\frac{1}{16}x^4 - \frac{1}{2}x^2\right) = x^2 - \frac{1}{16}x^4$

$d'(x) = 2x - \frac{1}{4}x^3 = 0 \Longrightarrow 8x = x^3$

$\Longrightarrow x = 0, 2\sqrt{2}$ (in $[0, 4]$)

The maximum distance is $d = 4$ when $x = 2\sqrt{2}$.

(c) $f'(x) = x$, Tangent line at $(2\sqrt{2}, 4)$ is

$$y - 4 = 2\sqrt{2}(x - 2\sqrt{2})$$

$$y = 2\sqrt{2}x - 4.$$

$g'(x) = \frac{1}{4}x^3 - x$, Tangent line at $(2\sqrt{2}, 0)$ is

$$y - 0 = \left(\tfrac{1}{4}(2\sqrt{2})^3 - 2\sqrt{2}\right)(x - 2\sqrt{2})$$

$$y = 2\sqrt{2}x - 8.$$

The tangent lines are parallel and 4 vertical units apart.

(d) The tangent lines will be parallel. If $d(x) = f(x) - g(x)$, then $d'(x) = 0 = f'(x) - g'(x)$ implies that $f'(x) = g'(x)$ at the point x where the distance is maximum.

45. Let F be the illumination at point P which is x units from source 1.

$$F = \frac{kI_1}{x^2} + \frac{kI_2}{(d - x)^2}$$

$$\frac{dF}{dx} = \frac{-2kI_1}{x^3} + \frac{2kI_2}{(d - x)^3}$$

$$= 0 \text{ when } \frac{2kI_1}{x^3} = \frac{2kI_2}{(d - x)^3}.$$

$$\frac{\sqrt[3]{I_1}}{\sqrt[3]{I_2}} = \frac{x}{d - x}$$

$$(d - x)\sqrt[3]{I_1} = x\sqrt[3]{I_2}$$

$$d\sqrt[3]{I_1} = x\left(\sqrt[3]{I_1} + \sqrt[3]{I_2}\right)$$

$$x = \frac{d\sqrt[3]{I_1}}{\sqrt[3]{I_1} + \sqrt[3]{I_2}}$$

$$\frac{d^2F}{dx^2} = \frac{6kI_1}{x^4} + \frac{6kI_2}{(d - x)^4}$$

$$> 0 \text{ when } x = \frac{d\sqrt[3]{I_1}}{\sqrt[3]{I_1} + \sqrt[3]{I_2}}.$$

This is the minimum point.

47. $T = \dfrac{\sqrt{x^2 + 4}}{v_1} + \dfrac{\sqrt{x^2 - 6x + 10}}{v_2}$

$$\frac{dT}{dx} = \frac{x}{v_1\sqrt{x^2 + 4}} + \frac{x - 3}{v_2\sqrt{x^2 - 6x + 10}} = 0$$

Since

$$\frac{x}{\sqrt{x^2 + 4}} = \sin\theta_1 \text{ and } \frac{x - 3}{\sqrt{x^2 - 6x + 10}} = -\sin\theta_2$$

we have

$$\frac{\sin\theta_1}{v_1} - \frac{\sin\theta_2}{v_2} = 0 \Longrightarrow \frac{\sin\theta_1}{v_1} = \frac{\sin\theta_2}{v_2}.$$

Since

$$\frac{d^2T}{dx^2} = \frac{4}{v_1(x^2 + 4)^{3/2}} + \frac{1}{v_2(x^2 - 6x + 10)^{3/2}} > 0$$

this condition yields a minimum time.

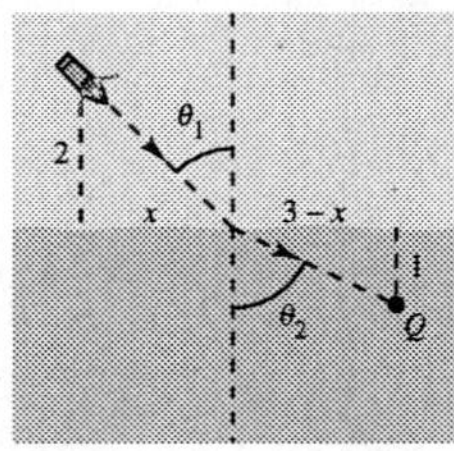

49. $f(x) = 2 - 2\sin x$

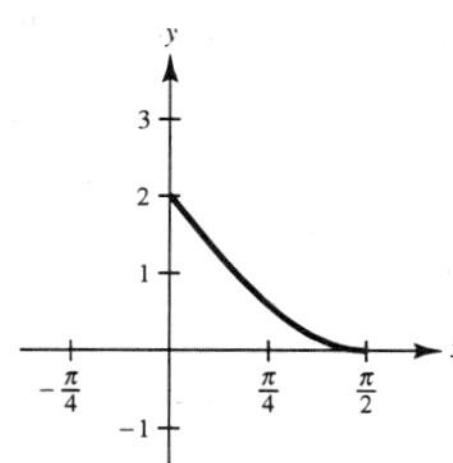

(a) Distance from origin to y-intercept is 2.
Distance from origin to x-intercept is $\pi/2 \approx 1.57$.

(b) $d = \sqrt{x^2 + y^2} = \sqrt{x^2 + (2 - 2\sin x)^2}$

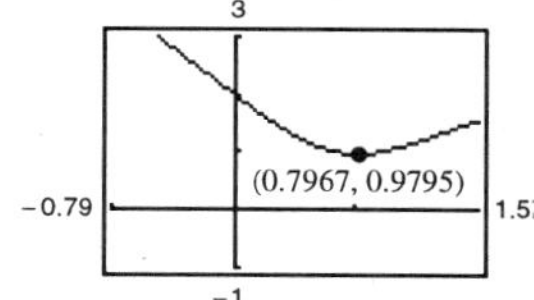

Minimum distance $= 0.9795$ at $x = 0.7967$.

(c) Let $f(x) = d^2(x) = x^2 - (2 - 2\sin x)^2$.

$f'(x) = 2x + 2(2 - 2\sin x)(-2\cos x)$

Setting $f'(x) = 0$, you obtain $x \approx 0.7967$, which corresponds to $d = 0.9795$.

51. $V = \frac{1}{3}\pi r^2 h = \frac{1}{3}\pi r^2\sqrt{144 - r^2}$

$$\frac{dV}{dr} = \frac{1}{3}\pi\left[r^2\left(\frac{1}{2}\right)(144 - r^2)^{-1/2}(-2r) + 2r\sqrt{144 - r^2}\right]$$

$$= \frac{1}{3}\pi\left[\frac{288r - 3r^3}{\sqrt{144 - r^2}}\right] = \pi\left[\frac{r(96 - r^2)}{\sqrt{144 - r^2}}\right] = 0 \text{ when } r = 0, 4\sqrt{6}.$$

By the First Derivative Test, V is maximum when $r = 4\sqrt{6}$ and $h = 4\sqrt{3}$.

Area of circle: $A = \pi(12)^2 = 144\pi$

Lateral surface area of cone: $S = \pi(4\sqrt{6})\sqrt{(4\sqrt{6})^2 + (4\sqrt{3})^2} = 48\sqrt{6}\pi$

Area of sector: $144\pi - 48\sqrt{6}\pi = \frac{1}{2}\theta r^2 = 72\theta$

$$\theta = \frac{144\pi - 48\sqrt{6}\pi}{72} = \frac{2\pi}{3}(3 - \sqrt{6}) \approx 1.153 \text{ radians or } 66°$$

53. $\sin\theta = \frac{2}{L_1} \Rightarrow L_1 = \frac{2}{\sin\theta}$

$\cos\theta = \frac{7}{L_2} \Rightarrow L_2 = \frac{7}{\cos\theta}$

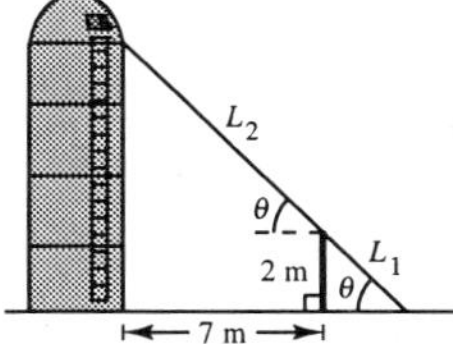

(a)

θ	L_1	L_2	$L_1 + L_2$
0.1	20.0	7.0	27.1
0.2	10.1	7.1	17.2
0.3	6.8	7.3	14.1
0.4	5.1	7.6	12.7
0.5	4.2	8.0	12.1
0.6	3.5	8.5	12.0

(b)

θ	L_1	L_2	$L_1 + L_2$
0.57	3.706	8.315	12.021
0.58	3.650	8.369	12.018
0.59	3.595	8.424	12.019

The minimum length of $L_1 + L_2$ is approximately 12.018 meters.

(c) $L = L_1 + L_2 = \frac{2}{\sin\theta} + \frac{7}{\cos\theta}$

(d)

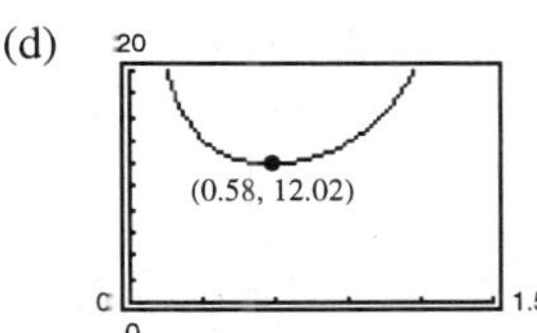

The minimum is approximately 12.018 meters.

—CONTINUED—

53. —CONTINUED—

(e) $L(\theta) = \dfrac{2}{\sin\theta} + \dfrac{7}{\cos\theta} = 2\csc\theta + 7\sec\theta$

$$L'(\theta) = -2\csc\theta\cot\theta + 7\sec\theta \cdot \tan\theta = 0$$

$$2\csc\theta\cot\theta = 7\sec\theta\tan\theta$$

$$2\frac{\cos\theta}{\sin^2\theta} = 7\frac{\sin\theta}{\cos^2\theta}$$

$$2\cos^3\theta = 7\sin^3\theta$$

$$\frac{2}{7} = \frac{\sin^3\theta}{\cos^3\theta} = \tan^3\theta$$

$$\tan\theta = \left(\frac{2}{7}\right)^{1/3} = \frac{\sqrt[3]{98}}{7} \Longrightarrow \theta \approx 0.5824 \text{ radians.}$$

$$L(0.5824) \approx 12.018 \text{ meters}$$

(f) height $= L \cdot \sin\theta = (12.018)(\sin(0.5824))$

≈ 6.61 meters

55. The slope of the line is $(-4/3)$, and the equation is

$$y - 0 = -\frac{4}{3}(x - 3)$$

$$y = -\frac{4}{3}x + 4$$

For the rectangle, let $(x, y) = (x, (-4/3)x + 4)$ be the vertex on the line. Then the area is

$$A = bh = x\left(-\frac{4}{3}x + 4\right) = -\frac{4}{3}x^2 + 4x.$$

$$A'(x) = -\frac{8}{3}x + 4 = 0 \Longrightarrow x = \frac{3}{2} \text{ and } y = 2. \text{ Base } = \frac{3}{2}, \text{ height } = 2.$$

For the circle, you can use geometry to note that the center of the inscribed circle is (r, r), and the distance from (r, r) to the line $(4/3)x + y - 4 = 0$ is also r:

$$\frac{|(4/3)r + r - 4|}{\sqrt{(16/9) + 1}} = r$$

$$|7r - 12| = 5r$$

$$7r - 12 = \pm 5r$$

$$r = 1, 6.$$

Clearly, we must choose $r = 1$. Hence, the center is $(1, 1)$ and the radius is 1.
For the semicircle, observe that the center must be of the form (r, r), and must lie on the line $y = (-4/3)x + 4$. Thus,

$$r = \frac{-4}{3}r + 4$$

$$3r = -4r + 12$$

$$7r = 12$$

$$r = \frac{12}{7}.$$

The center is $(12/7, 12/7)$ and the radius is $r = 12/7$.

Section 3.8 Newton's Method

1. $f(x) = x^2 - 3$

$f'(x) = 2x$

$x_1 = 1.7$

n	x_n	$f(x_n)$	$f'(x_n)$	$\frac{f(x_n)}{f'(x_n)}$	$x_n - \frac{f(x_n)}{f'(x_n)}$
1	1.7000	−0.1100	3.4000	−0.0324	1.7324
2	1.7324	0.0012	3.4648	0.0003	1.7321

3. $f(x) = \sin x$

$f'(x) = \cos x$

$x_1 = 3$

n	x_n	$f(x_n)$	$f'(x_n)$	$\frac{f(x_n)}{f'(x_n)}$	$x_n - \frac{f(x_n)}{f'(x_n)}$
1	3.0000	0.1411	−0.9900	−0.1425	3.1425
2	3.1425	−0.0009	−1.0000	0.0009	3.1416

5. $f(x) = x^3 + x - 1$

$f'(x) = 3x^2 + 1$

Approximation of the zero of f is 0.682.

n	x_n	$f(x_n)$	$f'(x_n)$	$\frac{f(x_n)}{f'(x_n)}$	$x_n - \frac{f(x_n)}{f'(x_n)}$
1	0.5000	−0.3750	1.7500	−0.2143	0.7143
2	0.7143	0.0788	2.5307	0.0311	0.6832
3	0.6832	0.0021	2.4003	0.0009	0.6823

7. $f(x) = 3\sqrt{x-1} - x$

$f'(x) = \dfrac{3}{2\sqrt{x-1}} - 1$

Approximation of the zero of f is 1.146.

Similarly, the other zero is approximately 7.854.

n	x_n	$f(x_n)$	$f'(x_n)$	$\frac{f(x_n)}{f'(x_n)}$	$x_n - \frac{f(x_n)}{f'(x_n)}$
1	1.2000	0.1416	2.3541	0.0602	1.1398
2	1.1398	−0.0181	3.0118	−0.0060	1.1458
3	1.1458	−0.0003	2.9284	−0.0001	1.1459

9. $f(x) = x^3 - 3.9x^2 + 4.79x - 1.881$

$f'(x) = 3x^2 - 7.8x + 4.79$

n	x_n	$f(x_n)$	$f'(x_n)$	$\frac{f(x_n)}{f'(x_n)}$	$x_n - \frac{f(x_n)}{f'(x_n)}$
1	0.5000	−0.3360	1.6400	−0.2049	0.7049
2	0.7049	−0.0921	0.7824	−0.1177	0.8226
3	0.8226	−0.0231	0.4037	−0.0573	0.8799
4	0.8799	−0.0045	0.2495	−0.0181	0.8980
5	0.8980	−0.0004	0.2048	−0.0020	0.9000
6	0.9000	0.0000	0.2000	0.0000	0.9000

Approximation of the zero of f is 0.900.

n	x_n	$f(x_n)$	$f'(x_n)$	$\frac{f(x_n)}{f'(x_n)}$	$x_n - \frac{f(x_n)}{f'(x_n)}$
1	1.1	0.0000	−0.1600	−0.0000	1.1000

Approximation of the zero of f is 1.100.

n	x_n	$f(x_n)$	$f'(x_n)$	$\frac{f(x_n)}{f'(x_n)}$	$x_n - \frac{f(x_n)}{f'(x_n)}$
1	1.9	0.0000	0.8000	0.0000	1.9000

Approximation of the zero of f is 1.900.

11. $f(x) = x + \sin(x + 1)$

$f'(x) = 1 + \cos(x + 1)$

n	x_n	$f(x_n)$	$f'(x_n)$	$\frac{f(x_n)}{f'(x_n)}$	$x_n - \frac{f(x_n)}{f'(x_n)}$
1	−0.5000	−0.0206	1.8776	−0.0110	−0.4890
2	−0.4890	0.0000	1.8723	0.0000	−0.4890

Approximation of the zero of f is -0.489.

13. $h(x) = f(x) - g(x) = 2x + 1 - \sqrt{x + 4}$

$$h'(x) = 2 - \frac{1}{2\sqrt{x + 4}}$$

n	x_n	$h(x_n)$	$h'(x_n)$	$\frac{h(x_n)}{h'(x_n)}$	$x_n - \frac{h(x_n)}{h'(x_n)}$
1	0.6000	0.0552	1.7669	0.0313	0.5687
2	0.5687	−0.0001	1.7661	0.0000	0.5687

Point of intersection of the graphs of f and g occurs when $x \approx 0.569$.

15. $h(x) = f(x) - g(x) = x - \tan x$

$h'(x) = 1 - \sec^2 x$

n	x_n	$h(x_n)$	$h'(x_n)$	$\frac{h(x_n)}{h'(x_n)}$	$x_n - \frac{h(x_n)}{h'(x_n)}$
1	4.5000	−0.1373	−21.5048	0.0064	4.4936
2	4.4936	−0.0039	−20.2271	0.0002	4.4934

Point of intersection of the graphs of f and g occurs when $x \approx 4.493$.

17. Let $g(x) = f(x) - x = \cos x - x$

$g'(x) = -\sin x - 1.$

n	x_n	$g(x_n)$	$g'(x_n)$	$\frac{g(x_n)}{g'(x_n)}$	$x_n - \frac{g(x_n)}{g'(x_n)}$
1	1.0000	−0.4597	−1.8415	0.2496	0.7504
2	0.7504	−0.0190	−1.6819	0.0113	0.7391
3	0.7391	0.0000	−1.6736	0.0000	0.7391

The fixed point is approximately 0.74.

19. $f(x) = x^3 - 3x^2 + 3,\ f'(x) = 3x^2 - 6x$

(a)

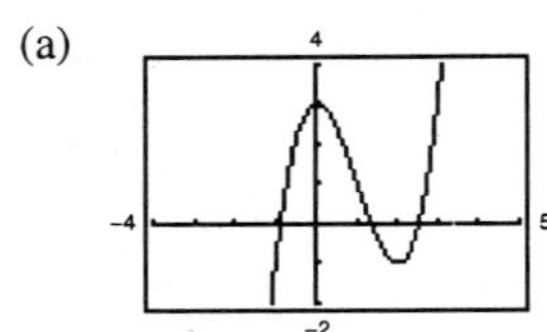

(b) $x_1 = 1$

$$x_2 = x_1 - \frac{f(x_1)}{f'(x_1)} \approx 1.333$$

Continuing, the zero is 1.347.

—CONTINUED—

19. —CONTINUED—

(c) $x_1 = \frac{1}{4}$

$$x_2 = x_1 - \frac{f(x_1)}{f'(x_1)} \approx 2.405$$

Continuing, the zero is 2.532.

(d)

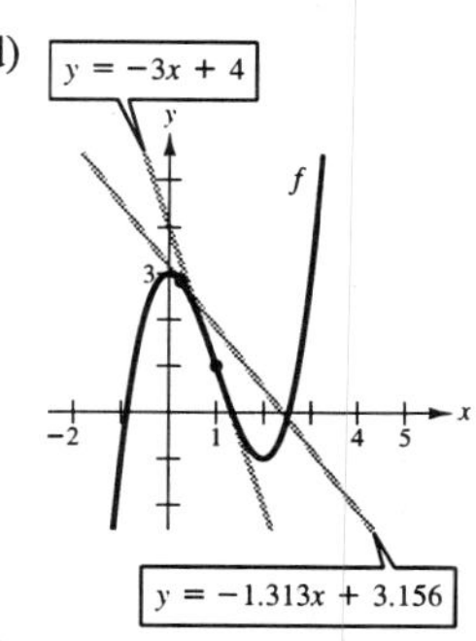

The x-intercepts correspond to the values resulting from the first iteration of Newton's Method.

(e) If the initial guess x_1 is not "close to" the desired zero of the function, the x-intercept of the tangent line may approximate another zero of the function.

21. $y = 2x^3 - 6x^2 + 6x - 1 = f(x)$

$y' = 6x^2 - 12x + 6 = f'(x)$

$x_1 = 1$

$f'(x) = 0$; therefore, the method fails.

n	x_n	$f(x_n)$	$f'(x_n)$
1	1	1	0

23. $y = -x^3 + 3x^2 - x + 1 = f(x)$

$y' = -3x^2 + 6x - 1 = f'(x)$

$x_1 = 1$

Fails to converge.

n	x_n	$f(x_n)$	$f'(x_n)$	$\frac{f(x_n)}{f'(x_n)}$	$x_n - \frac{f(x_n)}{f'(x_n)}$
1	1	2	2	1	0
2	0	1	-1	-1	1
3	1	2	2	1	0

25. $f(x) = x^2 - a = 0$

$f'(x) = 2x$

$$x_{i+1} = x_i - \frac{x_i^2 - a}{2x_i}$$

$$= \frac{2x_i^2 - x_i^2 + a}{2x_i} = \frac{x_i^2 + a}{2x_i} = \frac{x_i}{2} + \frac{a}{2x_i}$$

27. $x_{i+1} = \frac{x_i^2 + 7}{2x_i}$

i	1	2	3	4	5
x_i	2.0000	2.7500	2.6477	2.6458	2.6458

$\sqrt{7} \approx 2.646$

29. $x_{i+1} = \frac{3x_i^4 + 6}{4x_i^3}$

$\sqrt[4]{6} \approx 1.565$

i	1	2	3	4
x_i	1.5000	1.5694	1.5651	1.5651

31. $f(x) = \frac{1}{x} - a = 0$

$$f'(x) = -\frac{1}{x^2}$$

$$x_{n+1} = x_n - \frac{(1/x_n) - a}{-1/x_n^2} = x_n + x_n^2\left(\frac{1}{x_n} - a\right) = x_n + x_n - x_n^2a = 2x_n - x_n^2a = x_n(2 - ax_n)$$

33. $f(x) = 1 + \cos x$

$f'(x) = -\sin x$

n	x_n	$f(x_n)$	$f'(x_n)$	$\dfrac{f(x_n)}{f'(x_n)}$	$x_n - \dfrac{f(x_n)}{f'(x_n)}$
1	3.0000	0.0100	−0.1411	−0.0709	3.0709
2	3.0709	0.0025	−0.0706	−0.0354	3.1063
3	3.1063	0.0006	−0.0353	−0.0176	3.1239
4	3.1239	0.0002	−0.0177	−0.0088	3.1327
5	3.1327	0.0000	−0.0089	−0.0044	3.1371
6	3.1371	0.0000	−0.0045	−0.0022	3.1393
7	3.1393	0.0000	−0.0023	−0.0011	3.1404
8	3.1404	0.0000	−0.0012	−0.0006	3.1410

Approximation of the zero: 3.141

35. $f(x) = x \cos x$

$f'(x) = -x \sin x + \cos x = 0$

Letting $F(x) = f'(x)$, we can use Newton's Method as follows.

$[F'(x) = -2 \sin x + x \cos x]$

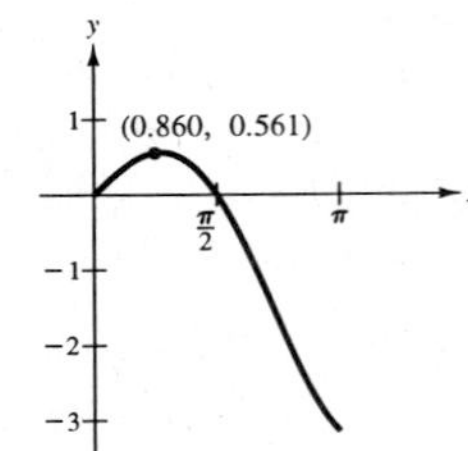

n	x_n	$F(x_n)$	$F'(x_n)$	$\dfrac{F(x_n)}{F'(x_n)}$	$x_n - \dfrac{F(x_n)}{F'(x_n)}$
1	0.9000	−0.0834	−2.1261	0.0392	0.8608
2	0.8608	−0.0010	−2.0778	0.0005	0.8603

Approximation to the critical number: 0.860

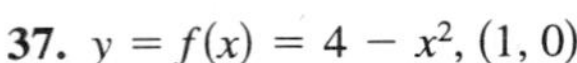

37. $y = f(x) = 4 - x^2, (1, 0)$

$d = \sqrt{(x-1)^2 + (y-0)^2} = \sqrt{(x-1)^2 + (4-x^2)^2} = \sqrt{x^4 - 7x^2 - 2x + 17}$

d is minimized when $D = x^4 - 7x^2 - 2x + 17$ is a minimum.

$g(x) = D' = 4x^3 - 14x - 2$

$g'(x) = 12x^2 - 14$

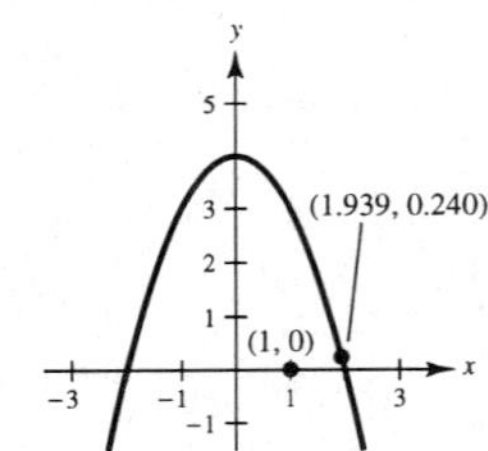

n	x_n	$g(x_n)$	$g'(x_n)$	$\dfrac{g(x_n)}{g'(x_n)}$	$x_n - \dfrac{g(x_n)}{g'(x_n)}$
1	2.0000	2.0000	34.0000	0.0588	1.9412
2	1.9412	0.0830	31.2191	0.0027	1.9385
3	1.9385	−0.0012	31.0934	0.0000	1.9385

$x \approx 1.939$

Point closest to $(1, 0)$ is $\approx (1.939, 0.240)$.

39.

$$\text{Minimize: } T = \frac{\text{Distance rowed}}{\text{Rate rowed}} + \frac{\text{Distance walked}}{\text{Rate walked}}$$

$$T = \frac{\sqrt{x^2 + 4}}{3} + \frac{\sqrt{x^2 - 6x + 10}}{4}$$

$$T' = \frac{x}{3\sqrt{x^2 + 4}} + \frac{x - 3}{4\sqrt{x^2 - 6x + 10}} = 0$$

$$4x\sqrt{x^2 - 6x + 10} = -3(x - 3)\sqrt{x^2 + 4}$$

$$16x^2(x^2 - 6x + 10) = 9(x - 3)^2(x^2 + 4)$$

$$7x^4 - 42x^3 + 43x^2 + 216x - 324 = 0$$

Let $f(x) = 7x^4 - 42x^3 + 43x^2 + 216x - 324$ and $f'(x) = 28x^3 - 126x^2 + 86x + 216$. Since $f(1) = -100$ and $f(2) = 56$, the solution is in the interval $(1, 2)$.

n	x_n	$f(x_n)$	$f'(x_n)$	$\frac{f(x_n)}{f'(x_n)}$	$x_n - \frac{f(x_n)}{f'(x_n)}$
1	1.7000	19.5887	135.6240	0.1444	1.5556
2	1.5556	−1.0480	150.2780	−0.0070	1.5626
3	1.5626	0.0014	49.5591	0.0000	1.5626

Approximation: $x \approx 1.563$ miles

41.

$$2{,}500{,}000 = -76x^3 + 4830x^2 - 320{,}000$$

$$76x^3 - 4830x^2 + 2{,}820{,}000 = 0$$

Let $f(x) = 76x^3 - 4830x^2 + 2{,}820{,}000$

$f'(x) = 228x^2 - 9660x.$

From the graph, choose $x_1 = 40$.

n	x_n	$f(x_n)$	$f'(x_n)$	$\frac{f(x_n)}{f'(x_n)}$	$x_n - \frac{f(x_n)}{f'(x_n)}$
1	40.0000	−44000.0000	−21600.0000	2.0370	37.9630
2	37.9630	17157.6209	−38131.4039	−0.4500	38.4130
3	38.4130	780.0914	−34642.2263	−0.0225	38.4355
4	38.4355	2.6308	−34465.3435	−0.0001	38.4356

The zero occurs when $x \approx 38.4356$ which corresponds to \$384,356.

43. False. Let $f(x) = (x^2 - 1)/(x - 1)$. $x = 1$ is a discontinuity. It is not a zero of $f(x)$. This statement would be true if $f(x) = p(x)/q(x)$ is given in **reduced** form.

45. True

47. $f(x) = \frac{1}{4}x^3 - 3x^2 + \frac{3}{4}x - 2$

$f'(x) = \frac{3}{4}x^2 - 6x + \frac{3}{4}$

Let $x_1 = 12$.

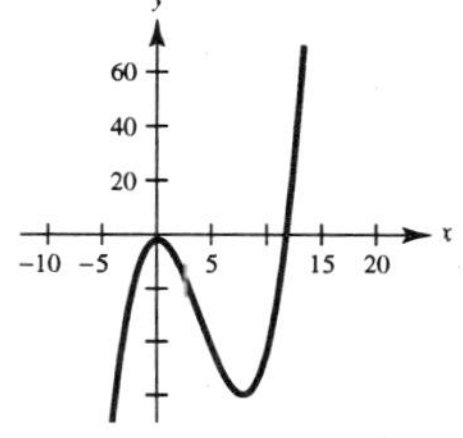

n	x_n	$f(x_n)$	$f'(x_n)$	$\frac{f(x_n)}{f'(x_n)}$	$x_n - \frac{f(x_n)}{f'(x_n)}$
1	12.0000	7.0000	36.7500	0.1905	11.8095
2	11.8095	0.2151	34.4912	0.0062	11.8033
3	11.8033	0.0015	34.4186	0.0000	11.8033

Approximation: $x \approx 11.803$

Section 3.9 Differentials

1. $f(x) = x^2$

$f'(x) = 2x$

Tangent line at $(2, 4)$: $y - f(2) = f'(2)(x - 2)$

$$y - 4 = 4(x - 2)$$

$$y = 4x - 4$$

x	1.9	1.99	2	2.01	2.1
$f(x) = x^2$	3.6100	3.9601	4	4.0401	4.4100
$T(x) = 4x - 4$	3.6000	3.9600	4	4.0400	4.4000

3. $f(x) = x^5$

$f'(x) = 5x^4$

Tangent line at $(2, 32)$: $y - f(2) = f'(2)(x - 2)$

$$y - 32 = 80(x - 2)$$

$$y = 80x - 128$$

x	1.9	1.99	2	2.01	2.1
$f(x) = x^5$	24.7610	31.2080	32	32.8080	40.8410
$T(x) = 80x - 128$	24.0000	31.2000	32	32.8000	40.0000

5. $f(x) = \sin x$

$f'(x) = \cos x$

Tangent line at $(2, \sin 2)$:

$$y - f(2) = f'(2)(x - 2)$$

$$y - \sin 2 = (\cos 2)(x - 2)$$

$$y = (\cos 2)(x - 2) + \sin 2$$

x	1.9	1.99	2	2.01	2.1
$f(x) = \sin x$	0.9463	0.9134	0.9093	0.9051	0.8632
$T(x) = (\cos 2)(x - 2) + \sin 2$	0.9509	0.9135	0.9093	0.9051	0.8677

7. $y = f(x) = x^3, f'(x) = 3x^2, x = 1, \Delta x = dx = 0.1$

$$\Delta y = f(x + \Delta x) - f(x) = f(1.1) - f(1) = (1.1)^3 - (1)^3 = 0.331$$

$$dy = f'(x)\,dx = f'(1)(0.1) = 3(0.1) = 0.3$$

9. $y = f(x) = x^4 + 1, f'(x) = 4x^3, x = -1, \Delta x = dx = 0.01$

$$\Delta y = f(x + \Delta x) - f(x) = f(-0.99) - f(-1) = [(-0.99)^4 + 1] - [(-1)^4 + 1] \approx -0.0394$$

$$dy = f'(x)\,dx = f'(-1)(0.01) = (-4)(0.01) = -0.04$$

11. $y = 3x^2 - 4$

$dy = 6x\,dx$

13. $y = \dfrac{x + 1}{2x - 1}$

$$dy = \frac{-3}{(2x - 1)^2}\,dx$$

15. $y = x\sqrt{1 - x^2}$

$$dy = \left(x\frac{-x}{\sqrt{1 - x^2}} + \sqrt{1 - x^2}\right)dx = \frac{1 - 2x^2}{\sqrt{1 - x^2}}\,dx$$

17. $y = \dfrac{\sec^2 x}{x^2 + 1}$

$$dy = \left[\frac{(x^2 + 1)2\sec^2 x \tan x - \sec^2 x(2x)}{(x^2 + 1)^2}\right]dx$$

$$= \left[\frac{2\sec^2 x(x^2 \tan x + \tan x - x)}{(x^2 + 1)^2}\right]dx$$

19. $y = \frac{1}{3}\cos\left(\frac{6\pi x - 1}{2}\right)$

$dy = -\pi \sin\left(\frac{6\pi x - 1}{2}\right) dx$

21. $A = x^2$

$x = 12$

$\Delta x = dx = \pm\frac{1}{64}$

$dA = 2x\, dx$

$\Delta A \approx dA = 2(12)\left(\pm\frac{1}{64}\right)$

$= \pm\frac{3}{8}$ square inches

23. $A = \pi r^2$

$r = 14$

$\Delta r = dr = \pm\frac{1}{4}$

$\Delta A \approx dA = 2\pi r\, dr = \pi(28)\left(\pm\frac{1}{4}\right)$

$= \pm 7\pi$ square inches

25. (a) $x = 15$ centimeter

$\Delta x = dx = \pm 0.05$ centimeters

$A = x^2$

$dA = 2x\, dx = 2(15)(\pm 0.05)$

$= \pm 1.5$ square centimeters

Percentage error:

$$\frac{dA}{A} = \frac{\pm 1.5}{(15)^2} = 0.00666\ldots = \frac{2}{3}\%$$

(b) $\frac{dA}{A} = \frac{2x\, dx}{x^2} = \frac{2\, dx}{x} \le 0.025$

$$\frac{dx}{x} \le \frac{0.025}{2} = 0.0125 = 1.25\%$$

27. $r = 6$ inches

$\Delta r = dr = \pm 0.02$ inches

(a) $V = \frac{4}{3}\pi r^3$

$dV = 4\pi r^2\, dr = 4\pi(6)^2(\pm 0.02) = \pm 2.88\pi$ cubic inches

(b) $S = 4\pi r^2$

$dS = 8\pi r\, dr = 8\pi(6)(\pm 0.02) = \pm 0.96\pi$ square inches

(c) Relative error: $\frac{dV}{V} = \frac{4\pi r^2\, dr}{(4/3)\pi r^3} = \frac{3dr}{r}$

$= \frac{3}{6}(0.02) = 0.01 = 1\%$

Relative error: $\frac{dS}{S} = \frac{8\pi r\, dr}{4\pi r^2} = \frac{2dr}{r}$

$= \frac{2(0.02)}{6} = 0.000666\ldots = \frac{2}{3}\%$

29. $V = \pi r^2 h = 40\pi r^2$, $r = 5$ cm, $h = 40$ cm, $dr = 0.2$ cm

$\Delta V \approx dV = 80\pi r\, dr = 80\pi(5)(0.2) = 80\pi\ \text{cm}^3$

31. (a) $T = 2\pi\sqrt{L/g}$

$dT = \frac{\pi}{g\sqrt{L/g}}\, dL$

Relative error:

$$\frac{dT}{T} = \frac{(\pi\, dL)/\left(g\sqrt{L/g}\right)}{2\pi\sqrt{L/g}}$$

$$= \frac{dL}{2L}$$

$$= \frac{1}{2}(\text{relative error in } L)$$

$$= \frac{1}{2}(0.005) = 0.0025$$

Percentage error: $\frac{dT}{T}(100) = 0.25\% = \frac{1}{4}\%$

(b) $(0.0025)(3600)(24) = 216$ seconds

$= 3.6$ minutes

33. $E = IR$

$R = \frac{E}{I}$

$dR = -\frac{E}{I^2}dI$

$\frac{dR}{R} = \frac{-(E/I^2)dI}{E/I} = -\frac{dI}{I}$

$\left|\frac{dR}{R}\right| = \left|-\frac{dI}{I}\right| = \left|\frac{dI}{I}\right|$

35. $\theta = 26°45' = 26.75°$

$d\theta = \pm 15' = \pm 0.25°$

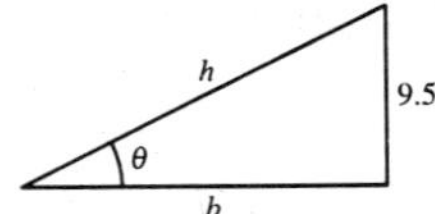

(a) $h = 9.5 \csc \theta$

$$dh = -9.5 \csc \theta \cot \theta \, d\theta$$

$$\frac{dh}{h} = -\cot \theta \, d\theta$$

$$\left|\frac{dh}{h}\right| = (\cot 26.75°)(0.25°)$$

Converting to radians, $(\cot 0.4669)(0.0044)$
$\approx 0.0087 = 0.87\%$ (in radians).

(b) $\left|\frac{dh}{h}\right| = \cot \theta \, d\theta \leq 0.02$

$$\frac{d\theta}{\theta} \leq \frac{0.02}{\theta(\cot \theta)} = \frac{0.02 \tan \theta}{\theta}$$

$$\frac{d\theta}{\theta} \leq \frac{0.02 \tan 26.75°}{26.75°} \approx \frac{0.02 \tan 0.4669}{0.4669}$$

$$\approx 0.0216 = 2.16\% \text{ (in radians)}$$

37. $r = \frac{v_0^2}{32}(\sin 2\theta)$

$v_0 = 2200$ ft/sec

θ changes from $10°$ to $11°$

$$dr = \frac{(2200)^2}{16}(\cos 2\theta)\, d\theta$$

$$\theta = 10\left(\frac{\pi}{180}\right)$$

$$d\theta = (11 - 10)\frac{\pi}{180}$$

$$\Delta r \approx dr$$

$$= \frac{(2200)^2}{16} \cos\left(\frac{20\pi}{180}\right)\left(\frac{\pi}{180}\right) \approx 4961 \text{ feet}$$

≈ 4961 feet

39. Let $f(x) = \sqrt{x}$, $x = 100$, $dx = -0.6$.

$$f(x + \Delta x) \approx f(x) + f'(x)dx$$

$$= \sqrt{x} + \frac{1}{2\sqrt{x}}dx$$

$$f(x + \Delta x) = \sqrt{99.4}$$

$$\approx \sqrt{100} + \frac{1}{2\sqrt{100}}(-0.6) = 9.97$$

Using a calculator: $\sqrt{99.4} \approx 9.96995$

41. Let $f(x) = \sqrt[4]{x}$, $x = 625$, $dx = -1$.

$$f(x + \Delta x) \approx f(x) + f'(x)\, dx = \sqrt[4]{x} + \frac{1}{4^4\sqrt{x^3}}\, dx$$

$$f(x + \Delta x) = \sqrt[4]{624} \approx \sqrt[4]{625} + \frac{1}{4(\sqrt[4]{625})^3}(-1)$$

$$= 5 - \frac{1}{500} = 4.998$$

Using a calculator, $\sqrt[4]{624} \approx 4.9980$.

43. Let $f(x) = x^4$, $x = 1$, $dx = -0.01$, $f'(x) = 4x^3$.

Then

$$f(0.99) \approx f(1) + f'(1)dx$$

$$(0.99)^4 \approx (1)^4 + 4(1)^3(-0.01) = 1 - 4(0.01).$$

45. Let $f(x) = \sec x$, $x = 0$, $dx = 0.03$, $f'(x) = \sec x \tan x$.

Then

$$f(0.03) \approx f(0) + f'(0)\, dx$$

$$\sec 0.03 \approx \sec 0 + (\sec 0 \tan 0)(0.03)$$

$$= 1 + 0(0.03).$$

47. True

49. True

51. $A(x) = x^2$

(a) $dA = 2x\, dx = 2x\, \Delta x$

$$\Delta A = (x + \Delta x)^2 - x^2$$

$$= 2x\, \Delta x + (\Delta x)^2$$

(b)

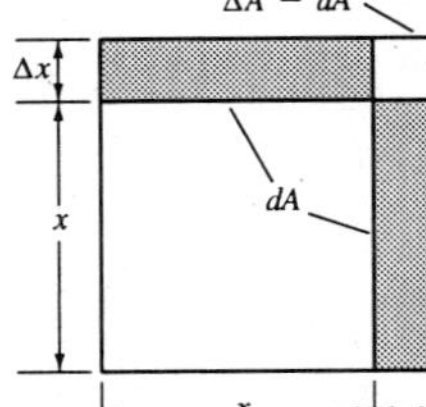

(c) $\Delta A - dA = (\Delta x)^2$

Section 3.10 Business and Economics Applications

1. (a) $C(0)$ represents the fixed costs.

(b)

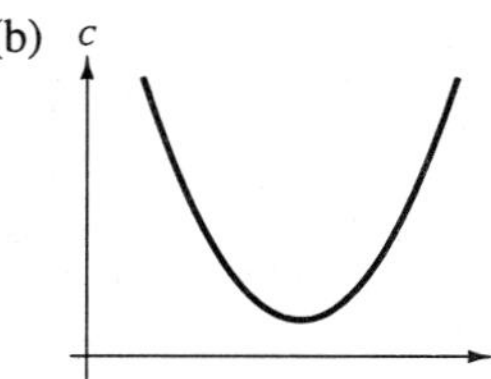

(c) The marginal cost function has a relative minimum. It occurs when production costs are increasing at their slowest rate.

3. $R = 900x - 0.1x^2$

$\dfrac{dR}{dx} = 900 - 0.2x = 0$ when $x = 4500$.

By the First Derivative Test, $x = 4500$ is a maximum.

5.

$$R = \frac{1{,}000{,}000x}{0.02x^2 + 1800}$$

$$\frac{dR}{dx} = 1{,}000{,}000\left[\frac{0.02x^2 + 1800 - x(0.04x)}{(0.02x^2 + 1800)^2}\right] = 0$$

$1800 - 0.02x^2 = 0$ when $x = 300$.

By the First Derivative Test, $x = 300$ is a maximum.

7. $\overline{C} = 0.125x + 20 + \dfrac{5000}{x}$

$\dfrac{d\overline{C}}{dx} = 0.125 - \dfrac{5000}{x^2} = 0$ when $x = 200$.

By the First Derivative Test, $x = 200$ yields the minimum average cost.

9. $\overline{C} = 3000 - x(300 - x)^{1/2}$

$$\frac{d\overline{C}}{dx} = -x\left(\frac{1}{2}\right)(300 - x)^{-1/2}(-1) - (300 - x)^{1/2}$$

$$= -\frac{3}{2}(300 - x)^{-1/2}(200 - x) = 0 \text{ when } x = 200.$$

By the First Derivative Test, $x = 200$ yields the minimum average cost.

11. $C = 100 + 30x$

$p = 90 - x$

$P = xp - C$

$= 90x - x^2 - 30x - 100$

$= -x^2 + 60x - 100$

$\dfrac{dP}{dx} = -2x + 60$

$= 0$ when $x = 30$, so $p = 60$.

By the First Derivative Test, $x = 30$ is a maximum.

13. $C = 4000 - 40x + 0.02x^2$

$p = 50 - 0.01x$

$P = xp - C = 50x - 0.01x^2 - 4000 + 40x - 0.02x^2$

$= -0.03x^2 + 90x - 4000$

$\dfrac{dP}{dx} = -0.06x + 90 = 0$ when $x = 1500$, so $p = 35$.

By the First Derivative Test, $x = 1500$ is a maximum.

15. $C = 2x^2 + 5x + 18$

Average cost $= \dfrac{C}{x} = \overline{C} = 2x + 5 + \dfrac{18}{x}$

$$\frac{d\overline{C}}{dx} = 2 - \frac{18}{x^2} = 0 \text{ when } x = 3.$$

$$\overline{C}(3) = 6 + 5 + 6 = 17$$

By the First Derivative Test, $x = 3$ is a minimum.

Marginal cost: $\dfrac{dC}{dx} = 4x + 5$

At $x = 3$: $\dfrac{dC}{dx} = 17 = \overline{C}(3)$

17. Average cost: $\overline{C}(x) = \dfrac{C(x)}{x}$

$$\frac{d\overline{C}}{dx} = \frac{xC'(x) - C(x)}{x^2} = 0 \Rightarrow xC'(x) - C(x) = 0 \text{ when } C'(x) = \frac{C(x)}{x} = \overline{C}(x).$$

Marginal cost = average cost
This condition will yield a minimum (if it exists).

19. (a)

Order size, x	Price	Profit
102	$90 - 2(0.15)$	$102[90 - 2(0.15)] - 102(60) = 3029.40$
104	$90 - 4(0.15)$	$104[90 - 4(0.15)] - 104(60) = 3057.60$
106	$90 - 6(0.15)$	$106[90 - 6(0.15)] - 106(60) = 3084.60$
108	$90 - 8(0.15)$	$108[90 - 8(0.15)] - 108(60) = 3110.40$
110	$90 - 10(0.15)$	$110[90 - 10(0.15)] - 110(60) = 3135.00$
112	$90 - 12(0.15)$	$112[90 - 12(0.15)] - 112(60) = 3158.40$

(b)

Order size, x	Profit
148	3374.40
149	3374.90
150	3375.00
151	3374.90
152	3374.40

The maximum profit is 3375.00 for $x = 150$.

(c) $P(x) = x[90 - (x - 100)(0.15)] - 60x$

$= x(45 - 0.15x),\ x \ge 100$

(d) $\dfrac{dP}{dx} = 45 - 0.30x = 0$ when $x = 150$.

Since $\dfrac{d^2P}{dx^2} < 0$, an order size of $x = 150$ units yields a maximum profit.

(e)

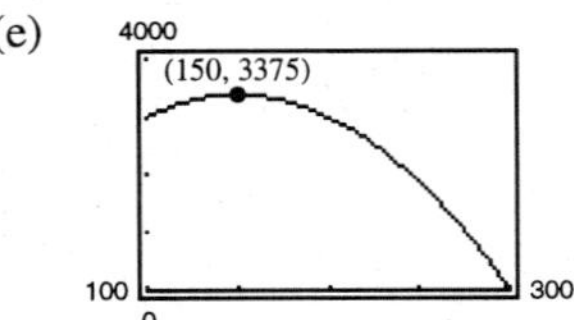

21. Total cost = (Cost per hour)(Number of hours)

$$T = \left(\frac{v^2}{600} + 5\right)\left(\frac{110}{v}\right) = \frac{11v}{60} + \frac{550}{v}$$

$$\frac{dT}{dv} = \frac{11}{60} - \frac{550}{v^2} = \frac{11v^2 - 33{,}000}{60v^2}$$

$$= 0 \text{ when } v = \sqrt{3000} = 10\sqrt{30} \approx 54.8 \text{ mph.}$$

$$\frac{d^2T}{dv^2} = \frac{1100}{v^3} > 0 \text{ when } v = 10\sqrt{30} \text{ so this value yields a minimum.}$$

23. The total cost $T(x)$ is

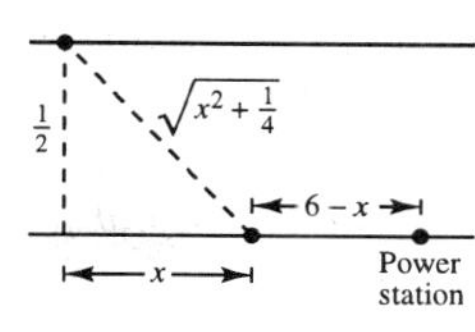

$$T(x) = 12(5280)(6 - x) + 16(5280)\sqrt{x^2 + \frac{1}{4}}$$

$$T'(x) = 5280\left[-12 + \frac{16x}{\sqrt{x^2 + (1/4)}}\right] = 0$$

$$12 = \frac{16x}{\sqrt{x^2 + (1/4)}}$$

$$3\sqrt{x^2 + (1/4)} = 4x$$

$$9\left(x^2 + \frac{1}{4}\right) = 16x^2$$

$$\frac{9}{4} = 7x^2 \Longrightarrow x = \frac{3}{2\sqrt{7}} \approx 0.57 \text{ miles.}$$

This is a minimum by the First Derivative Test.

25. $S_1 = (4m - 1)^2 + (5m - 6)^2 + (10m - 3)^2$

$$\frac{dS_1}{dm} = 2(4m - 1)(4) + 2(5m - 6)(5) + 2(10m - 3)(10) = 282m - 128 = 0 \text{ when } m = \frac{64}{141}.$$

Line: $y = \frac{64}{141}x$

$$S = \left|4\left(\frac{64}{141}\right) - 1\right| + \left|5\left(\frac{64}{141}\right) - 6\right| + \left|10\left(\frac{64}{141}\right) - 3\right| = \left|\frac{256}{141} - 1\right| + \left|\frac{320}{141} - 6\right| + \left|\frac{640}{141} - 3\right| = \frac{858}{141} \approx 6.1 \text{ mi}$$

27. $S_3 = \frac{|4m - 1|}{\sqrt{m^2 + 1}} + \frac{|5m - 6|}{\sqrt{m^2 + 1}} + \frac{|10m - 3|}{\sqrt{m^2 + 1}}$

Using a graphing utility, you can see that the minimum occurs when $x \approx 0.3$.

Line: $y \approx 0.3x$

$$S_3 = \frac{|4(0.3) - 1| + |5(0.3) - 6| + |10(0.3) - 3|}{\sqrt{(0.3)^2 + 1}}$$

$$\approx 4.5 \text{ mi.}$$

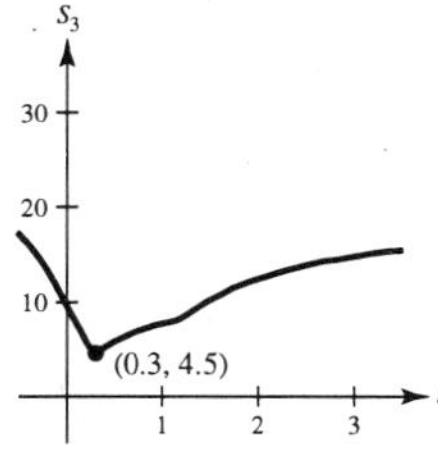

29. Let d be the amount deposited in the bank, i be the interest rate paid by the bank, and P be the profit.

$$P = (0.12)d - id$$

$$d = ki^2 \text{ (since } d \text{ is proportional to } i^2\text{)}$$

$$P = (0.12)(ki^2) - i(ki^2) = k(0.12i^2 - i^3)$$

$$\frac{dP}{di} = k(0.24i - 3i^2) = 0 \text{ when } i = \frac{0.24}{3} = 0.08.$$

$$\frac{d^2P}{di^2} = k(0.24 - 6i) < 0 \text{ when } i = 0.08 \text{ (**Note:** } k > 0\text{).}$$

The profit is a maximum when $i = 8\%$.

31. $C = 100\left(\frac{200}{x^2} + \frac{x}{x + 30}\right),\ 1 \le x$

$$C' = 100\left(-\frac{400}{x^3} + \frac{30}{(x + 30)^2}\right)$$

$$\overline{C}' = f(x) = 100\left(-\frac{400}{x^3} + \frac{30}{(x + 30)^2}\right)$$

$$f'(x) = 100\left(\frac{1200}{x^4} - \frac{60}{(x + 30)^3}\right)$$

Approximation: $x \approx 40$ units

n	x_n	$f(x_n)$	$f'(x_n)$	$\frac{f(x_n)}{f'(x_n)}$	$x_n - \frac{f(x_n)}{f'(x_n)}$
1	40.0000	−0.0128	0.0294	−0.4341	40.4341
2	40.4341	−0.0004	0.0277	−0.0131	40.4472
3	40.4472	0.0000	0.0277	0.0000	40.4472

33. $R = 900x - 0.1x^2$

$x = 3000$

$dx = 100$

$dR = (900 - 0.2x)dx$

$= [900 - 0.2(3000)](100)$

$= \$30{,}000$

35. $F = 100{,}000\left(1 + \sin\left[\frac{2\pi(t - 60)}{365}\right]\right)$

(a) $\frac{dF}{dt} = 100{,}000\left(\frac{2\pi}{365}\cos\left[\frac{2\pi(t - 60)}{365}\right]\right)$

$= 0$ when $\frac{2\pi(t - 60)}{365} = \frac{\pi}{2}$ or $\frac{3\pi}{2}$.

$t = 151.25$ or 333.75

The maximum occurs when $t = 151.25$ which corresponds to May 31.

(b) 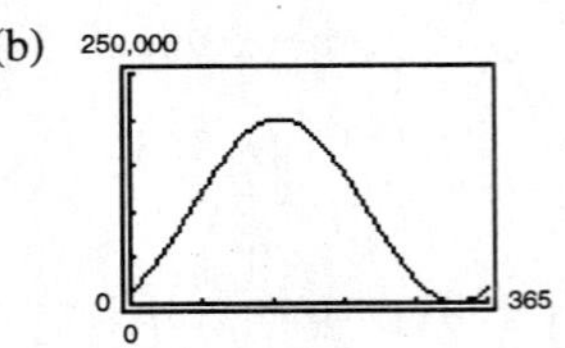

Sales are minimum when $t = 333.75$ which corresponds to November 30.

37. $R = 4.7t^4 - 193.5t^3 + 2941.7t^2 - 19{,}294.7t + 52{,}012$

(a) $R'(t) = 18.8t^3 - 580.5t^2 + 5883.4t - 19{,}294.7 = 0$

The real root is $t = 7.22$ (1987), which yields a minimum.

(b) The maximum occurs at $t = 14$ (1994).

(c) The minimum revenue was 5995 million dollars.

The maximum revenue was 8050.6 million dollars.

(d) 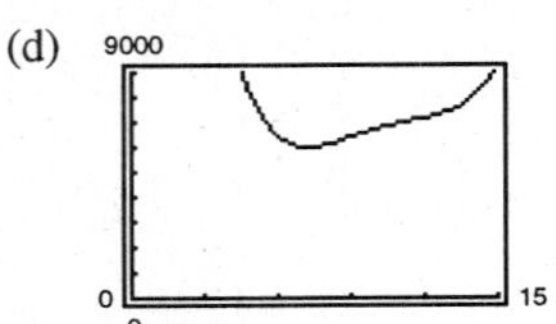

39. (a) Demand function

(b) Cost function

(c) Revenue function

(d) Profit function

41. $\eta = \frac{p/x}{dp/dx} = \frac{(400 - 3x)/x}{-3} = 1 - \frac{400}{3x}$

When $x = 20$, we have

$$\eta = 1 - \frac{400}{3(20)} = -\frac{17}{3}.$$

Since $|\eta| = \frac{17}{3} > 1$, the demand is elastic.

43. $\eta = \frac{p/x}{dp/dx} = \frac{(400 - 0.5x^2)/x}{-x} = \frac{1}{2} - \frac{400}{x^2}$

When $x = 20$, we have

$$\eta = \frac{1}{2} - \frac{400}{(20)^2} = -\frac{1}{2}.$$

Since $|\eta| = \frac{1}{2} < 1$, the demand is inelastic.

Review Exercises for Chapter 3

1. A number c in the domain of f is a critical number if $f'(c) = 0$ or f' is undefined at c.

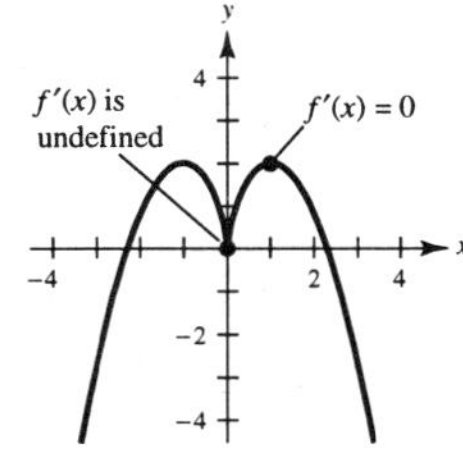

3. $g(x) = 2x + 5\cos x, [0, 2\pi]$

$$g'(x) = 2 - 5\sin x$$
$$= 0 \text{ when } \sin x = \tfrac{2}{5}.$$

Critical numbers: $x \approx 0.41, x \approx 2.73$

Left endpoint: $(0, 5)$

Critical number: $(0.41, 5.41)$

Critical number: $(2.73, 0.88)$ Minimum

Right endpoint: $(2\pi, 17.57)$ Maximum

5. $f(x) = 3 - |x - 4|$

(a)

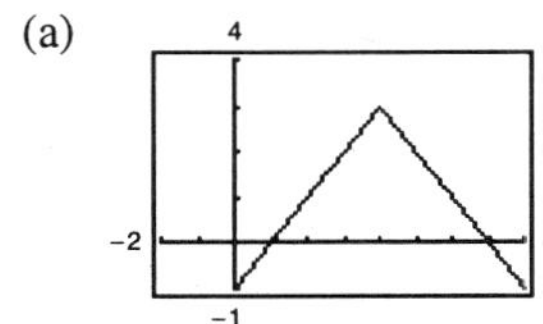

$f(1) = f(7) = 0$

(b) f is not differentiable at $x = 4$.

7.

$$f(x) = x^{2/3}, 1 \le x \le 8$$
$$f'(x) = \frac{2}{3}x^{-1/3}$$
$$\frac{f(b) - f(a)}{b - a} = \frac{4 - 1}{8 - 1} = \frac{3}{7}$$
$$f'(c) = \frac{2}{3}c^{-1/3} = \frac{3}{7}$$
$$c = \left(\frac{14}{9}\right)^3 = \frac{2744}{729} \approx 3.764$$

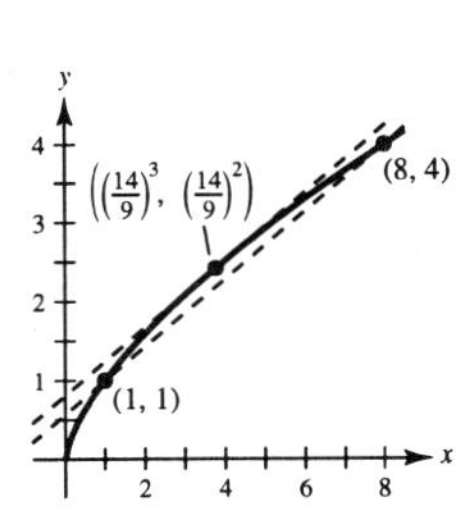

9.

$$f(x) = x - \cos x, -\frac{\pi}{2} \le x \le \frac{\pi}{2}$$
$$f'(x) = 1 + \sin x$$
$$\frac{f(b) - f(a)}{b - a} = \frac{(\pi/2) - (-\pi/2)}{(\pi/2) - (-\pi/2)} = 1$$
$$f'(c) = 1 + \sin c = 1$$
$$c = 0$$

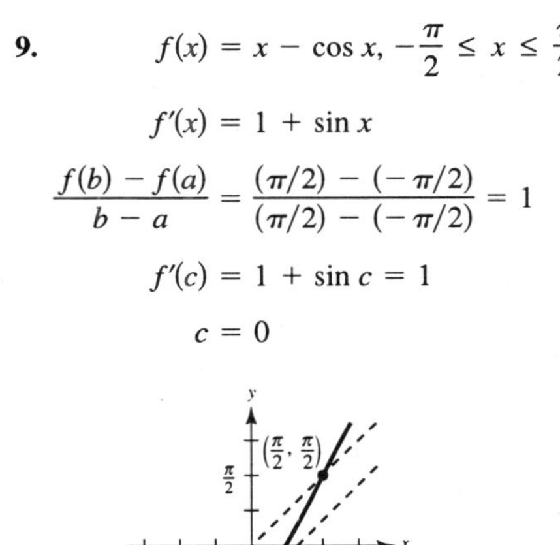

11.

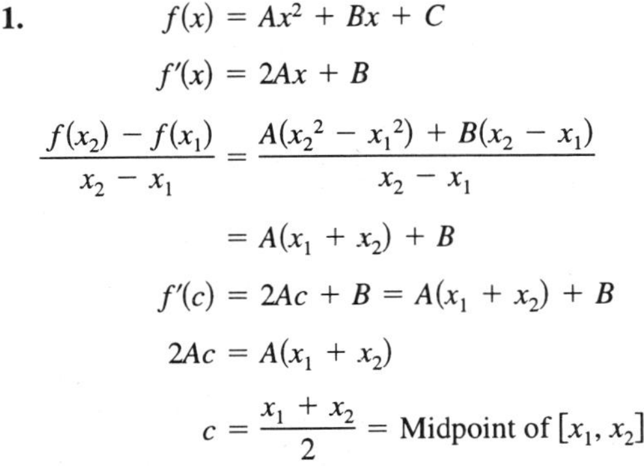

$$f(x) = Ax^2 + Bx + C$$
$$f'(x) = 2Ax + B$$
$$\frac{f(x_2) - f(x_1)}{x_2 - x_1} = \frac{A(x_2^2 - x_1^2) + B(x_2 - x_1)}{x_2 - x_1}$$
$$= A(x_1 + x_2) + B$$
$$f'(c) = 2Ac + B = A(x_1 + x_2) + B$$
$$2Ac = A(x_1 + x_2)$$
$$c = \frac{x_1 + x_2}{2} = \text{Midpoint of } [x_1, x_2]$$

13. $f(x) = (x-1)^2(x-3)$

$f'(x) = (x-1)^2(1) + (x-3)(2)(x-1)$

$= (x-1)(3x-7)$

Critical numbers: $x = 1$ and $x = \frac{7}{3}$

Interval	$-\infty < x < 1$	$1 < x < \frac{7}{3}$	$\frac{7}{3} < x < \infty$
Sign of $f'(x)$	$f'(x) > 0$	$f'(x) < 0$	$f'(x) > 0$
Conclusion	Increasing	Decreasing	Increasing

15. $h(x) = \sqrt{x}(x-3) = x^{3/2} - 3x^{1/2}$

Domain: $[0, \infty)$

$h'(x) = \frac{3}{2}x^{1/2} - \frac{3}{2}x^{-1/2}$

$= \frac{3}{2}x^{-1/2}(x-1) = \frac{3(x-1)}{2\sqrt{x}}$

Critical numbers: $x = 0, x = 1$

Interval	$0 < x < 1$	$1 < x < \infty$
Sign of $h'(x)$	$h'(x) < 0$	$h'(x) > 0$
Conclusion	Decreasing	Increasing

17. $h(t) = \frac{1}{4}t^4 - 8t$

$h'(t) = t^3 - 8 = 0$ when $t = 2$.

Relative minimum: $(2, -12)$

Test Interval	$-\infty < t < 2$	$2 < t < \infty$
Sign of $h'(t)$	$h'(t) < 0$	$h'(t) > 0$
Conclusion	Decreasing	Increasing

19. $f(x) = x + \cos x,\ 0 \le x \le 2\pi$

$f'(x) = 1 - \sin x$

$f''(x) = -\cos x = 0$ when $x = \frac{\pi}{2}, \frac{3\pi}{2}$.

Points of inflection: $\left(\frac{\pi}{2}, \frac{\pi}{2}\right), \left(\frac{3\pi}{2}, \frac{3\pi}{2}\right)$

Test Interval	$0 < x < \frac{\pi}{2}$	$\frac{\pi}{2} < x < \frac{3\pi}{2}$	$\frac{3\pi}{2} < x < 2\pi$
Sign of $f''(x)$	$f''(x) < 0$	$f''(x) > 0$	$f''(x) < 0$
Conclusion	Concave downward	Concave upward	Concave downward

21. $\lim_{x\to\infty} \frac{2x^2}{3x^2+5} = \lim_{x\to\infty} \frac{2}{3 + 5/x^2} = \frac{2}{3}$

23. $\lim_{x\to\infty} \frac{5\cos x}{x} = 0$, since $|5\cos x| \le 5$.

25. $h(x) = \frac{2x+3}{x-4}$

Discontinuity: $x = 4$

$$\lim_{x\to\infty} \frac{2x+3}{x-4} = \lim_{x\to\infty} \frac{2 + (3/x)}{1 - (4/x)} = 2$$

Vertical asymptote: $x = 4$

Horizontal asymptote: $y = 2$

27. $f(x) = \frac{3}{x} - 2$

Discontinuity: $x = 0$

$$\lim_{x\to\infty}\left(\frac{3}{x} - 2\right) = -2$$

Vertical asymptote: $x = 0$

Horizontal asymptote: $y = -2$

29. $f(x) = x^3 + \frac{243}{x}$

Relative minimum: $(3, 108)$

Relative maximum: $(-3, -108)$

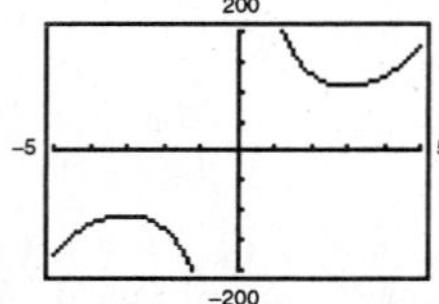

Vertical asymptote: $x = 0$

31. $f(x) = \frac{x-1}{1+3x^2}$

Relative minimum: $(-0.155, -1.077)$

Relative maximum: $(2.155, 0.077)$

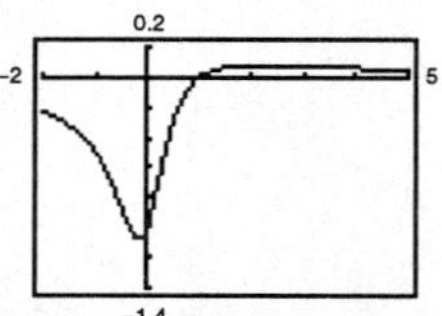

Horizontal asymptote: $y = 0$

33. $f(x) = 4x - x^2 = x(4 - x)$

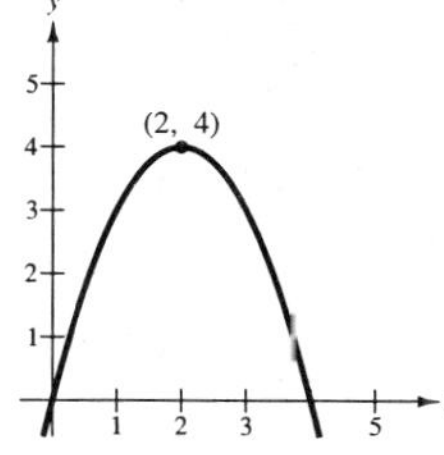

Domain: $(-\infty, \infty)$

Range: $(-\infty, 4)$

$f'(x) = 4 - 2x = 0$ when $x = 2$.

$f''(x) = -2$

Therefore, $(2, 4)$ is a relative maximum.

Intercepts: $(0, 0)$, $(4, 0)$

35. $f(x) = x\sqrt{16 - x^2}$, Domain: $[-4, 4]$, Range: $[-8, 8]$

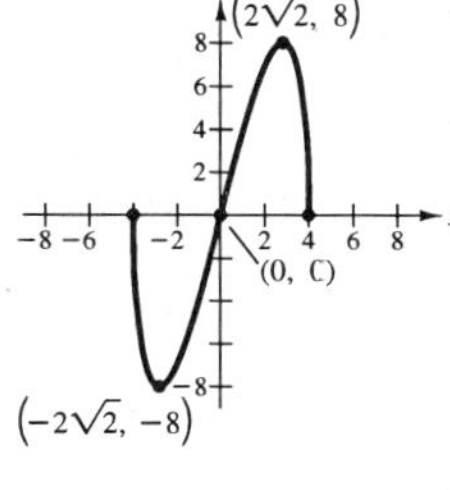

Domain: $[-4, 4]$

Range: $[-8, 8]$

$f'(x) = \dfrac{16 - 2x^2}{\sqrt{16 - x^2}} = 0$ when $x = \pm 2\sqrt{2}$ and undefined when $x = \pm 4$.

$f''(x) = \dfrac{2x(x^2 - 24)}{(16 - x^2)^{3/2}}$

$f'(-2\sqrt{2}) > 0$

Therefore, $(-2\sqrt{2}, -8)$ is a relative minimum.

$f''(2\sqrt{2}) < 0$

Therefore, $(2\sqrt{2}, 8)$ is a relative maximum.

Point of inflection: $(0, 0)$

Intercepts: $(-4, 0)$, $(0, 0)$, $(4, 0)$

Symmetry with respect to origin

37. $f(x) = (x - 1)^3(x - 3)^2$

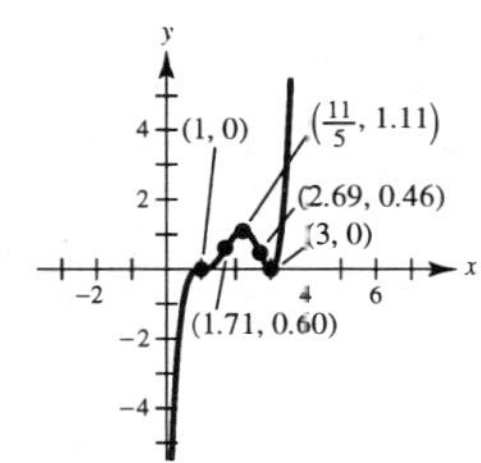

Domain: $(-\infty, \infty)$

Range: $(-\infty, \infty)$

$f'(x) = (x - 1)^2(x - 3)(5x - 11) = 0$ when $x = 1, \dfrac{11}{5}, 3$.

$f''(x) = 4(x - 1)(5x^2 - 22x + 23) = 0$ when $x = 1, \dfrac{11 \pm \sqrt{6}}{5}$.

$f''(3) > 0$

Therefore, $(3, 0)$ is a relative minimum.

$f''\left(\dfrac{11}{5}\right) < 0$

Therefore, $\left(\dfrac{11}{4}, \dfrac{3456}{3125}\right)$ is a relative maximum.

Points of inflection: $(1, 0)$, $\left(\dfrac{11 - \sqrt{6}}{5}, 0.60\right)$, $\left(\dfrac{11 + \sqrt{6}}{5}, 0.46\right)$

Intercepts: $(0, -9)$, $(1, 0)$, $(3, 0)$

39. $f(x) = x^{1/3}(x+3)^{2/3}$

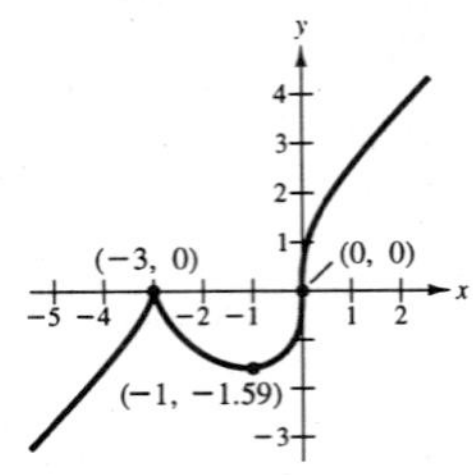

Domain: $(-\infty, \infty)$

Range: $(-\infty, \infty)$

$$f'(x) = \frac{x+1}{(x+3)^{1/3}x^{2/3}} = 0 \text{ when } x = -1 \text{ and undefined when } x = -3, 0.$$

$$f''(x) = \frac{-2}{x^{5/3}(x+3)^{4/3}} \text{ is undefined when } x = 0, -3.$$

By the First Derivative Test $(-3, 0)$ is a relative maximum and $\left(-1, -\sqrt[3]{4}\right)$ is a relative minimum. $(0, 0)$ is a point of inflection.

Intercepts: $(-3, 0), (0, 0)$

41. $f(x) = \dfrac{x+1}{x-1}$

Domain: $(-\infty, 1), (1, \infty)$

Range: $(-\infty, 1), (1, \infty)$

$$f'(x) = \frac{-2}{(x-1)^2} < 0 \text{ if } x \neq 1.$$

$$f''(x) = \frac{4}{(x-1)^3}$$

Horizontal asymptote: $y = 1$

Vertical asymptote: $x = 1$

Intercepts: $(-1, 0), (0, -1)$

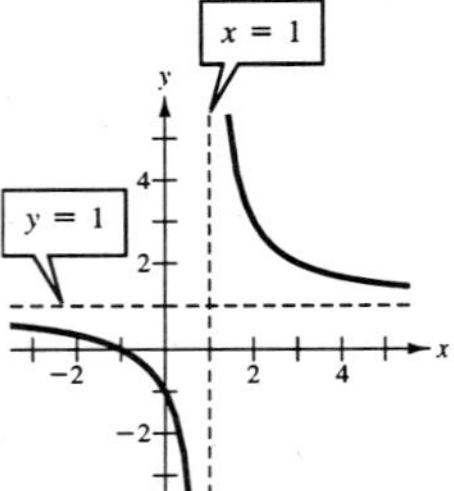

43. $f(x) = \dfrac{4}{1+x^2}$

Domain: $(-\infty, \infty)$

Range: $(0, 4]$

$$f'(x) = \frac{-8x}{(1+x^2)^2} = 0 \text{ when } x = 0.$$

$$f''(x) = \frac{-8(1-3x^2)}{(1+x^2)^3} = 0 \text{ when } x = \pm\frac{\sqrt{3}}{3}.$$

$f''(0) < 0$

Therefore, $(0, 4)$ is a relative maximum.

Points of inflection: $\left(\pm\sqrt{3}/3, 3\right)$

Intercept: $(0, 4)$

Symmetric to the y-axis

Horizontal asymptote: $y = 0$

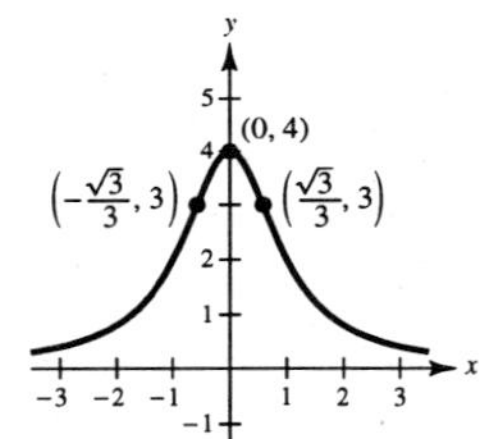

45. $f(x) = x^3 + x + \dfrac{4}{x}$

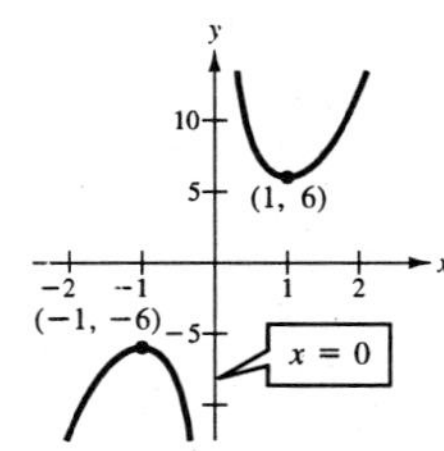

Domain: $(-\infty, 0), (0, \infty)$

Range: $(-\infty, -6], [6, \infty)$

$$f'(x) = 3x^2 + 1 - \frac{4}{x^2} = \frac{3x^4 + x^2 - 4}{x^2} = 0 \text{ when } x = \pm 1.$$

$$f''(x) = 6x + \frac{8}{x^3} = \frac{6x^4 + 8}{x^3} \neq 0$$

$f''(-1) < 0$

Therefore, $(-1, -6)$ is a relative maximum.

$f''(1) > 0$

Therefore, $(1, 6)$ is a relative minimum.

Vertical asymptote: $x = 0$

Symmetric with respect to origin

47. $f(x) = |x^2 - 9|$

Domain: $(-\infty, \infty)$

Range: $[0, \infty)$

$$f'(x) = \frac{2x(x^2 - 9)}{|x^2 - 9|} = 0 \text{ when } x = 0 \text{ and is undefined when } x = \pm 3.$$

$$f''(x) = \frac{2(x^2 - 9)}{|x^2 - 9|} \text{ is undefined at } x = \pm 3.$$

$f''(0) < 0$

Therefore, $(0, 9)$ is a relative maximum.

Relative minima: $(\pm 3, 0)$

Points of inflection: $(\pm 3, 0)$

Intercepts: $(\pm 3, 0)$, $(0, 9)$

Symmetric to the y-axis

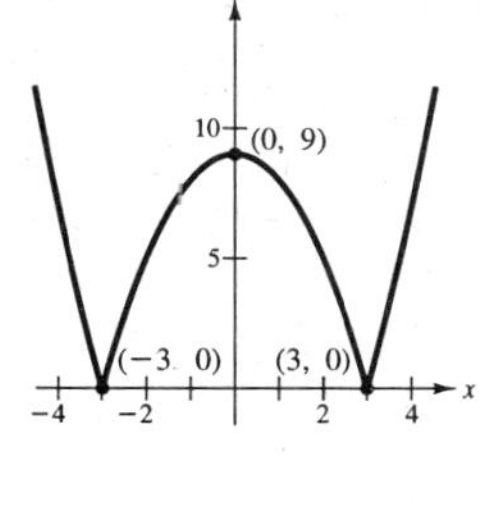

49. $f(x) = x + \cos x$

Domain: $[0, 2\pi]$

Range: $[1, 1 + 2\pi]$

$f'(x) = 1 - \sin x \geq 0$, f is increasing.

$$f''(x) = -\cos x = 0 \text{ when } x = \frac{\pi}{2}, \frac{3\pi}{2}.$$

Points of inflection: $\left(\frac{\pi}{2}, \frac{\pi}{2}\right), \left(\frac{3\pi}{2}, \frac{3\pi}{2}\right)$

Intercept: $(0, 1)$

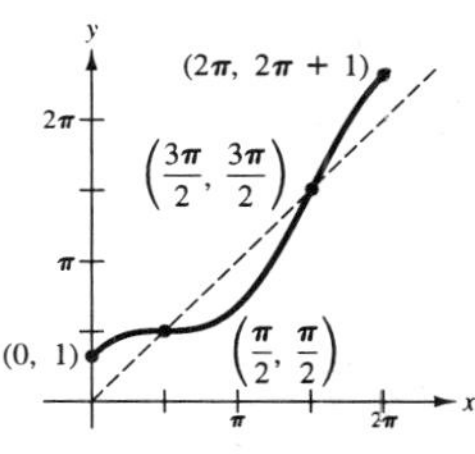

51. Let $t = 0$ at noon.

$$L = d^2 = (100 - 12t)^2 + (-10t)^2 = 10{,}000 - 2400t + 244t^2$$

$$\frac{dL}{dt} = -2400 + 488t = 0 \text{ when } t = \frac{300}{61} \approx 4.92 \text{ hr.}$$

Ship A at $(40.98, 0)$; Ship B at $(0, -49.18)$

$$d^2 = 10{,}000 - 2400t + 244t^2$$

$$\approx 4098.36 \text{ when } t \approx 4.92 \approx 4{:}55 \text{ P.M.}.$$

$d \approx 64$ km

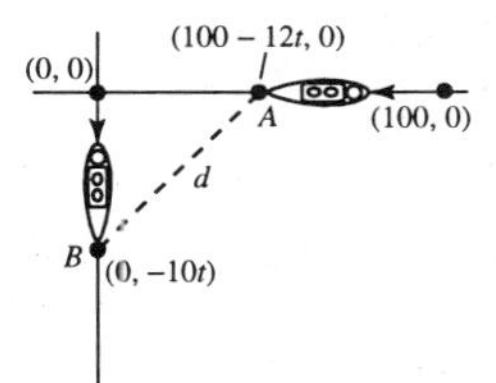

53. We have points $(0, y)$, $(x, 0)$, and $(1, 8)$. Thus,

$$m = \frac{y - 8}{0 - 1} = \frac{0 - 8}{x - 1} \text{ or } y = \frac{8x}{x - 1}.$$

Let $f(x) = L^2 = x^2 + \left(\frac{8x}{x - 1}\right)^2$.

$$f'(x) = 2x + 128\left(\frac{x}{x - 1}\right)\left[\frac{(x - 1) - x}{(x - 1)^2}\right] = 0$$

$$x - \frac{64x}{(x - 1)^3} = 0$$

$x[(x - 1)^3 - 64] = 0$ when $x = 0, 5$ (minimum).

Vertices of triangle: $(0, 0)$, $(5, 0)$, $(0, 10)$

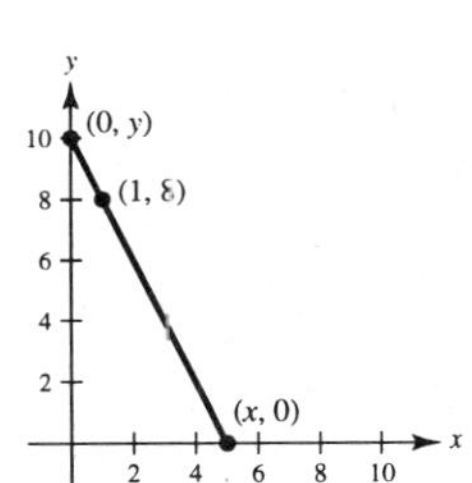

55. $A = (\text{Average of bases})(\text{Height})$

$$= \left(\frac{x+s}{2}\right)\frac{\sqrt{3s^2+2sx-x^2}}{2} \text{ (see figure)}$$

$$\frac{dA}{dx} = \frac{1}{4}\left[\frac{(s-x)(s+x)}{\sqrt{3s^2+2sx-x^2}} + \sqrt{3s^2+2sx-x^2}\right]$$

$$= \frac{2(2s-x)(s+x)}{4\sqrt{3s^2+2sx-x^2}} = 0 \text{ when } x = 2s.$$

A is a maximum when $x = 2s$.

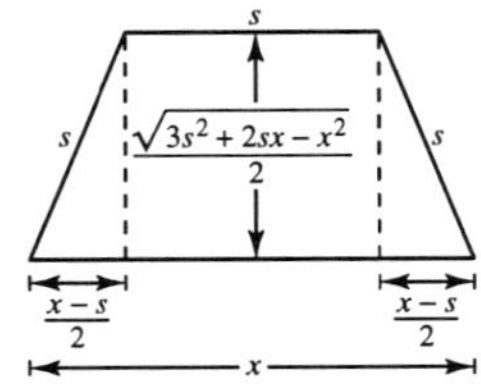

57. You can form a right triangle with vertices $(0, 0)$, $(x, 0)$ and $(0, y)$. Assume that the hypotenuse of length L passes through $(4, 6)$.

$$m = \frac{y-6}{0-4} = \frac{6-0}{4-x} \text{ or } y = \frac{6x}{x-4}$$

Let $f(x) = L^2 = x^2 + y^2 = x^2 + \left(\frac{6x}{x-4}\right)^2$.

$$f'(x) = 2x + 72\left(\frac{x}{x-4}\right)\left[\frac{-4}{(x-4)^2}\right] = 0$$

$$x[(x-4)^3 - 144] = 0 \text{ when } x = 0 \text{ or } x = 4 + \sqrt[3]{144}.$$

$$L \approx 14.05 \text{ feet}$$

59. $\csc\theta = \frac{L_1}{6}$ or $L_1 = 6\csc\theta$ (see figure)

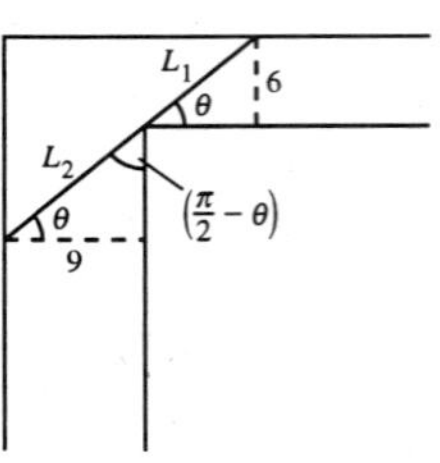

$$\csc\left(\frac{\pi}{2}-\theta\right) = \frac{L_2}{9} \text{ or } L_2 = 9\csc\left(\frac{\pi}{2}-\theta\right)$$

$$L = L_1 + L_2 = 6\csc\theta + 9\csc\left(\frac{\pi}{2}-\theta\right) = 6\csc\theta + 9\sec\theta$$

$$\frac{dL}{d\theta} = -6\csc\theta\cot\theta + 9\sec\theta\tan\theta = 0$$

$$\tan^3\theta = \frac{2}{3} \Rightarrow \tan\theta = \frac{\sqrt[3]{2}}{\sqrt[3]{3}}$$

$$\sec\theta = \sqrt{1+\tan^2\theta} = \sqrt{1+\left(\frac{2}{3}\right)^{2/3}} = \frac{\sqrt{3^{2/3}+2^{2/3}}}{3^{1/3}}$$

$$\csc\theta = \frac{\sec\theta}{\tan\theta} = \frac{\sqrt{3^{2/3}+2^{2/3}}}{2^{1/3}}$$

$$L = 6\frac{(3^{2/3}+2^{2/3})^{1/2}}{2^{1/3}} + 9\frac{(3^{2/3}+2^{2/3})^{1/2}}{3^{1/3}} = 3(3^{2/3}+2^{2/3})^{3/2} \text{ ft} \approx 21.07 \text{ ft}$$ (Compare to Exercise 58 using $a = 9$ and $b = 6$.)

61. $y = \frac{1}{3}\cos(12t) - \frac{1}{4}\sin(12t)$

$v = y' = -4\sin(12t) - 3\cos(12t)$

(a) When $t = \frac{\pi}{8}$, $y = \frac{1}{4}$ inch and $v = y' = 4$ inches/second.

(b) $y' = -4\sin(12t) - 3\cos(12t) = 0$ when $\frac{\sin(12t)}{\cos(12t)} = -\frac{3}{4} \Rightarrow \tan(12t) = -\frac{3}{4}$.

Therefore, $\sin(12t) = -\frac{3}{5}$ and $\cos(12t) = \frac{4}{5}$. The maximum displacement is

$$y = \left(\frac{1}{3}\right)\left(\frac{4}{5}\right) - \frac{1}{4}\left(-\frac{3}{5}\right) = \frac{5}{12} \text{ inch.}$$

(c) Period: $\frac{2\pi}{12} = \frac{\pi}{6}$

Frequency: $\frac{1}{\pi/6} = \frac{6}{\pi}$

63. $f(x) = x^3 - 3x - 1$

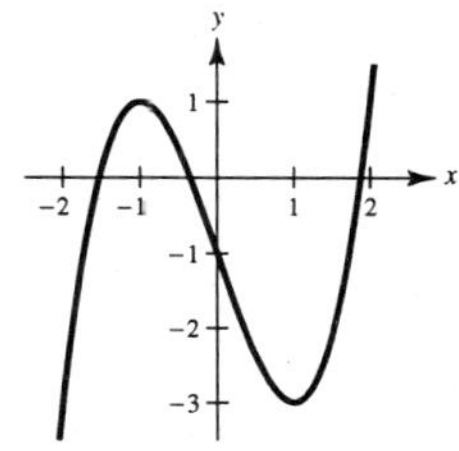

From the graph you can see that $f(x)$ has three real zeros.

$f'(x) = 3x^2 - 3$

n	x_n	$f(x_n)$	$f'(x_n)$	$\frac{f(x_n)}{f'(x_n)}$	$x_n - \frac{f(x_n)}{f'(x_n)}$
1	−1.5000	0.1250	3.7500	0.0333	−1.5333
2	−1.5333	−0.0049	4.0530	−0.0012	−1.5321

n	x_n	$f(x_n)$	$f'(x_n)$	$\frac{f(x_n)}{f'(x_n)}$	$x_n - \frac{f(x_n)}{f'(x_n)}$
1	−0.5000	0.3750	−2.2500	−0.1667	−0.3333
2	−0.3333	−0.0371	−2.6667	0.0139	−0.3472
3	−0.3472	−0.0003	−2.6384	0.0001	−0.3473

n	x_n	$f(x_n)$	$f'(x_n)$	$\frac{f(x_n)}{f'(x_n)}$	$x_n - \frac{f(x_n)}{f'(x_n)}$
1	−1.9000	0.1590	7.8300	0.0203	1.8797
2	1.8797	0.0024	7.5998	0.0003	1.8794

The three real zeros of $f(x)$ are $x \approx -1.532$, $x \approx -0.347$, and $x \approx 1.879$.

65. Find the zeros of $f(x) = x^4 - x - 3$.

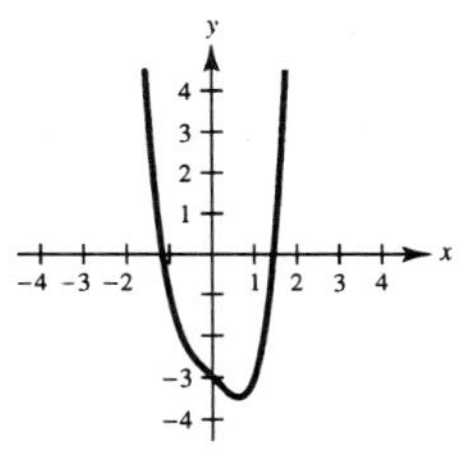

$f'(x) = 4x^3 - 1$

From the graph you can see that $f(x)$ has two real zeros.

f changes sign in $[-2, -1]$.

n	x_n	$f(x_n)$	$f'(x_n)$	$\frac{f(x_n)}{f'(x_n)}$	$x_n - \frac{f(x_n)}{f'(x_n)}$
1	−1.2000	0.2736	−7.9120	−0.0346	−1.1654
2	−1.1654	0.0100	−7.3312	−0.0014	−1.1640

On the interval $[-2, -1]$: $x \approx -1.164$.

f changes sign in $[1, 2]$.

n	x_n	$f(x_n)$	$f'(x_n)$	$\frac{f(x_n)}{f'(x_n)}$	$x_n - \frac{f(x_n)}{f'(x_n)}$
1	1.5000	0.5625	12.5000	0.0450	1.4550
2	1.4550	0.0268	11.3211	0.0024	1.4526
3	1.4526	−0.0003	11.2602	0.0000	1.4526

On the interval $[1, 2]$; $x \approx 1.453$.

67. $y = x(1 - \cos x) = x - x\cos x$

$$\frac{dy}{dx} = 1 + x\sin x - \cos x$$

$$dy = (1 + x\sin x - \cos x)\,dx$$

69. (a) $S = -0.0810t^3 + 2.9197t^2 - 15.5459t + 92.3247$

(b)

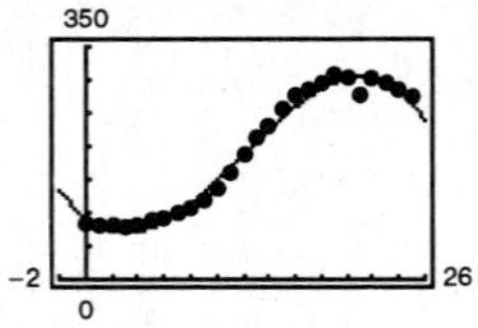

(c) Maximum is $S = 303.3$ at $t = 20.98$ (1990).
Note that this is not accurate!

(d) The derivative $S'(t)$ is greatest at $t = 12.0$ (1982).

71.

$$S = 4\pi r^2.\ dr = \Delta r = \pm 0.025$$

$$dS = 8\pi r\,dr = 8\pi(9)(\pm 0.025)$$

$$= \pm 1.8\pi \text{ square cm}$$

$$\frac{dS}{S}(100) = \frac{8\pi r\,dr}{4\pi r^2}(100) = \frac{2\,dr}{r}(100)$$

$$= \frac{2(\pm 0.025)}{9}(100) \approx \pm 0.56\%$$

$$V = \frac{4}{3}\pi r^3$$

$$dV = 4\pi r^2\,dr = 4\pi(9)^2(\pm 0.025)$$

$$= \pm 8.1\pi \text{ cubic cm}$$

$$\frac{dV}{V}(100) = \frac{4\pi r^2\,dr}{(4/3)\pi r^3}(100) = \frac{3\,dr}{r}(100)$$

$$= \frac{3(\pm 0.025)}{9}(100) \approx \pm 0.83\%$$

73.

$$p = 36 - 4x$$

$$C = 2x^2 + 6$$

$$R = xp$$

$$P = R - C$$

$$= x(36 - 4x) - (2x^2 + 6) = -6(x^2 - 6x + 1)$$

$$\frac{dP}{dx} = -6(2x - 6) = 0 \text{ when } x = 3.$$

$$\frac{d^2P}{dx^2} = -12 < 0,\ x = 3 \text{ is a maximum.}$$

For $x = 3$, the profit is $P = \$48$.

75. $R = (\text{Number of people})(\text{Rate per person})$

$$= n[8.00 - 0.05(n - 80)],\ n \geq 80$$

$$= 12n - 0.05n^2$$

$$\frac{dR}{dn} = 12 - 0.10n = 0 \text{ when } n = 120.$$

$$\frac{d^2R}{dn^2} = -0.10 < 0 \Longrightarrow \text{Revenue will be maximum when 120 people go on the bus.}$$

77. $C = \left(\frac{Q}{x}\right)s + \left(\frac{x}{2}\right)r$

$$\frac{dC}{dx} = -\frac{Qs}{x^2} + \frac{r}{2} = 0$$

$$\frac{Qs}{x^2} = \frac{r}{2}$$

$$x^2 = \frac{2Qs}{r}$$

$$x = \sqrt{\frac{2Qs}{r}}$$

79. $x^2 + 4y^2 - 2x - 16y + 13 = 0$

(a) $(x^2 - 2x + 1) + 4(y^2 - 4y + 4) = -13 + 1 + 16$

$$(x-1)^2 + 4(y-2)^2 = 4$$

$$\frac{(x-1)^2}{4} + \frac{(y-2)^2}{1} = 1$$

The graph is an ellipse:

Maximum: $(1, 3)$

Minimum: $(1, 1)$

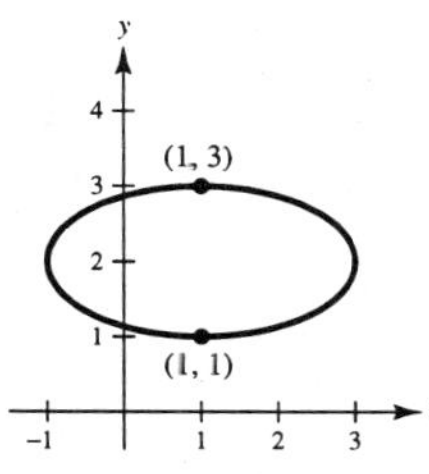

(b) $x^2 + 4y^2 - 2x - 16y + 13 = 0$

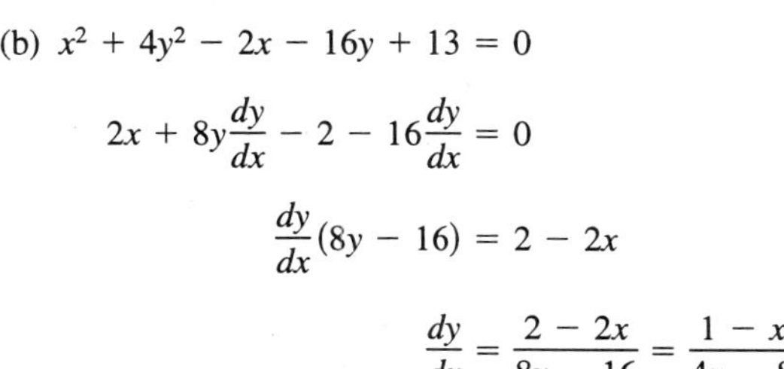

$$2x + 8y\frac{dy}{dx} - 2 - 16\frac{dy}{dx} = 0$$

$$\frac{dy}{dx}(8y - 16) = 2 - 2x$$

$$\frac{dy}{dx} = \frac{2 - 2x}{8y - 16} = \frac{1 - x}{4y - 8}$$

The critical numbers are $x = 1$ and $y = 2$. These correspond to the points $(1, 1)$, $(1, 3)$, $(2, -1)$, and $(2, 3)$. Hence, the maximum is $(1, 3)$ and the minimum is $(1, 1)$.

81. False

Let $f(x) = x^3$, $c = 0$.

83. The first derivative is positive and the second derivative is negative. The graph is increasing and is concave down.

CHAPTER 4
Integration

CHAPTER 4
Integration

Section 4.1 Antiderivatives and Indefinite Integration

Solutions to Odd-Numbered Exercises

1. $\dfrac{d}{dx}\left(\dfrac{3}{x^3} + C\right) = \dfrac{d}{dx}(3x^{-3} + C) = -9x^{-4} = \dfrac{-9}{x^4}$

3. $\dfrac{d}{dx}\left(\dfrac{1}{3}x^3 - 4x + C\right) = x^2 - 4 = (x - 2)(x + 2)$

	Given	*Rewrite*	*Integrate*	*Simplify*
5.	$\int \sqrt[3]{x}\,dx$	$\int x^{1/3}\,dx$	$\dfrac{x^{4/3}}{4/3} + C$	$\dfrac{3}{4}x^{4/3} + C$
7.	$\int \dfrac{1}{x\sqrt{x}}\,dx$	$\int x^{-3/2}\,dx$	$\dfrac{x^{-1/2}}{-1/2} + C$	$-\dfrac{2}{\sqrt{x}} + C$
9.	$\int \dfrac{1}{2x^3}\,dx$	$\dfrac{1}{2}\int x^{-3}\,dx$	$\dfrac{1}{2}\left(\dfrac{x^{-2}}{-2}\right) + C$	$-\dfrac{1}{4x^2} + C$

11. $\dfrac{dy}{dt} = 3t^2$

$y = t^3 + C$

Check: $\dfrac{d}{dt}[t^3 + C] = 3t^2$

13. $\dfrac{dy}{dx} = x^{3/2}$

$y = \dfrac{2}{5}x^{5/2} + C$

Check: $\dfrac{d}{dx}\left[\dfrac{2}{5}x^{5/2} + C\right] = x^{3/2}$

15. $\int (x^3 + 2)\,dx = \dfrac{1}{4}x^4 + 2x + C$

Check: $\dfrac{d}{dx}\left(\dfrac{1}{4}x^4 + 2x + C\right) = x^3 + 2$

17. $\int (x^{3/2} + 2x + 1)\,dx = \dfrac{2}{5}x^{5/2} + x^2 + x + C$

Check: $\dfrac{d}{dx}\left(\dfrac{2}{5}x^{5/2} + x^2 + x + C\right) = x^{3/2} + 2x + 1$

19. $\int \sqrt[3]{x^2}\,dx = \int x^{2/3}\,dx = \dfrac{x^{5/3}}{5/3} + C = \dfrac{3}{5}x^{5/3} + C$

Check: $\dfrac{d}{dx}\left(\dfrac{3}{5}x^{5/3} + C\right) = x^{2/3} = \sqrt[3]{x^2}$

21. $\int \dfrac{1}{x^3}\,dx = \int x^{-3}\,dx = \dfrac{x^{-2}}{-2} + C = -\dfrac{1}{2x^2} + C$

Check: $\dfrac{d}{dx}\left(-\dfrac{1}{2x^2} + C\right) = \dfrac{1}{x^3}$

23. $\int \dfrac{x^2 + x + 1}{\sqrt{x}}\,dx = \int (x^{3/2} + x^{1/2} + x^{-1/2})\,dx = \dfrac{2}{5}x^{5/2} + \dfrac{2}{3}x^{3/2} + 2x^{1/2} + C = \dfrac{2}{15}x^{1/2}(3x^2 - 5x + 15) + C$

Check: $\dfrac{d}{dx}\left(\dfrac{2}{5}x^{5/2} + \dfrac{2}{3}x^{3/2} + 2x^{1/2} + C\right) = x^{3/2} + x^{1/2} + x^{-1/2} = \dfrac{x^2 + x + 1}{\sqrt{x}}$

25. $\int (x + 1)(3x - 2)\,dx = \int (3x^2 + x - 2)\,dx = x^3 + \dfrac{1}{2}x^2 - 2x + C$

Check: $\dfrac{d}{dx}\left(x^3 + \dfrac{1}{2}x^2 - 2x + C\right) = 3x^2 + x - 2 = (x + 1)(3x - 2)$

27. $\int y^2\sqrt{y}\,dy = \int y^{5/2}\,dy = \frac{2}{7}y^{7/2} + C$

Check: $\frac{d}{dy}\left(\frac{2}{7}y^{7/2} + C\right) = y^{5/2} = y^2\sqrt{y}$

29. $\int dx = \int 1\,dx = x + C$

Check: $\frac{d}{dx}(x + C) = 1$

31. $\int (2\sin x + 3\cos x)\,dx = -2\cos x + 3\sin x + C$

Check: $\frac{d}{dx}(-2\cos x + 3\sin x + C) = 2\sin x + 3\cos x$

33. $\int (1 - \csc t\cot t)\,dt = t + \csc t + C$

Check: $\frac{d}{dt}(t + \csc t + C) = 1 - \csc t\cot t$

35. $\int (\sec^2\theta - \sin\theta)\,d\theta = \tan\theta + \cos\theta + C$

Check: $\frac{d}{d\theta}(\tan\theta + \cos\theta + C) = \sec^2\theta - \sin\theta$

37. $\int (\tan^2 y + 1)\,dy = \int \sec^2 y\,dy = \tan y + C$

Check: $\frac{d}{dy}(\tan y + C) = \sec^2 y = \tan^2 y + 1$

39. $f(x) = \cos x$

41. $f'(x) = 2$

$f(x) = 2x + C$

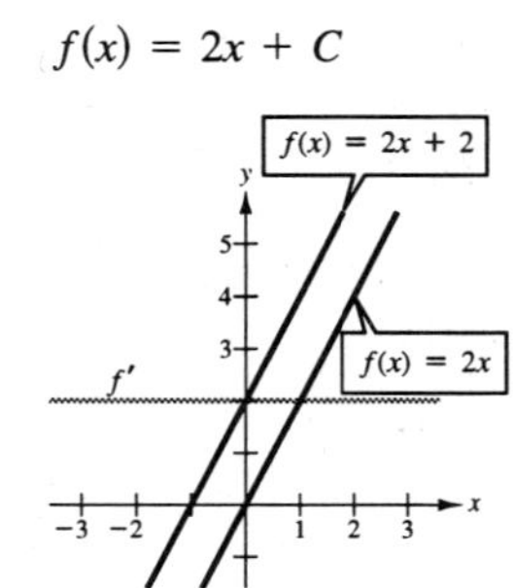

43. $f'(x) = 1 - x^2$

$f(x) = x - \frac{x^3}{3} + C$

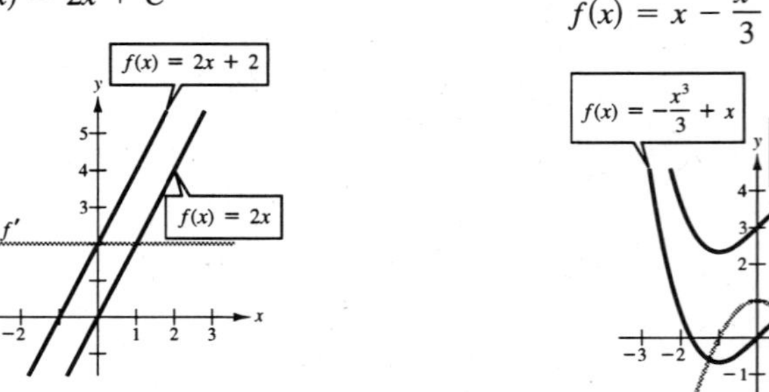

45. $\frac{dy}{dx} = 2x - 1,\ (1, 1)$

$y = \int (2x - 1)\,dx = x^2 - x + C$

$1 = (1)^2 - (1) + C \Rightarrow C = 1$

$y = x^2 - x + 1$

47. $\frac{dy}{dx} = \cos x,\ (0, 4)$

$y = \int \cos x\,dx = \sin x + C$

$4 = \sin 0 + C \Rightarrow C = 4$

$y = \sin x + 4$

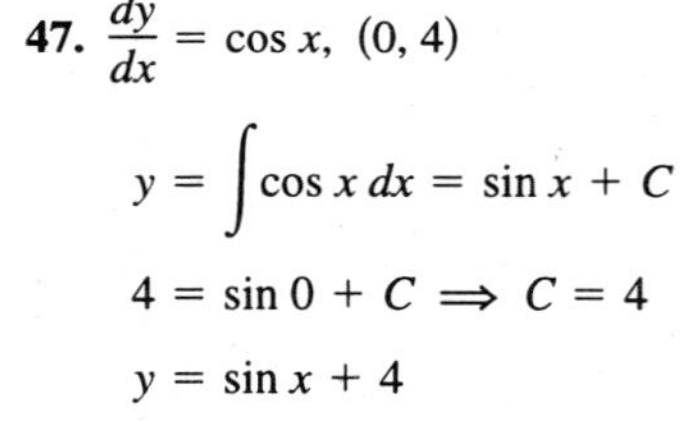

49. (a)

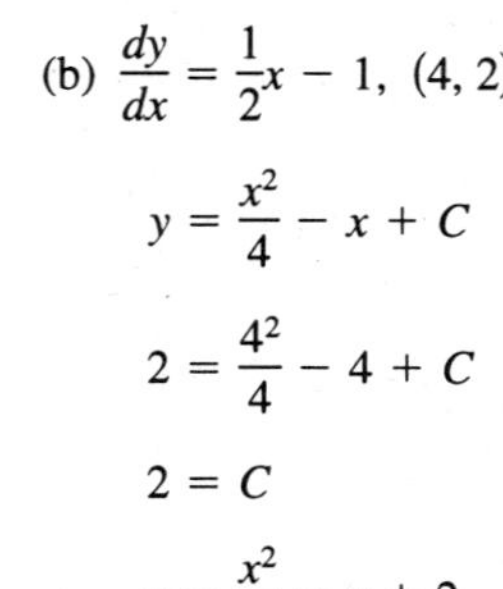

(b) $\frac{dy}{dx} = \frac{1}{2}x - 1,\ (4, 2)$

$y = \frac{x^2}{4} - x + C$

$2 = \frac{4^2}{4} - 4 + C$

$2 = C$

$y = \frac{x^2}{4} - x + 2$

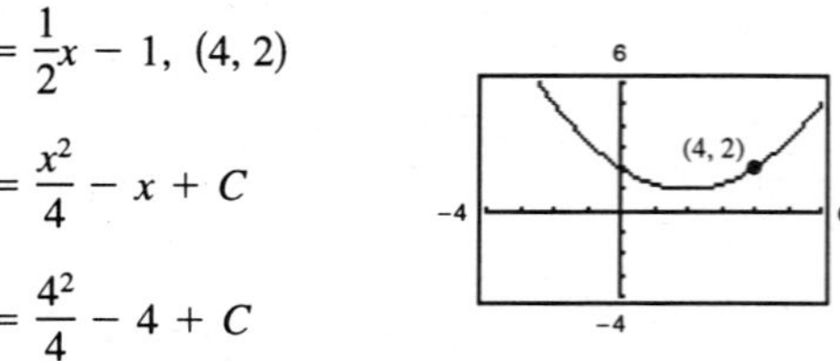

51. $f''(x) = 2$

$f'(2) = 5$

$f(2) = 10$

$f'(x) = \int 2\,dx = 2x + C_1$

$f'(2) = 4 + C_1 = 5 \Rightarrow C_1 = 1$

$f'(x) = 2x + 1$

$f(x) = \int (2x + 1)\,dx = x^2 + x + C_2$

$f(2) = 6 + C_2 = 10 \Rightarrow C_2 = 4$

$f(x) = x^2 + x + 4$

53. $f''(x) = x^{-3/2}$

$f'(4) = 2$

$f(0) = 0$

$f'(x) = \int x^{-3/2}\,dx = -2x^{-1/2} + C_1 = -\dfrac{2}{\sqrt{x}} + C_1$

$f'(4) = -\dfrac{2}{2} + C_1 = 2 \Rightarrow C_1 = 3$

$f'(x) = -\dfrac{2}{\sqrt{x}} + 3$

$f(x) = \int (-2x^{-1/2} + 3)\,dx = -4x^{1/2} + 3x + C_2$

$f(0) = 0 + 0 + C_2 = 0 \Rightarrow C_2 = 0$

$f(x) = -4x^{1/2} + 3x = -4\sqrt{x} + 3x$

55. (a) $h(t) = \int (1.5t + 5)\,dt = 0.75t^2 + 5t + C$

$h(0) = 0 + 0 + C = 12 \Rightarrow C = 12$

$h(t) = 0.75t^2 + 5t + 12$

(b) $h(6) = 0.75(6)^2 + 5(6) + 12 = 69$ cm

57. $a(t) = -32 \text{ ft/sec}^2$

$v(t) = \int -32\,dt = -32t + C_1$

$v(0) = 60 = C_1$

$v(t) = -32t + 60$

$s(t) = \int (-32t + 60)\,dt = -16t^2 + 60t + C_2$

$s(0) = 0 = C_2$

$s(t) = -16t^2 + 60t$

The ball reaches its maximum height when

$v(t) = 0$: $-32t + 60 = 0$

$32t = 60$

$t = \frac{15}{8}$.

$s\left(\frac{15}{8}\right) = -16\left(\frac{15}{8}\right)^2 + 60\left(\frac{15}{8}\right) = 56.26$ feet

59. From Exercise 58, we have:

$s(t) = -16t^2 + v_0 t$

$s'(t) = -32t + v_0 = 0$ when $t = \dfrac{v_0}{32}$ = time to reach maximum height.

$s\left(\dfrac{v_0}{32}\right) = -16\left(\dfrac{v_0}{32}\right)^2 + v_0\left(\dfrac{v_0}{32}\right) = 550$

$-\dfrac{v_0^2}{64} + \dfrac{v_0^2}{32} = 550$

$v_0^2 = 35{,}200$

$v_0 \approx 187.617$ ft/sec

61. $a(t) = -9.8$

$$v(t) = \int -9.8\,dt = -9.8t + C_1$$

$$v(0) = v_0 = C_1 \Rightarrow v(t) = -9.8t + v_0$$

$$f(t) = \int (-9.8t + v_0)\,dt = -4.9t^2 + v_0t + C_2$$

$$f(0) = s_0 = C_2 \Rightarrow f(t) = -4.9t^2 + v_0t + C_2$$

63. From Exercise 61, $f(t) = -4.9t^2 + 10t$.

$$v(t) = -9.8t + 10 = 0 \text{ (Maximum height when } v = 0.)$$

$$9.8t = 10$$

$$t = \frac{10}{9.8}$$

$$f\left(\frac{10}{9.8}\right) \approx 5.1 \text{ m}$$

65. $a = -1.6$

$$v(t) = \int -1.6\,dt = -1.6t + v_0 = -1.6t, \text{ since the stone was dropped, } v_0 = 0.$$

$$s(t) = \int (-1.6t)\,dt = -0.8t^2 + s_0$$

$$s(20) = 0 \Rightarrow -0.8(20)^2 + s_0 = 0$$

$$s_0 = 320$$

Thus, the height of the cliff is 320 meters.

$$v(t) = -1.6t$$

$$v(20) = -32 \text{ m/sec}$$

67. $x(t) = t^3 - 6t^2 + 9t - 2 \qquad 0 \le t \le 5$

(a) $v(t) = x'(t) = 3t^2 - 12t + 9$

$= 3(t^2 - 4t + 3) = 3(t - 1)(t - 3)$

$a(t) = v'(t) = 6t - 12 = 6(t - 2)$

(b) $v(t) > 0$ when $0 < t < 1$ or $3 < t < 5$.

(c) $a(t) = 6(t - 2) = 0$ when $t = 2$.

$v(2) = 3(1)(-1) = -3$

69. $v(t) = \dfrac{1}{\sqrt{t}} = t^{-1/2} \qquad t > 0$

$$x(t) = \int v(t)\,dt = 2t^{1/2} + C$$

$$x(1) = 4 = 2(1) + C \Rightarrow C = 2$$

$$x(t) = 2t^{1/2} + 2$$

$$a(t) = v'(t) = -\frac{1}{2}t^{-3/2} = \frac{-1}{2t^{3/2}}$$

71. (a) $v(0) = 25 \text{ km/hr} = 25 \cdot \dfrac{1000}{3600} = \dfrac{250}{36} \text{ m/sec}$

$$v(13) = 80 \text{ km/hr} = 80 \cdot \frac{1000}{3600} = \frac{800}{36} \text{ m/sec}$$

$$a(t) = a \text{ (constant acceleration)}$$

$$v(t) = at + C$$

$$v(0) = \frac{250}{36} \Rightarrow v(t) = at + \frac{250}{36}$$

$$v(13) = \frac{800}{36} = 13a + \frac{250}{36}$$

$$\frac{550}{36} = 13a$$

$$a = \frac{550}{468} = \frac{275}{234} \approx 1.175 \text{ m/sec}^2$$

(b) $s(t) = a\dfrac{t^2}{2} + \dfrac{250}{36}t \qquad (s(0) = 0)$

$$s(13) = \frac{275}{234}\frac{(13)^2}{2} + \frac{250}{36}(13) \approx 189.58 \text{ m}$$

73. Truck: $v(t) = 30$

$s(t) = 30t$ (Let $s(0) = 0$.)

Automobile: $a(t) = 6$

$v(t) = 6t$ (Let $v(0) = 0$.)

$s(t) = 3t^2$ (Let $s(0) = 0$.)

At the point where the automobile overtakes the truck:

$$30t = 3t^2$$

$$0 = 3t^2 - 30t$$

$$0 = 3t(t - 10) \text{ when } t = 10 \text{ sec.}$$

(a) $s(10) = 3(10)^2 = 300$ ft

(b) $v(10) = 6(10) = 60$ ft/sec ≈ 41 mph

75. $a(t) = k$

$v(t) = kt$

$s(t) = \frac{k}{2}t^2$ since $v(0) = s(0) = 0$.

At the time of lift-off, $kt = 160$ and $(k/2)t^2 = 0.7$. Since $(k/2)t^2 = 0.7$,

$$t = \sqrt{\frac{1.4}{k}}$$

$$v\left(\sqrt{\frac{1.4}{k}}\right) = k\sqrt{\frac{1.4}{k}} = 160$$

$$1.4k = 160^2 \implies k = \frac{160^2}{1.4}$$

$$\approx 18{,}285.714 \text{ mi/hr}^2$$

$$\approx 7.45 \text{ ft/sec}^2.$$

77. $\frac{dC}{dx} = 2x - 12$

$C(0) = 50$

$C(x) = x^2 - 12x + C$

$C(0) = 0 + C = 50 \implies C = 50$

Cost: $C(x) = x^2 - 12x + 50$

Average cost: $\overline{C}(x) = \frac{C(x)}{x} = x - 12 + \frac{50}{x}$

79. $\frac{dR}{dx} = 100 - 5x$

$R = 100x - \frac{5}{2}x^2 + C$

($C = 0$; if no units are sold, no revenue is generated.)

Revenue: $R = 100x - \frac{5}{2}x^2$

$$R = xp = x\left(100 - \frac{5}{2}x\right)$$

Demand: $p = 100 - \frac{5}{2}x$

81. True

83. True

85. $f(0) = -4$. Graph of f' is given.

(a) $f'(4) \approx -1.0$

(b) No. The slopes of the tangent lines are greater than 2 on $[0, 2]$. Therefore, f must increase more than 4 units on $[0, 4]$.

(c) No, $f(5) < f(4)$ because f is decreasing on $[4, 5]$.

(d) f is an maximum at $x = 3.5$ because $f'(3.5) \approx 0$ and the first derivative test.

(e) f is concave upward when f' is increasing on $(-\infty, 1)$ and $(5, \infty)$. f is concave downward on $(1, 5)$. Points of inflection at $x = 1, 5$.

(f) f'' is a minimum at $x = 3$.

(g)

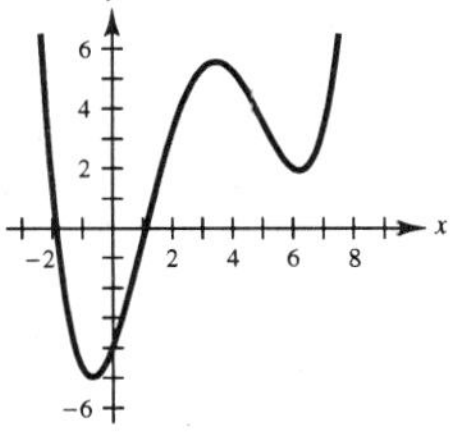

87. Let d be the distance traversed and a be the uniform acceleration. Furthermore, note that $v(0) = 0$ and let $s(0) = 0$.

$$a(t) = a$$

$$v(t) = at$$

$$s(t) = \frac{at^2}{2} = d \text{ when } t = \sqrt{\frac{2d}{a}}.$$

The highest speed is

$$v\left(\sqrt{\frac{2d}{a}}\right) = a\sqrt{\frac{2d}{a}} = \sqrt{2ad}$$

and the mean speed is

$$\frac{\sqrt{2ad}}{2} = \sqrt{\frac{ad}{2}}.$$

The time necessary to traverse the distance at the mean speed must satisfy the equation

$$\sqrt{\frac{ad}{2}}t = d \text{ or } t = \sqrt{\frac{2d}{a}}.$$

This is the same time traversed under uniform acceleration.

Section 4.2 Area

1. $\sum_{i=1}^{5}(2i+1) = 2\sum_{i=1}^{5} i + \sum_{i=1}^{5} 1 = 2(1+2+3+4+5)+5 = 35$

3. $\sum_{k=0}^{4}\frac{1}{k^2+1} = 1 + \frac{1}{2} + \frac{1}{5} + \frac{1}{10} + \frac{1}{17} = \frac{158}{85}$

5. $\sum_{k=1}^{4} c = c + c + c + c = 4c$

7. $\sum_{i=1}^{9}\frac{1}{3i}$

9. $\sum_{j=1}^{8}\left[2\left(\frac{j}{8}\right)+3\right]$

11. $\frac{2}{n}\sum_{i=1}^{n}\left[\left(\frac{2i}{n}\right)^3 - \left(\frac{2i}{n}\right)\right]$

13. $\frac{3}{n}\sum_{i=1}^{n}\left[2\left(1+\frac{3i}{n}\right)^2\right]$

15. $\sum_{i=1}^{20} 2i = 2\sum_{i=1}^{20} i = 2\left[\frac{20(21)}{2}\right] = 420$

17. $\sum_{i=1}^{20}(i-1)^2 = \sum_{i=1}^{19} i^2 = \left[\frac{19(20)(39)}{6}\right] = 2470$

19. $\sum_{i=1}^{15} i(i-1)^2 = \sum_{i=1}^{15} i^3 - 2\sum_{i=1}^{15} i^2 + \sum_{i=1}^{15} i$

$$= \frac{15^2(16)^2}{4} - 2\frac{15(16)(31)}{6} + \frac{15(16)}{2}$$

$$= 14{,}400 - 2{,}480 + 120$$

$$= 12{,}040$$

21. sum seq(x ^ 2 + 3, x, 1, 20, 1) = 2930 (*TI-82*)

$$\sum_{i=1}^{20}(i^2+3) = \frac{20(20+1)(2(20)+1)}{6} + 3(20)$$

$$= \frac{(20)(21)(41)}{6} + 60 = 2930$$

23. $\lim_{n\to\infty}\left[\left(\frac{4}{3n^3}\right)(2n^3+3n^2+n)\right] = \lim_{n\to\infty}\left[\frac{8}{3}+\frac{4}{n}+\frac{4}{3n^2}\right] = \frac{8}{3}$

25. $\lim_{n\to\infty}\left[\left(\frac{81}{n^4}\right)\frac{n^2(n+1)^2}{4}\right] = \frac{81}{4}\lim_{n\to\infty}\left[\frac{n^4+2n^3+n^2}{n^4}\right] = \frac{81}{4}(1) = \frac{81}{4}$

27. $\lim_{n\to\infty}\left[\left(\frac{18}{n^2}\right)\frac{n(n+1)}{2}\right] = \frac{18}{2}\lim_{n\to\infty}\left[\frac{n^2+n}{n^2}\right] = \frac{18}{2}(1) = 9$

29. $\lim_{n\to\infty}\sum_{i=1}^{n}\left(\frac{16i}{n^2}\right) = \lim_{n\to\infty}\frac{16}{n^2}\sum_{i=1}^{n} i = \lim_{n\to\infty}\frac{16}{n^2}\left(\frac{n(n+1)}{2}\right) = \lim_{n\to\infty}\left[8\left(\frac{n^2+n}{n^2}\right)\right] = 8\lim_{n\to\infty}\left(1+\frac{1}{n}\right) = 8$

31. $\lim_{n\to\infty}\sum_{i=1}^{n}\frac{1}{n^3}(i-1)^2 = \lim_{n\to\infty}\frac{1}{n^3}\sum_{i=1}^{n-1} i^2 = \lim_{n\to\infty}\frac{1}{n^3}\left[\frac{(n-1)(n)(2n-1)}{6}\right]$

$$= \lim_{n\to\infty}\frac{1}{6}\left[\frac{2n^3-3n^2+n}{n^3}\right] = \lim_{n\to\infty}\left[\frac{1}{6}\left(\frac{2-(3/n)+(1/n^2)}{1}\right)\right] = \frac{1}{3}$$

33. $\lim_{n\to\infty}\sum_{i=1}^{n}\left(1+\frac{i}{n}\right)\left(\frac{2}{n}\right) = 2\lim_{n\to\infty}\frac{1}{n}\left[\sum_{i=1}^{n} 1 + \frac{1}{n}\sum_{i=1}^{n} i\right] = 2\lim_{n\to\infty}\frac{1}{n}\left[n + \frac{1}{n}\left(\frac{n(n+1)}{2}\right)\right] = 2\lim_{n\to\infty}\left[1+\frac{n^2+n}{2n^2}\right] = 2\left(1+\frac{1}{2}\right) = 3$

35. $S(4) = \sqrt{\frac{1}{4}}\left(\frac{1}{4}\right) + \sqrt{\frac{1}{2}}\left(\frac{1}{4}\right) + \sqrt{\frac{3}{4}}\left(\frac{1}{4}\right) + \sqrt{1}\left(\frac{1}{4}\right) = \frac{1+\sqrt{2}+\sqrt{3}+2}{8} \approx 0.768$

$$s(4) = 0\left(\frac{1}{4}\right) + \sqrt{\frac{1}{4}}\left(\frac{1}{4}\right) + \sqrt{\frac{1}{2}}\left(\frac{1}{4}\right) + \sqrt{\frac{3}{4}}\left(\frac{1}{4}\right) = \frac{1+\sqrt{2}+\sqrt{3}}{8} \approx 0.518$$

37. $S(5) = 1\left(\frac{1}{5}\right) + \frac{1}{6/5}\left(\frac{1}{5}\right) + \frac{1}{7/5}\left(\frac{1}{5}\right) + \frac{1}{8/5}\left(\frac{1}{5}\right) + \frac{1}{9/5}\left(\frac{1}{5}\right) = \frac{1}{5} + \frac{1}{6} + \frac{1}{7} + \frac{1}{8} + \frac{1}{9} \approx 0.746$

$s(5) = \frac{1}{6/5}\left(\frac{1}{5}\right) + \frac{1}{7/5}\left(\frac{1}{5}\right) + \frac{1}{8/5}\left(\frac{1}{5}\right) + \frac{1}{9/5}\left(\frac{1}{5}\right) + \frac{1}{2}\left(\frac{1}{5}\right) = \frac{1}{6} + \frac{1}{7} + \frac{1}{8} + \frac{1}{9} + \frac{1}{10} \approx 0.646$

39. (a)

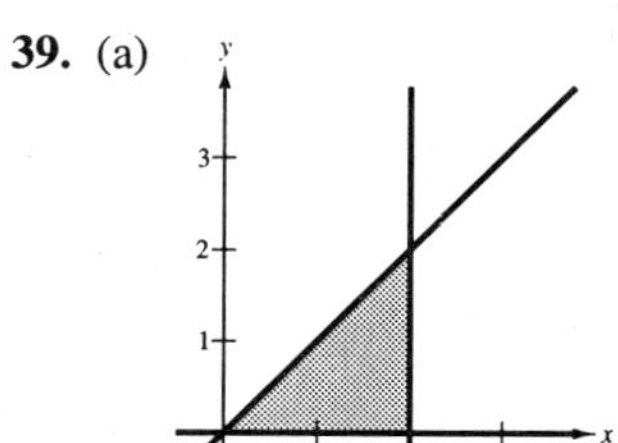

(b) 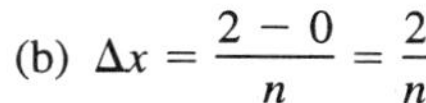
$\Delta x = \frac{2-0}{n} = \frac{2}{n}$

Endpoints:

$$0 < 1\left(\frac{2}{n}\right) < 2\left(\frac{2}{n}\right) < \cdots < (n-1)\left(\frac{2}{n}\right) < n\left(\frac{2}{n}\right) = 2$$

(c) Since $y = x$ is increasing, $f(m_i) = f(x_{i-1})$ on $[x_{i-1}, x_i]$.

$$s(n) = \sum_{i=1}^{n} f(x_{i-1})\,\Delta x$$

$$= \sum_{i=1}^{n} f\left(\frac{2i-2}{n}\right)\left(\frac{2}{n}\right) = \sum_{i=1}^{n}\left[(i-1)\left(\frac{2}{n}\right)\right]\left(\frac{2}{n}\right)$$

(d) $f(M_i) = f(x_i)$ on $[x_{i-1}, x_i]$

$$S(n) = \sum_{i=1}^{n} f(x_i)\,\Delta x = \sum_{i=1}^{n} f\left(\frac{2i}{n}\right)\frac{2}{n} = \sum_{i=1}^{n}\left[i\left(\frac{2}{n}\right)\right]\left(\frac{2}{n}\right)$$

(e)

x	5	10	50	100
$s(n)$	1.6	1.8	1.96	1.98
$S(n)$	2.4	2.2	2.04	2.02

(f) $\lim_{n\to\infty} \sum_{i=1}^{n}\left[(i-1)\left(\frac{2}{n}\right)\right]\left(\frac{2}{n}\right) = \lim_{n\to\infty} \frac{4}{n^2}\sum_{i=1}^{n}(i-1)$

$$= \lim_{n\to\infty} \frac{4}{n^2}\left[\frac{n(n+1)}{2} - n\right]$$

$$= \lim_{n\to\infty} \left[\frac{2(n+1)}{n} - \frac{4}{n}\right] = 2$$

$$\lim_{n\to\infty} \sum_{i=1}^{n}\left[i\left(\frac{2}{n}\right)\right]\left(\frac{2}{n}\right) = \lim_{n\to\infty} \frac{4}{n^2}\sum_{i=1}^{n} i$$

$$= \lim_{n\to\infty} \left(\frac{4}{n^2}\right)\frac{n(n+1)}{2}$$

$$= \lim_{n\to\infty} \frac{2(n+1)}{n} = 2$$

41. $y = -2x + 3$ on $[0, 1]$. $\left(\textbf{Note: } \Delta x = \frac{1-0}{n} = \frac{1}{n}\right)$

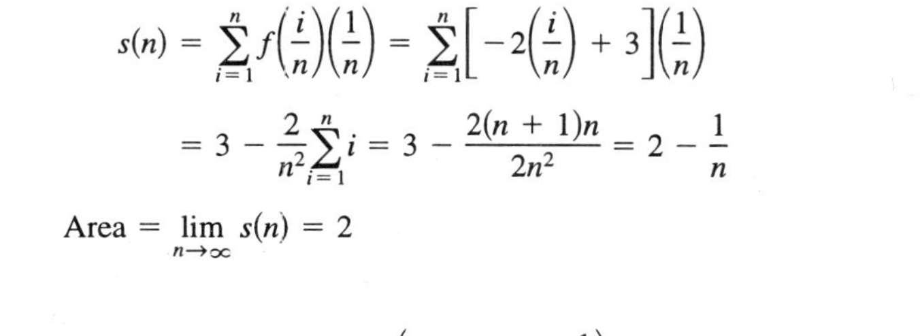

$$s(n) = \sum_{i=1}^{n} f\left(\frac{i}{n}\right)\left(\frac{1}{n}\right) = \sum_{i=1}^{n}\left[-2\left(\frac{i}{n}\right) + 3\right]\left(\frac{1}{n}\right)$$

$$= 3 - \frac{2}{n^2}\sum_{i=1}^{n} i = 3 - \frac{2(n+1)n}{2n^2} = 2 - \frac{1}{n}$$

$$\text{Area} = \lim_{n\to\infty} s(n) = 2$$

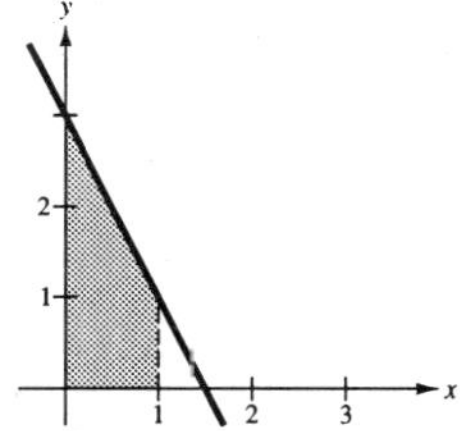

43. $y = x^2 + 2$ on $[0, 1]$. $\left(\textbf{Note: } \Delta x = \frac{1}{n}\right)$

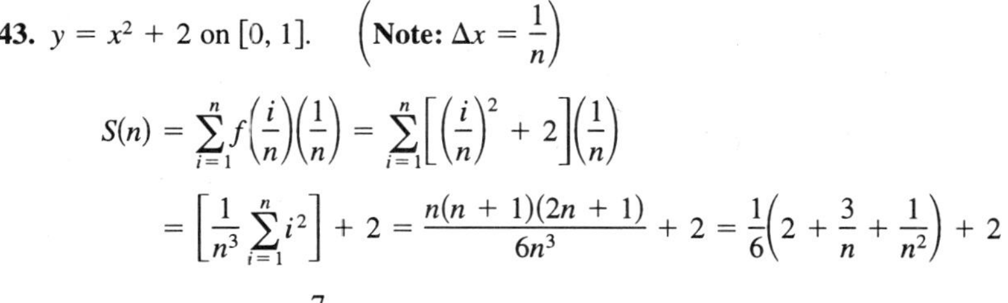

$$S(n) = \sum_{i=1}^{n} f\left(\frac{i}{n}\right)\left(\frac{1}{n}\right) = \sum_{i=1}^{n}\left[\left(\frac{i}{n}\right)^2 + 2\right]\left(\frac{1}{n}\right)$$

$$= \left[\frac{1}{n^3}\sum_{i=1}^{n} i^2\right] + 2 = \frac{n(n+1)(2n+1)}{6n^3} + 2 = \frac{1}{6}\left(2 + \frac{3}{n} + \frac{1}{n^2}\right) + 2$$

$$\text{Area} = \lim_{n\to\infty} S(n) = \frac{7}{3}$$

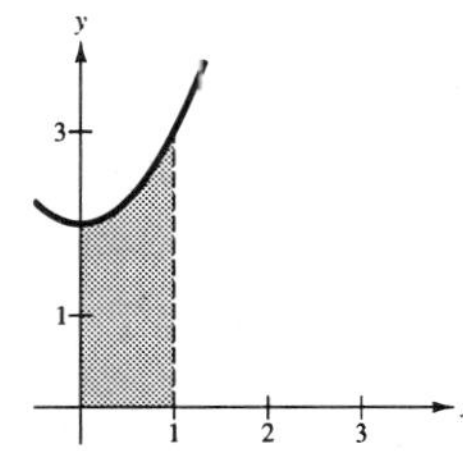

45. $y = 27 - x^3$ on $[1, 3]$. $\left(\textbf{Note: } \Delta x = \frac{3-1}{n} = \frac{2}{n}\right)$

$$s(n) = \sum_{i=1}^{n} f\left(1 + \frac{2i}{n}\right)\left(\frac{2}{n}\right) = \sum_{i=1}^{n}\left[27 - \left(1 + \frac{2i}{n}\right)^3\right]\left(\frac{2}{n}\right) = \frac{2}{n}\sum_{i=1}^{n}\left(26 - \frac{6i}{n} - \frac{12i^2}{n^2} - \frac{8i^3}{n^3}\right)$$

$$= \frac{2}{n}\left[26n - \frac{6}{n}\left(\frac{n(n+1)}{2}\right) - \frac{12}{n^2}\left(\frac{n(n+1)(2n+1)}{6}\right) - \frac{8}{n^3}\left(\frac{n^2(n+1)^2}{4}\right)\right]$$

$$= 34 - \frac{26}{n} - \frac{8}{n^2}$$

$$\text{Area} = \lim_{n\to\infty} s(n) = 34$$

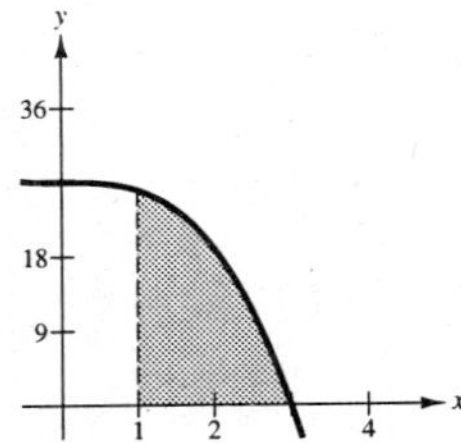

47. $y = x^2 - x^3$ on $[-1, 1]$. $\left(\textbf{Note: } \Delta x = \frac{1-(-1)}{n} = \frac{2}{n}\right)$

Again, $T(n)$ is neither an upper nor a lower sum.

$$T(n) = \sum_{i=1}^{n} f\left(-1 + \frac{2i}{n}\right)\left(\frac{2}{n}\right) = \sum_{i=1}^{n}\left[\left(-1 + \frac{2i}{n}\right)^2 - \left(-1 + \frac{2i}{n}\right)^3\right]\left(\frac{2}{n}\right)$$

$$= \sum_{i=1}^{n}\left[\left(1 - \frac{4i}{n} + \frac{4i^2}{n^2}\right) - \left(-1 + \frac{6i}{n} - \frac{12i^2}{n^2} + \frac{8i^3}{n^3}\right)\right]\left(\frac{2}{n}\right)$$

$$= \sum_{i=1}^{n}\left[2 - \frac{10i}{n} + \frac{16i^2}{n^2} - \frac{8i^3}{n^3}\right]\left(\frac{2}{n}\right) = \frac{4}{n}\sum_{i=1}^{n}1 - \frac{20}{n^2}\sum_{i=1}^{n}i + \frac{32}{n^3}\sum_{i=1}^{n}i^2 - \frac{16}{n^4}\sum_{i=1}^{n}i^3$$

$$= \frac{4}{n}(n) - \frac{20}{n^2}\cdot\frac{n(n+1)}{2} + \frac{32}{n^3}\cdot\frac{n(n+1)(2n+1)}{6} - \frac{16}{n^4}\cdot\frac{n^2(n+1)^2}{4}$$

$$= 4 - 10\left(1 + \frac{1}{n}\right) + \frac{16}{3}\left(2 + \frac{3}{n} + \frac{1}{n^2}\right) - 4\left(1 + \frac{2}{n} + \frac{1}{n^2}\right)$$

$$\text{Area} = \lim_{n\to\infty} T(n) = 4 - 10 + \frac{32}{3} - 4 = \frac{2}{3}$$

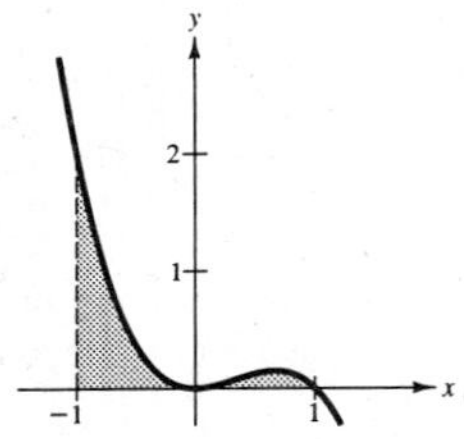

49. $f(y) = 3y$ on $[0, 2]$. $\left(\textbf{Note: } \Delta y = \frac{2-0}{n} = \frac{2}{n}\right)$

$$S(n) = \sum_{i=1}^{n} f(m_i)\,\Delta y = \sum_{i=1}^{n} f\left(\frac{2i}{n}\right)\left(\frac{2}{n}\right) = \sum_{i=1}^{n} 3\left(\frac{2i}{n}\right)\left(\frac{2}{n}\right)$$

$$= \frac{12}{n^2}\sum_{i=1}^{n} i = \left(\frac{12}{n^2}\right)\cdot\frac{n(n+1)}{2} = \frac{6(n+1)}{n} = 6 + \frac{6}{n}$$

$$\text{Area} = \lim_{n\to\infty} S(n) = \lim_{n\to\infty}\left(6 + \frac{6}{n}\right) = 6$$

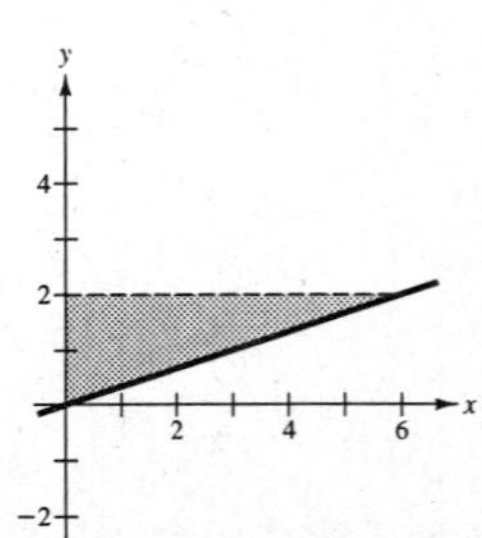

51. $f(x) = x^2 + 3, 0 \le x \le 2, n = 4$

Let $c_i = \dfrac{x_i + x_{i-1}}{2}$.

$$\Delta x = \frac{1}{2}, c_1 = \frac{1}{4}, c_2 = \frac{3}{4}, c_3 = \frac{5}{4}, c_4 = \frac{7}{4}$$

$$\text{Area} \approx \sum_{i=1}^{n} f(c_i)\,\Delta x = \sum_{i=1}^{4}[c_i^2 + 3]\left(\frac{1}{2}\right)$$

$$= \frac{1}{2}\left[\left(\frac{1}{16} + 3\right) + \left(\frac{9}{16} + 3\right) + \left(\frac{25}{16} + 3\right) + \left(\frac{49}{16} + 3\right)\right]$$

$$= \frac{69}{8}$$

53. $f(x) = \tan x, 0 \le x \le \dfrac{\pi}{4}, n = 4$

Let $c_i = \dfrac{x_i + x_{i-1}}{2}$.

$$\Delta x = \frac{\pi}{16}, c_1 = \frac{\pi}{32}, c_2 = \frac{3\pi}{32}, c_3 = \frac{5\pi}{32}, c_4 = \frac{7\pi}{32}$$

$$\text{Area} \approx \sum_{i=1}^{n} f(c_i)\,\Delta x = \sum_{i=1}^{4}(\tan c_i)\left(\frac{\pi}{16}\right)$$

$$= \frac{\pi}{16}\left(\tan\frac{\pi}{32} + \tan\frac{3\pi}{32} + \tan\frac{5\pi}{32} + \tan\frac{7\pi}{32}\right) \approx 0.345$$

55. $f(x) = \sqrt{x}$ on $[0, 4]$.

n	4	8	12	16	20
Approximate area	5.3838	5.3523	5.3439	5.3403	5.3384

(Exact value is 16/3)

57. $f(x) = \tan\left(\dfrac{\pi x}{8}\right)$ on $[1, 3]$.

n	4	8	12	16	20
Approximate area	2.2223	2.2387	2.2418	2.2430	2.2435

59. (a)

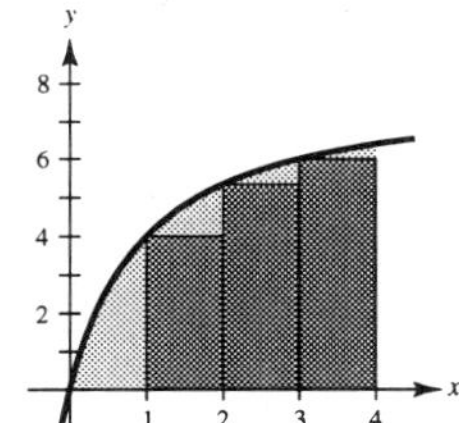

Lower sum:
$s(4) = 0 + 4 + 5\frac{1}{3} + 6 = 15\frac{1}{3} = \frac{46}{3} \approx 15.333$

(b)

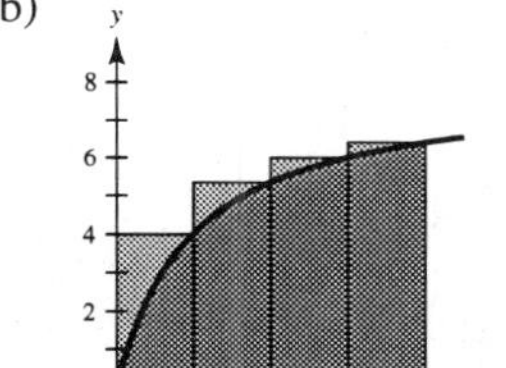

Upper sum:
$S(4) = 4 + 5\frac{1}{3} + 6 + 6\frac{2}{5} = 21\frac{11}{15} = \frac{326}{15} \approx 21.733$

(c)

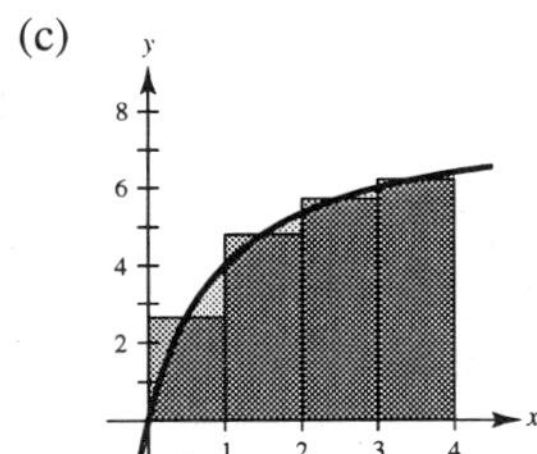

Midpoint Rule:
$M(4) = 2\frac{2}{3} + 4\frac{4}{5} + 5\frac{5}{7} + 6\frac{2}{9} = \frac{6112}{315} \approx 19.403$

(d) In each case, $\Delta x = 4/n$. The lower sum uses left endpoints, $(i - 1)(4/n)$. The upper sum uses right endpoints, $(i)(4/n)$. The Midpoint Rule uses midpoints, $\left(i - \frac{1}{2}\right)(4/n)$.

(e)

n	4	8	20	100	200
$s(n)$	15.333	17.368	18.459	18.995	19.06
$S(n)$	21.733	20.568	19.739	19.251	19.188
$M(n)$	19.403	19.201	19.137	19.125	19.125

(f) $s(n)$ increases because the lower sum approaches the exact value as n increases. $S(n)$ decreases because the upper sum approaches the exact value as n increases. Because of the shape of the graph, the lower sum is always smaller than the exact value, whereas the upper sum is always larger.

61.

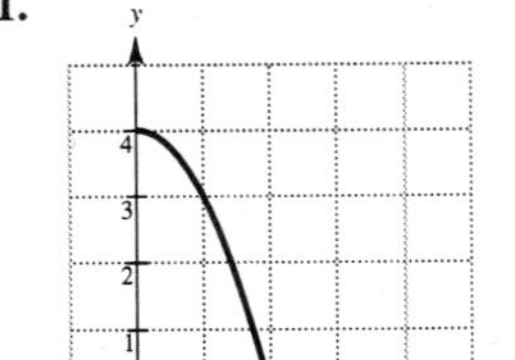

b. $A \approx 6$ square units

63. True. (Theorem 4.2 (2))

65. $f(x) = \sin x, \left[0, \frac{\pi}{2}\right]$

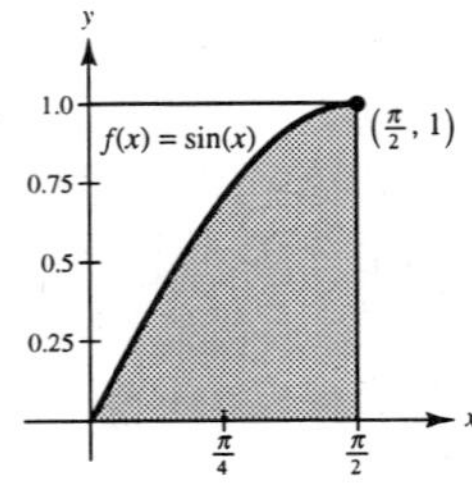

Let A_1 = area bounded by $f(x) = \sin x$, the x-axis, $x = 0$ and $x = \pi/2$. Let A_2 = area of the rectangle bounded by $y = 1$, $y = 0$, $x = 0$, and $x = \pi/2$. Thus, $A_2 = (\pi/2)(1) \approx 1.570796$. In this program, the computer is generating N_2 pairs of random points in the rectangle whose area is represented by A_2. It is keeping track of how many of these points, N_1, lie in the region whose area is represented by A_1. Since the points are randomly generated, we assume that

$$\frac{A_1}{A_2} \approx \frac{N_1}{N_2} \Rightarrow A_1 \approx \frac{N_1}{N_2} A_2.$$

The larger N_2 is the better the approximation to A_1.

67. Suppose there are n rows in the figure. The stars on the left total $1 + 2 + \cdots + n$, as do the stars on the right. There are $n(n + 1)$ stars in total, hence

$$2[1 + 2 + \cdots + n] = n(n + 1)$$
$$1 + 2 + \cdots + n = \tfrac{1}{2}(n)(n + 1).$$

69. (a) $y = (-4.09 \times 10^{-5})x^3 + 0.016x^2 - 2.67x + 452.9$

(b)

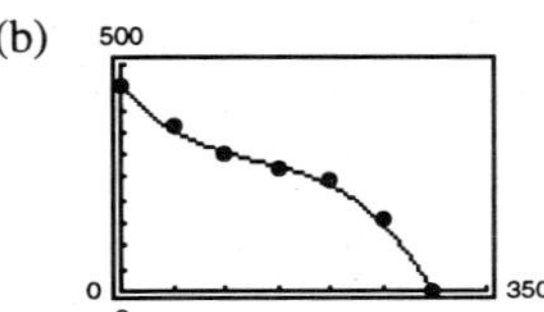

(c) Using the integration capability of a graphing utility, you obtain

$$A \approx 76{,}897.5 \text{ ft}^2.$$

Section 4.3 Riemann Sums and Definite Integrals

1. $\int_0^5 3\,dx$

3. $\int_{-4}^4 (4 - |x|)\,dx$

5. $\int_{-2}^2 (4 - x^2)\,dx$

7. $\int_0^\pi \sin x\,dx$

9. $\int_0^2 y^3\,dy$

11. Rectangle

$A = bh = 3(4)$

$A = \int_0^3 4\,dx = 12$

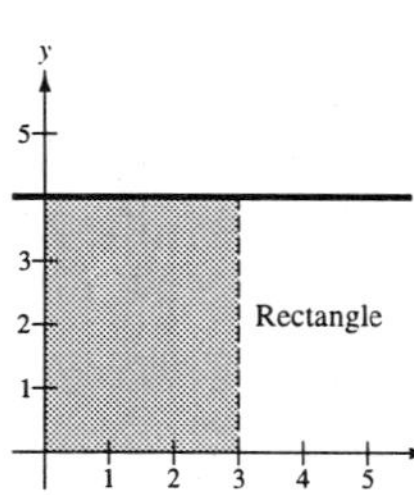

13. Triangle

$A = \frac{1}{2}bh = \frac{1}{2}(4)(4)$

$A = \int_0^4 x\,dx = 8$

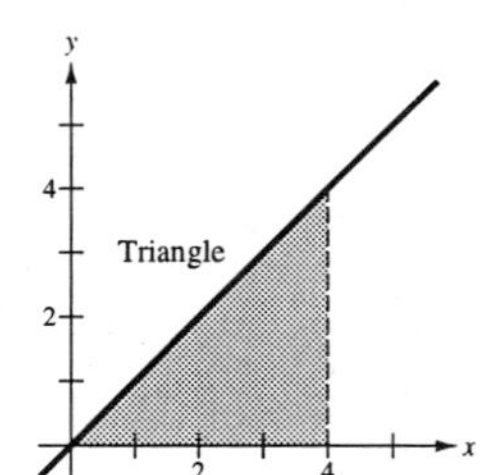

15. Trapezoid

$$A = \frac{b_1 + b_2}{2}h = \left(\frac{5 + 9}{2}\right)2$$

$$A = \int_0^2 (2x + 5)\, dx = 14$$

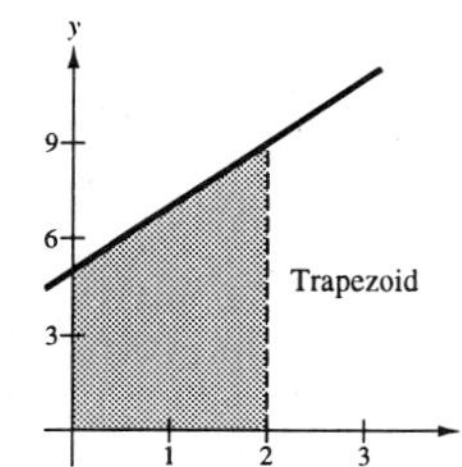

17. Triangle

$$A = \frac{1}{2}bh = \frac{1}{2}(2)(1)$$

$$A = \int_{-1}^{1} (1 - |x|)\, dx = 1$$

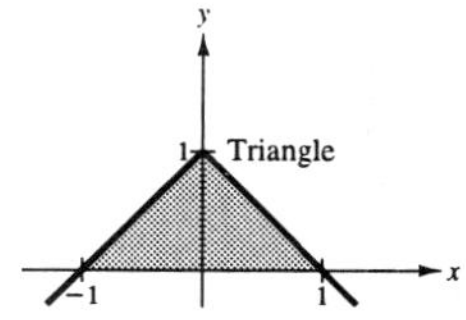

19. Semicircle

$$A = \frac{1}{2}\pi r^2 = \frac{1}{2}\pi(3)^2$$

$$A = \int_{-3}^{3} \sqrt{9 - x^2}\, dx = \frac{9\pi}{2}$$

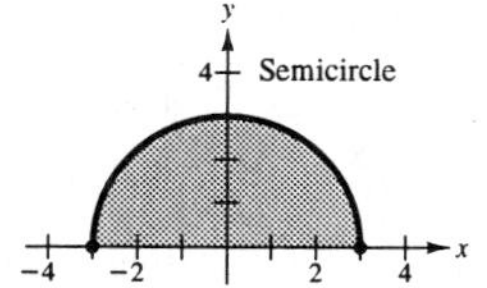

21. (a) $\displaystyle\int_0^7 f(x)\, dx = \int_0^5 f(x)\, dx + \int_5^7 f(x)\, dx = 10 + 3 = 13$

(b) $\displaystyle\int_5^0 f(x)\, dx = -\int_0^5 f(x)\, dx = -10$

(c) $\displaystyle\int_5^5 f(x)\, dx = 0$

(d) $\displaystyle\int_0^5 3f(x)\, dx = 3\int_0^5 f(x)\, dx = 3(10) = 30$

23. (a) $\displaystyle\int_2^6 [f(x) + g(x)]\, dx = \int_2^6 f(x)\, dx + \int_2^6 g(x)\, dx$

$= 10 + (-2) = 8$

(b) $\displaystyle\int_2^6 [g(x) - f(x)]\, dx = \int_2^6 g(x)\, dx - \int_2^6 f(x)\, dx$

$= -2 - 10 = -12$

(c) $\displaystyle\int_2^6 2g(x)\, dx = 2\int_2^6 g(x)\, dx = 2(-2) = -4$

(d) $\displaystyle\int_2^6 3f(x)\, dx = 3\int_2^6 f(x)\, dx = 3(10) = 30$

25. $y = 6$ on $[4, 10]$. $\left(\textbf{Note: } \Delta x = \dfrac{10 - 4}{n} = \dfrac{6}{n}, \|\Delta\| \to 0 \text{ as } n \to \infty\right)$

$$\sum_{i=1}^{n} f(c_i)\, \Delta x_i = \sum_{i=1}^{n} f\left(4 + \frac{6i}{n}\right)\left(\frac{6}{n}\right) = \sum_{i=1}^{n} 6\left(\frac{6}{n}\right) = \sum_{i=1}^{n} \frac{36}{n} = 36$$

$$\int_4^{10} 6\, dx = \lim_{n \to \infty} 36 = 36$$

27. $y = x^3$ on $[-1, 1]$. $\left(\textbf{Note: } \Delta x = \dfrac{1 - (-1)}{n} = \dfrac{2}{n}, \|\Delta\| \to 0 \text{ as } n \to \infty\right)$

$$\sum_{i=1}^{n} f(c_i)\, \Delta x_i = \sum_{i=1}^{n} f\left(-1 + \frac{2i}{n}\right)\left(\frac{2}{n}\right) = \sum_{i=1}^{n}\left(-1 + \frac{2i}{n}\right)^3\left(\frac{2}{n}\right) = \sum_{i=1}^{n}\left[-1 + \frac{6i}{n} - \frac{12i^2}{n^2} + \frac{8i^3}{n^3}\right]\left(\frac{2}{n}\right)$$

$$= -2 + \frac{12}{n^2}\sum_{i=1}^{n} i - \frac{24}{n^3}\sum_{i=1}^{n} i^2 + \frac{16}{n^4}\sum_{i=1}^{n} i^3$$

$$= -2 + 6\left(1 + \frac{1}{n}\right) - 4\left(2 + \frac{3}{n} + \frac{1}{n^2}\right) + 4\left(1 + \frac{2}{n} + \frac{1}{n^2}\right) = \frac{2}{n}$$

$$\int_{-1}^{1} x^3\, dx = \lim_{n \to \infty} \frac{2}{n} = 0$$

29. $y = x^2 + 1$ on $[1, 2]$. $\left(\textbf{Note: } \Delta x = \frac{2-1}{n} = \frac{1}{n}, \|\Delta\| \to 0 \text{ as } n \to \infty\right)$

$$\sum_{i=1}^{n} f(c_i)\,\Delta x_i = \sum_{i=1}^{n} f\left(1 + \frac{i}{n}\right)\left(\frac{1}{n}\right) = \sum_{i=1}^{n}\left[\left(1 + \frac{i}{n}\right)^2 + 1\right]\left(\frac{1}{n}\right) = \sum_{i=1}^{n}\left[1 + \frac{2i}{n} + \frac{i^2}{n^2} + 1\right]\left(\frac{1}{n}\right)$$

$$= 2 + \frac{2}{n^2}\sum_{i=1}^{n} i + \frac{1}{n^3}\sum_{i=1}^{n} i^2 = 2 + \left(1 + \frac{1}{n}\right) + \frac{1}{6}\left(2 + \frac{3}{n} + \frac{1}{n^2}\right) = \frac{10}{3} + \frac{3}{2n} + \frac{1}{6n^2}$$

$$\int_1^2 (x^2 + 1)\,dx = \lim_{n\to\infty}\left(\frac{10}{3} + \frac{3}{2n} + \frac{1}{6n^2}\right) = \frac{10}{3}$$

31. $\displaystyle \lim_{\|\Delta\|\to 0} \sum_{i=1}^{n} (3c_i + 10)\,\Delta x_i = \int_{-1}^{5} (3x + 10)\,dx$

on the interval $[-1, 5]$.

33. $\displaystyle \lim_{\|\Delta\|\to 0} \sum_{i=1}^{n} \sqrt{c_i^2 + 4}\,\Delta x_i = \int_0^3 \sqrt{x^2 + 4}\,dx$

on the interval $[0, 3]$.

35. $\displaystyle \int_0^3 x\sqrt{3 - x}\,dx$

n	4	8	12	16	20
$L(n)$	3.6830	3.9956	4.0707	4.1016	4.1177
$M(n)$	4.3082	4.2076	4.1838	4.1740	4.1690
$R(n)$	3.6830	3.9956	4.0707	4.1016	4.1177

37. $\displaystyle \int_0^{\pi/2} \sin^2 x\,dx$

n	4	8	12	16	20
$L(n)$	0.5890	0.6872	0.7199	0.7363	0.7461
$M(n)$	0.7854	0.7854	0.7854	0.7854	0.7854
$R(n)$	0.9817	0.8836	0.8508	0.8345	0.8247

39. The left endpoint approximation will be greater than the actual area: $>$

41. Because the curve is concave upward, the midpoint approximation will be less than the actual area: $<$

43. (a) Quarter circle below x-axis: $-\frac{1}{4}\pi r^2 = -\frac{1}{4}\pi(2)^2 = -\pi$

(b) Triangle: $\frac{1}{2}bh = \frac{1}{2}(4)(2) = 4$

(c) Triangle + Semicircle below x-axis: $-\frac{1}{2}(2)(1) - \frac{1}{2}\pi(2)^2 = -(1 + 2\pi)$

(d) Sum of parts (b) and (c): $4 - (1 + 2\pi) = 3 - 2\pi$

(e) Sum of absolute values of (b) and (c): $4 + (1 + 2\pi) = 5 + 2\pi$

(f) Answer to (d) plus $2(10) = 20$: $(3 - 2\pi) + 20 = 23 - 2\pi$

45.

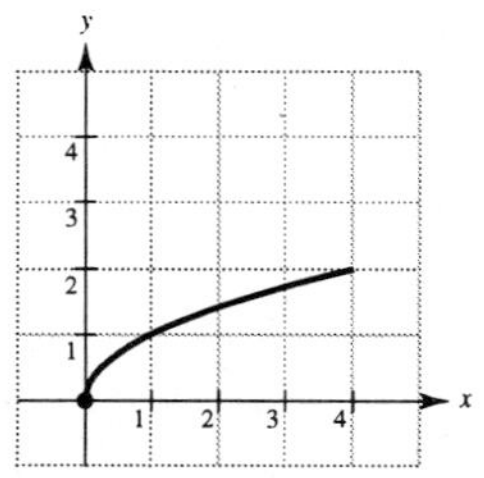

a. $A \approx 5$ square units

47. True

49. True

51. False

$$\int_0^2 (-x)\,dx = -2$$

53. $f(x) = x^2 + 3x$, $[0, 8]$

$x_0 = 0, x_1 = 1, x_2 = 3, x_3 = 7, x_4 = 8$

$\Delta x_1 = 1, \Delta x_2 = 2, \Delta x_3 = 4, \Delta x_4 = 1$

$c_1 = 1, c_2 = 2, c_3 = 5, c_4 = 8$

$$\sum_{i=1}^{4} f(c_i)\,\Delta x = f(1)\,\Delta x_1 + f(2)\,\Delta x_2 + f(5)\,\Delta x_3 + f(8)\,\Delta x_4 = (4)(1) + (10)(2) + (40)(4) + (88)(1) = 272$$

55. $f(x) = \sqrt{x}$, $y = 0$, $x = 0$, $x = 2$, $c_i = \dfrac{2i^2}{n^2}$

$$\Delta x_i = \frac{2i^2}{n^2} - \frac{2(i-1)^2}{n^2} = \frac{2(2i-1)}{n^2}$$

$$\begin{aligned}\lim_{n\to\infty}\sum_{i=1}^{n} f(c_i)\,\Delta x_i &= \lim_{n\to\infty}\sum_{i=1}^{n}\sqrt{\frac{2i^2}{n^2}}\left[\frac{2(2i-1)}{n^2}\right]\\ &= \lim_{n\to\infty}\frac{2\sqrt{2}}{n^3}\sum_{i=1}^{n}(2i^2 - i)\\ &= \lim_{n\to\infty}\frac{2\sqrt{2}}{n^3}\left[2\left(\frac{n(n+1)(2n+1)}{6}\right) - \frac{n(n+1)}{2}\right]\\ &= \lim_{n\to\infty}\frac{2\sqrt{2}}{n^3}\left[\frac{4n^3 + 3n^2 - n}{6}\right] = \lim_{n\to\infty}\sqrt{2}\left[\frac{4}{3} + \frac{1}{n} - \frac{2}{n^2}\right] = \frac{4\sqrt{2}}{3}\end{aligned}$$

57. $f(x) = \dfrac{1}{x-4}$

is not integrable on the interval $[3, 5]$ and f has a discontinuity at $x = 4$.

59. $f(x) = |x|/x$ is integrable on $[-1, 1]$, but is not contin-uous on $[-1, 1]$. There is discontinuity at $x = 0$. To see that

$$\int_{-1}^{1}\frac{|x|}{x}\,dx$$

is integrable, sketch a graph of the region bounded by $f(x) = |x|/x$ and the x-axis for $-1 \le x \le 1$. You see that the integral equals 0.

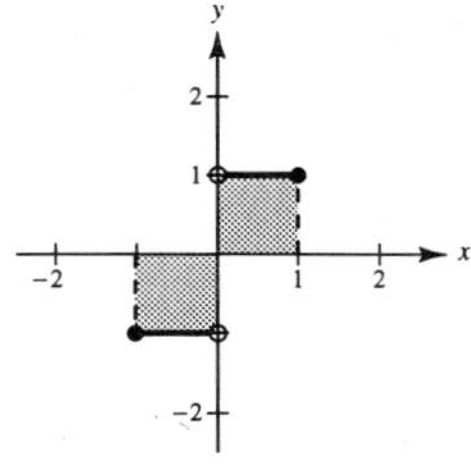

61.
$$\begin{aligned}\lim_{n\to\infty}\frac{1}{n^3}[1^2 + 2^2 + 3^2 + \cdots + n^2] &= \lim_{n\to\infty}\frac{1}{n^3}\cdot\frac{n(2n+1)(n+1)}{6}\\ &= \lim_{n\to\infty}\frac{2n^2 + 3n + 1}{6n^2} = \lim_{n\to\infty}\left(\frac{1}{3} + \frac{1}{2n} + \frac{1}{6n^2}\right) = \frac{1}{3}\end{aligned}$$

Let $f(x) = x^2$, $0 \le x \le 1$, and $\Delta x_i = 1/n$. The appropriate Riemann Sum is

$$\sum_{i=1}^{n} f(c_i)\Delta x_i = \sum_{i=1}^{n}\left(\frac{i}{n}\right)^2\frac{1}{n} = \frac{1}{n^3}\sum_{i=1}^{n} i^2.$$

63. Since $-|f(x)| \le f(x) \le |f(x)|$,

$$-\int_a^b |f(x)|\,dx \le \int_a^b f(x)\,dx \le \int_a^b |f(x)|\,dx \Rightarrow \left|\int_a^b f(x)\,dx\right| \le \int_a^b |f(x)|\,dx.$$

Section 4.4 The Fundamental Theorem of Calculus

1. $f(x) = \dfrac{4}{x^2+1}$

$\displaystyle\int_0^{\pi} \frac{4}{x^2+1}\,dx$ is positive.

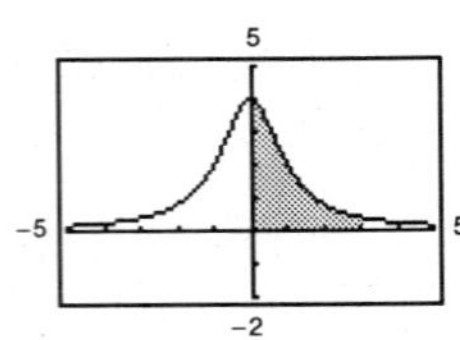

3. $f(x) = x\sqrt{x^2+1}$

$\displaystyle\int_{-2}^{2} x\sqrt{x^2+1}\,dx = 0$

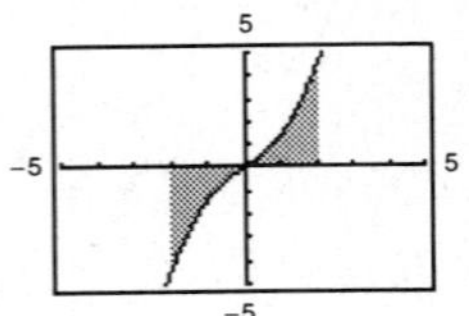

5. $\displaystyle\int_0^1 2x\,dx = \Big[x^2\Big]_0^1 = 1 - 0 = 1$

7. $\displaystyle\int_{-1}^{0} (x-2)\,dx = \left[\frac{x^2}{2} - 2x\right]_{-1}^{0} = 0 - \left(\frac{1}{2} + 2\right) = -\frac{5}{2}$

9. $\displaystyle\int_{-1}^{1} (t^2-2)\,dt = \left[\frac{t^3}{3} - 2t\right]_{-1}^{1} = \left(\frac{1}{3} - 2\right) - \left(-\frac{1}{3} + 2\right) = -\frac{10}{3}$

11. $\displaystyle\int_0^1 (2t-1)^2\,dt = \int_0^1 (4t^2 - 4t + 1)\,dt = \left[\frac{4}{3}t^3 - 2t^2 + t\right]_0^1 = \frac{4}{3} - 2 + 1 = \frac{1}{3}$

13. $\displaystyle\int_1^2 \left(\frac{3}{x^2} - 1\right) dx = \left[-\frac{3}{x} - x\right]_1^2 = \left(-\frac{3}{2} - 2\right) - (-3 - 1) = \frac{1}{2}$

15. $\displaystyle\int_1^4 \frac{u-2}{\sqrt{u}}\,du = \int_1^4 (u^{1/2} - 2u^{-1/2})\,du = \left[\frac{2}{3}u^{3/2} - 4u^{1/2}\right]_1^4 = \left[\frac{2}{3}(\sqrt{4})^3 - 4\sqrt{4}\right] - \left[\frac{2}{3} - 4\right] = \frac{2}{3}$

17. $\displaystyle\int_{-1}^{1} (\sqrt[3]{t} - 2)\,dt = \left[\frac{3}{4}t^{4/3} - 2t\right]_{-1}^{1} = \left(\frac{3}{4} - 2\right) - \left(\frac{3}{4} + 2\right) = -4$

19. $\displaystyle\int_0^1 \frac{x - \sqrt{x}}{3}\,dx = \frac{1}{3}\int_0^1 (x - x^{1/2})\,dx = \frac{1}{3}\left[\frac{x^2}{2} - \frac{2}{3}x^{3/2}\right]_0^1 = \frac{1}{3}\left(\frac{1}{2} - \frac{2}{3}\right) = -\frac{1}{18}$

21. $\displaystyle\int_{-1}^{0} (t^{1/3} - t^{2/3})\,dt = \left[\frac{3}{4}t^{4/3} - \frac{3}{5}t^{5/3}\right]_{-1}^{0} = 0 - \left(\frac{3}{4} + \frac{3}{5}\right) = -\frac{27}{20}$

23. $\displaystyle\int_0^3 |2x-3|\,dx = \int_0^{3/2} (3-2x)\,dx + \int_{3/2}^{3} (2x-3)\,dx$ $\left(\text{split up the integral at the zero } x = \frac{3}{2}\right)$

$\displaystyle = \Big[3x - x^2\Big]_0^{3/2} + \Big[x^2 - 3x\Big]_{3/2}^{3} = \left(\frac{9}{2} - \frac{9}{4}\right) - 0 + (9 - 9) - \left(\frac{9}{4} - \frac{9}{2}\right) = 2\left(\frac{9}{2} - \frac{9}{4}\right) = \frac{9}{2}$

25. $\displaystyle\int_0^{\pi} (1 + \sin x)\,dx = \Big[x - \cos x\Big]_0^{\pi} = (\pi + 1) - (0 - 1) = 2 + \pi$

27. $\displaystyle\int_{-\pi/6}^{\pi/6} \sec^2 x\,dx = \Big[\tan x\Big]_{-\pi/6}^{\pi/6} = \frac{\sqrt{3}}{3} - \left(-\frac{\sqrt{3}}{3}\right) = \frac{2\sqrt{3}}{3}$

29. $\displaystyle\int_{-\pi/3}^{\pi/3} 4\sec\theta\tan\theta\,d\theta = \Big[4\sec\theta\Big]_{-\pi/3}^{\pi/3} = 4(2) - 4(2) = 0$

31. $\displaystyle\int_0^3 10{,}000(t-6)\,dt = 10{,}000\left[\frac{t^2}{2} - 6t\right]_0^3 = -\$135{,}000$

33. $\displaystyle A = \int_0^1 (x - x^2)\,dx = \left[\frac{x^2}{2} - \frac{x^3}{3}\right]_0^1 = \frac{1}{6}$

35. $\displaystyle A = \int_0^3 (3-x)\sqrt{x}\,dx = \int_0^3 (3x^{1/2} - x^{3/2})\,dx = \left[2x^{3/2} - \frac{2}{5}x^{5/2}\right]_0^3 = \left[\frac{x\sqrt{x}}{5}(10 - 2x)\right]_0^3 = \frac{12\sqrt{3}}{5}$

37. $A = \int_0^{\pi/2} \cos x\, dx = \left[\sin x\right]_0^{\pi/2} = 1$

39. Since $y \geq 0$ on $[0, 2]$,

$$A = \int_0^2 (3x^2 + 1)\, dx = \left[x^3 + x\right]_0^2 = 8 + 2 = 10.$$

41. Since $y \geq 0$ on $[0, 2]$,

$$A = \int_0^2 (x^3 + x)\, dx = \left[\frac{x^4}{4} + \frac{x^2}{2}\right]_0^2 = 4 + 2 = 6.$$

43. $\int_0^2 \left(x - 2\sqrt{x}\right) dx = \left[\frac{x^2}{2} - \frac{4x^{3/2}}{3}\right]_0^2 = 2 - \frac{8\sqrt{2}}{3}$

$$f(c)(2 - 0) = \frac{6 - 8\sqrt{2}}{3}$$

$$c - 2\sqrt{c} = \frac{3 - 4\sqrt{2}}{3}$$

$$c - 2\sqrt{c} + 1 = \frac{3 - 4\sqrt{2}}{3} + 1$$

$$\left(\sqrt{c} - 1\right)^2 = \frac{6 - 4\sqrt{2}}{3}$$

$$\sqrt{c} - 1 = \pm\sqrt{\frac{6 - 4\sqrt{2}}{3}}$$

$$c = \left[1 \pm \sqrt{\frac{6 - 4\sqrt{2}}{3}}\right]^2$$

$$c \approx 0.4380 \text{ or } c \approx 1.7908$$

45. $\int_{-\pi/4}^{\pi/4} 2\sec^2 x\, dx = \left[2\tan x\right]_{-\pi/4}^{\pi/4} = 2(1) - 2(-1) = 4$

$$f(c)\left[\frac{\pi}{4} - \left(-\frac{\pi}{4}\right)\right] = 4$$

$$2\sec^2 c = \frac{8}{\pi}$$

$$\sec^2 c = \frac{4}{\pi}$$

$$\sec c = \pm\frac{2}{\sqrt{\pi}}$$

$$c = \pm\text{arcsec}\left(\frac{2}{\sqrt{\pi}}\right)$$

$$= \pm\arccos\frac{\sqrt{\pi}}{2} \approx \pm 0.4817$$

47. $\frac{1}{2 - (-2)}\int_{-2}^{2} (4 - x^2)\, dx = \frac{1}{4}\left[4x - \frac{1}{3}x^3\right]_{-2}^{2} = \frac{1}{4}\left[\left(8 - \frac{8}{3}\right) - \left(-8 + \frac{8}{3}\right)\right] = \frac{8}{3}$

Average value $= \frac{8}{3}$

$4 - x^2 = \frac{8}{3}$ when $x^2 = 4 - \frac{8}{3}$ or $x = \pm\frac{2\sqrt{3}}{3} \approx \pm 1.155.$

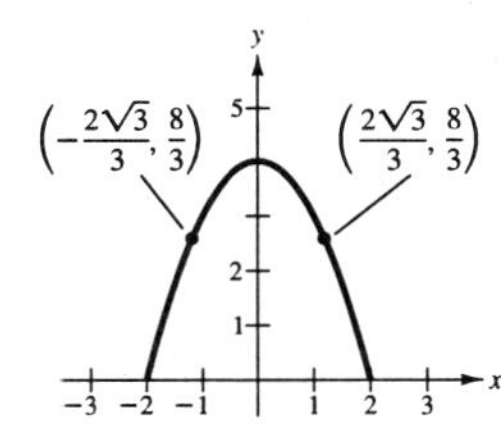

49. $\frac{1}{\pi - 0}\int_0^{\pi} \sin x\, dx = \left[-\frac{1}{\pi}\cos x\right]_0^{\pi} = \frac{2}{\pi}$

Average value $= \frac{2}{\pi}$

$$\sin x = \frac{2}{\pi}$$

$$x \approx 0.690, 2.451$$

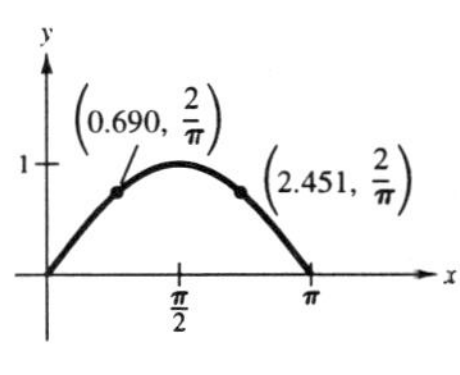

51. $\int_0^2 f(x)\, dx = -(\text{area of region } A) = -1.5$

53. $\int_0^6 |f(x)|\, dx = -\int_0^2 f(x)\, dx + \int_2^6 f(x)\, dx = 1.5 + 5.0 = 6.5$

55. $\int_0^6 [2 + f(x)]\, dx = \int_0^6 2\, dx + \int_0^6 f(x)\, dx = 12 + 3.5 = 15.5$

57. (a) $\int_1^7 f(x)\,dx = \text{Sum of the areas}$

$$= A_1 + A_2 + A_3 + A_4$$

$$= \frac{1}{2}(3+1) + \frac{1}{2}(1+2) + \frac{1}{2}(2+1) + (3)(1)$$

$$= 8$$

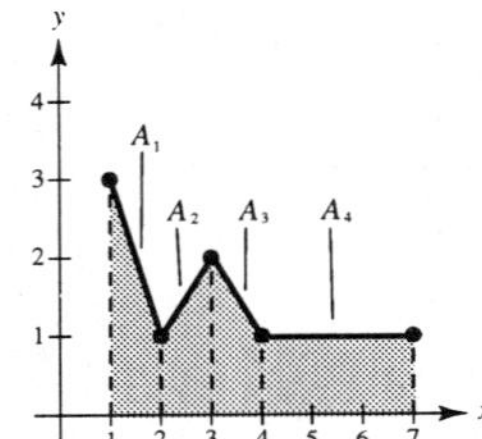

(b) Average value $= \dfrac{\int_1^7 f(x)\,dx}{7-1} = \dfrac{8}{6} = \dfrac{4}{3}$

(c) $A = 8 + (6)(2) = 20$

Average value $= \dfrac{20}{6} = \dfrac{10}{3}$

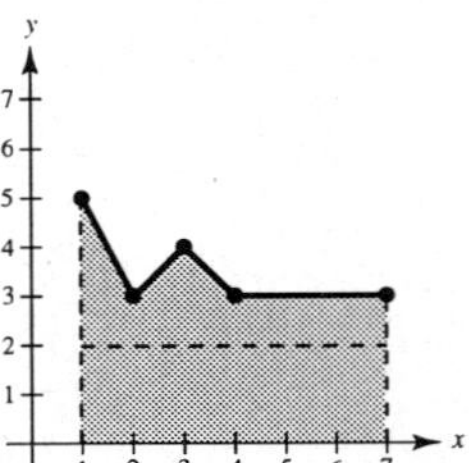

59. $R = -91.1 - 6.313x + 0.035x^2 + 45.794\sqrt{x},\ 20 \le x \le 60$

(a)

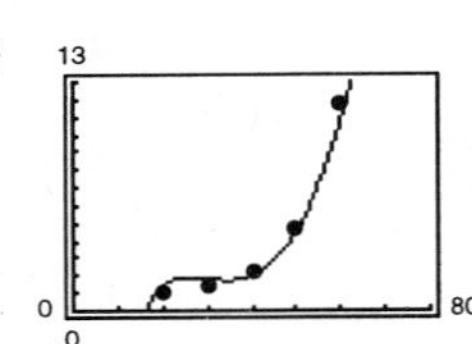

(b) $R'(x) = -6.313 + 0.07x + \dfrac{22.897}{\sqrt{x}}$

$R'(40) \approx 0.12$

$R'(50) \approx 0.43$

(c) $\dfrac{1}{10}\int_{30}^{40} R(x)\,dx \approx 1.80$ (between 30 and 40)

$\dfrac{1}{10}\int_{50}^{60} R(x)\,dx \approx 7.35$ (between 50 and 60)

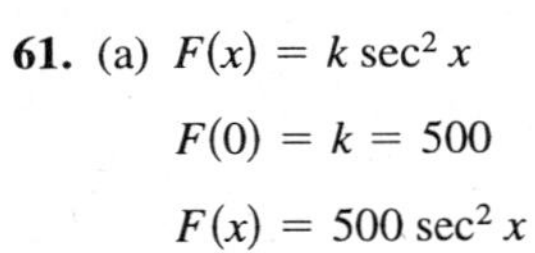

61. (a) $F(x) = k\sec^2 x$

$F(0) = k = 500$

$F(x) = 500\sec^2 x$

(b) $\dfrac{1}{\pi/3 - 0}\int_0^{\pi/3} 500\sec^2 x\,dx = \dfrac{1500}{\pi}\Big[\tan x\Big]_0^{\pi/3}$

$$= \frac{1500}{\pi}\left(\sqrt{3} - 0\right)$$

$$\approx 826.99 \text{ newtons}$$

$$\approx 827 \text{ newtons}$$

63. (a) histogram

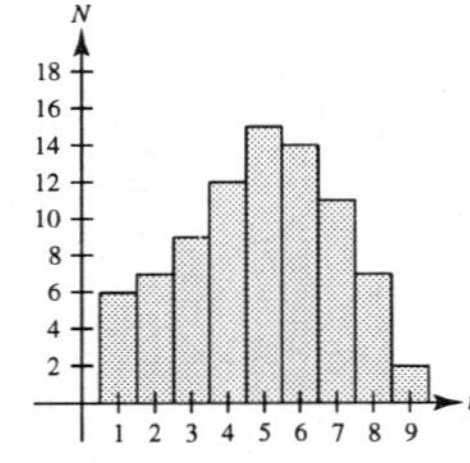

(b) $[6 + 7 + 9 + 12 + 15 + 14 + 11 + 7 + 2]60 = (83)60 = 4980$ customers

(c) Using a graphing utility, you obtain

$$N(t) = -0.084175t^3 + 0.63492t^2 + 0.79052 + 4.10317.$$

(d)

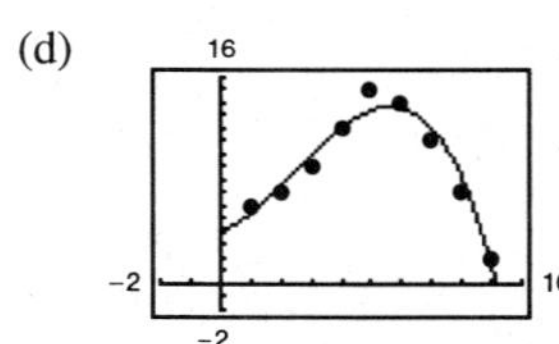

(e) $\int_0^9 N(t)\,dt \approx 85.162$

The estimated number of customers is $(85.162)(60) \approx 5110$.

(f) Between 3 P.M. and 7 P.M., the number of customers is approximately

$$\left(\int_3^7 N(t)\,dt\right)(60) \approx (50.28)(60) \approx 3017.$$

Hence, $3017/240 \approx 12.6$ per minute.

65. (a) $v = -8.61 \times 10^{-4}t^3 + 0.0782t^2 - 0.208t + 0.0952$

(b)

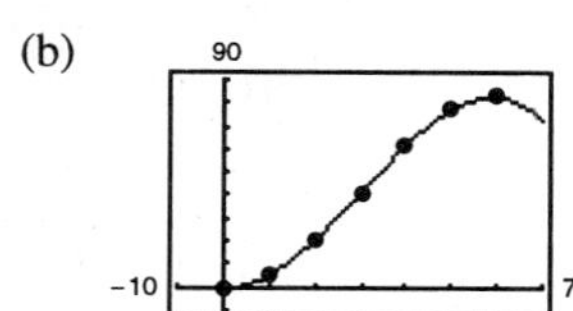

(c) $\displaystyle\int_0^{60} v(t)\,dt = \left[\frac{-8.61 \times 10^{-4}t^4}{4} + \frac{0.0782t^3}{3} - \frac{0.208t^2}{2} + 0.0952t\right]_0^{60} = 2472 \text{ meters}$

67. (a) $\displaystyle\int_0^x (t+2)\,dt = \left[\frac{t^2}{2} + 2t\right]_0^x = \frac{1}{2}x^2 + 2x$

(b) $\displaystyle\frac{d}{dx}\left[\frac{1}{2}x^2 + 2x\right] = x + 2$

69. (a) $\displaystyle\int_8^x \sqrt[3]{t}\,dt = \left[\frac{3}{4}t^{4/3}\right]_8^x = \frac{3}{4}(x^{4/3} - 16) = \frac{3}{4}x^{4/3} - 12$

(b) $\displaystyle\frac{d}{dx}\left[\frac{3}{4}x^{4/3} - 12\right] = x^{1/3} = \sqrt[3]{x}$

71. (a) $\displaystyle\int_{\pi/4}^x \sec^2 t\,dt = \Big[\tan t\Big]_{\pi/4}^x = \tan x - 1$

(b) $\displaystyle\frac{d}{dx}[\tan x - 1] = \sec^2 x$

73. $\displaystyle F(x) = \int_{-2}^x (t^2 - 2t)\,dt$

$F'(x) = x^2 - 2x$

75. $\displaystyle F(x) = \int_{-1}^x \sqrt{t^4 + 1}\,dt$

$F'(x) = \sqrt{x^4 + 1}$

77. $\displaystyle F(x) = \int_0^x t\cos t\,dt$

$F'(x) = x\cos x$

79. $\displaystyle F(x) = \int_x^{x+2} (4t + 1)\,dt$

$\displaystyle = \Big[2t^2 + t\Big]_x^{x+2}$

$= [2(x+2)^2 + (x+2)] - [2x^2 + x]$

$= 8x + 10$

$F'(x) = 8$

Alternate solution:

$\displaystyle F(x) = \int_x^{x+2} (4t+1)\,dt$

$\displaystyle = \int_x^0 (4t+1)\,dt + \int_0^{x+2} (4t+1)\,dt$

$\displaystyle = -\int_0^x (4t+1)\,dt + \int_0^{x+2} (4t+1)\,dt$

$F'(x) = -(4x + 1) + 4(x + 2) + 1 = 8$

81. $\displaystyle F(x) = \int_0^{\sin x} \sqrt{t}\,dt = \left[\frac{2}{3}t^{3/2}\right]_0^{\sin x} = \frac{2}{3}(\sin x)^{3/2}$

$F'(x) = (\sin x)^{1/2}\cos x = \cos x\sqrt{\sin x}$

Alternate solution

$\displaystyle F(x) = \int_0^{\sin x} \sqrt{t}\,dt$

$\displaystyle F'(x) = \sqrt{\sin x}\,\frac{d}{dx}(\sin x) = \sqrt{\sin x}(\cos x)$

83. $\displaystyle F(x) = \int_0^{x^3} \sin t^2\,dt$

$F'(x) = \sin(x^3)^2 \cdot 3x^2 = 3x^2 \sin x^6$

85. The extrema of F correspond to the zeros of f and the inflection point of F corresponds to the extrema of f.

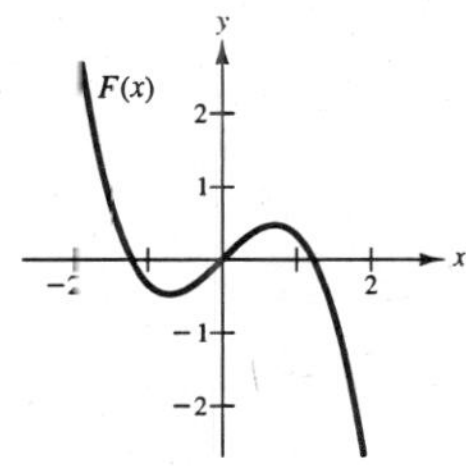

87. (a) $C(x) = 5000\left(25 + 3\int_0^x t^{1/4}\,dt\right)$

$= 5000\left(25 + 3\left[\frac{4}{5}t^{5/4}\right]_0^x\right)$

$= 5000\left(25 + \frac{12}{5}x^{5/4}\right) = 1000(125 + 12x^{5/4})$

(b) $C(1) = 1000(125 + 12(1)) = \$137,000$

$C(5) = 1000(125 + 12(5)^{5/4}) \approx \$214,721$

$C(10) = 1000(125 + 12(10)^{5/4}) \approx \$338,394$

89. True

91. False; $\int_{-1}^{1} x^{-2}\,dx = \int_{-1}^{0} x^{-2}\,dx + \int_{0}^{1} x^{-2}\,dx$

Each of these integrals is infinite. $f(x) = x^{-2}$ has a nonremovable discontinuity at $x = 0$.

93. $f(x) = \int_0^{1/x} \frac{1}{t^2+1}\,dt + \int_0^x \frac{1}{t^2+1}\,dt$

By the Second Fundamental Theorem of Calculus, we have

$$f'(x) = \frac{1}{(1/x)^2 + 1}\left(-\frac{1}{x^2}\right) + \frac{1}{x^2+1}$$

$$= -\frac{1}{1+x^2} + \frac{1}{x^2+1} = 0.$$

Since $f'(x) = 0$, $f(x)$ must be constant.

95. (a)

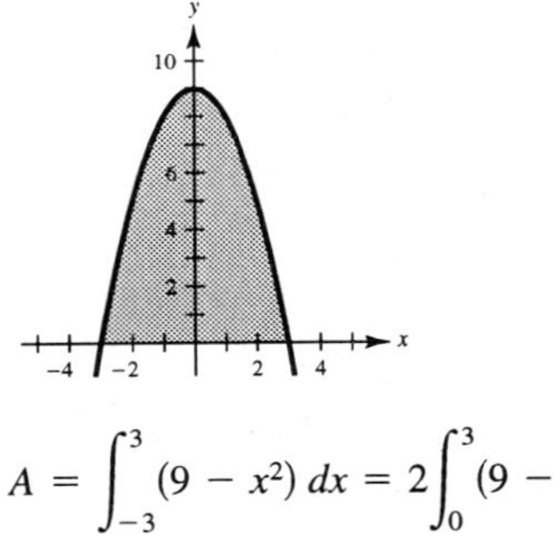

$$A = \int_{-3}^{3} (9 - x^2)\,dx = 2\int_0^3 (9 - x^2)\,dx$$

$$= 2\left[9x - \frac{x^3}{3}\right]_0^3 = 2\left[27 - \frac{27}{3}\right] = 36$$

(b) base $= 6$, height $= 9$, Area $= \frac{2}{3}(6)(9) = 36$

(c)

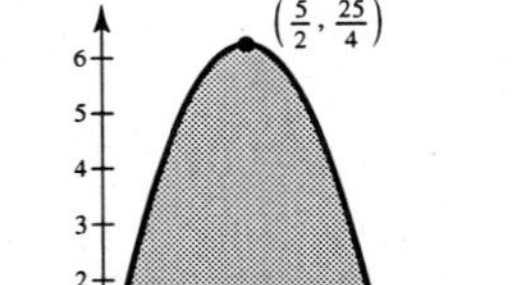

$$A = \int_0^5 (5x - x^2)\,dx = \left[\frac{5}{2}x^2 - \frac{x^3}{3}\right]_0^5 = \frac{125}{2} - \frac{125}{3} = \frac{125}{6}$$

Formula: $A = \frac{2}{3}(5)\left(\frac{25}{4}\right) = \frac{125}{6}$

97. $x(t) = (t - 1)(t - 3)^2 = t^3 - 7t^2 + 15t - 9$

$x'(t) = 3t^2 - 14t + 15$

Using a graphing utility,

$$\text{Total distance} = \int_0^5 |x'(t)|\,dt \approx 27.37 \text{ units}$$

Section 4.5 Integration by Substitution

$\int f(g(x))g'(x)\,dx$	$u = g(x)$	$du = g'(x)\,dx$
1. $\int (5x^2 + 1)^2(10x)\,dx$	$5x^2 + 1$	$10x\,dx$
3. $\int \frac{x}{\sqrt{x^2+1}}\,dx$	$x^2 + 1$	$2x\,dx$
5. $\int \tan^2 x \sec^2 x\,dx$	$\tan x$	$\sec^2 x\,dx$

7. $\int (1 + 2x)^4\, 2\,dx = \frac{(1+2x)^5}{5} + C$

Check: $\frac{d}{dx}\left[\frac{(1+2x)^5}{5} + C\right] = 2(1 + 2x)^4$

9. $\int (9 - x^2)^{1/2}(-2x)\,dx = \frac{(9-x^2)^{3/2}}{3/2} + C = \frac{2}{3}(9 - x^2)^{3/2} + C$

Check: $\frac{d}{dx}\left[\frac{2}{3}(9 - x^2)^{3/2} + C\right] = \frac{2}{3}\cdot\frac{3}{2}(9 - x^2)^{1/2}(-2x) = \sqrt{9 - x^2}(-2x)$

11. $\int x^2(x^3 - 1)^4\,dx = \frac{1}{3}\int (x^3 - 1)^4(3x^2)\,dx = \frac{1}{3}\left[\frac{(x^3-1)^5}{5}\right] + C = \frac{(x^3-1)^5}{15} + C$

Check: $\frac{d}{dx}\left[\frac{(x^3-1)^5}{15} + C\right] = \frac{5(x^3-1)^4(3x^2)}{15} = x^2(x^3 - 1)^4$

13. $\int 5x(1 - x^2)^{1/3}\,dx = -\frac{5}{2}\int (1 - x^2)^{1/3}(-2x)\,dx = -\frac{5}{2}\cdot\frac{(1-x^2)^{4/3}}{4/3} + C = -\frac{15}{8}(1 - x^2)^{4/3} + C$

Check: $\frac{d}{dx}\left[-\frac{15}{8}(1 - x^2)^{4/3} + C\right] = -\frac{15}{8}\cdot\frac{4}{3}(1 - x^2)^{1/3}(-2x) = 5x(1 - x^2)^{1/3} = 5x\sqrt[3]{1 - x^2}$

15. $\int \frac{x^2}{(1+x^3)^2}\,dx = \frac{1}{3}\int (1 + x^3)^{-2}(3x^2)\,dx = \frac{1}{3}\left[\frac{(1+x^3)^{-1}}{-1}\right] + C = -\frac{1}{3(1+x^3)} + C$

Check: $\frac{d}{dx}\left[-\frac{1}{3(1+x^3)} + C\right] = -\frac{1}{3}(-1)(1 + x^3)^{-2}(3x^2) = \frac{x^2}{(1+x^3)^2}$

17. $\int \left(1 + \frac{1}{t}\right)^3\left(\frac{1}{t^2}\right) dt = -\int \left(1 + \frac{1}{t}\right)^3\left(-\frac{1}{t^2}\right) dt = -\frac{[1 + (1/t)]^4}{4} + C$

Check: $\frac{d}{dt}\left[-\frac{[1 + (1/t)]^4}{4} + C\right] = -\frac{1}{4}(4)\left(1 + \frac{1}{t}\right)^3\left(-\frac{1}{t^2}\right) = \frac{1}{t^2}\left(1 + \frac{1}{t}\right)^3$

19. $\int \frac{1}{\sqrt{2x}}\,dx = \frac{1}{2}\int (2x)^{-1/2}\, 2\,dx = \frac{1}{2}\left[\frac{(2x)^{1/2}}{1/2}\right] + C = \sqrt{2x} + C$

Check: $\frac{d}{dx}\left[\sqrt{2x} + C\right] = \frac{1}{2}(2x)^{-1/2}(2) = \frac{1}{\sqrt{2x}}$

21. $\displaystyle\int \frac{x^2+3x+7}{\sqrt{x}}\,dx = \int (x^{3/2}+3x^{1/2}+7x^{-1/2})\,dx = \frac{2}{5}x^{5/2}+2x^{3/2}+14x^{1/2}+C = \frac{2}{5}\sqrt{x}(x^2+5x+35)+C$

Check: $\displaystyle\frac{d}{dx}\left[\frac{2}{5}x^{5/2}+2x^{3/2}+14x^{1/2}+C\right] = \frac{x^2+3x+7}{\sqrt{x}}$

23. $\displaystyle\int t^2\left(t-\frac{2}{t}\right)dt = \int (t^3-2t)\,dt = \frac{1}{4}t^4 - t^2 + C$

Check: $\displaystyle\frac{d}{dt}\left[\frac{1}{4}t^4 - t^2 + C\right] = t^3 - 2t = t^2\left(t-\frac{2}{t}\right)$

25. $\displaystyle\int (9-y)\sqrt{y}\,dy = \int (9y^{1/2}-y^{3/2})\,dy = 9\left(\frac{2}{3}y^{3/2}\right) - \frac{2}{5}y^{5/2} + C = \frac{2}{5}y^{3/2}(15-y)+C$

Check: $\displaystyle\frac{d}{dy}\left[\frac{2}{5}y^{3/2}(15-y)+C\right] = \frac{d}{dy}\left[6y^{3/2}-\frac{2}{5}y^{5/2}+C\right] = 9y^{1/2}-y^{3/2} = (9-y)\sqrt{y}$

27. $$\begin{aligned} y &= \int \left[4x + \frac{4x}{\sqrt{16-x^2}}\right]dx \\ &= 4\int x\,dx - 2\int (16-x^2)^{-1/2}(-2x)\,dx \\ &= 4\left(\frac{x^2}{2}\right) - 2\left[\frac{(16-x^2)^{1/2}}{1/2}\right] + C \\ &= 2x^2 - 4\sqrt{16-x^2} + C \end{aligned}$$

29. $$\begin{aligned} y &= \int \frac{x+1}{(x^2+2x-3)^2}\,dx \\ &= \frac{1}{2}\int (x^2+2x-3)^{-2}(2x+2)\,dx \\ &= \frac{1}{2}\left[\frac{(x^2+2x-3)^{-1}}{-1}\right] + C \\ &= -\frac{1}{2(x^2+2x-3)} + C \end{aligned}$$

31. (a)

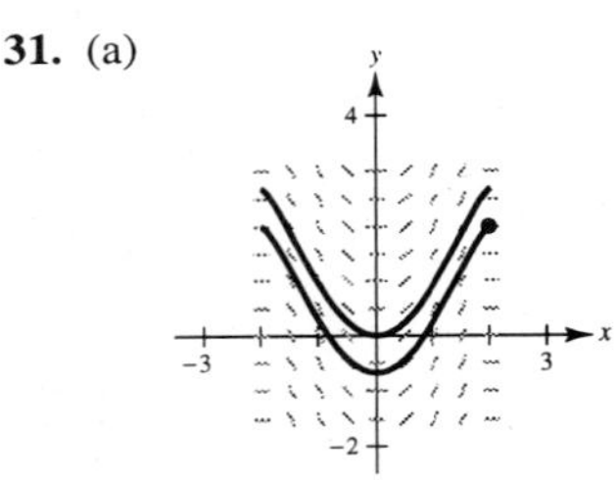

(b) $\displaystyle\frac{dy}{dx} = x\sqrt{4-x^2},\ (2, 2)$

$$\begin{aligned} y &= \int x\sqrt{4-x^2}\,dx = -\frac{1}{2}\int (4-x^2)^{1/2}(-2x\,dx) \\ &= -\frac{1}{2}\cdot\frac{2}{3}(4-x^2)^{3/2} + C = -\frac{1}{3}(4-x^2)^{3/2} + C \end{aligned}$$

$(2, 2)$: $2 = -\frac{1}{3}(4-2^2)^{3/2} + C \Rightarrow C = 2$

$$y = -\frac{1}{3}(4-x^2)^{3/2} + 2$$

33. $\displaystyle\int \sin 2x\,dx = \frac{1}{2}\int (\sin 2x)(2x)\,dx = -\frac{1}{2}\cos 2x + C$

35. $\displaystyle\int \frac{1}{\theta^2}\cos\frac{1}{\theta}\,d\theta = -\int \cos\frac{1}{\theta}\left(-\frac{1}{\theta^2}\right)d\theta = -\sin\frac{1}{\theta} + C$

37. $\displaystyle\int \sin 2x\cos 2x\,dx = \frac{1}{2}\int (\sin 2x)(2\cos 2x)\,dx = \frac{1}{2}\frac{(\sin 2x)^2}{2} + C = \frac{1}{4}\sin^2 2x + C$ OR

$\displaystyle\int \sin 2x\cos 2x\,dx = -\frac{1}{2}\int (\cos 2x)(-2\sin 2x)\,dx = -\frac{1}{2}\frac{(\cos 2x)^2}{2} + C_1 = -\frac{1}{4}\cos^2 2x + C_1$ OR

$\displaystyle\int \sin 2x\cos 2x\,dx = \frac{1}{2}\int 2\sin 2x\cos 2x\,dx = \frac{1}{2}\int \sin 4x\,dx = \frac{1}{8}\cos 4x + C_2$

39. $\displaystyle\int \tan^4 x\sec^2 x\,dx = \frac{\tan^5 x}{5} + C = \frac{1}{5}\tan^5 x + C$

41. $\displaystyle\int \frac{\csc^2 x}{\cot^3 x}\,dx = -\int (\cot x)^{-3}(-\csc^2 x)\,dx$

$$= -\frac{(\cot x)^{-2}}{-2} + C = \frac{1}{2\cot^2 x} + C = \frac{1}{2}\tan^2 x + C = \frac{1}{2}(\sec^2 x - 1) + C = \frac{1}{2}\sec^2 x + C_1$$

43. $\displaystyle\int \cot^2 x\,dx = \int (\csc^2 x - 1)\,dx = -\cot x - x + C$

45. $\displaystyle f(x) = \int \cos\frac{x}{2}\,dx = 2\sin\frac{x}{2} + C$

Since $f(0) = 3 = 2\sin 0 + C$, $C = 3$. Thus,

$$f(x) = 2\sin\frac{x}{2} + 3.$$

47. $u = x + 2,\ x = u - 2,\ dx = du$

$$\int x\sqrt{x+2}\,dx = \int (u-2)\sqrt{u}\,du = \int (u^{3/2} - 2u^{1/2})\,du$$

$$= \frac{2}{5}u^{5/2} - \frac{4}{3}u^{3/2} + C = \frac{2u^{3/2}}{15}(3u - 10) + C$$

$$= \frac{2}{15}(x+2)^{3/2}[3(x+2) - 10] + C = \frac{2}{15}(x+2)^{3/2}(3x - 4) + C$$

49. $u = 1 - x,\ x = 1 - u,\ dx = -du$

$$\int x^2\sqrt{1-x}\,dx = -\int (1-u)^2\sqrt{u}\,du = -\int (u^{1/2} - 2u^{3/2} + u^{5/2})\,du$$

$$= -\left(\frac{2}{3}u^{3/2} - \frac{4}{5}u^{5/2} + \frac{2}{7}u^{7/2}\right) + C = -\frac{2u^{3/2}}{105}(35 - 42u + 15u^2) + C$$

$$= -\frac{2}{105}(1-x)^{3/2}[35 - 42(1-x) + 15(1-x)^2] + C = -\frac{2}{105}(1-x)^{3/2}(15x^2 + 12x + 8) + C$$

51. $u = 2x - 1,\ x = \frac{1}{2}(u + 1),\ dx = \frac{1}{2}\,du$

$$\int \frac{x^2 - 1}{\sqrt{2x-1}}\,dx = \int \frac{[(1/2)(u+1)]^2 - 1}{\sqrt{u}}\,\frac{1}{2}\,du = \frac{1}{8}\int u^{-1/2}[(u^2 + 2u + 1) - 4]\,du = \frac{1}{8}\int (u^{3/2} + 2u^{1/2} - 3u^{-1/2})\,du$$

$$= \frac{1}{8}\left(\frac{2}{5}u^{5/2} + \frac{4}{3}u^{3/2} - 6u^{1/2}\right) + C = \frac{u^{1/2}}{60}(3u^2 + 10u - 45) + C$$

$$= \frac{\sqrt{2x-1}}{60}[3(2x-1)^2 + 10(2x-1) - 45] + C = \frac{1}{60}\sqrt{2x-1}(12x^2 + 8x - 52) + C$$

$$= \frac{1}{15}\sqrt{2x-1}(3x^2 + 2x - 13) + C$$

53. $u = x + 1,\ x = u - 1,\ dx = du$

$$\int \frac{-x}{(x+1) - \sqrt{x+1}}\,dx = \int \frac{-(u-1)}{u - \sqrt{u}}\,du = -\int \frac{(\sqrt{u}+1)(\sqrt{u}-1)}{\sqrt{u}(\sqrt{u}-1)}\,du = -\int (1 + u^{-1/2})\,du$$

$$= -(u + 2u^{1/2}) + C = -u - 2\sqrt{u} + C = -(x+1) - 2\sqrt{x+1} + C$$

$$= -x - 2\sqrt{x+1} - 1 + C = -\left(x + 2\sqrt{x+1}\right) + C_1$$

where $C_1 = -1 + C$.

55. Let $u = x^2 + 1$, $du = 2x\,dx$.

$$\int_{-1}^{1} x(x^2+1)^3\,dx = \frac{1}{2}\int_{-1}^{1}(x^2+1)^3(2x)\,dx = \left[\frac{1}{8}(x^2+1)^4\right]_{-1}^{1} = 0$$

57. Let $u = 2x + 1$, $du = 2\,dx$.

$$\int_0^4 \frac{1}{\sqrt{2x+1}}\,dx = \frac{1}{2}\int_0^4 (2x+1)^{-1/2}(2)\,dx = \left[\sqrt{2x+1}\right]_0^4 = \sqrt{9} - \sqrt{1} = 2$$

59. Let $u = 1 + \sqrt{x}$, $du = \dfrac{1}{2\sqrt{x}}\,dx$.

$$\int_1^9 \frac{1}{\sqrt{x}(1+\sqrt{x})^2}\,dx = 2\int_1^9 (1+\sqrt{x})^{-2}\left(\frac{1}{2\sqrt{x}}\right)dx = \left[-\frac{2}{1+\sqrt{x}}\right]_1^9 = -\frac{1}{2} + 1 = \frac{1}{2}$$

61. $u = 2 - x$, $x = 2 - u$, $dx = -du$

When $x = 1$, $u = 1$. When $x = 2$, $u = 0$.

$$\int_1^2 (x-1)\sqrt{2-x}\,dx = \int_1^0 -[(2-u)-1]\sqrt{u}\,du = \int_1^0 (u^{3/2} - u^{1/2})\,du = \left[\frac{2}{5}u^{5/2} - \frac{2}{3}u^{3/2}\right]_1^0 = -\left[\frac{2}{5} - \frac{2}{3}\right] = \frac{4}{15}$$

63. $$\int_0^{\pi/2} \cos\left(\frac{2}{3}x\right)dx = \left[\frac{3}{2}\sin\left(\frac{2}{3}x\right)\right]_0^{\pi/2} = \frac{3}{2}\left(\frac{\sqrt{3}}{2}\right) = \frac{3\sqrt{3}}{4}$$

65. $u = x + 1$, $x = u - 1$, $dx = du$

When $x = 0$, $u = 1$. When $x = 7$, $u = 8$.

$$\text{Area} = \int_0^7 x\sqrt[3]{x+1}\,dx = \int_1^8 (u-1)\sqrt[3]{u}\,du$$

$$= \int_1^8 (u^{4/3} - u^{1/3})\,du = \left[\frac{3}{7}u^{7/3} - \frac{3}{4}u^{4/3}\right]_1^8 = \left(\frac{384}{7} - 12\right) - \left(\frac{3}{7} - \frac{3}{4}\right) = \frac{1209}{28}$$

67. $$A = \int_0^{\pi} (2\sin x + \sin 2x)\,dx = -\left[2\cos x + \frac{1}{2}\cos 2x\right]_0^{\pi} = 4$$

69. $$\text{Area} = \int_{\pi/2}^{2\pi/3} \sec^2\left(\frac{x}{2}\right)dx = 2\int_{\pi/2}^{2\pi/3} \sec^2\left(\frac{x}{2}\right)\left(\frac{1}{2}\right)dx = \left[2\tan\left(\frac{x}{2}\right)\right]_{\pi/2}^{2\pi/3} = 2(\sqrt{3} - 1)$$

71. $$\int_0^4 \frac{x}{\sqrt{2x+1}}\,dx \approx 3.333 = \frac{10}{3}$$

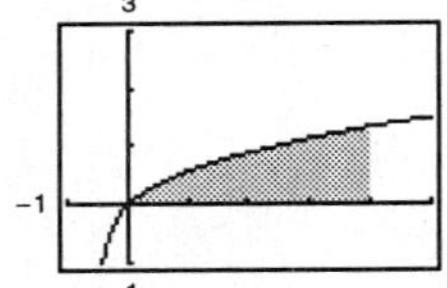

73. $$\int_3^7 x\sqrt{x-3}\,dx \approx 28.8$$

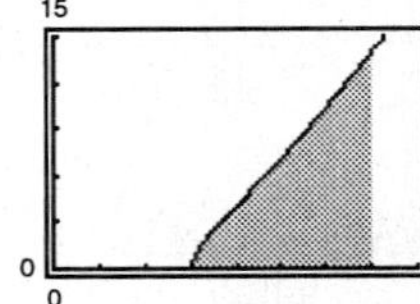

75. $$\int_0^3 \left(\theta + \cos\frac{\theta}{6}\right)d\theta \approx 7.377$$

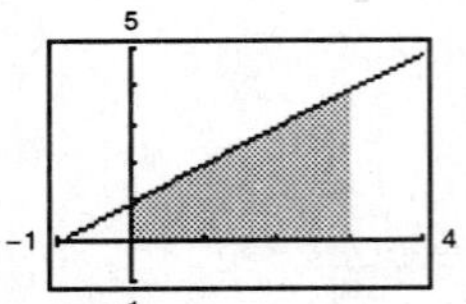

77. $$\int (2x-1)^2\,dx = \frac{1}{2}\int (2x-1)^2\,2\,dx = \frac{1}{6}(2x-1)^3 + C_1 = \frac{4}{3}x^3 - 2x^2 + x - \frac{1}{6} + C_1$$

$$\int (2x-1)^2\,dx = \int (4x^2 - 4x + 1)\,dx = \frac{4}{3}x^3 - 2x^2 + x + C_2$$

They differ by a constant: $C_2 = C_1 - \dfrac{1}{6}$.

79. $\int_0^2 x^2\,dx = \left[\frac{x^3}{3}\right]_0^2 = \frac{8}{3}$; the function x^2 is an even function.

(a) $\int_{-2}^0 x^2\,dx = \int_0^2 x^2\,dx = \frac{8}{3}$

(b) $\int_{-2}^2 x^2\,dx = 2\int_0^2 x^2\,dx = \frac{16}{3}$

(c) $\int_0^2 (-x^2)\,dx = -\int_0^2 x^2\,dx = -\frac{8}{3}$

(d) $\int_{-2}^0 3x^2\,dx = 3\int_0^2 x^2\,dx = 8$

81. $\int_{-4}^4 (x^3 + 6x^2 - 2x - 3)\,dx = \int_{-4}^4 (x^3 - 2x)\,dx + \int_{-4}^4 (6x^2 - 3)\,dx = 0 + 2\int_0^4 (6x^2 - 3)\,dx = 2\left[2x^3 - 3x\right]_0^4 = 232$

83. $\frac{dV}{dt} = \frac{k}{(t+1)^2}$

$V(t) = \int \frac{k}{(t+1)^2}\,dt = -\frac{k}{t+1} + C$

$V(0) = -k + C = 500{,}000$

$V(1) = -\frac{1}{2}k + C = 400{,}000$

Solving this system yields $k = -200{,}000$ and $C = 300{,}000$. Thus,

$$V(t) = \frac{200{,}000}{t+1} + 300{,}000.$$

When $t = 4$, $V(4) = \$340{,}000$.

85. $\frac{dC}{dx} = \frac{12}{\sqrt[3]{12x+1}}$

(a) $C(x) = \int \frac{12}{\sqrt[3]{12x+1}}\,dx$

$= \int (12x+1)^{-1/3}\,12\,dx = \frac{3}{2}(12x+1)^{2/3} + C_1$

$C(13) \approx 43.65 + C_1 = 100$

$C_1 = 56.35$

$C(x) = \frac{3}{2}(12x+1)^{2/3} + 56.35$

(b) 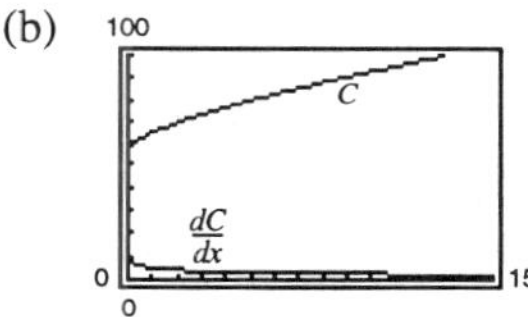

87. $\frac{1}{b-a}\int_a^b \left[74.50 + 43.75\sin\frac{\pi t}{6}\right] dt = \frac{1}{b-a}\left[74.50t - \frac{262.5}{\pi}\cos\frac{\pi t}{6}\right]_a^b$

(a) $\frac{1}{3}\left[74.50t - \frac{262.5}{\pi}\cos\frac{\pi t}{6}\right]_0^3 = \frac{1}{3}\left(223.5 + \frac{262.5}{\pi}\right) \approx 102.352$ thousand units

(b) $\frac{1}{3}\left[74.50t - \frac{262.5}{\pi}\cos\frac{\pi t}{6}\right]_3^6 = \frac{1}{3}\left(447 + \frac{262.5}{\pi} - 223.5\right) \approx 102.352$ thousand units

(c) $\frac{1}{12}\left[74.50t - \frac{262.5}{\pi}\cos\frac{\pi t}{6}\right]_0^{12} = \frac{1}{12}\left(894 - \frac{262.5}{\pi} + \frac{262.5}{\pi}\right) = 74.5$ thousand units

89. $\frac{1}{b-a}\int_a^b [2\sin(60\pi t) + \cos(120\pi t)]\,dt = \frac{1}{b-a}\left[-\frac{1}{30\pi}\cos(60\pi t) + \frac{1}{120\,\pi}\sin(120\pi t)\right]_a^b$

(a) $\frac{1}{(1/60) - 0}\left[-\frac{1}{30\pi}\cos(60\pi t) + \frac{1}{120\pi}\sin(120\pi t)\right]_0^{1/60} = 60\left[\left(\frac{1}{30\pi} + 0\right) - \left(-\frac{1}{30\pi}\right)\right] = \frac{4}{\pi} \approx 1.273$ amps

(b) $\frac{1}{(1/240) - 0}\left[-\frac{1}{30\pi}\cos(60\pi t) + \frac{1}{120\pi}\sin(120\pi t)\right]_0^{1/240} = 240\left[\left(-\frac{1}{30\sqrt{2}\pi} + \frac{1}{120\pi}\right) - \left(-\frac{1}{30\pi}\right)\right]$

$= \frac{2}{\pi}(5 - 2\sqrt{2}) \approx 1.382$ amps

(c) $\frac{1}{(1/30) - 0}\left[-\frac{1}{30\pi}\cos(60\pi t) + \frac{1}{120\pi}\sin(120\pi t)\right]_0^{1/30} = 30\left[\left(\frac{1}{30\pi}\right) - \left(-\frac{1}{30\pi}\right)\right] = 0$ amps

91. False

$$\int (2x+1)^2\,dx = \frac{1}{2}\int (2x+1)^2\,2\,dx = \frac{1}{6}(2x+1)^3 + C$$

93. True

$$\int_{-10}^{10} (ax^3 + bx^2 + cx + d)\,dx = \underbrace{\int_{-10}^{10} (ax^3 + cx)\,dx}_{\text{Odd}} + \underbrace{\int_{-10}^{10} (bx^2 + d)\,dx}_{\text{Even}} = 0 + 2\int_0^{10} (bx^2 + d)\,dx$$

95. True

$$4\int \sin x \cos x\,dx = 2\int \sin 2x\,dx = -\cos 2x + C$$

97. Let $u = x + h$, then $du = dx$. When $x = a$, $u = a + h$. When $x = b$, $u = b + h$. Thus,

$$\int_a^b f(x+h)\,dx = \int_{a+h}^{b+h} f(u)\,du = \int_{a+h}^{b+h} f(x)\,dx.$$

Section 4.6 Numerical Integration

1. Exact: $\displaystyle\int_0^2 x^2\,dx = \left[\frac{1}{3}x^3\right]_0^2 = \frac{8}{3} \approx 2.6667$

Trapezoidal: $\displaystyle\int_0^2 x^2\,dx \approx \frac{1}{4}\left[0 + 2\left(\frac{1}{2}\right)^2 + 2(1)^2 + 2\left(\frac{3}{2}\right)^2 + (2)^2\right] = \frac{11}{4} = 2.7500$

Simpson's: $\displaystyle\int_0^2 x^2\,dx \approx \frac{1}{6}\left[0 + 4\left(\frac{1}{2}\right)^2 + 2(1)^2 + 4\left(\frac{3}{2}\right)^2 + (2)^2\right] = \frac{8}{3} \approx 2.6667$

3. Exact: $\displaystyle\int_0^2 x^3\,dx = \left[\frac{x^4}{4}\right]_0^2 = 4.000$

Trapezoidal: $\displaystyle\int_0^2 x^3\,dx \approx \frac{1}{4}\left[0 + 2\left(\frac{1}{2}\right)^3 + 2(1)^3 + 2\left(\frac{3}{2}\right)^3 + (2)^3\right] = \frac{17}{4} = 4.2500$

Simpson's: $\displaystyle\int_0^2 x^3\,dx \approx \frac{1}{6}\left[0 + 4\left(\frac{1}{2}\right)^3 + 2(1)^3 + 4\left(\frac{3}{2}\right)^3 + (2)^3\right] = \frac{24}{6} = 4.0000$

5. Exact: $\displaystyle\int_0^2 x^3\,dx = \left[\frac{1}{4}x^4\right]_0^2 = 4.0000$

Trapezoidal: $\displaystyle\int_0^2 x^3\,dx \approx \frac{1}{8}\left[0 + 2\left(\frac{1}{4}\right)^3 + 2\left(\frac{2}{4}\right)^3 + 2\left(\frac{3}{4}\right)^3 + 2(1)^3 + 2\left(\frac{5}{4}\right)^3 + 2\left(\frac{6}{4}\right)^3 + 2\left(\frac{7}{4}\right)^3 + 8\right] = 4.0625$

Simpson's: $\displaystyle\int_0^2 x^3\,dx \approx \frac{1}{12}\left[0 + 4\left(\frac{1}{4}\right)^3 + 2\left(\frac{2}{4}\right)^3 + 4\left(\frac{3}{4}\right)^3 + 2(1)^3 + 4\left(\frac{5}{4}\right)^3 + 2\left(\frac{6}{4}\right)^3 + 4\left(\frac{7}{4}\right)^3 + 8\right] = 4.0000$

7. Exact: $\displaystyle\int_4^9 \sqrt{x}\,dx = \left[\frac{2}{3}x^{3/2}\right]_4^9 = 18 - \frac{16}{3} = \frac{38}{3} \approx 12.6667$

Trapezoidal: $\displaystyle\int_4^9 \sqrt{x}\,dx \approx \frac{5}{16}\left[2 + 2\sqrt{\frac{37}{8}} + 2\sqrt{\frac{21}{4}} + 2\sqrt{\frac{47}{8}} + 2\sqrt{\frac{26}{4}} + 2\sqrt{\frac{57}{8}} + 2\sqrt{\frac{31}{4}} + 2\sqrt{\frac{67}{8}} + 3\right]$

≈ 12.6640

Simpson's: $\displaystyle\int_4^9 \sqrt{x}\,dx \approx \frac{5}{24}\left[2 + 4\sqrt{\frac{37}{8}} + \sqrt{21} + 4\sqrt{\frac{47}{8}} + \sqrt{26} + 4\sqrt{\frac{57}{8}} + \sqrt{31} + 4\sqrt{\frac{67}{8}} + 3\right] \approx 12.6667$

9. Exact: $\int_1^2 \frac{1}{(x+1)^2}\,dx = \left[-\frac{1}{x+1}\right]_1^2 = -\frac{1}{3} + \frac{1}{2} = \frac{1}{6} \approx 0.1667$

Trapezoidal: $\int_1^2 \frac{1}{(x+1)^2}\,dx \approx \frac{1}{8}\left[\frac{1}{4} + 2\left(\frac{1}{((5/4)+1)^2}\right) + 2\left(\frac{1}{((3/2)+1)^2}\right) + 2\left(\frac{1}{((7/4)+1)^2}\right) + \frac{1}{9}\right]$

$= \frac{1}{8}\left(\frac{1}{4} + \frac{32}{81} + \frac{8}{25} + \frac{32}{121} + \frac{1}{9}\right) \approx 0.1676$

Simpson's: $\int_1^2 \frac{1}{(x+1)^2}\,dx \approx \frac{1}{12}\left[\frac{1}{4} + 4\left(\frac{1}{((5/4)+1)^2}\right) + 2\left(\frac{1}{((3/2)+1)^2}\right) + 4\left(\frac{1}{((7/4)+1)^2}\right) + \frac{1}{9}\right]$

$= \frac{1}{12}\left(\frac{1}{4} + \frac{64}{81} + \frac{8}{25} + \frac{64}{121} + \frac{1}{9}\right) \approx 0.1667$

11. Trapezoidal: $\int_0^2 \sqrt{1+x^3}\,dx \approx \frac{1}{4}[1 + 2\sqrt{1+(1/8)} + 2\sqrt{2} + 2\sqrt{1+(27/8)} + 3] \approx 3.283$

Simpson's: $\int_0^2 \sqrt{1+x^3}\,dx \approx \frac{1}{6}[1 + 4\sqrt{1+(1/8)} + 2\sqrt{2} + 4\sqrt{1+(27/8)} + 3] \approx 3.240$

Graphing utility: 3.241

13. $\int_0^1 \sqrt{x}\sqrt{1-x}\,dx = \int_0^1 \sqrt{x(1-x)}\,dx$

Trapezoidal: $\int_0^1 \sqrt{x(1-x)}\,dx \approx \frac{1}{8}\left[0 + 2\sqrt{\frac{1}{4}\left(1-\frac{1}{4}\right)} + 2\sqrt{\frac{1}{2}\left(1-\frac{1}{2}\right)} + 2\sqrt{\frac{3}{4}\left(1-\frac{3}{4}\right)}\right] \approx 0.342$

Simpson's: $\int_0^1 \sqrt{x(1-x)}\,dx \approx \frac{1}{12}\left[0 + 4\sqrt{\frac{1}{4}\left(1-\frac{1}{4}\right)} + 2\sqrt{\frac{1}{2}\left(1-\frac{1}{2}\right)} + 4\sqrt{\frac{3}{4}\left(1-\frac{3}{4}\right)}\right] \approx 0.372$

Graphing utility: 0.393

15. Trapezoidal: $\int_0^{\sqrt{\pi/2}} \cos(x^2)\,dx \approx \frac{\sqrt{\pi/2}}{8}\left[\cos 0 + 2\cos\left(\frac{\sqrt{\pi/2}}{4}\right)^2 + 2\cos\left(\frac{\sqrt{\pi/2}}{2}\right)^2 + 2\cos\left(\frac{\sqrt{\pi/2}}{4}\right)^2 + \cos\left(\sqrt{\frac{\pi}{2}}\right)^2\right]$

≈ 0.957

Simpson's: $\int_0^{\sqrt{\pi/2}} \cos(x^2)\,dx \approx \frac{\sqrt{\pi/2}}{12}\left[\cos 0 + 4\cos\left(\frac{\sqrt{\pi/2}}{4}\right)^2 + 2\cos\left(\frac{\sqrt{\pi/2}}{2}\right)^2 + 4\cos\left(\frac{\sqrt{\pi/2}}{4}\right)^2 + \cos\left(\sqrt{\frac{\pi}{2}}\right)^2\right]$

≈ 0.978

Graphing utility: 0.977

17. Trapezoidal: $\int_1^{1.1} \sin x^2\,dx \approx \frac{1}{80}[\sin(1) + 2\sin(1.025)^2 + 2\sin(1.05)^2 + 2\sin(1.075)^2 + \sin(1.1)^2] \approx 0.089$

Simpson's: $\int_1^{1.1} \sin x^2\,dx \approx \frac{1}{120}[\sin(1) + 4\sin(1.025)^2 + 2\sin(1.05)^2 + 4\sin(1.075)^2 + \sin(1.1)^2] \approx 0.089$

Graphing utility: 0.089

19. Trapezoidal: $\int_0^{\pi/4} x\tan x\,dx \approx \frac{\pi}{32}\left[0 + 2\left(\frac{\pi}{16}\right)\tan\left(\frac{\pi}{16}\right) + 2\left(\frac{2\pi}{16}\right)\tan\left(\frac{2\pi}{16}\right) + 2\left(\frac{3\pi}{16}\right)\tan\left(\frac{3\pi}{16}\right) + \frac{\pi}{4}\right] \approx 0.194$

Simpson's: $\int_0^{\pi/4} x\tan x\,dx \approx \frac{\pi}{48}\left[0 + 4\left(\frac{\pi}{16}\right)\tan\left(\frac{\pi}{16}\right) + 2\left(\frac{2\pi}{16}\right)\tan\left(\frac{2\pi}{16}\right) + 4\left(\frac{3\pi}{16}\right)\tan\left(\frac{3\pi}{16}\right) + \frac{\pi}{4}\right] \approx 0.186$

Graphing utility: 0.186

21. $f(x) = x^3$

$f'(x) = 3x^2$

$f''(x) = 6x$

$f'''(x) = 6$

$f^{(4)}(x) = 0$

(a) Trapezoidal: Error $\le \dfrac{(2-0)^3}{12(4^2)}(12) = 0.5$ since

$f''(x)$ is maximum in $[0, 2]$ when $x = 2$.

(b) Simpson's: Error $\le \dfrac{(2-0)^5}{180(4^4)}(0) = 0$ since

$f^{(4)}(x) = 0$.

23. $f''(x) = \dfrac{2}{x^3}$ in $[1, 3]$.

(a) $|f''(x)|$ is maximum when $x = 1$ and $|f''(1)| = 2$.

Trapezoidal: Error $\le \dfrac{2^3}{12n^2}(2) < 0.00001$, $n^2 > 133{,}333.33$, $n > 365.15$; let $n = 366$.

$f^{(4)}(x) = \dfrac{24}{x^5}$ in $[1, 3]$

(b) $|f^{(4)}(x)|$ is maximum when $x = 1$ and when $|f^{(4)}(1)| = 24$.

Simpson's: Error $\le \dfrac{2^5}{180n^4}(24) < 0.00001$, $n^4 > 426{,}666.67$, $n > 25.56$; let $n = 26$.

25. $f(x) = \sqrt{1+x}$

(a) $f''(x) = -\dfrac{1}{4(1+x)^{3/2}}$ in $[0, 2]$.

$|f''(x)|$ is maximum when $x = 0$ and $|f''(0)| = \dfrac{1}{4}$.

Trapezoidal: Error $\le \dfrac{8}{12n^2}\left(\dfrac{1}{4}\right) < 0.00001$, $n^2 > 16{,}666.67$, $n > 129.10$; let $n = 130$.

(b) $f^{(4)}(x) = \dfrac{-15}{16(1+x)^{7/2}}$ in $[0, 2]$

$|f^{(4)}(x)|$ is maximum when $x = 0$ and $|f^{(4)}(0)| = \dfrac{15}{16}$.

Simpson's: Error $\le \dfrac{32}{180n^4}\left(\dfrac{15}{16}\right) < 0.00001$, $n^4 > 16{,}666.67$, $n > 11.36$; let $n = 12$.

27. $f(x) = \tan(x^2)$

(a) $f''(x) = 2\sec^2(x^2)[1 + 4x^2\tan(x^2)]$ in $[0, 1]$.

$|f''(x)|$ is maximum when $x = 1$ and $|f''(1)| \approx 49.5305$.

Trapezoidal: Error $\le \dfrac{(1-0)^3}{12n^2}(49.5305) < 0.00001$, $n^2 > 412{,}754.17$, $n > 642.46$; let $n = 643$.

(b) $f^{(4)}(x) = 8\sec^2(x^2)[12x^2 + (3 + 32x^4)\tan(x^2) + 36x^2\tan^2(x^2) + 48x^4\tan^3(x^2)]$ in $[0, 1]$

$|f^{(4)}(x)|$ is maximum when $x = 1$ and $|f^{(4)}(1)| \approx 9184.4734$.

Simpson's: Error $\le \dfrac{(1-0)^5}{180n^4}(9184.4734) < 0.00001$, $n^4 > 5{,}102{,}485.22$, $n > 47.53$; let $n = 48$.

29. Let $f(x) = Ax^3 + Bx^2 + Cx + D$. Then $f^{(4)}(x) = 0$.

Simpson's: Error $\le \dfrac{(b-a)^5}{180n^4}(0) = 0$

Therefore, Simpson's Rule is exact when approximating the integral of a cubic polynomial.

Example: $\displaystyle\int_0^1 x^3\,dx = \frac{1}{6}\left[0 + 4\left(\frac{1}{2}\right)^3 + 1\right] = \frac{1}{4}$

This is the exact value of the integral.

31. $f(x) = \sqrt{2 + 3x^2}$ on $[0, 4]$.

n	$L(n)$	$M(n)$	$R(n)$	$T(n)$	$S(n)$
4	12.7771	15.3965	18.4340	15.6055	15.4845
8	14.0868	15.4480	16.9152	15.5010	15.4662
10	14.3569	15.4544	16.6197	15.4883	15.4658
12	14.5386	15.4578	16.4242	15.4814	15.4657
16	14.7674	15.4613	16.1816	15.4745	15.4657
20	14.9056	15.4628	16.0370	15.4713	15.4657

33. $f(x) = \sin\sqrt{x}$ on $[0, 4]$.

n	$L(n)$	$M(n)$	$R(n)$	$T(n)$	$S(n)$
4	2.8163	3.5456	3.7256	3.2709	3.3996
8	3.1809	3.5053	3.6356	3.4083	3.4541
10	3.2478	3.4990	3.6115	3.4296	3.4624
12	3.2909	3.4952	3.5940	3.4425	3.4674
16	3.3431	3.4910	3.5704	3.4568	3.4730
20	3.3734	3.4888	3.5552	3.4643	3.4759

35. $A = \displaystyle\int_0^{\pi/2} \sqrt{x}\cos x\,dx$

Simpson's Rule: $n = 14$

$$\int_0^{\pi/2} \sqrt{x}\cos x\,dx \approx \frac{\pi}{84}\left[\sqrt{0}\cos 0 + 4\sqrt{\frac{\pi}{28}}\cos\frac{\pi}{28} + 2\sqrt{\frac{\pi}{14}}\cos\frac{\pi}{14} + 4\sqrt{\frac{3\pi}{28}}\cos\frac{3\pi}{28} + \cdots + \sqrt{\frac{\pi}{2}}\cos\frac{\pi}{2}\right]$$
$$\approx 0.701$$

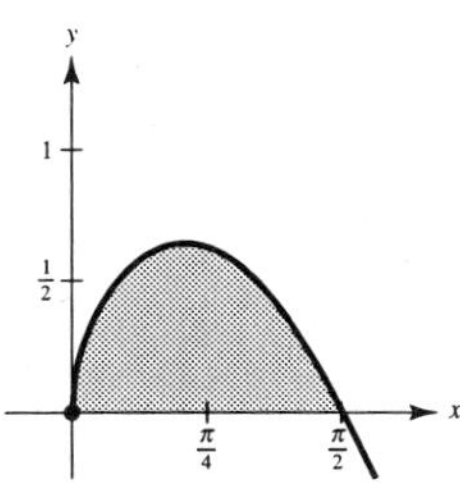

37. $W = \displaystyle\int_0^5 100x\sqrt{125 - x^3}\,dx$

Simpson's Rule: $n = 12$

$$\int_0^5 100x\sqrt{125 - x^3}\,dx \approx \frac{5}{3(12)}\left[0 + 400\left(\frac{5}{12}\right)\sqrt{125 - \left(\frac{5}{12}\right)^3} + 200\left(\frac{10}{12}\right)\sqrt{125 - \left(\frac{10}{12}\right)^3}\right.$$
$$\left. + 400\left(\frac{15}{12}\right)\sqrt{125 - \left(\frac{15}{12}\right)^3} + \cdots + 0\right] \approx 10{,}233.58 \text{ ft} \cdot \text{lb}$$

39. (a) Trapezoidal: Area $\approx \dfrac{160}{2(8)}[0 + 2(50) + 2(54) + 2(82) + 2(82) + 2(73) + 2(75) + 2(80) + 0] = 9920$ sq ft

(b) Simpson's: Area $\approx \dfrac{160}{3(8)}[0 + 4(50) + 2(54) + 4(82) + 2(82) + 4(73) + 2(75) + 4(80) + 0] = 10{,}413\frac{1}{3}$ sq ft

41. Area $\approx \dfrac{1000}{2(10)}[125 + 2(125) + 2(120) + 2(112) + 2(90) + 2(90) + 2(95) + 2(88) + 2(75) + 2(35)] = 89{,}250$ sq m

43. $\displaystyle\int_0^t \sin\sqrt{x}\,dx = 2,\ n = 10$

By trial and error, we obtain $t \approx 2.477$.

45. $\displaystyle L(x) = \int_1^x \frac{1}{t}\,dt$

(a) $\displaystyle L(1) = \int_1^1 \frac{1}{t}\,dt = 0$

(b) $L'(x) = \dfrac{1}{x}$

$L'(1) = 1$

(c) By trial and error, $x \approx 2.718$.

(d) $\displaystyle L(x_1x_2) = \int_1^{x_1x_2} \frac{1}{t}\,dt = \int_1^{x_1} \frac{1}{t}\,dt + \int_{x_1}^{x_1x_2} \frac{1}{t}\,dt$

But, this second integral is equal to $\int_1^{x_2}(1/t)\,dt$, by substitution $u = t/x_1$, $du = dt/x_1$. Hence,

$L(x_1x_2) = L(x_1) + L(x_2)$.

Review Exercises for Chapter 4

1.

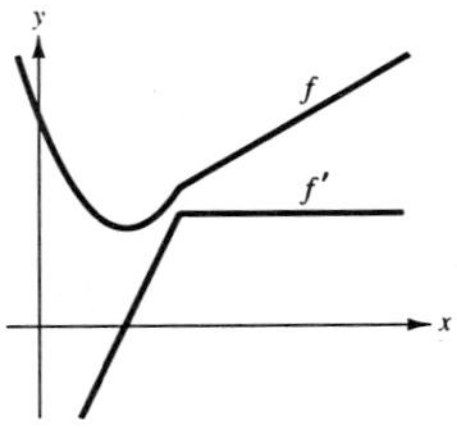

3. $\displaystyle\int (2x^2 + x - 1)\,dx = \frac{2}{3}x^3 + \frac{1}{2}x^2 - x + C$

5. $\displaystyle\int \frac{x^3+1}{x^2}\,dx = \int\left(x + \frac{1}{x^2}\right)dx = \frac{1}{2}x^2 - \frac{1}{x} + C$

7. $\displaystyle\int (4x - 3\sin x)\,dx = 2x^2 + 3\cos x + C$

9. $f'(x) = -2x,\ (-1, 1)$

$\displaystyle f(x) = \int -2x\,dx = -x^2 + C$

When $x = -1$:

$y = -1 + C = 1$

$C = 2$

$y = 2 - x^2$

11. $a(t) = a$

$\displaystyle v(t) = \int a\,dt = at + C_1$

$v(0) = 0 + C_1 = 0$ when $C_1 = 0$.

$v(t) = at$

$\displaystyle s(t) = \int at\,dt = \frac{a}{2}t^2 + C_2$

$s(0) = 0 + C_2 = 0$ when $C_2 = 0$.

$s(t) = \dfrac{a}{2}t^2$

$s(30) = \dfrac{a}{2}(30)^2 = 3600$ or

$a = \dfrac{2(3600)}{(30)^2} = 8$ ft/sec^2.

$v(30) = 8(30) = 240$ ft/sec

13. $a(t) = -32$

$v(t) = -32t + 96$

$s(t) = -16t^2 + 96t$

(a) $v(t) = -32t + 96 = 0$ when $t = 3$ sec.

(b) $s(3) = -144 + 288 = 144$ ft

(c) $v(t) = -32t + 96 = \frac{96}{2}$ when $t = \frac{3}{2}$ sec.

(d) $s\left(\frac{3}{2}\right) = -16\left(\frac{9}{4}\right) + 96\left(\frac{3}{2}\right) = 108$ ft

15. (a) $\sum_{i=1}^{10} (2i - 1)$ (b) $\sum_{i=1}^{n} i^3$ (c) $\sum_{i=1}^{10} (4i + 2)$

17. (a) $S = m\left(\frac{b}{4}\right)\left(\frac{b}{4}\right) + m\left(\frac{2b}{4}\right)\left(\frac{b}{4}\right) + m\left(\frac{3b}{4}\right)\left(\frac{b}{4}\right) + m\left(\frac{4b}{4}\right)\left(\frac{b}{4}\right) = \frac{mb^2}{16}(1 + 2 + 3 + 4) = \frac{5mb^2}{8}$

$s = m(0)\left(\frac{b}{4}\right) + m\left(\frac{b}{4}\right)\left(\frac{b}{4}\right) + m\left(\frac{2b}{4}\right)\left(\frac{b}{4}\right) + m\left(\frac{3b}{4}\right)\left(\frac{b}{4}\right) = \frac{mb^2}{16}(1 + 2 + 3) = \frac{3mb^2}{8}$

(b) $S(n) = \sum_{i=1}^{n} f\left(\frac{bi}{n}\right)\left(\frac{b}{n}\right) = \sum_{i=1}^{n} \left(\frac{mbi}{n}\right)\left(\frac{b}{n}\right) = m\left(\frac{b}{n}\right)^2 \sum_{i=1}^{n} i = \frac{mb^2}{n^2}\left(\frac{n(n + 1)}{2}\right) = \frac{mb^2(n + 1)}{2n}$

$s(n) = \sum_{i=0}^{n-1} f\left(\frac{bi}{n}\right)\left(\frac{b}{n}\right) = \sum_{i=0}^{n-1} m\left(\frac{bi}{n}\right)\left(\frac{b}{n}\right) = m\left(\frac{b}{n}\right)^2 \sum_{i=0}^{n-1} i = \frac{mb^2}{n^2}\left(\frac{(n - 1)n}{2}\right) = \frac{mb^2(n - 1)}{2n}$

(c) Area $= \lim_{n\to\infty} \frac{mb^2(n + 1)}{2n} = \lim_{n\to\infty} \frac{mb^2(n - 1)}{2n} = \frac{1}{2}mb^2 = \frac{1}{2}(b)(mb) = \frac{1}{2}$(base)(height)

(d) $\int_0^b mx\,dx = \left[\frac{1}{2}mx^2\right]_0^b = \frac{1}{2}mb^2$

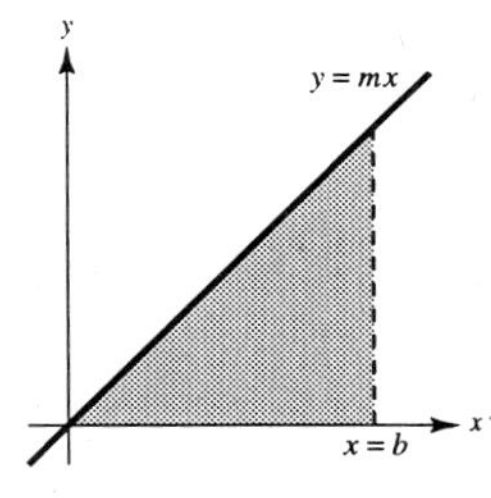

19. (a) $\int_2^6 [f(x) + g(x)]\,dx = \int_2^6 f(x)\,dx + \int_2^6 g(x)\,dx = 10 + 3 = 13$

(b) $\int_2^6 [f(x) - g(x)]\,dx = \int_2^6 f(x)\,dx - \int_2^6 g(x)\,dx = 10 - 3 = 7$

(c) $\int_2^6 [2f(x) - 3g(x)]\,dx = 2\int_2^6 f(x)\,dx - 3\int_2^6 g(x)\,dx = 2(10) - 3(3) = 11$

(d) $\int_2^6 5f(x)\,dx = 5\int_2^6 f(x)\,dx = 5(10) = 50$

21. $\int (x^2 + 1)^3\,dx = \int (x^6 + 3x^4 + 3x^2 + 1)\,dx = \frac{x^7}{7} + \frac{3}{5}x^5 + x^3 + x + C$

23. $u = x^3 + 3$, $du = 3x^2\,dx$

$\int \frac{x^2}{\sqrt{x^3 + 3}}\,dx = \int (x^3 + 3)^{-1/2}\,x^2\,dx = \frac{1}{3}\int (x^3 + 3)^{-1/2}\,3x^2\,dx = \frac{2}{3}(x^3 + 3)^{1/2} + C$

25. $u = 1 - 3x^2$, $du = -6x\,dx$

$\int x(1 - 3x^2)^4\,dx = -\frac{1}{6}\int (1 - 3x^2)^4(-6x\,dx) = -\frac{1}{30}(1 - 3x^2)^5 + C = \frac{1}{30}(3x^2 - 1)^5 + C$

27. $\int \sin^3 x \cos x\,dx = \frac{1}{4}\sin^4 x + C$

29. $\displaystyle\int \frac{\sin\theta}{\sqrt{1-\cos\theta}}\,dx = \int (1-\cos\theta)^{-1/2}\sin\theta\,d\theta = 2(1-\cos\theta)^{1/2} + C = 2\sqrt{1-\cos\theta} + C$

31. $\displaystyle\int \tan^n x\sec^2 x\,dx = \frac{\tan^{n+1} x}{n+1} + C, n \neq -1$

33. $\displaystyle\int (1+\sec\pi x)^2 \sec\pi x\tan\pi x\,dx = \frac{1}{\pi}\int (1+\sec\pi x)^2(\pi\sec\pi x\tan\pi x)\,dx = \frac{1}{3\pi}(1+\sec\pi x)^3 + C$

35. $\displaystyle\int_0^4 (2+x)\,dx = \left[2x + \frac{x^2}{2}\right]_0^4 = 8 + \frac{16}{2} = 16$

37. $\displaystyle\int_{-1}^1 (4t^3 - 2t)\,dt = \left[t^4 - t^2\right]_{-1}^1 = 0$

39. $\displaystyle\int_0^3 \frac{1}{\sqrt{1+x}}\,dx = \int_0^3 (1+x)^{-1/2}\,dx = \left[2(1+x)^{1/2}\right]_0^3 = 4 - 2 = 2$

41. $\displaystyle\int_4^9 x\sqrt{x}\,dx = \int_4^9 x^{3/2}\,dx = \left[\frac{2}{5}x^{5/2}\right]_4^9 = \frac{2}{5}\left[(\sqrt{9})^5 - (\sqrt{4})^5\right] = \frac{2}{5}(243 - 32) = \frac{422}{5}$

43. $u = 1 - y, y = 1 - u, dy = -du$

When $y = 0, u = 1$. When $y = 1, u = 0$.

$$2\pi\int_0^1 (y+1)\sqrt{1-y}\,dy = 2\pi\int_1^0 -[(1-u)+1]\sqrt{u}\,du = 2\pi\int_1^0 (u^{3/2} - 2u^{1/2})\,du = 2\pi\left[\frac{2}{5}u^{5/2} - \frac{4}{3}u^{3/2}\right]_1^0 = \frac{28\pi}{15}$$

45. $\displaystyle\int_0^\pi \cos\left(\frac{x}{2}\right)dx = 2\int_0^\pi \cos\left(\frac{x}{2}\right)\frac{1}{2}\,dx = \left[2\sin\left(\frac{x}{2}\right)\right]_0^\pi = 2$

47. $\displaystyle\int_1^3 (2x-1)\,dx = \left[x^2 - x\right]_1^3 = 6$

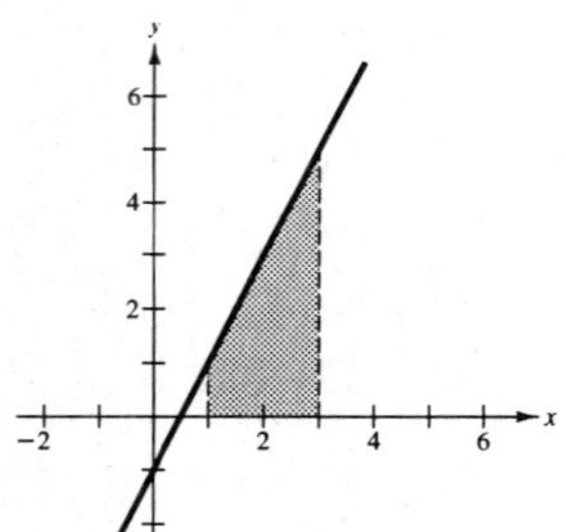

49. $\displaystyle\int_3^4 (x^2 - 9)\,dx = \left[\frac{x^3}{3} - 9x\right]_3^4$

$\displaystyle = \left(\frac{64}{3} - 36\right) - (9 - 27)$

$\displaystyle = \frac{64}{3} - \frac{54}{3} = \frac{10}{3}$

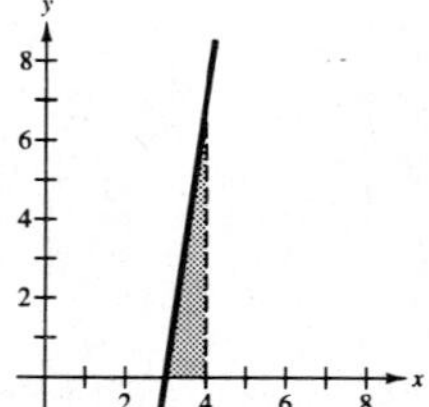

51. $\displaystyle\int_0^1 (x - x^3)\,dx = \left[\frac{x^2}{2} - \frac{x^4}{4}\right]_0^1 = \frac{1}{2} - \frac{1}{4} = \frac{1}{4}$

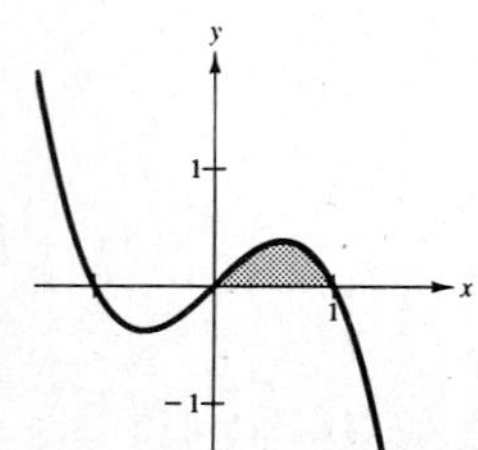

53. $\displaystyle\int_0^8 \frac{4}{\sqrt{x+1}}\,dx = 4\int_0^8 (x+1)^{-1/2}\,dx$

$\displaystyle = \Big[8\sqrt{x+1}\Big]_0^8 = 8(3-1) = 16$

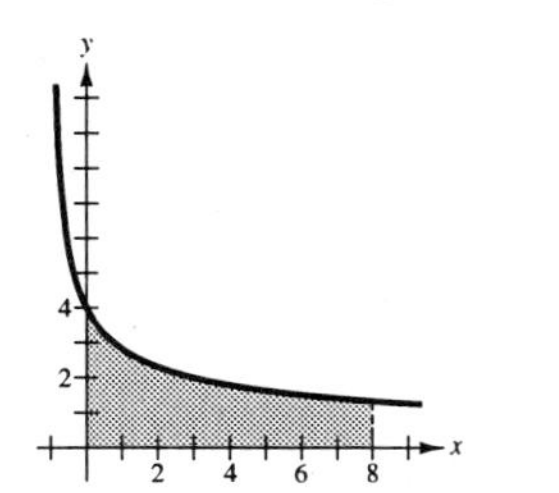

55. $\displaystyle\int_0^{\pi/3} \sec^2 x\,dx = \Big[\tan x\Big]_0^{\pi/3} = \sqrt{3}$

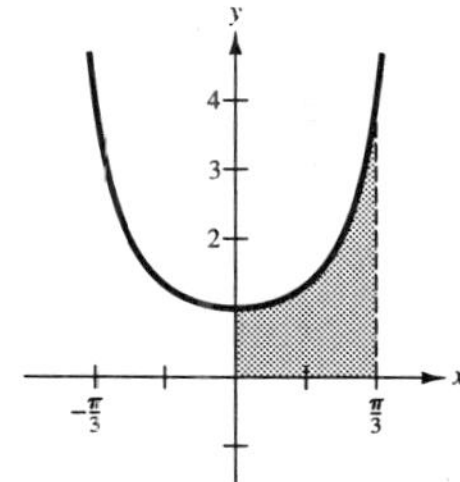

57. $\displaystyle\frac{1}{10-5}\int_5^{10} \frac{1}{\sqrt{x-1}}\,dx = \left[\frac{2}{5}\sqrt{x-1}\right]_5^{10} = \frac{2}{5}$

$\displaystyle\frac{1}{\sqrt{x-1}} = \frac{2}{5}$

$\displaystyle\sqrt{x-1} = \frac{5}{2}$

$\displaystyle x - 1 = \frac{25}{4}$

$\displaystyle x = \frac{29}{4}$

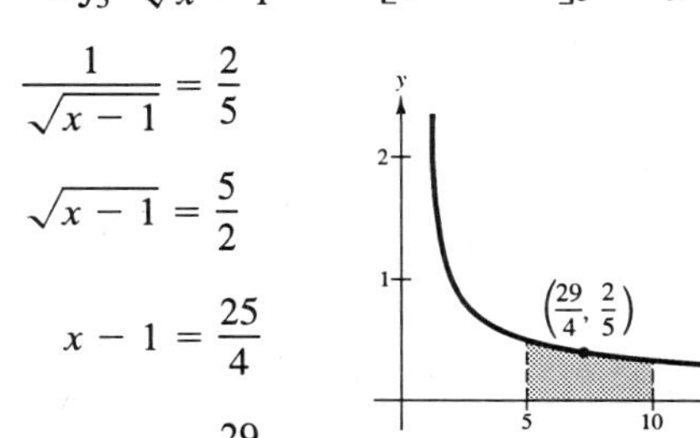

59. $\displaystyle\frac{1}{4-0}\int_0^4 x\,dx = \left[\frac{x^2}{8}\right]_0^4 = 2$

$x = 2$

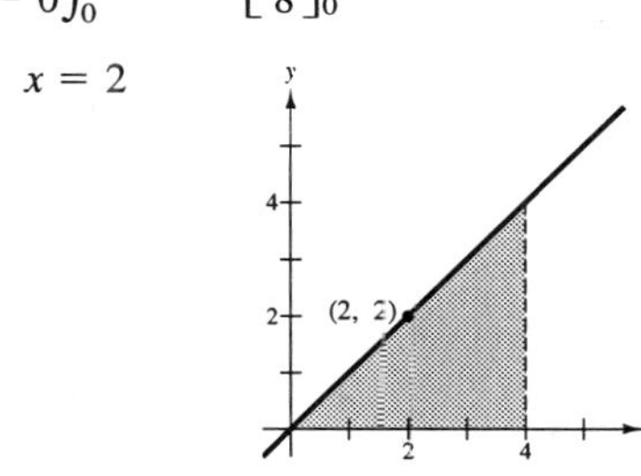

61. Trapezoidal Rule ($n = 4$): $\displaystyle\int_1^2 \frac{1}{1+x^3}\,dx \approx \frac{1}{8}\left[\frac{1}{1+1^3} + \frac{2}{1+(1.25)^3} + \frac{2}{1+(1.5)^3} + \frac{2}{1+(1.75)^3} + \frac{1}{1+2^3}\right] \approx 0.257$

Simpson's Rule ($n = 4$): $\displaystyle\int_1^2 \frac{1}{1+x^3}\,dx \approx \frac{1}{12}\left[\frac{1}{1+1^3} + \frac{4}{1+(1.25)^3} + \frac{2}{1+(1.5)^3} + \frac{4}{1+(1.75)^3} + \frac{1}{1+2^3}\right] \approx 0.254$

Graphing utility: 0.254

63. (a) $\displaystyle C = 0.1\int_8^{20}\left[12\sin\frac{\pi(t-8)}{12}\right]dt = \left[-\frac{14.4}{\pi}\cos\frac{\pi(t-8)}{12}\right]_8^{20} = \frac{-14.4}{\pi}(-1-1) \approx \$9.17$

(b) $\displaystyle C = 0.1\int_{10}^{18}\left[12\sin\frac{\pi(t-8)}{12} - 6\right]dt = \left[-\frac{14.4}{\pi}\cos\frac{\pi(t-8)}{12} - 0.6t\right]_{10}^{18}$

$\displaystyle = \left[\frac{-14.4}{\pi}\left(\frac{-\sqrt{3}}{2}\right) - 10.8\right] - \left[\frac{-14.4}{\pi}\left(\frac{\sqrt{3}}{2}\right) - 6\right] \approx \$3.14$

Savings $\approx 9.17 - 3.14 = \$6.03$.

65. $\displaystyle V = \int_0^3 0.85\sin\frac{\pi t}{3}\,dt = -\frac{3}{\pi}\left[0.85\cos\frac{\pi t}{3}\right]_0^3 = -\frac{2.55}{\pi}(-1-1) = \frac{5.1}{\pi} \approx 1.6234\,\text{liters}$

67. $p = 1.20 + 0.04t$

$\displaystyle C = \frac{15{,}000}{M}\int_t^{t+1} p\,ds$

(a) 2000 corresponds to $t = 10$.

$\displaystyle C = \frac{15{,}000}{M}\int_{10}^{11}[1.20 + 0.04t]\,dt$

$\displaystyle = \frac{15{,}000}{M}\Big[1.20t + 0.02t^2\Big]_{10}^{11} = \frac{24{,}300}{M}$

(b) 2005 corresponds to $t = 15$.

$\displaystyle C = \frac{15{,}000}{M}\Big[1.20t + 0.02t^2\Big]_{15}^{16} = \frac{27{,}300}{M}$

69. $u = 1 - x, x = 1 - u, dx = -du$

When $x = a, u = 1 - a$. When $x = b, u = 1 - b$.

$$P_{a,b} = \int_a^b \frac{1155}{32} x^3(1-x)^{3/2}\,dx = \frac{1155}{32}\int_{1-a}^{1-b} -(1-u)^3 u^{3/2}\,du$$

$$= \frac{1155}{32}\int_{1-a}^{1-b} (u^{9/2} - 3u^{7/2} + 3u^{5/2} - u^{3/2})\,du = \frac{1155}{32}\left[\frac{2}{11}u^{11/2} - \frac{2}{3}u^{9/2} + \frac{6}{7}u^{7/2} - \frac{2}{5}u^{5/2}\right]_{1-a}^{1-b}$$

$$= \frac{1155}{32}\left[\frac{2u^{5/2}}{1155}(105u^3 - 385u^2 + 495u - 231)\right]_{1-a}^{1-b} = \left[\frac{u^{5/2}}{16}(105u^3 - 385u^2 + 495u - 231)\right]_{1-a}^{1-b}$$

(a) $P_{0,0.25} = \left[\frac{u^{5/2}}{16}(105u^3 - 385u^2 + 495u - 231)\right]_1^{0.75} \approx 0.025 = 2.5\%$

(b) $P_{0.5,1} = \left[\frac{u^{5/2}}{16}(105u^3 - 385u^2 + 495u - 231)\right]_{0.5}^{0} \approx 0.736 = 73.6\%$

71. False

$$\int -\frac{1}{x^2}\,dx = \frac{1}{x} + C \text{ or } \frac{d}{dx}\left[-\frac{1}{x^2}\right] \neq \frac{1}{x}.$$

73. True

$$\frac{1}{2\pi - 0}\int_0^{2\pi} \sin x\,dx = \left[-\frac{1}{2\pi}\cos x\right]_0^{2\pi} = 0$$

CHAPTER 5
Logarithmic, Exponential, and Other Transcendental Functions

CHAPTER 5
Logarithmic, Exponential, and Other Transcendental Functions

Section 5.1 The Natural Logarithmic Function and Differentiation

Solutions to Odd-Numbered Exercises

1. Simpson's Rule: $n = 10$

x	0.5	1.5	2	2.5	3	3.5	4
$\int_1^x \frac{1}{t}\,dt$	-0.6932	0.4055	0.6932	0.9163	1.0987	1.2529	1.3865

Note: $\displaystyle\int_1^{0.5} \frac{1}{t}\,dt = -\int_{0.5}^1 \frac{1}{t}\,dt$

3. $f(x) = \ln x + 2$

Vertical shift 2 units upward

Matches (b)

5. $f(x) = \ln(x - 1)$

Horizontal shift 1 unit to the right

Matches (a)

7. $f(x) = 3 \ln x$

Domain: $x > 0$

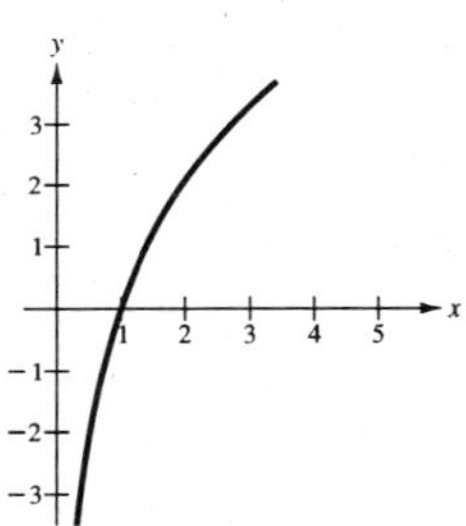

9. $f(x) = \ln 2x$

Domain: $x > 0$

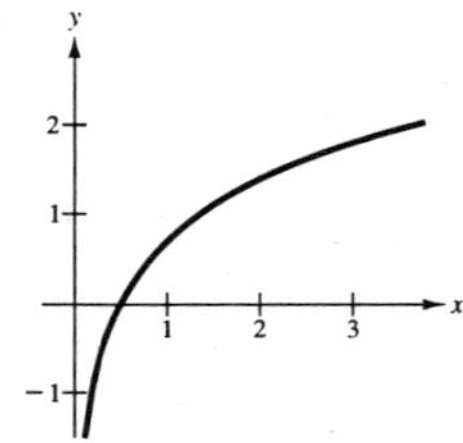

11. $f(x) = \ln(x - 1)$

Domain: $x > 1$

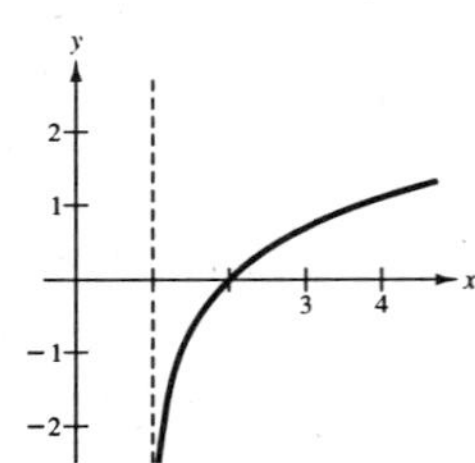

13. (a) $\ln 6 = \ln 2 + \ln 3 \approx 1.7917$

(b) $\ln \frac{2}{3} = \ln 2 - \ln 3 \approx -0.4055$

(c) $\ln 81 = \ln 3^4 = 4 \ln 3 \approx 4.3944$

(d) $\ln \sqrt{3} = \ln 3^{1/2} = \frac{1}{2} \ln 3 \approx 0.5493$

15. $\ln \dfrac{2}{3} = \ln 2 - \ln 3$

17. $\ln \dfrac{xy}{z} = \ln x + \ln y - \ln z$

19. $\ln \sqrt{2^3} = \ln 2^{3/2} = \dfrac{3}{2} \ln 2$

21. $$\ln\left(\frac{x^2 - 1}{x^3}\right)^3 = 3[\ln(x^2 - 1) - \ln x^3]$$
$$= 3[\ln(x + 1) + \ln(x - 1) - 3 \ln x]$$

23. $$\ln z(z - 1)^2 = \ln z + \ln(z - 1)^2$$
$$= \ln z + 2 \ln(z - 1)$$

25. $\ln(x - 2) - \ln(x + 2) = \ln \dfrac{x - 2}{x + 2}$

27. $\dfrac{1}{3}[2 \ln(x + 3) + \ln x - \ln(x^2 - 1)] = \dfrac{1}{3} \ln \dfrac{x(x + 3)^2}{x^2 - 1} = \ln \sqrt[3]{\dfrac{x(x + 3)^2}{x^2 - 1}}$

29. $2 \ln 3 - \frac{1}{2}\ln(x^2 + 1) = \ln 9 - \ln\sqrt{x^2+1} = \ln \frac{9}{\sqrt{x^2+1}}$

31.

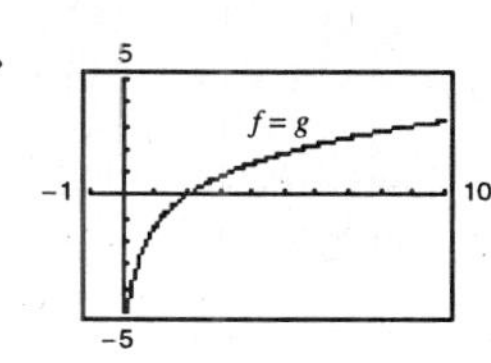

33. $\lim_{x \to 3^+} \ln(x - 3) = -\infty$

35. $\lim_{x \to 2^-} \ln[x^2(3 - x)] = \ln 4 \approx 1.3863$

37. $y = \ln x^3 = 3 \ln x$

$y' = \frac{3}{x}$

At $(1, 0)$, $y' = 3$.

39. $y = \ln x^2 = 2 \ln x$

$y' = \frac{2}{x}$

At $(1, 0)$, $y' = 2$.

41. $g(x) = \ln x^2 = 2 \ln x$

$g'(x) = \frac{2}{x}$

43. $y = (\ln x)^4$

$\frac{dy}{dx} = 4(\ln x)^3\left(\frac{1}{x}\right) = \frac{4(\ln x)^3}{x}$

45. $y = \ln x\sqrt{x^2 - 1} = \ln x + \frac{1}{2}\ln(x^2 - 1)$

$\frac{dy}{dx} = \frac{1}{x} + \frac{1}{2}\left(\frac{2x}{x^2 - 1}\right) = \frac{2x^2 - 1}{x(x^2 - 1)}$

47. $f(x) = \ln \frac{x}{x^2 + 1} = \ln x - \ln(x^2 + 1)$

$f'(x) = \frac{1}{x} - \frac{2x}{x^2 + 1} = \frac{1 - x^2}{x(x^2 + 1)}$

49. $g(t) = \frac{\ln t}{t^2}$

$g'(t) = \frac{t^2(1/t) - 2t \ln t}{t^4} = \frac{1 - 2 \ln t}{t^3}$

51. $y = \ln(\ln x^2)$

$\frac{dy}{dx} = \frac{1}{\ln x^2}\frac{d}{dx}(\ln x^2) = \frac{(2x/x^2)}{\ln x^2} = \frac{2}{x \ln x^2} = \frac{1}{x \ln x}$

53. $y = \ln\sqrt{\frac{x + 1}{x - 1}} = \frac{1}{2}[\ln(x + 1) - \ln(x - 1)]$

$\frac{dy}{dx} = \frac{1}{2}\left[\frac{1}{x + 1} - \frac{1}{x - 1}\right] = \frac{1}{1 - x^2}$

55. $f(x) = \ln \frac{\sqrt{4 + x^2}}{x} = \frac{1}{2}\ln(4 + x^2) - \ln x$

$f'(x) = \frac{x}{4 + x^2} - \frac{1}{x} = \frac{-4}{x(x^2 + 4)}$

57. $y = \frac{-\sqrt{x^2 + 1}}{x} + \ln\left(x + \sqrt{x^2 + 1}\right)$

$$\frac{dy}{dx} = \frac{-x\left(x/\sqrt{x^2 + 1}\right) + \sqrt{x^2 + 1}}{x^2} + \left(\frac{1}{x + \sqrt{x^2 + 1}}\right)\left(1 + \frac{x}{\sqrt{x^2 + 1}}\right)$$

$$= \frac{1}{x^2\sqrt{x^2 + 1}} + \left(\frac{1}{x + \sqrt{x^2 + 1}}\right)\left(\frac{\sqrt{x^2 + 1} + x}{\sqrt{x^2 + 1}}\right)$$

$$= \frac{1}{x^2\sqrt{x^2 + 1}} + \frac{1}{\sqrt{x^2 + 1}} = \frac{1 + x^2}{x^2\sqrt{x^2 + 1}} = \frac{\sqrt{x^2 + 1}}{x^2}$$

59. $y = \ln|\sin x|$

$\frac{dy}{dx} = \frac{\cos x}{\sin x} = \cot x$

61. $y = \ln\left|\frac{\cos x}{\cos x - 1}\right|$

$= \ln|\cos x| - \ln|\cos x - 1|$

$\frac{dy}{dx} = \frac{-\sin x}{\cos x} - \frac{-\sin x}{\cos x - 1} = -\tan x + \frac{\sin x}{\cos x - 1}$

63. $y = \ln\left|\dfrac{-1 + \sin x}{2 + \sin x}\right|$

$= \ln|-1 + \sin x| - \ln|2 + \sin x|$

$\dfrac{dy}{dx} = \dfrac{\cos x}{-1 + \sin x} - \dfrac{\cos x}{2 + \sin x}$

$= \dfrac{3\cos x}{(\sin x - 1)(\sin x + 2)}$

65. $f(x) = \sin 2x \ln x^2 = 2 \sin 2x \ln x$

$f'(x) = (2 \sin 2x)\left(\dfrac{1}{x}\right) + 4 \cos 2x \ln x$

$= \dfrac{2}{x}(\sin 2x + 2x \cos 2x \ln x)$

$= \dfrac{2}{x}(\sin 2x + x \cos 2x \ln x^2)$

67. (a) $y = 3x^2 - \ln x, \quad (1, 3)$

$\dfrac{dy}{dx} = 6x - \dfrac{1}{x}$

When $x = 1$, $\dfrac{dy}{dx} = 5$.

Tangent line: $y - 3 = 5(x - 1)$

$y = 5x - 2$

$0 = 5x - y - 2$

(b)

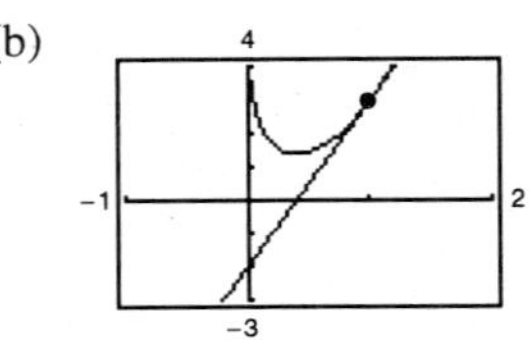

69. $x^2 - 3 \ln y + y^2 = 10$

$2x - \dfrac{3}{y}\dfrac{dy}{dx} + 2y\dfrac{dy}{dx} = 0$

$2x = \dfrac{dy}{dx}\left(\dfrac{3}{y} - 2y\right)$

$\dfrac{dy}{dx} = \dfrac{2x}{(3/y) - 2y} = \dfrac{2xy}{3 - 2y^2}$

71. $y = 2(\ln x) + 3$

$y' = \dfrac{2}{x}$

$y'' = -\dfrac{2}{x^2}$

$xy'' + y' = x\left(-\dfrac{2}{x^2}\right) + \dfrac{2}{x} = 0$

73. $y = \dfrac{x^2}{2} - \ln x$

Domain: $x > 0$

$y' = x - \dfrac{1}{x} = \dfrac{(x + 1)(x - 1)}{x} = 0$ when $x = 1$.

$y'' = 1 + \dfrac{1}{x^2} > 0$

Relative minimum: $\left(1, \frac{1}{2}\right)$

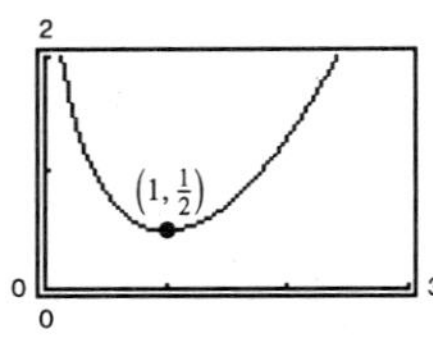

75. $y = x \ln x$

Domain: $x > 0$

$y' = x\left(\dfrac{1}{x}\right) + \ln x = 1 + \ln x = 0$ when $x = e^{-1}$.

$y'' = \dfrac{1}{x} > 0$

Relative minimum: $(e^{-1}, -e^{-1})$

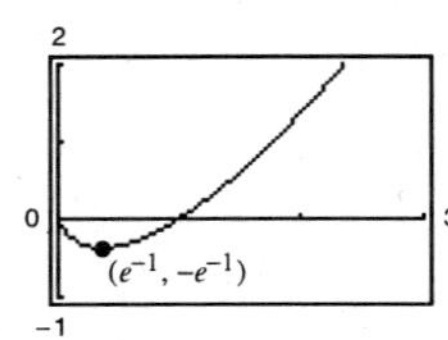

77. $y = \dfrac{x}{\ln x}$

Domain: $0 < x < 1, x > 1$

$$y' = \frac{(\ln x)(1) - (x)(1/x)}{(\ln x)^2} = \frac{\ln x - 1}{(\ln x)^2} = 0 \text{ when } x = e.$$

$$y'' = \frac{(\ln x)^2(1/x) - (\ln x - 1)(2/x)\ln x}{(\ln x)^4} = \frac{2 - \ln x}{x(\ln x)^3} = 0 \text{ when } x = e^2.$$

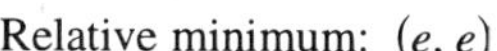

Relative minimum: (e, e)

Point of inflection: $(e^2, e^2/2)$

79. $f(x) = \ln x, \quad f(1) = 0$

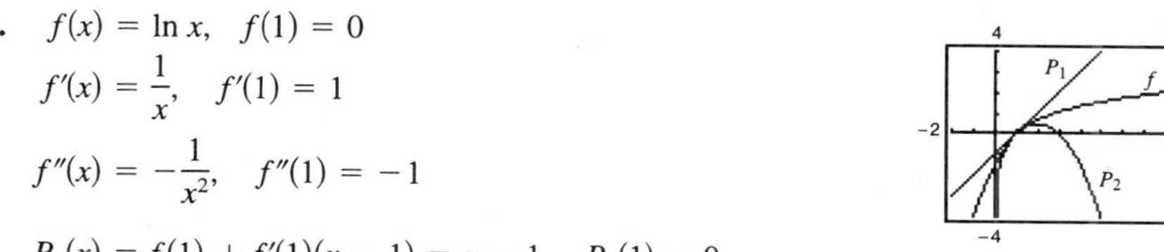

$f'(x) = \dfrac{1}{x}, \quad f'(1) = 1$

$f''(x) = -\dfrac{1}{x^2}, \quad f''(1) = -1$

$P_1(x) = f(1) + f'(1)(x - 1) = x - 1, \quad P_1(1) = 0$

$P_2(x) = f(1) + f'(1)(x - 1) + \dfrac{1}{2}f''(1)(x - 1)^2 = (x - 1) - \dfrac{1}{2}(x - 1)^2, \quad P_2(1) = 0$

$P_1'(x) = 1, \quad P_1'(1) = 1$

$P_2'(x) = 1 - (x - 1) = 2 - x, \quad P_2'(1) = 1$

$P_2''(x) = -1, \quad P_2''(1) = -1$

The values of f, P_1, P_2, and their first derivatives agree at $x = 1$. The values of the second derivatives of f and P_2 agree at $x = 1$.

81. Find x such that $\ln x = -x$.

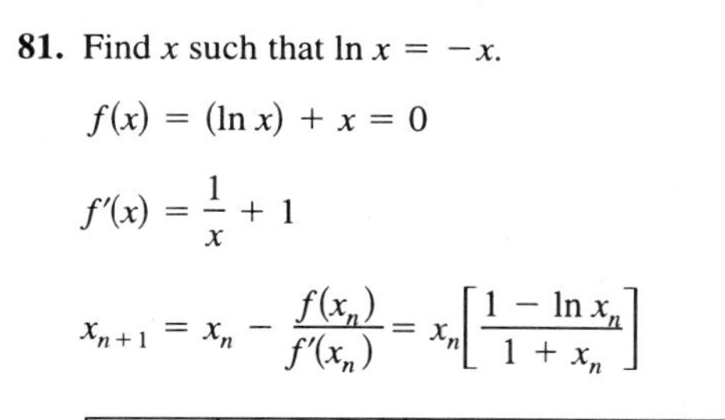

$f(x) = (\ln x) + x = 0$

$f'(x) = \dfrac{1}{x} + 1$

$$x_{n+1} = x_n - \frac{f(x_n)}{f'(x_n)} = x_n\left[\frac{1 - \ln x_n}{1 + x_n}\right]$$

n	1	2	3
x_n	0.5	0.5644	0.5671
$f(x_n)$	-0.1931	-0.0076	-0.0001

Approximate root: $x = 0.567$

83. $y = x\sqrt{x^2 - 1}$

$$\ln y = \ln x + \frac{1}{2}\ln(x^2 - 1)$$

$$\frac{1}{y}\left(\frac{dy}{dx}\right) = \frac{1}{x} + \frac{x}{x^2 - 1}$$

$$\frac{dy}{dx} = y\left[\frac{2x^2 - 1}{x(x^2 - 1)}\right] = \frac{2x^2 - 1}{\sqrt{x^2 - 1}}$$

85. $y = \dfrac{x^2\sqrt{3x - 2}}{(x - 1)^2}$

$$\ln y = 2\ln x + \frac{1}{2}\ln(3x - 2) - 2\ln(x - 1)$$

$$\frac{1}{y}\left(\frac{dy}{dx}\right) = \frac{2}{x} + \frac{3}{2(3x - 2)} - \frac{2}{x - 1}$$

$$\frac{dy}{dx} = y\left[\frac{3x^2 - 15x + 8}{2x(3x - 2)(x - 1)}\right]$$

$$= \frac{3x^3 - 15x^2 + 8x}{2(x - 1)^3\sqrt{3x - 2}}$$

87. $y = \dfrac{x(x - 1)^{3/2}}{\sqrt{x + 1}}$

$$\ln y = \ln x + \frac{3}{2}\ln(x - 1) - \frac{1}{2}\ln(x + 1)$$

$$\frac{1}{y}\left(\frac{dy}{dx}\right) = \frac{1}{x} + \frac{3}{2}\left(\frac{1}{x - 1}\right) - \frac{1}{2}\left(\frac{1}{x + 1}\right)$$

$$\frac{dy}{dx} = \frac{y}{2}\left[\frac{2}{x} + \frac{3}{x - 1} - \frac{1}{x + 1}\right]$$

$$= \frac{y}{2}\left[\frac{4x^2 + 4x - 2}{x(x^2 - 1)}\right] = \frac{(2x^2 + 2x - 1)\sqrt{x - 1}}{(x + 1)^{3/2}}$$

89. $$\beta = 10\log_{10}\left(\frac{I}{10^{-16}}\right) = \frac{10}{\ln 10}\ln\left(\frac{I}{10^{-16}}\right) = \frac{10}{\ln 10}[\ln I + 16\ln 10]$$

$$\beta(10^{-10}) = \frac{10}{\ln 10}[\ln 10^{-10} + 16\ln 10] = \frac{10}{\ln 10}[-10\ln 10 + 16\ln 10] = \frac{10}{\ln 10}[6\ln 10] = 60 \text{ decibels}$$

91. (a)

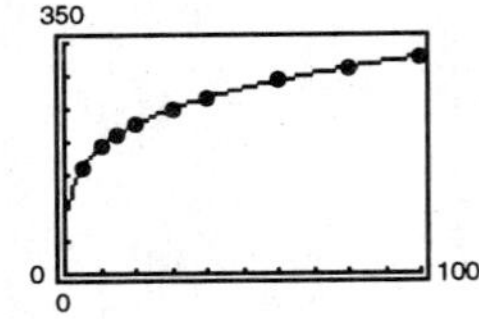

(b) $T'(p) = \dfrac{34.96}{p} + \dfrac{3.955}{\sqrt{p}}$

$T'(10) \approx 4.75 \text{ deg/lb/in}^2$

$T'(70) \approx 0.97 \text{ deg/lb/in}^2$

(c)

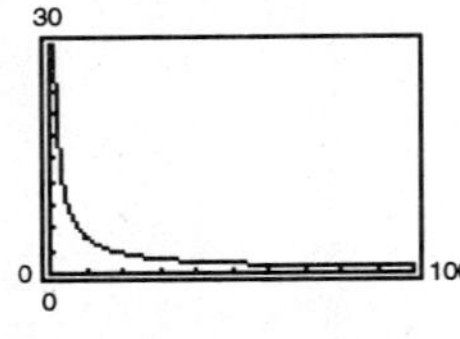

$\lim_{p\to\infty} T'(p) = 0$

93. (a) You get an error message because $\ln h$ does not exist for $h = 0$.

(b) Reversing the data, you obtain

$h = 0.8627 - 6.4474 \ln p.$

(c)

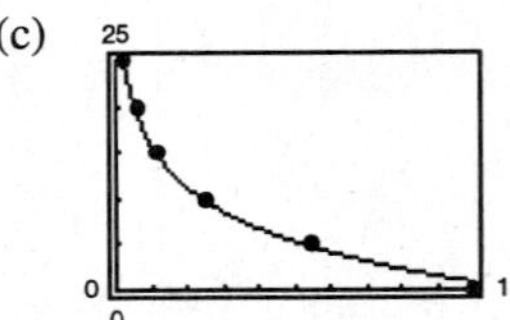

(d) If $p = 0.75$, $h \approx 2.72$ km.

(e) If $h = 13$ km, $p \approx 0.15$ atmosphere.

(f) $h = 0.8627 - 6.4474 \ln p$

$1 = -6.4474\dfrac{1}{p}\dfrac{dp}{dh}$ (implicit differentiation)

$\dfrac{dp}{dh} = \dfrac{p}{-6.4474}$

For $h = 5$, $p = 0.55$ and $dp/dh = -0.0853$ atmos/km.

For $h = 20$, $p = 0.06$ and $dp/dh = -0.00931$ atmos/km.

As the altitude increases, the rate of change of pressure decreases.

95. (a) $f(x) = \ln x$, $g(x) = \sqrt{x}$

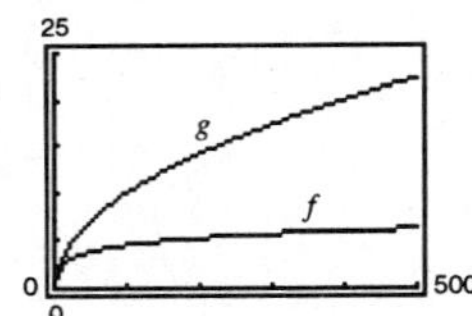

$f'(x) = \dfrac{1}{x}$, $g'(x) = \dfrac{1}{2\sqrt{x}}$

For $x > 4$, $g'(x) > f'(x)$. g is increasing at a higher rate than f for "large" values of x.

(b) $f(x) = \ln x$, $g(x) = \sqrt[4]{x}$

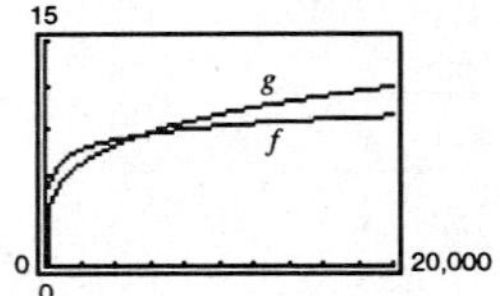

$f'(x) = \dfrac{1}{x}$, $g'(x) = \dfrac{1}{4\sqrt[4]{x^3}}$

For $x > 256$, $g'(x) > f'(x)$. g is increasing at a higher rate than f for "large" values of x. $f(x) = \ln x$ increases very slowly for "large" values of x.

97. False

$\ln x + \ln 25 = \ln(25x) \neq \ln(x + 25)$

Section 5.2 The Natural Logarithmic Function and Integration

1. $u = x + 1,\ du = dx$

$$\int \frac{1}{x+1}\,dx = \ln|x+1| + C$$

3. $u = 3 - 2x,\ du = -2\,dx$

$$\int \frac{1}{3-2x}\,dx = -\frac{1}{2}\int \frac{1}{3-2x}(-2)\,dx$$

$$= -\frac{1}{2}\ln|3-2x| + C$$

5. $u = x^2 + 1,\ du = 2x\,dx$

$$\int \frac{x}{x^2+1}\,dx = \frac{1}{2}\int \frac{1}{x^2+1}(2x)\,dx$$

$$= \frac{1}{2}\ln(x^2+1) + C = \ln\sqrt{x^2+1} + C$$

7. $$\int \frac{x^2-4}{x}\,dx = \int\left(x - \frac{4}{x}\right)dx$$

$$= \frac{x^2}{2} - 4\ln|x| + C$$

9. $u = x^3 + 3x^2 + 9x,\ du = 3(x^2 + 2x + 3)\,dx$

$$\int \frac{x^2+2x+3}{x^3+3x^2+9x}\,dx = \frac{1}{3}\int \frac{3(x^2+2x+3)}{x^3+3x^2+9x}\,dx$$

$$= \frac{1}{3}\ln|x^3+3x^2+9x| + C$$

11. $u = \ln x,\ du = \frac{1}{x}\,dx$

$$\int \frac{(\ln x)^2}{x}\,dx = \frac{1}{3}(\ln x)^3 + C$$

13. $u = x + 1,\ du = dx$

$$\int \frac{1}{\sqrt{x+1}}\,dx = \int (x+1)^{-1/2}\,dx$$

$$= 2(x+1)^{1/2} + C = 2\sqrt{x+1} + C$$

15. $u = \sqrt{x} - 3,\ du = \frac{1}{2\sqrt{x}}\,dx \Rightarrow 2(u+3)\,du = dx$

$$\int \frac{\sqrt{x}}{\sqrt{x}-3}\,dx = 2\int \frac{(u+3)^2}{u}\,du = 2\int \frac{u^2+6u+9}{u}\,du = 2\int\left(u + 6 + \frac{9}{u}\right)du$$

$$= 2\left[\frac{u^2}{2} + 6u + 9\ln|u|\right] + C_1 = u^2 + 12u + 18\ln|u| + C_1$$

$$= \left(\sqrt{x}-3\right)^2 + 12\left(\sqrt{x}-3\right) + 18\ln\left|\sqrt{x}-3\right| + C_1$$

$$= x + 6\sqrt{x} + 18\ln\left|\sqrt{x}-3\right| + C \text{ where } C = C_1 - 27.$$

17. $$\int \frac{2x}{(x-1)^2}\,dx = \int \frac{2x-2+2}{(x-1)^2}\,dx$$

$$= \int \frac{2(x-1)}{(x-1)^2}\,dx + 2\int \frac{1}{(x-1)^2}\,dx$$

$$= 2\int \frac{1}{x-1}\,dx + 2\int \frac{1}{(x-1)^2}\,dx$$

$$= 2\ln|x-1| - \frac{2}{(x-1)} + C$$

19. $$\int \frac{\cos\theta}{\sin\theta}\,d\theta = \ln|\sin\theta| + C$$

$(u = \sin\theta,\ du = \cos\theta\,d\theta)$

21. $\displaystyle\int \csc 2x\,dx = \frac{1}{2}\int (\csc 2x)(2)\,dx$

$\displaystyle= -\frac{1}{2}\ln|\csc 2x + \cot 2x| + C$

23. $\displaystyle\int \frac{\cos t}{1 + \sin t}\,dt = \ln|1 + \sin t| + C$

25. $\displaystyle\int \frac{\sec x \tan x}{\sec x - 1}\,dx = \ln|\sec x - 1| + C$

27. $\displaystyle y = \int \frac{3}{2 - x}\,dx$

$\displaystyle= -3\int \frac{1}{x - 2}\,dx$

$= -3\ln|x - 2| + C$

$(1, 0)$: $0 = -3\ln|1 - 2| + C \Rightarrow C = 0$

$y = -3\ln|x - 2|$

29. $\displaystyle s = \int \tan(2\theta)\,d\theta$

$\displaystyle= \frac{1}{2}\int \tan(2\theta)(2\,d\theta)$

$\displaystyle= -\frac{1}{2}\ln|\cos 2\theta| + C$

$(0, 2)$: $\displaystyle 2 = -\frac{1}{2}\ln|\cos(0)| + C \Rightarrow C = 2$

$\displaystyle s = -\frac{1}{2}\ln|\cos 2\theta| + 2$

31. $\displaystyle\frac{dy}{dx} = \frac{1}{x + 2}$, $(0, 1)$

(a)

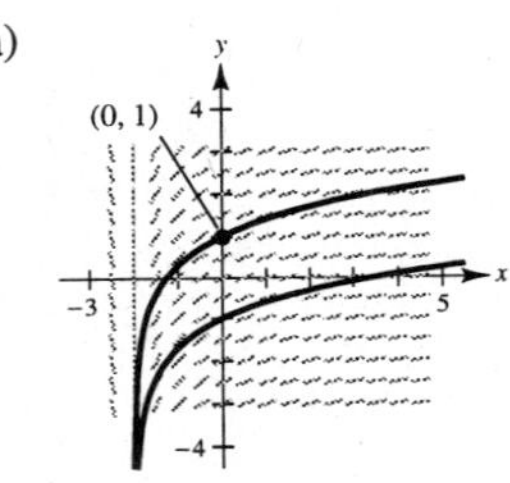

(b) $\displaystyle y = \int \frac{1}{x + 2}\,dx = \ln|x + 2| + C$

$y(0) = 1 \Rightarrow 1 = \ln 2 + C \Rightarrow C = 1 - \ln 2$

Hence, $\displaystyle y = \ln|x + 2| + 1 - \ln 2 = \ln\left|\frac{x + 2}{2}\right| + 1.$

33. $\displaystyle\int_0^4 \frac{5}{3x + 1}\,dx = \left[\frac{5}{3}\ln|3x + 1|\right]_0^4$

$\displaystyle= \frac{5}{3}\ln 13 \approx 4.275$

35. $\displaystyle u = 1 + \ln x,\ du = \frac{1}{x}\,dx$

$\displaystyle\int_1^e \frac{(1 + \ln x)^2}{x}\,dx = \left[\frac{1}{3}(1 + \ln x)^3\right]_1^e = \frac{7}{3}$

37. $\displaystyle\int_0^2 \frac{x^2 - 2}{x + 1}\,dx = \int_0^2 \left(x - 1 - \frac{1}{x + 1}\right)dx$

$\displaystyle= \left[\frac{1}{2}x^2 - x - \ln|x + 1|\right]_0^2 = -\ln 3$

39. $\displaystyle\int_1^2 \frac{1 - \cos\theta}{\theta - \sin\theta}\,d\theta = \Big[\ln|\theta - \sin\theta|\Big]_1^2$

$\displaystyle= \ln\left|\frac{2 - \sin 2}{1 - \sin 1}\right| \approx 1.929$

41. $\displaystyle\int \frac{1}{1 + \sqrt{x}}\,dx = 2\left(1 + \sqrt{x}\right) - 2\ln\left(1 + \sqrt{x}\right) + C_1$

$= 2\left[\sqrt{x} - \ln\left(1 + \sqrt{x}\right)\right] + C$ where $C = C_1 + 2$.

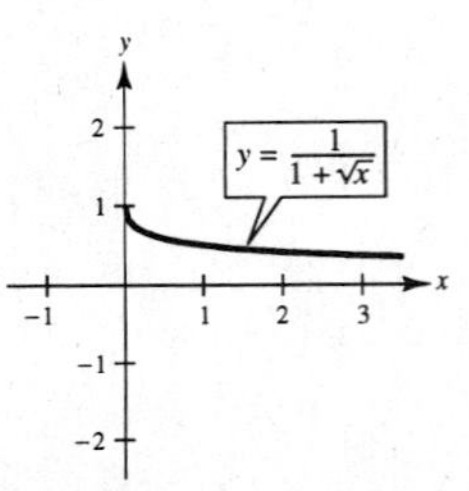

43. $\int \cos(1 - x)\, dx = -\sin(1 - x) + C$

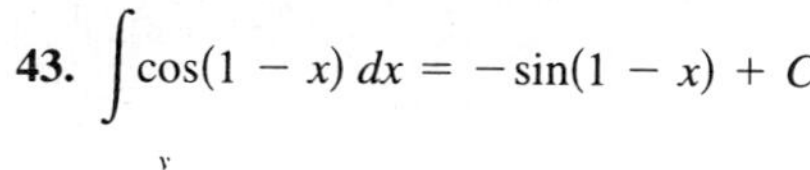

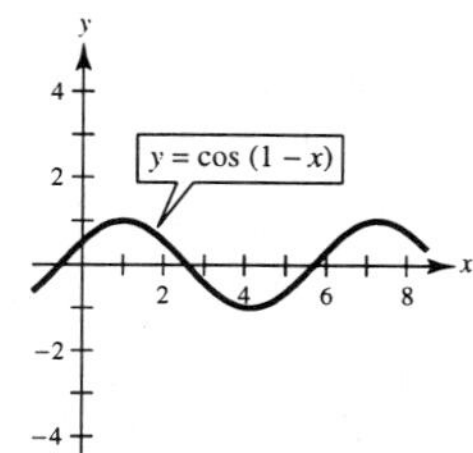

45. $\int_{\pi/4}^{\pi/2} (\csc x - \sin x)\, dx = \Big[-\ln|\csc x + \cot x| + \cos x\Big]_{\pi/4}^{\pi/2}$

$= \ln(\sqrt{2} + 1) - \dfrac{\sqrt{2}}{2} \approx 0.174$

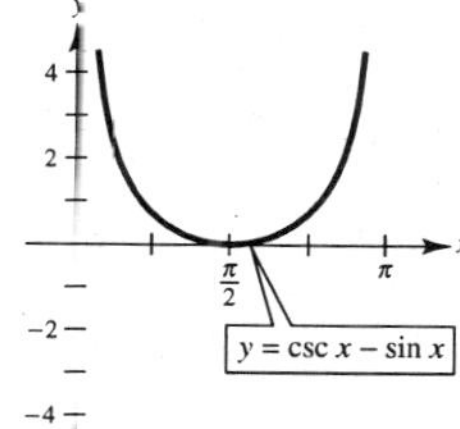

47. $-\ln|\cos x| + C = \ln\left|\dfrac{1}{\cos x}\right| + C = \ln|\sec x| + C$

49. $\ln|\sec x + \tan x| + C = \ln\left|\dfrac{(\sec x + \tan x)(\sec x - \tan x)}{(\sec x - \tan x)}\right| + C = \ln\left|\dfrac{\sec^2 x - \tan^2 x}{\sec x - \tan x}\right| + C$

$= \ln\left|\dfrac{1}{\sec x - \tan x}\right| + C = -\ln|\sec x - \tan x| - C$

Note: In Exercises 51 and 53, you can use the Second Fundamental Theorem of Calculus or integrate the function.

51. $F(x) = \int_1^x \dfrac{1}{t}\, dt$

$F'(x) = \dfrac{1}{x}$

53. $F(x) = \int_x^{3x} \dfrac{1}{t}\, dt$

$= \int_1^{3x} \dfrac{1}{t}\, dt - \int_1^x \dfrac{1}{t}\, dt$

$F'(x) = \dfrac{3}{3x} - \dfrac{1}{x} = 0$

55.

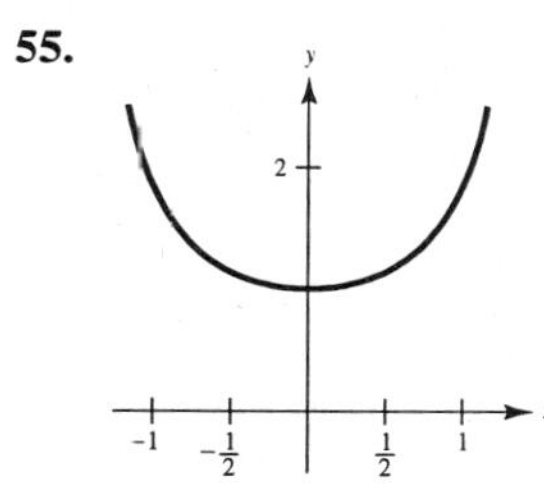

$A \approx 1.25$

Matches (d)

57. $A = \int_1^4 \dfrac{x^2 + 4}{x}\, dx = \int_1^4 \left(x + \dfrac{4}{x}\right) dx$

$= \left[\dfrac{x^2}{2} + 4 \ln x\right]_1^4 = (8 + 4 \ln 4) - \dfrac{1}{2}$

$= \dfrac{15}{2} + 8 \ln 2 \approx 13.045$ square units

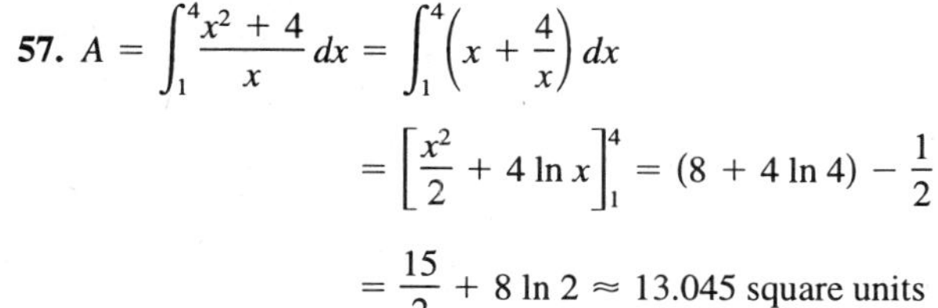

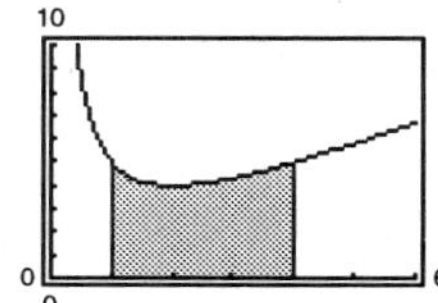

59. $\int_0^2 2 \sec \dfrac{\pi x}{6}\, dx = \dfrac{12}{\pi} \int_0^2 \sec\left(\dfrac{\pi x}{6}\right) \dfrac{\pi}{6}\, dx$

$= \left[\dfrac{12}{\pi} \ln\left|\sec \dfrac{\pi x}{6} + \tan \dfrac{\pi x}{6}\right|\right]_0^2$

$= \dfrac{12}{\pi} \ln\left|\sec \dfrac{\pi}{3} + \tan \dfrac{\pi}{3}\right| - \dfrac{12}{\pi} \ln|1 + 0|$

$= \dfrac{12}{\pi} \ln(2 + \sqrt{3}) \approx 5.03041$

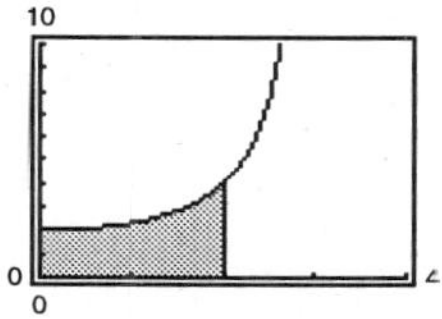

61. $P(t) = \int \frac{3000}{1 + 0.25t}\,dt = (3000)(4)\int \frac{0.25}{1 + 0.25t}\,dt = 12{,}000 \ln|1 + 0.25t| + C$

$P(0) = 12{,}000 \ln|1 + 0.25(0)| + C = 1000$

$C = 1000$

$P(t) = 12{,}000 \ln|1 + 0.25t| + 1000 = 1000[12 \ln|1 + 0.25t| + 1]$

$P(3) = 1000[12(\ln 1.75) + 1] \approx 7715$

63. $\frac{1}{50 - 40}\int_{40}^{50} \frac{90{,}000}{400 + 3x}\,dx = \Big[3000 \ln|400 + 3x|\Big]_{40}^{50} \approx \$168.27$

65. (a) $2x^2 - y^2 = 8$

$y^2 = 2x^2 - 8$

$y_1 = \sqrt{2x^2 - 8}$

$y_2 = -\sqrt{2x^2 - 8}$

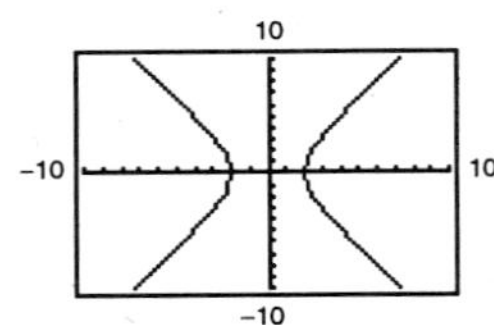

(b) $y^2 = e^{-\int(1/x)dx} = e^{-\ln x + C} = e^{\ln(1/x)}(e^C) = \frac{1}{x}k$

Let $k = 4$ and graph $y^2 = \frac{4}{x}\begin{pmatrix} y_1 = 2/\sqrt{x} \\ y_2 = -2/\sqrt{x}\end{pmatrix}$

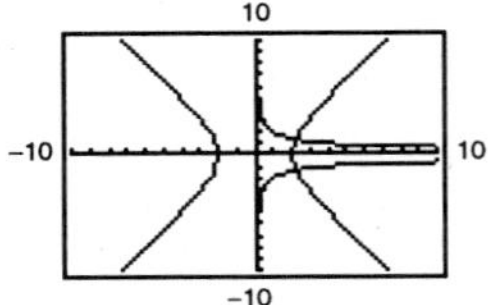

(c) In part (a), $2x^2 - y^2 = 8$

$4x - 2yy' = 0$

$y' = \frac{2x}{y}.$

In part (b), $y^2 = \frac{4}{x} = 4x^{-1}$

$2yy' = \frac{-4}{x^2}$

$y' = \frac{-2}{yx^2} = \frac{-2y}{y^2x^2} = \frac{-2y}{4x} = \frac{-y}{2x}.$

Using a graphing utility the graphs intersect at (2.214, 1.344). The slopes are 3.295 and $-0.304 = (-1)/3.295$, respectively.

67. False

$\frac{1}{2}(\ln x) = \ln(x^{1/2}) \neq (\ln x)^{1/2}$

69. True

$\int \frac{1}{x}\,dx = \ln|x| + C_1$

$= \ln|x| + \ln|C| = \ln|Cx|, \; C \neq 0$

Section 5.3 Inverse Functions

1. (a) $f(x) = 5x + 1$

$g(x) = \dfrac{x - 1}{5}$

$f(g(x)) = f\left(\dfrac{x-1}{5}\right) = 5\left(\dfrac{x-1}{5}\right) + 1 = x$

$g((f(x)) = g(5x + 1) = \dfrac{(5x + 1) - 1}{5} = x$

(b)

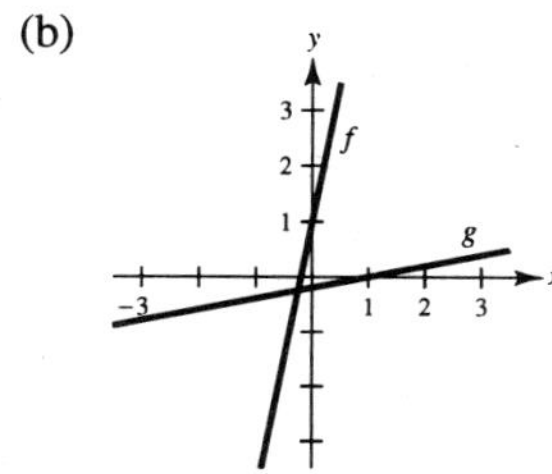

3. (a) $f(x) = x^3$

$g(x) = \sqrt[3]{x}$

$f(g(x)) = f(\sqrt[3]{x}) = (\sqrt[3]{x})^3 = x$

$g(f(x)) = g(x^3) = \sqrt[3]{x^3} = x$

(b)

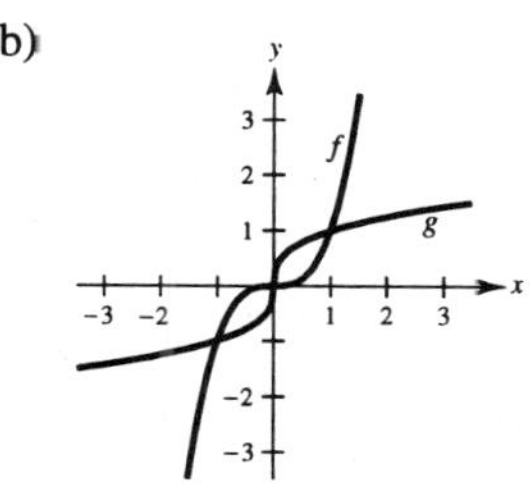

5. (a) $f(x) = \sqrt{x - 4}$

$g(x) = x^2 + 4,\ x \geq 0$

$f(g(x)) = f(x^2 + 4)$

$= \sqrt{(x^2 + 4) - 4} = \sqrt{x^2} = x$

$g(f(x)) = g(\sqrt{x - 4})$

$= (\sqrt{x - 4})^2 + 4 = x - 4 + 4 = x$

(b)

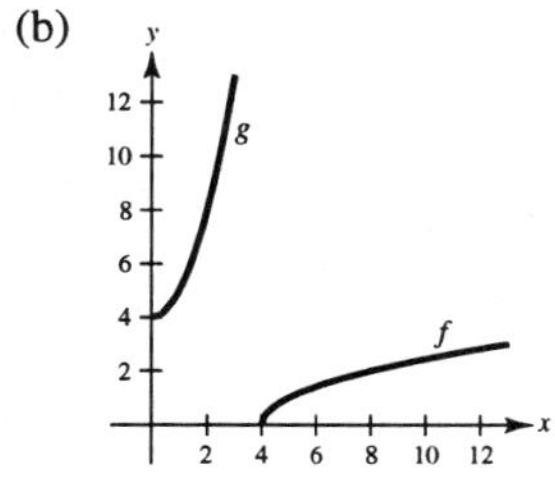

7. (a) $f(x) = \dfrac{1}{x}$

$g(x) = \dfrac{1}{x}$

$f(g(x)) = \dfrac{1}{1/x} = x$

$g(f(x)) = \dfrac{1}{1/x} = x$

(ɔ)

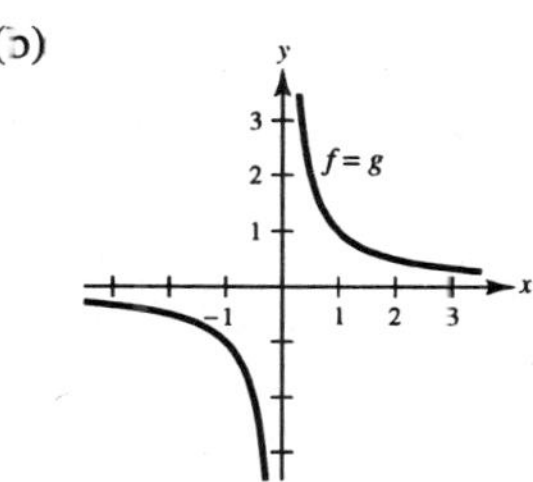

9. Matches (c)

11. Matches (a)

13. $f(x) = 2x - 3 = y$

$x = \dfrac{y + 3}{2}$

$y = \dfrac{x + 3}{2}$

$f^{-1}(x) = \dfrac{x + 3}{2}$

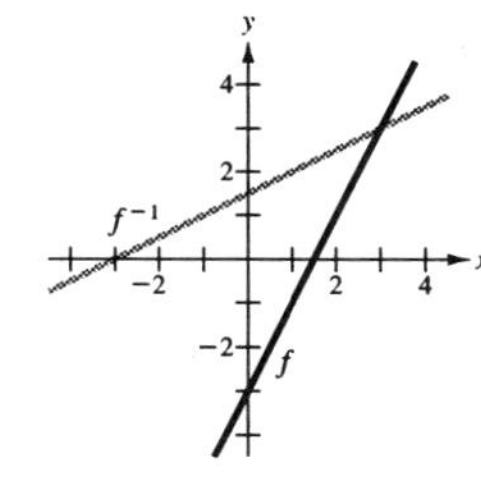

15. $f(x) = x^5 = y$

$x = \sqrt[5]{y}$

$y = \sqrt[5]{x}$

$f^{-1}(x) = \sqrt[5]{x} = x^{1/5}$

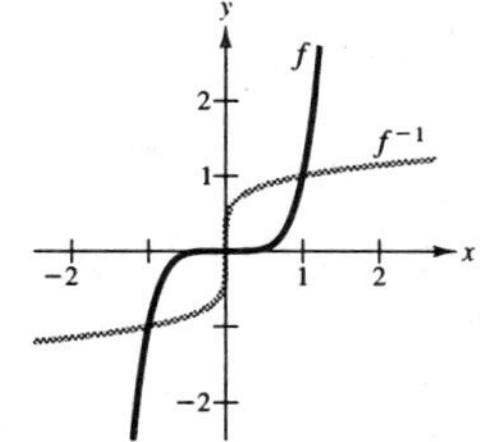

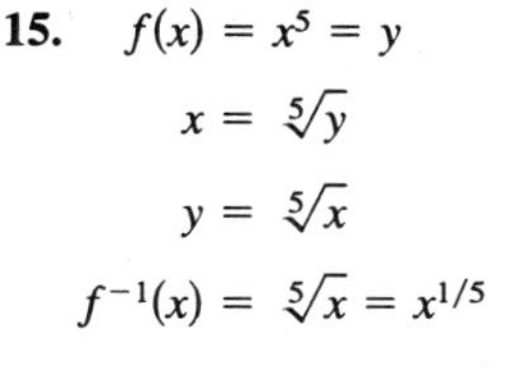

17. $f(x) = \sqrt{x} = y$

$x = y^2$

$y = x^2$

$f^{-1}(x) = x^2, \; x \geq 0$

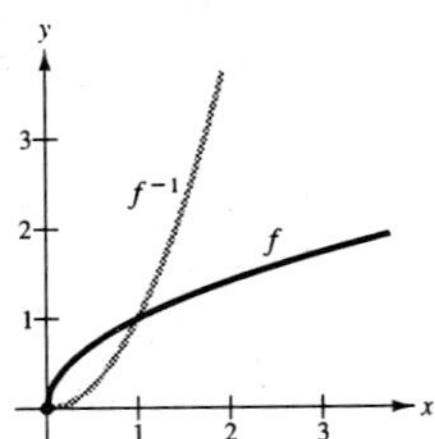

19. $f(x) = \sqrt{4 - x^2} = y, \; 0 \leq x \leq 2$

$x = \sqrt{4 - y^2}$

$y = \sqrt{4 - x^2}$

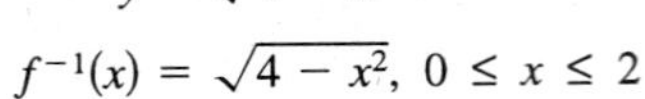

$f^{-1}(x) = \sqrt{4 - x^2}, \; 0 \leq x \leq 2$

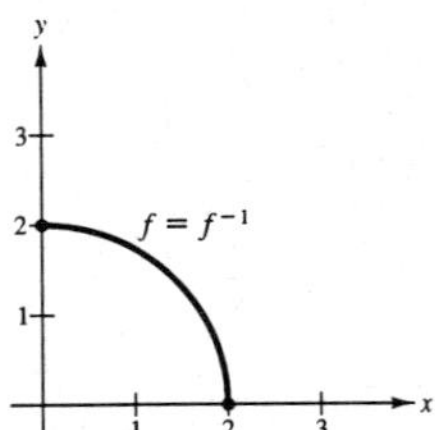

21. $f(x) = \sqrt[3]{x - 1} = y$

$x = y^3 + 1$

$y = x^3 + 1$

$f^{-1}(x) = x^3 + 1$

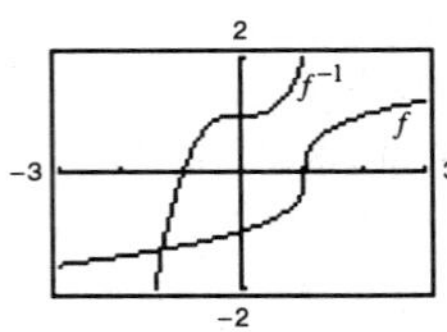

The graphs of f and f^{-1} are reflections of each other across the line $y = x$.

23. $f(x) = x^{2/3} = y, \; x \geq 0$

$x = y^{3/2}$

$y = x^{3/2}$

$f^{-1}(x) = x^{3/2}, \; x \geq 0$

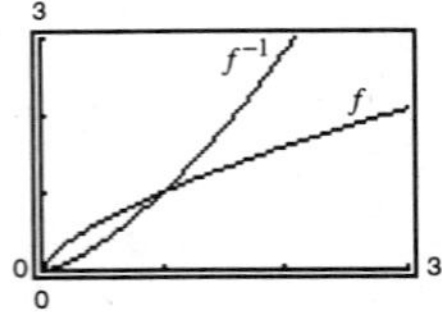

The graphs of f and f^{-1} are reflections of each other across the line $y = x$.

25. $f(x) = \dfrac{x}{\sqrt{x^2 + 7}} = y$

$x = \dfrac{\sqrt{7}y}{\sqrt{1 - y^2}}$

$y = \dfrac{\sqrt{7}x}{\sqrt{1 - x^2}}$

$f^{-1}(x) = \dfrac{\sqrt{7}x}{\sqrt{1 - x^2}}, \; -1 < x < 1$

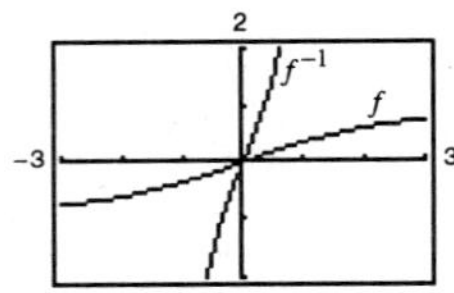

The graphs of f and f^{-1} are reflections of each other across the line $y = x$.

27. $f(x) = \dfrac{x}{x^2 - 4} = y$ on $(-2, 2)$

$x^2y - 4y = x$

$x^2y - x - 4y = 0$

$a = y, b = -1, c = -4y$

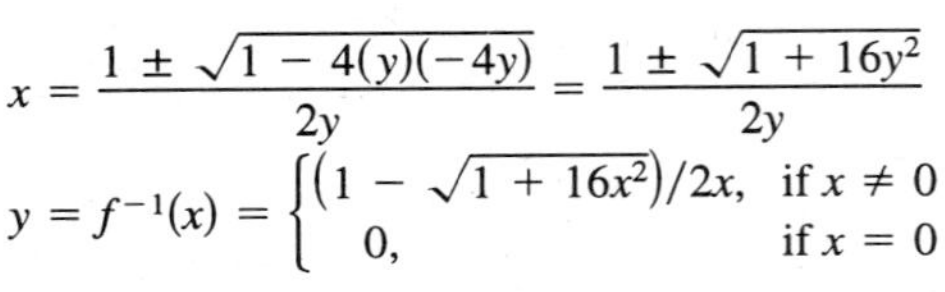

$x = \dfrac{1 \pm \sqrt{1 - 4(y)(-4y)}}{2y} = \dfrac{1 \pm \sqrt{1 + 16y^2}}{2y}$

$$y = f^{-1}(x) = \begin{cases} \left(1 - \sqrt{1 + 16x^2}\right)/2x, & \text{if } x \neq 0 \\ 0, & \text{if } x = 0 \end{cases}$$

Domain: all x

Range: $-2 < y < 2$

The graphs of f and f^{-1} are reflections of each other across the line $y = x$.

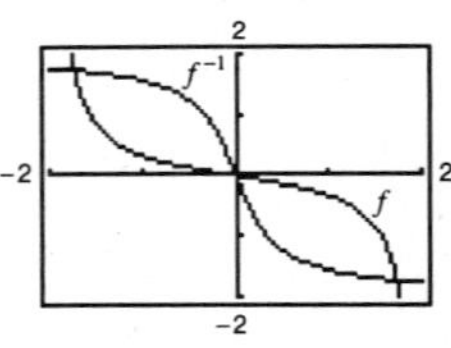

29. (a)

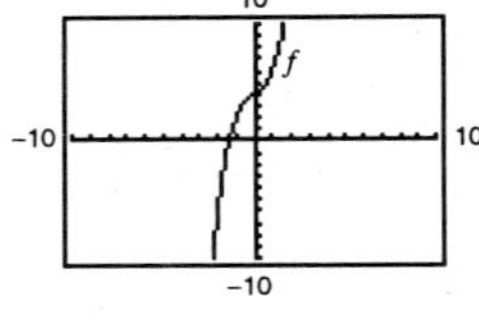

(b)

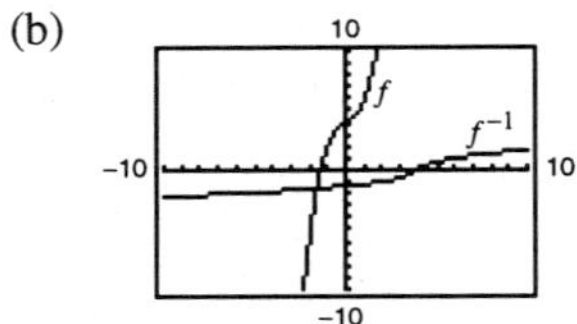

(c) Yes, f is one-to-one and has an inverse. The inverse relation is an inverse function.

31. (a)

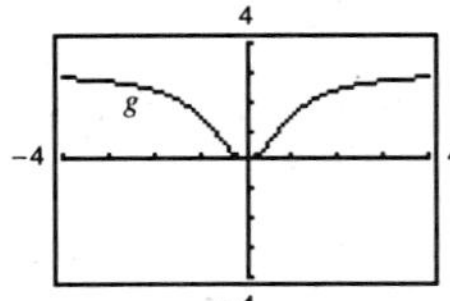

(b)

(c) g is not one-to-one and does not have an inverse. The inverse relation is not an inverse function.

33.

x	1	2	3	4
$f^{-1}(x)$	0	1	2	4

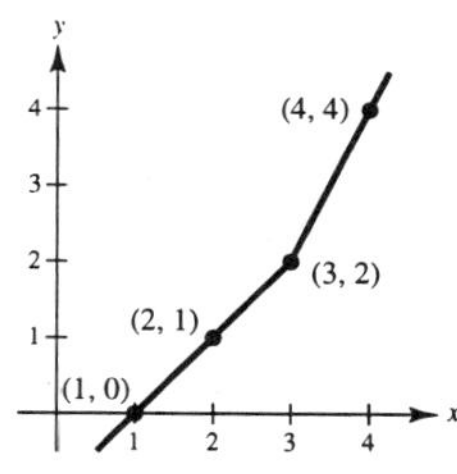

35. (a) Let x be the number of pounds of the commodity costing 1.25 per pound. Since there are 50 pounds total, the amount of the second commodity is $50 - x$. The total cost is

$$y = 1.25x + 1.60(50 - x)$$
$$= -0.35x + 80 \qquad 0 \le x \le 50.$$

(b) We find the inverse of the original function:

$$y = -0.35x + 80$$
$$0.35x = 80 - y$$
$$x = \tfrac{100}{35}(80 - y)$$

Inverse: $y = \frac{100}{35}(80 - x) = \frac{20}{7}(80 - x)$.

x represents cost and y represents pounds.

(c) Domain of inverse is $62.5 \le x \le 80$.

(d) If $x = 73$ in the inverse function,
$y = \frac{100}{35}(80 - 73) = \frac{100}{5} = 20$ pounds.

37. $f(x) = \dfrac{3}{4}x + 6$

One-to-one; has an inverse

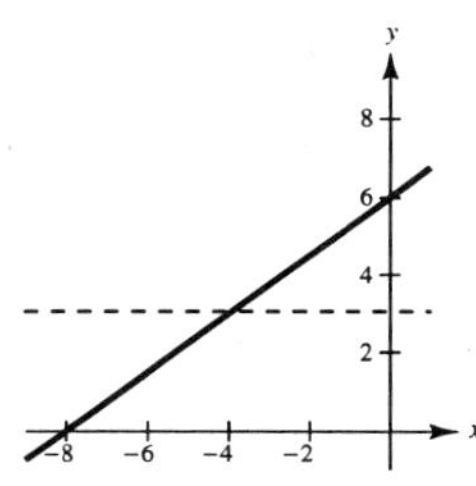

39. $f(\theta) = \sin \theta$

Not one-to-one; does not have an inverse

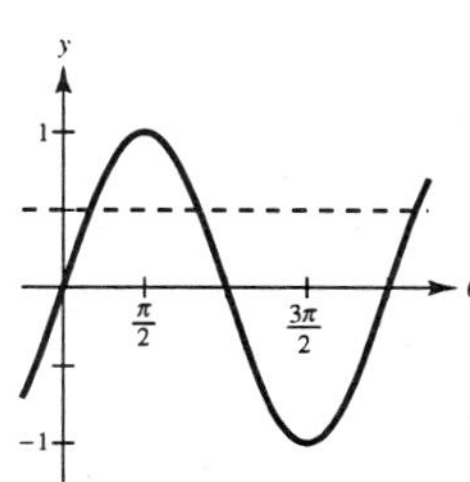

41. $h(s) = \dfrac{1}{s - 2} - 3$

One-to-one; has an inverse

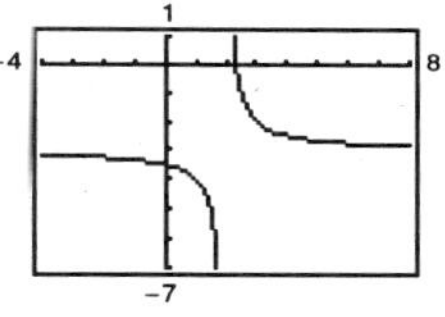

43. $f(x) = \ln x$

One-to-one; has an inverse

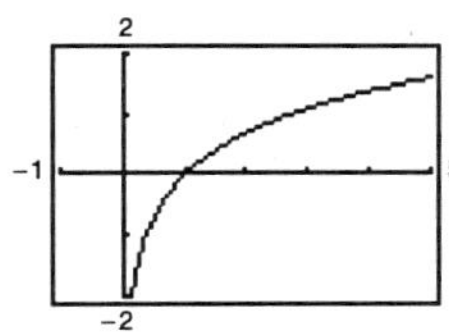

45. $g(x) = (x + 5)^3$

One-to-one; has an inverse

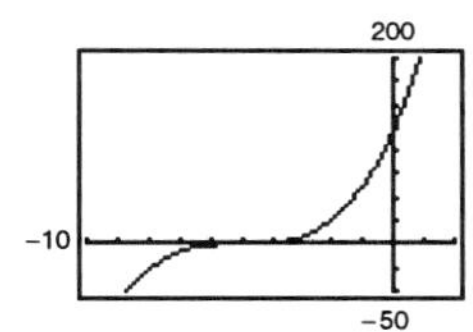

47. $f(x) = (x + a)^3 + b$

$f'(x) = 3(x + a)^2 \ge 0$ for all x.

f is increasing on $(-\infty, \infty)$. Therefore, f is strictly monotonic and has an inverse.

49. $f(x) = \dfrac{x^4}{4} - 2x^2$

$f'(x) = x^3 - 4x = 0$ when $x = 0, 2, -2$.

f is not strictly monotonic on $(-\infty, \infty)$. Therefore, f does not have an inverse.

51. $f(x) = 2 - x - x^3$

$f'(x) = -1 - 3x^2 < 0$ for all x.

f is decreasing on $(-\infty, \infty)$. Therefore, f is strictly monotonic and has an inverse.

53. $f(x) = (x - 4)^2$ on $[4, \infty)$

$f'(x) = 2(x - 4) > 0$ on $(4, \infty)$

f is increasing on $[4, \infty)$. Therefore, f is strictly monotonic and has an inverse.

55. $f(x) = \dfrac{4}{x^2}$ on $(0, \infty)$

$f'(x) = -\dfrac{8}{x^3} < 0$ on $(0, \infty)$

f is decreasing on $(0, \infty)$. Therefore, f is strictly monotonic and has an inverse.

57. $f(x) = \cos x$ on $[0, \pi]$

$f'(x) = -\sin x < 0$ on $(0, \pi)$

f is decreasing on $[0, \pi]$. Therefore, f is strictly monotonic and has an inverse.

59. f is not one-to-one because many different x-values yield the same y-value.

Example: $f(0) = f(\pi) = 0$

Not continuous at $\dfrac{(2n-1)\pi}{2}$.

61. $f(x) = \sqrt{x-2}$, Domain: $x \geq 2$

$f'(x) = \dfrac{1}{2\sqrt{x-2}} > 0$ for $x > 2$.

f is one-to-one; has an inverse

$$\sqrt{x-2} = y$$
$$x - 2 = y^2$$
$$x = y^2 + 2$$
$$y = x^2 + 2$$
$$f^{-1}(x) = x^2 + 2, x \geq 0$$

63. $f(x) = |x-2|, x \leq 2$

$$= -(x-2)$$
$$= 2 - x$$

f is one-to-one; has an inverse

$$2 - x = y$$
$$2 - y = x$$
$$f^{-1}(x) = 2 - x, \ x \geq 0$$

65. $f(x) = (x-3)^2$ is one-to-one for $x \geq 3$.

$$(x-3)^2 = y$$
$$x - 3 = \sqrt{y}$$
$$x = \sqrt{y} + 3$$
$$y = \sqrt{x} + 3$$
$$f^{-1}(x) = \sqrt{x} + 3, \ x \geq 0$$

67. $f(x) = |x+3|$ is one-to-one for $x \geq -3$.

$$x + 3 = y$$
$$x = y - 3$$
$$y = x - 3$$
$$f^{-1}(x) = x - 3, \ x \geq 0$$

69. Yes, the volume is an increasing function, and hence one-to-one. The inverse function gives the time t corresponding to the volume V.

71. No, $C(t)$ is not one-to-one because long distance costs are step functions. A call lasting 2.1 minutes costs the same as one lasting 2.2 minutes.

73.

$$f(x) = x^3 + 2x - 1, \ f(1) = 2 = a$$
$$f'(x) = 3x^2 + 2$$
$$(f^{-1})'(2) = \frac{1}{f'(f^{-1}(2))} = \frac{1}{f'(1)} = \frac{1}{3(1)^2 + 2} = \frac{1}{5}$$

75.

$$f(x) = \sin x, f\left(\frac{\pi}{6}\right) = \frac{1}{2} = a$$
$$f'(x) = \cos x$$
$$(f^{-1})'\left(\frac{1}{2}\right) = \frac{1}{f'(f^{-1}(1/2))} = \frac{1}{f'(\pi/6)} = \frac{1}{\cos(\pi/6)}$$
$$= \frac{1}{\sqrt{3}/2} = \frac{2\sqrt{3}}{3}$$

77.

$$f(x) = x^3 - \frac{4}{x}, \ f(2) = 6 = a$$
$$f'(x) = 3x^2 + \frac{4}{x^2}$$
$$(f^{-1})'(6) = \frac{1}{f'(f^{-1}(6))} = \frac{1}{f'(2)} = \frac{1}{3(2)^2 + (4/2^2)} = \frac{1}{13}$$

79. (a) Domain f = Domain $f^{-1} = (-\infty, \infty)$

(b) Range f = Range $f^{-1} = (-\infty, \infty)$

(c)

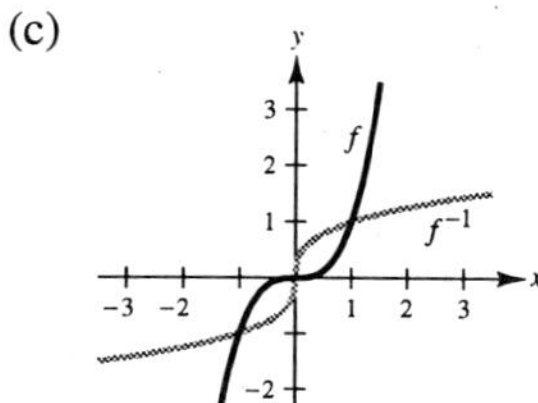

(d) $f(x) = x^3, \left(\frac{1}{2}, \frac{1}{8}\right)$

$f'(x) = 3x^2$

$f'\left(\frac{1}{2}\right) = \frac{3}{4}$

$f^{-1}(x) = \sqrt[3]{x}, \left(\frac{1}{8}, \frac{1}{2}\right)$

$(f^{-1})'(x) = \frac{1}{3\sqrt[3]{x^2}}$

$(f^{-1})'\left(\frac{1}{8}\right) = \frac{4}{3}$

81. (a) Domain $f = [4, \infty)$, Domain $f^{-1} = [0, \infty)$

(b) Range $f = [0, \infty)$, Range $f^{-1} = [4, \infty)$

(c)

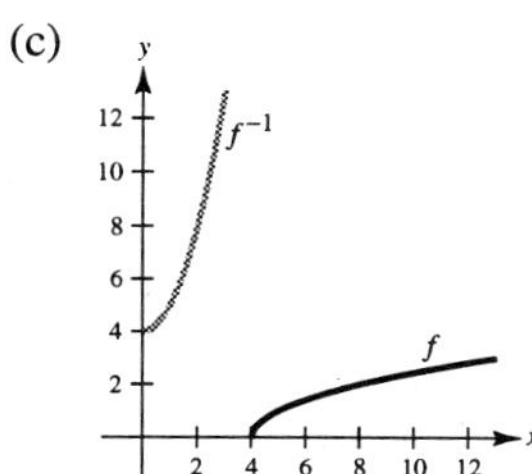

(d) $f(x) = \sqrt{x-4}, \ (5, 1)$

$f'(x) = \frac{1}{2\sqrt{x-4}}$

$f'(5) = \frac{1}{2}$

$f^{-1}(x) = x^2 + 4, \ (1, 5)$

$(f^{-1})'(x) = 2x$

$(f^{-1})'(1) = 2$

83. $x = y^3 - 7y^2 + 2$

$1 = 3y^2\frac{dy}{dx} - 14y\frac{dy}{dx}$

$\frac{dy}{dx} = \frac{1}{3y^2 - 14y}$. At $(-4, 1)$, $\frac{dy}{dx} = \frac{1}{3-14} = \frac{-1}{11}$.

Alternate solution: let $f(x) = x^3 - 7x^2 + 2$.

Then $f'(x) = 3x^2 - 14x$ and $f'(1) = -11$.

Hence, $\frac{dy}{dx} = \frac{1}{-11} = \frac{-1}{11}$.

In Exercises 85 and 87, use the following.

$$f(x) = \tfrac{1}{8}x - 3 \text{ and } g(x) = x^3$$

$$f^{-1}(x) = 8(x + 3) \text{ and } g^{-1}(x) = \sqrt[3]{x}$$

85. $(f^{-1} \circ g^{-1})(1) = f^{-1}(g^{-1}(1)) = f^{-1}(1) = 32$

87. $(f^{-1} \circ f^{-1})(6) = f^{-1}(f^{-1}(6)) = f^{-1}(72) = 600$

In Exercises 89 and 91, use the following.

$$f(x) = x + 4 \text{ and } g(x) = 2x - 5$$

$$f^{-1}(x) = x - 4 \text{ and } g^{-1}(x) = \frac{x+5}{2}$$

89. $(g^{-1} \circ f^{-1})(x) = g^{-1}(f^{-1}(x))$

$= g^{-1}(x - 4)$

$= \frac{(x-4)+5}{2}$

$= \frac{x+1}{2}$

91. $(f \circ g)(x) = f(g(x))$

$= f(2x - 5)$

$= (2x - 5) + 4$

$= 2x - 1$

Hence, $(f \circ g)^{-1}(x) = \frac{x+1}{2}$

(**Note:** $(f \circ g)^{-1} = g^{-1} \circ f^{-1}$)

93. Let $(f \circ g)(x) = y$ then $x = (f \circ g)^{-1}(y)$. Also,

$$\begin{aligned}(f \circ g)(x) &= y\\ f(g(x)) &= y\\ g(x) &= f^{-1}(y)\\ x &= g^{-1}(f^{-1}(y))\\ &= (g^{-1} \circ f^{-1})(y)\end{aligned}$$

Since f and g are one-to-one functions, $(f \circ g)^{-1} = g^{-1} \circ f^{-1}$.

95. Suppose $g(x)$ and $h(x)$ are both inverses of $f(x)$. Then the graph of $f(x)$ contains the point (a, b) if and only if the graphs of $g(x)$ and $h(x)$ contain the point (b, a). Since the graphs of $g(x)$ and $h(x)$ are the same, $g(x) = h(x)$. Therefore, the inverse of $f(x)$ is unique.

97. False

Let $f(x) = x^2$.

99. True

101. Not true

Let $f(x) = \begin{cases} x, & 0 \le x \le 1 \\ 1 - x, & 1 < x \le 2 \end{cases}$.

f is one-to-one, but not strictly monotonic.

103.
$$f(x) = \int_2^x \frac{dt}{\sqrt{1 + t^4}},\ f(2) = 0$$

$$f'(x) = \frac{1}{\sqrt{1 + x^4}}$$

$$(f^{-1})'(0) = \frac{1}{f'(2)} = \frac{1}{1/\sqrt{17}} = \sqrt{17}$$

Section 5.4 Exponential Functions: Differentiation and Integration

1. $e^0 = 1$

$\ln 1 = 0$

3. $\ln 2 = 0.6931$

$e^{0.6931\ldots} = 2$

5. $e^{\ln x} = 4$

$x = 4$

7. $\ln x = 2$

$x = e^2 \approx 7.3891$

9. $y = e^{-x}$

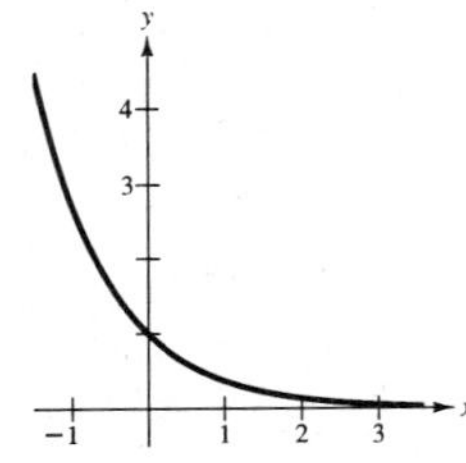

11. $y = e^{-x^2}$

Symmetric with respect to the y-axis

Horizontal asymptote: $y = 0$

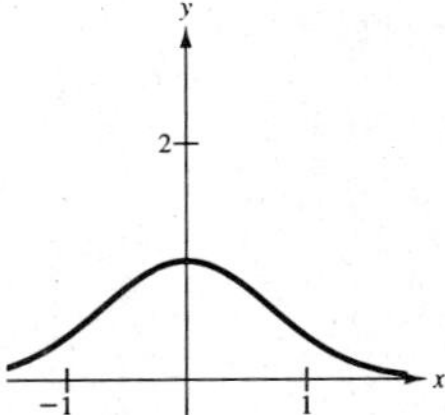

13. (a)

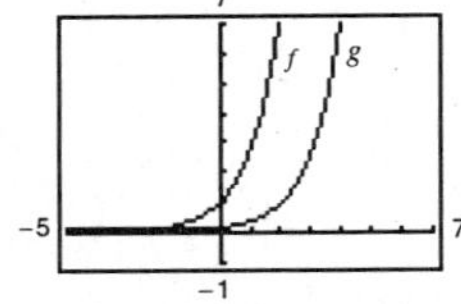

Horizontal shift 2 units to the right

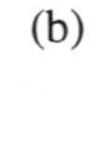
(b)

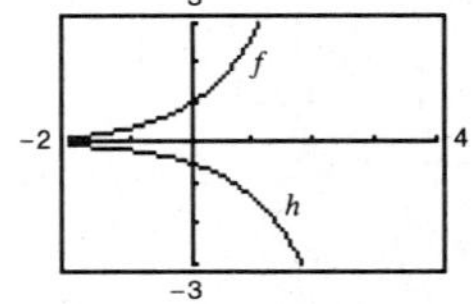

A reflection in the x-axis and a vertical shrink

(c)

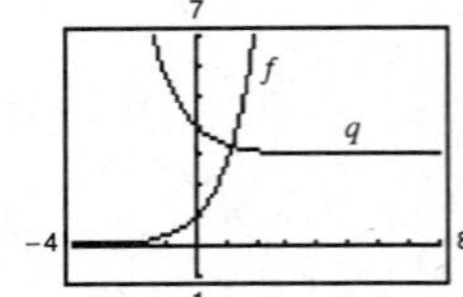

Vertical shift 3 units upward and a reflection in the y-axis

15. $f(x) = e^{2x}$

$g(x) = \ln\sqrt{x} = \dfrac{1}{2}\ln x$

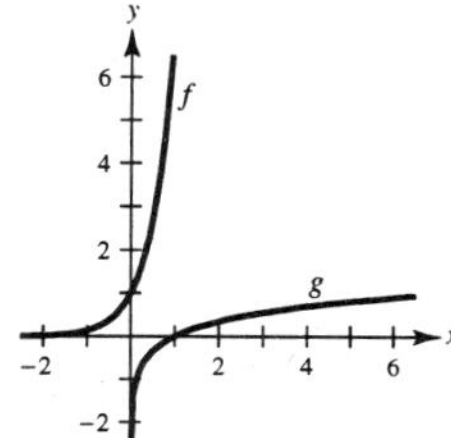

17. $f(x) = e^x - 1$

$g(x) = \ln(x + 1)$

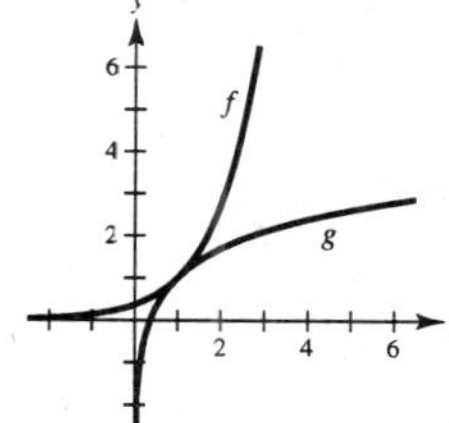

19. $y = Ce^{ax}$

Horizontal asymptote: $y = 0$

Matches (c)

21. $y = C(1 - e^{-ax})$

Vertical shift C units

Reflection in both the x- and y-axes

Matches (a)

23.

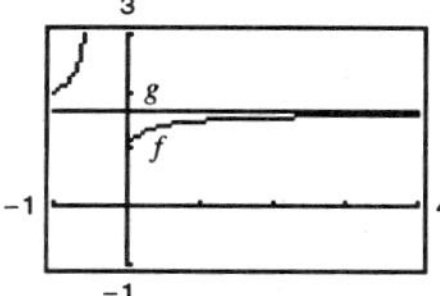

As $x \to \infty$, the graph of f approaches the graph of g.

$$\lim_{x \to \infty}\left(1 + \frac{0.5}{x}\right)^x = e^{0.5}$$

25. $\left(1 + \dfrac{1}{1{,}000{,}000}\right)^{1{,}000{,}000} \approx 2.718280469$

$e \approx 2.718281828$

$e > \left(1 + \dfrac{1}{1{,}000{,}000}\right)^{1{,}000{,}000}$

27. (a) $y = e^{3x}$

$y' = 3e^{3x}$

At $(0, 1)$, $y' = 3$.

(b) $y = e^{-3x}$

$y' = -3e^{-3x}$

At $(0, 1)$, $y' = -3$.

29. $f(x) = e^{2x}$

$f'(x) = 2e^{2x}$

31. $f(x) = e^{-2x+x^2}$

$\dfrac{dy}{dx} = 2(x - 1)e^{-2x+x^2}$

33. $y = e^{\sqrt{x}}$

$\dfrac{dy}{dx} = \dfrac{e^{\sqrt{x}}}{2\sqrt{x}}$

35. $g(t) = (e^{-t} + e^t)^3$

$g'(t) = 3(e^{-t} + e^t)^2(e^t - e^{-t})$

37. $y = \ln e^{x^2} = x^2$

$\dfrac{dy}{dx} = 2x$

39. $y = \ln(1 + e^{2x})$

$\dfrac{dy}{dx} = \dfrac{2e^{2x}}{1 + e^{2x}}$

41. $y = \dfrac{2}{e^x + e^{-x}} = 2(e^x + e^{-x})^{-1}$

$\dfrac{dy}{dx} = -2(e^x + e^{-x})^{-2}(e^x - e^{-x}) = \dfrac{-2(e^x - e^{-x})}{(e^x + e^{-x})^2}$

43. $y = x^2e^x - 2xe^x + 2e^x = e^x(x^2 - 2x + 2)$

$\dfrac{dy}{dx} = e^x(2x - 2) + e^x(x^2 - 2x + 2) = x^2e^x$

45. $f(x) = e^{-x}\ln x$

$f'(x) = e^{-x}\left(\dfrac{1}{x}\right) - e^{-x}\ln x = e^{-x}\left(\dfrac{1}{x} - \ln x\right)$

47. $y = e^x(\sin x + \cos x)$

$\dfrac{dy}{dx} = e^x(\cos x - \sin x) + (\sin x + \cos x)(e^x)$

$= e^x(2\cos x) = 2e^x\cos x$

49.

$$xe^y - 10x + 3y = 0$$

$$xe^y\frac{dy}{dx} + e^y - 10 + 3\frac{dy}{dx} = 0$$

$$\frac{dy}{dx}(xe^y + 3) = 10 - e^y$$

$$\frac{dy}{dx} = \frac{10 - e^y}{xe^y + 3}$$

51. $f(x) = (3 + 2x)e^{-3x}$

$$f'(x) = (3 + 2x)(-3e^{-3x}) + 2e^{-3x}$$

$$= (-7 - 6x)e^{-3x}$$

$$f''(x) = (-7 - 6x)(-3e^{-3x}) - 6e^{-3x}$$

$$= 3(6x + 5)e^{-3x}$$

53.

$$y = e^x(\cos\sqrt{2}x + \sin\sqrt{2}x)$$

$$y' = e^x(-\sqrt{2}\sin\sqrt{2}x + \sqrt{2}\cos\sqrt{2}x) + e^x(\cos\sqrt{2}x + \sin\sqrt{2}x)$$

$$= e^x\left[(1 + \sqrt{2})\cos\sqrt{2}x + (1 - \sqrt{2})\sin\sqrt{2}x\right]$$

$$y'' = e^x\left[-(\sqrt{2} + 2)\sin\sqrt{2}x + (\sqrt{2} - 2)\cos\sqrt{2}x\right] + e^x\left[(1 + \sqrt{2})\cos\sqrt{2}x + (1 - \sqrt{2})\sin\sqrt{2}x\right]$$

$$= e^x\left[(-1 - 2\sqrt{2})\sin\sqrt{2}x + (-1 + 2\sqrt{2})\cos\sqrt{2}x\right]$$

$$-2y' + 3y = -2e^x\left[(1 + \sqrt{2})\cos\sqrt{2}x + (1 - \sqrt{2})\sin\sqrt{2}x\right] + 3e^x\left[\cos\sqrt{2}x + \sin\sqrt{2}x\right]$$

$$= e^x\left[(1 - 2\sqrt{2})\cos\sqrt{2}x + (1 + 2\sqrt{2})\sin\sqrt{2}x\right] = -y''$$

Therefore, $-2y' + 3y = -y'' \Rightarrow y'' - 2y' + 3y = 0.$

55. $f(x) = \dfrac{1}{\sqrt{2\pi}}e^{-x^2/2}$

$$f'(x) = \frac{1}{\sqrt{2\pi}}(-xe^{-x^2/2}) = 0 \text{ when } x = 0.$$

$$f''(x) = \frac{1}{\sqrt{2\pi}}(x^2e^{-x^2/2} - e^{-x^2/2})$$

$$= \frac{e^{-x^2/2}}{\sqrt{2\pi}}(x^2 - 1) = 0 \text{ when } x = \pm 1.$$

Relative maximum: $\left(0, \dfrac{1}{\sqrt{2\pi}}\right)$

Points of inflection: $\left(\pm 1, \dfrac{1}{\sqrt{2e\pi}}\right)$

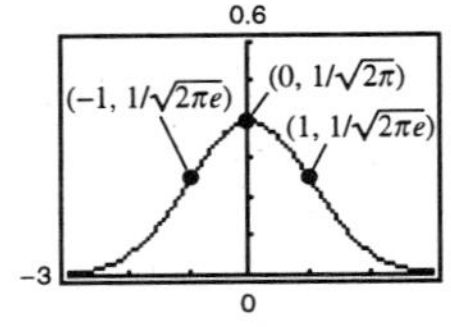

57. $f(x) = \dfrac{e^x + e^{-x}}{2}$

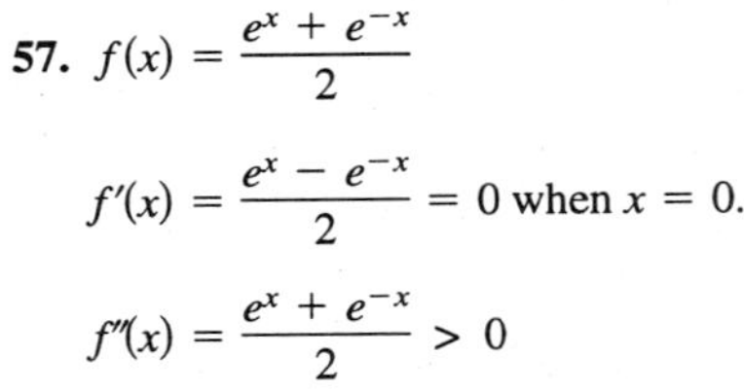

$$f'(x) = \frac{e^x - e^{-x}}{2} = 0 \text{ when } x = 0.$$

$$f''(x) = \frac{e^x + e^{-x}}{2} > 0$$

Relative minimum: $(0, 1)$

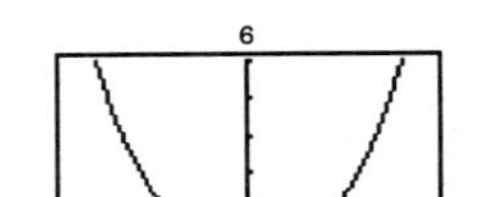

59. $f(x) = x^2e^{-x}$

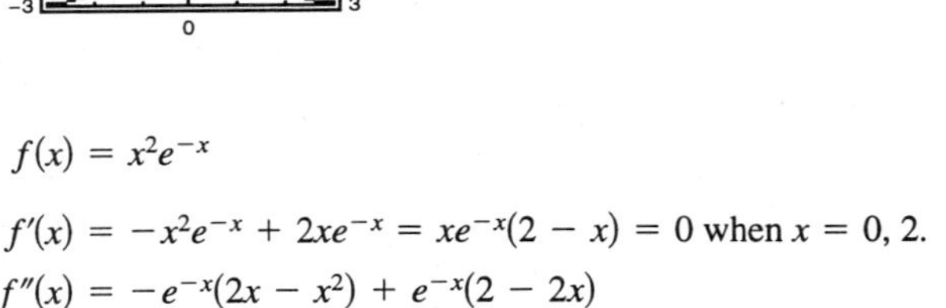

$$f'(x) = -x^2e^{-x} + 2xe^{-x} = xe^{-x}(2 - x) = 0 \text{ when } x = 0, 2.$$

$$f''(x) = -e^{-x}(2x - x^2) + e^{-x}(2 - 2x)$$

$$= e^{-x}(x^2 - 4x + 2) = 0 \text{ when } x = 2 \pm \sqrt{2}.$$

Relative minimum: $(0, 0)$

Relative maximum: $(2, 4e^{-2})$

$$x = 2 \pm \sqrt{2}$$

$$y = (2 \pm \sqrt{2})^2e^{-(2\pm\sqrt{2})}$$

Points of inflection: $(3.414, 0.384)$, $(0.586, 0.191)$

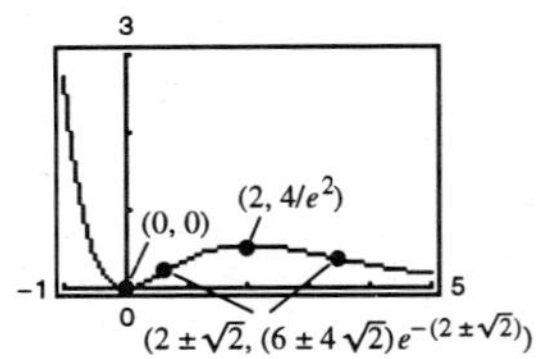

61. $A = (\text{base})(\text{height}) = 2xe^{-x^2}$

$$\frac{dA}{dx} = -4x^2e^{-x^2} + 2e^{-x^2}$$

$$= 2e^{-x^2}(1 - 2x^2) = 0 \text{ when } x = \frac{\sqrt{2}}{2}.$$

$A = \sqrt{2}e^{-1/2}$

63. $y = \dfrac{L}{1 + ae^{-x/b}}, a > 0, b > 0, L > 0$

$$y' = \frac{-L\left(-\frac{a}{b}e^{-x/b}\right)}{(1 + ae^{-x/b})^2} = \frac{\frac{aL}{b}e^{-x/b}}{(1 + ae^{-x/b})^2}$$

$$y'' = \frac{(1 + ae^{-x/b})^2\left(\frac{-aL}{b^2}e^{-x/b}\right) - \left(\frac{aL}{b}e^{-x/b}\right)2(1 + ae^{-x/b})\left(\frac{-a}{b}e^{-x/b}\right)}{(1 + ae^{-x/b})^4}$$

$$= \frac{(1 + ae^{-x/b})\left(\frac{-aL}{b^2}e^{-x/b}\right) + 2\left(\frac{aL}{b}e^{-x/b}\right)\left(\frac{a}{b}e^{-x/b}\right)}{(1 + ae^{-x/b})^3}$$

$$= \frac{Lae^{-x/b}[ae^{-x/b} - 1]}{(1 + ae^{-x/b})^3 b^2}$$

$y'' = 0$ if $ae^{-x/b} = 1 \Rightarrow \dfrac{-x}{b} = \ln\left(\dfrac{1}{a}\right) \Rightarrow x = b \ln a$

$$y(b \ln a) = \frac{L}{1 + ae^{-(b \ln a)/b}} = \frac{L}{1 + a(1/a)} = \frac{L}{2}$$

Therefore, the y-coordinate of the inflection point is $L/2$.

65. $e^{-x} = x \Rightarrow f(x) = x - e^{-x}$

$f'(x) = 1 + e^{-x}$

$$x_{n+1} = x_n - \frac{f(x_n)}{f'(x_n)} = x_n - \frac{x_n - e^{-x_n}}{1 + e^{-x_n}}$$

$x_1 = 1$

$$x_2 = x_1 - \frac{f(x_1)}{f'(x_1)} \approx 0.5379$$

$$x_3 = x_2 - \frac{f(x_2)}{f'(x_2)} \approx 0.5670$$

$$x_4 = x_3 - \frac{f(x_3)}{f'(x_3)} \approx 0.5671$$

We approximate the root of f to be $x = 0.567$.

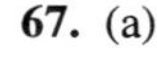

67. (a)

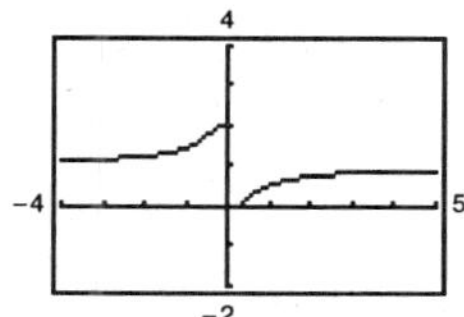

(b) When x increases without bound, $1/x$ approaches zero, and $e^{1/x}$ approaches 1. Therefore, $f(x)$ approaches $2/(1 + 1) = 1$. Thus, $f(x)$ has a horizontal asymptote at $y = 1$. As x approaches zero from the right, $1/x$ approaches ∞, $e^{1/x}$ approaches ∞ and $f(x)$ approaches zero. As x approaches zero from the left, $1/x$ approaches $-\infty$, $e^{1/x}$ approaches zero, and $f(x)$ approaches 2. The limit does not exist since the left limit does not equal the right limit. Therefore, $x = 0$ is a nonremovable discontinuity.

69.

h	0	5	10	15	20
P	10,332	5,583	2,376	1,240	517
$\ln P$	9.243	8.627	7.773	7.123	6.248

(a)

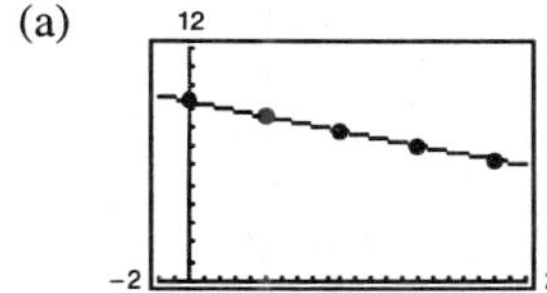

$y = -0.1499h + 9.3018$ is the regression line for data $(h, \ln P)$.

—CONTINUED—

69. —CONTINUED—

(b) $\ln P = ah + b$

$$P = e^{ah+b} = e^b e^{ah}$$

$$P = Ce^{ah},\ C = e^b$$

For our data, $a = -0.1499$ and $C = e^{9.3018} = 10{,}957.7$

$$P = 10{,}957.7e^{-0.1499h}$$

(c)

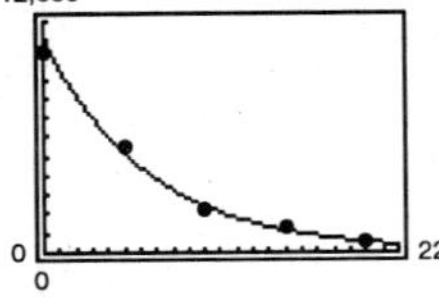

(d) $\dfrac{dP}{dh} = (10{,}957.71)(-0.1499)e^{-0.1499h}$

$$= -1642.56e^{-0.1499h}$$

For $h = 5$, $\dfrac{dP}{dh} = -776.3$. For $h = 18$, $\dfrac{dP}{dh} \approx -110.6$.

71. $f(x) = e^{x/2}, f(0) = 1$

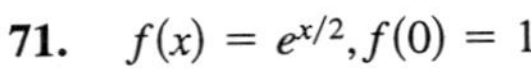

$f'(x) = \dfrac{1}{2}e^{x/2}, f'(0) = \dfrac{1}{2}$

$f''(x) = \dfrac{1}{4}e^{x/2}, f''(0) = \dfrac{1}{4}$

$P_1(x) = 1 + \dfrac{1}{2}(x - 0) = \dfrac{x}{2} + 1, P_1(0) = 1$

$P_1'(x) = \dfrac{1}{2}, P_1'(0) = \dfrac{1}{2}$

$P_2(x) = 1 + \dfrac{1}{2}(x - 0) + \dfrac{1}{8}(x - 0)^2 = \dfrac{x^2}{8} + \dfrac{x}{2} + 1, P_2(0) = 1$

$P_2'(x) = \dfrac{1}{4}x + \dfrac{1}{2}, P_2'(0) = \dfrac{1}{2}$

$P_2''(x) = \dfrac{1}{4}, P_2''(0) = \dfrac{1}{4}$

The values of f, P_1, P_2 and their first derivatives agree at $x = 0$. The values of the second derivatives of f and P_2 agree at $x = 0$.

73. (a) $y = e^x$

$y_1 = 1 + x$

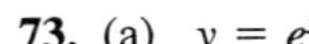

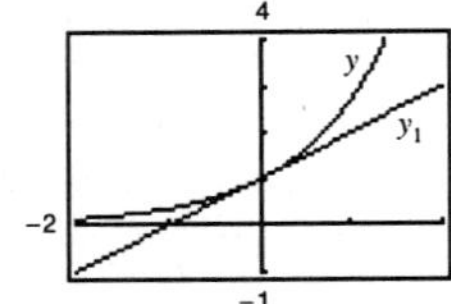

(b) $y = e^x$

$y_2 = 1 + x + \left(\dfrac{x^2}{2}\right)$

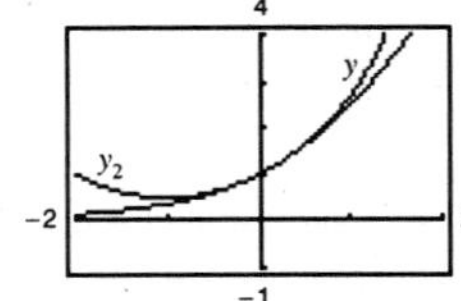

(c) $y = e^x$

$y_3 = 1 + x + \dfrac{x^2}{2} + \dfrac{x^3}{6}$

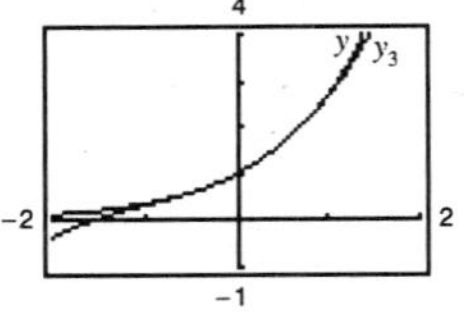

75. Let $u = 5x$, $du = 5\,dx$.

$$\int e^{5x}\,5dx = e^{5x} + C$$

77. Let $u = -2x$, $du = -2\,dx$.

$$\int_0^1 e^{-2x}\,dx = -\frac{1}{2}\int_0^1 e^{-2x}(-2)\,dx = \left[-\frac{1}{2}e^{-2x}\right]_0^1 = \frac{1}{2}(1 - e^{-2}) = \frac{e^2 - 1}{2e^2}$$

79. Let $u = 1 + e^{-x}$, $du = -e^{-x}\,dx$.

$$\int \frac{e^{-x}}{1 + e^{-x}}\,dx = -\int \frac{-e^{-x}}{1 + e^{-x}}\,dx$$

$$= -\ln(1 + e^{-x}) + C = \ln\left(\frac{e^x}{e^x + 1}\right) + C$$

$$= x - \ln(e^x + 1) + C$$

81. Let $u = \dfrac{3}{x}$, $du = -\dfrac{3}{x^2}\,dx$.

$$\int_1^3 \frac{e^{3/x}}{x^2}\,dx = -\frac{1}{3}\int_1^3 e^{3/x}\left(-\frac{3}{x^2}\right)dx$$

$$= \left[-\frac{1}{3}e^{3/x}\right]_1^3 = \frac{e}{3}(e^2 - 1)$$

83. Let $u = 1 - e^x$, $du = -e^x\,dx$.

$$\int e^x\sqrt{1 - e^x}\,dx = -\int (1 - e^x)^{1/2}(-e^x)\,dx$$

$$= -\frac{2}{3}(1 - e^x)^{3/2} + C$$

85. Let $u = e^x - e^{-x}$, $du = (e^x + e^{-x})\,dx$.

$$\int \frac{e^x + e^{-x}}{e^x - e^{-x}}\,dx = \ln|e^x - e^{-x}| + C$$

87. $$\int \frac{5 - e^x}{e^{2x}}\,dx = \int 5e^{-2x}\,dx - \int e^{-x}\,dx$$

$$= -\frac{5}{2}e^{-2x} + e^{-x} + C$$

89. $$\int e^{\sin \pi x}\cos \pi x\,dx = \frac{1}{\pi}\int e^{\sin \pi x}(\pi \cos \pi x)\,dx$$

$$= \frac{1}{\pi}e^{\sin \pi x} + C$$

91. $$\int e^{-x}\tan(e^{-x})\,dx = -\int [\tan(e^{-x})](-e^{-x})\,dx$$

$$= \ln|\cos(e^{-x})| + C$$

93. Let $u = ax^2$, $du = 2ax\,dx$. (Assume $a \neq 0$)

$$y = \int xe^{ax^2}\,dx = \frac{1}{2a}\int e^{ax^2}(2ax)\,dx = \frac{1}{2a}e^{ax^2} + C$$

95. $$f'(x) = \int \frac{1}{2}(e^x + e^{-x})\,dx = \frac{1}{2}(e^x - e^{-x}) + C_1$$

$$f'(0) = C_1 = 0$$

$$f(x) = \int \frac{1}{2}(e^x - e^{-x})\,dx = \frac{1}{2}(e^x + e^{-x}) + C_2$$

$$f(0) = 1 + C_2 = 1 \Rightarrow C_2 = 0$$

$$f(x) = \frac{1}{2}(e^x + e^{-x})$$

97. (a)

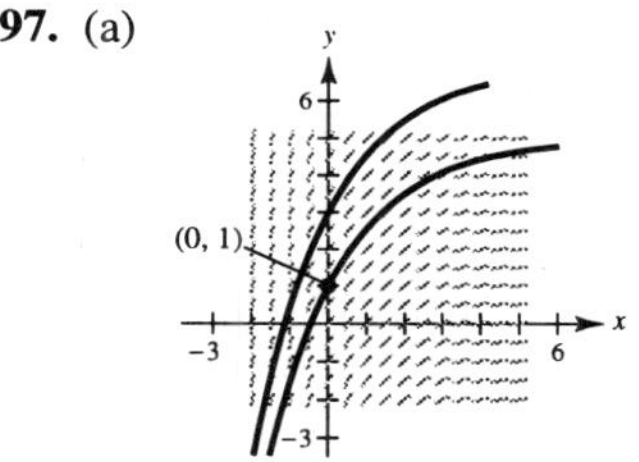

(b) $\dfrac{dy}{dx} = 2e^{-x/2}$, $(0, 1)$

$$y = \int 2e^{-x/2}\,dx = -4\int e^{-x/2}\left(-\frac{1}{2}dx\right)$$

$$= -4e^{-x/2} + C$$

$(0, 1)$: $1 = -4e^0 + C = -4 + C \Rightarrow C = 5$

$$y = -4e^{-x/2} + 5$$

99. $$\int_0^5 e^x\,dx = \Big[e^x\Big]_0^5 = e^5 - 1 \approx 147.413$$

101. $$\int_0^{\sqrt{2}} xe^{-(x^2/2)}dx = \Big[-e^{-(x^2/2)}\Big]_0^{\sqrt{2}}$$

$$= -e^{-1} + 1 \approx 0.632$$

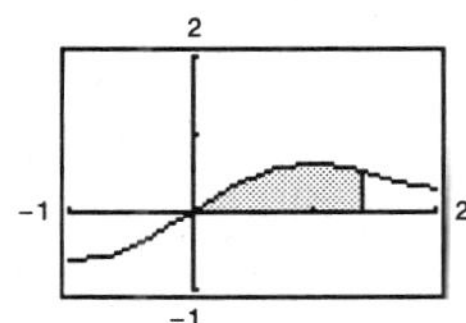

103. (a) $f(u - v) = e^{u-v} = (e^u)(e^{-v}) = \dfrac{e^u}{e^v} = \dfrac{f(u)}{f(v)}$

(b) $f(kx) = e^{kx} = (e^x)^k = [f(x)]^k$.

105. $0.0665 \displaystyle\int_{48}^{60} e^{-0.0139(t-48)^2}\,dt$

Graphing Utility: $0.4772 = 47.72\%$

107.

t	0	1	2	3	4
R	425	240	118	71	36
$\ln R$	6.052	5.481	4.771	4.263	3.584

(a) $\ln R = -0.6155t + 6.0609$

$R = e^{-0.6155t + 6.0609} = 428.78e^{-0.6155t}$

(c) $\displaystyle\int_0^4 R(t)\,dt = \int_0^4 428.78e^{-0.6155t}\,dt \approx 637.2$ liters

(b)

109. $f(x) = \dfrac{\ln x}{x}$

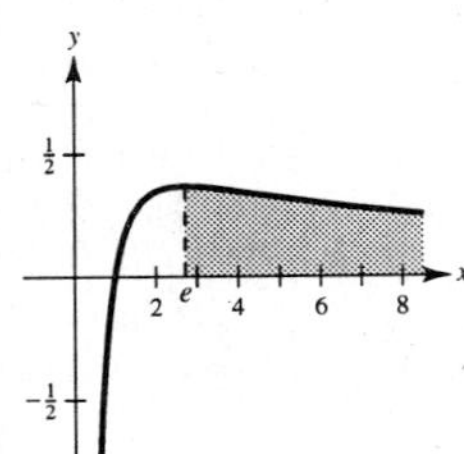

(a) $f'(x) = \dfrac{1 - \ln x}{x^2} = 0$ when $x = e$.

On $(0, e), f'(x) > 0 \Rightarrow f$ is increasing.

On $(e, \infty), f'(x) < 0 \Rightarrow f$ is decreasing.

(b) For $e \le A < B$, we have:

$$\frac{\ln A}{A} > \frac{\ln B}{B}$$

$$B \ln A > A \ln B$$

$$\ln A^B > \ln B^A$$

$$A^B > B^A.$$

(c) Since $e < \pi$, from part (b) we have $e^\pi > \pi^e$.

Section 5.5 Bases Other than e and Applications

1. $\log_2 \frac{1}{8} = \log_2 2^{-3} = -3$

3. $\log_7 1 = 0$

5. (a) $2^3 = 8$

$\log_2 8 = 3$

(b) $3^{-1} = \frac{1}{3}$

$\log_3 \frac{1}{3} = -1$

7. (a) $\log_{10} 0.01 = -2$

$10^{-2} = 0.01$

(b) $\log_{0.5} 8 = -3$

$0.5^{-3} = 8$

$\left(\frac{1}{2}\right)^{-3} = 8$

9. (a) $\log_{10} 1000 = x$

$10^x = 1000$

$x = 3$

(b) $\log_{10} 0.1 = x$

$10^x = 0.1$

$x = -1$

11. (a) $\log_3 x = -1$

$3^{-1} = x$

$x = \frac{1}{3}$

(b) $\log_2 x = -4$

$2^{-4} = x$

$x = \frac{1}{16}$

13. (a)

$$x^2 - x = \log_5 25$$
$$x^2 - x = \log_5 5^2 = 2$$
$$x^2 - x - 2 = 0$$
$$(x + 1)(x - 2) = 0$$
$$x = -1 \text{ OR } x = 2$$

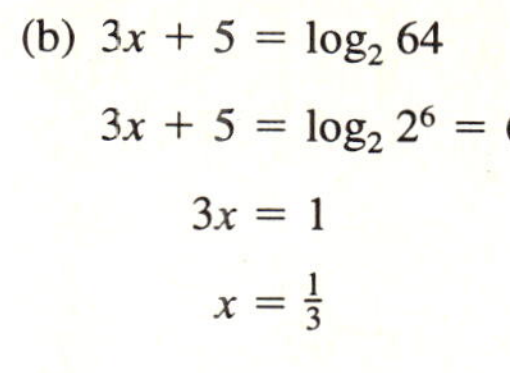

(b)

$$3x + 5 = \log_2 64$$
$$3x + 5 = \log_2 2^6 = 6$$
$$3x = 1$$
$$x = \tfrac{1}{3}$$

15. $y = 3^x$

x	-2	-1	0	1	2
y	$\frac{1}{9}$	$\frac{1}{3}$	1	3	9

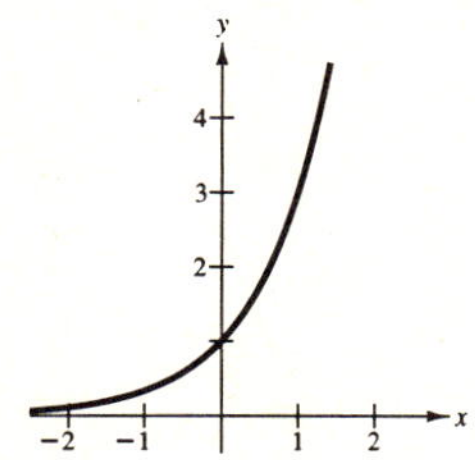

17. $y = \left(\frac{1}{3}\right)^x = 3^{-x}$

x	-2	-1	0	1	2
y	9	3	1	$\frac{1}{3}$	$\frac{1}{9}$

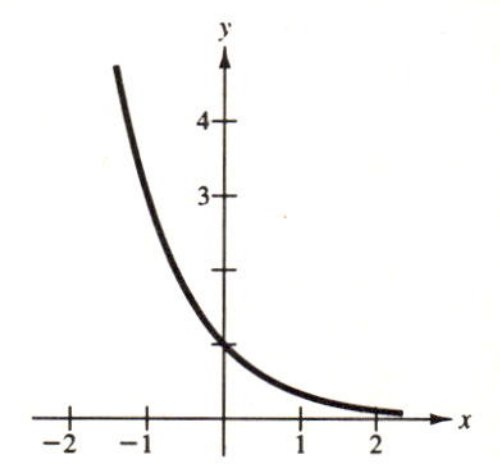

19. $h(x) = 5^{x-2}$

x	-1	0	1	2	3
y	$\frac{1}{125}$	$\frac{1}{25}$	$\frac{1}{5}$	1	5

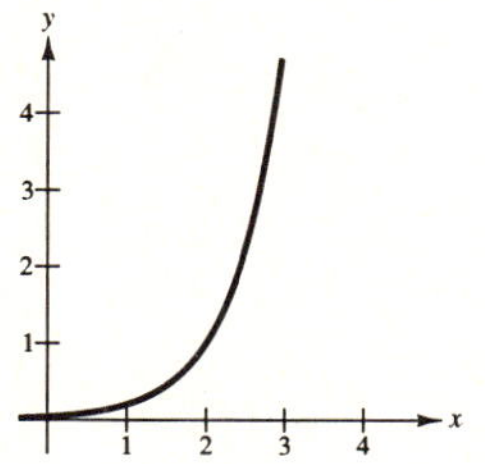

21. $g(x) = 6(2^{1-x}) - 25$

Zero: $x \approx -1.059$

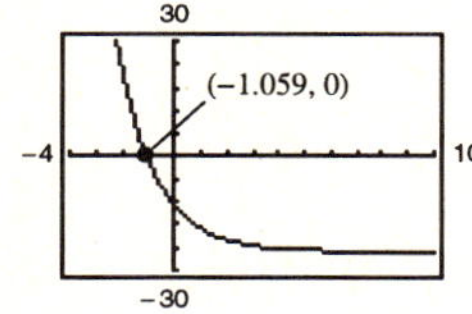

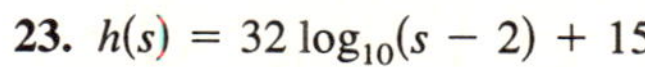

23. $h(s) = 32 \log_{10}(s - 2) + 15$

Zero: $s \approx 2.340$

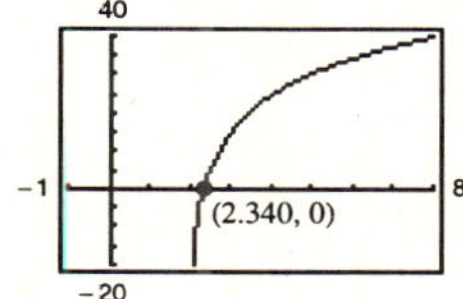

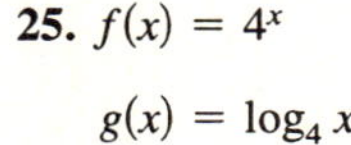

25. $f(x) = 4^x$

$g(x) = \log_4 x$

x	-2	-1	0	$\frac{1}{2}$	1
$f(x)$	$\frac{1}{16}$	$\frac{1}{4}$	1	2	4

x	$\frac{1}{16}$	$\frac{1}{4}$	1	2	4
$g(x)$	-2	-1	0	$\frac{1}{2}$	1

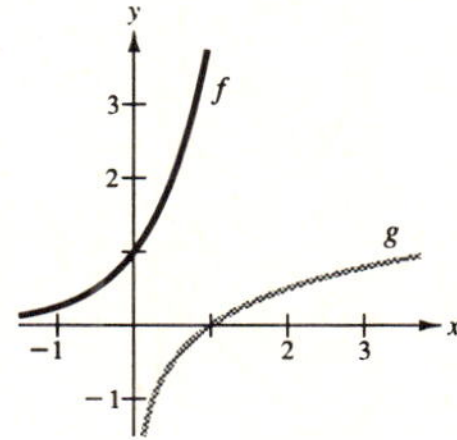

27.

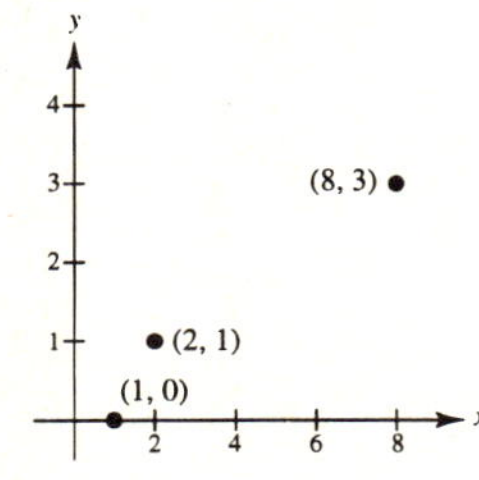

x	1	2	8
y	0	1	3

(a) y is an exponential function of x: False

(b) y is a logarithmic function of x: True; $y = \log_2 x$

(c) x is an exponential function of y: True, $2^y = x$

(d) y is a linear function of x: False

29. $f(x) = 4^x$

$$f'(x) = (\ln 4)\, 4^x$$

31. $y = 5^{x-2}$

$$\frac{dy}{dx} = (\ln 5)\, 5^{x-2}$$

33. $g(t) = t^2 2^t$

$$g'(t) = t^2 (\ln 2)\, 2^t + (2t)\, 2^t$$
$$= t2^t (t \ln 2 + 2)$$
$$= 2^t t(2 + t \ln 2)$$

35. $h(\theta) = 2^{-\theta}\cos \pi\theta$

$h'(\theta) = 2^{-\theta}(-\pi \sin \pi\theta) - (\ln 2)2^{-\theta}\cos \pi\theta$

$= -2^{-\theta}[(\ln 2)\cos \pi\theta + \pi \sin \pi\theta]$

37. $y = \log_3 x$

$\dfrac{dy}{dx} = \dfrac{1}{x \ln 3}$

39. $f(x) = \log_2 \dfrac{x^2}{x-1} = 2\log_2 x - \log_2(x-1)$

$f'(x) = \dfrac{2}{x \ln 2} - \dfrac{1}{(x-1)\ln 2} = \dfrac{x-2}{(\ln 2)x(x-1)}$

41. $y = \log_5 \sqrt{x^2-1} = \dfrac{1}{2}\log_5(x^2-1)$

$\dfrac{dy}{dx} = \dfrac{1}{2}\cdot\dfrac{2x}{(x^2-1)\ln 5} = \dfrac{x}{(x^2-1)\ln 5}$

43. $g(t) = \dfrac{10\log_4 t}{t} = \dfrac{10}{\ln 4}\left(\dfrac{\ln t}{t}\right)$

$g'(t) = \dfrac{10}{\ln 4}\left[\dfrac{t(1/t) - \ln t}{t^2}\right]$

$= \dfrac{10}{t^2 \ln 4}[1 - \ln t] = \dfrac{5}{t^2 \ln 2}(1 - \ln t)$

45. $y = x^{2/x}$

$\ln y = \dfrac{2}{x}\ln x$

$\dfrac{1}{y}\left(\dfrac{dy}{dx}\right) = \dfrac{2}{x}\left(\dfrac{1}{x}\right) + \ln x\left(-\dfrac{2}{x^2}\right) = \dfrac{2}{x^2}(1 - \ln x)$

$\dfrac{dy}{dx} = \dfrac{2y}{x^2}(1 - \ln x) = 2x^{(2/x)-2}(1 - \ln x)$

47. $y = (x-2)^{x+1}$

$\ln y = (x+1)\ln(x-2)$

$\dfrac{1}{y}\left(\dfrac{dy}{dx}\right) = (x+1)\left(\dfrac{1}{x-2}\right) + \ln(x-2)$

$\dfrac{dy}{dx} = y\left[\dfrac{x+1}{x-2} + \ln(x-2)\right]$

$= (x-2)^{x+1}\left[\dfrac{x+1}{x-2} + \ln(x-2)\right]$

49. $f(x) = \log_2 x \Rightarrow f'(x) = \dfrac{1}{x \ln 2}$

$g(x) = x^x \Rightarrow g'(x) = x^x(1 + \ln x)$

[**Note:** Let $y = g(x)$. Then: $\ln y = \ln x^x = x \ln x$

$\dfrac{1}{y}y' = x\cdot\dfrac{1}{x} + \ln x$

$y' = y(1 + \ln x)$

$y' = x^x(1 + \ln x) = g'(x)$.]

$h(x) = x^2 \Rightarrow h'(x) = 2x$

$k(x) = 2^x \Rightarrow k'(x) = (\ln 2)2^x$

From greatest to smallest rate of growth:
$g(x), k(x), h(x), f(x)$

51. $C(t) = P(1.05)^t$

(a) $C(10) = 24.95(1.05)^{10}$

$\approx \$40.64$

(b) $\dfrac{dC}{dt} = P(\ln 1.05)(1.05)^t$

When $t = 1$: $\dfrac{dC}{dt} \approx 0.051P$

When $t = 8$: $\dfrac{dC}{dt} \approx 0.072P$

(c) $\dfrac{dC}{dt} = (\ln 1.05)[P(1.05)^t]$

$= (\ln 1.05)C(t)$

The constant of proportionality is $\ln 1.05$.

53. $P = \$1000,\ r = 3\tfrac{1}{2}\% = 0.035,\ t = 10$

$A = 1000\left(1 + \dfrac{0.035}{n}\right)^{10n}$

$A = 1000e^{(0.035)(10)} = 1419.07$

n	1	2	4	12	365	Continuous
A	1410.60	1414.78	1416.91	1418.34	1419.04	1419.07

55. $P = \$1000,\ r = 5\% = 0.05,\ t = 30$

$A = 1000\left(1 + \dfrac{0.05}{n}\right)^{30n}$

$A = 1000e^{(0.05)30} = 4481.69$

n	1	2	4	12	365	Continuous
A	4321.94	4399.79	4440.21	4467.74	4481.23	4481.69

57. $100{,}000 = Pe^{0.05t} \Rightarrow P = 100{,}000e^{-0.05t}$

t	1	10	20	30	40	50
P	95,122.94	60,653.07	36,787.94	22,313.02	13,583.53	8208.50

59. $100{,}000 = P\left(1 + \dfrac{0.05}{12}\right)^{12t} \Rightarrow P = 100{,}000\left(1 + \dfrac{0.05}{12}\right)^{-12t}$

t	1	10	20	30	40	50
P	95,132.82	60,716.10	36,864.45	22,382.66	13,589.88	8251.24

61. (a) $A = 20{,}000\left(1 + \dfrac{0.06}{365}\right)^{(365)(8)} \approx \$32{,}320.21$

(b) $A = \$30{,}000$

(c) $A = 8000\left(1 + \dfrac{0.06}{365}\right)^{(365)(8)} + 20{,}000\left(1 + \dfrac{0.06}{365}\right)^{(365)(4)}$

$\approx \$12{,}928.09 + 25{,}424.48 = \$38{,}352.57$

(d) $A = 9000\left[\left(1 + \dfrac{0.06}{365}\right)^{(365)(8)} + \left(1 + \dfrac{0.06}{365}\right)^{(365)(4)} + 1\right]$

$\approx \$34{,}985.11$

Take option (c).

63. (a) $\lim_{t\to\infty} 6.7e^{(-48.1)/t} = 6.7e^0 = 6.7$ million ft^3

(b) $V' = \dfrac{322.27}{t^2}e^{-(48.1)/t}$

$V'(20) \approx 0.073$ million ft^3/yr

$V'(60) \approx 0.040$ million ft^3/yr

65. $y = \dfrac{300}{3 + 17e^{-0.0625x}}$

(a)

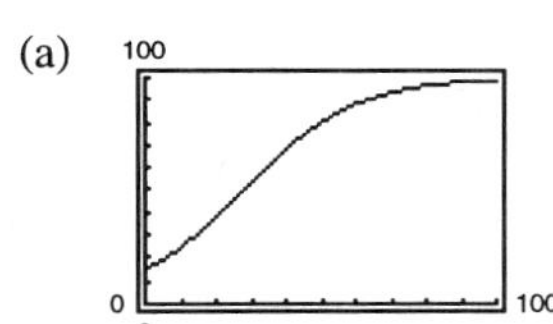

(b) If $x = 2$ (2000 egg masses), $y \approx 16.67 \approx 16.7\%$.

(c) If $y = 66.67\%$, then $x \approx 38.8$ or 38,800 egg masses.

(d) $y = 300(3 + 17e^{-0.0625x})^{-1}$

$y' = \dfrac{318.75e^{-0.0625x}}{(3 + 17e^{-0.0625x})^2}$

$y'' = \dfrac{19.921875e^{-0.0625x}(17e^{-0.0625x} - 3)}{(3 + 17e^{-0.0625x})^3}$

$17e^{-0.0625x} - 3 = 0 \Rightarrow x \approx 27.8$ or 27,800 egg masses.

67. (a) $y = (271.92)(1.096)^x = (271.92)e^{0.0918x}$

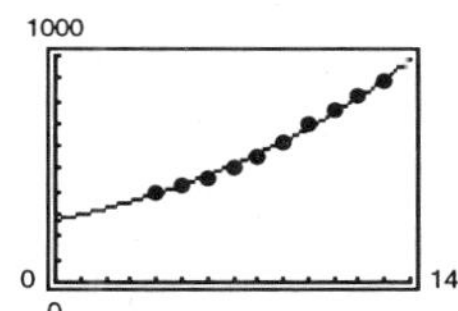

(b) $y = -254.08 + 418.41 \ln x$

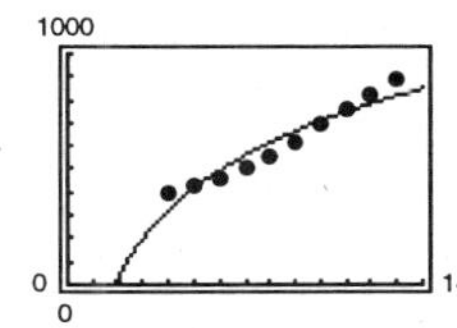

(c) The exponential model is better.

(d) Exponential model:

$\dfrac{dy}{dx} = (2171.92)(0.0918)(e^{0.0918(20)}) \approx 157$

Logarithmic model: $\dfrac{dy}{dx} = (418.41)\dfrac{1}{20} \approx 20.9$

Logarithmic model would be better.

69. $\displaystyle\int 3^x\,dx = \frac{3^x}{\ln 3} + C$

71. $\displaystyle\int_{-1}^{2} 2^x\,dx = \left[\frac{2^x}{\ln 2}\right]_{-1}^{2}$

$= \dfrac{1}{\ln 2}\left[4 - \dfrac{1}{2}\right] = \dfrac{7}{2\ln 2} = \dfrac{7}{\ln 4}$

73. $\displaystyle\int x5^{-x^2}\,dx = -\frac{1}{2}\int 5^{-x^2}(-2x)\,dx$

$\displaystyle= -\left(\frac{1}{2}\right)\frac{5^{-x^2}}{\ln 5} + C = \frac{-1}{2\ln 5}(5^{-x^2}) + C$

75. $\displaystyle\int \frac{3^{2x}}{1+3^{2x}}\,dx,\ u = 1 + 3^{2x},\ du = 2(\ln 3)3^{2x}\,dx$

$\displaystyle\frac{1}{2\ln 3}\int\frac{(2\ln 3)3^{2x}}{1+3^{2x}}\,dx = \frac{1}{2\ln 3}\ln(1+3^{2x}) + C$

77. $\displaystyle\frac{dy}{dx} = 0.4^{x/3},\ \left(0, \frac{1}{2}\right)$

$\displaystyle y = \int 0.4^{x/3}\,dx = 3\int 0.4^{x/3}\left(\frac{1}{3}\,dx\right)$

$\displaystyle= \frac{3}{\ln 0.4}0.4^{x/3} + C = 3(\ln 2.5)(0.4)^{x/3} + C$

$\displaystyle\left(0, \frac{1}{2}\right)\colon\ \frac{1}{2} = 3(\ln 2.5) + C \Rightarrow C = \frac{1}{2} - 3\ln 2.5$

$\displaystyle y = 3\ln 2.5(0.4)^{x/3} + \frac{1}{2} - 3\ln 2.5 = \frac{3(1-0.4^{x/3})}{\ln 2.5} + \frac{1}{2}$

79. (a) $\displaystyle\int_0^4 f(t)\,dt \approx 5.67$

$\displaystyle\int_0^4 g(t)\,dt \approx 5.67$

$\displaystyle\int_0^4 h(t)\,dt \approx 5.67$

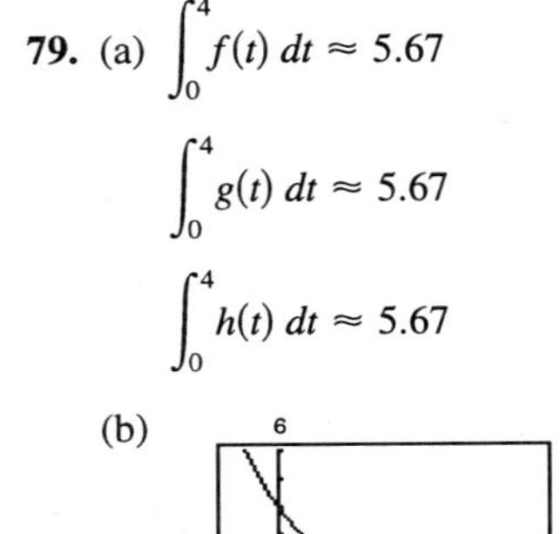

(b)

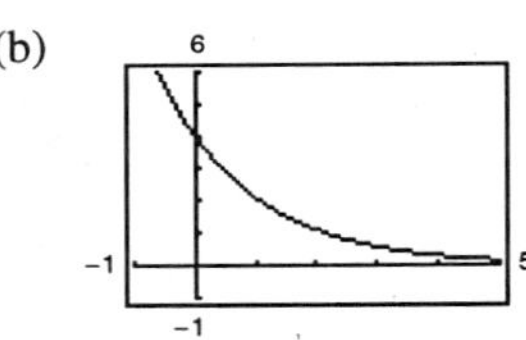

(c) The functions appear to be equal: $f(t) = g(t) = h(t)$

Analytically,

$\displaystyle f(t) = 4\left(\frac{3}{8}\right)^{2t/3} = 4\left[\left(\frac{3}{8}\right)^{2/3}\right]^t = 4\left(\frac{9^{1/3}}{4}\right)^t = g(t)$

$h(t) = 4e^{-0.653886t} = 4[e^{-0.653886}]^t = 4(0.52002)^t$

$\displaystyle g(t) = 4\left(\frac{9^{1/3}}{4}\right)^t = 4(0.52002)^t$

No. The definite integrals over a given interval may be equal when the functions are not equal.

81. $\displaystyle P = \int_0^{10} 2000e^{-0.06t}\,dt = \left[\frac{2000}{-0.06}e^{-0.06t}\right]_0^{10} \approx \$15{,}039.61$

83. $y = C(k^t)$

When $t = 0$, $y = 1200 \Rightarrow C = 1200$.

$y = 1200(k^t)$

$\displaystyle\frac{720}{1200} = 0.6,\ \frac{432}{720} = 0.6,\ \frac{259.20}{432} = 0.6,\ \frac{155.52}{259.20} = 0.6$

Let $k = 0.6$.

$y = 1200(0.6)^t$

t	0	1	2	3	4
y	1200	720	432	259.20	155.52

85. False. e is an irrational number.

87. True.

$f(g(x)) = 2 + e^{\ln(x-2)}$

$= 2 + x - 2 = x$

$g(f(x)) = \ln(2 + e^x - 2)$

$= \ln e^x = x$

89. True.

$\displaystyle\frac{d}{dx}[e^x] = e^x$ and $\displaystyle\frac{d}{dx}[e^{-x}] = -e^{-x}$

$e^x = e^{-x}$ when $x = 0$.

$(e^0)(-e^{-0}) = -1$

91. $\dfrac{dy}{dt} = \dfrac{8}{25}y\left(\dfrac{5}{4} - y\right), y(0) = 1$

$$\frac{dy}{y[(5/4) - y]} = \frac{8}{25}\,dt \Rightarrow \frac{4}{5}\int\left(\frac{1}{y} + \frac{1}{(5/4) - y}\right)dy = \int \frac{8}{25}\,dt \Rightarrow$$

$$\ln y - \ln\left(\frac{5}{4} - y\right) = \frac{2}{5}t + C$$

$$\ln\left(\frac{y}{(5/4) - y}\right) = \frac{2}{5}t + C$$

$$\frac{y}{(5/4) - y} = e^{(2/5)t + C} = C_1e^{(2/5)t}$$

$$y(0) = 1 \Rightarrow C_1 = 4 \Rightarrow 4e^{(2/5)t} = \frac{y}{(5/4) - y}$$

$$\Rightarrow 4e^{(2/5)t}\left(\frac{5}{4} - y\right) = y \Rightarrow 5e^{(2/5)t} = 4e^{(2/5)t}y + y = (4e^{(2/5)t} + 1)y$$

$$\Rightarrow y = \frac{5e^{(2/5)t}}{4e^{(2/5)t} + 1} = \frac{5}{4 + e^{-0.4t}} = \frac{1.25}{1 + 0.25e^{-0.4t}}$$

Section 5.6 Differential Equations: Growth and Decay

1. $y' = \dfrac{5x}{y}$

$$yy' = 5x$$

$$\int yy'\,dx = \int 5x\,dx$$

$$\int y\,dy = \int 5x\,dx$$

$$\frac{1}{2}y^2 = \frac{5}{2}x^2 + C_1$$

$$y^2 - 5x^2 = C$$

3. $y' = \sqrt{x}\,y$

$$\frac{y'}{y} = \sqrt{x}$$

$$\int \frac{y'}{y}\,dx = \int \sqrt{x}\,dx$$

$$\int \frac{dy}{y} = \int \sqrt{x}\,dx$$

$$\ln y = \frac{2}{3}x^{3/2} + C_1$$

$$y = e^{(2/3)x^{3/2} + C_1} = e^{C_1}e^{(2/3)x^{3/2}} = Ce^{(2/3)x^{3/2}}$$

5. $(1 + x^2)y' - 2xy = 0$

$$y' = \frac{2xy}{1 + x^2}$$

$$\frac{y'}{y} = \frac{2x}{1 + x^2}$$

$$\int \frac{y'}{y}\,dx = \int \frac{2x}{1 + x^2}\,dx$$

$$\int \frac{dy}{y} = \int \frac{2x}{1 + x^2}\,dx$$

$$\ln y = \ln(1 + x^2) + C_1$$

$$\ln y = \ln(1 + x^2) + \ln C$$

$$\ln y = \ln C(1 + x^2)$$

$$y = C(1 + x^2)$$

7. (a)

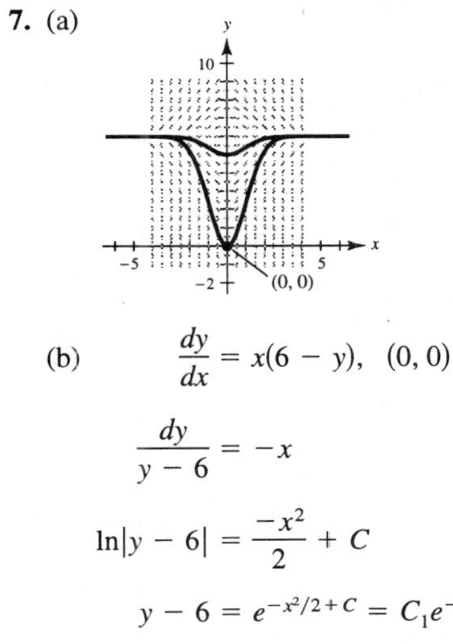

(b) $\dfrac{dy}{dx} = x(6 - y),\ (0, 0)$

$$\frac{dy}{y - 6} = -x$$

$$\ln|y - 6| = \frac{-x^2}{2} + C$$

$$y - 6 = e^{-x^2/2 + C} = C_1e^{-x^2/2}$$

$$y = 6 + C_1e^{-x^2/2}$$

$$(0, 0)\text{: } 0 = 6 + C_1 \Rightarrow C_1 = -6 \Rightarrow y = 6 - 6e^{-x^2/2}$$

9. $\dfrac{dQ}{dt} = \dfrac{k}{t^2}$

$$\int \frac{dQ}{dt}\,dt = \int \frac{k}{t^2}\,dt$$

$$\int dQ = -\frac{k}{t} + C$$

$$Q = -\frac{k}{t} + C$$

11. $\dfrac{dN}{ds} = k(250 - s)$

$$\int \frac{dN}{ds}\,ds = \int k(250 - s)\,ds$$

$$\int dN = -\frac{k}{2}(250 - s)^2 + C$$

$$N = -\frac{k}{2}(250 - s)^2 + C$$

13. $\dfrac{dy}{dt} = \dfrac{1}{2}t,\ (0, 10)$

$$\int dy = \int \frac{1}{2}t\,dt$$

$$y = \frac{1}{4}t^2 + C$$

$$10 = \frac{1}{4}(0)^2 + C \Rightarrow C = 10$$

$$y = \frac{1}{4}t^2 + 10$$

15. $\dfrac{dy}{dt} = -\dfrac{1}{2}y,\ (0, 10)$

$$\int \frac{dy}{y} = \int -\frac{1}{2}\,dt$$

$$\ln y = -\frac{1}{2}t + C_1$$

$$y = e^{-(t/2) + C_1} = e^{C_1}\,e^{-t/2} = Ce^{-t/2}$$

$$10 = Ce^0 \Rightarrow C = 10$$

$$y = 10e^{-t/2}$$

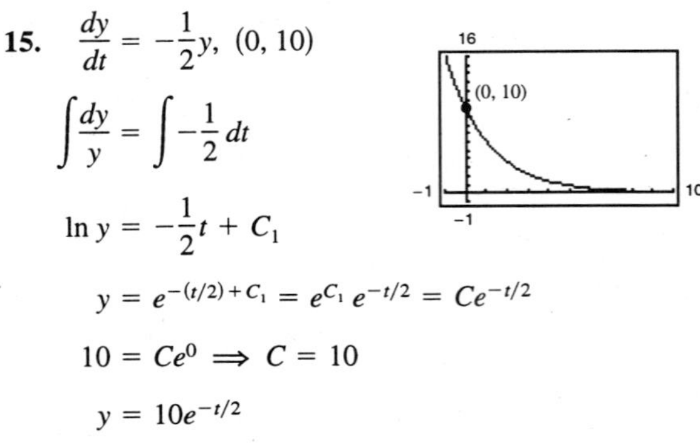

17. $y = Ce^{kt},\ \left(0, \dfrac{1}{2}\right),\ (5, 5)$

$$C = \frac{1}{2}$$

$$y = \frac{1}{2}e^{kt}$$

$$5 = \frac{1}{2}e^{5k}$$

$$k = \frac{\ln 10}{5} \approx 0.4605$$

$$y = \frac{1}{2}e^{0.4605t}$$

19. $y = Ce^{kt},\ (1, 1),\ (5, 5)$

$$1 = Ce^k$$

$$5 = Ce^{5k}$$

$$5Ce^k = Ce^{5k}$$

$$5e^k = e^{5k}$$

$$5 = e^{4k}$$

$$k = \frac{\ln 5}{4} \approx 0.4024$$

$$y = Ce^{0.4024t}$$

$$1 = Ce^{0.4024}$$

$$C \approx 0.6687$$

$$y = 0.6687e^{0.4024t}$$

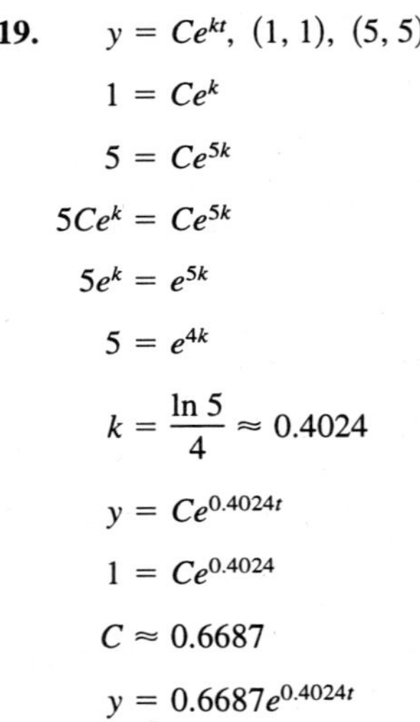

21. Since the intial quantity is 10 grams, $y = 10e^{[\ln(1/2)/1620]t}$. When $t = 1000$, $y = 10e^{[\ln(1/2)/1620](1000)} \approx 6.52$ grams. When $t = 10{,}000$, $y = 10e^{[\ln(1/2)/1620](10{,}000)} \approx 0.14$ gram.

23. Since $y = Ce^{[\ln(1/2)/5730]t}$, we have $2.0 = Ce^{[\ln(1/2)/5730](10{,}000)} \Rightarrow C \approx 6.70$ which implies that the initial quantity is 6.70 grams. When $t = 1000$, we have $y = 6.70e^{[\ln(1/2)/5730](1000)} \approx 5.94$ grams.

25. Since $y = Ce^{[\ln(1/2)/24{,}360]t}$, we have $2.1 = Ce^{[\ln(1/2)/24{,}360](1000)} \Rightarrow C \approx 2.16$. Thus, the initial quantity is 2.16 grams. When $t = 10{,}000$, $y = 2.16e^{[\ln(1/2)/24{,}360](10{,}000)} \approx 1.63$ grams.

27. Since $\frac{dy}{dx} = ky$, $y = Ce^{kt}$ or $y = y_0e^{kt}$.

$$\frac{1}{2}y_0 = y_0e^{1620k}$$

$$k = \frac{-\ln 2}{1620}$$

$$y = y_0e^{-(\ln 2)t/1620}.$$

When $t = 100$, $y = y_0e^{-(\ln 2)/16.2} \approx y_0(0.9581)$. Therefore, 95.81% of the present amount still exists.

29. Since $A = 1000e^{0.06t}$, the time to double is given by $2000 = 1000e^{0.06t}$ and we have

$$2 = e^{0.06t}$$

$$\ln 2 = 0.06t$$

$$t = \frac{\ln 2}{0.06} \approx 11.55 \text{ years.}$$

Amount after 10 years: $A = 1000e^{(0.06)(10)} \approx \$1822.12$

31. Since $A = 750e^{rt}$ and $A = 1500$ when $t = 7.75$, we have the following.

$$1500 = 750e^{7.75r}$$

$$r = \frac{\ln 2}{7.75} \approx 0.0894 = 8.94\%$$

Amount after 10 years: $A = 750e^{0.0894(10)} \approx \$1833.67$

33. Since $A = 500e^{rt}$ and $A = 1292.85$ when $t = 10$, we have the following.

$$1292.85 = 500e^{10r}$$

$$r = \frac{\ln(1292.85/500)}{10} \approx 0.0950 = 9.50\%$$

The time to double is given by

$$1000 = 500e^{0.0950t}$$

$$t = \frac{\ln 2}{0.095} \approx 7.30 \text{ years.}$$

35. $500{,}000 = P\left(1 + \frac{0.075}{12}\right)^{(12)(20)}$

$$P = 500{,}000\left(1 + \frac{0.075}{12}\right)^{-240}$$

$$\approx \$112{,}087.09$$

37. (a) $2000 = 1000(1 + 0.07)^t$

$$2 = 1.07^t$$

$$\ln 2 = t \ln 1.07$$

$$t = \frac{\ln 2}{\ln 1.07} \approx 10.24 \text{ years}$$

(b) $2000 = 1000\left(1 + \frac{0.07}{12}\right)^{12t}$

$$2 = \left(1 + \frac{0.007}{12}\right)^{12t}$$

$$\ln 2 = 12t \ln\left(1 + \frac{0.07}{12}\right)$$

$$t = \frac{\ln 2}{12 \ln(1 + (0.07/12))} \approx 9.93 \text{ years}$$

(c) $2000 = 1000\left(1 + \frac{0.07}{365}\right)^{365t}$

$$2 = \left(1 + \frac{0.07}{365}\right)^{365t}$$

$$\ln 2 = 365t \ln\left(1 + \frac{0.07}{365}\right)$$

$$t = \frac{\ln 2}{365 \ln(1 + (0.07/365))} \approx 9.90 \text{ years}$$

(d) $2000 = 1000e^{(0.07)t}$

$$2 = e^{0.07t}$$

$$\ln 2 = 0.07t$$

$$t = \frac{\ln 2}{0.07} \approx 9.90 \text{ years}$$

39. Let $t = 0$ represent 1990.

$$y = Ce^{kt},\ (0, 4.22),\ (10, 6.49)$$
$$C = 4.22$$
$$6.49 = 4.22e^{10k}$$
$$k = \frac{\ln(6.49/4.22)}{10} \approx 0.0430$$
$$y \approx 4.22e^{0.0430t}$$

When $t = 20$ (for 2010), $y \approx 9.97$ million.

41. Let $t = 0$ represent 1990.

$$y = Ce^{kt},\ (0, 3.0),\ (10, 2.74)$$
$$C = 3.0$$
$$2.74 = 3.0e^{10k}$$
$$k = \frac{\ln(2.74/3.0)}{10} \approx -0.0091$$
$$y \approx 3.0e^{-0.0091t}$$

When $t = 20$ (for 2010), $y \approx 2.51$ million.

43. (a) k; the larger the value of k, the greater the rate of growth of the population.

(b) k; if $k > 0$, there is growth. If $k < 0$, decrease.

45.
$$P = Ce^{kx},\ (0, 760),\ (1000, 672.71)$$
$$C = 760$$
$$672.71 = 760e^{1000x}$$
$$x = \frac{\ln(672.71/760)}{1000} \approx -0.000122$$
$$P \approx 760e^{-0.000122x}$$

When $x = 3000$, $P \approx 527.06$ mm Hg.

47. (a)
$$19 = 30(1 - e^{20k})$$
$$30e^{20k} = 11$$
$$k = \frac{\ln(11/30)}{20} \approx -0.0502$$
$$N \approx 30(1 - e^{-0.0502t})$$

(b)
$$25 = 30(1 - e^{-0.0502t})$$
$$e^{-0.0502t} = \frac{1}{6}$$
$$t = \frac{-\ln 6}{-0.0502} \approx 36 \text{ days}$$

49. $S = Ce^{k/t}$

(a)
$$S = 5 \text{ when } t = 1$$
$$5 = Ce^{k}$$
$$\lim_{t\to\infty} Ce^{k/t} = C = 30$$
$$5 = 30e^{k}$$
$$k = \ln \tfrac{1}{6} \approx -1.7918$$
$$S \approx 30e^{-1.7918/t}$$

(b) When $t = 5$, $S \approx 20.9646$ which is 20,965 units.

(c)

51. $A(t) = V(t)e^{-0.10t} = 100{,}000e^{0.8\sqrt{t}}\,e^{-0.10t} = 100{,}000e^{0.8\sqrt{t}-0.10t}$

$$\frac{dA}{dt} = 100{,}000\left(\frac{0.4}{\sqrt{t}} - 0.10\right)e^{0.8\sqrt{t}-0.10t} = 0 \text{ when } 16.$$

The timber should be harvested in the year 2014, (1998 + 16). **Note:** You could also use a graphing utility to graph $A(t)$ and find the maximum of $A(t)$. Use the viewing rectangle $0 \le x \le 30$ and $0 \le y \le 600{,}000$.

53. $\beta(I) = 10 \log_{10} \dfrac{I}{I_0}$, $I_0 = 10^{-16}$

(a) $\beta(10^{-14}) = 10 \log_{10} \dfrac{10^{-14}}{10^{-16}} = 20$ decibels

(b) $\beta(10^{-9}) = 10 \log_{10} \dfrac{10^{-9}}{10^{-16}} = 70$ decibels

(c) $\beta(10^{-6.5}) = 10 \log_{10} \dfrac{10^{-6.5}}{10^{-16}} = 95$ decibels

(d) $\beta(10^{-4}) = 10 \log_{10} \dfrac{10^{-4}}{10^{-16}} = 120$ decibels

55. $R = \dfrac{\ln I - 0}{\ln 10}, I = e^{R \ln 10} = 10^R$

(a) $8.3 = \dfrac{\ln I - 0}{\ln 10}$

$I = 10^{8.3} \approx 199{,}526{,}231.5$

(b) $2R = \dfrac{\ln I - 0}{\ln 10}$

$I = e^{2R \ln 10} = e^{2R \ln 10} = (e^{R \ln 10})^2 = (10^R)^2$

Increases by a factor of $e^{2R \ln 10}$ or 10^R.

(c) $\dfrac{dR}{dI} = \dfrac{1}{I \ln 10}$

57. Since $\dfrac{dy}{dt} = k(y - 20)$,

$$\int \frac{1}{y - 20}\,dy = \int k\,dt$$

$$\ln(y - 20) = kt + C$$

$$y = Ce^{kt} + 20.$$

When $t = 0, y = 72$. Therefore, $C = 52$.

When $t = 1, y = 48$. Therefore, $48 = 52e^k + 20$, $e^k = (28/52) = (7/13)$, and $k = \ln(7/13)$. Thus, $y = 52e^{[\ln(7/13)]t} + 20$.

When $t = 5, y = 52e^{5 \ln(7/13)} + 20 \approx 22.35°$.

59. (a) $u = 985.93 - \left(985.93 - \dfrac{(120{,}000)(0.095)}{12}\right)\left(1 + \dfrac{0.095}{12}\right)^{12t}$

$v = \left(985.93 - \dfrac{(120{,}000)(0.095)}{12}\right)\left(1 + \dfrac{0.095}{12}\right)^{12t}$

(b) The larger part goes for interest. The curves intersect when $t \approx 27.7$ years.

(c) The slopes are negatives of each other. Analytically,

$$u = 985.93 - v \implies \frac{du}{dt} = -\frac{dv}{dt}$$

$u'(15) = -v'(15) = -14.06$.

(d) $t = 12.7$ years

Again, the larger part goes for interest.

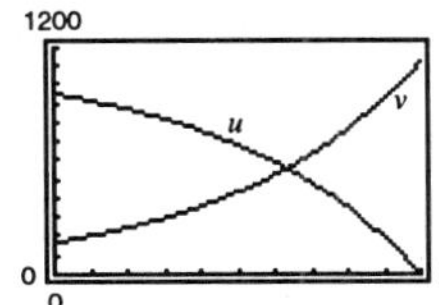

Section 5.7 Differential Equations: Separation of Variables

1. Differential equation: $y' = 4y$

Solution: $y = Ce^{4x}$

Check: $y' = 4Ce^{4x} = 4y$

3. Differential equation: $y'' + y = 0$

Solution: $y = C_1 \cos x + C_2 \sin x$

Check: $y' = -C_1 \sin x + C_2 \cos x$

$y'' = -C_1 \cos x - C_2 \sin x$

$y'' + y = -C_1 \cos x - C_2 \sin x + C_1 \cos x + C_2 \sin x = 0$

5. $y = -\cos x \ln|\sec x + \tan x|$

$$y' = (-\cos x)\frac{1}{\sec x + \tan x}(\sec x \cdot \tan x + \sec^2 x) + \sin x \ln|\sec x + \tan x|$$

$$= \frac{(-\cos x)}{\sec x + \tan x}(\sec x)(\tan x + \sec x) + \sin x \ln|\sec x + \tan x|$$

$$= -1 + \sin x \ln|\sec x + \tan x|$$

$$y'' = (\sin x)\frac{1}{\sec x + \tan x}(\sec x \cdot \tan x + \sec^2 x) + \cos x \ln|\sec x + \tan x|$$

$$= (\sin x)(\sec x) + \cos x \ln|\sec x + \tan x|$$

Substituting,

$$y'' + y = (\sin x)(\sec x) + \cos x \ln|\sec x + \tan x| - \cos x \ln|\sec x + \tan x|$$

$$= \tan x.$$

In Exercises 7–11, the differential equation is $y^{(4)} - 16y = 0$.

7. $y = 3\cos x$

$y^{(4)} = 3\cos x$

$y^{(4)} - 16y = -45\cos x \neq 0$, No.

9. $y = e^{-2x}$

$y^{(4)} = 16e^{-2x}$

$y^{(4)} - 16y = 16e^{-2x} - 16e^{-2x} = 0$, Yes.

11. $y = C_1e^{2x} + C_2e^{-2x} + C_3\sin 2x + C_4\cos 2x$

$y^{(4)} = 16C_1e^{2x} + 16C_2e^{-2x} + 16C_3\sin 2x + 16C_4\cos 2x$

$y^{(4)} - 16y = 0$, Yes.

In 13–17, the differential equation is $xy' - 2y = x^3e^x$.

13. $y = x^2,\ y' = 2x$

$xy' - 2y = x(2x) - 2(x^2) = 0 \neq x^3e^x$,

No.

15. $y = x^2(2 + e^x),\ \ y' = x^2(e^x) + 2x(2 + e^x)$

$xy' - 2y = x[x^2e^x + 2xe^x + 4x] - 2[x^2e^x + 2x^2] = x^3e^x$

Yes.

17. $y = \ln x,\ y' = \frac{1}{x}$

$xy' - 2y = x\left(\frac{1}{x}\right) - 2\ln x \neq x^3e^x$, No.

19. $y = Ce^{kx}$

$\frac{dy}{dx} = Cke^{kx}$

Since $dy/dx = 0.07y$, we have $Cke^{kx} = 0.07Ce^{kx}$.
Thus, $k = 0.07$.

21. $y^2 = Cx^3$ passes through $(4, 4)$

$16 = C(64) \Rightarrow C = \frac{1}{4}$

Particular solution: $y^2 = \frac{1}{4}x^3$ or $4y^2 = x^3$

23. Differential equation: $4yy' - x = 0$

General solution: $4y^2 - x^2 = C$

Particular solutions: $C = 0$, Two interesecting lines
$C = \pm 1,\ C = \pm 4$, Hyperbolas

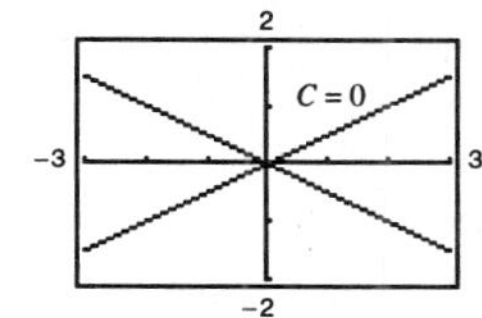

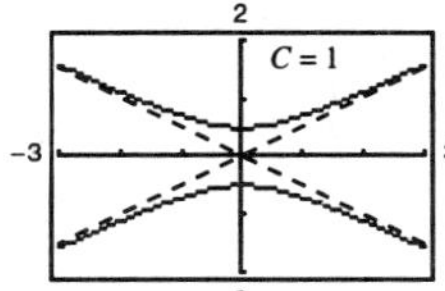

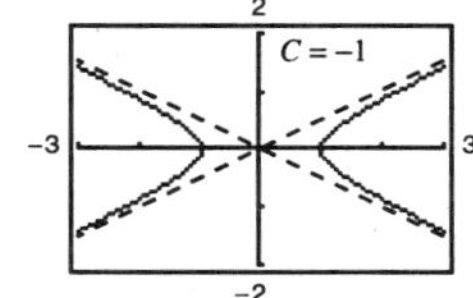

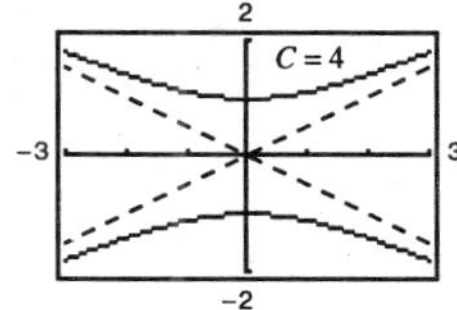

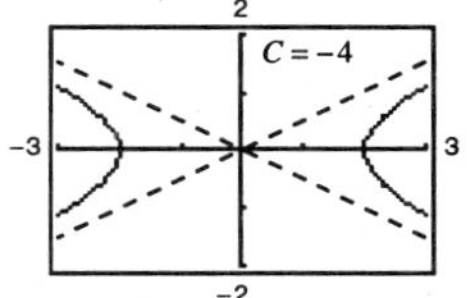

25. Differential equation: $y' + 2y = 0$

General Solution: $y = Ce^{-2x}$

$y' + 2y = C(-2)e^{-2x} + 2(Ce^{-2x}) = 0$

Initial condition: $y(0) = 3,\ 3 = Ce^0 = C$

Particular solution: $y = 3e^{-2x}$

27. Differential equation: $y'' + 9y = 0$

General solution: $y = C_1 \sin 3x + C_2 \cos 3x$

$y' = 3C_1 \cos 3x - 3C_2 \sin 3x,$

$y'' = -9C_1 \sin 3x - 9C_2 \cos 3x$

$$y'' + 9y = (-9C_1 \sin 3x - 9C_2 \cos 3x) + 9(C_1 \sin 3x + C_2 \cos 3x) = 0$$

Initial conditions: $y\left(\frac{\pi}{6}\right) = 2,\ y'\left(\frac{\pi}{6}\right) = 1$

$$2 = C_1 \sin\left(\frac{\pi}{2}\right) + C_2 \cos\left(\frac{\pi}{2}\right) \Rightarrow C_1 = 2$$

$$y' = 3C_1 \cos 3x - 3C_2 \sin 3x$$

$$1 = 3C_1 \cos\left(\frac{\pi}{2}\right) - 3C_2 \sin\left(\frac{\pi}{2}\right)$$

$$= -3C_2 \Rightarrow C_2 = -\frac{1}{3}$$

Particular solution: $y = 2 \sin 3x - \frac{1}{3} \cos 3x$

29. Differential equation: $x^2y'' - 3xy' + 3y = 0$

General solution: $y = C_1x + C_2x^3$

$y' = C_1 + 3C_2x^2,\ y'' = 6C_2x$

$$x^2y'' - 3xy' + 3y = x^2(6C_2x) - 3x(C_1 + 3C_2x^2) + 3(C_1x + C_2x^3) = 0$$

Initial conditions: $y(2) = 0,\ y'(2) = 4$

$$0 = 2C_1 + 8C_2$$

$$y' = C_1 + 3C_2x^2$$

$$4 = C_1 + 12C_2$$

$$\left.\begin{aligned} C_1 + 4C_2 &= 0 \\ C_1 + 12C_2 &= 4 \end{aligned}\right\} \quad C_2 = \frac{1}{2},\ C_1 = -2$$

Particular solution: $y = -2x + \frac{1}{2}x^3$

31. $\dfrac{dy}{dx} = 3x^2$

$$y = \int 3x^2\,dx = x^3 + C$$

33. $\dfrac{dy}{dx} = \dfrac{x-2}{x} = 1 - \dfrac{2}{x}$

$$y = \int \left[1 - \frac{2}{x}\right] dx$$

$$= x - 2\ln|x| + C = x - \ln x^2 + C$$

35. $\dfrac{dy}{dx} = \sin 2x$

$$y = \int \sin 2x\,dx = -\frac{1}{2}\cos 2x + C$$

$(u = 2x,\ du = 2dx)$

37. $\dfrac{dy}{dx} = x\sqrt{x-3}$ Let $u = \sqrt{x-3}$, then $x = u^2 + 3$ and $dx = 2u\,du$.

$$y = \int x\sqrt{x-3}\,dx = \int (u^2+3)(u)(2u)du$$

$$= 2\int (u^4 + 3u^2)\,du = 2\left(\frac{u^5}{5} + u^3\right) + C = \frac{2}{5}(x-3)^{5/2} + 2(x-3)^{3/2} + C$$

39. $\dfrac{dy}{dx} = \dfrac{x}{y}$

$$\int y\,dy = \int x\,dx$$

$$\frac{y^2}{2} = \frac{x^2}{2} + C_1$$

$$y^2 - x^2 = C$$

41. $\dfrac{dr}{ds} = 0.05r$

$$\int \frac{dr}{r} = \int 0.05\,ds$$

$$\ln|r| = 0.05s + C_1$$

$$r = e^{0.05s + C_1} = Ce^{0.05s}$$

43. $(2+x)y' = 3y$

$$\int \frac{dy}{y} = \int \frac{3}{2+x}\,dx$$

$$\ln y = 3\ln(2+x) + \ln C = \ln C(2+x)^3$$

$$y = C(x+2)^3$$

45. $yy' = \sin x$

$$\int y\,dy = \int \sin x\,dx$$

$$\frac{y^2}{2} = -\cos x + C_1$$

$$y^2 = -2\cos x + C$$

47. $y\ln x - xy' = 0$

$$\int \frac{dy}{y} = \int \frac{\ln x}{x}\,dx \quad \left(u = \ln x,\ du = \frac{dx}{x}\right)$$

$$\ln y = \frac{1}{2}(\ln x)^2 + C_1$$

$$y = e^{(1/2)(\ln x)^2 + C_1} = Ce^{(\ln x)^2/2}$$

49. $yy' - e^x = 0$

$$\int y\,dy = \int e^x\,dx$$

$$\frac{y^2}{2} = e^x + C_1$$

$$y^2 = 2e^x + C$$

Initial condition: $y(0) = 4,\ 16 = 2 + C,\ C = 14$

Particular solution: $y^2 = 2e^x + 14$

51. $y(x+1) + y' = 0$

$$\int \frac{dy}{y} = -\int (x+1)\,dx$$

$$\ln y = -\frac{(x+1)^2}{2} + C_1$$

$$y = Ce^{-(x+1)^2/2}$$

Initial condition: $y(-2) = 1,\ 1 = Ce^{-1/2},\ C = e^{1/2}$

Particular solution: $y = e^{[1-(x+1)^2]/2} = e^{-(x^2+2x)/2}$

53. $y(1+x^2)\dfrac{dy}{dx} = x(1+y^2)$

$$\frac{y}{1+y^2}\,dy = \frac{x}{1+x^2}\,dx$$

$$\frac{1}{2}\ln(1+y^2) = \frac{1}{2}\ln(1+x^2) + C_1$$

$$\ln(1+y^2) = \ln(1+x^2) + \ln C = \ln[C(1+x^2)]$$

$$1 + y^2 = C(1+x^2)$$

$$y(0) = \sqrt{3}:\ 1 + 3 = C \Rightarrow C = 4$$

$$1 + y^2 = 4(1+x^2)$$

$$y^2 = 3 + 4x^2$$

55. $\dfrac{du}{dv} = uv \sin v^2$

$$\int \frac{du}{u} = \int v \sin v^2 \, dv$$

$$\ln u = -\frac{1}{2}\cos v^2 + C_1$$

$$u = Ce^{-(\cos v^2)/2}$$

Initial condition: $u(0) = 1,\ C = \dfrac{1}{e^{-1/2}} = e^{1/2}$

Particular solution: $u = e^{(1 - \cos v^2)/2}$

57. $dP - kP\,dt = 0$

$$\int \frac{dP}{P} = k \int dt$$

$$\ln P = kt + C_1$$

$$P = Ce^{kt}$$

Initial condition: $P(0) = P_0,\ P_0 = Ce^0 = C$

Particular solution: $P = P_0 e^{kt}$

59. $\dfrac{dy}{dx} = \dfrac{-9x}{16y}$

$$\int 16y\,dy = -\int 9x\,dx$$

$$8y^2 = \frac{-9}{2}x^2 + C$$

Initial condition: $y(1) = 1,\ 8 = -\dfrac{9}{2} + C,\ C = \dfrac{25}{2}$

Particular solution: $8y^2 = \dfrac{-9}{2}x^2 + \dfrac{25}{2},$

$$16y^2 + 9x^2 = 25$$

61. $m = \dfrac{dy}{dx} = \dfrac{0 - y}{(x + 2) - x} = -\dfrac{y}{2}$

$$\int \frac{dy}{y} = \int -\frac{1}{2}\,dx$$

$$\ln y = -\frac{1}{2}x + C_1$$

$$y = Ce^{-x/2}$$

63. $f(x, y) = x^3 - 4xy^2 + y^3$

$$f(tx, ty) = t^3x^3 - 4tx t^2y^2 + t^3y^3$$

$$= t^3(x^3 - 4xy^2 + y^3)$$

Homogeneous of degree 3

65. $f(x, y) = 2 \ln xy$

$$f(tx, ty) = 2 \ln tx\,ty$$

$$= 2 \ln t^2 xy$$

$$= 2(\ln t^2 + \ln xy)$$

Not homogeneous

67. $f(x, y) = 2 \ln \dfrac{x}{y}$

$$f(tx, ty) = 2 \ln \frac{tx}{ty} = 2 \ln \frac{x}{y}$$

Homogeneous of degree 0

69. $y' = \dfrac{x + y}{2x},\ y = vx$

$$v + x\frac{dv}{dx} = \frac{x + vx}{2x}$$

$$x\frac{dv}{dx} = \frac{1 + v}{2} - v$$

$$2\int \frac{dv}{1 - v} = \int \frac{dx}{x}$$

$$-\ln(1 - v)^2 = \ln|x| + \ln C = \ln|Cx|$$

$$\frac{1}{(1 - v^2)} = |Cx|$$

$$\frac{1}{[1 - (y/x)]^2} = |Cx|$$

$$\frac{x^2}{(x - y)^2} = |Cx| = C|x|$$

$$|x| = C(x - y)^2$$

71. $y' = \dfrac{x - y}{x + y},\ y = vx$

$$v + x\frac{dv}{dx} = \frac{x - xv}{x + xv}$$

$$v\,dx + x\,dv = \frac{1 - v}{1 + v}dx$$

$$\int \frac{v + 1}{v^2 + 2v - 1}\,dv = -\int \frac{dx}{x}$$

$$\frac{1}{2}\ln|v^2 + 2v - 1| = -\ln|x| + \ln C_1 = \ln\left|\frac{C_1}{x}\right|$$

$$|v^2 + 2v - 1| = \frac{C}{x^2}$$

$$\left|\frac{y^2}{x^2} + 2\frac{y}{x} - 1\right| = \frac{C}{x^2}$$

$$|y^2 + 2xy - x^2| = C$$

73. $$y' = \frac{xy}{x^2 - y^2},\ y = vx$$

$$v + x\frac{dv}{dx} = \frac{x^2 v}{x^2 - x^2 v^2}$$

$$v\,dx + x\,dv = \frac{v}{1 - v^2}\,dx$$

$$\int \frac{1 - v^2}{v^3}\,dv = \int \frac{dx}{x}$$

$$-\frac{1}{2v^2} - \ln|v| = \ln|x| + \ln|C_1| = \ln|C_1 x|$$

$$\frac{-1}{2v^2} = \ln|C_1 x v|$$

$$\frac{-x^2}{2y^2} = \ln|C_1 y|$$

$$y = Ce^{-x^2/2y^2}$$

75. $$x\,dy - (2xe^{-y/x} + y)\,dx = 0,\ y = vx$$

$$x(v\,dx + x\,dv) - (2xe^{-v} + vx)\,dx = 0$$

$$\int e^v\,dv = \int \frac{2}{x}\,dx$$

$$e^v = \ln C_1 x^2$$

$$e^{y/x} = \ln C_1 + \ln x^2$$

$$e^{y/x} = C + \ln x^2$$

Initial condition: $y(1) = 0,\ 1 = C$

Particular solution: $e^{y/x} = 1 + \ln x^2$

77. $$\left(x \sec\frac{y}{x} + y\right)dx - x\,dy = 0,\ y = vx$$

$$(x \sec v + xv)\,dx - x(v\,dx + x\,dv) = 0$$

$$(\sec v + v)\,dx = v\,dx + x\,dv$$

$$\int \cos v\,dv = \int \frac{dx}{x}$$

$$\sin v = \ln x + \ln C_1$$

$$x = Ce^{\sin v}$$

$$= Ce^{\sin(y/x)}$$

Initial condition: $y(1) = 0,\ 1 = Ce^0 = C$

Particular solution: $x = e^{\sin(y/x)}$

79. $\dfrac{dy}{dx} = x$

(a)

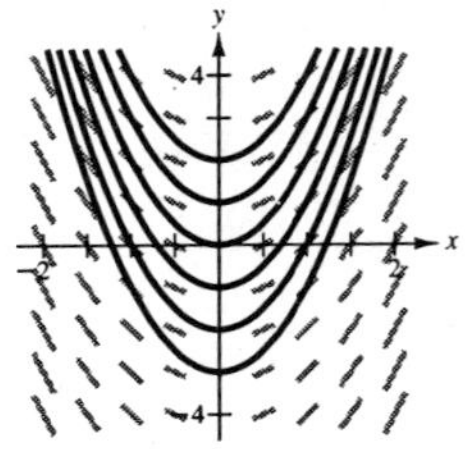

(b) $y = \displaystyle\int x\,dx = \frac{1}{2}x^2 + C$

81. $\dfrac{dy}{dx} = 4 - y$

(a)

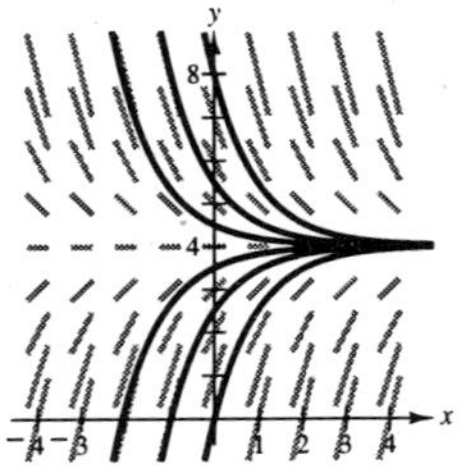

(b) $$\int \frac{dy}{4 - y} = \int dx$$

$$\ln|4 - y| = -x + C_1$$

$$4 - y = e^{-x + C_1}$$

$$y = 4 + Ce^{-x}$$

83. $\dfrac{dy}{dx} = k(y - 4)$

The direction field satisfies $(dy/dx) = 0$ along $y = 4$; but not along $y = 0$. Matches (a).

85. $\dfrac{dy}{dx} = ky(y - 4)$

The direction field satisfies $(dy/dx) = 0$ along $y = 0$ and $y = 4$. Matches (c).

87. (a) $\dfrac{dS}{dt} = k_1S(L - S)$

$S = \dfrac{L}{1 + Ce^{-kt}}$ is a solution because

$$\frac{dS}{dt} = -L(1 + Ce^{-kt})^{-2}(-C\,ke^{-kt})$$

$$= \frac{LC\,ke^{-kt}}{(1 + Ce^{-kt})^2}$$

$$= \left(\frac{k}{L}\right)\frac{L}{1 + Ce^{-kt}} \cdot \frac{C\,Le^{-kt}}{1 + Ce^{-kt}}$$

$$= \left(\frac{k}{L}\right)\frac{L}{1 + Ce^{-kt}} \cdot \left(L - \frac{L}{1 + Ce^{-kt}}\right)$$

$$= k_1S(L - S), \text{ where } k_1 = \frac{k}{L}.$$

$L = 100$. Also, $S = 10$ when $t = 0 \implies C = 9$. And, $S = 20$ when $t = 1 \implies k = -\ln(4/9)$.

Particular Solution. $S = \dfrac{100}{1 + 9e^{\ln(4/9)t}} = \dfrac{100}{1 + 9e^{-0.8109t}}$

(b) $\dfrac{dS}{dt} = \ln\left(\dfrac{4}{9}\right) S(100 - S)$

$$\frac{d^2S}{dt^2} = \ln\left(\frac{4}{9}\right)\left[S\left(-\frac{dS}{dt}\right) + (100 - S)\frac{dS}{dt}\right]$$

$$= \ln\left(\frac{4}{9}\right)(100 - 2S)\frac{dS}{dt}$$

$$= 0 \text{ when } S = 50 \text{ or } \frac{dS}{dt} = 0.$$

Choosing $S = 50$, we have:

$$50 = \frac{100}{1 + 9e^{\ln(4/9)t}}$$

$$2 = 1 + 9e^{\ln(4/9)t}$$

$$\frac{\ln(1/9)}{\ln(4/9)} = t$$

$$t \approx 2.7 \text{ months}$$

(c)

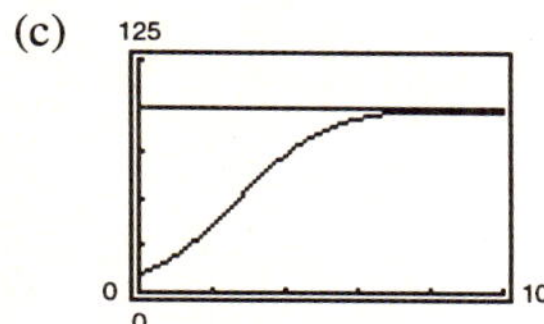

(d)

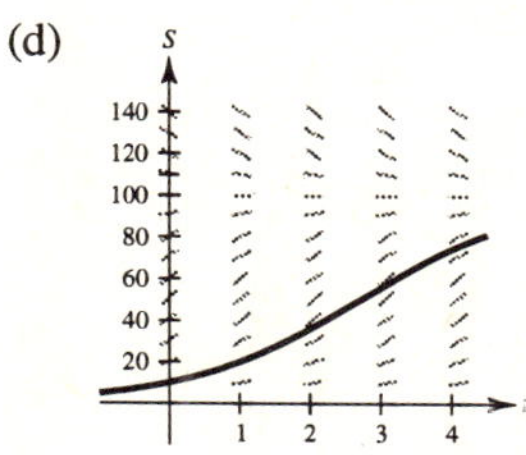

(e) Sales will decrease toward the line $S = L$.

89. $\dfrac{dy}{dt} = ky,\ y = Ce^{kt}$

Initial conditions: $y(0) = y_0$

$$y(1620) = \frac{y_0}{2}$$

$$C = y_0$$

$$\frac{y_0}{2} = y_0e^{1620k}$$

$$k = \frac{\ln(1/2)}{1620}$$

Particular solution: $y = y_0e^{-t(\ln 2)/1620}$

When $t = 25$, $y \approx 0.989y_0$, $y = 98.9\%$ of y_0.

91. (a) $\dfrac{dv}{dt} = k(W - v)$

$$\int \frac{dv}{W - v} = \int k\,dt$$

$$-\ln(W - v) = kt + C_1$$

$$v = W - Ce^{-kt}$$

Initial conditions:

$W = 20$, $v = 0$ when $t = 0$, and

$v = 5$ when $t = 1$.

$C = 20$, $k = -\ln(3/4)$

Particular solution:

$$v = 20(1 - e^{\ln(3/4)t}) \approx 20(1 - e^{-0.2877t})$$

(b) $s = \displaystyle\int 20(1 - e^{-0.2877t})\,dt$

$$\approx 20[t + 3.4761e^{-0.2877t}] + C$$

Since $S(0) = 0$, $C \approx -69.5$ and we have $s \approx 20t + 69.5(e^{-0.2877t} - 1)$.

93. Given family (circles): $x^2 + y^2 = C$

$$2x + 2yy' = 0$$

$$y' = -\frac{x}{y}$$

Orthogonal trajectory (lines): $y' = \frac{y}{x}$

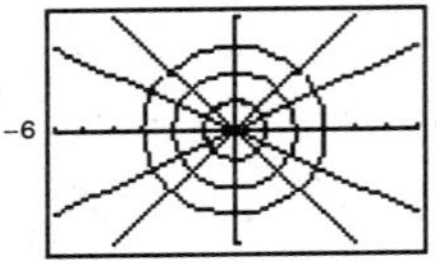

$$\int \frac{dy}{y} = \int \frac{dx}{x}$$

$$\ln y = \ln x + \ln K$$

$$y = Kx$$

95. Given family (parabolas): $x^2 = Cy$

$$2x = Cy'$$

$$y' = \frac{2x}{C} = \frac{2x}{x^2/y} = \frac{2y}{x}$$

Orthogonal trajectory (ellipses):

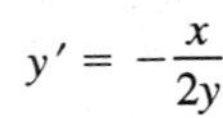

$$y' = -\frac{x}{2y}$$

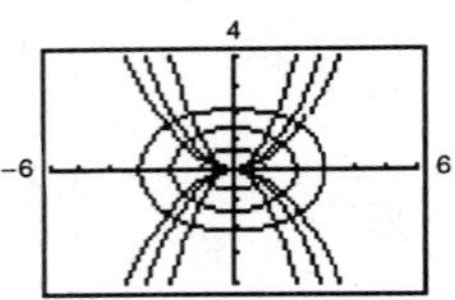

$$2\int y\,dy = -\int x\,dx$$

$$y^2 = -\frac{x^2}{2} + K_1$$

$$x^2 + 2y^2 = K$$

97. Given family: $y^2 = Cx^3$

$$2yy' = 3Cx^2$$

$$y' = \frac{3Cx^2}{2y} = \frac{3x^2}{2y}\left(\frac{y^2}{x^3}\right) = \frac{3y}{2x}$$

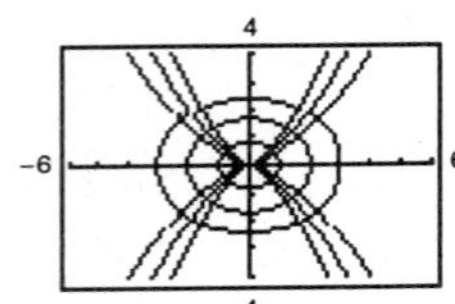

Orthogonal trajectory (ellipses): $y' = -\frac{2x}{3y}$

$$3\int y\,dy = -2\int x\,dx$$

$$\frac{3y^2}{2} = -x^2 + K_1$$

$$3y^2 + 2x^2 = K$$

99. False. Consider Example 2. $y = x^3$ is a solution to $xy' - 3y = 0$, but $y = x^3 + 1$ is not a solution.

101. False

$$f(tx, ty) = t^2x^2 + t^2xy + 2 \neq t^2 f(x, y)$$

Section 5.8 Inverse Trigonometric Functions and Differentiation

1. $y = \arcsin x$

(a)

x	-1	-0.8	-0.6	-0.4	-0.2	0	0.2	0.4	0.6	0.8	1
y	-1.571	-0.927	-0.644	-0.412	-0.201	0	0.201	0.412	0.644	0.927	1.571

(b)

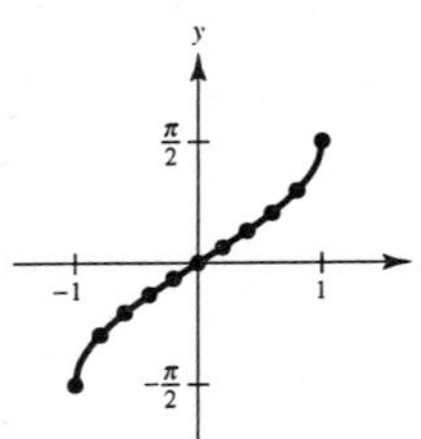

(c)

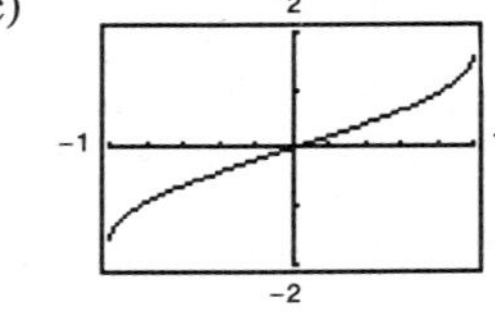

(d) Symmetric about origin: $\arcsin(-x) = -\arcsin x$

Intercept: $(0, 0)$

3. False.

$$\arccos \frac{1}{2} = \frac{\pi}{3}$$

since the range is $[0, \pi]$.

5. $\arcsin \frac{1}{2} = \frac{\pi}{6}$

7. $\arccos \frac{1}{2} = \frac{\pi}{3}$

9. $\arctan \dfrac{\sqrt{3}}{3} = \dfrac{\pi}{6}$

11. $\operatorname{arccsc}(-\sqrt{2}) = -\dfrac{\pi}{4}$

13. $\arccos(-0.8) \approx 2.50$

15. $\operatorname{arcsec}(1.269) = \arccos\left(\dfrac{1}{1.269}\right) \approx 0.66$

17. $\arctan 0 = 0$; π is not in the range of $y = \arctan x$.

19. (a) $\sin\left(\arctan \frac{3}{4}\right) = \frac{3}{5}$

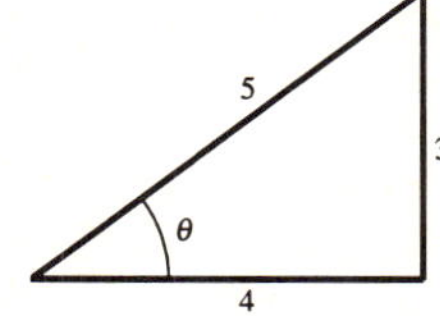

(b) $\sec\left(\arcsin \frac{4}{5}\right) = \frac{5}{3}$

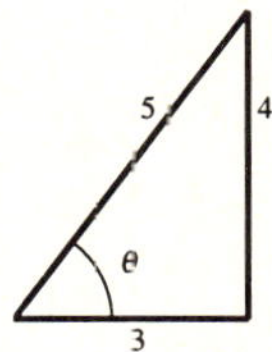

21. (a) $\cot\left[\arcsin\left(-\dfrac{1}{2}\right)\right] = \cot\left(-\dfrac{\pi}{6}\right) = -\sqrt{3}$

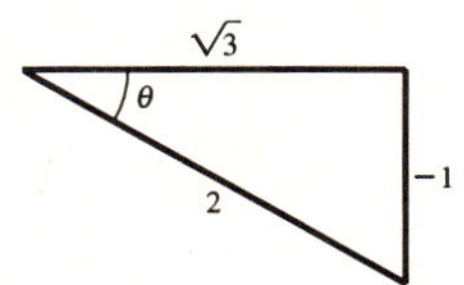

(b) $\csc\left[\arctan\left(-\dfrac{5}{12}\right)\right] = -\dfrac{13}{5}$

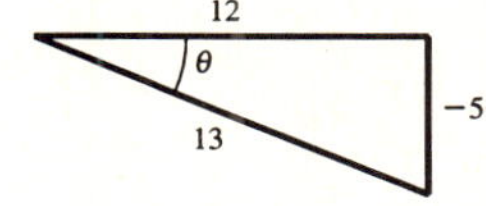

23. $y = \cos(\arcsin 2x)$

$\theta = \arcsin 2x$

$y = \cos \theta = \sqrt{1 - 4x^2}$

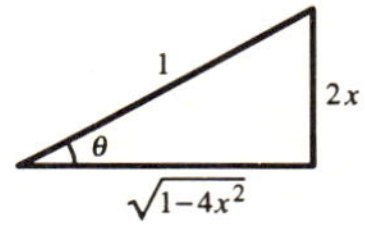

25. $y = \sin(\operatorname{arcsec} x)$

$\theta = \operatorname{arcsec} x,\ 0 \le \theta \le \pi,\ \theta \ne \dfrac{\pi}{2}$

$y = \sin \theta = \dfrac{\sqrt{x^2 - 1}}{|x|}$

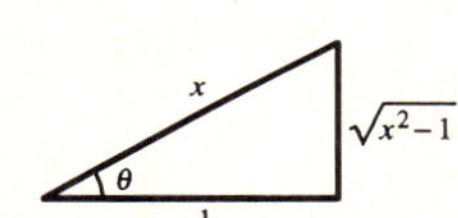

The absolute value bars on x are necessary because of the restriction $0 \le \theta \le \pi$, $\theta \ne \pi/2$, and $\sin \theta$ for this domain must always be nonnegative.

27. $y = \tan\left(\operatorname{arcsec} \dfrac{x}{3}\right)$

$\theta = \operatorname{arcsec} \dfrac{x}{3}$

$y = \tan \theta = \dfrac{\sqrt{x^2 - 9}}{3}$

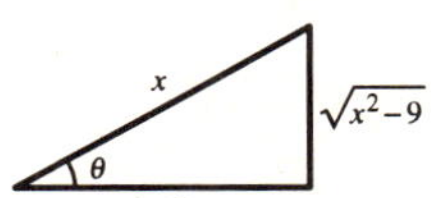

29. $y = \csc\left(\arctan \dfrac{x}{\sqrt{2}}\right)$

$\theta = \arctan \dfrac{x}{\sqrt{2}}$

$y = \csc \theta = \dfrac{\sqrt{x^2 + 2}}{x}$

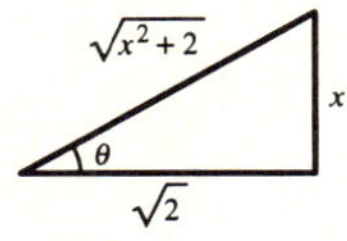

31. $\sin(\arctan 2x) = \dfrac{2x}{\sqrt{1 + 4x^2}}$

Asymptotes: $y = \pm 1$

$\arctan 2x = \theta$

$\tan \theta = 2x$

$\sin \theta = \dfrac{2x}{\sqrt{1 + 4x^2}}$

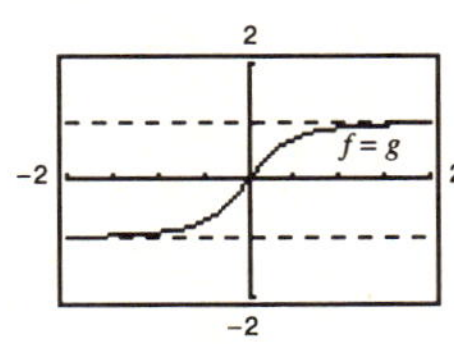

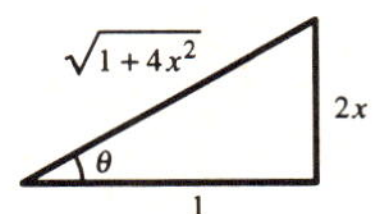

33. (a) $\operatorname{arccsc} x = \arcsin \dfrac{1}{x}, \ |x| \geq 1$

Let $y = \operatorname{arccsc} x$. Then for

$$-\frac{\pi}{2} \leq y < 0 \text{ and } 0 < y \leq \frac{\pi}{2},$$

$\csc y = x \Rightarrow \sin y = 1/x$. Thus, $y = \arcsin(1/x)$.
Therefore, $\operatorname{arccsc} x = \arcsin(1/x)$.

(b) $\arctan x + \arctan \dfrac{1}{x} = \dfrac{\pi}{2}, \ x > 0$

Let $y = \arctan x + \arctan(1/x)$. Then,

$$\tan y = \frac{\tan(\arctan x) + \tan[\arctan(1/x)]}{1 - \tan(\arctan x)\tan[\arctan(1/x)]}$$

$$= \frac{x + (1/x)}{1 - x(1/x)} = \frac{x + (1/x)}{0} \text{ (which is undefined).}$$

Thus, $y = \pi/2$. Therefore, $\arctan x + \arctan(1/x) = \pi/2$.

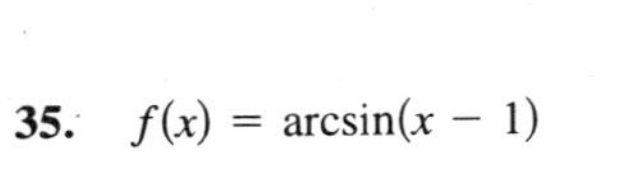

35. $f(x) = \arcsin(x - 1)$

$x - 1 = \sin y$

$x = 1 + \sin y$

Domain: $[0, 2]$

Range: $\left[-\dfrac{\pi}{2}, \dfrac{\pi}{2}\right]$

$f(x)$ is the graph of $\arcsin x$ shifted 1 unit to the right.

37. $f(x) = \operatorname{arcsec} 2x$

$2x = \sec y$

$x = \dfrac{1}{2} \sec y$

Domain: $\left(-\infty, -\dfrac{1}{2}\right], \left[\dfrac{1}{2}, \infty\right)$

Range: $\left[0, \dfrac{\pi}{2}\right), \left(\dfrac{\pi}{2}, \pi\right]$

39. $\arcsin(3x - \pi) = \frac{1}{2}$

$3x - \pi = \sin\left(\frac{1}{2}\right)$

$x = \frac{1}{3}\left[\sin\left(\frac{1}{2}\right) + \pi\right] \approx 1.207$

41. $\arcsin\sqrt{2x} = \arccos\sqrt{x}$

$\sqrt{2x} = \sin\left(\arccos\sqrt{x}\right)$

$\sqrt{2x} = \sqrt{1 - x}, \ 0 \leq x \leq 1$

$2x = 1 - x$

$3x = 1$

$x = \frac{1}{3}$

43. $f(x) = 2\arcsin(x - 1)$

$$f'(x) = \frac{2}{\sqrt{1 - (x - 1)^2}} = \frac{2}{\sqrt{2x - x^2}}$$

45. $g(x) = 3\arccos\dfrac{x}{2}$

$$g'(x) = \frac{-3(1/2)}{\sqrt{1 - (x^2/4)}} = \frac{-3}{\sqrt{4 - x^2}}$$

47. $f(x) = \arctan\dfrac{x}{a}$

$$f'(x) = \frac{1/a}{1 + (x^2/a^2)} = \frac{a}{a^2 + x^2}$$

49. $g(x) = \dfrac{\arcsin 3x}{x}$

$$g'(x) = \frac{x\left(3/\sqrt{1 - 9x^2}\right) - \arcsin 3x}{x^2} = \frac{3x - \sqrt{1 - 9x^2}\arcsin 3x}{x^2\sqrt{1 - 9x^2}}$$

51. $h(t) = \sin(\arccos t) = \sqrt{1 - t^2}$

$$h'(t) = \frac{1}{2}(1 - t^2)^{-1/2}(-2t) = \frac{-t}{\sqrt{1 - t^2}}$$

53. $y = \dfrac{1}{2}\left(\dfrac{1}{2}\ln\dfrac{x + 1}{x - 1} + \arctan x\right)$

$$= \frac{1}{4}[\ln(x + 1) - \ln(x - 1)] + \frac{1}{2}\arctan x$$

$$\frac{dy}{dx} = \frac{1}{4}\left(\frac{1}{x + 1} - \frac{1}{x - 1}\right) + \frac{1/2}{1 + x^2} = \frac{1}{1 - x^4}$$

55. $y = x\arcsin x + \sqrt{1 - x^2}$

$$\frac{dy}{dx} = x\left(\frac{1}{\sqrt{1 - x^2}}\right) + \arcsin x - \frac{x}{\sqrt{1 - x^2}} = \arcsin x$$

57. $f(x) = \arcsin x,\ a = \dfrac{1}{2}$

$$f'(x) = \frac{1}{\sqrt{1 - x^2}}$$

$$f''(x) = \frac{x}{(1 - x^2)^{3/2}}$$

$$P_1(x) = f\left(\frac{1}{2}\right) + f'\left(\frac{1}{2}\right)\left(x - \frac{1}{2}\right) = \frac{\pi}{6} + \frac{2\sqrt{3}}{3}\left(x - \frac{1}{2}\right)$$

$$P_2(x) = f\left(\frac{1}{2}\right) + f'\left(\frac{1}{2}\right)\left(x - \frac{1}{2}\right) + \frac{1}{2}f''\left(\frac{1}{2}\right)\left(x - \frac{1}{2}\right)^2 = \frac{\pi}{6} + \frac{2\sqrt{3}}{3}\left(x - \frac{1}{2}\right) + \frac{2\sqrt{3}}{9}\left(x - \frac{1}{2}\right)^2$$

59. $f(x) = \operatorname{arcsec} x - x$

$$f'(x) = \frac{1}{|x|\sqrt{x^2 - 1}} - 1 = 0 \text{ when } |x|\sqrt{x^2 - 1} = 1.$$

$$x^2(x^2 - 1) = 1$$

$$x^4 - x^2 - 1 = 0 \text{ when } x^2 = \frac{1 + \sqrt{5}}{2} \text{ or}$$

$$x = \pm\sqrt{\frac{1 + \sqrt{5}}{2}} = \pm 1.272$$

Relative maximum: $(1.272, -0.606)$

Relative minimum: $(-1.272, 3.747)$

61. (a) $h(t) = -16t^2 + 256$

$-16t^2 + 256 = 0$ when $t = 4$ sec.

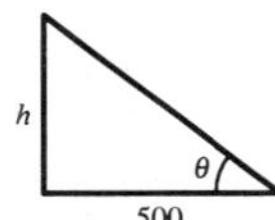

(b) $\tan\theta = \dfrac{h}{500} = \dfrac{-16t^2 + 256}{500}$

$$\theta = \arctan\left[\frac{16}{500}(-t^2 + 16)\right]$$

$$\frac{d\theta}{dt} = \frac{-8t/125}{1 + [(4/125)(-t^2 + 16)]^2} = \frac{-1000t}{15{,}625 + 16(16 - t^2)^2}$$

When $t = 1$, $d\theta/dt \approx -0.0520$ rad/sec.

When $t = 2$, $d\theta/dt \approx -0.1116$ rad/sec.

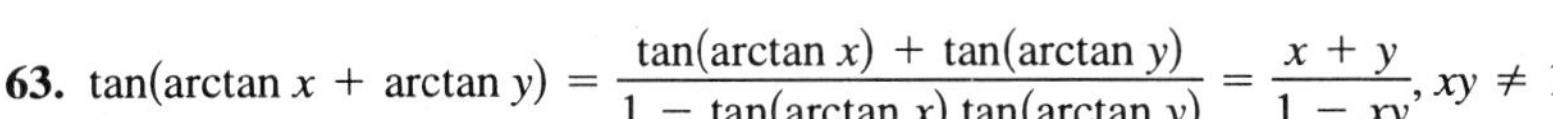

63. $\tan(\arctan x + \arctan y) = \dfrac{\tan(\arctan x) + \tan(\arctan y)}{1 - \tan(\arctan x)\tan(\arctan y)} = \dfrac{x + y}{1 - xy},\ xy \neq 1$

Therefore,

$$\arctan x + \arctan y = \arctan\left(\frac{x + y}{1 - xy}\right),\ xy \neq 1.$$

Let $x = \frac{1}{2}$ and $y = \frac{1}{3}$.

$$\arctan\left(\frac{1}{2}\right) + \arctan\left(\frac{1}{3}\right) = \arctan\frac{(1/2) + (1/3)}{1 - [(1/2)\cdot(1/3)]} = \arctan\frac{5/6}{1 - (1/6)} = \arctan\frac{5/6}{5/6} = \arctan 1 = \frac{\pi}{4}$$

65. $f(x) = kx + \sin x$

$f'(x) = k + \cos x \geq 0$ for $k \geq 1$

$f'(x) = k + \cos x \leq 0$ for $k \leq -1$

Therefore, $f(x) = kx + \sin x$ is strictly monotonic and has an inverse for $k \leq -1$ or $k \geq 1$.

67. True

$$\frac{d}{dx}[\arctan x] = \frac{1}{1 + x^2} > 0 \text{ for all } x.$$

69. True

$$\frac{d}{dx}[\arctan(\tan x)] = \frac{\sec^2 x}{1 + \tan^2 x} = \frac{\sec^2 x}{\sec^2 x} = 1$$

Section 5.9 Inverse Trigonometric Functions and Integration

1. Let $u = 3x$, $du = 3\,dx$.

$$\int_0^{1/6} \frac{1}{\sqrt{1 - 9x^2}}\,dx = \frac{1}{3}\int_0^{1/6} \frac{1}{\sqrt{1 - (3x)^2}}(3)\,dx = \left[\frac{1}{3}\arcsin(3x)\right]_0^{1/6} = \frac{\pi}{18}$$

3. Let $u = 2x$, $du = 2\,dx$.

$$\int_0^{\sqrt{3}/2} \frac{1}{1 + 4x^2}\,dx = \frac{1}{2}\int_0^{\sqrt{3}/2} \frac{2}{1 + (2x)^2}\,dx = \left[\frac{1}{2}\arctan(2x)\right]_0^{\sqrt{3}/2} = \frac{\pi}{6}$$

5. $\displaystyle\int \frac{1}{x\sqrt{4x^2 - 1}}\,dx = \int \frac{2}{2x\sqrt{(2x)^2 - 1}}\,dx = \operatorname{arcsec}|2x| + C$

7. $\displaystyle\int \frac{x^3}{x^2 + 1}\,dx = \int\left[x - \frac{x}{x^2 + 1}\right]dx = \int x\,dx - \frac{1}{2}\int \frac{2x}{x^2 + 1}\,dx = \frac{1}{2}x^2 - \frac{1}{2}\ln(x^2 + 1) + C$ (Use long division.)

9. $\displaystyle\int \frac{1}{\sqrt{1 - (x + 1)^2}}\,dx = \arcsin(x + 1) + C$

11. Let $u = t^2$, $du = 2t\,dt$.

$$\int \frac{t}{\sqrt{1 - t^4}}\,dt = \frac{1}{2}\int \frac{1}{\sqrt{1 - (t^2)^2}}(2t)\,dt = \frac{1}{2}\arcsin(t^2) + C$$

13. Let $u = \arcsin x$, $du = \dfrac{1}{\sqrt{1 - x^2}}\,dx$.

$$\int_0^{1/\sqrt{2}} \frac{\arcsin}{\sqrt{1 - x^2}}\,dx = \left[\frac{1}{2}\arcsin^2 x\right]_0^{1/\sqrt{2}} = \frac{\pi^2}{32} \approx 0.308$$

15. Let $u = 1 - x^2$, $du = -2x\,dx$.

$$\int_{-1/2}^{0} \frac{x}{\sqrt{1 - x^2}}\,dx = -\frac{1}{2}\int_{-1/2}^{0} (1 - x^2)^{-1/2}(-2x)\,dx$$

$$= \left[-\sqrt{1 - x^2}\right]_{-1/2}^{0} = \frac{\sqrt{3} - 2}{2}$$

$$\approx -0.134$$

17. Let $u = e^{2x}$, $du = 2e^{2x}\,dx$.

$$\int \frac{e^{2x}}{4 + e^{4x}}\,dx = \frac{1}{2}\int \frac{2e^{2x}}{4 + (e^{2x})^2}\,dx = \frac{1}{4}\arctan\frac{e^{2x}}{2} + C$$

19. Let $u = \cos x$, $du = -\sin x\,dx$.

$$\int_{\pi/2}^{\pi} \frac{\sin x}{1 + \cos^2 x}\,dx = -\int_{\pi/2}^{\pi} \frac{-\sin x}{1 + \cos^2 x}\,dx$$

$$= \left[-\arctan(\cos x)\right]_{\pi/2}^{\pi} = \frac{\pi}{4}$$

21. $\displaystyle\int_0^2 \frac{1}{x^2 - 2x + 2}\,dx = \int_0^2 \frac{1}{1 + (x - 1)^2}\,dx = \left[\arctan(x - 1)\right]_0^2 = \frac{\pi}{2}$

23. $\displaystyle\int \frac{2x}{x^2 + 6x + 13}\,dx = \int \frac{2x + 6}{x^2 + 6x + 13}\,dx - 6\int \frac{1}{x^2 + 6x + 13}\,dx = \int \frac{2x + 6}{x^2 + 6x + 13}\,dx - 6\int \frac{1}{4 + (x + 3)^2}\,dx$

$$= \ln|x^2 + 6x + 13| - 3\arctan\left(\frac{x + 3}{2}\right) + C$$

25. $\displaystyle\int \frac{1}{\sqrt{-x^2-4x}}\,dx = \int \frac{1}{\sqrt{4-(x+2)^2}}\,dx$

$\displaystyle = \arcsin\left(\frac{x+2}{2}\right) + C$

27. Let $u = -x^2 - 4x$, $du = (-2x - 4)\,dx$.

$\displaystyle\int \frac{x+2}{\sqrt{-x^2-4x}}\,dx = -\frac{1}{2}\int (-x^2-4x)^{-1/2}(-2x-4)\,dx$

$\displaystyle = -\sqrt{-x^2-4x} + C$

29. $\displaystyle\int_2^3 \frac{2x-3}{\sqrt{4x-x^2}}\,dx = \int_2^3 \frac{2x-4}{\sqrt{4x-x^2}}\,dx + \int_2^3 \frac{1}{\sqrt{4x-x^2}}\,dx = -\int_2^3 (4x-x^2)^{-1/2}(4-2x)\,dx + \int_2^3 \frac{1}{\sqrt{4-(x-2)^2}}\,dx$

$\displaystyle = \left[-2\sqrt{4x-x^2} + \arcsin\left(\frac{x-2}{2}\right)\right]_2^3 = 4 - 2\sqrt{3} + \frac{\pi}{6} \approx 1.059$

31. Let $u = x^2 + 1$, $du = 2x\,dx$.

$\displaystyle\int \frac{x}{x^4+2x^2+2}\,dx = \frac{1}{2}\int \frac{2x}{(x^2+1)^2+1}\,dx$

$\displaystyle = \frac{1}{2}\arctan(x^2+1) + C$

33. (a) $\displaystyle\int \frac{1}{\sqrt{1-x^2}}\,dx = \arcsin x + C$, $u = x$

(b) $\displaystyle\int \frac{x}{\sqrt{1-x^2}}\,dx = -\sqrt{1-x^2} + C$, $u = 1 - x^2$

(c) $\displaystyle\int \frac{1}{x\sqrt{1-x^2}}\,dx$ cannot be evaluated using the basic integration rules.

35. (a) $\displaystyle\int \sqrt{x-1}\,dx = \frac{2}{3}(x-1)^{3/2} + C$, $u = x - 1$

(b) Let $u = \sqrt{x-1}$. Then $x = u^2 + 1$ and $dx = 2u\,du$.

$\displaystyle\int x\sqrt{x-1}\,dx = \int (u^2+1)(u)(2u)\,du = 2\int (u^4+u^2)\,du = 2\left(\frac{u^5}{5} + \frac{u^3}{3}\right) + C$

$\displaystyle = \frac{2}{15}u^3(3u^2+5) + C = \frac{2}{15}(x-1)^{3/2}[3(x-1)+5] + C = \frac{2}{15}(x-1)^{3/2}(3x+2) + C$

(c) Let $u = \sqrt{x-1}$. Then $x = u^2 + 1$ and $dx = 2u\,du$.

$\displaystyle\int \frac{x}{\sqrt{x-1}}\,dx = \int \frac{u^2+1}{u}(2u)\,du = 2\int (u^2+1)\,du = 2\left(\frac{u^3}{3} + u\right) + C = \frac{2}{3}u(u^2+3) + C = \frac{2}{3}\sqrt{x-1}(x+2) + C$

Note: In (b) and (c), substitution was necessary *before* the basic integration rules could be used.

37. Let $u = \sqrt{e^t - 3}$. Then $u^2 + 3 = e^t$, $2u\,du = e^t\,dt$, and $\displaystyle\frac{2u\,du}{u^2+3} = dt$.

$\displaystyle\int \sqrt{e^t-3}\,dt = \int \frac{2u^2}{u^2+3}\,du = \int 2\,du - \int 6\frac{1}{u^2+3}\,du$

$\displaystyle = 2u - 2\sqrt{3}\arctan\frac{u}{\sqrt{3}} + C = 2\sqrt{e^t-3} - 2\sqrt{3}\arctan\sqrt{\frac{e^t-3}{3}} + C$

39. (a)

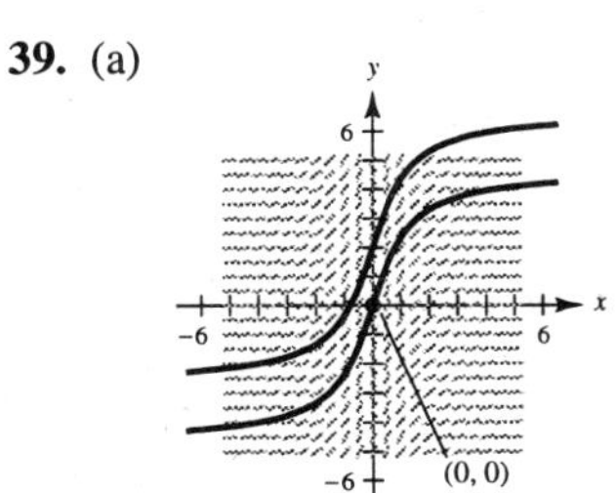

(b) $\displaystyle\frac{dy}{dx} = \frac{3}{1+x^2}$, $(0, 0)$

$\displaystyle y = 3\int \frac{dx}{1+x^2} = 3\arctan x + C$

$(0, 0)$: $0 = 3\arctan(0) + C \Rightarrow C = 0$

$y = 3\arctan x$

41. $A = \int_1^3 \frac{1}{x^2 - 2x + 1 + 4}\,dx = \int_1^3 \frac{1}{(x-1)^2 + 2^2}\,dx$

$= \left[\frac{1}{2}\arctan\left(\frac{x-1}{2}\right)\right]_1^3$

$= \frac{1}{2}\arctan(1) = \frac{\pi}{8} \approx 0.3927$

43. Area $\approx (1)(1) = 1$

Matches (c)

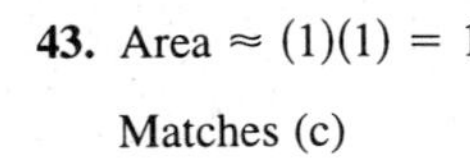

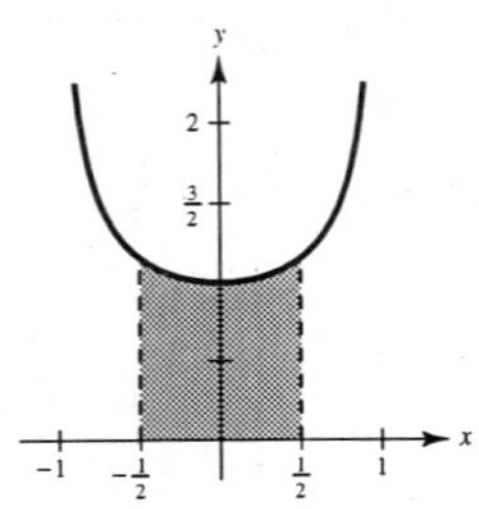

45. (a) $\int_0^1 \frac{4}{1+x^2}\,dx = \left[4\arctan x\right]_0^1 = 4\arctan 1 - 4\arctan 0 = 4\left(\frac{\pi}{4}\right) - 4(0) = \pi$

(b) Let $n = 6$.

$4\int_0^1 \frac{4}{1+x^2}\,dx \approx 4\left(\frac{1}{36}\right)\left[1 + \frac{4}{1+(1/36)} + \frac{2}{1+(1/9)} + \frac{4}{1+(1/4)} + \frac{2}{1+(4/9)} + \frac{4}{1+(25/36)} + \frac{1}{2}\right] \approx 3.1415918$

(c) 3.1415927

47. $\int \frac{1}{\sqrt{6x - x^2}}\,dx$

(a) $6x - x^2 = 9 - (x^2 - 6x + 9) = 9 - (x-3)^2$

$\int \frac{1}{\sqrt{6x-x^2}}\,dx = \int \frac{dx}{\sqrt{9-(x-3)^2}} = \arcsin\left(\frac{x-3}{3}\right) + C$

(b) $u = \sqrt{x},\ u^2 = x,\ 2u\,du = dx$

$\int \frac{1}{\sqrt{6u^2 - u^4}}(2u\,du) = \int \frac{2}{\sqrt{6-u^2}}\,du = 2\arcsin\left(\frac{u}{\sqrt{6}}\right) + C = 2\arcsin\left(\frac{\sqrt{x}}{\sqrt{6}}\right) + C$

(c)

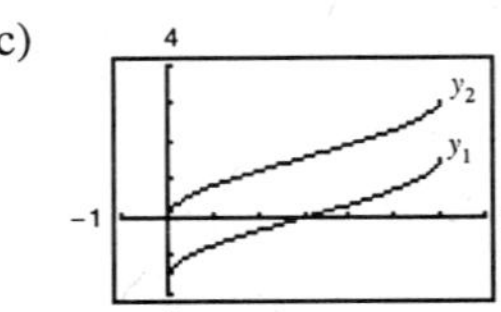

The antiderivatives differ by a constant, $\pi/2$.

Domain: $[0, 6]$

49. (a) $v(t) = -32t + 500$

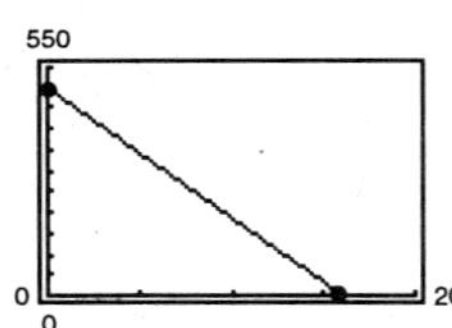

(b) $s(t) = \int v(t)\,dt = \int(-32t + 500)\,dt$

$= -16t^2 + 500t + C$

$s(0) = -16(0) + 500(0) + C = 0 \Rightarrow C = 0$

$s(t) = -16t^2 + 500t$

When the object reaches its maximum height, $v(t) = 0$.

$v(t) = -32t + 500 = 0$

$-32t = -500$

$t = 15.625$

$s(15.625) = -16(15.625)^2 + 500(15.625)$

$= 3906.25$ ft (Maximum height)

(c) $\int \frac{1}{32 + kv^2}\,dv = -\int dt$

$\frac{1}{\sqrt{32k}}\arctan\left(\sqrt{\frac{k}{32}}v\right) = -t + C_1$

$\arctan\left(\sqrt{\frac{k}{32}}v\right) = -\sqrt{32k}\,t + C$

$\sqrt{\frac{k}{32}}v = \tan\left(C - \sqrt{32k}\,t\right)$

$v = \sqrt{\frac{32}{k}}\tan\left(C - \sqrt{32k}\,t\right)$

When $t = 0$, $v = 500$, $C = \arctan\left(500\sqrt{k/32}\right)$, and we have

$v(t) = \sqrt{\frac{32}{k}}\tan\left[\arctan\left(500\sqrt{\frac{k}{32}}\right) - \sqrt{32k}\,t\right].$

—CONTINUED—

49. —CONTINUED—

(d) When $k = 0.001$, $v(t) = \sqrt{32{,}000}\tan\left[\arctan\left(500\sqrt{0.00003125}\right) - \sqrt{0.032}\,t\right]$.

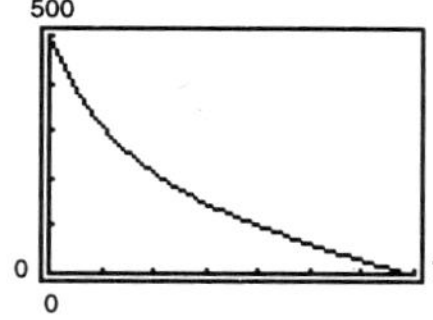

$v(t) = 0$ when $t_0 \approx 6.86$ sec.

(e) $h = \int_0^{6.86} \sqrt{32{,}000}\tan\left[\arctan\left(500\sqrt{0.00003125}\right) - \sqrt{0.032}\,t\right] dt$

Simpson's Rule: $n = 10$; $h \approx 1088$ feet

(f) Air resistance lowers the maximum height.

Section 5.10 Hyperbolic Functions

1. (a) $\sinh 3 = \dfrac{e^3 - e^{-3}}{2} \approx 10.018$

(b) $\tanh(-2) = \dfrac{\sinh(-2)}{\cosh(-2)} = \dfrac{e^{-2} - e^2}{e^{-2} + e^2} \approx -0.964$

3. (a) $\operatorname{csch}(\ln 2) = \dfrac{2}{e^{\ln 2} - e^{-\ln 2}} = \dfrac{2}{2 - (1/2)} = \dfrac{4}{3}$

(b) $\coth(\ln 5) = \dfrac{\cosh(\ln 5)}{\sinh(\ln 5)} = \dfrac{e^{\ln 5} + e^{-\ln 5}}{e^{\ln 5} - e^{-\ln 5}}$

$= \dfrac{5 + (1/5)}{5 - (1/5)} = \dfrac{13}{12}$

5. (a) $\cosh^{-1}(2) = \ln\left(2 + \sqrt{3}\right) \approx 1.317$

(b) $\operatorname{sech}^{-1}\left(\dfrac{2}{3}\right) = \ln\left(\dfrac{1 + \sqrt{1 - (4/9)}}{2/3}\right) \approx 0.962$

7. $\tanh^2 x + \operatorname{sech}^2 x = \left(\dfrac{e^x - e^{-x}}{e^x + e^{-x}}\right)^2 + \left(\dfrac{2}{e^x + e^{-x}}\right)^2$

$= \dfrac{e^{2x} - 2 + e^{-2x} + 4}{(e^x + e^{-x})^2}$

$= \dfrac{e^{2x} + 2 + e^{-2x}}{e^{2x} + 2 + e^{-2x}} = 1$

9. $\sinh x \cosh y + \cosh x \sinh y = \left(\dfrac{e^x - e^{-x}}{2}\right)\left(\dfrac{e^y + e^{-y}}{2}\right) + \left(\dfrac{e^x + e^{-x}}{2}\right)\left(\dfrac{e^y - e^{-y}}{2}\right)$

$= \dfrac{1}{4}[e^{x+y} - e^{-x+y} + e^{x-y} - e^{-(x+y)} + e^{x+y} + e^{-x+y} - e^{x-y} - e^{-(x+y)}]$

$= \dfrac{1}{4}[2(e^{x+y} - e^{-(x+y)})] = \dfrac{e^{(x+y)} - e^{-(x+y)}}{2} = \sinh(x + y)$

11. $3\sinh x + 4\sinh^3 x = \sinh x(3 + 4\sinh^2 x) = \left(\dfrac{e^x - e^{-x}}{2}\right)\left[3 + 4\left(\dfrac{e^x - e^{-x}}{2}\right)^2\right]$

$= \left(\dfrac{e^x - e^{-x}}{2}\right)[3 + e^{2x} - 2 + e^{-2x}] = \dfrac{1}{2}(e^x - e^{-x})(e^{2x} + e^{-2x} + 1)$

$= \dfrac{1}{2}[e^{3x} + e^{-x} + e^x - e^x - e^{-3x} - e^{-x}] = \dfrac{e^{3x} - e^{-3x}}{2} = \sinh(3x)$

13. $\sinh x = \dfrac{3}{2}$

$$\cosh^2 x - \left(\frac{3}{2}\right)^2 = 1 \Rightarrow \cosh^2 x = \frac{13}{4} \Rightarrow \cosh x = \frac{\sqrt{13}}{2}$$

$$\tanh x = \frac{3/2}{\sqrt{13}/2} = \frac{3\sqrt{13}}{13}$$

$$\text{csch } x = \frac{1}{3/2} = \frac{2}{3}$$

$$\text{sech } x = \frac{1}{\sqrt{13}/2} = \frac{2\sqrt{13}}{13}$$

$$\coth x = \frac{1}{3/\sqrt{13}} = \frac{\sqrt{13}}{3}$$

15. $y = \sinh(1 - x^2)$

$$y' = -2x\cosh(1 - x^2)$$

17. $f(x) = \ln(\sinh x)$

$$f'(x) = \frac{1}{\sinh}(\cosh x) = \coth x$$

19. $y = \ln\left(\tanh \dfrac{x}{2}\right)$

$$y' = \frac{1/2}{\tanh(x/2)}\text{sech}^2\left(\frac{x}{2}\right) = \frac{1}{2\sinh(x/2)\cosh(x/2)}$$

$$= \frac{1}{\sinh x} = \text{csch } x$$

21. $h(x) = \dfrac{1}{4}\sinh(2x) - \dfrac{x}{2}$

$$h'(x) = \frac{1}{2}\cosh(2x) - \frac{1}{2} = \frac{\cosh(2x) - 1}{2} = \sinh^2 x$$

23. $f(t) = \arctan(\sinh t)$

$$f'(t) = \frac{1}{1 + \sinh^2 t}(\cosh t) = \frac{\cosh t}{\cosh^2 t} = \text{sech } t$$

25. Let $y = g(x)$.

$$y = x^{\cosh x}$$

$$\ln y = \cosh x \ln x$$

$$\frac{1}{y}\left(\frac{dy}{dx}\right) = \frac{\cosh x}{x} + \sinh x \ln x$$

$$\frac{dy}{dx} = \frac{y}{x}[\cosh x + x(\sinh x)\ln x]$$

$$= \frac{x^{\cosh x}}{x}[\cosh x + x(\sinh x)\ln x]$$

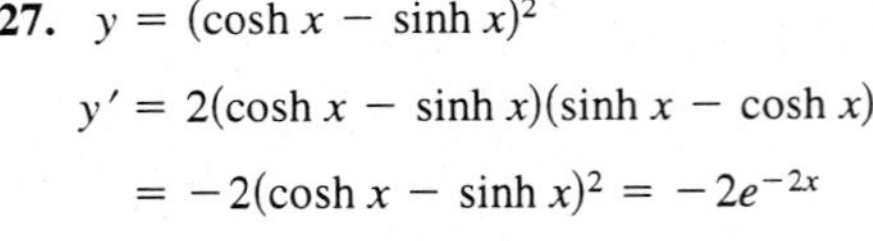

27. $y = (\cosh x - \sinh x)^2$

$$y' = 2(\cosh x - \sinh x)(\sinh x - \cosh x)$$

$$= -2(\cosh x - \sinh x)^2 = -2e^{-2x}$$

29. $f(x) = \sin x \sinh x - \cos x \cosh x, \ -4 \le x \le 4$

$$f'(x) = \sin x \cosh x + \cos x \sinh x - \cos x \sinh x + \sin x \cosh x$$

$$= 2\sin x \cosh x = 0 \text{ when } x = 0, \pm\pi.$$

Relative maxima: $(\pm\pi, \cosh \pi)$

Relative minimum: $(0, -1)$

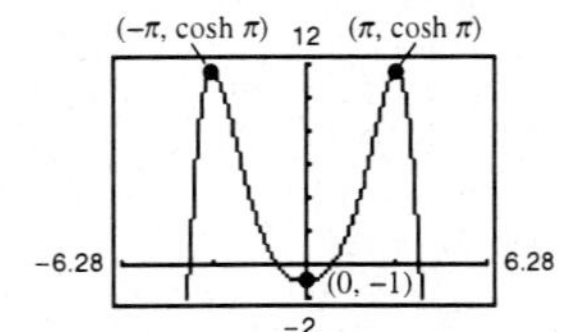

31. $g(x) = x \operatorname{sech} x = \dfrac{x}{\cosh x}$

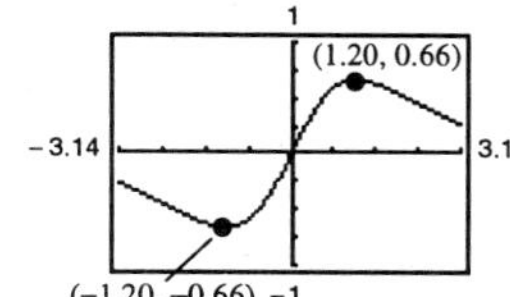

Relative maximum: $(1.20, 0.66)$

Relative minimum: $(-1.20, -0.66)$

33. $y = a \sinh x$

$y' = a \cosh x$

$y'' = a \sinh x$

$y''' = a \cosh x$

Therefore, $y''' - y' = 0$.

35. $f(x) = \tanh x$ $\qquad f(1) = \tanh(1) \approx 0.7616$

$f'(x) = \operatorname{sech}^2 x$ $\qquad f'(1) = \dfrac{1}{\cosh^2(1)} \approx 0.4200$

$f''(x) = -2 \operatorname{sech}^2 x \cdot \tanh x$ $\qquad f''(1) \approx -0.6397$

$P_1(x) = f(1) + f'(1)(x - 1) = 0.7616 + 0.42(x - 1)$

$P_2(x) = 0.7616 + 0.42(x - 1) - \dfrac{0.6397}{2}(x - 1)^2$

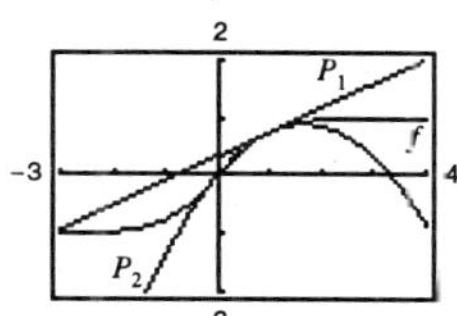

37. Let $u = 1 - 2x$, $du = -2\,dx$.

$$\int \sinh(1 - 2x)\,dx = -\frac{1}{2}\int \sinh(1 - 2x)(-2)\,dx$$
$$= -\frac{1}{2}\cosh(1 - 2x) + C$$

39. Let $u = \cosh(x - 1)$, $du = \sinh(x - 1)\,dx$.

$$\int \cosh^2(x - 1)\sinh(x - 1)\,dx = \frac{1}{3}\cosh^3(x - 1) + C$$

41. Let $u = \sinh x$, $du = \cosh x\,dx$.

$$\int \frac{\cosh x}{\sinh x}\,dx = \ln|\sinh x| + C$$

43. Let $u = \dfrac{x^2}{2}$, $du = x\,dx$.

$$\int x \operatorname{csch}^2 \frac{x^2}{2}\,dx = \int \left(\operatorname{csch}^2 \frac{x^2}{2}\right) x\,dx = -\coth \frac{x^2}{2} + C$$

45. Let $u = \dfrac{1}{x}$, $du = -\dfrac{1}{x^2}\,dx$.

$$\int \frac{\operatorname{csch}(1/x)\coth(1/x)}{x^2}\,dx = -\int \operatorname{csch}\frac{1}{x}\coth\frac{1}{x}\left(-\frac{1}{x^2}\right)dx = \operatorname{csch}\frac{1}{x} + C$$

47. $\displaystyle\int_0^4 \frac{1}{25 - x^2}\,dx = \left[\frac{1}{10}\ln\left|\frac{5 + x}{5 - x}\right|\right]_0^4 = \frac{1}{10}\ln 9 = \frac{1}{5}\ln 3$

49. Let $u = 2x$, $du = 2\,dx$.

$$\int_0^{\sqrt{2}/4} \frac{1}{\sqrt{1 - (2x)^2}}(2)\,dx = \Big[\arcsin(2x)\Big]_0^{\sqrt{2}/4} = \frac{\pi}{4}$$

51. Let $u = x^2$, $du = 2x\,dx$.

$$\int \frac{x}{x^4 + 1}\,dx = \frac{1}{2}\int \frac{2x}{(x^2)^2 + 1}\,dx = \frac{1}{2}\arctan(x^2) + C$$

53. $y = \cosh^{-1}(3x)$

$$y' = \frac{3}{\sqrt{9x^2 - 1}}$$

55. $y = \sinh^{-1}(\tan x)$

$$y' = \frac{1}{\sqrt{\tan^2 x + 1}}(\sec^2 x) = |\sec x|$$

57. $y = \coth^{-1}(\sin 2x)$

$$y' = \frac{1}{1 - \sin^2 2x}(2\cos 2x) = 2\sec 2x$$

59. $y = 2x \sinh^{-1}(2x) - \sqrt{1 + 4x^2}$

$$y' = 2x\left(\frac{2}{\sqrt{1 + 4x^2}}\right) + 2 \sinh^{-1}(2x) - \frac{4x}{\sqrt{1 + 4x^2}} = 2 \sinh^{-1}(2x)$$

61. $y = a \operatorname{sech}^{-1}\left(\frac{x}{a}\right) - \sqrt{a^2 - x^2}$

$$\frac{dy}{dx} = \frac{-1}{(x/a)\sqrt{1 - (x^2/a^2)}} + \frac{x}{\sqrt{a^2 - x^2}} = \frac{-a^2}{x\sqrt{a^2 - x^2}} + \frac{x}{\sqrt{a^2 - x^2}} = \frac{x^2 - a^2}{x\sqrt{a^2 - x^2}} = \frac{-\sqrt{a^2 - x^2}}{x}$$

63. $$\int \frac{1}{\sqrt{1 + e^{2x}}}\,dx = \int \frac{e^x}{e^x\sqrt{1 + (e^x)^2}}\,dx = -\operatorname{csch}^{-1}(e^x) + C = -\ln\left(\frac{1 + \sqrt{1 + e^{2x}}}{e^x}\right) + C$$

65. Let $u = \sqrt{x}$, $du = \frac{1}{2\sqrt{x}}\,dx$.

$$\int \frac{1}{\sqrt{x}\sqrt{1 + x}}\,dx = 2\int \frac{1}{\sqrt{1 + (\sqrt{x})^2}}\left(\frac{1}{2\sqrt{x}}\right)dx = 2 \sinh^{-1}\sqrt{x} + C = 2 \ln\left(\sqrt{x} + \sqrt{1 + x}\right) + C$$

67. $$\int \frac{-1}{4x - x^2}\,dx = \int \frac{1}{(x - 2)^2 - 4}\,dx = \frac{1}{4}\ln\left|\frac{(x - 2) - 2}{(x - 2) + 2}\right| = \frac{1}{4}\ln\left|\frac{x - 4}{x}\right| + C$$

69. $$\int \frac{1}{1 - 4x - 2x^2}\,dx = \int \frac{1}{3 - 2(x + 1)^2}\,dx = \frac{-1}{\sqrt{2}}\int \frac{\sqrt{2}}{\left[\sqrt{2}(x + 1)\right]^2 - \left(\sqrt{3}\right)^2}\,dx$$

$$= \frac{-1}{2\sqrt{6}}\ln\left|\frac{\sqrt{2}(x + 1) - \sqrt{3}}{\sqrt{2}(x + 1) + \sqrt{3}}\right| + C = \frac{1}{2\sqrt{6}}\ln\left|\frac{\sqrt{2}(x + 1) + \sqrt{3}}{\sqrt{2}(x + 1) - \sqrt{3}}\right| + C$$

71. Let $u = 4x - 1$, $du = 4\,dx$.

$$y = \int \frac{1}{\sqrt{80 + 8x - 16x^2}}\,dx = \frac{1}{4}\int \frac{4}{\sqrt{81 - (4x - 1)^2}}\,dx = \frac{1}{4}\arcsin\left(\frac{4x - 1}{9}\right) + C$$

73. $$y = \int \frac{x^3 - 21x}{5 + 4x - x^2}\,dx = \int\left(-x - 4 + \frac{20}{5 + 4x - x^2}\right)dx = \int(-x - 4)\,dx + 20\int \frac{1}{3^2 - (x - 2)^2}\,dx$$

$$= -\frac{x^2}{2} - 4x + \frac{20}{6}\ln\left|\frac{(x - 2) + 3}{(x - 2) - 3}\right| + C = -\frac{x^2}{2} - 4x + \frac{10}{3}\ln\left|\frac{x + 1}{x - 5}\right| + C = \frac{-x^2}{2} - 4x - \frac{10}{3}\ln\left|\frac{x - 5}{x + 1}\right| + C$$

75. $$A = 2\int_0^4 \operatorname{sech}\frac{x}{2}\,dx = 2\int_0^4 \frac{2}{e^{x/2} + e^{-x/2}}\,dx = 4\int_0^4 \frac{e^{x/2}}{(e^{x/2})^2 + 1}\,dx = \left[8 \arctan(e^{x/2})\right]_0^4 = 8 \arctan(e^2) - 2\pi \approx 5.207$$

77. $$A = \int_0^2 \frac{5x}{\sqrt{x^4 + 1}}\,dx = \frac{5}{2}\int_0^2 \frac{2x}{\sqrt{(x^2)^2 + 1}}\,dx = \left[\frac{5}{2}\ln\left(x^2 + \sqrt{x^4 + 1}\right)\right]_0^2 = \frac{5}{2}\ln\left(4 + \sqrt{17}\right) \approx 5.237$$

79. $$\int \frac{3k}{16}\,dt = \int \frac{1}{x^2 - 12x + 32}\,dx$$

$$\frac{3kt}{16} = \int \frac{1}{(x - 6)^2 - 4}\,dx = \frac{1}{2(2)}\ln\left|\frac{(x - 6) - 2}{(x - 6) + 2}\right| + C = \frac{1}{4}\ln\left|\frac{x - 8}{x - 4}\right| + C$$

When $x = 0$: $\quad t = 0$

$$C = -\frac{1}{4}\ln(2)$$

—CONTINUED—

79. —CONTINUED—

When $x = 1$: $t = 10$

$$\frac{30k}{16} = \frac{1}{4}\ln\left|\frac{-7}{-3}\right| - \frac{1}{4}\ln(2) = \frac{1}{4}\ln\left(\frac{7}{6}\right)$$

$$k = \frac{2}{15}\ln\left(\frac{7}{6}\right)$$

When $t = 20$: $\left(\frac{3}{16}\right)\left(\frac{2}{15}\right)\ln\left(\frac{7}{6}\right)(20) = \frac{1}{4}\ln\frac{x-8}{2x-8}$

$$\ln\left(\frac{7}{6}\right)^2 = \ln\frac{x-8}{2x-8}$$

$$\frac{49}{36} = \frac{x-8}{2x-8}$$

$$62x = 104$$

$$x = \frac{104}{62} = \frac{52}{31} \approx 1.677 \text{ kg}$$

81. As k increases, the time required for the object to reach the ground increases.

83. $y = \cosh x = \dfrac{e^x + e^{-x}}{2}$

$$y' = \frac{e^x - e^{-x}}{2} = \sinh x$$

85. $y = \cosh^{-1} x$

$$\cosh y = x$$

$$(\sinh y)(y') = 1$$

$$y' = \frac{1}{\sinh y} = \frac{1}{\sqrt{\cosh^2 y - 1}} = \frac{1}{\sqrt{x^2 - 1}}$$

87. $y = \operatorname{sech} x = \dfrac{2}{e^x + e^{-x}}$

$$y' = -2(e^x + e^{-x})^{-2}(e^x - e^{-x}) = \left(\frac{-2}{e^x + e^{-x}}\right)\left(\frac{e^x - e^{-x}}{e^x + e^{-x}}\right) = -\operatorname{sech} x \tanh x$$

Review Exercises for Chapter 5

1. $f(x) = \ln x + 3$

Vertical shift 3 units upward
Vertical asymptote: $x = 0$

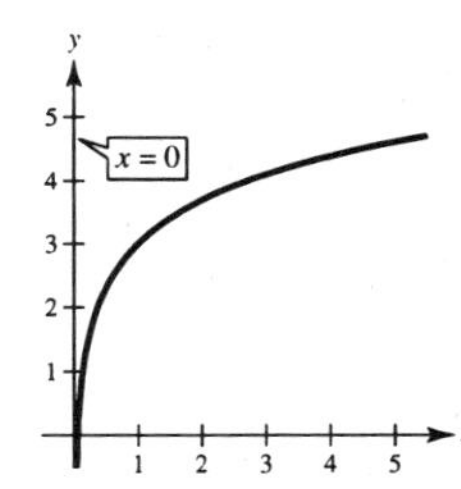

3. $\ln \sqrt[5]{\dfrac{4x^2 - 1}{4x^2 + 1}} = \dfrac{1}{5}\ln\dfrac{(2x-1)(2x+1)}{4x^2+1}$

$$= \frac{1}{5}[\ln(2x-1) + \ln(2x+1) - \ln(4x^2+1)]$$

5. $\ln 3 + \dfrac{1}{3}\ln(4 - x^2) - \ln x = \ln 3 + \ln\sqrt[3]{4 - x^2} - \ln x = \ln\left(\dfrac{3\sqrt[3]{4 - x^2}}{x}\right)$

7. False; the domain of $f(x) = \ln x$ is the set of all **positive** real numbers.

9. $\ln\sqrt{x+1} = 2$

$$\sqrt{x+1} = e^2$$

$$x + 1 = e^4$$

$$x = e^4 - 1 \approx 53.598$$

11. $g(x) = \ln\sqrt{x} = \dfrac{1}{2}\ln x$

$$g'(x) = \frac{1}{2x}$$

13. $f(x) = x\sqrt{\ln x}$

$$f'(x) = \left(\frac{x}{2}\right)(\ln x)^{-1/2}\left(\frac{1}{x}\right) + \sqrt{\ln x}$$

$$= \frac{1}{2\sqrt{\ln x}} + \sqrt{\ln x} = \frac{1 + 2\ln x}{2\sqrt{\ln x}}$$

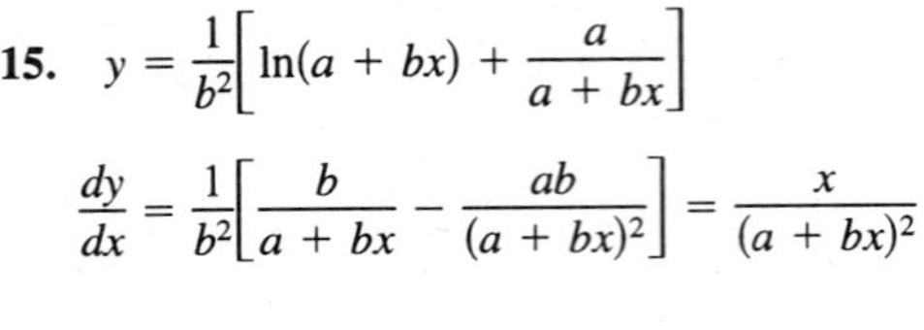

15. $y = \frac{1}{b^2}\left[\ln(a + bx) + \frac{a}{a + bx}\right]$

$$\frac{dy}{dx} = \frac{1}{b^2}\left[\frac{b}{a + bx} - \frac{ab}{(a + bx)^2}\right] = \frac{x}{(a + bx)^2}$$

17. $y = -\frac{1}{a}\ln\left(\frac{a + bx}{x}\right) = -\frac{1}{a}[\ln(a + bx) - \ln x]$

$$\frac{dy}{dx} = -\frac{1}{a}\left(\frac{b}{a + bx} - \frac{1}{x}\right) = \frac{1}{x(a + bx)}$$

19. $u = 7x - 2,\ du = 7dx$

$$\int \frac{1}{7x - 2}\,dx = \frac{1}{7}\int \frac{1}{7x - 2}(7)\,dx = \frac{1}{7}\ln|7x - 2| + C$$

21. $\displaystyle\int \frac{\sin x}{1 + \cos x}\,dx = -\int \frac{-\sin x}{1 + \cos x}\,dx = -\ln|1 + \cos x| + C$

23. $\displaystyle\int_1^4 \frac{x + 1}{x}\,dx = \int_1^4 \left(1 + \frac{1}{x}\right)dx = \Big[x + \ln|x|\Big]_1^4 = 3 + \ln 4$

25. $\displaystyle\int_0^{\pi/3} \sec\theta\,d\theta = \Big[\ln|\sec\theta + \tan\theta|\Big]_0^{\pi/3} = \ln\left(2 + \sqrt{3}\right)$

27. (a) $f(x) = \frac{1}{2}x - 3$

$y = \frac{1}{2}x - 3$

$2(y + 3) = x$

$2(x + 3) = y$

$f^{-1}(x) = 2x + 6$

(b)

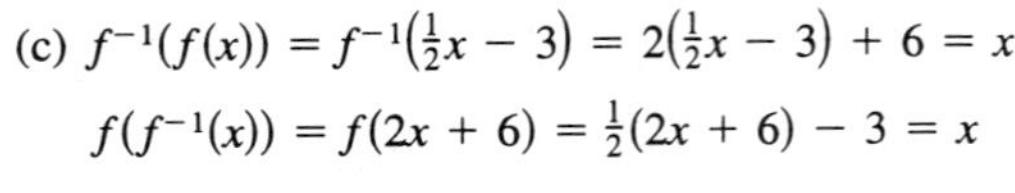

(c) $f^{-1}(f(x)) = f^{-1}\left(\frac{1}{2}x - 3\right) = 2\left(\frac{1}{2}x - 3\right) + 6 = x$

$f(f^{-1}(x)) = f(2x + 6) = \frac{1}{2}(2x + 6) - 3 = x$

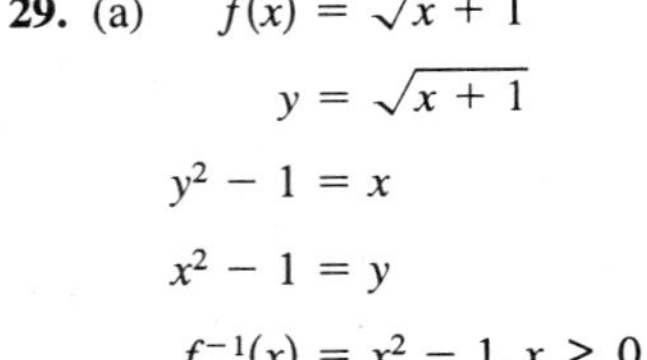

29. (a) $f(x) = \sqrt{x + 1}$

$y = \sqrt{x + 1}$

$y^2 - 1 = x$

$x^2 - 1 = y$

$f^{-1}(x) = x^2 - 1,\ x \geq 0$

(b)

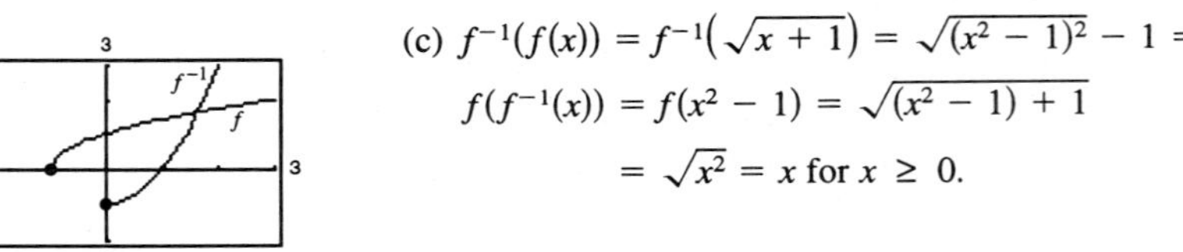

(c) $f^{-1}(f(x)) = f^{-1}\left(\sqrt{x + 1}\right) = \sqrt{(x^2 - 1)^2} - 1 = x$

$f(f^{-1}(x)) = f(x^2 - 1) = \sqrt{(x^2 - 1) + 1}$

$= \sqrt{x^2} = x$ for $x \geq 0$.

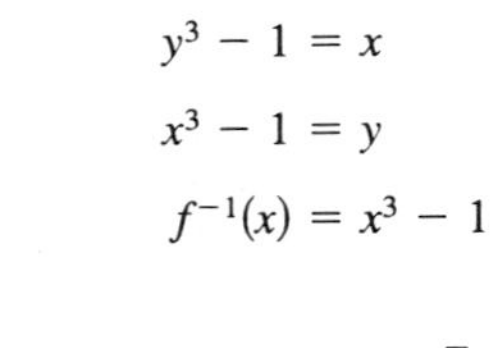

31. (a) $f(x) = \sqrt[3]{x + 1}$

$y = \sqrt[3]{x + 1}$

$y^3 - 1 = x$

$x^3 - 1 = y$

$f^{-1}(x) = x^3 - 1$

(b)

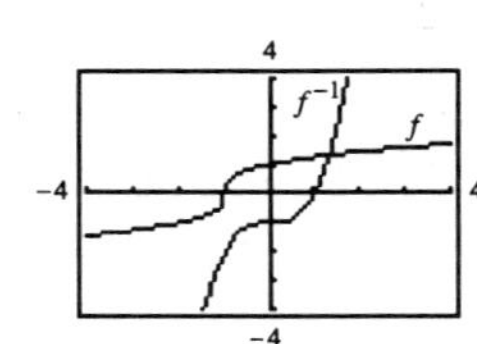

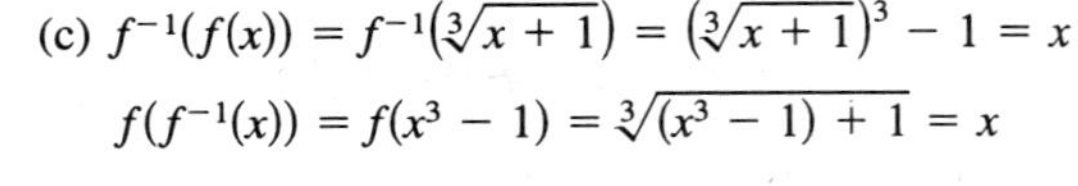

(c) $f^{-1}(f(x)) = f^{-1}\left(\sqrt[3]{x + 1}\right) = \left(\sqrt[3]{x + 1}\right)^3 - 1 = x$

$f(f^{-1}(x)) = f(x^3 - 1) = \sqrt[3]{(x^3 - 1) + 1} = x$

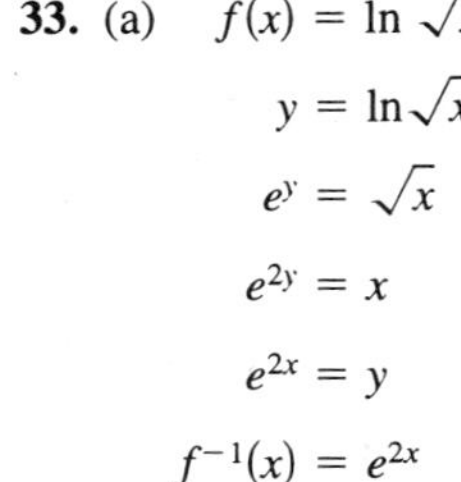

33. (a) $f(x) = \ln\sqrt{x}$

$y = \ln\sqrt{x}$

$e^y = \sqrt{x}$

$e^{2y} = x$

$e^{2x} = y$

$f^{-1}(x) = e^{2x}$

(b)

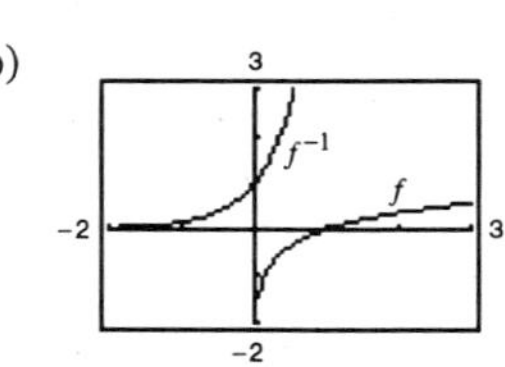

(c) $f^{-1}(f(x)) = f^{-1}\left(\ln\sqrt{x}\right) = e^{2\ln\sqrt{x}} = e^{\ln x} = x$

$f(f^{-1}(x)) = f(e^{2x}) = \ln\sqrt{e^{2x}} = \ln e^x = x$

35. $y = e^{-x/2}$

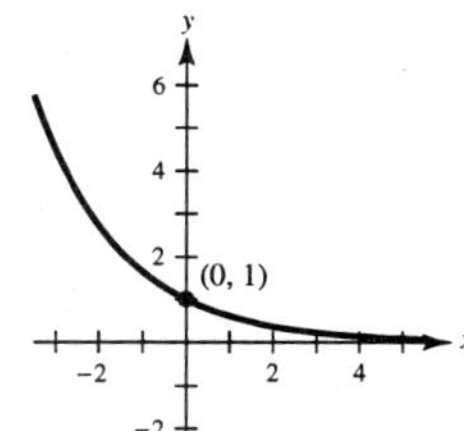

37. $h(x) = -3 \arcsin(2x)$

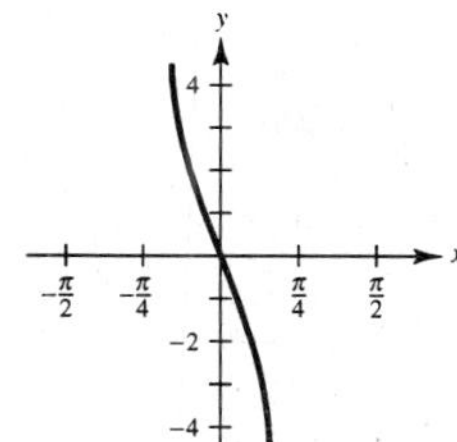

39. (a) Let $\theta = \arcsin \frac{1}{2}$

$$\sin \theta = \frac{1}{2}$$

$$\sin\left(\arcsin \frac{1}{2}\right) = \sin \theta = \frac{1}{2}.$$

(b) Let $\theta = \arcsin \frac{1}{2}$

$$\sin \theta = \frac{1}{2}$$

$$\cos\left(\arcsin \frac{1}{2}\right) = \cos \theta = \frac{\sqrt{3}}{2}.$$

41. $f(x) = \ln(e^{-x^2}) = -x^2$

$f'(x) = -2x$

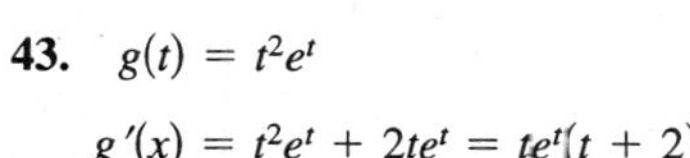

43. $g(t) = t^2e^t$

$g'(x) = t^2e^t + 2te^t = te^t(t + 2)$

45. $y = \sqrt{e^{2x} + e^{-2x}}$

$$y' = \frac{1}{2}(e^{2x} + e^{-2x})^{-1/2}(2e^{2x} - 2e^{-2x}) = \frac{e^{2x} - e^{-2x}}{\sqrt{e^{2x} + e^{-2x}}}$$

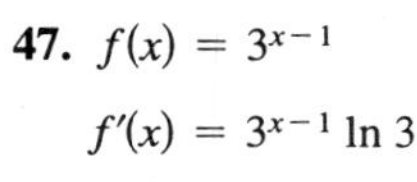

47. $f(x) = 3^{x-1}$

$f'(x) = 3^{x-1} \ln 3$

49. $g(x) = \dfrac{x^2}{e^x}$

$$g'(x) = \frac{e^x(2x) - x^2e^x}{e^{2x}} = \frac{x(2 - x)}{e^x}$$

51. $y = \tan(\arcsin x) = \dfrac{x}{\sqrt{1 - x^2}}$

$$y' = \frac{(1 - x^2)^{1/2} + x^2(1 - x^2)^{-1/2}}{1 - x^2}$$

$$= (1 - x^2)^{-3/2}$$

53. $y = x \operatorname{arcsec} x$

$$y' = \frac{x}{|x|\sqrt{x^2 - 1}} + \operatorname{arcsec} x$$

55. $y = x(\arcsin x)^2 - 2x + 2\sqrt{1 - x^2} \arcsin x$

$$y' = \frac{2x \arcsin x}{\sqrt{1 - x^2}} + (\arcsin x)^2 - 2 + \frac{2\sqrt{1 - x^2}}{\sqrt{1 - x^2}} - \frac{2x}{\sqrt{1 - x^2}} \arcsin x = (\arcsin x)^2$$

57. $y = 2x - \cosh\sqrt{x}$

$$y' = 2 - \frac{1}{2\sqrt{x}}(\sinh \sqrt{x}) = 2 - \frac{\sinh\sqrt{x}}{2\sqrt{x}}$$

59.

$$y(\ln x) + y^2 = 0$$

$$y\left(\frac{1}{x}\right) + (\ln x)\left(\frac{dy}{dx}\right) + 2y\left(\frac{dy}{dx}\right) = 0$$

$$(2y - \ln x)\frac{dy}{dx} = \frac{-y}{x}$$

$$\frac{dy}{dx} = \frac{-y}{x(2y + \ln x)}$$

61. (a) $y = x^a$

$y' = ax^{a-1}$

(b) $y = a^x$

$y' = (\ln a)a^x$

(c) $y = x^x$

$\ln y = x \ln x$

$\frac{1}{y}y' = x \cdot \frac{1}{x} + (1) \ln x$

$y' = y(1 + \ln x)$

$y' = x^x(1 + \ln x)$

(d) $y = a^a$

$y' = 0$

63. $2P = Pe^{10r}$

$2 = e^{10r}$

$\ln 2 = 10r$

$r = \dfrac{\ln 2}{10} \approx 6.93\%$

65. Let $u = -3x^2$, $du = -6x\,dx$.

$$\int xe^{-3x^2}\,dx = -\frac{1}{6}\int e^{-3x^2}(-6x)\,dx = -\frac{1}{6}e^{-3x^2} + C$$

67. $\displaystyle\int \frac{e^{4x} - e^{2x} + 1}{e^x}\,dx = \int (e^{3x} - e^x + e^{-x})\,dx$

$$= \frac{1}{3}e^{3x} - e^x - e^{-x} + C$$

$$= \frac{e^{4x} - 3e^{2x} - 3}{3e^x} + C$$

69. Let $u = e^x - 1$, $du = e^x\,dx$.

$$\int \frac{e^x}{e^x - 1}\,dx = \ln|e^x - 1| + C$$

71. Let $u = e^{2x}$, $du = 2e^{2x}\,dx$.

$$\int \frac{1}{e^{2x} + e^{-2x}}\,dx = \int \frac{e^{2x}}{1 + e^{4x}}\,dx = \frac{1}{2}\int \frac{1}{1 + (e^{2x})^2}(2e^{2x})\,dx = \frac{1}{2}\arctan(e^{2x}) + C$$

73. Let $u = x^2$, $du = 2x\,dx$.

$$\int \frac{x}{\sqrt{1 - x^4}}\,dx = \frac{1}{2}\int \frac{1}{\sqrt{1 - (x^2)^2}}(2x)\,dx = \frac{1}{2}\arcsin x^2 + C$$

75. Let $u = 16 + x^2$, $du = 2x\,dx$.

$$\int \frac{x}{16 + x^2}\,dx = \frac{1}{2}\int \frac{1}{16 + x^2}(2x)\,dx = \frac{1}{2}\ln(16 + x^2) + C$$

77. Let $u = \arctan\left(\dfrac{x}{2}\right)$, $du = \dfrac{2}{4 + x^2}\,dx$.

$$\int \frac{\arctan(x/2)}{4 + x^2}\,dx = \frac{1}{2}\int \left(\arctan\frac{x}{2}\right)\left(\frac{2}{4 + x^2}\right)dx = \frac{1}{4}\left(\arctan\frac{x}{2}\right)^2 + C$$

79. Let $u = x^2$, $du = 2x\,dx$.

$$\int \frac{x}{\sqrt{x^4 - 1}}\,dx = \frac{1}{2}\int \frac{1}{\sqrt{(x^2)^2 - 1}}(2x)\,dx = \frac{1}{2}\ln\left(x^2 + \sqrt{x^4 - 1}\right) + C$$

81. $\text{Area} = \displaystyle\int_0^4 xe^{-x^2}\,dx = \left[-\frac{1}{2}e^{-x^2}\right]_0^4 = -\frac{1}{2}(e^{-16} - 1) \approx 0.500$

83. $\dfrac{dy}{dx} = \dfrac{x^2 + 3}{x}$

$$\int dy = \int \left(x + \frac{3}{x}\right)dx$$

$$y = \frac{x^2}{2} + 3\ln|x| + C$$

85. $y' - 2xy = 0$

$$\frac{dy}{dx} = 2xy$$

$$\int \frac{1}{y}\,dy = \int 2x\,dx$$

$$\ln|y| = x^2 + C_1$$

$$e^{x^2 + C_1} = y$$

$$y = Ce^{x^2}$$

87. $\dfrac{dy}{dx} = \dfrac{x^2 + y^2}{2xy}$ (homogeneous differential equation)

$$(x^2 + y^2)\,dx - 2xy\,dy = 0$$

Let $y = vx$, $dy = x\,dv + v\,dx$.

$$(x^2 + v^2x^2)\,dx - 2x(vx)(x\,dv + v\,dx) = 0$$

$$(x^2 + v^2x^2 - 2x^2v^2)\,dx - 2x^3v\,dv = 0$$

$$(x^2 - x^2v^2)\,dx = 2x^3v\,dv$$

$$(1 - v^2)\,dx = 2x\ dv$$

$$\int \frac{dx}{x} = \int \frac{2v}{1 - v^2}\,dv$$

$$\ln|x| = -\ln|1 - v^2| + C_1 = -\ln|1 - v^2| + \ln C$$

$$x = \frac{C}{1 - v^2} = \frac{C}{1 - (y/x)^2} = \frac{Cx^2}{x^2 - y^2}$$

$$1 = \frac{Cx}{x^2 - y^2} \quad \text{or} \quad C_1 = \frac{x}{x^2 - y^2}$$

89. (a)

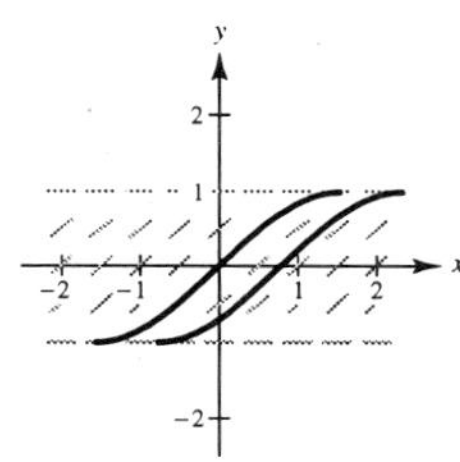

(b) Rate of change is greatest when $y = 0$ (slope lines are closest to vertical). Rate of change is least after $y = \pm 1$ (slope lines are horizontal).

(c)

$$\frac{dy}{dx} = \sqrt{1 - y^2}$$

$$\int \frac{1}{\sqrt{1 - y^2}}\,dy = \int dx$$

$$\arcsin y = x + C$$

$$y = \sin(x + C)$$

$$-\frac{\pi}{2} \le x + C \le \frac{\pi}{2}$$

91.

$$(x - C)^2 + y^2 = C^2 \quad \text{Circles}$$

$$x^2 - 2Cx + C^2 + y^2 = C^2$$

$$\frac{x^2 + y^2}{x} = 2C$$

$$\frac{x(2x + 2yy') - (x^2 + y^2)}{x^2} = 0$$

$$x^2 + 2xy\,y' - y^2 = 0$$

$$y' = \frac{y^2 - x^2}{2xy} \quad \text{slope of given curves}$$

Orthogonal family has negative reciprocal slope.

$\dfrac{dy}{dx} = \dfrac{2xy}{x^2 - y^2}$ (homogeneous differential equation)

—CONTINUED—

91. —CONTINUED—

Letting $x = vy$ and $dx = v\,dy + y\,dv$ (this is easier than the equivalent substitution $y = vx$), you obtain

$$(x^2 - y^2)\,dy = 2xy\,dx$$

$$(v^2y^2 - y^2)\,dy = 2(vy)y(v\,dy + y\,dv)$$

$$0 = (v^2y^2 + y^2)\,dy + 2vy^3\,dv$$

$$0 = (v^2 + 1)\,dy + 2vy\,dv$$

$$\int \frac{2v}{v^2+1}\,dv + \int \frac{dy}{y} = 0$$

$$\ln(1 + v^2) + \ln|y| = \ln k_1 \Rightarrow \left[1 + \left(\frac{x}{y}\right)^2\right]y = k_1$$

$y^2 + x^2 = k_1 y$ Circles $x^2 + (y - k)^2 = k^2$

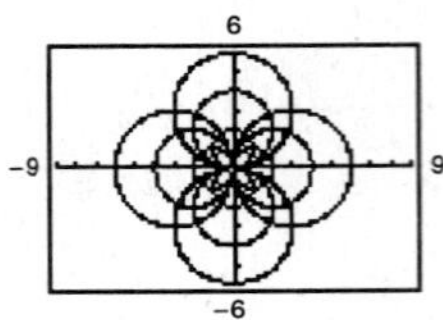

93. (a) $P = \frac{1}{100}\left(25 + \int_{2.5}^{10} \frac{25}{x}\,dx\right) = \frac{1}{100}\left(25 + \left[25 \ln x\right]_{2.5}^{10}\right) = \frac{1}{4}(1 + \ln 4) \approx 0.60$

(b) $P = \frac{1}{100}\left(50 + \int_{5}^{10} \frac{50}{x}\,dx\right) = \frac{1}{100}\left(50 + \left[50 \ln x\right]_{5}^{10}\right) = \frac{1}{2}(1 + \ln 2) \approx 0.85$

95. $$\int \frac{dy}{\sqrt{A^2 - y^2}} = \int \sqrt{\frac{k}{m}}\,dt$$

$$\arcsin\left(\frac{y}{A}\right) = \sqrt{\frac{k}{m}}\,t + C$$

Since $y = 0$ when $t = 0$, you have $C = 0$. Thus,

$$\sin\left(\sqrt{\frac{k}{m}}\,t\right) = \frac{y}{A}$$

$$y = A \sin\left(\sqrt{\frac{k}{m}}\,t\right)$$

97. (a) $$\frac{dy}{ds} = -0.012y,\; s > 50$$

$$\frac{-1}{0.012}\int \frac{dy}{y} = \int ds$$

$$\frac{-1}{0.012}\ln y = s + C_1$$

$$y = Ce^{-0.012s}$$

When $s = 50$, $y = 28 = Ce^{-0.012(50)} \Rightarrow C = 28e^{0.6}$

$$y = 28e^{0.6 - 0.012s},\; s > 50$$

(b)

Speed(s)	50	55	60	65	70
Miles per Gallon (y)	28	26.4	24.8	23.4	22.0

CHAPTER 6
Applications of Integration

CHAPTER 6
Applications of Integration

Section 6.1 Area of a Region Between Two Curves

Solutions to Odd-Numbered Exercises

1. $A = \int_0^6 [0 - (x^2 - 6x)]\,dx = -\int_0^6 (x^2 - 6x)\,dx$

3. $A = \int_0^3 [(-x^2 + 2x + 3) - (x^2 - 4x + 3)]\,dx = \int_0^3 (-2x^2 + 6x)\,dx$

5. $A = 2\int_{-1}^0 3(x^3 - x)\,dx = 6\int_{-1}^0 (x^3 - x)\,dx$ or $-6\int_0^1 (x^3 - x)\,dx$

7. $\int_0^4 \left[(x+1) - \frac{x}{2}\right] dx$

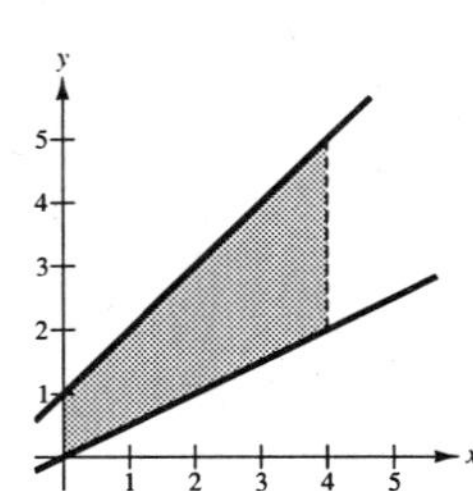

9. $\int_0^6 \left[4(2^{-x/3}) - \frac{x}{6}\right] dx$

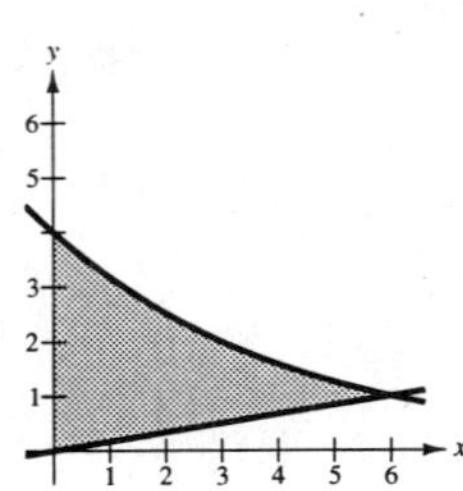

11. $f(x) = x + 1$

$g(x) = (x - 1)^2$

$A \approx 4$

Matches (d)

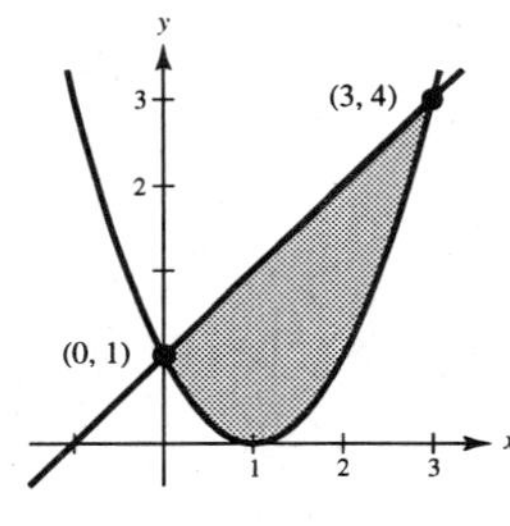

13. The points of intersection are given by:

$$x^2 - 4x = 0$$

$$x(x - 4) = 0 \quad \text{when} \quad x = 0, 4$$

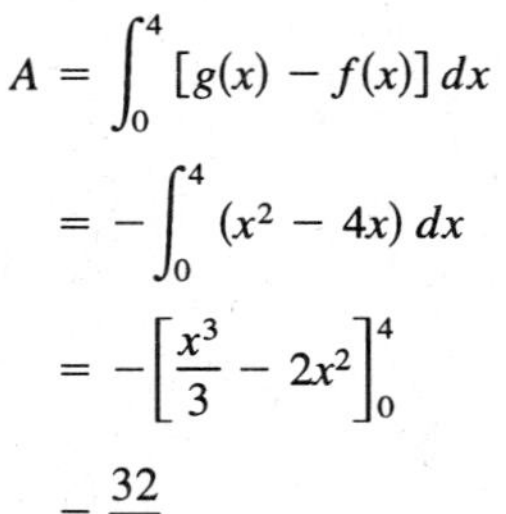

$$A = \int_0^4 [g(x) - f(x)]\,dx$$

$$= -\int_0^4 (x^2 - 4x)\,dx$$

$$= -\left[\frac{x^3}{3} - 2x^2\right]_0^4$$

$$= \frac{32}{3}$$

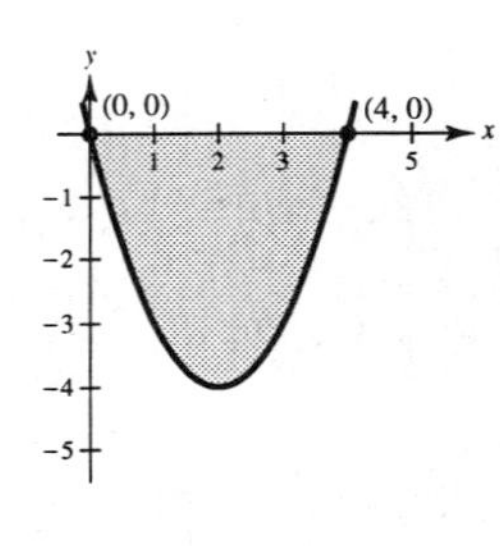

15. The points of intersection are given by:

$$x^2 + 2x + 1 = 3x + 3$$

$$(x - 2)(x + 1) = 0 \quad \text{when} \quad x = -1, 2$$

$$A = \int_{-1}^2 [g(x) - f(x)]\,dx$$

$$= \int_{-1}^2 [(3x + 3) - (x^2 + 2x + 1)]\,dx$$

$$= \int_{-1}^2 (2 + x - x^2)\,dx = \left[2x + \frac{x^2}{2} - \frac{x^3}{3}\right]_{-1}^2 = \frac{9}{2}$$

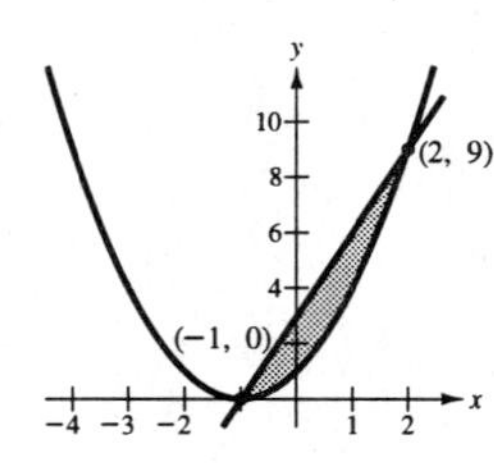

17. The points of intersection are given by:

$$x = 2 - x \quad \text{and} \quad x = 0 \quad \text{and} \quad 2 - x = 0$$

$$x = 1 \qquad\qquad x = 0 \qquad\qquad x = 2$$

$$A = \int_0^1 [(2 - y) - (y)]\, dy = \Big[2y - y^2\Big]_0^1 = 1$$

Note that if we integrate with respect to x, we need two integrals. Also, note that the region is a triangle.

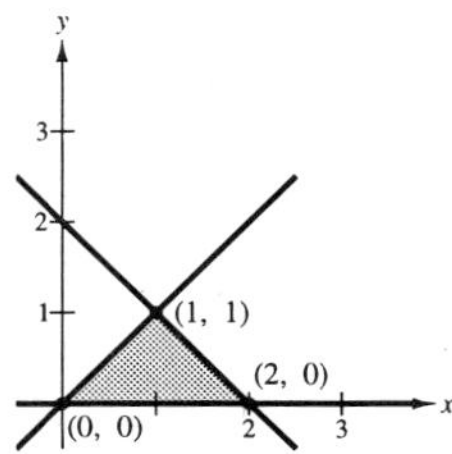

19. The points of intersection are given by:

$$\sqrt{3x} + 1 = x + 1$$

$$\sqrt{3x} = x \quad \text{when} \quad x = 0, 3$$

$$A = \int_0^3 [f(x) - g(x)]\, dx$$

$$= \int_0^3 \left[\left(\sqrt{3x} + 1\right) - (x + 1)\right] dx$$

$$= \int_0^3 [(3x)^{1/2} - x]\, dx$$

$$= \left[\frac{2}{9}(3x)^{3/2} - \frac{x^2}{2}\right]_0^3 = \frac{3}{2}$$

21. The points of intersection are given by:

$$y^2 = y + 2$$

$$(y - 2)(y + 1) = 0 \quad \text{when} \quad y = -1, 2$$

$$A = \int_{-1}^2 [g(y) - f(y)]\, dy$$

$$= \int_{-1}^2 [(y + 2) - y^2]\, dy$$

$$= \left[2y + \frac{y^2}{2} - \frac{y^3}{3}\right]_{-1}^2 = \frac{9}{2}$$

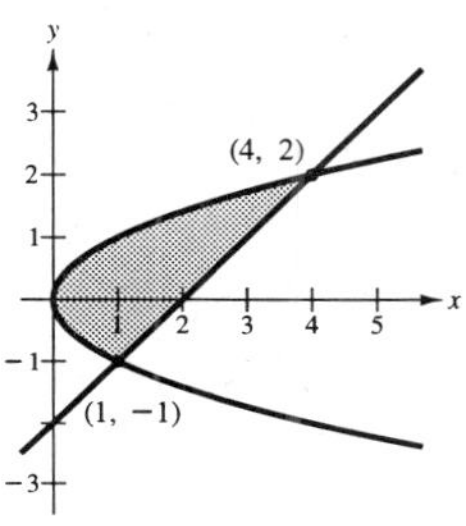

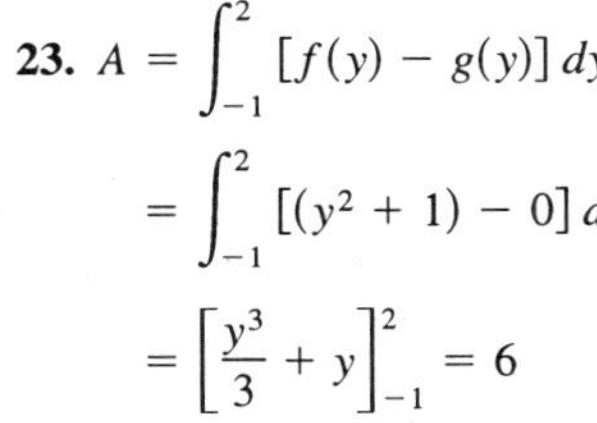

23. $$A = \int_{-1}^2 [f(y) - g(y)]\, dy$$

$$= \int_{-1}^2 [(y^2 + 1) - 0]\, dy$$

$$= \left[\frac{y^3}{3} + y\right]_{-1}^2 = 6$$

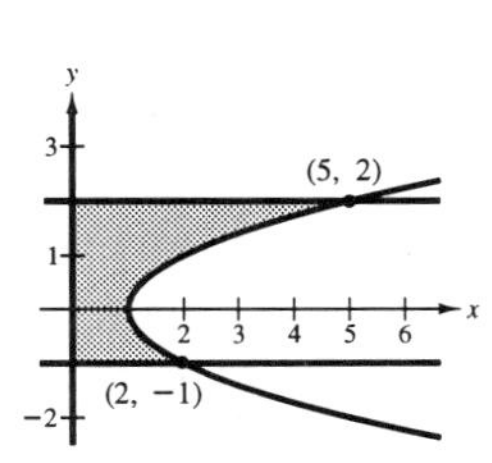

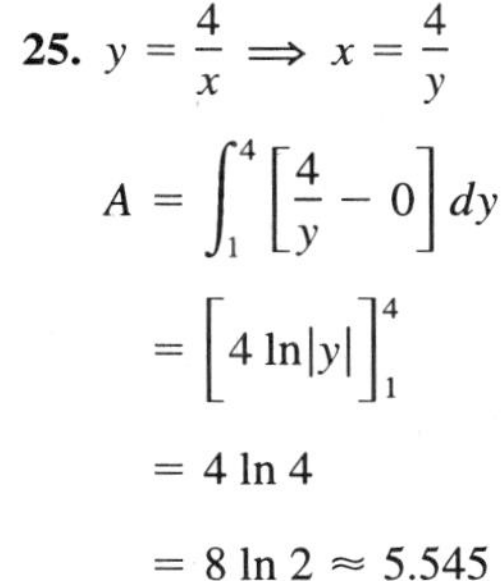

25. $$y = \frac{4}{x} \Rightarrow x = \frac{4}{y}$$

$$A = \int_1^4 \left[\frac{4}{y} - 0\right] dy$$

$$= \Big[4 \ln|y|\Big]_1^4$$

$$= 4 \ln 4$$

$$= 8 \ln 2 \approx 5.545$$

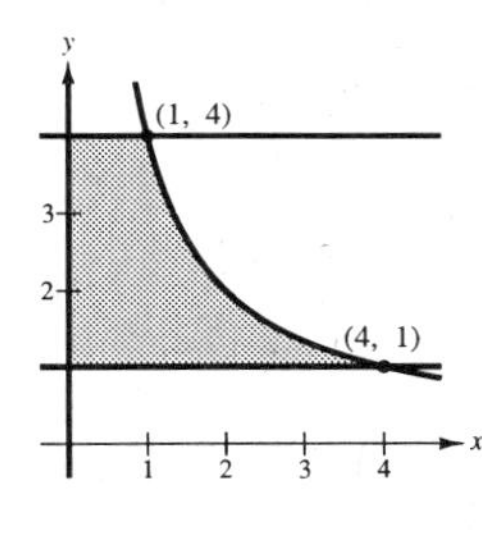

27. The points of intersection are given by:

$$x^3 - 3x^2 + 3x = x^2$$

$$x(x - 1)(x - 3) = 0 \quad \text{when} \quad x = 0, 1, 3$$

$$A = \int_0^1 [f(x) - g(x)]\, dx + \int_1^3 [g(x) - f(x)]\, dx$$

$$= \int_0^1 [(x^3 - 3x^2 + 3x) - x^2]\, dx + \int_1^3 [x^2 - (x^3 - 3x^2 + 3x)]\, dx$$

$$= \int_0^1 (x^3 - 4x^2 + 3x)\, dx + \int_1^3 (-x^3 + 4x^2 - 3x)\, dx = \left[\frac{x^4}{4} - \frac{4}{3}x^3 + \frac{3}{2}x^2\right]_0^1 + \left[\frac{-x^4}{4} + \frac{4}{3}x^3 - \frac{3}{2}x^2\right]_1^3 = \frac{37}{12}$$

Numerical approximation: $0.417 + 2.667 \approx 3.083$

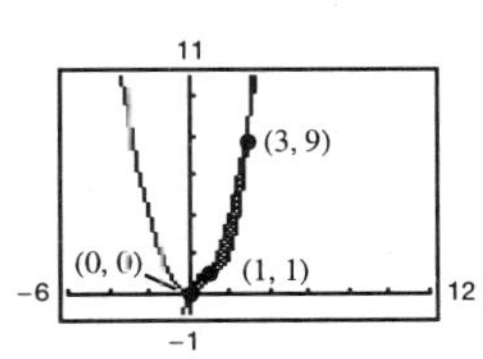

29. The points of intersection are given by:

$$x^2 - 4x + 3 = 3 + 4x - x^2$$

$$2x(x - 4) = 0 \quad \text{when} \quad x = 0, 4$$

$$A = \int_0^4 [(3 + 4x - x^2) - (x^2 - 4x + 3)]\,dx$$

$$= \int_0^4 (-2x^2 + 8x)\,dx$$

$$= \left[-\frac{2x^3}{3} + 4x^2\right]_0^4 = \frac{64}{3}$$

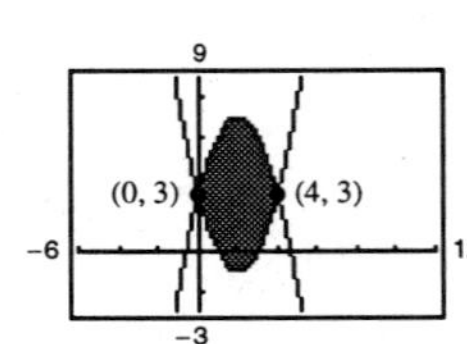

Numerical approximation: 21.333

31. $f(x) = x^4 - 4x^2, \quad g(x) = x^2 - 4$

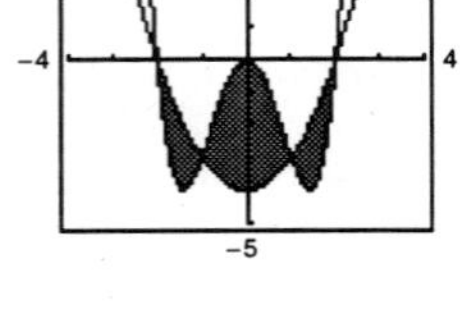

The points of intersection are given by:

$$x^4 - 4x^2 = x^2 - 4$$

$$x^4 - 5x^2 + 4 = 0$$

$$(x^2 - 4)(x^2 - 1) = 0 \quad \text{when} \quad x = \pm 2, \pm 1$$

By symmetry,

$$A = 2\int_0^1 [(x^4 - 4x^2) - (x^2 - 4)]\,dx + 2\int_1^2 [(x^2 - 4) - (x^4 - 4x^2)]\,dx$$

$$= 2\int_0^1 (x^4 - 5x^2 + 4)\,dx + 2\int_1^2 (-x^4 + 5x^2 - 4)\,dx$$

$$= 2\left[\frac{x^5}{5} - \frac{5x^3}{3} + 4x\right]_0^1 + 2\left[-\frac{x^5}{5} + \frac{5x^3}{3} - 4x\right]_1^2$$

$$= 2\left[\frac{1}{5} - \frac{5}{3} + 4\right] + 2\left[\left(-\frac{32}{5} + \frac{40}{3} - 8\right) - \left(-\frac{1}{5} + \frac{5}{3} - 4\right)\right] = 8.$$

Numerical approximation: $5.067 + 2.933 = 8.0$

33. The points of intersection are given by:

$$\frac{1}{1 + x^2} = \frac{x^2}{2}$$

$$x^4 + x^2 - 2 = 0$$

$$(x^2 + 2)(x^2 - 1) = 0$$

$$x = \pm 1$$

$$A = 2\int_0^1 [f(x) - g(x)]\,dx$$

$$= 2\int_0^1 \left[\frac{1}{1 + x^2} - \frac{x^2}{2}\right]dx$$

$$= 2\left[\arctan x - \frac{x^3}{6}\right]_0^1$$

$$= 2\left(\frac{\pi}{4} - \frac{1}{6}\right) = \frac{\pi}{2} - \frac{1}{3} \approx 1.237$$

Numerical approximation: 1.237

35. $\sqrt{1 + x^3} \le \frac{1}{2}x + 2$ on $[0, 2]$

Numerical approximation: 1.759

$$A = \int_0^2 \left[\frac{1}{2}x + 2 - \sqrt{1 + x^3}\right]dx \approx 1.759$$

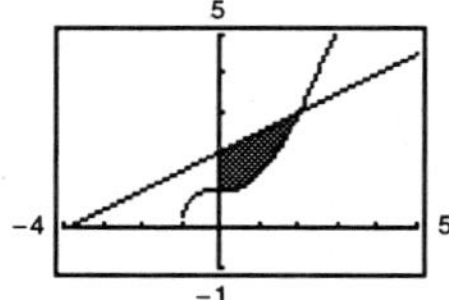

37. $A = 2\int_0^{\pi/3} [f(x) - g(x)]\,dx$

$= 2\int_0^{\pi/3} (2\sin x - \tan x)\,dx$

$= 2\Big[-2\cos x + \ln|\cos x|\Big]_0^{\pi/3}$

$= 2(1 - \ln 2) \approx 0.614$

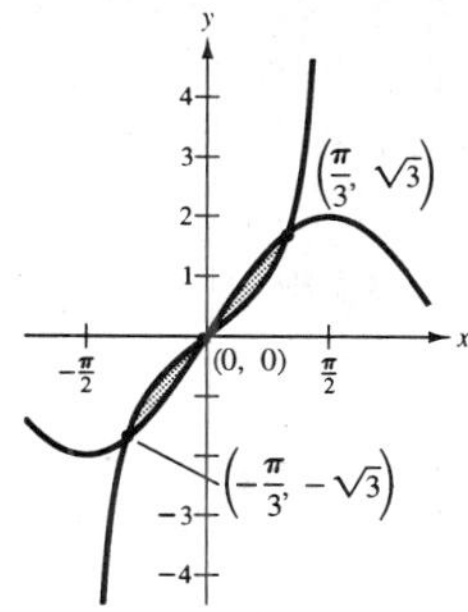

39. $A = \int_0^1 \left[xe^{-x^2} - 0\right] dx$

$= \left[-\frac{1}{2}e^{-x^2}\right]_0^1 = \frac{1}{2}\left(1 - \frac{1}{e}\right) \approx 0.316$

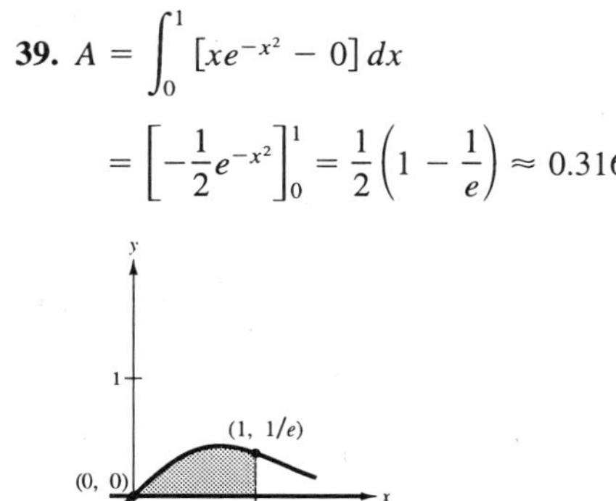

41. $A = \int_0^{\pi} [(2\sin x + \sin 2x) - 0]\,dx$

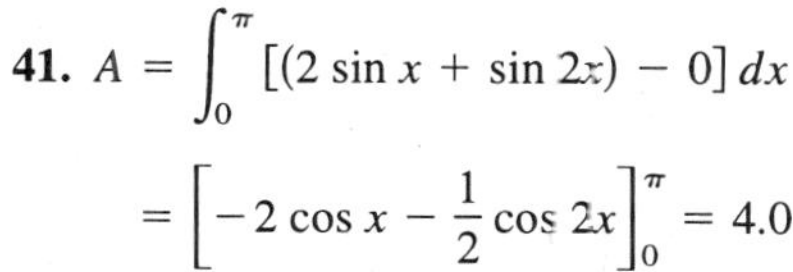

$= \left[-2\cos x - \frac{1}{2}\cos 2x\right]_0^{\pi} = 4.0$

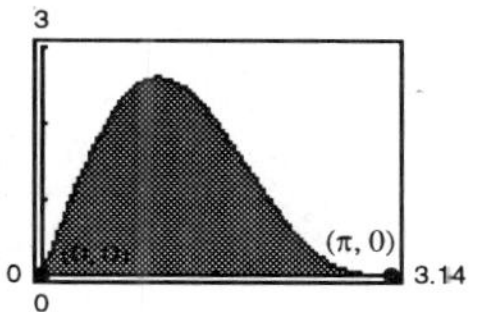

43. $A = \int_1^3 \left[\frac{1}{x^2}e^{1/x} - 0\right] dx$

$= \left[-e^{1/x}\right]_1^3 = e - e^{1/3} \approx 1.323$

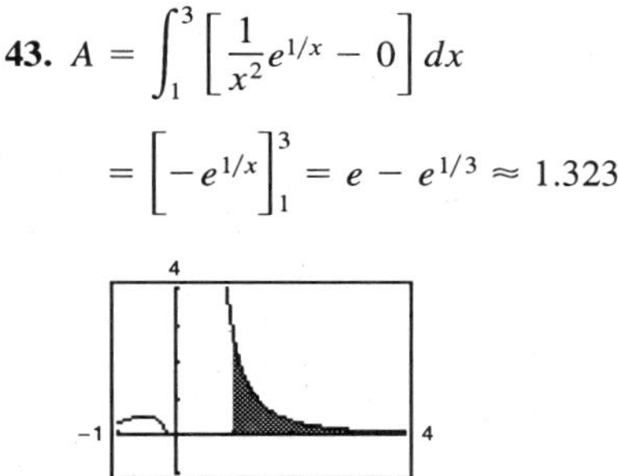

45. (a) $y = \sqrt{\dfrac{x^3}{4-x}}, \quad y = 0, \quad x = 3$

(b) $A = \int_0^3 \sqrt{\dfrac{x^3}{4-x}}\,dx,$

No, it cannot be evaluated by hand.

(c) 4.7721

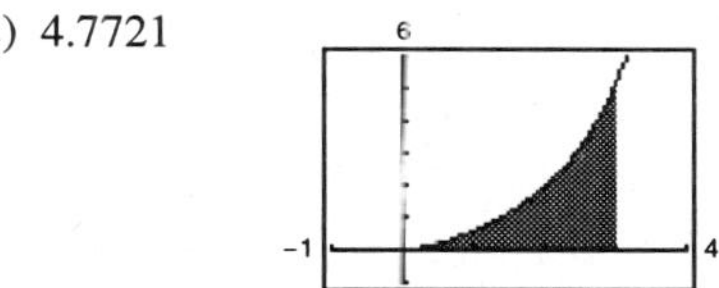

47. $A = \int_0^c \left[\left(\frac{b-a}{c}y + a\right) - \frac{b}{c}y\right] dy$

$= \int_0^c \left(-\frac{a}{c}y + a\right) dy$

$= \left[-\frac{a}{2c}y^2 + ay\right]_0^c$

$= -\frac{ac}{2} + ac = \frac{ac}{2} \quad \left(= \frac{1}{2}(\text{base})(\text{height})\right)$

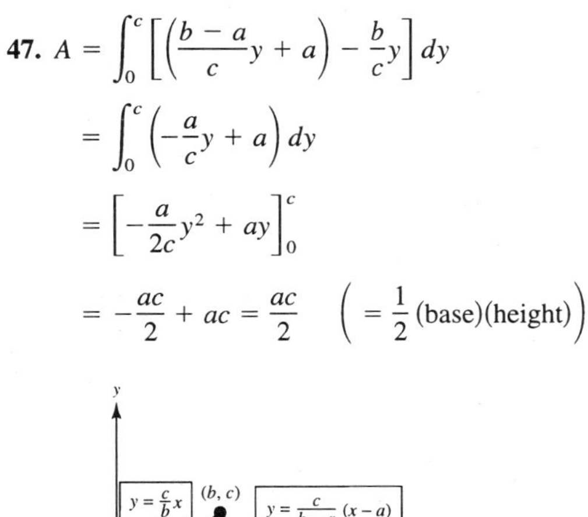

49. $f(x) = x^3$

$f'(x) = 3x^2$

At (1, 1), $f'(1) = 3$.

Tangent line:

$y - 1 = 3(x - 1)$ or $y = 3x - 2$

The tangent line intersects $f(x) = x^3$ at $x = -2$.

$$A = \int_{-2}^{1} [x^3 - (3x - 2)]\,dx = \left[\frac{x^4}{4} - \frac{3x^2}{2} + 2x\right]_{-2}^{1} = \frac{27}{4}$$

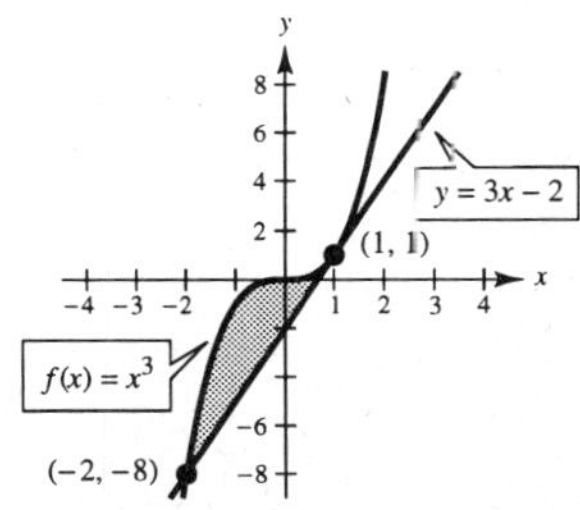

51. $x^4 - 2x^2 + 1 \le 1 - x^2$ on $[-1, 1]$

$$A = \int_{-1}^{1} [(1 - x^2) - (x^4 - 2x^2 + 1)]\, dx = \int_{-1}^{1} (x^2 - x^4)\, dx$$

$$= \left[\frac{x^3}{3} - \frac{x^5}{5}\right]_{-1}^{1} = \frac{4}{15}$$

You can use a single integral because $x^4 - 2x^2 + 1 \le 1 - x^2$ on $[-1, 1]$.

53. $$A = \int_{-3}^{3} (9 - x^2)\, dx = 36$$

$$\int_{-\sqrt{9-b}}^{\sqrt{9-b}} [(9 - x^2) - b]\, dx = 18$$

$$\int_{0}^{\sqrt{9-b}} [(9 - b) - x^2]\, dx = 9$$

$$\left[(9 - b)x - \frac{x^3}{3}\right]_0^{\sqrt{9-b}} = 9$$

$$\frac{2}{3}(9 - b)^{3/2} = 9$$

$$(9 - b)^{3/2} = \frac{27}{2}$$

$$9 - b = \frac{9}{\sqrt[3]{4}}$$

$$b = 9 - \frac{9}{\sqrt[3]{4}} \approx 3.330$$

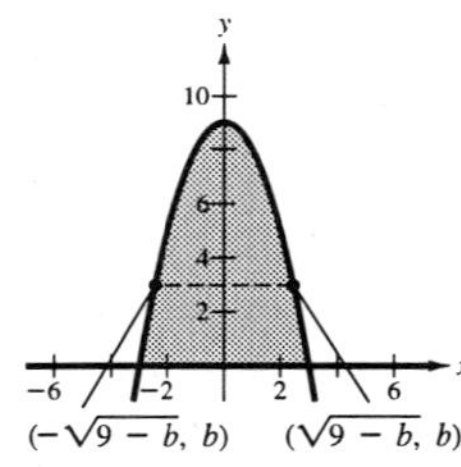

55. 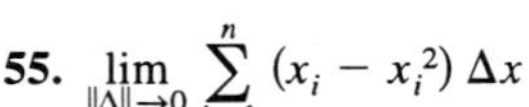

$$\lim_{\|\Delta\| \to 0} \sum_{i=1}^{n} (x_i - x_i^2)\, \Delta x$$

where $x_i = \dfrac{i}{n}$ and $\Delta x = \dfrac{1}{n}$ is the same as

$$\int_0^1 (x - x^2)\, dx = \left[\frac{x^2}{2} - \frac{x^3}{3}\right]_0^1 = \frac{1}{6}.$$

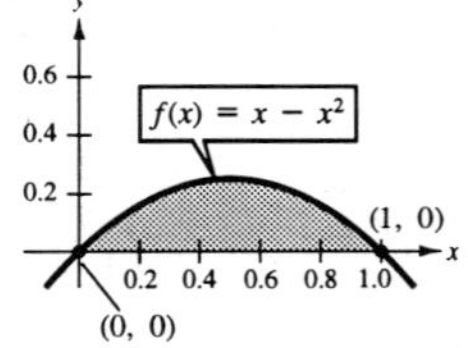

57. $$\int_0^5 [(7.21 + 0.58t) - (7.21 + 0.45t)]\, dt = \int_0^5 0.13t\, dt = \left[\frac{0.13t^2}{2}\right]_0^5 = \$1.625 \text{ billion}$$

59. $$f(t) = \begin{cases} 27.77 - 0.36t & 5 \le t \le 10 \\ 21.00 + 0.27t & 10 \le t \le 14 \end{cases}$$

(a)

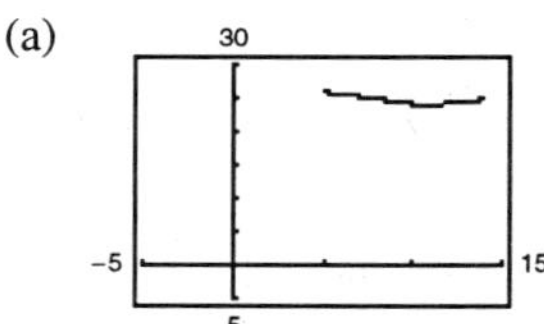

(b) $$\int_{10}^{14} [(21.00 + 0.27t) - (27.77 - 0.36t)]\, dt = \int_{10}^{14} [-6.77 + 0.63t]\, dt$$

$$= \left[-6.77t + 0.315t^2\right]_{10}^{14} = 3.16 \text{ billion pounds}$$

61. 5% : $P_1 = 893{,}000\, e^{(0.05)t}$

$3\frac{1}{2}\%$: $P_2 = 893{,}000\, e^{(0.035)t}$

Difference in profits over 5 years:

$$\int_0^5 [893{,}000e^{0.05t} - 893{,}000e^{0.035t}]\, dt = 893{,}000\left[\frac{e^{0.05t}}{0.05} - \frac{e^{0.035t}}{0.035}\right]_0^5$$

$$\approx 893{,}000[(25.6805 - 34.0356) - (20 - 28.5714)]$$

$$\approx 893{,}000(0.2163) \approx \$193{,}156$$

Note: Using a graphing utility you obtain $193,183.

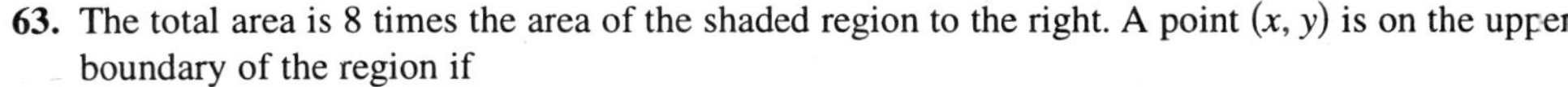

63. The total area is 8 times the area of the shaded region to the right. A point (x, y) is on the upper boundary of the region if

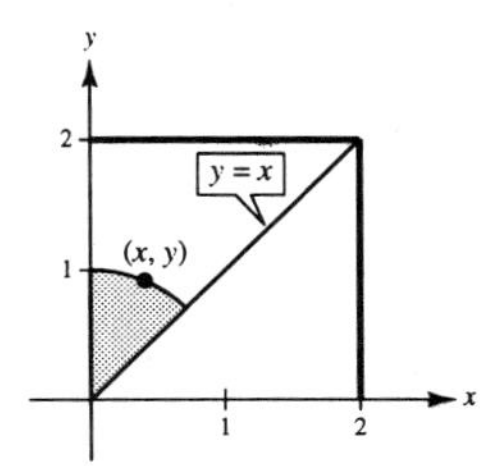

$$\sqrt{x^2 + y^2} = 2 - y$$

$$x^2 + y^2 = 4 - 4y + y^2$$

$$x^2 = 4 - 4y$$

$$4y = 4 - x^2$$

$$y = 1 - \frac{x^2}{4}.$$

We now determine where this curve intersects the line $y = x$.

$$x = 1 - \frac{x^2}{4}$$

$$x^2 + 4x - 4 = 0$$

$$x = \frac{-4 \pm \sqrt{16 + 16}}{2} = -2 \pm 2\sqrt{2} \Rightarrow x = -2 + 2\sqrt{2}$$

$$\text{Total area} = 8\int_0^{-2+2\sqrt{2}} \left(1 - \frac{x^2}{4} - x\right) dx = 8\left[x - \frac{x^3}{12} - \frac{x^2}{2}\right]_0^{-2+2\sqrt{2}} \approx 8(0.4379) = 3.503$$

65. (a) $$A = 2\left[\int_0^5 \left(1 - \frac{1}{3}\sqrt{5 - x}\right) dx + \int_5^{5.5} (1 - 0)\, dx\right]$$

$$= 2\left(\left[x + \frac{2}{9}(5 - x)^{3/2}\right]_0^5 + \left[x\right]_5^{5.5}\right) = 2\left(5 - \frac{10\sqrt{5}}{9} + 5.5 - 5\right) \approx 6.031 \text{ m}^2$$

(b) $V = 2A \approx 2(6.031) \approx 12.062 \text{ m}^3$

(c) $5000\, V \approx 5000(12.062) = 60{,}310$ pounds

67. $50 - 0.5x = 0.125x$

$$x = 80$$

$p_1(80) = p_2(80) = 10$

Point of equilibrium: (80, 10)

$$CS = \int_0^{80} [(50 - 0.5x) - 10]\, dx$$

$$= \left[-\frac{0.5x^2}{2} + 40x\right]_0^{80} = 1600$$

$$PS = \int_0^{80} [10 - 0.125x]\, dx$$

$$= \left[10x - \frac{0.125x^2}{2}\right]_0^{80} = 400$$

69. True

Section 6.2 Volume: The Disc Method

1. $V = \pi \int_0^1 (-x+1)^2\,dx = \pi \int_0^1 (x^2 - 2x + 1)\,dx = \pi \left[\frac{x^3}{3} - x^2 + x\right]_0^1 = \frac{\pi}{3}$

3. $V = \pi \int_1^4 (\sqrt{x})^2\,dx = \pi \int_1^4 x\,dx = \pi \left[\frac{x^2}{2}\right]_1^4 = \frac{15\pi}{2}$

5. $V = \pi \int_0^1 [(x^2)^2 - (x^3)^2]\,dx = \pi \int_0^1 (x^4 - x^6)\,dx = \pi \left[\frac{x^5}{5} - \frac{x^7}{7}\right]_0^1 = \frac{2\pi}{35}$

7. $y = x^2 \Rightarrow x = \sqrt{y}$

$V = \pi \int_0^4 (\sqrt{y})^2\,dy = \pi \int_0^4 y\,dy$

$= \pi \left[\frac{y^2}{2}\right]_0^4 = 8\pi$

9. $y = x^{2/3} \Rightarrow x = y^{3/2}$

$V = \pi \int_0^1 (y^{3/2})^2\,dy = \pi \int_0^1 y^3\,dy = \pi \left[\frac{y^4}{4}\right]_0^1 = \frac{\pi}{4}$

11. $y = \sqrt{x},\ y = 0,\ x = 4$

(a) $R(x) = \sqrt{x},\ r(x) = 0$

$V = \pi \int_0^4 (\sqrt{x})^2\,dx$

$= \pi \int_0^4 x\,dx = \left[\frac{\pi}{2}x^2\right]_0^4 = 8\pi$

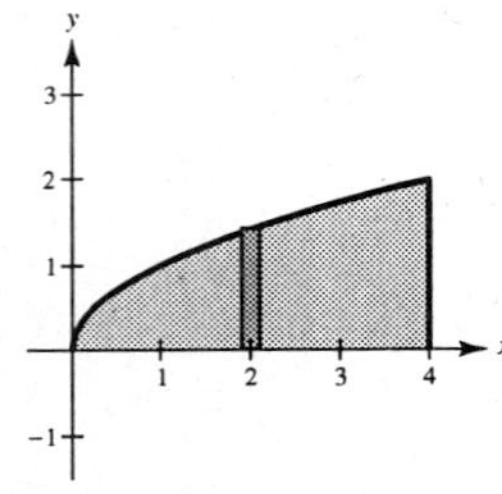

(b) $R(y) = 4,\ r(y) = y^2$

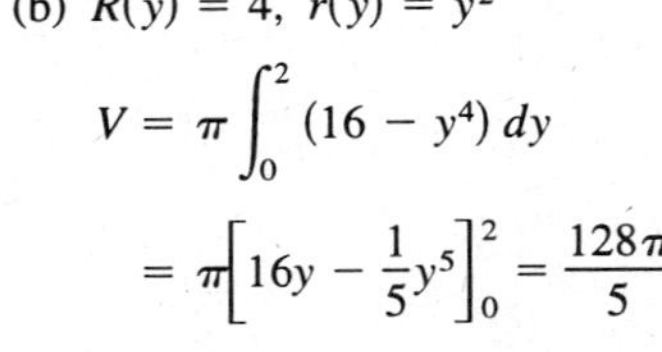

$V = \pi \int_0^2 (16 - y^4)\,dy$

$= \pi \left[16y - \frac{1}{5}y^5\right]_0^2 = \frac{128\pi}{5}$

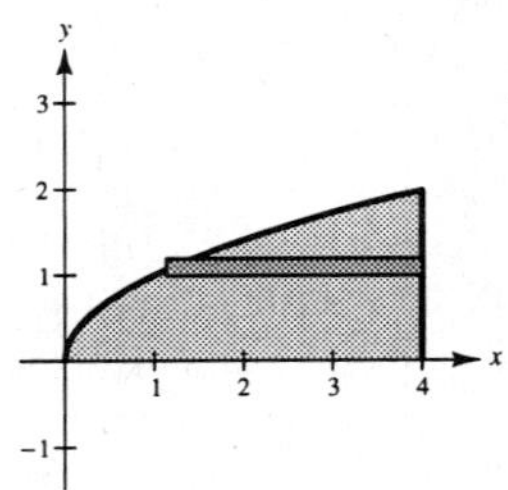

(c) $R(y) = 4 - y^2,\ r(y) = 0$

$V = \pi \int_0^2 (4 - y^2)^2\,dy$

$= \pi \int_0^2 (16 - 8y^2 + y^4)\,dy$

$= \pi \left[16y - \frac{8}{3}y^3 + \frac{1}{5}y^5\right]_0^2 = \frac{256\pi}{15}$

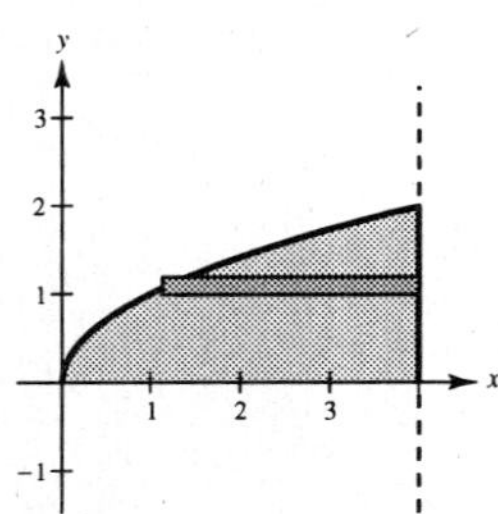

(d) $R(y) = 6 - y^2,\ r(y) = 2$

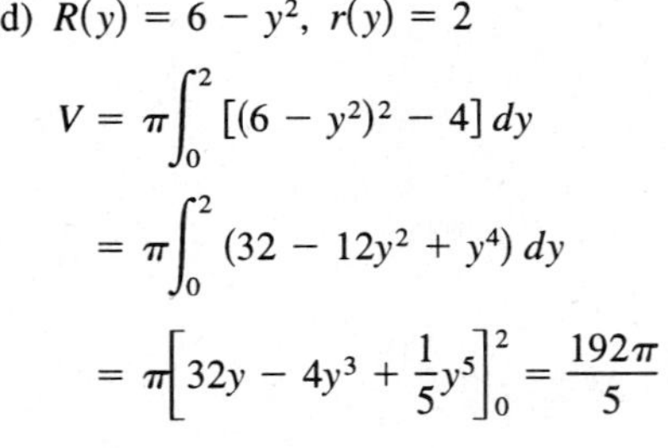

$V = \pi \int_0^2 [(6 - y^2)^2 - 4]\,dy$

$= \pi \int_0^2 (32 - 12y^2 + y^4)\,dy$

$= \pi \left[32y - 4y^3 + \frac{1}{5}y^5\right]_0^2 = \frac{192\pi}{5}$

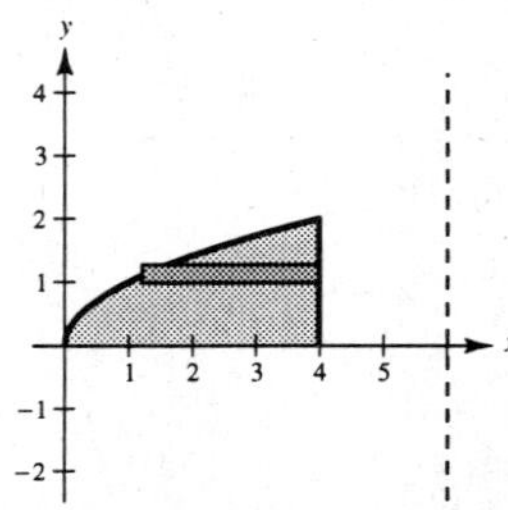

13. $y = x^2$, $y = 4x - x^2$ intersect at $(0, 0)$ and $(2, 4)$.

(a) $R(x) = 4x - x^2$ $r(x) = x^2$

$$V = \pi\int_0^2 [(4x - x^2)^2 - x^4]\,dx$$

$$= \pi\int_0^2 (16x^2 - 8x^3)\,dx$$

$$= \pi\left[\frac{16}{3}x^3 - 2x^4\right]_0^2 = \frac{32\pi}{3}$$

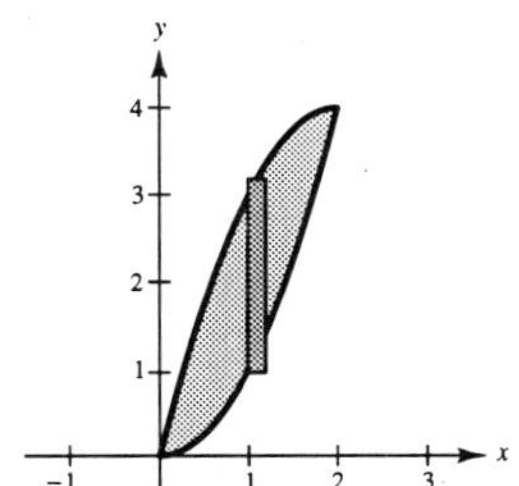

(b) $R(x) = 6 - x^2$, $r(x) = 6 - (4x - x^2)$

$$V = \pi\int_0^2 [(6 - x^2)^2 - (6 - 4x + x^2)^2]\,dx$$

$$= 8\pi\int_0^2 (x^3 - 5x^2 + 6x)\,dx$$

$$= 8\pi\left[\frac{x^4}{4} - \frac{5}{3}x^3 + 3x^2\right]_0^2 = \frac{64\pi}{3}$$

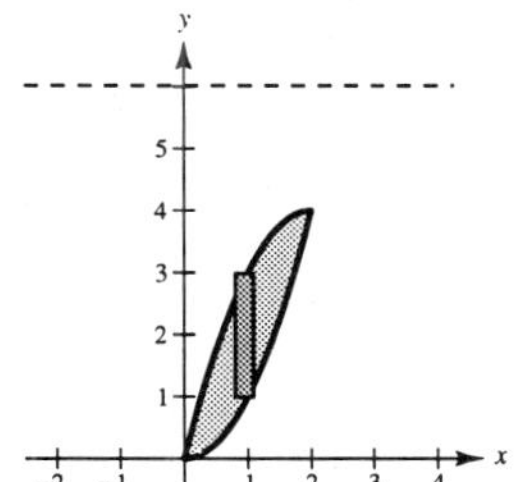

15. $R(x) = 4 - x$, $r(x) = 1$

$$V = \pi\int_0^3 [(4 - x)^2 - (1)^2]\,dx$$

$$= \pi\int_0^3 (x^2 - 8x + 15)\,dx$$

$$= \pi\left[\frac{x^3}{3} - 4x^2 + 15x\right]_0^3 = 18\pi$$

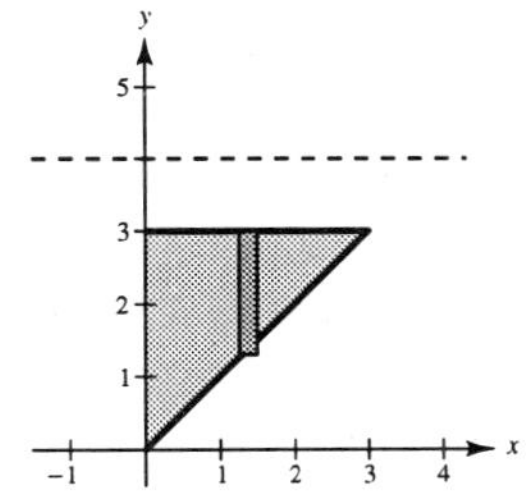

17. $R(x) = 4$, $r(x) = 4 - \dfrac{1}{x}$

$$V = \pi\int_1^4 \left[(4)^2 - \left(4 - \frac{1}{x}\right)^2\right] dx$$

$$= \pi\int_1^4 \left(\frac{8}{x} - \frac{1}{x^2}\right) dx$$

$$= \pi\left[8\ln|x| + \frac{1}{x}\right]_1^4$$

$$= \pi\left(8\ln 4 - \frac{3}{4}\right) \approx 32.49$$

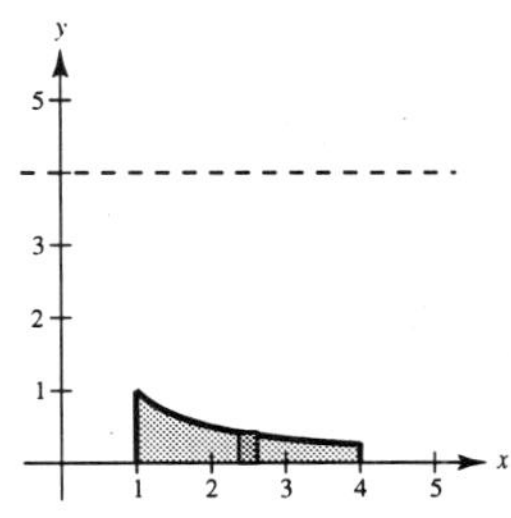

19. $R(y) = 6 - y$, $r(y) = 0$

$$V = \pi\int_0^4 (6 - y)^2\,dy$$

$$= \pi\int_0^4 (y^2 - 12y + 36)\,dy$$

$$= \pi\left[\frac{y^3}{3} - 6y^2 + 36y\right]_0^4$$

$$= \frac{208\pi}{3}$$

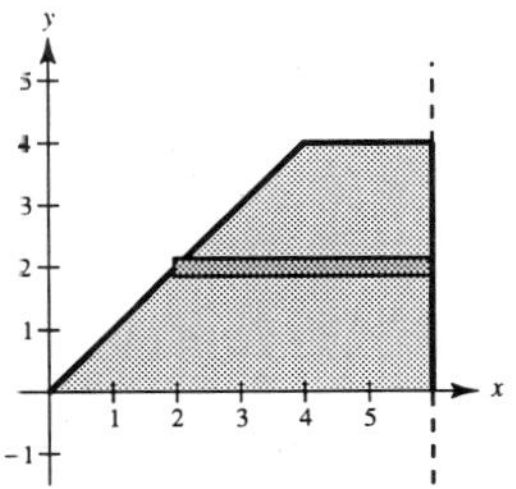

21. $R(y) = 6 - y^2$, $r(y) = 2$

$$V = \pi\int_{-2}^2 [(6 - y^2)^2 - (2)^2]\,dy$$

$$= 2\pi\int_0^2 (y^4 - 12y^2 + 32)\,dy$$

$$= 2\pi\left[\frac{y^5}{5} - 4y^3 + 32y\right]_0^2$$

$$= \frac{384\pi}{5}$$

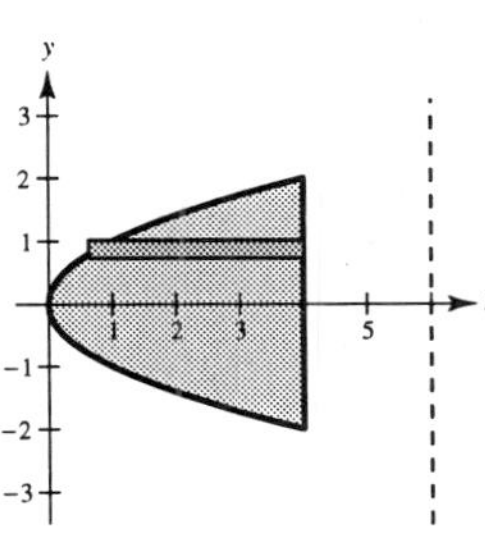

23. $R(x) = \dfrac{1}{\sqrt{x+1}}, \; r(x) = 0$

$$V = \pi\int_0^3 \left(\frac{1}{\sqrt{x+1}}\right)^2 dx$$

$$= \pi\int_0^3 \frac{1}{x+1}\,dx$$

$$= \Big[\pi \ln|x+1|\Big]_0^3 = \pi \ln 4$$

25. $R(x) = \dfrac{1}{x}, \; r(x) = 0$

$$V = \pi\int_1^4 \left(\frac{1}{x}\right)^2 dx$$

$$= \pi\left[-\frac{1}{x}\right]_1^4$$

$$= \frac{3\pi}{4}$$

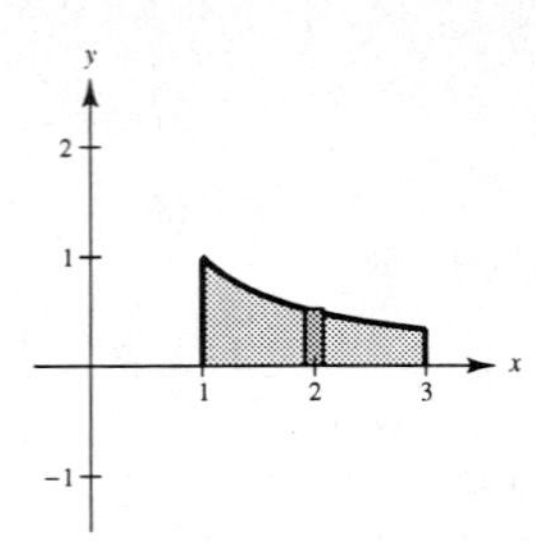

27. $R(x) = e^{-x}, \; r(x) = 0$

$$V = \pi\int_0^1 (e^{-x})^2\,dx$$

$$= \pi\int_0^1 e^{-2x}\,dx$$

$$= \left[-\frac{\pi}{2}e^{-2x}\right]_0^1$$

$$= \frac{\pi}{2}(1 - e^{-2}) \approx 1.358$$

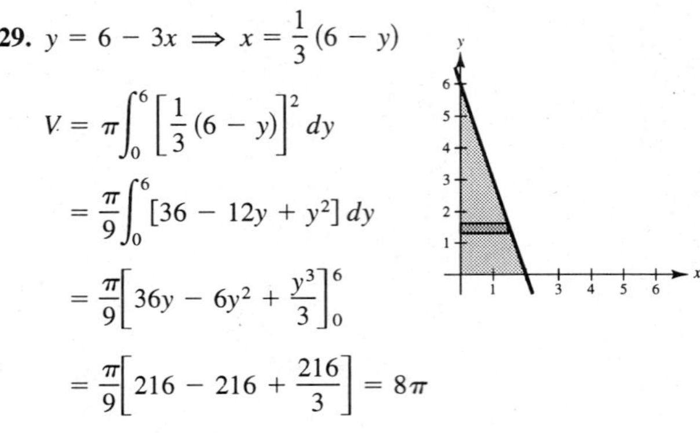

29. $y = 6 - 3x \Rightarrow x = \dfrac{1}{3}(6 - y)$

$$V = \pi\int_0^6 \left[\frac{1}{3}(6-y)\right]^2 dy$$

$$= \frac{\pi}{9}\int_0^6 [36 - 12y + y^2]\,dy$$

$$= \frac{\pi}{9}\left[36y - 6y^2 + \frac{y^3}{3}\right]_0^6$$

$$= \frac{\pi}{9}\left[216 - 216 + \frac{216}{3}\right] = 8\pi$$

31. $V = \pi\displaystyle\int_0^{\pi} [\sin x]^2\,dx \approx 4.9348$

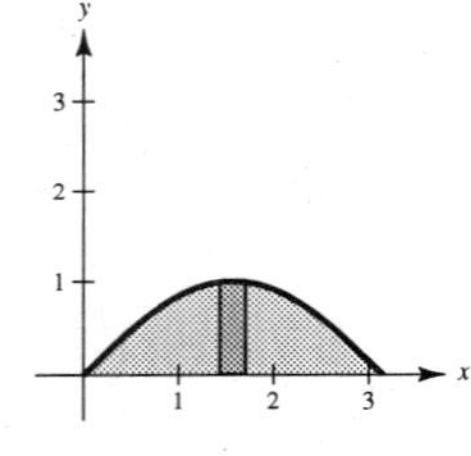

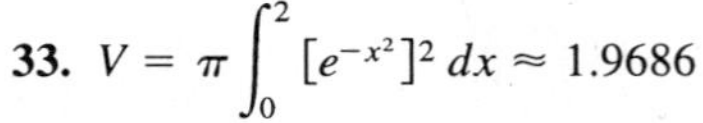

33. $V = \pi\displaystyle\int_0^2 [e^{-x^2}]^2\,dx \approx 1.9686$

35. $V = \pi\displaystyle\int_{-1}^2 [e^{x/2} + e^{-x/2}]^2\,dx$

≈ 49.0218

37. $A \approx 3$

Matches (a)

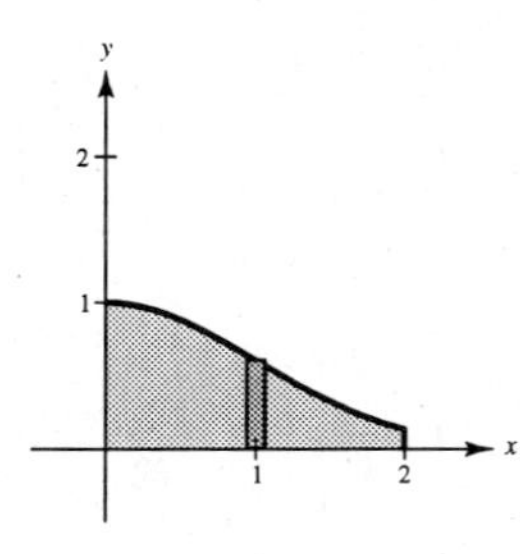

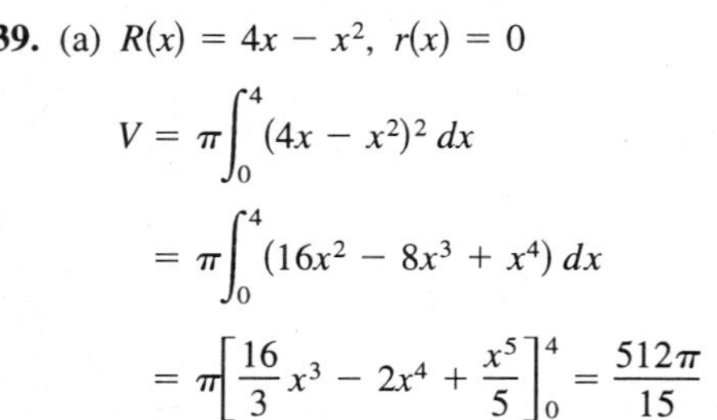

39. (a) $R(x) = 4x - x^2, \; r(x) = 0$

$$V = \pi\int_0^4 (4x - x^2)^2\,dx$$

$$= \pi\int_0^4 (16x^2 - 8x^3 + x^4)\,dx$$

$$= \pi\left[\frac{16}{3}x^3 - 2x^4 + \frac{x^5}{5}\right]_0^4 = \frac{512\pi}{15}$$

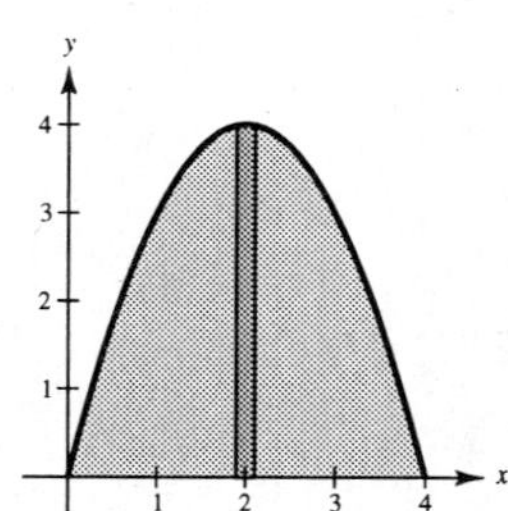

(b) Completing the square we have $4x - x^2 = 4 - (x^2 - 4x + 4) = 4 - (x - 2)^2$. Thus, $y = 4 - x^2$ has the same volume as in part (a) since the solid has been translated only horizontally.

41. $R(x) = \frac{1}{2}x, \; r(x) = 0$

$$V = \pi\int_0^6 \frac{1}{4}x^2\,dx$$

$$= \left[\frac{\pi}{12}x^3\right]_0^6 = 18\pi$$

Note: $V = \frac{1}{3}\pi r^2 h$

$$= \frac{1}{3}\pi(3^2)6$$

$$= 18\pi$$

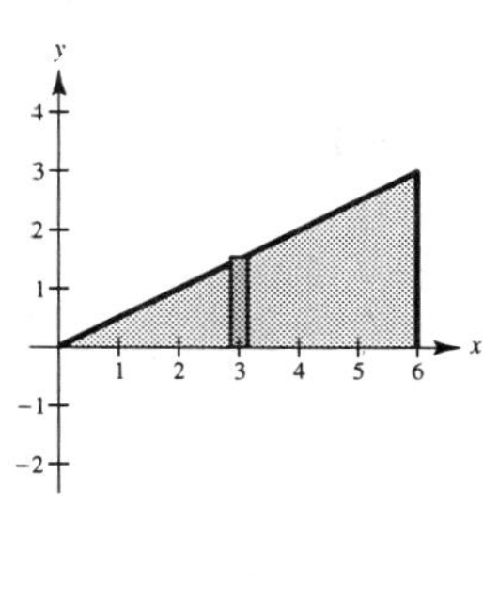

43. $R(x) = \sqrt{r^2 - x^2}, \; r(x) = 0$

$$V = \pi\int_{-r}^{r}(r^2 - x^2)\,dx$$

$$= 2\pi\int_0^r (r^2 - x^2)\,dx$$

$$= 2\pi\left[r^2x - \frac{1}{3}x^3\right]_0^r$$

$$= 2\pi\left(r^3 - \frac{1}{3}r^3\right)$$

$$= \frac{4}{3}\pi r^3$$

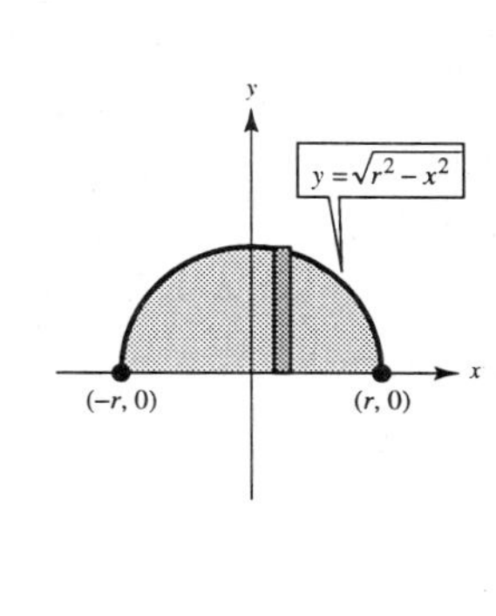

45. $x = r - \frac{r}{H}y = r\left(1 - \frac{y}{H}\right), \; R(y) = r\left(1 - \frac{y}{H}\right), \; r(y) = 0$

$$V = \pi\int_0^h \left[r\left(1 - \frac{y}{H}\right)\right]^2 dy = \pi r^2\int_0^h\left(1 - \frac{2}{H}y + \frac{1}{H^2}y^2\right)dy$$

$$= \pi r^2\left[y - \frac{1}{H}y^2 + \frac{1}{3H^2}y^3\right]_0^h$$

$$= \pi r^2\left(h - \frac{h^2}{H} + \frac{h^3}{3H^2}\right)$$

$$= \pi r^2 h\left(1 - \frac{h}{H} + \frac{h^2}{3H^2}\right)$$

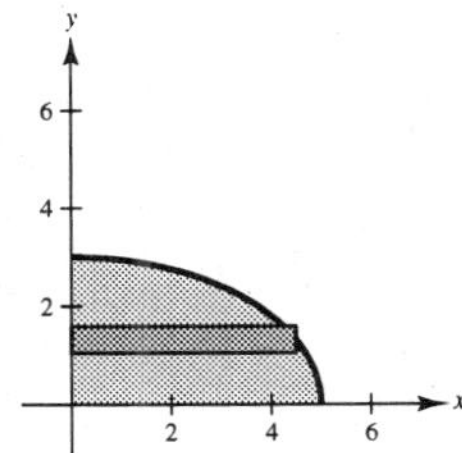

47. $V = \pi\int_0^2\left(\frac{1}{8}x^2\sqrt{2-x}\right)^2 dx = \frac{\pi}{64}\int_0^2 x^4(2-x)\,dx = \frac{\pi}{64}\left[\frac{2x^5}{5} - \frac{x^6}{6}\right]_0^2 = \frac{\pi}{30}$

49. (a) $R(x) = \frac{3}{5}\sqrt{25 - x^2}, \; r(x) = 0$

$$V = \frac{9\pi}{25}\int_{-5}^{5}(25 - x^2)\,dx$$

$$= \frac{18\pi}{25}\int_0^5 (25 - x^2)\,dx$$

$$= \frac{18\pi}{25}\left[25x - \frac{x^3}{3}\right]_0^5 = 60\pi$$

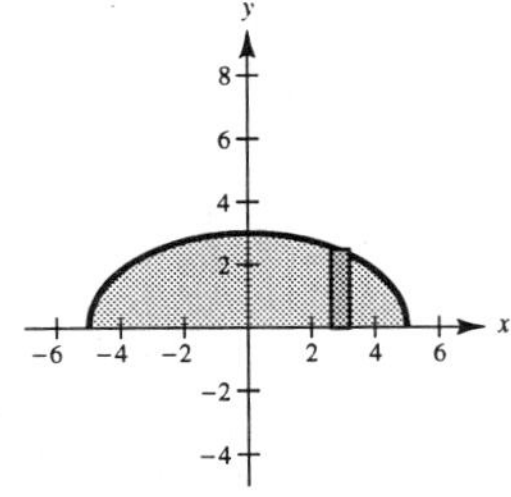

(b) $R(y) = \frac{5}{3}\sqrt{9 - y^2}, \; r(y) = 0, \; x \geq 0$

$$V = \frac{25\pi}{9}\int_0^3 (9 - y^2)\,dy$$

$$= \frac{25\pi}{9}\left[9y - \frac{y^3}{3}\right]_0^3 = 50\pi$$

51. Total volume: $V = \dfrac{4\pi(50)^3}{3} = \dfrac{500{,}000}{3}$ ft^3

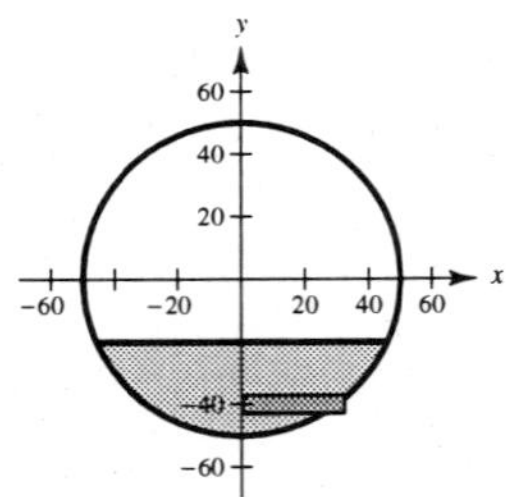

Volume of water in the tank:

$$\pi\int_{-50}^{y_0}\left(\sqrt{2500 - y^2}\right)^2 dy = \pi\int_{-50}^{y_0}(2500 - y^2)\,dy$$

$$= \pi\left[2500y - \frac{y^3}{3}\right]_{-50}^{y_0}$$

$$= \pi\left(2500y_0 - \frac{y_0^3}{3} + \frac{250{,}000}{3}\right)$$

When the tank is one-fourth of its capacity:

$$\frac{1}{4}\left(\frac{500{,}000\pi}{3}\right) = \pi\left(2500y_0 - \frac{y_0^3}{3} + \frac{250{,}00}{3}\right)$$

$$125{,}000 = 7500y_0 - y_0^3 + 250{,}000$$

$$y_0^3 - 7500y_0 - 125{,}000 = 0$$

$$y_0 \approx -17.36$$

Depth: $-17.36 - (-50) = 32.64$ feet

When the tank is three-fourths of its capacity the depth is $100 - 32.64 = 67.36$ feet.

53. (a) $\pi\displaystyle\int_0^h r^2\,dx$ (ii)

is the volume of a right circular cylinder with radius r and height h.

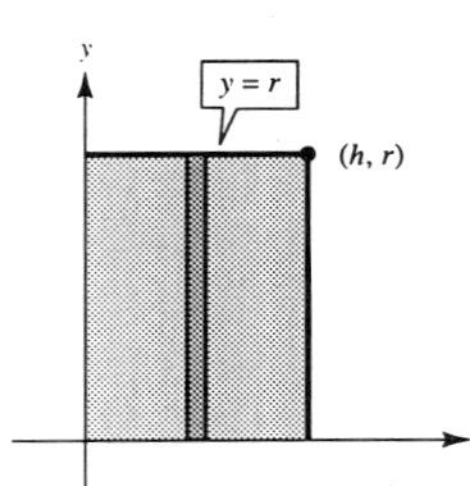

(b) $\pi\displaystyle\int_{-b}^{b}\left(a\sqrt{1 - \frac{x^2}{b^2}}\right)^2 dx$ (iv)

is the volume of an ellipsoid with axes $2a$ and $2b$.

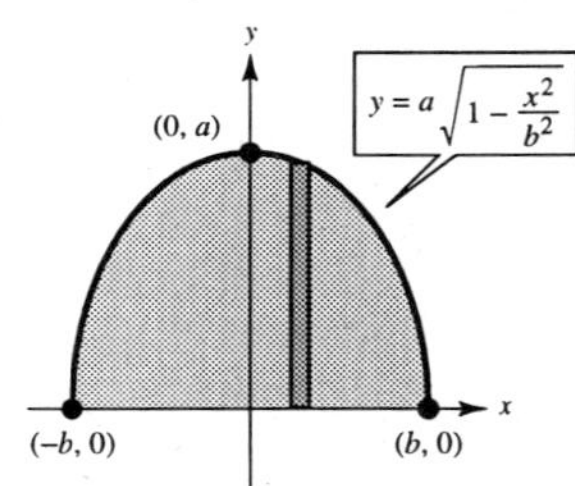
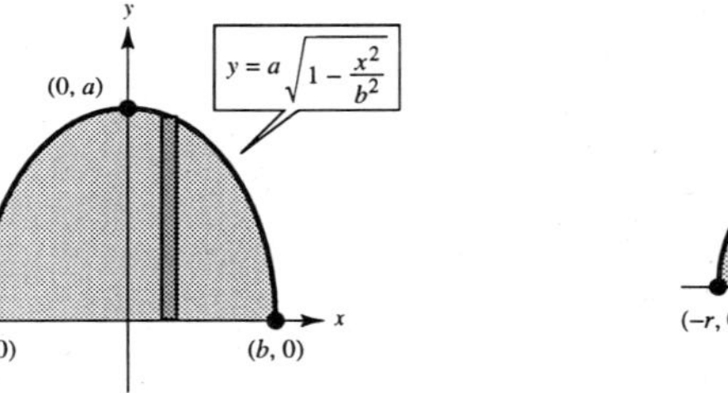

(c) $\pi\displaystyle\int_{-r}^{r}\left(\sqrt{r^2 - x^2}\right)^2 dx$ (iii)

is the volume of a sphere with radius r.

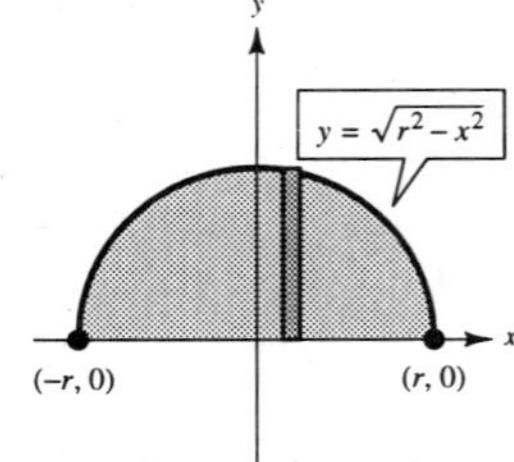

(d) $\pi\displaystyle\int_0^h\left(\frac{rx}{h}\right)^2 dx$ (i)

is the volume of a right circular cone with the radius of the base as r and height h.

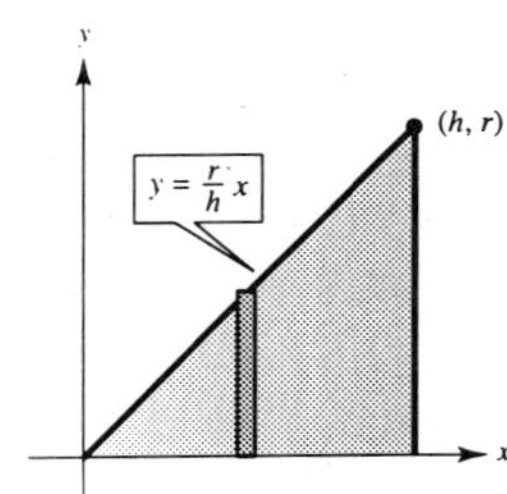

(e) $\pi\displaystyle\int_{-r}^{r}\left[\left(R + \sqrt{r^2 - x^2}\right)^2 - \left(R - \sqrt{r^2 - x^2}\right)^2\right]dx$ (v)

is the volume of a torus with the radius of its circular cross section as r and the distance from the axis of the torus to the center of its cross section as R.

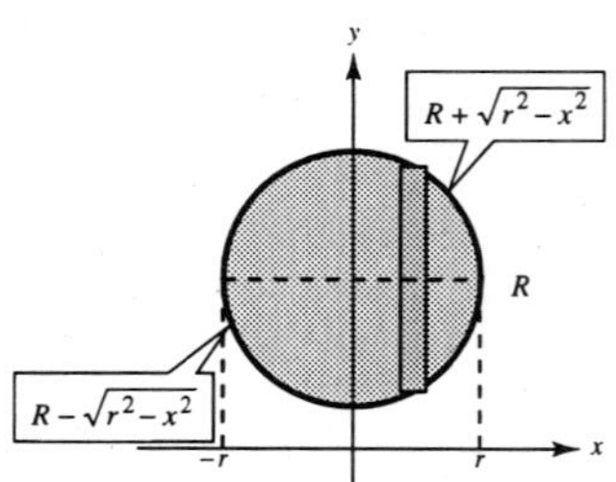

55.

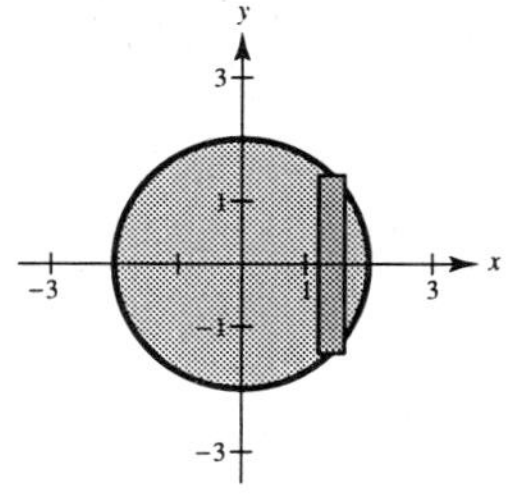

Base of Cross Section $= 2\sqrt{4 - x^2}$

(a) $A(x) = b^2 = \left(2\sqrt{4 - x^2}\right)^2$

$$V = \int_{-2}^{2} 4(4 - x^2)\,dx = 4\left[4x - \frac{x^3}{3}\right]_{-2}^{2} = \frac{128}{3}$$

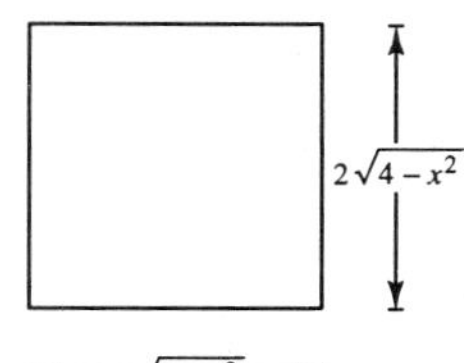

(c) $A(x) = \dfrac{1}{2}\pi r^2 = \dfrac{\pi}{2}\left(\sqrt{4 - x^2}\right)^2 = \dfrac{\pi}{2}(4 - x^2)$

$$V = \frac{\pi}{2}\int_{-2}^{2} (4 - x^2)\,dx = \frac{\pi}{2}\left[4x - \frac{x^3}{3}\right]_{-2}^{2} = \frac{16\pi}{3}$$

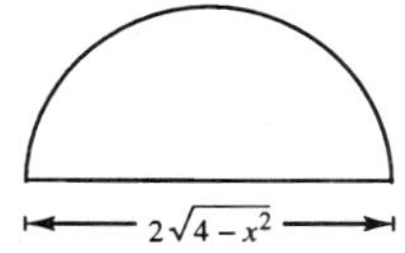

(b) $A(x) = \dfrac{1}{2}bh = \dfrac{1}{2}\left(2\sqrt{4 - x^2}\right)\left(\sqrt{3}\sqrt{4 - x^2}\right) = \sqrt{3}\,(4 - x^2)$

$$V = \sqrt{3}\int_{-2}^{2} (4 - x^2)\,dx = \sqrt{3}\left[4x - \frac{x^3}{3}\right]_{-2}^{2} = \frac{32\sqrt{3}}{3}$$

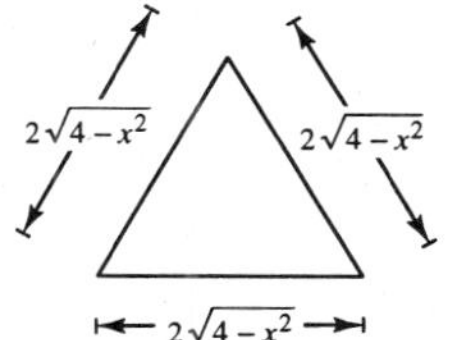

(d) $A(x) = \dfrac{1}{2}bh = \dfrac{1}{2}\left(2\sqrt{4 - x^2}\right)\left(\sqrt{4 - x^2}\right) = 4 - x^2$

$$V = \int_{-2}^{2} (4 - x^2)\,dx = \left[4x - \frac{x^3}{3}\right]_{-2}^{2} = \frac{32}{3}$$

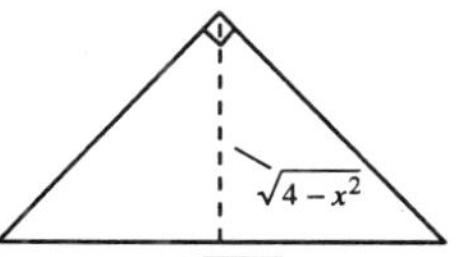

57.

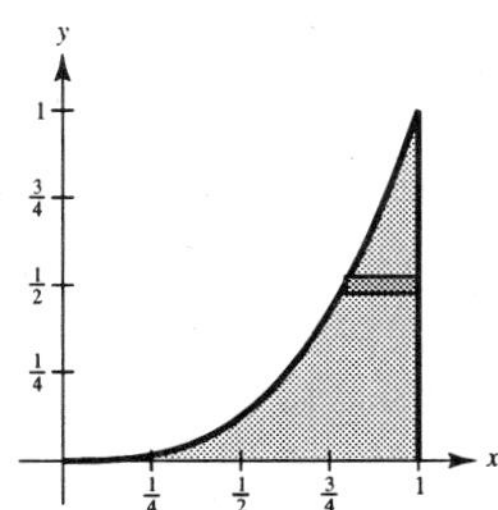

Base of Cross Section $= 1 - \sqrt[3]{y}$

—CONTINUED—

(a) $A(y) = b^2 = \left(1 - \sqrt[3]{y}\right)^2$

$$V = \int_0^1 \left(1 - \sqrt[3]{y}\right)^2 dy$$

$$= \int_0^1 \left(1 - 2y^{1/3} + y^{2/3}\right) dy$$

$$= \left[y - \frac{3}{2}y^{4/3} + \frac{3}{5}y^{5/3}\right]_0^1 = \frac{1}{10}$$

(b) $A(y) = \dfrac{1}{2}\pi r^2 = \dfrac{1}{2}\pi\left(\dfrac{1 - \sqrt[3]{y}}{2}\right)^2 = \dfrac{1}{8}\pi\left(1 - \sqrt[3]{y}\right)^2$

$$V = \frac{1}{8}\pi\int_0^1 \left(1 - \sqrt[3]{y}\right)^2 dy = \frac{\pi}{8}\left(\frac{1}{10}\right) = \frac{\pi}{80}$$

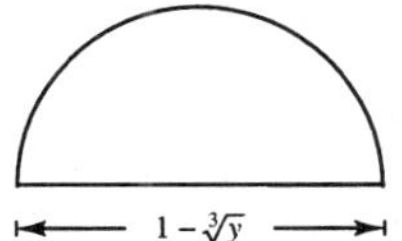

57. —CONTINUED—

(c) $A(y) = \frac{1}{2}bh = \frac{1}{2}(1 - \sqrt[3]{y})\left(\frac{\sqrt{3}}{2}\right)(1 - \sqrt[3]{y})$

$= \frac{\sqrt{3}}{4}(1 - \sqrt[3]{y})^2$

$V = \frac{\sqrt{3}}{4}\int_0^1 (1 - \sqrt[3]{y})^2\, dy = \frac{\sqrt{3}}{4}\left(\frac{1}{10}\right) = \frac{\sqrt{3}}{40}$

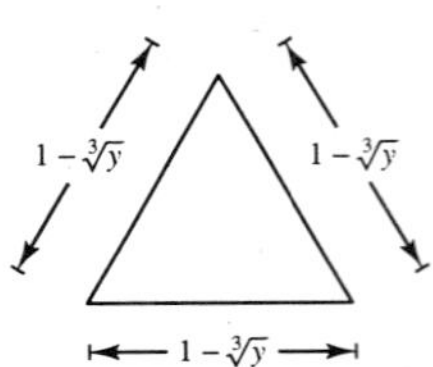

(d) $A(y) = \frac{1}{2}\pi ab = \frac{\pi}{2}(2)(1 - \sqrt[3]{y})\frac{1 - \sqrt[3]{y}}{2}$

$= \frac{\pi}{2}(1 - \sqrt[3]{y})^2$

$V = \frac{\pi}{2}\int_0^1 (1 - \sqrt[3]{y})^2\, dy = \frac{\pi}{2}\left(\frac{1}{10}\right) = \frac{\pi}{20}$

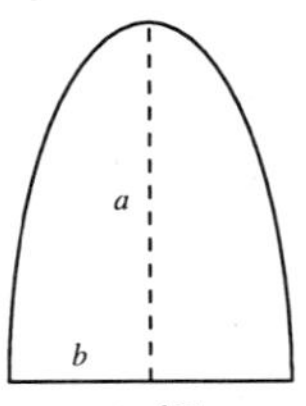

59. Assume that the oil just hits the top of the cylinder, which means that the volume of oil is one-half of the volume of the cylinder. Then we have

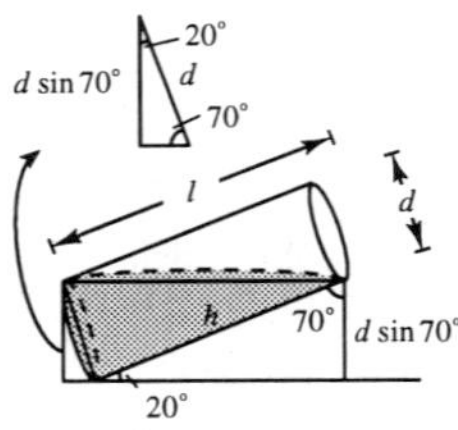

$$\sin 20° = \frac{d \sin 70°}{h}$$

$$h = \frac{d \sin 70°}{\sin 20°}$$

$$\text{Volume} = \frac{1}{2}(\pi r^2 h) = \frac{\pi}{2}\left(\frac{d}{2}\right)^2\left(\frac{d \sin 70°}{\sin 20°}\right) = \frac{\pi\, d^3 \sin 70°}{8 \sin 20°} = \pi d^3 \cot 20°.$$

61. (a) Since the cross sections are isosceles right triangles:

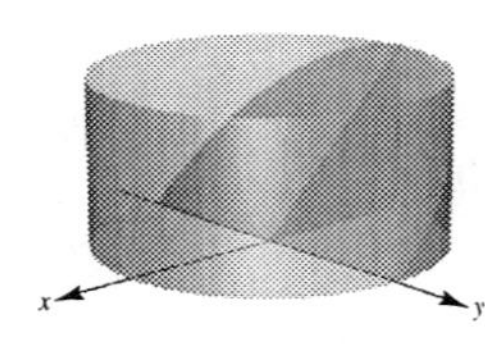

$$A(x) = \frac{1}{2}bh = \frac{1}{2}\left(\sqrt{r^2 - y^2}\right)\left(\sqrt{r^2 - y^2}\right) = \frac{1}{2}(r^2 - y^2)$$

$$V = \frac{1}{2}\int_{-r}^{r} (r^2 - y^2)\, dy = \int_0^r (r^2 - y^2)\, dy = \left[r^2 y - \frac{y^3}{3}\right]_0^r = \frac{2}{3}r^3$$

(b) $A(x) = \frac{1}{2}bh = \frac{1}{2}\sqrt{r^2 - y^2}\left(\sqrt{r^2 - y^2}\tan\theta\right) = \frac{\tan\theta}{2}(r^2 - y^2)$

$$V = \frac{\tan\theta}{2}\int_{-r}^{r} (r^2 - y^2)\, dy = \tan\theta \int_0^r (r^2 - y^2)\, dy = \tan\theta\left[r^2 y - \frac{y^3}{3}\right]_0^r = \frac{2}{3}r^3 \tan\theta$$

As $\theta \to 90°$, $V \to \infty$.

63. $\frac{4}{3}\pi(25 - r^2)^{3/2} = \frac{1}{2}\left(\frac{4}{3}\right)\pi(125)$

$$(25 - r^2)^{3/2} = \frac{125}{2}$$

$$25 - r^2 = \left(\frac{125}{2}\right)^{2/3}$$

$$25 - \frac{25}{(2^{2/3})} = r^2$$

$$25(1 - 2^{-2/3}) = r^2$$

$$r = 5\sqrt{1 - 2^{-2/3}} \approx 3.0415$$

Section 6.3 Volume: The Shell Method

1. $p(x) = x$

$h(x) = x$

$V = 2\pi\int_0^2 x(x)\,dx = \left[\frac{2\pi x^3}{3}\right]_0^2 = \frac{16\pi}{3}$

3. $p(x) = x$

$h(x) = \sqrt{x}$

$V = 2\pi\int_0^4 x\sqrt{x}\,dx$

$= 2\pi\int_0^4 x^{3/2}\,dx = \left[\frac{4\pi}{5}x^{5/2}\right]_0^4 = \frac{128\pi}{5}$

5. $p(x) = x$

$h(x) = x^2$

$V = 2\pi\int_0^2 x^3\,dx$

$= \left[\frac{\pi}{2}x^4\right]_0^2 = 8\pi$

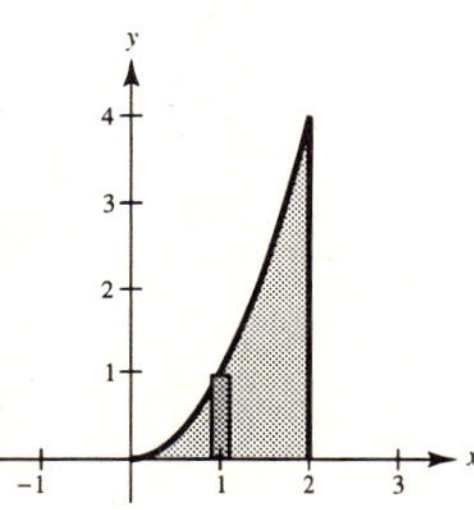

7. $p(x) = x$

$h(x) = (4x - x^2) - x^2 = 4x - 2x^2$

$V = 2\pi\int_0^2 x(4x - 2x^2)\,dx$

$= 4\pi\int_0^2 (2x^2 - x^3)\,dx$

$= 4\pi\left[\frac{2}{3}x^3 - \frac{1}{4}x^4\right]_0^2 = \frac{16\pi}{3}$

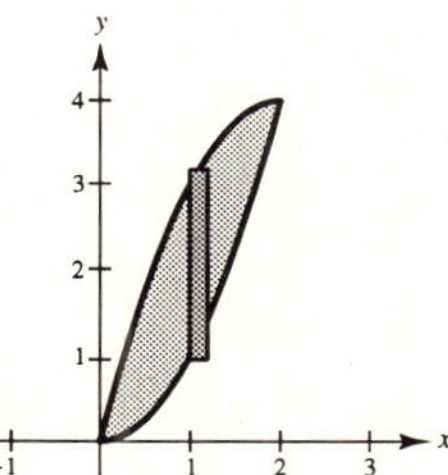

9. $p(x) = x$

$h(x) = 4 - (4x - x^2) = x^2 - 4x + 4$

$V = 2\pi\int_0^2 (x^3 - 4x^2 + 4x)\,dx$

$= 2\pi\left[\frac{x^4}{4} - \frac{4}{3}x^3 + 2x^2\right]_0^2 = \frac{8\pi}{3}$

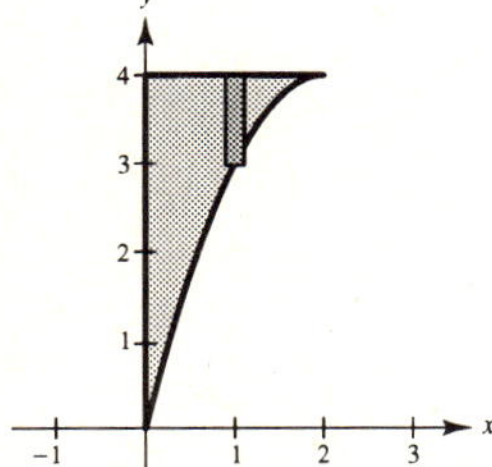

11. $p(x) = x$

$h(x) = \frac{1}{\sqrt{2\pi}}e^{-x^2/2}$

$V = 2\pi\int_0^1 x\left(\frac{1}{\sqrt{2\pi}}e^{-x^2/2}\right)dx$

$= \sqrt{2\pi}\int_0^1 e^{-x^2/2}x\,dx$

$= \left[-\sqrt{2\pi}\,e^{-x^2/2}\right]_0^1 = \sqrt{2\pi}\left(1 - \frac{1}{\sqrt{e}}\right) \approx 0.986$

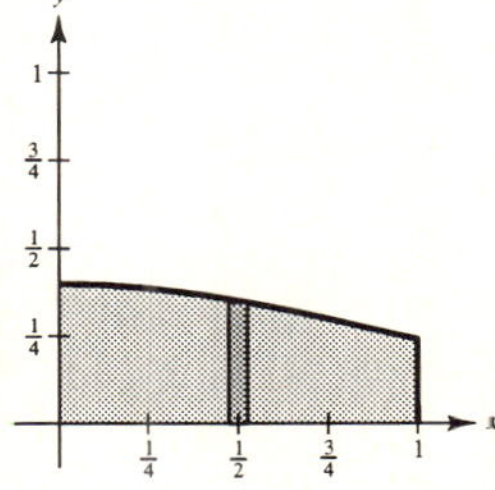

13. $p(y) = y$

$h(y) = 2 - y$

$$V = 2\pi \int_0^2 y(2 - y)\,dy$$

$$= 2\pi \int_0^2 (2y - y^2)\,dy$$

$$= 2\pi \left[y^2 - \frac{y^3}{3}\right]_0^2 = \frac{8\pi}{3}$$

15. $p(y) = y$ and $h(y) = 1$ if $0 \le y < \frac{1}{2}$.

$p(y) = y$ and $h(y) = \frac{1}{y} - 1$ if $\frac{1}{2} \le y \le 1$.

$$V = 2\pi \int_0^{1/2} y\,dy + 2\pi \int_{1/2}^1 (1 - y)\,dy$$

$$= 2\pi \left[\frac{y^2}{2}\right]_0^{1/2} + 2\pi \left[y - \frac{y^2}{2}\right]_{1/2}^1 = \frac{\pi}{4} + \frac{\pi}{4} = \frac{\pi}{2}$$

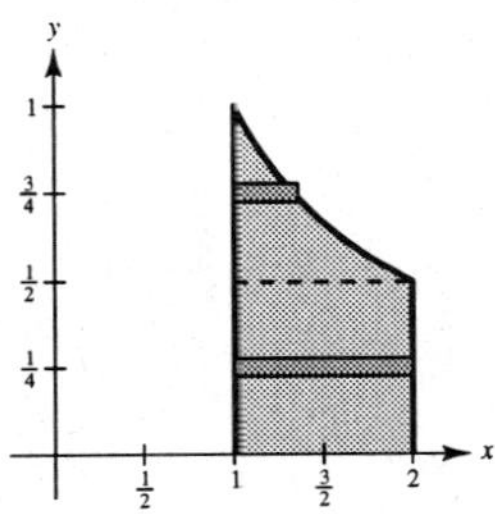

17. $p(x) = 4 - x$

$h(x) = 4x - x^2 - x^2 = 4x - 2x^2$

$$V = 2\pi \int_0^2 (4 - x)(4x - 2x^2)\,dx$$

$$= 2\pi(2) \int_0^2 (x^3 - 6x^2 + 8x)\,dx$$

$$= 4\pi \left[\frac{x^4}{4} - 2x^3 + 4x^2\right]_0^2 = 16\pi$$

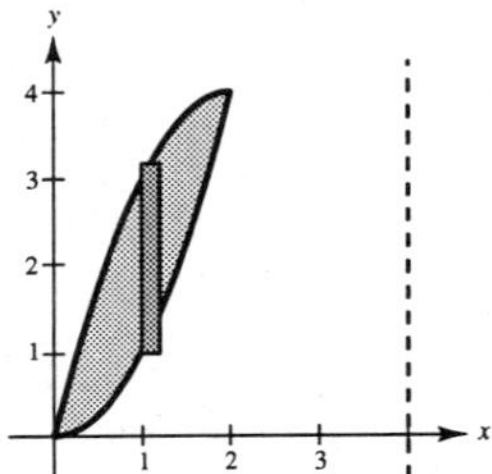

19. $p(x) = 5 - x$

$h(x) = 4x - x^2$

$$V = 2\pi \int_0^4 (5 - x)(4x - x^2)\,dx$$

$$= 2\pi \int_0^4 (x^3 - 9x^2 + 20x)\,dx$$

$$= 2\pi \left[\frac{x^4}{4} - 3x^3 + 10x^2\right]_0^4 = 64\pi$$

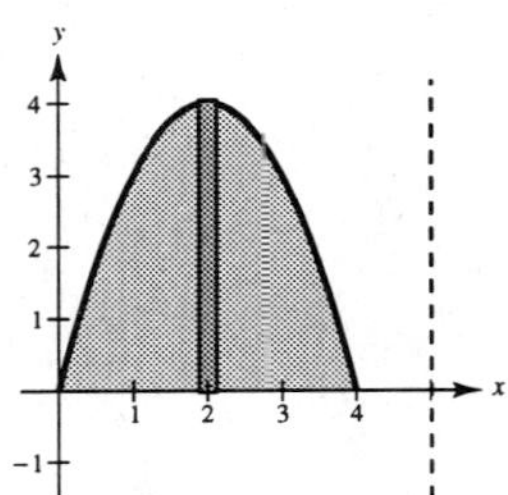

21. (a) **Disc**

$R(x) = x^3$

$r(x) = 0$

$$V = \pi \int_0^2 x^6\,dx = \pi \left[\frac{x^7}{7}\right]_0^2 = \frac{128\pi}{7}$$

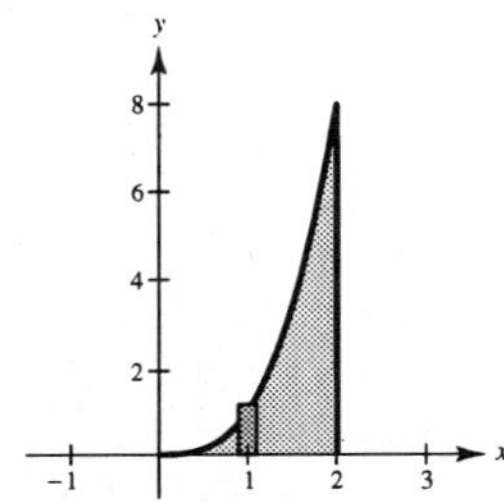

(b) **Shell**

$p(x) = x$

$h(x) = x^3$

$$V = 2\pi \int_0^2 x^4\,dx = 2\pi \left[\frac{x^5}{5}\right]_0^2 = \frac{64\pi}{5}$$

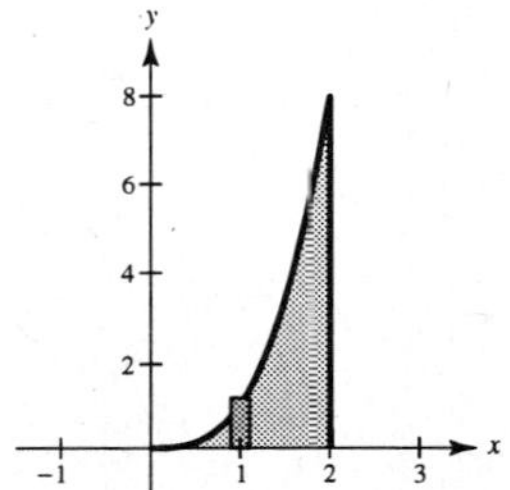

—CONTINUED—

21. —CONTINUED—

(c) **Shell**

$p(x) = 4 - x$

$h(x) = x^3$

$$V = 2\pi \int_0^2 (4 - x)x^3\,dx = 2\pi \int_0^2 (4x^3 - x^4)\,dx = 2\pi \left[x^4 - \frac{1}{5}x^5\right]_0^2 = \frac{96\pi}{5}$$

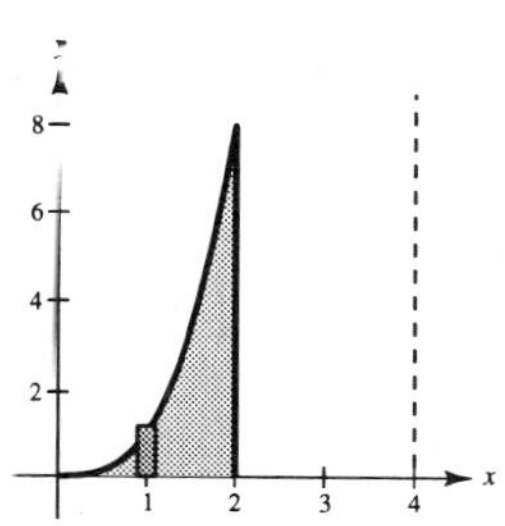

23. (a) **Shell**

$p(y) = y$

$h(y) = (a^{1/2} - y^{1/2})^2$

$$V = 2\pi \int_0^a y(a - 2a^{1/2}y^{1/2} + y)\,dy$$

$$= 2\pi \int_0^a (ay - 2a^{1/2}y^{3/2} + y^2)\,dy$$

$$= 2\pi \left[\frac{a}{2}y^2 - \frac{4a^{1/2}}{5}y^{5/2} + \frac{y^3}{3}\right]_0^a$$

$$= 2\pi \left[\frac{a^3}{2} - \frac{4a^3}{5} + \frac{a^3}{3}\right] = \frac{\pi a^3}{15}$$

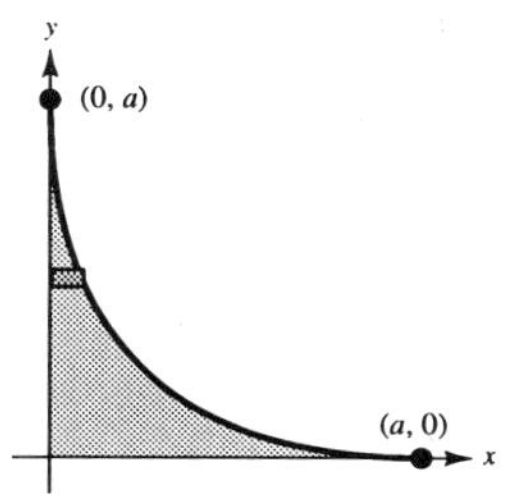

(b) Same as part a by symmetry

(c) **Shell**

$p(x) = a - x$

$h(x) = (a^{1/2} - x^{1/2})^2$

$$V = 2\pi \int_0^a (a - x)(a^{1/2} - x^{1/2})^2\,dx$$

$$= 2\pi \int_0^a (a^2 - 2a^{3/2}x^{1/2} + 2a^{1/2}x^{3/2} - x^2)\,dx$$

$$= 2\pi \left[a^2x - \frac{4}{3}a^{3/2}x^{3/2} + \frac{4}{5}a^{1/2}x^{5/2} - \frac{1}{3}x^3\right]_0^a = \frac{4\pi a^3}{15}$$

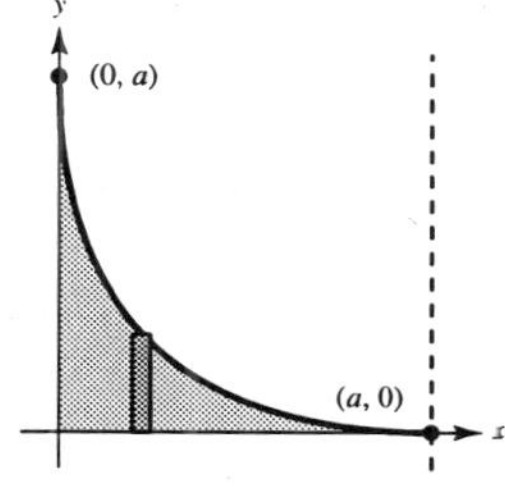

25. (a)

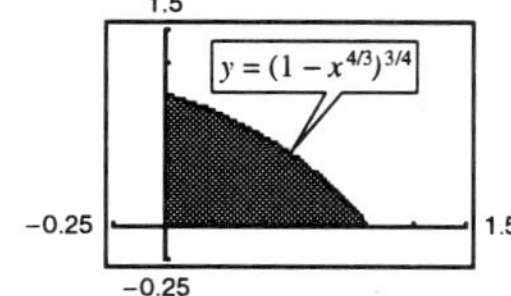

(b) $x^{4/3} + y^{4/3} = 1,\ x = 0,\ y = 0$

$y = (1 - x^{4/3})^{3/4}$

$$V = 2\pi \int_0^1 x(1 - x^{4/3})^{3/4}\,dx \approx 1.5056$$

27. (a)

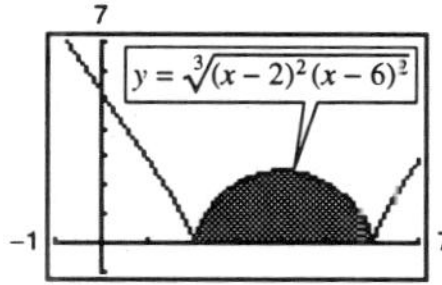

(b) $V = 2\pi \int_2^6 x\sqrt[3]{(x - 2)^2(x - 6)^2}\,dx \approx 187.249$

29. (a)

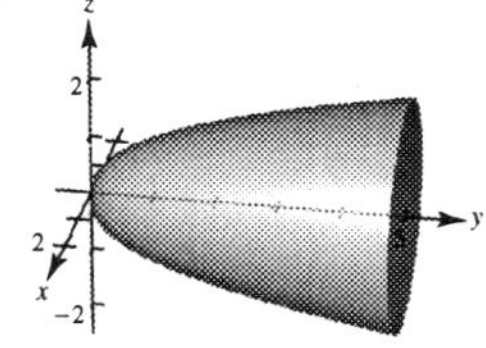

(b)

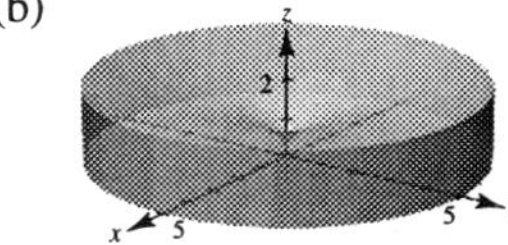

(c)

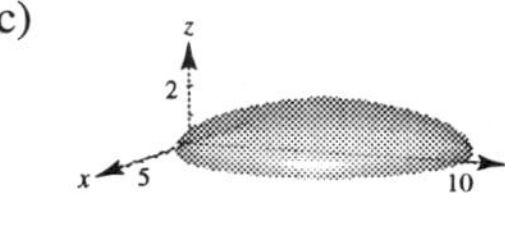

$a < c < b$

31. $y = 2e^{-x}$, $y = 0$, $x = 0$, $x = 2$

Volume ≈ 7.5

Matches (d)

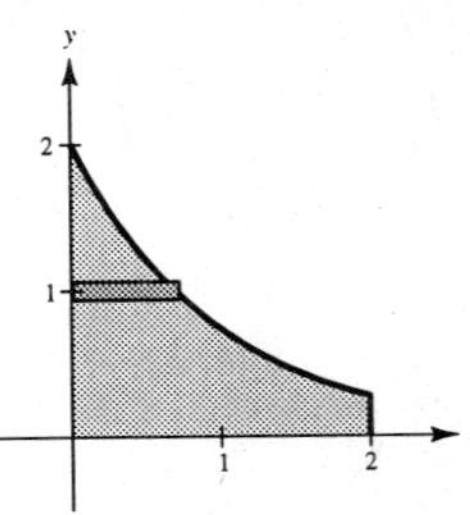

33. $p(x) = x$

$h(x) = 2 - \frac{1}{2}x^2$

$$V = 2\pi \int_0^2 x\left(2 - \frac{1}{2}x^2\right) dx = 2\pi \int_0^2 \left(2x - \frac{1}{2}x^3\right) dx = 2\pi\left[x^2 - \frac{1}{8}x^4\right]_0^2 = 4\pi \text{ (total volume)}$$

Now find x_0 such that

$$\pi = 2\pi \int_0^{x_0} \left(2x - \frac{1}{2}x^3\right) dx$$

$$1 = 2\left[x^2 - \frac{1}{8}x^4\right]_0^{x_0}$$

$$1 = 2x_0^2 - \frac{1}{4}x_0^4$$

$$x_0^4 - 8x_0^2 + 4 = 0$$

$$x_0^2 = 4 \pm 2\sqrt{3} \quad \text{(Quadratic Formula)}$$

Take $x_0 = \sqrt{4 - 2\sqrt{3}}$ since the other root is too large.
Diameter: $2\sqrt{4 - 2\sqrt{3}} \approx 1.464$

35. $x = \sqrt{r^2 - \left(\frac{h}{2}\right)^2} = \frac{\sqrt{4r^2 - h^2}}{2}$

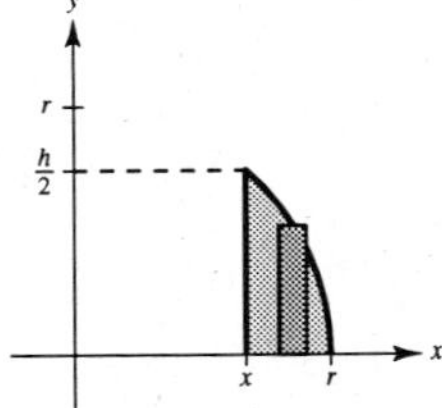

$$V = 4\pi \int_{\sqrt{4r^2-h^2}/2}^{r} x\sqrt{r^2 - x^2}\, dx$$

$$= \left[-2\pi\left(\frac{2}{3}\right)(r^2 - x^2)^{3/2}\right]_{\sqrt{4r^2-h^2}/2}^{r}$$

$$= 0 + \frac{4\pi}{3}\left[r^2 - \frac{4r^2 - h^2}{4}\right]^{3/2}$$

$$= \frac{4\pi}{3}\left(\frac{h^2}{4}\right)^{3/2}$$

$$= \frac{4\pi}{3} \cdot \frac{h^3}{8} = \frac{\pi h^3}{6} \quad \text{(independent of } r\text{!)}$$

37. $V = 4\pi \int_{-r}^{r} (R - x)\sqrt{r^2 - x^2}\, dx$

$$= 4\pi R \int_{-r}^{r} \sqrt{r^2 - x^2}\, dx - 4\pi \int_{-r}^{r} x\sqrt{r^2 - x^2}\, dx$$

$$= 4\pi R\left(\frac{\pi r^2}{2}\right) + \left[2\pi\left(\frac{2}{3}\right)(r^2 - x^2)^{3/2}\right]_{-r}^{r}$$

$$= 2\pi^2 r^2 R$$

39. $\pi\int_1^5 (x-1)\,dx = \pi\int_1^5 \left(\sqrt{x-1}\right)^2 dx$

This integral represents the volume of the solid generated by revolving the region bounded by $y = \sqrt{x-1}$, $y = 0$, and $x = 5$ about the x-axis by using the Disc Method.

$$2\pi\int_0^2 y[5-(y^2+1)]\,dy$$

represents this same volume by using the Shell Method.

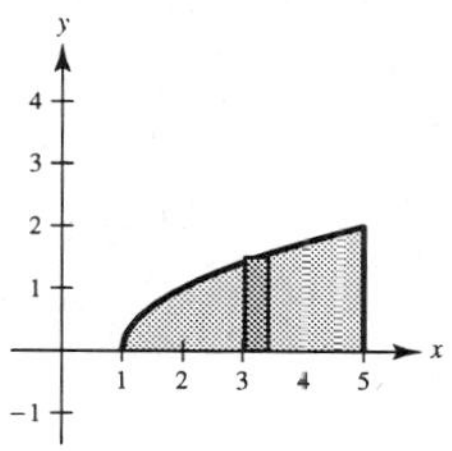

Disc Method

41. (a) $2\pi\int_0^r hx\left(1-\dfrac{x}{r}\right)dx$ (ii)

is the volume of a right circular cone with the radius of the base as r and height h.

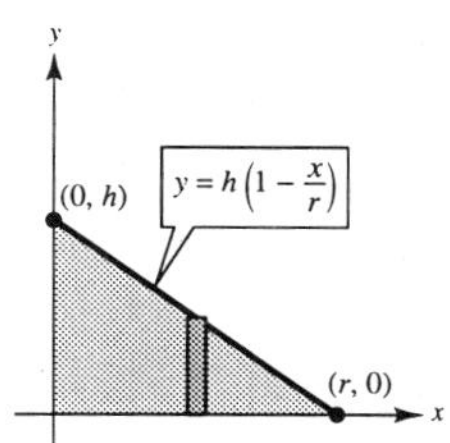

(b) $2\pi\int_{-r}^r (R-x)\left(2\sqrt{r^2-x^2}\right)dx$ (v)

is the volume of a torus with the radius of its circular cross section as r and the distance from the axis of the torus to the center of its cross section as R.

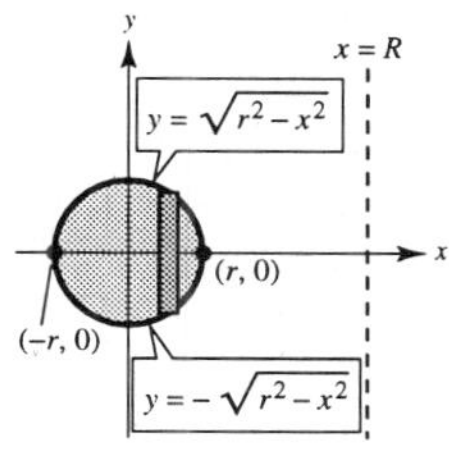

(c) $2\pi\int_0^r 2x\sqrt{r^2-x^2}\,dx$ (iii) is the volume of a sphere with radius r.

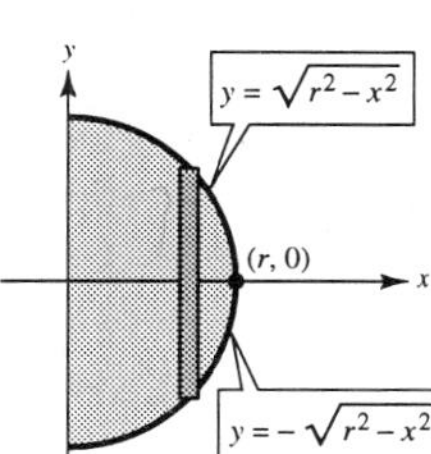

(d) $2\pi\int_0^r hx\,dx$ (i) is the volume of a right circular cylinder with a radius of r and a height of h.

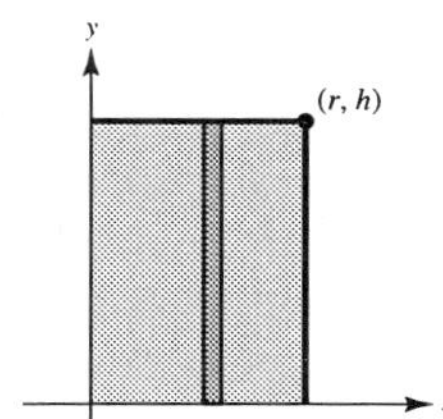

(e) $2\pi\int_0^b 2ax\sqrt{1-(x^2/b^2)}\,dx$ (iv) is the volume of an ellipsoid with axes $2a$ and $2b$.

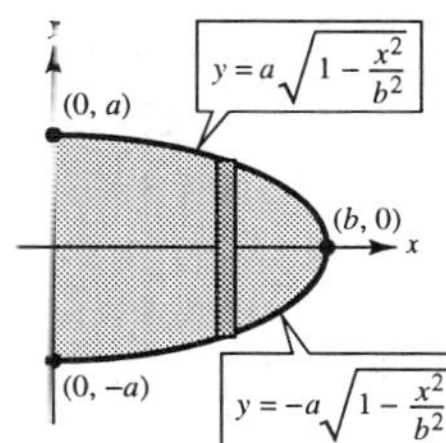

43. (a) $V = 2\pi\int_0^{200} xf(x)\,dx$

$$\approx \frac{2\pi(200)}{3(8)}[0 + 4(25)(19) + 2(50)(19) + 4(75)(17) + 2(100)15 + 4(125)(14) + 2(150)(10) + 4(175)(6) + 0]$$

$\approx$ 1,366,593 cubic feet

(b) $d = -0.000561x^2 + 0.0189x + 19.39$

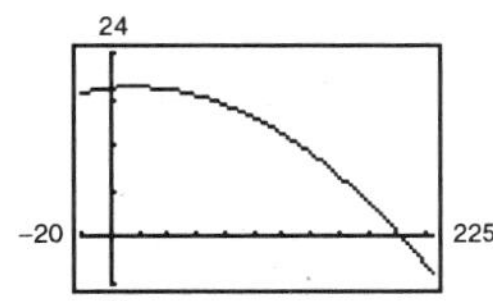

(c) $V \approx 2\pi\int_0^{200} xd(x)\,dx \approx 2\pi(213{,}800) = 1{,}343{,}345$ cubic feet

(d) Number gallons $\approx V(7.48) = 10{,}048{,}221$ gallons

Section 6.4 Arc Length and Surfaces of Revolution

1. $(0, 0), (5, 12)$

(a) $d = \sqrt{(5-0)^2 + (12-0)^2} = 13$

(b) $y = \frac{12}{5}x$

$y' = \frac{12}{5}$

$s = \int_0^5 \sqrt{1 + \left(\frac{12}{5}\right)^2}\,dx = \left[\frac{13}{5}x\right]_0^5 = 13$

3. $y = \frac{2}{3}x^{3/2} + 1$

$y' = x^{1/2}, [0, 1]$

$s = \int_0^1 \sqrt{1+x}\,dx$

$= \left[\frac{2}{3}(1+x)^{3/2}\right]_0^1$

$= \frac{2}{3}\left(\sqrt{8} - 1\right) \approx 1.219$

5. $y = \frac{3}{2}x^{2/3}$

$y' = \frac{1}{x^{1/3}}, [1, 8]$

$s = \int_1^8 \sqrt{1 + \left(\frac{1}{x^{1/3}}\right)^2}\,dx$

$= \int_1^8 \sqrt{\frac{x^{2/3}+1}{x^{2/3}}}\,dx$

$= \frac{3}{2}\int_1^8 \sqrt{x^{2/3}+1}\left(\frac{2}{3x^{1/3}}\right)dx$

$= \frac{3}{2}\left[\frac{2}{3}(x^{2/3}+1)^{3/2}\right]_1^8$

$= 5\sqrt{5} - 2\sqrt{2} \approx 8.352$

7. $y = \frac{x^4}{8} + \frac{1}{4x^2}$

$y' = \frac{1}{2}x^3 - \frac{1}{2x^3}$

$1 + (y')^2 = \left(\frac{1}{2}x^3 + \frac{1}{2x^3}\right)^2, [1, 2]$

$s = \int_a^b \sqrt{1 + (y')^2}\,dx$

$= \int_1^2 \left(\frac{1}{2}x^3 + \frac{1}{2x^3}\right)dx$

$= \left[\frac{1}{8}x^4 - \frac{1}{4x^2}\right]_1^2 = \frac{33}{16} \approx 2.063$

9. (a) $y = 4 - x^2, 0 \le x \le 2$

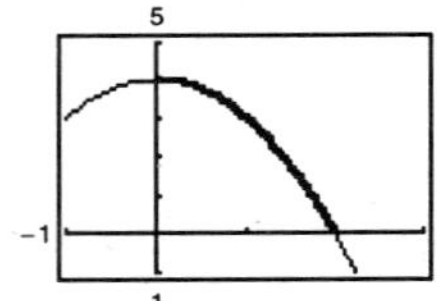

(b) $y' = -2x$

$1 + (y')^2 = 1 + 4x^2$

$L = \int_0^2 \sqrt{1 + 4x^2}\,dx$

(c) $L \approx 4.647$

11. (a) $y = \frac{1}{x}, 1 \le x \le 3$

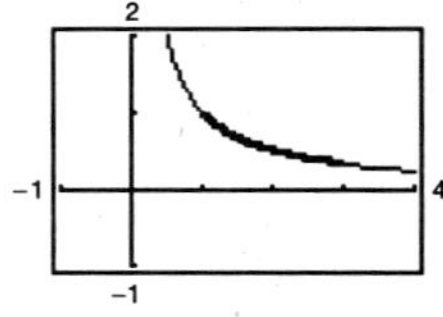

(b) $y' = -\frac{1}{x^2}$

$1 + (y')^2 = 1 + \frac{1}{x^4}$

$L = \int_1^3 \sqrt{1 + \frac{1}{x^4}}\,dx$

(c) $L \approx 2.147$

13. (a) $y = \sin x, 0 \le x \le \pi$

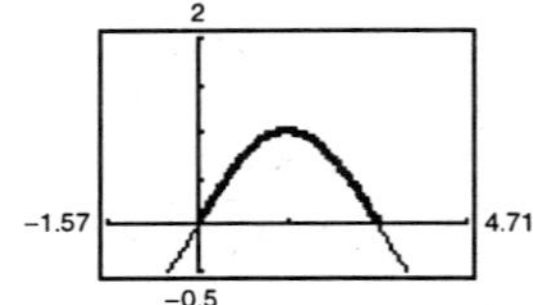

(b) $y' = \cos x$

$1 + (y')^2 = 1 + \cos^2 x$

$L = \int_0^\pi \sqrt{1 + \cos^2 x}\,dx$

(c) $L \approx 3.820$

15. (a) $x = e^{-y}, 0 \le y \le 2$

$y = -\ln x$

$1 \ge x \ge e^{-2} \approx 0.135$

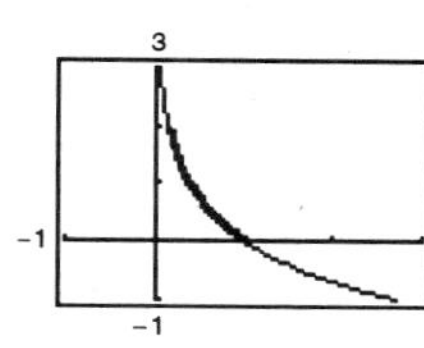

(b) $y' = -\dfrac{1}{x}$

$1 + (y')^2 = 1 + \dfrac{1}{x^2}$

$L = \displaystyle\int_{e^{-2}}^{1} \sqrt{1 + \frac{1}{x^2}}\, dx$

(c) $L \approx 2.221$

Alternatively, you can do all the computations with respect to y.

(a) $x = e^{-y}\ 0 \le y \le 2$

(b) $\dfrac{dx}{dy} = -e^{-y}$

$1 + \left(\dfrac{dx}{dy}\right)^2 = 1 + e^{-2y}$

$L = \displaystyle\int_0^2 \sqrt{1 + e^{-2y}}\, dy$

(c) $L \approx 2.221$

17. (a) $y = 2 \arctan x, 0 \le x \le 1$

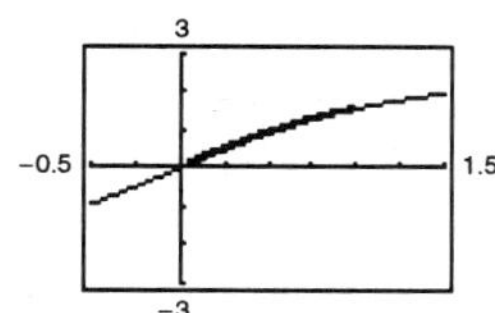

(b) $y' = \dfrac{2}{1 + x^2}$

$L = \displaystyle\int_0^1 \sqrt{1 + \frac{4}{(1 + x^2)^2}}\, dx$

(c) $L \approx 1.871$

19. $\displaystyle\int_0^2 \sqrt{1 + \left[\frac{d}{dx}\left(\frac{5}{x^2 + 1}\right)\right]^2}\, dx$

$s \approx 5$

Matches (b)

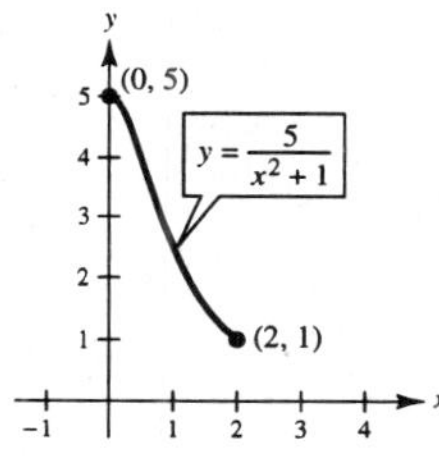

21. $y = x^3, [0, 4]$

(a) $d = \sqrt{(4 - 0)^2 + (64 - 0)^2} \approx 64.125$

(b) $d = \sqrt{(1 - 0)^2 + (1 - 0)^2} + \sqrt{(2 - 1)^2 + (8 - 1)^2} + \sqrt{(3 - 2)^2 + (27 - 8)^2} + \sqrt{(4 - 3)^2 + (64 - 27)^2}$

≈ 64.525

(c) $s = \displaystyle\int_0^4 \sqrt{1 + (3x^2)^2}\, dx = \int_0^4 \sqrt{1 + 9x^4}\, dx \approx 64.666$

(d) 64.672

23. (a)

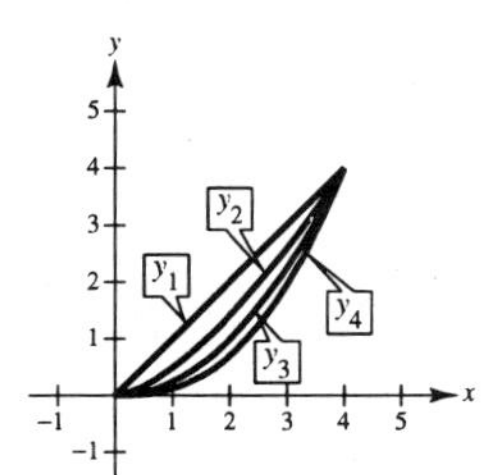

(b) y_1, y_2, y_3, y_4

(c) $y_1' = 1,\ L_1 = \displaystyle\int_0^4 \sqrt{2}\, dx \approx 5.657$

$y_2' = \dfrac{3}{4}x^{1/2},\ L_2 = \displaystyle\int_0^4 \sqrt{1 + \frac{9x}{16}}\, dx \approx 5.759$

$y_3' = \dfrac{1}{2}x,\ L_3 = \displaystyle\int_0^4 \sqrt{1 + \frac{x^2}{4}}\, dx \approx 5.916$

$y_4' = \dfrac{5}{16}x^{3/2},\ L_4 = \displaystyle\int_0^4 \sqrt{1 + \frac{25}{256}x^3}\, dx \approx 6.063$

25. $y = \frac{1}{3}[x^{3/2} - 3x^{1/2} + 2]$

When $x = 0$, $y = \frac{2}{3}$. Thus, the fleeting object has traveled $\frac{2}{3}$ units when it is caught.

$$y' = \frac{1}{3}\left[\frac{3}{2}x^{1/2} - \frac{3}{2}x^{-1/2}\right] = \left(\frac{1}{2}\right)\frac{x-1}{x^{1/2}}$$

$$1 + (y')^2 = 1 + \frac{(x-1)^2}{4x} = \frac{(x+1)^2}{4x}$$

$$s = \int_0^1 \frac{x+1}{2x^{1/2}}\,dx = \frac{1}{2}\int_0^1 (x^{1/2} + x^{-1/2})\,dx = \frac{1}{2}\left[\frac{2}{3}x^{3/2} + 2x^{1/2}\right]_0^1 = \frac{4}{3} = 2\left(\frac{2}{3}\right)$$

The pursuer has traveled twice the distance that the fleeing object has traveled when it is caught.

27. $y = 20\cosh\frac{x}{20},\ -20 \le x \le 20$

$$y' = \sinh\frac{x}{20}$$

$$1 + (y')^2 = 1 + \sinh^2\frac{x}{20} = \cosh^2\frac{x}{20}$$

$$L = \int_{-20}^{20}\cosh\frac{x}{20}\,dx = 2\int_0^{20}\cosh\frac{x}{20}\,dx = 2(20)\sinh\left[\frac{x}{20}\right]_0^{20} = 40\sinh(1) \approx 47.008 \text{ m.}$$

29. $y = \sqrt{9 - x^2}$

$$y' = \frac{-x}{\sqrt{9 - x^2}}$$

$$1 + (y')^2 = \frac{9}{9 - x^2}$$

$$s = \int_0^2 \sqrt{\frac{9}{9 - x^2}}\,dx = \int_0^2 \frac{3}{\sqrt{9 - x^2}}\,dx$$

$$= \left[3\arcsin\frac{x}{3}\right]_0^2 = 3\left(\arcsin\frac{2}{3} - \arcsin 0\right)$$

$$= 3\arcsin\frac{2}{3} \approx 2.1892$$

31. $y = \frac{x^3}{3}$

$$y' = x^2,\ [0, 3]$$

$$S = 2\pi\int_0^3 \frac{x^3}{3}\sqrt{1 + x^4}\,dx$$

$$= \frac{\pi}{6}\int_0^3 (1 + x^4)^{1/2}(4x^3)\,dx$$

$$= \left[\frac{\pi}{9}(1 + x^4)^{3/2}\right]_0^3$$

$$= \frac{\pi}{9}\left(82\sqrt{82} - 1\right) \approx 258.85$$

33. $y = \frac{x^3}{6} + \frac{1}{2x}$

$$y' = \frac{x^2}{2} - \frac{1}{2x^2}$$

$$1 + (y')^2 = \left(\frac{x^2}{2} + \frac{1}{2x^2}\right)^2,\ [1, 2]$$

$$S = 2\pi\int_1^2 \left(\frac{x^3}{6} + \frac{1}{2x}\right)\left(\frac{x^2}{2} + \frac{1}{2x^2}\right)dx$$

$$= 2\pi\int_1^2 \left(\frac{x^5}{12} + \frac{x}{3} + \frac{1}{4x^3}\right)dx$$

$$= 2\pi\left[\frac{x^6}{72} + \frac{x^2}{6} - \frac{1}{8x^2}\right]_1^2 = \frac{47\pi}{16}$$

35. $y = \sqrt[3]{x} + 2$

$$y' = \frac{1}{3x^{2/3}},\ [1, 8]$$

$$S = 2\pi\int_1^8 x\sqrt{1 + \frac{1}{9x^{4/3}}}\,dx$$

$$= \frac{2\pi}{3}\int_1^8 x^{1/3}\sqrt{9x^{4/3} + 1}\,dx$$

$$= \frac{\pi}{18}\int_1^8 (9x^{4/3} + 1)^{1/2}(12x^{1/3})\,dx$$

$$= \left[\frac{\pi}{27}(9x^{4/3} + 1)^{3/2}\right]_1^8$$

$$= \frac{\pi}{27}\left(145\sqrt{145} - 10\sqrt{10}\right) \approx 199.48$$

37. $y = \sin x$

$y' = \cos x, [0, \pi]$

$$S = 2\pi \int_0^{\pi} \sin x\sqrt{1 + \cos^2 x}\, dx$$

$$\approx 14.4236$$

39.

$$y = \frac{hx}{r}$$

$$y' = \frac{h}{r}$$

$$1 + (y')^2 = \frac{r^2 + h^2}{r^2}$$

$$S = 2\pi \int_0^r x\sqrt{\frac{r^2 + h^2}{r^2}}\, dx$$

$$= \left[\frac{2\pi\sqrt{r^2 + h^2}}{r}\left(\frac{x^2}{2}\right)\right]_0^r = \pi r\sqrt{r^2 + h^2}$$

41.

$$y = \sqrt{9 - x^2}$$

$$y' = \frac{-x}{\sqrt{9 - x^2}}$$

$$\sqrt{1 + (y')^2} = \frac{3}{\sqrt{9 - x^2}}$$

$$S = 2\pi \int_0^2 \frac{3x}{\sqrt{9 - x^2}}\, dx = -3\pi \int_0^2 \frac{-2x}{\sqrt{9 - x^2}}\, dx = \left[-6\pi\sqrt{9 - x^2}\right]_0^2 = 6\pi(3 - \sqrt{5}) \approx 14.40$$

See figure in Exercise 42.

43.

$$y = \frac{1}{3}x^{1/2} - x^{3/2}$$

$$y' = \frac{1}{6}x^{-1/2} - \frac{3}{2}x^{1/2} = \frac{1}{6}(x^{-1/2} - 9x^{1/2})$$

$$1 + (y')^2 = 1 + \frac{1}{36}(x^{-1} - 18 + 81x) = \frac{1}{36}(x^{-1/2} + 9x^{1/2})^2$$

$$S = 2\pi \int_0^{1/3} \left(\frac{1}{3}x^{1/2} - x^{3/2}\right)\sqrt{\frac{1}{36}(x^{-1/2} + 9^{1/2})^2}\, dx = \frac{2\pi}{6}\int_0^{1/3}\left(\frac{1}{3}x^{1/2} - x^{3/2}\right)(x^{-1/2} + 9x^{1/2})\, dx$$

$$= \frac{\pi}{3}\int_0^{1/3}\left(\frac{1}{3} + 2x - 9x^2\right) dx = \frac{\pi}{3}\left[\frac{1}{3}x + x^2 - 3x^3\right]_0^{1/3} = \frac{\pi}{27}\text{ ft}^2 \approx 0.1164 \text{ ft}^2 \approx 16.8 \text{ in}^2$$

Amount of glass needed: $V = \frac{\pi}{27}\left(\frac{0.015}{12}\right) \approx 0.00015 \text{ ft}^3 \approx 0.25 \text{ in}^3$

45. Individual project, see Exercise 44.

47. (a) $\frac{ds}{dx} = \sqrt{1 + [f'(x)]^2}$ (Second Fundamental Theorem of Calculus)

(b) $ds = \sqrt{1 + [f'(x)]^2}\, dx$

$$(ds)^2 = \left[1 + [f'(x)]^2\right](dx)^2 = \left[1 + \left(\frac{dy}{dx}\right)^2\right](dx)^2 = (dx)^2 + (dy)^2$$

(c) $s(x) = \int_1^x \sqrt{1 + \left(\frac{3}{2}t^{1/2}\right)^2}\, dt = \int_1^x \sqrt{1 + \frac{9}{4}t}\, dt = \left[\frac{4}{9}\left(\frac{2}{3}\right)\left(1 + \frac{9}{4}t\right)^{3/2}\right]_1^x$

$$= \frac{8}{27}\left[\left(1 + \frac{9}{4}x\right)^{3/2} - \left(\frac{13}{4}\right)^{3/2}\right] = \frac{8}{27}\left[\frac{(4 + 9x)^{3/2}}{8} - \frac{13\sqrt{13}}{8}\right] = \frac{1}{27}\left[(4 + 9x)^{3/2} - 13\sqrt{13}\right]$$

$$s(2) = \frac{1}{27}\left[22\sqrt{22} - 13\sqrt{13}\right] = \frac{22}{27}\sqrt{22} - \frac{13}{27}\sqrt{13} \approx 2.086$$

49. (a) Area of circle with radius L: $A = \pi L^2$

Area of sector with central angle θ (in radians)

$$S = \frac{\theta}{2\pi}A = \frac{\theta}{2\pi}(\pi L^2) = \frac{1}{2}L^2\theta$$

(b) Let s be the arc length of the sector, which is the circumference of the base of the cone. Here, $s = L\theta = 2\pi r$, and you have

$$S = \frac{1}{2}L^2\theta = \frac{1}{2}L^2\left(\frac{s}{L}\right) = \frac{1}{2}Ls = \frac{1}{2}L(2\pi r) = \pi rL$$

(c) The lateral surface area of the frustum is the difference of the large cone and the small one.

$$S = \pi r_2(L + L_1) - \pi r_1 L_1$$
$$= \pi r_2 L + \pi L_1(r_2 - r_1)$$

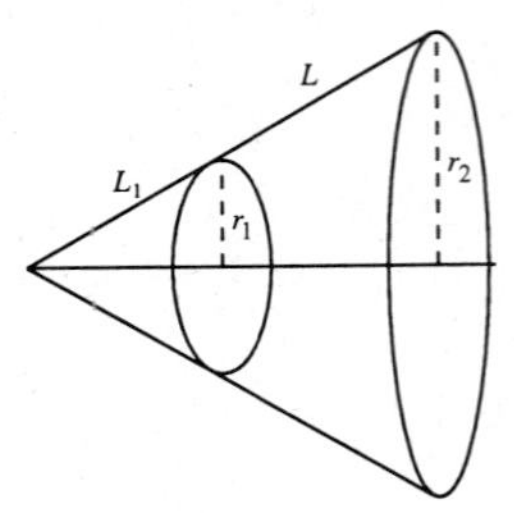

By similar triangles, $\dfrac{L + L_1}{r_2} = \dfrac{L_1}{r_1} \Rightarrow Lr_1 = L_1(r_2 - r_1)$

Hence,

$$S = \pi r_2 L + \pi L_1(r_2 - r_1) = \pi r_2 L + \pi L r_1$$
$$= \pi L(r_1 + r_2).$$

Section 6.5 Work

1. $W = Fd = (100)(10) = 1000$ ft · lb

3. $W = Fd = (112)(4) = 448$ joules (newton-meters)

5. Since the work equals the area under the force function, you have $(c) < (d) < (a) < (b)$.

7. $F(x) = kx$

$5 = k(4)$

$k = \frac{5}{4}$

$$W = \int_0^7 \frac{5}{4}x\,dx = \left[\frac{5}{8}x^2\right]_0^7$$

$= \frac{245}{8}$ in · lb

$= 30.625$ in · lb ≈ 2.55 ft · lb

9. $F(x) = kx$

$250 = k(30) \Rightarrow k = \frac{25}{3}$

$$W = \int_{20}^{50} F(x)\,dx = \int_{20}^{50} \frac{25}{3}x\,dx = \frac{25x^2}{6}\bigg]_{20}^{50}$$

$= 8750$ n · cm $= 87.5$ joules or Nm

11. $F(x) = kx$

$15 = 6k$

$k = \frac{5}{2}$

$$W = \int_0^{12} \frac{5}{2}x\,dx = \left[\frac{5}{4}x^2\right]_0^{12} = 180 \text{ in} \cdot \text{lb} = 15 \text{ ft} \cdot \text{lb}$$

13. $W = 18 = \displaystyle\int_0^{1/3} kx\,dx = \frac{kx^2}{2}\bigg]_0^{1/3} = \frac{k}{18} \Rightarrow k = 324$

$$W = \int_{1/3}^{7/12} 324x\,dx = 162x^2\bigg]_{1/3}^{7/12} = 37.125 \text{ ft} \cdot \text{lbs}$$

[**Note:** 4 inches $= \frac{1}{3}$ foot]

15. Assume that the earth has a radius of 4000 miles.

$$F(x) = \frac{k}{x^2}$$

$$4 = \frac{k}{(4000)^2}$$

$$k = 64{,}000{,}000$$

$$F(x) = \frac{64{,}000{,}000}{x^2}$$

(a) $W = \int_{4000}^{4200} \frac{64{,}000{,}000}{x^2}\,dx = \left[-\frac{64{,}000{,}000}{x}\right]_{4000}^{4200} \approx -15{,}238.095 + 16{,}000$

$= 761.905 \text{ mi} \cdot \text{ton} \approx 8.05 \times 10^9 \text{ ft} \cdot \text{lb}$

(b) $W = \int_{4000}^{4400} \frac{64{,}000{,}000}{x^2}\,dx = \left[-\frac{64{,}000{,}000}{x}\right]_{4000}^{4400} \approx -14{,}545.455 + 16{,}000$

$= 1454.545 \text{ mi} \cdot \text{ton} \approx 1.54 \times 10^{10} \text{ ft} \cdot \text{lb}$

17. Assume that the earth has a radius of 4000 miles.

$$F(x) = \frac{k}{x^2}$$

$$10 = \frac{k}{(4000)^2}$$

$$k = 160{,}000{,}000$$

$$F(x) = \frac{160{,}000{,}000}{x^2}$$

(a) $W = \int_{4000}^{15{,}000} \frac{160{,}000{,}000}{x^2}\,dx = \left[-\frac{160{,}000{,}000}{x}\right]_{4000}^{15{,}000} \approx -10{,}666.667 + 40{,}000$

$= 29{,}333.333 \text{ mi} \cdot \text{ton}$

$\approx 2.93 \times 10^4 \text{ mi} \cdot \text{ton}$

$\approx 3.10 \times 10^{11} \text{ ft} \cdot \text{lb}$

(b) $W = \int_{4000}^{26{,}000} \frac{160{,}000{,}000}{x^2}\,dx = \left[-\frac{160{,}000{,}000}{x}\right]_{4000}^{26{,}000} \approx -6{,}153.846 + 40{,}000$

$= 33{,}846.154 \text{ mi} \cdot \text{ton}$

$\approx 3.38 \times 10^4 \text{ mi} \cdot \text{ton}$

$\approx 3.57 \times 10^{11} \text{ ft} \cdot \text{lb}$

19. Weight of each layer: $62.4(20)\,\Delta y$

Distance: $4 - y$

(a) $W = \int_2^4 62.4(20)(4 - y)\,dy = \left[4992y - 624y^2\right]_2^4 = 2496 \text{ ft} \cdot \text{lb}$

(b) $W = \int_0^4 62.4(20)(4 - y)\,dy = \left[4992y - 624y^2\right]_0^4 = 9984 \text{ ft} \cdot \text{lb}$

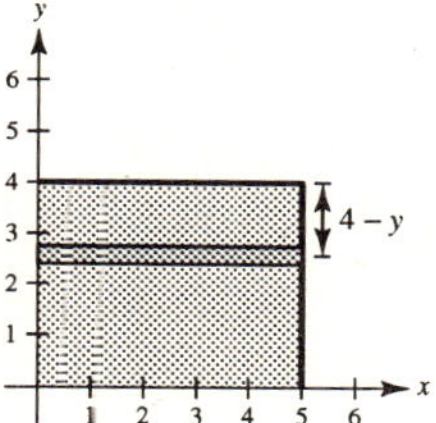

21. Volume of disc of water: $\pi(2)^2\,\Delta y = 4\pi\,\Delta y$

Weight of disc of water: $(9800)(4\pi\,\Delta y)$

Distance the disc of water is moved: $5 - y$

$$W = \int_0^4 (5 - y)(39{,}200\pi)\,dy = 39{,}200\pi \int_0^4 (5 - y)\,dy$$

$$= 39{,}200\pi\left[5y - \frac{y^2}{2}\right]_0^4 = 39{,}200\pi[12] = 470{,}400\pi \text{ Newton} \cdot \text{m}$$

23. Volume of disc: $\pi\left(\frac{2}{3}y\right)^2 \Delta y$

Weight of disc: $62.4\pi\left(\frac{2}{3}y\right)^2 \Delta y$

Distance: $6 - y$

$$W = \frac{4(62.4)\pi}{9}\int_0^6 (6 - y)y^2\,dy = \frac{4}{9}(62.4)\pi\left[2y^3 - \frac{1}{4}y^4\right]_0^6 = 2995.2\pi \text{ ft} \cdot \text{lb}$$

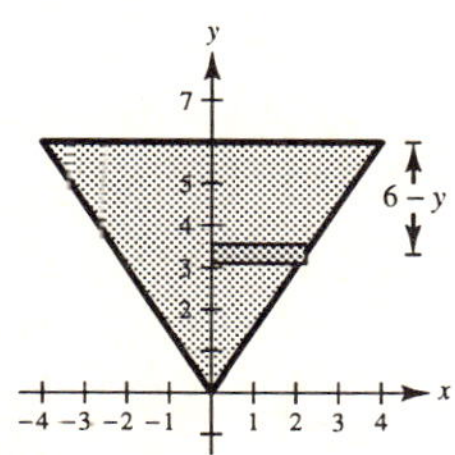

25. Volume of disc: $\pi\left(\sqrt{36 - y^2}\right)^2 \Delta y$

Weight of disc: $62.4\pi(36 - y^2)\, \Delta y$

Distance: y

$$W = 62.4\pi \int_0^6 y(36 - y^2)\, dy$$

$$= 62.4\pi \int_0^6 (36y - y^3)\, dy = 62.4\pi \left[18y^2 - \frac{1}{4}y^4\right]_0^6 = 20{,}217.6\pi \text{ ft} \cdot \text{lb}$$

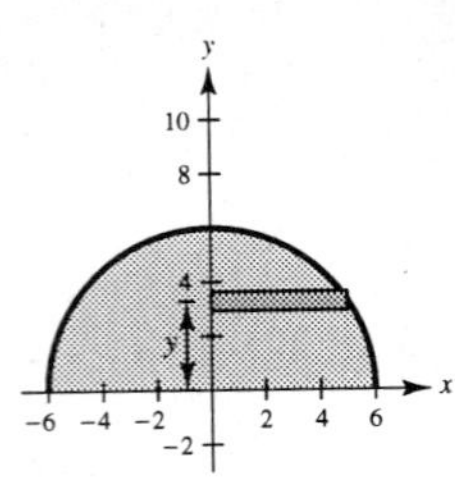

27. Volume of layer: $V = lwh = 4(2)\sqrt{(9/4) - y^2}\, \Delta y$

Weight of layer: $W = 42(8)\sqrt{(9/4) - y^2}\, \Delta y$

Distance: $\dfrac{13}{2} - y$

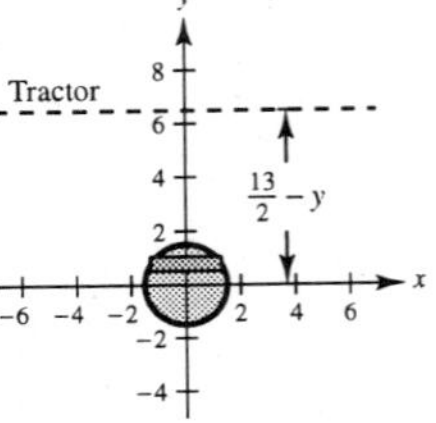

$$W = \int_{-1.5}^{1.5} 42(8)\sqrt{(9/4) - y^2}\left(\frac{13}{2} - y\right) dy$$

$$= 336\left[\frac{13}{2}\int_{-1.5}^{1.5} \sqrt{(9/4) - y^2}\, dy - \int_{-1.5}^{1.5} \sqrt{(9/4) - y^2}\, y\, dy\right]$$

The second integral is zero since the integrand is odd and the limits of integration are symmetric to the origin. The first integral represents the area of a semicircle of radius $\frac{3}{2}$. Thus, the work is

$$W = 336\left(\frac{13}{2}\right)\pi\left(\frac{3}{2}\right)^2\left(\frac{1}{2}\right) = 2457\pi \text{ ft} \cdot \text{lb}$$

29. Weight of section of chain: $3\, \Delta y$

Distance: $15 - y$

$$W = 3\int_0^{15} (15 - y)\, dy = \left[-\frac{3}{2}(15 - y)^2\right]_0^{15} = 337.5 \text{ ft} \cdot \text{lb}$$

31. The lower 5 feet of chain are raised 10 feet with a constant force.

$$W_1 = 3(5)(10) = 150 \text{ ft} \cdot \text{lb}$$

The top 10 feet of chain are raised with a variable force.

Weight per section: $3\, \Delta y$

Distance: $10 - y$

$$W_2 = 3\int_0^{10} (10 - y)\, dy$$

$$= \left[-\frac{3}{2}(10 - y)^2\right]_0^{10} = 150 \text{ ft} \cdot \text{lb}$$

$$W = W_1 + W_2 = 300 \text{ ft} \cdot \text{lb}$$

33. Weight of section of chain: $3\, \Delta y$

Distance: $15 - 2y$

$$W = 3\int_0^{7.5} (15 - 2y)\, dy$$

$$= \left[-\frac{3}{4}(15 - 2y)^2\right]_0^{7.5} = \frac{3}{4}(15)^2 = 168.75 \text{ ft} \cdot \text{lb}$$

35. Work to pull up the ball: $W_1 = 500(15) = 7500 \text{ ft} \cdot \text{lb}$

Work to wind up the top 15 feet of cable: force is variable

Weight per section: $1\, \Delta y$

Distance: $15 - x$

$$W_2 = \int_0^{15} (15 - x)\, dx = \left[-\frac{1}{2}(15 - x)^2\right]_0^{15}$$

$$= 112.5 \text{ ft} \cdot \text{lb}$$

Work to lift the lower 25 feet of cable with a constant force:

$$W_3 = (1)(25)(15) = 375 \text{ ft} \cdot \text{lb}$$

$$W = W_1 + W_2 + W_3 = 7500 + 112.5 + 375$$

$$= 7987.5 \text{ ft} \cdot \text{lb}$$

37. $p = \dfrac{k}{V}$

$1000 = \dfrac{k}{2}$

$k = 2000$

$W = \displaystyle\int_2^3 \frac{2000}{V}\,dV = \Big[2000 \ln |V|\Big]_2^3$

$= 2000 \ln\left(\dfrac{3}{2}\right) \approx 810.93 \text{ ft} \cdot \text{lb}$

39. $F(x) = \dfrac{k}{(2-x)^2}$

$W = \displaystyle\int_{-2}^{1} \frac{k}{(2-x)^2}\,dx$

$= \left[\dfrac{k}{2-x}\right]_{-2}^{1} = k\left(1 - \dfrac{1}{4}\right) = \dfrac{3k}{4}$ (units of work)

41. $W = \displaystyle\int_0^5 1000[1.8 - \ln(x+1)]\,dx \approx 3249.44$

43. $W = \displaystyle\int_0^5 100x\sqrt{125 - x^3}\,dx \approx 10{,}330.3 \text{ ft} \cdot \text{lb}$

Section 6.6 Moments, Centers of Mass, and Centroids

1. $\bar{x} = \dfrac{6(-5) + 3(1) + 5(3)}{6 + 3 + 5} = -\dfrac{6}{7}$

3. $\bar{x} = \dfrac{1(7) + 1(8) + 1(12) + 1(15) + 1(18)}{1 + 1 + 1 + 1 + 1} = 12$

5. (a) $\bar{x} = \dfrac{(7+5) + (8+5) + (12+5) + (15+5) + (18+5)}{5} = 17 = 12 + 5$

(b) $\bar{x} = \dfrac{12(-3-3) + 1(-2-3) + 6(-1-3) + 3(0-3) + 11(4-3)}{12 + 1 + 6 + 3 + 11} = -3 = 0 - 3$

7. $50x = 75(L - x) = 75(10 - x)$

$50x = 750 - 75x$

$125x = 750$

$x = 6$ feet

9. $\bar{x} = \dfrac{5(2) + 1(-3) + 3(1)}{5 + 1 + 3} = \dfrac{10}{9}$

$\bar{y} = \dfrac{5(2) + 1(1) + 3(-4)}{5 + 1 + 3} = -\dfrac{1}{9}$

$(\bar{x}, \bar{y}) = \left(\dfrac{10}{9}, -\dfrac{1}{9}\right)$

11. $\bar{x} = \dfrac{3(-2) + 4(-1) + 2(7) + 1(0) + 6(-3)}{3 + 4 + 2 + 1 + 6} = -\dfrac{7}{8}$

$\bar{y} = \dfrac{3(-3) + 4(0) + 2(1) + 1(0) + 6(0)}{3 + 4 + 2 + 1 + 6} = -\dfrac{7}{16}$

$(\bar{x}, \bar{y}) = \left(-\dfrac{7}{8}, -\dfrac{7}{16}\right)$

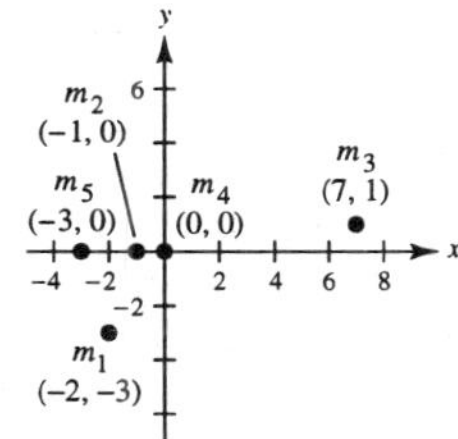

13. $m = \rho \int_0^4 \sqrt{x}\,dx = \left[\frac{2\rho}{3}x^{3/2}\right]_0^4 = \frac{16\rho}{3}$

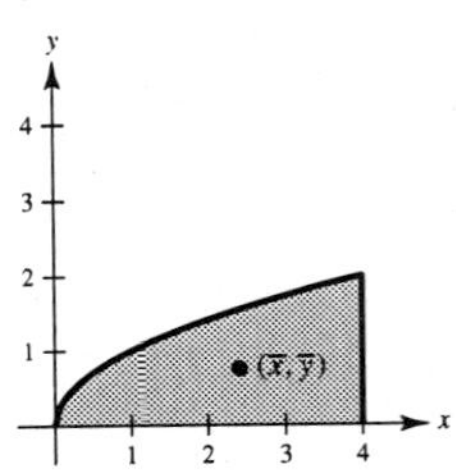

$M_x = \rho \int_0^4 \frac{\sqrt{x}}{2}(\sqrt{x})\,dx = \left[\rho\frac{x^2}{4}\right]_0^4 = 4\rho$

$\bar{y} = \frac{M_x}{m} = 4\rho\left(\frac{3}{16\rho}\right) = \frac{3}{4}$

$M_y = \rho \int_0^4 x\sqrt{x}\,dx = \left[\rho\frac{2}{5}x^{5/2}\right]_0^4 = \frac{64\rho}{5}$

$\bar{x} = \frac{M_y}{m} = \frac{64\rho}{5}\left(\frac{3}{16\rho}\right) = \frac{12}{5}$

$(\bar{x}, \bar{y}) = \left(\frac{12}{5}, \frac{3}{4}\right)$

15.

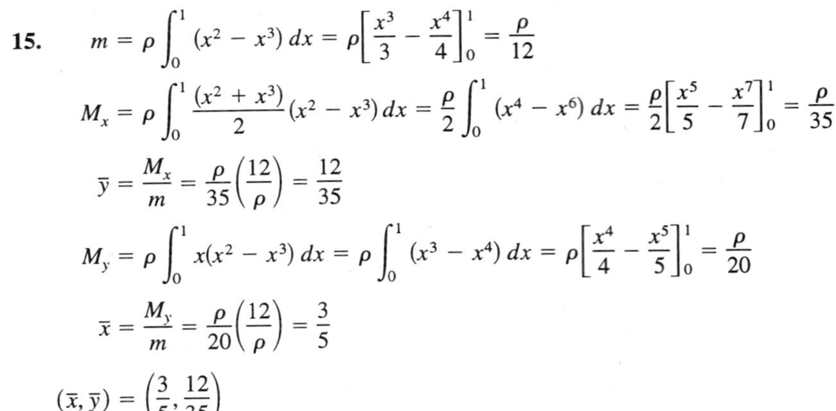

$m = \rho \int_0^1 (x^2 - x^3)\,dx = \rho\left[\frac{x^3}{3} - \frac{x^4}{4}\right]_0^1 = \frac{\rho}{12}$

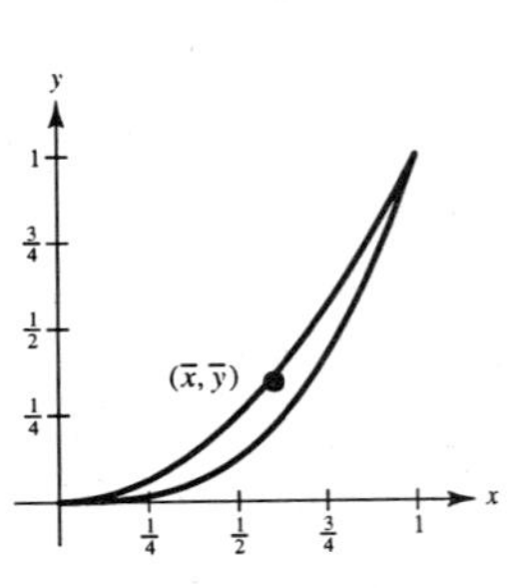

$M_x = \rho \int_0^1 \frac{(x^2 + x^3)}{2}(x^2 - x^3)\,dx = \frac{\rho}{2}\int_0^1 (x^4 - x^6)\,dx = \frac{\rho}{2}\left[\frac{x^5}{5} - \frac{x^7}{7}\right]_0^1 = \frac{\rho}{35}$

$\bar{y} = \frac{M_x}{m} = \frac{\rho}{35}\left(\frac{12}{\rho}\right) = \frac{12}{35}$

$M_y = \rho \int_0^1 x(x^2 - x^3)\,dx = \rho \int_0^1 (x^3 - x^4)\,dx = \rho\left[\frac{x^4}{4} - \frac{x^5}{5}\right]_0^1 = \frac{\rho}{20}$

$\bar{x} = \frac{M_y}{m} = \frac{\rho}{20}\left(\frac{12}{\rho}\right) = \frac{3}{5}$

$(\bar{x}, \bar{y}) = \left(\frac{3}{5}, \frac{12}{35}\right)$

17.

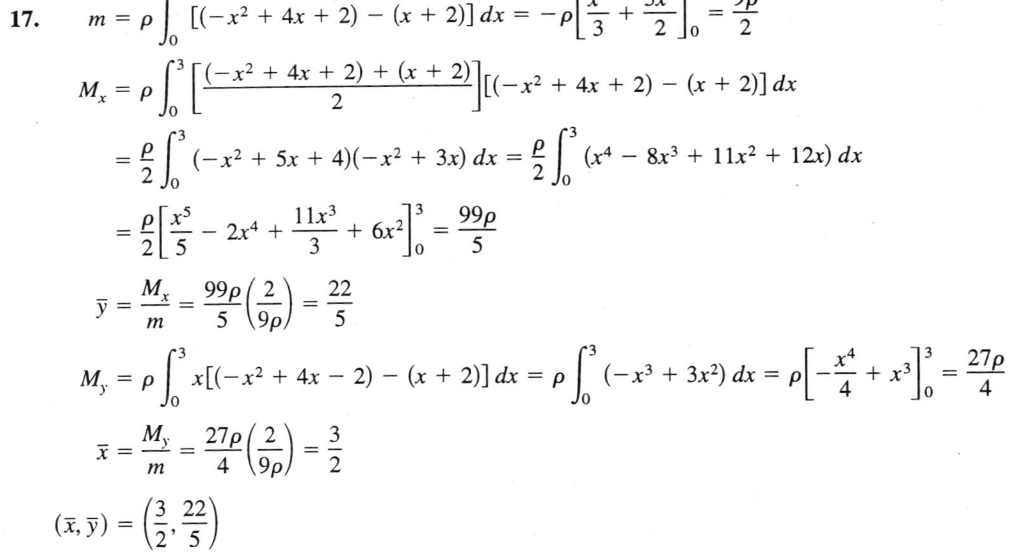

$m = \rho \int_0^3 [(-x^2 + 4x + 2) - (x + 2)]\,dx = -\rho\left[\frac{x^3}{3} + \frac{3x^2}{2}\right]_0^3 = \frac{9\rho}{2}$

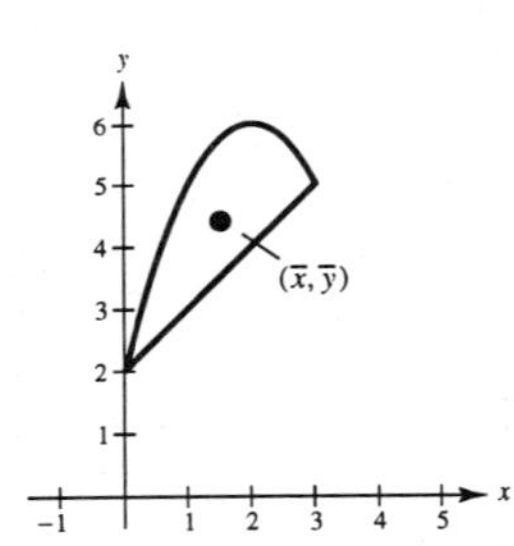

$M_x = \rho \int_0^3 \left[\frac{(-x^2 + 4x + 2) + (x + 2)}{2}\right][(-x^2 + 4x + 2) - (x + 2)]\,dx$

$= \frac{\rho}{2}\int_0^3 (-x^2 + 5x + 4)(-x^2 + 3x)\,dx = \frac{\rho}{2}\int_0^3 (x^4 - 8x^3 + 11x^2 + 12x)\,dx$

$= \frac{\rho}{2}\left[\frac{x^5}{5} - 2x^4 + \frac{11x^3}{3} + 6x^2\right]_0^3 = \frac{99\rho}{5}$

$\bar{y} = \frac{M_x}{m} = \frac{99\rho}{5}\left(\frac{2}{9\rho}\right) = \frac{22}{5}$

$M_y = \rho \int_0^3 x[(-x^2 + 4x - 2) - (x + 2)]\,dx = \rho \int_0^3 (-x^3 + 3x^2)\,dx = \rho\left[-\frac{x^4}{4} + x^3\right]_0^3 = \frac{27\rho}{4}$

$\bar{x} = \frac{M_y}{m} = \frac{27\rho}{4}\left(\frac{2}{9\rho}\right) = \frac{3}{2}$

$(\bar{x}, \bar{y}) = \left(\frac{3}{2}, \frac{22}{5}\right)$

19. $m = \rho\int_0^8 x^{2/3}\,dx = \rho\left[\frac{3}{5}x^{5/3}\right]_0^8 = \frac{96\rho}{5}$

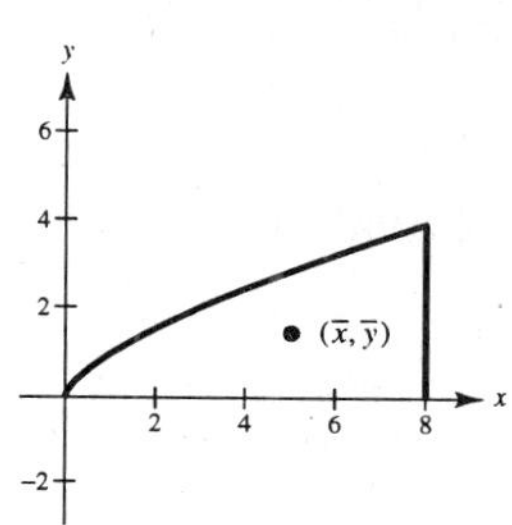

$M_x = \rho\int_0^8 \frac{x^{2/3}}{2}(x^{2/3})\,dx = \frac{\rho}{2}\left[\frac{3}{7}x^{7/3}\right]_0^8 = \frac{192\rho}{7}$

$\bar{y} = \frac{M_x}{m} = \frac{192\rho}{7}\left(\frac{5}{96\rho}\right) = \frac{10}{7}$

$M_y = \rho\int_0^8 x(x^{2/3})\,dx = \rho\left[\frac{3}{8}x^{8/3}\right]_0^8 = 96\rho$

$\bar{x} = \frac{M_y}{m} = 96\rho\left(\frac{5}{96\rho}\right) = 5$

$(\bar{x}, \bar{y}) = \left(5, \frac{10}{7}\right)$

21. 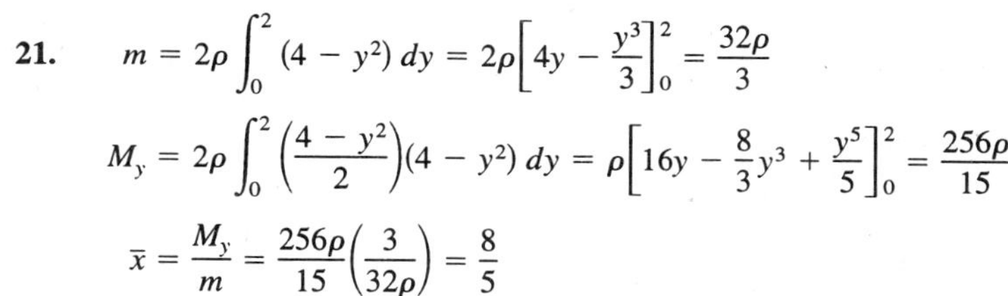

$m = 2\rho\int_0^2 (4 - y^2)\,dy = 2\rho\left[4y - \frac{y^3}{3}\right]_0^2 = \frac{32\rho}{3}$

$M_y = 2\rho\int_0^2 \left(\frac{4 - y^2}{2}\right)(4 - y^2)\,dy = \rho\left[16y - \frac{8}{3}y^3 + \frac{y^5}{5}\right]_0^2 = \frac{256\rho}{15}$

$\bar{x} = \frac{M_y}{m} = \frac{256\rho}{15}\left(\frac{3}{32\rho}\right) = \frac{8}{5}$

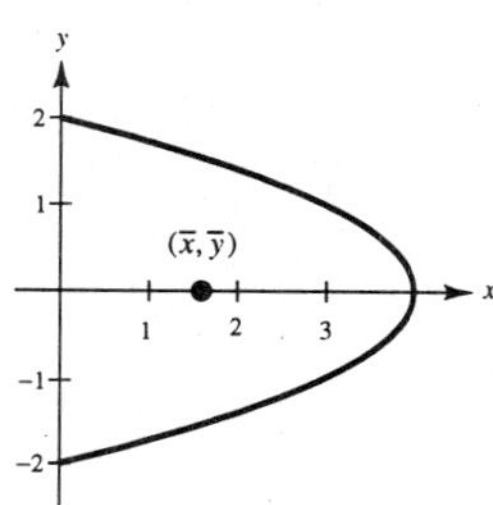

By symmetry, M_x and $\bar{y} = 0$.

$(\bar{x}, \bar{y}) = \left(\frac{8}{5}, 0\right)$

23. $m = \rho\int_0^3 [(2y - y^2) - (-y)]\,dy = \rho\left[\frac{3y^2}{2} - \frac{y^3}{3}\right]_0^3 = \frac{9\rho}{2}$

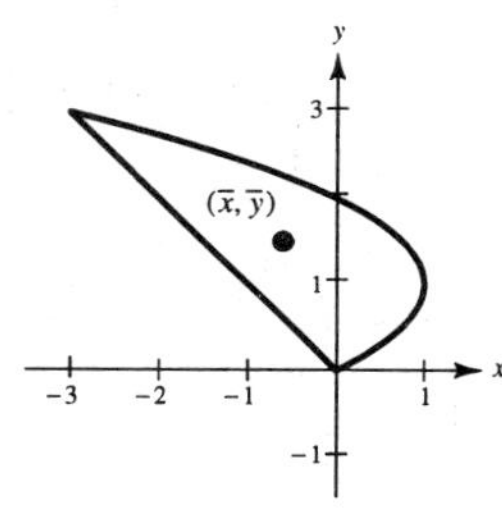

$M_y = \rho\int_0^3 \frac{[(2y - y^2) + (-y)]}{2}[(2y - y^2) - (-y)]\,dy = \frac{\rho}{2}\int_0^3 (y - y^2)(3y - y^2)\,dy$

$= \frac{\rho}{2}\int_0^3 (y^4 - 4y^3 + 3y^2)\,dy = \frac{\rho}{2}\left[\frac{y^5}{5} - y^4 + y^3\right]_0^3 = -\frac{27\rho}{10}$

$\bar{x} = \frac{M_y}{m} = -\frac{27\rho}{10}\left(\frac{2}{9\rho}\right) = -\frac{3}{5}$

$M_x = \rho\int_0^3 y[(2y - y^2) - (-y)]\,dy = \rho\int_0^3 (3y^2 - y^3)\,dy = \rho\left[y^3 - \frac{y^4}{4}\right]_0^3 = \frac{27\rho}{4}$

$\bar{y} = \frac{M_x}{m} = \frac{27\rho}{4}\left(\frac{2}{9\rho}\right) = \frac{3}{2}$

$(\bar{x}, \bar{y}) = \left(-\frac{3}{5}, \frac{3}{2}\right)$

25. $A = \int_0^1 (x - x^2)\,dx = \left[\frac{1}{2}x^2 - \frac{x^3}{3}\right]_0^1 = \frac{1}{6}$

$M_x = \frac{1}{2}\int_0^1 (x^2 - x^4)\,dx$

$= \frac{1}{2}\left[\frac{x^3}{3} - \frac{x^5}{5}\right]_0^1 = \frac{1}{2}\left(\frac{1}{3} - \frac{1}{5}\right) = \frac{1}{15}$

$M_y = \int_0^1 (x^2 - x^3)\,dx = \left[\frac{x^3}{3} - \frac{x^4}{4}\right]_0^1 = \left(\frac{1}{3} - \frac{1}{4}\right) = \frac{1}{12}$

27. $A = \int_0^3 (2x + 4)\,dx = \left[x^2 + 4x\right]_0^3 = 9 + 12 = 21$

$M_x = \frac{1}{2}\int_0^3 (2x + 4)^2\,dx = \int_0^3 (2x^2 + 8x + 8)\,dx$

$= \left[\frac{2x^3}{3} + 4x^2 + 8x\right]_0^3 = 18 + 36 + 24 = 78$

$M_y = \int_0^3 (2x^2 + 4x)\,dx = \left[\frac{2x^3}{3} + 2x^2\right]_0^3 = 18 + 18 = 36$

29. $m = \rho \int_0^5 10x\sqrt{125 - x^3}\, dx \approx 1033.0\rho$

$$M_x = \rho \int_0^5 \left(\frac{10x\sqrt{125 - x^3}}{2}\right)\left(10x\sqrt{125 - x^3}\right) dx = 50\rho \int_0^5 x^2(125 - x^3)\, dx = \frac{3{,}124{,}375\rho}{24} \approx 130{,}208\rho$$

$$M_y = \rho \int_0^5 10x^2\sqrt{125 - x^3}\, dx = -\frac{10\rho}{3}\int_0^5 \sqrt{125 - x^3}(-3x^2)\, dx = \frac{12{,}500\sqrt{5}\rho}{9} \approx 3105.6\rho$$

$$\bar{x} = \frac{M_y}{m} \approx 3.0$$

$$\bar{y} = \frac{M_x}{m} \approx 126.0$$

Therefore, the centroid is (3.0, 126.0).

31. $m = \rho \int_{-20}^{20} 5\sqrt[3]{400 - x^2}\, dx \approx 1239.76\rho$

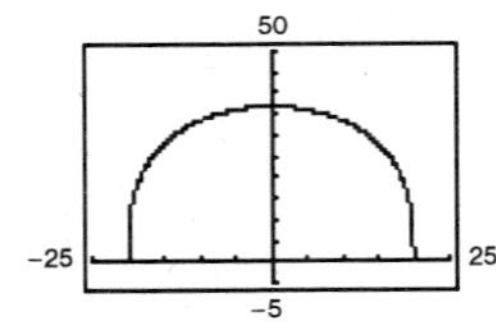

$$M_x = \rho \int_{-20}^{20} \frac{5\sqrt[3]{400 - x^2}}{2}\left(5\sqrt[3]{400 - x^2}\right) dx$$

$$= \frac{25\rho}{2}\int_{-20}^{20} (400 - x^2)^{2/3}\, dx \approx 20064.27$$

$$\bar{y} = \frac{M_x}{m} \approx 16.18$$

$\bar{x} = 0$ by symmetry. Therefore, the centroid is (0, 16.2).

33. $A = \frac{1}{2}(2a)c = ac$

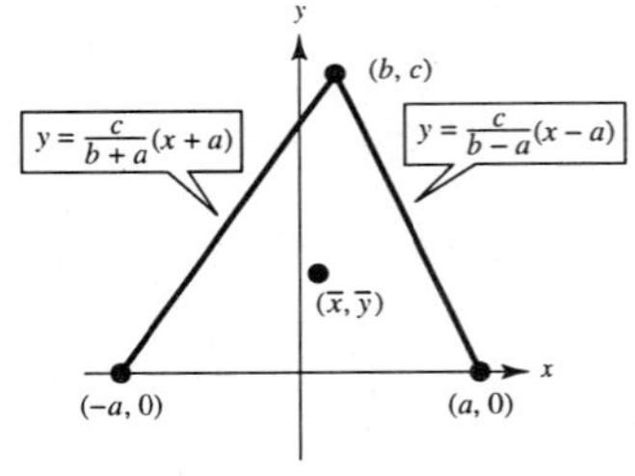

$$\frac{1}{A} = \frac{1}{ac}$$

$$\bar{x} = \left(\frac{1}{ac}\right)\frac{1}{2}\int_0^c \left[\left(\frac{b - a}{c}y + a\right)^2 - \left(\frac{b + a}{c}y - a\right)^2\right] dy$$

$$= \frac{1}{2ac}\int_0^c \left[\frac{4ab}{c}y - \frac{4ab}{c^2}y^2\right] dy$$

$$= \frac{1}{2ac}\left[\frac{2ab}{c}y^2 - \frac{4ab}{3c^2}y^3\right]_0^c = \frac{1}{2ac}\left(\frac{2}{3}abc\right) = \frac{b}{3}$$

$$\bar{y} = \frac{1}{ac}\int_0^c y\left[\left(\frac{b - a}{c}y + a\right) - \left(\frac{b + a}{c}y - a\right)\right] dy$$

$$= \frac{1}{ac}\int_0^c y\left(-\frac{2a}{c}y + 2a\right) dy = \frac{2}{c}\int_0^c \left(y - \frac{y^2}{c}\right) dy$$

$$= \frac{2}{c}\left[\frac{y^2}{2} - \frac{y^3}{3c}\right]_0^c = \frac{c}{3}$$

$$(\bar{x}, \bar{y}) = \left(\frac{b}{3}, \frac{c}{3}\right)$$

In Exercise 566 of Section P.2, you found that $(b/3, c/3)$ is the point of intersection of the medians.

35. $A = \frac{c}{2}(a + b)$

$$\frac{1}{A} = \frac{2}{c(a + b)}$$

$$\bar{x} = \frac{2}{c(a + b)}\int_0^c x\left(\frac{b - a}{c}x + a\right)dx = \frac{2}{c(a + b)}\int_0^c \left(\frac{b - a}{c}x^2 + ax\right)dx = \frac{2}{c(a + b)}\left[\frac{b - a}{c}\frac{x^3}{3} + \frac{ax^2}{2}\right]_0^c$$

$$= \frac{2}{c(a + b)}\left[\frac{(b - a)c^2}{3} + \frac{ac^2}{2}\right] = \frac{2}{c(a + b)}\left[\frac{2bc^2 - 2ac^2 + 3ac^2}{6}\right] = \frac{c(2b + a)}{3(a + b)} = \frac{(a + 2b)c}{3(a + b)}$$

$$\bar{y} = \frac{2}{c(a + b)}\frac{1}{2}\int_0^c \left(\frac{b - a}{c}x + a\right)^2 dx = \frac{1}{c(a + b)}\int_0^c \left[\left(\frac{b - a}{c}\right)^2 x^2 + \frac{2a(b - a)}{c}x + a^2\right]dx$$

$$= \frac{1}{c(a + b)}\left[\left(\frac{b - a}{c}\right)^2\frac{x^3}{3} + \frac{2a(b - a)}{c}\frac{x^2}{2} + a^2x\right]_0^c = \frac{1}{c(a + b)}\left[\frac{(b - a)^2c}{3} + ac(b - a) + a^2c\right]$$

$$= \frac{1}{3c(a + b)}[(b^2 - 2ab + a^2)c + 3ac(b - a) + 3a^2c]$$

$$= \frac{1}{3(a + b)}[b^2 - 2ab + a^2 + 3ab - 3a^2 + 3a^2] = \frac{a^2 + ab + b^2}{3(a + b)}$$

Thus,

$$(\bar{x}, \bar{y}) = \left(\frac{(a + 2b)c}{3(a + b)}, \frac{a^2 + ab + b^2}{3(a + b)}\right).$$

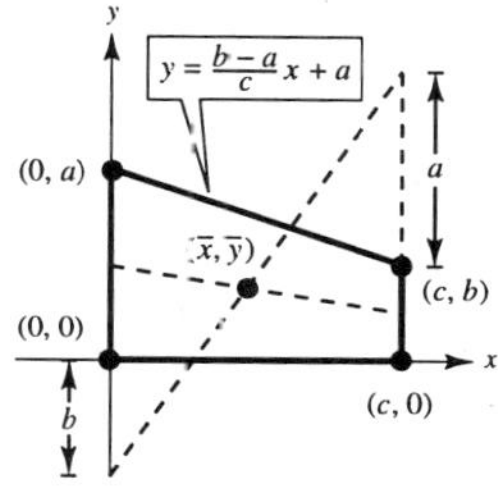

The one line passes through $(0, a/2)$ and $(c, b/2)$. It's equation is

$$y = \frac{b - a}{2c}x + \frac{a}{2}.$$

The other line passes through $(0, -b)$ and $(c, a + b)$. It's equation is

$$y = \frac{a + 2b}{c}x - b.$$

$(\bar{x}, \bar{y})$ is the point of intersection of these two lines.

37. $\bar{x} = 0$ by symmetry

$$A = \frac{1}{2}\pi ab$$

$$\frac{1}{A} = \frac{2}{\pi ab}$$

$$\bar{y} = \frac{2}{\pi ab}\frac{1}{2}\int_{-a}^a \left(\frac{b}{a}\sqrt{a^2 - x^2}\right)^2 dx = \frac{1}{\pi ab}\left(\frac{b^2}{a^2}\right)\left[a^2x - \frac{x^3}{3}\right]_{-a}^a = \frac{b}{\pi a^3}\left[\frac{4a^3}{3}\right] = \frac{4b}{3\pi}$$

$$(\bar{x}, \bar{y}) = \left(0, \frac{4b}{3\pi}\right)$$

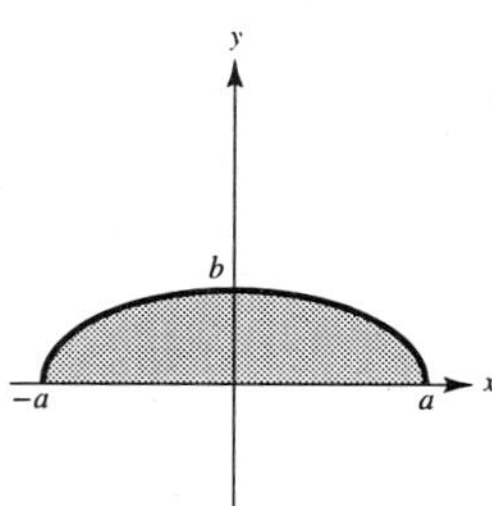

39. (a)

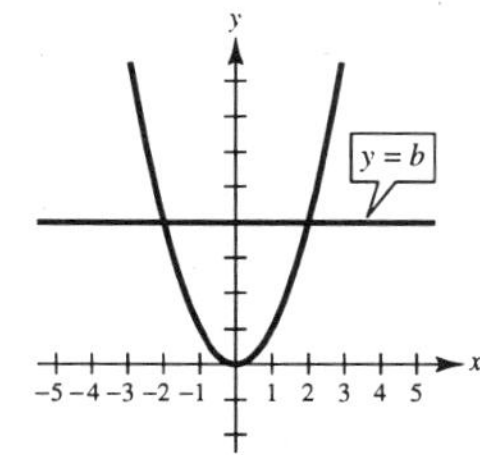

(b) $\bar{x} = 0$ by symmetry

(c) $M_y = \int_{-\sqrt{b}}^{\sqrt{b}} x(b - x^2)\,dx = 0$ because $bx - x^3$ is odd

(d) $\bar{y} > \frac{b}{2}$ since there is more area above $y = \frac{b}{2}$ than below

—CONTINUED—

39. —CONTINUED—

(e) $M_x = \int_{-\sqrt{b}}^{\sqrt{b}} \frac{(b + x^2)(b - x^2)}{2}\,dx = \int_{-\sqrt{b}}^{\sqrt{b}} \frac{b^2 - x^4}{2}\,dx = \frac{1}{2}\left[b^2x - \frac{x^5}{5}\right]_{-\sqrt{b}}^{\sqrt{b}} = b^2\sqrt{b} - \frac{b^2\sqrt{b}}{5} = \frac{4b^2\sqrt{b}}{5}$

$A = \int_{-\sqrt{b}}^{\sqrt{b}} (b - x^2)\,dx = \left[bx - \frac{x^3}{3}\right]_{-\sqrt{b}}^{\sqrt{b}} = \left(b\sqrt{b} - \frac{b\sqrt{b}}{3}\right)2 = 4\frac{b\sqrt{b}}{3}$

$\bar{y} = \frac{M_x}{A} = \frac{4b^2\sqrt{b}/5}{4b\sqrt{b}/3} = \frac{3}{5}b.$

41. (a) $\bar{x} = 0$ by symmetry

$A = 2\int_0^{40} f(x)\,dx = \frac{2(40)}{3(4)}[30 + 4(29) + 2(26) + 4(20) + 0] = \frac{20}{3}(278) = \frac{5560}{3}$

$M_x = \int_{-40}^{40} \frac{f(x)^2}{2}\,dx = \frac{40}{3(4)}[30^2 + 4(29)^2 + 2(26)^2 + 4(20)^2 + 0] = \frac{10}{3}(7216) = \frac{72160}{3}$

$\bar{y} = \frac{M_x}{A} = \frac{72160/3}{5560/3} = \frac{72160}{5560} \approx 12.98$

$(\bar{x}, \bar{y}) = (0, 12.98)$

(b) $y = (-1.02 \times 10^{-5})x^4 - 0.0019x^2 + 29.28$

(c) $\bar{y} = \frac{M_x}{A} \approx \frac{23697.68}{1843.54} \approx 12.85$

$(\bar{x}, \bar{y}) = (0, 12.85)$

43. Centroids of the given regions: $(1, 0)$ and $(3, 0)$

Area: $A = 4 + \pi$

$\bar{x} = \frac{4(1) + \pi(3)}{4 + \pi} = \frac{4 + 3\pi}{4 + \pi}$

$\bar{y} = \frac{4(0) + \pi(0)}{4 + \pi} = 0$

$(\bar{x}, \bar{y}) = \left(\frac{4 + 3\pi}{4 + \pi}, 0\right) \approx (1.88, 0)$

45. Centroids of the given regions: $\left(0, \frac{3}{2}\right)$, $(0, 5)$, and $\left(0, \frac{15}{2}\right)$

Area: $A = 15 + 12 + 7 = 34$

$\bar{x} = \frac{15(0) + 12(0) + 7(0)}{34} = 0$

$\bar{y} = \frac{15(3/2) + 12(5) + 7(15/2)}{34} = \frac{135}{34}$

$(\bar{x}, \bar{y}) = \left(0, \frac{135}{34}\right)$

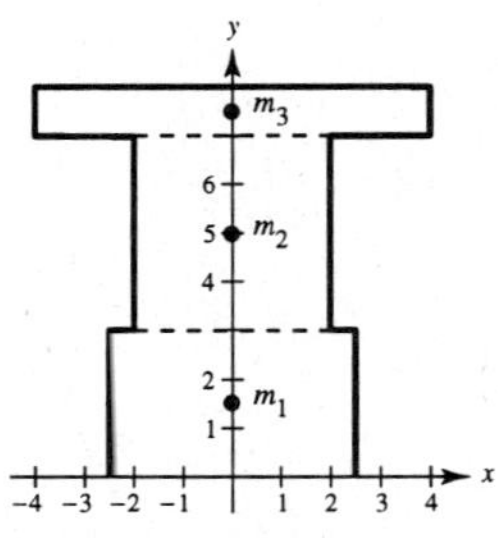

47. Centroids of the given regions: $(1, 0)$ and $(3, 0)$

Mass: $4 + 2\pi$

$\bar{x} = \frac{4(1) + 2\pi(3)}{4 + 2\pi} = \frac{2 + 3\pi}{2 + \pi}$

$\bar{y} = 0$

$(\bar{x}, \bar{y}) = \left(\frac{2 + 3\pi}{2 + \pi}, 0\right) \approx (2.22, 0)$

49. $V = 2\pi rA = 2\pi(5)(16\pi) = 160\pi^2 \approx 1579.14$

51. $A = \frac{1}{2}(4)(4) = 8$

$$\bar{y} = \left(\frac{1}{8}\right)\frac{1}{2}\int_0^4 (4 + x)(4 - x)\,dx = \frac{1}{16}\left[16x - \frac{x^3}{3}\right]_0^4 = \frac{8}{3}$$

$$r = \bar{y} = \frac{8}{3}$$

$$V = 2\pi rA = 2\pi\left(\frac{8}{3}\right)(8) = \frac{128\pi}{3} \approx 134.04$$

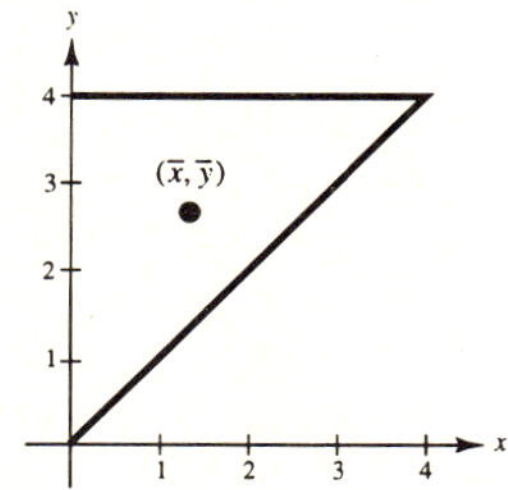

53. The surface area of the sphere is $S = 4\pi r^2$. The arc length of C is $s = \pi r$. The distance traveled by the centroid is

$$d = \frac{S}{s} = \frac{4\pi r^2}{\pi r} = 4r.$$

This distance is also the circumference of the circle of radius y.

$$d = 2\pi y$$

Thus, $2\pi y = 4r$ and we have $y = 2r/\pi$. Therefore, the centroid of the semicircle $y = \sqrt{r^2 - x^2}$ is $(0, 2r/\pi)$.

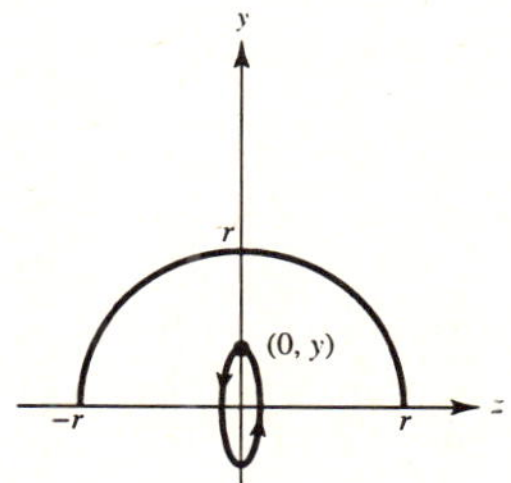

55. $A = \int_0^1 x^n\,dx = \left[\frac{x^{n+1}}{n+1}\right]_0^1 = \frac{1}{n+1}$

$$m = \rho A = \frac{\rho}{n+1}$$

$$M_x = \frac{\rho}{2}\int_0^1 (x^n)^2\,dx = \left[\frac{\rho}{2}\cdot\frac{x^{2n+1}}{2n+1}\right]_0^1 = \frac{\rho}{2(2n+1)}$$

$$M_y = \rho\int_0^1 x(x^n)\,dx = \left[\rho\cdot\frac{x^{n+2}}{n+2}\right]_0^1 = \frac{\rho}{n+2}$$

$$\bar{x} = \frac{M_y}{m} = \frac{n+1}{n+2}$$

$$\bar{y} = \frac{M_x}{m} = \frac{n+1}{2(2n+1)} = \frac{n+1}{4n+2}$$

Centroid: $\left(\frac{n+1}{n+2}, \frac{n+1}{4n+2}\right)$

As $n \to \infty$, $(\bar{x}, \bar{y}) \to \left(1, \frac{1}{4}\right)$. The graph approaches the x-axis and the line $x = 1$ as $n \to \infty$.

Section 6.7 Fluid Pressure and Fluid Force

1. $F = PA = [62.4(5)](3) = 936$ lb

3. $F = 62.4(h + 2)(6) - (62.4)(h)(6)$

$= 62.4(2)(6) = 748.8$ lb

5. $h(y) = 3 - y$

$L(y) = 4$

$$F = 62.4\int_0^3 (3 - y)(4)\,dy = 249.6\int_0^3 (3 - y)\,dy$$

$$= 249.6\left[3y - \frac{y^2}{2}\right]_0^3 = 1123.2 \text{ lb}$$

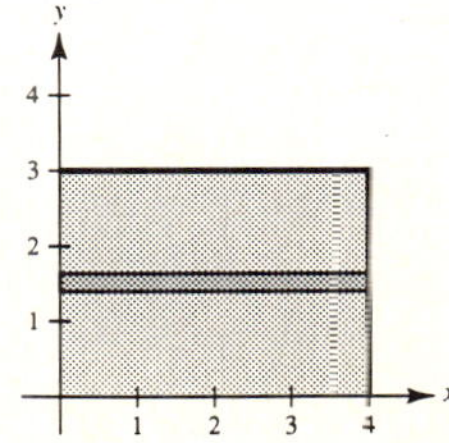

7. $h(y) = 3 - y$

$L(y) = 2\left(\frac{y}{3} + 1\right)$

$$F = 2(62.4)\int_0^3 (3 - y)\left(\frac{y}{3} + 1\right) dy$$

$$= 124.8\int_0^3 \left(3 - \frac{y^2}{3}\right) dy$$

$$= 124.8\left[3y - \frac{y^3}{9}\right]_0^3 = 748.8 \text{ lb}$$

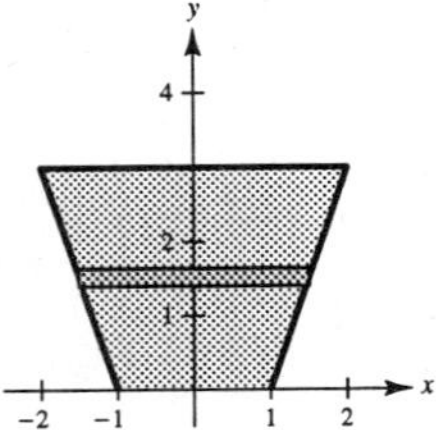

9. $h(y) = 4 - y$

$L(y) = 2\sqrt{y}$

$$F = 2(62.4)\int_0^4 (4 - y)\sqrt{y}\, dy$$

$$= 124.8\int_0^4 (4y^{1/2} - y^{3/2})\, dy$$

$$= 124.8\left[\frac{8y^{3/2}}{3} - \frac{2y^{5/2}}{5}\right]_0^4 = 1064.96 \text{ lb}$$

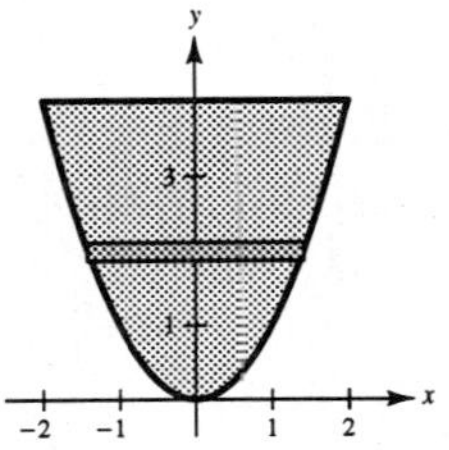

11. $h(y) = 4 - y$

$L(y) = 2$

$$F = 1000\int_0^2 2(4 - y)\, dy$$

$$= 1000\left[8y - y^2\right]_0^2 = 12{,}000 \text{ Newtons}$$

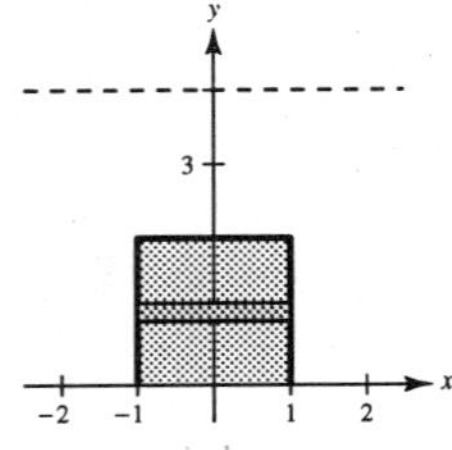

13. $h(y) = 12 - y$

$L(y) = 6 - \frac{2y}{3}$

$$F = 1000\int_0^9 (12 - y)\left(6 - \frac{2y}{3}\right) dy$$

$$= 1000\left[72y - 7y^2 + \frac{2y^3}{9}\right]_0^9 = 243{,}000 \text{ Newtons}$$

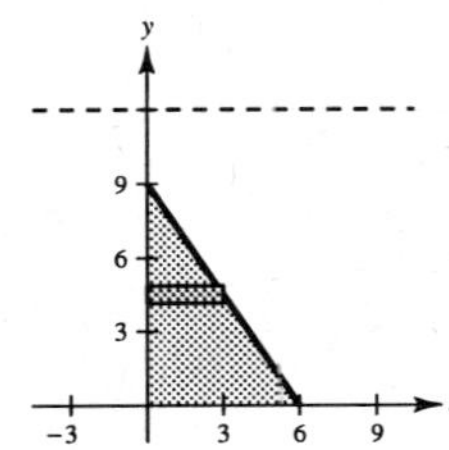

15. $h(y) = 2 - y$

$L(y) = 10$

$$F = 140.7\int_0^2 (2 - y)(10)\, dy$$

$$= 1407\int_0^2 (2 - y)\, dy$$

$$= 1407\left[2y - \frac{y^2}{2}\right]_0^2 = 2814 \text{ lb}$$

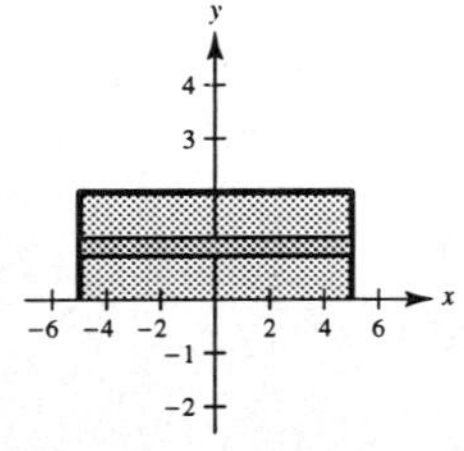

17. $h(y) = 4 - y$

$L(y) = 6$

$$F = 140.7\int_0^4 (4 - y)(6)\, dy$$

$$= 844.2\int_0^4 (4 - y)\, dy$$

$$= 844.2\left[4y - \frac{y^2}{2}\right]_0^4 = 6753.6 \text{ lb}$$

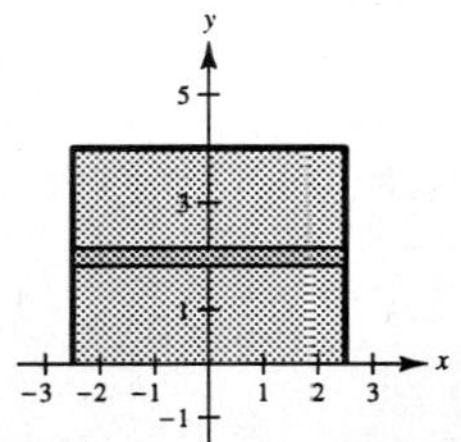

19. $h(y) = -y$

$$L(y) = 2\left(\frac{1}{2}\right)\sqrt{9 - 4y^2}$$

$$F = 42\int_{-3/2}^{0} (-y)\sqrt{9 - 4y^2}\,dy$$

$$= \frac{42}{8}\int_{-3/2}^{0} (9 - 4y^2)^{1/2}(-8y)\,dy$$

$$= \left[\left(\frac{21}{4}\right)\left(\frac{2}{3}\right)(9 - 4y^2)^{3/2}\right]_{-3/2}^{0} = 94.5 \text{ lb}$$

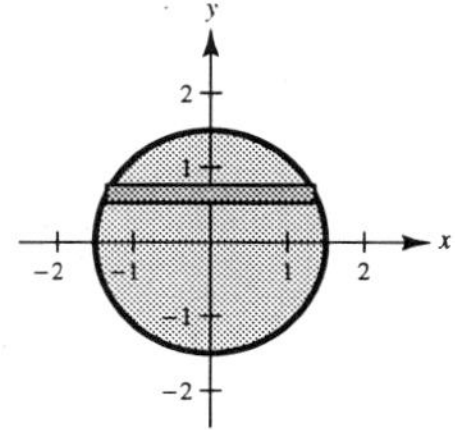

21. $h(y) = k - y$

$$L(y) = 2\sqrt{r^2 - y^2}$$

$$F = w\int_{-r}^{r} (k - y)\sqrt{r^2 - y^2}\,(2)\,dy$$

$$= w\left[2k\int_{-r}^{r} \sqrt{r^2 - y^2}\,dy + \int_{-r}^{r} \sqrt{r^2 - y^2}\,(-2y)\,dy\right]$$

The second integral is zero since its integrand is odd and the limits of integration are symmetric to the origin. The first integral is the area of a semicircle with radius r.

$$F = w\left[(2k)\frac{\pi r^2}{2} + 0\right] = wk\pi r^2$$

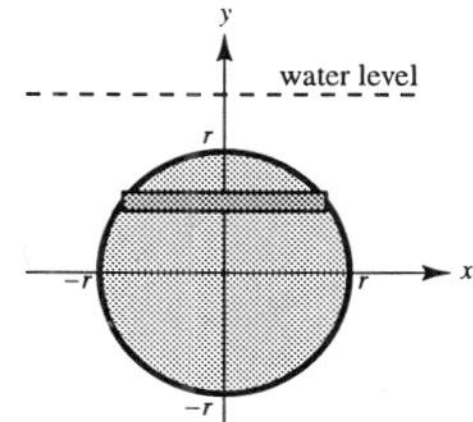

23. From Exercise 22:

$$F = 64(15)(1)(1) = 960 \text{ lb}$$

25. $h(y) = 4 - y$

$$F = 62.4\int_{0}^{4} (4 - y)L(y)\,dy$$

Using Simpson's Rule with $n = 8$ we have:

$$F \approx 62.4\left(\frac{4 - 0}{3(8)}\right)[0 + 4(3.5)(3) + 2(3)(5) + 4(2.5)(8) + 2(2)(9) + 4(1.5)(10) + 2(1)(10.25) + 4(0.5)(10.5) + 0]$$

$$= 3010.8 \text{ lb}$$

27. $h(y) = 12 - y$

$$L(y) = 2(4^{2/3} - y^{2/3})^{3/2}$$

$$F = 62.4\int_{0}^{4} 2(12 - y)(4^{2/3} - y^{2/3})^{3/2}\,dy$$

$$\approx 6448.73 \text{ lb}$$

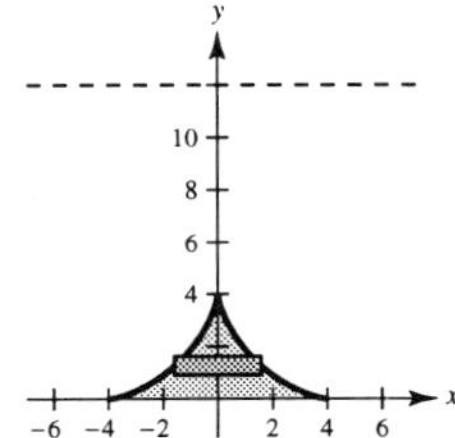

29. (a) If the fluid force is one half of 1123.2 lb, and the height of the water is b, then

$$h(y) = b - y$$

$$L(y) = 4$$

$$F = 62.4\int_{0}^{b} (b - y)(4)\,dy = \frac{1}{2}(1123.2)$$

$$\int_{0}^{b} (b - y)\,dy = 2.25$$

$$\left[by - \frac{y^2}{2}\right]_{0}^{b} = 2.25$$

$$b^2 - \frac{b^2}{2} = 2.25$$

$$b^2 = 4.5 \implies b \approx 2.12 \text{ ft.}$$

(b) The pressure increases with increasing depth.

Review Exercises for Chapter 6

1. $A = \int_1^5 \frac{1}{x^2}\,dx = \left[-\frac{1}{x}\right]_1^5 = \frac{4}{5}$

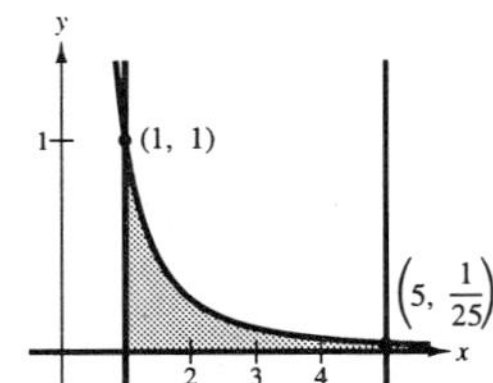

3. $A = \int_{-1}^1 \frac{1}{x^2+1}\,dx$

$= \left[\arctan x\right]_{-1}^1$

$= \frac{\pi}{4} - \left(-\frac{\pi}{4}\right) = \frac{\pi}{2}$

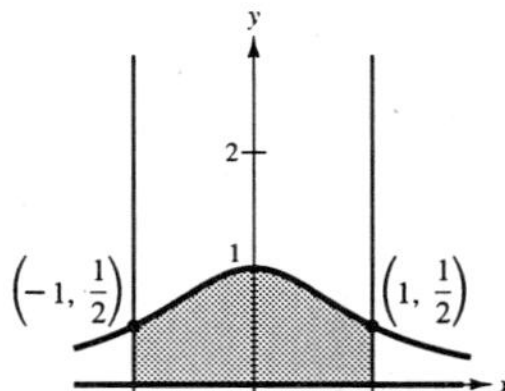

5. $A = 2\int_0^1 (x - x^3)\,dx$

$= 2\left[\frac{1}{2}x^2 - \frac{1}{4}x^4\right]_0^1$

$= \frac{1}{2}$

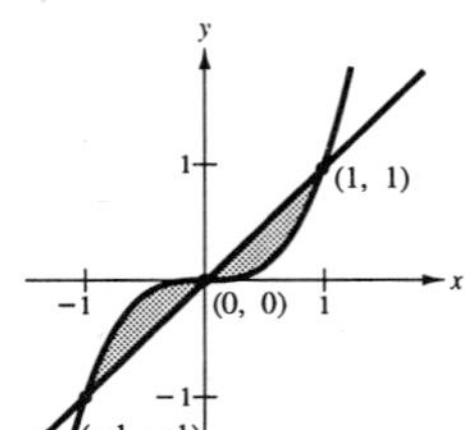

7. $A = \int_0^2 (e^2 - e^x)\,dx$

$= \left[xe^2 - e^x\right]_0^2$

$= e^2 + 1$

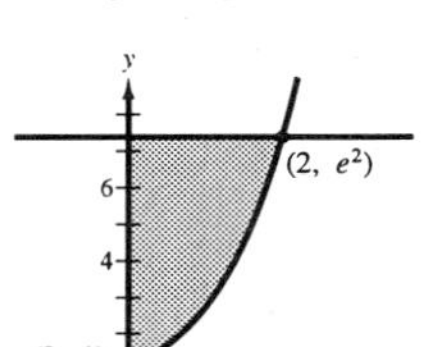

9. $A = \int_{\pi/4}^{5\pi/4} (\sin x - \cos x)\,dx$

$= \left[-\cos x - \sin x\right]_{\pi/4}^{5\pi/4}$

$= \left(\frac{1}{\sqrt{2}} + \frac{1}{\sqrt{2}}\right) - \left(-\frac{1}{\sqrt{2}} - \frac{1}{\sqrt{2}}\right) = \frac{4}{\sqrt{2}} = 2\sqrt{2}$

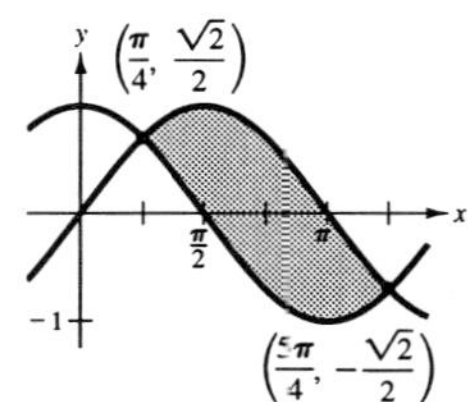

11. $A = \int_0^8 [(3 + 8x - x^2) - (x^2 - 8x + 3)]\,dx$

$= \int_0^8 (16x - 2x^2)\,dx$

$= \left[8x^2 - \frac{2}{3}x^3\right]_0^8 = \frac{512}{3} \approx 170.667$

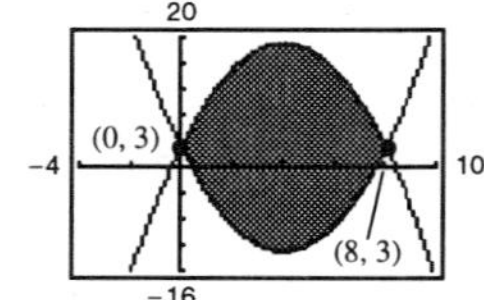

13. $y = (1 - \sqrt{x})^2$

$A = \int_0^1 (1 - \sqrt{x})^2\,dx = \int_0^1 (1 - 2x^{1/2} + x)\,dx$

$= \left[x - \frac{4}{3}x^{3/2} + \frac{1}{2}x^2\right]_0^1 = \frac{1}{6} \approx 0.1667$

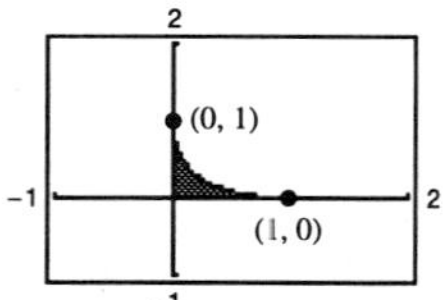

15. $x = y^2 - 2y \Rightarrow x + 1 = (y - 1)^2 \Rightarrow y = 1 \pm \sqrt{x+1}$

$A = \int_{-1}^0 \left[\left(1 + \sqrt{x+1}\right) - \left(1 - \sqrt{x+1}\right)\right]dx = \int_{-1}^0 2\sqrt{x+1}\,dx$

$A = \int_0^2 [0 - (y^2 - 2y)]\,dy = \int_0^2 (2y - y^2)\,dy = \left[y^2 - \frac{1}{3}y^3\right]_0^2 = \frac{4}{3}$

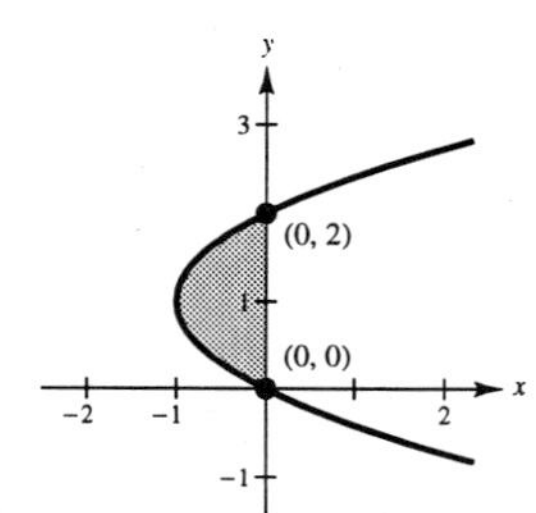

17. $A = \int_0^2 \left[1 - \left(1 - \frac{x}{2}\right)\right] dx + \int_2^3 [1 - (x - 2)]\, dx = \int_0^2 \frac{x}{2}\, dx + \int_2^3 (3 - x)\, dx$

$y = 1 - \frac{x}{2} \Rightarrow x = 2 - 2y$

$y = x - 2 \Rightarrow x = y + 2,\ y = 1$

$A = \int_0^1 [(y + 2) - (2 - 2y)]\, dy = \int_0^1 3y\, dy = \left[\frac{3}{2}y^2\right]_0^1 = \frac{3}{2}$

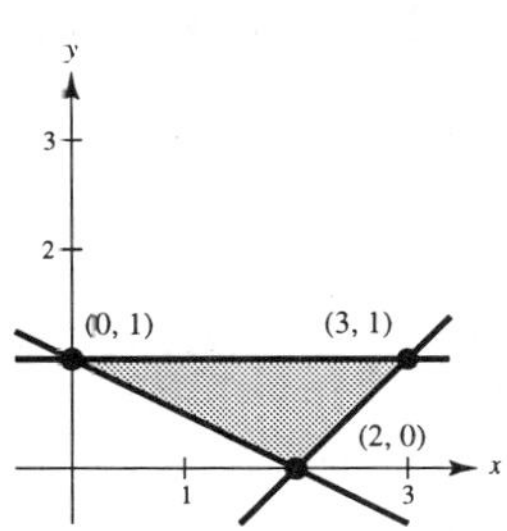

19. Job 1 is better. The salary for Job 1 is greater than the salary for Job 2 for all the years except the first and 10th years.

21. (a) **Disc**

$V = \pi \int_0^4 x^2\, dx = \left[\frac{\pi x^3}{3}\right]_0^4 = \frac{64\pi}{3}$

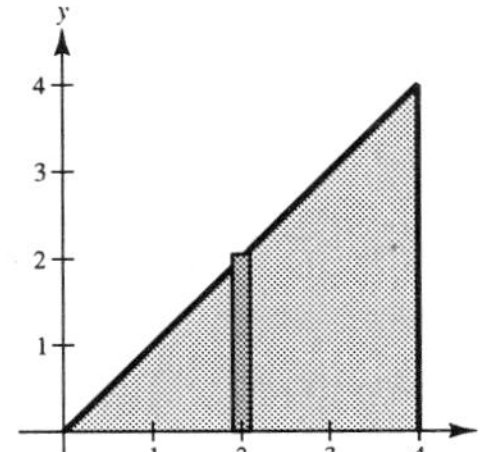

(b) **Shell**

$V = 2\pi \int_0^4 x^2\, dx = \left[\frac{2\pi}{3}x^3\right]_0^4 = \frac{128\pi}{3}$

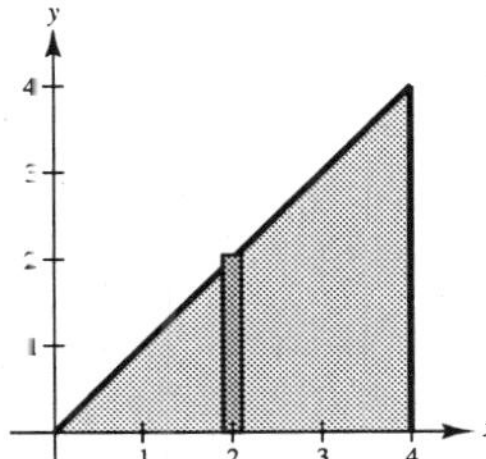

(c) **Shell**

$$V = 2\pi \int_0^4 (4 - x)x\, dx$$

$$= 2\pi \int_0^4 (4x - x^2)\, dx$$

$$= 2\pi\left[2x^2 - \frac{x^3}{3}\right]_0^4 = \frac{64\pi}{3}$$

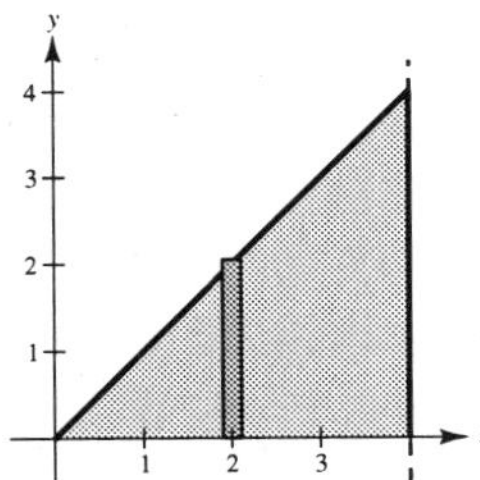

(d) **Shell**

$$V = 2\pi \int_0^4 (6 - x)x\, dx$$

$$= 2\pi \int_0^4 (6x - x^2)\, dx$$

$$= 2\pi\left[3x^2 - \frac{1}{3}x^3\right]_0^4 = \frac{160\pi}{3}$$

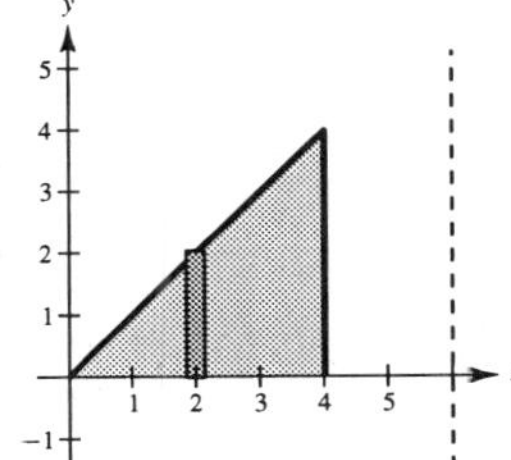

23. (a) **Shell**

$$V = 4\pi \int_0^4 x\left(\frac{3}{4}\right)\sqrt{16 - x^2}\, dx$$

$$= \left[3\pi\left(-\frac{1}{2}\right)\left(\frac{2}{3}\right)(16 - x^2)^{3/2}\right]_0^4 = 64\pi$$

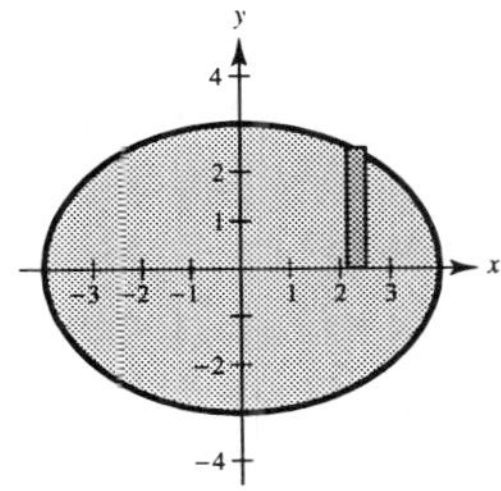

—CONTINUED—

23. —CONTINUED—

(b) **Disc**

$$V = 2\pi\int_0^4 \left[\frac{3}{4}\sqrt{16 - x^2}\right]^2 dx$$

$$= \frac{9\pi}{8}\left[16x - \frac{x^3}{3}\right]_0^4 = 48\pi$$

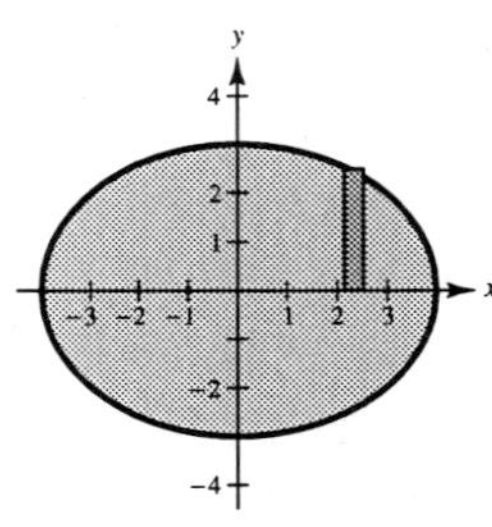

25. Shell

$$V = 2\pi\int_0^1 \frac{x}{x^4 + 1}\,dx = \pi\int_0^1 \frac{(2x)}{(x^2)^2 + 1}\,dx$$

$$= \left[\pi\arctan(x^2)\right]_0^1$$

$$= \pi\left[\frac{\pi}{4} - 0\right] = \frac{\pi^2}{4}$$

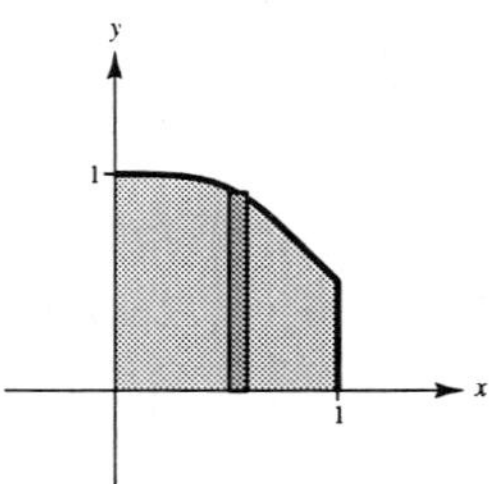

27. Shell

$u = \sqrt{x - 2}$

$x = u^2 + 2$

$dx = 2u\,du$

$$V = 2\pi\int_2^6 \frac{x}{1 + \sqrt{x - 2}}\,dx = 4\pi\int_0^2 \frac{(u^2 + 2)u}{1 + u}\,du$$

$$= 4\pi\int_0^2 \frac{u^3 + 2u}{1 + u}\,du = 4\pi\int_0^2 \left(u^2 - u + 3 - \frac{3}{1 + u}\right) du$$

$$= 4\pi\left[\frac{1}{3}u^3 - \frac{1}{2}u^2 + 3u - 3\ln(1 + u)\right]_0^2 = \frac{4\pi}{3}(20 - 9\ln 3) \approx 42.359$$

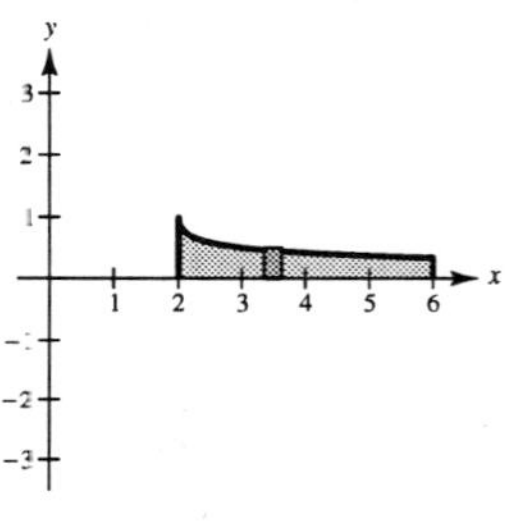

29. Since $y \le 0$, $A = -\int_{-1}^0 x\sqrt{x + 1}\,dx$.

$u = x + 1$

$x = u - 1$

$dx = du$

$$A = -\int_0^1 (u - 1)\sqrt{u}\,du = -\int_0^1 (u^{3/2} - u^{1/2})\,du$$

$$= -\left[\frac{2}{5}u^{5/2} - \frac{2}{3}u^{3/2}\right]_0^1 = \frac{4}{15}$$

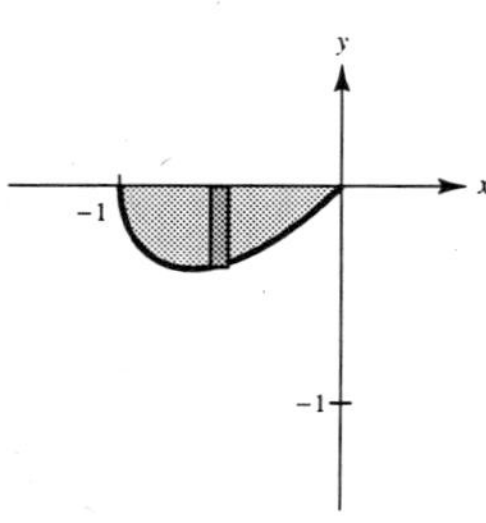

31. From Exercise 23(a) we have: $V = 64\pi\text{ ft}^3$

$$\frac{1}{4}V = 16\pi$$

Disc: $$\pi\int_{-3}^{y_0} \frac{16}{9}(9 - y^2)\,dy = 16\pi$$

$$\frac{1}{9}\int_{-3}^{y_0} (9 - y^2)\,dy = 1$$

$$\left[9y - \frac{1}{3}y^3\right]_{-3}^{y_0} = 9$$

$$\left(9y_0 - \frac{1}{3}y_0^3\right) - (-27 + 9) = 9$$

$$y_0^3 - 27y_0 - 27 = 0$$

By Newton's Method, $y_0 \approx -1.042$ and the depth of the gasoline is $3 - 1.042 = 1.958$ ft.

33. $f(x) = \frac{4}{5}x^{5/4}$

$f'(x) = x^{1/4}$

$1 + [f'(x)]^2 = 1 + \sqrt{x}$

$u = 1 + \sqrt{x}$

$x = (u - 1)^2$

$dx = 2(u - 1)\,du$

$$s = \int_0^4 \sqrt{1 + \sqrt{x}}\,dx = 2\int_1^3 \sqrt{u}(u - 1)\,du$$

$$= 2\int_1^3 (u^{3/2} - u^{1/2})\,du$$

$$= 2\left[\frac{2}{5}u^{5/2} - \frac{2}{3}u^{3/2}\right]_1^3 = \frac{4}{15}\left[u^{3/2}(3u - 5)\right]_1^3$$

$$= \frac{8}{15}(1 + 6\sqrt{3}) \approx 6.076$$

35. $y = 300\cosh\left(\frac{x}{2000}\right) - 280, \ -2000 \le x \le 2000$

$y' = \frac{3}{20}\sinh\left(\frac{x}{2000}\right)$

$$s = \int_{-2000}^{2000} \sqrt{1 + \left[\frac{3}{20}\sinh\left(\frac{x}{2000}\right)\right]^2}\,dx$$

$$= \frac{1}{20}\int_{-2000}^{2000} \sqrt{400 + 9\sinh^2\left(\frac{x}{2000}\right)}\,dx$$

≈ 4018.2 ft (by Simpson's Rule or graphing utility)

37. $y = \frac{3}{4}x$

$y' = \frac{3}{4}$

$1 + (y')^2 = \frac{25}{16}$

$$S = 2\pi\int_0^4 \left(\frac{3}{4}x\right)\sqrt{\frac{25}{16}}\,dx = \left[\left(\frac{15\pi}{8}\right)\frac{x^2}{2}\right]_0^4 = 15\pi$$

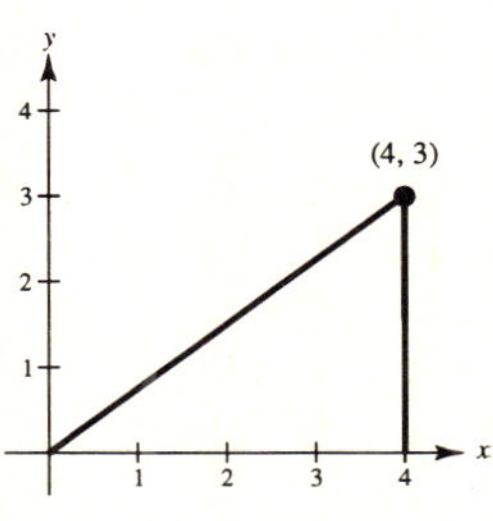

39. $F = kx$

$4 = k(1)$

$F = 4x$

$$W = \int_0^5 4x\,dx = \left[2x^2\right]_0^5$$

$= 50$ in $\cdot$ lb ≈ 4.167 ft $\cdot$ lb

41. Volume of disc: $\pi\left(\frac{1}{3}\right)^2 \Delta y$

Weight of disc: $62.4\pi\left(\frac{1}{3}\right)^2 \Delta y$

Distance: $175 - y$

$$W = \frac{62.4\pi}{9}\int_0^{150} (175 - y)\,dy = \frac{62.4\pi}{9}\left[175y - \frac{y^2}{2}\right]_0^{150}$$

$= 104{,}000\pi$ ft $\cdot$ lb ≈ 163.4 ft $\cdot$ ton

43. Weight of section of chain: $5\,\Delta x$

Distance moved: $10 - x$

$$W = 5\int_0^{10} (10 - x)\,dx$$

$$= \left[-\frac{5}{2}(10 - x)^2\right]_0^{10} = 250 \text{ ft} \cdot \text{lb}$$

45. $W = \int_a^b F(x)\,dx$

$$80 = \int_0^4 ax^2\,dx = \frac{ax^3}{3}\Big]_0^4 = \frac{64}{3}a$$

$$a = \frac{3(80)}{64} = \frac{15}{4} = 3.75$$

47. $A = \int_0^a (\sqrt{a} - \sqrt{x})^2\,dx = \int_0^a (a - 2\sqrt{a}\,x^{1/2} + x)\,dx = \left[ax - \frac{4}{3}\sqrt{a}\,x^{3/2} + \frac{1}{2}x^2\right]_0^a = \frac{a^2}{6}$

$$\frac{1}{A} = \frac{6}{a^2}$$

$$\bar{x} = \frac{6}{a^2}\int_0^a x(\sqrt{a} - \sqrt{x})^2\,dx = \frac{6}{a^2}\int_0^a (ax - 2\sqrt{a}\,x^{3/2} + x^2)\,dx$$

$$\bar{y} = \left(\frac{6}{a^2}\right)\frac{1}{2}\int_0^a (\sqrt{a} - \sqrt{x})^4\,dx$$

$$= \frac{3}{a^2}\int_0^a (a^2 - 4a^{3/2}x^{1/2} + 6ax - 4a^{1/2}x^{3/2} + x^2)\,dx$$

$$= \frac{3}{a^2}\left[a^2x - \frac{8}{3}a^{3/2}x^{3/2} + 3ax^2 - \frac{8}{5}a^{1/2}x^{5/2} + \frac{1}{3}x^3\right]_0^a = \frac{a}{5}$$

$$(\bar{x}, \bar{y}) = \left(\frac{a}{5}, \frac{a}{5}\right)$$

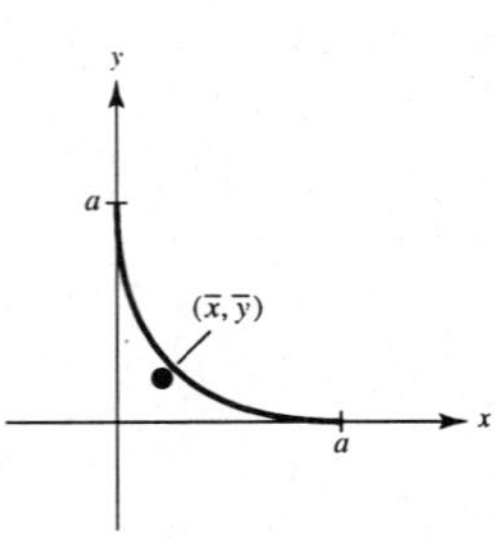

49. By symmetry, $x = 0$.

$$A = 2\int_0^1 (a^2 - x^2)\,dx = 2\left[a^2x - \frac{x^3}{3}\right]_0^a = \frac{4a^3}{3}$$

$$\frac{1}{A} = \frac{3}{4a^3}$$

$$\bar{y} = \left(\frac{3}{4a^3}\right)\frac{1}{2}\int_{-a}^a (a^2 - x^2)^2\,dx = \frac{6}{8a^3}\int_0^a (a^4 - 2a^2x^2 + x^4)\,dx$$

$$= \frac{6}{8a^3}\left[a^4x - \frac{2a^2}{3}x^3 + \frac{1}{5}x^5\right]_0^a = \frac{6}{8a^3}\left(a^5 - \frac{2}{3}a^5 + \frac{1}{5}a^5\right) = \frac{2a^2}{5}$$

$$(\bar{x}, \bar{y}) = \left(0, \frac{2a^2}{5}\right)$$

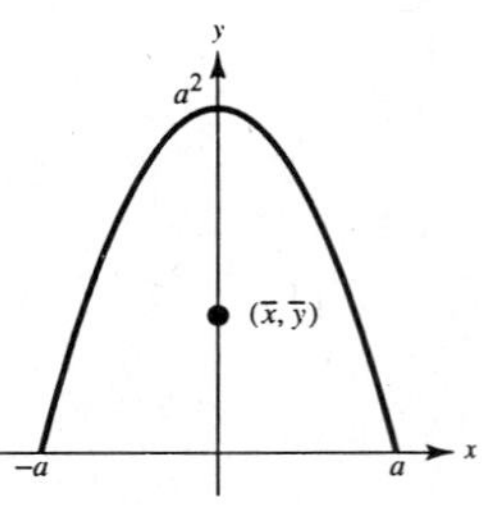

51. $\bar{y} = 0$ by symmetry

For the trapezoid:

$$m = [(4)(6) - (1)(6)]\rho = 18\rho$$

$$M_y = \rho\int_0^6 x\left[\left(\frac{1}{6}x + 1\right) - \left(-\frac{1}{6}x - 1\right)\right]dx$$

$$= \rho\int_0^6 \left(\frac{1}{3}x^2 + 2x\right)dx = \rho\left[\frac{x^3}{9} + x^2\right]_0^6 = 60\rho$$

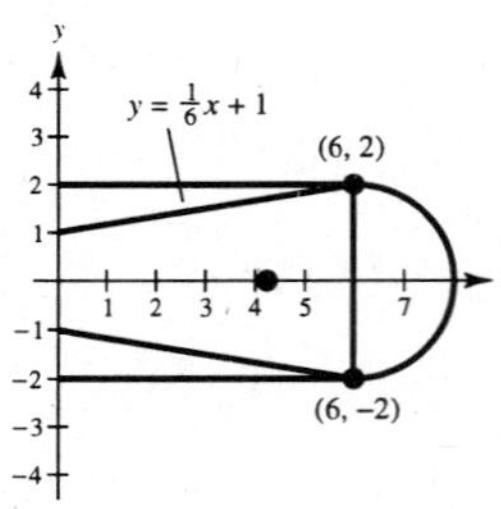

For the semicircle:

$$m = \left(\frac{1}{2}\right)(\pi)(2)^2\rho = 2\pi\rho$$

$$M_y = \rho\int_6^8 x\left[\sqrt{4 - (x - 6)^2} - \left(-\sqrt{4 - (x - 6)^2}\right)\right]dx = 2\rho\int_6^8 x\sqrt{4 - (x - 6)^2}\,dx$$

—CONTINUED—

51. —CONTINUED—

Let $u = x - 6$, then $x = u + 6$ and $dx = du$. When $x = 6$, $u = 0$. When $x = 8$, $u = 2$.

$$M_y = 2\rho\int_0^2 (u + 6)\sqrt{4 - u^2}\,du = 2\rho\int_0^2 u\sqrt{4 - u^2}\,du + 12\rho\int_0^2 \sqrt{4 - u^2}\,du$$

$$= 2\rho\left[\left(-\frac{1}{2}\right)\left(\frac{2}{3}\right)(4 - u^2)^{3/2}\right]_0^2 + 12\rho\left[\frac{\pi(2)^2}{4}\right] = \frac{16\rho}{3} + 12\pi\rho = \frac{4\rho(4 + 9\pi)}{3}$$

Thus, we have:

$$\bar{x}(18\rho + 2\pi\rho) = 60\rho + \frac{4\rho(4 + 9\pi)}{3}$$

$$\bar{x} = \frac{180\rho + 4\rho(4 + 9\pi)}{3}\cdot\frac{1}{2\rho(9 + \pi)} = \frac{2(9\pi + 49)}{3(\pi + 9)}$$

The centroid of the blade is $\left(\dfrac{2(9\pi + 49)}{3(\pi + 9)}, 0\right)$.

53. Wall at shallow end:

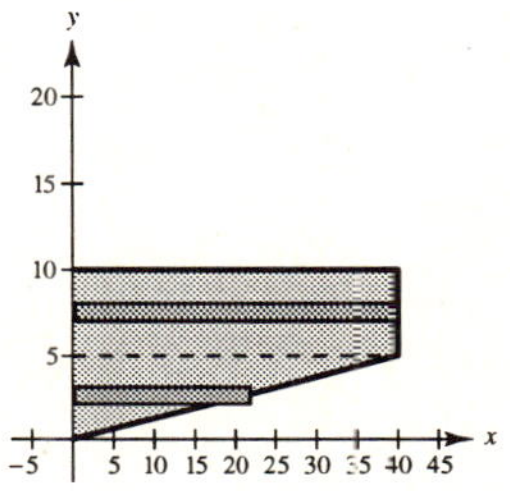

$$F = 62.4\int_0^5 y(20)\,dy = \left[(1248)\frac{y^2}{2}\right]_0^5 = 15{,}600 \text{ lb}$$

Wall at deep end:

$$F = 62.4\int_0^{10} y(20)\,dy = \left[(624)y^2\right]_0^{10} = 62{,}400 \text{ lb}$$

Side wall:

$$F_1 = 62.4\int_0^5 y(40)\,dy = \left[(1248)y^2\right]_0^5 = 31{,}200 \text{ lb} \quad F_2 = 62.4\int_0^5 (10 - y)8y\,dy = 62.4\int_0^5 (80y - 8y^2)\,dy$$

$$F = F_1 + F_2 = 72{,}800 \text{ lb}$$

55. $F = 62.4(16\pi)5 = 4992\pi$ lb

CHAPTER 7
Integration Techniques, L'Hôpital's Rule, and Improper Integrals

CHAPTER 7
Integration Techniques, L'Hôpital's Rule, and Improper Integrals

Section 7.1 Basic Integration Formulas

Solutions to Odd-Numbered Exercises

1. (a) $\frac{d}{dx}\left[2\sqrt{x^2+1}+C\right] = 2\left(\frac{1}{2}\right)(x^2+1)^{-1/2}(2x) = \frac{2x}{\sqrt{x^2+1}}$

(b) $\frac{d}{dx}\left[\sqrt{x^2+1}+C\right] = \frac{1}{2}(x^2+1)^{-1/2}(2x) = \frac{x}{\sqrt{x^2+1}}$

(c) $\frac{d}{dx}\left[\frac{1}{2}\sqrt{x^2+1}+C\right] = \frac{1}{2}\left(\frac{1}{2}\right)(x^2+1)^{-1/2}(2x) = \frac{x}{2\sqrt{x^2+1}}$

(d) $\frac{d}{dx}\left[\ln(x^2+1)+C\right] = \frac{2x}{x^2+1}$

$\int \frac{x}{\sqrt{x^2+1}}\,dx$ matches (b).

3. (a) $\frac{d}{dx}\left[\ln\sqrt{x^2+1}+C\right] = \frac{1}{2}\left(\frac{2x}{x^2+1}\right) = \frac{x}{x^2+1}$

(b) $\frac{d}{dx}\left[\frac{2x}{(x^2+1)^2}+C\right] = \frac{(x^2+1)^2(2)-(2x)(2)(x^2+1)(2x)}{(x^2+1)^4} = \frac{2(1-3x^2)}{(x^2+1)^3}$

(c) $\frac{d}{dx}\left[\arctan x + C\right] = \frac{1}{1+x^2}$

(d) $\frac{d}{dx}\left[\ln(x^2+1)+C\right] = \frac{2x}{x^2+1}$

$\int \frac{1}{x^2+1}\,dx$ matches (c).

5. $\int (3x-2)^4\,dx$

$u = 3x-2,\ du = 3\,dx,\ n = 4$

Use $\int u^n\,du$.

7. $\int \frac{1}{\sqrt{x}(1-2\sqrt{x})}\,dx$

$u = 1-2\sqrt{x},\ du = -\frac{1}{\sqrt{x}}\,dx$

Use $\int \frac{du}{u}$.

9. $\int \frac{3}{\sqrt{1-t^2}}\,dt$

$u = t,\ du = dt,\ a = 1$

Use $\int \frac{du}{\sqrt{a^2-u^2}}$

11. $\int t\sin t^2\,dt$

$u = t^2,\ du = 2t\,dt$

Use $\int \sin u\,du$.

13. $\int \cos x e^{\sin x}\,dx$

$u = \sin x,\ du = \cos x\,dx$

Use $\int e^u\,du$.

15. Let $u = -2x+5,\ du = -2\,dx$.

$$\int (-2x+5)^{3/2}\,dx = -\frac{1}{2}\int (-2x+5)^{3/2}(-2)\,dx = -\frac{1}{5}(-2x+5)^{5/2}+C$$

269

17. $$\int\left[v + \frac{1}{(3v-1)^3}\right]dv = \int v\,dv + \frac{1}{3}\int(3v-1)^{-3}(3)dv$$
$$= \frac{1}{2}v^2 - \frac{1}{6(3v-1)^2} + C$$

19. Let $u = -t^3 + 9t + 1$,
$du = (-3t^2 + 9)\,dt = -3(t^2 - 3)\,dt$.
$$\int\frac{t^2-3}{-t^3+9t+1}\,dt = -\frac{1}{3}\int\frac{-3(t^2-3)}{-t^3+9t+1}\,dt$$
$$= -\frac{1}{3}\ln\left|-t^3 + 9t + 1\right| + C$$

21. $$\int\frac{x^2}{x-1}\,dx = \int(x+1)\,dx + \int\frac{1}{x-1}\,dx$$
$$= \frac{1}{2}x^2 + x + \ln|x-1| + C$$

23. Let $u = 1 + e^x$, $du = e^x\,dx$.
$$\int\frac{e^x}{1+e^x}\,dx = \ln(1+e^x) + C$$

25. $$\int(1+2x^2)^2\,dx = \int(4x^4 + 4x^2 + 1)dx = \frac{4}{5}x^5 + \frac{4}{3}x^3 + x + C = \frac{x}{15}(12x^4 + 20x^2 + 15) + C$$

27. Let $u = 2\pi x^2$, $du = 4\pi x\,dx$.
$$\int x(\cos 2\pi x^2)\,dx = \frac{1}{4\pi}\int(\cos 2\pi x^2)(4\pi x)\,dx$$
$$= \frac{1}{4\pi}\sin 2\pi x^2 + C$$

29. Let $u = \pi x$, $du = \pi\,dx$.
$$\int\csc(\pi x)\cot(\pi x)\,dx = \frac{1}{\pi}\int\csc(\pi x)\cot(\pi x)\pi\,dx$$
$$= -\frac{1}{\pi}\csc(\pi x) + C$$

31. Let $u = 5x$, $du = 5\,dx$.
$$\int e^{5x}\,dx = \frac{1}{5}\int e^{5x}(5)\,dx = \frac{1}{5}e^{5x} + C$$

33. Let $u = 1 + e^x$, $du = e^x\,dx$.
$$\int\frac{2}{e^{-x}+1}\,dx = 2\int\left(\frac{1}{e^{-x}+1}\right)\left(\frac{e^x}{e^x}\right)dx$$
$$= 2\int\frac{e^x}{1+e^x}\,dx = 2\ln(1+e^x) + C$$

35. $$\int\frac{1+\sin x}{\cos x}\,dx = \int(\sec x + \tan x)\,dx$$
$$= \ln|\sec x + \tan x| + \ln|\sec x| + C$$
$$= \ln|\sec x(\sec x + \tan x)| + C$$

37. $$\int\frac{2t-1}{t^2+4}\,dt = \int\frac{2t}{t^2+4}\,dt - \int\frac{1}{4+t^2}\,dt$$
$$= \ln(t^2+4) - \frac{1}{2}\arctan\frac{t}{2} + C$$

39. Let $u = 2t - 1$, $du = 2\,dt$.
$$\int\frac{-1}{\sqrt{1-(2t-1)^2}}\,dt = -\frac{1}{2}\int\frac{2}{\sqrt{1-(2t-1)^2}}\,dt$$
$$= -\frac{1}{2}\arcsin(2t-1) + C$$

41. Let $u = \cos\left(\frac{2}{t}\right)$, $du = \frac{2\sin(2/t)}{t^2}\,dt$.
$$\int\frac{\tan(2/t)}{t^2}\,dt = \frac{1}{2}\int\frac{1}{\cos(2/t)}\left[\frac{2\sin(2/t)}{t^2}\right]dt$$
$$= \frac{1}{2}\ln\left|\cos\left(\frac{2}{t}\right)\right| + C$$

43. $$\int\frac{3}{\sqrt{6x-x^2}}\,dx = 3\int\frac{1}{\sqrt{9-(x-3)^2}}\,dx$$
$$= 3\arcsin\left(\frac{x-3}{3}\right) + C$$

45. $$\int\frac{4}{4x^2+4x+65}\,dx = \int\frac{1}{[x+(1/2)]^2+16}\,dx$$
$$= \frac{1}{4}\arctan\left[\frac{x+(1/2)}{4}\right] + C$$
$$= \frac{1}{4}\arctan\left(\frac{2x+1}{8}\right) + C$$

47. $\frac{ds}{dt} = \frac{t}{\sqrt{1-t^4}}, \left(0, -\frac{1}{2}\right)$

(a)

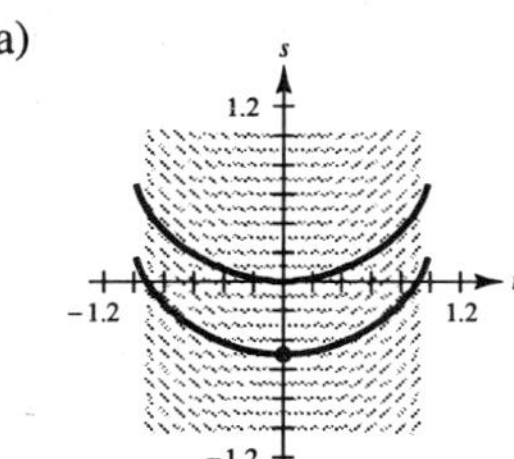

(b) $u = t^2, du = 2t\,dt$

$$\int \frac{t}{\sqrt{1-t^4}}\,dt = \frac{1}{2}\int \frac{2t}{\sqrt{1-(t^2)^2}}\,dt = \frac{1}{2}\arcsin t^2 + C$$

$$\left(0, -\frac{1}{2}\right): -\frac{1}{2} = \frac{1}{2}\arcsin 0 + C \Rightarrow C = -\frac{1}{2}$$

$$s = \frac{1}{2}\arcsin t^2 - \frac{1}{2}$$

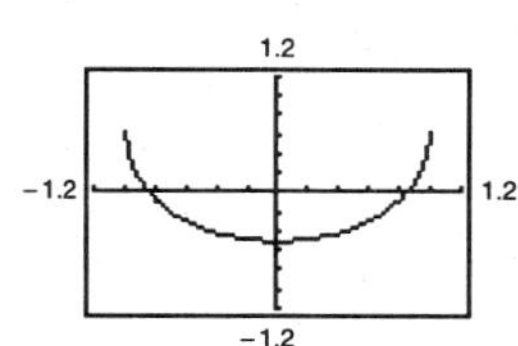

49. $y = \int (1 + e^x)^2\,dx = \int (e^{2x} + 2e^x + 1)\,dx$

$$= \frac{1}{2}e^{2x} + 2e^x + x + C$$

51. $\frac{dy}{dx} = \frac{\sec^2 x}{4 + \tan^2 x}$

Let $u = \tan x, du = \sec^2 x\,dx.$

$$y = \int \frac{\sec^2 x}{4 + \tan^2 x}\,dx = \frac{1}{2}\arctan\left(\frac{\tan x}{2}\right) + C$$

53. Let $u = 2x, du = 2\,dx.$

$$\int_0^{\pi/4} \cos 2x\,dx = \frac{1}{2}\int_0^{\pi/4} \cos 2x(2)\,dx$$

$$= \left[\frac{1}{2}\sin 2x\right]_0^{\pi/4} = \frac{1}{2}$$

55. Let $u = -x^2, du = -2x\,dx.$

$$\int_0^1 xe^{-x^2}\,dx = -\frac{1}{2}\int_0^1 e^{-x^2}(-2x)\,dx = \left[-\frac{1}{2}e^{-x^2}\right]_0^1$$

$$= \frac{1}{2}(1 - e^{-1}) \approx 0.316$$

57. Let $u = x^2 + 9, du = 2x\,dx.$

$$\int_0^4 \frac{2x}{\sqrt{x^2+9}}\,dx = \int_0^4 (x^2+9)^{-1/2}(2x)\,dx$$

$$= \left[2\sqrt{x^2+9}\right]_0^4 = 4$$

59. Let $u = 3x, du = 3\,dx.$

$$\int_0^{2/\sqrt{3}} \frac{1}{4 + 9x^2}\,dx = \frac{1}{3}\int_0^{2/\sqrt{3}} \frac{3}{4 + (3x)^2}\,dx$$

$$= \left[\frac{1}{6}\arctan\left(\frac{3x}{2}\right)\right]_0^{2/\sqrt{3}}$$

$$= \frac{\pi}{18} \approx 0.175$$

61. $\int \frac{1}{x^2 + 4x + 13}\,dx = \frac{1}{3}\arctan\left(\frac{x+2}{3}\right) + C$

The antiderivatives are vertical translations of each other.

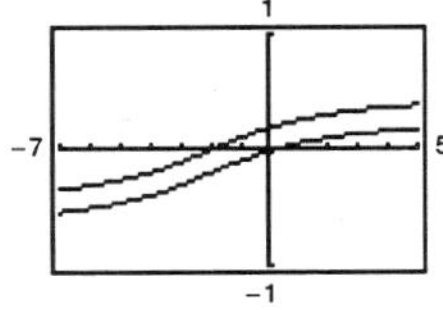

63. $\int \frac{1}{1 + \sin\theta}\,d\theta = \tan\theta - \sec\theta + C\left(\text{or } \frac{-2}{1 + \tan(\theta/2)}\right)$

The antiderivatives are vertical translations of each other.

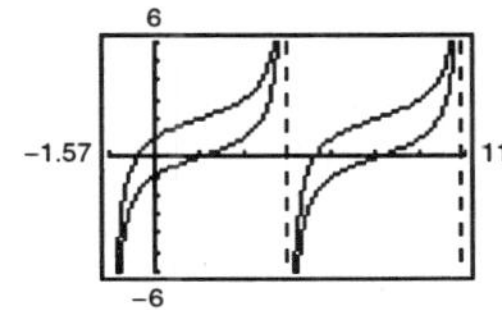

65. $\sin x + \cos x = a\sin(x + b)$

$\sin x + \cos x = a\sin x\cos b + a\cos x\sin b$

$\sin x + \cos x = (a\cos b)\sin x + (a\sin b)\cos x$

Equate coefficients of like terms to obtain the following.

$1 = a\cos b$ and $1 = a\sin b$

—CONTINUED—

65. —CONTINUED—

Thus, $a = 1/\cos b$. Now, substitute for a in $1 = a \sin b$.

$$1 = \left(\frac{1}{\cos b}\right)\sin b$$

$$1 = \tan b \Rightarrow b = \frac{\pi}{4}$$

Since $b = \frac{\pi}{4}$, $a = \frac{1}{\cos(\pi/4)} = \sqrt{2}$. Thus, $\sin x + \cos x = \sqrt{2}\sin\left(x + \frac{\pi}{4}\right)$.

$$\int \frac{dx}{\sin x + \cos x} = \int \frac{dx}{\sqrt{2}\sin(x + (\pi/4))} = \frac{1}{\sqrt{2}}\int \csc\left(x + \frac{\pi}{4}\right)dx = -\frac{1}{\sqrt{2}}\ln\left|\csc\left(x + \frac{\pi}{4}\right) + \cot\left(x + \frac{\pi}{4}\right)\right| + C$$

67. $\displaystyle\int_0^2 \frac{4x}{x^2 + 1}\,dx \approx 3$

Matches (a).

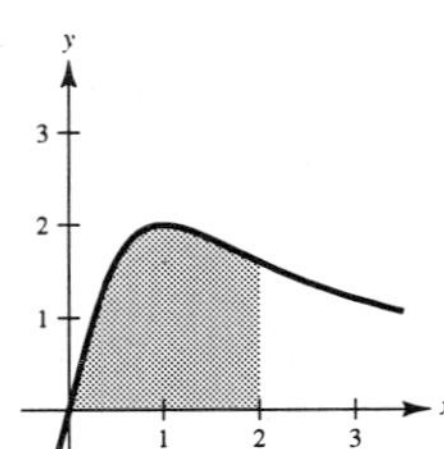

69. Let $u = 1 - x^2$, $du = -2x\,dx$.

$$A = 4\int_0^1 x\sqrt{1 - x^2}\,dx$$

$$= -2\int_0^1 (1 - x^2)^{1/2}(-2x)\,dx$$

$$= \left[-\frac{4}{3}(1 - x^2)^{3/2}\right]_0^1 = \frac{4}{3}$$

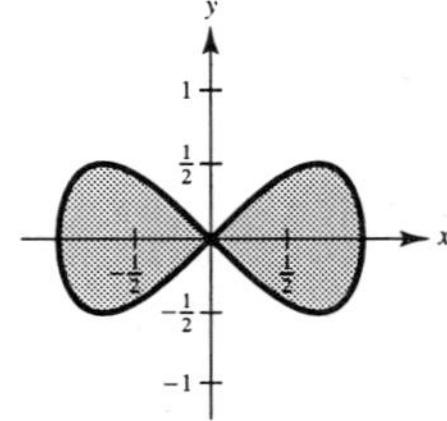

71. $\displaystyle\int_0^{1/a} (x - ax^2)\,dx = \left[\frac{1}{2}x^2 - \frac{a}{3}x^3\right]_0^{1/a}$

$$= \frac{1}{6a^2}$$

Let $\frac{1}{6a^2} = \frac{2}{3}$, $12a^2 = 3$, $a = \frac{1}{2}$.

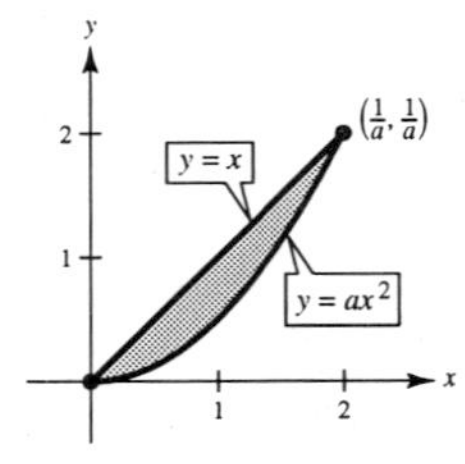

73. (a) **Shell Method:**

Let $u = -x^2$, $du = -2x\,dx$.

$$V = 2\pi\int_0^1 xe^{-x^2}\,dx$$

$$= -\pi\int_0^1 e^{-x^2}(-2x)\,dx$$

$$= \left[-\pi e^{-x^2}\right]_0^1$$

$$= \pi(1 - e^{-1}) \approx 1.986$$

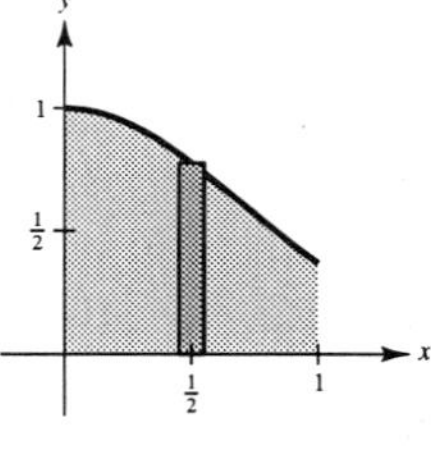

(b) **Shell Method:**

$$V = 2\pi\int_0^b xe^{-x^2}\,dx$$

$$= \left[-\pi e^{-x^2}\right]_0^b$$

$$= \pi(1 - e^{-b^2}) = \frac{4}{3}$$

$$e^{-b^2} = \frac{3\pi - 4}{3\pi}$$

$$b = \sqrt{\ln\left(\frac{3\pi}{3\pi - 4}\right)} \approx 0.743$$

75. $\displaystyle A = \int_0^4 \frac{5}{\sqrt{25 - x^2}}\,dx = \left[5\arcsin\frac{x}{5}\right]_0^4 = 5\arcsin\frac{4}{5}$

$$\bar{x} = \frac{1}{A}\int_0^4 x\left(\frac{5}{\sqrt{25 - x^2}}\right)dx = \frac{1}{5\arcsin(4/5)}\left(-\frac{5}{2}\right)\int_0^4 (25 - x^2)^{-1/2}(-2x)\,dx$$

$$= \frac{1}{5\arcsin(4/5)}(-5)\left[(25 - x^2)^{1/2}\right]_0^4$$

$$= -\frac{1}{\arcsin(4/5)}[3 - 5] = \frac{2}{\arcsin(4/5)} \approx 2.157$$

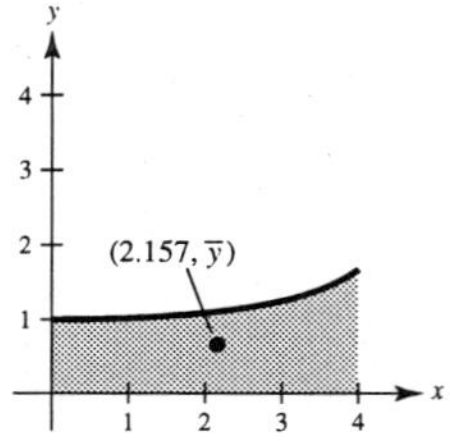

77. $$y = \tan(\pi x)$$
$$y' = \pi \sec^2(\pi x)$$
$$1 + (y')^2 = 1 + \pi^2 \sec^4(\pi x)$$
$$s = \int_0^{1/4} \sqrt{1 + \pi^2 \sec^4(\pi x)}\, dx$$
$$\approx 1.0320$$

79. True. If $u = x^4 - 1$, then $du = 4x^3\, dx$.

$$\int \frac{x^3}{x^4 - 1}\, dx = \frac{1}{4}\int \frac{4x^3\, dx}{x^4 - 1} = \frac{1}{4}\int \frac{du}{u}$$

Section 7.2 Integration by Parts

1. (a) $\dfrac{d}{dx}[\sin x - x\cos x] = \cos x - (-x\sin x + \cos x) = x\sin x$. Matches (ii)

(b) $\dfrac{d}{dx}[x^2\sin x + 2x\cos x - 2\sin x] = x^2\cos x + 2x\sin x - 2x\sin x + 2\cos x - 2\cos x = x^2\cos x$. Matches (iv)

(c) $\dfrac{d}{dx}[x^2e^x - 2xe^x + 2e^x] = x^2e^x + 2xe^x - 2xe^x - 2e^x + 2e^x = x^2e^x$. Matches (iii)

(d) $\dfrac{d}{dx}[-x + x\ln x] = -1 + x\left(\dfrac{1}{x}\right) + \ln x = \ln x$. Matches (i)

3. $\displaystyle\int xe^{2x}\, dx$

$u = x,\ dv = e^{2x}\, dx$

5. $\displaystyle\int (\ln x)^2\, dx$

$u = (\ln x)^2,\ dv = dx$

7. $\displaystyle\int x\sec^2 x\, dx$

$u = x,\ dv = \sec^2 x\, dx$

9. $dv = e^{-2x}\, dx \implies v = \displaystyle\int e^{-2x}\, dx = -\frac{1}{2}e^{-2x}$

$u = x \implies du = dx$

$$\int xe^{-2x}\, dx = -\frac{1}{2}xe^{-2x} - \int -\frac{1}{2}e^{-2x}\, dx = -\frac{1}{2}xe^{-2x} - \frac{1}{4}e^{-2x} + C = \frac{-1}{4e^{2x}}(2x + 1) + C$$

11. Use integration by parts three times.

(1) $dv = e^x\, dx \implies v = \displaystyle\int e^x\, dx = e^x$

$u = x^3 \implies du = 3x^2\, dx$

(2) $dv = e^x\, dx \implies v = \displaystyle\int e^x\, dx = e^x$

$u = x^2 \implies du = 2x\, dx$

(3) $dv = e^x\, dx \implies v = \displaystyle\int e^x\, dx = e^x$

$u = x \implies du = dx$

$$\int x^3e^x\, dx = x^3e^x - 3\int x^2e^x\, dx = x^3e^x - 3x^2e^x + 6\int xe^x\, dx$$
$$= x^3e^x - 3x^2e^x + 6xe^x - 6e^x + C = e^x(x^3 - 3x^2 + 6x - 6) + C$$

13. $$\int x^2 e^{x^3}\, dx = \frac{1}{3}\int e^{x^3}(3x^2)\, dx = \frac{1}{3}e^{x^3} + C$$

15. $dv = t\, dt \implies v = \displaystyle\int t\, dt = \frac{t^2}{2}$

$u = \ln(t + 1) \implies du = \dfrac{1}{t + 1}\, dt$

$$\int t\ln(t + 1)\, dt = \frac{t^2}{2}\ln(t + 1) - \frac{1}{2}\int \frac{t^2}{t + 1}\, dt = \frac{t^2}{2}\ln(t + 1) - \frac{1}{2}\int\left(t - 1 + \frac{1}{t + 1}\right) dt$$
$$= \frac{t^2}{2}\ln(t + 1) - \frac{1}{2}\left[\frac{t^2}{2} - t + \ln(t + 1)\right] + C = \frac{1}{4}[2(t^2 - 1)\ln|t + 1| - t^2 + 2t] + C$$

17. Let $u = \ln x$, $du = \dfrac{1}{x}\,dx$.

$$\int \frac{(\ln x)^2}{x}\,dx = \int (\ln x)^2\left(\frac{1}{x}\right)dx = \frac{(\ln x)^3}{3} + C$$

19. $dv = \dfrac{1}{(2x+1)^2}\,dx \implies v = \displaystyle\int (2x+1)^{-2}\,dx = -\frac{1}{2(2x+1)}$

$u = xe^{2x} \implies du = (2xe^{2x} + e^{2x})\,dx = e^{2x}(2x+1)\,dx$

$$\int \frac{xe^{2x}}{(2x+1)^2}\,dx = -\frac{xe^{2x}}{2(2x+1)} + \int \frac{e^{2x}}{2}\,dx = \frac{-xe^{2x}}{2(2x+1)} + \frac{e^{2x}}{4} + C = \frac{e^{2x}}{4(2x+1)} + C$$

21. Use integration by parts twice.

(1) $dv = e^x\,dx \implies v = \displaystyle\int e^x\,dx = e^x$

$u = x^2 \implies du = 2x\,dx$

(2) $dv = e^x\,dx \implies v = \displaystyle\int e^x\,dx = e^x$

$u = x \implies du = dx$

$$\int (x^2 - 1)e^x\,dx = \int x^2e^x\,dx - \int e^x\,dx = x^2e^x - 2\int xe^x\,dx - e^x$$

$$= x^2e^x - 2\left[xe^x - \int e^x\,dx\right] - e^x = x^2e^x - 2xe^x + e^x + C = (x-1)^2e^x + C$$

23. $dv = \sqrt{x-1}\,dx \implies v = \displaystyle\int (x-1)^{1/2}\,dx = \frac{2}{3}(x-1)^{3/2}$

$u = x \implies du = dx$

$$\int x\sqrt{x-1}\,dx = \frac{2}{3}x(x-1)^{3/2} - \frac{2}{3}\int (x-1)^{3/2}\,dx$$

$$= \frac{2}{3}x(x-1)^{3/2} - \frac{4}{15}(x-1)^{5/2} + C = \frac{2(x-1)^{3/2}}{15}(3x+2) + C$$

25. $dv = \cos x\,dx \implies v = \displaystyle\int \cos x\,dx = \sin x$

$u = x \implies du = dx$

$$\int x\cos x\,dx = x\sin x - \int \sin x\,dx = x\sin x + \cos x + C$$

27. $dv = dx \implies v = \displaystyle\int dx = x$

$u = \arctan x \implies du = \dfrac{1}{1+x^2}\,dx$

$$\int \arctan x\,dx = x\arctan x - \int \frac{x}{1+x^2}\,dx$$

$$= x\arctan x - \frac{1}{2}\ln(1+x^2) + C$$

29. Use integration by parts twice.

(1) $dv = e^{2x}dx \implies v = \displaystyle\int e^{2x}\,dx = \frac{1}{2}e^{2x}$

$u = \sin x \implies du = \cos x\,dx$

(2) $dv = e^{2x}\,dx \implies v = \displaystyle\int e^{2x}\,dx = \frac{1}{2}e^{2x}$

$u = \cos x \implies du = -\sin x\,dx$

$$\int e^{2x}\sin x\,dx = \frac{1}{2}e^{2x}\sin x - \frac{1}{2}\int e^{2x}\cos x\,dx = \frac{1}{2}e^{2x}\sin x - \frac{1}{2}\left(\frac{1}{2}e^{2x}\cos x + \frac{1}{2}\int e^{2x}\sin x\,dx\right)$$

$$\frac{5}{4}\int e^{2x}\sin x\,dx = \frac{1}{2}e^{2x}\sin x - \frac{1}{4}e^{2x}\cos x$$

$$\int e^{2x}\sin x\,dx = \frac{1}{5}e^{2x}(2\sin x - \cos x) + C$$

31. $y' = xe^{x^2}$

$$y = \int xe^{x^2}\,dx = \frac{1}{2}e^{x^2} + C$$

33. Use integration by parts twice.

(1) $dv = \dfrac{1}{\sqrt{2+3t}}\,dt \implies v = \displaystyle\int (2+3t)^{-1/2}\,dt = \frac{2}{3}\sqrt{2+3t}$

$u = t^2 \implies du = 2t\,dt$

(2) $dv = \sqrt{2+3t}\,dt \implies v = \displaystyle\int (2+3t)^{1/2}\,dt = \frac{2}{9}(2+3t)^{3/2}$

$u = t \implies du = dt$

$$y = \int \frac{t^2}{\sqrt{2+3t}}\,dt = \frac{2t^2\sqrt{2+3t}}{3} - \frac{4}{3}\int t\sqrt{2+3t}\,dt = \frac{2t^2\sqrt{2+3t}}{3} - \frac{4}{3}\left[\frac{2t}{9}(2+3t)^{3/2} - \frac{2}{9}\int (2+3t)^{3/2}\,dt\right]$$

$$= \frac{2t^2\sqrt{2+3t}}{3} - \frac{8t}{27}(2+3t)^{3/2} + \frac{16}{405}(2+3t)^{5/2} + C = \frac{2\sqrt{2+3t}}{405}(27t^2 - 24t + 32) + C$$

35. $(\cos y)y' = 2x$

$$\int \cos y\,dy = \int 2x\,dx$$

$$\sin y = x^2 + C$$

37. (a)

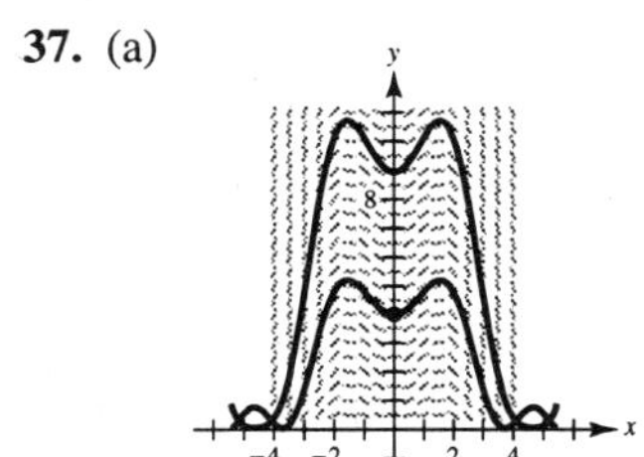

(b) $\dfrac{dy}{dx} = x\sqrt{y}\cos x,\ (0, 4)$

$$\int \frac{dy}{\sqrt{y}} = \int x\cos x\,dx$$

$$\int y^{-1/2}\,dy = \int x\cos x\,dx \qquad (u = x,\ du = dx,\ dv = \cos x\,dx,\ v = \sin x)$$

$$2y^{1/2} = x\sin x - \int \sin x\,dx = x\sin x + \cos x + C$$

$(0, 4)$: $2(4)^{1/2} = 0 - 1 + C \implies C = 3$

$2\sqrt{y} = x\sin x + \cos x + 3$

39. $dv = \sin 2x\,dx \implies v = \displaystyle\int \sin 2x\,dx = -\frac{1}{2}\cos 2x$

$u = x \implies du = dx$

$$\int x\sin 2x\,dx = \frac{-1}{2}x\cos 2x + \frac{1}{2}\int \cos 2x\,dx = \frac{-1}{2}x\cos 2x + \frac{1}{4}\sin 2x + C = \frac{1}{4}(\sin 2x - 2x\cos 2x) + C$$

Thus, $\displaystyle\int_0^{\pi} x\sin 2x\,dx = \left[\frac{1}{4}(\sin 2x - 2x\cos 2x)\right]_0^{\pi} = -\frac{\pi}{2}.$

41. Use integration by parts twice.

(1) $dv = e^x\,dx \implies v = \displaystyle\int e^x\,dx = e^x$

$u = \sin x \implies du = \cos x\,dx$

(2) $dv = e^x\,dx \implies v = \displaystyle\int e^x\,dx = e^x$

$u = \cos x \implies du = -\sin x\,dx$

—CONTINUED—

41. —CONTINUED—

$$\int e^x \sin x\,dx = e^x \sin x - \int e^x \cos x\,dx = e^x \sin x - e^x \cos x - \int e^x \sin x\,dx$$

$$2\int e^x \sin x\,dx = e^x(\sin x - \cos x)$$

$$\int e^x \sin x\,dx = \frac{e^x}{2}(\sin x - \cos x) + C$$

Thus, $\displaystyle\int_0^1 e^x \sin x\,dx = \left[\frac{e^x}{2}(\sin x - \cos x)\right]_0^1 = \frac{e}{2}(\sin 1 - \cos 1) + \frac{1}{2} = \frac{e(\sin 1 - \cos 1) + 1}{2} \approx 0.909.$

43. See Exercise 25.

$$\int_0^{\pi/2} x \cos x\,dx = \Big[x \sin x + \cos x\Big]_0^{\pi/2} = \frac{\pi}{2} - 1$$

45. $\displaystyle\int x^2e^{2x}\,dx = x^2\left(\frac{1}{2}e^{2x}\right) - (2x)\left(\frac{1}{4}e^{2x}\right) + 2\left(\frac{1}{8}e^{2x}\right) + C$

$$= \frac{1}{2}x^2e^{2x} - \frac{1}{2}xe^{2x} + \frac{1}{4}e^{2x} + C$$

$$= \frac{1}{4}e^{2x}(2x^2 - 2x + 1) + C$$

Alternate signs	u and its derivatives	v' and its antiderivatives
+	x^2	e^{2x}
−	$2x$	$\frac{1}{2}e^{2x}$
+	2	$\frac{1}{4}e^{2x}$
−	0	$\frac{1}{8}e^{2x}$

47. $\displaystyle\int x^3 \sin x\,dx = x^3(-\cos x) - 3x^2(-\sin x) + 6x \cos x - 6 \sin x + C$

$$= -x^3 \cos x + 3x^2 \sin x + 6x \cos x - 6 \sin x + C$$

$$= (3x^2 - 6) \sin x - (x^3 - 6x) \cos x + C$$

Alternate signs	u and its derivatives	v' and its antiderivatives
+	x^3	$\sin x$
−	$3x^2$	$-\cos x$
+	$6x$	$-\sin x$
−	6	$\cos x$
+	0	$\sin x$

49. $\displaystyle\int x \sec^2 x\,dx = x \tan x + \ln|\cos x| + C$

Alternate signs	u and its derivatives	v' and its antiderivatives
+	x	$\sec^2 x$
−	1	$\tan x$
+	0	$-\ln\|\cos x\|$

51. $\displaystyle\int t^3e^{-4t}\,dt = -\frac{e^{-4t}}{128}(32t^3 + 24t^2 + 12t + 3) + C$

53. $\displaystyle\int_0^{\pi/2} e^{-2x} \sin 3x\,dx = \left[\frac{e^{-2x}(-2 \sin 3x - 3 \cos 3x)}{13}\right]_0^{\pi/2}$

$$= \frac{1}{13}(2e^{-\pi} + 3) \approx 0.2374$$

55. (a) $dv = \sqrt{2x-3}\,dx \implies v = \int (2x-3)^{1/2}\,dx = \frac{1}{3}(2x-3)^{3/2}$

$u = 2x \implies du = 2\,dx$

$$\int 2x\sqrt{2x-3}\,dx = \frac{2}{3}x(2x-3)^{3/2} - \frac{2}{3}\int (2x-3)^{3/2}\,dx$$

$$= \frac{2}{3}x(2x-3)^{3/2} - \frac{2}{15}(2x-3)^{5/2} + C$$

$$= \frac{2}{15}(2x-3)^{3/2}(3x+3) + C = \frac{2}{5}(2x-3)^{3/2}(x+1) + C$$

(b) $u = 2x-3 \implies x = \dfrac{u+3}{2}$ and $dx = \dfrac{1}{2}\,du$

$$\int 2x\sqrt{2x-3}\,dx = \int 2\left(\frac{u+3}{2}\right)u^{1/2}\left(\frac{1}{2}\right)du = \frac{1}{2}\int (u^{3/2} + 3u^{1/2})\,du = \frac{1}{2}\left[\frac{2}{5}u^{5/2} + 2u^{3/2}\right] + C$$

$$= \frac{1}{5}u^{3/2}(u+5) + C = \frac{1}{5}(2x-3)^{3/2}[(2x-3)+5] + C = \frac{2}{5}(2x-3)^{3/2}(x+1) + C$$

57. (a) $dv = \dfrac{x}{\sqrt{4+x^2}}\,dx \implies v = \int (4+x^2)^{-1/2}x\,dx = \sqrt{4+x^2}$

$u = x^2 \implies du = 2x\,dx$

$$\int \frac{x^3}{\sqrt{4+x^2}}\,dx = x^2\sqrt{4+x^2} - 2\int x\sqrt{4+x^2}\,dx$$

$$= x^2\sqrt{4+x^2} - \frac{2}{3}(4+x^2)^{3/2} + C = \frac{1}{3}\sqrt{4+x^2}\,(x^2-8) + C$$

(b) $u = 4+x^2 \implies x^2 = u-4$ and $2x\,dx = du \implies x\,dx = \dfrac{1}{2}\,du$

$$\int \frac{x^3}{\sqrt{4+x^2}}\,dx = \int \frac{x^2}{\sqrt{4+x^2}}\,x\,dx = \int \frac{u-4}{\sqrt{u}}\,\frac{1}{2}\,du$$

$$= \frac{1}{2}\int (u^{1/2} - 4u^{-1/2})\,du = \frac{1}{2}\left(\frac{2}{3}u^{3/2} - 8u^{1/2}\right) + C$$

$$= \frac{1}{3}u^{1/2}(u-12) + C = \frac{1}{3}\sqrt{4+x^2}\,[(4+x^2)-12] + C = \frac{1}{3}\sqrt{4+x^2}\,(x^2-8) + C$$

59. $n = 0$: $\displaystyle\int \ln x\,dx = x(\ln x - 1) + C$

$n = 1$: $\displaystyle\int x\ln x\,dx = \frac{x^2}{4}(2\ln x - 1) + C$

$n = 2$: $\displaystyle\int x^2\ln x\,dx = \frac{x^3}{9}(3\ln x - 1) + C$

$n = 3$: $\displaystyle\int x^3\ln x\,dx = \frac{x^4}{16}(4\ln x - 1) + C$

$n = 4$: $\displaystyle\int x^4\ln x\,dx = \frac{x^5}{25}(5\ln x - 1) + C$

In general,

$$\int x^n\ln x\,dx = \frac{x^{n+1}}{(n+1)^2}[(n+1)\ln x - 1] + C.$$

(See Exercise 63)

61. $dv = \sin x\,dx \implies v = -\cos x$

$u = x^n \implies du = nx^{n-1}\,dx$

$$\int x^n\sin x\,dx = -x^n\cos x + n\int x^{n-1}\cos x\,dx$$

63. $dv = x^n\,dx \implies v = \dfrac{x^{n+1}}{n+1}$

$u = \ln x \implies du = \dfrac{1}{x}\,dx$

$$\int x^n \ln x\,dx = \frac{x^{n+1}}{n+1}\ln x - \int \frac{x^n}{n+1}\,dx$$

$$= \frac{x^{n+1}}{n+1}\ln x - \frac{x^{n+1}}{(n+1)^2} + C = \frac{x^{n+1}}{(n+1)^2}[(n+1)\ln x - 1] + C$$

65. Use integration by parts twice.

(1) $dv = e^{ax}\,dx \implies v = \dfrac{1}{a}e^{ax}$

$u = \sin bx \implies du = b\cos bx\,dx$

(2) $dv = e^{ax}\,dx \implies v = \dfrac{1}{a}e^{ax}$

$u = \cos bx \implies du = -b\sin bx\,dx$

$$\int e^{ax}\sin bx\,dx = \frac{e^{ax}\sin bx}{a} - \frac{b}{a}\int e^{ax}\cos bx\,dx$$

$$= \frac{e^{ax}\sin bx}{a} - \frac{b}{a}\left[\frac{e^{ax}\cos bx}{a} + \frac{b}{a}\int e^{ax}\sin bx\,dx\right] = \frac{e^{ax}\sin bx}{a} - \frac{b^2}{a^2}\int e^{ax}\sin bx\,dx$$

Therefore, $\left(1 + \dfrac{b^2}{a^2}\right)\displaystyle\int e^{ax}\sin bx\,dx = \dfrac{e^{ax}(a\sin bx - b\cos bx)}{a^2}$

$$\int e^{ax}\sin bx\,dx = \frac{e^{ax}(a\sin bx - b\cos bx)}{a^2+b^2} + C.$$

67. $n = 3$ (Use formula in Exercise 63.)

$$\int x^3\ln x\,dx = \frac{x^4}{16}[4\ln x - 1] + C$$

69. $a = 2, b = 3$ (Use formula in Exercise 66.)

$$\int e^{2x}\cos 3x\,dx = \frac{e^{2x}(2\cos 3x + 3\sin 3x)}{13} + C$$

71. $dv = e^{-x}\,dx \implies v = -e^{-x}$

$u = x \implies du = dx$

$$A = \int_0^4 xe^{-x}\,dx = \Big[-xe^{-x}\Big]_0^4 + \int_0^4 e^{-x}\,dx = \frac{-4}{e^4} - \Big[e^{-x}\Big]_0^4$$

$$= 1 - \frac{5}{e^4} \approx 0.908$$

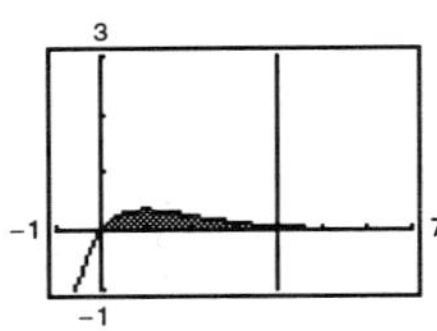

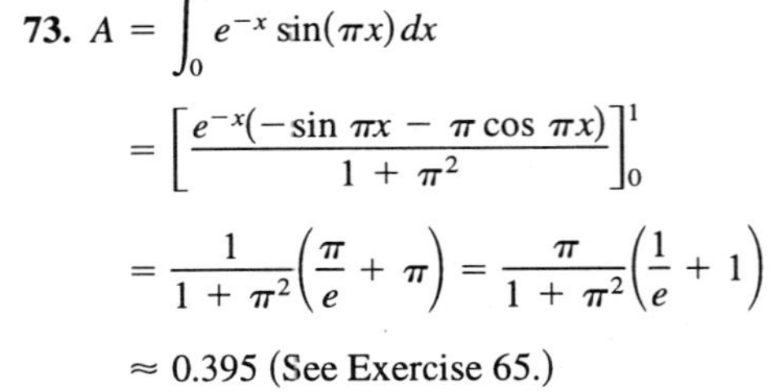

73. $A = \displaystyle\int_0^1 e^{-x}\sin(\pi x)\,dx$

$$= \left[\frac{e^{-x}(-\sin\pi x - \pi\cos\pi x)}{1+\pi^2}\right]_0^1$$

$$= \frac{1}{1+\pi^2}\left(\frac{\pi}{e} + \pi\right) = \frac{\pi}{1+\pi^2}\left(\frac{1}{e} + 1\right)$$

≈ 0.395 (See Exercise 65.)

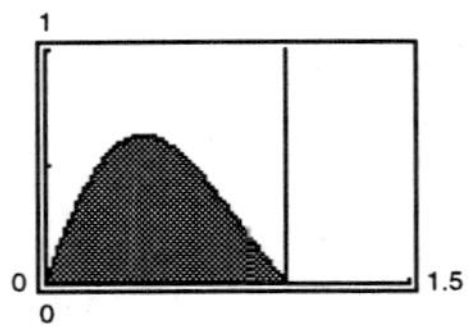

75. (a) $A = \displaystyle\int_1^e \ln x\,dx = \Big[-x + x\ln x\Big]_1^e = 1$ (See Exercise 1.)

(b) $R(x) = \ln x,\ r(x) = 0$

$$V = \pi\int_1^e (\ln x)^2\,dx$$

$$= \pi\Big[x(\ln x)^2 - 2x\ln x + 2x\Big]_1^e \text{ (Use integration by parts twic, see Exercise 5.)}$$

$$= \pi(e-2) \approx 2.257$$

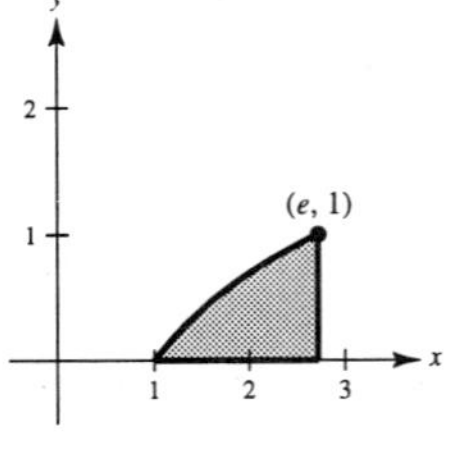

—CONTINUED—

75. —CONTINUED—

(c) $p(x) = x, h(x) = \ln x$

$$V = 2\pi\int_1^e x\ln x\,dx = 2\pi\left[\frac{x^2}{4}(-1 + 2\ln x)\right]_1^e = \frac{(e^2+1)\pi}{2} \approx 13.177 \text{ (See Exercise 63.)}$$

(d) $$\bar{x} = \frac{\int_1^e x\ln x\,dx}{1} = \frac{e^2+1}{4} \approx 2.097$$

$$\bar{y} = \frac{\frac{1}{2}\int_1^e (\ln x)^2 dx}{1} = \frac{e-2}{2} \approx 0.359$$

$$(\bar{x}, \bar{y}) = \left(\frac{e^2+1}{4}, \frac{e-2}{2}\right) \approx (2.097, 0.359)$$

77. Average value $= \dfrac{1}{\pi}\displaystyle\int_0^\pi e^{-4t}(\cos 2t + 5\sin 2t)\,dt$

$$= \frac{1}{\pi}\left[e^{-4t}\left(\frac{-4\cos 2t + 2\sin 2t}{20}\right) + 5e^{-4t}\left(\frac{-4\sin 2t - 2\cos 2t}{20}\right)\right]_0^\pi \text{ (From Exercises 65 and 66)}$$

$$= \frac{7}{10\pi}(1 - e^{-4\pi}) \approx 0.223$$

79. $c(t) = 100{,}000 + 4000t, r = 5\%, t_1 = 10$

$$P = \int_0^{10}(100{,}000 + 4000t)e^{-0.05t}\,dt = 4000\int_0^{10}(25 + t)e^{-0.05t}\,dt$$

Let $u = 25 + t, dv = e^{-0.05t}dt, du = dt, v = -\dfrac{100}{5}e^{-0.05t}$

$$P = 4000\left\{\left[(25+t)\left(-\frac{100}{5}e^{-0.05t}\right)\right]_0^{10} + \frac{100}{5}\int_0^{10} e^{-0.05t}\,dt\right\}$$

$$= 4000\left\{\left[(25+t)\left(-\frac{100}{5}e^{-0.05t}\right)\right]_0^{10} - \left[\frac{10{,}000}{25}e^{-0.05t}\right]_0^{10}\right\} \approx \$931{,}265$$

81. $$\int_{-\pi}^{\pi} x\sin nx\,dx = \left[-\frac{x}{n}\cos nx + \frac{1}{n^2}\sin nx\right]_{-\pi}^{\pi}$$

$$= -\frac{\pi}{n}\cos \pi n - \frac{\pi}{n}\cos(-\pi n)$$

$$= -\frac{2\pi}{n}\cos\pi n = \begin{cases} -(2\pi/n), & \text{if } n \text{ is even} \\ (2\pi/n), & \text{if } n \text{ is odd}\end{cases}$$

83. Let $u = x, dv = \sin\left(\dfrac{n\pi}{2}x\right)dx, du = dx, v = -\dfrac{2}{n\pi}\cos\left(\dfrac{n\pi}{2}x\right).$

$$I_1 = \int_0^1 x\sin\left(\frac{n\pi}{2}x\right)dx = \left[\frac{-2x}{n\pi}\cos\left(\frac{n\pi}{2}x\right)\right]_0^1 + \frac{2}{n\pi}\int_0^1 \cos\left(\frac{n\pi}{2}x\right)dx$$

$$= -\frac{2}{n\pi}\cos\left(\frac{n\pi}{2}\right) + \left[\left(\frac{2}{n\pi}\right)^2\sin\left(\frac{n\pi}{2}x\right)\right]_0^1$$

$$= -\frac{2}{n\pi}\cos\left(\frac{n\pi}{2}\right) + \left(\frac{2}{n\pi}\right)^2\sin\left(\frac{n\pi}{2}\right)$$

—CONTINUED—

83. —CONTINUED—

Let $u = (-x + 2)$, $dv = \sin\left(\frac{n\pi}{2}x\right)dx$, $du = -dx$, $v = -\frac{2}{n\pi}\cos\left(\frac{n\pi}{2}x\right)$.

$$I_2 = \int_1^2 (-x + 2)\sin\left(\frac{n\pi}{2}x\right)dx = \left[\frac{-2(-x + 2)}{n\pi}\cos\left(\frac{n\pi}{2}x\right)\right]_1^2 - \frac{2}{n\pi}\int_1^2 \cos\left(\frac{n\pi}{2}x\right)dx$$

$$= \frac{2}{n\pi}\cos\left(\frac{n\pi}{2}\right) - \left[\left(\frac{2}{n\pi}\right)^2 \sin\left(\frac{n\pi}{2}x\right)\right]_1^2$$

$$= \frac{2}{n\pi}\cos\left(\frac{n\pi}{2}\right) + \left(\frac{2}{n\pi}\right)^2 \sin\left(\frac{n\pi}{2}\right)$$

$$h(I_1 + I_2) = b_n = h\left[\left(\frac{2}{n\pi}\right)^2 \sin\left(\frac{n\pi}{2}\right) + \left(\frac{2}{n\pi}\right)^2 \sin\left(\frac{n\pi}{2}\right)\right] = \frac{8h}{(n\pi)^2}\sin\left(\frac{n\pi}{2}\right)$$

85. Shell Method:

$$V = 2\pi\int_a^b x f(x)\, dx$$

$$dv = x\, dx \implies v = \frac{x^2}{2}$$

$$u = f(x) \implies du = f'(x)\, dx$$

$$V = 2\pi\left[\frac{x^2}{2}f(x) - \int \frac{x^2}{2}f'(x)\, dx\right]_a^b$$

$$= \pi\left[(b^2f(b) - a^2f(a)) - \int_a^b x^2 f'(x)\, dx\right]$$

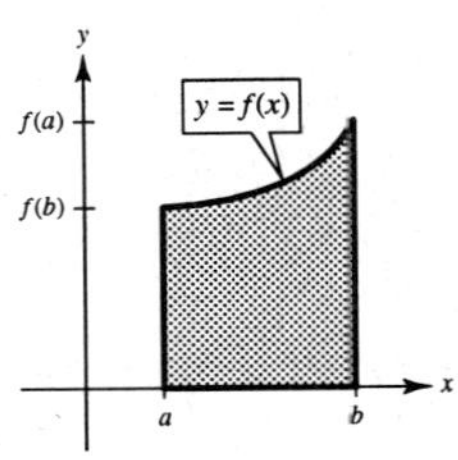

Disc Method:

$$V = \pi\int_0^{f(a)} (b^2 - a^2)\, dy + \pi\int_{f(a)}^{f(b)} [b^2 - [f^{-1}(y)]^2]\, dy$$

$$= \pi(b^2 - a^2)f(a) + \pi b^2(f(b) - f(a)) - \pi\int_{f(a)}^{f(b)} [f^{-1}(y)]^2\, dy$$

$$= \pi\left[(b^2f(b) - a^2f(a)) - \int_{f(a)}^{f(b)} [f^{-1}(y)]^2\, dy\right]$$

Since $x = f^{-1}(y)$, we have $f(x) = y$ and $f'(x)dx = dy$. When $y = f(a)$, $x = a$. When $y = f(b)$, $x = b$. Thus,

$$\int_{f(a)}^{f(b)} [f^{-1}(y)]^2\, dy = \int_a^b x^2 f'(x)\, dx$$

and the volumes are the same.

87. $f'(x) = \cos\sqrt{x}$, $f(0) = 2$

(a) It cannot be solved by integration.

(b) You obtain the points

n	x_n	y_n
0	0	0
1	0.05	2.05
2	0.10	2.098755
3	0.15	2.146276
⋮	⋮	⋮
80	4.0	2.8403565

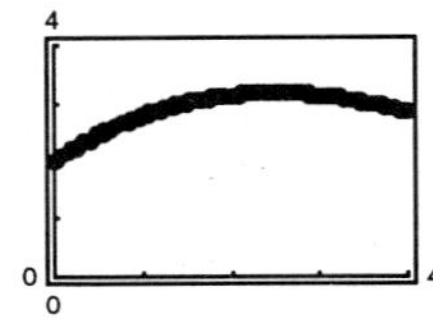

Section 7.3 Trigonometric Integrals

1. $f(x) = \sin^4 x + \cos^4 x$

(a) $\sin^4 x + \cos^4 x = \left(\dfrac{1 - \cos 2x}{2}\right)^2 + \left(\dfrac{1 + \cos 2x}{2}\right)^2$

$$= \frac{1}{4}[1 - 2\cos 2x + \cos^2 2x + 1 + 2\cos 2x + \cos^2 2x]$$

$$= \frac{1}{4}\left[2 + 2\,\frac{1 + \cos 4x}{2}\right]$$

$$= \frac{1}{4}[3 + \cos 4x]$$

(b) $\sin^4 x + \cos^4 x = (\sin^2 x)^2 + \cos^4 x$

$$= (1 - \cos^2 x)^2 + \cos^4 x$$

$$= 1 - 2\cos^2 x + 2\cos^4 x$$

(c) $\sin^4 x + \cos^4 x = \sin^4 x + 2\sin^2 x\cos^2 x + \cos^4 x - 2\sin^2 x\cos^2 x$

$$= (\sin^2 x + \cos^2 x)^2 - 2\sin^2 x\cos^2 x$$

$$= 1 - 2\sin^2 x\cos^2 x$$

(d) $1 - 2\sin^2 x\cos^2 x = 1 - (2\sin x\cos x)(\sin x\cos x)$

$$= 1 - (\sin 2x)\left(\frac{1}{2}\sin 2x\right)$$

$$= 1 - \frac{1}{2}\sin^2(2x)$$

(e) Four ways. There is often more than one way to rewrite a trigonometric expression.

3. Let $u = \cos x$, $du = -\sin x\,dx$.

$$\int \cos^3 x \sin x\,dx = -\int \cos^3 x(-\sin x)\,dx$$

$$= -\frac{1}{4}\cos^4 x + C$$

5. Let $u = \sin 2x$, $du = 2\cos 2x\,dx$.

$$\int \sin^5 2x\cos 2x\,dx = \frac{1}{2}\int \sin^5 2x(2\cos 2x)dx$$

$$= \frac{1}{12}\sin^6 2x + C$$

7. Let $u = \cos x$, $du = -\sin x\,dx$.

$$\int \sin^5 x\cos^2 x\,dx = \int \sin x(1 - \cos^2 x)^2\cos^2 x\,dx$$

$$= -\int (\cos^2 x - 2\cos^4 x + \cos^6 x)(-\sin x)\,dx = \frac{-1}{3}\cos^3 x + \frac{2}{5}\cos^5 x - \frac{1}{7}\cos^7 x + C$$

9. $\displaystyle\int \cos^2 3x\,dx = \int \frac{1 + \cos 6x}{2}\,dx$

$$= \frac{1}{2}\left(x + \frac{1}{6}\sin 6x\right) + C$$

$$= \frac{1}{12}(6x + \sin 6x) + C$$

11. Integration by parts.

$$dv = \sin^2 x\, dx = \frac{1 - \cos 2x}{2} \implies v = \frac{x}{2} - \frac{\sin 2x}{4} = \frac{1}{4}(2x - \sin 2x)$$

$$u = x \implies du = dx$$

$$\int x \sin^2 x\, dx = \frac{1}{4}x(2x - \sin 2x) - \frac{1}{4}\int (2x - \sin 2x)\, dx$$

$$= \frac{1}{4}x(2x - \sin 2x) - \frac{1}{4}\left(x^2 + \frac{1}{2}\cos 2x\right) + C = \frac{1}{8}(2x^2 - 2x\sin 2x - \cos 2x) + C$$

13. Let $u = \sin x$, $du = \cos x\, dx$.

$$\int_0^{\pi/2} \cos^3 x\, dx = \int_0^{\pi/2} (1 - \sin^2 x)\cos x\, dx = \left[\sin x - \frac{1}{3}\sin^3 x\right]_0^{\pi/2} = \frac{2}{3}$$

15. Let $u = \sin x$, $du = \cos x\, dx$.

$$\int_0^{\pi/2} \cos^7 x\, dx = \int_0^{\pi/2} (1 - \sin^2 x)^3 \cos x\, dx = \int_0^{\pi/2} (1 - 3\sin^2 x + 3\sin^4 x - \sin^6 x)\cos x\, dx$$

$$= \left[\sin x - \sin^3 x + \frac{3}{5}\sin^5 x - \frac{1}{7}\sin^7 x\right]_0^{\pi/2} = \frac{16}{35}$$

17. $$\int \sec(3x)\, dx = \frac{1}{3}\ln|\sec 3x + \tan 3x| + C$$

19. $$\int \sec^4 5x\, dx = \int (1 + \tan^2 5x)\sec^2 5x\, dx$$

$$= \frac{1}{5}\left(\tan 5x + \frac{\tan^3 5x}{3}\right) + C$$

$$= \frac{\tan 5x}{15}(3 + \tan^2 5x) + C$$

21. $$dv = \sec^2 \pi x\, dx \implies v = \frac{1}{\pi}\tan \pi x$$

$$u = \sec \pi x \implies du = \pi \sec \pi x \tan \pi x\, dx$$

$$\int \sec^3 \pi x\, dx = \frac{1}{\pi}\sec \pi x \tan \pi x - \int \sec \pi x \tan^2 \pi x\, dx = \frac{1}{\pi}\sec \pi x \tan \pi x - \int \sec \pi x(\sec^2 \pi x - 1)\, dx$$

$$2\int \sec^3 \pi x\, dx = \frac{1}{\pi}(\sec \pi x \tan \pi x + \ln|\sec \pi x + \tan \pi x|) + C_1$$

$$\int \sec^3 \pi x\, dx = \frac{1}{2\pi}(\sec \pi x \tan \pi x + \ln|\sec \pi x + \tan \pi x|) + C$$

23. $$\int \tan^5 \frac{x}{4}\, dx = \int \left(\sec^2 \frac{x}{4} - 1\right)\tan^3 \frac{x}{4}\, dx$$

$$= \int \tan^3 \frac{x}{4}\sec^2 \frac{x}{4}\, dx - \int \tan^3 \frac{x}{4}\, dx$$

$$= \tan^4 \frac{x}{4} - \int \left(\sec^2 \frac{x}{4} - 1\right)\tan \frac{x}{4}\, dx$$

$$= \tan^4 \frac{x}{4} - 2\tan^2 \frac{x}{4} - 4\ln\left|\cos \frac{x}{4}\right| + C$$

25. $u = \tan x$, $du = \sec^2 x\, dx$

$$\int \sec^2 x \tan x\, dx = \frac{1}{2}\tan^2 x + C$$

27. $\displaystyle\int \tan^2 x \sec^2 x\,dx = \frac{\tan^3 x}{3} + C$

29. $\displaystyle\int \sec^6 4x \tan 4x\,dx = \frac{1}{4}\int \sec^5 4x(4 \sec 4x \tan 4x)\,dx$

$\displaystyle= \frac{\sec^6 4x}{24} + C$

31. Let $u = \sec x$, $du = \sec x \tan x\,dx$.

$\displaystyle\int \sec^3 x \tan x\,dx = \int \sec^2 x(\sec x \tan x)\,dx$

$\displaystyle= \frac{1}{3}\sec^3 x + C$

33. $\displaystyle r = \int \sin^4(\pi\theta)\,d\theta = \frac{1}{4}\int [1 - \cos(2\pi\theta)]^2\,d\theta$

$\displaystyle= \frac{1}{4}\int [1 - 2\cos(2\pi\theta) + \cos^2(2\pi\theta)]\,d\theta$

$\displaystyle= \frac{1}{4}\int \left[1 - 2\cos(2\pi\theta) + \frac{1 + \cos(4\pi\theta)}{2}\right] d\theta$

$\displaystyle= \frac{1}{4}\left[\theta - \frac{1}{\pi}\sin(2\pi\theta) + \frac{\theta}{2} + \frac{1}{8\pi}\sin(4\pi\theta)\right] + C$

$\displaystyle= \frac{1}{32\pi}[12\pi\theta - 8\sin(2\pi\theta) + \sin(4\pi\theta)] + C$

35. $\displaystyle y = \int \tan^3 3x \sec 3x\,dx$

$\displaystyle= \int (\sec^2 3x - 1)\sec 3x \tan 3x\,dx$

$\displaystyle= \frac{1}{3}\int \sec^2 3x(3\sec 3x \tan 3x)\,dx - \frac{1}{3}\int 3\sec 3x \tan 3x\,dx$

$\displaystyle= \frac{1}{9}\sec^3 3x - \frac{1}{3}\sec 3x + C$

37. (a)

(b) $\displaystyle\frac{dy}{dx} = \sin^2 x,\ (0, 0)$

$\displaystyle y = \int \sin^2 x\,dx = \int \frac{1 - \cos 2x}{2}\,dx$

$\displaystyle= \frac{1}{2}x - \frac{\sin 2x}{4} + C$

$\displaystyle(0, 0):\ 0 = C,\ y = \frac{1}{2}x - \frac{\sin 2x}{4}$

39. $\displaystyle\int \sin 3x \cos 2x\,dx = \frac{1}{2}\int (\sin 5x + \sin x)\,dx$

$\displaystyle= \frac{-1}{2}\left(\frac{1}{5}\cos 5x + \cos x\right) + C$

$\displaystyle= \frac{-1}{10}(\cos 5x + 5\cos x) + C$

41. $\displaystyle\int \sin\theta \sin 3\theta\,d\theta = \frac{1}{2}\int (\cos 2\theta - \cos 4\theta)\,d\theta$

$\displaystyle= \frac{1}{2}\left(\frac{1}{2}\sin 2\theta - \frac{1}{4}\sin 4\theta\right) + C$

$\displaystyle= \frac{1}{8}(2\sin 2\theta - \sin 4\theta) + C$

43. $\displaystyle\int \cot^3 2x\,dx = \int (\csc^2 2x - 1)\cot 2x\,dx$

$\displaystyle= -\frac{1}{2}\int \cot 2x(-2\csc^2 2x)\,dx - \frac{1}{2}\int \frac{2\cos 2x}{\sin 2x}\,dx$

$\displaystyle= -\frac{1}{4}\cot^2 2x - \frac{1}{2}\ln|\sin 2x| + C$

$\displaystyle= \frac{1}{4}(\ln|\csc^2 2x| - \cot^2 2x) + C$

45. Let $u = \cot\theta$, $du = -\csc^2\theta\,d\theta$.

$\displaystyle\int \csc^4\theta\,d\theta = \int \csc^2\theta(1 + \cot^2\theta)\,d\theta$

$\displaystyle= \int \csc^2\theta\,d\theta + \int \csc^2\theta \cot^2\theta\,d\theta$

$\displaystyle= -\cot\theta - \frac{1}{3}\cot^3\theta + C$

47. $\displaystyle\int \frac{\cot^2 t}{\csc t}\,dt = \int \frac{\csc^2 t - 1}{\csc t}\,dt = \int (\csc t - \sin t)\,dt = \ln|\csc t - \cot t| + \cos t + C$

49. $\displaystyle\int \frac{1}{\sec x \tan x}\,dx = \int \frac{\cos^2 x}{\sin x}\,dx = \int \frac{1-\sin^2 x}{\sin x}\,dx = \int (\csc x - \sin x)\,dx = \ln|\csc x - \cot x| + \cos x + C$

51. $\displaystyle\int (\tan^4 t - \sec^4 t)\,dt = \int (\tan^2 t + \sec^2 t)(\tan^2 t - \sec^2 t)\,dt \qquad (\tan^2 t - \sec^2 t = -1)$

$$= -\int (\tan^2 t + \sec^2 t)\,dt = -\int (2\sec^2 t - 1)\,dt = -2\tan t + t + C$$

53. $\displaystyle\int_{-\pi}^{\pi} \sin^2 x\,dx = 2\int_0^{\pi} \frac{1-\cos 2x}{2}\,dx$

$$= \left[x - \frac{1}{2}\sin 2x\right]_0^{\pi} = \pi$$

55. $\displaystyle\int_0^{\pi/4} \tan^3 x\,dx = \int_0^{\pi/4} (\sec^2 x - 1)\tan x\,dx$

$$= \int_0^{\pi/4} \sec^2 x \tan x\,dx - \int_0^{\pi/4} \frac{\sin x}{\cos x}\,dx$$

$$= \left[\frac{1}{2}\tan^2 x + \ln|\cos x|\right]_0^{\pi/4} = \frac{1}{2}(1 - \ln 2)$$

57. Let $u = 1 + \sin t$, $du = \cos t\,dt$.

$$\int_0^{\pi/2} \frac{\cos t}{1+\sin t}\,dt = \Big[\ln|1+\sin t|\Big]_0^{\pi/2} = \ln 2$$

59. Let $u = \sin x$, $du = \cos x\,dx$.

$$\int_{-\pi/2}^{\pi/2} \cos^3 x\,dx = 2\int_0^{\pi/2} (1-\sin^2 x)\cos x\,dx$$

$$= 2\left[\sin x - \frac{1}{3}\sin^3 x\right]_0^{\pi/2} = \frac{4}{3}$$

61. $\displaystyle\int \cos^4 \frac{x}{2}\,dx = \frac{1}{16}[6x + 8\sin x + \sin 2x] + C$

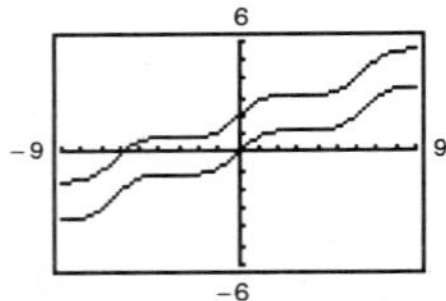

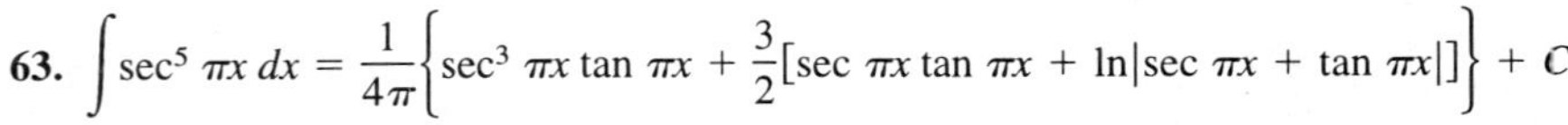

63. $\displaystyle\int \sec^5 \pi x\,dx = \frac{1}{4\pi}\left\{\sec^3 \pi x \tan \pi x + \frac{3}{2}[\sec \pi x \tan \pi x + \ln|\sec \pi x + \tan \pi x|]\right\} + C$

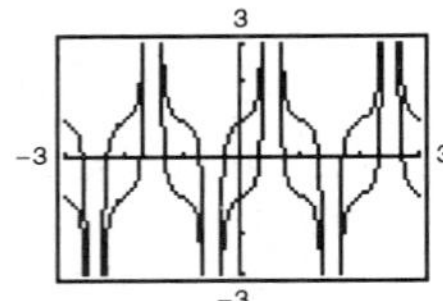

65. $\displaystyle\int \sec^5 \pi x \tan \pi x\,dx = \frac{1}{5\pi}\sec^5 \pi x + C$

67. $\displaystyle\int_0^{\pi/4} \sin 2\theta \sin 3\theta\,d\theta = \frac{1}{2}\left[\sin\theta - \frac{1}{5}\sin 5\theta\right]_0^{\pi/4} = \frac{3\sqrt{2}}{10}$

69. $\displaystyle\int_0^{\pi/2} \sin^4 x\,dx = \frac{1}{4}\left[\frac{3x}{2} - \sin 2x + \frac{1}{8}\sin 4x\right]_0^{\pi/2} = \frac{3\pi}{16}$

71. (a) Let $u = \tan 3x$, $du = 3\sec^2 3x\,dx$.

$$\int \sec^4 3x \tan^3 3x\,dx = \int \sec^2 3x \tan^3 3x \sec^2 3x\,dx$$
$$= \frac{1}{3}\int (\tan^2 3x + 1)\tan^3 3x(3\sec^2 3x)\,dx$$
$$= \frac{1}{3}\int (\tan^5 3x + \tan^3 3x)(3\sec^2 3x)\,dx$$
$$= \frac{\tan^6 3x}{18} + \frac{\tan^4 3x}{12} + C_1$$

Or let $u = \sec 3x$, $du = 3\sec 3x \tan 3x\,dx$.

$$\int \sec^4 3x \tan^3 3x\,dx = \int \sec^3 3x \tan^2 3x \sec 3x \tan 3x\,dx$$
$$= \frac{1}{3}\int \sec^3 3x(\sec^2 3x - 1)(3\sec 3x \tan 3x)\,dx$$
$$= \frac{\sec^6 3x}{18} - \frac{\sec^4 3x}{12} + C$$

(b)

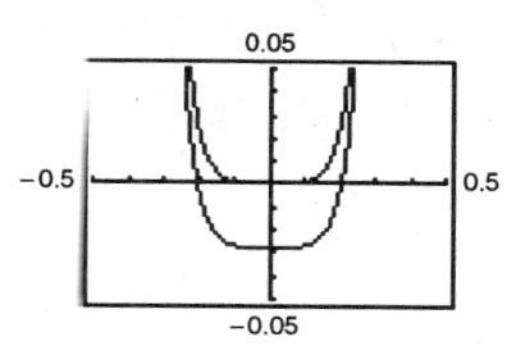

(c) $$\frac{\sec^6 3x}{18} - \frac{\sec^4 3x}{12} + C = \frac{(1+\tan^2 3x)^3}{18} - \frac{(1+\tan^2 3x)^2}{12} + C$$
$$= \frac{1}{18}\tan^6 3x + \frac{1}{6}\tan^4 3x + \frac{1}{6}\tan^2 3x + \frac{1}{18} - \frac{1}{12}\tan^4 3x - \frac{1}{6}\tan^2 3x - \frac{1}{12} + C$$
$$= \frac{\tan^6 3x}{18} + \frac{\tan^4 3x}{12} + \left(\frac{1}{18} - \frac{1}{12}\right) + C$$
$$= \frac{\tan^6 3x}{18} + \frac{\tan^4 3x}{12} + C_2$$

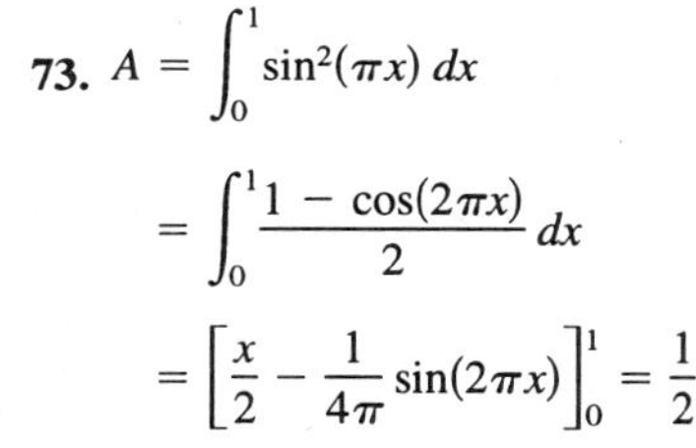

73. $$A = \int_0^1 \sin^2(\pi x)\,dx$$
$$= \int_0^1 \frac{1 - \cos(2\pi x)}{2}\,dx$$
$$= \left[\frac{x}{2} - \frac{1}{4\pi}\sin(2\pi x)\right]_0^1 = \frac{1}{2}$$

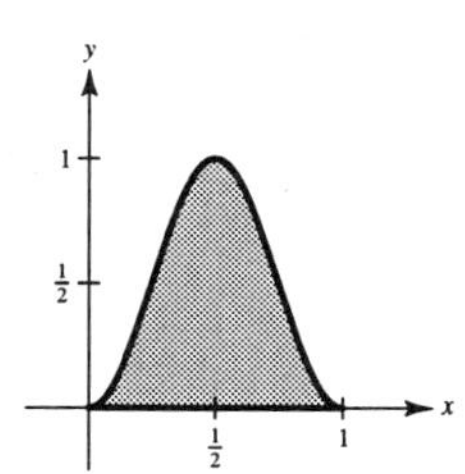

75. (a) $$V = \pi\int_0^\pi \sin^2 x\,dx = \frac{\pi}{2}\int_0^\pi (1 - \cos 2x)\,dx = \frac{\pi}{2}\left[x - \frac{1}{2}\sin 2x\right]_0^\pi = \frac{\pi^2}{2}$$

(b) $$A = \int_0^\pi \sin x\,dx = \left[-\cos x\right]_0^\pi = 1 + 1 = 2$$

Let $u = x$, $dv = \sin x\,dx$, $du = dx$, $v = -\cos x$.

$$\bar{x} = \frac{1}{A}\int_0^\pi x\sin x\,dx = \frac{1}{2}\left[\left[-x\cos x\right]_0^\pi + \int_0^\pi \cos x\,dx\right] = \frac{1}{2}\left[-x\cos x + \sin x\right]_0^\pi = \frac{\pi}{2}$$

$$\bar{y} = \frac{1}{2A}\int_0^\pi \sin^2 x\,dx = \frac{1}{8}\int_0^\pi (1 - \cos 2x)\,dx$$
$$= \frac{1}{8}\left[x - \frac{1}{2}\sin 2x\right]_0^\pi = \frac{\pi}{8}$$

$$(\bar{x}, \bar{y}) = \left(\frac{\pi}{2}, \frac{\pi}{8}\right)$$

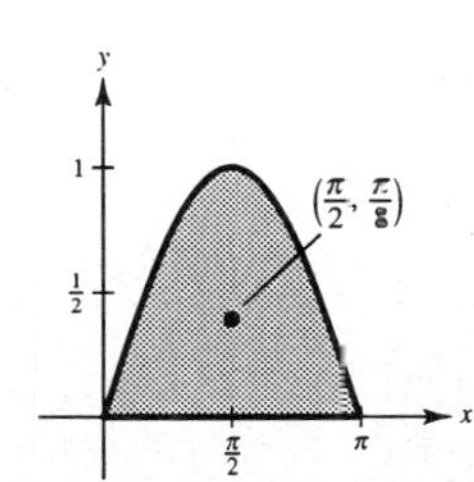

77. $dv = \sin x\,dx \implies v = -\cos x$

$u = \sin^{n-1} x \implies du = (n-1)\sin^{n-2} x \cos x\,dx$

$$\int \sin^n x\,dx = -\sin^{n-1} x \cos x + (n-1)\int \sin^{n-2} x \cos^2 x\,dx$$

$$= -\sin^{n-1} x \cos x + (n-1)\int \sin^{n-2} x(1 - \sin^2 x)\,dx$$

$$= -\sin^{n-1} x \cos x + (n-1)\int \sin^{n-2} x\,dx - (n-1)\int \sin^n x\,dx$$

Therefore, $n\displaystyle\int \sin^n x\,dx = -\sin^{n-1} x \cos x + (n-1)\int \sin^{n-2} x\,dx$

$$\int \sin^n x\,dx = \frac{-\sin^{n-1} x \cos x}{n} + \frac{n-1}{n}\int \sin^{n-2} x\,dx.$$

79. Let $u = \sin^{n-1} x$, $du = (n-1)\sin^{n-2} x \cos x\,dx$, $dv = \cos^m x \sin x\,dx$, $v = \dfrac{-\cos^{m+1} x}{m+1}$.

$$\int \cos^m x \sin^n x\,dx = \frac{-\sin^{n-1} x \cos^{m+1} x}{m+1} + \frac{n-1}{m+1}\int \sin^{n-2} x \cos^{m+2} x\,dx$$

$$= \frac{-\sin^{n-1} x \cos^{m+1} x}{m+1} + \frac{n-1}{m+1}\int \sin^{n-2} x \cos^m x(1 - \sin^2 x)\,dx$$

$$= \frac{-\sin^{n-1} x \cos^{m+1} x}{m+1} + \frac{n-1}{m+1}\int \sin^{n-2} x \cos^m x\,dx - \frac{n-1}{m+1}\int \sin^n x \cos^m x\,dx$$

$$\frac{m+n}{m+1}\int \cos^m x \sin^n x\,dx = \frac{-\sin^{n-1} x \cos^{m+1} x}{m+1} + \frac{n-1}{m+1}\int \sin^{n-2} x \cos^m x\,dx$$

$$\int \cos^m x \sin^n x\,dx = \frac{-\cos^{m+1} x \sin^{n-1}}{m+n} + \frac{n-1}{m+n}\int \cos^m x \sin^{n-2} x\,dx$$

81. $$\int \sin^5 x\,dx = -\frac{\sin^4 x \cos x}{5} + \frac{4}{5}\int \sin^3 x\,dx$$

$$= -\frac{\sin^4 x \cos x}{5} + \frac{4}{5}\left[-\frac{\sin^2 x \cos x}{3} + \frac{2}{3}\int \sin x\,dx\right]$$

$$= -\frac{1}{5}\sin^4 x \cos x - \frac{4}{15}\sin^2 x \cos x - \frac{8}{15}\cos x + C$$

$$= -\frac{\cos x}{15}[3\sin^4 x + 4\sin^2 x + 8] + C$$

83. $$\int \sec^4\left(\frac{2\pi x}{5}\right)dx = \frac{5}{2\pi}\int \sec^4\left(\frac{2\pi x}{5}\right)\frac{2\pi}{5}\,dx$$

$$= \frac{5}{2\pi}\left[\frac{1}{3}\sec^2\left(\frac{2\pi x}{5}\right)\tan\left(\frac{2\pi x}{5}\right) + \frac{2}{3}\int \sec^2\left(\frac{2\pi x}{5}\right)\frac{2\pi}{5}\,dx\right]$$

$$= \frac{5}{6\pi}\left[\sec^2\left(\frac{2\pi x}{5}\right)\tan\left(\frac{2\pi x}{5}\right) + 2\tan\left(\frac{2\pi x}{5}\right)\right] + C$$

$$= \frac{5}{6\pi}\tan\left(\frac{2\pi x}{5}\right)\left[\sec^2\left(\frac{2\pi x}{5}\right) + 2\right] + C$$

85. (a) n is odd and $n \geq 3$.

$$\int_0^{\pi/2} \cos^n x\,dx = \left[\frac{\cos^{n-1} x \sin x}{n}\right]_0^{\pi/2} + \frac{n-1}{n}\int_0^{\pi/2} \cos^{n-2} x\,dx$$

$$= \frac{n-1}{n}\left[\left[\frac{\cos^{n-3} x \sin x}{n-2}\right]_0^{\pi/2} + \frac{n-3}{n-2}\int_0^{\pi/2} \cos^{n-4} x\,dx\right]$$

$$= \frac{n-1}{n}\cdot\frac{n-3}{n-2}\left[\left[\frac{\cos^{n-5} x \sin x}{n-4}\right]_0^{\pi/2} + \frac{n-5}{n-4}\int_0^{\pi/2} \cos^{n-6} x\,dx\right]$$

$$= \frac{n-1}{n}\cdot\frac{n-3}{n-2}\cdot\frac{n-5}{n-4}\int_0^{\pi/2} \cos^{n-6} x\,dx$$

$$= \frac{n-1}{n}\cdot\frac{n-3}{n-2}\cdot\frac{n-5}{n-4}\cdots\int_0^{\pi/2} \cos x\,dx$$

$$= \left[\frac{n-1}{n}\cdot\frac{n-3}{n-2}\cdot\frac{n-5}{n-4}\cdots(\sin x)\right]_0^{\pi/2}$$

$$= \frac{n-1}{n}\cdot\frac{n-3}{n-2}\cdot\frac{n-5}{n-4}\cdots 1 \quad \text{(Reverse the order)}$$

$$= (1)\left(\frac{2}{3}\right)\left(\frac{4}{5}\right)\left(\frac{6}{7}\right)\cdots\left(\frac{n-1}{n}\right)$$

$$= \left(\frac{2}{3}\right)\left(\frac{4}{5}\right)\left(\frac{6}{7}\right)\cdots\left(\frac{n-1}{n}\right)$$

(b) n is even and $n \geq 2$.

$$\int_0^{\pi/2} \cos^n x\,dx = \frac{n-1}{n}\cdot\frac{n-3}{n-2}\cdot\frac{n-5}{n-4}\cdots\int_0^{\pi/2} \cos^2 x\,dx \quad \text{(From part (a).)}$$

$$= \left[\frac{n-1}{n}\cdot\frac{n-3}{n-2}\cdot\frac{n-5}{n-4}\cdots\left(\frac{x}{2} + \frac{1}{4}\sin 2x\right)\right]_0^{\pi/2}$$

$$= \frac{n-1}{n}\cdot\frac{n-3}{n-2}\cdot\frac{n-5}{n-4}\cdots\frac{\pi}{4} \quad \text{(Reverse the order)}$$

$$= \left(\frac{\pi}{2}\cdot\frac{1}{2}\right)\left(\frac{3}{4}\right)\left(\frac{5}{6}\right)\cdots\left(\frac{n-1}{n}\right)$$

$$= \left(\frac{1}{2}\right)\left(\frac{3}{4}\right)\left(\frac{5}{6}\right)\cdots\left(\frac{n-1}{n}\right)\left(\frac{\pi}{2}\right)$$

87. (a) $f(t) = a_0 + a_1 \cos\frac{\pi t}{6} + b_1 \sin\frac{\pi t}{6}$ where:

$$a_0 = \frac{1}{12}\int_0^{12} f(t)\,dt$$

$$a_1 = \frac{1}{6}\int_0^{12} f(t)\cos\frac{\pi t}{6}\,dt$$

$$b_1 = \frac{1}{6}\int_0^{12} f(t)\sin\frac{\pi t}{6}\,dt$$

—CONTINUED—

87. —CONTINUED—

$$a_0 \approx \frac{12-0}{3(12)^2}[30.9 + 4(32.2) + 2(41.1) + 4(53.7) + 2(64.6) + 4(74.0) + 2(78.2) + 4(77.0) + 2(71.0) + 4(60.1) + 2(47.1) + 4(35.7) + 30.9] \approx 55.46$$

$$a_1 \approx \frac{12-0}{6(3)(12)}\left[30.9\cos 0 + 4\left(32.2\cos\frac{\pi}{6}\right) + 2\left(41.1\cos\frac{\pi}{3}\right) + 4\left(53.7\cos\frac{\pi}{2}\right) + 2\left(64.6\cos\frac{2\pi}{3}\right) + 4\left(74.0\cos\frac{5\pi}{6}\right) + 2(78.2\cos\pi) + 4\left(77.0\cos\frac{7\pi}{6}\right) + 2\left(71.0\cos\frac{4\pi}{3}\right) + 4\left(60.1\cos\frac{3\pi}{2}\right) + 2\left(47.1\cos\frac{5\pi}{3}\right) + 4\left(35.7\cos\frac{11\pi}{6}\right) + 30.9\cos 2\pi\right] \approx -23.88$$

$$b_1 \approx \frac{12-0}{6(3)(12)}\left[30.9\sin 0 + 4\left(32.2\sin\frac{\pi}{6}\right) + 2\left(41.1\sin\frac{\pi}{3}\right) + 4\left(53.7\sin\frac{\pi}{2}\right) + 2\left(64.6\sin\frac{2\pi}{3}\right) + 4\left(74.0\sin\frac{5\pi}{6}\right) + 2(78.2\sin\pi) + 4\left(77.0\sin\frac{7\pi}{6}\right) + 2\left(71.0\sin\frac{4\pi}{3}\right) + 4\left(60.1\sin\frac{3\pi}{2}\right) + 2\left(47.1\sin\frac{5\pi}{3}\right) + 4\left(35.7\sin\frac{11\pi}{6}\right) + 30.9\sin 2\pi\right] \approx -3.34$$

$$H(t) \approx 55.46 - 23.88\cos\frac{\pi t}{6} - 3.34\sin\frac{\pi t}{6}$$

(b) $$a_0 \approx \frac{12-0}{3(12)^2}[18.0 + 4(17.7) + 2(25.8) + 4(36.1) + 2(45.4) + 4(55.2) + 2(59.9) + 4(59.4) + 2(53.1) + 4(43.2) + 2(34.3) + 4(24.2) + 18.0] \approx 39.34$$

$$a_1 \approx \frac{12-0}{6(3)(12)}\left[18.0\cos 0 + 4\left(17.7\cos\frac{\pi}{6}\right) + 2\left(25.8\cos\frac{\pi}{3}\right) + 4\left(36.1\cos\frac{\pi}{2}\right) + 2\left(45.4\cos\frac{2\pi}{3}\right) + 4\left(55.2\cos\frac{5\pi}{6}\right) + 2(59.9\cos\pi) + 4\left(59.4\cos\frac{7\pi}{6}\right) + 2\left(53.1\cos\frac{4\pi}{3}\right) + 4\left(43.2\cos\frac{3\pi}{2}\right) + 2\left(34.3\cos\frac{5\pi}{3}\right) + 4\left(24.2\cos\frac{11\pi}{6}\right) + 18\cos 2\pi\right] \approx -20.78$$

$$b_1 \approx \frac{12-0}{6(3)(12)}\left[18.0\sin 0 + 4\left(17.7\sin\frac{\pi}{6}\right) + 2\left(25.8\sin\frac{\pi}{3}\right) + 4\left(36.1\sin\frac{\pi}{2}\right) + 2\left(45.4\sin\frac{2\pi}{3}\right) + 4\left(55.2\sin\frac{5\pi}{6}\right) + 2(59.9\sin\pi) + 4\left(59.4\sin\frac{7\pi}{6}\right) + 2\left(53.1\sin\frac{4\pi}{3}\right) + 4\left(43.2\sin\frac{3\pi}{2}\right) + 2\left(34.3\sin\frac{5\pi}{3}\right) + 4\left(24.2\sin\frac{11\pi}{6}\right) + 18\sin 2\pi\right] \approx -4.33$$

$$L(t) \approx 39.34 - 20.78\cos\frac{\pi t}{6} - 4.33\sin\frac{\pi t}{6}$$

(c) The difference between the maximum and minimum temperatures is greatest in the summer.

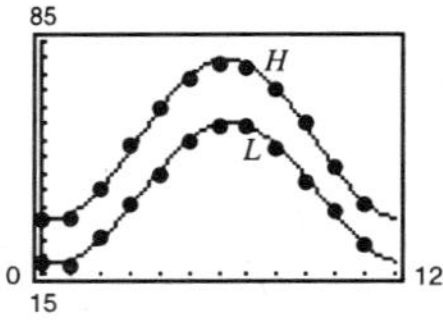

Section 7.4 Trigonometric Substitution

1. $$\frac{d}{dx}\left[4\ln\left|\frac{\sqrt{x^2+16}-4}{x}\right| + \sqrt{x^2+16} + C\right] = \frac{d}{dx}\left[4\ln\left|\sqrt{x^2+16}-4\right| - 4\ln|x| + \sqrt{x^2+16} + C\right]$$

$$= 4\left[\frac{x/\sqrt{x^2+16}}{\sqrt{x^2+16}-4}\right] - \frac{4}{x} + \frac{x}{\sqrt{x^2+16}}$$

$$= \frac{4x}{\sqrt{x^2+16}\left(\sqrt{x^2+16}-4\right)} - \frac{4}{x} + \frac{x}{\sqrt{x^2+16}}$$

$$= \frac{4x^2 - 4\sqrt{x^2+16}\left(\sqrt{x^2+16}-4\right) + x^2\left(\sqrt{x^2-16}-4\right)}{x\sqrt{x^2+16}\left(\sqrt{x^2+16}-4\right)}$$

$$= \frac{4x^2 - 4(x^2+16) + 16\sqrt{x^2+16} + x^2\sqrt{x^2+16} - 4x^2}{x\sqrt{x^2+16}\left(\sqrt{x^2+16}-4\right)}$$

$$= \frac{\sqrt{x^2+16}(x^2+16) - 4(x^2+16)}{x\sqrt{x^2+16}\left(\sqrt{x^2+16}-4\right)}$$

$$= \frac{(x^2+16)\left(\sqrt{x^2+16}-4\right)}{x\sqrt{x^2+16}\left(\sqrt{x^2+16}-4\right)} = \frac{\sqrt{x^2+16}}{x}$$

Indefinite integral: $\displaystyle\int \frac{\sqrt{x^2+16}}{x}\,dx$ Matches (b)

3. $$\frac{d}{dx}\left[8\arcsin\frac{x}{4} - \frac{x\sqrt{16-x^2}}{2} + C\right] = 8\frac{1/4}{\sqrt{1-(x/4)^2}} - \frac{x(1/2)(16-x^2)^{-1/2}(-2x) + \sqrt{16-x^2}}{2}$$

$$= \frac{8}{\sqrt{16-x^2}} + \frac{x^2}{2\sqrt{16-x^2}} - \frac{\sqrt{16-x^2}}{2}$$

$$= \frac{16}{2\sqrt{16-x^2}} + \frac{x^2}{2\sqrt{16-x^2}} - \frac{(16-x^2)}{2\sqrt{16-x^2}} = \frac{x^2}{\sqrt{16-x^2}}$$

Matches (a)

5. Let $x = 5\sin\theta$, $dx = 5\cos\theta\,d\theta$, $\sqrt{25-x^2} = 5\cos\theta$.

$$\int \frac{1}{(25-x^2)^{3/2}}\,dx = \int \frac{5\cos\theta}{(5\cos\theta)^3}\,d\theta = \frac{1}{25}\int \sec^2\theta\,d\theta$$

$$= \frac{1}{25}\tan\theta + C$$

$$= \frac{x}{25\sqrt{25-x^2}} + C$$

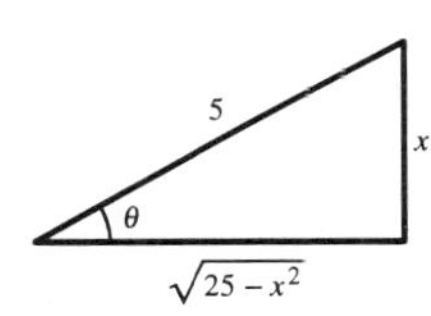

7. Same substitution as in Exercise 5

$$\int \frac{\sqrt{25-x^2}}{x}\,dx = \int \frac{25\cos^2\theta\,d\theta}{5\sin\theta} = 5\int \frac{1-\sin^2\theta}{\sin\theta}\,d\theta = 5\int(\csc\theta - \sin\theta)\,d\theta$$

$$= 5[\ln|\csc\theta - \cot\theta| + \cos\theta] + C = 5\ln\left|\frac{5-\sqrt{25-x^2}}{x}\right| + \sqrt{25-x^2} + C$$

9. Let $x = 2\sec\theta$, $dx = 2\sec\theta\tan\theta\,d\theta$, $\sqrt{x^2-4} = 2\tan\theta$.

$$\int \frac{1}{\sqrt{x^2-4}}\,dx = \int \frac{2\sec\theta\tan\theta\,d\theta}{2\tan\theta} = \int \sec\theta\,d\theta = \ln|\sec\theta + \tan\theta| + C_1$$

$$= \ln\left|\frac{x}{2} + \frac{\sqrt{x^2-4}}{2}\right| + C_1$$

$$= \ln\left|x + \sqrt{x^2-4}\right| - \ln 2 + C_1 = \ln\left|x + \sqrt{x^2-4}\right| + C$$

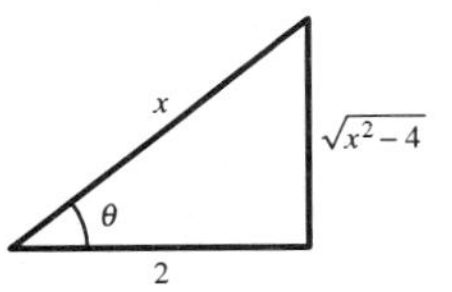

11. Same substitution as in Exercise 9

$$\int x^3\sqrt{x^2-4}\,dx = \int (8\sec^3\theta)(2\tan\theta)(2\sec\theta\tan\theta)\,d\theta = 32\int \tan^2\theta\sec^4\theta\,d\theta$$

$$= 32\int \tan^2\theta(1+\tan^2\theta)\sec^2\theta\,d\theta = 32\left(\frac{\tan^3\theta}{3}+\frac{\tan^5\theta}{5}\right)+C$$

$$= \frac{32}{15}\tan^3\theta[5+3\tan^2\theta]+C = \frac{32}{15}\frac{(x^2-4)^{3/2}}{8}\left[5+3\frac{(x^2-4)}{4}\right]+C$$

$$= \frac{1}{15}(x^2-4)^{3/2}[20+3(x^2-4)]+C = \frac{1}{15}(x^2-4)^{3/2}(3x^2+8)+C$$

13. Let $x = \tan\theta$, $dx = \sec^2\theta\,d\theta$, $\sqrt{1+x^2} = \sec\theta$.

$$\int x\sqrt{1+x^2}\,dx = \int \tan\theta(\sec\theta)\sec^2\theta\,d\theta = \frac{\sec^3\theta}{3}+C = \frac{1}{3}(1+x^2)^{3/2}+C$$

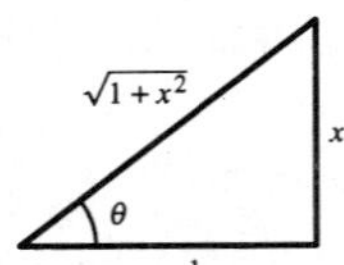

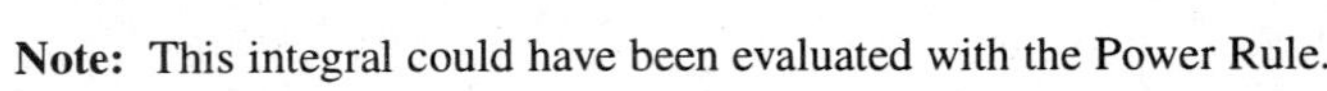

Note: This integral could have been evaluated with the Power Rule.

15. Same substitution as in Exercise 13

$$\int \frac{1}{(1+x^2)^2}\,dx = \int \frac{1}{(\sqrt{1+x^2})^4}\,dx$$

$$= \int \frac{\sec^2\theta\,d\theta}{\sec^4\theta}$$

$$= \int \cos^2\theta\,d\theta = \frac{1}{2}\int(1+\cos 2\theta)\,d\theta$$

$$= \frac{1}{2}\left[\theta+\frac{\sin 2\theta}{2}\right]$$

$$= \frac{1}{2}[\theta+\sin\theta\cos\theta]+C$$

$$= \frac{1}{2}\left[\arctan x+\left(\frac{x}{\sqrt{1+x^2}}\right)\left(\frac{1}{\sqrt{1+x^2}}\right)\right]+C$$

$$= \frac{1}{2}\left[\arctan x+\frac{x}{1+x^2}\right]+C$$

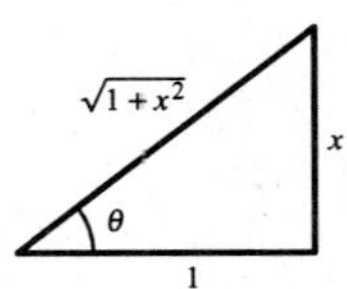

17. Let $u = 3x$, $a = 2$, and $du = 3\,dx$.

$$\int \sqrt{4+9x^2}\,dx = \frac{1}{3}\int \sqrt{(2)^2+(3x)^2}\,3\,dx$$

$$= \frac{1}{3}\left(\frac{1}{2}\right)\left(3x\sqrt{4+9x^2}+4\ln\left|3x+\sqrt{4+9x^2}\right|\right)+C$$

$$= \frac{1}{2}x\sqrt{4+9x^2}+\frac{2}{3}\ln\left|3x+\sqrt{4+9x^2}\right|+C$$

19. $$\int \frac{x}{\sqrt{x^2+9}}\,dx = \frac{1}{2}\int (x^2+9)^{-1/2}(2x)\,dx$$

$$= \sqrt{x^2+9}+C$$

(Power Rule)

21. Let $x = 2\sin\theta$, $dx = 2\cos\theta\,d\theta$, $\sqrt{4-x^2} = 2\cos\theta$.

$$\int_0^2 \sqrt{16-4x^2}\,dx = 2\int_0^2 \sqrt{4-x^2}\,dx$$
$$= 2\int_0^{\pi/2} 2\cos\theta(2\cos\theta\,d\theta)$$
$$= 8\int_0^{\pi/2} \cos^2\theta\,d\theta$$
$$= 4\int_0^{\pi/2} (1+\cos 2\theta)\,d\theta$$
$$= 4\left[\theta + \frac{1}{2}\sin 2\theta\right]_0^{\pi/2} = 2\pi$$

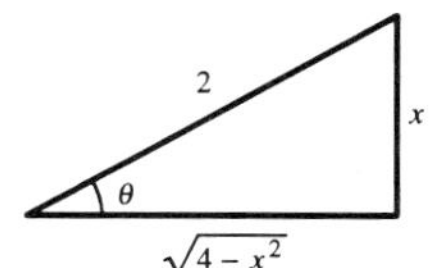

23. Let $x = 3\sec\theta$, $dx = 3\sec\theta\tan\theta\,d\theta$, $\sqrt{x^2-9} = 3\tan\theta$.

$$\int \frac{1}{\sqrt{x^2-9}}\,dx = \int \frac{3\sec\theta\tan\theta\,d\theta}{3\tan\theta}$$
$$= \int \sec\theta\,d\theta$$
$$= \ln\left|\sec\theta + \tan\theta\right| + C_1$$
$$= \ln\left|\frac{x}{3} + \frac{\sqrt{x^2-9}}{3}\right| + C_1$$
$$= \ln\left|x + \sqrt{x^2-9}\right| + C$$

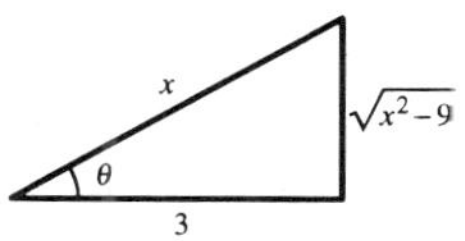

25. Let $x = \sin\theta$, $dx = \cos\theta\,d\theta$, $\sqrt{1-x^2} = \cos\theta$.

$$\int \frac{\sqrt{1-x^2}}{x^4}\,dx = \int \frac{\cos\theta(\cos\theta\,d\theta)}{\sin^4\theta}$$
$$= \int \cot^2\theta\csc^2\theta\,d\theta$$
$$= -\frac{1}{3}\cot^3\theta + C$$
$$= \frac{-(1-x^2)^{3/2}}{3x^3} + C$$

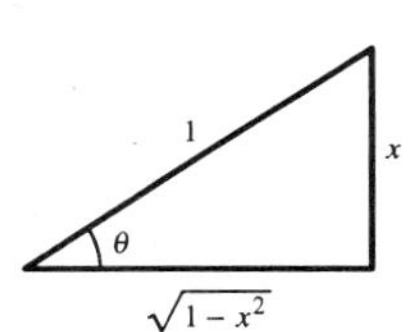

27. Same substitutions as in Exercise 26

$$\int \frac{1}{x\sqrt{4x^2+9}}\,dx = \int \frac{(3/2)\sec^2\theta\,d\theta}{(3/2)\tan\theta\,3\sec\theta}$$
$$= \frac{1}{3}\int \csc\theta\,d\theta$$
$$= -\frac{1}{3}\ln\left|\csc\theta + \cot\theta\right| + C$$
$$= -\frac{1}{3}\ln\left|\frac{\sqrt{4x^2+9}+3}{2x}\right| + C$$

29. Same substitutions as in Exercise 28

$$\int \frac{x}{(x^2+3)^{3/2}}\,dx = \int \frac{\sqrt{3}\tan\theta\left(\sqrt{3}\sec^2\theta\,d\theta\right)}{3\sqrt{3}\sec^3\theta} = \frac{\sqrt{3}}{3}\int \sin\theta\,d\theta = -\frac{\sqrt{3}}{3}\cos\theta + C = \frac{-1}{\sqrt{x^2+3}} + C$$

Note: This integral could have been evaluated with the Power Rule: $u = x^2 + 3$, $du = 2x\,dx$

31. Let $u = 1 + e^{2x}$, $du = 2e^{2x}\,dx$.

$$\int e^{2x}\sqrt{1+e^{2x}}\,dx = \frac{1}{2}\int (1+e^{2x})^{1/2}(2e^{2x})dx = \frac{1}{3}(1+e^{2x})^{3/2} + C$$

33. Let $e^x = \sin\theta$, $e^x\,dx = \cos\theta\,d\theta$, $\sqrt{1-e^{2x}} = \cos\theta$.

$$\int e^x\sqrt{1-e^{2x}}\,dx = \int \cos^2\theta\,d\theta = \frac{1}{2}\int (1+\cos 2\theta)\,d\theta$$
$$= \frac{1}{2}\left[\theta + \frac{\sin 2\theta}{2}\right]$$
$$= \frac{1}{2}(\theta + \sin\theta\cos\theta) + C = \frac{1}{2}\left(\arcsin e^x + e^x\sqrt{1-e^{2x}}\right) + C$$

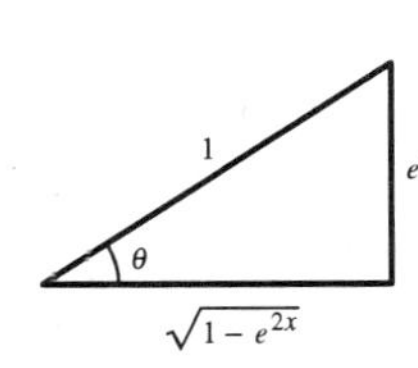

35. Let $x = \sqrt{2}\tan\theta$, $dx = \sqrt{2}\sec^2\theta\,d\theta$, $x^2 + 2 = 2\sec^2\theta$.

$$\int \frac{1}{4 + 4x^2 + x^4}\,dx = \int \frac{1}{(x^2+2)^2}\,dx$$
$$= \int \frac{\sqrt{2}\sec^2\theta\,d\theta}{4\sec^4\theta}$$
$$= \frac{\sqrt{2}}{4}\int \cos^2\theta\,d\theta$$
$$= \frac{\sqrt{2}}{4}\left(\frac{1}{2}\right)\int (1 + \cos 2\theta)\,d\theta$$
$$= \frac{\sqrt{2}}{8}\left(\theta + \frac{1}{2}\sin 2\theta\right) + C$$
$$= \frac{\sqrt{2}}{8}(\theta + \sin\theta\cos\theta) + C$$
$$= \frac{1}{4}\left[\frac{x}{x^2+2} + \frac{1}{\sqrt{2}}\arctan\frac{x}{\sqrt{2}}\right] + C$$

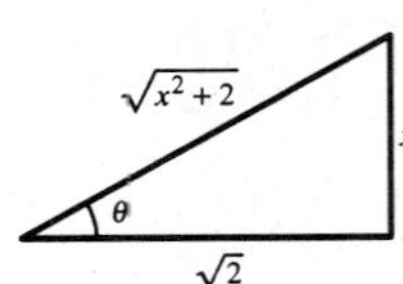

37. Assume $x > 0$.

$$u = \operatorname{arcsec} 2x, \implies du = \frac{1}{x\sqrt{4x^2-1}}\,dx,\ dv = dx \implies v = x$$

$$\int \operatorname{arcsec} 2x\,dx = x\operatorname{arcsec} 2x - \int \frac{1}{\sqrt{4x^2-1}}\,dx$$

$$2x = \sec\theta,\ dx = \frac{1}{2}\sec\theta\tan\theta\,d\theta,\ \sqrt{4x^2-1} = \tan\theta$$

$$\int \operatorname{arcsec} 2x\,dx = x\operatorname{arcsec} 2x - \int \frac{(1/2)\sec\theta\tan\theta\,d\theta}{\tan\theta} = x\operatorname{arcsec} 2x - \frac{1}{2}\int \sec\theta\,d\theta$$
$$= x\operatorname{arcsec} 2x - \frac{1}{2}\ln|\sec\theta + \tan\theta| + C = x\operatorname{arcsec} 2x - \frac{1}{2}\ln\left|2x + \sqrt{4x^2-1}\right| + C$$

A similar argument applies if $x < 0$.

39. $$\int \frac{1}{\sqrt{4x - x^2}}\,dx = \int \frac{1}{\sqrt{4-(x-2)^2}}\,dx = \arcsin\left(\frac{x-2}{2}\right) + C$$

41. Let $x + 2 = 2\tan\theta$, $dx = 2\sec^2\theta\,d\theta$, $\sqrt{(x+2)^2 + 4} = 2\sec\theta$.

$$\int \frac{x}{\sqrt{x^2+4x+8}}\,dx = \int \frac{x}{\sqrt{(x+2)^2+4}}\,dx = \int \frac{(2\tan\theta - 2)(2\sec^2\theta)\,d\theta}{2\sec\theta}$$
$$= 2\int (\tan\theta - 1)(\sec\theta)\,d\theta$$
$$= 2[\sec\theta - \ln|\sec\theta + \tan\theta|] + C_1$$
$$= 2\left[\frac{\sqrt{(x+2)^2+4}}{2} - \ln\left|\frac{\sqrt{(x+2)^2+4}}{2} + \frac{x+2}{2}\right|\right] + C_1$$
$$= \sqrt{x^2+4x+8} - 2\left[\ln\left|\sqrt{x^2+4x+8} + (x+2)\right| - \ln 2\right] + C_1$$
$$= \sqrt{x^2+4x+8} - 2\ln\left|\sqrt{x^2+4x+8} + (x+2)\right| + C$$

43. Let $t = \sin\theta$, $dt = \cos\theta\, d\theta$, $1 - t^2 = \cos^2\theta$.

(a) $$\int \frac{t^2}{(1-t^2)^{3/2}}\,dt = \int \frac{\sin^2\theta\cos\theta\,d\theta}{\cos^3\theta}$$
$$= \int \tan^2\theta\,d\theta$$
$$= \int (\sec^2\theta - 1)\,d\theta$$
$$= \tan\theta - \theta + C$$
$$= \frac{t}{\sqrt{1-t^2}} - \arcsin t + C$$

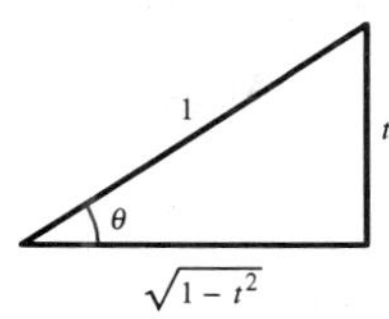

Thus, $$\int_0^{\sqrt{3}/2} \frac{t^2}{(1-t^2)^{3/2}}\,dt = \left[\frac{t}{\sqrt{1-t^2}} - \arcsin t\right]_0^{\sqrt{3}/2} = \frac{\sqrt{3}/2}{\sqrt{1/4}} - \arcsin\frac{\sqrt{3}}{2} = \sqrt{3} - \frac{\pi}{3} \approx 0.685.$$

(b) When $t = 0$, $\theta = 0$. When $t = \sqrt{3}/2$, $\theta = \pi/3$. Thus,
$$\int_0^{\sqrt{3}/2} \frac{t^2}{(1-t^2)^{3/2}}\,dt = \left[\tan\theta - \theta\right]_0^{\pi/3} = \sqrt{3} - \frac{\pi}{3} \approx 0.685.$$

45. (a) Let $x = 3\tan\theta$, $dx = 3\sec^2\theta\,d\theta$, $\sqrt{x^2+9} = 3\sec\theta$.
$$\int \frac{x^3}{\sqrt{x^2+9}}\,dx = \int \frac{(27\tan^3\theta)(3\sec^2\theta\,d\theta)}{3\sec\theta}$$
$$= 27\int (\sec^2\theta - 1)\sec\theta\tan\theta\,d\theta$$
$$= 27\left[\frac{1}{3}\sec^3\theta - \sec\theta\right] + C = 9[\sec^3\theta - 3\sec\theta] + C$$
$$= 9\left[\left(\frac{\sqrt{x^2+9}}{3}\right)^3 - 3\left(\frac{\sqrt{x^2+9}}{3}\right)\right] + C = \frac{1}{3}(x^2+9)^{3/2} - 9\sqrt{x^2+9} + C$$

Thus, $$\int_0^3 \frac{x^3}{\sqrt{x^2+9}}\,dx = \left[\frac{1}{3}(x^2+9)^{3/2} - 9\sqrt{x^2+9}\right]_0^3$$
$$= \left(\frac{1}{3}\left(54\sqrt{2}\right) - 27\sqrt{2}\right) - (9 - 27)$$
$$= 18 - 9\sqrt{2} = 9\left(2 - \sqrt{2}\right) \approx 5.272.$$

(b) When $x = 0$, $\theta = 0$. When $x = 3$, $\theta = \pi/4$. Thus,
$$\int_0^3 \frac{x^3}{\sqrt{x^2+9}}\,dx = 9\left[\sec^3\theta - 3\sec\theta\right]_0^{\pi/4} = 9\left(2\sqrt{2} - 3\sqrt{2}\right) - 9(1-3) = 9\left(2 - \sqrt{2}\right) \approx 5.272.$$

47. $$\int \frac{x^2}{\sqrt{x^2+10x+9}}\,dx = \frac{1}{2}\sqrt{x^2+10x+9}\,(x-15) + 33\ln\left|(x+5) + \sqrt{x^2+10x+9}\right| + C$$

49. $$\int \frac{x^2}{\sqrt{x^2-1}}\,dx = \frac{1}{2}\left(x\sqrt{x^2-1} + \ln\left|x + \sqrt{x^2-1}\right|\right) + C$$

51. $A = 4\int_0^a \frac{b}{a}\sqrt{a^2 - x^2}\,dx$

$$= \frac{4b}{a}\int_0^a \sqrt{a^2 - x^2}\,dx$$

$$= \left[\frac{4b}{a}\left(\frac{1}{2}\right)\left(a^2 \arcsin\frac{x}{a} + x\sqrt{a^2 - x^2}\right)\right]_0^a$$

$$= \frac{2b}{a}\left(a^2\left(\frac{\pi}{2}\right)\right)$$

$$= \pi ab$$

Note: See Theorem 7.2 for $\int \sqrt{a^2 - x^2}\,dx$.

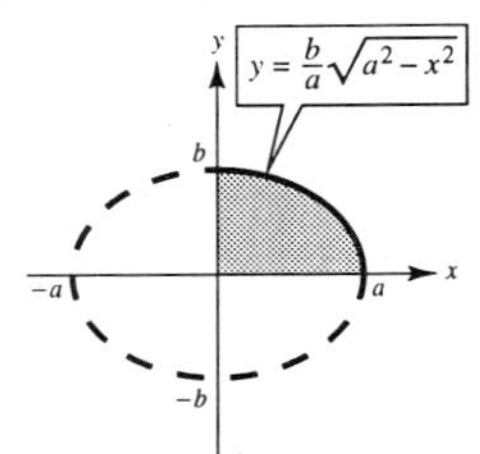

53. Let $x - 3 = \sin\theta$, $dx = \cos\theta\,d\theta$, $\sqrt{1 - (x-3)^2} = \cos\theta$.

Shell Method:

$$V = 4\pi\int_2^4 x\sqrt{1 - (x-3)^2}\,dx$$

$$= 4\pi\int_{-\pi/2}^{\pi/2} (3 + \sin\theta)\cos^2\theta\,d\theta$$

$$= 4\pi\left[\frac{3}{2}\int_{-\pi/2}^{\pi/2} (1 + \cos 2\theta)\,d\theta + \int_{-\pi/2}^{\pi/2} \cos^2\theta\sin\theta\,d\theta\right]$$

$$= 4\pi\left[\frac{3}{2}\left(\theta + \frac{1}{2}\sin 2\theta\right) - \frac{1}{3}\cos^3\theta\right]_{-\pi/2}^{\pi/2} = 6\pi^2$$

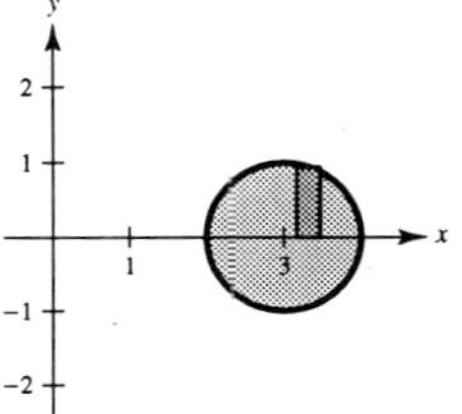

55. $y = \ln x$, $y' = \frac{1}{x}$, $1 + (y')^2 = 1 + \frac{1}{x^2} = \frac{x^2 + 1}{x^2}$

Let $x = \tan\theta$, $dx = \sec^2\theta\,d\theta$, $\sqrt{x^2 + 1} = \sec\theta$.

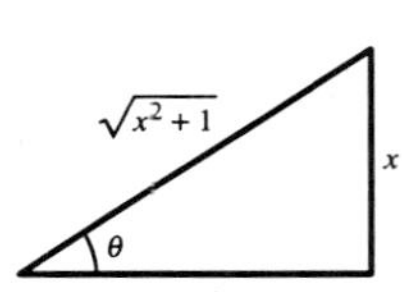

$$s = \int_1^5 \sqrt{\frac{x^2 + 1}{x^2}}\,dx = \int_1^5 \frac{\sqrt{x^2 + 1}}{x}\,dx$$

$$= \int_a^b \frac{\sec\theta}{\tan\theta}\sec^2\theta\,d\theta = \int_a^b \frac{\sec\theta}{\tan\theta}(1 + \tan^2\theta)\,d\theta$$

$$= \int_a^b (\csc\theta + \sec\theta\tan\theta)\,d\theta$$

$$= \Big[-\ln|\csc\theta + \cot\theta| + \sec\theta\Big]_a^b$$

$$= \left[-\ln\left|\frac{\sqrt{x^2 + 1}}{x} + \frac{1}{x}\right| + \sqrt{x^2 + 1}\right]_1^5$$

$$= \left[-\ln\left(\frac{\sqrt{26} + 1}{5}\right) + \sqrt{26}\right] - \left[-\ln(\sqrt{2} + 1) + \sqrt{2}\right]$$

$$= \ln\left[\frac{5(\sqrt{2} + 1)}{\sqrt{26} + 1}\right] + \sqrt{26} - \sqrt{2} \approx 4.367 \text{ or } \ln\left[\frac{\sqrt{26} - 1}{5(\sqrt{2} - 1)}\right] + \sqrt{26} - \sqrt{2}$$

57. Length of one arch of sine curve: $y = \sin x$, $y' = \cos x$

$$L_1 = \int_0^\pi \sqrt{1 + \cos^2 x}\,dx$$

Length of one arch of cosine curve: $y = \cos x$, $y' = -\sin x$

$$L_2 = \int_{-\pi/2}^{\pi/2} \sqrt{1 + \sin^2 x}\,dx$$

$$= \int_{-\pi/2}^{\pi/2} \sqrt{1 + \cos^2\left(x - \frac{\pi}{2}\right)}\,dx \qquad u = x - \frac{\pi}{2},\ du = dx$$

$$= \int_{-\pi}^{0} \sqrt{1 + \cos^2 u}\,du = \int_0^\pi \sqrt{1 + \cos^2 u}\,du = L_1$$

59. (a)

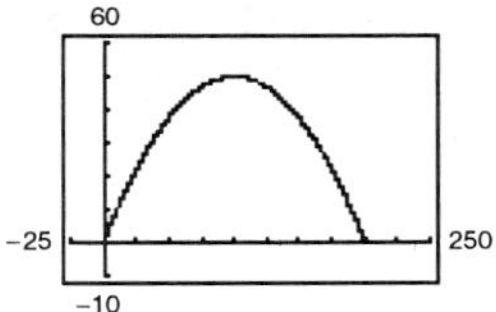

(b) $y = 0$ for $x = 200$ (range)

(c) $y = x - 0.005x^2$, $y' = 1 - 0.01x$, $1 + (y')^2 = 1 + (1 - 0.01x)^2$

Let $u = 1 - 0.01x$, $du = -0.01\,dx$, $a = 1$. (See Theorem 7.2.)

$$s = \int_0^{200} \sqrt{1 + (1 - 0.01x)^2}\,dx = -100\int_0^{200} \sqrt{(1 - 0.01x)^2 + 1}\,(-0.01)\,dx$$

$$= -50\left[(1 - 0.01x)\sqrt{(1 - 0.01x)^2 + 1} + \ln\left|(1 - 0.01x) + \sqrt{(1 - 0.01x)^2 + 1}\right|\right]_0^{200}$$

$$= -50\left[\left(-\sqrt{2} + \ln\left|-1 + \sqrt{2}\right|\right) - \left(\sqrt{2} + \ln\left|1 + \sqrt{2}\right|\right)\right]$$ (b) $y = 0$ for $x = 200$ (range)

$$= 100\sqrt{2} + 50\ln\left(\frac{\sqrt{2} + 1}{\sqrt{2} - 1}\right) \approx 229.559$$

61. $y = x^2$, $\quad y' = 2x$, $\quad 1 + (y') = 1 + 4x^2$

$2x = \tan\theta$, $dx = \dfrac{1}{2}\sec^2\theta\,d\theta$, $\sqrt{1 + 4x^2} = \sec\theta$

(For $\int \sec^5\theta\,d\theta$ and $\int \sec^3\theta\,d\theta$, see Exercise 80 in Section 7.3)

$$S = 2\pi\int_0^{\sqrt{2}} x^2\sqrt{1 + 4x^2}\,dx$$

$$= 2\pi\int_a^b \left(\frac{\tan\theta}{2}\right)^2(\sec\theta)\left(\frac{1}{2}\sec^2\theta\right)d\theta$$

$$= \frac{\pi}{4}\int_a^b \sec^3\theta\tan^2\theta\,d\theta = \frac{\pi}{4}\left[\int_a^b \sec^5\theta\,d\theta - \int_a^b \sec^3\theta\,d\theta\right]$$

$$= \frac{\pi}{4}\left\{\frac{1}{4}\left[\sec^3\theta\tan\theta + \frac{3}{2}(\sec\theta\tan\theta + \ln|\sec\theta + \tan\theta|)\right] - \frac{1}{2}(\sec\theta\tan\theta + \ln|\sec\theta + \tan\theta|)\right\}_a^b$$

$$= \frac{\pi}{4}\left[\frac{1}{4}[(1 + 4x^2)^{3/2}(2x)] - \frac{1}{8}[(1 + 4x^2)^{1/2}(2x) + \ln\left|\sqrt{1 + 4x^2} + 2x\right|]\right]_0^{\sqrt{2}}$$

$$= \frac{\pi}{4}\left[\frac{54\sqrt{2}}{4} - \frac{6\sqrt{2}}{6} = \frac{1}{8}\ln\left(3 + 2\sqrt{2}\right)\right]$$

$$= \frac{\pi}{4}\left(\frac{51\sqrt{2}}{4} - \frac{\ln\left(3 + 2\sqrt{2}\right)}{8}\right) = \frac{\pi}{32}\left[102\sqrt{2} - \ln\left(3 + 2\sqrt{2}\right)\right] \approx 13.989$$

63. First find where the curves intersect.

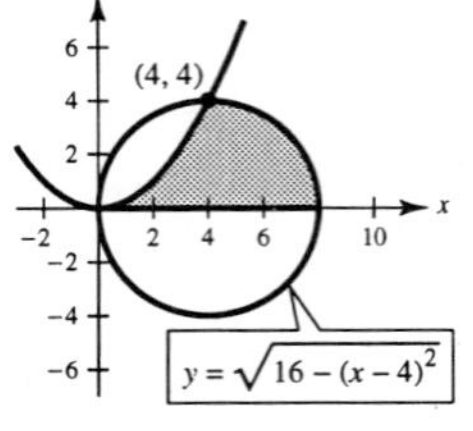

$$y^2 = 16 - (x-4)^2 = \frac{1}{16}x^4$$

$$16^2 - 16(x-4)^2 = x^4$$

$$16^2 - 16x^2 + 128x - 16^2 = x^4$$

$$x^4 + 16x^2 - 128x = 0$$

$$x(x-4)(x^2+4x+32) \Rightarrow x = 0, 4$$

$$A = \int_0^4 \frac{1}{4}x^2 dx + \frac{1}{4}\pi(4)^2 = \frac{1}{12}x^3\Big]_0^4 + 4\pi = \frac{16}{3} + 4\pi$$

$$M_y = \int_0^4 x\left[\frac{1}{4}x^2\right]dx + \int_4^8 x\sqrt{16-(x-4)^2}\,dx$$

$$= \frac{x^4}{16}\Big]_0^4 + \int_4^8 (x-4)\sqrt{16-(x-4)^2}\,dx + \int_4^8 4\sqrt{16-(x-4)^2}\,dx$$

$$= 16 + \left[\frac{-1}{3}(16-(x-4)^2)^{3/2}\right]_4^8 + 2\left[16\arcsin\frac{x-4}{4} + (x-4)\sqrt{16-(x-4)^2}\right]_4^8$$

$$= 16 + \frac{1}{3}16^{3/2} + 2\left[16\left(\frac{\pi}{2}\right)\right] = 16 + \frac{64}{3} + 16\pi = \frac{112}{3} + 16\pi$$

$$M_x = \int_0^4 \frac{1}{2}\left(\frac{1}{4}x^2\right)^2 dx + \int_4^8 \frac{1}{2}(16-(x-4)^2)\,dx$$

$$= \left[\frac{1}{32}\cdot\frac{x^5}{5}\right]_0^4 + \left[8x - \frac{(x-4)^3}{6}\right]_4^8$$

$$= \frac{32}{5} + \left(64 - \frac{64}{6}\right) - 32 = \frac{416}{15}$$

$$\bar{x} = \frac{M_y}{A} = \frac{112/3 + 16\pi}{16/3 + 4\pi} = \frac{112 + 48\pi}{16 + 12\pi} = \frac{28 + 12\pi}{4 + 3\pi} \approx 4.89$$

$$\bar{y} = \frac{M_x}{A} = \frac{416/15}{(16/3) + 4\pi} = \frac{104}{5(4+3\pi)} \approx 1.55$$

$(\bar{x}, \bar{y}) \approx (4.89, 1.55)$

65. (a) Area of representative rectangle: $2\sqrt{1-y^2}\,\Delta y$

Pressure: $2(62.4)(3-y)\sqrt{1-y^2}\,\Delta y$

$$F = 124.8\int_{-1}^{1} (3-y)\sqrt{1-y^2}\,dy$$

$$= 124.8\left[3\int_{-1}^{1}\sqrt{1-y^2}\,dy - \int_{-1}^{1} y\sqrt{1-y^2}\,dy\right]$$

$$= 124.8\left[\frac{3}{2}\left(\arcsin y + y\sqrt{1-y^2}\right) + \frac{1}{2}\left(\frac{2}{3}\right)(1-y^2)^{3/2}\right]_{-1}^{1}$$

$$= (62.4)3[\arcsin 1 - \arcsin(-1)] = 187.2\pi \text{ lb}$$

(b) $$F = 124.8\int_{-1}^{1}(d-y)\sqrt{1-y^2}\,dy = 124.8d\int_{-1}^{1}\sqrt{1-y^2}\,dy - 124.8\int_{-1}^{1} y\sqrt{1-y^2}\,dy$$

$$= 124.8\left(\frac{d}{2}\right)\left[\arcsin y + y\sqrt{1-y^2}\right]_{-1}^{1} - 124.8(0) = 62.4\pi d \text{ lb}$$

67. (a) $m = \dfrac{dy}{dx} = \dfrac{y - \left(y + \sqrt{144 - x^2}\right)}{x - 0}$

$= -\dfrac{\sqrt{144 - x^2}}{x}$

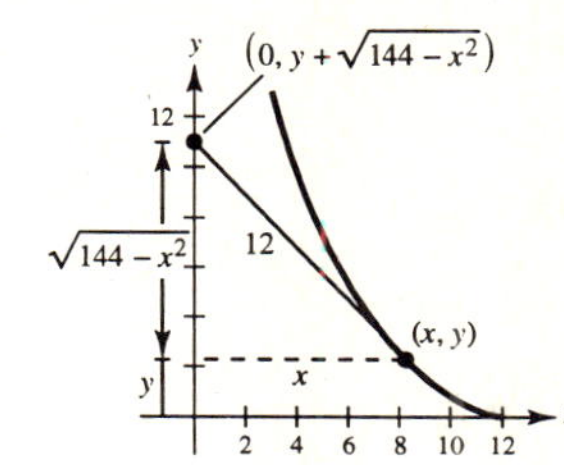

(b) $y = -\displaystyle\int \frac{\sqrt{144 - x^2}}{x}\,dx$

Let $x = 12 \sin\theta$, $dx = 12\cos\theta\,d\theta$, $\sqrt{144 - x^2} = 12\cos\theta$.

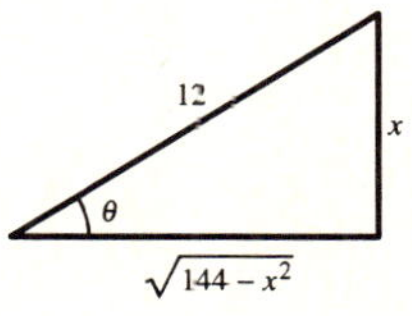

$$y = -\int \frac{12\cos\theta}{12\sin\theta}\,12\cos\theta\,d\theta = -12\int \frac{1 - \sin^2\theta}{\sin\theta}\,d\theta$$

$$= -12\int (\csc\theta - \sin\theta)\,d\theta = -12\ln|\csc\theta - \cot\theta| - 12\cos\theta + C$$

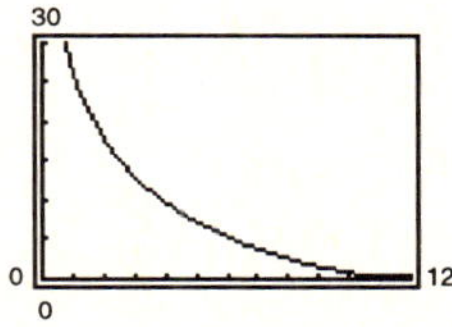

$$= -12\ln\left|\frac{12}{x} - \frac{\sqrt{144 - x^2}}{x}\right| - 12\left(\frac{\sqrt{144 - x^2}}{12}\right) + C$$

$$= -12\ln\left|\frac{12 - \sqrt{144 - x^2}}{x}\right| - \sqrt{144 - x^2} + C$$

When $x = 12$, $y = 0 \Rightarrow C = 0$. Thus, $y = -12\ln\left(\dfrac{12 - \sqrt{144 - x^2}}{x}\right) - \sqrt{144 - x^2}$.

Note: $\dfrac{12 - \sqrt{144 - x^2}}{x} > 0$ for $0 < x \le 12$

(c) Vertical asymptote: $x = 0$

(d) $y + \sqrt{144 - x^2} = 12 \Rightarrow y = 12 - \sqrt{144 - x^2}$

Thus,

$$12 - \sqrt{144 - x^2} = -12\ln\left(\frac{12 - \sqrt{144 - x^2}}{x}\right) - \sqrt{144 - x^2}$$

$$-1 = \ln\left(\frac{12 - \sqrt{144 - x^2}}{x}\right)$$

$$xe^{-1} = 12 - \sqrt{144 - x^2}$$

$$(xe^{-1} - 12)^2 = \left(-\sqrt{144 - x^2}\right)^2$$

$$x^2e^{-2} - 24xe^{-1} + 144 = 144 - x^2$$

$$x^2(e^{-2} + 1) - 24xe^{-1} = 0$$

$$x[x(e^{-2} + 1) - 24e^{-1}] = 0$$

$$x = 0 \text{ or } x = \frac{24e^{-1}}{e^{-2} + 1} \approx 7.77665.$$

Therefore,

$$s = \int_{7.77665}^{12} \sqrt{1 + \left(-\frac{\sqrt{144 - x^2}}{x}\right)^2}\,dx = \int_{7.77665}^{12} \sqrt{\frac{x^2 + (144 - x^2)}{x^2}}\,dx$$

$$= \int_{7.77665}^{12} \frac{12}{x}\,dx = \Big[12\ln|x|\Big]_{7.77665}^{12} = 12(\ln 12 - \ln 7.77665) \approx 5.2 \text{ meters.}$$

69. True

$$\int \frac{dx}{\sqrt{1-x^2}} = \int \frac{\cos\theta\,d\theta}{\cos\theta} = \int d\theta$$

71. False

$$\int_0^{\sqrt{3}} \frac{dx}{\left(\sqrt{1+x^2}\right)^3} = \int_0^{\pi/3} \frac{\sec^2\theta\,d\theta}{\sec^3\theta} = \int_0^{\pi/3} \cos\theta\,d\theta$$

73. Let $u = a\sin\theta$, $du = a\cos\theta\,d\theta$, $\sqrt{a^2-u^2} = a\cos\theta$.

$$\int \sqrt{a^2-u^2}\,du = \int a^2\cos^2\theta\,d\theta = a^2\int \frac{1+\cos 2\theta}{2}\,d\theta$$

$$= \frac{a^2}{2}\left(\theta + \frac{1}{2}\sin 2\theta\right) + C = \frac{a^2}{2}(\theta + \sin\theta\cos\theta) + C$$

$$= \frac{a^2}{2}\left[\arcsin\frac{u}{a} + \left(\frac{u}{a}\right)\left(\frac{\sqrt{a^2+u^2}}{a}\right)\right] + C = \frac{1}{2}\left[a^2\arcsin\frac{u}{a} + u\sqrt{a^2-u^2}\right] + C$$

Let $u = a\sec\theta$, $du = a\sec\theta\tan\theta\,d\theta$, $\sqrt{u^2-a^2} = a\tan\theta$.

$$\int \sqrt{u^2-a^2}\,du = \int a\tan\theta(a\sec\theta\tan\theta)\,d\theta = a^2\int \tan^2\theta\sec\theta\,d\theta$$

$$= a^2\int(\sec^2\theta - 1)\sec\theta\,d\theta = a^2\int(\sec^3\theta - \sec\theta)\,d\theta$$

$$= a^2\left[\frac{1}{2}\sec\theta\tan\theta + \frac{1}{2}\int\sec\theta\,d\theta\right] - a^2\int\sec\theta\,d\theta = a^2\left[\frac{1}{2}\sec\theta\tan\theta - \frac{1}{2}\ln|\sec\theta + \tan\theta|\right]$$

$$= \frac{a^2}{2}\left[\frac{u}{a}\cdot\frac{\sqrt{u^2-a^2}}{a} - \ln\left|\frac{u}{a} + \frac{\sqrt{u^2-a^2}}{a}\right|\right] + C_1 = \frac{1}{2}\left[u\sqrt{u^2-a^2} - a^2\ln\left|u + \sqrt{u^2-a^2}\right|\right] + C$$

Let $u = a\tan\theta$, $du = a\sec^2\theta\,d\theta$, $\sqrt{u^2+a^2} = a\sec\theta\,d\theta$.

$$\int \sqrt{u^2+a^2}\,du = \int(a\sec\theta)(a\sec^2\theta)\,d\theta$$

$$= a^2\int\sec^3\theta\,d\theta = a^2\left[\frac{1}{2}\sec\theta\tan\theta + \frac{1}{2}\ln|\sec\theta + \tan\theta|\right] + C_1$$

$$= \frac{a^2}{2}\left[\frac{\sqrt{u^2+a^2}}{a}\cdot\frac{u}{a} + \ln\left|\frac{\sqrt{u^2+a^2}}{a} + \frac{u}{a}\right|\right] + C_1 = \frac{1}{2}\left[u\sqrt{u^2+a^2} + a^2\ln\left|u + \sqrt{u^2+a^2}\right|\right] + C$$

Section 7.5 Partial Fractions

1. $\dfrac{5}{x^2-10x} = \dfrac{5}{x(x-10)} = \dfrac{A}{x} + \dfrac{B}{x-10}$

3. $\dfrac{2x-3}{x^3+10x} = \dfrac{2x-3}{x(x^2+10)} = \dfrac{A}{x} + \dfrac{Bx+C}{x^2+10}$

5. $\dfrac{16x}{x^3-10x^2} = \dfrac{16x}{x^2(x-10)} = \dfrac{A}{x} + \dfrac{B}{x^2} + \dfrac{C}{x-10}$

7. $\dfrac{1}{x^2-1} = \dfrac{1}{(x+1)(x-1)} = \dfrac{A}{x+1} + \dfrac{B}{x-1}$

$$1 = A(x-1) + B(x+1)$$

When $x = -1$, $1 = -2A$, $A = -\frac{1}{2}$.

When $x = 1$, $1 = 2B$, $B = \frac{1}{2}$.

$$\int\frac{1}{x^2-1}\,dx = -\frac{1}{2}\int\frac{1}{x+1}\,dx + \frac{1}{2}\int\frac{1}{x-1}\,dx$$

$$= -\frac{1}{2}\ln|x+1| + \frac{1}{2}\ln|x-1| + C$$

$$= \frac{1}{2}\ln\left|\frac{x-1}{x+1}\right| + C$$

9. $\dfrac{3}{x^2+x-2} = \dfrac{3}{(x-1)(x+2)} = \dfrac{A}{x-1} + \dfrac{B}{x+2}$

$$3 = (x+2) + B(x-1)$$

When $x = 1$, $3 = 3A$, $A = 1$.

When $x = -2$, $3 = -3B$, $B = -1$.

$$\int\frac{3}{x^2+x-2}\,dx = \int\frac{1}{x-1}\,dx - \int\frac{1}{x+2}\,dx$$

$$= \ln|x-1| - \ln|x+2| + C$$

$$= \ln\left|\frac{x-1}{x+2}\right| + C$$

11. $\dfrac{5-x}{2x^2+x-1} = \dfrac{5-x}{(2x-1)(x+1)} = \dfrac{A}{2x-1} + \dfrac{B}{x+1}$

$$5 - x = A(x+1) + B(2x-1)$$

When $x = \frac{1}{2}, \frac{9}{2} = \frac{3}{2}A, A = 3$. When $x = -1, 6 = -3B, B = -2$.

$$\int \frac{5-x}{2x^2+x-1}\,dx = 3\int \frac{1}{2x-1}\,dx - 2\int \frac{1}{x+1}\,dx$$

$$= \frac{3}{2}\ln|2x-1| - 2\ln|x+1| + C$$

13. $\dfrac{x^2+12x+12}{x(x+2)(x-2)} = \dfrac{A}{x} + \dfrac{B}{x+2} + \dfrac{C}{x-2}$

$$x^2 + 12x + 12 = A(x+2)(x-2) + Bx(x-2) + Cx(x+2)$$

When $x = 0, 12 = -4A, A = -3$. When $x = -2, -8 = 8B, B = -1$. When $x = 2, 40 = 8C, C = 5$.

$$\int \frac{x^2+12x+12}{x^3-4x}\,dx = 5\int \frac{1}{x-2}\,dx - \int \frac{1}{x+2}\,dx - 3\int \frac{1}{x}\,dx$$

$$= 5\ln|x-2| - \ln|x+2| - 3\ln|x| + C$$

15. $\dfrac{2x^3-4x^2-15x+5}{x^2-2x-8} = 2x + \dfrac{x+5}{(x-4)(x+2)} = 2x + \dfrac{A}{x-4} + \dfrac{B}{x+2}$

$$x + 5 = A(x+2) + B(x-4)$$

When $x = 4, 9 = 6A, A = \frac{3}{2}$. When $x = -2, 3 = -6B, B = -\frac{1}{2}$.

$$\int \frac{2x^3-4x^2-15x+5}{x^2-2x-8}\,dx = \int \left[2x + \frac{3/2}{x-4} - \frac{1/2}{x+2}\right] dx$$

$$= x^2 + \frac{3}{2}\ln|x-4| - \frac{1}{2}\ln|x+2| + C$$

17. $\dfrac{4x^2+2x-1}{x^2(x+1)} = \dfrac{A}{x} + \dfrac{B}{x^2} + \dfrac{C}{x+1}$

$$4x^2 + 2x - 1 = Ax(x+1) + B(x+1) + Cx^2$$

When $x = 0, B = -1$. When $x = -1, C = 1$.
When $x = 1, A = 3$.

$$\int \frac{4x^2+2x-1}{x^3+x^2}\,dx = \int \left[\frac{3}{x} - \frac{1}{x^2} + \frac{1}{x+1}\right] dx$$

$$= 3\ln|x| + \frac{1}{x} + \ln|x+1| + C$$

$$= \frac{1}{x} + \ln|x^4 + x^3| + C$$

19. $\dfrac{x^2-1}{x(x^2+1)} = \dfrac{A}{x} + \dfrac{Bx+C}{x^2+1}$

$$x^2 - 1 = A(x^2+1) + (Bx+C)x$$

When $x = 0, A = -1$. When $x = 1, 0 = -2 + B + C$.
When $x = -1, 0 = -2 + B + C$. Solving these equations we have $A = -1, B = 2, C = 0$.

$$\int \frac{x^2-1}{x^3+x}\,dx = -\int \frac{1}{x}\,dx + \int \frac{2x}{x^2+1}\,dx$$

$$= \ln|x^2+1| - \ln|x| + C$$

$$= \ln\left|\frac{x^2+1}{x}\right| + C$$

21. $\dfrac{x^2}{x^4-2x^2-8} = \dfrac{A}{x-2} + \dfrac{B}{x+2} + \dfrac{Cx+D}{x^2+2}$

$$x^2 = A(x+2)(x^2+2) + B(x-2)(x^2+2) + (Cx+D)(x+2)(x-2)$$

When $x = 2, 4 = 24A$. When $x = -2, 4 = -24B$. When $x = 0, 0 = 4A - 4B - 4D$, and when $x = 1$, $1 = 9A - 3B - 3C - 3D$. Solving these equations we have $A = \frac{1}{6}, B = -\frac{1}{6}, C = 0, D = \frac{1}{3}$.

$$\int \frac{x^2}{x^4-2x^2-8}\,dx = \frac{1}{6}\left[\int \frac{1}{x-2}\,dx - \int \frac{1}{x+2}\,dx + 2\int \frac{1}{x^2+2}\,dx\right] = \frac{1}{6}\left[\ln\left|\frac{x-2}{x+2}\right| + \sqrt{2}\arctan\frac{x}{\sqrt{2}}\right] + C$$

23. $\dfrac{x}{(2x-1)(2x+1)(4x^2+1)} = \dfrac{A}{2x-1} + \dfrac{B}{2x+1} + \dfrac{Cx+D}{4x^2+1}$

$$x = A(2x+1)(4x^2+1) + B(2x-1)(4x^2+1) + (Cx+D)(2x-1)(2x+1)$$

When $x = \frac{1}{2}, \frac{1}{2} = 4A$. When $x = -\frac{1}{2}, -\frac{1}{2} = -4B$. When $x = 0, 0 = A - B - D$, and when $x = 1$, $1 = 15A + 5B + 3C + 3D$. Solving these equations we have $A = \frac{1}{8}, B = \frac{1}{8}, C = -\frac{1}{2}, D = 0$.

$$\int \frac{x}{16x^4-1}\,dx = \frac{1}{8}\left[\int \frac{1}{2x-1}\,dx + \int \frac{1}{2x+1}\,dx - 4\int \frac{x}{4x^2+1}\,dx\right]$$

$$= \frac{1}{16}\ln\left|\frac{4x^2-1}{4x^2+1}\right| + C$$

25. $\dfrac{x^2+5}{(x+1)(x^2-2x+3)} = \dfrac{A}{x+1} + \dfrac{Bx+C}{x^2-2x+3}$

$$x^2+5 = A(x^2-2x+3) + (Bx+C)(x+1)$$

$$= (A+B)x^2 + (-2A+B+C)x + (3A+C)$$

When $x = -1, A = 1$. By equating coefficients of like terms, we have $A + B = 1, -2A + B + C = 0, 3A + C = 5$. Solving these equations we have $A = 1, B = 0, C = 2$.

$$\int \frac{x^2+5}{x^3-x^2+x+3}\,dx = \int \frac{1}{x+1}\,dx + 2\int \frac{1}{(x-1)^2+2}\,dx$$

$$= \ln|x+1| + \sqrt{2}\arctan\left(\frac{x-1}{\sqrt{2}}\right) + C$$

27. $\dfrac{3}{(2x+1)(x+2)} = \dfrac{A}{2x+1} + \dfrac{B}{x+2}$

$$3 = A(x+2) + B(2x+1)$$

When $x = -\frac{1}{2}, A = 2$. When $x = -2, B = -1$.

$$\int_0^1 \frac{3}{2x^2+5x+2}\,dx = \int_0^1 \frac{2}{2x+1}\,dx - \int_0^1 \frac{1}{x+2}\,dx$$

$$= \Big[\ln|2x-1| - \ln|x+2|\Big]_0^1 = \ln 2$$

29. $\dfrac{x+1}{x(x^2+1)} = \dfrac{A}{x} + \dfrac{Bx+C}{x^2+1}$

$$x+1 = A(x^2+1) + (Bx+C)x$$

When $x = 0, A = 1$. When $x = 1, 2 = 2A + B + C$. When $x = -1, 0 = 2A + B - C$. Solving these equations we have $A = 1, B = -1, C = 1$.

$$\int_1^2 \frac{x+1}{x(x^2+1)}\,dx = \int_1^2 \frac{1}{x}\,dx - \int_1^2 \frac{x}{x^2+1}\,dx + \int_1^2 \frac{1}{x^2+1}\,dx$$

$$= \left[\ln|x| - \frac{1}{2}\ln(x^2+1) + \arctan x\right]_1^2$$

$$= \frac{1}{2}\ln\frac{8}{5} - \frac{\pi}{4} + \arctan 2 \approx 0.557$$

31. $\displaystyle\int \frac{3x\,dx}{x^2-6x+9} = 3\ln|x-3| - \frac{9}{x-3} + C$

$(4, 0)$: $\quad 3\ln|4-3| - \dfrac{9}{4-3} + C = 0 \Rightarrow C = 9$

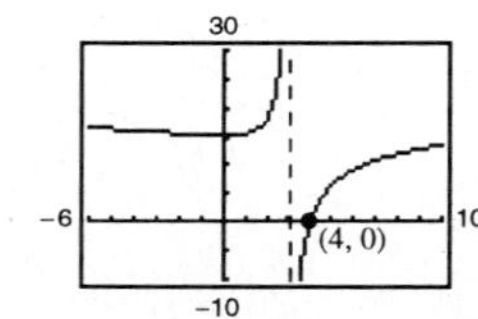

33. $\displaystyle\int \frac{x^2 + x + 2}{(x^2 + 2)^2}\,dx = \frac{\sqrt{2}}{2}\arctan\frac{x}{\sqrt{2}} - \frac{1}{2(x^2 + 2)} + C$

$(0, 1)$: $0 - \dfrac{1}{4} + C = 1 \Longrightarrow C = \dfrac{5}{4}$

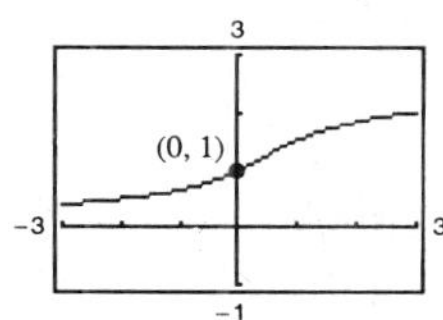

35. $\displaystyle\int \frac{2x^2 - 2x + 3}{x^3 - x^2 - x - 2}\,dx = \ln|x - 2| + \frac{1}{2}\ln|x^2 + x + 1| - \sqrt{3}\arctan\left(\frac{2x + 1}{\sqrt{3}}\right) + C$

$(3, 10)$: $0 + \dfrac{1}{2}\ln 13 - \sqrt{3}\arctan\dfrac{7}{\sqrt{3}} + C = 10 \Longrightarrow C = 10 - \dfrac{1}{2}\ln 13 + \sqrt{3}\arctan\dfrac{7}{\sqrt{3}}$

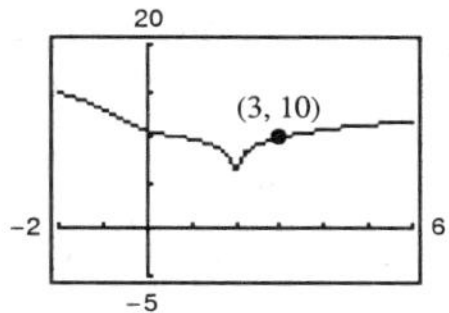

37. $\displaystyle\int \frac{x^2 - x + 2}{x^3 - x^2 + x - 1}\,dx = -\arctan x + \ln|x - 1| + C$

$(2, 6)$: $-\arctan 2 + 0 + C = 6 \Longrightarrow C = 6 + \arctan 2$

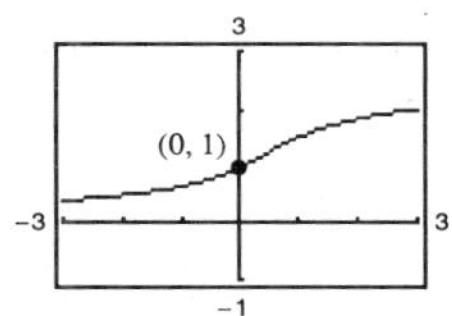

39. Let $u = \cos x$ $du = -\sin x\,dx$.

$$\frac{1}{u(u - 1)} = \frac{A}{u} + \frac{B}{u - 1}$$

$$1 = A(u - 1) + Bu$$

When $u = 0, A = -1$. When $u = 1, B = 1, u = \cos x$, $du = -\sin x\,dx$.

$$\begin{aligned}\int \frac{\sin x}{\cos x(\cos x - 1)}\,dx &= -\int \frac{1}{u(u - 1)}\,du \\ &= \int \frac{1}{u}\,du - \int \frac{1}{u - 1}\,du \\ &= \ln|u| - \ln|u - 1| + C \\ &= \ln\left|\frac{u}{u - 1}\right| + C \\ &= \ln\left|\frac{\cos x}{\cos x - 1}\right| + C\end{aligned}$$

41. $$\begin{aligned}\int \frac{3\cos x}{\sin^2 x + \sin x - 2}\,dx &= 3\int \frac{1}{u^2 + u - 2}\,du \\ &= \ln\left|\frac{u - 1}{u + 2}\right| + C \\ &= \ln\left|\frac{-1 + \sin x}{2 + \sin x}\right| + C\end{aligned}$$

(From Exercise 9 with $u = \sin x$, $du = \cos x\,dx$)

43. Let $u = e^x$, $du = e^x\,dx$.

$$\frac{1}{(u - 1)(u + 4)} = \frac{A}{u - 1} + \frac{B}{u + 4}$$

$$1 = A(u + 4) + B(u - 1)$$

When $u = 1, A = \frac{1}{5}$.

When $u = -4, B = -\frac{1}{5}, u = e^x, du = e^x\,dx$.

$$\begin{aligned}\int \frac{e^x}{(e^x - 1)(e^x + 4)}\,dx &= \int \frac{1}{(u - 1)(u + 4)}\,du \\ &= \frac{1}{5}\left(\int \frac{1}{u - 1}\,du - \int \frac{1}{u + 4}\,du\right) \\ &= \frac{1}{5}\ln\left|\frac{u - 1}{u + 4}\right| + C \\ &= \frac{1}{5}\ln\left|\frac{e^x - 1}{e^x + 4}\right| + C\end{aligned}$$

45. $$\frac{1}{x(a + bx)} = \frac{A}{x} + \frac{B}{a + bx}$$

$$1 = A(a + bx) + Bx$$

When $x = 0$, $1 = aA \Longrightarrow A = 1/a$.
When $x = -a/b$, $1 = -(a/b)B \Longrightarrow B = -b/a$.

$$\begin{aligned}\int \frac{1}{x(a + bx)}\,dx &= \frac{1}{a}\int \left(\frac{1}{x} - \frac{b}{a + bx}\right)dx \\ &= \frac{1}{a}\left(\ln|x| - \ln|a + bx|\right) + C \\ &= \frac{1}{a}\ln\left|\frac{x}{a + bx}\right| + C\end{aligned}$$

47. $\dfrac{x}{(a+bx)^2} = \dfrac{A}{a+bx} + \dfrac{B}{(a+bx)^2}$

$$x = A(a+bx) + B$$

When $x = -a/b$, $B = -a/b$.
When $x = 0$, $0 = aA + B \Rightarrow A = 1/b$.

$$\int \frac{x}{(a+bx)^2}\,dx = \int \left(\frac{1/b}{a+bx} + \frac{-a/b}{(a+bx)^2}\right)dx$$

$$= \frac{1}{b}\int \frac{1}{a+bx}\,dx - \frac{a}{b}\int \frac{1}{(a+bx)^2}\,dx$$

$$= \frac{1}{b^2}\ln|a+bx| + \frac{a}{b^2}\left(\frac{1}{a+bx}\right) + C$$

$$= \frac{1}{b^2}\left(\frac{a}{a+bx} + \ln|a+bx|\right) + C$$

49. $A = \displaystyle\int_1^3 \frac{10}{x(x^2+1)}\,dx \approx 3$

Matches (c)

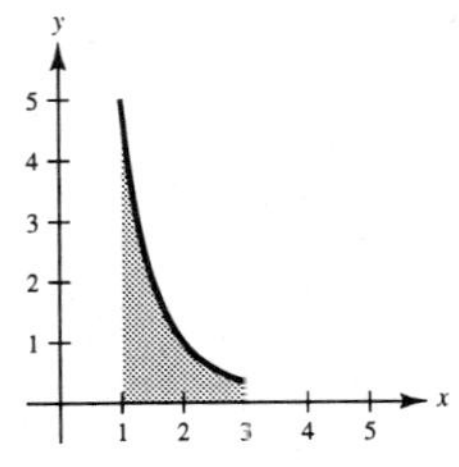

51. (a) $V = \pi\displaystyle\int_0^3 \left(\frac{2x}{x^2+1}\right)^2 dx = 4\pi\int_0^3 \frac{x^2}{(x^2+1)^2}\,dx$

$$= 4\pi\int_0^3 \left(\frac{1}{x^2+1} - \frac{1}{(x^2+1)^2}\right)dx \qquad \text{(partial fractions)}$$

$$= 4\pi\left[\arctan x - \frac{1}{2}\left(\arctan x + \frac{x}{x^2+1}\right)\right]_0^3 \qquad \text{(trigonometric substitution)}$$

$$= 2\pi\left[\arctan x - \frac{x}{x^2+1}\right]_0^3 = 2\pi\left[\arctan 3 - \frac{3}{10}\right] \approx 5.963$$

(b) $A = \displaystyle\int_0^3 \frac{2x}{x^2+1}\,dx = \Big[\ln(x^2+1)\Big]_0^3 = \ln 10$

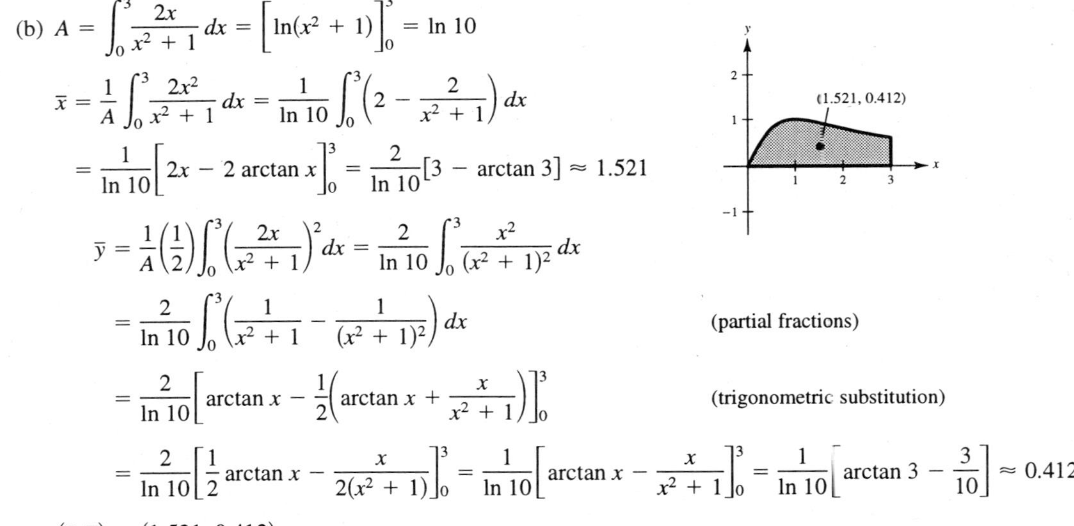

$$\bar{x} = \frac{1}{A}\int_0^3 \frac{2x^2}{x^2+1}\,dx = \frac{1}{\ln 10}\int_0^3 \left(2 - \frac{2}{x^2+1}\right)dx$$

$$= \frac{1}{\ln 10}\Big[2x - 2\arctan x\Big]_0^3 = \frac{2}{\ln 10}[3 - \arctan 3] \approx 1.521$$

$$\bar{y} = \frac{1}{A}\left(\frac{1}{2}\right)\int_0^3 \left(\frac{2x}{x^2+1}\right)^2 dx = \frac{2}{\ln 10}\int_0^3 \frac{x^2}{(x^2+1)^2}\,dx$$

$$= \frac{2}{\ln 10}\int_0^3 \left(\frac{1}{x^2+1} - \frac{1}{(x^2+1)^2}\right)dx \qquad \text{(partial fractions)}$$

$$= \frac{2}{\ln 10}\left[\arctan x - \frac{1}{2}\left(\arctan x + \frac{x}{x^2+1}\right)\right]_0^3 \qquad \text{(trigonometric substitution)}$$

$$= \frac{2}{\ln 10}\left[\frac{1}{2}\arctan x - \frac{x}{2(x^2+1)}\right]_0^3 = \frac{1}{\ln 10}\left[\arctan x - \frac{x}{x^2+1}\right]_0^3 = \frac{1}{\ln 10}\left[\arctan 3 - \frac{3}{10}\right] \approx 0.412$$

$(\bar{x}, \bar{y}) \approx (1.521, 0.412)$

53. $$\frac{1}{(x+1)(n-x)} = \frac{A}{x+1} + \frac{B}{n-x},\ A = B = \frac{1}{n+1}$$

$$\frac{1}{n+1}\int\left(\frac{1}{x+1} + \frac{1}{n-x}\right)dx = kt + C$$

$$\frac{1}{n+1}\ln\left|\frac{x+1}{n-x}\right| = kt + C$$

When $t = 0, x = 0, C = \frac{1}{n+1}\ln\frac{1}{n}$.

$$\frac{1}{n+1}\ln\left|\frac{x+1}{n-x}\right| = kt + \frac{1}{n+1}\ln\frac{1}{n}$$

$$\frac{1}{n+1}\left[\ln\left|\frac{x+1}{n-x}\right| - \ln\frac{1}{n}\right] = kt$$

$$\ln\frac{nx+n}{n-x} = (n+1)kt$$

$$\frac{nx+n}{n-x} = e^{(n+1)kt}$$

$$x = \frac{n[e^{(n+1)kt} - 1]}{n + e^{(n+1)kt}} \qquad \textbf{Note: } \lim_{t\to\infty} x = n$$

55. $$\frac{x}{1+x^4} = \frac{Ax+B}{x^2+\sqrt{2}x+1} + \frac{Cx+D}{x^2-\sqrt{2}x+1}$$

$$\begin{aligned} x &= (Ax+B)(x^2-\sqrt{2}x+1) + (Cx+D)(x^2+\sqrt{2}x+1) \\ &= (A+C)x^3 + (B+D-\sqrt{2}A+\sqrt{2}C)x^2 + (A+C-\sqrt{2}B+\sqrt{2}D)x + (B+D) \end{aligned}$$

$0 = A + C \Rightarrow C = -A$

$0 = B + D - \sqrt{2}A + \sqrt{2}C$ $\quad -2\sqrt{2}A = 0 \Rightarrow A = 0$ and $C = 0$

$1 = A + C - \sqrt{2}B + \sqrt{2}D$ $\quad -2\sqrt{2}B = 1 \Rightarrow B = -\frac{\sqrt{2}}{4}$ and $D = \frac{\sqrt{2}}{4}$

$0 = B + D \Rightarrow D = -B$

Thus,

$$\begin{aligned} \int_0^1 \frac{x}{1+x^4}\,dx &= \int_0^1 \left[\frac{-\sqrt{2}/4}{x^2+\sqrt{2}x+1} + \frac{\sqrt{2}/4}{x^2-\sqrt{2}x+1}\right]dx \\ &= \frac{\sqrt{2}}{4}\int_0^1\left[\frac{-1}{[x+(\sqrt{2}/2)]^2+(1/2)} + \frac{1}{[x-(\sqrt{2}/2)]^2+(1/2)}\right]dx \\ &= \frac{\sqrt{2}}{4}\cdot\frac{1}{1/\sqrt{2}}\left[-\arctan\left(\frac{x+(\sqrt{2}/2)}{1/\sqrt{2}}\right) + \arctan\left(\frac{x-(\sqrt{2}/2)}{1/\sqrt{2}}\right)\right]_0^1 \\ &= \frac{1}{2}\left[-\arctan(\sqrt{2}x+1) + \arctan(\sqrt{2}x-1)\right]_0^1 \\ &= \frac{1}{2}\left[(-\arctan(\sqrt{2}+1) + \arctan(\sqrt{2}-1)) - (-\arctan 1 + \arctan(-1))\right] \\ &= \frac{1}{2}\left[\arctan(\sqrt{2}-1) - \arctan(\sqrt{2}+1) + \frac{\pi}{4} + \frac{\pi}{4}\right]. \end{aligned}$$

Since $\arctan x - \arctan y = \arctan[(x-y)/(1+xy)]$, we have:

$$\int_0^1 \frac{x}{1+x^4}\,dx = \frac{1}{2}\left[\arctan\left(\frac{(\sqrt{2}-1)-(\sqrt{2}+1)}{1+(\sqrt{2}-1)(\sqrt{2}+1)}\right) + \frac{\pi}{2}\right] = \frac{1}{2}\left[\arctan\left(\frac{-2}{2}\right) + \frac{\pi}{2}\right] = \frac{1}{2}\left[-\frac{\pi}{4} + \frac{\pi}{2}\right] = \frac{\pi}{8}$$

57. $\dfrac{x^3 - 3x^2 + 1}{x^4 - 13x^2 + 12x} = \dfrac{P_1}{x} + \dfrac{P_2}{x-1} + \dfrac{P_3}{x+4} + \dfrac{P_4}{x-3} \Rightarrow c_1 = 0,\ c_2 = 1,\ c_3 = -4,\ c_4 = 3$

$N(x) = x^3 - 3x^2 + 1$

$D'(x) = 4x^3 - 26x + 12$

$P_1 = \dfrac{N(0)}{D'(0)} = \dfrac{1}{12}$

$P_2 = \dfrac{N(1)}{D'(1)} = \dfrac{-1}{-10} = \dfrac{1}{10}$

$P_3 = \dfrac{N(-4)}{D'(-4)} = \dfrac{-111}{-140} = \dfrac{111}{140}$

$P_4 = \dfrac{N(3)}{D'(3)} = \dfrac{1}{42}$

Thus, $\dfrac{x^3 - 3x^2 + 1}{x^4 - 13x^2 + 12x} = \dfrac{1/12}{x} + \dfrac{1/10}{x-1} + \dfrac{111/140}{x+4} + \dfrac{1/42}{x-3}$.

Section 7.6 Integration by Tables and Other Integration Techniques

1. By Formula 6: $\displaystyle\int \frac{x^2}{1+x}\,dx = -\frac{x}{2}(2 - x) + \ln|1 + x| + C$

3. By Formula 26: $\displaystyle\int e^x\sqrt{1 + e^{2x}}\,dx = \frac{1}{2}\left[e^x\sqrt{e^{2x} + 1} + \ln\left|e^x + \sqrt{e^{2x} + 1}\right|\right] + C$

$u = e^x,\ du = e^x\,dx$

5. By Formula 44: $\displaystyle\int \frac{1}{x^2\sqrt{1 - x^2}}\,dx = -\frac{\sqrt{1 - x^2}}{x} + C$

7. By Formulas 50 and 48:
$$\begin{aligned}\int \sin^4(2x)\,dx &= \frac{1}{2}\int \sin^4(2x)(2)\,dx\\ &= \frac{1}{2}\left[\frac{-\sin^3(2x)\cos(2x)}{4} + \frac{3}{4}\int \sin^2(2x)(2)\,dx\right]\\ &= \frac{1}{2}\left[\frac{-\sin^3(2x)\cos(2x)}{4} + \frac{3}{8}(2x - \sin 2x\cos 2x)\right] + C\\ &= \frac{1}{16}(6x - 3\sin 2x\cos 2x - 2\sin^3 2x\cos 2x) + C\end{aligned}$$

9. By Formula 57:
$$\begin{aligned}\int \frac{1}{\sqrt{x}\left(1 - \cos\sqrt{x}\right)}\,dx &= 2\int \frac{1}{1 - \cos\sqrt{x}}\left(\frac{1}{2\sqrt{x}}\right)dx\\ &= -2\left(\cot\sqrt{x} + \csc\sqrt{x}\right) + C\end{aligned}$$

$u = \sqrt{x},\ du = \dfrac{1}{2\sqrt{x}}\,dx$

11. By Formula 84:

$$\int \frac{1}{1 + e^{2x}}\,dx = x - \frac{1}{2}\ln(1 + e^{2x}) + C$$

13. By Formula 89:

$$\int x^3\ln x\,dx = \frac{x^4}{16}(4\ln|x| - 1) + C$$

15. (a) By Formulas 83 and 82: $\displaystyle\int x^2e^x\,dx = x^2e^x - 2\int xe^x\,dx$

$$= x^2e^x - 2[(x-1)e^x + C_1]$$

$$= x^2e^x - 2xe^x + 2e^x + C$$

(b) Integration by parts: $u = x^2, du = 2x\,dx, dv = e^x\,dx, v = e^x$

$$\int x^2e^x\,dx = x^2e^x - \int 2xe^x\,dx$$

Parts again: $u = 2x, du = 2\,dx, dv = e^x\,dx, v = e^x$

$$\int x^2e^x\,dx = x^2e^x - \left[2xe^x - \int 2e^x\,dx\right] = x^2e^x - 2xe^x + 2e^x + C$$

17. (a) By Formula: 12, $a = b = 1, u = x$, and

$$\int \frac{1}{x^2(x+1)}\,dx = \frac{-1}{1}\left(\frac{1}{x} + \frac{1}{1}\ln\left|\frac{x}{1+x}\right|\right) + C$$

$$= \frac{-1}{x} - \ln\left|\frac{x}{1+x}\right| + C$$

$$= \frac{-1}{x} + \ln\left|\frac{x+1}{x}\right| + C$$

(b) Partial fractions:

$$\frac{1}{x^2(x+1)} = \frac{A}{x} + \frac{B}{x^2} + \frac{C}{x+1}$$

$$1 = Ax(x+1) - B(x+1) + Cx^2$$

$x = 0$: $1 = B$

$x = -1$: $1 = C$

$x = 1$: $1 = 2A + 2 + 1 \Rightarrow A = -1$

$$\int \frac{1}{x^2(x+1)}\,dx = \int\left[\frac{-1}{x} + \frac{1}{x^2} + \frac{1}{x+1}\right]dx$$

$$= -\ln|x| - \frac{1}{x} + \ln|x+1| + C$$

$$= -\frac{1}{x} - \ln\left|\frac{x}{x+1}\right| + C$$

19. By Formula 81: $\displaystyle\int xe^{x^2} = \frac{1}{2}e^{x^2} + C$

21. By Formula 79: $\displaystyle\int x\,\text{arcsec}(x^2+1)\,dx = \frac{1}{2}\int \text{arcsec}(x^2+1)(2x)\,dx$

$$= \frac{1}{2}\left[(x^2+1)\,\text{arcsec}(x^2+1) - \ln\left((x^2+1) + \sqrt{x^4+2x^2}\right)\right] + C$$

$u = x^2 + 1, du = 2x\,dx$

23. By Formula 89:

$$\int x^2\ln x\,dx = \frac{x^3}{9}\left(-1 + 3\ln|x|\right) + C$$

25. By Formula 35: $\displaystyle\int \frac{1}{x^2\sqrt{x^2-4}}\,dx = \frac{\sqrt{x^2-4}}{4x} + C$

27. By Formula 4: $\displaystyle\int \frac{2x}{(1-3x)^2}\,dx = 2\int \frac{x}{(1-3x)^2}\,dx = \frac{2}{9}\left(\ln|1-3x| + \frac{1}{1-3x}\right) + C$

29. By Formula 76:

$$\int e^x \arccos e^x\,dx = e^x \arccos e^x - \sqrt{1-e^{2x}} + C$$

$u = e^x, du = e^x\,dx$

31. By Formula 73:

$$\int \frac{x}{1-\sec x^2}\,dx = \frac{1}{2}\int \frac{2x}{1-\sec x^2}\,dx$$

$$= \frac{1}{2}(x^2 + \cot x^2 + \csc x^2) + C$$

33. By Formula 23:

$$\int \frac{\cos x}{1 + \sin^2 x}\, dx = \arctan(\sin x) + C$$

$u = \sin x,\ du = \cos x\, dx$

35. By Formula 14: $\displaystyle\int \frac{\cos\theta}{3 + 2\sin\theta + \sin^2\theta}\, d\theta = \frac{\sqrt{2}}{2}\arctan\left(\frac{1 + \sin\theta}{\sqrt{2}}\right) + C$

$u = \sin\theta,\ du = \cos\theta\, d\theta$

37. By Formula 35: $\displaystyle\int \frac{1}{x^2\sqrt{2 + 9x^2}}\, dx = 3\int \frac{3}{(3x)^2\sqrt{\left(\sqrt{2}\right)^2 + (3x)^2}}\, dx$

$$= -\frac{3\sqrt{2 + 9x^2}}{6x} + C = -\frac{\sqrt{2 + 9x^2}}{2x} + C$$

39. By Formulas 55 and 54: $\displaystyle\int t^4 \cos t\, dt = t^4 \sin t - 4\int t^3 \sin t\, dt$

$$= t^4 \sin t - 4\left[-t^3 \cos t + 3\int t^2 \cos t\, dt\right]$$

$$= t^4 \sin t + 4t^3 \cos t - 12\left[t^2 \sin t - 2\int t \sin t\, dt\right]$$

$$= t^4 \sin t + 4t^3 \cos t - 12t^2 \sin t + 24(-t\cos t + \sin t) + C$$

$$= (t^4 - 12t^2 + 24)\sin t + (4t^3 - 24t)\cos t + C$$

41. By Formula 3:

$$\int \frac{\ln x}{x(3 + 2\ln x)}\, dx = \frac{1}{4}\left(2\ln|x| - 3\ln|3 + 2\ln|x||\right) + C$$

$u = \ln x,\ du = \dfrac{1}{x}\, dx$

43. By Formulas 1, 25, and 33: $\displaystyle\int \frac{x}{(x^2 - 6x + 10)^2}\, dx = \frac{1}{2}\int \frac{2x - 6 + 6}{(x^2 - 6x + 10)^2}\, dx$

$$= \frac{1}{2}\int (x^2 - 6x + 10)^{-2}(2x - 6)\, dx + 3\int \frac{1}{[(x - 3)^2 + 1]^2}\, dx$$

$$= -\frac{1}{2(x^2 - 6x + 10)} + \frac{3}{2}\left[\frac{x - 3}{x^2 - 6x + 10} + \arctan(x - 3)\right] + C$$

$$= \frac{3x - 10}{2(x^2 - 6x + 10)} + \frac{3}{2}\arctan(x - 3) + C$$

45. By Formula 31: $\displaystyle\int \frac{x}{\sqrt{x^4 - 6x^2 + 5}}\, dx = \frac{1}{2}\int \frac{2x}{\sqrt{(x^2 - 3)^2 - 4}}\, dx$

$$= \frac{1}{2}\ln\left|x^2 - 3 + \sqrt{x^4 - 6x^2 + 5}\right| + C$$

$u = x^2 - 3,\ du = 2x\, dx$

47. $$\int \frac{x^3}{\sqrt{4-x^2}}\,dx = \int \frac{8\sin^3\theta(2\cos\theta\,d\theta)}{2\cos\theta}$$

$$= 8\int (1-\cos^2\theta)\sin\theta\,d\theta$$

$$= 8\int [\sin\theta - \cos^2\theta(\sin\theta)]\,d\theta$$

$$= -8\cos\theta + \frac{8\cos^3\theta}{3} + C$$

$$= \frac{-\sqrt{4-x^2}}{3}(x^2+8) + C$$

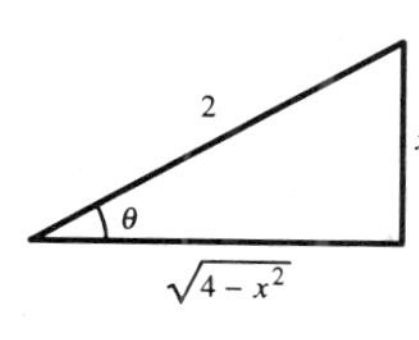

$x = 2\sin\theta,\ dx = 2\cos\theta\,d\theta,\ \sqrt{4-x^2} = 2\cos\theta$

49. By Formula 8:

$$\int \frac{e^{3x}}{(1+e^x)^3}\,dx = \int \frac{(e^x)^2}{(1+e^x)^3}(e^x)\,dx = \frac{2}{1+e^x} - \frac{1}{2(1+e^x)^2} + \ln|1+e^x| + C$$

$u = e^x,\ du = e^x\,dx$

51. $$\frac{u^2}{(a+bu)^2} = \frac{1}{b^2} - \frac{(2a/b)u + (a^2/b^2)}{(a+bu)^2} = \frac{1}{b^2} + \frac{A}{a+bu} + \frac{B}{(a+bu)^2}$$

$$-\frac{2a}{b}u - \frac{a^2}{b^2} = A(a+bu) + B = (aA+B) + bAu$$

Equating the coefficients of like terms we have $aA + B = -a^2/b^2$ and $bA = -2a/b$. Solving these equations we have $A = -2a/b^2$ and $B = a^2/b^2$.

$$\int \frac{u^2}{(a+bu)^2}\,du = \frac{1}{b^2}\int du - \frac{2a}{b^2}\left(\frac{1}{b}\right)\int \frac{1}{a+bu}b\,du + \frac{a^2}{b^2}\left(\frac{1}{b}\right)\int \frac{1}{(a+bu)^2}b\,du$$

$$= \frac{1}{b^2}u - \frac{2a}{b^3}\ln|a+bu| - \frac{a^2}{b^3}\left(\frac{1}{a+bu}\right) + C$$

$$= \frac{1}{b^3}\left(bu - \frac{a^2}{a+bu} - 2a\ln|a+bu|\right) + C$$

53. When we have $u^2 + a^2$:

$$u = a\tan\theta$$

$$du = a\sec^2\theta\,d\theta$$

$$u^2 + a^2 = a^2\sec^2\theta$$

$$\int \frac{1}{(u^2+a^2)^{3/2}}\,du = \int \frac{a\sec^2\theta\,d\theta}{a^3\sec^3\theta}$$

$$= \frac{1}{a^2}\int \cos\theta\,d\theta$$

$$= \frac{1}{a^2}\sin\theta + C$$

$$= \frac{u}{a^2\sqrt{u^2+a^2}} + C$$

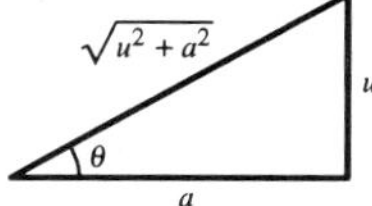

When we have $u^2 - a^2$:

$$u = a\sec\theta$$

$$du = a\sec\theta\tan\theta\,d\theta$$

$$u^2 - a^2 = a^2\tan^2\theta$$

$$\int \frac{1}{(u^2-a^2)^{3/2}}\,du = \int \frac{a\sec\theta\tan\theta\,d\theta}{a^3\tan^3\theta}$$

$$= \frac{1}{a^2}\int \frac{\cos\theta}{\sin^2\theta}\,d\theta$$

$$= -\frac{1}{a^2}\csc\theta + C$$

$$= \frac{-u}{a^2\sqrt{u^2-a^2}} + C$$

55. $\displaystyle\int (\arctan u)\, du = u \arctan u - \frac{1}{2}\int \frac{2u}{1+u^2}\, du$

$\displaystyle= u \arctan u - \frac{1}{2}\ln(1+u^2) + C$

$\displaystyle= u \arctan u - \ln\sqrt{1+u^2} + C$

$\displaystyle w = \arctan u,\ dv = du,\ dw = \frac{du}{1+u^2},\ v = u$

57. $\displaystyle\int \frac{1}{x^{3/2}\sqrt{1-x}}\, dx = \frac{-2\sqrt{1-x}}{\sqrt{x}} + C$

$\displaystyle\left(\frac{1}{2}, 5\right):\ \frac{-2\sqrt{1/2}}{\sqrt{1/2}} + C = 5 \Rightarrow C = 7$

$\displaystyle y = \frac{-2\sqrt{1-x}}{\sqrt{x}} + 7$

8

$(\frac{1}{2}, 5)$

−0.5 1.5

−2

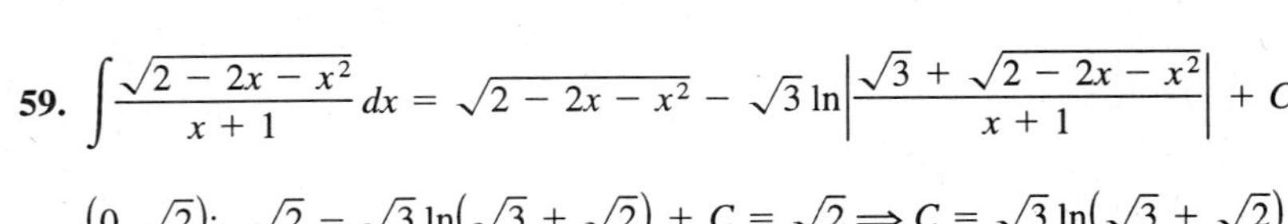

59. $\displaystyle\int \frac{\sqrt{2-2x-x^2}}{x+1}\, dx = \sqrt{2-2x-x^2} - \sqrt{3}\ln\left|\frac{\sqrt{3}+\sqrt{2-2x-x^2}}{x+1}\right| + C$

$(0, \sqrt{2})$: $\sqrt{2} - \sqrt{3}\ln(\sqrt{3}+\sqrt{2}) + C = \sqrt{2} \Rightarrow C = \sqrt{3}\ln(\sqrt{3}+\sqrt{2})$

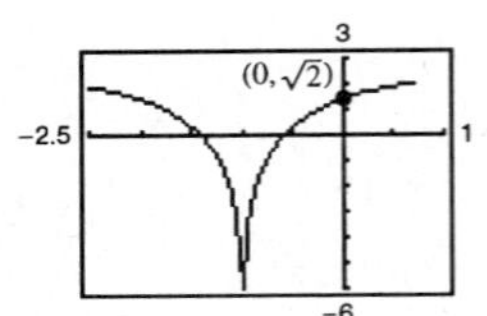

61. $\displaystyle\int \frac{1}{\sin\theta\tan\theta}\, d\theta = -\csc\theta + C$

$\displaystyle\left(\frac{\pi}{4}, 2\right):\ -\frac{2}{\sqrt{2}} + C = 2 \Rightarrow C = 2 + \sqrt{2}$

$y = -\csc\theta + 2 + \sqrt{2}$

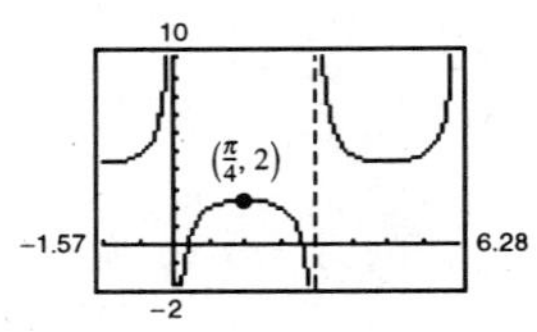

63. $\displaystyle\int \frac{1}{2-3\sin\theta}\, d\theta = \int\left[\frac{\frac{2\,du}{1+u^2}}{2-3\left(\frac{2u}{1+u^2}\right)}\right]$

$\displaystyle= \int 2\frac{2}{(1+u^2)-6u}\, du$

$\displaystyle= \int \frac{1}{u^2-3u+1}\, du$

$\displaystyle= \int \frac{1}{\left(u-\frac{3}{2}\right)^2 - \frac{5}{4}}\, du$

$\displaystyle= \frac{1}{\sqrt{5}}\ln\left|\frac{\left(u-\frac{3}{2}\right)-\frac{\sqrt{5}}{2}}{\left(u-\frac{3}{2}\right)+\frac{\sqrt{5}}{2}}\right| + C$

$\displaystyle= \frac{1}{\sqrt{5}}\ln\left|\frac{2u-3-\sqrt{5}}{2u-3+\sqrt{5}}\right| + C$

$\displaystyle= \frac{1}{\sqrt{5}}\ln\left|\frac{2\tan\left(\frac{\theta}{2}\right)-3-\sqrt{5}}{2\tan\left(\frac{\theta}{2}\right)-3+\sqrt{5}}\right| + C$

$\displaystyle u = \tan\frac{\theta}{2}$

65. $\displaystyle\int_0^{\pi/2} \frac{1}{1+\sin\theta+\cos\theta}\, d\theta = \int_0^1\left[\frac{\frac{2\,du}{1+u^2}}{1+\frac{2u}{1+u^2}+\frac{1-u^2}{1+u^2}}\right]$

$\displaystyle= \int_0^1 \frac{1}{1+u}\, du$

$\displaystyle= \Big[\ln|1+u|\Big]_0^1$

$= \ln 2$

$\displaystyle u = \tan\frac{\theta}{2}$

67. $$\int \frac{\sin\theta}{3-2\cos\theta}\,d\theta = \frac{1}{2}\int \frac{2\sin\theta}{3-2\cos\theta}\,d\theta$$
$$= \frac{1}{2}\ln|u| + C$$
$$= \frac{1}{2}\ln(3-2\cos\theta) + C$$

$u = 3 - 2\cos\theta,\ du = 2\sin\theta\,d\theta$

69. $$\int \frac{\cos\sqrt{\theta}}{\sqrt{\theta}}\,d\theta = 2\int \cos\sqrt{\theta}\left(\frac{1}{2\sqrt{\theta}}\right)d\theta$$
$$= 2\sin\sqrt{\theta} + C$$

$u = \sqrt{\theta},\ du = \dfrac{1}{2\sqrt{\theta}}\,d\theta$

71. $$A = \int_0^8 \frac{x}{\sqrt{x+1}}\,dx$$
$$= \left[\frac{-2(2-x)}{3}\sqrt{x+1}\right]_0^8$$
$$= 12 - \left(-\frac{4}{3}\right)$$
$$= \frac{40}{3} \approx 13.333 \text{ square units}$$

73. $$W = \int_0^5 2000xe^{-x}\,dx$$
$$= -2000\int_0^5 -xe^{-x}\,dx$$
$$= 2000\int_0^5 (-x)e^{-x}(-1)\,dx$$
$$= 2000\Big[(-x)e^{-x} - e^{-x}\Big]_0^5$$
$$= 2000\left(-\frac{6}{e^5} + 1\right)$$
$$\approx 1919.145 \text{ ft} \cdot \text{lbs}$$

75. (a) $$V = 20(2)\int_0^3 \frac{2}{\sqrt{1+y^2}}\,dy$$
$$= \Big[80\ln\left|y + \sqrt{1+y^2}\right|\Big]_0^3$$
$$= 80\ln\left(3 + \sqrt{10}\right)$$
$$\approx 145.5 \text{ cubic feet}$$

$$W = 148\left(80\ln\left(3+\sqrt{10}\right)\right)$$
$$= 11{,}840\ln\left(3+\sqrt{10}\right)$$
$$\approx 21{,}530.4 \text{ lb}$$

(b) By symmetry, $\bar{x} = 0$.

$$M = \rho(2)\int_0^3 \frac{2}{\sqrt{1+y^2}}\,dy = \Big[4\rho\ln\left|y+\sqrt{1+y^2}\right|\Big]_0^3 = 4\rho\ln\left(3+\sqrt{10}\right)$$

$$M_x = 2\rho\int_0^3 \frac{2y}{\sqrt{1+y^2}}\,dy = \Big[4\rho\sqrt{1+y^2}\Big]_0^3 = 4\rho\left(\sqrt{10}-1\right)$$

$$\bar{y} = \frac{M_x}{M} = \frac{4\rho\left(\sqrt{10}-1\right)}{4\rho\ln\left(3+\sqrt{10}\right)} \approx 1.19$$

Centroid: $(\bar{x}, \bar{y}) \approx (0, 1.19)$

77. (a) $$\int_0^4 \frac{k}{2+3x}\,dx = 10$$

$$k = \frac{10}{\displaystyle\int_0^4 \frac{1}{2+3x}\,dx} \approx \frac{10}{0.6486}$$
$$= 15.417\ \left(= \frac{30}{\ln 7}\right)$$

(b) $$\int_0^4 \frac{15.417}{2+3x}\,dx$$

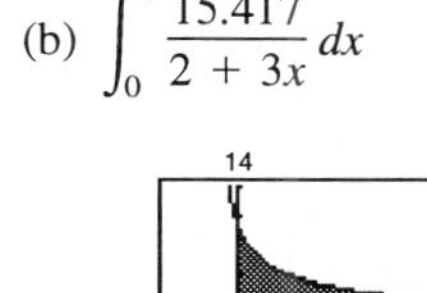
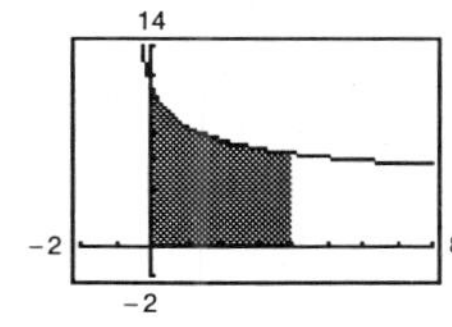

Section 7.7 Indeterminate Forms and L'Hôpital's Rule

1. $\lim_{x\to 0} \dfrac{\sin 5x}{\sin 2x} \approx 2.5 \left(\text{exact: } \dfrac{5}{2}\right)$

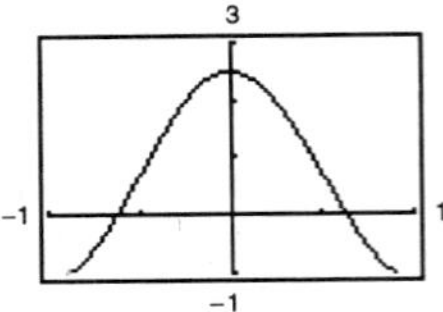

x	-0.1	-0.01	-0.001	0.001	0.01	0.1
$f(x)$	2.4132	2.4991	2.500	2.500	2.4991	2.4132

3. $\lim_{x\to\infty} x^5 e^{-x/100} \approx 0$

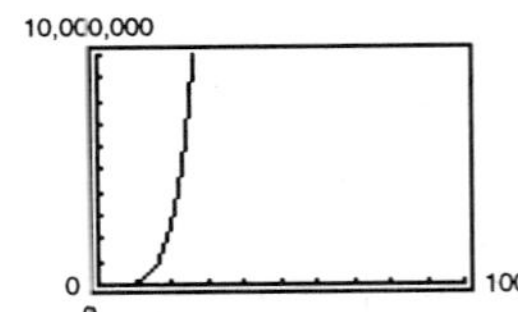

x	1	10	10^2	10^3	10^4	10^5
$f(x)$	0.9901	90,484	3.7×10^9	4.5×10^{10}	0	0

5. (a) $\lim_{x\to 3} \dfrac{2(x-3)}{x^2-9} = \lim_{x\to 3} \dfrac{2(x-3)}{(x+3)(x-3)} = \lim_{x\to 3} \dfrac{2}{x+3} = \dfrac{1}{3}$

(b) $\lim_{x\to 3} \dfrac{2(x-3)}{x^2-9} = \lim_{x\to 3} \dfrac{(d/dx)[2(x-3)]}{(d/dx)[x^2-9]} = \lim_{x\to 3} \dfrac{2}{2x} = \dfrac{2}{6} = \dfrac{1}{3}$

7. (a) $\lim_{x\to 3} \dfrac{\sqrt{x+1}-2}{x-3} = \lim_{x\to 3} \dfrac{\sqrt{x+1}-2}{x-3} \cdot \dfrac{\sqrt{x+1}+2}{\sqrt{x+1}+2} = \lim_{x\to 3} \dfrac{(x+1)-4}{(x-3)\left[\sqrt{x+1}+2\right]} = \lim_{x\to 3} \dfrac{1}{\sqrt{x+1}+2} = \dfrac{1}{4}$

(b) $\lim_{x\to 3} \dfrac{\sqrt{x+1}-2}{x-3} = \lim_{x\to 3} \dfrac{(d/dx)\left[\sqrt{x+1}-2\right]}{(d/dx)[x-3]} = \lim_{x\to 3} \dfrac{1/\left(2\sqrt{x+1}\right)}{1} = \dfrac{1}{4}$

9. (a) $\lim_{x\to\infty} \dfrac{5x^2-3x+1}{3x^2-5} = \lim_{x\to\infty} \dfrac{5-(3/x)+(1/x^2)}{3-(5/x^2)} = \dfrac{5}{3}$

(b) $\lim_{x\to\infty} \dfrac{5x^2-3x+1}{3x^2-5} = \lim_{x\to\infty} \dfrac{(d/dx)[5x^2-3x+1]}{(d/dx)[3x^2-5]} = \lim_{x\to\infty} \dfrac{10x-3}{6x} = \lim_{x\to\infty} \dfrac{(d/dx)[10x-3]}{(d/dx)[6x]} = \lim_{x\to\infty} \dfrac{10}{6} = \dfrac{5}{3}$

11. $\lim_{x\to 2} \dfrac{x^2-x-2}{x-2} = \lim_{x\to 2} \dfrac{2x-1}{1} = 3$

13. $\lim_{x\to 0} \dfrac{\sqrt{4-x^2}-2}{x} = \lim_{x\to 0} \dfrac{-x/\sqrt{4-x^2}}{1} = 0$

15. $\lim_{x\to 0} \dfrac{e^x-(1-x)}{x} = \lim_{x\to 0} \dfrac{e^x+1}{1} = 2$

17. Case 1: $n = 1$

$$\lim_{x\to 0^+} \frac{e^x-(1+x)}{x} = \lim_{x\to 0^+} \frac{e^x-1}{1} = 0$$

Case 2: $n = 2$

$$\lim_{x\to 0^+} \frac{e^x-(1+x)}{x^2} = \lim_{x\to 0^+} \frac{e^x-1}{2x} = \lim_{x\to 0^+} \frac{e^x}{2} = \frac{1}{2}$$

Case 3: $n \geq 3$

$$\lim_{x\to 0^+} \frac{e^x-(1+x)}{x^n} = \lim_{x\to 0^+} \frac{e^x-1}{nx^{n-1}} = \lim_{x\to 0^+} \frac{e^x}{n(n-1)x^{n-2}} = \infty$$

19. $\lim_{x\to 0} \frac{\sin 2x}{\sin 3x} = \lim_{x\to 0} \frac{2\cos 2x}{3\cos 3x} = \frac{2}{3}$

21. $\lim_{x\to 0} \frac{\arcsin x}{x} = \lim_{x\to 0} \frac{1/\sqrt{1-x^2}}{1} = 1$

23. $\lim_{x\to\infty} \frac{3x^2 - 2x + 1}{2x^2 + 3} = \lim_{x\to\infty} \frac{6x - 2}{4x} = \lim_{x\to\infty} \frac{6}{4} = \frac{3}{2}$

25. $\lim_{x\to\infty} \frac{x^2 + 2x + 3}{x - 1} = \lim_{x\to\infty} \frac{2x + 2}{1} = \infty$

27. $\lim_{x\to\infty} \frac{x}{\sqrt{x^2+1}} = \lim_{x\to\infty} \frac{1}{\sqrt{1 + (1/x^2)}} = 1$

Note: L'Hôpital's Rule does not work on this limit. See Exercise 67.

29. $\lim_{x\to\infty} \frac{\ln x}{x} = \lim_{x\to\infty} \frac{1/x}{1} = 0$

31. (a) $\lim_{x\to 0^+} (-x\ln x) = (-0)(-\infty) = (0)(\infty)$

(b) $\lim_{x\to 0^+} (-x\ln x) = \lim_{x\to 0^+} \frac{\ln x}{-1/x}$

$= \lim_{x\to 0^+} \frac{1/x}{1/x^2}$

$= \lim_{x\to 0^+} x = 0$

(c)

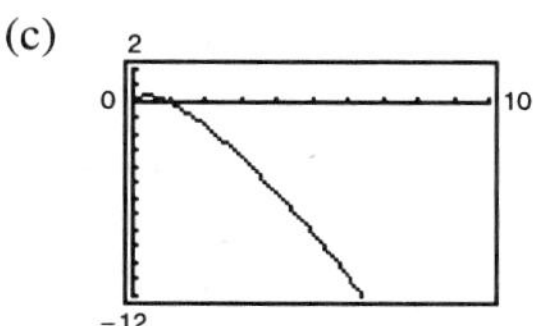

33. (a) $\lim_{x\to\infty} \left(x\sin\frac{1}{x}\right) = (\infty)(0)$

(b) $\lim_{x\to\infty} x\sin\frac{1}{x} = \lim_{x\to\infty} \frac{\sin(1/x)}{1/x}$

$= \lim_{x\to\infty} \frac{(-1/x^2)\cos(1/x)}{-1/x^2}$

$= \lim_{x\to\infty} \cos\left(\frac{1}{x}\right) = 1$

(c)

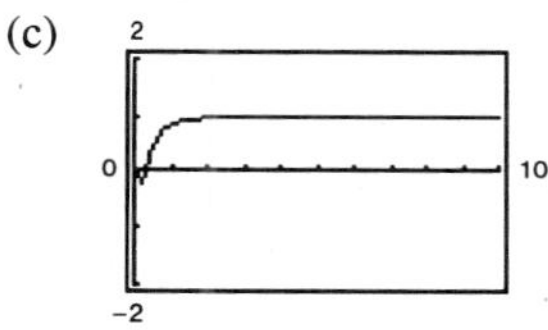

35. (a) $\lim_{x\to 0^+} x^{1/x} = 0^\infty = 0$, not indeterminant

(See Exercise 83)

(b) Let $y = x^{1/x}$

$\ln y = \ln x^{1/x} = \frac{1}{x}\ln x.$

Since $x \to 0^+$, $\frac{1}{x}\ln x \to (\infty)(-\infty) = -\infty$. Hence,

$\ln y \to -\infty \Rightarrow y \to 0^+$.

Therefore, $\lim_{x\to 0^+} x^{1/x} = 0$.

(c)

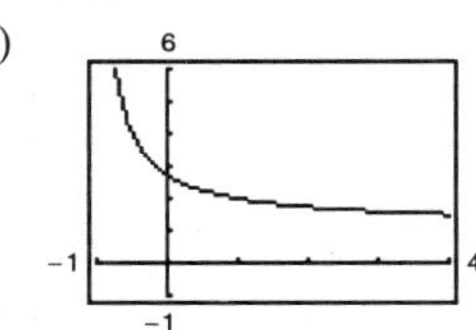

37. (a) $\lim_{x\to\infty} x^{1/x} = \infty^0$

(b) Let $y = \lim_{x\to\infty} x^{1/x}$.

$\ln y = \lim_{x\to\infty} \frac{\ln x}{x} = \lim_{x\to\infty} \left(\frac{1/x}{1}\right) = 0$

Thus, $\ln y = 0 \Rightarrow y = e^0 = 1$. Therefore,

$\lim_{x\to\infty} x^{1/x} = 1$.

(c)

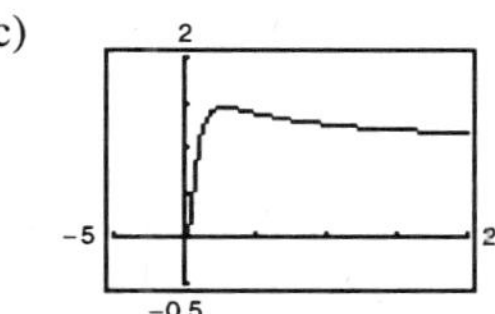

39. (a) $\lim_{x\to 0^+} (1 + x)^{1/x} = 1^\infty$

(c)

(b) Let $y = \lim_{x\to 0^+} (1 + x)^{1/x}$.

$\ln y = \lim_{x\to 0^+} \frac{\ln(1+x)}{x}$

$= \lim_{x\to 0^+} \left(\frac{1/(1+x)}{1}\right) = 1$

Thus, $\ln y = 1 \Rightarrow y = e^1 = e$.

Therefore, $\lim_{x\to 0^+} (1 + x)^{1/x} = e$.

41. (a) $\lim_{x \to 2^+} \left(\dfrac{8}{x^2 - 4} - \dfrac{x}{x - 2} \right) = \infty - \infty$

(b) $$\lim_{x \to 2^+} \left(\frac{8}{x^2 - 4} - \frac{x}{x - 2} \right) = \lim_{x \to 2^+} \frac{8 - x(x + 2)}{x^2 - 4}$$
$$= \lim_{x \to 2^+} \frac{(2 - x)(4 + x)}{(x + 2)(x - 2)}$$
$$= \lim_{x \to 2^+} \frac{-(x + 4)}{x + 2} = \frac{-3}{2}$$

(c)

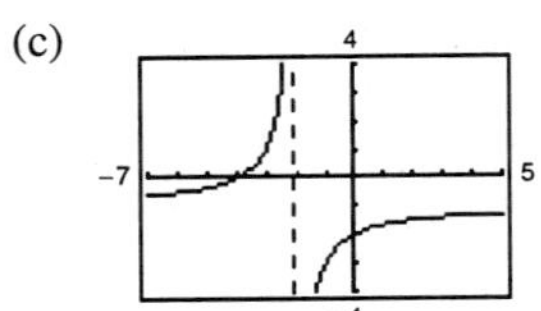

43. (a) $\lim_{x \to 1^+} \left(\dfrac{3}{\ln x} - \dfrac{2}{x - 1} \right) = \infty - \infty$

(b) $$\lim_{x \to 1^+} \left(\frac{3}{\ln x} - \frac{2}{x - 1} \right) = \lim_{x \to 1^+} \frac{3x - 3 - 2 \ln x}{(x - 1)\ln x}$$
$$= \lim_{x \to 1^+} \frac{3 - (2/x)}{[(x - 1)/x] + \ln x} = \infty$$

(c)

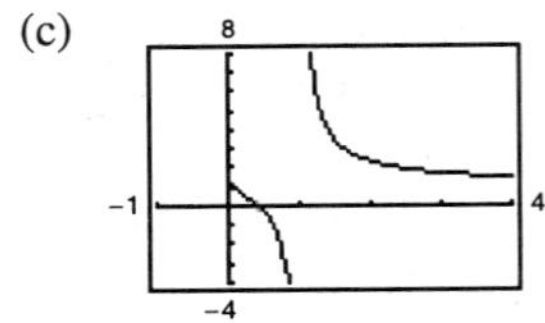

45. (a)

(b) $$\lim_{x \to 3} \frac{x - 3}{\ln(2x - 5)} = \lim_{x \to 3} \frac{1}{2/(2x - 5)}$$
$$= \lim_{x \to 3} \frac{2x - 5}{2} = \frac{1}{2}$$

47. (a)

(b) $$\lim_{x \to \infty} \left(\sqrt{x^2 + 5x + 2} - x \right) = \lim_{x \to \infty} \left(\sqrt{x^2 + 5x + 2} - x \right) \frac{\left(\sqrt{x^2 + 5x + 2} + x \right)}{\left(\sqrt{x^2 + 5x + 2} + x \right)}$$
$$= \lim_{x \to \infty} \frac{(x^2 + 5x + 2) - x^2}{\sqrt{x^2 + 5x + 2} + x}$$
$$= \lim_{x \to \infty} \frac{5x + 2}{\sqrt{x^2 + 5x + 2} + x}$$
$$= \lim_{x \to \infty} \frac{5 + (2/x)}{\sqrt{1 + (5/x) + (2/x^2)} + 1} = \frac{5}{2}$$

49. (a) Let $f(x) = x^2 - 25$ and $g(x) = x - 5$.

(b) Let $f(x) = (x - 5)^2$ and $g(x) = x^2 - 25$.

(c) Let $f(x) = x^2 - 25$ and $g(x) = (x - 5)^3$.

51. $\lim_{x \to \infty} \dfrac{x^2}{e^{5x}} = \lim_{x \to \infty} \dfrac{2x}{5e^{5x}} = \lim_{x \to \infty} \dfrac{2}{25e^{5x}} = 0$

53. $$\lim_{x \to \infty} \frac{(\ln x)^3}{x} = \lim_{x \to \infty} \frac{3(\ln x)^2(1/x)}{1}$$
$$= \lim_{x \to \infty} \frac{3(\ln x)^2}{x}$$
$$= \lim_{x \to \infty} \frac{6(\ln x)(1/x)}{1}$$
$$= \lim_{x \to \infty} \frac{6(\ln x)}{x} = \lim_{x \to \infty} \frac{6}{x} = 0$$

55. $$\lim_{x \to \infty} \frac{(\ln x)^n}{x^m} = \lim_{x \to \infty} \frac{n(\ln x)^{n-1}/x}{mx^{m-1}}$$
$$= \lim_{x \to \infty} \frac{n(\ln x)^{n-1}}{mx^m}$$
$$= \lim_{x \to \infty} \frac{n(n - 1)(\ln x)^{n-2}}{m^2x^m}$$
$$= \cdots = \lim_{x \to \infty} \frac{n!}{m^n x^m} = 0$$

57.

x	10	10^2	10^4	10^6	10^8	10^{10}
$\dfrac{(\ln x)^4}{x}$	2.811	4.498	0.720	0.036	0.001	0.000

59. $y = x^{1/x}, x > 0$

Horizontal asymptote: $y = 1$ (See Exercise 37)

$$\ln y = \frac{1}{x}\ln x$$

$$\frac{1}{y}\frac{dy}{dx} = \frac{1}{x}\left(\frac{1}{x}\right) + (\ln x)\left(-\frac{1}{x^2}\right)$$

$$\frac{dy}{dx} = x^{1/x}\left(\frac{1}{x^2}\right)(1 - \ln x) = x^{(1/x)-2}(1 - \ln x) = 0$$

Critical number: $x = e$

Intervals:	$(0, e)$	(e, ∞)
Sign of dy/dx:	$+$	$-$
$y = f(x)$:	Increasing	Decreasing

Relative maximum: $(e, e^{1/e})$

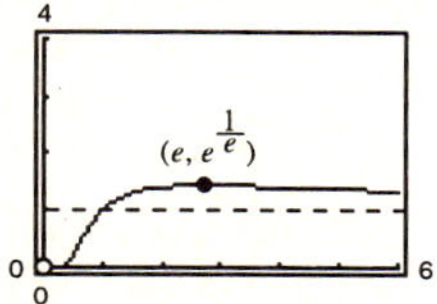

61. $y = 2xe^{-x}$

$$\lim_{x\to\infty} \frac{2x}{e^x} = \lim_{x\to\infty} \frac{2}{e^x} = 0$$

Horizontal asymptote: $y = 0$

$$\frac{dy}{dx} = 2x(-e^{-x}) + 2e^{-x}$$

$$= 2e^{-x}(1 - x) = 0$$

Critical number: $x = 1$

Intervals:	$(-\infty, 1)$	$(1, \infty)$
Sign of dy/dx:	$+$	$-$
$y = f(x)$:	Increasing	Decreasing

Relative maximum: $\left(1, \frac{2}{e}\right)$

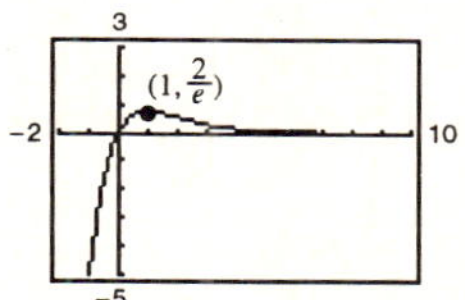

63. $\lim_{x\to 0} \frac{e^{2x} - 1}{e^x} = \frac{0}{1} = 0$

Limit is not of the form $0/0$ or ∞/∞.
L'Hôpital's Rule does not apply.

65. $\lim_{x\to\infty} x \cos\frac{1}{x} = \infty(1) = \infty$

Limit is not of the form $0/0$ or ∞/∞.
L'Hôpital's Rule does not apply.

67. (a) $$\lim_{x\to\infty} \frac{x}{\sqrt{x^2+1}} = \lim_{x\to\infty} \frac{x/x}{\sqrt{x^2+1}/x}$$

$$= \lim_{x\to\infty} \frac{1}{\sqrt{x^2+1}/\sqrt{x^2}}$$

$$= \lim_{x\to\infty} \frac{1}{\sqrt{1 + (1/x^2)}}$$

$$= \frac{1}{\sqrt{1+0}} = 1$$

(b) $$\lim_{x\to\infty} \frac{x}{\sqrt{x^2+1}} = \lim_{x\to\infty} \frac{1}{x/\sqrt{x^2+1}}$$

$$= \lim_{x\to\infty} \frac{\sqrt{x^2+1}}{x} = \lim_{x\to\infty} \frac{x/\sqrt{x^2+1}}{1}$$

$$= \lim_{x\to\infty} \frac{x}{\sqrt{x^2+1}}$$

Applying L'Hôpital's rule twice results in the original limit, so L'Hôpital's rule fails.

(c)

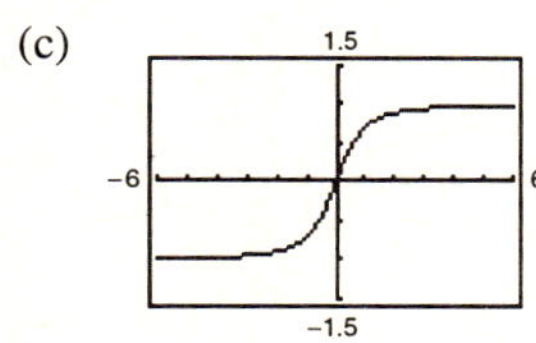

69. $$\lim_{k\to 0} \frac{32\left(1 - e^{-kt} + \frac{v_0ke^{-kt}}{32}\right)}{k} = \lim_{k\to 0} \frac{32(1 - e^{-kt})}{k} + \lim_{k\to 0}(v_0e^{-kt})$$

$$= \lim_{k\to 0} \frac{32(0 + te^{-kt})}{1} + \lim_{k\to 0}\left(\frac{v_0}{e^{kt}}\right) = 32t + v_0$$

71. Area of triangle: $\frac{1}{2}(2x)(1 - \cos x) = x - x\cos x$

Shaded area: Area of rectangle − Area under curve

$$2x(1 - \cos x) - 2\int_0^x (1 - \cos t)\,dt = 2x(1 - \cos x) - 2\Big[t - \sin t\Big]_0^x$$

$$= 2x(1 - \cos x) - 2(x - \sin x) = 2\sin x - 2x\cos x$$

Ratio: $$\lim_{x\to 0} \frac{x - x\cos x}{2\sin x - 2x\cos x} = \lim_{x\to 0} \frac{1 + x\sin x - \cos x}{2\cos x + 2x\sin x - 2\cos x}$$

$$= \lim_{x\to 0} \frac{1 + x\sin x - \cos x}{2x\sin x}$$

$$= \lim_{x\to 0} \frac{x\cos x + \sin x + \sin x}{2x\cos x + 2\sin x}$$

$$= \lim_{x\to 0} \frac{x\cos x + 2\sin x}{2x\cos x + 2\sin x}\cdot\frac{1/\cos x}{1/\cos x}$$

$$= \lim_{x\to 0} \frac{x + 2\tan x}{2x + 2\tan x}$$

$$= \lim_{x\to 0} \frac{1 + 2\sec^2 x}{2 + 2\sec^2 x} = \frac{3}{4}$$

73. $f(x) = x^3,\ g(x) = x^2 + 1,\ [0, 1]$

$$\frac{f(b) - f(a)}{g(b) - g(a)} = \frac{f'(c)}{g'(c)}$$

$$\frac{f(1) - f(0)}{g(1) - g(0)} = \frac{3c^2}{2c}$$

$$\frac{1}{1} = \frac{3c}{2}$$

$$c = \frac{2}{3}$$

75. $f(x) = \sin x,\ g(x) = \cos x,\ \left[0, \frac{\pi}{2}\right]$

$$\frac{f(\pi/2) - f(0)}{g(\pi/2) - g(0)} = \frac{f'(c)}{g'(c)}$$

$$\frac{1}{-1} = \frac{\cos c}{-\sin c}$$

$$-1 = -\cot c$$

$$c = \frac{\pi}{4}$$

77. False. L'Hôpital's Rule does not apply since

$$\lim_{x\to 0} (x^2 + x + 1) \neq 0.$$

$$\lim_{x\to 0} \frac{x^2 + x + 1}{x} = \lim_{x\to 0}\left(x + 1 + \frac{1}{x}\right) = 1 + \infty = \infty$$

79. True

81. (a) $\sin\theta = BD$

$\cos\theta = DO \Rightarrow AD = 1 - \cos\theta$

Area $\triangle ABD = \frac{1}{2}bh = \frac{1}{2}(1 - \cos\theta)\sin\theta = \frac{1}{2}\sin\theta - \frac{1}{2}\sin\theta\cos\theta$

(b) Area of sector: $\frac{1}{2}\theta$

Shaded area: $\frac{1}{2}\theta$ − Area $\triangle OBD = \frac{1}{2}\theta - \frac{1}{2}(\cos\theta)(\sin\theta) = \frac{1}{2}\theta - \frac{1}{2}\sin\theta\cos\theta$

(c) $$R = \frac{(1/2)\sin\theta - (1/2)\sin\theta\cos\theta}{(1/2)\theta - (1/2)\sin\theta\cos\theta} = \frac{\sin\theta - \sin\theta\cos\theta}{\theta - \sin\theta\cos\theta}$$

(d) $$\lim_{\theta\to 0} R = \lim_{\theta\to 0} \frac{\sin\theta - (1/2)\sin 2\theta}{\theta - (1/2)\sin 2\theta}$$

$$= \lim_{\theta\to 0} \frac{\cos\theta - \cos 2\theta}{1 - \cos 2\theta} = \lim_{\theta\to 0} \frac{-\sin\theta + 2\sin 2\theta}{2\sin 2\theta} = \lim_{\theta\to 0} \frac{-\cos\theta + 4\cos 2\theta}{4\cos 2\theta} = \frac{3}{4}$$

83. $\lim_{x \to a} f(x)^{g(x)}$

$$y = f(x)^{g(x)}$$

$$\ln y = g(x) \ln f(x)$$

$$\lim_{x \to a} g(x) \ln f(x) = (\infty)(-\infty) = -\infty$$

As $x \to a$, $\ln y \Rightarrow -\infty$, and hence $y = 0$. Thus,

$$\lim_{x \to a} f(x)^{g(x)} = 0.$$

85. $f'(a)(b-a) - \int_a^b f''(t)(t-b)\,dt = f'(a)(b-a) - \left\{\left[f'(t)(t-b)\right]_a^b - \int_a^b f'(t)\,dt\right\}$

$$= f'(a)(b-a) + f'(a)(a-b) + \left[f(t)\right]_a^b = f(b) - f(a)$$

$$dv = f''(t)dt \Rightarrow v = f'(t)$$

$$u = t - b \Rightarrow du = dt$$

Section 7.8 Improper Integrals

1. Infinite discontinuity at $x = 0$.

$$\int_0^4 \frac{1}{\sqrt{x}}\,dx = \lim_{b \to 0^+} \int_b^4 \frac{1}{\sqrt{x}}\,dx$$

$$= \lim_{b \to 0^+} \left[2\sqrt{x}\right]_b^4$$

$$= \lim_{b \to 0^+} \left(4 - 2\sqrt{b}\right) = 4$$

Converges

3. Infinite discontinuity at $x = 1$.

$$\int_0^2 \frac{1}{(x-1)^2}\,dx = \int_0^1 \frac{1}{(x-1)^2}\,dx + \int_1^2 \frac{1}{(x-1)^2}\,dx$$

$$= \lim_{b \to 1^-} \int_0^b \frac{1}{(x-1)^2}\,dx + \lim_{c \to 1^+} \int_c^2 \frac{1}{(x-1)^2}\,dx$$

$$= \lim_{b \to 1^-} \left[-\frac{1}{x-1}\right]_0^b + \lim_{c \to 1^+} \left[-\frac{1}{x-1}\right]_c^2 = (\infty - 1) + (-1 + \infty)$$

Diverges

5. Infinite limit of integration.

$$\int_0^\infty e^{-x}\,dx = \lim_{b \to \infty} \int_0^b e^{-x}\,dx$$

$$= \lim_{b \to \infty} \left[-e^{-x}\right]_0^b = 0 + 1 = 1$$

Converges

7. $\int_{-1}^1 \frac{1}{x^2}\,dx \neq -2$

because the integrand is not defined at $x = 0$.
Diverges

9. $\int_{-\infty}^0 xe^{-2x}\,dx = \lim_{b \to -\infty} \int_b^0 xe^{-2x}\,dx = \lim_{b \to -\infty} \frac{1}{4}\left[(-2x-1)e^{-2x}\right]_b^0 = \lim_{b \to -\infty} \frac{1}{4}\left[-1 + (2b+1)e^{-2b}\right] = -\infty$ (Integration by parts)

Diverges

11. $\displaystyle\int_0^{\infty} x^2e^{-x}\,dx = \lim_{b\to\infty}\int_0^b x^2e^{-x}\,dx = \lim_{b\to\infty}\Big[-e^{-x}(x^2+2x+2)\Big]_0^b = \lim_{b\to\infty}\left(-\frac{b^2+2b+2}{e^b}+2\right) = 2$

Since $\displaystyle\lim_{b\to\infty}\left(-\frac{b^2+2b+2}{e^b}\right) = 0$ by L'Hôpital's Rule.

13. $\displaystyle\int_1^{\infty}\frac{1}{x^2}\,dx = \lim_{b\to\infty}\int_1^b\frac{1}{x^2}\,dx$

$\displaystyle= \lim_{b\to\infty}\left[-\frac{1}{x}\right]_1^b = 1$

15. $\displaystyle\int_0^{\infty} e^{-x}\cos x\,dx = \lim_{b\to\infty}\frac{1}{2}\Big[e^{-x}(-\cos x+\sin x)\Big]_0^b$

$\displaystyle= \frac{1}{2}[0-(-1)] = \frac{1}{2}$

17. $\displaystyle\int_{-\infty}^{\infty}\frac{1}{1+x^2}\,dx = \int_{-\infty}^{0}\frac{1}{1+x^2}\,dx + \int_0^{\infty}\frac{1}{1+x^2}\,dx$

$\displaystyle= \lim_{b\to-\infty}\int_b^0\frac{1}{1+x^2}\,dx + \lim_{c\to\infty}\int_0^c\frac{1}{1+x^2}\,dx$

$\displaystyle= \lim_{b\to-\infty}\Big[\arctan x\Big]_b^0 + \lim_{c\to\infty}\Big[\arctan x\Big]_0^c$

$\displaystyle= \frac{\pi}{2}+\frac{\pi}{2} = \pi$

19. $\displaystyle\int_0^{\infty}\frac{1}{e^x+e^{-x}}\,dx = \lim_{b\to\infty}\int_0^b\frac{e^x}{1+e^{2x}}\,dx$

$\displaystyle= \lim_{b\to\infty}\Big[\arctan(e^x)\Big]_0^b$

$\displaystyle= \frac{\pi}{2}-\frac{\pi}{4} = \frac{\pi}{4}$

21. $\displaystyle\int_0^{\infty}\cos\pi x\,dx = \lim_{b\to\infty}\left[\frac{1}{\pi}\sin\pi x\right]_0^b$

Diverges since $\sin\pi x$ does not approach a limit as $x\to\infty$.

23. $\displaystyle\int_0^1\frac{1}{x^2}\,dx = \lim_{b\to0^+}\left[\frac{-1}{x}\right]_b^1 = -1+\infty$

Diverges

25. $\displaystyle\int_0^8\frac{1}{\sqrt[3]{8-x}}\,dx = \lim_{b\to8^-}\int_0^b\frac{1}{\sqrt[3]{8-x}}\,dx = \lim_{b\to8^-}\left[\frac{-3}{2}(8-x)^{2/3}\right]_0^b = 6$

27. $\displaystyle\int_0^1 x\ln x\,dx = \lim_{b\to0^+}\left[\frac{x^2}{2}\ln|x| - \frac{x^2}{4}\right]_b^1 = \lim_{b\to0^+}\left[\frac{-1}{4}-\frac{b^2\ln b}{2}+\frac{b^2}{4}\right] = \frac{-1}{4}$ since $\displaystyle\lim_{b\to0^+}(b^2\ln b) = 0$ by L'Hopital's Rule.

29. $\displaystyle\int_0^{\pi/2}\tan\theta\,d\theta = \lim_{b\to(\pi/2)^-}\Big[\ln|\sec\theta|\Big]_0^b = \infty,$

Diverges

31. $\displaystyle\int_2^4\frac{1}{\sqrt{x^2-4}}\,dx = \lim_{b\to2^+}\left[\ln\left|x+\sqrt{x^2-4}\right|\right]_b^4$

$= \ln\left(4+2\sqrt{3}\right) - \ln 2$

$= \ln\left(2+\sqrt{3}\right) \approx 1.317$

33. $\displaystyle\int_0^2\frac{1}{\sqrt[3]{x-1}}\,dx = \int_0^1\frac{1}{\sqrt[3]{x-1}}\,dx + \int_1^2\frac{1}{\sqrt[3]{x-1}}\,dx$

$\displaystyle= \lim_{b\to1^-}\left[\frac{3}{2}(x-1)^{2/3}\right]_0^b + \lim_{c\to1^+}\left[\frac{3}{2}(x-1)^{2/3}\right]_c^2 = \frac{-3}{2}+\frac{3}{2} = 0$

35. If $p=1$, $\displaystyle\int_1^{\infty}\frac{1}{x}\,dx = \lim_{b\to\infty}\int_1^b\frac{1}{x}\,dx = \lim_{b\to\infty}\ln x\Big]_1^b.$

Diverges. For $p\neq1$,

$$\int_1^{\infty}\frac{1}{x^p}\,dx = \lim_{b\to\infty}\left[\frac{x^{1-p}}{1-p}\right]_1^b = \lim_{b\to\infty}\left[\frac{b^{1-p}}{1-p}-\frac{1}{1-p}\right].$$

This converges to $\dfrac{1}{p-1}$ if $1-p<0$ or $p>1$.

37. For $n = 1$ we have $\int_0^{\infty} xe^{-x}\,dx = 1$ (see Exercise 10). Assume $\int_0^{\infty} x^n e^{-x}\,dx$ converges, then for $n + 1$ we have $\int_0^{\infty} x^{n+1}e^{-x}\,dx$. Using integration by parts, $u = x^{n+1}$, $du = (n+1)x^n\,dx$, $dv = e^{-x}\,dx$, $v = -e^{-x}$,

$$\int_0^{\infty} x^{n+1}e^{-x}\,dx = \lim_{b\to\infty}\Big[-x^{n+1}e^{-x}\Big]_0^b + (n+1)\int_0^{\infty} x^n e^{-x} dx = (n+1)\int_0^{\infty} x^n e^{-x}\,dx \text{ which converges.}$$

39. $\displaystyle\int_0^1 \frac{1}{x^3}\,dx$ diverges.

(See Exercise 36, $p = 3 \not< 1$.)

41. $\displaystyle\int_1^{\infty} \frac{1}{x^3}\,dx = \frac{1}{3-1} = \frac{1}{2}$ converges.

(See Exercise 35, $p = 3$.)

43. Since $\dfrac{1}{x^2+5} \le \dfrac{1}{x^2}$ on $[1, \infty)$ and $\displaystyle\int_1^{\infty} \frac{1}{x^2}\,dx$ converges by Exercise 35, $\displaystyle\int_1^{\infty} \frac{1}{x^2+5}\,dx$ converges.

45. Since $\dfrac{1}{\sqrt[3]{x(x-1)}} \ge \dfrac{1}{\sqrt[3]{x^2}}$ on $[2, \infty)$ and $\displaystyle\int_2^{\infty} \frac{1}{\sqrt[3]{x^2}}\,dx$ diverges by Exercise 35, $\displaystyle\int_2^{\infty} \frac{1}{\sqrt[3]{x(x-1)}}\,dx$ diverges.

47. Since $e^{-x^2} \le e^{-x}$ on $[1, \infty)$ and $\displaystyle\int_0^{\infty} e^{-x}\,dx$ converges (see Exercise 5), $\displaystyle\int_0^{\infty} e^{-x^2}\,dx$ converges.

49. $f(t) = 1$

$$F(s) = \int_0^{\infty} e^{-st}\,dx = \lim_{b\to\infty}\left[-\frac{1}{s}e^{-st}\right]_0^b = \frac{1}{s},\ s > 0$$

51. $f(t) = t^2$

$$F(s) = \int_0^{\infty} t^2e^{-st}\,dx = \lim_{b\to\infty}\left[\frac{1}{s^3}(-s^2t^2 - 2st - 2)e^{-st}\right]_0^b$$

$$= \frac{2}{s^3},\ s > 0$$

53. $f(t) = \cos at$

$$F(s) = \int_0^{\infty} e^{-st}\cos at\,dt$$

$$= \lim_{b\to\infty}\left[\frac{e^{-st}}{s^2+a^2}(-s\cos at + a\sin at)\right]_0^b = 0 + \frac{s}{s^2+a^2} = \frac{s}{s^2+a^2},\ s > 0$$

55. $f(t) = \cosh at$

$$F(s) = \int_0^{\infty} e^{-st}\cosh at\,dt = \int_0^{\infty} e^{-st}\left(\frac{e^{at}+e^{-at}}{2}\right)dt = \frac{1}{2}\int_0^{\infty}\left[e^{t(-s+a)} + e^{t(-s-a)}\right]dt$$

$$= \lim_{b\to\infty}\frac{1}{2}\left[\frac{1}{(-s+a)}e^{t(-s+a)} + \frac{1}{(-s-a)}e^{t(-s-a)}\right]_0^b = 0 - \frac{1}{2}\left[\frac{1}{(-s+a)} + \frac{1}{(-s-a)}\right]$$

$$= \frac{-1}{2}\left[\frac{1}{(-s+a)} + \frac{1}{(-s-a)}\right] = \frac{s}{s^2-a^2},\ s > |a|$$

57. (a) $\displaystyle A = \int_0^{\infty} e^{-x}\,dx$

$$= \lim_{b\to\infty}\Big[-e^{-x}\Big]_0^b$$

$$= 0 - (-1) = 1$$

(b) **Disc:**

$$V = \pi\int_0^{\infty} (e^{-x})^2\,dx$$

$$= \lim_{b\to\infty} \pi\left[-\frac{1}{2}e^{-2x}\right]_0^b = \frac{\pi}{2}$$

(c) **Shell:**

$$V = 2\pi\int_0^{\infty} xe^{-x}\,dx$$

$$= \lim_{b\to\infty}\left\{2\pi\Big[-e^{-x}(x+1)\Big]_0^b\right\}$$

$$= 2\pi$$

59. $x^{2/3} + y^{2/3} = 1$

$$\frac{2}{3}x^{-1/3} + \frac{2}{3}y^{-1/3}y' = 0$$

$$y' = \frac{-y^{1/3}}{x^{1/3}}$$

$$\sqrt{1 + (y')^2} = \sqrt{1 + \frac{y^{2/3}}{x^{2/3}}} = \sqrt{\frac{1}{x^{2/3}}} = \frac{1}{x^{1/3}}$$

$$s = 4\int_0^1 \frac{1}{x^{1/3}}\,dx = \lim_{b\to 0^+}\left[4\left(\frac{3}{2}x^{2/3}\right)\right]_b^1 = 6$$

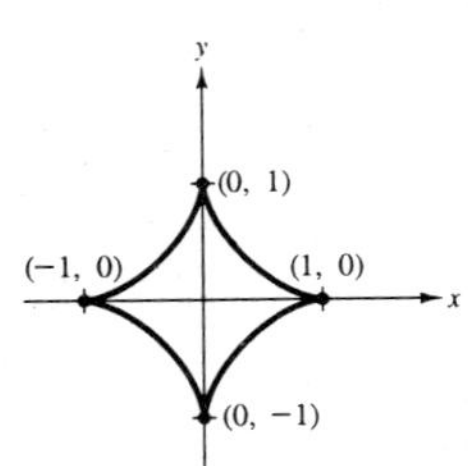

61. $\Gamma(n) = \int_0^\infty x^{n-1}e^{-x}\,dx$

(a) $\Gamma(1) = \int_0^\infty e^{-x}\,dx = \lim_{b\to\infty}\left[-e^{-x}\right]_0^b = 1$

$\Gamma(2) = \int_0^\infty xe^{-x}\,dx = \lim_{b\to\infty}\left[-e^{-x}(x+1)\right]_0^b = 1$

$\Gamma(3) = \int_0^\infty x^2e^{-x}\,dx = \lim_{b\to\infty}\left[-x^2e^{-x} - 2xe^{-x} - 2e^{-x}\right]_0^b = 2$

(b) $\Gamma(n+1) = \int_0^\infty x^ne^{-x}\,dx = \lim_{b\to\infty}\left[-x^ne^{-x}\right]_0^b + \lim_{b\to\infty} n\int_0^b x^{n-1}e^{-x}\,dx = 0 + n\Gamma(n)$ $(u = x^n, dv = e^{-x}\,dx)$

(c) $\Gamma(n) = (n-1)!$

63. (a) $\int_{-\infty}^\infty \frac{1}{7}e^{-t/7}\,dt = \int_0^\infty \frac{1}{7}e^{-t/7}\,dt = \lim_{b\to\infty}\left[-e^{-t/7}\right]_0^b = 1$

(b) $\int_0^4 \frac{1}{7}e^{-t/7}\,dt = \left[-e^{-t/7}\right]_0^4 = -e^{-4/7} + 1 \approx 0.4353 = 43.53\%$

(c) $\int_0^\infty t\left[\frac{1}{7}e^{-t/7}\right]dt = \lim_{b\to\infty}\left[-te^{-t/7} - 7e^{-t/7}\right]_0^b = 0 + 7 = 7$

65. (a) $C = 650{,}000 + \int_0^5 25{,}000\,e^{-0.06t}\,dt = 650{,}000 - \left[\frac{25{,}000}{0.06}e^{-0.06t}\right]_0^5 \approx \$757{,}992.41$

(b) $C = 650{,}000 + \int_0^{10} 25{,}000e^{-0.06t}\,dt \approx \$837{,}995.15$

(c) $C = 650{,}000 + \int_0^\infty 25{,}000e^{-0.06t}\,dt = 650{,}000 - \lim_{b\to\infty}\left[\frac{25{,}000}{0.06}e^{-0.06t}\right]_0^b \approx \$1{,}066{,}666.67$

67. Let $x = a\tan\theta$, $dx = a\sec^2\theta\,d\theta$, $\sqrt{a^2 + x^2} = a\sec\theta$.

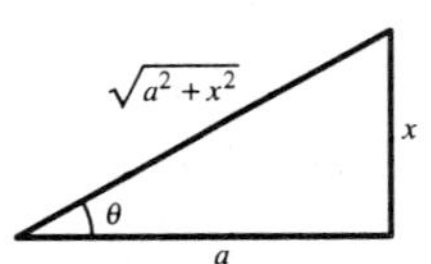

$$\int \frac{1}{(a^2+x^2)^{3/2}}\,dx = \int \frac{a\sec^2\theta\,d\theta}{a^3\sec^3\theta} = \frac{1}{a^2}\int\cos\theta\,d\theta$$

$$= \frac{1}{a^2}\sin\theta = \frac{1}{a^2}\frac{x}{\sqrt{a^2+x^2}}$$

Hence,

$$P = k\int_1^\infty \frac{1}{(a^2+x^2)^{3/2}}\,dx = \frac{k}{a^2}\lim_{b\to\infty}\left[\frac{x}{\sqrt{a^2+x^2}}\right]_1^b$$

$$= \frac{k}{a^2}\left[1 - \frac{1}{\sqrt{a^2+x^2}}\right] = \frac{k\sqrt{a^2+x^2} - 1}{a^2\sqrt{a^2+x^2}}$$

69. For $n = 1$,

$$I_1 = \int_0^{\infty} \frac{x}{(x^2+1)^4}\,dx = \lim_{b\to\infty} \frac{1}{2}\int_0^b (x^2+1)^{-4}(2x\,dx) = \lim_{b\to\infty}\left[-\frac{1}{6}\frac{1}{(x^2+1)^3}\right]_0^b = \frac{1}{6}.$$

For $n > 1$,

$$I_n = \int_0^{\infty} \frac{x^{2n-1}}{(x^2+1)^{n+3}}\,dx = \lim_{b\to\infty}\left[\frac{-x^{2n-2}}{2(n+2)(x^2+1)^n} + 2\right]_0^b + \frac{n-1}{n+2}\int_0^{\infty}\frac{x^{2n-3}}{(x^2+1)^{n+2}}\,dx = 0 + \frac{n-1}{n+2}(I_{n-1})$$

$u = x^{2n-2}$, $du = (2n-2)x^{2n-3}\,dx$, $dv = \dfrac{x}{(x^2+1)^{n+3}}\,dx$, $v = \dfrac{-1}{2(n+2)(x^2+1)^{n+2}}$

(a) $\displaystyle\int_0^{\infty}\frac{x}{(x^2+1)^4}\,dx = \lim_{b\to\infty}\left[-\frac{1}{6(x^2+1)^3}\right]_0^b = \frac{1}{6}$

(b) $\displaystyle\int_0^{\infty}\frac{x^3}{(x^2+1)^5}\,dx = \frac{1}{4}\int_0^{\infty}\frac{x}{(x^2+1)^4}\,dx = \frac{1}{4}\left(\frac{1}{6}\right) = \frac{1}{24}$

(c) $\displaystyle\int_0^{\infty}\frac{x^5}{(x^2+1)^6} = \frac{2}{5}\int_0^{\infty}\frac{x^3}{(x^2+1)^5}\,dx = \frac{2}{5}\left(\frac{1}{24}\right) = \frac{1}{60}$

71. False. $f(x) = 1/(x+1)$ is continuous on $[0, \infty)$, $\displaystyle\lim_{x\to\infty} 1/(x+1) = 0$, but

$$\int_0^{\infty}\frac{1}{x+1}\,dx = \lim_{b\to\infty}\Big[\ln|x+1|\Big]_0^b = \infty.$$

Diverges

73. True

Review Exercises for Chapter 7

1.
$$\int e^{2x}\sin 3x\,dx = -\frac{1}{3}e^{2x}\cos 3x + \frac{2}{3}\int e^{2x}\cos 3x\,dx$$
$$= -\frac{1}{3}e^{2x}\cos 3x + \frac{2}{3}\left(\frac{1}{3}e^{2x}\sin 3x - \frac{2}{3}\int e^{2x}\sin 3x\,dx\right)$$
$$\frac{13}{9}\int e^{2x}\sin 3x\,dx = -\frac{1}{3}e^{2x}\cos 3x + \frac{2}{9}e^{2x}\sin 3x$$
$$\int e^{2x}\sin 3x\,dx = \frac{e^{2x}}{13}(2\sin 3x - 3\cos 3x) + C$$

(1) $dv = \sin 3x\,dx \Rightarrow v = -\dfrac{1}{3}\cos 3x$

$u = e^{2x} \Rightarrow du = 2e^{2x}\,dx$

(2) $dv = \cos 3x\,dx \Rightarrow v = \dfrac{1}{3}\sin 3x$

$u = e^{2x} \Rightarrow du = 2e^{2x}\,dx$

3.
$$\int\cos^3(\pi x - 1)\,dx = \int[1 - \sin^2(\pi x - 1)]\cos(\pi x - 1)\,dx$$
$$= \frac{1}{\pi}\left[\sin(\pi x - 1) - \frac{1}{3}\sin^3(\pi x - 1)\right] + C$$
$$= \frac{1}{3\pi}\sin(\pi x - 1)[3 - \sin^2(\pi x - 1)] + C$$
$$= \frac{1}{3\pi}\sin(\pi x - 1)[3 - (1 - \cos^2(\pi x - 1))] + C$$
$$= \frac{1}{3\pi}\sin(\pi x - 1)[2 + \cos^2(\pi x - 1)] + C$$

5. $$\int \sec^4\left(\frac{x}{2}\right) dx = \int \left[\tan^2\left(\frac{x}{2}\right) + 1\right] \sec^2\left(\frac{x}{2}\right) dx$$
$$= \int \tan^2\left(\frac{x}{2}\right) \sec^2\left(\frac{x}{2}\right) dx + \int \sec^2\left(\frac{x}{2}\right) dx$$
$$= \frac{2}{3}\tan^3\left(\frac{x}{2}\right) + 2\tan\left(\frac{x}{2}\right) + C = \frac{2}{3}\left[\tan^3\left(\frac{x}{2}\right) + 3\tan\left(\frac{x}{2}\right)\right] + C$$

7. $$\int \frac{-12}{x^2\sqrt{4-x^2}}\, dx = \int \frac{-24\cos\theta\, d\theta}{(4\sin^2\theta)(2\cos\theta)}$$
$$= -3\int \csc^2\theta\, d\theta$$
$$= 3\cot\theta + C$$
$$= \frac{3\sqrt{4-x^2}}{x} + C$$

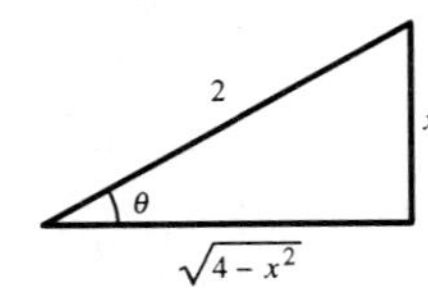

$x = 2\sin\theta,\ dx = 2\cos\theta\, d\theta,\ \sqrt{4-x^2} = 2\cos\theta$

9. $$\frac{x^2+2x}{(x-1)(x^2+1)} = \frac{A}{x-1} + \frac{Bx+C}{x^2+1}$$
$$x^2 + 2x = A(x^2+1) + (Bx+C)(x-1)$$

Let $x = 1$: $3 = 2A \implies A = \frac{3}{2}$

Let $x = 0$: $0 = A - C \implies C = \frac{3}{2}$

Let $x = 2$: $8 = 5A + 2B + C \implies B = -\frac{1}{2}$

$$\int \frac{x^2+2x}{x^3-x^2+x-1}\, dx = \frac{3}{2}\int \frac{1}{x-1}\, dx - \frac{1}{2}\int \frac{x-3}{x^2+1}\, dx$$
$$= \frac{3}{2}\int \frac{1}{x-1}\, dx - \frac{1}{4}\int \frac{2x}{x^2+1}\, dx + \frac{3}{2}\int \frac{1}{x^2+1}\, dx$$
$$= \frac{3}{2}\ln|x-1| - \frac{1}{4}\ln|x^2+1| + \frac{3}{2}\arctan x + C$$
$$= \frac{1}{4}[6\ln|x-1| - \ln(x^2+1) + 6\arctan x] + C$$

11. $$\frac{x^2}{x^2+2x-15} = 1 + \frac{15-2x}{x^2+2x-15}$$
$$\frac{15-2x}{(x-3)(x+5)} = \frac{A}{x-3} + \frac{B}{x+5}$$
$$15 - 2x = A(x+5) + B(x-3)$$

Let $x = 3$: $9 = 8A \implies A = \frac{9}{8}$

Let $x = -5$: $25 = -8B \implies B = -\frac{25}{8}$

$$\int \frac{x^2}{x^2+2x-15}\, dx = \int dx + \frac{9}{8}\int \frac{1}{x-3}\, dx - \frac{25}{8}\int \frac{1}{x+5}\, dx$$
$$= x + \frac{9}{8}\ln|x-3| - \frac{25}{8}\ln|x+5| + C$$

13. $\displaystyle\int \frac{1}{1 - \sin\theta}\,d\theta = \int \frac{1 + \sin\theta}{\cos^2\theta}\,d\theta = \int (\sec^2\theta + \sec\theta\tan\theta)\,d\theta = \tan\theta - \sec\theta + C$

15. $\displaystyle\int \frac{\ln 2x}{x^2}\,dx = \frac{-\ln 2x}{x} + \int \frac{1}{x^2}\,dx$

$\displaystyle= \frac{-\ln 2x}{x} - \frac{1}{x} + C$

$\displaystyle= -\frac{1}{x}(1 + \ln 2x) + C$

$\displaystyle dv = \frac{1}{x^2}\,dx \Rightarrow v = -\frac{1}{x}$

$\displaystyle u = \ln 2x \Rightarrow du = \frac{1}{x}\,dx$

17. $\displaystyle\int \sqrt{4 - x^2}\,dx = \int (2\cos\theta)(2\cos\theta)\,d\theta$

$\displaystyle= 2\int (1 + \cos 2\theta)\,d\theta$

$\displaystyle= 2\left(\theta + \frac{1}{2}\sin 2\theta\right) + C$

$= 2(\theta + \sin\theta\cos\theta) + C$

$\displaystyle= 2\left[\arcsin\left(\frac{x}{2}\right) + \frac{x}{2}\left(\frac{\sqrt{4 - x^2}}{2}\right)\right] + C$

$\displaystyle= \frac{1}{2}\left[4\arcsin\left(\frac{x}{2}\right) + x\sqrt{4 - x^2}\right] + C$

$x = 2\sin\theta,\ dx = 2\cos\theta\,d\theta,\ \sqrt{4 - x^2} = 2\cos\theta$

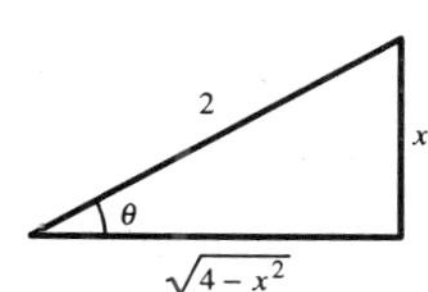

19. $\displaystyle\frac{3x^3 + 4x}{(x^2 + 1)^2} = \frac{Ax + B}{x^2 + 1} + \frac{Cx + D}{(x^2 + 1)^2}$

$3x^3 + 4x = (Ax + B)(x^2 + 1) + Cx + D$

$= Ax^3 + Bx^2 + (A + C)x + (B + D)$

$A = 3, B = 0, A + C = 4 \Rightarrow C = 1,$

$B + D = 0 \Rightarrow D = 0$

$\displaystyle\int \frac{3x^3 + 4x}{(x^2 + 1)^2}\,dx = 3\int \frac{x}{x^2 + 1}\,dx + \int \frac{x}{(x^2 + 1)^2}\,dx$

$\displaystyle= \frac{3}{2}\ln(x^2 + 1) - \frac{1}{2(x^2 + 1)} + C$

21. $\displaystyle\int \frac{16}{\sqrt{16 - x^2}}\,dx = 16\arcsin\left(\frac{x}{4}\right) + C$

23. $\displaystyle\int \frac{x}{x^2 + 4x + 8}\,dx = \frac{1}{2}\int \frac{2x + 4 - 4}{x^2 + 4x + 8}\,dx$

$\displaystyle= \frac{1}{2}\int \frac{2x + 4}{x^2 + 4x + 8}\,dx - 2\int \frac{1}{(x + 2)^2 + 4}\,dx = \frac{1}{2}\ln|x^2 + 4x + 8| - \arctan\left(\frac{x + 2}{2}\right) + C$

25. $\displaystyle\int \theta\sin\theta\cos\theta\,d\theta = \frac{1}{2}\int \theta\sin 2\theta\,d\theta$

$\displaystyle= -\frac{1}{4}\theta\cos 2\theta + \frac{1}{4}\int \cos 2\theta\,d\theta = -\frac{1}{4}\theta\cos 2\theta + \frac{1}{8}\sin 2\theta + C = \frac{1}{8}(\sin 2\theta - 2\theta\cos 2\theta) + C$

$\displaystyle dv = \sin 2\theta\,d\theta \Rightarrow v = -\frac{1}{2}\cos 2\theta$

$u = \theta \Rightarrow du = d\theta$

27. $\displaystyle\int (\sin\theta + \cos\theta)^2\, d\theta = \int (\sin^2\theta + 2\sin\theta\cos\theta + \cos^2\theta)\, d\theta$

$$= \int (1 + \sin 2\theta)\, d\theta = \theta - \frac{1}{2}\cos 2\theta + C = \frac{1}{2}(2\theta - \cos 2\theta) + C$$

29. $\displaystyle\int \sqrt{1 + \cos x}\, dx = \int \frac{\sin x}{\sqrt{1 - \cos x}}\, dx$

$$= \int (1 - \cos x)^{-1/2}(\sin x)\, dx$$

$$= 2\sqrt{1 - \cos x} + C$$

$u = 1 - \cos x,\ du = \sin x\, dx$

31. $\displaystyle\int \cos x \ln(\sin x)\, dx = \sin x \ln(\sin x) - \int \cos x\, dx$

$$= \sin x \ln(\sin x) - \sin x + C$$

$dv = \cos x\, dx \implies v = \sin x$

$u = \ln(\sin x) \implies du = \dfrac{\cos x}{\sin x}\, dx$

33. $\displaystyle\int x \arcsin 2x\, dx = \frac{x^2}{2}\arcsin 2x - \int \frac{x^2}{\sqrt{1 - 4x^2}}\, dx$

$$= \frac{x^2}{2}\arcsin 2x - \frac{1}{8}\int \frac{2(2x)^2}{\sqrt{1 - (2x)^2}}\, dx$$

$$= \frac{x^2}{2}\arcsin 2x - \frac{1}{8}\left(\frac{1}{2}\right)\left[-(2x)\sqrt{1 - 4x^2} + \arcsin 2x\right] + C \quad \text{(by Formula 43 of Integration Tables)}$$

$$= \frac{1}{16}\left[(8x^2 - 1)\arcsin 2x + 2x\sqrt{1 - 4x^2}\right] + C$$

$dv = x\, dx \implies v = \dfrac{x^2}{2}$

$u = \arcsin 2x \implies du = \dfrac{2}{\sqrt{1 - 4x^2}}\, dx$

35. $\displaystyle\int \frac{x^{1/4}}{1 + x^{1/2}}\, dx = 4\int \frac{u(u^3)}{1 + u^2}\, du = 4\int \left(u^2 - 1 + \frac{1}{u^2 + 1}\right) du$

$$= 4\left(\frac{1}{3}u^3 - u + \arctan u\right) + C = \frac{4}{3}\left[x^{3/4} - 3x^{1/4} + 3\arctan(x^{1/4})\right] + C$$

$y = \sqrt[4]{x},\ x = u^4,\ dx = 4u^3\, du$

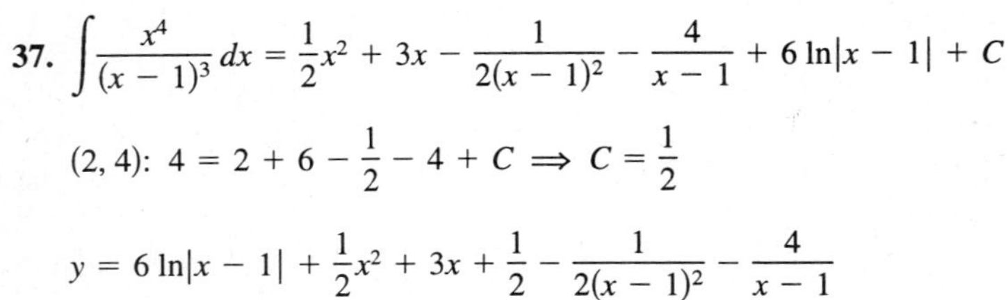

37. $\displaystyle\int \frac{x^4}{(x - 1)^3}\, dx = \frac{1}{2}x^2 + 3x - \frac{1}{2(x - 1)^2} - \frac{4}{x - 1} + 6\ln|x - 1| + C$

$(2, 4)$: $4 = 2 + 6 - \dfrac{1}{2} - 4 + C \implies C = \dfrac{1}{2}$

$y = 6\ln|x - 1| + \dfrac{1}{2}x^2 + 3x + \dfrac{1}{2} - \dfrac{1}{2(x - 1)^2} - \dfrac{4}{x - 1}$

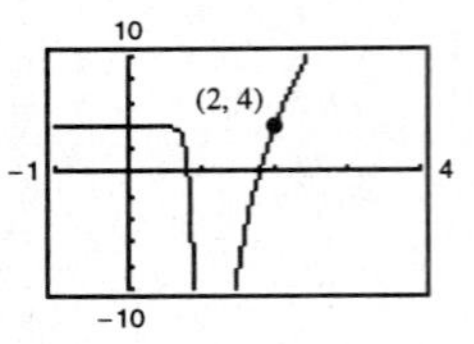

39. $\displaystyle\int \frac{6x^2 - 3x + 14}{x^3 - 2x^2 + 4x - 8}\, dx = 4\ln|x - 2| + \ln(x^2 + 4) + \frac{1}{2}\arctan\left(\frac{1}{2}x\right) + C$

$(3, 2)$: $\ln 13 + \dfrac{1}{2}\arctan\dfrac{3}{2} + C = 2 \implies C = 2 - \ln 13 - \dfrac{1}{2}\arctan\left(\dfrac{3}{2}\right)$

$y = \dfrac{1}{2}\arctan\dfrac{x}{2} + \ln(x^2 + 4) + 4\ln|x - 2| - \dfrac{1}{2}\arctan\dfrac{3}{2} - \ln 13 + 2$

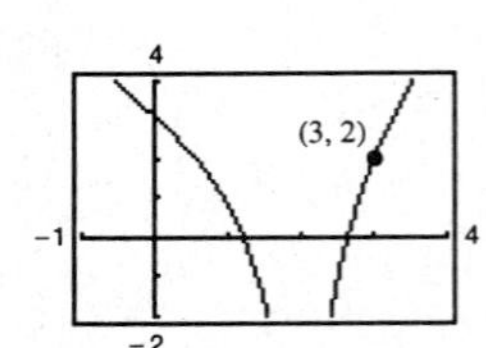

41. $\displaystyle\int \frac{d\theta}{2 - 3\sin\theta} = \frac{\sqrt{5}}{5}\ln\left|\frac{\cos\theta - \sqrt{5}(\sin\theta - 1)}{\cos\theta + \sqrt{5}(\sin\theta - 1)}\right| + C$

$(1, 1)$: $\displaystyle\frac{\sqrt{5}}{5}\ln\left|\frac{\cos(1) - \sqrt{5}(\sin(1) - 1)}{\cos(1) + \sqrt{5}(\sin(1) - 1)}\right| + C = 1 \Rightarrow C \approx 0.297$

$\displaystyle\frac{\sqrt{5}}{5}\ln\left|\frac{\cos\theta - \sqrt{5}(\sin\theta - 1)}{\cos\theta + \sqrt{5}(\sin\theta - 1)}\right| + 0.297$

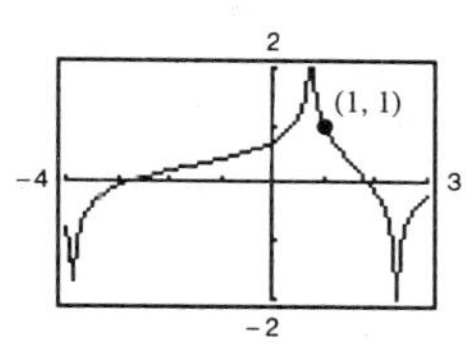

43. $\displaystyle\int \frac{d\theta}{1 + \sin\theta + \cos\theta} = \ln\left|2 + 2\tan\left(\frac{\theta}{2}\right)\right| + C$

or $\displaystyle\ln\left|\frac{1 + \sin\theta + \cos\theta}{1 + \cos\theta}\right| + C$

$(0, 0)$: $\ln 1 + C = 0 \Rightarrow C = 0$

$\displaystyle y = \ln\left|\frac{1 + \sin\theta + \cos\theta}{1 + \cos\theta}\right|$

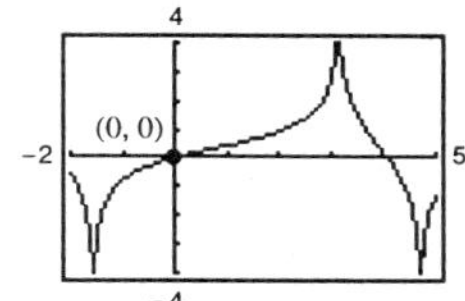

45. $\displaystyle\int \frac{\sin\theta}{3 - 2\cos\theta}\,d\theta = \frac{1}{2}\ln|3 - 2\cos\theta| + C$

$(0, 0)$: $\displaystyle\frac{1}{2}\ln 1 + C = 0 \Rightarrow C = 0$

$\displaystyle y = \frac{1}{2}\ln(3 - 2\cos\theta)$

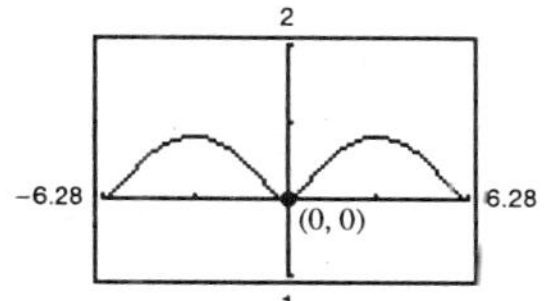

47. $\displaystyle y = \int \frac{9}{x^2 - 9}\,dx = \frac{3}{2}\ln\left|\frac{x - 3}{x + 3}\right| + C$ (by Formula 24 of Integration Tables)

49. $\displaystyle y = \int \ln(x^2 + x)dx = x\ln|x^2 + x| - \int \frac{2x^2 + x}{x^2 + x}\,dx = x\ln|x^2 + x| - \int \frac{2x + 1}{x + 1}\,dx$

$\displaystyle = x\ln|x^2 + x| - \int 2dx + \int \frac{1}{x + 1}\,dx$

$= x\ln|x^2 + x| - 2x + \ln|x + 1| + C$

$dv = dx \Rightarrow v = x$

$\displaystyle u = \ln(x^2 + x) \Rightarrow du = \frac{2x + 1}{x^2 + x}\,dx$

51. (a) $\displaystyle\int \frac{1}{x^2\sqrt{4 + x^2}}\,dx = \frac{1}{4}\int \frac{\cos\theta}{\sin^2\theta}\,d\theta$

$\displaystyle = \frac{-1}{4\sin\theta} + C$

$\displaystyle = -\frac{1}{4}\csc\theta + C$

$\displaystyle = \frac{-\sqrt{4 + x^2}}{4x} + C$

$x = 2\tan\theta$, $dx = 2\sec^2\theta\,d\theta$, $\sqrt{4 + x^2} = 2\sec\theta$

(b) $\displaystyle\int \frac{1}{x^2\sqrt{4 + x^2}}\,dx = -\frac{1}{4}\int \frac{u}{\sqrt{1 + u^2}}\,du$

$\displaystyle = -\frac{1}{4}\sqrt{1 + u^2} + C$

$\displaystyle = -\frac{1}{4}\sqrt{1 + \frac{4}{x^2}}$

$\displaystyle = \frac{-\sqrt{4 + x^2}}{4x} + C$

$\displaystyle x = \frac{2}{u}$, $\displaystyle dx = -\frac{2du}{u^2}$

53. (a) $\displaystyle\int \frac{x^3}{\sqrt{4+x^2}}\,dx = 8\int \frac{\sin^3\theta}{\cos^4\theta}\,d\theta$

$\displaystyle = 8\int (\cos^{-4}\theta - \cos^{-2}\theta)\sin\theta\,d\theta$

$\displaystyle = \frac{8}{3}\sec\theta(\sec^2\theta - 3) + C$

$\displaystyle = \frac{\sqrt{4+x^2}}{3}(x^2-8) + C$

$x = 2\tan\theta,\ dx = 2\sec^2\theta\,d\theta$

(b) $\displaystyle\int \frac{x^3}{\sqrt{4+x^2}}\,dx = \int (u^2-4)\,du$

$\displaystyle = \frac{1}{3}u^3 - 4u + C$

$\displaystyle = \frac{u}{3}(u^2-12) + C$

$\displaystyle = \frac{\sqrt{4+x^2}}{3}(x^2-8) + C$

$u^2 = 4 + x^2,\ 2u\,du = 2x\,dx$

(c) $\displaystyle\int \frac{x^3}{\sqrt{4+x^2}}\,dx = x^2\sqrt{4+x^2} - \int 2x\sqrt{4+x^2}\,dx$

$\displaystyle = x^2\sqrt{4+x^2} - \frac{2}{3}(4+x^2)^{3/2} + C = \frac{\sqrt{4+x^2}}{3}(x^2-8) + C$

$\displaystyle dv = \frac{x}{\sqrt{4+x^2}}\,dx \Rightarrow v = \sqrt{4+x^2}$

$u = x^2 \Rightarrow du = 2x\,dx$

55. $\displaystyle\int_2^{\sqrt{5}} x(x^2-4)^{3/2}\,dx = \left[\frac{1}{5}(x^2-4)^{5/2}\right]_2^{\sqrt{5}} = \frac{1}{5}$

57. $\displaystyle\int_1^4 \frac{\ln x}{x}\,dx = \left[\frac{1}{2}(\ln x)^2\right]_1^4 = \frac{1}{2}(\ln 4)^2 = 2(\ln 2)^2 \approx 0.961$

59. $\displaystyle\int_0^{\pi} x\sin x\,dx = \Big[-x\cos x + \sin x\Big]_0^{\pi} = \pi$

61. $\displaystyle A = \int_0^4 x\sqrt{4-x}\,dx = \int_2^0 (4-u^2)u(-2u)\,du$

$\displaystyle = \int_2^0 2(u^4 - 4u^2)\,du$

$\displaystyle = \left[2\left(\frac{u^5}{5} - \frac{4u^3}{3}\right)\right]_2^0 = \frac{128}{15}$

$u = \sqrt{4-x},\ x = 4 - u^2,\ dx = -2u\,du$

63. By symmetry, $\bar{x} = 0$, $A = \dfrac{1}{2}\pi$.

$\displaystyle \bar{y} = \frac{2}{\pi}\left(\frac{1}{2}\right)\int_{-1}^{1}\left(\sqrt{1-x^2}\right)^2 dx = \frac{1}{\pi}\left[x - \frac{1}{3}x^3\right]_{-1}^{1} = \frac{4}{3\pi}$

$\displaystyle (\bar{x}, \bar{y}) = \left(0, \frac{4}{3\pi}\right)$

65. (a) $\displaystyle\int_0^1 e^x\,dx = \Big[e^x\Big]_0^1 = e - 1 \approx 1.72$

(b) $\displaystyle\int_0^1 xe^x\,dx = \Big[e^x(x-1)\Big]_0^1 = 1.00$

(c) $\displaystyle\int_0^1 xe^{x^2}\,dx = \left[\frac{1}{2}e^{x^2}\right]_0^1 = \frac{1}{2}(e-1) \approx 0.86$

(d) Simpson's Rule ($n = 8$)

$$\int_0^1 e^{x^2}\,dx = \frac{1}{24}\left[1 + 4e^{(1/8)^2} + 2e^{(1/4)^2} + 4e^{(3/8)^2} + 2e^{(1/2)^2} + 4e^{(5/8)^2} + 2e^{(3/4)^2} + 4e^{(7/8)^2} + e\right] \approx 1.46$$

67. $\displaystyle s = \int_0^{\pi}\sqrt{1+\cos^2 x}\,dx \approx 3.82$

69. $\displaystyle\lim_{x\to 1}\left[\frac{(\ln x)^2}{x-1}\right] = \lim_{x\to 1}\left[\frac{2(1/x)\ln x}{1}\right] = 0$

71. $\displaystyle\lim_{x\to\infty}\frac{e^{2x}}{x^2} = \lim_{x\to\infty}\frac{2e^{2x}}{2x} = \lim_{x\to\infty}\frac{4e^{2x}}{2} = \infty$

73. $\displaystyle y = \lim_{x\to\infty}(\ln x)^{2/x}$

$\displaystyle \ln y = \lim_{x\to\infty}\frac{2\ln(\ln x)}{x} = \lim_{x\to\infty}\left[\frac{2/(x\ln x)}{1}\right] = 0$

Since $\ln y = 0$, $y = 1$.

75. $\lim_{n\to\infty} 1000\left(1+\frac{0.09}{n}\right)^n = 1000 \lim_{n\to\infty}\left(1+\frac{0.09}{n}\right)^n$

Let $y = \lim_{n\to\infty}\left(1+\frac{0.09}{n}\right)^n$.

$$\ln y = \lim_{n\to\infty} n\ln\left(1+\frac{0.09}{n}\right) = \lim_{n\to\infty}\frac{\ln\left(1+\frac{0.09}{n}\right)}{\frac{1}{n}} = \lim_{n\to\infty}\left(\frac{\frac{-0.09/n^2}{1+(0.09/n)}}{-\frac{1}{n^2}}\right) = \lim_{n\to\infty}\frac{0.09}{1+\left(\frac{0.09}{n}\right)} = 0.09$$

Thus, $\ln y = 0.09 \Rightarrow y = e^{0.09}$ and $\lim_{n\to\infty} 1000\left(1+\frac{0.09}{n}\right)^n = 1000e^{0.09} \approx 1094.17$.

77. $\int_0^{16}\frac{1}{\sqrt[4]{x}}\,dx = \lim_{b\to 0^+}\left[\frac{4}{3}x^{3/4}\right]_b^{16} = \frac{32}{3}$

Converges

79. $\int_1^{\infty} x^2\ln x\,dx = \lim_{b\to\infty}\left[\frac{x^3}{9}(-1+3\ln x)\right]_1^b = \infty$

Diverges

81. $\int_0^{t_0} 500{,}000e^{-0.05t}\,dt = \left[\frac{500{,}000}{-0.05}e^{-0.05t}\right]_0^{t_0} = \frac{-500{,}000}{0.05}(e^{-0.05t_0}-1) = 10{,}000{,}000(1-e^{-0.05t_0})$

(a) $t_0 = 20$: \$6,321,205.59

(b) $t_0 \to \infty$: \$10,000,000

83. (a) $P(13 \le x < \infty) = \frac{1}{0.95\sqrt{2\pi}}\int_{13}^{\infty} e^{-(x-12.9)^2/2(0.95)^2}\,dx \approx 0.4581$

(b) $P(15 \le x < 20) = \frac{1}{0.95\sqrt{2\pi}}\int_{15}^{\infty} e^{-(x-12.9)^2/2(0.95)^2}\,dx \approx 0.0135$

85. $dv = dx \quad \Rightarrow \quad v = x$

$u = (\ln x)^n \Rightarrow du = n(\ln x)^{n-1}\frac{1}{x}dx$

$$\int(\ln x)^n\,dx = x(\ln x)^n - n\int(\ln x)^{n-1}\,dx$$

87. False

$u = \ln x^2 = 2\ln x \Rightarrow du = \frac{2}{x}dx$

$$\int\frac{\ln x^2}{x}\,dx = \frac{1}{2}\int(\ln x^2)\frac{2}{x}\,dx = \frac{1}{2}\int u\,du$$

89. False

$$\int_{-1}^{1}\sqrt{x^2-x^3}\,dx = \int_{-1}^{1}\sqrt{x^2(1-x)}\,dx = \int_{-1}^{1}|x|\sqrt{1-x}\,dx$$

91. $\int_x^1\frac{1}{1+t^2}\,dt = \left[\arctan t\right]_x^1 = \frac{\pi}{4} - \arctan x$

$\int_1^{1/x}\frac{1}{1+t^2}\,dt = \left[\arctan t\right]_1^{1/x} = \arctan\frac{1}{x} - \frac{\pi}{4}$

Since $\arctan x + \arctan\frac{1}{x} = \frac{\pi}{2}$, $x > 0$, we have:

$$\arctan\frac{1}{x} = \frac{\pi}{4} + \frac{\pi}{4} - \arctan x$$

$$\arctan\frac{1}{x} - \frac{\pi}{4} = \frac{\pi}{4} - \arctan x$$

Therefore, $\int_1^{1/x}\frac{1}{1+t^2}\,dt = \int_x^1\frac{1}{1+t^2}\,dt$.

CHAPTER 8
Infinite Series

CHAPTER 8
Infinite Series

Section 8.1 Sequences

Solutions to Odd-Numbered Exercises

1. $a_n = 2^n$

$a_1 = 2^1 = 2$

$a_2 = 2^2 = 4$

$a_3 = 2^3 = 8$

$a_4 = 2^4 = 16$

$a_5 = 2^5 = 32$

3. $a_n = \left(-\frac{1}{2}\right)^n$

$a_1 = \left(-\frac{1}{2}\right)^1 = -\frac{1}{2}$

$a_2 = \left(-\frac{1}{2}\right)^2 = \frac{1}{4}$

$a_3 = \left(-\frac{1}{2}\right)^3 = -\frac{1}{8}$

$a_4 = \left(-\frac{1}{2}\right)^4 = \frac{1}{16}$

$a_5 = \left(-\frac{1}{2}\right)^5 = -\frac{1}{32}$

5. $a_n = \dfrac{(-1)^{n(n+1)/2}}{n^2}$

$a_1 = \dfrac{(-1)^1}{1^2} = -1$

$a_2 = \dfrac{(-1)^3}{2^2} = -\dfrac{1}{4}$

$a_3 = \dfrac{(-1)^6}{3^2} = \dfrac{1}{9}$

$a_4 = \dfrac{(-1)^{10}}{4^2} = \dfrac{1}{16}$

$a_5 = \dfrac{(-1)^{15}}{5^2} = -\dfrac{1}{25}$

7. $a_n = \dfrac{3^n}{n!}$

$a_1 = \dfrac{3}{1!} = 3$

$a_2 = \dfrac{3^2}{2!} = \dfrac{9}{2}$

$a_3 = \dfrac{3}{3!} = \dfrac{27}{6}$

$a_4 = \dfrac{3^4}{4!} = \dfrac{81}{24}$

$a_5 = \dfrac{3^5}{5!} = \dfrac{243}{120}$

9. $a_1 = 3, a_{k+1} = 2(a_k - 1)$

$a_2 = 2(a_1 - 1)$

$= 2(3 - 1) = 4$

$a_3 = 2(a_2 - 1)$

$= 2(4 - 1) = 6$

$a_4 = 2(a_3 - 1)$

$= 2(6 - 1) = 10$

$a_5 = 2(a_4 - 1)$

$= 2(10 - 1) = 18$

11. $a_1 = 32, a_{k+1} = \frac{1}{2}a_k$

$a_2 = \frac{1}{2}a_1 = \frac{1}{2}(32) = 16$

$a_3 = \frac{1}{2}a_2 = \frac{1}{2}(16) = 8$

$a_4 = \frac{1}{2}a_3 = \frac{1}{2}(8) = 4$

$a_5 = \frac{1}{2}a_4 = \frac{1}{2}(4) = 2$

13. Because $a_1 = 8/(1 + 1) = 4$ and $a_2 = 8/(2 + 1) = \frac{8}{3}$, the sequence matches graph (d).

15. This sequence decreases and $a_1 = 4$, $a_2 = 4(0.5) = 2$. Matches (c).

17.

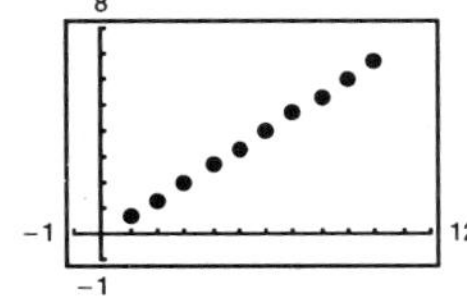

19.

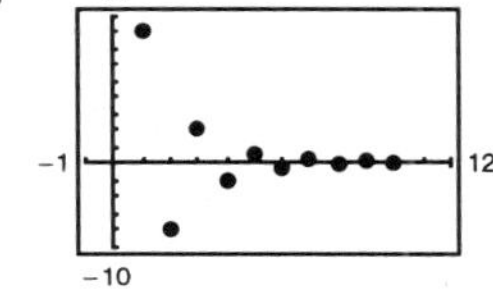

21.

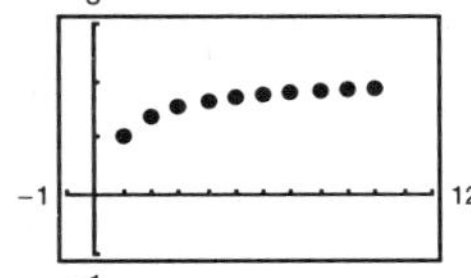

23. $a_n = 3n - 1$

$a_5 = 3(5) - 1 = 14$

$a_6 = 3(6) - 1 = 17$

25. $a_n = \dfrac{3}{(-2)^{n-1}}$

$a_n = \dfrac{3}{(-2)^4} = \dfrac{3}{16}$

$a_6 = \dfrac{3}{(-2)^5} = -\dfrac{3}{32}$

27. $\dfrac{10!}{8!} = \dfrac{8!(9)(10)}{8!}$

$= (9)(10) = 90$

29. $\dfrac{(n+1)!}{n!} = \dfrac{n!(n+1)}{n!}$

$= n + 1$

31. $\dfrac{(2n-1)!}{(2n+1)!} = \dfrac{(2n-1)!}{(2n-1)!(2n)(2n+1)}$

$= \dfrac{1}{2n(2n+1)}$

33. $a_n = 3n - 2$

35. $a_n = n^2 - 2$

37. $a_n = \dfrac{n+1}{n+2}$

39. $a_n = \dfrac{(-1)^{n-1}}{2^{n-2}}$

41. $a_n = 1 + \dfrac{1}{n} = \dfrac{n+1}{n}$

43. $a_n = \dfrac{n}{(n+1)(n+2)}$

45. $a_n = \dfrac{(-1)^{n-1}}{1 \cdot 3 \cdot 5 \cdots (2n-1)}$

$= \dfrac{(-1)^{n-1}2^n n!}{(2n)!}$

47.

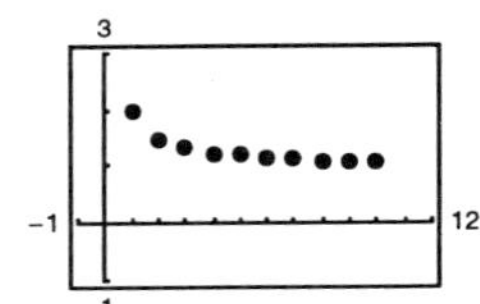

The graph seems to indicate that the sequence converges to 1. Analytically,

$$\lim_{n\to\infty} a_n = \lim_{n\to\infty} \frac{n+1}{n} = \lim_{x\to\infty} \frac{x+1}{x} = \lim_{x\to\infty} 1 = 1.$$

49.

2

-1 12

-2

The graph seems to indicate that the sequence diverges. Analytically, the sequence is

$$\{a_n\} = \{0, -1, 0, 1, 0, -1, \ldots\}.$$

Hence, $\lim_{n\to\infty} a_n$ does not exist.

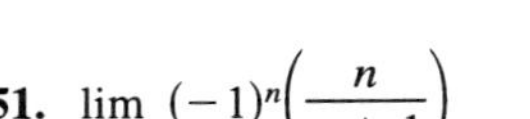

51. $\lim_{n\to\infty} (-1)^n\left(\dfrac{n}{n+1}\right)$

does not exist (oscillates between -1 and 1), diverges.

53. $\lim_{n\to\infty} \dfrac{3n^2 - n + 4}{2n^2 + 1} = \dfrac{3}{2}$, converges

55. $\lim_{n\to\infty} \dfrac{1 + (-1)^n}{n} = 0$ converges

57. $\lim_{n\to\infty} \left(\dfrac{3}{4}\right)^n = 0$, converges

59. $\lim_{n\to\infty} \dfrac{(n+1)!}{n!} = \lim_{n\to\infty} (n+1) = \infty$, diverges

61. $\lim_{n\to\infty} \left(\dfrac{n-1}{n} - \dfrac{n}{n-1}\right) = \lim_{n\to\infty} \dfrac{(n-1)^2 - n^2}{n(n-1)} = \lim_{n\to\infty} \dfrac{1-2n}{n^2-n} = 0$, converges

63. $\lim_{n\to\infty} \dfrac{n^p}{e^n} = 0$, converges

$(p > 0, n \geq 2)$

65. $a_n = \left(1 + \dfrac{k}{n}\right)^n$

$$\lim_{n\to\infty} \left(1 + \frac{k}{n}\right)^n = \lim_{u\to 0} [(1+u)^{1/u}]^k = e^k$$

where $u = \dfrac{k}{n}$, converges

67. $a_n = 4 - \dfrac{1}{n} < 4 - \dfrac{1}{n+1} = a_{n+1}$,

monotonic; $|a_n| < 4$ bounded.

69. $a_n = \dfrac{\cos n}{n}$

$a_1 = 0.5403$

$a_2 = -0.2081$

$a_3 = -0.3230$

$a_4 = -0.1634$

Not monotonic; $|a_n| \le 1$, bounded

71. $a_n = (-1)^n\left(\dfrac{1}{n}\right)$

$a_1 = -1$

$a_2 = \dfrac{1}{2}$

$a_3 = -\dfrac{1}{3}$

Not monotonic; $|a_n| \le 1$, bounded

73. $a_n = \left(\frac{2}{3}\right)^n > \left(\frac{2}{3}\right)^{n+1} = a_{n+1}$

Monotonic; $|a_n| \le \frac{2}{3}$, bounded

75. $a_n = \sin\left(\dfrac{n\pi}{6}\right)$

$a_1 = 0.500$

$a_2 = 0.8660$

$a_3 = 1.000$

$a_4 = 0.8660$

Not monotonic; $|a_n| \le 1$, bounded

77. (a) $a_n = 5 + \dfrac{1}{n}$

$\left|5 + \dfrac{1}{n}\right| \le 6 \Rightarrow \{a_n\}$ bounded

$a_n = 5 + \dfrac{1}{n} > 5 + \dfrac{1}{n+1}$

$= a_{n+1} \Rightarrow \{a_n\}$ monotonic

Therefore, $\{a_n\}$ converges.

(b)

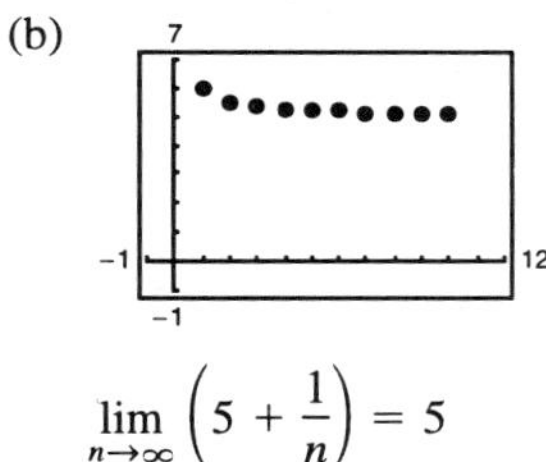

$\lim_{n\to\infty}\left(5 + \dfrac{1}{n}\right) = 5$

79. (a) $a_n = \dfrac{1}{3}\left(1 - \dfrac{1}{3^n}\right)$

$\left|\dfrac{1}{3}\left(1 - \dfrac{1}{3^n}\right)\right| < \dfrac{1}{3} \Rightarrow \{a_n\}$ bounded

$a_n = \dfrac{1}{3}\left(1 - \dfrac{1}{3^n}\right) < \dfrac{1}{3}\left(1 - \dfrac{1}{3^{n+1}}\right)$

$= a_{n+1} \Rightarrow \{a_n\}$ monotonic

Therefore, $\{a_n\}$ converges.

(b)

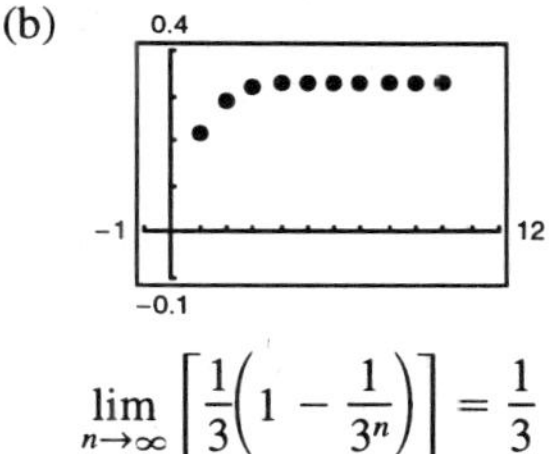

$\lim_{n\to\infty}\left[\dfrac{1}{3}\left(1 - \dfrac{1}{3^n}\right)\right] = \dfrac{1}{3}$

81. (a) $a_n = 10 - \dfrac{1}{n}$

(b) Not possible; a bounded monotonic sequence must converge—see Theorem 8.5.

(c) $a_n = \dfrac{3n}{4n + 1}$

(d) Not possible; an unbounded sequence does not converge.

83. $A_n = P\left[1 + \dfrac{r}{12}\right]^n$

(a) $\lim_{n\to\infty} A_n = \infty$, divergent. The amount will grow arbitrarily large over time.

(b) $A_n = 9000\left[1 + \dfrac{0.115}{12}\right]^n$

$A_1 = \$9086.25$	$A_6 = \$9530.06$
$A_2 = \$9173.33$	$A_7 = \$9621.39$
$A_3 = \$9261.24$	$A_8 = \$9713.59$
$A_4 = \$9349.99$	$A_9 = \$9806.68$
$A_5 = \$9439.60$	$A_{10} = \$9900.66$

85. (a) $A_n = (0.8)^n(2.5)$ billion

(b) $A_1 = \$2$ billion

$A_2 = \$1.6$ billion

$A_3 = \$1.28$ billion

$A_4 = \$1.024$ billion

(c) $\lim_{n\to\infty} (0.8)^n(2.5) = 0$

87. (a) $a_n = 59.69n + 697.32$

(b) For the year 2000, $n = 10$ and $a_{10} \approx \$1294$

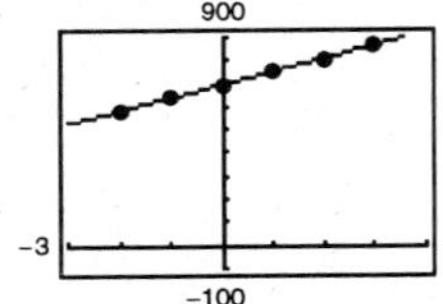

89. $S_6 = 130 + 70 + 40 = 240$

$S_7 = 240 + 130 + 70 = 440$

$S_8 = 440 + 240 + 130 = 810$

$S_9 = 810 + 440 + 240 = 1490$

$S_{10} = 1490 + 810 + 440 = 2740$

91. $a_n = \dfrac{10^n}{n!}$

(a) $a_9 = a_{10} = \dfrac{10^9}{9!}$

$= \dfrac{1{,}000{,}000{,}000}{362{,}880}$

$= \dfrac{1{,}562{,}500}{567}$

(b) Decreasing

(c) Factorials increase more rapidly than exponentials.

93. $\{a_n\} = \left\{\sqrt[n]{n}\right\} = \{n^{1/n}\}$

$a_1 = 1^{1/1} = 1$

$a_2 = \sqrt{2} \approx 1.4142$

$a_3 = \sqrt[3]{3} \approx 1.4422$

$a_4 = \sqrt[4]{4} \approx 1.4142$

$a_5 = \sqrt[5]{5} \approx 1.3797$

$a_6 = \sqrt[6]{6} \approx 1.3480$

Let $y = \lim_{n\to\infty} n^{1/n}$.

$$\ln y = \lim_{n\to\infty}\left(\frac{1}{n}\ln n\right) = \lim_{n\to\infty}\frac{\ln n}{n} = \lim_{n\to\infty}\frac{1/n}{1} = 0$$

Since $\ln y = 0$, we have $y = e^0 = 1$. Therefore,

$$\lim_{n\to\infty}\sqrt[n]{n} = 1.$$

95. $a_{n+2} = a_n + a_{n+1}$

(a) $a_1 = 1$ $\quad a_7 = 8 + 5 = 13$

$a_2 = 1$ $\quad a_8 = 13 + 8 = 21$

$a_3 = 1 + 1 = 2$ $\quad a_9 = 21 + 13 = 34$

$a_4 = 2 + 1 = 3$ $\quad a_{10} = 34 + 21 = 55$

$a_5 = 3 + 2 = 5$ $\quad a_{11} = 55 + 34 = 89$

$a_6 = 5 + 3 = 8$ $\quad a_{12} = 89 + 55 = 144$

(b) $b_n = \dfrac{a_{n+1}}{a_n},\ n \geq 1$

$b_1 = \dfrac{1}{1} = 1$ $\quad b_6 = \dfrac{13}{8}$

$b_2 = \dfrac{2}{1} = 2$ $\quad b_7 = \dfrac{21}{13}$

$b_3 = \dfrac{3}{2}$ $\quad b_8 = \dfrac{34}{21}$

$b_4 = \dfrac{5}{3}$ $\quad b_9 = \dfrac{55}{34}$

$b_5 = \dfrac{8}{5}$ $\quad b_{10} = \dfrac{89}{55}$

(c) $1 + \dfrac{1}{b_{n-1}} = 1 + \dfrac{1}{a_n/a_{n-1}}$

$= 1 + \dfrac{a_{n-1}}{a_n}$

$= \dfrac{a_n + a_{n-1}}{a_n} = \dfrac{a_{n+1}}{a_n} = b_n$

(d) If $\lim_{n\to\infty} b_n = \rho$, then $\lim_{n\to\infty}\left(1 + \dfrac{1}{b_{n-1}}\right) = \rho$.

Since $\lim_{n\to\infty} b_n = \lim_{n\to\infty} b_{n-1}$ we have,

$1 + (1/\rho) = \rho.$

$\rho + 1 = \rho^2$

$0 = \rho^2 - \rho - 1$

$\rho = \dfrac{1 \pm \sqrt{1 + 4}}{2} = \dfrac{1 \pm \sqrt{5}}{2}$

Since a_n, and thus b_n, is positive,

$\rho = \left(1 + \sqrt{5}\right)/2 \approx 1.6180.$

97. True

99. True

101. $a_1 = \sqrt{2} \approx 1.4142$

$a_2 = \sqrt{2 + \sqrt{2}} \approx 1.8478$

$a_3 = \sqrt{2 + \sqrt{2 + \sqrt{2}}} \approx 1.9616$

$a_4 = \sqrt{2 + \sqrt{2 + \sqrt{2 + \sqrt{2}}}} \approx 1.9904$

$a_5 = \sqrt{2 + \sqrt{2 + \sqrt{2 + \sqrt{2 + \sqrt{2}}}}} \approx 1.9976$

$\{a_n\}$ is increasing and bounded by 2, and hence converges to L. Letting $\lim_{n\to\infty} a_n = L$ implies that $\sqrt{2 + L} = L \Rightarrow L = 2$. Hence, $\lim_{n\to\infty} a_n = 2$.

Section 8.2 Series and Convergence

1. $S_1 = 1$

$S_2 = 1 + \frac{1}{4} = 1.2500$

$S_3 = 1 + \frac{1}{4} + \frac{1}{9} \approx 1.3611$

$S_4 = 1 + \frac{1}{4} + \frac{1}{9} + \frac{1}{16} \approx 1.4236$

$S_5 = 1 + \frac{1}{4} + \frac{1}{9} + \frac{1}{16} + \frac{1}{25} \approx 1.4636$

3. $S_1 = 3$

$S_2 = 3 - \frac{9}{2} = -1.5$

$S_3 = 3 - \frac{9}{2} + \frac{27}{4} = 5.25$

$S_4 = 3 - \frac{9}{2} + \frac{27}{4} - \frac{81}{8} = -4.875$

$S_5 = 3 - \frac{9}{2} + \frac{27}{4} - \frac{81}{8} + \frac{243}{16} = 10.3125$

5. $S_1 = 3$

$S_2 = 3 + \frac{3}{2} = 4.5$

$S_3 = 3 + \frac{3}{2} + \frac{3}{4} = 5.250$

$S_4 = 3 + \frac{3}{2} + \frac{3}{4} + \frac{3}{8} = 5.625$

$S_5 = 3 + \frac{3}{2} + \frac{3}{4} + \frac{3}{8} + \frac{3}{16} = 5.8125$

7. $\sum_{n=1}^{\infty} \frac{n}{n+1}$

$\lim_{n\to\infty} \frac{n}{n+1} = 1 \neq 0$

Diverges by Theorem 8.9

9. $\sum_{n=1}^{\infty} \frac{n^2}{n^2+1}$

$\lim_{n\to\infty} \frac{n^2}{n^2+1} = 1 \neq 0$

Diverges by Theorem 8.9

11. $\sum_{n=0}^{\infty} 3\left(\frac{3}{2}\right)^n$ Geometric series

$r = \frac{3}{2} > 1$

Diverges by Theorem 8.6

13. $\sum_{n=0}^{\infty} 1000(1.055)^n$ Geometric series

$r = 1.055 > 1$

Diverges by Theorem 8.6

15. $\sum_{n=0}^{\infty} \frac{2^n + 1}{2^{n+1}}$

$\lim_{n\to\infty} \frac{2^n + 1}{2^{n+1}} = \lim_{n\to\infty} \frac{1 + 2^{-n}}{2} = \frac{1}{2} \neq 0$

Diverges by Theorem 8.9

17. $\sum_{n=0}^{\infty} 2\left(\frac{3}{4}\right)^n$

Geometric series with $r = \frac{3}{4} < 1$.

Converges by Theorem 8.6

19. $\sum_{n=0}^{\infty} (0.9)^n$

Geometric series with $r = 0.9 < 1$.

Converges by Theorem 8.6

21. $\sum_{n=1}^{\infty} \frac{1}{n(n+1)} = \sum_{n=1}^{\infty} \left(\frac{1}{n} - \frac{1}{n+1}\right) = \left(1 - \frac{1}{2}\right) + \left(\frac{1}{2} - \frac{1}{3}\right) + \left(\frac{1}{3} - \frac{1}{4}\right) + \left(\frac{1}{4} - \frac{1}{5}\right) + \cdots$

$\sum_{n=1}^{\infty} \frac{1}{n(n+1)} = \lim_{n\to\infty} S_n = \lim_{n\to\infty} \left(1 - \frac{1}{n+1}\right) = 1$

23. $\sum_{n=0}^{\infty} \frac{9}{4}\left(\frac{1}{4}\right)^n = \frac{9}{4}\left[1 + \frac{1}{4} + \frac{1}{16} + \cdots\right]$

$S_0 = \frac{9}{4}, S_1 = \frac{9}{4} \cdot \frac{5}{4} = \frac{45}{16}, S_2 = \frac{9}{4} \cdot \frac{21}{16} \approx 2.95, \ldots$

Matches graph (c).

Analytically, the series is geometric:

$$\sum_{n=0}^{\infty} \left(\frac{9}{4}\right)\left(\frac{1}{4}\right)^n = \frac{9/4}{1 - 1/4} = \frac{9/4}{3/4} = 3$$

25. $\sum_{n=0}^{\infty} \frac{15}{4}\left(-\frac{1}{4}\right)^n = \frac{15}{4}\left[1 - \frac{1}{4} + \frac{1}{16} - \cdots\right]$

$S_0 = \frac{15}{4}, S_1 = \frac{45}{16}, S_2 \approx 3.05, \ldots$

Matches graph (a).

Analytically, the series is geometric:

$$\sum_{n=0}^{\infty} \frac{15}{4}\left(-\frac{1}{4}\right)^n = \frac{15/4}{1 - (-1/4)} = \frac{15/4}{5/4} = 3$$

27. (a) $\sum_{n=1}^{\infty} 2(0.9)^{n-1} = \sum_{n=0}^{\infty} 2(0.9)^n = \frac{2}{1 - 0.9} = 20$

(b)

n	5	10	20	50	100
S_n	8.1902	13.0264	17.5685	19.8969	19.9995

(c)

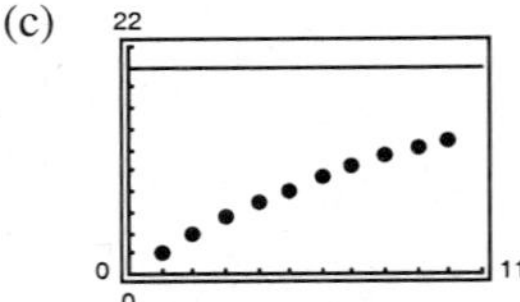

(d) The terms of the series decrease in magnitude slowly. Thus, the sequence of partial sums approaches the sum slowly.

29. (a) $\sum_{n=1}^{\infty} 10(0.25)^{n-1} = \frac{10}{1 - 0.25} = \frac{40}{3} \approx 13.3333$

(b)

n	5	10	20	50	100
S_n	13.3203	13.3333	13.3333	13.3333	13.3333

(c)

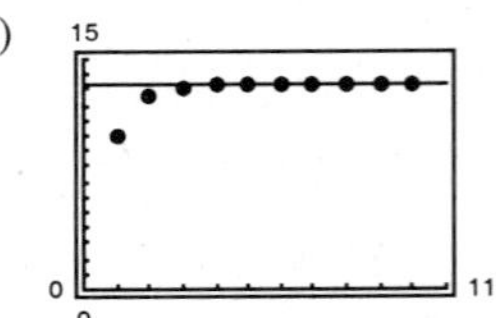

(d) The terms of the series decrease in magnitude rapidly. Thus, the sequence of partial sums approaches the sum rapidly.

31. $\sum_{n=0}^{\infty} \left(\frac{1}{2}\right)^n = \frac{1}{1 - (1/2)} = 2$

33. $\sum_{n=0}^{\infty} \left(-\frac{1}{2}\right)^n = \frac{1}{1 - (-1/2)} = \frac{2}{3}$

35. $\sum_{n=0}^{\infty} \left(\frac{1}{10}\right)^n = \frac{1}{1 - (1/10)} = \frac{10}{9}$

37. $\sum_{n=0}^{\infty} 3\left(-\frac{1}{3}\right)^n = \frac{3}{1 - (-1/3)} = \frac{9}{4}$

39. $$\sum_{n=2}^{\infty} \frac{1}{n^2 - 1} = \sum_{n=2}^{\infty} \left(\frac{1/2}{n-1} - \frac{1/2}{n+1}\right) = \frac{1}{2}\sum_{n=2}^{\infty}\left(\frac{1}{n-1} - \frac{1}{n+1}\right)$$

$$= \frac{1}{2}\left[\left(1 - \frac{1}{3}\right) + \left(\frac{1}{2} - \frac{1}{4}\right) + \left(\frac{1}{3} - \frac{1}{5}\right) + \left(\frac{1}{4} - \frac{1}{6}\right) + \cdots\right]$$

$$= \frac{1}{2}\left(1 + \frac{1}{2}\right) = \frac{3}{4}$$

41. $$\sum_{n=1}^{\infty} \frac{4}{n(n+2)} = 2\sum_{n=1}^{\infty}\left(\frac{1}{n} - \frac{1}{n+2}\right) = 2\left[\left(1 - \frac{1}{3}\right) + \left(\frac{1}{2} - \frac{1}{4}\right) + \left(\frac{1}{3} - \frac{1}{5}\right) + \cdots\right] = 2\left(1 + \frac{1}{2}\right) = 3$$

43. $$\sum_{n=0}^{\infty}\left(\frac{1}{2^n} - \frac{1}{3^n}\right) = \sum_{n=0}^{\infty}\left(\frac{1}{2}\right)^n - \sum_{n=0}^{\infty}\left(\frac{1}{3}\right)^n = \frac{1}{1 - (1/2)} - \frac{1}{1 - (1/3)} = 2 - \frac{3}{2} = \frac{1}{2}$$

45. $0.\overline{4} = \sum_{n=0}^{\infty} \frac{4}{10}\left(\frac{1}{10}\right)^n$

Geometric series with $a = \frac{4}{10}$ and $r = \frac{1}{10}$

$$S = \frac{a}{1 - r} = \frac{4/10}{1 - (1/10)} = \frac{4}{9}$$

47. $0.0\overline{75} = \sum_{n=0}^{\infty} \frac{3}{40}\left(\frac{1}{100}\right)^n$

Geometric series with $a = \frac{3}{40}$ and $r = \frac{1}{100}$

$$S = \frac{a}{1 - r} = \frac{3/40}{99/100} = \frac{5}{66}$$

49. $\sum_{n=1}^{\infty} \frac{n+10}{10n+1}$

$$\lim_{n\to\infty} \frac{n+10}{10n+1} = \frac{1}{10} \neq 0$$

Diverges by Theorem 8.9

51. $\sum_{n=1}^{\infty} \left(\frac{1}{n} - \frac{1}{n+2}\right) = \left(1 - \frac{1}{3}\right) + \left(\frac{1}{2} - \frac{1}{4}\right) + \left(\frac{1}{3} - \frac{1}{5}\right) + \left(\frac{1}{4} - \frac{1}{6}\right) + \cdots = 1 + \frac{1}{2} = \frac{3}{2}$, converges

53. $\sum_{n=1}^{\infty} \frac{3n-1}{2n+1}$

$$\lim_{n\to\infty} \frac{3n-1}{2n+1} = \frac{3}{2} \neq 0$$

Diverges by Theorem 8.9

55. $\sum_{n=0}^{\infty} \frac{4}{2^n} = 4\sum_{n=0}^{\infty} \left(\frac{1}{2}\right)^n$

Geometric series with $r = \frac{1}{2}$

Converges by Theorem 8.6

57. $\sum_{n=0}^{\infty} (1.075)^n$

Geometric series with $r = 1.075$

Diverges by Theorem 8.6

59. $\sum_{n=2}^{\infty} \frac{n}{\ln n}$

$$\lim_{n\to\infty} \frac{n}{\ln n} = \lim_{n\to\infty} \frac{1}{1/n} = \infty$$

(by L'Hôpital's Rule) Diverges by Theorem 8.9

61. (a) x is the common ratio.

(b) $1 + x + x^2 + \cdots = \sum_{n=0}^{\infty} x^n = \frac{1}{1-x}, \ |x| < 1$

Geometric series: $a = 1, r = x, |x| < 1$

(c) $y_1 = \frac{1}{1-x}$

$y_2 = 1 + x$

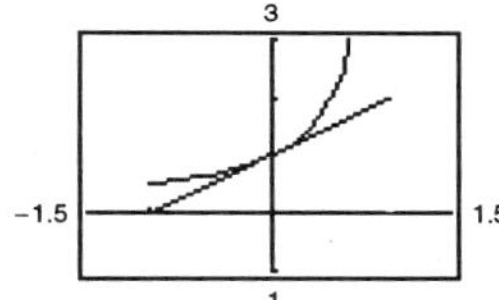

63. $f(x) = 3\left[\frac{1 - 0.5^x}{1 - 0.5}\right]$

Horizontal asymptote: $y = 6$

$$\sum_{n=0}^{\infty} 3\left(\frac{1}{2}\right)^n$$

$$S = \frac{3}{1 - (1/2)} = 6$$

The horizontal asymptote is the sum of the series. $f(n)$ is the nth partial sum.

65. $\frac{1}{n(n+1)} < 0.001$

$10{,}000 < n^2 + n$

$0 < n^2 + n - 10{,}000$

$$n = \frac{-1 \pm \sqrt{1^2 - 4(1)(-10{,}000)}}{2}$$

Choosing the positive value for n we have $n \approx 99.5012$. The first *term* that is less than 0.001 is $n = 100$.

$$\left(\frac{1}{8}\right)^n < 0.001$$

$10{,}000 < 8^n$

This inequality is true when $n = 5$. This series converges at a faster rate.

67. $\sum_{i=0}^{n-1} 8000(0.9)^i = \frac{8000[1 - (0.9)^{(n-1)+1}]}{1 - 0.9} = 80{,}000(1 - 0.9^n), \ n > 0$

69. $D_1 = 16$

$$D_2 = \underbrace{0.81(16)}_{\text{up}} + \underbrace{0.81(16)}_{\text{down}} = 32(0.81)$$

$$D_3 = 16(0.81)^2 + 16(0.81)^2 = 32(0.81)^2$$

$$\vdots$$

$$D = 16 + 32(0.81) + 32(0.81)^2 + \ldots = -16 + \sum_{n=0}^{\infty} 32(0.81)^n = -16 + \frac{32}{1 - 0.81} = 152.42 \text{ ft}$$

71. $P(n) = \frac{1}{2}\left(\frac{1}{2}\right)^n$

$$P(2) = \frac{1}{2}\left(\frac{1}{2}\right)^2 = \frac{1}{8}$$

$$\sum_{n=0}^{\infty} \frac{1}{2}\left(\frac{1}{2}\right)^n = \frac{1/2}{1 - (1/2)} = 1$$

73. (a) $64 + 32 + 16 + 8 + 4 + 2 = 126 \text{ in.}^2$

(b) $\displaystyle\sum_{n=0}^{\infty} 64\left(\frac{1}{2}\right)^n = \frac{64}{1 - (1/2)} = 128 \text{ in.}^2$

Note: This is one-half of the area of the original square!

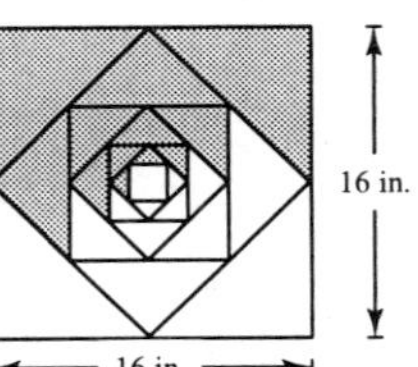

75. $w = \displaystyle\sum_{i=0}^{n-1} 0.01(2)^i = \frac{0.01(1 - 2^n)}{1 - 2} = 0.01(2^n - 1)$

(a) When $n = 29$: $w = \$5{,}368{,}709.11$

(b) When $n = 30$: $w = \$10{,}737{,}418.23$

(c) When $n = 31$: $w = \$21{,}474{,}836.47$

77. $P = 50, r = 0.03, t = 20$

(a) $A = 50\left(\frac{12}{0.03}\right)\left[\left(1 + \frac{0.03}{12}\right)^{12(20)} - 1\right] \approx \$16{,}415.10$

(b) $A = \dfrac{50 - (e^{0.03(20)} - 1)}{e^{0.03/12} - 1} \approx \$16{,}421.83$

79. $P = 100, r = 0.04, t = 40$

(a) $A = 100\left(\frac{12}{0.04}\right)\left[\left(1 + \frac{0.04}{12}\right)^{12(40)} - 1\right] \approx \$118{,}196.13$

(b) $A = \dfrac{100(e^{0.04(40)} - 1)}{e^{0.04/12} - 1} \approx \$118{,}393.43$

81. $x = 0.749999\ldots = 0.74 + \displaystyle\sum_{n=0}^{\infty} 0.009(0.1)^n$

$$= 0.74 + \frac{0.009}{1 - 0.1}$$

$$= 0.74 + 0.01 = 0.75$$

83. By letting $S_0 = 0$, we have $a_n = \displaystyle\sum_{k=1}^{n} a_k - \sum_{k=1}^{n-1} a_k = S_n - S_{n-1}$. Thus,

$$\sum_{n=1}^{\infty} a_n = \sum_{n=1}^{\infty} (S_n - S_{n-1}) = \sum_{n=1}^{\infty} (S_n - S_{n-1} + c - c) = \sum_{n=1}^{\infty} [(c - S_{n-1}) - (c - S_n)].$$

85. $\displaystyle\sum_{n=0}^{39} 30{,}000(1.05)^n = \frac{30{,}000(1 - 1.05^{40})}{1 - 1.05} \approx \$3{,}623{,}993.23$

87. Let $\sum a_n = \displaystyle\sum_{n=0}^{\infty} 1$ and $\sum b_n = \displaystyle\sum_{n=0}^{\infty} (-1)$.

Both are divergent series.

$$\sum (a_n + b_n) = \sum_{n=0}^{\infty} [1 + (-1)] = \sum_{n=0}^{\infty} [1 - 1] = 0$$

89. False. $\displaystyle\lim_{n\to\infty} \frac{1}{n} = 0$, but $\displaystyle\sum_{n=1}^{\infty} \frac{1}{n}$ diverges.

91. False

$$\sum_{n=1}^{\infty} ar^n = \left(\frac{a}{1 - r}\right) - a$$

The formula requires that the geometric series begins with $n = 0$.

93. Let H represent the half-life of the drug. If a patient receives n equal doses of P units each of this drug, administered at equal time interval of length t, the total amount of the drug in the patient's system at the time the last dose is administered is given by

$$T_n = P + Pe^{kt} + Pe^{2kt} + \cdots + Pe^{(n-1)kt}$$

where $k = -(\ln 2)/H$. One time interval *after* the last dose is administered is given by

$$T_{n+1} = Pe^{kt} + Pe^{2kt} + Pe^{3kt} + \cdots + Pe^{nkt}.$$

Two time intervals *after* the last dose is administered is given by

$$T_{n+1} = Pe^{2kt} + Pe^{3kt} + Pe^{4kt} + \cdots + Pe^{(n+1)kt}$$

and so on. Since $k < 0$, $T_n \to 0$ as $n \to \infty$.

Section 8.3 The Integral Test and p-series

1. $\displaystyle\sum_{n=1}^{\infty} \frac{1}{n+1}$

Let $f(x) = \dfrac{1}{x+1}$.

f is positive, continuous and decreasing for $x \ge 1$.

$$\int_1^{\infty} \frac{1}{x+1}\,dx = \Big[\ln(x+1)\Big]_1^{\infty} = \infty$$

Diverges by Theorem 8.10

3. $\displaystyle\sum_{n=1}^{\infty} e^{-n}$

Let $f(x) = e^{-x}$.

f is positive, continuous, and decreasing for $x \ge 1$.

$$\int_1^{\infty} e^{-x}\,dx = \Big[-e^{-x}\Big]_1^{\infty} = \frac{1}{e}$$

Converges by Theorem 8.10

5. $\displaystyle\sum_{n=1}^{\infty} \frac{1}{n^2+1}$

Let $f(x) = \dfrac{1}{x^2+1}$.

f is positive, continuous, and decreasing for $x \ge 1$.

$$\int_1^{\infty} \frac{1}{x^2+1}\,dx = \Big[\arctan x\Big]_1^{\infty} = \frac{\pi}{4}$$

Converges by Theorem 8.10

7. $\displaystyle\sum_{n=1}^{\infty} \frac{\ln(n+1)}{n+1}$

Let $f(x) = \dfrac{\ln(x+1)}{x+1}$.

f is positive, continuous, and decreasing for $x \ge 2$ since

$$f'(x) = \frac{1 - \ln(x+1)}{(x+1)^2} < 0 \text{ for } x \ge 2.$$

$$\int_1^{\infty} \frac{\ln(x+1)}{x+1}\,dx = \left[\frac{\ln^2(x+1)}{2}\right]_1^{\infty} = \infty$$

Diverges by Theorem 8.10

9. $\displaystyle\sum_{n=1}^{\infty} \frac{n^{k-1}}{n^k+c}$

Let $f(x) = \dfrac{x^{k-1}}{x^k+c}$.

f is positive, continuous, and decreasing for $x > \sqrt[k]{c(k-1)}$ since

$$f'(x) = \frac{x^{k-2}[c(k-1) - x^k]}{(x^k+c)^2} < 0$$

for $x > \sqrt[k]{c(k-1)}$.

$$\int_1^{\infty} \frac{x^{k-1}}{x^k+c}\,dx = \left[\frac{1}{k}\ln(x^k+c)\right]_1^{\infty} = \infty$$

Diverges by Theorem 8.10

11. $\displaystyle\sum_{n=1}^{\infty} \frac{1}{n^3}$

Let $f(x) = \dfrac{1}{x^3}$.

f is positive, continuous, and decreasing for $x \ge 1$.

$$\int_1^{\infty} \frac{1}{x^3}\,dx = \left[-\frac{1}{2x^2}\right]_1^{\infty} = \frac{1}{2}$$

Converges by Theorem 8.10

13. $\displaystyle\sum_{n=2}^{\infty} \frac{1}{n(\ln n)^p}$

If $p = 1$, then the series diverges by the Integral Test. If $p \neq 1$,

$$\int_2^{\infty} \frac{1}{x(\ln x)^p}\, dx = \int_2^{\infty} (\ln x)^{-p} \frac{1}{x}\, dx = \left[\frac{(\ln x)^{-p+1}}{-p+1}\right]_2^{\infty}.$$

Converges for $-p + 1 < 0$ or $p > 1$.

15. $\displaystyle\sum_{n=1}^{\infty} \frac{1}{\sqrt[5]{n}} = \sum_{n=1}^{\infty} \frac{1}{n^{1/5}}$

Divergent p-series with $p = \frac{1}{5} < 1$

17. $\displaystyle\sum_{n=1}^{\infty} \frac{1}{n^{1/2}}$

Divergent p-series with $p = \frac{1}{2} < 1$

19. $\displaystyle\sum_{n=1}^{\infty} \frac{1}{n^{3/2}}$

Convergent p-series with $p = \frac{3}{2} > 1$

21. $\displaystyle\sum_{n=1}^{\infty} \frac{1}{n^{1.04}}$

Convergent p-series with $p = 1.04 > 1$

23. $\displaystyle\sum_{n=1}^{\infty} \frac{2}{\sqrt[4]{n^3}} = \frac{2}{1} + \frac{2}{2^{3/4}} + \frac{2}{3^{3/4}} + \cdots$

$S_1 = 2$

$S_2 \approx 3.189$

$S_3 \approx 4.067$

Matches (a)

Diverges—p-series with $p = \frac{3}{4} < 1$

25. $\displaystyle\sum_{n=1}^{\infty} \frac{2}{n\sqrt{n}} = 2 + 2/2^{3/2} + 2/3^{3/2} + \cdots$

$S_1 = 2$

$S_2 \approx 2.707$

$S_3 \approx 3.092$

Matches (b)

Converges—p-series with $p = \frac{3}{2} > 1$

27. No. Theorem 8.9 says that if the series converges, then the terms a_n tend to zero. Some of the series in Exercises 23-26 converge because the terms tend to 0 very rapidly.

29. $\displaystyle\sum_{n=1}^{N} \frac{1}{n} = 1 + \frac{1}{2} + \frac{1}{3} + \frac{1}{4} + \cdots + \frac{1}{N} > M$

(a)

M	2	4	6	8
N	4	31	227	1674

(b) No. Since the terms are decreasing (approaching zero), more and more terms are required to increase the partial sum by 2.

31. Since f is positive, continuous, and decreasing for $x \geq 1$ and $a_n = f(n)$, we have,

$$R_N = S - S_N = \sum_{n=1}^{\infty} a_n - \sum_{n=1}^{N} a_n = \sum_{n=N+1}^{\infty} a_n > 0.$$

Also, $R_N = S - S_N = \displaystyle\sum_{n=N+1}^{\infty} a_n \leq a_{N+1} + \int_{N+1}^{\infty} f(x)\, dx \leq \int_N^{\infty} f(x)\, dx$. Thus,

$$0 \leq R_N \leq \int_N^{\infty} f(x)\, dx.$$

33. $S_6 = 1 + \dfrac{1}{2^4} + \dfrac{1}{3^4} + \dfrac{1}{4^4} + \dfrac{1}{5^4} + \dfrac{1}{6^4} \approx 1.0811$

$$R_6 \leq \int_6^{\infty} \frac{1}{x^4}\, dx = \left[-\frac{1}{3x^3}\right]_6^{\infty} \approx 0.0015$$

$$1.0811 \leq \sum_{n=1}^{\infty} \frac{1}{n^4} \leq 1.0811 + 0.0015 = 1.0826$$

35. $S_{10} = \frac{1}{2} + \frac{1}{5} + \frac{1}{10} + \frac{1}{17} + \frac{1}{26} + \frac{1}{37} + \frac{1}{50} + \frac{1}{65} + \frac{1}{82} + \frac{1}{101} \approx 0.9818$

$$R_{10} = \leq \int_{10}^{\infty} \frac{1}{x^2+1}\,dx = \left[\arctan x\right]_{10}^{\infty} = \frac{\pi}{2} - \arctan 10 \approx 0.0997$$

$$0.9818 \leq \sum_{n=1}^{\infty} \frac{1}{n^5} \leq 0.9818 + 0.0997 = 1.0815$$

37. $S_4 = \frac{1}{e} + \frac{2}{e^4} + \frac{3}{e^9} + \frac{4}{e^{16}} \approx 0.4049$

$$R_4 \leq \int_4^{\infty} xe^{-x^2}\,dx = \left[-\frac{1}{2}e^{-x^2}\right]_4^{\infty} = 5.6 \times 10^{-8}$$

$$0.4049 \leq \sum_{n=1}^{\infty} ne^{-n^2} \leq 0.4049 + 5.6 \times 10^{-8}$$

39. $0 < R_N < \int_N^{\infty} \frac{1}{x^4}\,dx = \left[-\frac{1}{3x^3}\right]_N^{\infty} = \frac{1}{3N^3} < 0.001$

$\frac{1}{N^3} < 0.003$

$N^3 > 333.33$

$N > 6.93$

$N \geq 7$

41. $R_N < \int_N^{\infty} e^{-5x}\,dx = \left[-\frac{1}{5}e^{-5x}\right]_N^{\infty} = \frac{e^{-5N}}{5} < 0.001$

$\frac{1}{e^{5N}} < 0.005$

$e^{5N} > 200$

$5N > \ln 200$

$N > \frac{\ln 200}{5}$

$N > 1.0597$

$N \geq 2$

43. Your friend is not correct. The series

$$\sum_{n=10,000}^{\infty} \frac{1}{n} = \frac{1}{10,000} + \frac{1}{10,001} + \cdots$$

is the harmonic series, starting with the 10,000th term, and hence diverges.

45. (a) $\sum_{n=2}^{\infty} \frac{1}{n^{1.1}}$. This is a convergent p-series with $p = 1.1 > 1$.

$\sum_{n=2}^{\infty} \frac{1}{n \ln n}$ is a divergent series. Use the Integral Test.

$$\int_2^{\infty} \frac{1}{x \ln x}\,dx = \left[\ln|\ln x|\right]_2^{\infty} = \infty$$

(b) $\sum_{n=2}^{6} \frac{1}{n^{1.1}} = \frac{1}{2^{1.1}} + \frac{1}{3^{1.1}} + \frac{1}{4^{1.1}} + \frac{1}{5^{1.1}} + \frac{1}{6^{1.1}} \approx 0.4665 + 0.2987 + 0.2176 + 0.1703 + 0\ 1393$

$\sum_{n=2}^{6} \frac{1}{n \ln n} = \frac{1}{2 \ln 2} + \frac{1}{3 \ln 3} + \frac{1}{4 \ln 4} + \frac{1}{5 \ln 5} + \frac{1}{6 \ln 6} \approx 0.7213 + 0.3034 + 0.1803 - 0.1243 + 0.0930$

The terms of the convergent series **seem** to be larger than those of the divergent series!

(c) $\frac{1}{n^{1.1}} < \frac{1}{n \ln n}$

$n \ln n < n^{1.1}$

$\ln n < n^{0.1}$

This inequality holds when $n \geq 3.5 \times 10^{15}$. Or, $n > e^{40}$. Then $\ln e^{40} = 40 < (e^{40})^{0.1} = e^4 \approx 55$.

47. The harmonic series $\sum_{n=1}^{\infty} \frac{1}{n}$.

49. $\sum_{n=1}^{\infty} \frac{1}{2n-1}$

Let $f(x) = \frac{1}{2x-1}$.

f is positive, continuous, and decreasing for $x \geq 1$.

$$\int_1^{\infty} \frac{1}{2x-1}\,dx = \Big[\ln\sqrt{2x-1}\Big]_1^{\infty} = \infty$$

Diverges by Theorem 8.10

51. $\sum_{n=1}^{\infty} \frac{1}{n\sqrt[4]{n}} = \sum_{n=1}^{\infty} \frac{1}{n^{5/4}}$

p-series with $p = \frac{5}{4}$

Converges by Theorem 8.11

53. $\sum_{n=0}^{\infty} \left(\frac{2}{3}\right)^n$

Geometric series with $r = \frac{2}{3}$

Converges by Theorem 8.6

55. $\sum_{n=1}^{\infty} \frac{n}{\sqrt{n^2+1}}$

$$\lim_{n\to\infty} \frac{n}{\sqrt{n^2+1}} = \lim_{n\to\infty} \frac{1}{\sqrt{1+(1/n^2)}}$$

$$= 1 \neq 0$$

Diverges by Theorem 8.9

57. $\sum_{n=1}^{\infty} \left(1+\frac{1}{n}\right)^n$

$$\lim_{n\to\infty} \left(1+\frac{1}{n}\right)^n = e \neq 0$$

Fails nth Term Test

Diverges by Theorem 8.9

59. $\sum_{n=2}^{\infty} \frac{1}{n(\ln n)^3}$

Let $f(x) = \frac{1}{x(\ln x)^3}$.

f is positive, continuous and decreasing for $x \geq 2$.

$$\int_2^{\infty} \frac{1}{x(\ln x)^3}\,dx = \int_2^{\infty} (\ln x)^{-3}\frac{1}{x}\,dx = \left[\frac{(\ln x)^{-2}}{-2}\right]_2^{\infty} = \left[-\frac{1}{2(\ln x)^2}\right]_2^{\infty} = \frac{1}{2(\ln 2)^2}$$

Converges by Theorem 8.10. See Exercise 13.

Section 8.4 Comparisons of Series

1. (a) $\sum_{n=1}^{\infty} \frac{6}{n^{3/2}} = \frac{6}{1} + \frac{6}{2^{3/2}} + \cdots \quad S_1 = 6$

$$\sum_{n=1}^{\infty} \frac{6}{n^{3/2}+3} = \frac{6}{4} + \frac{6}{2^{3/2}+3} + \cdots \quad S_1 = \frac{3}{2}$$

$$\sum_{n=1}^{\infty} \frac{6}{n\sqrt{n^2+0.5}} = \frac{6}{1\sqrt{1.5}} + \frac{6}{2\sqrt{4.5}} + \cdots \quad S_1 = \frac{6}{\sqrt{1.5}} \approx 4.9$$

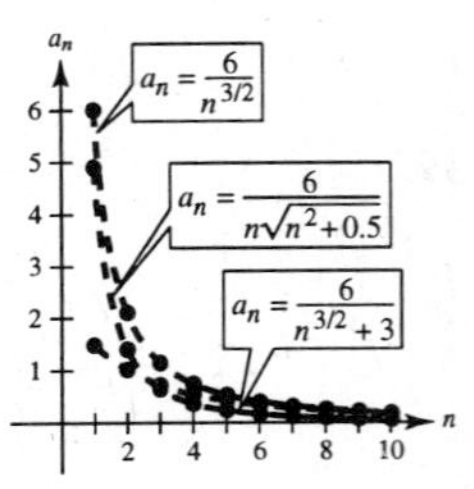

(b) The first series is a p-series. It converges $(p = 3/2 > 1)$.

(c) The magnitude of the terms of the other two series are less than the corresponding terms at the convergent p-series. Hence, the other two series converge.

(d) The smaller the magnitude of the terms, the smaller the magnitude of the terms of the sequence of partial sums.

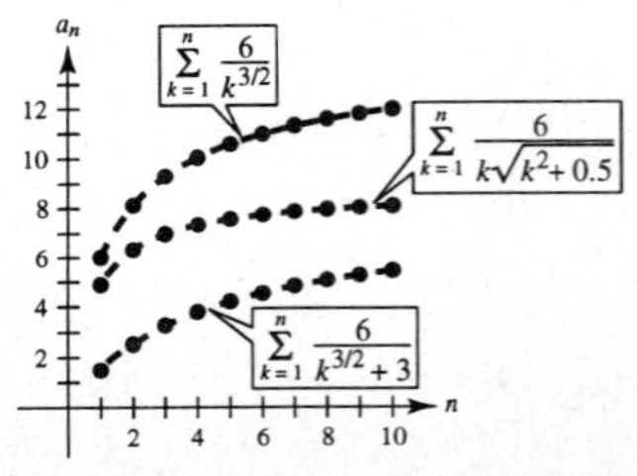

3. $\dfrac{1}{n^2+1} < \dfrac{1}{n^2}$

Therefore,

$$\sum_{n=1}^{\infty} \frac{1}{n^2+1}$$

converges by comparison with the convergent p-series

$$\sum_{n=1}^{\infty} \frac{1}{n^2}.$$

5. $\dfrac{1}{n-1} > \dfrac{1}{n}$ for $n \ge 2$

Therefore,

$$\sum_{n=2}^{\infty} \frac{1}{n-1}$$

diverges by comparison with the divergent p-series

$$\sum_{n=2}^{\infty} \frac{1}{n}.$$

7. $\dfrac{1}{3^n+1} < \dfrac{1}{3^n}$

Therefore,

$$\sum_{n=0}^{\infty} \frac{1}{3^n+1}$$

converges by comparison with the convergent geometric series

$$\sum_{n=0}^{\infty} \left(\frac{1}{3}\right)^n.$$

9. For $n \ge 3$, $\dfrac{\ln n}{n+1} > \dfrac{1}{n+1}$.

Therefore,

$$\sum_{n=1}^{\infty} \frac{\ln n}{n+1}$$

diverges by comparison with the divergent series

$$\sum_{n=1}^{\infty} \frac{1}{n+1}.$$

Note: $\displaystyle\sum_{n=1}^{\infty} \frac{1}{n+1}$ diverges by the integral test.

11. For $n > 3$, $\dfrac{1}{n^2} > \dfrac{1}{n!}$.

Therefore,

$$\sum_{n=0}^{\infty} \frac{1}{n!}$$

converges by comparison with the convergent p-series

$$\sum_{n=1}^{\infty} \frac{1}{n^2}.$$

13. $\dfrac{1}{e^{n^2}} \le \dfrac{1}{e^n}$

Therefore,

$$\sum_{n=0}^{\infty} \frac{1}{e^{n^2}}$$

converges by comparison with the convergent geometric series

$$\sum_{n=0}^{\infty} \left(\frac{1}{e}\right)^n.$$

15. $\displaystyle\lim_{n\to\infty} \frac{n/(n^2+1)}{1/n} = \lim_{n\to\infty} \frac{n^2}{n^2+1} = 1$

Therefore,

$$\sum_{n=1}^{\infty} \frac{n}{n^2+1}$$

diverges by a limit comparison with the divergent p-series

$$\sum_{n=1}^{\infty} \frac{1}{n}.$$

17. $\displaystyle\lim_{n\to\infty} \frac{1/\sqrt{n^2+1}}{1/n} = \lim_{n\to\infty} \frac{n}{\sqrt{n^2+1}} = 1$

Therefore,

$$\sum_{n=0}^{\infty} \frac{1}{\sqrt{n^2+1}}$$

diverges by a limit comparison with the divergent p-series

$$\sum_{n=1}^{\infty} \frac{1}{n}.$$

19. $\displaystyle\lim_{n\to\infty} \frac{\dfrac{2n^2-1}{3n^5+2n+1}}{1/n^3} = \lim_{n\to\infty} \frac{2n^5-n^3}{3n^5+2n+1} = \frac{2}{3}$

Therefore,

$$\sum_{n=1}^{\infty} \frac{2n^2-1}{3n^5+2n+1}$$

converges by a limit comparison with the convergent p-series

$$\sum_{n=1}^{\infty} \frac{1}{n^3}.$$

21. $\displaystyle\lim_{n\to\infty} \frac{\dfrac{n+3}{n(n+2)}}{1/n} = \lim_{n\to\infty} \frac{n^2+3n}{n^2+2n} = 1$

Therefore,

$$\sum_{n=1}^{\infty} \frac{n+3}{n(n+2)}$$

diverges by a limit comparison with the divergent p-series

$$\sum_{n=1}^{\infty} \frac{1}{n}.$$

23. $\lim_{n\to\infty} \frac{1/(n\sqrt{n^2+1})}{1/n^2} = \lim_{n\to\infty} \frac{n^2}{n\sqrt{n^2+1}} = 1$

Therefore,

$$\sum_{n=1}^{\infty} \frac{1}{n\sqrt{n^2+1}}$$

converges by a limit comparison with the convergent p-series

$$\sum_{n=1}^{\infty} \frac{1}{n^2}.$$

25. $\lim_{n\to\infty} \frac{(n^{k-1})/(n^k+1)}{1/n} = \lim_{n\to\infty} \frac{n^k}{n^k+1} = 1$

Therefore,

$$\sum_{n=1}^{\infty} \frac{n^{k-1}}{n^k+1}$$

diverges by a limit comparison with the divergent p-series

$$\sum_{n=1}^{\infty} \frac{1}{n}.$$

27. $\lim_{n\to\infty} \frac{\sin(1/n)}{1/n} = \lim_{n\to\infty} \frac{(-1/n^2)\cos(1/n)}{-1/n^2} = \lim_{n\to\infty} \cos\left(\frac{1}{n}\right) = 1$

Therefore, $\sum_{n=1}^{\infty} \sin\left(\frac{1}{n}\right)$ diverges by a limit comparison with the divergent p-series $\sum_{n=1}^{\infty} \frac{1}{n}$.

29. $\sum_{n=1}^{\infty} \frac{\sqrt{n}}{n} = \sum_{n=1}^{\infty} \frac{1}{\sqrt{n}}$

Diverges

p-series with $p = \frac{1}{2}$

31. $\sum_{n=1}^{\infty} \frac{1}{3^n+2}$

Converges

Direct comparison with $\sum_{n=1}^{\infty} \left(\frac{1}{3}\right)^n$

33. $\sum_{n=1}^{\infty} \frac{n}{2n+3}$

Diverges; nth Term Test

$$\lim_{n\to\infty} \frac{n}{2n+3} = \frac{1}{2} \neq 0$$

35. $\sum_{n=1}^{\infty} \frac{n}{(n^2+1)^2}$

Converges; integral test

37. $\lim_{n\to\infty} \frac{a_n}{1/n} = \lim_{n\to\infty} na_n \neq 0$

Therefore,

$\sum_{n=1}^{\infty} a_n$ diverges by a limit comparison with the p-series $\sum_{n=1}^{\infty} \frac{1}{n}$.

39. $\frac{1}{2} + \frac{2}{5} + \frac{3}{10} + \frac{4}{17} + \frac{5}{26} + \cdots = \sum_{n=1}^{\infty} \frac{n}{n^2+1}$,

which diverges since the degree of the numerator is only one less than the degree of the denominator.

41. $\sum_{n=1}^{\infty} \frac{1}{n^3+1}$

converges since the degree of the numerator is three less than the degree of the denominator.

43. $\lim_{n\to\infty} n\left(\frac{n^3}{5n^4+3}\right) = \lim_{n\to\infty} \frac{n^4}{5n^4+3} = \frac{1}{5} \neq 0$

Therefore, $\sum_{n=1}^{\infty} \frac{n^3}{5n^4+3}$ diverges.

45. (a) $\sum_{n=1}^{\infty} \frac{1}{(2n-1)^2} = \sum_{n=1}^{\infty} \frac{1}{4n^2 - 4n + 1}$

converges since the degree of the numerator is two less than the degree of the denominator. (See Exercise 38.)

(b)

n	5	10	20	50	100
S_n	1.1839	1.02087	1.2212	1.2287	1.2312

(c) $\sum_{n=3}^{\infty} \frac{1}{(2n-1)^2} = \frac{\pi^2}{8} - S_2 \approx 0.1226$

(d) $\sum_{n=10}^{\infty} \frac{1}{(2n-1)^2} = \frac{\pi^2}{8} - S_9 \approx 0.0277$

47. False. Let $a_n = 1/n^3$ and $b_n = 1/n^2$. $0 < a_n \le b_n$ and both

$$\sum_{n=1}^{\infty} \frac{1}{n^3} \text{ and } \sum_{n=1}^{\infty} \frac{1}{n^2}$$

converge.

49. True

51. Since $\sum_{n=1}^{\infty} b_n$ converges, $\lim_{n\to\infty} b_n = 0$. There exists N such that $b_n < 1$ for $n > N$. Thus,

$$a_n b_n < a_n \text{ for } n > N \text{ and } \sum_{n=1}^{\infty} a_n b_n$$

converges by comparison to the convergent series $\sum_{i=1}^{\infty} a_n$.

53. $\sum \frac{1}{n^2}$ and $\sum \frac{1}{n^3}$ both converge, and hence so does

$$\sum\left(\frac{1}{n^2}\right)\left(\frac{1}{n^3}\right) = \sum \frac{1}{n^5}$$

55. (a) Suppose Σb_n converges and Σa_n diverges. Then there exists N such that $0 < b_n < a_n$ for $n \ge N$. This means that $1 < a_n/b_n$ for $n \ge N$. Therefore, $\lim_{n\to\infty} a_n/b_n \ne 0$. Thus, Σa_n must also converge.

(b) Suppose Σb_n diverges and Σa_n converges. Then there exists N such that $0 < a_n < b_n$ for $n \ge N$. This means that $0 < a_n/b_n < 1$ for $n \ge N$. Therefore, $\lim_{n\to\infty} a_n/b_n \ne \infty$. Thus, Σa_n must also diverge.

57. Since $0 < a_n < 1$, $0 < a_n^2 < a_n < 1$. The squared terms will be below the others.

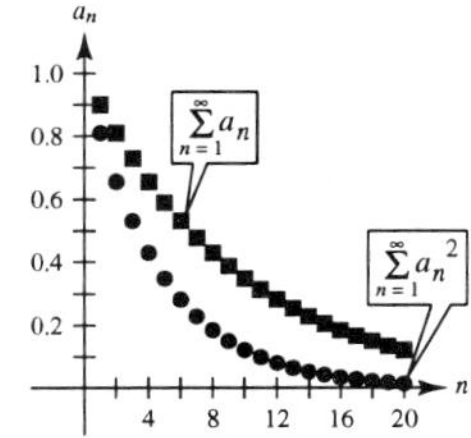

Section 8.5 Alternating Series

1. $\sum_{n=1}^{\infty} \frac{6}{n^2} = \frac{6}{1} + \frac{6}{4} + \frac{6}{9} + \cdots$

$S_1 = 6, S_2 = 7.5$

Matches (b)

3. $\sum_{n=1}^{\infty} \frac{10}{n2^n} = \frac{10}{2} + \frac{10}{8} + \cdots$

$S_1 = 5, S_2 = 6.25$

Matches (c)

5. $\sum_{n=1}^{\infty} \frac{(-1)^{n-1}}{2n-1} = \frac{\pi}{4} \approx 0.7854$

(a)

n	1	2	3	4	5	6	7	8	9	10
S_n	1	0.6667	0.8667	0.7238	0.8349	0.7440	0.8209	0.7543	0.8131	0.7605

(b)

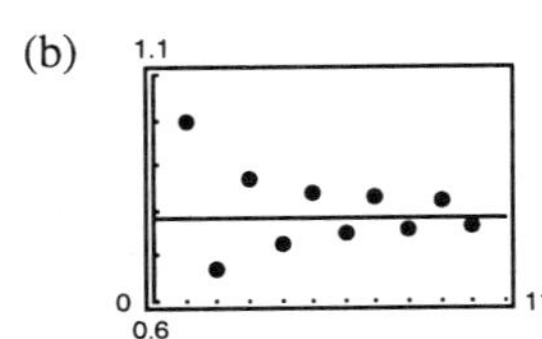

(c) The points alternate sides of the horizontal line that represents the sum of the series. The distance between successive points and the line decreases.

(d) The distance in part (c) is always less than the magnitude of the next term of the series.

7. $\sum_{n=1}^{\infty} \frac{(-1)^{n-1}}{n^2} = \frac{\pi^2}{12} \approx 0.8225$

(a)

n	1	2	3	4	5	6	7	8	9	10
S_n	1	0.75	0.8611	0.7986	0.8386	0.8108	0.8312	0.8156	0.8280	0.8180

(b)

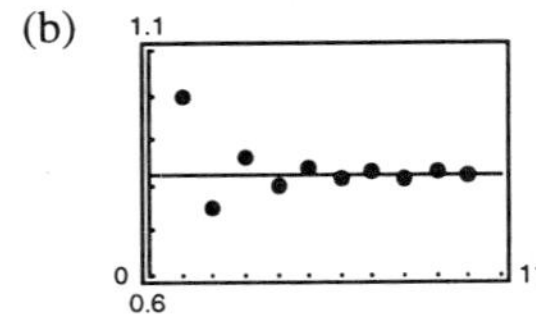

(c) The points alternate sides of the horizontal line that represents the sum of the series. The distance between successive points and the line decreases.

(d) The distance in part (c) is always less than the magnitude of the next series.

9. $\sum_{n=1}^{\infty} \frac{(-1)^{n+1}}{n}$

$a_{n+1} = \frac{1}{n+1} < \frac{1}{n} = a_n$

$\lim_{n\to\infty} \frac{1}{n} = 0$

Converges by Theorem 8.14.

11. $\sum_{n=1}^{\infty} \frac{(-1)^{n+1}}{2n-1}$

$a_{n+1} = \frac{1}{2(n+1)-1} < \frac{1}{2n-1} = a_n$

$\lim_{n\to\infty} \frac{1}{2n-1} = 0$

Converges by Theorem 8.14

13. $\sum_{n=1}^{\infty} \frac{(-1)^n n^2}{n^2+1}$

$\lim_{n\to\infty} \frac{n^2}{n^2+1} = 1$

Diverges by the nth Term Test

15. $\sum_{n=1}^{\infty} \frac{(-1)^n}{\sqrt{n}}$

$a_{n+1} = \frac{1}{\sqrt{n+1}} < \frac{1}{\sqrt{n}} = a_n$

$\lim_{n\to\infty} \frac{1}{\sqrt{n}} = 0$

Converges by Theorem 8.14

17. $\sum_{n=1}^{\infty} \frac{(-1)^{n+1}(n+1)}{\ln(n+1)}$

$$\lim_{n\to\infty} \frac{n+1}{\ln(n+1)} = \lim_{n\to\infty} \frac{1}{1/(n+1)} = \lim_{n\to\infty} (n+1) = \infty$$

Diverges by the nth Term Test

19. $\sum_{n=1}^{\infty} \sin\left[\frac{(2n-1)\pi}{2}\right] = \sum_{n=1}^{\infty} (-1)^{n+1}$

Diverges by the nth Term Test

21. $\sum_{n=1}^{\infty} \frac{1}{n} \sin\left[\frac{(2n-1)\pi}{2}\right] = \sum_{n=1}^{\infty} \frac{(-1)^{n+1}}{n}$

Converges; (see Exercise 9)

23. $\sum_{n=0}^{\infty} \frac{(-1)^n}{n!}$

$$a_{n+1} = \frac{1}{(n+1)!} < \frac{1}{n!} = a_n$$

$$\lim_{n\to\infty} \frac{1}{n!} = 0$$

Converges by Theorem 8.14

25. $\sum_{n=1}^{\infty} \frac{(-1)^{n+1}\sqrt{n}}{n+2}$

$$a_{n+1} = \frac{\sqrt{n+1}}{(n+1)+2} < \frac{\sqrt{n}}{n+2} \text{ for } n \geq 2$$

$$\lim_{n\to\infty} \frac{\sqrt{n}}{n+2} = 0$$

Converges by Theorem 8.14

27. $\sum_{n=1}^{\infty} \frac{(-1)^{n+1}(2)}{e^n - e^{-n}} = \sum_{n=1}^{\infty} \frac{(-1)^{n+1}(2e^n)}{e^{2n}-1}$

Let $f(x) = \frac{2e^x}{e^{2x}-1}$. Then

$$f'(x) = \frac{-2e^x(e^{2x}+1)}{(e^{2x}-1)^2} < 0.$$

Thus, $f(x)$ is decreasing. Therefore, $a_{n+1} < a_n$, and

$$\lim_{n\to\infty} \frac{2e^n}{e^{2n}-1} = \lim_{n\to\infty} \frac{2e^n}{2e^{2n}} = \lim_{n\to\infty} \frac{1}{e^n} = 0.$$

The series converges by Theorem 8.14.

29. $\sum_{n=0}^{\infty} \frac{(-1)^n}{n!}$

(a) By Theorem 8.15,

$$|R_N| \leq a_{N+1} = \frac{1}{(N+1)!} < 0.001.$$

This inequality is valid when $N = 6$.

(b) We may approximate the series by

$$\sum_{n=0}^{6} \frac{(-1)^n}{n!} = 1 - 1 + \frac{1}{2} - \frac{1}{6} + \frac{1}{24} - \frac{1}{120} + \frac{1}{720}$$

$$\approx 0.368.$$

(7 terms. Note that the sum begins with $n = 0$.)

31. $\sum_{n=0}^{\infty} \frac{(-1)^n}{(2n+1)!}$

(a) By Theorem 8.15,

$$|R_N| \leq a_{N+1} = \frac{1}{[2(N+1)+1]!} < 0.001.$$

This inequality is valid when $N = 2$.

(b) We may approximate the series by

$$\sum_{n=0}^{2} \frac{(-1)^n}{(2n+1)!} = 1 - \frac{1}{6} + \frac{1}{120} \approx 0.842.$$

(3 terms. Note that the sum begins with $n = 0$.)

33. $\sum_{n=1}^{\infty} \frac{(-1)^{n+1}}{n}$

(a) By Theorem 8.15,

$$|R_N| \leq a_{N+1} = \frac{1}{N+1} < 0.001.$$

This inequality is valid when $N = 1000$.

(b) We may approximate the series by

$$\sum_{n=1}^{1000} \frac{(-1)^{n+1}}{n} = 1 - \frac{1}{2} + \frac{1}{3} - \frac{1}{4} + \cdots - \frac{1}{1000}$$

$$\approx 0.693.$$

(1000 terms)

35. $\sum_{n=1}^{\infty} \frac{(-1)^{n+1}}{2n^3 - 1}$

By Theorem 8.15,

$$|R_N| \leq a_{N+1} = \frac{1}{2(N+1)^3 - 1} < 0.001.$$

This inequality is valid when $N = 7$.

37. $\sum_{n=1}^{\infty} \frac{(-1)^{n+1}}{(n+1)^2}$

$\sum_{n=1}^{\infty} \frac{1}{(n+1)^2}$ converges by comparison to the p-series

$$\sum_{n=1}^{\infty} \frac{1}{n^2}.$$

Therefore, the given series converge absolutely.

39. $\sum_{n=1}^{\infty} \frac{(-1)^{n+1}}{\sqrt{n}}$

The given series converges by the Alternating Series Test, but does not converge absolutely since

$$\sum_{n=1}^{\infty} \frac{1}{\sqrt{n}}$$

is a divergent p-series. Therefore, the series converges conditionally.

41. $\sum_{n=1}^{\infty} \frac{(-1)^{n+1} n^2}{(n+1)^2}$

$\lim_{n\to\infty} \frac{n^2}{(n+1)^2} = 1$ Therefore, the series diverges by the nth Term Test.

43. $\sum_{n=2}^{\infty} \frac{(-1)}{\ln(n)}$

The given series converges by the Alternating Series Test, but does not converge absolutely since the series

$$\sum_{n=2}^{\infty} \frac{1}{\ln n}$$

diverges by comparison to the harmonic series

$$\sum_{n=1}^{\infty} \frac{1}{n}.$$

Therefore, the series converges conditionally.

45. $\sum_{n=2}^{\infty} \frac{(-1)^n n}{n^3 - 1}$

$$\sum_{n=2}^{\infty} \frac{n}{n^3 - 1}$$

converges by a limit comparison to the convergent p-series

$$\sum_{n=2}^{\infty} \frac{1}{n^2}.$$

Therefore, the given series converges absolutely.

47. $\sum_{n=0}^{\infty} \frac{(-1)^n}{(2n+1)!}$

$$\sum_{n=0}^{\infty} \frac{1}{(2n+1)!}$$

is convergent by comparison to the convergent geometric series

$$\sum_{n=0}^{\infty} \left(\frac{1}{2}\right)^n$$

since

$$\frac{1}{(2n+1)!} < \frac{1}{2^n} \text{ for } n > 0.$$

Therefore, the given series converges absolutely.

49. $\sum_{n=0}^{\infty} \frac{\cos n\pi}{n+1} = \sum_{n=0}^{\infty} \frac{(-1)^n}{n+1}$

The given series converges by the Alternating Series Test, but

$$\sum_{n=0}^{\infty} \frac{|\cos n\pi|}{n+1} = \sum_{n=0}^{\infty} \frac{1}{n+1}$$

diverges by a limit comparison to the divergent harmonic series,

$$\sum_{n=1}^{\infty} \frac{1}{n}.$$

$\lim_{n\to\infty} \frac{|\cos n\pi|/(n+1)}{1/n} = 1$, therefore the series converges conditionally.

51. $\sum_{n=1}^{\infty} \frac{\cos n\pi}{n^2} = \sum_{n=1}^{\infty} \frac{(-1)^n}{n^2}$

$\sum_{n=1}^{\infty} \frac{1}{n^2}$ is a convergent p-series. Therefore, the given series converges absolutely.

53. $\sum_{n=1}^{\infty} \frac{(-1)^n}{n^p}$

If $p = 0$, then

$$\lim_{n\to\infty} \frac{1}{n^p} = 1$$

and the series diverges. If $p > 0$, then

$$\lim_{n\to\infty} \frac{1}{n^p} = 0 \text{ and } \frac{1}{(n+1)^p} < \frac{1}{n^p}.$$

Therefore, the series converge by the Alternating Series Test.

55. $\sum_{n=1}^{\infty} \frac{(-1)^{n-1}}{n}$ converges, but $\sum_{n=1}^{\infty} \frac{1}{n}$ diverges

57. (a) $\sum_{n=1}^{\infty} \frac{x^n}{n}$

converges absolutely (by comparison) for

$$-1 < x < 1,$$

since

$$\left|\frac{x^n}{n}\right| < |x^n| \text{ and } \sum x^n$$

is a convergent geometric series for $-1 < x < 1$.

(b) When $x = -1$, we have the convergent alternating series

$$\sum_{n=1}^{\infty} \frac{(-1)^n}{n}.$$

When $x = 1$, we have the divergent harmonic series $\frac{1}{n}$.

Therefore,

$$\sum_{n=1}^{\infty} \frac{x^n}{n}$$

converges conditionally for $x = -1$.

59. False

Let $a_n = \frac{(-1)^n}{n}$.

Section 8.6 The Ratio and Root Tests

1. $\frac{(n+1)!}{(n-2)!} = \frac{(n+1)(n)(n-1)(n-2)!}{(n-2)!} = (n+1)(n)(n-1)$

3. Use the Principle of Mathematical Induction. When $k = 1$, the formula is valid since $1 = \frac{(2(1))!}{2^1 \cdot 1!}$. Assume that

$$1 \cdot 3 \cdot 5 \cdots (2n-1) = \frac{(2n)!}{2^n n!}$$

and show that

$$1 \cdot 3 \cdot 5 \cdots (2n-1)(2n+1) = \frac{(2n+2)!}{2^{n+1}(n+1)!}.$$

To do this, note that:

$$\begin{aligned} 1 \cdot 3 \cdot 5 \cdots (2n-1)(2n+1) &= [1 \cdot 3 \cdot 5 \cdots (2n-1)](2n+1) \\ &= \frac{(2n)!}{2^n n!} \cdot (2n+1) \\ &= \frac{(2n)!(2n+1)}{2^n n!} \cdot \frac{(2n+2)}{2(n+1)} \\ &= \frac{(2n)!(2n+1)(2n+2)}{2^{n+1} n!(n+1)} = \frac{(2n+2)!}{2^{n+1}(n+1)} \end{aligned}$$

The formula is valid for all $n \geq 1$.

5. $\sum_{n=1}^{\infty} n\left(\frac{3}{4}\right)^n = 1\left(\frac{3}{4}\right) + 2\left(\frac{9}{16}\right) + \cdots$

$S_1 = \frac{3}{4}, S_2 \approx 1.875$

Matches (d)

7. $\sum_{n=1}^{\infty} \frac{(-3)^{n+1}}{n!} = 9 - \frac{3^3}{2} + \cdots$

$S_1 = 9$

Matches (a)

9. (a) Ratio Test: $\lim_{n\to\infty}\left|\frac{a_{n+1}}{a_n}\right| = \lim_{n\to\infty}\frac{(n+1)^2(5/8)^{n+1}}{n^2(5/8)^n} = \lim_{n\to\infty}\left(\frac{n+1}{n}\right)^2\frac{5}{8} = \frac{5}{8} < 1$. Converges

(b)

n	5	10	15	20	25
S_n	9.2104	16.7598	18.8016	19.1878	19.2491

(c)

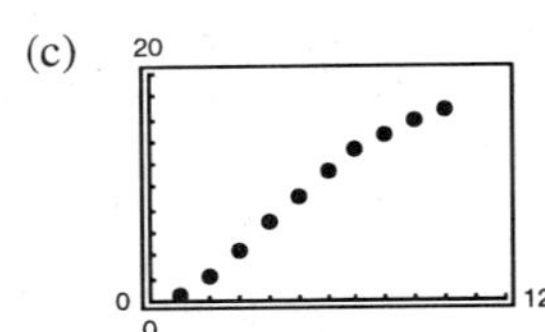

(d) The sum is approximately 19.26.

(e) The more rapidly the terms of the series approach 0, the more rapidly the sequence of the partial sums approaches the sum of the series.

11. $\sum_{n=0}^{\infty} \frac{n!}{3^n}$

$$\lim_{n\to\infty}\left|\frac{a_{n+1}}{a_n}\right| = \lim_{n\to\infty}\left|\frac{(n+1)!}{3^{n+1}}\cdot\frac{3^n}{n!}\right| = \lim_{n\to\infty}\frac{n+1}{3} = \infty$$

Therefore, by the Ratio Test, the series diverges.

13. $\sum_{n=0}^{\infty} \frac{3^n}{n!}$

$$\lim_{n\to\infty}\left|\frac{a_{n+1}}{a_n}\right| = \lim_{n\to\infty}\left|\frac{3^{n+1}}{(n+1)!}\cdot\frac{n!}{3^n}\right| = \lim_{n\to\infty}\frac{3}{n+1} = 0$$

Therefore, by the Ratio Test, the series converges.

15. $\sum_{n=1}^{\infty} \frac{n}{2^n}$

$$\lim_{n\to\infty}\left|\frac{a_{n+1}}{a_n}\right| = \lim_{n\to\infty}\left|\frac{n+1}{2^{n+1}}\cdot\frac{2^n}{n}\right| = \lim_{n\to\infty}\frac{n+1}{2n} = \frac{1}{2}$$

Therefore, by the Ratio Test, the series converges.

17. $\sum_{n=1}^{\infty} \frac{2^n}{n^2}$

$$\lim_{n\to\infty}\left|\frac{a_{n+1}}{a_n}\right| = \lim_{n\to\infty}\left|\frac{2^{n+1}}{(n+1)^2}\cdot\frac{n^2}{2^n}\right| = \lim_{n\to\infty}\frac{2n^2}{(n+1)^2} = 2$$

Therefore, by the Ratio Test, the series diverges.

19. $\sum_{n=0}^{\infty} \frac{(-1)^n 2^n}{n!}$

$$\lim_{n\to\infty}\left|\frac{a_{n+1}}{a_n}\right| = \lim_{n\to\infty}\left|\frac{2^{n+1}}{(n+1)!}\cdot\frac{n!}{2^n}\right| = \lim_{n\to\infty}\frac{2}{n+1} = 0$$

Therefore, by the Ratio Test, the series converges.

21. $\sum_{n=1}^{\infty} \frac{n!}{n3^n}$

$$\lim_{n\to\infty}\left|\frac{a_{n+1}}{a_n}\right| = \lim_{n\to\infty}\left|\frac{(n+1)!}{(n+1)3^{n+1}}\cdot\frac{n3^n}{n!}\right| = \lim_{n\to\infty}\frac{n}{3} = \infty$$

Therefore, by the Ratio Test, the series diverges.

23. $\sum_{n=0}^{\infty} \frac{4^n}{n!}$

$$\lim_{n\to\infty}\left|\frac{a_{n+1}}{a_n}\right| = \lim_{n\to\infty}\left|\frac{4^{n+1}}{(n+1)!}\cdot\frac{n!}{4^n}\right| = \lim_{n\to\infty}\frac{4}{n+1} = 0$$

Therefore, by the Ratio Test, the series converges.

25. $\displaystyle\sum_{n=0}^{\infty} \frac{3^n}{(n+1)^n}$

$$\lim_{n\to\infty}\left|\frac{a_{n+1}}{a_n}\right| = \lim_{n\to\infty}\left|\frac{3^{n+1}}{(n+2)^{n+1}}\cdot\frac{(n+1)^n}{3^n}\right| = \lim_{n\to\infty}\frac{3(n+1)^n}{(n+2)^{n+1}} = \lim_{n\to\infty}\frac{3}{n+2}\left(\frac{n+1}{n+2}\right)^n = (0)\left(\frac{1}{e}\right) = 0$$

To find $\displaystyle\lim_{n\to\infty}\left(\frac{n+1}{n+2}\right)^n$, let $y = \displaystyle\lim_{n\to\infty}\left(\frac{n+1}{n+2}\right)^n$. Then,

$$\ln y = \lim_{n\to\infty} n\ln\left(\frac{n+1}{n+2}\right) = \lim_{n\to\infty}\frac{\ln[(n+1)/(n+2)]}{1/n} = \frac{0}{0}$$

$$\ln y = \lim_{n\to\infty}\frac{[(1)/(n+1)] - [(1)/(n+2)]}{-(1/n^2)} = -1 \text{ by L'Hôpital's Rule}$$

$$y = e^{-1} = \frac{1}{e}.$$

Therefore, by the Ratio Test, the series converges.

27. $\displaystyle\sum_{n=0}^{\infty} \frac{4^n}{3^n+1}$

$$\lim_{n\to\infty}\left|\frac{a_{n+1}}{a_n}\right| = \lim_{n\to\infty}\left|\frac{4^{n+1}}{3^{n+1}+1}\cdot\frac{3^n+1}{4^n}\right| = \lim_{n\to\infty}\frac{4(3^n+1)}{3^{n+1}+1} = \lim_{n\to\infty}\frac{4(1+1/3^n)}{3+1/3^n} = \frac{4}{3}$$

Therefore, by the Ratio Test, the series diverges.

29. $\displaystyle\sum_{n=0}^{\infty} \frac{(-1)^{n+1}n!}{1\cdot 3\cdot 5\cdots(2n+1)}$

$$\lim_{n\to\infty}\left|\frac{a_{n+1}}{a_n}\right| = \lim_{n\to\infty}\left|\frac{(n+1)!}{1\cdot 3\cdot 5\cdots(2n+1)(2n+3)}\cdot\frac{1\cdot 3\cdot 5\cdots(2n+1)}{n!}\right| = \lim_{n\to\infty}\frac{n+1}{2n+3} = \frac{1}{2}$$

Therefore, by the Ratio Test, the series converges.

Note: The first few terms of this series are $-1 + \dfrac{1}{1\cdot 3} - \dfrac{2!}{1\cdot 3\cdot 5} + \dfrac{3!}{1\cdot 3\cdot 5\cdot 7} - \cdots$

31. (a) $\displaystyle\sum_{n=1}^{\infty} \frac{1}{n^{3/2}}$

$$\lim_{n\to\infty}\left|\frac{a_{n+1}}{a_n}\right| = \lim_{n\to\infty}\left|\frac{1}{(n+1)^{3/2}}\cdot\frac{n^{3/2}}{1}\right| = \lim_{n\to\infty}\left(\frac{n}{n+1}\right)^{3/2} = 1$$

(b) $\displaystyle\sum_{n=1}^{\infty} \frac{1}{n^{1/2}}$

$$\lim_{n\to\infty}\left|\frac{a_{n+1}}{a_n}\right| = \lim_{n\to\infty}\left|\frac{1}{(n+1)^{1/2}}\cdot\frac{n^{1/2}}{1}\right| = \lim_{n\to\infty}\left(\frac{n}{n+1}\right)^{1/2} = 1$$

33. $\displaystyle\sum_{n=1}^{\infty} \left(\frac{n}{2n+1}\right)^n$

$$\lim_{n\to\infty}\sqrt[n]{|a_n|} = \lim_{n\to\infty}\sqrt[n]{\left(\frac{n}{2n+1}\right)^n}$$

$$= \lim_{n\to\infty}\frac{n}{2n+1} = \frac{1}{2}$$

Therefore, by the Root Test, the series converges.

35. $\displaystyle\sum_{n=2}^{\infty} \frac{(-1)^n}{(\ln n)^n}$

$$\lim_{n\to\infty}\sqrt[n]{|a_n|} = \lim_{n\to\infty}\sqrt[n]{\left|\frac{(-1)^n}{(\ln n)^n}\right|}$$

$$= \lim_{n\to\infty}\frac{1}{|\ln n|} = 0$$

Therefore, by the Root Test, the series converges.

37. $\sum_{n=1}^{\infty} (2\sqrt[n]{n} + 1)^n$

$$\lim_{n\to\infty} \sqrt[n]{|a_n|} = \lim_{n\to\infty} \sqrt[n]{(2\sqrt[n]{n} + 1)^n} = \lim_{n\to\infty} (2\sqrt[n]{n} + 1)$$

To find $\lim_{n\to\infty} \sqrt[n]{n}$, let $y = \lim_{n\to\infty} \sqrt[x]{x}$. Then

$$\ln y = \lim_{n\to\infty} \left(\ln \sqrt[x]{x}\right) = \lim_{n\to\infty} \frac{1}{x} \ln x = \lim_{n\to\infty} \frac{\ln x}{x} = \lim_{n\to\infty} \frac{1/x}{1} = 0.$$

Thus, $\ln y = 0$, so $y = e^0 = 1$ and $\lim_{n\to\infty} \left(2\sqrt[n]{n} + 1\right) = 2(1) + 1 = 3$. Therefore, by the Root Test, the series diverges.

39. $\sum_{n=3}^{\infty} \frac{1}{(\ln n)^n}$

$$\lim_{n\to\infty} \sqrt[n]{|a_n|} = \lim_{n\to\infty} \sqrt[n]{\frac{1}{(\ln n)^n}} = \lim_{n\to\infty} \frac{1}{\ln n} = 0$$

Therefore, by the Root Test, the series converges.

41. $\sum_{n=1}^{\infty} \frac{(-1)^{n+1}5}{n}$

$$a_{n+1} = \frac{5}{n+1} < \frac{5}{n} = a_n$$

$$\lim_{n\to\infty} \frac{5}{n} = 0$$

Therefore, by the Alternating Series Test, the series converges (conditional convergence).

43. $\sum_{n=1}^{\infty} \frac{3}{n\sqrt{n}} = 3\sum_{n=1}^{\infty} \frac{1}{n^{3/2}}$

This is convergent p-series.

45. $\sum_{n=1}^{\infty} \frac{2n}{n+1}$

$$\lim_{n\to\infty} \frac{2n}{n+1} = 2 \neq 0$$

This diverges by the nth Term Test for Divergence.

47. $\sum_{n=1}^{\infty} \frac{(-1)^n 3^{n-2}}{2^n} = \sum_{n=1}^{\infty} \frac{(-1)^n 3^n 3^{-2}}{2^n} = \sum_{n=1}^{\infty} \frac{1}{9}\left(-\frac{3}{2}\right)^n$

Since $|r| = \frac{3}{2} > 1$, this is a divergent geometric series.

49. $\sum_{n=1}^{\infty} \frac{10n+3}{n2^n}$

$$\lim_{n\to\infty} \frac{(10n+3)/n2^n}{1/2^n} = \lim_{n\to\infty} \frac{10n+3}{n} = 10$$

Therefore, the series converges by a limit comparison test with the geometric series

$$\sum_{n=0}^{\infty} \left(\frac{1}{2}\right)^n.$$

51. $\sum_{n=1}^{\infty} \frac{\cos(n)}{2^n}$

$$\left|\frac{\cos(n)}{2^n}\right| \leq \frac{1}{2^n}$$

Therefore, the series $\sum_{n=1}^{\infty} \left|\frac{\cos(n)}{2^n}\right|$ converges by comparison with the geometric series $\sum_{n=0}^{\infty} \left(\frac{1}{2}\right)^n$.

53. $\sum_{n=1}^{\infty} \frac{n7^n}{n!}$

$$\lim_{n\to\infty} \left|\frac{a_{n+1}}{a_n}\right| = \lim_{n\to\infty} \left|\frac{(n+1)7^{n+1}}{(n+1)!} \cdot \frac{n!}{n7^n}\right| = \lim_{n\to\infty} \frac{7}{n} = 0$$

Therefore, by the Ratio Test, the series converges.

55. $\sum_{n=1}^{\infty} \frac{(-1)^n 3^{n-1}}{n!}$

$$\lim_{n\to\infty} \left|\frac{a_{n+1}}{a_n}\right| = \lim_{n\to\infty} \left|\frac{3^n}{(n+1)!} \cdot \frac{n!}{3^{n-1}}\right| = \lim_{n\to\infty} \frac{3}{n+1} = 0$$

Therefore, by the Ratio Test, the series converges.

57. $\sum_{n=1}^{\infty} \frac{(-3)^n}{3 \cdot 5 \cdot 7 \cdots (2n+1)}$

$$\lim_{n\to\infty} \left|\frac{a_{n+1}}{a_n}\right| = \lim_{n\to\infty} \left|\frac{(-3)^{n+1}}{3 \cdot 5 \cdot 7 \cdots (2n+1)(2n+3)} \cdot \frac{3 \cdot 5 \cdot 7 \cdots (2n+1)}{(-3)^n}\right| = \lim_{n\to\infty} \frac{3}{2n+3} = 0$$

Therefore, by the Ratio Test, the series converges.

59. (a) and (c)

$$\sum_{n=1}^{\infty} \frac{n5^n}{n!} = \sum_{n=0}^{\infty} \frac{(n+1)5^{n+1}}{(n+1)!}$$

$$= 5 + \frac{(2)(5)^2}{2!} + \frac{(3)(5)^3}{3!} + \frac{(4)(5)^4}{4!} + \cdots$$

61. (a) and (b) are the same.

63. Replace n with $n + 1$.

$$\sum_{n=1}^{\infty} \frac{n}{4^n} = \sum_{n=0}^{\infty} \frac{n+1}{4^{n+1}}$$

65. Since

$$\frac{3^{10}}{2^{10}\,10!} = 1.59 \times 10^{-5},$$

use 9 terms.

$$\sum_{k=1}^{9} \frac{(-3)^k}{2^k\,k!} \approx -0.7769$$

67. No. Let $a_n = \dfrac{1}{n + 10{,}000}$.

The series $\sum_{n=1}^{\infty} \frac{1}{n + 10{,}000}$ diverges.

69. First, let

$$\lim_{n\to\infty} \sqrt[n]{|a_n|} = r < 1$$

and choose R such that $0 \le r < R < 1$. There must exist some $N > 0$ such that $\sqrt[n]{|a_n|} < R$ for all $n > N$. Thus, for $n > N$, we $|a_n| < R^n$ and since the geometric series

$$\sum_{n=0}^{\infty} R^n$$

converges, we can apply the Comparison Test to conclude that

$$\sum_{n=1}^{\infty} |a_n|$$

converges which in turn implies that $\sum_{n=1}^{\infty} a_n$ converges.

Second, let

$$\lim_{n\to\infty} {}^n\!\sqrt{|a_n|} = r > R > 1.$$

Then there must exist some $M > 0$ such that $\sqrt[n]{|a_n|} > R$ for all $n > M$. Thus, for $n > M$, we have $|a_n| > R_n > 1$ which implies that $\lim_{n\to\infty} a_n \neq 0$ which in turn implies that

$\sum_{n=1}^{\infty} a_n$ diverges.

Section 8.7 Taylor Polynomials and Approximations

1. $y = -\frac{1}{2}x^2 + 1$

Parabola

Matches (d)

3. $y = e^{-1/2}[(x + 1) + 1]$

Linear

Matches (a)

5. $f(x) = \cos x$

$P_2(x) = 1 - \frac{1}{2}x^2$

$P_4(x) = 1 - \frac{1}{2}x^2 + \frac{1}{24}x^4$

$P_6(x) = 1 - \frac{1}{2}x^2 + \frac{1}{24}x^4 - \frac{1}{720}x^6$

(a)

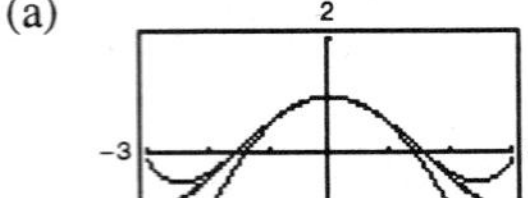

(c) In general, $f^{(n)}(0) = P_n^{(n)}(0)$ for all n.

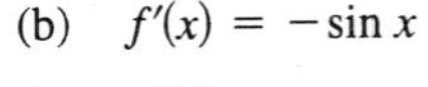

(b) $f'(x) = -\sin x$ $\quad P_2'(x) = -x$

$f''(x) = -\cos x$ $\quad P_2''(x) = -1$

$f''(0) = P_2''(0) = -1$

$f'''(x) = \sin x$ $\quad P_4'''(x) = x$

$f^{(4)}(x) = \cos x$ $\quad P_4^{(4)}(x) = 1$

$f^{(4)}(0) = 1 = P_4^{(4)}(0)$

$f^{(5)}(x) = -\sin x$ $\quad P_6^{(5)}(x) = -x$

$f^{(6)}(x) = -\cos x$ $\quad P^{(6)}(x) = -1$

$f^{(6)}(0) = -1 = P_6^{(6)}(0)$

7. $f(x) = e^{-x}$ $\quad f(0) = 1$

$f'(x) = -e^{-x}$ $\quad f'(0) = -1$

$f''(x) = e^{-x}$ $\quad f''(0) = 1$

$f'''(x) = -e^{-x}$ $\quad f'''(0) = -1$

$$P_3(x) = f(0) + f'(0)x + \frac{f''(0)}{2!}x^2 + \frac{f'''(0)}{3!}x^3$$

$$= 1 - x + \frac{x^2}{2} - \frac{x^3}{6}$$

9. $f(x) = e^{2x}$ $\quad f(0) = 1$

$f'(x) = 2e^{2x}$ $\quad f'(0) = 2$

$f''(x) = 4e^{2x}$ $\quad f''(0) = 4$

$f'''(x) = 8e^{2x}$ $\quad f'''(0) = 8$

$f^{(4)}(x) = 16^{2x}$ $\quad f^{(4)}(0) = 16$

$$P_4(x) = 1 + 2x + \frac{4}{2!}x^2 + \frac{8}{3!}x^3 + \frac{16}{4!}x^4$$

$$= 1 + 2x + 2x^2 + \frac{4}{3}x^3 + \frac{2}{3}x^4$$

11. $f(x) = \sin x$ $\quad f(0) = 0$

$f'(x) = \cos x$ $\quad f'(0) = 1$

$f''(x) = -\sin x$ $\quad f''(0) = 0$

$f'''(x) = -\cos x$ $\quad f'''(0) = -1$

$f^{(4)}(x) = \sin x$ $\quad f^{(4)}(0) = 0$

$f^{(5)}(x) = \cos x$ $\quad f^{(5)}(0) = 1$

$$P_5(x) = 0 + (1)x + \frac{0}{2!}x^2 + \frac{-1}{3!}x^3 + \frac{0}{4!}x^4 + \frac{1}{5!}x^5$$

$$= x - \frac{1}{6}x^3 + \frac{1}{120}x^5$$

13. $f(x) = xe^x$ $\quad f(0) = 0$

$f'(x) = xe^x + e^x$ $\quad f'(0) = 1$

$f''(x) = xe^x + 2e^x$ $\quad f''(0) = 2$

$f'''(x) = xe^x + 3e^x$ $\quad f'''(0) = 3$

$f^{(4)}(x) = xe^x + 4e^x$ $\quad f^{(4)}(0) = 4$

$$P_4(x) = 0 + x + \frac{2}{2!}x^2 + \frac{3}{3!}x^3 + \frac{4}{4!}x^4$$

$$= x + x^2 + \frac{1}{2}x^3 + \frac{1}{6}x^4$$

15. $f(x) = \dfrac{1}{x+1}$ $\quad f(0) = 1$

$f'(x) = -\dfrac{1}{(x+1)^2}$ $\quad f'(0) = -1$

$f''(x) = \dfrac{2}{(x+1)^2}$ $\quad f''(0) = 2$

$f'''(x) = \dfrac{-6}{(x+1)^4}$ $\quad f'''(0) = -6$

$f^{(4)}(x) = \dfrac{24}{(x+1)^5}$ $\quad f^{(4)}(0) = 24$

$$P_4(x) = 1 - x + \frac{2}{2!}x^2 + \frac{-6}{3!}x^3 + \frac{24}{4!}x^4 = 1 - x + x^2 - x^3 + x^4$$

17. $f(x) = \dfrac{1}{x}$ $\qquad f(1) = 1$

$f'(x) = -\dfrac{1}{x^2}$ $\qquad f'(1) = -1$

$f''(x) = \dfrac{2}{x^3}$ $\qquad f''(1) = 2$

$f'''(x) = -\dfrac{6}{x^4}$ $\qquad f'''(1) = -6$

$f^{(4)}(x) = \dfrac{24}{x^5}$ $\qquad f^{(4)}(1) = 24$

$$P_4(x) = 1 - (x-1) + \frac{2}{2!}(x-1)^2 + \frac{-6}{3!}(x-1)^3 + \frac{24}{4!}(x-1)^4 = 1 - (x-1) + (x-1)^2 - (x-1)^3 + (x-1)^4$$

19. $f(x) = \ln x$ $\qquad f(1) = 0$

$f'(x) = \dfrac{1}{x}$ $\qquad f'(1) = 1$

$f''(x) = -\dfrac{1}{x^2}$ $\qquad f''(1) = -1$

$f'''(x) = \dfrac{2}{x^3}$ $\qquad f'''(1) = 2$

$f^{(4)}(x) = -\dfrac{6}{x^4}$ $\qquad f^{(4)}(1) = -6$

$$P_4(x) = 0 + (x-1) - \frac{1}{2}(x-1)^2 + \frac{1}{3}(x-1)^3 - \frac{1}{4}(x-1)^4$$

21. $f(x) = \tan x$

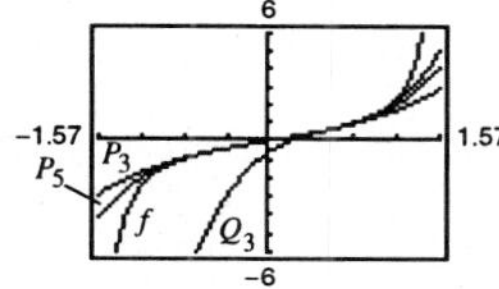

$f'(x) = \sec^2 x$

$f''(x) = 2\sec^2 x \tan x$

$f'''(x) = 4\sec^2 x \tan^2 x + 2\sec^4 x$

$f^{(4)}(x) = 8\sec^2 x \tan^3 x + 16\sec^4 x \tan x$

$f^{(5)}(x) = 16\sec^2 x \tan^4 x + 88\sec^4 x \tan^2 x + 16\sec^6 x$

(a) $n = 3, c = 0$

$$P_3(x) = 0 + x + \frac{0}{2!}x^2 + \frac{2}{3!}x^3 = x + \frac{1}{3}x^3$$

(b) $n = 5, c = 0$

$$P_5(x) = 0 + x + \frac{0}{2!}x^2 + \frac{2}{3!}x^3 + \frac{0}{4!}x^4 + \frac{16}{5!}x^5 = x + \frac{1}{3}x^3 + \frac{2}{15}x^5$$

(c) $n = 3, c = \dfrac{\pi}{4}$

$$Q_3(x) = 1 + 2\left(x - \frac{\pi}{4}\right) + \frac{4}{2!}\left(x - \frac{\pi}{4}\right)^2 + \frac{16}{3!}\left(x - \frac{\pi}{4}\right)^3 = 1 + 2\left(x - \frac{\pi}{4}\right) + 2\left(x - \frac{\pi}{4}\right)^2 + \frac{8}{3}\left(x - \frac{\pi}{4}\right)^3$$

23. $f(x) = \sin x$

$P_1(x) = x$

$P_3(x) = x - \frac{1}{6}x^3$

$P_5(x) = x - \frac{1}{6}x^3 + \frac{1}{120}x^5$

$P_7(x) = x - \frac{1}{6}x^3 + \frac{1}{120}x^5 - \frac{1}{5040}x^7$

(a)

x	0.00	0.25	0.50	0.75	1.00
$\sin x$	0.0000	0.2474	0.4794	0.6816	0.8415
$P_1(x)$	0.0000	0.2500	0.5000	0.7500	1.0000
$P_3(x)$	0.0000	0.2474	0.4792	0.6797	0.8333
$P_5(x)$	0.0000	0.2474	0.4794	0.6817	0.8417
$P_7(x)$	0.0000	0.2474	0.4794	0.6816	0.8415

(b)

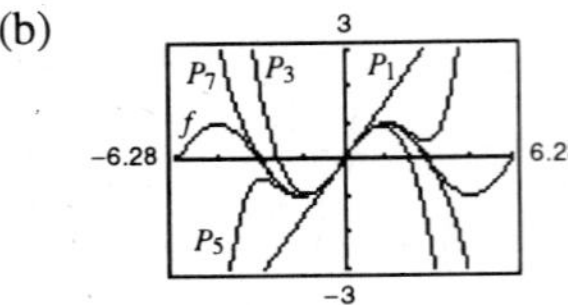

(c) As the distance increases, the accuracy decreases

25. $f(x) = \arcsin x$

(a) $P_3(x) = x + \dfrac{x^3}{6}$

(b)

x	−0.75	−0.50	−0.25	0	0.25	0.50	0.75
$f(x)$	−0.848	−0.524	−0.253	0	0.253	0.524	0.848
$P_3(x)$	−0.820	−0.521	−0.253	0	0.253	0.521	0.820

(c)

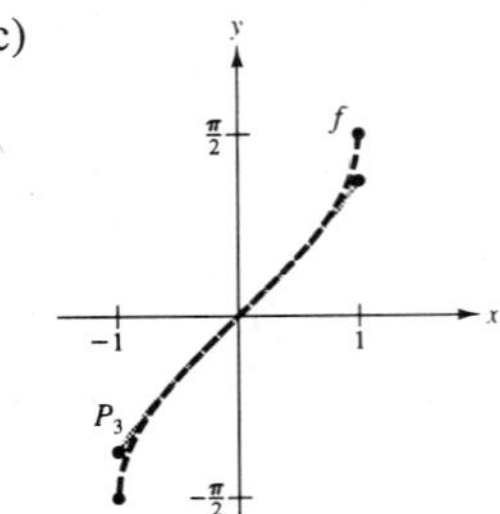

27. $f(x) = \cos x$

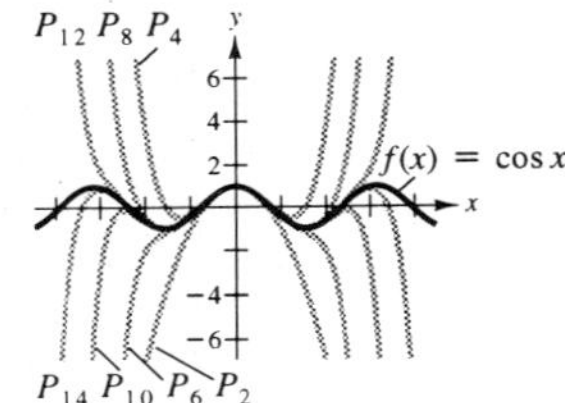

29. $f(x) = e^{-x} \approx 1 - x + \dfrac{x^2}{2} - \dfrac{x^3}{6}$

$f\left(\dfrac{1}{2}\right) \approx 0.6042$

31. $f(x) = \ln x \approx (x-1) - \frac{1}{2}(x-1)^2 + \frac{1}{3}(x-1)^3 - \frac{1}{4}(x-1)^4$

$f(1.2) \approx 0.1823$

33. $f(x) = \cos x; f^{(5)}(x) = -\sin x \Rightarrow$ Max on $[0, 0.3]$ is 1.

$R_4(x) \le \dfrac{1}{5!}(0.3)^5 = 2.025 \times 10^{-5}$

35. $f(x) = \arcsin x; f^{(4)}(x) = \dfrac{x(6x^2+9)}{(1-x^2)^{7/2}} \Rightarrow$ Max on $[0, 0.4]$ is $f^{(4)}(0.4) \approx 7.3340$.

$R_3(x) \le \dfrac{7.3340}{4!}(0.4)^4 \approx 0.00782 = 7.82 \times 10^{-3}$

37. $g(x) = \sin x$

$g^{(n+1)}(x) \le 1$ for all x

$$R_n(x) \le \frac{1}{(n+1)!}(0.3)^{n+1} < 0.001$$

By trial and error, $n = 3$.

39. $f(x) = \ln(x + 1)$

$$f^{(n+1)}(x) = \frac{(-1)^{n+1}n!}{(x+1)^{n+1}} \Rightarrow \text{Max on } [0, 0.5] \text{ is } n!.$$

$$R_n \le \frac{n!}{(n+1)!}(0.5)^{n+1} = \frac{(0.5)^{n+1}}{n+1} < 0.0001$$

By trial and error, $n = 9$. (See Example 9.) Using 9 terms, $\ln(1.5) \approx 0.4055$.

41. $f(x) = e^x \approx 1 + x + \frac{x^2}{2} + \frac{x^3}{6}, \; x < 0$

$$R_3(x) = \frac{e^z}{4!}x^4 < 0.001$$

$$e^z x^4 < 0.024$$

$$xe^{z/4} < 0.3936$$

$$x < \frac{0.3936}{e^{z/4}} < 0.3936, \; z < 0$$

$$-0.3936 < x < 0$$

43. (a) $f(x) = e^x$

$$P_4(x) = 1 + x + \frac{1}{2}x^2 + \frac{1}{6}x^3 + \frac{1}{24}x^4$$

$g(x) = xe^x$

$$Q_5(x) = x + x^2 + \frac{1}{2}x^3 + \frac{1}{6}x^4 + \frac{1}{24}x^5$$

$$Q_5(x) = x\,P_4(x)$$

(b) $f(x) = \sin x$

$$P_5(x) = x - \frac{x^3}{3!} + \frac{x^5}{5!}$$

$g(x) = x \sin x$

$$Q_6(x) = x\,P_5(x) = x^2 - \frac{x^4}{3!} + \frac{x^6}{5!}$$

(c) $g(x) = \frac{\sin x}{x} = \frac{1}{x}P_5(x) = 1 - \frac{x^2}{3!} + \frac{x^4}{5!}$

45. Let f be an odd function and P_n be the nth Maclaurin polynomial for f. Since f is odd, f' is even:

$$f'(-x) = \lim_{h\to 0}\frac{f(-x+h) - f(-x)}{h} = \lim_{h\to 0}\frac{-f(x-h) + f(x)}{h} = \lim_{h\to 0}\frac{f(x+(-h)) - f(x)}{-h} = f'(x).$$

Similarly, f'' is odd, f''' is even, etc. Therefore, $f, f'', f^{(4)}$, etc. are all odd functions, which implies that $f(0) = f''(0) = \ldots = 0$. Hence, in the formula

$$P_n(x) = f(0) + f'(0)x + \frac{f''(0)x^2}{2!} + \cdots$$ all the coefficients of the even power of x are zero.

47. Let $P_n(x) = a_0 + a_1(x - c) + a_2(x - c)^2 + \cdots + a_n(x - c)^n$ where $a_i = \frac{f^{(i)}(c)}{i!}$.

$$P_n(c) = a_0 = f(c)$$

For $1 \le k \le n$, $P_n^{(k)}(c) = a_n k! = \left(\frac{f^{(k)}(c)}{k!}\right)k! = f^{(k)}(c)$.

Section 8.8 Power Series

1. $\sum_{n=0}^{\infty} (-1)^n \frac{x^n}{n+1}$

$$L = \lim_{n\to\infty} \left|\frac{u_{n+1}}{u_n}\right| = \lim_{n\to\infty} \left|\frac{(-1)^{n+1}x^{n+1}}{n+2} \cdot \frac{n+1}{(-1)^n x^n}\right|$$

$$= \lim_{n\to\infty} \left|\frac{n+1}{n+2}\right||x| = |x|$$

$|x| < 1 \Rightarrow R = 1$

3. $\sum_{n=1}^{\infty} \frac{(2x)^n}{n^2}$

$$L = \lim_{n\to\infty} \left|\frac{u_{n+1}}{u_n}\right| = \lim_{n\to\infty} \left|\frac{(2x)^{n+1}}{(n+1)^2} \cdot \frac{n^2}{(2x)^n}\right|$$

$$= \lim_{n\to\infty} \left|\frac{2n^2x}{(n+1)^2}\right| = 2|x|$$

$2|x| < 1 \Rightarrow R = \frac{1}{2}$

5. $\sum_{n=0}^{\infty} \frac{(2x)^n}{n!}$

$$L = \lim_{n\to\infty} \left|\frac{u_{n+1}}{u_n}\right| = \lim_{n\to\infty} \left|\frac{(2x)^{n+1}}{(n+1)!} \cdot \frac{n!}{(2x)^n}\right|$$

$$= \lim_{n\to\infty} \left|\frac{2x}{n+1}\right| = 0$$

Thus, the series converges for all x. R is infinite.

$R = \infty$

7. $\sum_{n=0}^{\infty} \left(\frac{x}{2}\right)^n$

Since the series is geometric, it converges only if $|x/2| < 1$ or $-2 < x < 2$.

9. $\sum_{n=1}^{\infty} \frac{(-1)^n x^n}{n}$

$$\lim_{n\to\infty} \left|\frac{u_{n+1}}{u_n}\right| = \lim_{n\to\infty} \left|\frac{(-1)^{n+1}x^{n+1}}{n+1} \cdot \frac{n}{(-1)^n x^n}\right|$$

$$= \lim_{n\to\infty} \left|\frac{nx}{n+1}\right| = |x|$$

Interval: $-1 < x < 1$

When $x = 1$, the alternating series $\sum_{n=1}^{\infty} \frac{(-1)^n}{n}$ converges.

When $x = 1$, the p-series $\sum_{n=1}^{\infty} \frac{1}{n}$ diverges.

Therefore, the interval of convergence is $-1 < x \le 1$.

11. $\sum_{n=0}^{\infty} \frac{x^n}{n!}$

$$\lim_{n\to\infty} \left|\frac{u_{n+1}}{u_n}\right| = \lim_{n\to\infty} \left|\frac{x^{n+1}}{(n+1)!} \cdot \frac{n!}{x^n}\right|$$

$$= \lim_{n\to\infty} \left|\frac{x}{n+1}\right| = 0$$

The series converges for all x. Therefore, the interval of convergence is $-\infty < x < \infty$.

13. $\sum_{n=0}^{\infty} (2n)!\left(\frac{x}{2}\right)^n$

$$\lim_{n\to\infty} \left|\frac{u_{n+1}}{u_n}\right| = \lim_{n\to\infty} \left|\frac{(2n+2)!x^{n+1}}{2^{n+1}} \cdot \frac{2^n}{(2n)!x^n}\right| = \lim_{n\to\infty} \left|\frac{(2n+2)(2n+1)x}{2}\right| = \infty$$

Therefore, the series converges only for $x = 0$.

15. $\sum_{n=1}^{\infty} \frac{(-1)^{n+1}x^n}{4^n}$

Since the series is geometric, it converges only if $|x/4| < 1$ or $-4 < x < 4$.

17. $\displaystyle\sum_{n=1}^{\infty} \frac{(-1)^{n+1}(x-5)^n}{n5^n}$

$$\lim_{n\to\infty}\left|\frac{u_{n+1}}{u_n}\right| = \lim_{n\to\infty}\left|\frac{(-1)^{n+2}(x-5)^{n+1}}{(n+1)5^{n+1}} \cdot \frac{n5^n}{(-1)^{n+1}(x-5)^n}\right| = \lim_{n\to\infty}\left|\frac{n(x-5)}{5(n+1)}\right| = \frac{1}{5}|x-5|$$

$R = 5$

Center: $x = 5$

Interval: $-5 < x - 5 < 5$ or $0 < x < 10$

When $x = 0$, the p-series $\displaystyle\sum_{n=1}^{\infty} \frac{-1}{n}$ diverges.

When $x = 10$, the alternating series $\displaystyle\sum_{n=1}^{\infty} \frac{(-1)^{n+1}}{n}$ converges.

Therefore, the interval of convergence is $0 < x \le 10$.

19. $\displaystyle\sum_{n=0}^{\infty} \frac{(-1)^{n+1}(x-1)^{n+1}}{n+1}$

$$\lim_{n\to\infty}\left|\frac{u_{n+1}}{u_n}\right| = \lim_{n\to\infty}\left|\frac{(-1)^{n+2}(x-1)^{n+2}}{n+2} \cdot \frac{n+1}{(-1)^{n+1}(x-1)^{n+1}}\right| = \lim_{n\to\infty}\left|\frac{(n+1)(x-1)}{n+2}\right| = |x-1|$$

$R = 1$

Center: $x = 1$

Interval: $-1 < x - 1 < 1$ or $0 < x < 2$

When $x = 0$, the series $\displaystyle\sum_{n=0}^{\infty} \frac{1}{n+1}$ diverges by the integral test.

When $x = 2$, the alternating series $\displaystyle\sum_{n=0}^{\infty} \frac{(-1)^{n+1}}{n+1}$ converges.

Therefore, the interval of convergence is $0 < x \le 2$.

21. $\displaystyle\sum_{n=1}^{\infty} \frac{(x-c)^{n-1}}{c^{n-1}}$

$$\lim_{n\to\infty}\left|\frac{u_{n+1}}{u_n}\right| = \lim_{n\to\infty}\left|\frac{(x-c)^n}{c^n} \cdot \frac{c^{n-1}}{(x-c)^{n-1}}\right| = \frac{1}{c}|x-c|$$

$R = c$

Center: $x = c$

Interval: $-c < x - c < c$ or $0 < x < 2c$

When $x = 0$, the series $\displaystyle\sum_{n=1}^{\infty} (-1)^{n-1}$ diverges.

When $x = 2c$, the series $\displaystyle\sum_{n=1}^{\infty} 1$ diverges.

Therefore, the interval of convergence is $0 < x < 2c$.

23. $\displaystyle\sum_{n=1}^{\infty} \frac{n}{n+1}(-2x)^{n-1}$

$$\lim_{n\to\infty}\left|\frac{u_{n+1}}{u_n}\right| = \lim_{n\to\infty}\left|\frac{(n+1)(-2x)^n}{n+2} \cdot \frac{n+1}{n(-2x)^{n-1}}\right|$$

$$= \lim_{n\to\infty}\left|\frac{(-2x)(n+1)^2}{n(n+2)}\right| = 2|x|$$

$R = \dfrac{1}{2}$

Interval: $-\dfrac{1}{2} < x < \dfrac{1}{2}$

When $x = -\dfrac{1}{2}$, the series $\displaystyle\sum_{n=1}^{\infty} \frac{n}{n+1}$ diverges by the nth Term Test.

When $x = \dfrac{1}{2}$, the alternating series $\displaystyle\sum_{n=1}^{\infty} \frac{(-1)^{n-1}n}{n+1}$ diverges.

Therefore, the interval of convergence is $-\dfrac{1}{2} < x < \dfrac{1}{2}$.

25. $\displaystyle\sum_{n=0}^{\infty} \frac{x^{2n+1}}{(2n+1)!}$

$$\lim_{n\to\infty}\left|\frac{u_{n+1}}{u_n}\right| = \lim_{n\to\infty}\left|\frac{x^{2n+3}}{(2n+3)!} \cdot \frac{(2n+1)!}{x^{2n+1}}\right| = \lim_{n\to\infty}\left|\frac{x^2}{(2n+2)(2n+3)}\right| = 0$$

Therefore, the interval of convergence is $-\infty < x < \infty$.

27. $\displaystyle\sum_{n=1}^{\infty} \frac{k(k+1)\cdots(k+n-1)x^n}{n!}$

$$\lim_{n\to\infty}\left|\frac{u_{n+1}}{u_n}\right| = \lim_{n\to\infty}\left|\frac{k(k+1)\cdots(k+n-1)(k+n)x^{n+1}}{(n+1)!}\cdot\frac{n!}{k(k+1)\cdots(k+n-1)x^n}\right| = \lim_{n\to\infty}\left|\frac{(k+n)x}{n+1}\right| = |x|$$

$R = 1$

When $x = \pm 1$, the series diverges and the interval of convergence is $-1 < x < 1$.

$$\left[\frac{k(k+1)\cdots(k+n-1)}{1\cdot 2\cdots n} \geq 1\right]$$

29. $\displaystyle\sum_{n=1}^{\infty} \frac{(-1)^{n+1}3\cdot 7\cdot 11\cdots(4n-1)(x-3)^n}{4^n}$

$$\lim_{n\to\infty}\left|\frac{u_{n+1}}{u_n}\right| = \lim_{n\to\infty}\left|\frac{(-1)^{n+2}\cdot 3\cdot 7\cdot 11\cdots(4n-1)(4n+3)(x-3)^{n+1}}{4^{n+1}}\cdot\frac{4^n}{(-1)^{n+1}\cdot 3\cdot 7\cdot 11\cdots(4n-1)(x-3)^n}\right|$$

$$= \lim_{n\to\infty}\left|\frac{(4n+3)(x-3)}{4}\right| = \infty$$

$R = 0$

Center: $x = 3$

Therefore, the series converges only for $x = 3$.

31. (a) $f(x) = \displaystyle\sum_{n=0}^{\infty}\left(\frac{x}{2}\right)^n$, $-2 < x < 2$ (Geometric)

(b) $f'(x) = \displaystyle\sum_{n=1}^{\infty}\left(\frac{n}{2}\right)\left(\frac{x}{2}\right)^{n-1}$, $-2 < x < 2$

(c) $f''(x) = \displaystyle\sum_{n=2}^{\infty}\left(\frac{n}{2}\right)\left(\frac{n-1}{2}\right)\left(\frac{x}{2}\right)^{n-2}$, $-2 < x < 2$

(d) $\displaystyle\int f(x)\,dx = \sum_{n=0}^{\infty}\frac{2}{n+1}\left(\frac{x}{2}\right)^{n+1}$, $-2 \leq x < 2$

33. (a) $f(x) = \displaystyle\sum_{n=0}^{\infty}\frac{(-1)^{n+1}(x-1)^{n+1}}{n+1}$, $0 < x \leq 2$

(b) $f'(x) = \displaystyle\sum_{n=0}^{\infty}(-1)^{n+1}(x-1)^n$, $0 < x < 2$

(c) $f''(x) = \displaystyle\sum_{n=1}^{\infty}(-1)^{n+1}n(x-1)^{n-1}$, $0 < x < 2$

(d) $\displaystyle\int f(x)\,dx = \sum_{n=1}^{\infty}\frac{(-1)^{n+1}(x-1)^{n+2}}{(n+1)(n+2)}$, $0 \leq x \leq 2$

35. $g(1) = \displaystyle\sum_{n=0}^{\infty}\left(\frac{1}{3}\right)^n = 1 + \frac{1}{3} + \frac{1}{9} + \cdots$

$S_1 = 1, S_2 = 1.33.$

Matches (c)

37. $g(3.1) = \displaystyle\sum_{n=0}^{\infty}\left(\frac{3.1}{3}\right)^n$ diverges.

Matches (b)

39. (a) $f(x) = \displaystyle\sum_{n=0}^{\infty}\frac{(-1)^n x^{2n+1}}{(2n+1)!}$, $-\infty < x < \infty$ (See Exercise 25)

$g(x) = \displaystyle\sum_{n=0}^{\infty}\frac{(-1)^n x^{2n}}{(2n)!}$, $-\infty < x < \infty$

(b) $f'(x) = \displaystyle\sum_{n=0}^{\infty}\frac{(-1)^n x^{2n}}{(2n)!} = g(x)$

(c) $g'(x) = \displaystyle\sum_{n=1}^{\infty}\frac{(-1)^n x^{2n-1}}{(2n-1)!} = \sum_{n=0}^{\infty}\frac{(-1)^{n+1}x^{2n+1}}{(2n+1)!} = -\sum_{n=0}^{\infty}\frac{(-1)^n x^{2n+1}}{(2n+1)!} = -f(x)$

(d) $f(x) = \sin x$ and $g(x) = \cos x$

41.

$$y = \sum_{n=0}^{\infty} \frac{x^{2n}}{2^n n!}$$

$$y' = \sum_{n=1}^{\infty} \frac{2nx^{2n-1}}{2^n n!}$$

$$y'' = \sum_{n=1}^{\infty} \frac{2n(2n-1)x^{2n-2}}{2^n n!}$$

$$y'' - xy' - y = \sum_{n=1}^{\infty} \frac{2n(2n-1)x^{2n-2}}{2^n n!} - \sum_{n=1}^{\infty} \frac{2nx^{2n}}{2^n n!} - \sum_{n=0}^{\infty} \frac{x^{2n}}{2^n n!} = \sum_{n=1}^{\infty} \frac{2n(2n-1)x^{2n-2}}{2^n n!} - \sum_{n=0}^{\infty} \frac{(2n+1)x^{2n}}{2^n n!}$$

$$= \sum_{n=0}^{\infty} \left[\frac{(2n+2)(2n+1)x^{2n}}{2^{n+1}(n+1)!} - \frac{(2n+1)x^{2n}}{2^n n!} \cdot \frac{2(n+1)}{2(n+1)}\right]$$

$$= \sum_{n=0}^{\infty} \frac{2(n+1)x^{2n}[(2n+1) - (2n+1)]}{2^{n+1}(n+1)!} = 0$$

43. $J_0(x) = \displaystyle\sum_{k=0}^{\infty} \frac{(-1)^k x^{2k}}{2^{2k}(k!)^2}$

(a) $\displaystyle\lim_{k\to\infty} \left|\frac{u_{k+1}}{u_k}\right| = \lim_{k\to\infty} \left|\frac{(-1)^{k+1}x^{2k+2}}{2^{2k+2}[(k+1)!]^2} \cdot \frac{2^{2k}(k!)^2}{(-1)^k x^{2k}}\right| = \lim_{k\to\infty} \left|\frac{(-1)x^2}{2^2(k+1)^2}\right| = 0$

Therefore, the interval of convergence is $-\infty < x < \infty$.

(b)

$$J_0 = \sum_{k=0}^{\infty} (-1)^k \frac{x^{2k}}{4^k (k!)^2}$$

$$J_0' = \sum_{k=1}^{\infty} (-1)^k \frac{2kx^{2k-1}}{4^k (k!)^2} = \sum_{k=0}^{\infty} (-1)^{k+1} \frac{(2k+2)x^{2k+1}}{4^{k+1}[(k+1)!]^2}$$

$$J_0'' = \sum_{k=1}^{\infty} (-1)^k \frac{2k(2k-1)x^{2k-2}}{4^k (k!)^2} = \sum_{k=0}^{\infty} (-1)^{k+1} \frac{(2k+2)(2k+1)x^{2k}}{4^{k+1}[(k+1)!]^2}$$

$$x^2 J_0'' + xJ_0' + x^2 J_0 = \sum_{k=0}^{\infty} (-1)^{k+1} \frac{2(2k+1)x^{2k+2}}{4^{k+1}(k+1)!k!} + \sum_{k=0}^{\infty} (-1)^{k+1} \frac{2x^{2k+2}}{4^{k+1}(k+1)!k!} + \sum_{k=0}^{\infty} (-1)^k \frac{x^{2k+2}}{4^k (k!)^2}$$

$$= \sum_{k=0}^{\infty} \frac{(-1)^k x^{2k+2}}{4^k (k!)^2} \left[(-1)\frac{2(2k+1)}{4(k+1)} + (-1)\frac{2}{4(k+1)} + 1\right]$$

$$= \sum_{k=0}^{\infty} \frac{(-1)^k x^{2k+2}}{4^k (k!)^2} \left[\frac{-4k-2}{4k+4} - \frac{2}{4k+4} + \frac{4k+4}{4k+4}\right] = 0$$

(c) $P_6(x) = 1 - \dfrac{x^2}{4} + \dfrac{x^4}{64} - \dfrac{x^6}{2304}$

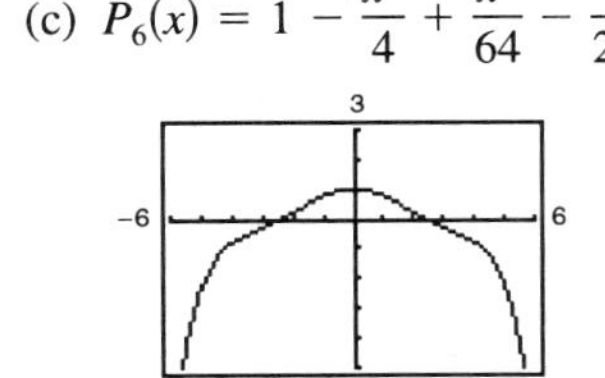

(d) $\displaystyle\int_0^1 J_0 dx = \int_0^1 \sum_{k=0}^{\infty} \frac{(-1)^k x^{2k}}{4^k (k!)^2} dx = \left[\sum_{k=0}^{\infty} \frac{(-1)^k x^{2k+1}}{4^k (k!)^2 (2k+1)}\right]_0^1$

$$= \sum_{k=0}^{\infty} \frac{(-1)^k}{4^k (k!)^2 (2k+1)} = 1 - \frac{1}{12} + \frac{1}{320} \approx 0.92$$

(exact integral is 0.9197304101)

45. $f(x) = \displaystyle\sum_{n=0}^{\infty} (-1)^n \frac{x^{2n}}{(2n)!} = \cos x$

(See Exercise 39.)

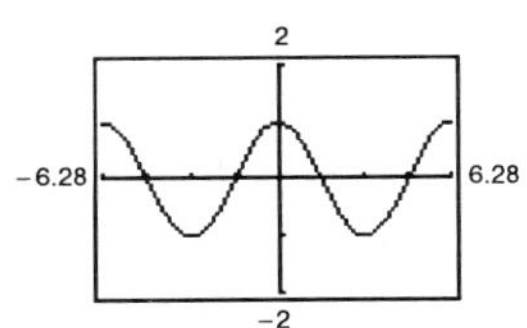

47. $f(x) = \displaystyle\sum_{n=0}^{\infty} (-1)^n x^n = \sum_{n=0}^{\infty} (-x)^n$

$$= \frac{1}{1-(-x)} = \frac{1}{1+x} \text{ for } -1 < x < 1$$

3

-1 1

0

49. $\sum_{n=0}^{\infty}\left(\frac{x}{2}\right)^n$

(a) $\sum_{n=0}^{\infty}\left(\frac{3/4}{2}\right)^n = \sum_{n=0}^{\infty}\left(\frac{3}{8}\right)^n$

$= \frac{1}{1-(3/8)} = \frac{8}{5} = 1.6$

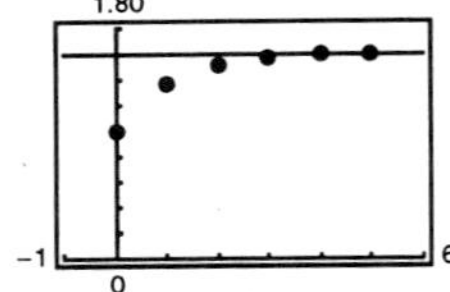

(b) $\sum_{n=0}^{\infty}\left(\frac{-3/4}{2}\right)^n = \sum_{n=0}^{\infty}\left(-\frac{3}{8}\right)^n$

$= \frac{1}{1-(-3/8)} = \frac{8}{11} \approx 0.7272$

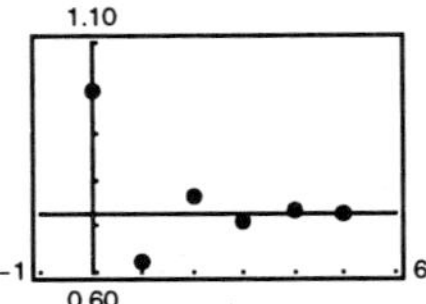

(c) The alternating series converges more rapidly. The partial sums of the series of positive terms approach the sum from below. The partial sums of the alternating series alternate sides of the horizontal line representing the sum.

(d) $\sum_{n=0}^{N}\left(\frac{3}{2}\right)^n > M$

M	10	100	1000	10,000
N	4	9	15	21

51. False;

$$\sum_{n=0}^{\infty}\frac{(-1)^n x^n}{n2^n}$$

converges for $x = 2$ but diverges for $x = -2$.

53. True; the radius of convergence is $R = 1$ for both series.

Section 8.9 Representation of Functions by Power Series

1. (a) $\frac{1}{2-x} = \frac{1/2}{1-(x/2)} = \frac{a}{1-r}$

$= \sum_{n=0}^{\infty}\frac{1}{2}\left(\frac{x}{2}\right)^n = \sum_{n=0}^{\infty}\frac{x^n}{2^{n+1}}$

This series converges on $(-2, 2)$.

(b)
$$\begin{array}{r|l} & \frac{1}{2}+\frac{x}{4}+\frac{x^2}{8}+\frac{x^3}{16}+\cdots \\ 2-x & 1 \\ & 1-\frac{x}{2} \\ & \frac{x}{2} \\ & \frac{x}{2}-\frac{x^2}{4} \\ & \frac{x^2}{4} \\ & \frac{x^2}{4}-\frac{x^3}{8} \\ & \frac{x^3}{8} \\ & \frac{x^3}{8}-\frac{x^4}{16} \\ & \vdots \end{array}$$

3. (a) $\frac{1}{2+x} = \frac{1/2}{1-(-x/2)} = \frac{a}{1-r}$

$= \sum_{n=0}^{\infty}\frac{1}{2}\left(-\frac{x}{2}\right)^n = \sum_{n=0}^{\infty}\frac{(-1)^n x^n}{2^{n+1}}$

This series converges on $(-2, 2)$.

(b)
$$\begin{array}{r|l} & \frac{1}{2}-\frac{x}{4}+\frac{x^2}{8}-\frac{x^3}{16}+\cdots \\ 2+x & 1 \\ & 1+\frac{x}{2} \\ & -\frac{x}{2} \\ & -\frac{x}{2}-\frac{x^2}{4} \\ & \frac{x^2}{4} \\ & \frac{x^2}{4}+\frac{x^3}{8} \\ & -\frac{x^3}{8} \\ & -\frac{x^3}{8}-\frac{x^4}{16} \\ & \vdots \end{array}$$

5. Writing $f(x)$ in the form $a/(1-r)$, we have

$$\frac{1}{2-x} = \frac{1}{-3-(x-5)} = \frac{-1/3}{1+(1/3)(x-5)}$$

which implies that $a = -1/3$ and $r = (-1/3)(x-5)$.

Therefore, the power series for $f(x)$ is given by

$$\frac{1}{2-x} = \sum_{n=0}^{\infty} ar^n = \sum_{n=0}^{\infty} -\frac{1}{3}\left[-\frac{1}{3}(x-5)\right]^n$$

$$= \sum_{n=0}^{\infty} \frac{(x-5)^n}{(-3)^{n+1}},\ |x-5| < 3 \text{ or } 2 < x < 8.$$

7. Writing $f(x)$ in the form $a/(1-r)$, we have

$$\frac{3}{2x-1} = \frac{-3}{1-2x} = \frac{a}{1-r}$$

which implies that $a = -3$ and $r = 2x$.

Therefore, the power series for $f(x)$ is given by

$$\frac{3}{2x-1} = \sum_{n=0}^{\infty} ar^n = \sum_{n=0}^{\infty} (-3)(2x)^n$$

$$= -3\sum_{n=0}^{\infty} (2x)^n,\ |2x| < 1 \text{ or } -\frac{1}{2} < x < \frac{1}{2}.$$

9. Writing $f(x)$ in the form $a/(1-r)$, we have

$$\frac{1}{2x-5} = \frac{-1}{11-2(x+3)}$$

$$= \frac{-1/11}{1-(2/11)(x+3)} = \frac{a}{1-r}$$

which implies that $a = -1/11$ and $r = (2/11)(x+3)$. Therefore, the power series for $f(x)$ is given by

$$\frac{1}{2x-5} = \sum_{n=0}^{\infty} ar^n = \sum_{n=0}^{\infty}\left(-\frac{1}{11}\right)\left[\frac{2}{11}(x+3)\right]^n$$

$$= -\sum_{n=0}^{\infty} \frac{2^n(x+3)^n}{11^{n+1}},$$

$$|x+3| < \frac{11}{2} \text{ or } -\frac{17}{2} < x < \frac{5}{2}.$$

11. Writing $f(x)$ in the form $a/(1-r)$, we have

$$\frac{3}{x+2} = \frac{3}{2+x} = \frac{3/2}{1+(1/2)x} = \frac{a}{1-r}$$

which implies that $a = 3/2$ and $r = (-1/2)x$. Therefore, the power series for $f(x)$ is given by

$$\frac{3}{x+2} = \sum_{n=0}^{\infty} ar^n = \sum_{n=0}^{\infty} \frac{3}{2}\left(-\frac{1}{2}x\right)^n$$

$$= 3\sum_{n=0}^{\infty} \frac{(-1)^n x^n}{2^{n+1}} = \frac{3}{2}\sum_{n=0}^{\infty}\left(-\frac{x}{2}\right)^n,$$

$$|x| < 2 \text{ or } -2 < x < 2.$$

13. $$\frac{3x}{x^2+x-2} = \frac{2}{x+2} + \frac{1}{x-1} = \frac{2}{2+x} + \frac{1}{-1+x} = \frac{1}{1+(1/2)x} + \frac{-1}{1-x}$$

Writing $f(x)$ as a sum of two geometric series, we have

$$\frac{3x}{x^2+x-2} = \sum_{n=0}^{\infty}\left(-\frac{1}{2}x\right)^n + \sum_{n=0}^{\infty}(-1)(x)^n = \sum_{n=0}^{\infty}\left[\frac{1}{(-2)^n} - 1\right]x^n.$$

The interval of convergence is $-1 < x < 1$ since

$$\lim_{n\to\infty}\left|\frac{u_{n+1}}{u_n}\right| = \lim_{n\to\infty}\left|\frac{(1-(-2)^{n+1})x^{n+1}}{(-2)^{n+1}} \cdot \frac{(-2)^n}{(1-(-2)^n)x^n}\right| = \lim_{n\to\infty}\left|\frac{(1-(-2)^{n+1})x}{-2-(-2)^{n+1}}\right| = |x|.$$

15. $$\frac{2}{1-x^2} = \frac{1}{1-x} + \frac{1}{1+x}$$

Writing $f(x)$ as a sum of two geometric series, we have

$$\frac{2}{1-x^2} = \sum_{n=0}^{\infty} x^n + \sum_{n=0}^{\infty}(-x)^n = \sum_{n=0}^{\infty}(1+(-1)^n)x^n = \sum_{n=0}^{\infty} 2x^{2n}.$$

The interval of convergence is $|x^2| < 1$ or $-1 < x < 1$ since $\displaystyle\lim_{n\to\infty}\left|\frac{u_{n+1}}{u_n}\right| = \lim_{n\to\infty}\left|\frac{2x^{2n+2}}{2x^2}\right| = |x^2|$.

17. $\dfrac{1}{1+x} = \displaystyle\sum_{n=0}^{\infty} (-1)^n x^n$

$\dfrac{1}{1-x} = \displaystyle\sum_{n=0}^{\infty} (-1)^n(-x)^n = \sum_{n=0}^{\infty} (-1)^{2n} x^n = \sum_{n=0}^{\infty} x^n$

$h(x) = \dfrac{-2}{x^2-1} = \dfrac{1}{1+x} + \dfrac{1}{1-x} = \displaystyle\sum_{n=0}^{\infty} (-1)^n x^n + \sum_{n=0}^{\infty} x^n = \sum_{n=0}^{\infty} [(-1)^n + 1] x^n$

$= 2 + 0x + 2x^2 + 0x^3 + 2x^4 + 0x^5 + 2x^6 + \cdots = \displaystyle\sum_{n=0}^{\infty} 2x^{2n}, \; -1 < x < 1$ (See Exercise 15.)

19. By taking the first derivative, we have $\dfrac{d}{dx}\left[\dfrac{1}{x+1}\right] = \dfrac{-1}{(x+1)^2}$. Therefore,

$$\frac{-1}{(x+1)^2} = \frac{d}{dx}\left[\sum_{n=0}^{\infty} (-1)^n x^n\right] = \sum_{n=1}^{\infty} (-1)^n n x^{n-1}$$

$$= \sum_{n=0}^{\infty} (-1)^{n+1}(n+1)x^n, \; -1 < x < 1.$$

21. By integrating, we have $\displaystyle\int \frac{1}{x+1}\,dx = \ln(x+1)$. Therefore,

$$\ln(x+1) = \int\left[\sum_{n=0}^{\infty} (-1)^n x^n\right] dx = C + \sum_{n=0}^{\infty} \frac{(-1)^n x^{n+1}}{n+1}, \; -1 < x \le 1.$$

To solve for C, let $x = 0$ and conclude that $C = 0$. Therefore,

$$\ln(x+1) = \sum_{n=0}^{\infty} \frac{(-1)^n x^{n+1}}{n+1}, \; -1 < x \le 1.$$

23. $\dfrac{1}{x^2+1} = \displaystyle\sum_{n=0}^{\infty} (-1)^n (x^2)^n = \sum_{n=0}^{\infty} (-1)^n x^{2n}, \; -1 < x < 1$

25. Since, $\dfrac{1}{x+1} = \displaystyle\sum_{n=0}^{\infty} (-1)^n x^n$, we have $\dfrac{1}{4x^2+1} = \displaystyle\sum_{n=0}^{\infty} (-1)^n (4x^2)^n = \sum_{n=0}^{\infty} (-1)^n 4^n x^{2n} = \sum_{n=0}^{\infty} (-1)^n (2x)^{2n}, \; -\frac{1}{2} < x < \frac{1}{2}.$

27. $x - \dfrac{x^2}{2} \le \ln(x+1) \le x - \dfrac{x^2}{2} + \dfrac{x^3}{3}$

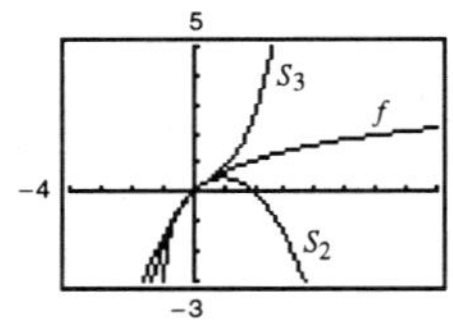

x	0.0	0.2	0.4	0.6	0.8	1.0
$x - \dfrac{x^2}{2}$	0.000	0.180	0.320	0.420	0.480	0.500
$\ln(x+1)$	0.000	0.180	0.336	0.470	0.588	0.693
$x - \dfrac{x^2}{2} + \dfrac{x^3}{3}$	0.000	0.183	0.341	0.492	0.651	0.833

29. $g(x) = x$, line, Matches (c)

31. $g(x) = x - \dfrac{x^3}{3} + \dfrac{x^5}{5}$, Matches (a)

33. $f(x) = \arctan x$ is an odd function (symmetric to the origin)

In Exercises 35 and 37, $\arctan x = \sum_{n=0}^{\infty} (-1)^n \frac{x^{2n+1}}{2n+1}$.

35. $\arctan \frac{1}{4} = \sum_{n=0}^{\infty} (-1)^n \frac{(1/4)^{2n+1}}{2n+1} = \sum_{n=0}^{\infty} \frac{(-1)^n}{(2n+1)4^{2n+1}} = \frac{1}{4} - \frac{1}{192} + \frac{1}{5120} + \cdots$

Since $\frac{1}{5120} < 0.001$, we can approximate the series by its first two terms: $\arctan \frac{1}{4} \approx \frac{1}{4} - \frac{1}{192} \approx 0.245$.

37. $$\frac{\arctan x^2}{x} = \sum_{n=0}^{\infty} (-1)^n \frac{x^{4n+1}}{2n+1}$$

$$\int \frac{\arctan x^2}{x}\,dx = \sum_{n=0}^{\infty} (-1)^n \frac{x^{4n+2}}{(4n+2)(2n+1)}$$

$$\int_0^{1/2} \frac{\arctan x^2}{x}\,dx = \sum_{n=0}^{\infty} (-1)^n \frac{1}{(4n+2)(2n+1)2^{4n+2}} = \frac{1}{8} - \frac{1}{1152} + \cdots$$

Since $\frac{1}{1152} < 0.001$, we can approximate the series by its first term: $\int_0^{1/2} \frac{\arctan x^2}{x}\,dx \approx 0.125$

In Exercises 39 and 41, $\frac{1}{1-x} = \sum_{n=0}^{\infty} x^n$.

39. $\frac{1}{(1-x)^2} = \frac{d}{dx}\left[\frac{1}{1-x}\right] = \frac{d}{dx}\left[\sum_{n=0}^{\infty} x^n\right] = \sum_{n=1}^{\infty} nx^{n-1}, \; -1 < x < 1$

41. $P(n) = \left(\frac{1}{2}\right)^n$

$$E(n) = \sum_{n=1}^{\infty} nP(n) = \sum_{n=1}^{\infty} n\left(\frac{1}{2}\right)^n = \frac{1}{2}\sum_{n=1}^{\infty} n\left(\frac{1}{2}\right)^{n-1} = \frac{1}{2}\,\frac{1}{[1-(1/2)]^2} = 2 \quad \text{(Exercise 39)}$$

Since the probability of obtaining a head on a single toss is $\frac{1}{2}$, it is expected that, on average, a head will be obtained in two tosses.

43. Let $\arctan x + \arctan y = \theta$. Then,

$$\tan(\arctan x + \arctan y) = \tan \theta$$

$$\frac{\tan(\arctan x) + \tan(\arctan y)}{1 - \tan(\arctan x)\tan(\arctan y)} = \tan \theta$$

$$\frac{x+y}{1-xy} = \tan \theta$$

$$\arctan\left(\frac{x+y}{1-xy}\right) = \theta.$$

Therefore, $\arctan x + \arctan y = \arctan\left(\frac{x+y}{1-xy}\right)$ for $xy \neq 1$.

45. (a) $2\arctan \frac{1}{2} = \arctan \frac{1}{2} + \arctan \frac{1}{2} = \arctan\left[\frac{2(1/2)}{1-(1/2)^2}\right] = \arctan \frac{4}{3}$

$$2\arctan \frac{1}{2} - \arctan \frac{1}{7} = \arctan \frac{4}{3} + \arctan\left(-\frac{1}{7}\right) = \arctan\left[\frac{(4/3)-(1/7)}{1+(4/3)(1/7)}\right] = \arctan \frac{25}{25} = \arctan 1 = \frac{\pi}{4}$$

(b) $\pi = 8\arctan \frac{1}{2} - 4\arctan \frac{1}{7} \approx 8\left[\frac{1}{2} - \frac{(0.5)^3}{3} + \frac{(0.5)^5}{5} - \frac{(0.5)^7}{7}\right] - 4\left[\frac{1}{7} - \frac{(1/7)^3}{3} + \frac{(1/7)^5}{5} - \frac{(1/7)^7}{7}\right] \approx 3.14$

47. From Exercise 21, we have

$$\ln(x+1) = \sum_{n=0}^{\infty} \frac{(-1)^n x^{n+1}}{n+1} = \sum_{n=1}^{\infty} \frac{(-1)^{n-1} x^n}{n} = \sum_{n=1}^{\infty} \frac{(-1)^{n+1} x^n}{n}.$$

Thus, $\displaystyle\sum_{n=1}^{\infty} (-1)^{n+1} \frac{1}{2^n n} = \sum_{n=1}^{\infty} \frac{(-1)^{n+1}(1/2)^n}{n} = \ln\left(\frac{1}{2}+1\right) = \ln\frac{3}{2} \approx 0.4055$

49. From Exercise 47, we have

$$\sum_{n=1}^{\infty} (-1)^{n+1} \frac{2^n}{5^n n} = \sum_{n=1}^{\infty} \frac{(-1)^{n+1}(2/5)^n}{n}$$

$$= \ln\left(\frac{2}{5}+1\right) = \ln\frac{7}{5} \approx 0.3365.$$

51. From Exercise 50, we have

$$\sum_{n=0}^{\infty} (-1)^n \frac{1}{2^{2n+1}(2n+1)} = \sum_{n=0}^{\infty} (-1)^n \frac{(1/2)^{2n+1}}{2n+1}$$

$$= \arctan\frac{1}{2} \approx 0.4636.$$

53. The series in Exercise 50 converges to its sum at a slower rate because its terms approach 0 at a much slower rate.

55. $\displaystyle f(x) = \sum_{n=1}^{\infty} (-1)^{n+1} \frac{(x-1)^n}{n}, \quad 0 < x \le 2$

$$f(0.5) = \sum_{n=1}^{\infty} (-1)^{n+1} \frac{(-0.5)^n}{n} = \sum_{n=1}^{\infty} -\frac{(1/2)^n}{n}$$

$$\sum_{n=1}^{\infty} -\frac{(1/2)^n}{n} = -0.6931$$

Section 8.10 Taylor and Maclaurin Series

1. For $c = 0$, we have:

$$f(x) = e^{2x}$$

$$f^{(n)}(x) = 2^n e^{2x} \Rightarrow f^{(n)}(0) = 2^n$$

$$e^{2x} = 1 + 2x + \frac{4x^2}{2!} + \frac{8x^3}{3!} + \frac{16x^4}{4!} + \cdots = \sum_{n=0}^{\infty} \frac{(2x)^n}{n!}$$

3. For $c = \pi/4$, we have:

$$f(x) = \cos(x) \qquad f\left(\frac{\pi}{4}\right) = \frac{\sqrt{2}}{2}$$

$$f'(x) = -\sin(x) \qquad f'\left(\frac{\pi}{4}\right) = -\frac{\sqrt{2}}{2}$$

$$f''(x) = -\cos(x) \qquad f''\left(\frac{\pi}{4}\right) = -\frac{\sqrt{2}}{2}$$

$$f'''(x) = \sin(x) \qquad f'''\left(\frac{\pi}{4}\right) = \frac{\sqrt{2}}{2}$$

$$f^{(4)}(x) = \cos(x) \qquad f^{(4)}\left(\frac{\pi}{4}\right) = \frac{\sqrt{2}}{2}$$

and so on. Therefore we have

$$\cos x = \sum_{n=0}^{\infty} \frac{f^{(n)}(\pi/4)[x-(\pi/4)]^n}{n!}$$

$$= \frac{\sqrt{2}}{2}\left[1 - \left(x - \frac{\pi}{4}\right) - \frac{[x-(\pi/4)]^2}{2!} + \frac{[x-(\pi/4)]^3}{3!} + \frac{[x-(\pi/4)]^4}{4!} - \cdots\right]$$

$$= \frac{\sqrt{2}}{2} \sum_{n=0}^{\infty} \frac{(-1)^{n(n+1)/2}[x-(\pi/4)]^n}{n!}.$$

[**Note:** $(-1)^{n(n+1)/2} = 1, -1, -1, 1, 1, -1, -1, 1, \ldots$]

5. For $c = 1$, we have,

$$\begin{aligned} f(x) &= \ln x & f(1) &= 0 \\ f'(x) &= \frac{1}{x} & f'(1) &= 1 \\ f''(x) &= -\frac{1}{x^2} & f''(1) &= -1 \\ f'''(x) &= \frac{2}{x^3} & f'''(1) &= 2 \\ f^{(4)}(x) &= -\frac{6}{x^4} & f^{(4)}(1) &= -6 \\ f^{(5)}(x) &= \frac{24}{x^5} & f^{(5)}(1) &= 24 \end{aligned}$$

and so on. Therefore, we have:

$$\begin{aligned} \ln x &= \sum_{n=0}^{\infty} \frac{f^{(n)}(1)(x-1)^n}{n!} \\ &= 0 + (x-1) - \frac{(x-1)^2}{2!} + \frac{2(x-1)^3}{3!} - \frac{6(x-1)^4}{4!} + \frac{24(x-1)^5}{5!} - \cdots \\ &= (x-1) - \frac{(x-1)^2}{2} + \frac{(x-1)^3}{3} - \frac{(x-1)^4}{4} + \frac{(x-1)^5}{5} - \cdots = \sum_{n=0}^{\infty} (-1)^n \frac{(x-1)^{n+1}}{n+1} \end{aligned}$$

7. For $c = 0$, we have:

$$\begin{aligned} f(x) &= \sin 2x & f(0) &= 0 \\ f'(x) &= 2\cos 2x & f'(0) &= 2 \\ f''(x) &= -4\sin 2x & f''(0) &= 0 \\ f'''(x) &= -8\cos 2x & f'''(0) &= -8 \\ f^{(4)}(x) &= 16\sin 2x & f^{(4)}(0) &= 0 \\ f^{(5)}(x) &= 32\cos 2x & f^{(5)}(0) &= 32 \\ f^{(6)}(x) &= -64\sin 2x & f^{(6)}(0) &= 0 \\ f^{(7)}(x) &= -128\cos 2x & f^{(7)}(0) &= -128 \end{aligned}$$

and so on. Therefore, we have:

$$\begin{aligned} \sin 2x &= \sum_{n=0}^{\infty} \frac{f^{(n)}(0)x^n}{n!} = 0 + 2x + \frac{0x^2}{2!} - \frac{8x^3}{3!} + \frac{0x^4}{4!} + \frac{32x^5}{5!} + \frac{0x^6}{6!} - \frac{128x^7}{7!} + \cdots \\ &= 2x - \frac{8x^3}{3!} + \frac{32x^5}{5!} - \frac{128x^7}{7!} + \cdots = \sum_{n=0}^{\infty} \frac{(-1)^n(2x)^{2n+1}}{(2n+1)!} \end{aligned}$$

9. For $c = 0$, we have:

$$\begin{aligned} f(x) &= \sec(x) & f(0) &= 1 \\ f'(x) &= \sec(x)\tan(x) & f'(0) &= 0 \\ f''(x) &= \sec^3(x) + \sec(x)\tan^2(x) & f''(0) &= 1 \\ f'''(x) &= 5\sec^3(x)\tan(x) + \sec(x)\tan^3(x) & f'''(0) &= 0 \\ f^{(4)}(x) &= 5\sec^5(x) + 18\sec^3(x)\tan^2(x) + \sec(x)\tan^4(x) & f^{(4)}(0) &= 5 \end{aligned}$$

$$\sec(x) = \sum_{n=0}^{\infty} \frac{f^{(n)}(0)x^n}{n!} = 1 + \frac{x^2}{2!} + \frac{5x^4}{4!} + \cdots$$

11. Since $(1+x)^{-k} = 1 - kx + \frac{k(k+1)x^2}{2!} - \frac{k(k+1)(k+2)x^3}{3!} + \cdots$, we have

$$(1+x)^{-2} = 1 - 2x + \frac{2(3)x^2}{2!} - \frac{2(3)(4)x^3}{3!} + \frac{2(3)(4)(5)x^4}{5!} - \cdots = 1 - 2x + 3x^2 - 4x^3 + 5x^4 - \ldots$$

$$= \sum_{n=0}^{\infty} (-1)^n (n+1)x^n.$$

13. $\frac{1}{\sqrt{4+x^2}} = \left(\frac{1}{2}\right)\left[1 + \left(\frac{x}{2}\right)^2\right]^{-1/2}$ and since $(1+x)^{-1/2} = 1 + \sum_{n=1}^{\infty} \frac{(-1)^n 1 \cdot 3 \cdot 5 \cdots (2n-1)x^n}{2^n n!}$, we have

$$\frac{1}{\sqrt{4+x^2}} = \frac{1}{2}\left[1 + \sum_{n=1}^{\infty} \frac{(-1)^n 1 \cdot 3 \cdot 5 \cdots (2n-1)(x/2)^{2n}}{2^n n!}\right] = \frac{1}{2} + \sum_{n=1}^{\infty} \frac{(-1)^n 1 \cdot 3 \cdot 5 \cdots (2n-1)x^{2n}}{2^{3n+1} n!}.$$

15. Since $(1+x)^{1/2} = 1 + \frac{x}{2} + \sum_{n=2}^{\infty} \frac{(-1)^{n+1} 1 \cdot 3 \cdot 5 \cdots (2n-3)x^n}{2^n n!}$ (Exercise 14)

we have $(1+x^2)^{1/2} = 1 + \frac{x^2}{2} + \sum_{n=2}^{\infty} \frac{(-1)^{n+1} 1 \cdot 3 \cdot 5 \cdots (2n-3)x^{2n}}{2^n n!}$.

17.

$$e^x = \sum_{n=0}^{\infty} \frac{x^n}{n!} = 1 + x + \frac{x^2}{2!} + \frac{x^3}{3!} + \frac{x^4}{4!} + \frac{x^5}{5!} + \cdots$$

$$e^{x^2/2} = \sum_{n=0}^{\infty} \frac{(x^2/2)^n}{n!} = \sum_{n=0}^{\infty} \frac{x^{2n}}{2^n n!} = 1 + \frac{x^2}{2} + \frac{x^4}{2^2 2!} + \frac{x^6}{2^3 3!} + \frac{x^8}{2^4 4!} + \cdots$$

19.

$$\sin x = \sum_{n=0}^{\infty} \frac{(-1)^n x^{2n+1}}{(2n+1)!} = x - \frac{x^3}{3!} + \frac{x^5}{5!} - \frac{x^7}{7!} + \cdots$$

$$\sin 2x = \sum_{n=0}^{\infty} \frac{(-1)^n (2x)^{2n+1}}{(2n+1)!} = \sum_{n=0}^{\infty} \frac{(-1)^n 2^{2n+1} x^{2n+1}}{(2n+1)!} = 2x - \frac{8x^3}{3!} + \frac{32x^5}{5!} - \frac{128x^7}{7!} + \cdots$$

21.

$$\cos x = \sum_{n=0}^{\infty} \frac{(-1)^n x^{2n}}{(2n)!} = 1 - \frac{x^2}{2!} + \frac{x^4}{4!} - \cdots$$

$$\cos x^{3/2} = \sum_{n=0}^{\infty} \frac{(-1)^n (x^{3/2})^{2n}}{(2n)!} = \sum_{n=0}^{\infty} \frac{(-1)^n x^{3n}}{(2n)!} = 1 - \frac{x^3}{2!} + \frac{x^6}{4!} - \ldots$$

23.

$$\frac{\sin x}{x} = \frac{1}{x} \sum_{n=0}^{\infty} \frac{(-1)^n x^{2n+1}}{(2n+1)!} = \sum_{n=0}^{\infty} \frac{(-1)^n x^{2n}}{(2n+1)!}$$

$$= 1 - \frac{x^2}{3!} + \frac{x^4}{5!} - \frac{x^6}{7!} + \frac{x^8}{9!} - \cdots$$

25.

$$e^x = 1 + x + \frac{x^2}{2!} + \frac{x^3}{3!} + \frac{x^4}{4!} + \frac{x^5}{5!} + \cdots$$

$$e^{-x} = 1 - x + \frac{x^2}{2!} - \frac{x^3}{3!} + \frac{x^4}{4!} - \frac{x^5}{5!} + \cdots$$

$$e^x - e^{-x} = 2x + \frac{2x^3}{3!} + \frac{2x^5}{5!} + \frac{2x^7}{7!} + \cdots$$

$$\sinh(x) = \frac{1}{2}(e^x - e^{-x}) = x + \frac{x^3}{3!} + \frac{x^5}{5!} + \frac{x^7}{7!} + \cdots = \sum_{n=0}^{\infty} \frac{x^{2n+1}}{(2n+1)!}$$

27. $\cos^2(x) = \frac{1}{2}[1 + \cos(2x)] = \frac{1}{2}\left[1 + 1 - \frac{(2x)^2}{2!} + \frac{(2x)^4}{4!} - \frac{(2x)^6}{6!} - \cdots\right] = \frac{1}{2}\left[1 + \sum_{n=0}^{\infty} \frac{(-1)^n(2x)^{2n}}{(2n)!}\right]$

29.

$$e^{ix} = 1 + ix + \frac{(ix)^2}{2!} + \frac{(ix)^3}{3!} + \frac{(ix)^4}{4!} + \cdots = 1 + ix - \frac{x^2}{2!} - \frac{ix^3}{3!} + \frac{x^4}{4!} + \frac{ix^5}{5!} - \frac{x^6}{6!} - \cdots$$

$$e^{-ix} = 1 - ix + \frac{(-ix)^2}{2!} + \frac{(-ix)^3}{3!} + \frac{(-ix)^4}{4!} + \cdots = 1 - ix - \frac{x^2}{2!} + \frac{ix^3}{3!} + \frac{x^4}{4!} - \frac{ix^5}{5!} - \frac{x^6}{6!} + \cdots$$

$$e^{ix} - e^{-ix} = 2ix - \frac{2ix^3}{3!} + \frac{2ix^5}{5!} - \frac{2ix^7}{7!} + \cdots$$

$$\frac{e^{ix} - e^{-ix}}{2i} = x - \frac{x^3}{3!} + \frac{x^5}{5!} - \frac{x^7}{7!} + \cdots = \sum_{n=0}^{\infty} \frac{(-1)^n x^{2n+1}}{(2n+1)!} = \sin(x)$$

31. $f(x) = e^x \sin x$

$$= \left(1 + x + \frac{x^2}{2} + \frac{x^3}{6} + \frac{x^4}{24} + \cdots\right)\left(x - \frac{x^3}{6} + \frac{x^5}{120} - \cdots\right)$$

$$= x + x^2 + \left(\frac{x^3}{2} - \frac{x^3}{6}\right) + \left(\frac{x^4}{6} - \frac{x^4}{6}\right) + \left(\frac{x^5}{120} - \frac{x^5}{12} + \frac{x^5}{24}\right) + \cdots$$

$$= x + x^2 + \frac{x^3}{3} - \frac{x^5}{30} + \cdots$$

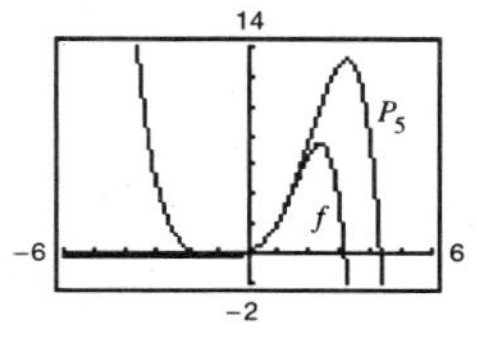

33. $h(x) = \cos x \ln(1 + x)$

$$= \left(1 - \frac{x^2}{2} + \frac{x^4}{24} - \cdots\right)\left(x - \frac{x^2}{2} + \frac{x^3}{3} - \frac{x^4}{4} + \frac{x^5}{5} - \cdots\right)$$

$$= x - \frac{x^2}{2} + \left(\frac{x^3}{3} - \frac{x^3}{2}\right) + \left(\frac{x^4}{4} - \frac{x^4}{4}\right) + \left(\frac{x^5}{5} - \frac{x^5}{6} + \frac{x^5}{24}\right) + \cdots$$

$$= x - \frac{x^2}{2} - \frac{x^3}{6} + \frac{3x^5}{40} + \cdots$$

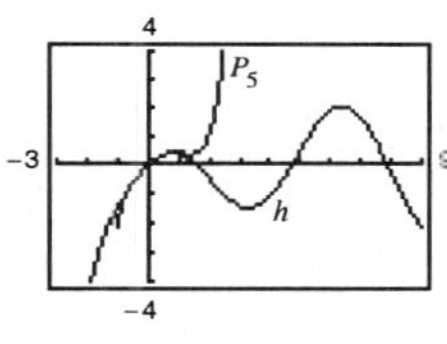

35. $g(x) = \frac{\sin x}{1 + x}$. Divide the series for $\sin x$ by $(1 + x)$.

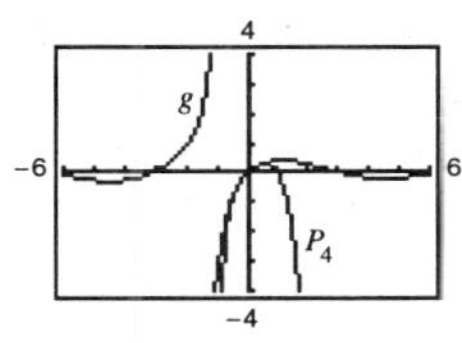

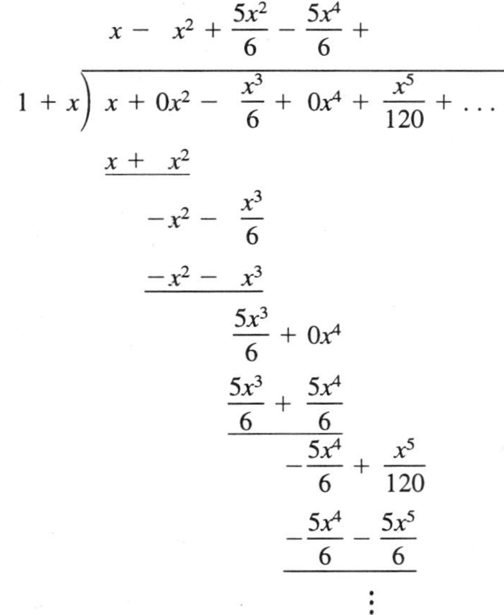

$$\begin{array}{r|l}
 & x - x^2 + \frac{5x^2}{6} - \frac{5x^4}{6} + \\
1 + x & x + 0x^2 - \frac{x^3}{6} + 0x^4 + \frac{x^5}{120} + \cdots \\
 & x + x^2 \\
 & -x^2 - \frac{x^3}{6} \\
 & -x^2 - x^3 \\
 & \frac{5x^3}{6} + 0x^4 \\
 & \frac{5x^3}{6} + \frac{5x^4}{6} \\
 & -\frac{5x^4}{6} + \frac{x^5}{120} \\
 & -\frac{5x^4}{6} - \frac{5x^5}{6} \\
 & \vdots
\end{array}$$

$$g(x) = x - x^2 + \frac{5x^3}{6} - \frac{5x^4}{6} + \cdots$$

37. $y = x^2 - \dfrac{x^4}{3!} = x\left(x - \dfrac{x^3}{3!}\right) \approx x \sin x.$

Matches (a)

39. $y = x + x^2 + \dfrac{x^3}{2!} = x\left(1 + x + \dfrac{x^2}{2!}\right) \approx xe^x.$

Matches (c)

41. $\displaystyle\int_0^x (e^{-t^2} - 1)dt = \int_0^x \left[\left(\sum_{n=0}^{\infty} \frac{(-1)^n t^{2n}}{n!}\right) - 1\right] dt$

$\displaystyle = \int_0^x \left[\sum_{n=0}^{\infty} \frac{(-1)^{n+1} t^{2n+2}}{(n+1)!}\right] dt = \left[\sum_{n=0}^{\infty} \frac{(-1)^{n+1} t^{2n+3}}{(2n+3)(n+1)!}\right]_0^x = \sum_{n=0}^{\infty} \frac{(-1)^{n+1} x^{2n+3}}{(2n+3)(n+1)!}$

43. Since $\displaystyle \ln x = \sum_{n=0}^{\infty} \frac{(-1)^n (x-1)^{n+1}}{n+1} = (x-1) - \frac{(x-1)^2}{2} + \frac{(x-1)^3}{3} - \frac{(x-1)^4}{4} + \cdots$, we have

$$\ln 2 = 1 - \frac{1}{2} + \frac{1}{3} - \frac{1}{4} + \cdots = \sum_{n=1}^{\infty} (-1)^{n+1} \frac{1}{n} \approx 0.6931. \quad \text{(10,001 terms)}$$

45. Since $\displaystyle e^x = \sum_{n=0}^{\infty} \frac{x^n}{n!} = 1 + x + \frac{x^2}{2!} + \frac{x^3}{3!} + \cdots$,, we have

$$e^2 = 1 + 2 + \frac{2^2}{2!} + \frac{2^3}{3!} + \cdots = \sum_{n=0}^{\infty} \frac{2^n}{n!} \approx 7.3891. \quad \text{(12 terms)}$$

47. Since

$$\cos x = \sum_{n=0}^{\infty} \frac{(-1)^n x^{2n}}{(2n)!} = 1 - \frac{x^2}{2!} + \frac{x^4}{4!} - \frac{x^6}{6!} + \frac{x^8}{8!} - \cdots$$

$$1 - \cos x = \frac{x^2}{2!} - \frac{x^4}{4!} + \frac{x^6}{6!} - \frac{x^8}{8!} + \cdots = \sum_{n=0}^{\infty} \frac{(-1)^n x^{2n+2}}{(2n+2)!}$$

$$\frac{1 - \cos}{x} = \frac{x}{2!} - \frac{x^3}{4!} + \frac{x^5}{6!} - \frac{x^7}{8!} + \cdots = \sum_{n=0}^{\infty} \frac{(-1)^n x^{2n+1}}{(2n+2)!}, \text{ we have } \lim_{x\to 0} \frac{1 - \cos x}{x} = \lim_{x \to 0} \sum_{n=0}^{\infty} \frac{(-1)x^{2n+1}}{(2n+2)!} = 0.$$

49. $\displaystyle\int_0^1 \frac{\sin x}{x}\, dx = \int_0^1 \left[\sum_{n=0}^{\infty} \frac{(-1)^n x^{2n}}{(2n+1)!}\right] dx = \left[\sum_{n=0}^{\infty} \frac{(-1)^n x^{2n+1}}{(2n+1)(2n+1)!}\right]_0^1 = \sum_{n=0}^{\infty} \frac{(-1)^n}{(2n+1)(2n+1)!}$

Since $1/(7 \cdot 7!) < 0.0001$, we have

$$\int_0^1 \frac{\sin x}{x}\, dx = 1 - \frac{1}{3 \cdot 3!} + \frac{1}{5 \cdot 5!} - \cdots \approx 0.9461.$$

Note: We are using $\displaystyle\lim_{x \to 0^+} \frac{\sin x}{x} = 1.$

51. $\displaystyle\int_0^{\pi/2} \sqrt{x} \cos x\, dx = \int_0^{\pi/2} \left[\sum_{n=0}^{\infty} \frac{(-1)^n x^{(4n+1)/2}}{(2n)!}\right] dx = \left[\sum_{n=0}^{\infty} \frac{(-1)^n x^{(4n+3)/2}}{\left(\frac{4n+3}{2}\right)(2n)!}\right]_0^{\pi/2} = \left[\sum_{n=0}^{\infty} \frac{(-1)^n 2x^{(4n+3)/2}}{(4n+3)(2n)!}\right]_0^{\pi/2}$

Since $(\pi/2)^{19/2}/766{,}080 < 0.0001$, we have

$$\int_0^1 \sqrt{x} \cos x\, dx = 2\left[\frac{(\pi/2)^{3/2}}{3} - \frac{(\pi/2)^{7/2}}{14} + \frac{(\pi/2)^{11/2}}{264} - \frac{(\pi/2)^{15/2}}{10{,}800} + \frac{(\pi/2)^{19/2}}{766{,}080}\right] \approx 0.7040.$$

53. $\displaystyle\int_{0.1}^{0.3} \sqrt{1 + x^3}\, dx = \int_{0.1}^{0.3} \left(1 + \frac{x^3}{2} - \frac{x^6}{8} + \frac{x^9}{16} - \frac{5x^{12}}{128} + \cdots\right) dx = \left[x + \frac{x^4}{8} - \frac{x^7}{56} + \frac{x^{10}}{160} - \frac{5x^{13}}{1664} + \cdots\right]_{0.1}^{0.3}$

Since $\frac{1}{56}(0.3^7 - 0.1^7) < 0.0001$, we have

$$\int_{0.1}^{0.3} \sqrt{1 + x^3}\, dx = \left[(0.3 - 0.1) + \frac{1}{8}(0.3^4 - 0.1^4) - \frac{1}{56}(0.3^7 - 0.1^7)\right] \approx 0.2010.$$

55. From Exercise 17, we have

$$\frac{1}{\sqrt{2\pi}}\int_0^1 e^{-x^2/2}\,dx = \frac{1}{\sqrt{2\pi}}\int_0^1 \sum_{n=0}^{\infty}\frac{(-1)^n x^{2n}}{2^n n!}\,dx = \frac{1}{\sqrt{2\pi}}\left[\sum_{n=0}^{\infty}\frac{(-1)^n x^{2n+1}}{2^n n!(2n+1)}\right]_0^1$$

$$= \frac{1}{\sqrt{2\pi}}\sum_{n=0}^{\infty}\frac{(-1)^n}{2^n n!(2n+1)}$$

$$\approx \frac{1}{\sqrt{2\pi}}\left[1 - \frac{1}{2\cdot 1\cdot 3} + \frac{1}{2^2\cdot 2!\cdot 5} - \frac{1}{2^3\cdot 3!\cdot 7}\right] \approx 0.3414.$$

57. $f(x) = x\cos 2x = \displaystyle\sum_{n=0}^{\infty}\frac{(-1)^n 4^n x^{2n+1}}{(2n)!}$

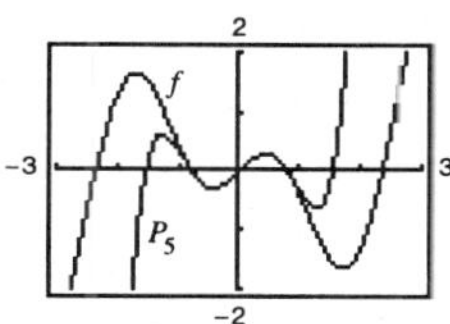

$P_5(x) = x - 2x^3 + \dfrac{2x^5}{3}$

The polynomial is a reasonable approximation on the interval $\left[-\frac{3}{4}, \frac{3}{4}\right]$.

59. $f(x) = \sqrt{x}\ln x,\ c = 1$

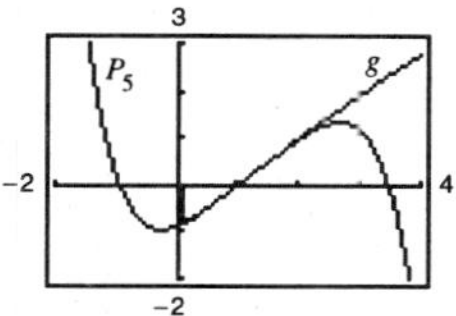

$P_5(x) = (x-1) - \dfrac{(x-1)^3}{24} + \dfrac{(x-1)^4}{24} - \dfrac{71(x-1)^5}{1920}$

The polynomial is a reasonable approximation on the interval $\left[\frac{1}{4}, 2\right]$.

61. $y = \left(\tan\theta - \dfrac{g}{kv_0\cos\theta}\right)x - \dfrac{g}{k^2}\ln\left(1 - \dfrac{kx}{v_0\cos\theta}\right)$

$$= (\tan\theta)x - \frac{gx}{kv_0\cos\theta} - \frac{g}{k^2}\left[-\frac{kx}{v_0\cos\theta} - \frac{1}{2}\left(\frac{kx}{v_0\cos\theta}\right)^2 - \frac{1}{3}\left(\frac{kx}{v_0\cos\theta}\right)^3 - \frac{1}{4}\left(\frac{kx}{v_o\cos\theta}\right)^4 - \cdots\right]$$

$$= (\tan\theta)x - \frac{gx}{kv_0\cos\theta} + \frac{gx}{kv_0\cos\theta} + \frac{gx^2}{2v_0^2\cos^2\theta} + \frac{gkx^3}{3v_0^3\cos^3\theta} + \frac{gk^2x^4}{4v_0^4\cos^4\theta} + \cdots\Bigg]$$

$$= (\tan\theta)x + \frac{gx^2}{2v_0^2\cos^2\theta} + \frac{kgx^3}{3v_0^3\cos^3\theta} + \frac{k^2gx^4}{4v_0^4\cos^4\theta} + \cdots$$

63. $f(x) = \begin{cases} e^{-1/x^2}, & x \neq 0 \\ 0, & x = 0 \end{cases}$

(a)

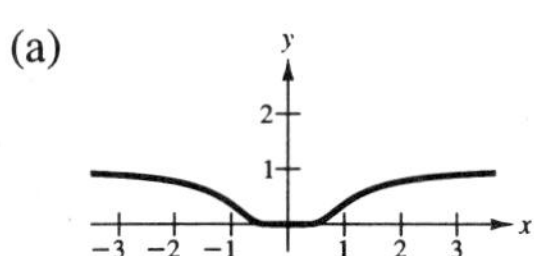

(b) $f'(0) = \displaystyle\lim_{x\to 0}\frac{f(x) - f(0)}{x - 0} = \lim_{x\to 0}\frac{e^{-1/x^2} - 0}{x}$

Let $y = \displaystyle\lim_{x\to 0}\frac{e^{-1/x^2}}{x}$. Then

$$\ln y = \lim_{x\to 0}\ln\left(\frac{e^{-1/x^2}}{x}\right) = \lim_{x\to 0^+}\left[-\frac{1}{x^2} - \ln x\right]$$

$$= \lim_{x\to 0^+}\left[\frac{-1 - x^2\ln x}{x^2}\right] = -\infty.$$

Thus, $y = e^{-\infty} = 0$ and we have $f'(0) = 0$.

(c) $\displaystyle\sum_{n=0}^{\infty}\frac{f^{(n)}(0)}{n!}x^n = f(0) + \frac{f'(0)x}{1!} + \frac{f''(0)x^2}{2!} + \cdots$

$= 0 \neq f(x)$

This series converges to f at $x = 0$ only.

65. By the Ratio Test:

$$\lim_{n\to\infty}\left|\frac{x^{n+1}}{(n+1)!}\cdot\frac{n!}{x^n}\right| = \lim_{n\to\infty}\frac{|x|}{n+1} = 0 \text{ which shows that } \sum_{n=0}^{\infty}\frac{x^n}{n!} \text{ converges for all } x.$$

Review Exercises for Chapter 8

1. $a_n = \dfrac{1}{n!}$

3. $a_n = 4 + \dfrac{2}{n}$: 6, 5, 4.67, . . .

Matches (a)

5. $a_n = 10(0.3)^{n-1}$: 10, 3, . . .

Matches (d)

7. $a_n = \dfrac{5n + 2}{n}$

The sequence seems to converge to 5.

$$\lim_{n\to\infty} a_n = \lim_{n\to\infty} \frac{5n + 2}{n} = \lim_{n\to\infty} \left(5 + \frac{2}{n}\right) = 5$$

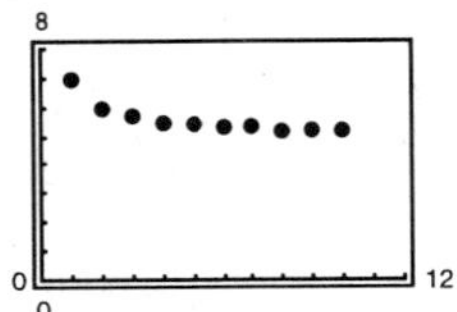

9. $\displaystyle\lim_{n\to\infty} \frac{n + 1}{n^2} = 0$

Converges

11. $\displaystyle\lim_{n\to\infty} \frac{n^3}{n^2 + 1} = \infty$

Diverges

13. $\displaystyle\lim_{n\to\infty} \left(\sqrt{n + 1} - \sqrt{n}\right) = \lim_{n\to\infty} \left(\sqrt{n + 1} - \sqrt{n}\right)\frac{\sqrt{n + 1} + \sqrt{n}}{\sqrt{n + 1} + \sqrt{n}} = \lim_{n\to\infty} \frac{1}{\sqrt{n + 1} + \sqrt{n}} = 0$ Converges

15. $\displaystyle\lim_{n\to\infty} \frac{\sin(n)}{\sqrt{n}} = 0$

Converges

17. $A_n = 5000\left(1 + \dfrac{0.05}{4}\right)^n = 5000(1.0125)^n$

$n = 1, 2, 3$

(a) $A_1 = 5062.50$ $\quad A_5 \approx 5320.41$

$A_2 \approx 5125.78$ $\quad A_6 \approx 5386.92$

$A_3 \approx 5189.85$ $\quad A_7 \approx 5454.25$

$A_4 \approx 5254.73$ $\quad A_8 \approx 5522.43$

(b) $A_{40} \approx 8218.10$

19. (a)

k	5	10	15	20	25
S_k	13.2	113.3	873.8	6448.5	50,500.3

(b)

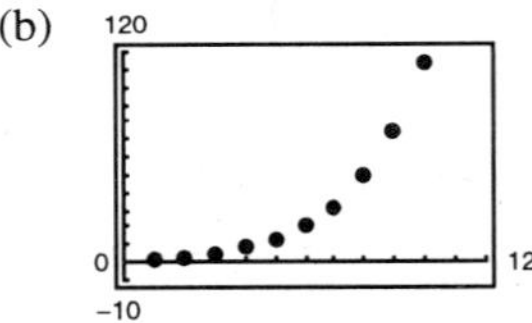

(c) The series diverges $\left(\text{geometric } r = \frac{3}{2} > 1\right)$

21. (a)

k	5	10	15	20	25
S_k	0.4597	0.4597	0.4597	0.4597	0.4597

(b)

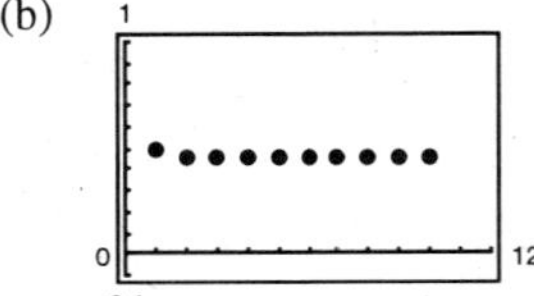

(c) The series converges by the Alternating Series Test.

23. 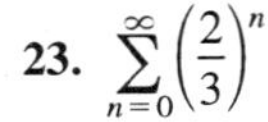 $\displaystyle\sum_{n=0}^{\infty} \left(\frac{2}{3}\right)^n$

Geometric series with $a = 1$ and $r = \frac{2}{3}$.

$$S = \frac{a}{1 - r} = \frac{1}{1 - (2/3)} = \frac{1}{1/3} = 3$$

25. $\displaystyle\sum_{n=0}^{\infty} \left(\frac{1}{2^n} - \frac{1}{3^n}\right) = \sum_{n=0}^{\infty} \left(\frac{1}{2}\right)^n - \sum_{n=0}^{\infty} \left(\frac{1}{3}\right)^n$

$$= \frac{1}{1 - (1/2)} - \frac{1}{1 - (1/3)} = 2 - \frac{3}{2} = \frac{1}{2}$$

27. $0.\overline{09} = 0.09 + 0.0009 + 0.000009 + \cdots = 0.09(1 + 0.01 + 0.0001 + \cdots) = \sum_{n=0}^{\infty} (0.09)(0.01)^n = \dfrac{0.09}{1 - 0.01} = \dfrac{1}{11}$

29. $D_1 = 8$

$D_2 = 0.7(8) + 0.7(8) = 16(0.7)$

$\vdots$

$D = 8 + 16(0.7) + 16(0.7)^2 + \cdots + 16(0.7)^n + \cdots$

$= -8 + \sum_{n=0}^{\infty} 16(0.7)^n = -8 + \dfrac{16}{1 - 0.7} = 45\frac{1}{3}$ meters

31. See Exercise 76 in Section 8.2.

$$A = \frac{P(e^{rt} - 1)}{e^{r/12} - 1}$$

$$= \frac{200(e^{(0.06)(2)} - 1)}{e^{0.06/12} - 1}$$

$$\approx \$5087.14$$

33. (a) Ratio Test: $\lim_{n\to\infty} \left|\dfrac{a_{n+1}}{a_n}\right| = \lim_{n\to\infty} \dfrac{(n + 1)(3/5)^{n+1}}{n(3/5)^n}$

$$= \lim_{n\to\infty} \left(\frac{n + 1}{n}\right)\left(\frac{3}{5}\right) = \frac{3}{5} < 1$$

Converges

(b)

x	5	10	15	20	25
S_n	2.8752	3.6366	3.7377	3.7488	3.7499

(c) 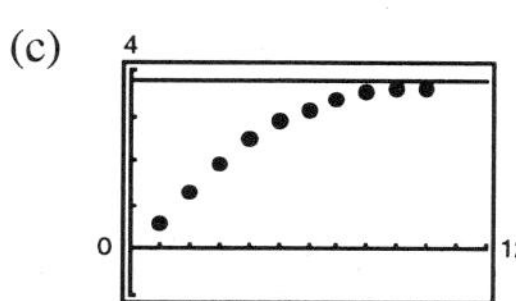

(d) The sum is approximately 3.75.

35. $\displaystyle\sum_{n=1}^{\infty} \frac{2^n}{n^3}$

$$\lim_{n\to\infty} \left|\frac{a_{n+1}}{a_n}\right| = \lim_{n\to\infty} \left|\frac{2^{n+1}}{(n + 1)^3} \cdot \frac{n^3}{2^n}\right|$$

$$= \lim_{n\to\infty} \frac{2n^3}{(n + 1)^3} = 2$$

Therefore, by the Ratio Test, the series diverges.

37. $\displaystyle\sum_{n=1}^{\infty} \frac{1}{\sqrt{n^3 + 2n}}$

$$\lim_{n\to\infty} \frac{1/\sqrt{n^3 + 2n}}{1/(n^{3/2})} = \lim_{n\to\infty} \frac{n^{3/2}}{\sqrt{n^3 + 2n}} = 1$$

By a limit comparison test with the convergent p-series

$\displaystyle\sum_{n=1}^{\infty} \frac{1}{n^{3/2}}$, the series converges.

39. $\displaystyle\sum_{n=1}^{\infty} \frac{n}{e^{n^2}}$

$$\lim_{n\to\infty} \left|\frac{a_{n+1}}{a_n}\right| = \lim_{n\to\infty} \left|\frac{n + 1}{e^{(n+1)^2}} \cdot \frac{e^{n^2}}{n}\right| = \lim_{n\to\infty} \left|\frac{e^{n^2}(n + 1)}{e^{n^2 + 2n + 1}n}\right|$$

$$= \lim_{n\to\infty} \left(\frac{1}{e^{2n+1}}\right)\left(\frac{n + 1}{n}\right)$$

$$= (0)(1) = 0 < 1$$

By the Ratio Test, the series converges.

41. $\displaystyle\sum_{n=1}^{\infty} \frac{(-1)^n n}{\ln(n)}$

$$\lim_{n\to\infty} \frac{n}{\ln(n)} = \infty \neq 0$$

The series diverges by the nth term test.

43. $\displaystyle\sum_{n=1}^{\infty} \left(\frac{1}{n^2} - \frac{1}{n}\right) = \sum_{n=1}^{\infty} \frac{1}{n^2} - \sum_{n=1}^{\infty} \frac{1}{n}$

Since the second series is a divergent p-series while the first series is a convergent p-series, the difference diverges.

45. $\displaystyle\sum_{n=1}^{\infty} \frac{1 \cdot 3 \cdot 5 \cdots (2n-1)}{2 \cdot 4 \cdot 6 \cdots (2n)}$

$$a_n = \frac{1 \cdot 3 \cdot 5 \cdots (2n-1)}{2 \cdot 4 \cdot 6 \cdots (2n)} = \left(\frac{3}{2} \cdot \frac{5}{4} \cdots \frac{2n-1}{2n-2}\right)\frac{1}{2n} > \frac{1}{2n}$$

Since $\displaystyle\sum_{n=1}^{\infty} \frac{1}{2n} = \frac{1}{2}\sum_{n=1}^{\infty} \frac{1}{n}$ diverges (harmonic series), so does the original series.

47. (a) $\displaystyle\int_N^{\infty} \frac{1}{x^2}\,dx = \left[-\frac{1}{x}\right]_N^{\infty} = \frac{1}{N}$

N	5	10	20	30	40
$\displaystyle\sum_{n=1}^{N} \frac{1}{n^2}$	1.4636	1.5498	1.5962	1.6122	1.6202
$\displaystyle\int_N^{\infty} \frac{1}{x^2}\,dx$	0.2000	0.1000	0.0500	0.0333	0.0250

(b) $\displaystyle\int_N^{\infty} \frac{1}{x^5}\,dx = \left[-\frac{1}{4x^4}\right]_N^{\infty} = \frac{1}{4N^4}$

N	5	10	20	30	40
$\displaystyle\sum_{n=1}^{N} \frac{1}{n^5}$	1.0367	1.0369	1.0369	1.0369	1.0369
$\displaystyle\int_N^{\infty} \frac{1}{x^5}\,dx$	0.0004	0.0000	0.0000	0.0000	0.0000

The series in part (b) converges more rapidly. The integral values represent the remainders of the partial sums.

49. $\displaystyle\sum_{n=0}^{\infty} \left(\frac{x}{10}\right)^n$

Geometric series which converges only if $|x/10| < 1$ or $-10 < x < 10$.

51. $\displaystyle\sum_{n=0}^{\infty} \frac{(-1)^n (x-2)^n}{(n+1)^2}$

$$\lim_{n\to\infty}\left|\frac{u_{n+1}}{u_n}\right| = \lim_{n\to\infty}\left|\frac{(-1)^{n+1}(x-2)^{n+1}}{(n+2)^2} \cdot \frac{(n+1)^2}{(-1)^n(x-2)^n}\right| = |x-2|$$

$R = 1$

Center: 2

Since the series converges when $x = 1$ and when $x = 3$, the interval of convergence is $1 \le x \le 3$.

53. $\displaystyle\sum_{n=0}^{\infty} n!(x-2)^n$

$$\lim_{n\to\infty}\left|\frac{u_{n+1}}{u_n}\right| = \lim_{n\to\infty}\left|\frac{(n+1)!(x-2)^{n+1}}{n!(x-2)^n}\right| = \infty$$

which implies that the series converges only at the center $x = 2$.

55. $f(x) = \sin(x)$

$f'(x) = \cos(x)$

$f''(x) = -\sin(x)$

$f'''(x) = -\cos(x), \ldots$

$$\sin(x) = \sum_{n=0}^{\infty} \frac{f^{(n)}(x)[x-(3\pi/4)]^n}{n!} = \frac{\sqrt{2}}{2} - \frac{\sqrt{2}}{2}\left(x - \frac{3\pi}{4}\right) - \frac{\sqrt{2}}{2 \cdot 2!}\left(x - \frac{3\pi}{4}\right)^2 + \cdots = \frac{\sqrt{2}}{2}\sum_{n=0}^{\infty} \frac{(-1)^{n(n+1)/2}[x-(3\pi/4)]^n}{n!}$$

57. $3^x = (e^{\ln(3)})^x = e^{x\ln(3)}$ and since $e^x = \displaystyle\sum_{n=0}^{\infty} \frac{x^n}{n!}$, we have

$$3^x = \sum_{n=0}^{\infty} \frac{(x \ln 3)^n}{n!} = 1 + x \ln 3 + \frac{x^2 \ln^2 3}{2!} + \frac{x^3 \ln^3 3}{3!} + \frac{x^4 \ln^4 3}{4!} + \cdots.$$

59. $f(x) = \frac{1}{x}$

$f'(x) = -\frac{1}{x^2}$

$f''(x) = \frac{2}{x^3}$

$f'''(x) = -\frac{6}{x^4}, \dots$

$$\frac{1}{x} = \sum_{n=0}^{\infty} \frac{f^{(n)}(-1)(x+1)^n}{n!} = \sum_{n=0}^{\infty} \frac{-n!(x+1)^n}{n!} = -\sum_{n=0}^{\infty} (x+1)^n$$

61. $g(x) = \frac{2}{3-x} = \frac{2/3}{1-(x/3)} = \sum_{n=0}^{\infty} \frac{2}{3}\left(\frac{x}{3}\right)^n = \sum_{n=0}^{\infty} \frac{2x^n}{3^{n+1}}$

63. $\ln x = \sum_{n=1}^{\infty} (-1)^{n+1} \frac{(x-1)^n}{n}, \quad 0 < x \le 2$

$$\ln\left(\frac{5}{4}\right) = \sum_{n=1}^{\infty} (-1)^{n+1}\left(\frac{(5/4)-1}{n}\right)^n$$

$$= \sum_{n=1}^{\infty} (-1)^{n+1} \frac{1}{4^n n} \approx 0.2231$$

65. $e^x = \sum_{n=0}^{\infty} \frac{x^n}{n!}, \quad -\infty < x < \infty$

$$e^{1/2} = \sum_{n=0}^{\infty} \frac{1}{2^n n!} \approx 1.6487$$

67. $\cos x = \sum_{n=0}^{\infty} (-1)^n \frac{x^{2n}}{(2n)!}, \quad -\infty < x < \infty$

$$\cos\left(\frac{2}{3}\right) = \sum_{n=0}^{\infty} (-1)^n \frac{2^{2n}}{3^{2n}(2n)!} \approx 0.7859$$

69. The series for Exercise 37 converges very slowly because the terms approach 0 at a slow rate.

71. (a) $f(x) = e^{2x} \qquad f(0) = 1$

$f'(x) = 2e^{2x} \qquad f'(0) = 2$

$f''(x) = 4e^{2x} \qquad f''(0) = 4$

$f'''(x) = 8e^{2x} \qquad f'''(0) = 8$

$$e^{2x} = 1 + 2x + \frac{4x^2}{2!} + \frac{8x^3}{3!} + \cdots$$

$$= 1 + 2x + 2x^2 + \frac{4}{3}x^3 + \cdots$$

(b) $e^x = \sum_{n=0}^{\infty} \frac{x^n}{n!}$

$$e^{2x} = \sum_{n=0}^{\infty} \frac{(2x)^n}{n!} = 1 + 2x + \frac{4x^2}{2!} + \frac{8x^3}{3!} + \cdots$$

$$= 1 + 2x + 2x^2 + \frac{4}{3}x^3 + \cdots$$

(c) $e^{2x} = e^x \cdot e^x = \left(1 + x + \frac{x^2}{2} + \frac{x^3}{6} + \cdots\right)\left(1 + x + \frac{x^2}{2} + \frac{x^3}{6} + \cdots\right)$

$$= 1 + (x + x) + \left(x^2 + \frac{x^2}{2} + \frac{x^2}{2}\right) + \left(\frac{x^3}{6} + \frac{x^3}{6} + \frac{x^3}{2} + \frac{x^3}{2}\right) + \cdots = 1 + 2x + 2x^2 + \frac{4}{3}x^3 + \cdots$$

73. $1 + \frac{2}{3}x + \frac{4}{9}x^2 + \frac{8}{27}x^3 + \cdots = \sum_{n=0}^{\infty}\left(\frac{2x}{3}\right)^n = \frac{1}{1-(2x/3)} = \frac{3}{3-2x}, \quad -\frac{3}{2} < x < \frac{3}{2}$

75. $\sin t = \sum_{n=0}^{\infty} \frac{(-1)^n t^{2n+1}}{(2n+1)!}$

$$\frac{\sin t}{t} = \sum_{n=0}^{\infty} \frac{(-1)^n t^{2n}}{(2n+1)!}$$

$$\int_0^x \frac{\sin t}{t}\,dt = \left[\sum_{n=0}^{\infty} \frac{(-1)^n t^{2n+1}}{(2n+1)(2n+1)!}\right]_0^x = \sum_{n=0}^{\infty} \frac{(-1)^n x^{2n+1}}{(2n+1)(2n+1)!}$$

77. $\frac{1}{1+t} = \sum_{n=0}^{\infty} (-1)^n t^n$

$$\ln(1+t) = \int \frac{1}{1+t}\,dt = \sum_{n=0}^{\infty} \frac{(-1)^n t^{n+1}}{n+1}$$

$$\frac{\ln(t+1)}{t} = \sum_{n=0}^{\infty} \frac{(-1)^n t^n}{n+1}$$

$$\int_0^x \frac{\ln(t+1)}{t}\,dt = \left[\sum_{n=0}^{\infty} \frac{(-1)^n t^{n+1}}{(n+1)^2}\right]_0^x = \sum_{n=0}^{\infty} \frac{(-1)^n x^{n+1}}{(n+1)^2}$$

79. $\arctan x = x - \frac{x^3}{3} + \frac{x^5}{5} - \frac{x^7}{7} + \frac{x^9}{9} - \cdots$

$$\frac{\arctan x}{\sqrt{x}} = \sqrt{x} - \frac{x^{5/2}}{3} + \frac{x^{9/2}}{5} - \frac{x^{13/2}}{7} + \frac{x^{17/2}}{9} - \cdots$$

$$\lim_{x\to 0} \frac{\arctan x}{\sqrt{x}} = 0$$

By L'Hôpital's Rule, $\lim_{x\to 0} \frac{\arctan x}{\sqrt{x}} = \lim_{x\to 0} \frac{\left(\frac{1}{1+x^2}\right)}{\left(\frac{1}{2\sqrt{x}}\right)} = \lim_{x\to 0} \frac{2\sqrt{x}}{1+x^2} = 0.$

81. $y = \sum_{n=0}^{\infty} (-1)^n \frac{x^{2n}}{4^n(n!)^2}$

$$y' = \sum_{n=1}^{\infty} \frac{(-1)^n(2n)x^{2n-1}}{4^n(n!)^2} = \sum_{n=0}^{\infty} \frac{(-1)^{n+1}(2n+2)x^{2n+1}}{4^{n+1}[(n+1)!]^2}$$

$$y'' = \sum_{n=0}^{\infty} \frac{(-1)^{n+1}(2n+2)(2n+1)x^{2n}}{4^{n+1}[(n+1)!]^2}$$

$$x^2y'' + xy' + x^2y = \sum_{n=0}^{\infty} \frac{(-1)^{n+1}(2n+2)(2n+1)x^{2n+2}}{4^{n+1}[(n+1)!]^2} + \sum_{n=0}^{\infty} \frac{(-1)^{n+1}(2n+2)x^{2n+2}}{4^{n+1}[(n+1)!]^2} + \sum_{n=0}^{\infty} (-1)^n \frac{x^{2n+1}}{4^n(n!)^2}$$

$$= \sum_{n=0}^{\infty}\left[(-1)^{n+1}\frac{(2n+2)(2n+1)}{4^{n+1}[(n+1)!]^2} + \frac{(-1)^{n+1}(2n+2)}{4^{n+1}[(n+1)!]^2} + \frac{(-1)^n}{4^n(n!)^2}\right]x^{2n+2}$$

$$= \sum_{n=0}^{\infty}\left[\frac{(-1)^{n+1}(2n+2)(2n+1+1)}{4^{n+1}[(n+1)!]^2} + (-1)^n\frac{1}{4^n(n!)^2}\right]x^{2n+2}$$

$$= \sum_{n=0}^{\infty}\left[\frac{(-1)^{n+1}4(n+1)^2}{4^{n+1}[(n+1)!]^2} + (-1)^n\frac{1}{4^n(n!)^2}\right]x^{2n+2} = \sum_{n=0}^{\infty}\left[\frac{(-1)^{n+1}1}{4^n(n!)^2} + (-1)^n\frac{1}{4^n(n!)^2}\right]x^{2n+2} = 0$$

83. $\sin(95°) = \sin\left(\frac{95\pi}{180}\right) \approx \frac{95\pi}{180} - \frac{(95\pi)^3}{180^3 3!} + \frac{(95\pi)^5}{180^5 5!} - \frac{(95\pi)^7}{180^7 7!} + \frac{(95\pi)^9}{180^9 9!} \approx 0.996$

85. $\ln(1.75) \approx (0.75) - \frac{(0.75)^2}{2} + \frac{(0.75)^3}{3} - \frac{(0.75)^4}{4} + \frac{(0.75)^5}{5} - \frac{(0.75)^6}{6} + \cdots + \frac{(0.75)^{15}}{15} \approx 0.560$

87. $f(x) = \cos x,\ c = 0$

$$R_n(x) = \frac{f^{(n+1)}(z)}{(n+1)!}x^{n+1}$$

$$|f^{(n+1)}(z)| \le 1 \Rightarrow R_n(x) \le \frac{x^{n+1}}{(n+1)!}$$

(a) $R_n(x) \le \frac{(0.5)^{n+1}}{(n+1)!} < 0.001$

This inequality is true for $n = 4$.

(b) $R_n(x) \le \frac{(1)^{n+1}}{(n+1)!} < 0.001$

This inequality is true for $n = 6$.

(c) $R_n(x) \le \frac{(0.5)^{n+1}}{(n+1)!} < 0.0001$

This inequality is true for $n = 5$.

(d) $R_n(x) \le \frac{2^{n+1}}{(n+1)!} < 0.0001$

This inequality is true for $n = 10$.

CHAPTER 9
Conics, Parametric Equations, and Polar Coordinates

CHAPTER 9
Conics, Parametric Equations, and Polar Coordinates

Section 9.1 Conics and Calculus

Solutions to Odd-Numbered Exercises

1. $y^2 = 4x$

Vertex: $(0, 0)$

$p = 1 > 0$

Opens to the right
Matches graph (h).

3. $(x + 3)^2 = -2(y - 2)$

Vertex: $(-3, 2)$

$p = -\frac{1}{2} < 0$

Opens downward
Matches graph (e).

5. $\dfrac{x^2}{9} + \dfrac{y^2}{4} = 1$

Center: $(0, 0)$
Ellipse
Matches (f)

7. $\dfrac{y^2}{16} - \dfrac{x^2}{1} = 1$

Hyperbola

Center: $(0, 0)$

Vertical transverse axis.
Matches (c)

9. $y^2 = -6x = 4\left(-\frac{3}{2}\right)x$

Vertex: $(0, 0)$

Focus: $\left(-\frac{3}{2}, 0\right)$

Directrix: $x = \frac{3}{2}$

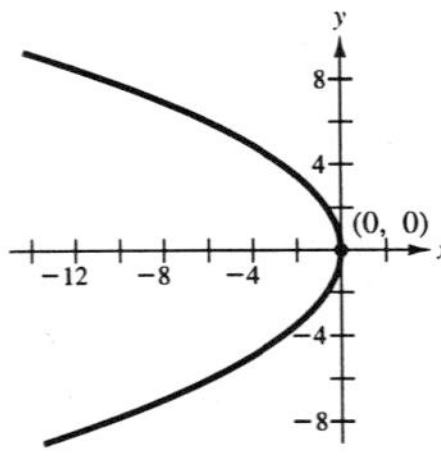

11. $(x + 3) + (y - 2)^2 = 0$

$$(y - 2)^2 = 4\left(-\tfrac{1}{4}\right)(x + 3)$$

Vertex: $(-3, 2)$

Focus: $(-3.25, 2)$

Directrix: $x = -2.75$

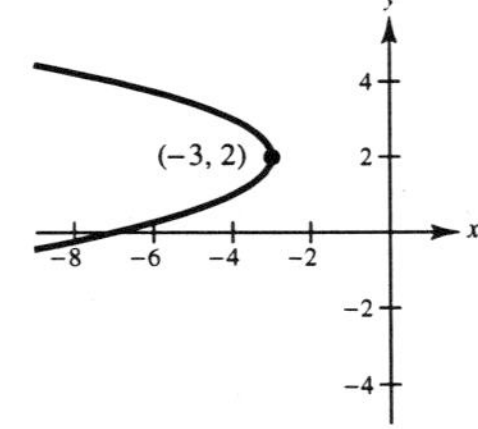

13. $y^2 - 4y - 4x = 0$

$$y^2 - 4y + 4 = 4x + 4$$

$$(y - 2)^2 = 4(1)(x + 1)$$

Vertex: $(-1, 2)$

Focus: $(0, 2)$

Directrix: $x = -2$

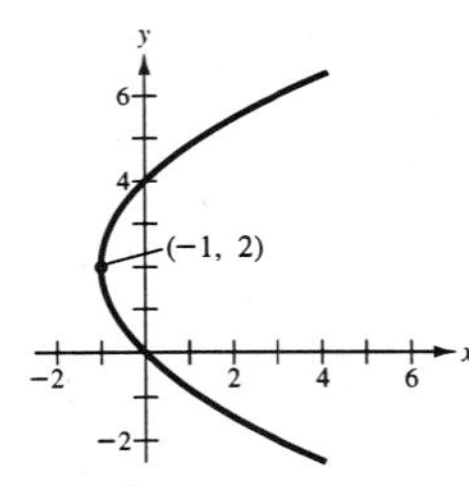

15. $x^2 + 4x + 4y - 4 = 0$

$$x^2 + 4x + 4 = -4y + 4 + 4$$

$$(x + 2)^2 = 4(-1)(y - 2)$$

Vertex: $(-2, 2)$

Focus: $(-2, 1)$

Directrix: $y = 3$

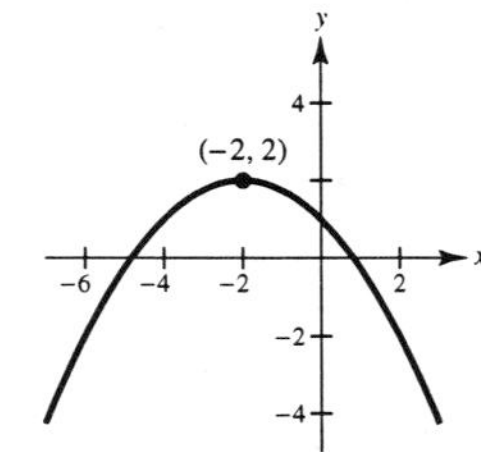

17. $y^2 + x + y = 0$

$$y^2 + y + \tfrac{1}{4} = -x + \tfrac{1}{4}$$

$$\left(y + \tfrac{1}{2}\right)^2 = 4\left(-\tfrac{1}{4}\right)\left(x - \tfrac{1}{4}\right)$$

Vertex: $\left(\frac{1}{4}, -\frac{1}{2}\right)$

Focus: $\left(0, -\frac{1}{2}\right)$

Directrix: $x = \frac{1}{2}$

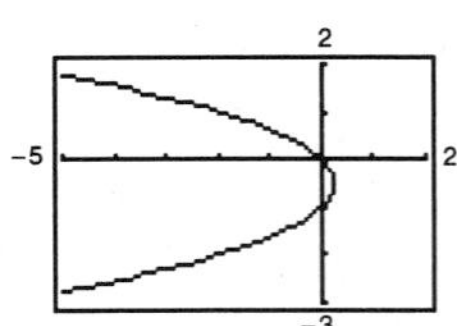

19. $y^2 - 4x - 4 = 0$

$$y^2 = 4x + 4$$
$$= 4(1)(x + 1)$$

Vertex: $(-1, 0)$

Focus: $(0, 0)$

Directrix: $x = -2$

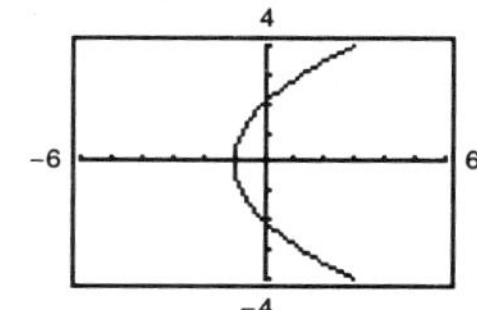

21. $(y - 2)^2 = 4(-2)(x - 3)$

$$y^2 - 4y + 8x - 20 = 0$$

23. $(x - h)^2 = 4p(y - k)$

$$x^2 = 4(6)(y - 4)$$
$$x^2 - 24y + 96 = 0$$

25. $y = 4 - x^2$

$$x^2 + y - 4 = 0$$

27. Since the axis of the parabola is vertical, the form of the equation is $y = ax^2 + bx + c$. Now, substituting the values of the given coordinates into this equation, we obtain

$$3 = c,\ 4 = 9a + 3b + c,\ 11 = 16a + 4b + c.$$

Solving this system, we have $a = \frac{5}{3}, b = -\frac{14}{3}, c = 3$. Therefore, $y = \frac{5}{3}x^2 - \frac{14}{3}x + 3$ or $5x^2 - 14x - 3y + 9 = 0$.

29. Assume that the vertex is at the origin.

$$x^2 = 4py$$
$$(3)^2 = 4p(1)$$
$$\frac{9}{4} = p$$

The pipe is located $\frac{9}{4}$ meters from the vertex.

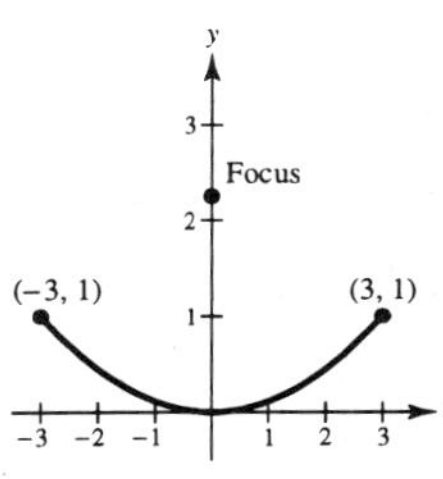

31. $y = ax^2$

$$y' = 2ax$$

The equation of the tangent line is

$$y - ax_0^2 = 2ax_0(x - x_0) \text{ or } y = 2ax_0x - ax_0^2.$$

Let $y = 0$. Then:

$$-ax_0^2 = 2ax_0x - 2ax_0^2$$
$$ax_0^2 = 2ax_0x$$

Therefore, $x_0/2 = x$ is the x-intercept.

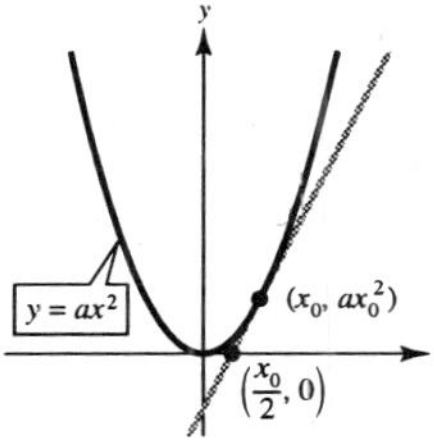

33. (a) Consider the parabola $x^2 = 4py$. Let m_0 be the slope of the one tangent line at (x_1, y_1) and therefore, $-1/m_0$ is the slope of the second at (x_2, y_2). From the derivative given in Exercise 32 we have:

$$m_0 = \frac{1}{2p}x_1 \text{ or } x_1 = 2pm_0$$

$$\frac{-1}{m_0} = \frac{1}{2p}x_2 \text{ or } x_2 = \frac{-2p}{m_0}$$

Substituting these values of x into the equation $x^2 = 4py$, we have the coordinates of the points of tangency $(2pm_0, pm_0^2)$ and $(-2p/m_0, p/m_0^2)$ and the equations of the tangent lines are

$$(y - pm_0^2) = m_0(x - 2pm_0) \quad \text{and} \quad \left(y - \frac{p}{m_0^2}\right) = \frac{-1}{m_0}\left(x + \frac{2p}{m_0}\right).$$

The point of intersection of these lines is $\left(\frac{p(m_0^2 - 1)}{m_0}, -p\right)$ and is on the directrix, $y = -p$.

—CONTINUED—

33. —CONTINUED—

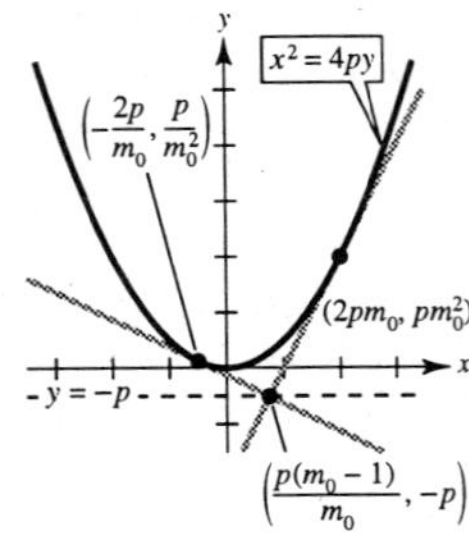

(b) $x^2 - 4x - 4y + 8 = 0$

$$(x - 2)^2 = 4(y - 1). \text{ Vertex } (2, 1)$$

$$2x - 4 - 4\frac{dy}{dx} = 0$$

$$\frac{dy}{dx} = \frac{1}{2}x - 1$$

At $(-2, 5)$, $dy/dx = -2$. At $\left(3, \frac{5}{4}\right)$, $dy/dx = \frac{1}{2}$.

Tangent line at $(-2, 5)$: $y - 5 = -2(x + 2) \implies 2x + y - 1 = 0.$

Tangent line at $\left(3, \frac{5}{4}\right)$: $y - \frac{5}{4} = \frac{1}{2}(x - 3) \implies 2x - 4y - 1 = 0.$

Since $m_1 m_2 = (-2)\left(\frac{1}{2}\right) = -1$, the lines are perpendicular.

Point of intersection: $-2x + 1 = \frac{1}{2}x - \frac{1}{4}$

$$-\frac{5}{2}x = -\frac{5}{4}$$

$$x = \frac{1}{2}$$

$$y = 0$$

Directrix: $y = 0$ and the point of intersection $\left(\frac{1}{2}, 0\right)$ lies on this line.

35. $y = x - x^2$

$$\frac{dy}{dx} = 1 - 2x$$

At (x_1, y_1) on the mountain, $m = 1 - 2x_1$. Also, $m = \dfrac{y_1 - 1}{x_1 + 1}$.

$$\frac{y_1 - 1}{x_1 + 1} = 1 - 2x_1$$

$$(x_1 - x_1^2) - 1 = (1 - 2x_1)(x_1 + 1)$$

$$-x_1^2 + x_1 - 1 = -2x_1^2 - x_1 + 1$$

$$x_1^2 + 2x_1 - 2 = 0$$

$$x_1 = \frac{-2 \pm \sqrt{2^2 - 4(1)(-2)}}{2(1)} = \frac{-2 \pm 2\sqrt{3}}{2} = -1 \pm \sqrt{3}$$

Choosing the positive value for x_1, we have $x_1 = -1 + \sqrt{3}$.

$$m = 1 - 2\left(-1 + \sqrt{3}\right) = 3 - 2\sqrt{3}$$

$$m = \frac{0 - 1}{x_0 + 1} = -\frac{1}{x_0 + 1}$$

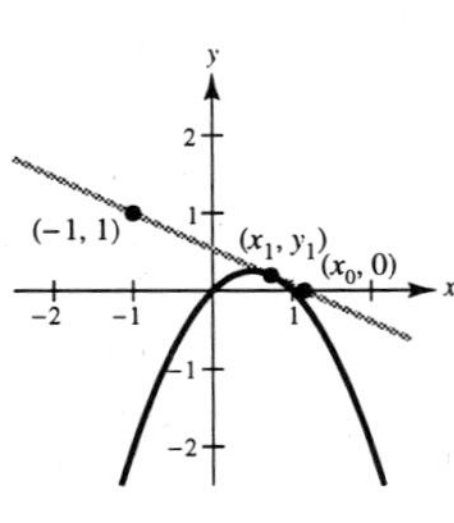

Thus, $$-\frac{1}{x_0 + 1} = 3 - 2\sqrt{3}$$

$$\frac{-1}{3 - 2\sqrt{3}} = x_0 + 1$$

$$\frac{3 + 2\sqrt{3}}{3} - 1 = x_0$$

$$\frac{2\sqrt{3}}{3} = x_0.$$

The closest the receiver can be to the hill is $\left(2\sqrt{3}/3\right) - 1 \approx 0.155$.

37. Parabola

Vertex: $(0, 4)$

$$x^2 = 4p(y - 4)$$
$$4^2 = 4p(0 - 4)$$
$$p = -1$$
$$x^2 = -4(y - 4)$$
$$y = 4 - \frac{x^2}{4}$$

Circle

Center: $(0, k)$

Radius: 8

$$x^2 + (y - k)^2 = 64$$
$$4^2 + (0 - k)^2 = 64$$
$$k^2 = 48$$
$$k = -4\sqrt{3} \quad \text{(Center is on the negative } y\text{-axis.)}$$
$$x^2 + \left(y + 4\sqrt{3}\right)^2 = 64$$
$$y = -4\sqrt{3} \pm \sqrt{64 - x^2}$$

Since the y value is positive when $x = 0$, we have $y = -4\sqrt{3} + \sqrt{64 - x^2}$.

$$A = 2\int_0^4 \left[\left(4 - \frac{x^2}{4}\right) - \left(-4\sqrt{3} + \sqrt{64 - x^2}\right)\right] dx$$
$$= 2\left[4x - \frac{x^3}{12} + 4\sqrt{3}x - \frac{1}{2}\left(x\sqrt{64 - x^2} + 64 \arcsin\frac{x}{8}\right)\right]_0^4$$
$$= 2\left[16 - \frac{64}{12} + 16\sqrt{3} - 2\sqrt{48} - 32 \arcsin\frac{1}{2}\right]$$
$$= \frac{16\left(4 + 3\sqrt{3} - 2\pi\right)}{3} \approx 15.536 \text{ square feet}$$

39. (a) Assume that $y = ax^2$.

$$20 = a(60)^2 \implies a = \frac{2}{360} = \frac{1}{180} \implies y = \frac{1}{180}x^2$$

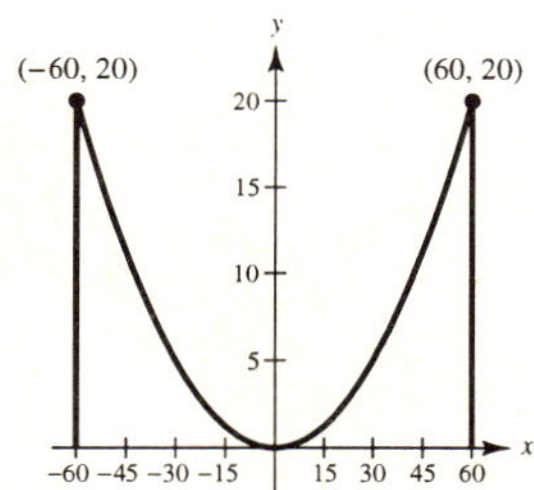

(b) $f(x) = \frac{1}{180}x^2, f'(x) = \frac{1}{90}x$

$$S = 2\int_0^{60} \sqrt{1 + \left(\frac{1}{90}x\right)^2}\, dx = \frac{2}{90}\int_0^{60} \sqrt{90^2 + x^2}\, dx$$
$$= \frac{2}{90}\frac{1}{2}\left[x\sqrt{90^2 + x^2} + 90^2 \ln\left|x + \sqrt{90^2 + x^2}\right|\right]_0^{60} \quad \text{(formula 26)}$$
$$= \frac{1}{90}\left[60\sqrt{11{,}700} + 90^2 \ln\left(60 + \sqrt{11{,}700}\right) - 90^2 \ln 90\right]$$
$$= \frac{1}{90}\left[1800\sqrt{13} + 90^2 \ln\left(60 + 30\sqrt{13}\right) - 90^2 \ln 90\right]$$
$$= 20\sqrt{13} + 90 \ln\left(\frac{60 + 30\sqrt{13}}{90}\right)$$
$$= 10\left[2\sqrt{13} + 9 \ln\left(\frac{2 + \sqrt{13}}{3}\right)\right] \approx 128.4 \text{ m}$$

41. $x^2 + 4y^2 = 4$

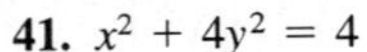

$$\frac{x^2}{4} + \frac{y^2}{1} = 1$$

$a^2 = 4,\ b^2 = 1,\ c^2 = 3$

Center: $(0, 0)$

Foci: $\left(\pm\sqrt{3}, 0\right)$

Vertices: $(\pm 2, 0)$

$$e = \frac{\sqrt{3}}{2}$$

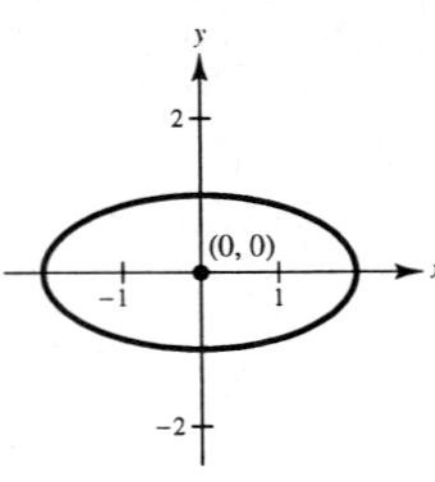

43. $\dfrac{(x-1)^2}{9} + \dfrac{(y-5)^2}{25} = 1$

$a^2 = 25,\ b^2 = 9,\ c^2 = 16$

Center: $(1, 5)$

Foci: $(1, 9), (1, 1)$

Vertices: $(1, 10), (1, 0)$

$$e = \frac{4}{5}$$

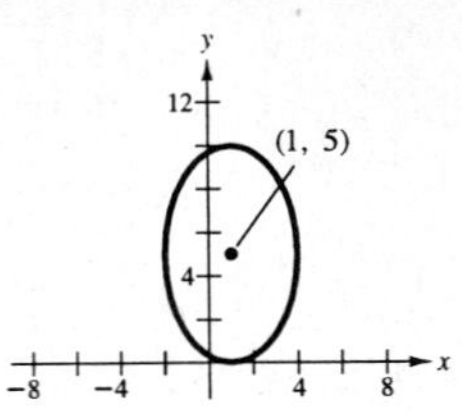

45. $9x^2 + 4y^2 + 36x - 24y + 36 = 0$

$$9(x^2 + 4x + 4) + 4(y^2 - 6y + 9) = -36 + 36 + 36$$
$$= 36$$
$$\frac{(x+2)^2}{4} + \frac{(y-3)^2}{9} = 1$$

$a^2 = 9,\ b^2 = 4,\ c^2 = 5$

Center: $(-2, 3)$

Foci: $\left(-2, 3 \pm \sqrt{5}\right)$

Vertices: $(-2, 6), (-2, 0)$

$$e = \frac{\sqrt{5}}{3}$$

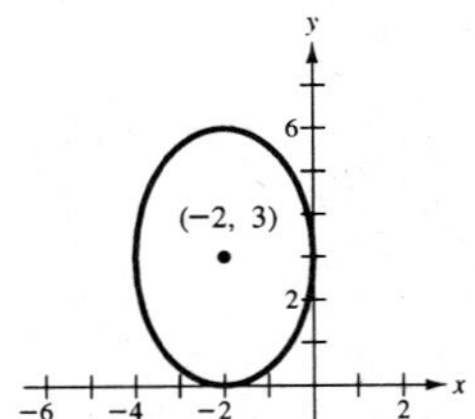

47. $12x^2 + 20y^2 - 12x + 40y - 37 = 0$

$$12\left(x^2 - x + \frac{1}{4}\right) + 20(y^2 + 2y + 1) = 37 + 3 + 20$$
$$= 60$$
$$\frac{[x - (1/2)]^2}{5} + \frac{(y+1)^2}{3} = 1$$

$a^2 = 5,\ b^2 = 3,\ c^2 = 2$

Center: $\left(\frac{1}{2}, -1\right)$

Foci: $\left(\frac{1}{2} \pm \sqrt{2}, -1\right)$

Vertices: $\left(\frac{1}{2} \pm \sqrt{5}, -1\right)$

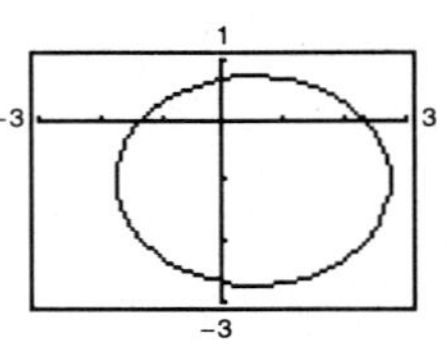

Solve for y:

$$20(y^2 + 2y + 1) = -12x^2 + 12x + 37 + 20$$
$$(y+1)^2 = \frac{57 + 12x - 12x^2}{20}$$
$$y = -1 \pm \sqrt{\frac{57 + 12x - 12x^2}{20}}$$

(Graph each of these separately.)

49. $x^2 + 2y^2 - 3x + 4y + 0.25 = 0$

$$\left(x^2 - 3x + \frac{9}{4}\right) + 2(y^2 + 2y + 1) = -\frac{1}{4} + \frac{9}{4} + 2 = 4$$
$$\frac{[x - (3/2)]^2}{4} + \frac{(y+1)^2}{2} = 1$$

$a^2 = 4,\ b^2 = 2,\ c^2 = 2$

Center: $\left(\frac{3}{2}, -1\right)$

Foci: $\left(\frac{3}{2} \pm \sqrt{2}, -1\right)$

Vertices: $\left(-\frac{1}{2}, -1\right), \left(\frac{7}{2}, -1\right)$

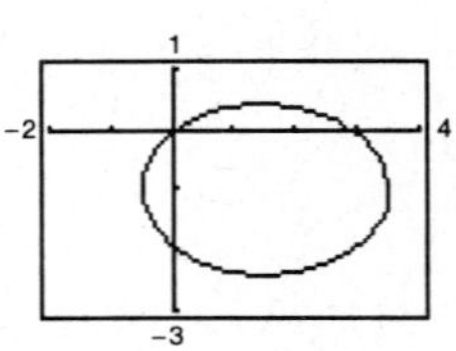

Solve for y: $2(y^2 + 2y + 1) = -x^2 + 3x - \dfrac{1}{4} + 2$

$$(y+1)^2 = \frac{1}{2}\left(\frac{7}{4} + 3x - x^2\right)$$
$$y = -1 \pm \sqrt{\frac{7 + 12x - 4x^2}{8}}$$

(Graph each of these separately.)

51. Center: (0, 0)
Focus: (2, 0)
Vertex: (3, 0)
Horizontal major axis

$a = 3, c = 2 \Rightarrow b = \sqrt{5}$

$$\frac{x^2}{9} + \frac{y^2}{5} = 1$$

53. Vertices: (3, 1), (3, 9)
Minor axis length: 6
Vertical major axis
Center: (3, 5)

$a = 4, b = 3$

$$\frac{(x-3)^2}{9} + \frac{(y-5)^2}{16} = 1$$

55. Center: (0, 0)
Horizontal major axis
Points on ellipse: (3, 1), (4, 0)

Since the major axis is horizontal,

$$\left(\frac{x^2}{a^2}\right) + \left(\frac{y^2}{b^2}\right) = 1.$$

Substituting the values of the coordinates of the given points into this equation, we have

$$\left(\frac{9}{a^2}\right) + \left(\frac{1}{b^2}\right) = 1, \text{ and } \frac{16}{a^2} = 1.$$

The solution to this system is $a^2 = 16$, $b^2 = \frac{16}{7}$. Therefore,

$$\frac{x^2}{16} + \frac{y^2}{16/7} = 1, \frac{x^2}{16} + \frac{7y^2}{16} = 1.$$

57. 

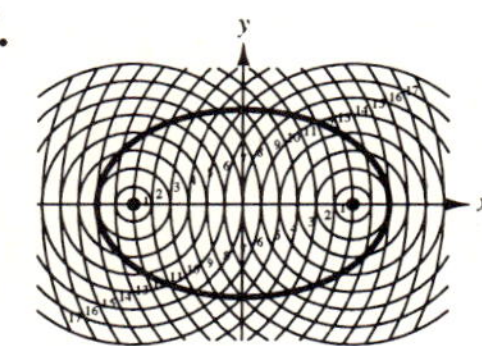

59. $a = \frac{5}{2}, b = 2, c = \sqrt{\left(\frac{5}{2}\right)^2 - (2)^2} = \frac{3}{2}$

The tacks should be placed 1.5 feet from the center. The string should be $2a = 5$ feet long.

61. $e = \frac{c}{a}$

$A + P = 2a$

$a = \frac{A + P}{2}$

$c = a - P = \frac{A + P}{2} - P = \frac{A - P}{2}$

$e = \frac{c}{a} = \frac{(A - P)/2}{(A + P)/2} = \frac{A - P}{A + P}$

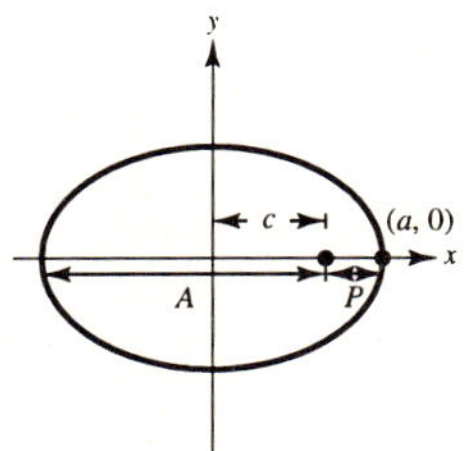

63. $e = \frac{A - P}{A + P} = \frac{35.34au - 0.59au}{35.34au + 0.59au} \approx 0.9672$

65. $16x^2 + 9y^2 + 96x + 36y + 36 = 0$

$$32x + 18yy' + 96 + 36y' = 0$$

$$y'(18y + 36) = -(32x + 96)$$

$$y' = \frac{-(32x + 96)}{18y + 36}$$

$y' = 0$ when $x = -3$. y' is undefined when $y = -2$.

At $x = -3$, $y = 2$ or -6.

Endpoints of major axis: $(-3, 2), (-3, -6)$

At $y = -2$, $x = 0$ or -6.

Endpoints of minor axis: $(0, -2), (-6, -2)$

Note: Equation of ellipse is $\dfrac{(x+3)^2}{9} + \dfrac{(y+2)^2}{16} = 1$

67. $\dfrac{x^2}{10^2} + \dfrac{y^2}{5^2} = 1$

$$\frac{2x}{10^2} + \frac{2yy'}{5^2} = 0$$

$$y' = \frac{-5^2x}{10^2y} = \frac{-x}{4y}$$

At $(-8, 3)$: $y' = \dfrac{8}{12} = \dfrac{2}{3}$

The equation of the tangent line is $y - 3 = \frac{2}{3}(x + 8)$. It will cross the y-axis when $x = 0$ and $y = \frac{2}{3}(8) + 3 = \frac{25}{3}$.

69. (a) $A = 4\displaystyle\int_0^2 \frac{1}{2}\sqrt{4 - x^2}\,dx = \left[x\sqrt{4 - x^2} + 4\arcsin\left(\frac{x}{2}\right)\right]_0^2 = 2\pi$ [or, $A = \pi ab = \pi(2)(1) = 2\pi$]

(b) **Disc:**

$$V = 2\pi\int_0^2 \frac{1}{4}(4 - x^2)\,dx = \frac{1}{2}\pi\left[4x - \frac{1}{3}x^3\right]_0^2 = \frac{8\pi}{3}$$

$$y = \frac{1}{2}\sqrt{4 - x^2}$$

$$y' = \frac{-x}{2\sqrt{4 - x^2}}$$

$$\sqrt{1 + (y')^2} = \sqrt{1 + \frac{x^2}{16 - 4x^2}} = \sqrt{\frac{16 - 3x^2}{4y}}$$

$$S = 2(2\pi)\int_0^2 y\left(\frac{\sqrt{16 - 3x^2}}{4y}\right)dx$$

$$= \frac{\pi}{2\sqrt{3}}\left[\sqrt{3}x\sqrt{16 - 3x^2} + 16\arcsin\left(\frac{\sqrt{3}x}{4}\right)\right]_0^2 = \frac{2\pi}{9}(9 + 4\sqrt{3}\pi) \approx 21.48$$

(c) **Shell:**

$$V = 2\pi\int_0^2 x\sqrt{4 - x^2}\,dx = -\pi\int_0^2 -2x(4 - x^2)^{1/2}\,dx = -\frac{2\pi}{3}\left[(4 - x^2)^{3/2}\right]_0^2 = \frac{16\pi}{3}$$

$$x = 2\sqrt{1 - y^2}$$

$$x' = \frac{-2y}{\sqrt{1 - y^2}}$$

$$\sqrt{1 + (x')^2} = \sqrt{1 + \frac{4y^2}{1 - y^2}} = \frac{\sqrt{1 + 3y^2}}{\sqrt{1 - y^2}}$$

$$S = 2(2\pi)\int_0^1 2\sqrt{1 - y^2}\frac{\sqrt{1 + 3y^2}}{\sqrt{1 - y^2}}\,dy = 8\pi\int_0^1 \sqrt{1 + 3y^2}\,dy$$

$$= \frac{8\pi}{2\sqrt{3}}\left[\sqrt{3}y\sqrt{1 + 3y^2} + \ln\left|\sqrt{3}y + \sqrt{1 + 3y^2}\right|\right]_0^1$$

$$= \frac{4\pi}{3}\left|6 + \sqrt{3}\ln\left(2 + \sqrt{3}\right)\right| \approx 34.69$$

71. From Example 5 we have

$$C = 4a\int_0^{\pi/2} \sqrt{1 - e^2 \sin^2 \theta}\, d\theta.$$

For $(x^2/9) + (y^2/16) = 1$ we have

$$a = 4,\ b = 3,\ c = \sqrt{7},\ e = \frac{\sqrt{7}}{4},\ C = 4(4)\int_0^{\pi/2} \sqrt{1 - \left(\frac{7}{16}\right)\sin^2 \theta}\, d\theta.$$

Applying Simpson's Rule with $n = 12$, or the integration capability of a graphing utility, produces $C \approx 22.10$.

73. Area circle $= \pi r^2 = 100\pi$

Area ellipse $= \pi ab = \pi a(10)$

$$2(100\pi) = 10\pi a \Rightarrow a = 20$$

Hence, the length of the major axis is $2a = 40$.

75. $\dfrac{y^2}{1} - \dfrac{x^2}{4} = 1$

$a = 1, b = 2, c = \sqrt{5}$

Center: $(0, 0)$

Vertices: $(0, \pm 1)$

Foci: $\left(0, \pm\sqrt{5}\right)$

Asymptotes: $y = \pm\dfrac{1}{2}x$

77. $\dfrac{(x-1)^2}{4} - \dfrac{(y+2)^2}{1} = 1$

$a = 2, b = 1, c = \sqrt{5}$

Center: $(1, -2)$

Vertices: $(-1, -2), (3, -2)$

Foci: $\left(1 \pm \sqrt{5}, -2\right)$

Asymptotes: $y = -2 \pm \dfrac{1}{2}(x - 1)$

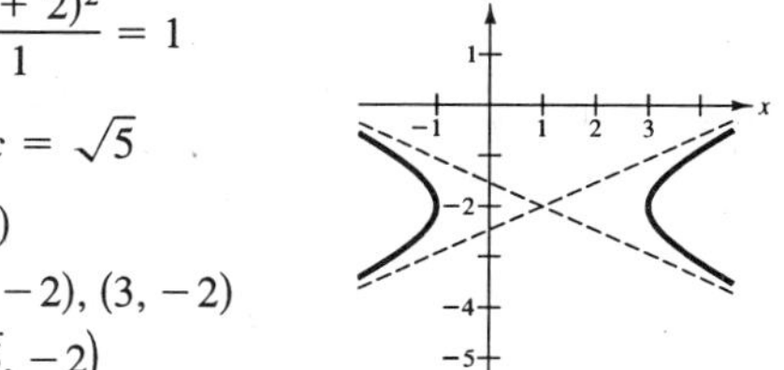

79.

$$9x^2 - y^2 - 36x - 6y + 18 = 0$$
$$9(x^2 - 4x + 4) - (y^2 + 6y + 9) = -18 + 36 - 9$$
$$\frac{(x-2)^2}{1} - \frac{(y+3)^2}{9} = 1$$

$a = 1, b = 3, c = \sqrt{10}$

Center: $(2, -3)$

Vertices: $(1, -3), (3, -3)$

Foci: $\left(2 \pm \sqrt{10}, -3\right)$

Asymptotes: $y = -3 \pm 3(x - 2)$

81.

$$x^2 - 9y^2 + 2x - 54y - 80 = 0$$
$$(x^2 + 2x + 1) - 9(y^2 + 6y + 9) = 80 + 1 - 81 = 0$$
$$(x+1)^2 - 9(y+3)^2 = 0$$
$$y + 3 = \pm\frac{1}{3}(x + 1)$$

Degenerate hyperbola is two lines intersecting at $(-1, -3)$.

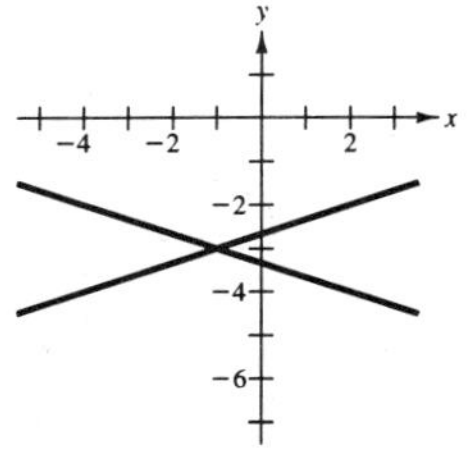

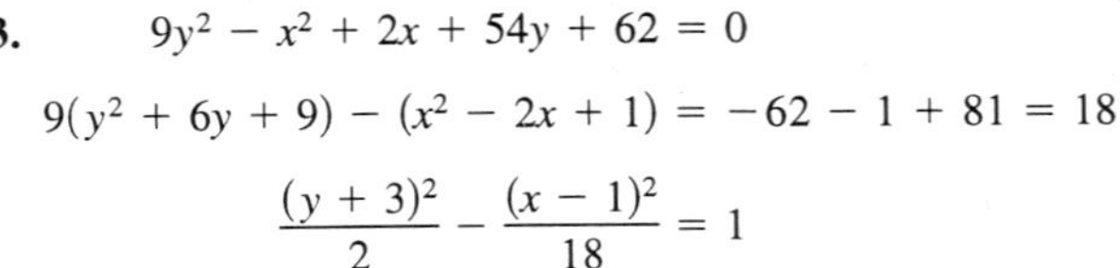

83. $$9y^2 - x^2 + 2x + 54y + 62 = 0$$

$$9(y^2 + 6y + 9) - (x^2 - 2x + 1) = -62 - 1 + 81 = 18$$

$$\frac{(y+3)^2}{2} - \frac{(x-1)^2}{18} = 1$$

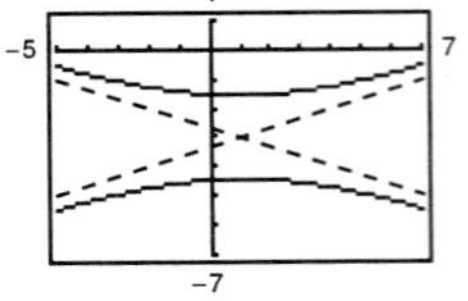

$a = \sqrt{2}, b = 3\sqrt{2}, c = 2\sqrt{5}$

Center: $(1, -3)$

Vertices: $\left(1, -3 \pm \sqrt{2}\right)$

Foci: $\left(1, -3 \pm 2\sqrt{5}\right)$

Solve for y:

$$9(y^2 + 6y + 9) = x^2 - 2x - 62 + 81$$

$$(y+3)^2 = \frac{x^2 - 2x + 19}{9}$$

$$y = -3 \pm \frac{1}{3}\sqrt{x^2 - 2x + 19}$$

(Graph each curve separately.)

85. $$3x^2 - 2y^2 - 6x - 12y - 27 = 0$$

$$3(x^2 - 2x + 1) - 2(y^2 + 6y + 9) = 27 + 3 - 18 = 12$$

$$\frac{(x-1)^2}{4} - \frac{(y+3)^2}{6} = 1$$

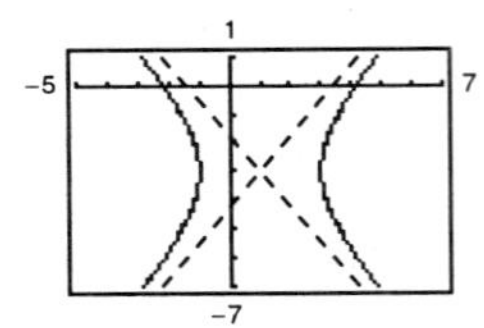

$a = 2, b = \sqrt{6}, c = \sqrt{10}$

Center: $(1, -3)$

Vertices: $(-1, -3), (3, -3)$

Foci: $\left(1 \pm \sqrt{10}, -3\right)$

Solve for y:

$$2(y^2 + 6y + 9) = 3x^2 - 6x - 27 + 18$$

$$(y+3)^2 = \frac{3x^2 - 6x - 9}{2}$$

$$y = -3 \pm \sqrt{\frac{3(x^2 - 2x - 3)}{2}}$$

(Graph each curve separately.)

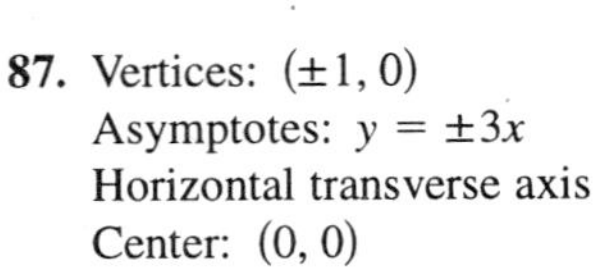

87. Vertices: $(\pm 1, 0)$
Asymptotes: $y = \pm 3x$
Horizontal transverse axis
Center: $(0, 0)$

$a = 1, \pm\frac{b}{a} = \pm\frac{b}{1} = \pm 3 \Rightarrow b = 3$

Therefore, $\frac{x^2}{1} - \frac{y^2}{9} = 1.$

89. Vertices: $(2, \pm 3)$
Point on graph: $(0, 5)$
Vertical transverse axis
Center: $(2, 0)$

$a = 3$

Therefore, the equation is of the form

$$\frac{y^2}{9} - \frac{(x-2)^2}{b^2} = 1.$$

Substituting the coordinates of the point $(0, 5)$, we have

$$\frac{25}{9} - \frac{4}{b^2} = 1 \quad \text{or} \quad b^2 = \frac{9}{4}.$$

Therefore, the equation is $\frac{y^2}{9} - \frac{(x-2)^2}{9/4} = 1.$

91. Center: $(0, 0)$
Vertex: $(0, 2)$
Focus: $(0, 4)$
Vertical transverse axis

$a = 2, c = 4, b^2 = c^2 - a^2 = 12$

Therefore, $\dfrac{y^2}{4} - \dfrac{x^2}{12} = 1$.

93. Vertices: $(0, 2), (6, 2)$

Asymptotes: $y = \dfrac{2}{3}x,\ y = 4 - \dfrac{2}{3}x$

Horizontal transverse axis

Center: $(3, 2)$

$a = 3$

Slopes of asymptotes: $\pm\dfrac{b}{a} = \pm\dfrac{2}{3}$

Thus, $b = 2$. Therefore, $\dfrac{(x-3)^2}{9} - \dfrac{(y-2)^2}{4} = 1$.

95. The transverse axis is horizontal since $(2, 2)$ and $(10, 2)$ are the foci (see definition of hyperbola).

Center: $(6, 2)$

$c = 4, 2a = 6, b^2 = c^2 - a^2 = 7$

Therefore, the equation is $\dfrac{(x-6)^2}{9} - \dfrac{(y-2)^2}{7} = 1$.

97. The transverse axis is vertical since $(-3, 0)$ and $(-3, 3)$ are the foci.

Center: $\left(-3, \dfrac{3}{2}\right)$

$c = \dfrac{3}{2},\ 2a = 2,\ b^2 = c^2 - a^2 = \dfrac{5}{4}$

Therefore, the equation is $\dfrac{[y - (3/2)]^2}{1} - \dfrac{(x+3)^2}{5/4} = 1$.

99. Time for sound of bullet hitting target to reach (x, y): $\dfrac{2c}{v_m} + \dfrac{\sqrt{(x-c)^2 + y^2}}{v_s}$

Time for sound of rifle to reach (x, y): $\dfrac{\sqrt{(x+c)^2 + y^2}}{v_s}$

Since the times are the same, we have:

$$\frac{2c}{v_m} + \frac{\sqrt{(x-c)^2 + y^2}}{v_s} = \frac{\sqrt{(x+c)^2 + y^2}}{v_s}$$

$$\frac{4c^2}{v_m^2} + \frac{4c}{v_m v_s}\sqrt{(x-c)^2 + y^2} + \frac{(x-c)^2 + y^2}{v_s^2} = \frac{(x+c)^2 + y^2}{v_s^2}$$

$$\sqrt{(x-c)^2 + y^2} = \frac{v_m^2 x - v_s^2 c}{v_s v_m}$$

$$\left(1 - \frac{v_m^2}{v_s^2}\right)x^2 + y^2 = \left(\frac{v_s^2}{v_m^2} - 1\right)c^2$$

$$\frac{x^2}{c^2 v_s^2/v_m^2} - \frac{y^2}{c^2(v_m^2 - v_s^2)/v_m^2} = 1$$

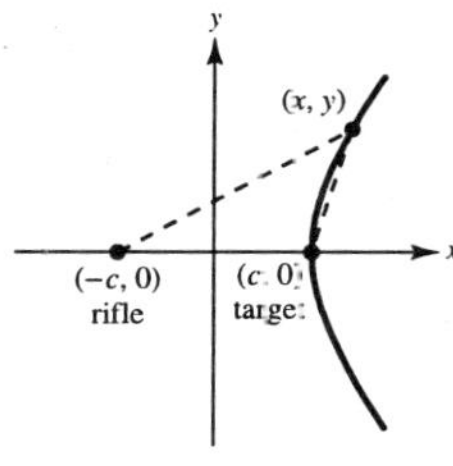

101. The point (x, y) lies on the line between $(0, 10)$ and $(10, 0)$. Thus, $y = 10 - x$. The point also lies on the hyperbola $(x^2/36) - (y^2/64) = 1$. Using substitution, we have:

$$\frac{x^2}{36} - \frac{(10-x)^2}{64} = 1$$

$$16x^2 - 9(10-x)^2 = 576$$

$$7x^2 + 180x - 1476 = 0$$

$$x = \frac{-180 \pm \sqrt{180^2 - 4(7)(-1476)}}{2(7)}$$

$$= \frac{-180 \pm 192\sqrt{2}}{14} = \frac{-90 \pm 96\sqrt{2}}{7}$$

Choosing the positive value for x we have: $x = \dfrac{-90 + 96\sqrt{2}}{7} \approx 6.538$ and $y = \dfrac{160 - 96\sqrt{2}}{7} \approx 3.462$

103. (a) $\frac{y^2}{4} - \frac{x^2}{2} = 1$, $y^2 - 2x^2 = 4$, $2yy' - 4x = 0$,

$y' = \frac{4x}{2y} = \frac{2x}{y}$

At $x = 4$: $y = \pm 6$, $y' = \frac{\pm 2(4)}{6} = \pm\frac{4}{3}$

At $(4, 6)$: $y - 6 = -\frac{4}{3}(x - 4)$ or $4x - 3y + 2 = 0$

At $(4, -6)$: $y + 6 = -\frac{4}{3}(x - 4)$ or $4x + 3y + 2 = 0$

(b) From part (a) we know that the slopes of the normal lines must be $\mp\frac{3}{4}$.

At $(4, 6)$: $y - 6 = -\frac{3}{4}(x - 4)$ or $3x + 4y - 36 = 0$

At $(4, -6)$: $y + 6 = \frac{3}{4}(x - 4)$ or $3x - 4y - 36 = 0$

105.

$$\frac{x^2}{a^2} + \frac{2y^2}{b^2} = 1 \Rightarrow \frac{2y^2}{b^2} = 1 - \frac{x^2}{a^2},\ c^2 = a^2 - b^2$$

$$\frac{x^2}{a^2 - b^2} - \frac{2y^2}{b^2} = 1 \Rightarrow \frac{2y^2}{b^2} = \frac{x^2}{a^2 - b^2} - 1$$

$$1 - \frac{x^2}{a^2} = \frac{x^2}{a^2 - b^2} - 1 \Rightarrow 2 = x^2\left(\frac{1}{a^2} + \frac{1}{a^2 - b^2}\right)$$

$$x^2 = \frac{2a^2(a^2 - b^2)}{2a^2 - b^2} \Rightarrow x = \pm\frac{\sqrt{2}a\sqrt{a^2 - b^2}}{\sqrt{2a^2 - b^2}} = \pm\frac{\sqrt{2}ac}{\sqrt{2a^2 - b^2}}$$

$$\frac{2y^2}{b^2} = 1 - \frac{1}{a^2}\left(\frac{2a^2c^2}{2a^2 - b^2}\right) \Rightarrow \frac{2y^2}{b^2} = \frac{b^2}{2a^2 - b^2}$$

$$y^2 = \frac{b^4}{2(2a^2 - b^2)} \Rightarrow y = \pm\frac{b^2}{\sqrt{2}\sqrt{2a^2 - b^2}}$$

There are four points of intersection: $\left(\frac{\sqrt{2}ac}{\sqrt{2a^2 - b^2}}, \pm\frac{b^2}{\sqrt{2}\sqrt{2a^2 - b^2}}\right)$, $\left(-\frac{\sqrt{2}ac}{\sqrt{2a^2 - b^2}}, \pm\frac{b^2}{\sqrt{2}\sqrt{2a^2 - b^2}}\right)$

$$\frac{x^2}{a^2} + \frac{2y^2}{b^2} = 1 \Rightarrow \frac{2x}{a^2} + \frac{4yy'}{b^2} = 0 \Rightarrow y'_e = -\frac{b^2x}{2a^2y}$$

$$\frac{x^2}{a^2 - b^2} - \frac{2y^2}{b^2} = 1 \Rightarrow \frac{2x}{c^2} - \frac{4yy'}{b^2} = 0 \Rightarrow y'_h = \frac{b^2x}{2c^2y}$$

At $\left(\frac{\sqrt{2}ac}{\sqrt{2a^2 - b^2}}, \frac{b^2}{\sqrt{2}\sqrt{2a^2 - b^2}}\right)$, the slopes of the tangent lines are:

$$y'_e = \frac{-b^2\left(\frac{\sqrt{2}ac}{\sqrt{2a^2 - b^2}}\right)}{2a^2\left(\frac{b^2}{\sqrt{2}\sqrt{2a^2 - b^2}}\right)} = -\frac{c}{a} \quad \text{and} \quad y'_h = \frac{b^2\left(\frac{\sqrt{2}ac}{\sqrt{2a^2 - b^2}}\right)}{2c^2\left(\frac{b^2}{\sqrt{2}\sqrt{2a^2 - b^2}}\right)} = \frac{a}{c}$$

Since the slopes are negative reciprocals, the tangent lines are perpendicular. Similarly, the curves are perpendicular at the other three points of intersection.

107. True

109. False. The y^4 term should be y^2.

111. True

113. $Ax^2 + Cy^2 + Dx + Ey + F = 0$ (Assume $A \neq 0$ and $C \neq 0$; see (b) below)

$$A\left(x^2 + \frac{D}{A}x\right) + C\left(y^2 + \frac{E}{C}y\right) = -F$$

$$A\left(x^2 + \frac{D}{A}x + \frac{D^2}{4A^2}\right) + C\left(y^2 + \frac{E}{C}y + \frac{E^2}{4C^2}\right) = -F + \frac{D^2}{4A} + \frac{E^2}{4C} = R$$

$$\frac{\left[x + \left(\frac{D}{2A}\right)\right]^2}{C} + \frac{\left[y + \left(\frac{E}{2C}\right)\right]^2}{A} = \frac{R}{AC}$$

(a) If $A = C$, we have

$$\left(x + \frac{D}{2A}\right)^2 + \left(y + \frac{E}{2C}\right)^2 = \frac{R}{A}$$

which is the standard equation of a circle.

(b) If $C = 0$, we have

$$A\left(x + \frac{D}{2A}\right)^2 = -F - Ey + \frac{D^2}{4A}.$$

If $A = 0$, we have

$$C\left(y + \frac{E}{2C}\right)^2 = -F - Dx + \frac{E^2}{4C}.$$

These are the equations of parabolas.

(c) If $AC > 0$, we have

$$\frac{\left[x + \left(\frac{D}{2A}\right)\right]^2}{\left|\frac{R}{A}\right|} + \frac{\left[y + \left(\frac{E}{2C}\right)\right]^2}{\left|\frac{R}{C}\right|} = 1$$

which is the equation of an ellipse.

(d) If $AC < 0$, we have

$$\frac{\left[x + \left(\frac{D}{2A}\right)\right]^2}{\left|\frac{R}{A}\right|} - \frac{\left[y + \left(\frac{E}{2C}\right)\right]^2}{\left|\frac{R}{C}\right|} = \pm 1$$

which is the equation of a hyperbola.

115. $4x^2 - y^2 - 4x - 3 = 0$

$A = 4, C = -1$

$AC < 0$

Hyperbola

117. $25x^2 - 10x - 200y - 119 = 0$

$A = 25, C = 0$

Parabola

119. $y^2 - x - 4y - 5 = 0$

$A = 0, C = 1$

Parabola**116.** $y^2 - 4y - 4x = 0$

$A = 0, C = 1$

Parabola

121. $2x^2 - 2xy = 3y - y^2 - 2xy$

$2x^2 + y^2 - 3y = 0$

$A = 2, C = 1, AC > 0$

Ellipse

123. $9x^2 + 54x + 81 = 36 - 4(y^2 - 4y + 4)$

$9x^2 + 4y^2 + 54x - 16y + 61 = 0$

$A = 9, C = 4, AC > 0$

Ellipse

Section 9.2 Plane Curves and Parametric Equations

1. $x = \sqrt{t},\ y = 1 - t$

(a)

t	0	1	2	3	4
x	0	1	$\sqrt{2}$	$\sqrt{3}$	2
y	1	0	-1	-2	-3

(b)

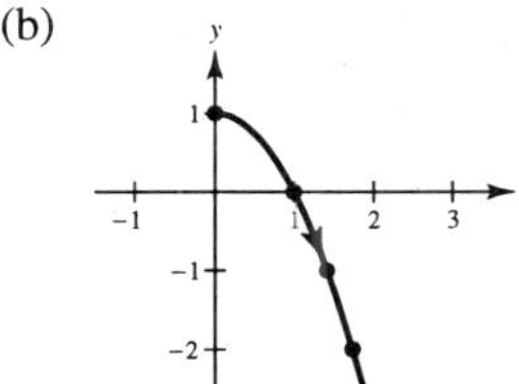

(c)

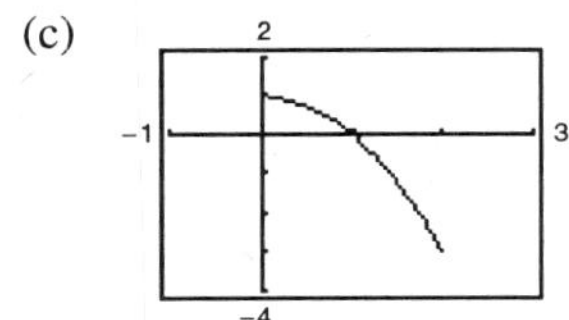

(d) $x^2 = t$

$y = 1 - x^2, x \geq 0$

3. $x = 3t - 1$

$y = 2t + 1$

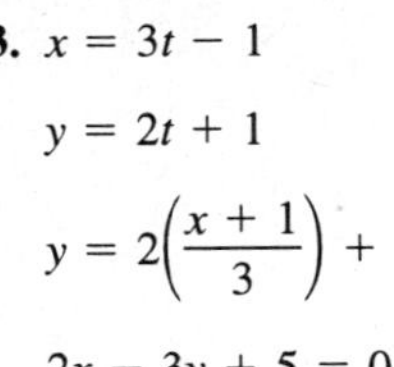

$y = 2\left(\dfrac{x+1}{3}\right) + 1$

$2x - 3y + 5 = 0$

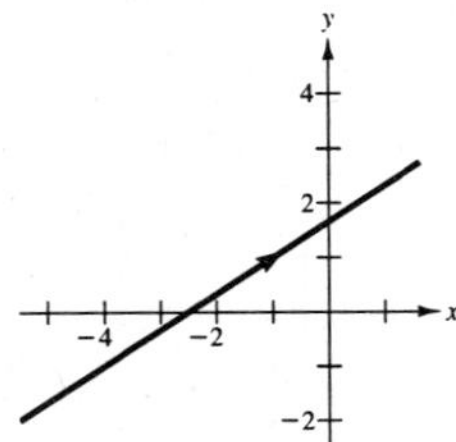

5. $x = t + 1$

$y = t^2$

$y = (x - 1)^2$

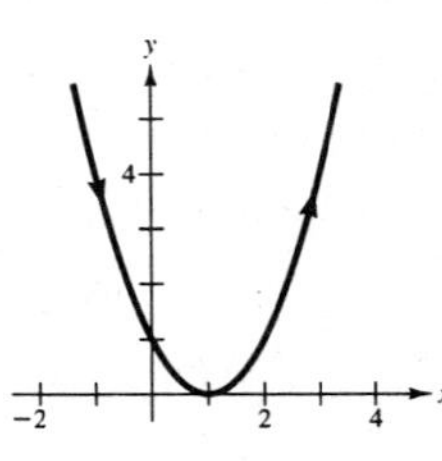

7. $x = t^3$

$y = \frac{1}{2}t^2$

$x = t^3$ implies $t = x^{1/3}$

$y = \frac{1}{2}x^{2/3}$

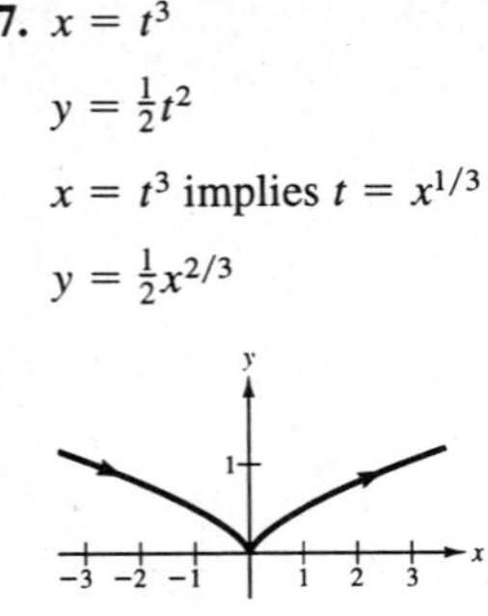

9. $x = t - 1$

$y = \dfrac{t}{t-1}$

$y = \dfrac{x+1}{x}$

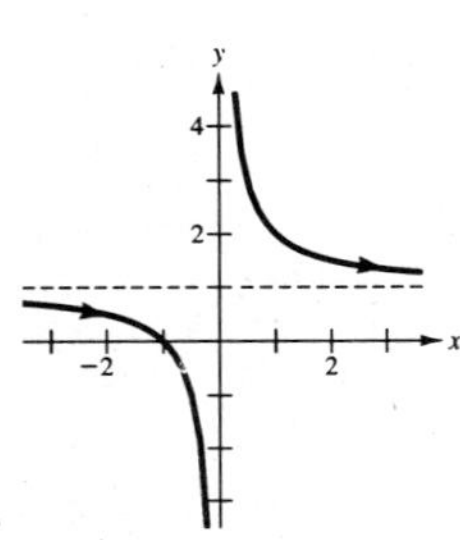

11. $x = 2t$

$y = |t - 2|$

$y = \left|\dfrac{x}{2} - 2\right| = \dfrac{|x-4|}{2}$

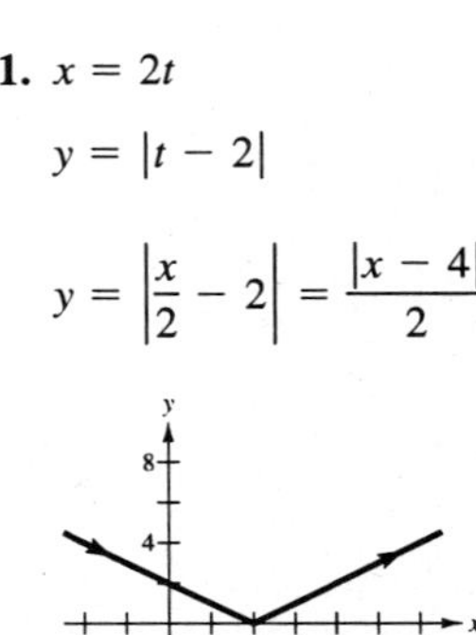

13. $x = \sec\theta$

$y = \cos\theta$

$0 \le \theta < \dfrac{\pi}{2}, \dfrac{\pi}{2} < \theta \le \pi$

$xy = 1$

$y = \dfrac{1}{x}$

$|x| \ge 1, \ |y| \le 1$

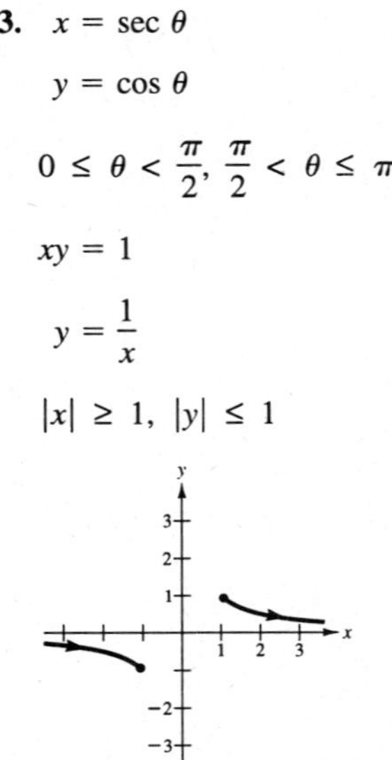

15. $x = 3\cos\theta, \ y = 3\sin\theta$

Squaring both equations and adding, we have

$x^2 + y^2 = 9.$

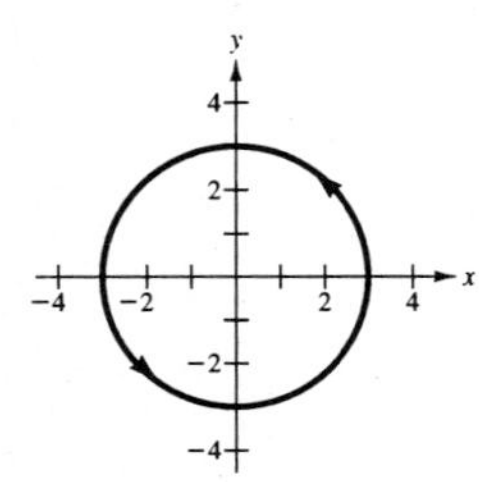

17. $x = 4\sin 2\theta$

$y = 2\cos 2\theta$

$\dfrac{x^2}{16} = \sin^2 2\theta$

$\dfrac{y^2}{4} = \cos^2 2\theta$

$\dfrac{x^2}{16} + \dfrac{y^2}{4} = 1$

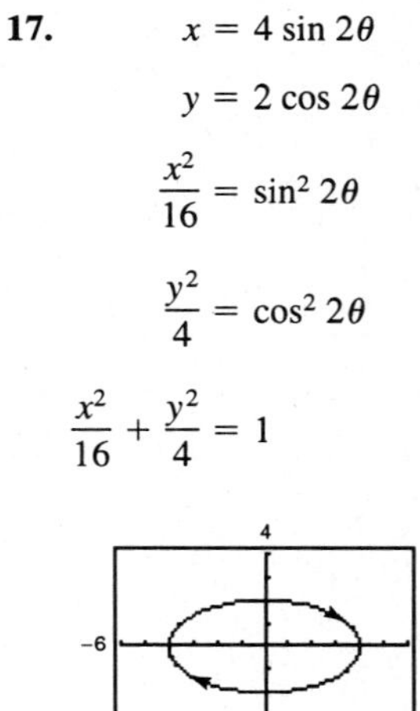

19. $x = 4 + 2\cos\theta$

$y = -1 + \sin\theta$

$\dfrac{(x-4)^2}{4} = \cos^2\theta$

$\dfrac{(y+1)^2}{1} = \sin^2\theta$

$\dfrac{(x-4)^2}{4} + \dfrac{(y+1)^2}{1} = 1$

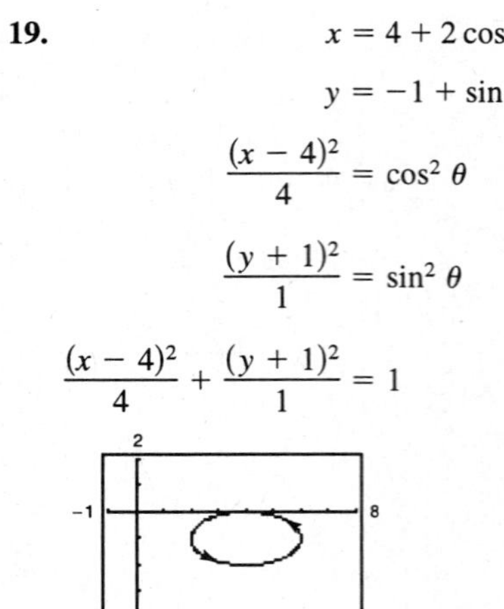

21.

$$x = 4 + 2\cos\theta$$

$$y = -1 + 4\sin\theta$$

$$\frac{(x-4)^2}{4} = \cos^2\theta$$

$$\frac{(y+1)^2}{16} = \sin^2\theta$$

$$\frac{(x-4)^2}{4} + \frac{(y+1)^2}{16} = 1$$

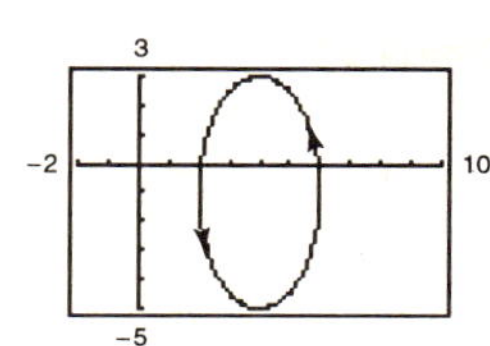

23.

$$x = 4\sec\theta$$

$$y = 3\tan\theta$$

$$\frac{x^2}{16} = \sec^2\theta$$

$$\frac{y^2}{9} = \tan^2\theta$$

$$\frac{x^2}{16} - \frac{y^2}{9} = 1$$

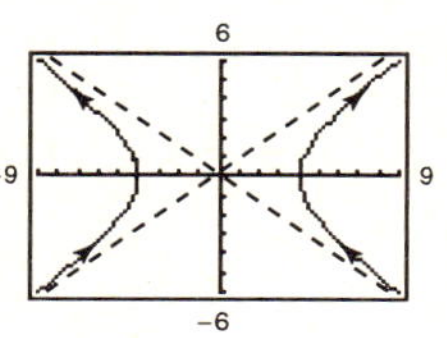

25. $x = t^3$

$y = 3\ln t$

$y = 3\ln\sqrt[3]{x} = \ln x$

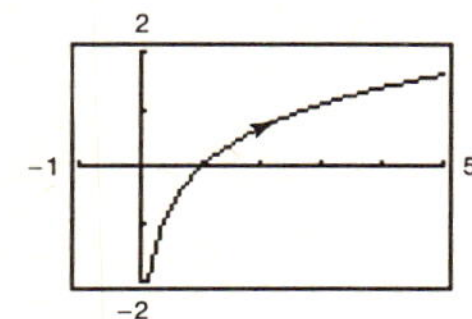

27.

$$x = e^{-t}$$

$$y = e^{3t}$$

$$e^t = \frac{1}{x}$$

$$e^t = \sqrt[3]{y}$$

$$\sqrt[3]{y} = \frac{1}{x}$$

$$y = \frac{1}{x^3}$$

$$x > 0$$

$$y > 0$$

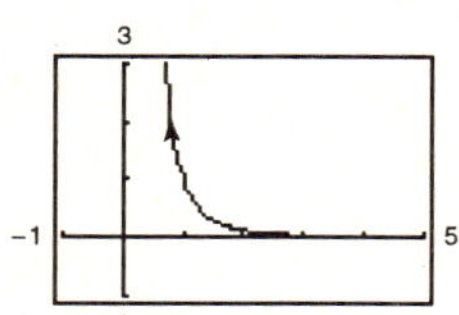

29. By eliminating the parameters in (a) – (d), we get $y = 2x + 1$. They differ from each other in orientation and in restricted domains. These curves are all smooth except for (b).

(a) $x = t,\ y = 2t + 1$

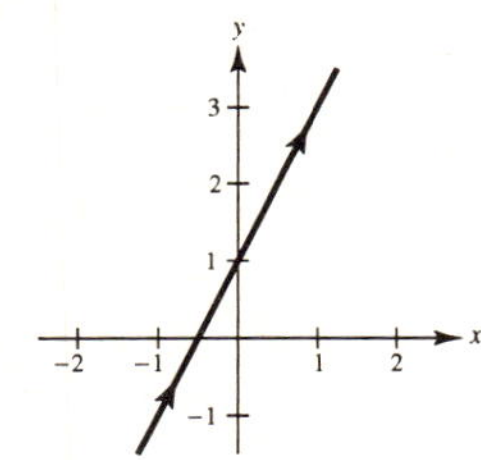

(b) $x = \cos\theta \qquad y = 2\cos\theta + 1$

$-1 \le x \le 1 \qquad -1 \le y \le 3$

$$\frac{dx}{d\theta} = \frac{dy}{d\theta} = 0 \text{ when } \theta = 0, \pm\pi, \pm 2\pi, \ldots$$

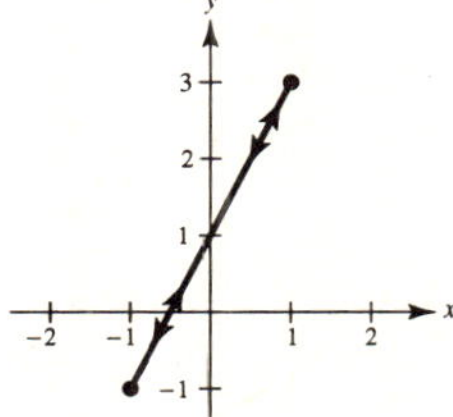

(c) $x = e^{-t} \qquad y = 2e^{-t} + 1$

$x > 0 \qquad y > 1$

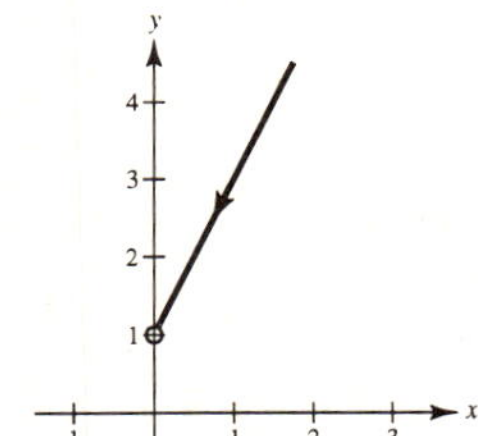

(d) $x = e^t \qquad y = 2e^t + 1$

$x > 0 \qquad y > 1$

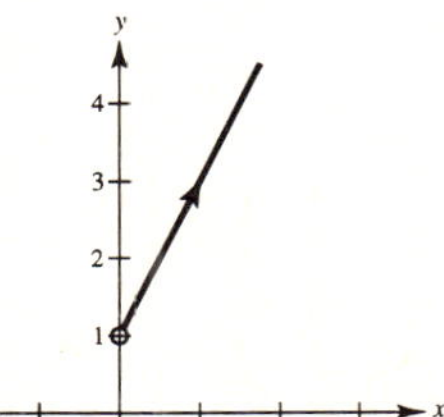

31. The curves are identical on $0 < \theta < \pi$. They are both smooth.

33. (a)

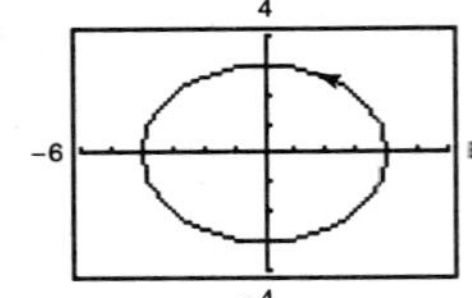

(b) The orientation of the second curve is reversed.

(c) The orientation will be reversed.

(d) Many answers possible. For example, $x = 1 + t$, $y = 1 + 2t$, and $x = 1 - t$, $x = 1 - 2t$.

35.

$$x = x_1 + t(x_2 - x_1)$$
$$y = y_1 + t(y_2 - y_1)$$
$$\frac{x - x_1}{x_2 - x_1} = t$$
$$y = y_1 + \left(\frac{x - x_1}{x_2 - x_1}\right)(y_2 - y_1)$$
$$y - y_1 = \frac{y_2 - y_1}{x_2 - x_1}(x - x_1)$$
$$y - y_1 = m(x - x_1)$$

37.

$$x = h + a\cos\theta$$
$$y = k + b\sin\theta$$
$$\frac{x - h}{a} = \cos\theta$$
$$\frac{y - k}{b} = \sin\theta$$
$$\frac{(x - h)^2}{a^2} + \frac{(y - k)^2}{b^2} = 1$$

39. From Exercise 35 we have

$x = 5t$

$y = -2t.$

Solution not unique

41. From Exercise 36 we have

$x = 2 + 4\cos\theta$

$y = 1 + 4\sin\theta.$

Solution not unique

43. From Exercise 37 we have

$a = 5, c = 4 \Rightarrow b = 3$

$x = 5\cos\theta$

$y = 3\sin\theta.$

Center: $(0, 0)$
Solution not unique

45. From Exercise 38 we have

$a = 4, c = 5 \Rightarrow b = 3$

$x = 4\sec\theta$

$y = 3\tan\theta.$

Center: $(0, 0)$
Solution not unique

47. $y = 3x - 2$

Example

$x = t, \quad y = 3t - 2$

$x = t - 3, \quad y = 3t - 11$

49. $y = x^3$

Example

$x = t, \quad y = t^3$

$x = \sqrt[3]{t}, \quad y = t$

$x = \tan t, \quad y = \tan^3 t$

51. $x = 2(\theta - \sin\theta)$

$y = 2(1 - \cos\theta)$

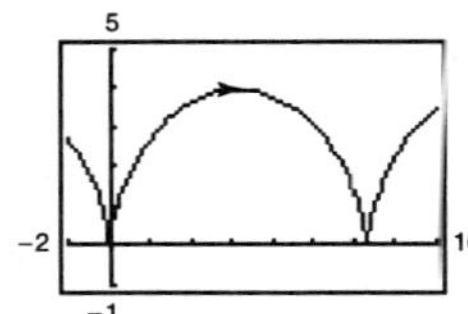

Not smooth at $\theta = 2n\pi$

53. $x = \theta - \frac{3}{2}\sin\theta$

$y = 1 - \frac{3}{2}\cos\theta$

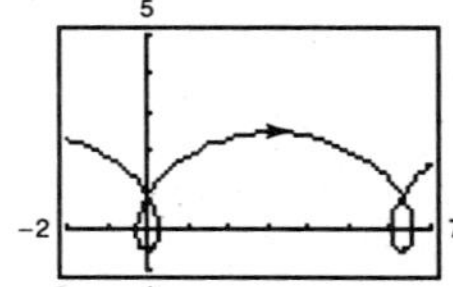

55. $x = 3\cos^3\theta$

$y = 3\sin^3\theta$

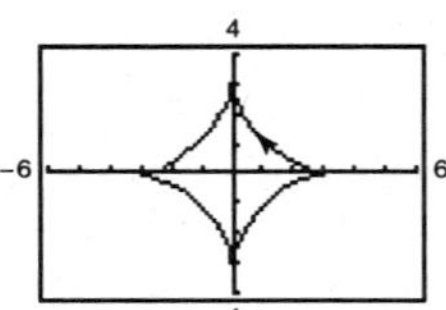

Not smooth at $(x, y) = (\pm 3, 0)$ and $(0, \pm 3)$, or $\theta = \frac{1}{2}n\pi$.

57. $x = 2 \cot \theta$

$y = 2 \sin^2 \theta$

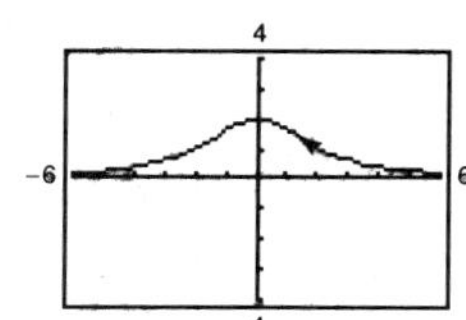

Smooth everywhere

59. $x = 4 \cos \theta$

$y = 2 \sin 2\theta$

Matches (d)

61. $x = \cos \theta + \theta \sin \theta$

$y = \sin \theta - \theta \cos \theta$

Matches (b)

63. When the circle has rolled θ radians, we know that the center is at $(a\theta, a)$.

$$\sin \theta = \sin(180° - \theta) = \frac{|AC|}{b} = \frac{|BD|}{b} \quad \text{or} \quad |BD| = b \sin \theta$$

$$\cos \theta = -\cos(180° - \theta) = \frac{|AP|}{-b} \quad \text{or} \quad |AP| = -b \cos \theta$$

Therefore, $x = a\theta - b \sin \theta$ and $y = a - b \cos \theta$.

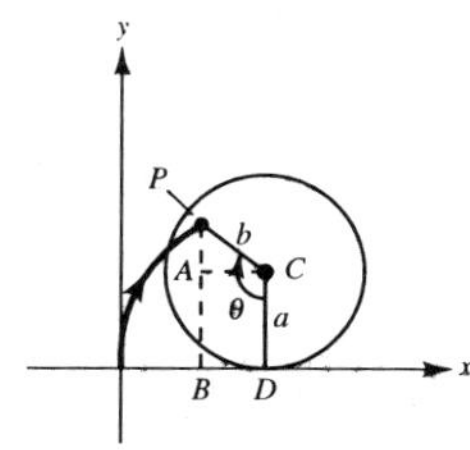

65. True

67. False. Let $x = t^2$ and $y = t$. Then $x = y^2$ and y is not a function of x.

69. (a) $100 \text{ mi/hr} = \dfrac{(100)(5280)}{3600} = \dfrac{440}{3} \text{ ft/sec}$

$$x = (v_0 \cos \theta)t = \left(\frac{440}{3} \cos \theta\right)t$$

$$y = h + (v_0 \sin \theta)t - 16t^2$$

$$= 3 + \left(\frac{440}{3} \sin \theta\right)t - 16t^2$$

(b)

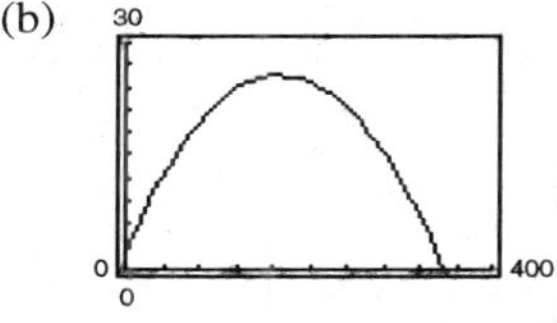

It is not a home run—when $x = 400$, $y \le 20$.

(c)

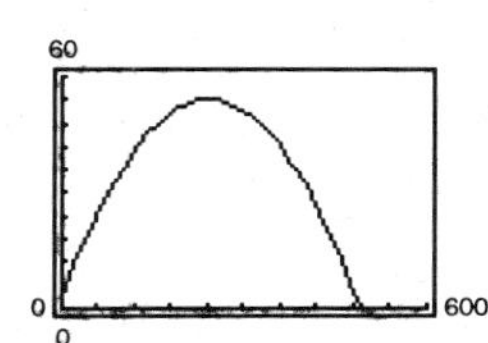

Yes, it's a home run when $x = 400$, $y > 10$.

(d) We need to find the angle θ (and time t) such that

$$x = \left(\frac{440}{3} \cos \theta\right)t = 400$$

$$y = 3 + \left(\frac{440}{3} \sin \theta\right)t - 16t^2 = 10.$$

From the first equation $t = 1200/440 \cos \theta$. Substituting into the second equation,

$$10 = 3 + \left(\frac{440}{3} \sin \theta\right)\left(\frac{1200}{440 \cos \theta}\right) - 16\left(\frac{1200}{440 \cos \theta}\right)^2$$

$$7 = 400 \tan \theta - 16\left(\frac{120}{44}\right)^2 \sec^2 \theta = 400 \tan \theta - 16\left(\frac{120}{44}\right)^2(\tan^2 \theta + 1).$$

We now solve the quadratic for $\tan \theta$:

$$16\left(\frac{120}{44}\right)^2 \tan^2 \theta - 400 \tan \theta + 7 + 16\left(\frac{120}{44}\right)^2 = 0$$

$\tan \theta \approx 0.35185 \implies \theta \approx 19.4°$

71. $x = \dfrac{1 - t^2}{1 + t^2}$ and $y = \dfrac{2t}{1 + t^2}$

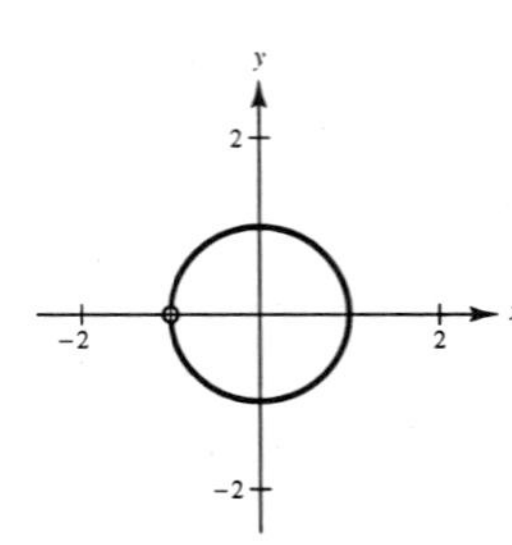

The graph is the circle $x^2 + y^2 = 1$ except the point $(-1, 0)$. Thus,

$$\left(\frac{1 - t^2}{1 + t^2}\right)^2 + \left(\frac{2t}{1 + t^2}\right)^2 = 1$$

$$(1 - t^2)^2 + (2t)^2 = (1 + t^2)^2.$$

When $t = 2$: $(-3)^2 + (4)^2 = (5)^2$

When $t = 3$: $(-8)^2 + (6)^2 = (10)^2$

Section 9.3 Parametric Equations and Calculus

1. $x = 2t,\ y = 3t - 1$

$$\frac{dy}{dx} = \frac{dy/dt}{dx/dt} = \frac{3}{2}$$

$$\frac{d^2y}{dx^2} = 0$$

3. $x = t + 1,\ y = t^2 + 3t$

$$\frac{dy}{dx} = \frac{2t + 3}{1} = 1 \text{ when } t = -1.$$

$$\frac{d^2y}{dx^2} = 2$$

5. $x = 2\cos\theta,\ y = 2\sin\theta$

$$\frac{dy}{dx} = \frac{2\cos\theta}{-2\sin\theta} = -\cot\theta = -1 \text{ when } \theta = \frac{\pi}{4}.$$

$$\frac{d^2y}{dx^2} = \frac{\csc^2\theta}{-2\sin\theta} = \frac{-\csc^3\theta}{2} = -\sqrt{2} \text{ when } \theta = \frac{\pi}{4}.$$

7. $x = 2 + \sec\theta,\ y = 1 + 2\tan\theta$

$$\frac{dy}{dx} = \frac{2\sec^2\theta}{\sec\theta\tan\theta}$$

$$= \frac{2\sec\theta}{\tan\theta} = 2\csc\theta = 4 \text{ when } \theta = \frac{\pi}{6}.$$

$$\frac{d^2y}{dx^2} = \frac{-2\csc\theta\cot\theta}{\sec\theta\tan\theta}$$

$$= -2\cot^3\theta = -6\sqrt{3} \text{ when } \theta = \frac{\pi}{6}.$$

9. $x = \cos^3\theta,\ y = \sin^3\theta$

$$\frac{dy}{dx} = \frac{3\sin^2\theta\cos\theta}{-3\cos^2\theta\sin\theta}$$

$$= -\tan\theta = -1 \text{ when } \theta = \frac{\pi}{4}.$$

$$\frac{d^2y}{dx^2} = \frac{-\sec^2\theta}{-3\cos^2\theta\sin\theta} = \frac{1}{3\cos^4\theta\sin\theta}$$

$$= \frac{\sec^4\theta\csc\theta}{3} = \frac{4\sqrt{2}}{3} \text{ when } \theta = \frac{\pi}{4}.$$

11. $x = 2\cot\theta,\ y = 2\sin^2\theta$

$$\frac{dy}{dx} = \frac{4\sin\theta\cos\theta}{-2\csc^2\theta} = -2\sin^3\theta\cos\theta$$

At $\left(-\dfrac{2}{\sqrt{3}}, \dfrac{3}{2}\right)$, $\theta = \dfrac{2\pi}{3}$, and $\dfrac{dy}{dx} = \dfrac{3\sqrt{3}}{8}$.

Tangent line: $y - \dfrac{3}{2} = \dfrac{3\sqrt{3}}{8}\left(x + \dfrac{2}{\sqrt{3}}\right)$

$$3\sqrt{3}x - 8y + 18 = 0$$

At $(0, 2)$, $\theta = \dfrac{\pi}{2}$, and $\dfrac{dy}{dx} = 0$.

Tangent line: $y - 2 = 0$

At $\left(2\sqrt{3}, \dfrac{1}{2}\right)$, $\theta = \dfrac{\pi}{6}$, and $\dfrac{dy}{dx} = -\dfrac{\sqrt{3}}{8}$.

Tangent line: $y - \dfrac{1}{2} = -\dfrac{\sqrt{3}}{8}(x - 2\sqrt{3})$

$$\sqrt{3}x + 8y - 10 = 0$$

13. $x = 2t,\ y = t^2 - 1,\ t = 2$

(a)

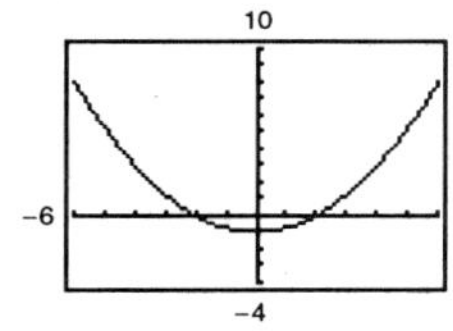

(b) At $t = 2$, $(x, y) = (4, 3)$, and $\dfrac{dx}{dt} = 2,\ \dfrac{dy}{dt} = 4,\ \dfrac{dy}{dx} = 2$

(c) $\dfrac{dy}{dx} = 2$. At $(4, 3)$, $y - 3 = 2(x - 4)$

$$y = 2x - 5$$

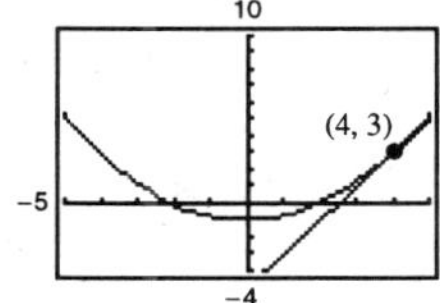

15. $x = t^2 - t + 2,\ y = t^3 - 3t,\ t = -1$

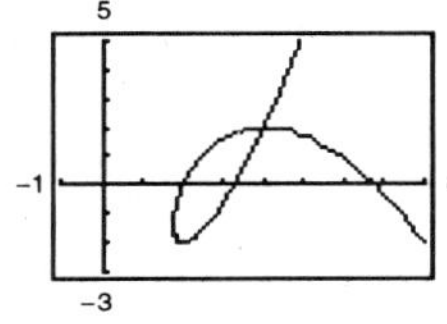

(b) At $t = -1$, $(x, y) = (4, 2)$, and

$$\frac{dx}{dt} = -3,\ \frac{dy}{dt} = 0,\ \frac{dy}{dx} = 0$$

(c) $\dfrac{dy}{dx} = 0$. At $(4, 2)$, $y - 2 = 0(x - 4)$

$$y = 2$$

(d)

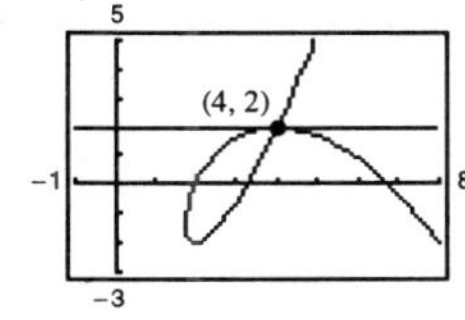

17. $x = \cos\theta + \theta\sin\theta,\ y = \sin\theta - \theta\cos\theta$

Horizontal tangents: $\dfrac{dy}{d\theta} = \theta\sin\theta = 0$ when $\theta = 0,\ \pi,\ 2\pi,\ 3\pi, \ldots.$

Points: $(-1, [2n - 1]\pi),\ (1, 2n\pi)$ where n is an integer.

Points shown: $(1, 0),\ (-1, \pi),\ (1, -2\pi)$

Vertical tangents: $\dfrac{dx}{d\theta} = \theta\cos\theta = 0$ when $\theta = \dfrac{\pi}{2}, \dfrac{3\pi}{2}, \dfrac{5\pi}{2}, \ldots.$

Points: $\left(\dfrac{(-1)^{n+1}(2n - 1)\pi}{2}, (-1)^{n+1}\right)$

Points shown: $\left(\dfrac{\pi}{2}, 1\right),\ \left(-\dfrac{3\pi}{2}, -1\right),\ \left(\dfrac{5\pi}{2}, 1\right)$

19. $x = 1 - t,\ y = t^2$

Horizontal tangents: $\dfrac{dy}{dt} = 2t = 0$ when $t = 0$.

Point: $(1, 0)$

Vertical tangents: $\dfrac{dx}{dt} = -1 \neq 0$; none

21. $x = 1 - t,\ y = t^3 - 3t$

Horizontal tangents: $\dfrac{dy}{dt} = 3t^2 - 3 = 0$ when $t = \pm 1$.

Points: $(0, -2),\ (2, 2)$

Vertical tangents: $\dfrac{dx}{dt} = -1 \neq 0$; none

23. $x = 3\cos\theta,\ y = 3\sin\theta$

Horizontal tangents: $\dfrac{dy}{d\theta} = 3\cos\theta = 0$ when $\theta = \dfrac{\pi}{2}, \dfrac{3\pi}{2}$.

Points: $(0, 3),\ (0, -3)$

Vertical tangents: $\dfrac{dx}{d\theta} = -3\sin\theta = 0$ when $\theta = 0, \pi$.

Points: $(3, 0),\ (-3, 0)$

25. $x = 4 + 2\cos\theta,\ y = -1 + \sin\theta$

Horizontal tangents: $\dfrac{dy}{d\theta} = \cos\theta = 0$ when $\theta = \dfrac{\pi}{2}, \dfrac{3\pi}{2}$.

Points: $(4, 0),\ (4, -2)$

Vertical tangents: $\dfrac{dx}{d\theta} = -2\sin\theta = 0$ when $x = 0, \pi$.

Points: $(6, -1),\ (2, -1)$

27. $x = \sec\theta,\ y = \tan\theta$

Horizontal tangents: $\dfrac{dy}{d\theta} = \sec^2\theta \neq 0$; none

Vertical tangents: $\dfrac{dx}{d\theta} = \sec\theta\tan\theta = 0$ when $x = 0, \pi$.

Points: $(1, 0), (-1, 0)$

29. One possible answer is the graph given by

$x = t,\ y = -t.$

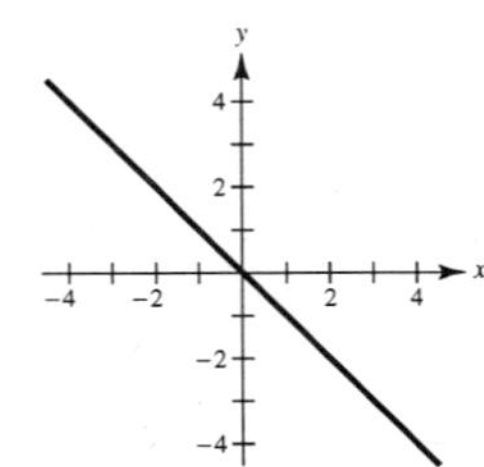

31. $x = e^{-t}\cos t,\ y = e^{-t}\sin t,\ 0 \le t \le \dfrac{\pi}{2}$

$$\frac{dx}{dt} = -e^{-t}(\sin t + \cos t),\ \frac{dy}{dt} = e^{-t}(\cos t - \sin t)$$

$$s = \int_0^{\pi/2}\sqrt{\left(\frac{dx}{dt}\right)^2 + \left(\frac{dy}{dt}\right)^2}\,dt$$

$$= \int_0^{\pi/2}\sqrt{2e^{-2t}}\,dt = -\sqrt{2}\int_0^{\pi/2} e^{-t}(-1)\,dt$$

$$= \left[-\sqrt{2}e^{-t}\right]_0^{\pi/2} = \sqrt{2}(1 - e^{-\pi/2}) \approx 1.12$$

33. $x = t^2,\ y = 2t,\ 0 \le t \le 2$

$$\frac{dx}{dt} = 2t,\ \frac{dy}{dt} = 2,\ \left(\frac{dx}{dt}\right)^2 + \left(\frac{dy}{dt}\right)^2 = 4t^2 + 4 = 4(t^2 + 1)$$

$$s = 2\int_0^2 \sqrt{t^2 + 1}\,dt$$

$$= \left[t\sqrt{t^2 + 1} + \ln\left|t + \sqrt{t^2 + 1}\right|\right]_0^2$$

$$= 2\sqrt{5} + \ln\left(2 + \sqrt{5}\right) \approx 5.916$$

35. $x = \sqrt{t},\ y = 3t - 1,\ \dfrac{dx}{dt} = \dfrac{1}{2\sqrt{t}},\ \dfrac{dy}{dt} = 3$

$$S = \int_0^1\sqrt{\frac{1}{4t} + 9}\,dt = \frac{1}{2}\int_0^1 \frac{\sqrt{1 + 36t}}{\sqrt{t}}\,dt$$

$$= \frac{1}{6}\int_0^6 \sqrt{1 + u^2}\,du$$

$$= \frac{1}{12}\left[\ln\left(\sqrt{1 + u^2} + u\right) + u\sqrt{1 + u^2}\right]_0^6$$

$$= \frac{1}{12}\left[\ln\left(\sqrt{37} + 6\right) + 6\sqrt{37}\right] \approx 3.249$$

$u = 6\sqrt{t},\ du = \dfrac{3}{\sqrt{t}}\,dt$

37. $x = a\cos^3\theta,\ y = a\sin^3\theta,\ \dfrac{dx}{d\theta} = -3a\cos^2\theta\sin\theta,$

$$\frac{dy}{d\theta} = 3a\sin^2\theta\cos\theta$$

$$S = 4\int_0^{\pi/2}\sqrt{9a^2\cos^4\theta\sin^2 + 9a^2\sin^4\theta\cos^2\theta}\,d\theta$$

$$= 12a\int_0^{\pi/2}\sin\theta\cos\theta\sqrt{\cos^2\theta + \sin^2\theta}\,d\theta$$

$$= 6a\int_0^{\pi/2}\sin 2\theta\,d\theta = \left[-3a\cos 2\theta\right]_0^{\pi/2} = 6a$$

39. $x = a(\theta - \sin\theta),\ y = a(1 - \cos\theta),\ \dfrac{dx}{d\theta} = a(1 - \cos\theta),$

$$\frac{dy}{d\theta} = a\sin\theta$$

$$S = 2\int_0^{\pi}\sqrt{a^2(1 - \cos\theta)^2 + a^2\sin^2\theta}\,d\theta$$

$$= 2\sqrt{2}a\int_0^{\pi}\sqrt{1 - \cos\theta}\,d\theta$$

$$= 2\sqrt{2}a\int_0^{\pi}\frac{\sin\theta}{\sqrt{1 + \cos\theta}}\,d\theta$$

$$= \left[-4\sqrt{2}a\sqrt{1 + \cos\theta}\right]_0^{\pi} = 8a$$

41. $x = (90\cos 30°)t,\ y = (90\sin 30°)t - 16t^2$

(a) 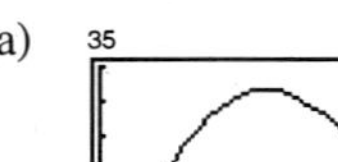

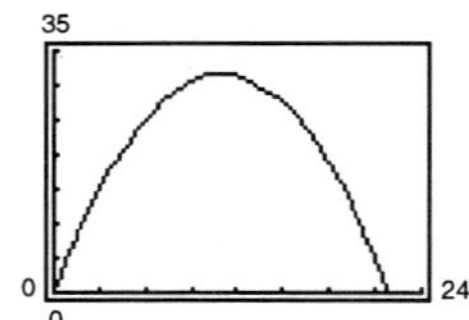

(b) Range: 219.2 ft

(c) $\dfrac{dx}{dt} = 90\cos 30°,\ \dfrac{dy}{dt} = 90\sin 30° - 32t.\ y = 0$

for $t = \dfrac{45}{16}.$

$$s = \int_0^{45/16}\sqrt{(90\cos 30°)^2 + (90\sin 30° - 32t)^2}\,dt$$

$$= 230.8 \text{ ft}$$

43. (a) $x = t - \sin t$ $\quad$ $x = 2t - \sin(2t)$

$y = 1 - \cos t$ $\quad$ $y = 1 - \cos(2t)$

$0 \le t \le 2\pi$ $\quad$ $0 \le t \le \pi$

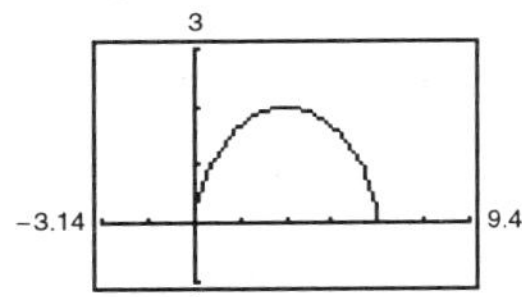

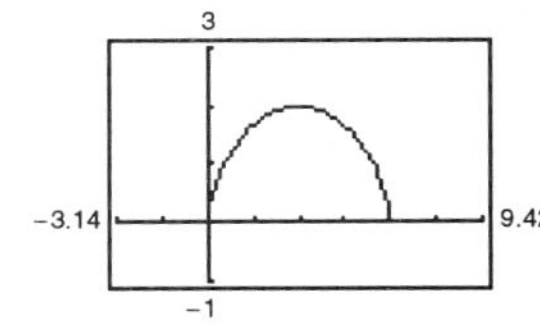

(b) The average speed of the particle on the second path is twice the average speed of a particle on the first path.

(c) $x = \frac{1}{2}t - \sin\left(\frac{1}{2}t\right)$

$y = 1 - \cos\left(\frac{1}{2}t\right)$

The time required for the particle to traverse the same path is $t = 4\pi$.

45. $x = t,\ y = 2t,\ \dfrac{dx}{dt} = 1,\ \dfrac{dy}{dt} = 2$

(a) $S = 2\pi\displaystyle\int_0^4 2t\sqrt{1+4}\,dt = 4\sqrt{5}\pi\int_0^4 t\,dt$

$= \left[2\sqrt{5}\pi t^2\right]_0^4 = 32\pi\sqrt{5}$

(b) $S = 2\pi\displaystyle\int_0^4 t\sqrt{1+4}\,dt = 2\sqrt{5}\pi\int_0^4 t\,dt$

$= \left[\sqrt{5}\pi t^2\right]_0^4 = 16\pi\sqrt{5}$

47. $x = 4\cos\theta,\ y = 4\sin\theta,\ \dfrac{dx}{d\theta} = -4\sin\theta,\ \dfrac{dy}{d\theta} = 4\cos\theta$

$$S = 2\pi\int_0^{\pi/2} 4\cos\theta\sqrt{(-4\sin\theta)^2 + (4\cos\theta)^2}\,d\theta$$

$$= 32\pi\int_0^{\pi/2}\cos\theta\,d\theta = \left[32\pi\sin\theta\right]_0^{\pi/2} = 32\pi$$

49. $x = a\cos^3\theta,\ y = a\sin^3\theta,\ \dfrac{dx}{d\theta} = -3a\cos^2\theta\sin\theta,\ \dfrac{dy}{d\theta} = 3a\sin^2\theta\cos\theta$

$$S = 4\pi\int_0^{\pi/2} a\sin^3\theta\sqrt{9a^2\cos^4\theta\sin^2\theta + 9a^2\sin^4\theta\cos^2\theta}\,d\theta = 12a^2\pi\int_0^{\pi/2}\sin^4\theta\cos\theta\,d\theta = \frac{12\pi a^2}{5}\left[\sin^5\theta\right]_0^{\pi/2} = \frac{12}{5}\pi a^2$$

51. $x = r\cos\phi,\ y = r\sin\phi$

$$S = 2\pi\int_0^{\theta} r\sin\phi\sqrt{r^2\sin^2\phi + r^2\cos^2\phi}\,d\phi$$

$$= 2\pi r^2\int_0^{\theta}\sin\phi\,d\phi$$

$$= \left[-2\pi r^2\cos\phi\right]_0^{\theta} = 2\pi r^2(1-\cos\theta)$$

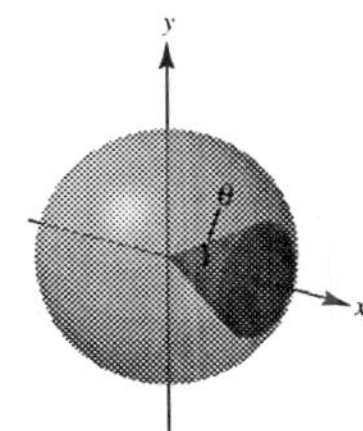

53. $x = \sqrt{t},\ y = 4 - t,\ 0 \le t \le 4$

$$A = \int_0^4 (4-t)\frac{1}{2\sqrt{t}}\,dt = \frac{1}{2}\int_0^4 (4t^{-1/2} - t^{1/2})\,dt = \left[\frac{1}{2}\left(8\sqrt{t} - \frac{2}{3}t\sqrt{t}\right)\right]_0^4 = \frac{16}{3}$$

$$\bar{x} = \frac{3}{16}\int_0^4 (4-t)\sqrt{t}\left(\frac{1}{2\sqrt{t}}\right)dt = \frac{3}{32}\int_0^4 (4-t)\,dt = \left[\frac{3}{32}\left(4t - \frac{t^2}{2}\right)\right]_0^4 = \frac{3}{4}$$

$$\bar{y} = \frac{3}{32}\int_0^4 (4-t)^2\frac{1}{2\sqrt{t}}\,dt = \frac{3}{64}\int_0^4 \left[16t^{-1/2} - 8t^{1/2} + t^{3/2}\right]dt = \frac{3}{64}\left[32\sqrt{t} - \frac{16}{3}t\sqrt{t} + \frac{2}{5}t^2\sqrt{t}\right]_0^4 = \frac{8}{5}$$

$$(\bar{x}, \bar{y}) = \left(\frac{3}{4}, \frac{8}{5}\right)$$

55. $x = 3\cos\theta,\ y = 3\sin\theta,\ \dfrac{dx}{d\theta} = -3\sin\theta$

$$V = 2\pi\int_{\pi/2}^{0}(3\sin\theta)^2(-3\sin\theta)\,d\theta = -54\pi\int_{\pi/2}^{0}\sin^3\theta\,d\theta = -54\pi\int_{\pi/2}^{0}(1-\cos^2\theta)\sin\theta\,d\theta$$

$$= -54\pi\left[-\cos\theta + \frac{\cos^3\theta}{3}\right]_{\pi/2}^{0} = 36\pi$$

57. $x = 2\sin^2\theta$

$y = 2\sin^2\theta\tan\theta$

$\dfrac{dx}{d\theta} = 4\sin\theta\cos\theta$

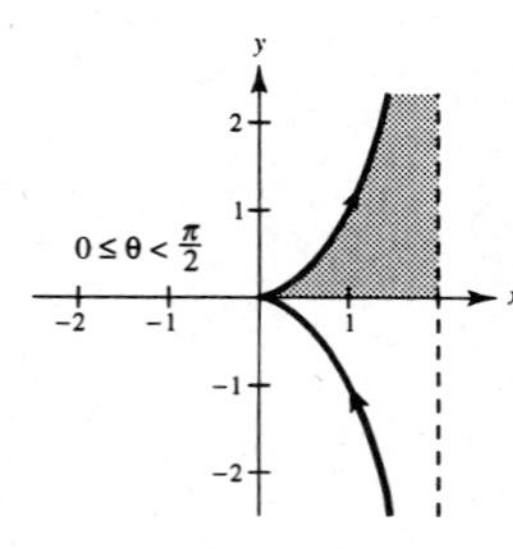

$$A = \int_0^{\pi/2} 2\sin^2\theta\tan\theta(4\sin\theta\cos\theta)\,d\theta = 8\int_0^{\pi/2}\sin^4\theta\,d\theta$$

$$= 8\left[\frac{-\sin^3\theta\cos\theta}{4} - \frac{3}{8}\sin\theta\cos\theta + \frac{3}{8}\theta\right]_0^{\pi/2} = \frac{3\pi}{2}$$

59. πab is area of ellipse (d).

61. $6\pi a^2$ is area of cardioid (f).

63. $\frac{8}{3}ab$ is area of hourglass (a).

65. (a) $x = \dfrac{1-t^2}{1+t^2},\ y = \dfrac{2t}{1+t^2},\ -20 \le t \le 20$

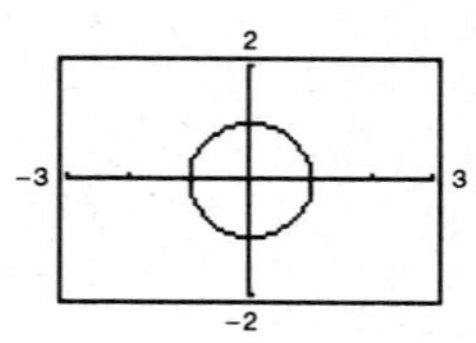

The graph is the circle $x^2 + y^2 = 1$, except the point $(-1, 0)$.

Verify: $x^2 + y^2 = \left(\dfrac{1-t^2}{1+t^2}\right)^2 + \left(\dfrac{2t}{1+t^2}\right)^2 = \dfrac{1 - 2t^2 + t^4 + 4t^2}{(1+t^2)^2} = \dfrac{(1+t^2)^2}{(1+t^2)^2} = 1$

(b) As t increases from -20 to 0, the speed increases, and as t increases from 0 to 20, the speed decreases.

67. (a) The first plane makes an angle of 70° with the positive x-axis, and is 150 miles from P:

$x_1 = \cos 70°(150 - 375t)$

$y_1 = \sin 70°(150 - 375t)$

Similarly for the second plane,

$x_2 = \cos 135°(190 - 450t)$

$\quad = \cos 45°(-190 + 450t)$

$y_2 = \sin 135°(190 - 450t)$

$\quad = \sin 45°(190 - 450t)$

(b) $d = \sqrt{(x_2 - x_1)^2 + (y_2 - y_1)^2}$

$= [[\cos 45(-190 + 450t) - \cos 70(150 - 375t)]^2 + [\sin 45(190 - 450t) - \sin 70(150 - 375t)]^2]^{1/2}$

(c)

The minimum distance is 7.59 miles when $t = 0.4145$.

69. False

$$\frac{d^2y}{dx^2} = \frac{\dfrac{d}{dt}\left[\dfrac{g'(t)}{f'(t)}\right]}{f'(t)} = \frac{f'(t)g''(t) - g'(t)f''(t)}{[f'(t)]^3}$$

Section 9.4 Polar Coordinates and Polar Graphs

1. $\left(4, \frac{\pi}{2}\right)$

$x = 4\cos\left(\frac{\pi}{2}\right) = 0$

$y = 4\sin\left(\frac{\pi}{2}\right) = 4$

$(x, y) = (0, 4)$

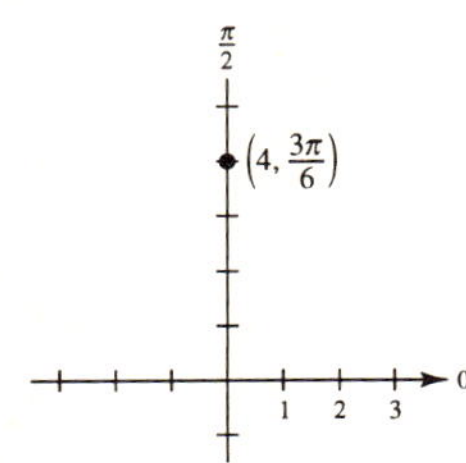

3. $\left(-4, -\frac{\pi}{3}\right)$

$x = -4\cos\left(-\frac{\pi}{3}\right) = -2$

$y = -4\sin\left(-\frac{\pi}{3}\right) = 2\sqrt{3}$

$(x, y) = \left(-2, 2\sqrt{3}\right)$

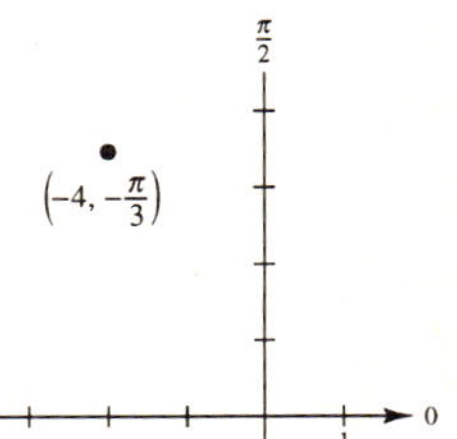

5. $\left(\sqrt{2}, 2.36\right)$

$x = \sqrt{2}\cos(2.36) \approx -1.004$

$y = \sqrt{2}\sin(2.36) \approx 0.996$

$(x, y) = (-1.004, 0.996)$

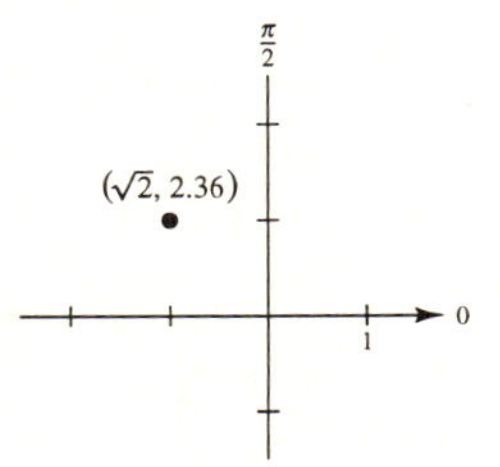

7. $(r, \theta) = \left(5, \frac{3\pi}{4}\right)$

$(x, y) = (-3.5355, 3.5355)$

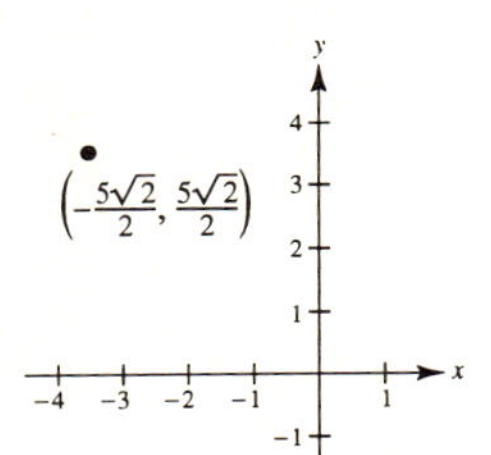

9. $(r, \theta) = (-3.5, 2.5)$

$(x, y) = (2.804, -2.095)$

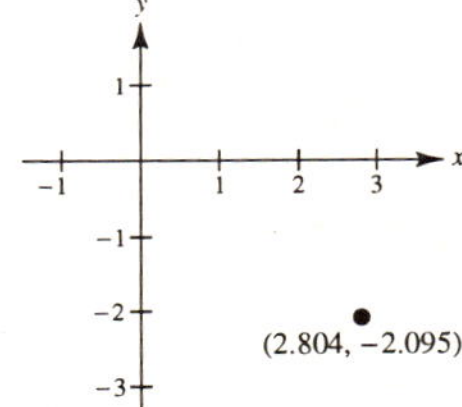

11. $(x, y) = (1, 1)$

$r = \pm\sqrt{2}$

$\tan\theta = 1$

$\theta = \frac{\pi}{4}, \frac{5\pi}{4}, \left(\sqrt{2}, \frac{\pi}{4}\right), \left(-\sqrt{2}, \frac{5\pi}{4}\right)$

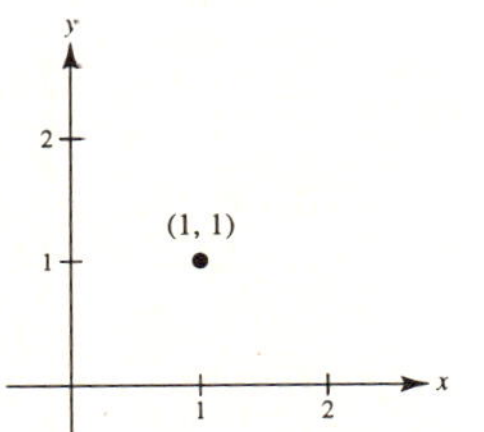

13. $(x, y) = (-3, 4)$

$r = \pm\sqrt{9 + 16} = \pm 5$

$\tan\theta = -\frac{4}{3}$

$\theta \approx 2.214, 5.356, (5, 2.214), (-5, 5.356)$

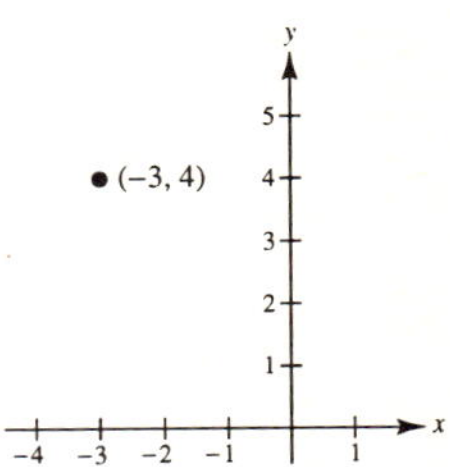

15. $(x, y) = (3, -2)$

$(r, \theta) = (3.606, -0.588)$

17. $(x, y) = \left(\frac{5}{2}, \frac{4}{3}\right)$

$(r, \theta) = (2.833, 0.490)$

19. (a) $(x, y) = (4, 3.5)$

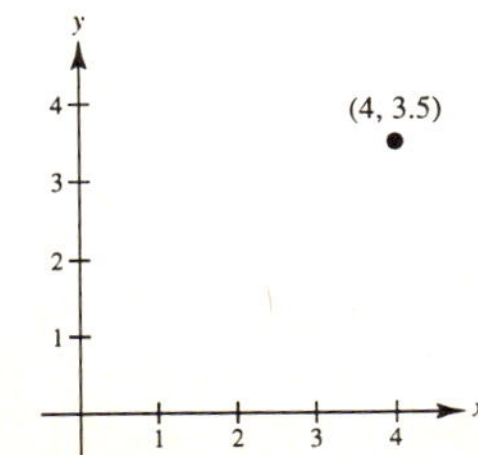

(b) $(r, \theta) = (4, 3.5)$

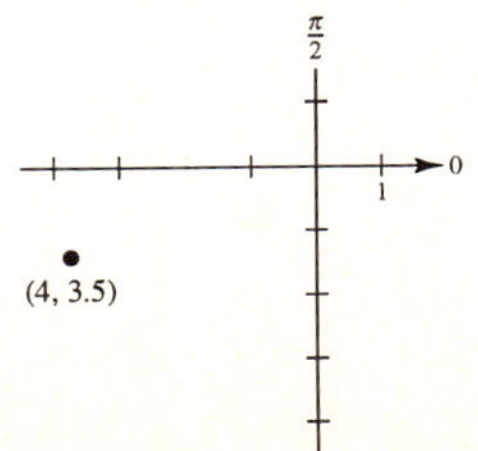

21. $x^2 + y^2 = a^2$

$r = a$

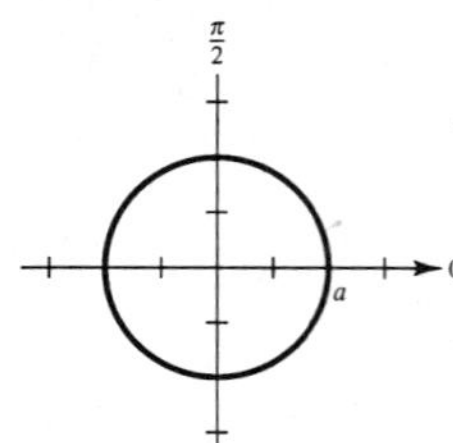

23. $y = 4$

$r \sin \theta = 4$

$r = 4 \csc \theta$

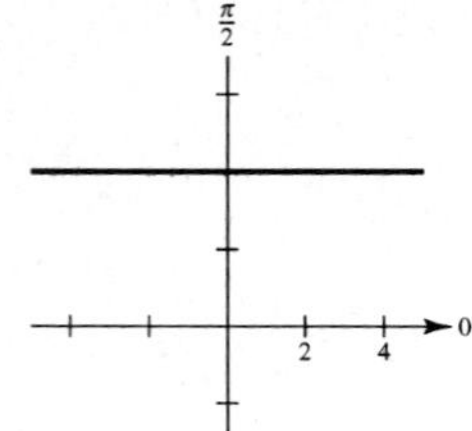

25.

$$3x - y + 2 = 0$$
$$3r \cos \theta - r \sin \theta + 2 = 0$$
$$r(3 \cos \theta - \sin \theta) = -2$$
$$r = \frac{-2}{3 \cos \theta - \sin \theta}$$

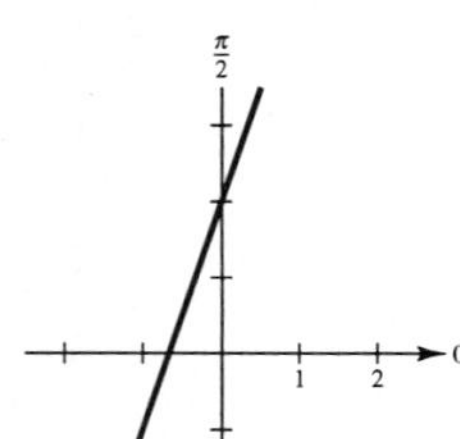

27.

$$y^2 = 9x$$
$$r^2 \sin^2 \theta = 9r \cos \theta$$
$$r = \frac{9 \cos \theta}{\sin^2 \theta}$$
$$r = 9 \csc^2 \theta \cos \theta$$

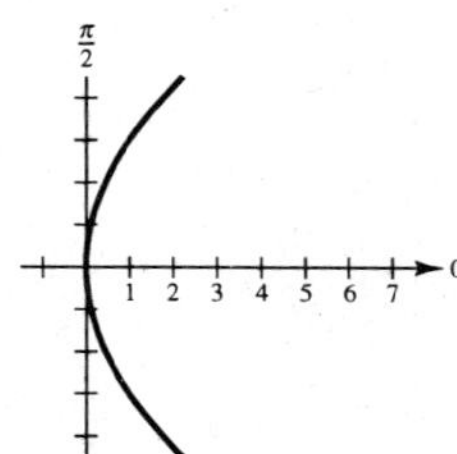

29.

$$r = 3$$
$$r^2 = 9$$
$$x^2 + y^2 = 9$$

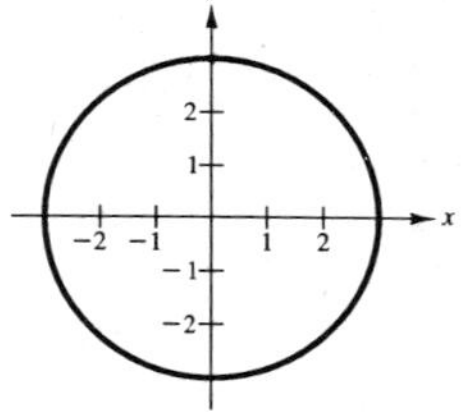

31.

$$r = \sin \theta$$
$$r^2 = r \sin \theta$$
$$x^2 + y^2 = y$$
$$x^2 + \left(y - \frac{1}{2}\right)^2 = \frac{1}{4}$$
$$x^2 + y^2 - y = 0$$

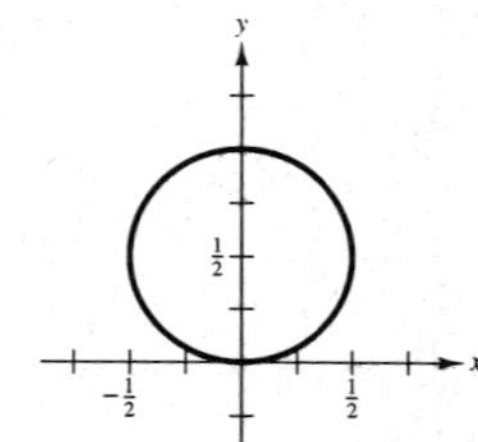

33.

$$r = \theta$$
$$\tan r = \tan \theta$$
$$\tan \sqrt{x^2 + y^2} = \frac{y}{x}$$
$$\sqrt{x^2 + y^2} = \arctan \frac{y}{x}$$

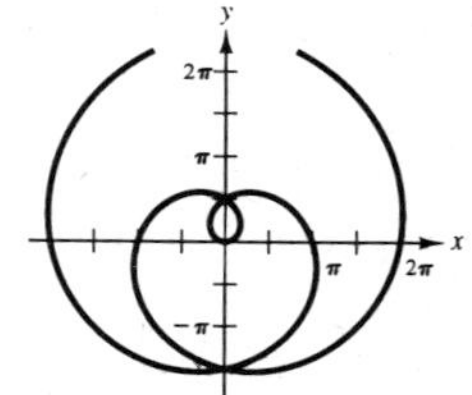

35.

$$r = 3 \sec \theta$$
$$r \cos \theta = 3$$
$$x = 3$$
$$x - 3 = 0$$

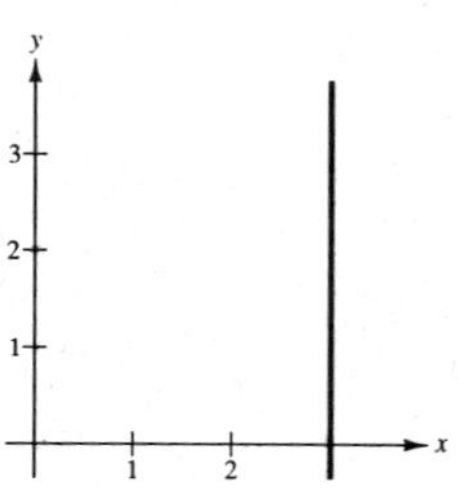

37.

$$r = 2(h \cos \theta + k \sin \theta)$$
$$r^2 = 2r(h \cos \theta + k \sin \theta)$$
$$r^2 = 2[h(r \cos \theta) + k(r \sin \theta)]$$
$$x^2 + y^2 = 2(hx + ky)$$
$$x^2 + y^2 - 2hx - 2ky = 0$$
$$(x^2 - 2hx + h^2) + (y^2 - 2ky + k^2) = 0 + h^2 + k^2$$
$$(x - h)^2 + (y - k)^2 = h^2 + k^2$$

Radius: $\sqrt{h^2 + k^2}$

Center: (h, k)

39. $\left(4, \frac{2\pi}{3}\right), \left(2, \frac{\pi}{6}\right)$

$$d = \sqrt{4^2 + 2^2 - 2(4)(2)\cos\left(\frac{2\pi}{3} - \frac{\pi}{6}\right)}$$

$$= \sqrt{20 - 16\cos\frac{\pi}{2}} = 2\sqrt{5} \approx 4.5$$

41. $(2, 0.5), (7, 1.2)$

$$d = \sqrt{2^2 + 7^2 - 2(2)(7)\cos(0.5 - 1.2)}$$

$$= \sqrt{53 - 28\cos(-0.7)} \approx 5.6$$

43. $r = 2 + 3\sin\theta$

$$\frac{dy}{dx} = \frac{3\cos\theta\sin\theta + \cos\theta(2 + 3\sin\theta)}{3\cos\theta\cos\theta - \sin\theta(2 + 3\sin\theta)}$$

$$= \frac{2\cos\theta(3\sin\theta + 1)}{3\cos 2\theta - 2\sin\theta} = \frac{2\cos\theta(3\sin\theta + 1)}{6\cos^2\theta - 2\sin\theta - 3}$$

At $\left(5, \frac{\pi}{2}\right)$, $\frac{dy}{dx} = 0$.

At $(2, \pi)$, $\frac{dy}{dx} = -\frac{2}{3}$.

At $\left(-1, \frac{3\pi}{2}\right)$, $\frac{dy}{dx} = 0$.

45. (a), (b) $r = 3(1 - \cos\theta)$

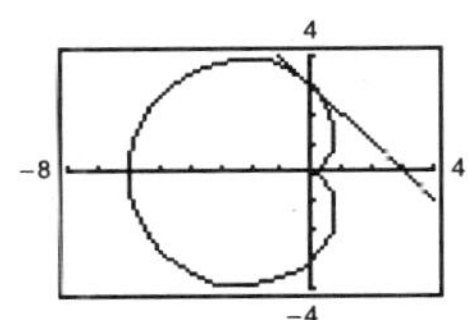

$(r, \theta) = \left(3, \frac{\pi}{2}\right) \Rightarrow (x, y) = (0, 3)$

Tangent line: $y - 3 = -1(x - 0)$

$$y = -x + 3$$

(c) At $\theta = \frac{\pi}{2}, \frac{dy}{dx} = -1.0$.

47. (a), (b) $r = 3\sin\theta$

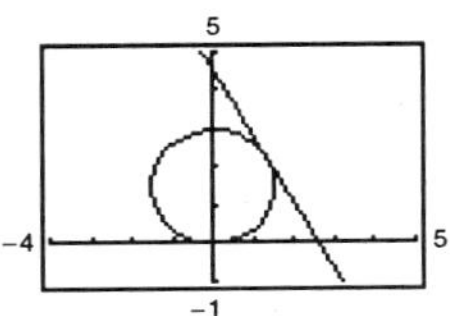

$(r, \theta) = \left(\frac{3\sqrt{3}}{2}, \frac{\pi}{3}\right) \Rightarrow (x, y) = \left(\frac{3\sqrt{3}}{4}, \frac{9}{4}\right)$

Tangent line: $y - \frac{9}{4} = -\sqrt{3}\left(x - \frac{3\sqrt{3}}{4}\right)$

$$y = -\sqrt{3}x + \frac{9}{2}$$

(c) At $\theta = \frac{\pi}{3}, \frac{dy}{dx} = -\sqrt{3} \approx 1.732$.

49. $r = 1 + \sin\theta$

$$\frac{dy}{d\theta} = (1 + \sin\theta)\cos\theta + \cos\theta\sin\theta$$

$$= \cos\theta(1 + 2\sin\theta) = 0$$

$\cos\theta = 0,\ \sin\theta = -\frac{1}{2},\ \theta = \frac{\pi}{2}, \frac{3\pi}{2},\ \theta = \frac{7\pi}{6}, \frac{11\pi}{6}$

Horizontal: $\left(2, \frac{\pi}{2}\right), \left(\frac{1}{2}, \frac{7\pi}{6}\right), \left(\frac{1}{2}, \frac{11\pi}{6}\right)$

$$\frac{dx}{d\theta} = -(1 + \sin\theta)\sin\theta + \cos^2\theta$$

$$= (1 + \sin\theta)(1 - 2\sin\theta) = 0$$

$\sin\theta = -1,\ \sin\theta = \frac{1}{2},\ \theta = \frac{3\pi}{2},\ \theta = \frac{\pi}{6}, \frac{5\pi}{6}$

Vertical: $\left(\frac{3}{2}, \frac{\pi}{6}\right), \left(\frac{3}{2}, \frac{5\pi}{6}\right)$

51. $r = 2\csc\theta + 3$

$$\frac{dy}{d\theta} = (2\csc\theta + 3)\cos\theta + (-2\csc\theta\cot\theta)\sin\theta$$

$$= 3\cos\theta = 0$$

$$\theta = \frac{\pi}{2}, \frac{3\pi}{2}$$

Horizontal: $\left(5, \frac{\pi}{2}\right), \left(1, \frac{3\pi}{2}\right)$

53. $r = 4\sin\theta\cos^2\theta$

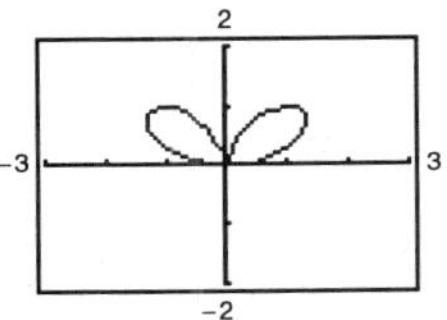

Horizontal tangents:

$(0, 0), (1.4142, 0.7854), (1.4142, 2.3562)$

55. $r = 2 \csc \theta + 5$

Horizontal tangents: $\left(7, \frac{\pi}{2}\right), \left(3, \frac{3\pi}{2}\right)$

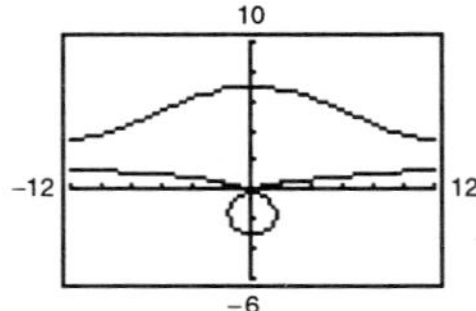

57.
$$r = 3 \sin \theta$$
$$r^2 = 3r \sin \theta$$
$$x^2 + y^2 = 3y$$
$$x^2 + \left(y - \tfrac{3}{2}\right)^2 = \tfrac{9}{4}$$

Circle $r = \frac{3}{2}$

Center: $\left(0, \frac{3}{2}\right)$

Tangent at the pole: $\theta = 0$

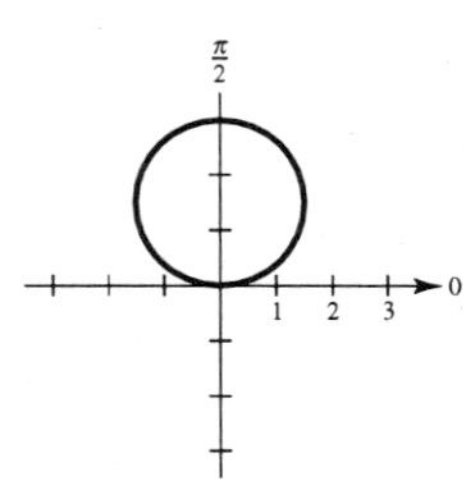

59. $r = 2 \cos(3\theta)$

Rose curve with three petals

Symmetric to the polar axis

Relative extrema: $(2, 0), \left(-2, \frac{\pi}{3}\right), \left(2, \frac{2\pi}{3}\right)$

Tangents at the pole: $\theta = \frac{\pi}{6}, \frac{\pi}{2}, \frac{5\pi}{6}$

θ	0	$\frac{\pi}{6}$	$\frac{\pi}{4}$	$\frac{\pi}{3}$	$\frac{\pi}{2}$	$\frac{2\pi}{3}$	$\frac{5\pi}{6}$	π
r	2	0	$-\sqrt{2}$	-2	0	2	0	-2

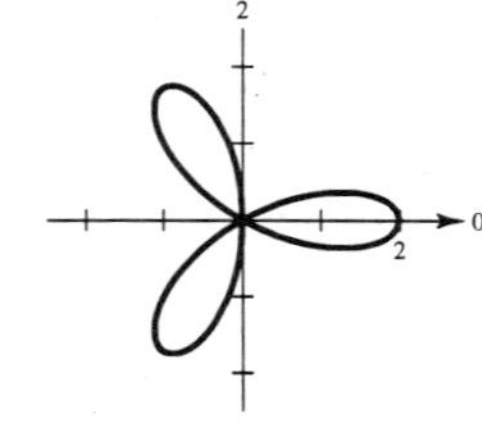

61. $r = 3 \sin 2\theta$

Rose curve with four petals

Symmetric to the polar axis, $\theta = \frac{\pi}{2}$, and pole

Relative extrema: $\left(\pm 3, \frac{\pi}{4}\right), \left(\pm 3, \frac{5\pi}{4}\right)$

Tangents at the pole: $\theta = 0, \frac{\pi}{2}$

($\theta = \pi$, $3\pi/2$ give the same tangents.)

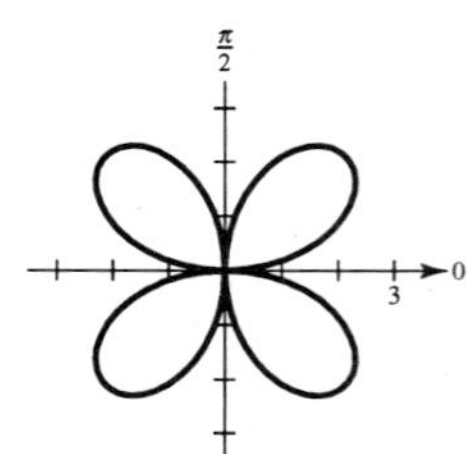

63. $r = 3 - 2 \cos \theta$

Limaçon

Symmetric to polar axis

θ	0	$\frac{\pi}{3}$	$\frac{\pi}{2}$	$\frac{2\pi}{3}$	π
r	1	2	3	4	5

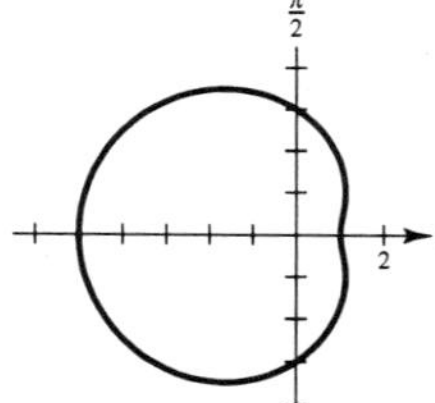

65.
$$r = 3 \csc \theta$$
$$r \sin \theta = 3$$
$$y = 3$$

Horizontal line

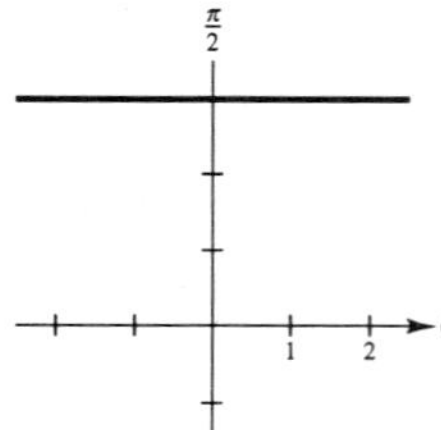

67. $r = 2\theta$

Spiral of Archimedes

Symmetric to $\theta = \frac{\pi}{2}$

Tangent at the pole: $\theta = 0$

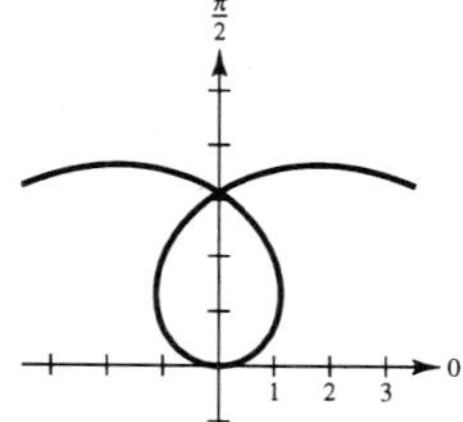

θ	0	$\frac{\pi}{4}$	$\frac{\pi}{2}$	$\frac{3\pi}{4}$	π	$\frac{5\pi}{4}$	$\frac{3\pi}{2}$
r	0	$\frac{\pi}{2}$	π	$\frac{3\pi}{2}$	2π	$\frac{5\pi}{2}$	3π

69. $r^2 = 4 \cos(2\theta)$

Lemniscate

Symmetric to the polar axis, $\theta = \frac{\pi}{2}$, and pole

Relative extrema: $(\pm 2, 0)$

Tangents at the pole: $\theta = \frac{\pi}{4}, \frac{3\pi}{4}$

θ	0	$\frac{\pi}{6}$	$\frac{\pi}{4}$
r	± 2	$\pm\sqrt{2}$	0

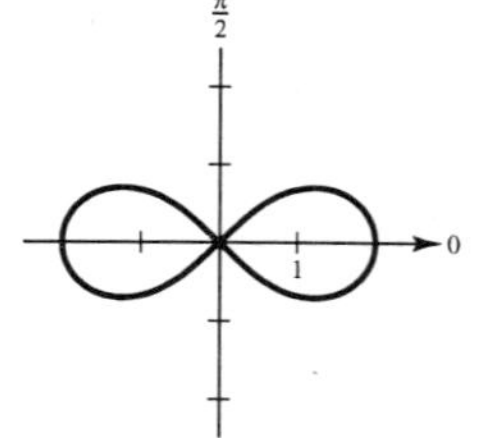

71. $r = 3 - 4\cos\theta$

$0 \le \theta < 2\pi$

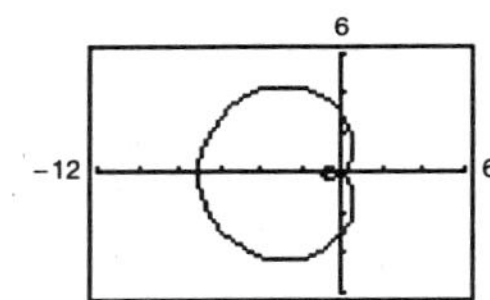

73. $r = 2 + \sin\theta$

$0 \le \theta < 2\pi$

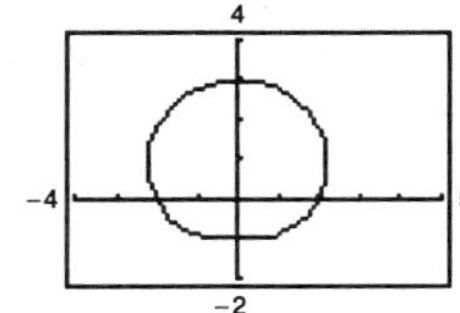

75. $r = \dfrac{2}{1 + \cos\theta}$

Traced out once on $-\pi < \theta < \pi$

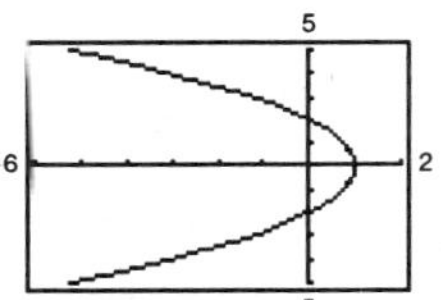

77. $r = 2\cos\left(\dfrac{3\theta}{2}\right)$

$0 \le \theta < 4\pi$

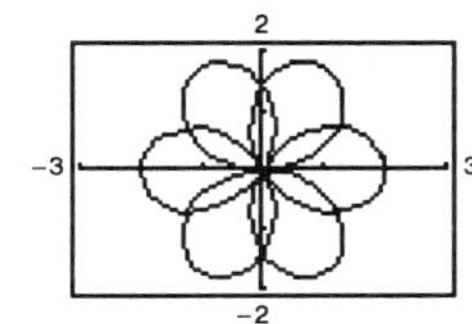

79. $r^2 = 4\sin 2\theta$

$0 \le \theta < \dfrac{\pi}{2}$

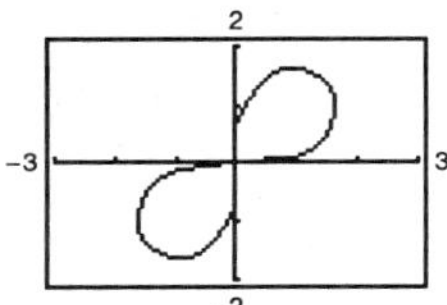

81. Since

$$r = 2 - \sec\theta = 2 - \frac{1}{\cos\theta},$$

the graph has polar axis symmetry and the lengths at the pole are

$$\theta = \frac{\pi}{3}, \frac{-\pi}{3}.$$

Furthermore,

$$r \Longrightarrow -\infty \text{ as } \theta \Longrightarrow \frac{\pi}{2^-}$$

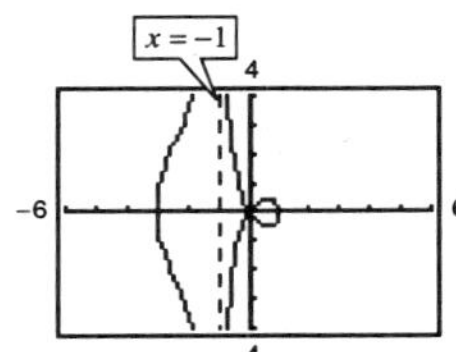

$$r \Longrightarrow \infty \text{ as } \theta \Longrightarrow -\frac{\pi}{2^+}.$$

Also, $r = 2 - \dfrac{1}{\cos\theta} = 2 - \dfrac{r}{r\cos\theta} = 2 - \dfrac{r}{x}$

$$rx = 2x - r$$

$$r = \frac{2x}{1 + x}.$$

Thus, $r \Longrightarrow \pm\infty$ as $x \Longrightarrow -1$.

83. $r = \dfrac{2}{\theta}$

Hyperbolic spiral

$r \Longrightarrow \infty$ as $\theta \Longrightarrow 0$

$$r = \frac{2}{\theta} \Longrightarrow \theta = \frac{2}{r} = \frac{2\sin\theta}{r\sin\theta} = \frac{2\sin\theta}{y}$$

$$y = \frac{2\sin\theta}{\theta}$$

$$\lim_{\theta\to 0}\frac{2\sin\theta}{\theta} = \lim_{\theta\to 0}\frac{2\cos\theta}{1} = 2$$

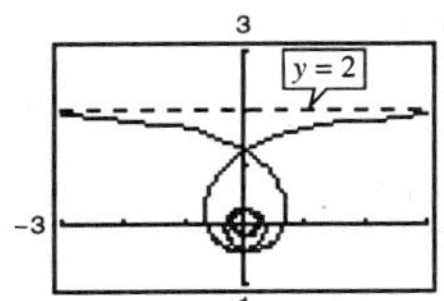

85. $r = 4\sin\theta$

(a) $0 \le \theta \le \dfrac{\pi}{2}$

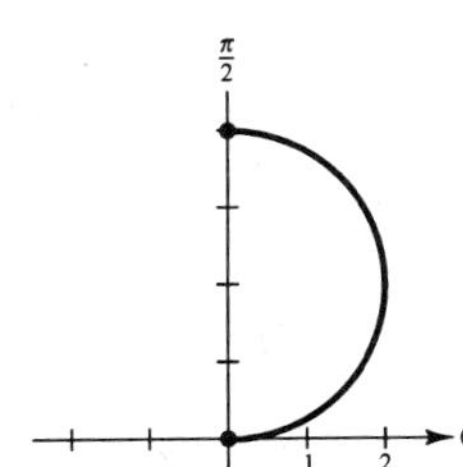

(b) $\dfrac{\pi}{2} \le \theta \le \pi$

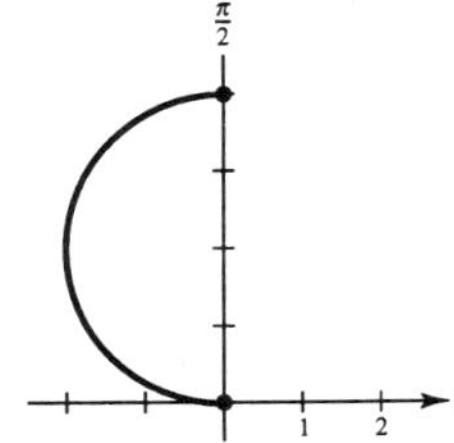

(c) $-\dfrac{\pi}{2} \le \theta \le \dfrac{\pi}{2}$

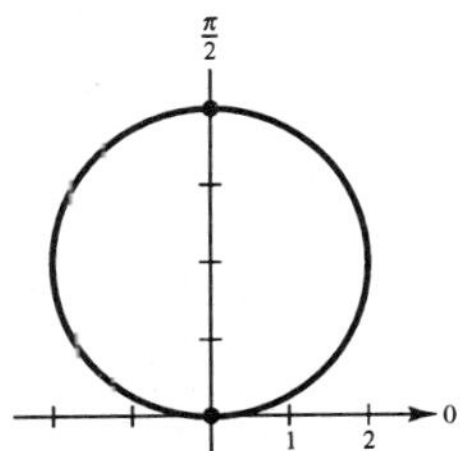

87. Let the curve $r = f(\theta)$ be rotated by ϕ to form the curve $r = g(\theta)$. If (r_1, θ_1) is a point on $r = f(\theta)$, then $(r_1, \theta_1 + \phi)$ is on $r = g(\theta)$. That is,

$$g(\theta_1 + \phi) = r_1 = f(\theta_1).$$

Letting $\theta = \theta_1 + \phi$, or $\theta_1 = \theta - \phi$, we see that

$$g(\theta) = g(\theta_1 + \phi) = f(\theta_1) = f(\theta - \phi).$$

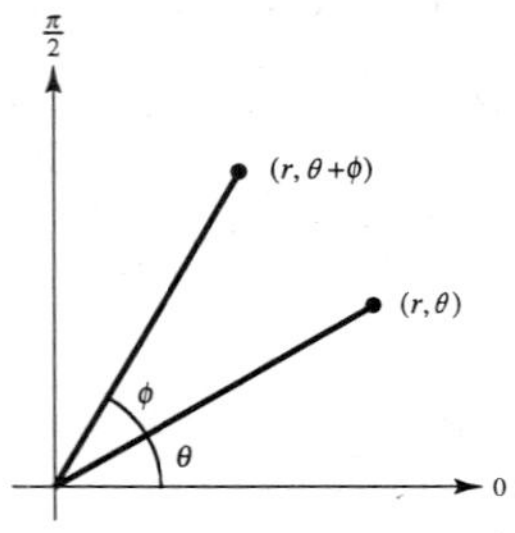

89. $r = 2 - \sin\theta$

(a) $r = 2 - \sin\left(\theta - \dfrac{\pi}{4}\right) = 2 - \dfrac{\sqrt{2}}{2}(\sin\theta - \cos\theta)$

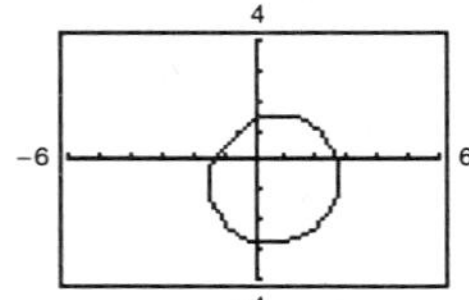

(b) $r = 2 - (-\cos\theta) = 2 + \cos\theta$

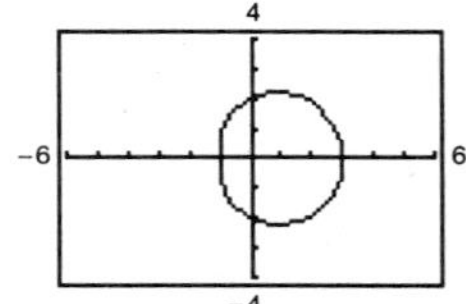

(c) $r = 2 - (-\sin\theta) = 2 + \sin\theta$

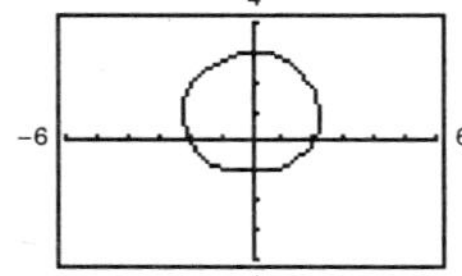

(d) $r = 2 - \cos\theta$

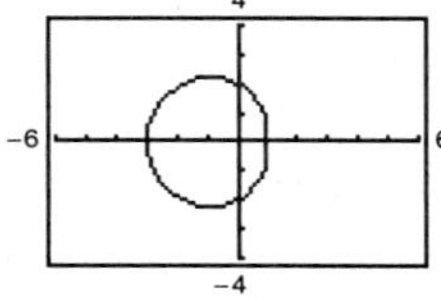

91. (a) $r = 1 - \sin\theta$

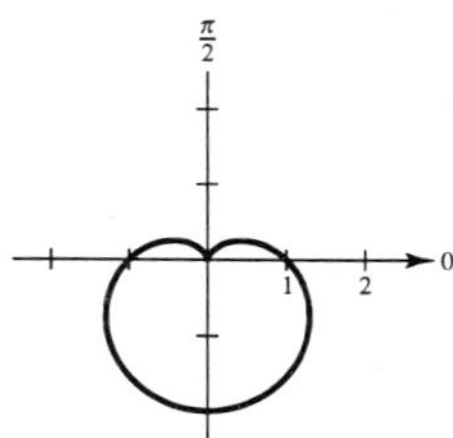

(b) $r = 1 - \sin\left(\theta - \dfrac{\pi}{4}\right)$

Rotate the graph of

$r = 1 - \sin\theta$

through the angle $\pi/4$.

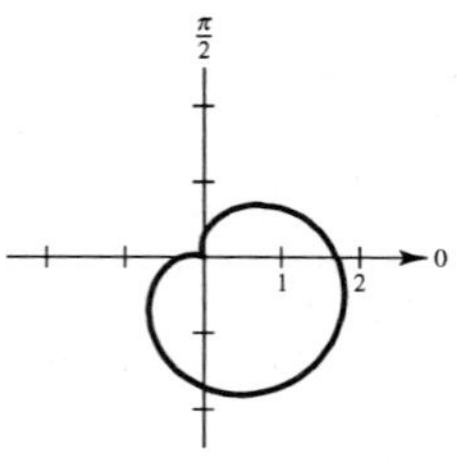

93. $\tan\psi = \dfrac{r}{dr/d\theta} = \dfrac{2(1 - \cos\theta)}{2\sin\theta}$

At $\theta = \pi$, $\tan\psi$ is undefined $\Rightarrow \psi = \dfrac{\pi}{2}$.

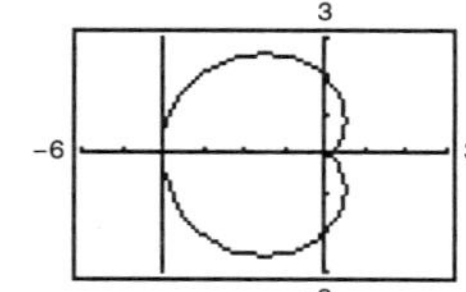

95. $\tan\psi = \dfrac{r}{dr/d\theta} = \dfrac{2\cos 3\theta}{-6\sin 3\theta}$

At $\theta = \dfrac{\pi}{6}$, $\tan\psi = 0 \Rightarrow \psi = 0$.

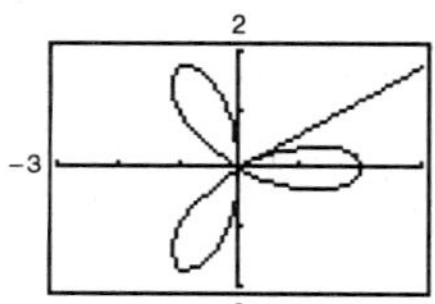

97. $r = \dfrac{6}{1 - \cos\theta} = 6(1 - \cos\theta)^{-1} \Rightarrow \dfrac{dr}{d\theta} = \dfrac{6\sin\theta}{(1 - \cos\theta)^2}$

$$\tan\psi = \frac{r}{\frac{dr}{d\theta}} = \frac{\frac{6}{(1 - \cos\theta)}}{\frac{6\sin\theta}{(1 - \cos\theta)^2}} = \frac{1 - \cos\theta}{\sin\theta}$$

At $\theta = \dfrac{2\pi}{3}$, $\tan\psi = \dfrac{1 - \left(-\frac{1}{2}\right)}{\frac{\sqrt{3}}{2}} = \sqrt{3}$.

$\psi = \dfrac{\pi}{3}\ (60°)$

99. The curve is produced over the interval

$0 \le \theta \le 9\pi$.

101. True

103. True

Section 9.5 Area and Arc Length in Polar Coordinates

1. (a) $r = 8\sin\theta$

$A = \pi(4)^2 = 16\pi$

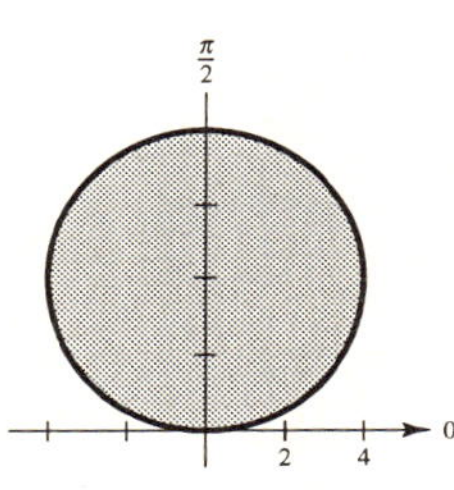

(b) $A = 2\left(\dfrac{1}{2}\right)\displaystyle\int_0^{\pi/2}\left[8\sin\theta\right]^2 d\theta$

$= 64\displaystyle\int_0^{\pi/2}\sin^2\theta\, d\theta$

$= 32\displaystyle\int_0^{\pi/2}(1 - \cos 2\theta)\, d\theta$

$= 32\left[\theta - \dfrac{\sin 2\theta}{2}\right]_0^{\pi/2} = 16\pi$

3. $A = 2\left[\dfrac{1}{2}\displaystyle\int_0^{\pi/6}(2\cos 3\theta)^2\, d\theta\right] = 2\left[\theta + \dfrac{1}{6}\sin 6\theta\right]_0^{\pi/6} = \dfrac{\pi}{3}$

5. $A = 2\left[\dfrac{1}{2}\displaystyle\int_0^{\pi/4}(\cos 2\theta)^2\, d\theta\right]$

$= \dfrac{1}{2}\left[\theta + \dfrac{1}{4}\sin 4\theta\right]_0^{\pi/4} = \dfrac{\pi}{8}$

7. $A = 2\left[\dfrac{1}{2}\displaystyle\int_{-\pi/2}^{\pi/2}(1 - \sin\theta)^2\, d\theta\right]$

$= \left[\dfrac{3}{2}\theta + 2\cos\theta - \dfrac{1}{4}\sin 2\theta\right]_{-\pi/2}^{\pi/2} = \dfrac{3\pi}{2}$

9. $A = 2\left[\dfrac{1}{2}\displaystyle\int_{2\pi/3}^{\pi}(1 + 2\cos\theta)^2\, d\theta\right]$

$= \left[3\theta + 4\sin\theta + \sin 2\theta\right]_{2\pi/3}^{\pi} = \dfrac{2\pi - 3\sqrt{3}}{2}$

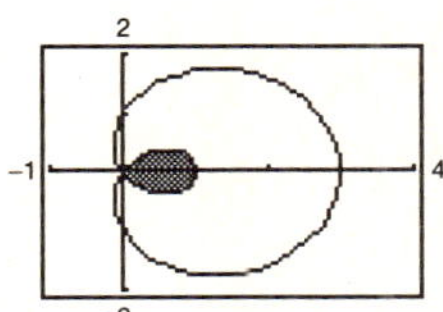

11. The area inside the outer loop is

$2\left[\dfrac{1}{2}\displaystyle\int_0^{2\pi/3}(1 + 2\cos\theta)^2\, d\theta\right] = \left[3\theta + 4\sin\theta + \sin 2\theta\right]_0^{2\pi/3} = \dfrac{4\pi + 3\sqrt{3}}{2}.$

From the result of Exercise 9, the area between the loops is

$A = \left(\dfrac{4\pi + 3\sqrt{3}}{2}\right) - \left(\dfrac{2\pi - 3\sqrt{3}}{2}\right) = \pi + 3\sqrt{3}.$

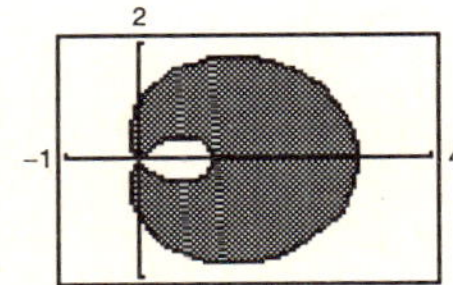

13. $r = 1 + \cos\theta$

$r = 1 - \cos\theta$

Solving simultaneously,

$$1 + \cos\theta = 1 - \cos\theta$$

$$2\cos\theta = 0$$

$$\theta = \frac{\pi}{2}, \frac{3\pi}{2}.$$

Replacing r by $-r$ and θ by $\theta + \pi$ in the first equation and solving, $-1 + \cos\theta = 1 - \cos\theta$, $\cos\theta = 1$, $\theta = 0$. Both curves pass through the pole, $(0, \pi)$, and $(0, 0)$, respectively.

Points of intersection: $\left(1, \frac{\pi}{2}\right), \left(1, \frac{3\pi}{2}\right), (0, 0)$

15. $r = 1 + \cos\theta$

$r = 1 - \sin\theta$

Solving simultaneously,

$$1 + \cos\theta = 1 - \sin\theta$$

$$\cos\theta = -\sin\theta$$

$$\tan\theta = -1$$

$$\theta = \frac{3\pi}{4}, \frac{7\pi}{4}.$$

Replacing r by $-r$ and θ by $\theta + \pi$ in the first equation and solving, $-1 + \cos\theta = 1 - \sin\theta$, $\sin\theta + \cos\theta = 2$, which has no solution. Both curves pass through the pole, $(0, \pi)$, and $(0, \pi/2)$, respectively.

Points of intersection: $\left(\frac{2 - \sqrt{2}}{2}, \frac{3\pi}{4}\right), \left(\frac{2 + \sqrt{2}}{2}, \frac{7\pi}{4}\right), (0, 0)$

17. $r = 4 - 5\sin\theta$

$r = 3\sin\theta$

Solving simultaneously,

$$4 - 5\sin\theta = 3\sin\theta$$

$$\sin\theta = \frac{1}{2}$$

$$\theta = \frac{\pi}{6}, \frac{5\pi}{6}.$$

Both curves pass through the pole, $(0, \arcsin 4/5)$, and $(0, 0)$, respectively.

Points of intersection: $\left(\frac{3}{2}, \frac{\pi}{6}\right), \left(\frac{3}{2}, \frac{5\pi}{6}\right), (0, 0)$

19. $r = \frac{\theta}{2}$

$r = 2$

Solving simultaneously, we have

$\theta/2 = 2$, $\theta = 4$.

Points of intersection:

$(2, 4), (-2, -4)$

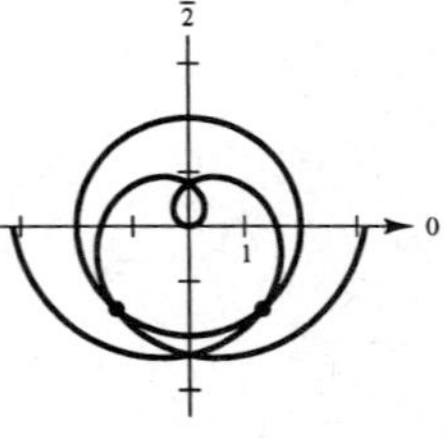

21. $r = 4\sin 2\theta$

$r = 2$

$r = 4\sin 2\theta$ is the equation of a rose curve with four petals and is symmetric to the polar axis, $\theta = \pi/2$, and the pole. Also, $r = 2$ is the equation of a circle of radius 2 centered at the pole. Solving simultaneously,

$$4\sin 2\theta = 2$$

$$2\theta = \frac{\pi}{6}, \frac{5\pi}{6}$$

$$\theta = \frac{\pi}{12}, \frac{5\pi}{12}.$$

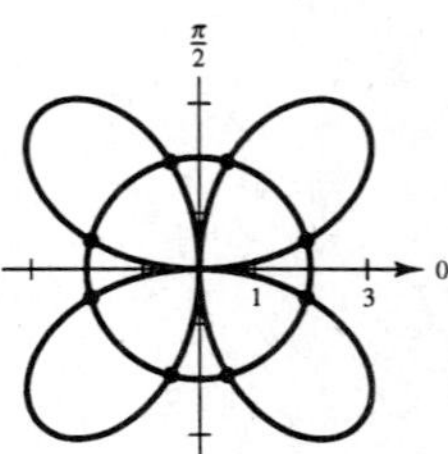

Therefore, the points of intersection for one petal are $(2, \pi/12)$ and $(2, 5\pi/12)$. By symmetry, the other points of intersection are $(2, 7\pi/12)$, $(2, 11\pi/12)$, $(2, 13\pi/12)$, $(2, 17\pi/12)$, $(2, 19\pi/12)$, and $(2, 23\pi/12)$.

23. $r = 2 + 3\cos\theta$

$r = \dfrac{\sec\theta}{2}$

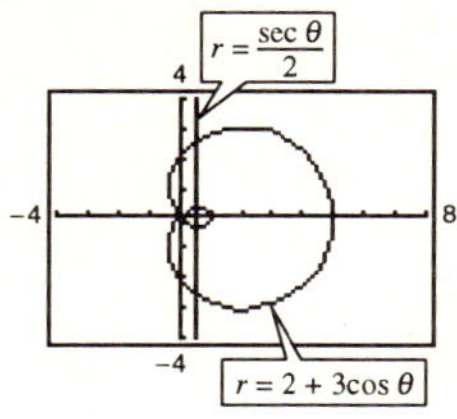

The graph of $r = 2 + 3\cos\theta$ is a limaçon with an inner loop $(b > a)$ and is symmetric to the polar axis. The graph of $r = (\sec\theta)/2$ is the vertical line $x = 1/2$. Therefore, there are four points of intersection. Solving simultaneously,

$$2 + 3\cos\theta = \frac{\sec\theta}{2}$$

$$6\cos^2\theta + 4\cos\theta - 1 = 0$$

$$\cos\theta = \frac{-2 \pm \sqrt{10}}{6}$$

$$\theta = \arccos\left(\frac{-2+\sqrt{10}}{6}\right) \approx 1.376$$

$$\theta = \arccos\left(\frac{-2-\sqrt{10}}{6}\right) \approx 2.6068.$$

Points of intersection: $(-0.581, \pm 2.607)$, $(2.581, \pm 1.376)$

25. $r = \cos\theta$

$r = 2 - 3\sin\theta$

Points of intersection:

$(0, 0)$, $(0.935, 0.363)$, $(0.535, -1.006)$

The graphs reach the pole at different times (θ values).

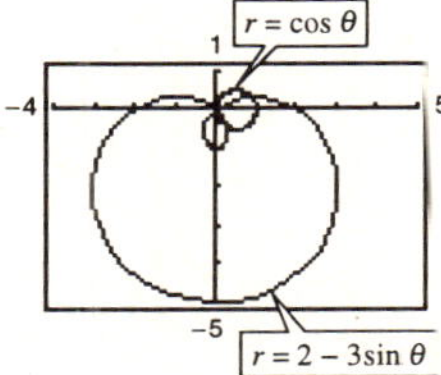

27. From Exercise 21, the points of intersection for one petal are $(2, \pi/12)$ and $(2, 5\pi/12)$. The area within one petal is

$$A = \frac{1}{2}\int_0^{\pi/12} (4\sin 2\theta)^2\,d\theta + \frac{1}{2}\int_{\pi/12}^{5\pi/12} (2)^2\,d\theta + \frac{1}{2}\int_{5\pi/12}^{\pi/2} (4\sin 2\theta)^2\,d\theta$$

$$= 16\int_0^{\pi/12} \sin^2(2\theta)\,d\theta + 2\int_{\pi/12}^{5\pi/12} d\theta \text{ (by symmetry of the petal)}$$

$$= 8\left[\theta - \frac{1}{4}\sin 4\theta\right]_0^{\pi/12} + \left[2\theta\right]_{\pi/12}^{5\pi/12} = \frac{4\pi}{3} - \sqrt{3}.$$

$$\text{Total area} = 4\left(\frac{4\pi}{3} - \sqrt{3}\right) = \frac{16\pi}{3} - 4\sqrt{3} = \frac{4}{3}(4\pi - 3\sqrt{3})$$

29. $$A = 4\left[\frac{1}{2}\int_0^{\pi/2} (3 - 2\sin\theta)^2\,d\theta\right]$$

$$= 2\left[11\theta + 12\cos\theta - \sin(2\theta)\right]_0^{\pi/2} = 11\pi - 24$$

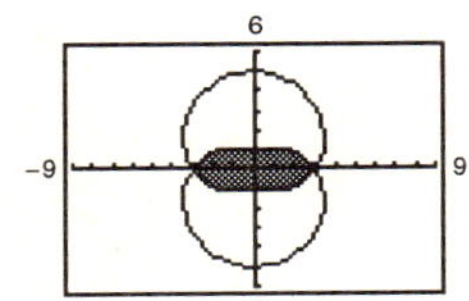

31. $$A = 2\left[\frac{1}{2}\int_0^{\pi/6} (4\sin\theta)^2\,d\theta + \frac{1}{2}\int_{\pi/6}^{\pi/2} (2)^2\,d\theta\right]$$

$$= 16\left[\frac{1}{2}\theta - \frac{1}{4}\sin(2\theta)\right]_0^{\pi/6} + \left[4\theta\right]_{\pi/6}^{\pi/2}$$

$$= \frac{8\pi}{3} - 2\sqrt{3} = \frac{2}{3}(4\pi - 3\sqrt{3})$$

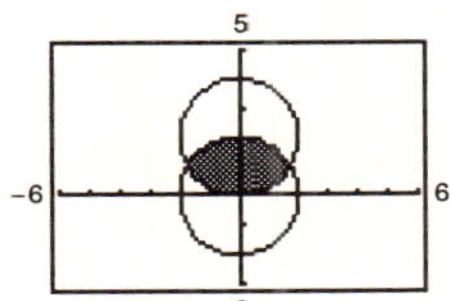

33. $A = 2\left[\frac{1}{2}\int_0^{\pi} [a(1 + \cos\theta)]^2\, d\theta\right] - \frac{a^2\pi}{4} = a^2\left[\frac{3}{2}\theta + 2\sin\theta + \frac{\sin 2\theta}{4}\right]_0^{\pi} - \frac{a^2\pi}{4} = \frac{3a^2\pi}{2} - \frac{a^2\pi}{4} = \frac{5a^2\pi}{4}$

35. $A = \frac{\pi a^2}{8} + \frac{1}{2}\int_{\pi/2}^{\pi} [a(1 + \cos\theta)]^2\, d\theta$

$= \frac{\pi a^2}{8} + \frac{a^2}{2}\int_{\pi/2}^{\pi}\left(\frac{3}{2} + 2\cos\theta + \frac{\cos 2\theta}{2}\right) d\theta$

$= \frac{\pi a^2}{8} + \frac{a^2}{2}\left[\frac{3}{2}\theta + 2\sin\theta + \frac{\sin 2\theta}{4}\right]_{\pi/2}^{\pi}$

$= \frac{\pi a^2}{8} + \frac{a^2}{2}\left[\frac{3\pi}{2} - \frac{3\pi}{4} - 2\right] = \frac{a^2}{2}[\pi - 2]$

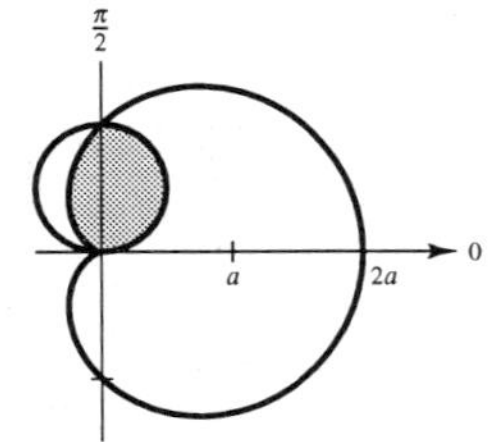

37. (a) $r = a\cos^2\theta$

$r^3 = ar^2\cos^2\theta$

$(x^2 + y^2)^{3/2} = ax^2$

(b)

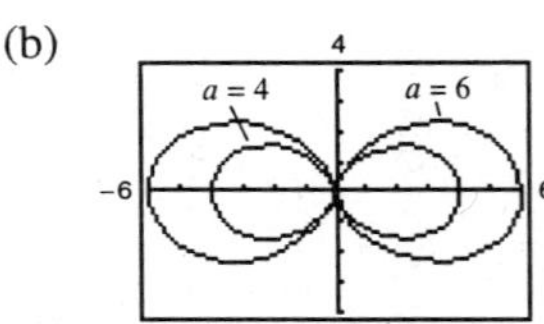

(c) $A = 4\left(\frac{1}{2}\right)\int_0^{\pi/2} [(6\cos^2\theta)^2 - (4\cos^2\theta)^2]\, d\theta$

$= 40\int_0^{\pi/2} \cos^4\theta\, d\theta = 10\int_0^{\pi/2} (1 + \cos 2\theta)^2\, d\theta$

$= 10\int_0^{\pi/2}\left(1 + 2\cos 2\theta + \frac{1 - \cos 4\theta}{2}\right) d\theta$

$= 10\left[\frac{3}{2}\theta + \sin 2\theta + \frac{1}{8}\sin 4\theta\right]_0^{\pi/2} = \frac{15\pi}{2}$

39. $r = a\cos(n\theta)$

For $n = 1$:

$r = a\cos\theta$

$A = \pi\left(\frac{a}{2}\right)^2 = \frac{\pi a^2}{4}$

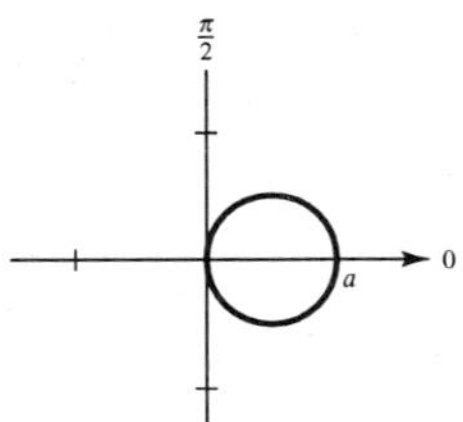

For $n = 2$:

$r = a\cos 2\theta$

$A = 8\left(\frac{1}{2}\right)\int_0^{\pi/4} (a\cos 2\theta)^2\, d\theta = \frac{\pi a^2}{2}$

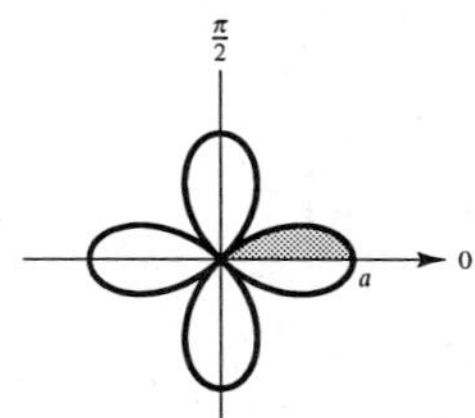

For $n = 3$:

$r = a\cos 3\theta$

$A = 6\left(\frac{1}{2}\right)\int_0^{\pi/6} (a\cos 3\theta)^2\, d\theta = \frac{\pi a^2}{4}$

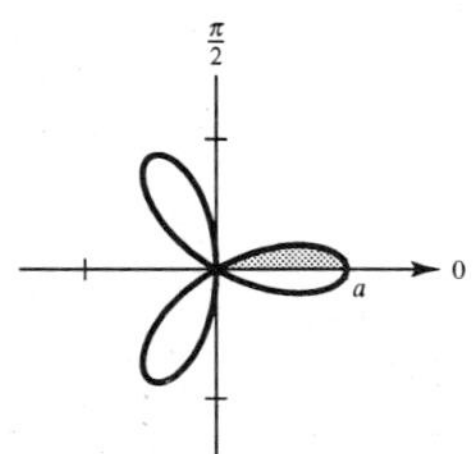

For $n = 4$:

$r = a\cos 4\theta$

$A = 16\left(\frac{1}{2}\right)\int_0^{\pi/8} (a\cos 4\theta)^2\, d\theta = \frac{\pi a^2}{2}$

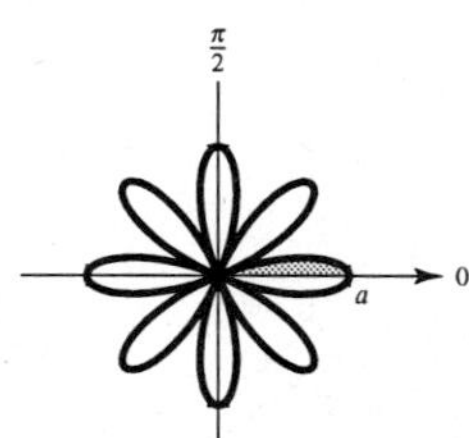

In general, the area of the region enclosed by $r = a\cos(n\theta)$ for $n = 1, 2, 3, \ldots$ is $(\pi a^2)/4$ if n is odd and is $(\pi a^2)/2$ if n is even.

41. $r = a$

$r' = 0$

$$s = \int_0^{2\pi} \sqrt{a^2 + 0^2}\, d\theta = \Big[a\theta\Big]_0^{2\pi} = 2\pi a$$

(circumference of circle of radius a)

43. $r = 1 + \sin\theta$

$r' = \cos\theta$

$$s = 2\int_{\pi/2}^{3\pi/2} \sqrt{(1 + \sin\theta)^2 + (\cos\theta)^2}\, d\theta$$

$$= 2\sqrt{2}\int_{\pi/2}^{3\pi/2} \sqrt{1 + \sin\theta}\, d\theta$$

$$= 2\sqrt{2}\int_{\pi/2}^{3\pi/2} \frac{-\cos\theta}{\sqrt{1 - \sin\theta}}\, d\theta$$

$$= \Big[4\sqrt{2}\sqrt{1 - \sin\theta}\Big]_{\pi/2}^{3\pi/2}$$

$$= 4\sqrt{2}\left(\sqrt{2} - 0\right) = 8$$

45. $r = 2\theta,\ 0 \le \theta \le \dfrac{\pi}{2}$

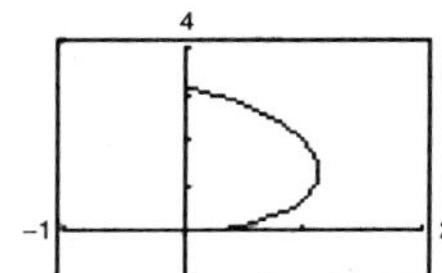

Length ≈ 4.16

47. $r = \dfrac{1}{\theta},\ \pi \le \theta \le 2\pi$

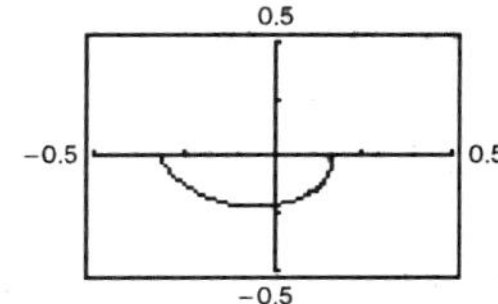

Length ≈ 0.71

49. $r = \sin(3\cos\theta),\ 0 \le \theta \le \pi$

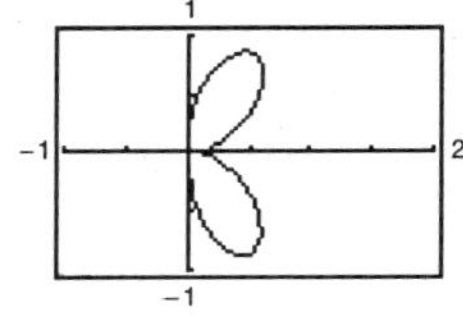

Length ≈ 4.39

51. $r = 2\cos\theta$

$r' = -2\sin\theta$

$$S = 2\pi\int_0^{\pi/2} 2\cos\theta\sin\theta\sqrt{4\cos^2\theta + 4\sin^2\theta}\, d\theta = 8\pi\int_0^{\pi/2} \sin\theta\cos\theta\, d\theta = \Big[4\pi\sin^2\theta\Big]_0^{\pi/2} = 4\pi$$

53. $r = e^{a\theta}$

$r' = ae^{a\theta}$

$$S = 2\pi\int_0^{\pi/2} e^{a\theta}\cos\theta\sqrt{(e^{a\theta})^2 + (ae^{a\theta})^2}\, d\theta$$

$$= 2\pi\sqrt{1 + a^2}\int_0^{\pi/2} e^{2a\theta}\cos\theta\, d\theta = 2\pi\sqrt{1 + a^2}\left[\frac{e^{2a\theta}}{4a^2 + 1}(2a\cos\theta + \sin\theta)\right]_0^{\pi/2} = \frac{2\pi\sqrt{1 + a^2}}{4a^2 + 1}(e^{\pi a} - 2a)$$

55. $r = 4\cos 2\theta$

$r' = -8\sin 2\theta$

$$S = 2\pi\int_0^{\pi/4} 4\cos 2\theta\sin\theta\sqrt{16\cos^2 2\theta + 64\sin^2 2\theta}\, d\theta = 32\pi\int_0^{\pi/4} \cos 2\theta\sin\theta\sqrt{\cos^2 2\theta + 4\sin^2 2\theta}\, d\theta \approx 21.87$$

57. Revolve $r = a$ about the line $r = b\sec\theta$ where $b > a > 0$.

$f(\theta) = a$

$f'(\theta) = 0$

$$S = 2\pi\int_0^{2\pi} [b - a\cos\theta]\sqrt{a^2 + 0^2}\, d\theta$$

$$= 2\pi a\Big[b\theta - a\sin\theta\Big]_0^{2\pi} = 2\pi a(2\pi b) = 4\pi^2 ab$$

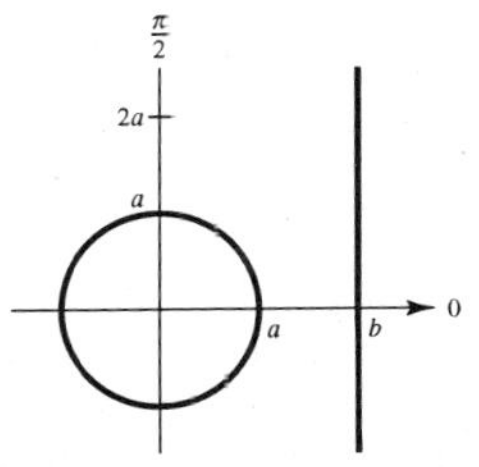

59. $r = 8\cos\theta,\ 0 \le \theta \le \pi$

(a) $A = \dfrac{1}{2}\displaystyle\int_0^{\pi} r^2\,d\theta = \dfrac{1}{2}\int_0^{\pi} 64\cos^2\theta\,d\theta = 32\int_0^{\pi}\dfrac{1+\cos 2\theta}{2}\,d\theta = 16\left[\theta + \dfrac{\sin 2\theta}{2}\right]_0^{\pi} = 16\pi$

(Area circle $= \pi r^2 = \pi 4^2 = 16\pi$)

(b)

θ	0.2	0.4	0.6	0.8	1.0	1.2	1.4
A	6.32	12.14	17.06	20.80	23.27	24.60	25.08

(c), (d) For $\frac{1}{4}$ of area $(4\pi \approx 12.57)$: 0.42
For $\frac{1}{2}$ of area $(8\pi \approx 25.13)$: 1.57 $(\pi/2)$
For $\frac{3}{4}$ of area $(12\pi \approx 37.70)$: 2.73

(e) No, it does not depend on the radius.

61. False. $f(\theta) = 1$ and $g(\theta) = -1$ have the same graphs.

63. True. The area enclosed by $r = \sin(n\theta)$ is $\pi/2$ and the area enclosed by $r = \sin[(n+1)\theta]$ is $\pi/4$.

Section 9.6 Polar Equations of Conics and Kepler's Laws

1. $r = \dfrac{2e}{1 + e\cos\theta}$

(a) $e = 1,\ r = \dfrac{2}{1+\cos\theta}$, parabola

(b) $e = 0.5,\ r = \dfrac{1}{1+0.5\cos\theta} = \dfrac{2}{2+\cos\theta}$, ellipse

(c) $e = 1.5,\ r = \dfrac{3}{1+1.5\cos\theta} = \dfrac{6}{2+3\cos\theta}$, hyperbola

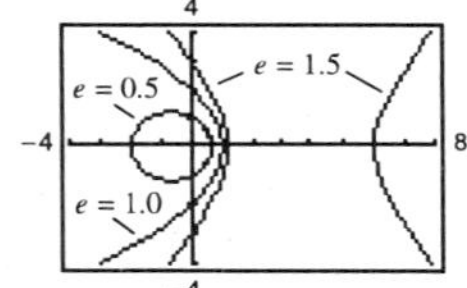

3. $r = \dfrac{2e}{1 - e\sin\theta}$

(a) $e = 1,\ r = \dfrac{2}{1-\sin\theta}$, parabola

(b) $e = 0.5,\ r = \dfrac{1}{1-0.5\sin\theta} = \dfrac{2}{2-\sin\theta}$, ellipse

(c) $e = 1.5,\ r = \dfrac{3}{1-1.5\sin\theta} = \dfrac{6}{2-3\sin\theta}$, hyperbola

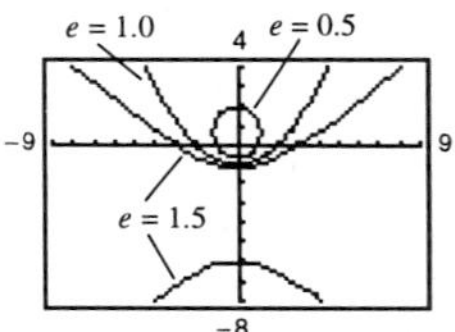

5. $r = \dfrac{4}{1 + e\sin\theta}$

(a)

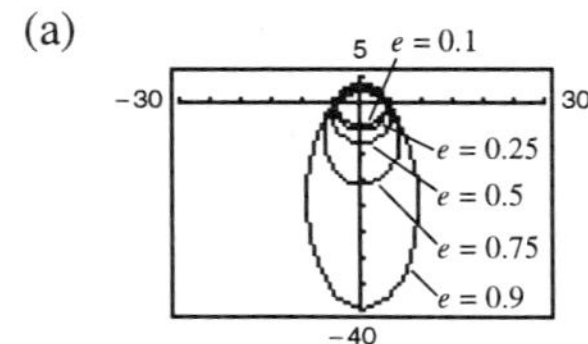

The conic is an ellipse. As $e \to 1^-$, the ellipse becomes more elliptical, and as $e \to 0^+$, it becomes more circular.

(b)

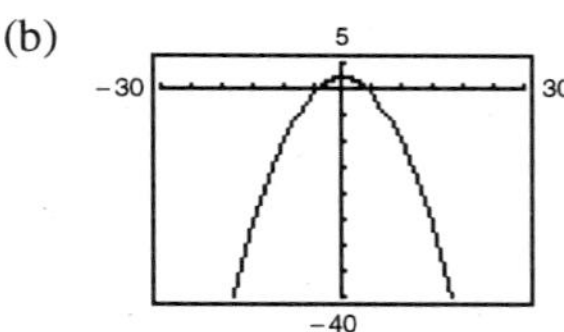

The conic is a parabola.

(c)

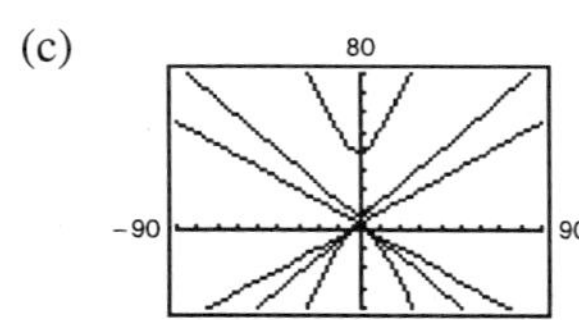

The conic is a hyperbola. As $e \to 1^+$, the hyperbolas opens more slowly, and as $e \to \infty$, they open more rapidly.

7. Parabola; Matches (c)

9. Hyperbola; Matches (a)

11. Ellipse; Matches (b)

13. $r = \dfrac{-1}{1 - \sin \theta}$

Parabola since $e = 1$

Vertex: $\left(-\dfrac{1}{2}, \dfrac{3\pi}{2}\right)$

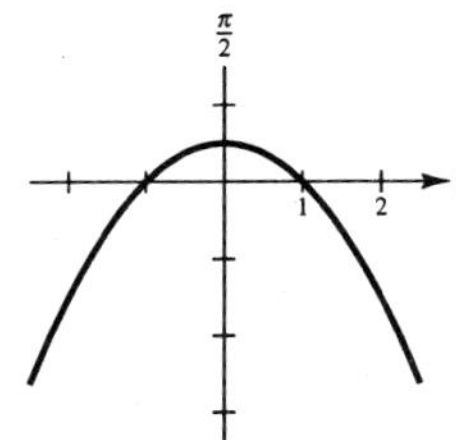

15. $r = \dfrac{6}{2 + \cos \theta}$

$= \dfrac{3}{1 + (1/2) \cos \theta}$

Ellipse since $e = \dfrac{1}{2} < 1$

Vertices: $(2, 0), (6, \pi)$

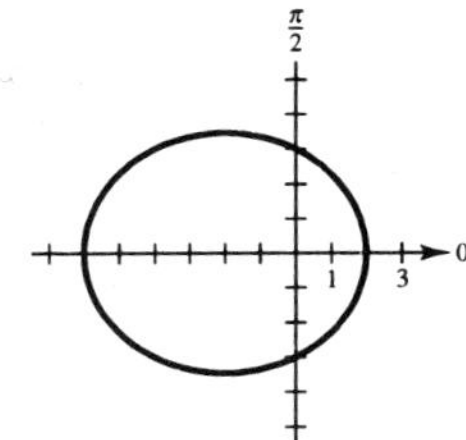

17. $r(2 + \sin \theta) = 4$

$r = \dfrac{4}{2 + \sin \theta}$

$= \dfrac{2}{1 + (1/2) \sin \theta}$

Ellipse since $e = \dfrac{1}{2} < 1$

Vertices: $\left(\dfrac{4}{3}, \dfrac{\pi}{2}\right), \left(4, \dfrac{3\pi}{2}\right)$

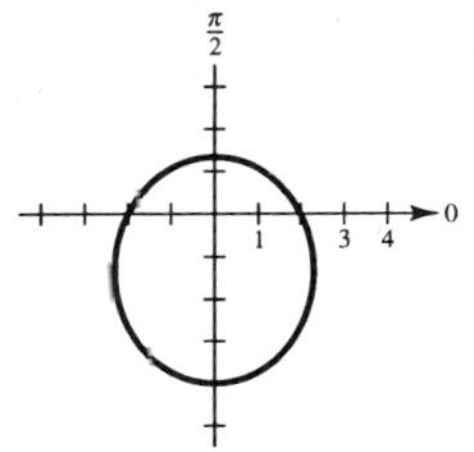

19. $r = \dfrac{5}{-1 + 2 \cos \theta} = \dfrac{-5}{1 - 2 \cos \theta}$

Hyperbola since $e = 2 > 1$

Vertices: $(5, 0), \left(-\dfrac{5}{3}, \pi\right)$

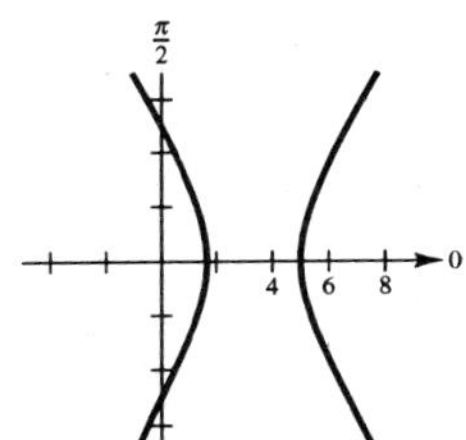

21. $r = \dfrac{3}{2 + 6 \sin \theta} = \dfrac{3/2}{1 + 3 \sin \theta}$

Hyperbola since $e = 3 > 1$

Vertices: $\left(\dfrac{3}{8}, \dfrac{\pi}{2}\right), \left(-\dfrac{3}{4}, \dfrac{3\pi}{2}\right)$

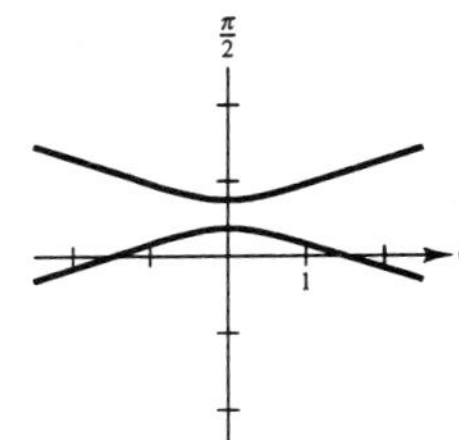

23.

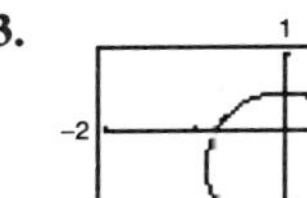

Ellipse

25.

Parabola

27. $r = \dfrac{-1}{1 - \sin\left(\theta - \dfrac{\pi}{4}\right)}$

Rotate the graph of

$r = \dfrac{-1}{1 - \sin \theta}$

counterclockwise through the angle $\pi/4$.

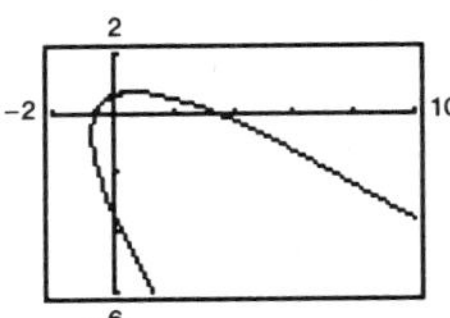

29. $r = \dfrac{6}{2 + \cos\left(\theta + \dfrac{\pi}{6}\right)}$

Rotate the graph of

$r = \dfrac{6}{2 + \cos \theta}$

clockwise through the angle $\pi/6$.

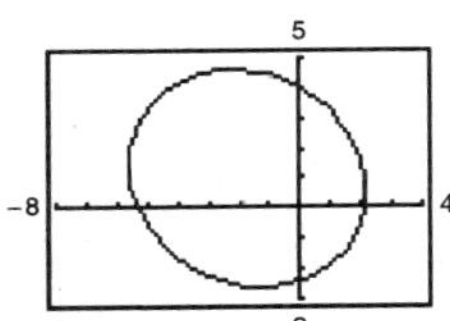

31. Change θ to $\theta + \frac{\pi}{4}$:

$$r = \frac{5}{5 + 3\cos\left(\theta + \frac{\pi}{4}\right)}.$$

33. Parabola

$e = 1, x = -1, d = 1$

$$r = \frac{ed}{1 - e\cos\theta} = \frac{1}{1 - \cos\theta}$$

35. Ellipse

$e = \frac{1}{2}, y = 1, d = 1$

$$r = \frac{ed}{1 + e\sin\theta}$$

$$= \frac{1/2}{1 + (1/2)\sin\theta} = \frac{1}{2 + \sin\theta}$$

37. Hyperbola

$e = 2, x = 1, d = 1$

$$r = \frac{ed}{1 + e\cos\theta} = \frac{2}{1 + 2\cos\theta}$$

39. Parabola

Vertex: $\left(1, -\frac{\pi}{2}\right)$

$$e = 1, d = 2, r = \frac{2}{1 - \sin\theta}$$

41. Ellipse

Vertices: $(2, 0), (8, \pi)$

$e = \frac{3}{5}, d = \frac{16}{3}$

$$r = \frac{ed}{1 + e\cos\theta}$$

$$= \frac{16/5}{1 + (3/5)\cos\theta} = \frac{16}{5 + 3\cos\theta}$$

43. Hyperbola

Vertices: $\left(1, \frac{3\pi}{2}\right), \left(9, \frac{3\pi}{2}\right)$

$e = \frac{5}{4}, d = \frac{9}{5}$

$$r = \frac{ed}{1 - e\sin\theta}$$

$$= \frac{9/4}{1 - (5/4)\sin\theta}$$

$$= \frac{9}{4 - 5\sin\theta}$$

45.

$$\frac{x^2}{a^2} + \frac{y^2}{b^2} = 1$$

$$x^2b^2 + y^2a^2 = a^2b^2$$

$$b^2r^2\cos^2\theta + a^2r^2\sin^2\theta = a^2b^2$$

$$r^2[b^2\cos^2\theta + a^2(1 - \cos^2\theta)] = a^2b^2$$

$$r^2[a^2 + \cos^2\theta(b^2 - a^2)] = a^2b^2$$

$$r^2 = \frac{a^2b^2}{a^2 + (b^2 - a^2)\cos^2\theta} = \frac{a^2b^2}{a^2 - c^2\cos^2\theta}$$

$$= \frac{b^2}{1 - (c/a)^2\cos^2\theta} = \frac{b^2}{1 - e^2\cos^2\theta}$$

47. $a = 5, c = 4, e = \frac{4}{5}, b = 3$

$$r^2 = \frac{9}{1 - (16/25)\cos^2\theta}$$

49. $a = 3, b = 4, c = 5, e = \frac{5}{3}$

$$r^2 = \frac{-16}{1 - (25/9)\cos^2\theta}$$

51. $A = 2\left[\frac{1}{2}\int_0^{\pi}\left(\frac{3}{2 - \cos\theta}\right)^2 d\theta\right]$

$$= 9\int_0^{\pi}\frac{1}{(2 - \cos\theta)^2}\,d\theta \approx 10.88$$

53. Vertices: $(126{,}000, 0), (4119, \pi)$

$$a = \frac{126{,}000 + 4119}{2} = 65{,}059.5,\ c = 65{,}059.5 - 4119 = 60{,}940.5,\ e = \frac{c}{a} = \frac{40{,}627}{43{,}373},\ d = 4119\left(\frac{84{,}000}{40{,}627}\right)$$

$$r = \frac{ed}{1 - e\cos\theta} = \frac{4119(84{,}000/43{,}373)}{1 - (40{,}627/43{,}373)\cos\theta} = \frac{345{,}996{,}000}{43{,}373 - 40{,}627\cos\theta}$$

When $\theta = 60°$, $r = \frac{345{,}996{,}000}{23{,}059.5} \approx 15{,}004.49$.

Distance between the surface of the earth and the satellite is $r - 4000 = 11{,}004.49$ miles.

55. $a = 92.957 \times 10^6$ mi, $e = 0.0167$

$$r = \frac{(1 - e^2)a}{1 - e\cos\theta} = \frac{92{,}931{,}075.2223}{1 - 0.0167\cos\theta}$$

Perihelion distance: $a(1 - e) \approx 91{,}404{,}618$ mi

Aphelion distance: $a(1 + e) \approx 94{,}509{,}382$ mi

57. $a = 5.900 \times 10^9$ km, $e = 0.2481$

$$r = \frac{(1 - e^2)a}{1 - e\cos\theta} \approx \frac{5.537 \times 10^9}{1 - 0.2481\cos\theta}$$

Perihelion distance: $a(1 - e) = 4.436 \times 10^9$ km

Aphelion distance: $a(1 + e) = 7.364 \times 10^9$ km

59. $r = \dfrac{5.537 \times 10^9}{1 - 0.2481\cos\theta}$

(a) $A = \dfrac{1}{2}\displaystyle\int_0^{\pi/9} \left[\frac{5.537 \times 10^9}{1 - 0.2481\cos\theta}\right]^2 d\theta \approx 9.341 \times 10^{18}\ \text{km}^2$

$$248\left[\frac{\dfrac{1}{2}\displaystyle\int_0^{\pi/9}\left[\frac{5.537 \times 10^9}{1 - 0.2481\cos\theta}\right]^2 d\theta}{\dfrac{1}{2}\displaystyle\int_0^{2\pi}\left[\frac{5.537 \times 10^9}{1 - 0.2481\cos\theta}\right]^2 d\theta}\right] \approx 21.867\ \text{yr}$$

(b) $\dfrac{1}{2}\displaystyle\int_\pi^{\alpha - \pi}\left[\frac{5.537 \times 10^9}{1 - 0.2481\cos\theta}\right]^2 d\theta = 9.341 \times 10^{18}$

$\alpha \approx \pi + 0.8995$ rad

In part (a) the ray swept through a smaller angle to generate the same area since the length of the ray is longer than in part (b).

(c) $r' = \dfrac{(-5.537 \times 10^9)(0.2481\sin\theta)}{(1 - 0.2481\cos\theta)^2}$

$$s = \int_0^{\pi/9}\sqrt{\left(\frac{5.537 \times 10^9}{1 - 0.2481\cos\theta}\right)^2 + \left[\frac{-1.3737297 \times 10^9 \sin\theta}{(1 - 0.2481\cos\theta)^2}\right]^2}\, d\theta \approx 2.559 \times 10^9\ \text{km}$$

$$\frac{2.559 \times 10^9\ \text{km}}{21.867\ \text{yr}} \approx 1.17 \times 10^8\ \text{km/yr}$$

$$s = \int_\pi^{\pi + 0.899}\sqrt{\left(\frac{5.537 \times 10^9}{1 - 0.2481\cos\theta}\right)^2 + \left[\frac{-1.3737297 \times 10^9 \sin\theta}{(1 - 0.2481\cos\theta)^2}\right]^2}\, d\theta \approx 4.119 \times 10^9\ \text{km}$$

$$\frac{4.119 \times 10^9\ \text{km}}{21.867\ \text{yr}} \approx 1.88 \times 10^8\ \text{km/yr}$$

61. $r_1 = \dfrac{ed}{1 + \sin\theta}$ and $r_2 = \dfrac{ed}{1 - \sin\theta}$

Points of intersection: $(ed, 0)$, (ed, π)

$$r_1\text{:}\quad \frac{dy}{dx} = \frac{\left(\dfrac{ed}{1 + \sin\theta}\right)(\cos\theta) + \left(\dfrac{-ed\cos\theta}{(1 + \sin\theta)^2}\right)(\sin\theta)}{\left(\dfrac{-ed}{1 + \sin\theta}\right)(\sin\theta) + \left(\dfrac{-ed\cos\theta}{(1 + \sin\theta)^2}\right)(\cos\theta)}$$

At $(ed, 0)$, $\dfrac{dy}{dx} = -1$. At (ed, π), $\dfrac{dy}{dx} = 1$.

$$r_2\text{:}\quad \frac{dy}{dx} = \frac{\left(\dfrac{ed}{1 - \sin\theta}\right)(\cos\theta) + \left(\dfrac{ed\cos\theta}{(1 - \sin\theta)^2}\right)(\sin\theta)}{\left(\dfrac{-ed}{1 - \sin\theta}\right)(\sin\theta) + \left(\dfrac{ed\cos\theta}{(1 - \sin\theta)^2}\right)(\cos\theta)}$$

At $(ed, 0)$, $\dfrac{dy}{dx} = 1$. At (ed, π), $\dfrac{dy}{dx} = -1$.

Therefore, at $(ed, 0)$ we have $m_1m_2 = (-1)(1) = -1$, and at (ed, π) we have $m_1m_2 = 1(-1) = -1$. The curves intersect at right angles.

Review Exercises for Chapter 9

1. Matches (d) - ellipse

3. Matches (a) - parabola

5. $$16x^2 + 16y^2 - 16x + 24y - 3 = 0$$

$$\left(x^2 - x + \frac{1}{4}\right) + \left(y^2 + \frac{3}{2}y + \frac{9}{16}\right) = \frac{3}{16} + \frac{1}{4} + \frac{9}{16}$$

$$\left(x - \frac{1}{2}\right)^2 + \left(y + \frac{3}{4}\right)^2 = 1$$

Circle

Center: $\left(\frac{1}{2}, -\frac{3}{4}\right)$

Radius: 1

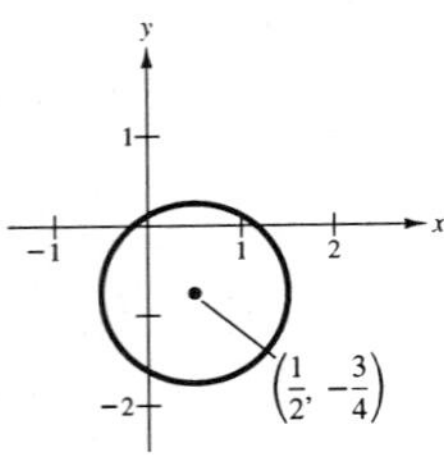

7. $$3x^2 - 2y^2 + 24x + 12y + 24 = 0$$

$$3(x^2 + 8x + 16) - 2(y^2 - 6y + 9) = -24 + 48 - 18$$

$$\frac{(x+4)^2}{2} - \frac{(y-3)^2}{3} = 1$$

Hyperbola

Center: $(-4, 3)$

Vertices: $(-4 \pm \sqrt{2}, 3)$

Asymptotes: $y = 3 \pm \sqrt{\frac{3}{2}}(x + 4)$

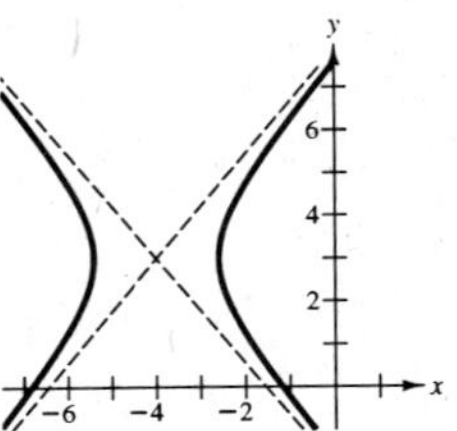

9. $$3x^2 + 2y^2 - 12x + 12y + 29 = 0$$

$$3(x^2 - 4x + 4) + 2(y^2 + 6y + 9) = -29 + 12 + 18$$

$$\frac{(x-2)^2}{1/3} + \frac{(y+3)^2}{1/2} = 1$$

Ellipse

Center: $(2, -3)$

Vertices: $\left(2, -3 \pm \frac{\sqrt{2}}{2}\right)$

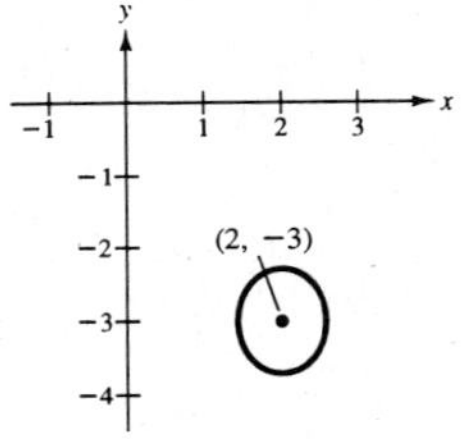

11. Vertex: $(0, 2)$

Directrix: $x = -3$

Parabola opens to the right

$p = 3$

$(y - 2)^2 = 4(3)(x - 0)$

$y^2 - 4y - 12x + 4 = 0$

13. Vertices: $(-3, 0), (7, 0)$

Foci: $(0, 0), (4, 0)$

Horizontal major axis

Center: $(2, 0)$

$a = 5, c = 2, b = \sqrt{21}$

$$\frac{(x-2)^2}{25} + \frac{y^2}{21} = 1$$

15. $\frac{x^2}{9} + \frac{y^2}{4} = 1, a = 3, b = 2, c = \sqrt{5}, e = \frac{\sqrt{5}}{3}$

By Example 5 of Section 9.1,

$$C = 12\int_0^{\pi/2} \sqrt{1 - \left(\frac{5}{9}\right)\sin^2\theta}\, d\theta \approx 15.87.$$

17. $y = x - 2$ has a slope of 1. The perpendicular slope is -1.

$$y = x^2 - 2x + 2$$

$\frac{dy}{dx} = 2x - 2 = -1$ when $x = \frac{1}{2}$ and $y = \frac{5}{4}$.

Perpendicular line: $y - \frac{5}{4} = -1\left(x - \frac{1}{2}\right)$

$$4x + 4y - 7 = 0$$

19. (a) $V = (\pi ab)(\text{Length}) = 12\pi(16) = 192\pi \text{ ft}^3$

(b) $F = 2(62.4)\int_{-3}^{3}(3-y)\frac{4}{3}\sqrt{9-y^2}\,dy = \frac{8}{3}(62.4)\left[3\int_{-3}^{3}\sqrt{9-y^2}\,dy - \int_{-3}^{3}y\sqrt{9-y^2}\,dy\right]$

$$= \frac{8}{3}(62.4)\left[\frac{3}{2}\left(y\sqrt{9-y^2} + 9\arcsin\frac{y}{3}\right) + \frac{1}{3}(9-y^2)^{3/2}\right]_{-3}^{3}$$

$$= \frac{8}{3}(62.4)\left[\frac{3}{2}\left(\frac{9\pi}{2}\right) - \frac{3}{2}\left(-\frac{9\pi}{2}\right)\right] = \frac{8}{3}(62.4)\left(\frac{27\pi}{2}\right) \approx 7057.274$$

(c) You want $\frac{3}{4}$ of the total area of 12π covered. Find h so that

$$2\int_0^h \frac{4}{3}\sqrt{9-y^2}\,dy = 3\pi$$

$$\int_0^h \sqrt{9-y^2}\,dy = \frac{9\pi}{8}$$

$$\frac{1}{2}\left[y\sqrt{9-y^2} + 9\arcsin\left(\frac{y}{3}\right)\right]_0^h = \frac{9\pi}{8}$$

$$h\sqrt{9-h^2} + 9\arcsin\left(\frac{h}{3}\right) = \frac{9\pi}{4}.$$

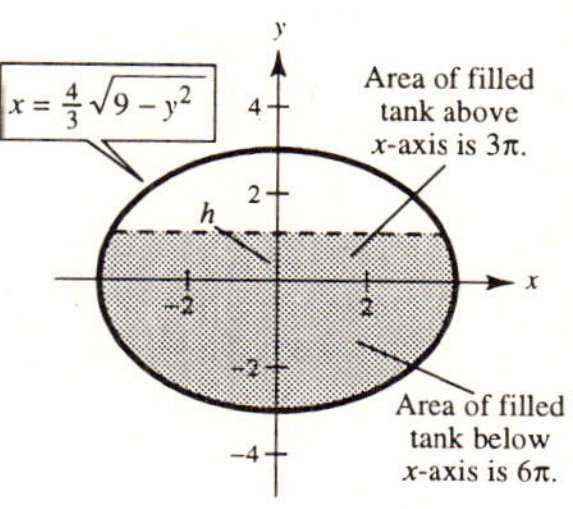

By Newton's Method, $h \approx 1.212$. Therefore, the total height of the water is $1.212 + 3 = 4.212$ ft.

(d) Area of ends $= 2(12\pi) = 24\pi$

Area of sides $=$ (Perimeter)(Length)

$$= 16\int_0^{\pi/2}\left(\sqrt{1-\left(\frac{7}{16}\right)\sin^2\theta}\right)d\theta(16) \text{ [from Example 5 of Section 9.1]}$$

$$\approx 256\left(\frac{\pi/2}{12}\right)\left[\sqrt{1-\left(\frac{7}{16}\right)\sin^2(0)} + 4\sqrt{1-\left(\frac{7}{16}\right)\sin^2\left(\frac{\pi}{8}\right)} + 2\sqrt{1-\left(\frac{7}{16}\right)\sin^2\left(\frac{\pi}{4}\right)}\right.$$
$$\left. + 4\sqrt{1-\left(\frac{7}{16}\right)\sin^2\left(\frac{3\pi}{8}\right)} + \sqrt{1-\left(\frac{7}{16}\right)\sin^2\left(\frac{\pi}{2}\right)}\right] \approx 353.65$$

Total area $= 24\pi + 353.65 \approx 429.05$

21. $x = 1 + 4t$

$y = 2 - 3t$

(a) $\frac{dy}{dx} = -\frac{3}{4}$

No horizontal tangents

(b) $t = \frac{x-1}{4}$

$y = 2 - \frac{3}{4}(x-1) = \frac{-3x+11}{4}$

(c)

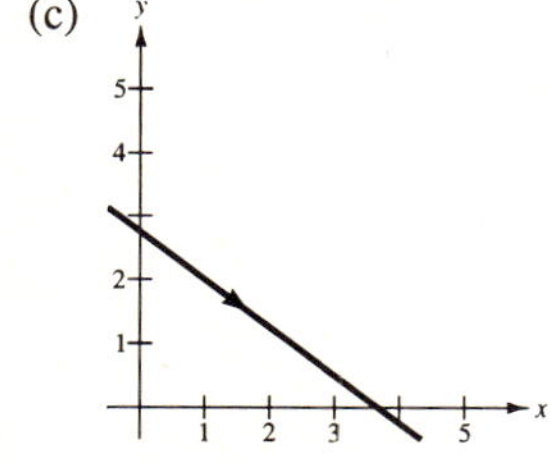

23. $x = \frac{1}{t}$

$y = 2t + 3$

(a) $\frac{dy}{dx} = \frac{2}{-1/t^2} = -2t^2$

No horizontal tangents $(t \neq 0)$

(b) $t = \frac{1}{x}$

$y = \frac{2}{x} + 3$

(c)

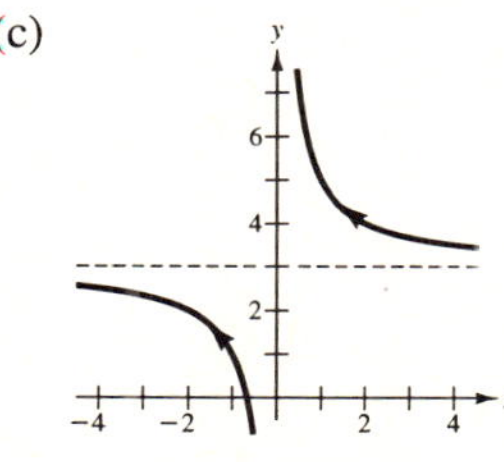

25. $x = \dfrac{1}{2t+1}$

$y = \dfrac{1}{t^2 - 2t}$

(a) $\dfrac{dy}{dx} = \dfrac{\dfrac{-(2t-2)}{(t^2-2t)^2}}{\dfrac{-2}{(2t+1)^2}} = \dfrac{(t-1)(2t+1)^2}{t^2(t-2)^2} = 0$ when $t = 1$.

Point of horizontal tangency: $\left(\frac{1}{3}, -1\right)$

(b) $2t + 1 = \dfrac{1}{x} \Rightarrow t = \dfrac{1}{2}\left(\dfrac{1}{x} - 1\right)$

$$y = \frac{1}{\dfrac{1}{2}\left(\dfrac{1-x}{x}\right)\left[\dfrac{1}{2}\left(\dfrac{1-x}{x}\right) - 2\right]}$$

$$= \frac{4x^2}{(1-x)^2 - 4x(1-x)} = \frac{4x^2}{(5x-1)(x-1)}$$

(c)

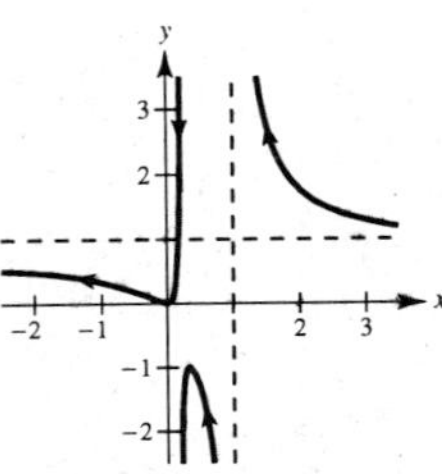

27. $x = 3 + 2\cos\theta$

$y = 2 + 5\sin\theta$

(a) $\dfrac{dy}{dx} = \dfrac{5\cos\theta}{-2\sin\theta} = -2.5\cot\theta = 0$ when $\theta = \dfrac{\pi}{2}, \dfrac{3\pi}{2}$.

Points of horizontal tangency: $(3, 7), (3, -3)$

(b) $\dfrac{(x-3)^2}{4} + \dfrac{(y-2)^2}{25} = 1$

(c)

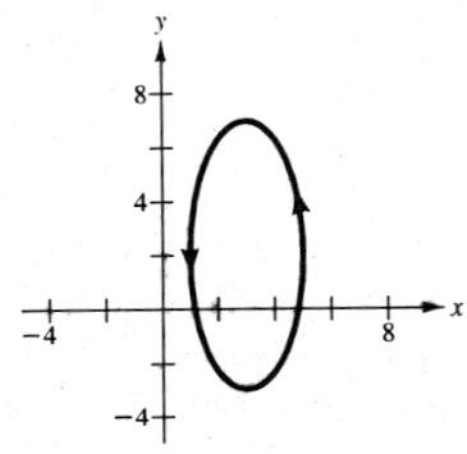

29. $x = \cos^3\theta$

$y = 4\sin^3\theta$

(a) $\dfrac{dy}{dx} = \dfrac{12\sin^2\theta\cos\theta}{3\cos^2\theta(-\sin\theta)}$

$= \dfrac{-4\sin\theta}{\cos\theta} = -4\tan\theta = 0$ when $\theta = 0, \pi$.

But, $\dfrac{dy}{dt} = \dfrac{dx}{dt} = 0$ at $\theta = 0, \pi$. Hence no points of horizontal tangency.

(b) $x^{2/3} + \left(\dfrac{y}{4}\right)^{2/3} = 1$

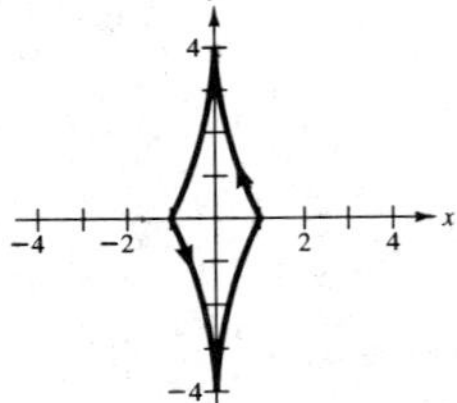

31. $x = \cot\theta$

$y = \sin 2\theta = 2\sin\theta\cos\theta$

(a), (c)

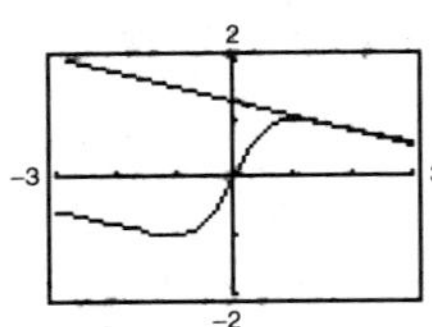

(b) At $\theta = \dfrac{\pi}{6}, \dfrac{dx}{d\theta} = -4, \dfrac{dy}{d\theta} = 1$, and $\dfrac{dy}{dx} = -\dfrac{1}{4}$

33. $x = 3 + (3 - (-2))t = 3 + 5t$

$y = 2 + (2 - 6)t = 2 - 4t$

(other answers possible)

35. $\dfrac{(x+3)^2}{16} + \dfrac{(y-4)^2}{9} = 1$

Let $\dfrac{(x+3)^2}{16} = \cos^2\theta$ and $\dfrac{(y-4)^2}{9} = \sin^2\theta$.

Then $x = -3 + 4\cos\theta$ and $y = 4 + 3\sin\theta$.

37. $x = \cos 3\theta + 5\cos\theta$

$y = \sin 3\theta + 5\sin\theta$

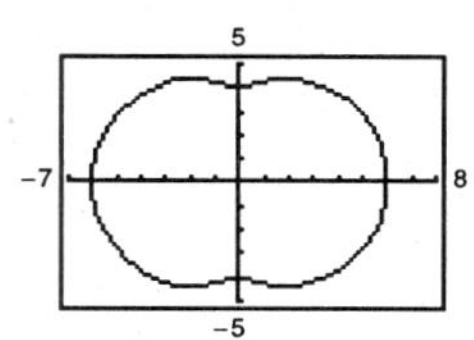

39. $x = a(\theta - \sin\theta)$

$y = a(1 - \cos\theta)$

$$\cos\theta = \frac{a-y}{a}$$

$$\theta = \arccos\left(\frac{a-y}{a}\right)$$

$$x = a\arccos\left(\frac{a-y}{a}\right) - a\sin\left[\arccos\left(\frac{a-y}{a}\right)\right]$$

$$= a\arccos\left(\frac{a-y}{a}\right) \pm \sqrt{2ay - y^2}$$

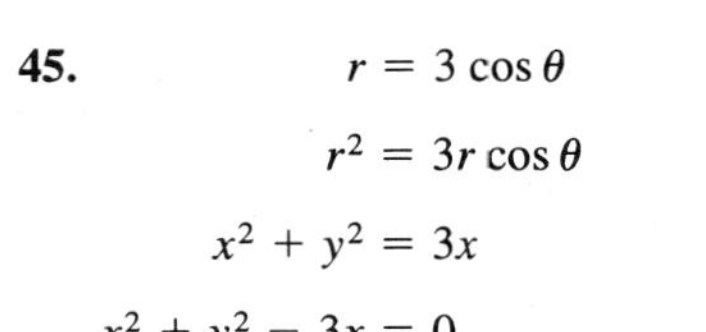

41. $x = r(\cos\theta + \theta\sin\theta)$

$y = r(\sin\theta - \theta\cos\theta)$

$$\frac{dx}{d\theta} = r\theta\cos\theta$$

$$\frac{dy}{d\theta} = r\theta\sin\theta$$

$$s = r\int_0^{\pi} \sqrt{\theta^2\cos^2\theta + \theta^2\sin^2\theta}\, d\theta$$

$$= r\int_0^{\pi} \theta\, d\theta = \frac{r}{2}\Big[\theta^2\Big]_0^{\pi} = \frac{1}{2}\pi^2 r$$

43. $(x, y) = (4, -4)$

$$r = \sqrt{4^2 + (-4)^2} = 4\sqrt{2}$$

$$\theta = 7\frac{\pi}{4}$$

$$(r, \theta) = \left(4\sqrt{2}, \frac{7\pi}{4}\right), \left(-4\sqrt{2}, \frac{3\pi}{4}\right)$$

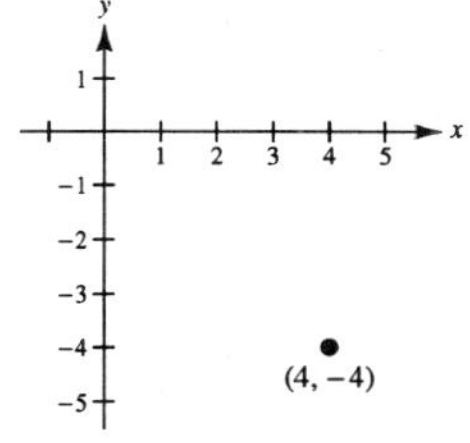

45.

$$r = 3\cos\theta$$
$$r^2 = 3r\cos\theta$$
$$x^2 + y^2 = 3x$$
$$x^2 + y^2 - 3x = 0$$

47.

$$r = -2(1 + \cos\theta)$$
$$r^2 = -2r(1 + \cos\theta)$$
$$x^2 + y^2 = -2\left(\pm\sqrt{x^2 + y^2}\right) - 2x$$
$$(x^2 + y^2 + 2x)^2 = 4(x^2 + y^2)$$

49.

$$r^2 = \cos 2\theta = \cos^2\theta - \sin^2\theta$$
$$r^4 = r^2\cos^2\theta - r^2\sin^2\theta$$
$$(x^2 + y^2)^2 = x^2 - y^2$$

51.

$$r = 4\cos 2\theta\sec\theta$$
$$= 4(2\cos^2\theta - 1)\left(\frac{1}{\cos\theta}\right)$$
$$r\cos\theta = 8\cos^2\theta - 4$$
$$x = 8\left(\frac{x^2}{x^2 + y^2}\right) - 4$$
$$x^3 + xy^2 = 4x^2 - 4y^2$$
$$y^2 = x^2\left(\frac{4 - x}{4 + x}\right)$$

53. $(x^2 + y^2)^2 = ax^2y$

$$r^4 = a(r^2 \cos^2 \theta)(r \sin \theta)$$

$$r = a \cos^2 \theta \sin \theta$$

55. $x^2 + y^2 = a^2\left(\arctan \dfrac{y}{x}\right)^2$

$$r^2 = a^2\theta^2$$

57. $r = 4$

Circle of radius 4

Centered at the pole

Symmetric to polar axis,

$\theta = \pi/2$, and pole

59. $r = -\sec \theta = \dfrac{-1}{\cos \theta}$

$r \cos \theta = -1, x = -1$

Vertical line

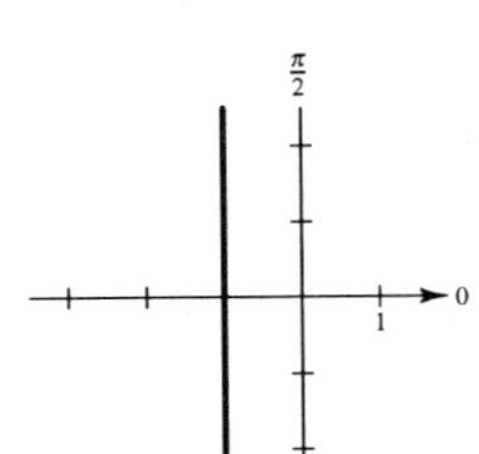

61. $r = -2(1 + \cos \theta)$

Cardioid

Symmetric to polar axis

63. $r = 4 - 3 \cos \theta$

Limaçon

Symmetric to polar axis

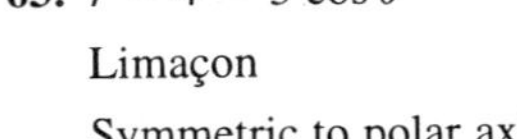

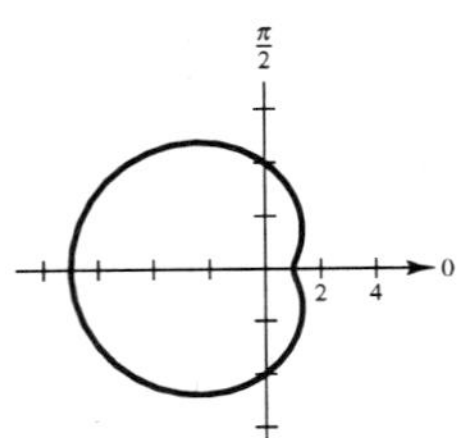

θ	0	$\frac{\pi}{3}$	$\frac{\pi}{2}$	$\frac{2\pi}{3}$	π
r	1	$\frac{5}{2}$	4	$\frac{11}{2}$	7

65. $r = -3 \cos(2\theta)$

Rose curve with four petals

Symmetric to polar axis, $\theta = \dfrac{\pi}{2}$, and pole

Relative extrema: $(-3, 0), \left(3, \dfrac{\pi}{2}\right), (-3, \pi), \left(3, \dfrac{3\pi}{2}\right)$

Tangents at the pole: $\theta = \dfrac{\pi}{4}, \dfrac{3\pi}{4}$

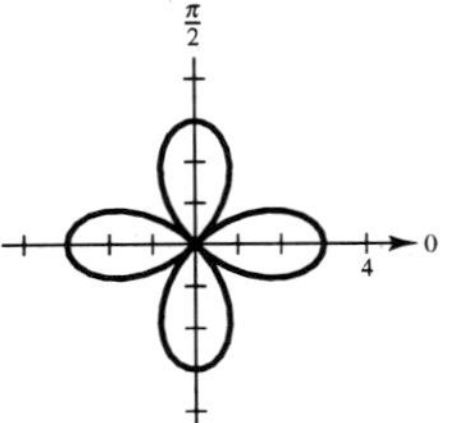

67. $r^2 = 4 \sin^2 (2\theta)$

$r = \pm 2 \sin(2\theta)$

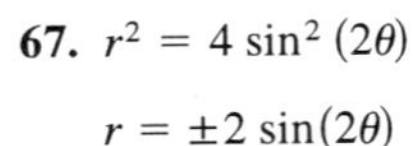

Rose curve with four petals

Symmetric to the polar axis, $\theta = \dfrac{\pi}{2}$, and pole

Relative extrema: $\left(\pm 2, \dfrac{\pi}{4}\right), \left(\pm 2, \dfrac{3\pi}{4}\right)$

Tangents at the pole: $\theta = 0, \dfrac{\pi}{2}$

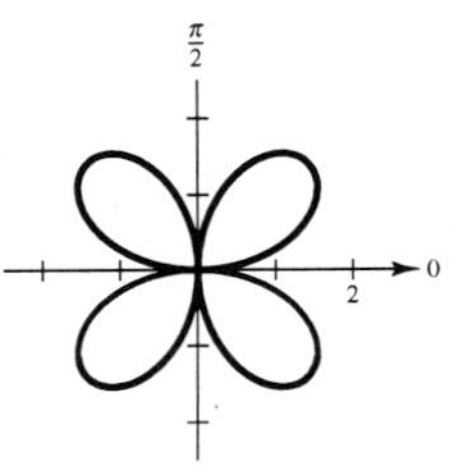

69. $r = \dfrac{2}{1 - \sin \theta}$

Parabola

Focus at the pole

Vertex: $\left(1, \dfrac{3\pi}{2}\right)$

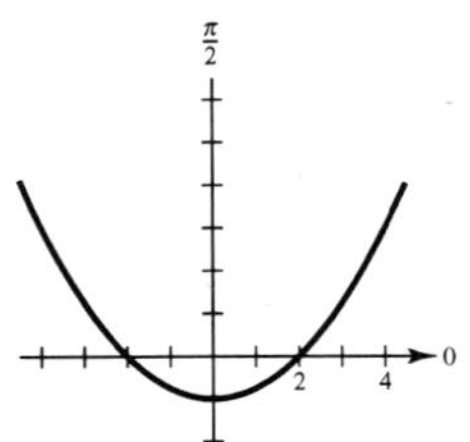

71. $r = \dfrac{3}{\cos[\theta - (\pi/4)]}$

Graph of $r = 3 \sec \theta$ rotated through an angle of $\pi/4$

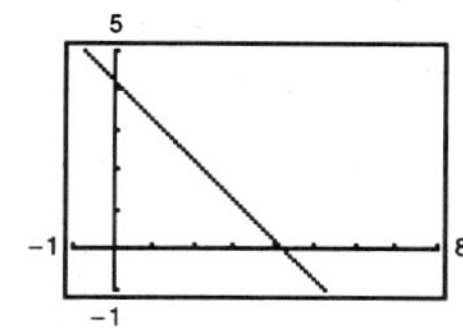

73. $r = 4 \cos 2\theta \sec \theta$

Strophoid

Symmetric to the polar axis

$r \Rightarrow -\infty$ as $\theta \Rightarrow \dfrac{\pi^-}{2}$

$r \Rightarrow -\infty$ as $\theta \Rightarrow \dfrac{-\pi^+}{2}$

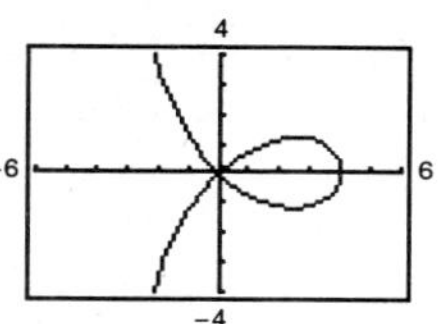

75. $r = 1 - 2 \cos \theta$

(a) The graph has polar symmetry and the tangents at the pole are

$$\theta = \frac{\pi}{3}, -\frac{\pi}{3}.$$

(b) $\dfrac{dy}{dx} = \dfrac{2 \sin^2 \theta + (1 - 2 \cos \theta) \cos \theta}{2 \sin \theta \cos \theta - (1 - 2 \cos \theta) \sin \theta}$

Horizontal tangents: $-4 \cos^2 \theta + \cos \theta + 2 = 0, \cos \theta = \dfrac{-1 \pm \sqrt{1 + 32}}{-8} = \dfrac{1 \pm \sqrt{33}}{8}$

When $\cos \theta = \dfrac{1 \pm \sqrt{33}}{8}, r = 1 - 2\left(\dfrac{1 + \sqrt{33}}{8}\right) = \dfrac{3 \mp \sqrt{33}}{4},$

$$\left[\frac{3 - \sqrt{33}}{4}, \arccos\left(\frac{1 + \sqrt{33}}{8}\right)\right] \approx (-0.686, 0.568)$$

$$\left[\frac{3 - \sqrt{33}}{4}, -\arccos\left(\frac{1 + \sqrt{33}}{8}\right)\right] \approx (-0.686, -0.568)$$

$$\left[\frac{3 + \sqrt{33}}{4}, \arccos\left(\frac{1 - \sqrt{33}}{8}\right)\right] \approx (2.186, 2.206)$$

$$\left[\frac{3 + \sqrt{33}}{4}, -\arccos\left(\frac{1 - \sqrt{33}}{8}\right)\right] \approx (2.186, -2.206).$$

Vertical tangents:

$$\sin \theta(4 \cos \theta - 1) = 0, \sin \theta = 0, \cos \theta = \frac{1}{4},$$

$$\theta = 0, \pi, \theta = \pm\arccos\left(\frac{1}{4}\right), (-1, 0), (3, \pi)$$

$$\left(\frac{1}{2}, \pm\arccos \frac{1}{4}\right) \approx (0.5, \pm 1.318)$$

(c)

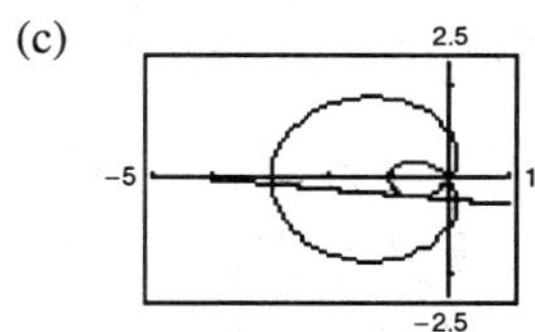

77. $r = 1 + \cos\theta,\ r = 1 - \cos\theta$

The points $(1, \pi/2)$ and $(1, 3\pi/2)$ are the two points of intersection (other than the pole). The slope of the graph of $r = 1 + \cos\theta$ is

$$m_1 = \frac{dy}{dx} = \frac{r'\sin\theta + r\cos\theta}{r'\cos\theta - r\sin\theta} = \frac{-\sin^2\theta + \cos\theta(1 + \cos\theta)}{-\sin\theta\cos\theta - \sin\theta(1 + \cos\theta)}.$$

At $(1, \pi/2)$, $m_1 = -1/-1 = 1$ and at $(1, 3\pi/2)$, $m_1 = -1/1 = -1$. The slope of the graph of $r = 1 - \cos\theta$ is

$$m_2 = \frac{dy}{dx} = \frac{\sin^2\theta + \cos\theta(1 - \cos\theta)}{\sin\theta\cos\theta - \sin\theta(1 - \cos\theta)}.$$

At $(1, \pi/2)$, $m_2 = 1/-1 = -1$ and at $(1, 3\pi/2)$, $m_2 = 1/1 = 1$. In both cases, $m_1 = -1/m_2$ and we conclude that the graphs are orthogonal at $(1, \pi/2)$ and $(1, 3\pi/2)$.

79. Circle: $r = 3\sin\theta$

$$\frac{dy}{dx} = \frac{3\cos\theta\sin\theta + 3\sin\theta\cos\theta}{3\cos\theta\cos\theta - 3\sin\theta\sin\theta} = \frac{\sin 2\theta}{\cos^2\theta - \sin^2\theta} = \tan 2\theta \text{ at } \theta = \frac{\pi}{6},\ \frac{dy}{dx} = \sqrt{3}$$

Limaçon: $r = 4 - 5\sin\theta$

$$\frac{dy}{dx} = \frac{-5\cos\theta\sin\theta + (4 - 5\sin\theta)\cos\theta}{-5\cos\theta\cos\theta - (4 - 5\sin\theta)\sin\theta} \text{ at } \theta = \frac{\pi}{6},\ \frac{dy}{dx} = \frac{\sqrt{3}}{9}$$

Let α be the angle between the curves:

$$\tan\alpha = \frac{\sqrt{3} - (\sqrt{3}/9)}{1 + (1/3)} = \frac{2\sqrt{3}}{3}.$$

Therefore, $\alpha = \arctan\left(\dfrac{2\sqrt{3}}{3}\right) \approx 49.1°$.

81. $r = 2 + \cos\theta$

$$A = 2\left[\frac{1}{2}\int_0^{\pi} (2 + \cos\theta)^2\, d\theta\right] \approx 14.14 \quad \left(\frac{9\pi}{2}\right)$$

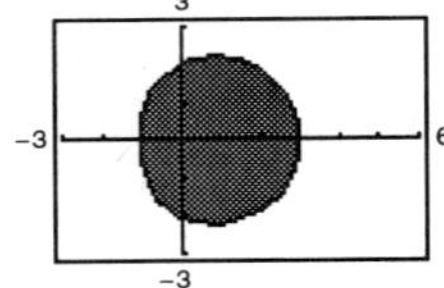

83. $r = \sin\theta \cdot \cos^2\theta$

$$A = 2\left[\frac{1}{2}\int_0^{\pi/2} (\sin\theta\cos^2\theta)^2\, d\theta\right]$$

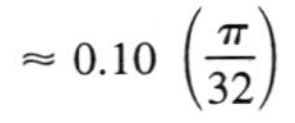

$$\approx 0.10 \quad \left(\frac{\pi}{32}\right)$$

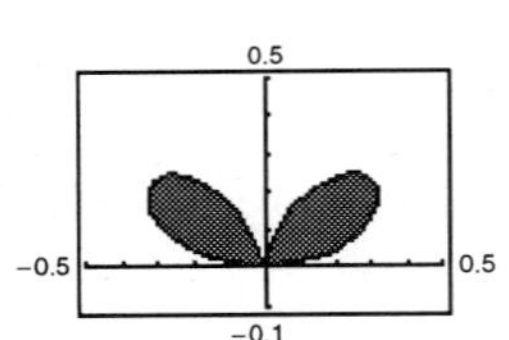

85. $r^2 = 4\sin 2\theta$

$$A = 2\left[\frac{1}{2}\int_0^{\pi/2} 4\sin 2\theta\, d\theta\right] = 4$$

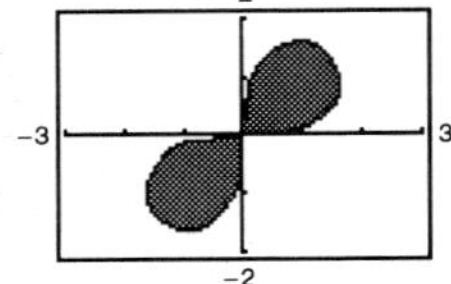

87. $r = 4\cos\theta,\ r = 2$

$$A = 2\left[\frac{1}{2}\int_0^{\pi/3} 4\, d\theta + \frac{1}{2}\int_{\pi/3}^{\pi/2} (4\cos\theta)^2\, d\theta\right] \approx 4.91$$

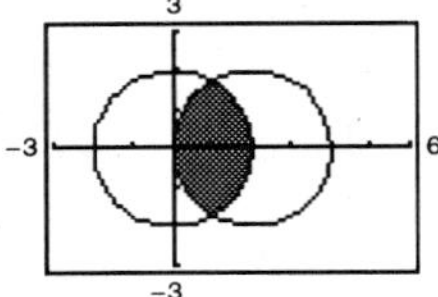

89. $s = 2\displaystyle\int_0^{\pi} \sqrt{a^2(1 - \cos\theta)^2 + a^2\sin^2\theta}\, d\theta$

$$= 2\sqrt{2}\,a\int_0^{\pi} \sqrt{1 - \cos\theta}\, d\theta = 2\sqrt{2}\,a\int_0^{\pi} \frac{\sin\theta}{\sqrt{1 + \cos\theta}}\, d\theta = \left[-4\sqrt{2}\,a(1 + \cos\theta)^{1/2}\right]_0^{\pi} = 8a$$

91. Circle

Center: $\left(5, \frac{\pi}{2}\right) = (0, 5)$ in rectangular coordinates

Solution point: $(0, 0)$

$$x^2 + (y - 5)^5 = 25$$

$$x^2 + y^2 - 10y = 0$$

$$r^2 - 10r \sin \theta = 0$$

$$r = 10 \sin \theta$$

93. Parabola

Vertex: $(2, \pi)$

Focus: $(0, 0)$

$e = 1, d = 4$

$$r = \frac{4}{1 - \cos \theta}$$

95. Ellipse

Vertices: $(5, 0), (1, \pi)$

Focus: $(0, 0)$

$a = 3, c = 2, e = \frac{2}{3}, d = \frac{5}{2}$

$$r = \frac{\left(\frac{2}{3}\right)\left(\frac{5}{2}\right)}{1 - \left(\frac{2}{3}\right) \cos \theta} = \frac{5}{3 - 2 \cos \theta}$$

CHAPTER 10
Vectors and the Geometry of Space

CHAPTER 10
Vectors and the Geometry of Space

Section 10.1 Vectors in the Plane

Solutions to Odd-Numbered Exercises

1. (a) $\mathbf{v} = \langle 5-1, 3-1\rangle = \langle 4, 2\rangle$

(b)

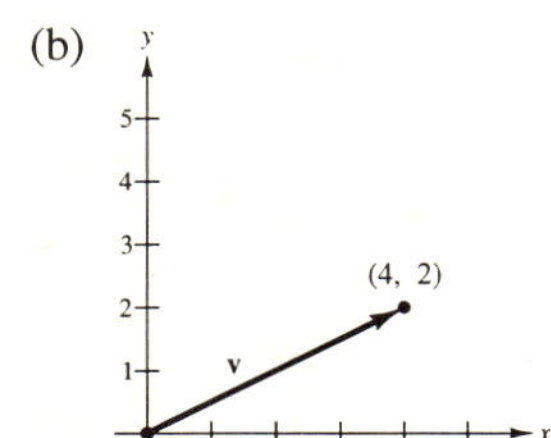

3. (a) $\mathbf{v} = \langle -4-3, -2-(-2)\rangle$
$= \langle -7, 0\rangle$

(b)

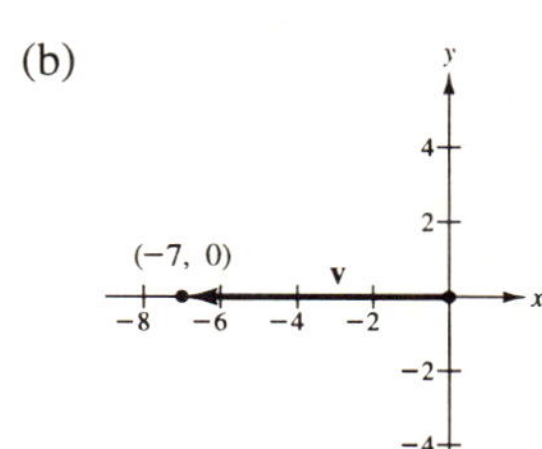

5. (b) $\mathbf{v} = \langle 5-1, 5-2\rangle = \langle 4, 3\rangle$

(a) and (c)

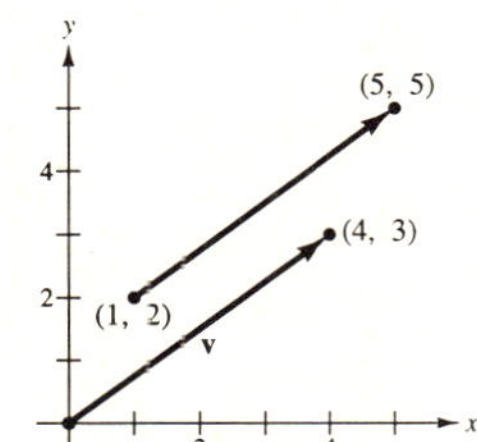

7. (b) $\mathbf{v} = \langle 6-10, -1-2\rangle$
$= \langle -4, -3\rangle$

(a) and (c).

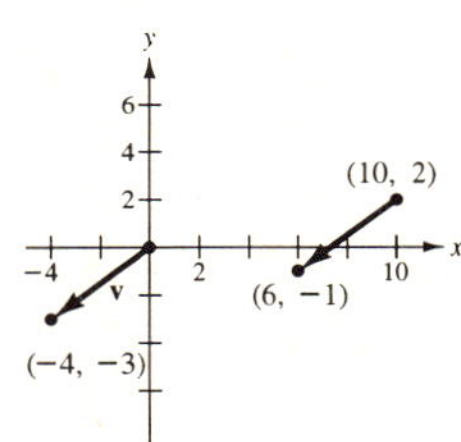

9. (b) $\mathbf{v} = \langle 6-6, 6-2\rangle = \langle 0, 4\rangle$

(a) and (c).

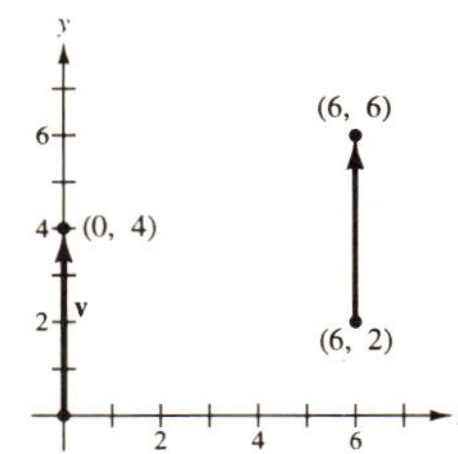

11. (b) $\mathbf{v} = \left\langle \frac{1}{2}-\frac{3}{2}, 3-\frac{4}{3}\right\rangle = \left\langle -1, \frac{5}{3}\right\rangle$

(a) and (c).

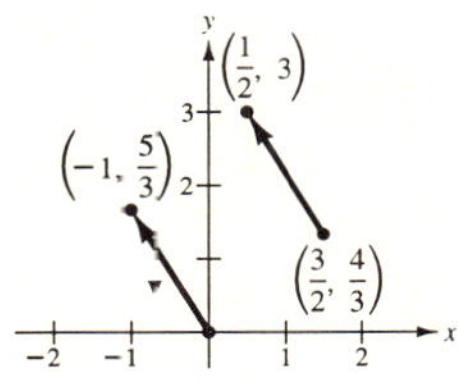

13. (a) $2\mathbf{v} = \langle 4, 6\rangle$

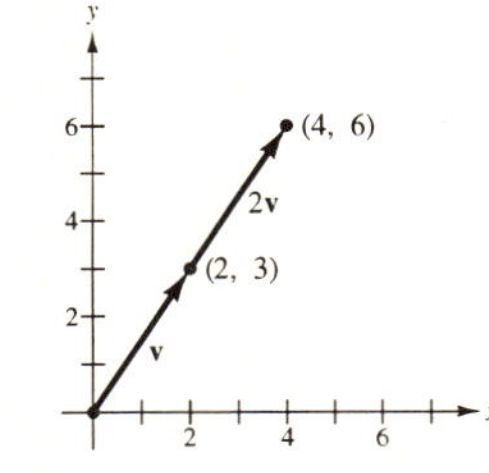

(b) $-3\mathbf{v} = \langle -6, -9\rangle$

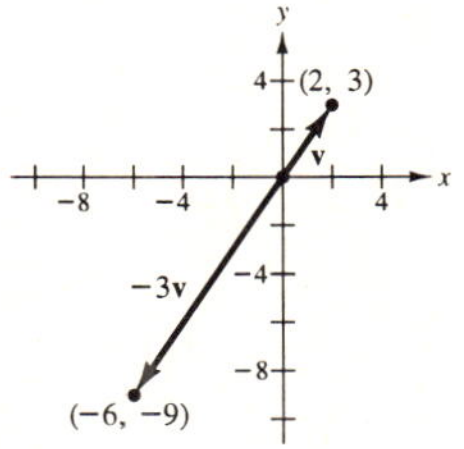

(c) $\frac{7}{2}\mathbf{v} = \left\langle 7, \frac{21}{2}\right\rangle$

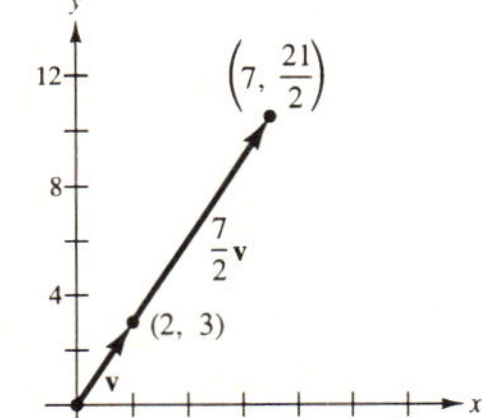

(d) $\frac{2}{3}\mathbf{v} = \left\langle \frac{4}{3}, 2\right\rangle$

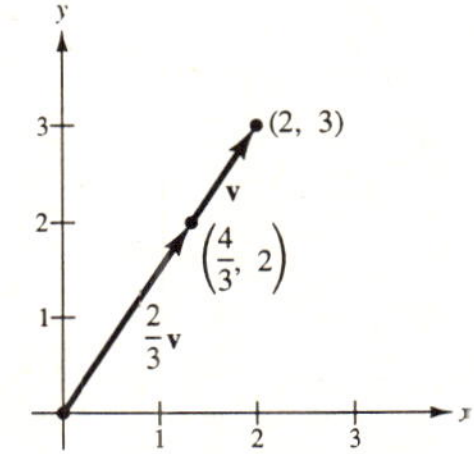

15.

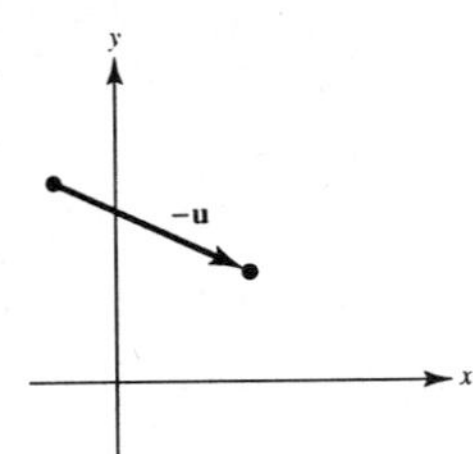

17.

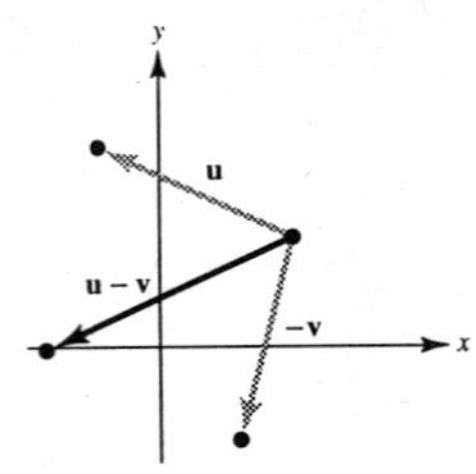

19. $\mathbf{v} = \frac{3}{2}(2\mathbf{i} - \mathbf{j}) = 3\mathbf{i} - \frac{3}{2}\mathbf{j}$

$= \left\langle 3, -\frac{3}{2} \right\rangle$

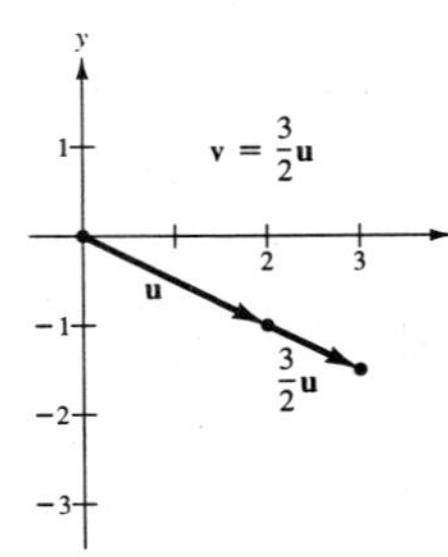

21. $\mathbf{v} = (2\mathbf{i} - \mathbf{j}) + 2(\mathbf{i} + 2\mathbf{j})$

$= 4\mathbf{i} + 3\mathbf{j} = \langle 4, 3 \rangle$

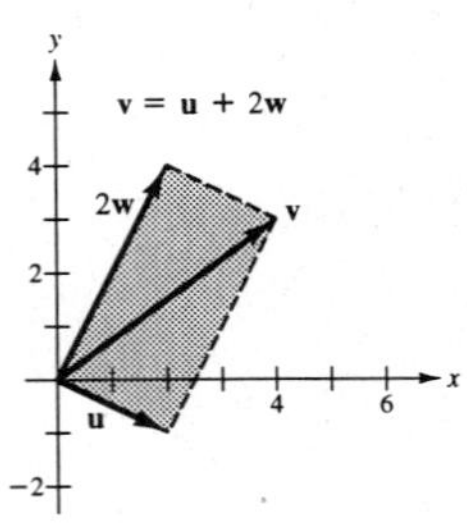

For Exercises 23–27, $a\mathbf{u} + b\mathbf{w} = a(\mathbf{i} + 2\mathbf{j}) + b(\mathbf{i} - \mathbf{j}) = (a + b)\mathbf{i} + (2a - b)\mathbf{j}$.

23. $\mathbf{v} = 2\mathbf{i} + \mathbf{j}$. Therefore, $a + b = 2$, $2a - b = 1$. Solving simultaneously, we have $a = 1, b = 1$.

25. $\mathbf{v} = 3\mathbf{i}$. Therefore, $a + b = 3$, $2a - b = 0$. Solving simultaneously, we have $a = 1, b = 2$.

27. $\mathbf{v} = \mathbf{i} + \mathbf{j}$. Therefore, $a + b = 1$, $2a - b = 1$. Solving simultaneously, we have $a = \frac{2}{3}, b = \frac{1}{3}$.

29. $u_1 - 4 = -1$

$u_2 - 2 = 3$

$u_1 = 3$

$u_2 = 5$

$Q = (3, 5)$

31. $\|\mathbf{v}\| = \sqrt{16 + 9} = 5$

33. $\|\mathbf{v}\| = \sqrt{36 + 25}$

$= \sqrt{61}$

35. $\|\mathbf{v}\| = \sqrt{0 + 16} = 4$

37. $\|\mathbf{u}\| = \langle 1, -1 \rangle, \mathbf{v} = \langle -1, 2 \rangle$

(a) $\|\mathbf{u}\| = \sqrt{1 + 1} = \sqrt{2}$

(b) $\|\mathbf{v}\| = \sqrt{1 + 4} = \sqrt{5}$

(c) $\mathbf{u} + \mathbf{v} = \langle 0, 1 \rangle$

$\|\mathbf{u} + \mathbf{v}\| = \sqrt{0 + 1} = 1$

(d) $\dfrac{\mathbf{u}}{\|\mathbf{u}\|} = \dfrac{1}{\sqrt{2}}\langle 1, -1 \rangle$

$\left\| \dfrac{\mathbf{u}}{\|\mathbf{u}\|} \right\| = 1$

(e) $\dfrac{\mathbf{v}}{\|\mathbf{v}\|} = \dfrac{1}{\sqrt{5}}\langle -1, 2 \rangle$

$\left\| \dfrac{\mathbf{v}}{\|\mathbf{v}\|} \right\| = 1$

(f) $\dfrac{\mathbf{u} + \mathbf{v}}{\|\mathbf{u} + \mathbf{v}\|} = \langle 0, 1 \rangle$

$\left\| \dfrac{\mathbf{u} + \mathbf{v}}{\|\mathbf{u} + \mathbf{v}\|} \right\| = 1$

39. $\mathbf{u} = \left\langle 1, \dfrac{1}{2} \right\rangle, \mathbf{v} = \langle 2, 3 \rangle$

(a) $\|\mathbf{u}\| = \sqrt{1 + \dfrac{1}{4}} = \dfrac{\sqrt{5}}{2}$

(b) $\|\mathbf{v}\| = \sqrt{4 + 9} = \sqrt{13}$

(c) $\mathbf{u} + \mathbf{v} = \left\langle 3, \dfrac{7}{2} \right\rangle$

$\|\mathbf{u} + \mathbf{v}\| = \sqrt{9 + \dfrac{49}{4}} = \dfrac{\sqrt{85}}{2}$

(d) $\dfrac{\mathbf{u}}{\|\mathbf{u}\|} = \dfrac{2}{\sqrt{5}}\left\langle 1, \dfrac{1}{2} \right\rangle$

$\left\| \dfrac{\mathbf{u}}{\|\mathbf{u}\|} \right\| = 1$

(e) $\dfrac{\mathbf{v}}{\|\mathbf{v}\|} = \dfrac{1}{\sqrt{13}}\langle 2, 3 \rangle$

$\left\| \dfrac{\mathbf{v}}{\|\mathbf{v}\|} \right\| = 1$

(f) $\dfrac{\mathbf{u} + \mathbf{v}}{\|\mathbf{u} + \mathbf{v}\|} = \dfrac{2}{\sqrt{85}}\left\langle 3, \dfrac{7}{2} \right\rangle$

$\left\| \dfrac{\mathbf{u} + \mathbf{v}}{\|\mathbf{u} + \mathbf{v}\|} \right\| = 1$

41.
$$\mathbf{u} = \langle 2, 1\rangle$$
$$\|\mathbf{u}\| = \sqrt{5} \approx 2.236$$
$$\mathbf{v} = \langle 5, 4\rangle$$
$$\|\mathbf{v}\| = \sqrt{41} \approx 6.403$$
$$\mathbf{u} + \mathbf{v} = \langle 7, 5\rangle$$
$$\|\mathbf{u} + \mathbf{v}\| = \sqrt{74} \approx 8.602$$
$$\|\mathbf{u} + \mathbf{v}\| \le \|\mathbf{u}\| + \|\mathbf{v}\|$$

43.
$$\frac{\mathbf{u}}{\|\mathbf{u}\|} = \frac{1}{\sqrt{2}}\langle 1, 1\rangle$$
$$4\left(\frac{\mathbf{u}}{\|\mathbf{u}\|}\right) = 2\sqrt{2}\langle 1, 1\rangle$$
$$\mathbf{v} = \langle 2\sqrt{2}, 2\sqrt{2}\rangle$$

45.
$$\frac{\mathbf{u}}{\|\mathbf{u}\|} = \frac{1}{2\sqrt{3}}\langle \sqrt{3}, 3\rangle$$
$$2\left(\frac{\mathbf{u}}{\|\mathbf{u}\|}\right) = \frac{1}{\sqrt{3}}\langle \sqrt{3}, 3\rangle$$
$$\mathbf{v} = \langle 1, \sqrt{3}\rangle$$

47. $y = x^3, y' = 3x^2 = 3$ at $x = 1$.

(a) $m = 3$. Let $\mathbf{w} = \langle 1, 3\rangle$, then
$$\frac{\mathbf{w}}{\|\mathbf{w}\|} = \pm\frac{1}{\sqrt{10}}\langle 1, 3\rangle.$$

(b) $m = -\frac{1}{3}$. Let $\mathbf{w} = \langle 3, -1\rangle$, then
$$\frac{\mathbf{w}}{\|\mathbf{w}\|} = \pm\frac{1}{\sqrt{10}}\langle 3, -1\rangle.$$

49. $f(x) = \sqrt{25 - x^2}$
$$f'(x) = \frac{-x}{\sqrt{25 - x^2}} = \frac{-3}{4} \text{ at } x = 3.$$

(a) $m = -\frac{3}{4}$. Let $\mathbf{w} = \langle -4, 3\rangle$, then
$$\frac{\mathbf{w}}{\|\mathbf{w}\|} = \pm\frac{1}{5}\langle -4, 3\rangle.$$

(b) $m = \frac{4}{3}$. Let $\mathbf{w} = \langle 3, 4\rangle$, then
$$\frac{\mathbf{w}}{\|\mathbf{w}\|} = \pm\frac{1}{5}\langle 3, 4\rangle$$

51. $\mathbf{v} = 3[(\cos 0°)\mathbf{i} + (\sin 0°)\mathbf{j}] = 3\mathbf{i} = \langle 3, 0\rangle$

53. $\mathbf{v} = 2[(\cos 150°)\mathbf{i} + (\sin 150°)\mathbf{j}]$
$$= -\sqrt{3}\mathbf{i} + \mathbf{j} = \langle -\sqrt{3}, 1\rangle$$

55.
$$\mathbf{u} = \mathbf{i}$$
$$\mathbf{v} = \frac{3\sqrt{2}}{2}\mathbf{i} + \frac{3\sqrt{2}}{2}\mathbf{j}$$
$$\mathbf{u} + \mathbf{v} = \left(\frac{2 + 3\sqrt{2}}{2}\right)\mathbf{i} + \frac{3\sqrt{2}}{2}\mathbf{j}$$

57.
$$\mathbf{u} = 2(\cos 4)\mathbf{i} + 2(\sin 4)\mathbf{j}$$
$$\mathbf{v} = (\cos 2)\mathbf{i} + (\sin 2)\mathbf{j}$$
$$\mathbf{u} + \mathbf{v} = (2\cos 4 + \cos 2)\mathbf{i} + (2\sin 4 + \sin 2)\mathbf{j}$$

59.
$$\mathbf{u} = \frac{\sqrt{2}}{2}\mathbf{i} + \frac{\sqrt{2}}{2}\mathbf{j}$$
$$\mathbf{u} + \mathbf{v} = \sqrt{2}\mathbf{j}$$
$$\mathbf{v} = (\mathbf{u} + \mathbf{v}) - \mathbf{u} = -\frac{\sqrt{2}}{2}\mathbf{i} + \frac{\sqrt{2}}{2}\mathbf{j}$$

61. Programs will vary.

63. $\|\mathbf{F}_1\| = 2, \theta_{\mathbf{F}_1} = 33°$
$$\|\mathbf{F}_2\| = 3, \theta_{\mathbf{F}_2} = -125°$$
$$\|\mathbf{F}_3\| = 2.5, \theta_{\mathbf{F}_3} = 110°$$
$$\|\mathbf{R}\| = \|\mathbf{F}_1 + \mathbf{F}_2 + \mathbf{F}_3\| \approx 1.33$$
$$\theta_{\mathbf{R}} = \theta_{\mathbf{F}_1 + \mathbf{F}_2 + \mathbf{F}_3} \approx 132.5°$$

65. (a) The forces act along the same direction. $\theta = 0°$.

(b) The forces cancel out each other. $\theta = 180°$.

(c) No, the magnitude of the resultant can not be greater than the sum.

67. (a) $180(\cos 30\mathbf{i} + \sin 30\mathbf{j}) + 275\mathbf{i} = 430.88\mathbf{i} + 90\mathbf{j}$

Direction: $\alpha = \arctan\left(\dfrac{90}{430.88}\right) = 0.206(= 11.8°)$

Magnitude: $\sqrt{430.88^2 + 90^2} = 440.18$ newtons

—CONTINUED—

(b) $M = \sqrt{(275 + 180\cos\theta)^2 + (180\sin\theta)^2}$
$$\alpha = \arctan\left[\frac{180\sin\theta}{275 + 180\cos\theta}\right]$$

67. —CONTINUED—

(c)

θ	0°	30°	60°	90°	120°	150°	180°
M	455	440.2	396.9	328.7	241.9	149.3	95
α	0°	11.8°	23.1°	33.2°	40.1°	37.1°	0

(d)

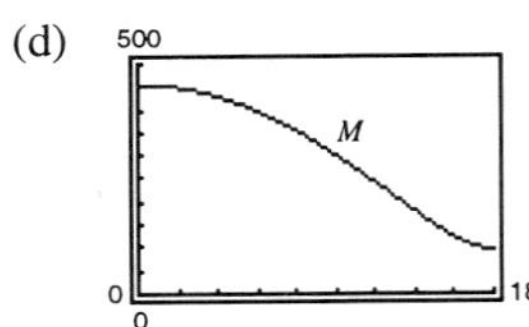

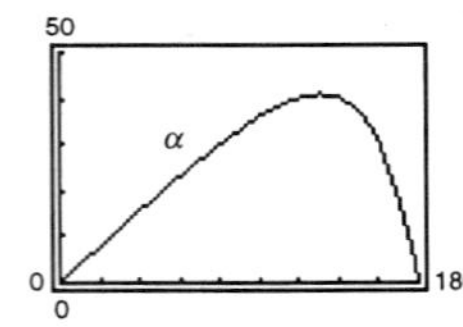

(e) M decreases because the forces change from acting in the same direction to acting in the opposite direction as θ increases from 0° to 180°.

69. $\mathbf{F}_1 + \mathbf{F}_2 + \mathbf{F}_3 = (75\cos 30°\mathbf{i} + 75\sin 30°\mathbf{j}) + (100\cos 45°\mathbf{i} + 100\sin 45°\mathbf{j}) + (125\cos 120°\mathbf{i} + 125\sin 120°\mathbf{j})$

$$= \left(\frac{75}{2}\sqrt{3} + 50\sqrt{2} - \frac{125}{2}\right)\mathbf{i} + \left(\frac{75}{2} + 50\sqrt{2} + \frac{125}{2}\sqrt{3}\right)\mathbf{j}$$

$$\|\mathbf{R}\| = \|\mathbf{F}_1 + \mathbf{F}_2 + \mathbf{F}_3\| \approx 228.5 \text{ lb}$$

$$\theta_{\mathbf{R}} = \theta_{\mathbf{F}_1+\mathbf{F}_2+\mathbf{F}_3} \approx 71.3°$$

71. $(-4, -1), (6, 5), (10, 3)$

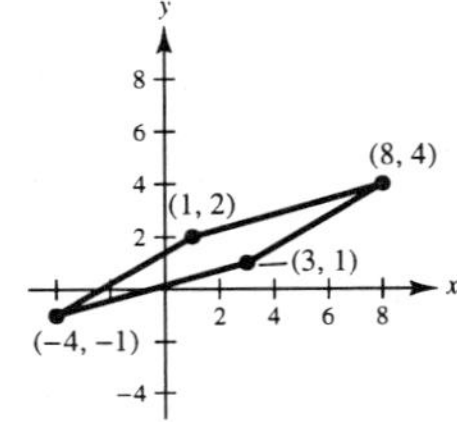

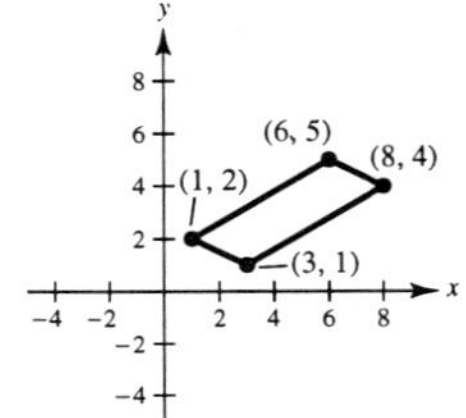

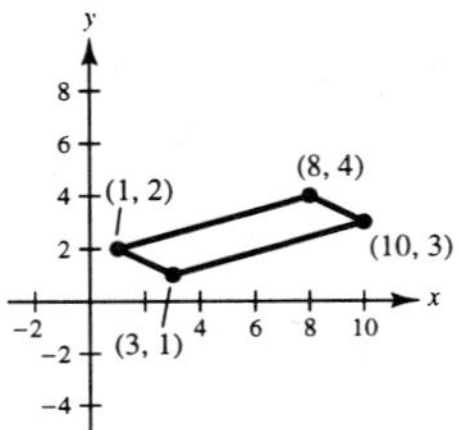

73. (a) $\mathbf{u} = \|\mathbf{u}\|(\cos 0\,\mathbf{i} + \sin 0\,\mathbf{j}) = \|\mathbf{u}\|\mathbf{i}$

Downward force $\mathbf{w} = -\mathbf{j}$

$$\mathbf{T} = \|\mathbf{T}\|(\cos(90° + \theta)\mathbf{i} + \sin(90° + \theta)\mathbf{j})$$

$$= \|\mathbf{T}\|(-\sin\theta\,\mathbf{i} + \cos\theta\,\mathbf{j})$$

$$\mathbf{0} = \mathbf{u} + \mathbf{w} + \mathbf{T} = \|\mathbf{u}\|\mathbf{i} - \mathbf{j} + \|\mathbf{T}\|(-\sin\theta\,\mathbf{i} + \cos\theta\,\mathbf{j})$$

$$\|\mathbf{u}\| = \sin\theta\,\|\mathbf{T}\|$$

$$1 = \cos\theta\,\|\mathbf{T}\|$$

If $\theta = 30°$, $\|\mathbf{u}\| = (1/2)\|\mathbf{T}\|$ and $1 = (\sqrt{3}/2)\|\mathbf{T}\|$

$$\Rightarrow \|\mathbf{T}\| = \frac{2}{\sqrt{3}} \approx 1.1547 \text{ lb}$$

and

$$\|\mathbf{u}\| = \frac{1}{2}\left(\frac{2}{\sqrt{3}}\right) \approx 0.5774 \text{ lb}$$

(b) From part (a), $\|\mathbf{u}\| = \tan\theta$ and $\|\mathbf{T}\| = \sec\theta$.

Domain: $0 \le \theta \le 90°$

(c)

θ	0°	10°	20°	30°	40°	50°	60°
T	1	1.0154	1.0642	1.1547	1.3054	1.5557	2
$\|\mathbf{u}\|$	0	0.1763	0.3640	0.5774	0.8391	1.1918	1.7321

(d)

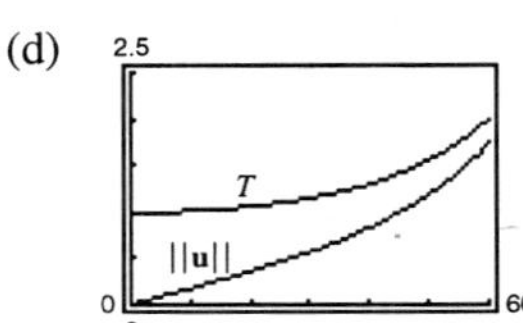

(e) Both are increasing functions.

(f) $\lim_{\theta\to\pi/2^-} T = \infty$ and $\lim_{\theta\to\pi/2^-} \|\mathbf{u}\| = \infty$.

75. Horizontal component $= \|\mathbf{v}\| \cos\theta = 1200\cos 6° \approx 1193.43$ ft/sec

Vertical component $= \|\mathbf{v}\| \sin\theta = 1200\sin 6° \approx 125.43$ ft/sec

77.
$$\mathbf{u} = 900[\cos 148°\,\mathbf{i} + \sin 148°\,\mathbf{j}]$$
$$\mathbf{v} = 100[\cos 45°\,\mathbf{i} + \sin 45°\,\mathbf{j}]$$
$$\mathbf{u} + \mathbf{v} = [900\cos 148° + 100\cos 45°]\mathbf{i} + [900\sin 148° + 100\sin 45°]\mathbf{j}$$
$$\approx -692.53\,\mathbf{i} + 547.64\,\mathbf{j}$$
$$\theta \approx \arctan\left(\frac{547.64}{-692.53}\right) \approx -38.34°. \quad 38.34° \text{ North of West.}$$
$$\|\mathbf{u} + \mathbf{v}\| = \sqrt{(-692.53)^2 + (547.64)^2} \approx 882.9 \text{ km/hr.}$$

79. $\mathbf{F}_1 = \mathbf{F}_2 + \mathbf{F}_3 = 0$

$-3600\mathbf{j} + T_2(\cos 35°\mathbf{i} - \sin 35°\,\mathbf{j}) + T_3(\cos 92°\mathbf{i} + \sin 92°\mathbf{j}) = 0$

$T_2\cos 35° + T_3\cos 92° = 0$

$-T_2\cos 35° + T_3\sin 92° = 3600$

$T_2 = \dfrac{-T_3\cos 92°}{\cos 35°} \Rightarrow \dfrac{T_3\cos 92°}{\cos 35°}\sin 35° + T_3\sin 92° = 3600$ and $T_3(0.97495) = 3600 \Rightarrow T_3 \approx 3692.48$

Finally, $T_2 = 157.32$

81. Let the triangle have vertices at $(0, 0)$, $(a, 0)$, and (b, c). Let $\mathbf{u}$ be the vector joining $(0, 0)$ and (b, c), as indicated in the figure. Then $\mathbf{v}$, the vector joining the midpoints, is

$$\mathbf{v} = \left(\frac{a+b}{2} - \frac{a}{2}\right)\mathbf{i} + \frac{c}{2}\mathbf{j}$$
$$= \frac{b}{2}\mathbf{i} + \frac{c}{2}\mathbf{j} = \frac{1}{2}(b\mathbf{i} + c\mathbf{j}) = \frac{1}{2}\mathbf{u}$$

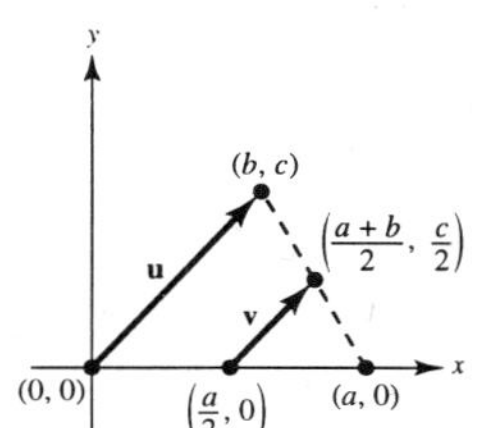

83. $\mathbf{w} = \|\mathbf{u}\|\mathbf{v} + \|\mathbf{v}\|\mathbf{u}$

$$= \|\mathbf{u}\|\big[\|\mathbf{v}\|\cos\theta_{\mathbf{v}}\mathbf{i} + \|\mathbf{v}\|\sin\theta_{\mathbf{v}}\mathbf{j}\big] + \|\mathbf{v}\|\big[\|\mathbf{u}\|\cos\theta_{\mathbf{u}}\mathbf{i} + \|\mathbf{u}\|\sin\theta_{\mathbf{u}}\mathbf{j}\big] = \|\mathbf{u}\|\,\|\mathbf{v}\|\big[(\cos\theta_{\mathbf{u}} + \cos\theta_{\mathbf{v}})\mathbf{i} + (\sin\theta_{\mathbf{u}} + \sin\theta_{\mathbf{v}})\mathbf{j}\big]$$
$$= 2\|\mathbf{u}\|\,\|\mathbf{v}\|\left[\cos\left(\frac{\theta_{\mathbf{u}} + \theta_{\mathbf{v}}}{2}\right)\cos\left(\frac{\theta_{\mathbf{u}} - \theta_{\mathbf{v}}}{2}\right)\mathbf{i} + \sin\left(\frac{\theta_{\mathbf{u}} + \theta_{\mathbf{v}}}{2}\right)\cos\left(\frac{\theta_{\mathbf{u}} - \theta_{\mathbf{v}}}{2}\right)\mathbf{j}\right]$$
$$\tan\theta_{\mathbf{w}} = \frac{\sin\left(\frac{\theta_{\mathbf{u}} + \theta_{\mathbf{v}}}{2}\right)\cos\left(\frac{\theta_{\mathbf{u}} - \theta_{\mathbf{v}}}{2}\right)}{\cos\left(\frac{\theta_{\mathbf{u}} + \theta_{\mathbf{v}}}{2}\right)\cos\left(\frac{\theta_{\mathbf{u}} - \theta_{\mathbf{v}}}{2}\right)} = \tan\left(\frac{\theta_{\mathbf{u}} + \theta_{\mathbf{v}}}{2}\right)$$

Thus, $\theta_{\mathbf{w}} = (\theta_{\mathbf{u}} + \theta_{\mathbf{v}})/2$ and $\mathbf{w}$ bisects the angle between $\mathbf{u}$ and $\mathbf{v}$.

85. True

87. False

$a = b = 0$

89. True

Section 10.2 Space Coordinates and Vectors in Space

1.

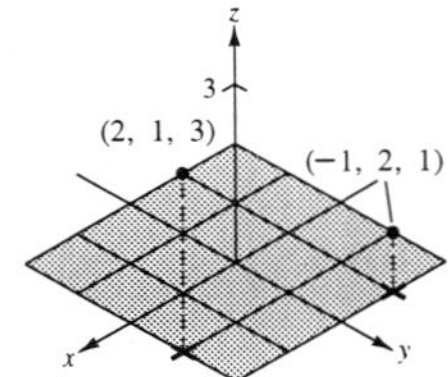

3.

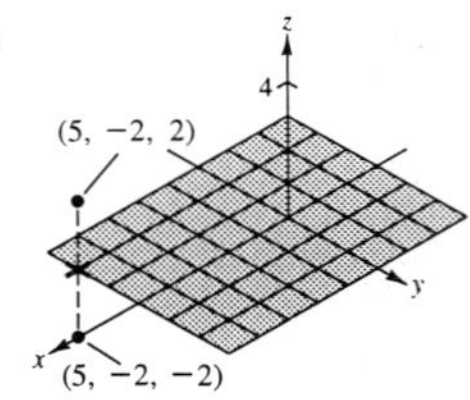

5. $A(2, 3, 4)$

$B(-1, -2, 2)$

7. $x = -3, y = 4, z = 5$: $(-3, 4, 5)$

9. $y = z = 0, x = 10$: $(10, 0, 0)$

11. The z-coordinate is 0.

13. The point (x, y, z) is below the xy-plane, and below either quadrant I or III.

15. The point could be above the xy-plane and thus above quadrants II or IV, or below the xy-plane, and thus below quadrants I or III.

17. $A(0, 0, 0), B(2, 2, 1), C(2, -4, 4)$

$|AB| = \sqrt{4 + 4 + 1} = 3$

$|AC| = \sqrt{4 + 16 + 16} = 6$

$|BC| = \sqrt{0 + 36 + 9} = 3\sqrt{5}$

$|BC|^2 = |AB|^2 + |AC|^2$

Right triangle

19. $A(1, -3, -2), B(5, -1, 2), C(-1, 1, 2)$

$|AB| = \sqrt{16 + 4 + 16} = 6$

$|AC| = \sqrt{4 + 16 + 16} = 6$

$|BC| = \sqrt{36 + 4 + 0} = 2\sqrt{10}$

Since $|AB| = |AC|$, the triangle is isosceles.

21. The z-coordinate is changed by 5 units:

$(0, 0, 5), (2, 2, 6), (2, -4, 9)$

23. $\left(\frac{5 + (-2)}{2}, \frac{-9 + 3}{2}, \frac{7 + 3}{2}\right) = \left(\frac{3}{2}, -3, 5\right)$

25. Center: $(0, 2, 5)$

Radius: 2

$(x - 0)^2 + (y - 2)^2 + (z - 5)^2 = 4$

$x^2 + y^2 + z^2 - 4y - 10z + 25 = 0$

27. Center: $\frac{(2, 0, 0) + (0, 6, 0)}{2} = (1, 3, 0)$

Radius: $\sqrt{10}$

$(x - 1)^2 + (y - 3)^2 + (z - 0)^2 = 10$

$x^2 + y^2 + z^2 - 2x - 6y = 0$

29.

$$x^2 + y^2 + z^2 - 2x + 6y + 8z + 1 = 0$$

$$(x^2 - 2x + 1) + (y^2 + 6y + 9) + (z^2 + 8z + 16) = -1 + 1 + 9 + 16$$

$$(x - 1)^2 + (y + 3)^2 + (z + 4)^2 = 25$$

Center: $(1, -3, -4)$

Radius: 5

31.

$$9x^2 + 9y^2 + 9z^2 - 6x + 18y + 1 = 0$$

$$x^2 + y^2 + z^2 - \frac{2}{3}x + 2y + \frac{1}{9} = 0$$

$$\left(x^2 - \frac{2}{3}x + \frac{1}{9}\right) + (y^2 + 2y + 1) + z^2 = -\frac{1}{9} + \frac{1}{9} + 1$$

$$\left(x - \frac{1}{3}\right)^2 + (y + 1)^2 + (z - 0)^2 = 1$$

Center: $\left(\frac{1}{3}, -1, 0\right)$; Radius: 1

33. (a) $\mathbf{v} = (2 - 4)\mathbf{i} + (4 - 2)\mathbf{j} + (3 - 1)\mathbf{k}$

$= -2\mathbf{i} + 2\mathbf{j} + 2\mathbf{k} = \langle -2, 2, 2\rangle$

(b)

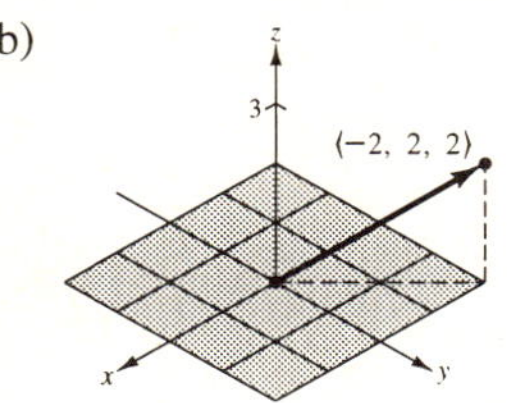

35. (a) $\mathbf{v} = (0 - 3)\mathbf{i} + (3 - 3)\mathbf{j} + (3 - 0)\mathbf{k}$

$= -3\mathbf{i} + 3\mathbf{k} = \langle -3, 0, 3\rangle$

(b)

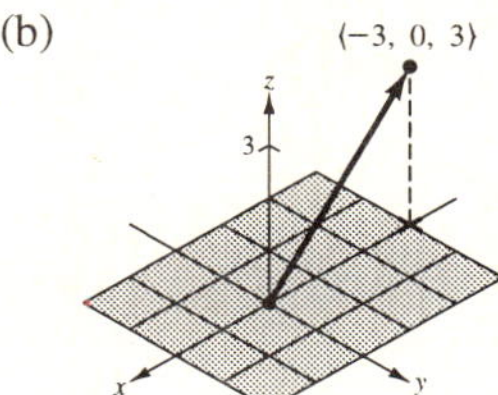

37. (a) and (c)

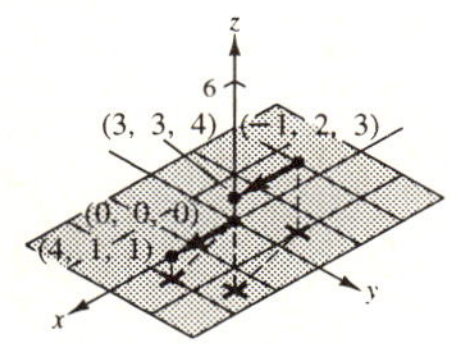

(b) $\mathbf{v} = (3 + 1)\mathbf{i} + (3 - 2)\mathbf{j} + (4 - 3)\mathbf{k}$

$= 4\mathbf{i} + \mathbf{j} + \mathbf{k} = \langle 4, 1, 1\rangle$

39. $(q_1, q_2, q_3) - (0, 6, 2) = (3, -5, 6)$

$Q = (3, 1, 8)$

41. (a) $2\mathbf{v} = \langle 2, 4, 4\rangle$

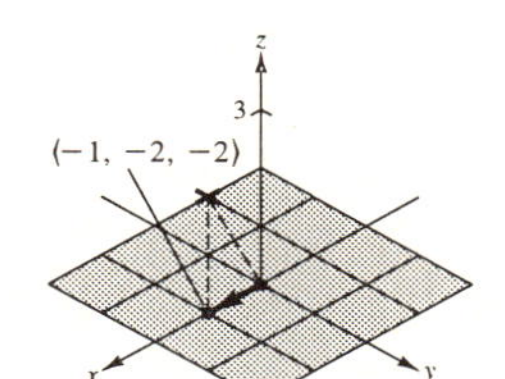

(b) $-\mathbf{v} = \langle -1, -2, -2\rangle$

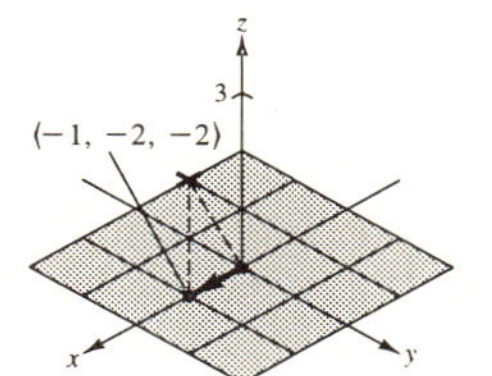

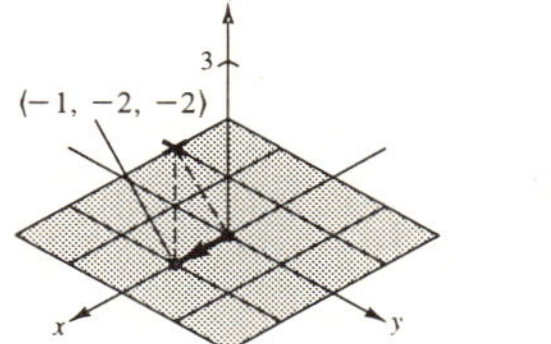

(c) $\frac{3}{2}\mathbf{v} = \left\langle \frac{3}{2}, 3, 3\right\rangle$

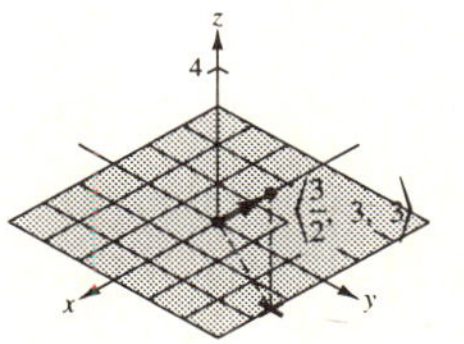

(d) $0\mathbf{v} = \langle 0, 0, 0\rangle$

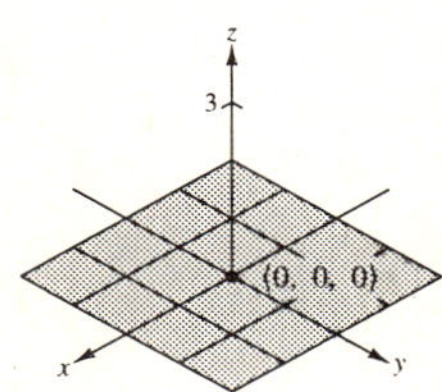

43. $\mathbf{z} = \mathbf{u} - \mathbf{v} = \langle 1,$ $2, 3\rangle - \langle 2$

45. $\mathbf{z} = 2\mathbf{u} + 4\mathbf{v} - \mathbf{w} = \langle 2, 4, 6\rangle + \langle 8, 8, -4\rangle - \langle 4, 0, -4\rangle = \langle 6, 12, 6\rangle$

47. $2\mathbf{z} - 3\mathbf{u} = 2\langle z_1, z_2, z_3\rangle - 3\langle 1, 2, 3\rangle = \langle 4, 0, -4\rangle$

$2z_1 - 3 = 4 \implies z_1 = \frac{7}{2}$

$2z_2 - 6 = 0 \implies z_2 = 3$

$2z_3 - 9 = -4 \implies z_3 = \frac{5}{2}$

$\mathbf{z} = \left\langle \frac{7}{2}, 3, \frac{5}{2}\right\rangle$

49. (a) and (b) are parallel since

$\langle -6, -4, 10\rangle = -2\langle 3, 2, -5\rangle$

and

$\left\langle 2, \frac{4}{3}, -\frac{10}{3}\right\rangle = \frac{2}{3}\langle 3, 2, -5\rangle.$

51. $\mathbf{z} = -3\mathbf{i} + 4\mathbf{j} + 2\mathbf{k}$

(a) is parallel since $-6\mathbf{i} + 8\mathbf{j} + 4\mathbf{k} = 2\mathbf{z}$.

53. $P(0, -2, -5), Q(3, 4, 4), R(2, 2, 1)$

$\overrightarrow{PQ} = \langle 3, 6, 9\rangle$

$\overrightarrow{PR} = \langle 2, 4, 6\rangle$

$\langle 3, 6, 9\rangle = \frac{3}{2}\langle 2, 4, 6\rangle$

Therefore, $\overrightarrow{PQ}$ and $\overrightarrow{PR}$ are parallel. The points are collinear.

55. $P(1, 2, 4), Q(2, 5, 0), R(0, 1, 5)$

$\overrightarrow{PQ} = \langle 1, 3, -4\rangle$

$\overrightarrow{PR} = \langle -1, -1, 1\rangle$

Since $\overrightarrow{PQ}$ and $\overrightarrow{PR}$ are not parallel, the points are not collinear.

57. $A(2, 9, 1), B(3, 11, 4), C(0, 10, 2), D(1, 12, 5)$

$\overrightarrow{AB} = \langle 1, 2, 3\rangle$

$\overrightarrow{CD} = \langle 1, 2, 3\rangle$

$\overrightarrow{AC} = \langle -2, 1, 1\rangle$

$\overrightarrow{BD} = \langle -2, 1, 1\rangle$

Since $\overrightarrow{AB} = \overrightarrow{CD}$ and $\overrightarrow{AC} = \overrightarrow{BD}$, the given points form the vertices of a parallelogram.

59. $\|\mathbf{v}\| = 0$

61. $\mathbf{v} = \langle 1, -2, -3\rangle$

$\|\mathbf{v}\| = \sqrt{1 + 4 + 9} = \sqrt{14}$

63. $\mathbf{v} = \langle 0, 3, -5\rangle$

$\|\mathbf{v}\| = \sqrt{0 + 9 + 25} = \sqrt{34}$

65. $\mathbf{u} = \langle 2, -1, 2\rangle$

$\|\mathbf{u}\| = \sqrt{4 + 1 + 4} = 3$

(a) $\dfrac{\mathbf{u}}{\|\mathbf{u}\|} = \dfrac{1}{3}\langle 2, -1, 2\rangle$

(b) $-\dfrac{\mathbf{u}}{\|\mathbf{u}\|} = -\dfrac{1}{3}\langle 2, -1, 2\rangle$

67. $\mathbf{u} = \langle 3, 2, -5\rangle$

$\|\mathbf{u}\| = \sqrt{9 + 4 + 25} = \sqrt{38}$

(a) $\dfrac{\mathbf{u}}{\|\mathbf{u}\|} = \dfrac{1}{\sqrt{38}}\langle 3, 2, -5\rangle$

(b) $-\dfrac{\mathbf{u}}{\|\mathbf{u}\|} = -\dfrac{1}{\sqrt{38}}\langle 3, 2, -5\rangle$

69. Programs will vary.

71. $c\mathbf{v} = \langle 2c, 2c, -c\rangle$

$\|c\mathbf{v}\| = \sqrt{4c^2 + 4c^2 + c^2} = 5$

$9c^2 = 25$

$c = \pm\dfrac{5}{3}$

73. $\mathbf{v} = 10\dfrac{\mathbf{u}}{\|\mathbf{u}\|} = 10\left\langle 0, \dfrac{1}{\sqrt{2}}, \dfrac{1}{\sqrt{2}}\right\rangle = \left\langle 0, \dfrac{5}{\sqrt{2}}, \dfrac{5}{\sqrt{2}}\right\rangle$

$= \left\langle 0, \dfrac{5\sqrt{2}}{2}, \dfrac{5\sqrt{2}}{2}\right\rangle$

75. $\mathbf{v} = \dfrac{3}{2}\dfrac{\mathbf{u}}{\|\mathbf{u}\|} = \dfrac{3}{2}\left\langle \dfrac{2}{3}, \dfrac{-2}{3}, \dfrac{1}{3}\right\rangle = \left\langle 1, -1, \dfrac{1}{2}\right\rangle$

77. $\mathbf{v} = 2[\cos(\pm 30°)\mathbf{j} + \sin(\pm 30°)\mathbf{k}]$

$= \sqrt{3}\mathbf{j} \pm \mathbf{k} = \langle 0, \sqrt{3}, \pm 1\rangle$

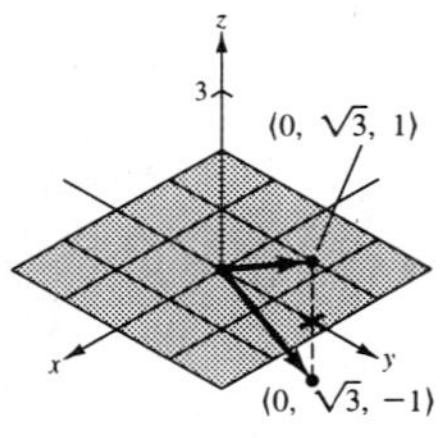

79. $\mathbf{v} = \langle -3, -6, 3\rangle$

$\frac{2}{3}\mathbf{v} = \langle -2, -4, 2\rangle$

$(4, 3, 0) + (-2, -4, 2) = (2, -1, 2)$

81. (a)

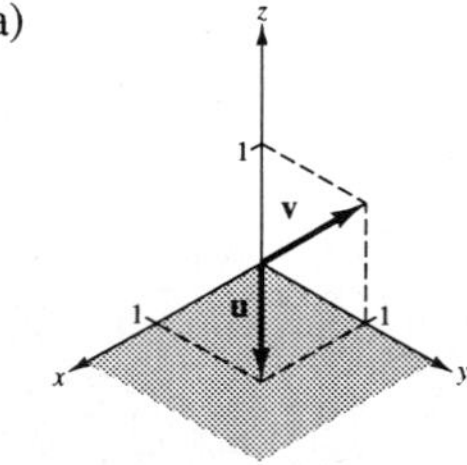

(b) $\mathbf{w} = a\mathbf{u} + b\mathbf{v}$

$= a\mathbf{i} + (a + b)\mathbf{j} + b\mathbf{k} = \mathbf{0}$

$a = 0, a + b = 0, b = 0$

Thus, a and b are both zero.

(c) $a\mathbf{i} + (a + b)\mathbf{j} + b\mathbf{k} = \mathbf{i} + 2\mathbf{j} + \mathbf{k}$

$a = 1, b = 1$

$\mathbf{w} = \mathbf{u} + \mathbf{v}$

(d) $a\mathbf{i} + (a + b)\mathbf{j} + b\mathbf{k} = \mathbf{i} + 2\mathbf{j} + 3\mathbf{k}$

$a = 1, a + b = 2, b = 3$

Not possible

83. (a) The height of the right triangle is $h = \sqrt{L^2 - 18^2}$.
The vector $\overrightarrow{PQ}$ is given by

$$\overrightarrow{PQ} = \langle 0, -18, h \rangle.$$

The tension vector $\mathbf{T}$ in each wire is

$$\mathbf{T} = c\langle 0, -18, h \rangle \text{ where } ch = \frac{24}{3} = 8.$$

Hence, $\mathbf{T} = \frac{8}{h}\langle 0, -18, h \rangle$ and

$$T = \|\mathbf{T}\| = \frac{8}{h}\sqrt{18^2 + h^2} = \frac{8}{\sqrt{L^2 - 18^2}}\sqrt{18^2 + (L^2 - 18^2)} = \frac{8L}{\sqrt{L^2 - 18^2}}$$

(b)

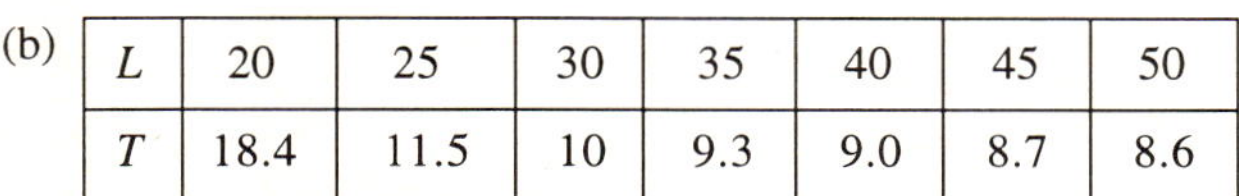

L	20	25	30	35	40	45	50
T	18.4	11.5	10	9.3	9.0	8.7	8.6

(c)

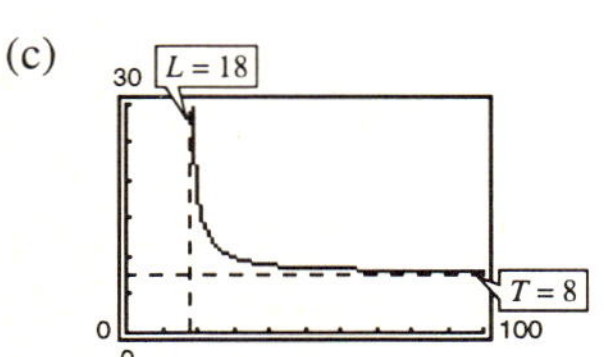

$x = 18$ is a vertical asymptote and $y = 8$ is a horizontal asymptote.

(d) $\lim_{L \to 18^+} \frac{8L}{\sqrt{L^2 - 18^2}} = \infty$

$$\lim_{L \to \infty} \frac{8L}{\sqrt{L^2 - 18^2}} = \lim_{L \to \infty} \frac{8}{\sqrt{1 - (18/L)^2}} = 8$$

(e) From the table, $T = 10$ implies $L = 30$ inches.

85. Let α be the angle between $\mathbf{v}$ and the coordinate axes.

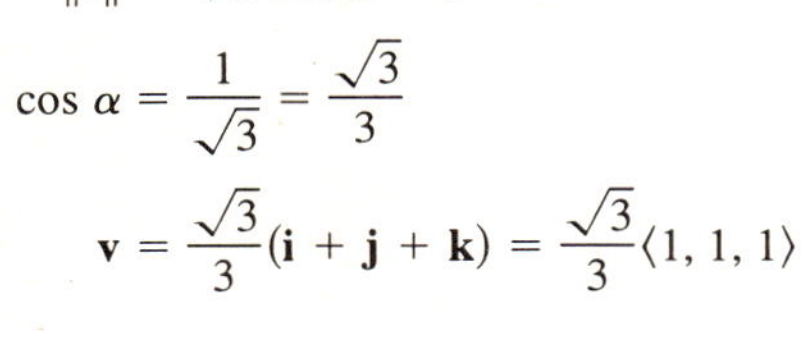

$$\mathbf{v} = (\cos\alpha)\mathbf{i} + (\cos\alpha)\mathbf{j} + (\cos\alpha)\mathbf{k}$$

$$\|\mathbf{v}\| = \sqrt{3}\cos\alpha = 1$$

$$\cos\alpha = \frac{1}{\sqrt{3}} = \frac{\sqrt{3}}{3}$$

$$\mathbf{v} = \frac{\sqrt{3}}{3}(\mathbf{i} + \mathbf{j} + \mathbf{k}) = \frac{\sqrt{3}}{3}\langle 1, 1, 1 \rangle$$

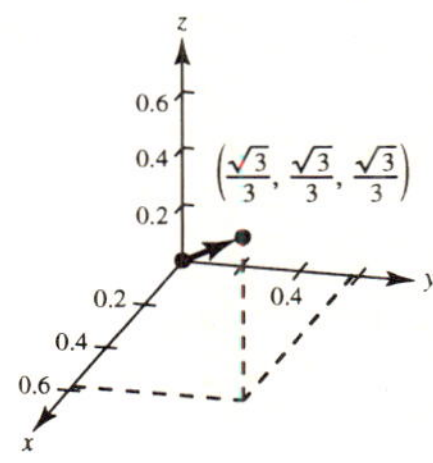

87. $\overrightarrow{AB} = \langle 0, 70, 115 \rangle, \mathbf{F}_1 = C_1\langle 0, 70, 115 \rangle$

$\overrightarrow{AC} = \langle -60, 0, 115 \rangle, \mathbf{F}_2 = C_2\langle -60, 0, 115 \rangle$

$\overrightarrow{AD} = \langle 45, -65, 115 \rangle, \mathbf{F}_3 = C_3\langle 45, -65, 115 \rangle$

$\mathbf{F} = \mathbf{F}_1 + \mathbf{F}_2 + \mathbf{F}_3 = \langle 0, 0, -500 \rangle$

Thus:

$$\begin{aligned} -60C_2 + 45C_3 &= 0 \\ 70C_1 \qquad - 65C_3 &= 0 \\ 115(C_1 + C_2 + C_3) &= -500 \end{aligned}$$

Solving this system yields $C_1 = -\frac{104}{69}$, $C_2 = -\frac{28}{23}$ and$C_3 = -\frac{112}{69}$. Thus:

$\|\mathbf{F}_1\| \approx 202.919N$

$\|\mathbf{F}_2\| \approx 157.909N$

$\|\mathbf{F}_3\| \approx 226.521N$

89. $d(AP) = 2d(BP)$

$$\sqrt{x^2 + (y+1)^2 + (z-1)^2} = 2\sqrt{(x-1)^2 + (y-2)^2 + z^2}$$

$$x^2 + y^2 + z^2 + 2y - 2z + 2 = 4(x^2 + y^2 + z^2 - 2x - 4y + 5)$$

$$0 = 3x^2 + 3y^2 + 3z^2 - 8x - 18y + 2z + 18$$

$$-6 + \frac{16}{9} + 9 + \frac{1}{9} = \left(x^2 - \frac{8}{3}x + \frac{16}{9}\right) + (y^2 - 6y + 9) + \left(z^2 + \frac{2}{3}z + \frac{1}{9}\right)$$

$$\frac{44}{9} = \left(x - \frac{4}{3}\right)^2 + (y-3)^2 + \left(z + \frac{1}{3}\right)^2$$

Sphere; center: $\left(\frac{4}{3}, 3, -\frac{1}{3}\right)$, radius: $\frac{2\sqrt{11}}{3}$

Section 10.3 The Dot Product of Two Vectors

1. $\mathbf{u} = \langle 3, 4\rangle, \mathbf{v} = \langle 2, -3\rangle$

(a) $\mathbf{u} \cdot \mathbf{v} = 3(2) + 4(-3) = -6$

(b) $\mathbf{u} \cdot \mathbf{u} = 3(3) + 4(4) = 25$

(c) $\|\mathbf{u}\|^2 = 25$

(d) $(\mathbf{u} \cdot \mathbf{v})\mathbf{v} = -6\langle 2, -3\rangle = \langle -12, 18\rangle$

(e) $\mathbf{u} \cdot (2\mathbf{v}) = 2(\mathbf{u} \cdot \mathbf{v}) = 2(-6) = -12$

3. $\mathbf{u} = \langle 2, -3, 4\rangle, \mathbf{v} = \langle 0, 6, 5\rangle$

(a) $\mathbf{u} \cdot \mathbf{v} = 2(0) + (-3)(6) + (4)(5) = 2$

(b) $\mathbf{u} \cdot \mathbf{u} = 2(2) + (-3)(-3) + 4(4) = 29$

(c) $\|\mathbf{u}\|^2 = 29$

(d) $(\mathbf{u} \cdot \mathbf{v})\mathbf{v} = 2\langle 0, 6, 5\rangle = \langle 0, 12, 10\rangle$

(e) $\mathbf{u} \cdot (2\mathbf{v}) = 2(\mathbf{u} \cdot \mathbf{v}) = 2(2) = 4$

5. $\mathbf{u} = 2\mathbf{i} - \mathbf{j} + \mathbf{k}, \mathbf{v} = \mathbf{i} - \mathbf{k}$

(a) $\mathbf{u} \cdot \mathbf{v} = 2(1) + (-1)(0) + 1(-1) = 1$

(b) $\mathbf{u} \cdot \mathbf{u} = 2(2) + (-1)(-1) + (1)(1) = 6$

(c) $\|\mathbf{u}\|^2 = 6$

(d) $(\mathbf{u} \cdot \mathbf{v})\mathbf{v} = \mathbf{v} = \mathbf{i} - \mathbf{k}$

(e) $\mathbf{u} \cdot (2\mathbf{v}) = 2(\mathbf{u} \cdot \mathbf{v}) = 2$

7. $\mathbf{u} = \langle 3240, 1450, 2235\rangle$

$\mathbf{v} = \langle 2.22, 1.85, 3.25\rangle$

$\mathbf{u} \cdot \mathbf{v} = \$17{,}139.05$

This gives the total amount that the person earned on his products.

9. $\frac{\mathbf{u} \cdot \mathbf{v}}{\|\mathbf{u}\|\,\|\mathbf{v}\|} = \cos\theta$

$$\mathbf{u} \cdot \mathbf{v} = (8)(5)\cos\frac{\pi}{3} = 20$$

11. $\mathbf{u} = \langle 1, 1\rangle, \mathbf{v} = \langle 2, -2\rangle$

$$\cos\theta = \frac{\mathbf{u} \cdot \mathbf{v}}{\|\mathbf{u}\|\,\|\mathbf{v}\|} = \frac{0}{\sqrt{2}\sqrt{8}} = 0$$

$$\theta = \frac{\pi}{2}$$

13. $\mathbf{u} = 3\mathbf{i} + \mathbf{j}, \mathbf{v} = -2\mathbf{i} + 4\mathbf{j}$

$$\cos\theta = \frac{\mathbf{u} \cdot \mathbf{v}}{\|\mathbf{u}\|\,\|\mathbf{v}\|} = \frac{-2}{\sqrt{10}\sqrt{20}} = \frac{-1}{5\sqrt{2}}$$

$$\theta = \arccos\left(-\frac{1}{5\sqrt{2}}\right) \approx 98.1^\circ$$

15. $\mathbf{u} = \langle 1, 1, 1\rangle, \mathbf{v} = \langle 2, 1, -1\rangle$

$$\cos\theta = \frac{\mathbf{u} \cdot \mathbf{v}}{\|\mathbf{u}\|\,\|\mathbf{v}\|} = \frac{2}{\sqrt{3}\sqrt{6}} = \frac{\sqrt{2}}{3}$$

$$\theta = \arcos\frac{\sqrt{2}}{3} \approx 61.9^\circ$$

17. $\mathbf{u} = 3\mathbf{i} + 4\mathbf{j}, \mathbf{v} = -2\mathbf{j} + 3\mathbf{k}$

$$\cos\theta = \frac{\mathbf{u} \cdot \mathbf{v}}{\|\mathbf{u}\|\,\|\mathbf{v}\|} = \frac{-8}{5\sqrt{13}} = \frac{-8\sqrt{13}}{65}$$

$$\theta = \arccos\left(-\frac{8\sqrt{13}}{65}\right) \approx 116.3^\circ$$

19. Programs will vary.

21. $\mathbf{u} = \langle 4, 0 \rangle, \mathbf{v} = \langle 1, 1 \rangle$

$\mathbf{u} \neq c\mathbf{v} \Rightarrow$ not parallel

$\mathbf{u} \cdot \mathbf{v} = 4 \neq 0 \Rightarrow$ not orthogonal

Neither

23. $\mathbf{u} = \langle 4, 3 \rangle, \mathbf{v} = \left\langle \frac{1}{2}, -\frac{2}{3} \right\rangle$

$\mathbf{u} \neq c\mathbf{v} \Rightarrow$ not parallel

$\mathbf{u} \cdot \mathbf{v} = 0 \Rightarrow$ orthogonal

25. $\mathbf{u} = \mathbf{j} + 6\mathbf{k}, \mathbf{v} = \mathbf{i} - 2\mathbf{j} - \mathbf{k}$

$\mathbf{u} \neq c\mathbf{v} \Rightarrow$ not parallel

$\mathbf{u} \cdot \mathbf{v} = -8 \neq 0 \Rightarrow$ not orthogonal

Neither

27. $\mathbf{u} = \langle 2, -3, 1 \rangle, \mathbf{v} = \langle -1, -1, -1 \rangle$

$\mathbf{u} \neq c\mathbf{v} \Rightarrow$ not parallel

$\mathbf{u} \cdot \mathbf{v} = 0 \Rightarrow$ orthogonal

29. In a rhombus, $\|\mathbf{u}\| = \|\mathbf{v}\|$. The diagonals are $\mathbf{u} + \mathbf{v}$ and $\mathbf{u} - \mathbf{v}$.

$$\begin{aligned}(\mathbf{u} + \mathbf{v}) \cdot (\mathbf{u} - \mathbf{v}) &= (\mathbf{u} + \mathbf{v}) \cdot \mathbf{u} - (\mathbf{u} + \mathbf{v}) \cdot \mathbf{v} \\ &= \mathbf{u} \cdot \mathbf{u} + \mathbf{v} \cdot \mathbf{u} - \mathbf{u} \cdot \mathbf{v} - \mathbf{v} \cdot \mathbf{v} \\ &= \|\mathbf{u}\|^2 - \|\mathbf{v}\|^2 = 0\end{aligned}$$

Therefore, the diagonals are orthogonal.

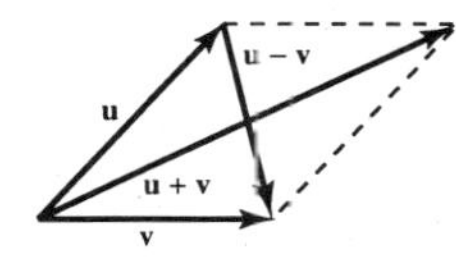

31. (a) $\mathbf{u} \cdot \mathbf{v} = 0 \Rightarrow \mathbf{u}$ and $\mathbf{v}$ are orthogonal and $\theta = \dfrac{\pi}{2}$.

(b) $\mathbf{u} \cdot \mathbf{v} > 0 \Rightarrow \cos\theta > 0 \Rightarrow 0 < \theta < \dfrac{\pi}{2}$

(c) $\mathbf{u} \cdot \mathbf{v} < 0 \Rightarrow \cos\theta < 0 \Rightarrow \dfrac{\pi}{2} < \theta < \pi$

33. $\mathbf{u} = \langle \cos\alpha, \sin\alpha, 0 \rangle, \mathbf{v} = \langle \cos\beta, \sin\beta, 0 \rangle$

The angle between $\mathbf{u}$ and $\mathbf{v}$ is $\alpha - \beta$. (Assuming that $\alpha > \beta$). Also,

$$\begin{aligned}\cos(\alpha - \beta) &= \frac{\mathbf{u} \cdot \mathbf{v}}{\|\mathbf{u}\|\,\|\mathbf{v}\|} \\ &= \frac{\cos\alpha\cos\beta + \sin\alpha\sin\beta}{(1)(1)} \\ &= \cos\alpha\cos\beta + \sin\alpha\sin\beta.\end{aligned}$$

35. $\mathbf{u} = \mathbf{i} + 2\mathbf{j} + 2\mathbf{k}, \|\mathbf{u}\| = 3$

$\cos\alpha = \dfrac{1}{3}$

$\cos\beta = \dfrac{2}{3}$

$\cos\gamma = \dfrac{2}{3}$

$\cos^2\alpha + \cos^2\beta + \cos^2\gamma = \dfrac{1}{9} + \dfrac{4}{9} + \dfrac{4}{9} = 1$

37. $\mathbf{u} = \langle 0, 6, -4 \rangle, \|\mathbf{u}\| = \sqrt{52} = 2\sqrt{13}$

$\cos\alpha = 0$

$\cos\beta = \dfrac{3}{\sqrt{13}}$

$\cos\gamma = -\dfrac{2}{\sqrt{13}}$

$\cos^2\alpha + \cos^2\beta + \cos^2\gamma = 0 + \dfrac{9}{13} + \dfrac{4}{13} = 1$

39. $\mathbf{F}_1$: $C_1 = \dfrac{50}{\|\mathbf{F}_1\|} \approx 4.3193$

$\mathbf{F}_2$: $C_2 = \dfrac{80}{\|\mathbf{F}_2\|} \approx 5.4183$

$$\begin{aligned}\mathbf{F} &= \mathbf{F}_1 + \mathbf{F}_2 \\ &\approx 4.3193\langle 10, 5, 3 \rangle + 5.4183\langle 12, 7, -5 \rangle \\ &= \langle 108.2126, 59.5246, -14.1336 \rangle\end{aligned}$$

$\|\mathbf{F}\| \approx 124.310$ lb

$\cos\alpha \approx \dfrac{108.2126}{\|\mathbf{F}\|} \Rightarrow \alpha \approx 29.48°$

$\cos\beta \approx \dfrac{59.5246}{\|\mathbf{F}\|} \Rightarrow \beta \approx 61.39°$

$\cos\gamma \approx \dfrac{-14.1336}{\|\mathbf{F}\|} \Rightarrow \gamma \approx 96.53°$

41. Let s = length of a side.

$$\mathbf{v} = \langle s, s, s \rangle$$

$$\|\mathbf{v}\| = s\sqrt{3}$$

$$\cos\alpha = \cos\beta = \cos\gamma = \frac{s}{s\sqrt{3}} = \frac{1}{\sqrt{3}}$$

$$\alpha = \beta = \gamma = \arccos\left(\frac{1}{\sqrt{3}}\right) \approx 54.7°$$

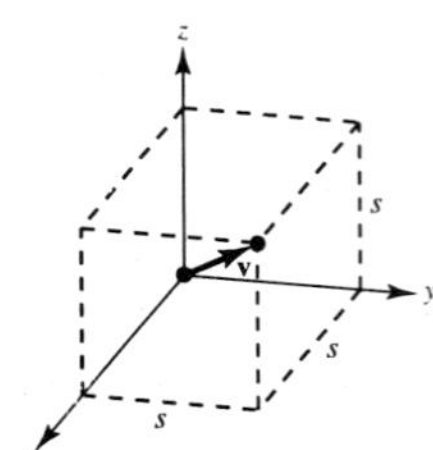

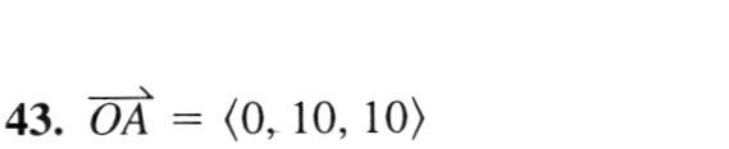

43. $\overrightarrow{OA} = \langle 0, 10, 10 \rangle$

$$\cos\alpha = \frac{0}{\sqrt{0^2 + 10^2 + 10^2}} = 0 \Rightarrow \alpha = 90°$$

$$\cos\beta = \cos\gamma = \frac{10}{\sqrt{0^2 + 10^2 + 10^2}}$$

$$= \frac{1}{\sqrt{2}} \Rightarrow \beta = \gamma = 45°$$

45. $\mathbf{u} = \langle 2, 3 \rangle, \mathbf{v} = \langle 5, 1 \rangle$

(a) $\mathbf{w}_1 = \left(\frac{\mathbf{u} \cdot \mathbf{v}}{\|\mathbf{v}\|^2}\right)\mathbf{v} = \frac{13}{26}\langle 5, 1 \rangle = \left\langle \frac{5}{2}, \frac{1}{2} \right\rangle$

(b) $\mathbf{w}_2 = \mathbf{u} - \mathbf{w}_1 = \left\langle -\frac{1}{2}, \frac{5}{2} \right\rangle$

47. $\mathbf{u} = \langle 2, 1, 2 \rangle, \mathbf{v} = \langle 0, 3, 4 \rangle$

(a) $\mathbf{w}_1 = \left(\frac{\mathbf{u} \cdot \mathbf{v}}{\|\mathbf{v}\|^2}\right)\mathbf{v} = \frac{11}{25}\langle 0, 3, 4 \rangle = \left\langle 0, \frac{33}{25}, \frac{44}{25} \right\rangle$

(b) $\mathbf{w}_2 = \mathbf{u} - \mathbf{w}_1 = \left\langle 2, -\frac{8}{25}, \frac{6}{25} \right\rangle$

49. Programs will vary.

51. Because $\mathbf{u}$ appears to be perpendicular to $\mathbf{v}$, the projection of $\mathbf{u}$ onto $\mathbf{v}$ is $\mathbf{0}$. Analytically,

$$\text{proj}_{\mathbf{v}}\mathbf{u} = \frac{\mathbf{u} \cdot \mathbf{v}}{\|\mathbf{v}\|^2}\mathbf{v} = \frac{\langle 2, -3 \rangle \cdot \langle 6, 4 \rangle}{\|\langle 6, 4 \rangle\|^2}\langle 6, 4 \rangle = 0\langle 6, 4 \rangle = \mathbf{0}.$$

53. (a) $\left(\frac{\mathbf{u} \cdot \mathbf{v}}{\|\mathbf{v}\|^2}\right)\mathbf{v} = \mathbf{u} \Rightarrow \mathbf{u} = c\mathbf{v} \Rightarrow$ $\mathbf{u}$ and $\mathbf{v}$ are parallel.

(b) $\left(\frac{\mathbf{u} \cdot \mathbf{v}}{\|\mathbf{v}\|^2}\right)\mathbf{v} = \mathbf{0} \Rightarrow \mathbf{u} \cdot \mathbf{v} = 0 \Rightarrow$ $\mathbf{u}$ and $\mathbf{v}$ are orthogonal.

55. $\mathbf{u} = \frac{1}{2}\mathbf{i} - \frac{2}{3}\mathbf{j}$. Want $\mathbf{u} \cdot \mathbf{v} = 0$.

$\mathbf{v} = 8\mathbf{i} + 6\mathbf{j}$ and $-\mathbf{v} = -8\mathbf{i} - 6\mathbf{j}$ are orthogonal to $\mathbf{u}$.

57. $\mathbf{u} = \langle 3, 1, -2 \rangle$. Want $\mathbf{u} \cdot \mathbf{v} = 0$.

$\mathbf{v} = \langle 0, 2, 1 \rangle$ and $-\mathbf{v} = \langle 0, -2, -1 \rangle$ are orthogonal to $\mathbf{u}$.

59. (a) Gravitational force $\mathbf{F} = -32{,}000\mathbf{j}$

$$\mathbf{v} = \cos 15°\mathbf{i} + \sin 15°\mathbf{j}$$

$$\mathbf{w}_1 = \frac{\mathbf{F} \cdot \mathbf{v}}{\|\mathbf{v}\|^2}\mathbf{v} = (\mathbf{F} \cdot \mathbf{v})\mathbf{v} = (-32{,}000)(\sin 15°)\mathbf{v} \approx -8282.2(\cos 15°\mathbf{i} + \sin 15°\mathbf{j})$$

$\|\mathbf{w}_1\| \approx 8282.2$ lb

(b) $\mathbf{w}_2 = \mathbf{F} - \mathbf{w}_1 = -32{,}000\mathbf{j} + 8282.2(\cos 15°\mathbf{i} + \sin 15°\mathbf{j})$

$\|\mathbf{w}_2\| \approx 30{,}909.6$ lb

61. $\mathbf{F} = 85\left(\frac{1}{2}\mathbf{i} + \frac{\sqrt{3}}{2}\mathbf{j}\right)$

$\mathbf{v} = 10\mathbf{i}$

$W = \mathbf{F} \cdot \mathbf{v} = 425 \text{ ft} \cdot \text{lb}$

63. $\overrightarrow{PQ} = \langle 4, 7, 5 \rangle$

$\mathbf{v} = \langle 1, 4, 8 \rangle$

$W = \overrightarrow{PQ} \cdot \mathbf{v} = 72$

65.
$$\begin{aligned}\|\mathbf{u} - \mathbf{v}\|^2 &= (\mathbf{u} - \mathbf{v}) \cdot (\mathbf{u} - \mathbf{v}) \\ &= (\mathbf{u} - \mathbf{v}) \cdot \mathbf{u} - (\mathbf{u} - \mathbf{v}) \cdot \mathbf{v} \\ &= \mathbf{u} \cdot \mathbf{u} - \mathbf{v} \cdot \mathbf{u} - \mathbf{u} \cdot \mathbf{v} + \mathbf{v} \cdot \mathbf{v} \\ &= \|\mathbf{u}\|^2 - \mathbf{u} \cdot \mathbf{v} - \mathbf{u} \cdot \mathbf{v} + \|\mathbf{v}\|^2 \\ &= \|\mathbf{u}\|^2 + \|\mathbf{v}\|^2 - 2\mathbf{u} \cdot \mathbf{v}\end{aligned}$$

67.
$$\begin{aligned}\|\mathbf{u} + \mathbf{v}\|^2 &= (\mathbf{u} + \mathbf{v}) \cdot (\mathbf{u} + \mathbf{v}) \\ &= (\mathbf{u} + \mathbf{v}) \cdot \mathbf{u} + (\mathbf{u} + \mathbf{v}) \cdot \mathbf{v} \\ &= \mathbf{u} \cdot \mathbf{u} + \mathbf{v} \cdot \mathbf{u} + \mathbf{u} \cdot \mathbf{v} + \mathbf{v} \cdot \mathbf{v} \\ &= \|\mathbf{u}\|^2 + 2\mathbf{u} \cdot \mathbf{v} + \|\mathbf{v}\|^2 \\ &\le \|\mathbf{u}\|^2 + 2\|\mathbf{u}\|\,\|\mathbf{v}\| - \|\mathbf{v}\|^2 \text{ from Exercise 66} \\ &\le (\|\mathbf{u}\| + \|\mathbf{v}\|)^2\end{aligned}$$

Therefore, $\|\mathbf{u} + \mathbf{v}\| \le \|\mathbf{u}\| + \|\mathbf{v}\|$.

Section 10.4 The Cross Product of Two Vectors in Space

1. $\mathbf{j} \times \mathbf{i} = \begin{vmatrix} \mathbf{i} & \mathbf{j} & \mathbf{k} \\ 0 & 1 & 0 \\ 1 & 0 & 0 \end{vmatrix} = -\mathbf{k}$

3. $\mathbf{j} \times \mathbf{k} = \begin{vmatrix} \mathbf{i} & \mathbf{j} & \mathbf{k} \\ 0 & 1 & 0 \\ 0 & 0 & 1 \end{vmatrix} = \mathbf{i}$

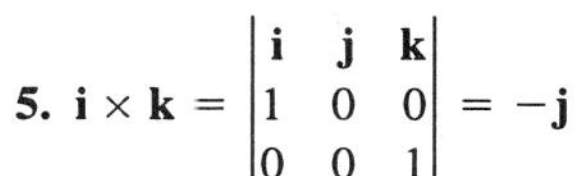

5. $\mathbf{i} \times \mathbf{k} = \begin{vmatrix} \mathbf{i} & \mathbf{j} & \mathbf{k} \\ 1 & 0 & 0 \\ 0 & 0 & 1 \end{vmatrix} = -\mathbf{j}$

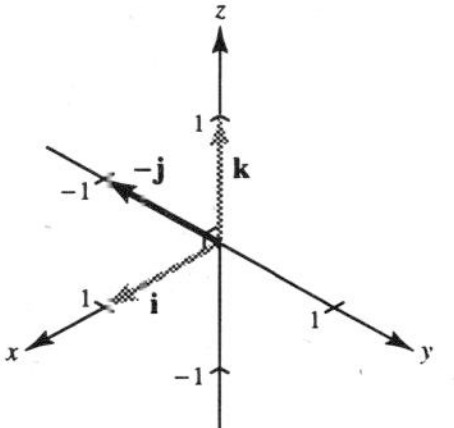

7. $\mathbf{u} = \langle 2, -3, 1 \rangle, \mathbf{v} = \langle 1, -2, 1 \rangle$

$$\mathbf{u} \times \mathbf{v} = \begin{vmatrix} \mathbf{i} & \mathbf{j} & \mathbf{k} \\ 2 & -3 & 1 \\ 1 & -2 & 1 \end{vmatrix} = -\mathbf{i} - \mathbf{j} - \mathbf{k} = \langle -1, -1, -1 \rangle$$

$\mathbf{u} \cdot (\mathbf{u} \times \mathbf{v}) = 2(-1) + (-3)(-1) + (1)(-1) = 0 \Rightarrow \mathbf{u} \perp \mathbf{u} \times \mathbf{v}$

$\mathbf{v} \cdot (\mathbf{u} \times \mathbf{v}) = 1(-1) + (-2)(-1) + (1)(-1) = 0 \Rightarrow \mathbf{v} \perp \mathbf{u} \times \mathbf{v}$

9. $\mathbf{u} = \langle 12, -3, 0 \rangle, \mathbf{v} = \langle -2, 5, 0 \rangle$

$$\mathbf{u} \times \mathbf{v} = \begin{vmatrix} \mathbf{i} & \mathbf{j} & \mathbf{k} \\ 12 & -3 & 0 \\ -2 & 5 & 0 \end{vmatrix} = 54\mathbf{k} = \langle 0, 0, 54 \rangle$$

$$\begin{aligned}\mathbf{u} \cdot (\mathbf{u} \times \mathbf{v}) &= 12(0) + (-3)(0) + 0(54) \\ &= 0 \Rightarrow \mathbf{u} \perp \mathbf{u} \times \mathbf{v}\end{aligned}$$

$$\begin{aligned}\mathbf{v} \cdot (\mathbf{u} \times \mathbf{v}) &= -2(0) + 5(0) + 0(54) \\ &= 0 \Rightarrow \mathbf{v} \perp \mathbf{u} \times \mathbf{v}\end{aligned}$$

11. $\mathbf{u} = \mathbf{i} + \mathbf{j} + \mathbf{k}, \mathbf{v} = 2\mathbf{i} + \mathbf{j} - \mathbf{k}$

$$\mathbf{u} \times \mathbf{v} = \begin{vmatrix} \mathbf{i} & \mathbf{j} & \mathbf{k} \\ 1 & 1 & 1 \\ 2 & 1 & -1 \end{vmatrix} = -2\mathbf{i} + 3\mathbf{j} - \mathbf{k} = \langle -2, 3, -1 \rangle$$

$$\begin{aligned}\mathbf{u} \cdot (\mathbf{u} \times \mathbf{v}) &= 1(-2) + 1(3) + 1(-1) \\ &= 0 \Rightarrow \mathbf{u} \perp \mathbf{u} \times \mathbf{v}\end{aligned}$$

$$\begin{aligned}\mathbf{v} \cdot (\mathbf{u} \times \mathbf{v}) &= 2(-2) + 1(3) + (-1)(-1) \\ &= 0 \Rightarrow \mathbf{v} \perp \mathbf{u} \times \mathbf{v}\end{aligned}$$

13.

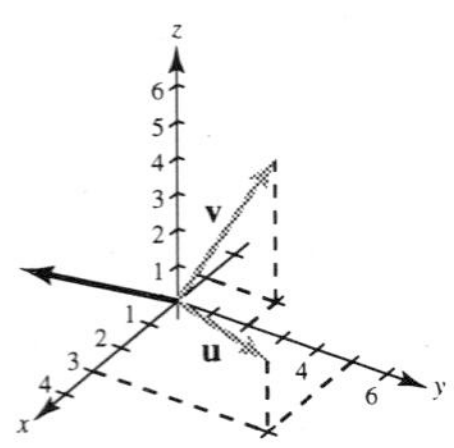

15.

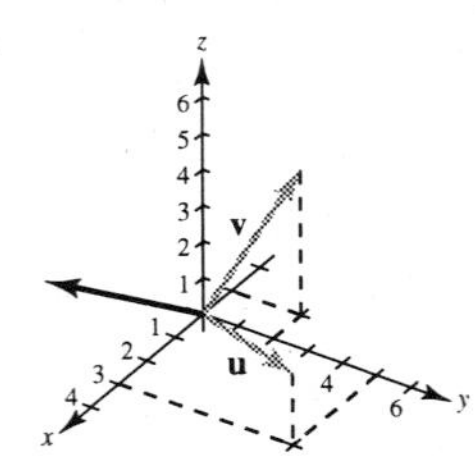

17. $\mathbf{u} = \langle 4, -3.5, 7\rangle$

$\mathbf{v} = \langle -1, 8, 4\rangle$

$\mathbf{u} \times \mathbf{v} = \left\langle -70, -23, \frac{57}{2}\right\rangle$

$\frac{\mathbf{u} \times \mathbf{v}}{\|\mathbf{u} \times \mathbf{v}\|} = \left\langle \frac{-140}{\sqrt{24{,}965}}, \frac{-46}{\sqrt{24{,}965}}, \frac{57}{\sqrt{24{,}965}}\right\rangle$

19. $\mathbf{u} = -3\mathbf{i} + 2\mathbf{j} - 5\mathbf{k}$

$\mathbf{v} = \frac{1}{2}\mathbf{i} - \frac{3}{4}\mathbf{j} + \frac{1}{10}\mathbf{k}$

$\mathbf{u} \times \mathbf{v} = \left\langle -\frac{71}{20}, -\frac{11}{5}, \frac{5}{4}\right\rangle$

$\frac{\mathbf{u} \times \mathbf{v}}{\|\mathbf{u} \times \mathbf{v}\|} = \frac{20}{\sqrt{7602}}\left\langle -\frac{71}{20}, -\frac{11}{5}, \frac{5}{4}\right\rangle$

$= \left\langle -\frac{71}{\sqrt{7602}}, -\frac{44}{\sqrt{7602}}, \frac{25}{\sqrt{7602}}\right\rangle$

21. Programs will vary.

23. $\mathbf{u} = \mathbf{j}$

$\mathbf{v} = \mathbf{j} + \mathbf{k}$

$\mathbf{u} \times \mathbf{v} = \begin{vmatrix} \mathbf{i} & \mathbf{j} & \mathbf{k} \\ 0 & 1 & 0 \\ 0 & 1 & 1 \end{vmatrix} = \mathbf{i}$

$A = \|\mathbf{u} \times \mathbf{v}\| = \|\mathbf{i}\| = 1$

25. $\mathbf{u} = \langle 3, 2, -1\rangle$

$\mathbf{v} = \langle 1, 2, 3\rangle$

$\mathbf{u} \times \mathbf{v} = \begin{vmatrix} \mathbf{i} & \mathbf{j} & \mathbf{k} \\ 3 & 2 & -1 \\ 1 & 2 & 3 \end{vmatrix} = \langle 8, -10, 4\rangle$

$A = \|\mathbf{u} \times \mathbf{v}\| = \|\langle 8, -10, 4\rangle\| = \sqrt{180} = 6\sqrt{5}$

27. $A(1, 1, 1,), B(2, 3, 4), C(6, 5, 2), D(7, 7, 5)$

$\overrightarrow{AB} = \langle 1, 2, 3\rangle, \overrightarrow{AC} = \langle 5, 4, 1\rangle, \overrightarrow{CD} = \langle 1, 2, 3\rangle,$
$\overrightarrow{BD} = \langle 5, 4, 1\rangle$

Since $\overrightarrow{AB} = \overrightarrow{CD}$ and $\overrightarrow{AC} = \overrightarrow{BD}$, the figure is a parallelogram. $\overrightarrow{AB}$ and $\overrightarrow{AC}$ are adjacent sides and

$\overrightarrow{AB} \times \overrightarrow{AC} = \begin{vmatrix} \mathbf{i} & \mathbf{j} & \mathbf{k} \\ 1 & 2 & 3 \\ 5 & 4 & 1 \end{vmatrix} = -10\mathbf{i} + 14\mathbf{j} - 6\mathbf{k}.$

$A = \|\overrightarrow{AB} \times \overrightarrow{AC}\| = \sqrt{332} = 2\sqrt{83}$

29. $A(0, 0, 0), B(1, 2, 3), C(-3, 0, 0)$

$\overrightarrow{AB} = \langle 1, 2, 3\rangle, \overrightarrow{AC} = \langle -3, 0, 0\rangle$

$\overrightarrow{AB} \times \overrightarrow{AC} = \begin{vmatrix} \mathbf{i} & \mathbf{j} & \mathbf{k} \\ 1 & 2 & 3 \\ -3 & 0 & 0 \end{vmatrix} = -9\mathbf{j} + 6\mathbf{k}$

$A = \frac{1}{2}\|\overrightarrow{AB} \times \overrightarrow{AC}\| = \frac{1}{2}\sqrt{117} = \frac{3}{2}\sqrt{13}$

31. $A(1, 3, 5), B(3, 3, 0), C(-2, 0, 5)$

$\overrightarrow{AB} = \langle 2, 0, -5\rangle, \overrightarrow{AC} = \langle -3, -3, 0\rangle$

$\overrightarrow{AB} \times \overrightarrow{AC} = \begin{vmatrix} \mathbf{i} & \mathbf{j} & \mathbf{k} \\ 2 & 0 & -5 \\ -3 & -3 & 0 \end{vmatrix} = -15\mathbf{i} + 15\mathbf{j} - 6\mathbf{k}$

$A = \frac{1}{2}\|\overrightarrow{AB} \times \overrightarrow{AC}\| = \frac{1}{2}\sqrt{486} = \frac{9}{2}\sqrt{6}$

33. $\mathbf{u} \cdot (\mathbf{v} \times \mathbf{w}) = \begin{vmatrix} 1 & 0 & 0 \\ 0 & 1 & 0 \\ 0 & 0 & 1 \end{vmatrix} = 1$

35. $\mathbf{u} \cdot (\mathbf{v} \times \mathbf{w}) = \begin{vmatrix} 2 & 0 & 1 \\ 0 & 3 & 0 \\ 0 & 0 & 1 \end{vmatrix} = 6$

37. $\mathbf{u} \cdot (\mathbf{v} \times \mathbf{w}) = \begin{vmatrix} 1 & 1 & 0 \\ 0 & 1 & 1 \\ 1 & 0 & 1 \end{vmatrix} = 2$

$V = |\mathbf{u} \cdot (\mathbf{v} \times \mathbf{w})| = 2$

39. $\mathbf{u} = \langle 3, 0, 0 \rangle$

$\mathbf{v} = \langle 0, 5, 1 \rangle$

$\mathbf{w} = \langle 2, 0, 5 \rangle$

$\mathbf{u} \cdot (\mathbf{v} \times \mathbf{w}) = \begin{vmatrix} 3 & 0 & 0 \\ 0 & 5 & 1 \\ 2 & 0 & 5 \end{vmatrix} = 75$

$V = |\mathbf{u} \cdot (\mathbf{v} \times \mathbf{w})| = 75$

41. $\mathbf{F} = -20\mathbf{k}$

$\overrightarrow{PQ} = \frac{1}{2}(\cos 40°\mathbf{j} + \sin 40°\mathbf{k})$

$\overrightarrow{PQ} \times \mathbf{F} = \begin{vmatrix} \mathbf{i} & \mathbf{j} & \mathbf{k} \\ 0 & \cos 40°/2 & \sin 40°/2 \\ 0 & 0 & -20 \end{vmatrix} = -10 \cos 40°\mathbf{i}$

$\|\overrightarrow{PQ} \times \mathbf{F}\| = 10 \cos 40° \approx 7.65 \text{ ft} \cdot \text{lb}$

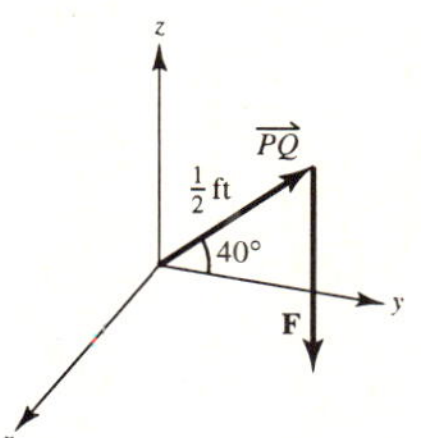

43. (a) B is $-\frac{15}{12} = -\frac{5}{4}$ to the left of A, and one foot upwards:

$\overrightarrow{AB} = \frac{-5}{4}\mathbf{j} + \mathbf{k}$

$\mathbf{F} = -200(\cos\theta\mathbf{j} + \sin\theta\mathbf{k})$

(b) $\overrightarrow{AB} \times \mathbf{F} = \begin{vmatrix} \mathbf{i} & \mathbf{j} & \mathbf{k} \\ 0 & -5/4 & 1 \\ 0 & -200\cos\theta & -200\sin\theta \end{vmatrix}$

$= (250 \sin\theta + 200 \cos\theta)\mathbf{i}$

$\|\overrightarrow{AB} \times \mathbf{F}\| = |250 \sin\theta + 200 \cos\theta|$

$= 25(10 \sin\theta + 8 \cos\theta)$

(c) For $\theta = 30°$,

$\|\overrightarrow{AB} \times \mathbf{F}\| = 25\left(10\left(\frac{1}{2}\right) + 8\left(\frac{\sqrt{3}}{2}\right)\right)$

$= 25(5 + 4\sqrt{3}) \approx 298.2.$

(d) If $T = \|\overrightarrow{AB} \times \mathbf{F}\|$,

$\frac{dT}{d\theta} = 25(10\cos\theta - 8\sin\theta) = 0 \Rightarrow \tan\theta = \frac{5}{4}$

$\Rightarrow \theta \approx 51.34°.$

The vectors are orthogonal.

(e) The zero is $\theta \approx 141.34°$, the angle making $\overrightarrow{AB}$ parallel to $\mathbf{F}$.

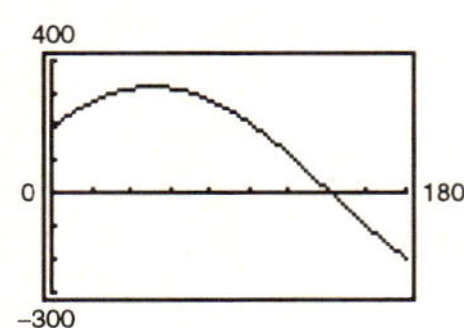

45. $\mathbf{u} = \langle u_1, u_2, u_3 \rangle$, $\mathbf{v} = \langle v_1, v_2, v_3 \rangle$, $\mathbf{w} = \langle w_1, w_2, w_3 \rangle$

$\mathbf{u} \times (\mathbf{v} + \mathbf{w}) = \begin{vmatrix} \mathbf{i} & \mathbf{j} & \mathbf{k} \\ u_1 & u_2 & u_3 \\ v_1 + w_1 & v_2 + w_2 & v_3 + w_3 \end{vmatrix}$

$= [u_2(v_3 + w_3) - u_3(v_2 + w_2)]\mathbf{i} - [u_1(v_3 + w_3) - u_3(v_1 + w_1)]\mathbf{j} + [u_1(v_2 + w_2) - u_2(v_1 + w_1)]\mathbf{k}$

$= (u_2v_3 - u_3v_2)\mathbf{i} - (u_1v_3 - u_3v_1)\mathbf{j} + (u_1v_2 - u_2v_1)\mathbf{k} + (u_2w_3 - u_3w_2)\mathbf{i} - (u_1w_3 - u_3w_1)\mathbf{j} + (u_1w_2 - u_2w_1)\mathbf{k}$

$= (\mathbf{u} \times \mathbf{v}) + (\mathbf{u} \times \mathbf{w})$

47. $\mathbf{u} = \langle u_1, u_2, u_3 \rangle$

$\mathbf{u} \times \mathbf{u} = \begin{vmatrix} \mathbf{i} & \mathbf{j} & \mathbf{k} \\ u_1 & u_2 & u_3 \\ u_1 & u_2 & u_3 \end{vmatrix} = (u_2u_3 - u_3u_2)\mathbf{i} - (u_1u_3 - u_3u_1)\mathbf{j} + (u_1u_2 - u_2u_1)\mathbf{k} = \mathbf{0}$

49. $\mathbf{u} \times \mathbf{v} = (u_2v_3 - u_3v_2)\mathbf{i} - (u_1v_3 - u_3v_1)\mathbf{j} + (u_1v_2 - u_2v_1)\mathbf{k}$

$$(\mathbf{u} \times \mathbf{v}) \cdot \mathbf{u} = (u_2v_3 - u_3v_2)u_1 + (u_3v_1 - u_1v_3)u_2 + (u_1v_2 - u_2v_1)u_3 = 0$$

$$(\mathbf{u} \times \mathbf{v}) \cdot \mathbf{v} = (u_2v_3 - u_3v_2)v_1 + (u_3v_1 - u_1v_3)v_2 + (u_1v_2 - u_2v_1)v_3 = 0$$

Thus, $\mathbf{u} \times \mathbf{v} \perp \mathbf{u}$ and $\mathbf{u} \times \mathbf{v} \perp \mathbf{v}$.

51. $\|\mathbf{u} \times \mathbf{v}\| = \|\mathbf{u}\|\,\|\mathbf{v}\| \sin\theta$

If $\mathbf{u}$ and $\mathbf{v}$ are orthogonal, $\theta = \pi/2$ and $\sin\theta = 1$. Therefore, $\|\mathbf{u} \times \mathbf{v}\| = \|\mathbf{u}\|\,\|\mathbf{v}\|$.

53. $\mathbf{u} = \langle u_1, u_2, u_3 \rangle$, $\mathbf{v} = \langle v_1, v_2, v_3 \rangle$, $\mathbf{w} = \langle w_1, w_2, w_3 \rangle$

$$\mathbf{u} = u_1\mathbf{i} + u_2\mathbf{j} + u_3\mathbf{k}$$

$$\mathbf{v} \times \mathbf{w} = (v_2w_3 - v_3w_2)\mathbf{i} - (v_1w_3 - v_3w_1)\mathbf{j} + (v_1w_2 - v_2w_1)\mathbf{k}$$

$$\mathbf{u} \cdot (\mathbf{v} + \mathbf{w}) = u_1(v_2w_3 - v_3w_2) - u_2(v_1w_3 - v_3w_1) + u_3(v_1w_2 - v_2w_1) = \begin{vmatrix} u_1 & u_2 & u_3 \\ v_1 & v_2 & v_3 \\ w_1 & w_2 & w_3 \end{vmatrix}$$

55. Form the vectors for two sides of the triangle, and compute their cross product:

$$\langle x_2 - x_1, y_2 - y_1, z_2 - z_1 \rangle \times \langle x_3 - x_1, y_3 - y_1, z_3 - z_1 \rangle$$

57. Let $\theta = \alpha - \beta$, the angle between $\mathbf{u}$ and $\mathbf{v}$. Then

$$\sin(\alpha - \beta) = \frac{\|\mathbf{u} \times \mathbf{v}\|}{\|\mathbf{u}\|\,\|\mathbf{v}\|} = \frac{\|\mathbf{v} \times \mathbf{u}\|}{\|\mathbf{u}\|\,\|\mathbf{v}\|}.$$

For $\mathbf{u} = \langle \cos\alpha, \sin\alpha, 0 \rangle$ and $\mathbf{v} = \langle \cos\beta, \sin\beta, 0 \rangle$, $\|\mathbf{u}\| = \|\mathbf{v}\| = 1$ and

$$\mathbf{v} \times \mathbf{u} = \begin{vmatrix} \mathbf{i} & \mathbf{j} & \mathbf{k} \\ \cos\beta & \sin\beta & 0 \\ \cos\alpha & \sin\alpha & 0 \end{vmatrix} = (\sin\alpha\cos\beta - \cos\alpha\sin\beta)\mathbf{k}.$$

Thus, $\sin(\alpha - \beta) = \|\mathbf{v} \times \mathbf{u}\| = \sin\alpha\cos\beta - \cos\alpha\sin\beta$.

Section 10.5 Lines and Planes in Space

1. $x = 1 + 3t$, $y = 2 - t$, $z = 2 + 5t$

(a)

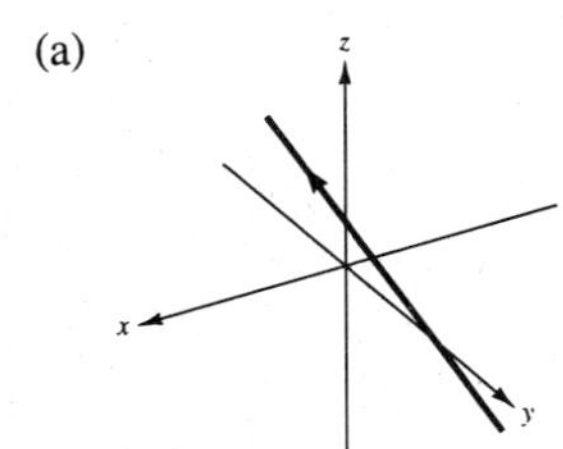

(b) When $t = 0$ we have $P = (1, 2, 2)$. When $t = 3$ we have $Q = (10, -1, 17)$.

$$\overrightarrow{PQ} = \langle 9, -3, 15 \rangle$$

The components of the vector and the coefficients of t are proportional since the line is parallel to $\overrightarrow{PQ}$.

(c) $y = 0$ when $t = 2$. Thus, $x = 7$ and $z = 12$.
Point: $(7, 0, 12)$

$x = 0$ when $t = -\frac{1}{3}$. Point: $\left(0, \frac{7}{3}, \frac{1}{3}\right)$

$z = 0$ when $t = -\frac{2}{5}$. Point: $\left(-\frac{1}{5}, \frac{12}{5}, 0\right)$

3. Point: $(0, 0, 0)$

Direction vector: $\mathbf{v} = \langle 1, 2, 3 \rangle$

Direction numbers: $1, 2, 3$

(a) Parametric: $x = t, y = 2t, z = 3t$

(b) Symmetric: $x = \dfrac{y}{2} = \dfrac{z}{3}$

5. Point: $(-2, 0, 3)$

Direction vector: $\mathbf{v} = \langle 2, 4, -2 \rangle$

Direction numbers: $2, 4, -2$

(a) Parametric: $x = -2 + 2t, y = 4t, z = 3 - 2t$

(b) Symmetric: $\dfrac{x+2}{2} = \dfrac{y}{4} = \dfrac{z-3}{-2}$

7. Point: $(1, 0, 1)$

Direction vector: $\mathbf{v} = 3\mathbf{i} - 2\mathbf{j} + \mathbf{k}$

Direction numbers: $3, -2, 1$

(a) Parametric: $x = 1 + 3t, y = -2t, z = 1 + t$

(b) Symmetric: $\dfrac{x-1}{3} = \dfrac{y}{-2} = \dfrac{z-1}{1}$

9. Points: $(5, -3, -2), \left(\dfrac{-2}{3}, \dfrac{2}{3}, 1\right)$

Direction vector: $\mathbf{v} = \dfrac{17}{3}\mathbf{i} - \dfrac{11}{3}\mathbf{j} - 3\mathbf{k}$

Direction numbers: $17, -11, -9$

(a) Parametric: $x = 5 + 17t, y = -3 - 11t, z = -2 - 9t$

(b) Symmetric: $\dfrac{x-5}{17} = \dfrac{y+3}{-11} = \dfrac{z+2}{-9}$

11. Point: $(2, 3, 4)$

Direction vector: $\mathbf{v} = \mathbf{k}$

Direction numbers: $0, 0, 1$

Parametric: $x = 2, y = 3, z = 4 + t$

13. Point: $(-2, 3, 1)$

Direction vector: $\mathbf{v} = 4\mathbf{i} - \mathbf{k}$

Direction numbers: $4, 0, -1$

Parametric: $x = -2 + 4t, y = 3, z = 1 - t$

Symmetric: $\dfrac{x+2}{4} = \dfrac{z-1}{-1}, y = 3$

(a) On line

(b) On line

(c) Not on line $(y \neq 3)$

(d) Not on line $\left(\dfrac{6+2}{4} \neq \dfrac{-2-1}{-1}\right)$

15. At the point of intersection, the coordinates for one line equal the corresponding coordinates for the other line. Thus,

(i) $4t + 2 = 2s + 2$, (ii) $3 = 2s + 3$, and (iii) $-t + 1 = s + 1$.

From (ii), we find that $s = 0$ and consequently, from (iii), $t = 0$. Letting $s = t = 0$, we see that equation (i) is satisfied and therefore the two lines intersect. Substituting zero for s or for t, we obtain the point $(2, 3, 1)$.

$\mathbf{u} = 4\mathbf{i} - \mathbf{k}$ (First line)

$\mathbf{v} = 2\mathbf{i} + 2\mathbf{j} + \mathbf{k}$ (Second line)

$$\cos\theta = \frac{|\mathbf{u}\cdot\mathbf{v}|}{\|\mathbf{u}\|\,\|\mathbf{v}\|} = \frac{8-1}{\sqrt{17}\sqrt{9}} = \frac{7}{3\sqrt{17}} = \frac{7\sqrt{17}}{51}$$

17. Writing the equations of the lines in parametric form we have

$x = 3t \qquad y = 2 - t \qquad z = -1 + t$

$x = 1 + 4s \qquad y = -2 + s \qquad z = -3 - 3s.$

For the coordinates to be equal, $3t = 1 + 4s$ and $2 - t = -2 + s$. Solving this system yields $t = \frac{17}{7}$ and $s = \frac{11}{7}$. When using these values for s and t, the z coordinates are not equal. The lines do not intersect.

19. $x = 2t + 3 \qquad x = -2s + 7$

$y = 5t - 2 \qquad y = s + 8$

$z = -t + 1 \qquad z = 2s - 1$

Point of intersection: $(7, 8, -1)$

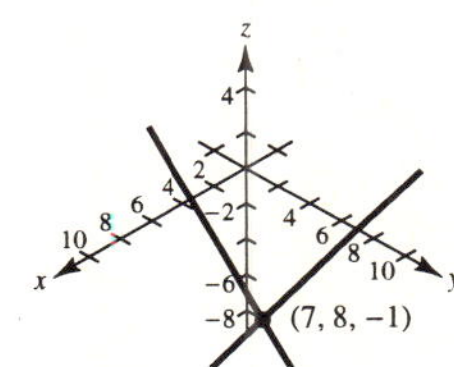

21. $4x - 3y - 6z = 6$

(a) $P = (0, 0, -1), Q = (0, -2, 0), R = (3, 4, -1)$

$\overrightarrow{PQ} = \langle 0, -2, 1\rangle, \overrightarrow{PR} = \langle 3, 4, 0\rangle$

(b) $\overrightarrow{PQ} \times \overrightarrow{PR} = \begin{vmatrix} \mathbf{i} & \mathbf{j} & \mathbf{k} \\ 0 & -2 & 1 \\ 3 & 4 & 0 \end{vmatrix} = \langle -4, 3, 6\rangle$

The components of the cross product are proportional to the coefficients of the variables in the equation. The cross product is parallel to the normal vector.

23. Point: $(2, 1, 2)$

$\mathbf{n} = \mathbf{i} = \langle 1, 0, 0\rangle$

$1(x - 2) + 0(y - 1) + 0(z - 2) = 0$

$x - 2 = 0$

25. Point: $(3, 2, 2)$

Normal vector: $\mathbf{n} = 2\mathbf{i} + 3\mathbf{j} - \mathbf{k}$

$2(x - 3) + 3(y - 2) - 1(z - 2) = 0$

$2x + 3y - z = 10$

27. Point: $(0, 0, 6)$

Normal vector: $\mathbf{n} = -\mathbf{i} + \mathbf{j} - 2\mathbf{k}$

$-1(x - 0) + 1(y - 0) - 2(z - 6) = 0$

$-x + y - 2z + 12 = 0$

$x - y + 2z = 12$

29. Let $\mathbf{u}$ be the vector from $(0, 0, 0)$ to $(1, 2, 3)$:

$\mathbf{u} = \mathbf{i} + 2\mathbf{j} + 3\mathbf{k}$

Let $\mathbf{v}$ be the vector from $(0, 0, 0)$ to $(-2, 3, 3)$:

$\mathbf{v} = -2\mathbf{i} + 3\mathbf{j} + 3\mathbf{k}$

Normal vector: $\mathbf{u} \times \mathbf{v} = \begin{vmatrix} \mathbf{i} & \mathbf{j} & \mathbf{k} \\ 1 & 2 & 3 \\ -2 & 3 & 3 \end{vmatrix}$

$= -3\mathbf{i} + (-9)\mathbf{j} + 7\mathbf{k}$

$-3(x - 0) - 9(y - 0) + 7(z - 0) = 0$

$3x + 9y - 7z = 0$

31. Let $\mathbf{u}$ be the vector from $(1, 2, 3)$ to $(3, 2, 1)$: $\mathbf{u} = 2\mathbf{i} - 2\mathbf{k}$

Let $\mathbf{v}$ be the vector from $(1, 2, 3)$ to $(-1, -2, 2)$: $\mathbf{v} = -2\mathbf{i} - 4\mathbf{j} - \mathbf{k}$

Normal vector: $\left(\frac{1}{2}\mathbf{u}\right) \times (-\mathbf{v}) = \begin{vmatrix} \mathbf{i} & \mathbf{j} & \mathbf{k} \\ 1 & 0 & -1 \\ 2 & 4 & 1 \end{vmatrix} = 4\mathbf{i} - 3\mathbf{j} + 4\mathbf{k}$

$4(x - 1) - 3(y - 2) + 4(z - 3) = 0$

$4x - 3y + 4z = 10$

33. $(1, 2, 3)$, Normal vector: $\mathbf{v} = \mathbf{k}$, $1(z - 3) = 0, z = 3$

35. The direction vectors for the lines are $\mathbf{u} = -2\mathbf{i} + \mathbf{j} + \mathbf{k}, \mathbf{v} = -3\mathbf{i} + 4\mathbf{j} - \mathbf{k}$.

Normal vector: $\mathbf{u} \times \mathbf{v} = \begin{vmatrix} \mathbf{i} & \mathbf{j} & \mathbf{k} \\ -2 & 1 & 1 \\ -3 & 4 & -1 \end{vmatrix} = -5(\mathbf{i} + \mathbf{j} + \mathbf{k})$

Point of intersection of the lines: $(-1, 5, 1)$

$(x + 1) + (y - 5) + (z - 1) = 0$

$x + y + z = 5$

37. Let $\mathbf{v}$ be the vector from $(-1, 1, -1)$ to $(2, 2, 1)$: $\mathbf{v} = 3\mathbf{i} + \mathbf{j} + 2\mathbf{k}$

Let $\mathbf{n}$ be a vector normal to the plane $2x - 3y + z = 3$: $\mathbf{n} = 2\mathbf{i} - 3\mathbf{j} + \mathbf{k}$

Since v and n both lie in the plane p, the normal vector to p is

$\mathbf{v} \times \mathbf{n} = \begin{vmatrix} \mathbf{i} & \mathbf{j} & \mathbf{k} \\ 3 & 1 & 2 \\ 2 & -3 & 1 \end{vmatrix} = 7\mathbf{i} + \mathbf{j} - 11\mathbf{k}$

$7(x - 2) + 1(y - 2) - 11(z - 1) = 0$

$7x + y - 11z = 5$

39. Let $\mathbf{u} = \mathbf{i}$ and let $\mathbf{v}$ be the vector from $(1, -2, -1)$ to $(2, 5, 6)$: $\mathbf{v} = \mathbf{i} + 7\mathbf{j} + 7\mathbf{k}$

Since $\mathbf{u}$ and $\mathbf{v}$ both lie in the plane P, the normal vector to P is:

$$\mathbf{u} \times \mathbf{v} = \begin{vmatrix} \mathbf{i} & \mathbf{j} & \mathbf{k} \\ 1 & 0 & 0 \\ 1 & 7 & 7 \end{vmatrix} = -7\mathbf{j} + 7\mathbf{k} = -7(\mathbf{j} - \mathbf{k})$$

$$[y - (-2)] - [z - (-1)] = 0$$

$$y - z = -1$$

41. The normal vectors to the planes are

$$\mathbf{n}_1 = \langle 5, -3, 1\rangle, \mathbf{n}_2 = \langle 1, 4, 7\rangle, \cos\theta = \frac{|\mathbf{n}_1 \cdot \mathbf{n}_2|}{\|\mathbf{n}_1\|\,\|\mathbf{n}_2\|} = 0.$$

Thus, $\theta = \pi/2$ and the planes are orthogonal.

43. The normal vectors to the planes are

$$\mathbf{n}_1 = \mathbf{i} - 3\mathbf{j} + 6\mathbf{k},\ \mathbf{n}_2 = 5\mathbf{i} + \mathbf{j} - \mathbf{k},$$

$$\cos\theta = \frac{|\mathbf{n}_1 \cdot \mathbf{n}_2|}{\|\mathbf{n}_1\|\,\|\mathbf{n}_2\|} = \frac{|5 - 3 - 6|}{\sqrt{46}\sqrt{27}} = \frac{4\sqrt{138}}{414}.$$

Therefore, $\theta = \arccos\left(\dfrac{4\sqrt{138}}{414}\right) \approx 83.5°$.

45. The normal vectors to the planes are $\mathbf{n}_1 = \langle 1, -5, -1\rangle$ and $\mathbf{n}_2 = \langle 5, -25, -5\rangle$. Since $\mathbf{n}_2 = 5\mathbf{n}_1$, the planes are parallel, but not equal.

47. $4x + 2y + 6z = 12$

49. $2x - y + 3z = 4$

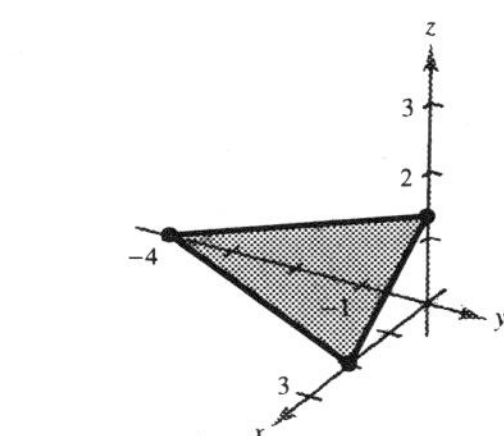

51. $y + z = 5$

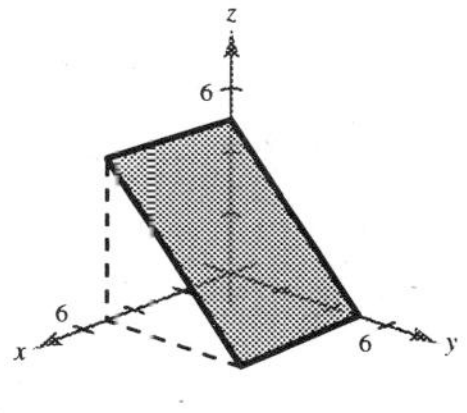

53. $2x + y - z = 6$

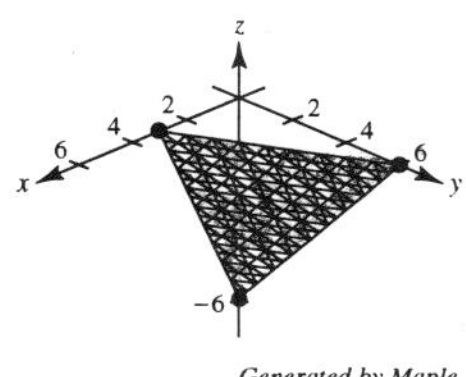

55. $-5x + 4y - 6z + 8 = 0$

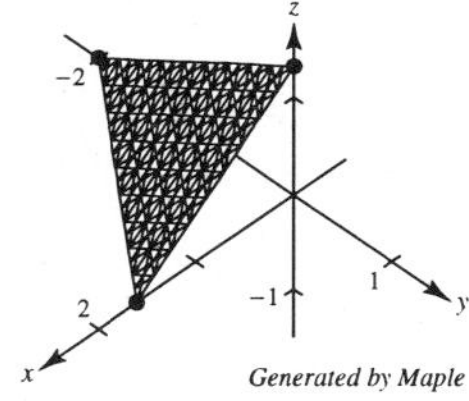

57. The normals to the planes are $\mathbf{n}_1 = 3\mathbf{i} + 2\mathbf{j} - \mathbf{k}$ and $\mathbf{n}_2 = \mathbf{i} - 4\mathbf{j} + 2\mathbf{k}$. The direction vector for the line is

$$\mathbf{n}_2 \times \mathbf{n}_1 = \begin{vmatrix} \mathbf{i} & \mathbf{j} & \mathbf{k} \\ 1 & -4 & 2 \\ 3 & 2 & -1 \end{vmatrix} = 7(\mathbf{j} + 2\mathbf{k}).$$

Now find a point of intersection of the planes.

$$\begin{aligned} 6x + 4y - 2y &= 14 \\ x - 4y + 2z &= 0 \\ \hline 7x \qquad &= 14 \\ x &= 2 \end{aligned}$$

Substituting 2 for x in the second equation, we have $-4y + 2z = -2$ or $z = 2y - 1$. Letting $y = 1$, a point of intersection is $(2, 1, 1)$.

$$x = 2, y = 1 + t, z = 1 + 2t$$

59. Writing the equation of the line in parametric form and substituting into the equation of the plane we have:

$$x = \frac{1}{2} + t,\ y = \frac{-3}{2} - t,\ z = -1 + 2t$$

$$2\left(\frac{1}{2} + t\right) - 2\left(\frac{-3}{2} - t\right) + (-1 + 2t) = 12,\ t = \frac{3}{2}$$

Substituting $t = 3/2$ into the parametric equations for the line we have the point of intersection $(2, -3, 2)$. The line does not lie in the plane.

61. Writing the equation of the line in parametric form and substituting into the equation of the plane we have:

$$x = 1 + 3t,\ y = -1 - 2t,\ z = 3 + t$$

$$2(1 + 3t) + 3(-1 - 2t) = 10,\ -1 = 10, \text{ contradiction}$$

Therefore, the line does not intersect the plane.

63. Point: $Q(0, 0, 0)$

Plane: $2x + 3y + z - 12 = 0$

Normal to plane: $\mathbf{n} = \langle 2, 3, 1\rangle$

Point in plane: $P(6, 0, 0)$

Vector $\overrightarrow{PQ} = \langle -6, 0\ 0\rangle$

$$D = \frac{|\overrightarrow{PQ} \cdot \mathbf{n}|}{\|\mathbf{n}\|} = \frac{|-12|}{\sqrt{14}} = \frac{6\sqrt{14}}{7}$$

65. The normal vectors to the planes are $\mathbf{n}_1 = \langle 1, -3, 4\rangle$ and $\mathbf{n}_2 = \langle 1, -3, 4\rangle$. Since $\mathbf{n}_1 = \mathbf{n}_2$, the planes are parallel. Choose a point in each plane.

$P = (10, 0, 0)$ is a point in $x - 3y + 4z = 10$.
$Q = (6, 0, 0)$ is a point in $x - 3y + 4z = 6$.

$$\overrightarrow{PQ} = \langle -4, 0, 0\rangle,\ D = \frac{|\overrightarrow{PQ} \cdot \mathbf{n}_1\|}{\|\mathbf{n}_1\|} = \frac{4}{\sqrt{26}} = \frac{2\sqrt{26}}{13}$$

67. $\mathbf{u} = \langle 4, 0, -1\rangle$ is the direction vector for the line. $Q(1, 5, -2)$ is the given point, and $P(-2, 3, 1)$ is on the line. Hence, $\overrightarrow{PQ} = \langle 3, 2, -3\rangle$ and

$$\overrightarrow{PQ} \times \mathbf{u} = \begin{vmatrix} \mathbf{i} & \mathbf{j} & \mathbf{k} \\ 3 & 2 & -3 \\ 4 & 0 & -1 \end{vmatrix} = \langle -2, -9, -8\rangle$$

$$D = \frac{\|\overrightarrow{PQ} \times \mathbf{u}\|}{\|\mathbf{u}\|} = \frac{|4 + 81 + 64|}{\sqrt{16 + 1}} = \frac{149}{\sqrt{17}} = \frac{149\sqrt{17}}{17}$$

69. (a) $z = 276 - 0.987x - 1.71y$

Year	1970	1975	1980	1985	1990	1991	1992	1993
z (approx.)	213.6	173.7	144.7	121.1	85.3	81.9	81.4	83.6

(b) An increase in x or y will cause a decrease in z. In fact, any increase in two variables will cause a decrease in the third.

(c)

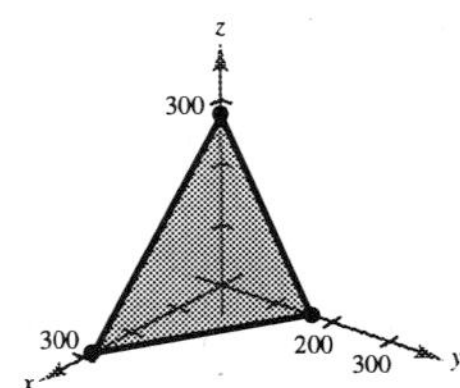

71. (a) Sphere

$$(x - 3)^2 + (y + 2)^2 + (z - 5)^2 = 16$$

$$x^2 + y^2 + z^2 - 6x + 4y - 10z + 22 = 0$$

(b) Parallel planes

$$4x - 3y + z = 10 \pm 4\|\mathbf{n}\| = 10 \pm 4\sqrt{26}$$

73. On one side we have the points $(0, 0, 0)$, $(6, 0, 0)$, and $(-1, -1, 8)$.

$$\mathbf{n}_1 = \begin{vmatrix} \mathbf{i} & \mathbf{j} & \mathbf{k} \\ 6 & 0 & 0 \\ -1 & -1 & 8 \end{vmatrix} = -48\mathbf{j} - 6\mathbf{k}$$

On the adjacent side we have the points $(0, 0, 0)$, $(0, 6, 0)$, and $(-1, -1, 8)$.

$$\mathbf{n}_2 = \begin{vmatrix} \mathbf{i} & \mathbf{j} & \mathbf{k} \\ 0 & 6 & 0 \\ -1 & -1 & 8 \end{vmatrix} = 48\mathbf{i} + 6\mathbf{k}$$

$$\cos\theta = \frac{|\mathbf{n}_1 \cdot \mathbf{n}_2|}{\|\mathbf{n}_1\|\,\|\mathbf{n}_2\|} = \frac{36}{2340} = \frac{1}{65}$$

$$\theta = \arccos\frac{1}{65} \approx 89.1°$$

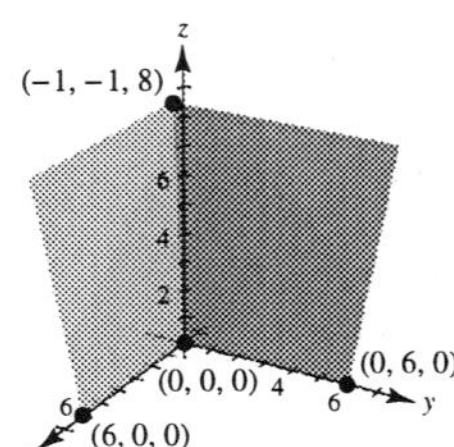

75. True

77. From Theorem 10.13 and Theorem 10.7 (6) we have

$$D = \frac{|\overrightarrow{PQ} \cdot \mathbf{n}|}{\|\mathbf{n}\|} = \frac{|\mathbf{w} \cdot (\mathbf{u} \times \mathbf{v})|}{\|\mathbf{u} \times \mathbf{v}\|} = \frac{|(\mathbf{u} \times \mathbf{v}) \cdot \mathbf{w}|}{\|\mathbf{u} \times \mathbf{v}\|} = \frac{|\mathbf{u} \cdot (\mathbf{v} \times \mathbf{w})|}{\|\mathbf{u} \times \mathbf{v}\|}.$$

Section 10.6 Surfaces in Space

1. Ellipsoid

Matches graph (c)

3. Hyperboloid of one sheet

Matches graph (f)

5. Elliptic paraboloid

Matches graph (d)

7. $z = 3$

Plane parallel to the xy-coordinate plane

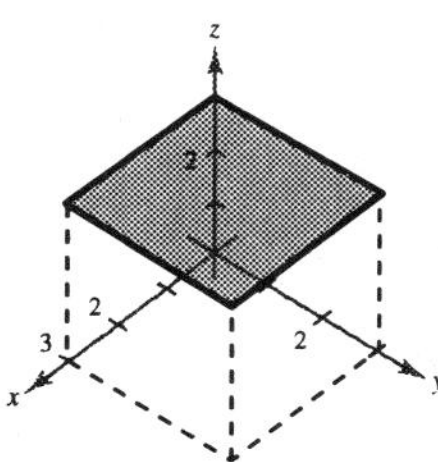

9. $y^2 + z^2 = 9$

The x-coordinate is missing so we have a cylindrical surface with rulings parallel to the x-axis. The generating curve is a circle.

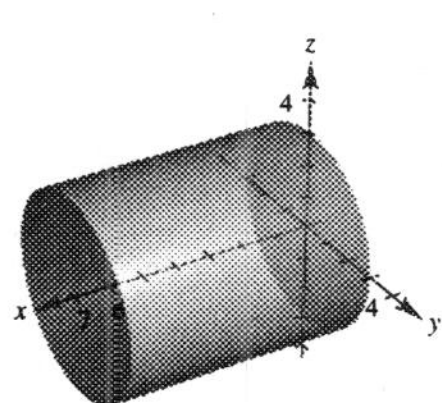

11. $y = x^2$

The z-coordinate is missing so we have a cylindrical surface with rulings parallel to the z-axis. The generating curve is a parabola.

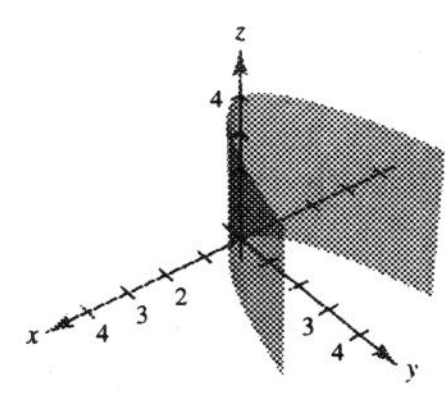

13. $4x^2 + y^2 = 4$

$$\frac{x^2}{1} + \frac{y^2}{4} = 1$$

The z-coordinate is missing so we have a cylindrical surface with rulings parallel to the z-axis. The generating curve is an ellipse.

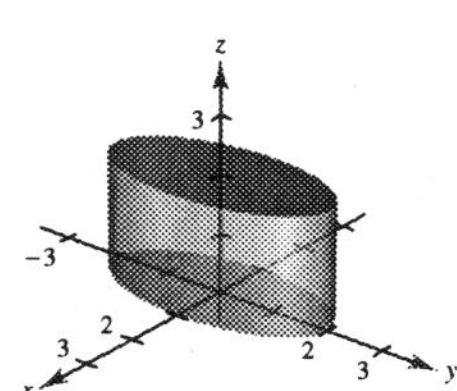

15. $z = \sin y$

The x-coordinate is missing so we have a cylindrical surface with rulings parallel to the x-axis. The generating curve is the sine curve.

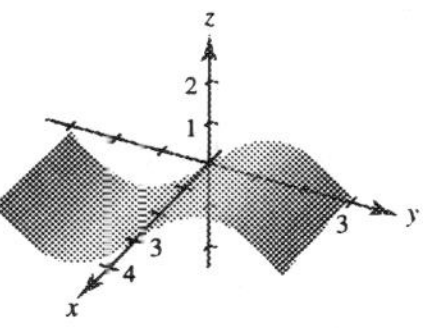

17. $x = x^2 + y^2$

(a) You are viewing the paraboloid from the x-axis: $(20, 0, 0)$

(b) You are viewing the paraboloid from above, but not on the z-axis: $(10, 10, 20)$

(c) You are viewing the paraboloid from the z-axis: $(0, 0, 20)$

(d) You are viewing the paraboloid from the y-axis: $(0, 20, 0)$

19. $\dfrac{x^2}{1} + \dfrac{y^2}{4} + \dfrac{z^2}{1} = 1$

Ellipsoid

xy-trace: $\dfrac{x^2}{1} + \dfrac{y^2}{4} = 1$

xz-trace: $x^2 + z^2 = 1$

yz-trace: $\dfrac{y^2}{4} + \dfrac{z^2}{1} = 1$

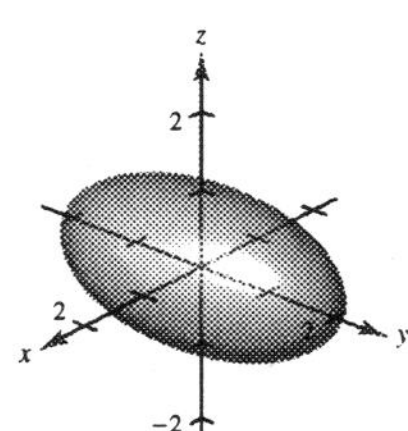

21. $16x^2 - y^2 + 16z^2 = 4$

$$4x^2 - \frac{y^2}{4} + 4z^2 = 1$$

Hyperboloid on one sheet

xy-trace: $4x^2 - \dfrac{y^2}{4} = 1$

xz-trace: $4(x^2 + z^2) = 1$

yz-trace: $\dfrac{-y^2}{4} + 4z^2 = 1$

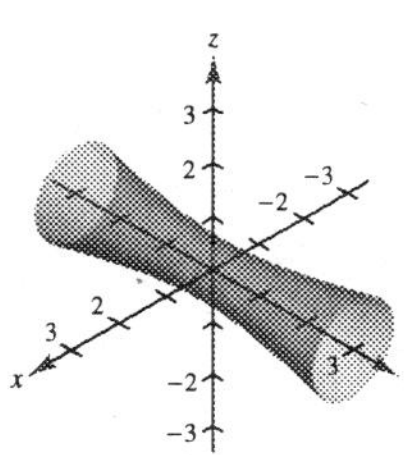

23. $x^2 - y + z^2 = 0$

Elliptic paraboloid

xy-trace: $y = x^2$

xz-trace: $x^2 + z^2 = 0$, point $(0, 0, 0)$

yz-trace: $y = z^2$

$y = 1$: $x^2 + z^2 = 1$

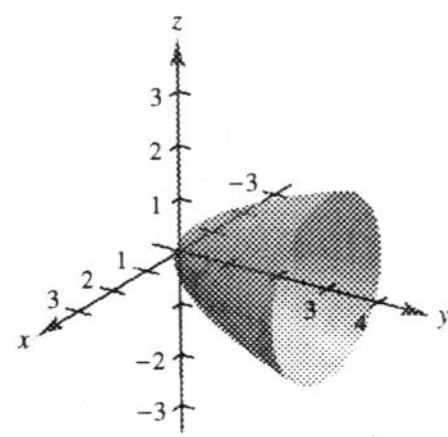

25. $x^2 - y^2 + z = 0$

Hyperbolic paraboloid

xy-trace: $y = \pm x$

xz-trace: $z = -x^2$

yz-trace: $z = y^2$

$y = \pm 1$: $z = 1 - x^2$

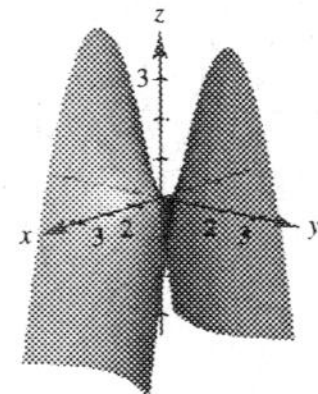

27. $z^2 = x^2 + \dfrac{y^2}{4}$

Elliptic Cone

xy-trace: point $(0, 0, 0)$

xz-trace: $z = \pm x$

yz-trace: $z = \dfrac{\pm 1}{2} y$

$z = \pm 1$: $x^2 + \dfrac{y^2}{4} = 1$

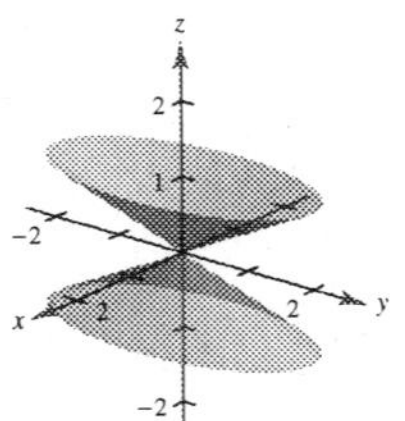

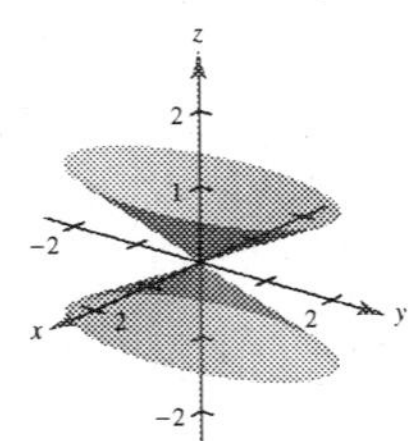

29.
$$16x^2 + 9y^2 + 16z^2 - 32x - 36y + 36 = 0$$
$$16(x^2 - 2x + 1) + 9(y^2 - 4y + 4) + 16z^2 = -36 + 16 + 36$$
$$16(x - 1)^2 + 9(y - 2)^2 + 16z^2 = 16$$
$$\frac{(x - 1)^2}{1} + \frac{(y - 2)^2}{16/9} + \frac{z^2}{1} = 1$$

Ellipsoid with center $(1, 2, 0)$.

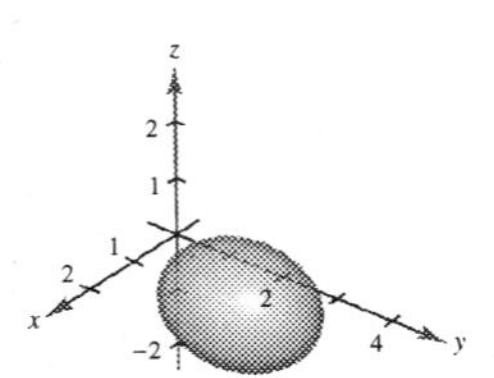

31. $z = 2 \sin x$

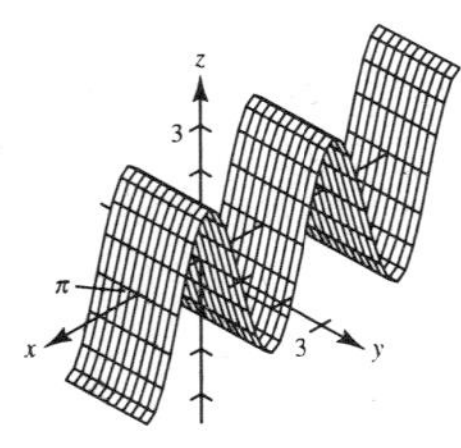

33. $z^2 = x^2 + 4y^2$

$z = \pm\sqrt{x^2 + 4y^2}$

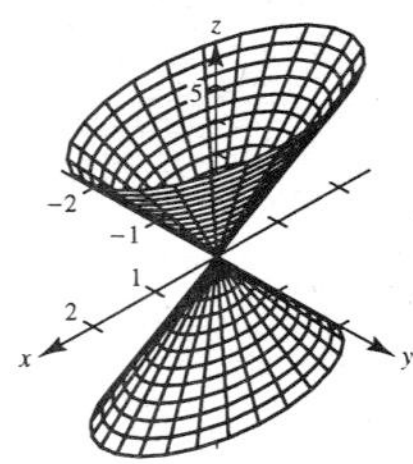

35. $x^2 + y^2 = \left(\dfrac{2}{z}\right)^2$

$y = \pm\sqrt{\dfrac{4}{z^2} - x^2}$

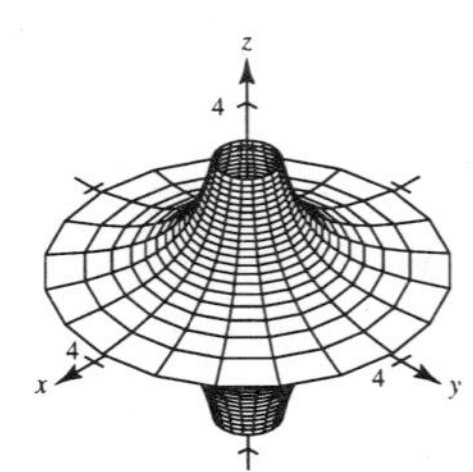

37. $z = 4 - \sqrt{|xy|}$

39. $4x^2 - y^2 + 4z^2 = -16$

$z = \pm\sqrt{\dfrac{y^2}{4} - x^2 - 4}$

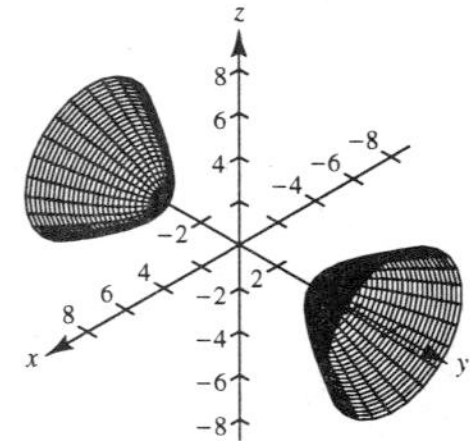

41. $z = 2\sqrt{x^2 + y^2}$

$z = 2$

$2\sqrt{x^2 + y^2} = 2$

$x^2 + y^2 = 1$

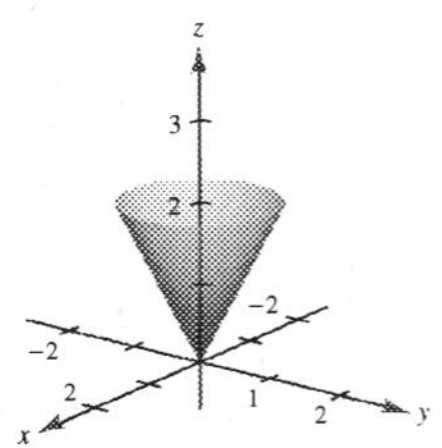

43. $x^2 + y^2 = 1$

$x + z = 2$

$z = 0$

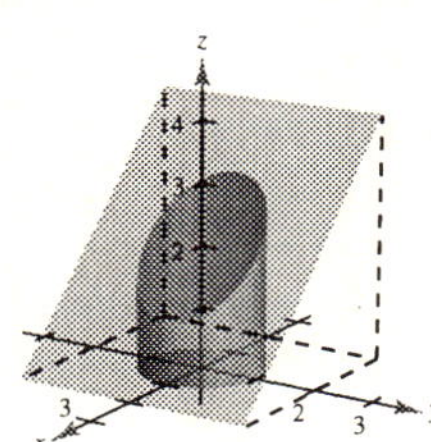

45. $x^2 + z^2 = [r(y)]^2$ and $z = r(y) = \pm 2\sqrt{y}$; therefore, $x^2 + z^2 = 4y$.

47. $x^2 + y^2 = [r(z)]^2$ and $y = r(z) = \dfrac{z}{2}$; therefore,

$$x^2 + y^2 = \frac{z^2}{4},\ 4x^2 + 4y^2 = z^2.$$

49. $y^2 + z^2 = [r(x)]^2$ and $y = r(x) = \dfrac{2}{x}$; therefore,

$$y^2 + z^2 = \left(\frac{2}{x}\right)^2,\ y^2 + z^2 = \frac{4}{x^2}.$$

51. $x^2 + y^2 - 2z = 0$

$x^2 + y^2 = \left(\sqrt{2z}\right)^2$

Equation of generating curve: $y = \sqrt{2z}$ or $x = \sqrt{2z}$

53. $V = 2\pi \displaystyle\int_0^4 x(4x - x^2)\,dx$

$$= 2\pi\left[\frac{4x^3}{3} - \frac{x^4}{4}\right]_0^4 = \frac{128\pi}{3}$$

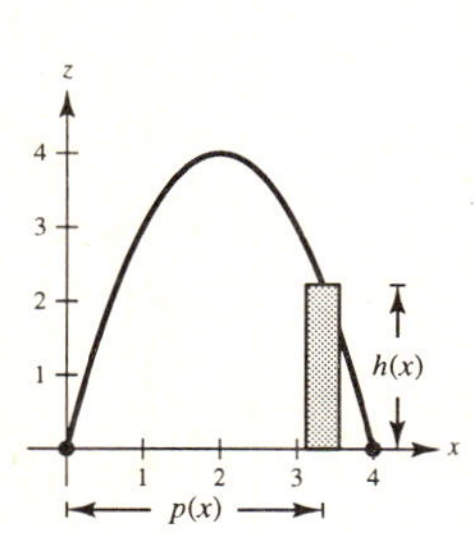

55. $z = \dfrac{x^2}{2} + \dfrac{y^2}{4}$

(a) When $z = 2$ we have $2 = \dfrac{x^2}{2} + \dfrac{y^2}{4}$, or $1 = \dfrac{x^2}{4} + \dfrac{y^2}{8}$

Major axis: $2\sqrt{8} = 4\sqrt{2}$ or 1

Minor axis: $2\sqrt{4} = 4$

$c^2 = a^2 - b^2, c^2 = 4, c = 2$

Foci: $(0, \pm 2, 2)$

(b) When $z = 8$ we have $8 = \dfrac{x^2}{2} + \dfrac{y^2}{4}$, or $1 = \dfrac{x^2}{16} + \dfrac{y^2}{32}$.

Major axis: $2\sqrt{32} = 8\sqrt{2}$

Minor axis: $2\sqrt{16} = 8$

$c^2 = 32 - 16 = 16, c = 4$

Foci: $(0, \pm 4, 8)$

57. $\dfrac{x^2}{3963^2} + \dfrac{y^2}{3963^2} + \dfrac{z^2}{3942^2} = 1$

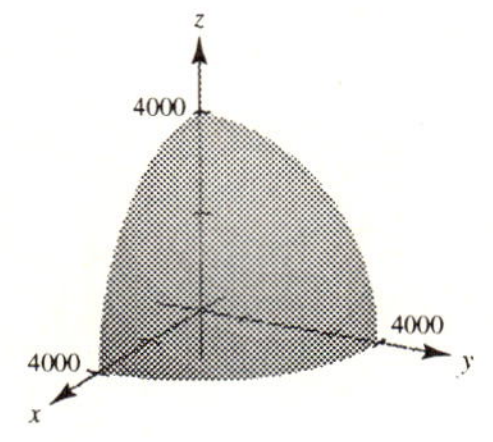

59. $z = \dfrac{y^2}{b^2} - \dfrac{x^2}{a^2},\ z = bx + ay$

$$bx + ay = \frac{y^2}{b^2} - \frac{x^2}{a^2}$$

$$\frac{1}{a^2}\left(x^2 + a^2bx + \frac{a^4b^2}{4}\right) = \frac{1}{b^2}\left(y^2 - ab^2y + \frac{a^2b^4}{4}\right)$$

$$\frac{\left(x + \frac{a^2b}{2}\right)^2}{a^2} = \frac{\left(y - \frac{ab^2}{2}\right)^2}{b^2}$$

$$y = \pm\frac{b}{a}\left(x + \frac{a^2b}{2}\right) + \frac{ab^2}{2}$$

Letting $x = at$, you obtain the two intersecting lines $x = at,\ y = -bt,\ z = 0$ and $x = at,\ y = bt + ab^2$ $z = 2abt + a^2b^2$.

Section 10.7 Cylindrical and Spherical Coordinates

1. $(0, 5, 1)$, rectangular

$r = \sqrt{(0)^2 + (5)^2} = 5$

$\theta = \arctan \frac{5}{0} = \frac{\pi}{2}$

$z = 1$

$\left(5, \frac{\pi}{2}, 1\right)$, cylindrical

3. $\left(1, \sqrt{3}, 4\right)$, rectangular

$r = \sqrt{1^2 + \left(\sqrt{3}\right)^2} = 2$

$\theta = \arctan \sqrt{3} = \frac{\pi}{3}$

$z = 4$

$\left(2, \frac{\pi}{3}, 4\right)$, cylindrical

5. $(2, -2, -4)$, rectangular

$r = \sqrt{2^2 + (-2)^2} = 2\sqrt{2}$

$\theta = \arctan(-1) = -\frac{\pi}{4}$

$z = -4$

$\left(2\sqrt{2}, \frac{-\pi}{4}, -4\right)$, cylindrical

7. $(5, 0, 2)$, cylindrical

$x = 5 \cos 0 = 5$

$y = 5 \sin 0 = 0$

$z = 2$

$(5, 0, 2)$, rectangular

9. $\left(2, \frac{\pi}{3}, 2\right)$, cylindrical

$x = 2 \cos \frac{\pi}{3} = 1$

$y = 2 \sin \frac{\pi}{3} = \sqrt{3}$

$z = 2$

$\left(1, \sqrt{3}, 2\right)$, rectangular

11. $\left(4, \frac{7\pi}{6}, 3\right)$, cylindrical

$x = 4 \cos \frac{7\pi}{6} = -2\sqrt{3}$

$y = 4 \sin \frac{7\pi}{6} = -2$

$z = 3$

$\left(-2\sqrt{3}, -2, 3\right)$, rectangular

13. $r = 2$

$\sqrt{x^2 + y^2} = 2$

$x^2 + y^2 = 4$

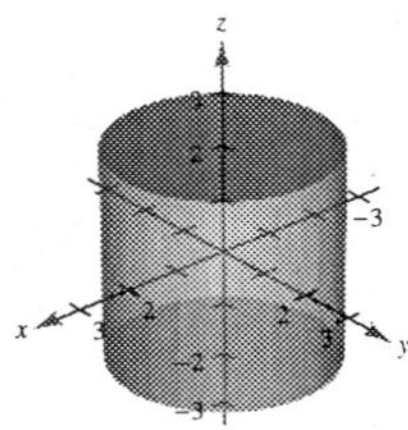

15. $\theta = \frac{\pi}{6}$

$\tan \frac{\pi}{6} = \frac{y}{x}$

$\frac{1}{\sqrt{3}} = \frac{y}{x}$

$x = \sqrt{3}y$

$x - \sqrt{3}y = 0$

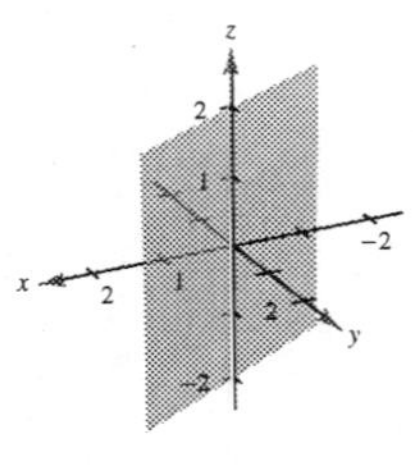

17. $r = 2 \sin \theta$

$r^2 = 2r \sin \theta$

$x^2 + y^2 = 2y$

$x^2 + y^2 - 2y = 0$

$x^2 + (y - 1)^2 = 1$

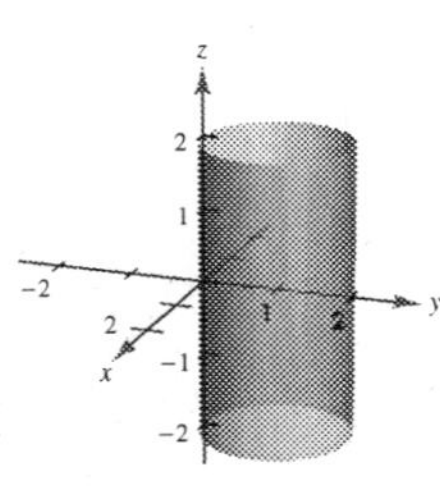

19. $r^2 + z^2 = 4$

$x^2 + y^2 + z^2 = 4$

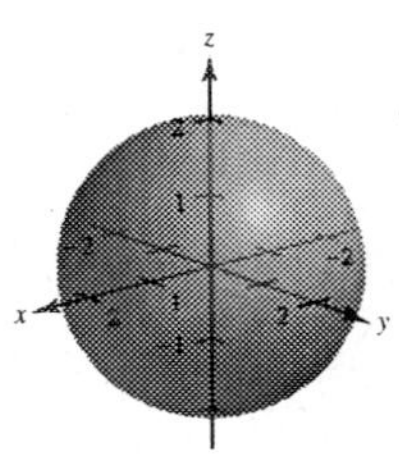

21. $(4, 0, 0)$, rectangular

$\rho = \sqrt{4^2 + 0^2 + 0^2} = 4$

$\theta = \arctan 0 = 0$

$\phi = \arccos 0 = \frac{\pi}{2}$

$\left(4, 0, \frac{\pi}{2}\right)$, spherical

23. $\left(-2, 2\sqrt{3}, 4\right)$, rectangular

$\rho = \sqrt{(-2)^2 + \left(2\sqrt{3}\right)^2 + 4^2} = 4\sqrt{2}$

$\theta = \arctan\left(-\sqrt{3}\right) = \frac{2\pi}{3}$

$\phi = \arccos \frac{1}{\sqrt{2}} = \frac{\pi}{4}$

$\left(4\sqrt{2}, \frac{2\pi}{3}, \frac{\pi}{4}\right)$, spherical

25. $\left(\sqrt{3}, 1, 2\sqrt{3}\right)$, rectangular

$\rho = \sqrt{3 + 1 + 12} = 4$

$\theta = \arctan \frac{1}{\sqrt{3}} = \frac{\pi}{6}$

$\phi = \arccos \frac{\sqrt{3}}{2} = \frac{\pi}{6}$

$\left(4, \frac{\pi}{6}, \frac{\pi}{6}\right)$, spherical

27. $\left(4, \frac{\pi}{6}, \frac{\pi}{4}\right)$, spherical

$x = 4 \sin \frac{\pi}{4} \cos \frac{\pi}{6} = \sqrt{6}$

$y = 4 \sin \frac{\pi}{4} \sin \frac{\pi}{6} = \sqrt{2}$

$z = 4 \cos \frac{\pi}{4} = 2\sqrt{2}$

$\left(\sqrt{6}, \sqrt{2}, 2\sqrt{2}\right)$, rectangular

29. $\left(12, \frac{-\pi}{4}, 0\right)$, spherical

$x = 12 \sin 0 \cos\left(\frac{-\pi}{4}\right) = 0$

$y = 12 \sin 0 \sin\left(\frac{-\pi}{4}\right) = 0$

$z = 12 \cos 0 = 12$

$(0, 0, 12)$, rectangular

31. $\left(5, \frac{\pi}{4}, \frac{3\pi}{4}\right)$, spherical

$x = 5 \sin \frac{3\pi}{4} \cos \frac{\pi}{4} = \frac{5}{2}$

$y = 5 \sin \frac{3\pi}{4} \sin \frac{\pi}{4} = \frac{5}{2}$

$z = 5 \cos \frac{3\pi}{4} = -\frac{5\sqrt{2}}{2}$

$\left(\frac{5}{2}, \frac{5}{2}, -\frac{5\sqrt{2}}{2}\right)$, rectangular

33. (a) Programs will vary.

(b) $(x, y, z) = (3, -4, 2)$

$(\rho, \theta, \phi) = (5.385, -0.927, 1.190)$

35. $\rho = 2$

$x^2 + y^2 + z^2 = 4$

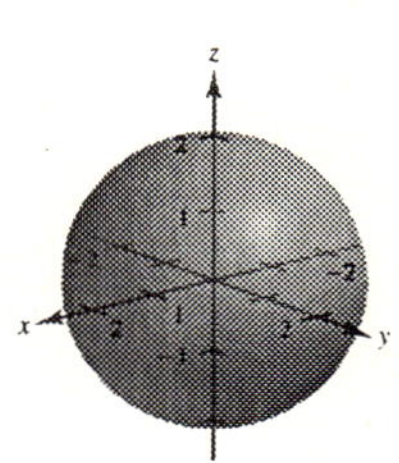

37. $\phi = \frac{\pi}{6}$

$\cos \phi = \frac{z}{\sqrt{x^2 + y^2 + z^2}}$

$\frac{\sqrt{3}}{2} = \frac{z}{\sqrt{x^2 + y^2 + z^2}}$

$\frac{3}{4} = \frac{z^2}{x^2 + y^2 + z^2}$

$3x^2 + 3y^2 - z^2 = 0$

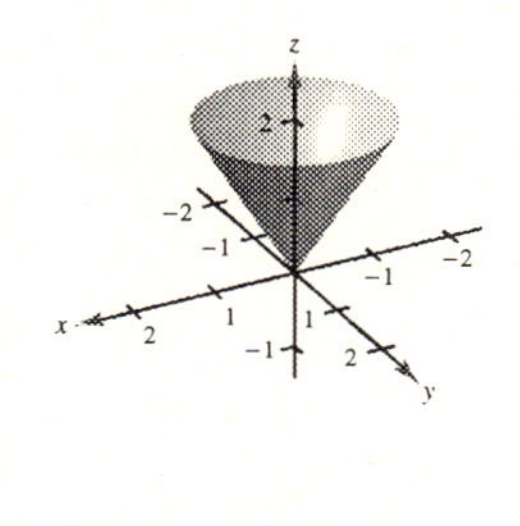

39. $\rho = 4 \cos \phi$

$\sqrt{x^2 + y^2 + z^2} = \frac{4z}{\sqrt{x^2 + y^2 + z^2}}$

$x^2 + y^2 + z^2 - 4z = 0$

$x^2 + y^2 + (z - 2)^2 = 4$

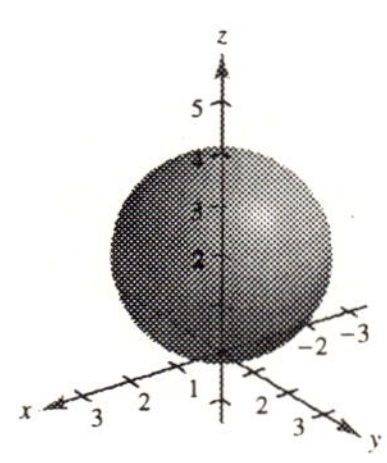

41. $\rho = \csc \phi$

$\rho \sin \phi = 1$

$\sqrt{x^2 + y^2} = 1$

$x^2 + y^2 = 1$

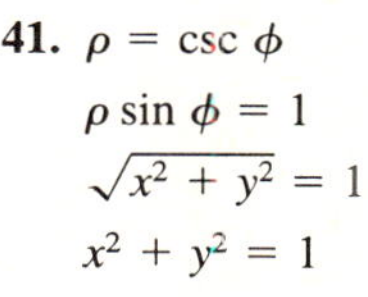

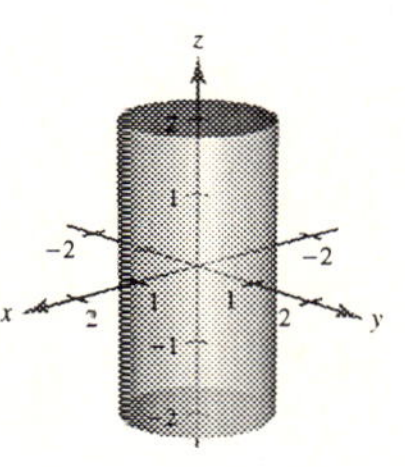

43. $\left(4, \frac{\pi}{4}, 0\right)$, cylindrical

$\rho = \sqrt{4^2 + 0^2} = 4$

$\theta = \frac{\pi}{4}$

$\phi = \arccos 0 = \frac{\pi}{2}$

$\left(4, \frac{\pi}{4}, \frac{\pi}{2}\right)$, spherical

45. $\left(4, \frac{-\pi}{6}, 6\right)$, cylindrical

$\rho = \sqrt{4^2 + 6^2} = 2\sqrt{13}$

$\theta = \frac{-\pi}{6}$

$\phi = \arccos \frac{3}{\sqrt{13}}$

$\left(2\sqrt{13}, \frac{-\pi}{6}, \arccos \frac{3}{\sqrt{13}}\right)$, spherical

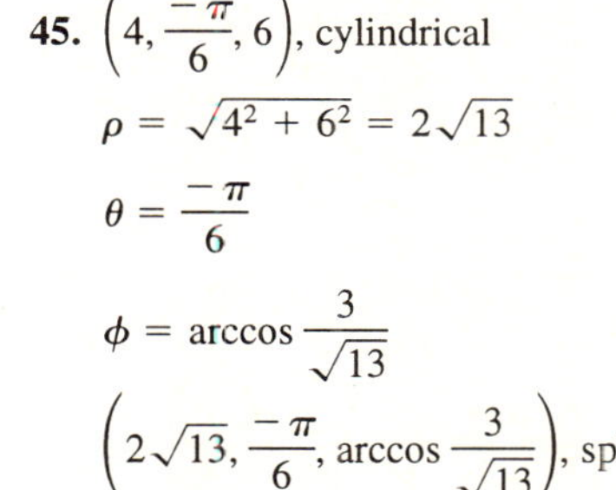

47. $(12, \pi, 5)$, cylindrical

$\rho = \sqrt{12^2 + 5^2} = 13$

$\theta = \pi$

$\phi = \arccos \frac{5}{13}$

$\left(13, \pi, \arccos \frac{5}{13}\right)$, spherical

49. $\left(10, \frac{\pi}{6}, \frac{\pi}{2}\right)$, spherical

$r = 10 \sin \frac{\pi}{2} = 10$

$\theta = \frac{\pi}{6}$

$z = 10 \cos \frac{\pi}{2} = 0$

$\left(10, \frac{\pi}{6}, 0\right)$, cylindrical

51. $\left(6, -\frac{\pi}{6}, \frac{\pi}{3}\right)$, spherical

$r = 6 \sin \frac{\pi}{3} = 3\sqrt{3}$

$\theta = -\frac{\pi}{6}$

$z = 6 \cos \frac{\pi}{3} = 3$

$\left(3\sqrt{3}, -\frac{\pi}{6}, 3\right)$, cylindrical

53. $\left(8, \frac{7\pi}{6}, \frac{\pi}{6}\right)$, spherical

$r = 8 \sin \frac{\pi}{6} = 4$

$\theta = \frac{7\pi}{6}$

$z = 8 \cos \frac{\pi}{6} = \frac{8\sqrt{3}}{2}$

$\left(4, \frac{7\pi}{6}, 4\sqrt{3}\right)$, cylindrical

	Rectangular	Cylindrical	Spherical
55.	(4, 6, 3)	(7.211, 0.983, 3)	(7.810, 0.983, 1.177)
57.	(4.698, 1.710, 8)	$\left(5, \frac{\pi}{9}, 8\right)$	(9.434, 0.349, 0.559)
59.	(−7.071, 12.247, 14.142)	(14.142, 2.094, 14.142)	$\left(20, \frac{2\pi}{3}, \frac{\pi}{4}\right)$
61.	(3, −2, 2)	(3.606, −0.588, 2)	(4.123, −0.588, 1.064)
63.	$\left(\frac{5}{2}, \frac{4}{3}, \frac{-3}{2}\right)$	(2.833, 0.490, −1.5)	(3.206, 0.490, 2.058)
65.	(−3.536, 3.536, −5)	$\left(5, \frac{3\pi}{4}, -5\right)$	(7.071, 2.356, 2.356)
67.	(2.804, −2.095, 6)	(−3.5, 2.5, 6)	(6.946, 5.642, 0.528)

[**Note:** Use the cylindrical coordinates (3.5, 5.642, 6)]

69. $r = 5$

Cylinder

Matches graph (d)

71. $\rho = 5$

Sphere

Matches graph (c)

73. $r^2 = z, x^2 + y^2 = z$

Paraboloid

Matches graph (f)

75. $x^2 + y^2 + z^2 = 16$

(a) $r^2 + z^2 = 16$

(b) $\rho^2 = 16, \rho = 4$

77. $x^2 + y^2 + z^2 - 2z = 0$

(a) $r^2 + z^2 - 2z = 0, r^2 + (z - 1)^2 = 1$

(b) $\rho^2 - 2\rho \cos \phi = 0, \rho(\rho - 2 \cos \phi) = 0,$

$\rho = 2 \cos \phi$

79. $x^2 + y^2 = 4y$

(a) $r^2 = 4r \sin \theta,\ r = 4 \sin \theta$

(b) $\rho^2 \sin^2 \phi = 4\rho \sin \phi \sin \theta,$

$\rho \sin \phi(\rho \sin \phi - 4 \sin \theta) = 0,$

$\rho = \frac{4 \sin \theta}{\sin \phi},\ \rho = 4 \sin \theta \csc \phi$

81. $x^2 - y^2 = 9$

(a) $r^2 \cos^2 \theta - r^2 \sin^2 \theta = 9,$

$r^2 = \frac{9}{\cos^2 \theta - \sin^2 \theta}$

(b) $\rho^2 \sin^2 \phi \cos^2 \theta - \rho^2 \sin^2 \phi \sin^2 \theta = 9,$

$\rho^2 \sin^2 \phi = \frac{9}{\cos^2 \theta - \sin^2 \theta},$

$\rho^2 = \frac{9 \csc^2 \phi}{\cos^2 \theta - \sin^2 \theta}$

83. $0 \le \theta \le \frac{\pi}{2}$

$0 \le r \le 2$

$0 \le z \le 4$

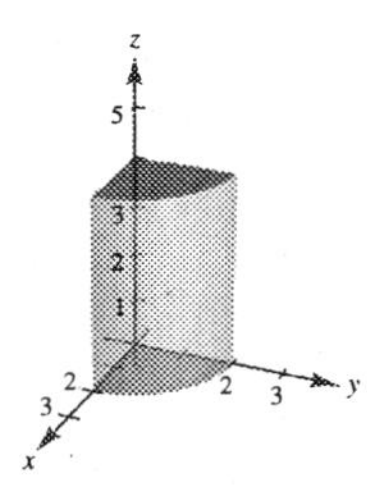

85. $0 \le \theta \le 2\pi$

$0 \le r \le a$

$r \le z \le a$

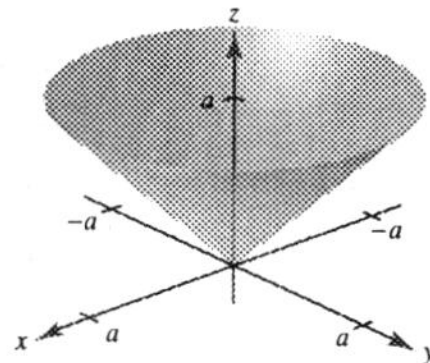

87. $0 \le \theta \le 2\pi$

$0 \le \phi \le \frac{\pi}{6}$

$0 \le \rho \le a \sec \phi$

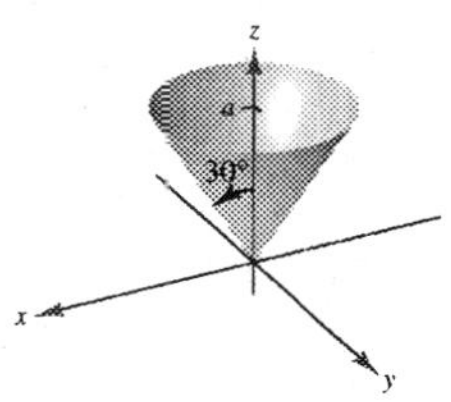

89. Rectangular

$0 \le x \le 10$

$0 \le y \le 10$

$0 \le z \le 10$

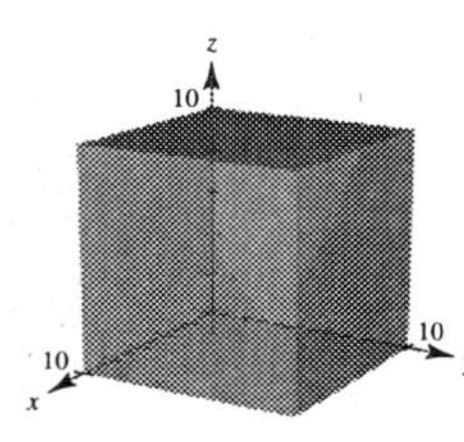

91. Spherical

$4 \le \rho \le 6$

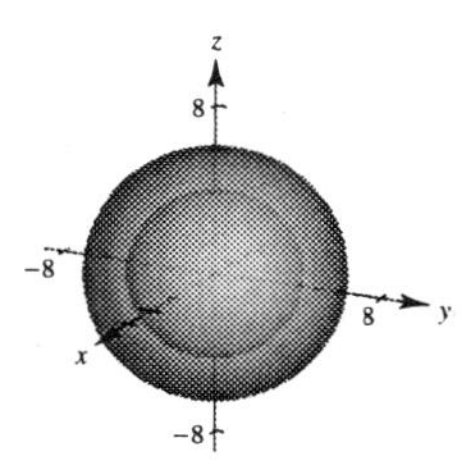

93. $z = \sin \theta, r = 1$

$$z = \frac{y}{r} = \frac{y}{1} = y$$

The curve of intersection is the ellipse formed by the intersection of the plane $z = y$ and the cylinder $r = 1$.

Review Exercises for Chapter 10

1. $P = (1, 2),\ Q = (4, 1),\ R = (5, 4)$

(a) $\mathbf{u} = \overrightarrow{PQ} = \langle 3, -1 \rangle = 3\mathbf{i} - \mathbf{j}$,

$\mathbf{v} = \overrightarrow{PR} = \langle 4, 2 \rangle = 4\mathbf{i} + 2\mathbf{j}$

(b) $\|\mathbf{v}\| = \sqrt{4^2 + 2^2} = 2\sqrt{5}$

(c) $\mathbf{u} \cdot \mathbf{v} = 3(4) + (-1)(2) = 10$

(d) $2\mathbf{u} + \mathbf{v} = \langle 6, -2 \rangle + \langle 4, 2 \rangle = \langle 10, 0 \rangle = 10\mathbf{i}$

(e) $\mathbf{w}_1 = \text{proj}_{\mathbf{v}}(\mathbf{u}) = \left(\frac{\mathbf{u} \cdot \mathbf{v}}{\|\mathbf{v}\|^2}\right)\mathbf{v} = \frac{10}{(2\sqrt{5})^2}(4\mathbf{i} + 2\mathbf{j}) = 2\mathbf{i} + \mathbf{j}$

(f) $\mathbf{w}_2 = \mathbf{u} - \mathbf{w}_1 = (3\mathbf{i} - \mathbf{j}) - (2\mathbf{i} - \mathbf{j}) = \mathbf{i} - 2\mathbf{j}$

3. $z = 0,\ y = 4,\ x = -5{:}\ (-5, 4, 0)$

5. Looking down from the positive x-axis towards the yz-plane, the point is either in the first quadrant $(y > 0, z > 0)$ or in the third quadrant $(y < 0, z < 0)$. The x-coordinate can be any number.

7. $(x - 3)^2 + (y + 2)^2 + (z - 6)^2 = \left(\frac{15}{2}\right)^2$

9. $(x^2 - 4x + 4) + (y^2 - 6y + 9) + z^2 = -4 + 4 + 9$

$(x - 2)^2 + (y - 3)^2 + z^2 = 9$

Center: $(2, 3, 0)$

Radius: 3

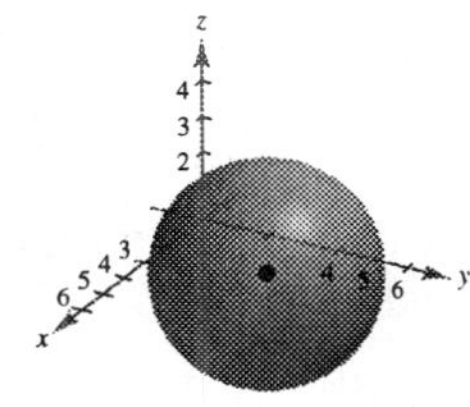

11. $P = (5, 0, 0),\ Q = (4, 4, 0),\ R = (2, 0, 6)$

(a) $\mathbf{u} = \overrightarrow{PQ} = \langle -1, 4, 0\rangle = -\mathbf{i} + 4\mathbf{j},$

$\mathbf{v} = \overrightarrow{PR} = \langle -3, 0, 6\rangle = -3\mathbf{i} + 6\mathbf{k}$

(b) $\mathbf{u} \cdot \mathbf{v} = (-1)(-3) + 4(0) + 0(6) = 3$

(c) $\mathbf{u} \times \mathbf{v} = \begin{vmatrix} \mathbf{i} & \mathbf{j} & \mathbf{k} \\ -1 & 4 & 0 \\ -3 & 0 & 6 \end{vmatrix} = 24\mathbf{i} + 6\mathbf{j} + 12\mathbf{k}$

(d) $\mathbf{n} = 24\mathbf{i} + 6\mathbf{j} + 12\mathbf{k},\ 24(x - 5) + 6y + 12z = 0,$

$4x + y + 2z - 20 = 0$

(e) $x = 5 - t,\ y = 4t,\ z = 0,$ or $x = 4 - t,$

$y = 4 + 4t,\ z = 0$

13. $\mathbf{u} = \langle 7, -2, 3\rangle,\ \mathbf{v} = \langle -1, 4, 5\rangle$

Since $\mathbf{u} \cdot \mathbf{v} = 0$, the vectors are orthogonal.

15. $\mathbf{u} = 5\left(\cos \frac{3\pi}{4}\mathbf{i} + \sin \frac{3\pi}{4}\mathbf{j}\right) = \frac{5\sqrt{2}}{2}[-\mathbf{i} + \mathbf{j}]$

$\mathbf{v} = 2\left(\cos \frac{2\pi}{3}\mathbf{i} + \sin \frac{2\pi}{3}\mathbf{j}\right) = -\mathbf{i} + \sqrt{3}\mathbf{j}$

$\mathbf{u} \cdot \mathbf{v} = \frac{5\sqrt{2}}{2}(1 + \sqrt{3})$

$\|\mathbf{u}\| = 5$

$\|\mathbf{v}\| = 2$

$\cos \theta = \frac{|\mathbf{u} \cdot \mathbf{v}|}{\|\mathbf{u}\|\ \|\mathbf{v}\|} = \frac{(5\sqrt{2}/2)(1 + \sqrt{3})}{5(2)} = \frac{\sqrt{2} + \sqrt{6}}{4}$

$\theta = \arccos \frac{\sqrt{2} + \sqrt{6}}{4} = 15°$

17. $\mathbf{u} = \langle 10, -5, 15\rangle,\ \mathbf{v} = \langle -2, 1, -3\rangle$

$\mathbf{u} = -5\mathbf{v} \Rightarrow \mathbf{u}$ is parallel to $\mathbf{v}$ and in the opposite direction.

$\theta = \pi$

19. $\mathbf{u} = 4[(\cos 135°)\mathbf{i} + (\sin 135°)\mathbf{j}]$

$= 4\left[-\frac{\sqrt{2}}{2}\mathbf{i} + \frac{\sqrt{2}}{2}\mathbf{j}\right] = -2\sqrt{2}\mathbf{i} + 2\sqrt{2}\mathbf{j}$

21. $\mathbf{v} = \mathbf{i} - 3\mathbf{j} + 4\mathbf{k}$

$\mathbf{u} = 3\left(\frac{\mathbf{v}}{\|\mathbf{v}\|}\right) = 3\left(\frac{\mathbf{i} - 3\mathbf{j} + 4\mathbf{k}}{\sqrt{26}}\right) = \frac{3}{\sqrt{26}}\mathbf{i} - \frac{9}{\sqrt{26}}\mathbf{j} + \frac{12}{\sqrt{26}}\mathbf{k}$

In Exercises 23–31, $\mathbf{u} = \langle 3, -2, 1\rangle,\ \mathbf{v} = \langle 2, -4, -3\rangle,\ \mathbf{w} = \langle -1, 2, 2\rangle$.

23. $\|\mathbf{u}\| = \sqrt{3^2 + (-2)^2 + 1^2} = \sqrt{14}$

25. $\mathbf{u} \cdot \mathbf{u} = 3(3) + (-2)(-2) + (1)(1) = 14 = \left(\sqrt{14}\right)^2 = \|\mathbf{u}\|^2$

27. $\text{proj}_{\mathbf{u}}\mathbf{w} = \left(\frac{\mathbf{u} \cdot \mathbf{w}}{\|\mathbf{u}\|^2}\right)\mathbf{u} = -\frac{5}{14}\langle 3, -2, 1\rangle = \left\langle -\frac{15}{14}, \frac{10}{14}, -\frac{5}{14}\right\rangle$

29. $\mathbf{u} \cdot (\mathbf{v} + \mathbf{w}) = \langle 3, -2, 1\rangle \cdot \langle 1, -2, -1\rangle = 6$

$\mathbf{u} \cdot \mathbf{v} + \mathbf{u} \cdot \mathbf{w} = \langle 3, -2, 1\rangle \cdot \langle 2, -4, -3\rangle + \langle 3, -2, 1\rangle \cdot \langle -1, 2, 2\rangle = 11 + (-5) = 6$

Thus, $\mathbf{u} \cdot (\mathbf{v} + \mathbf{w}) = \mathbf{u} \cdot \mathbf{v} + \mathbf{u} \cdot \mathbf{w}$.

31. $V = |\mathbf{u} \cdot (\mathbf{v} \times \mathbf{w})|$

$= |\langle 3, -2, 1\rangle \cdot \langle -2, -1, 0\rangle| = |-4| = 4$

33. $\|AC\| \cos 40° - \|BC\| \cos 20° = 0$

$\|AC\| \sin 40° + \|BC\| \sin 20° = 20$

$$\|AC\| = \frac{\|BC\| \cos 20°}{\cos 40°}$$

By substitution, we have:

$\|BC\|(\cos 20° \tan 40° + \sin 20°) = 20$

$\|BC\| \approx 17.7 \text{ N}$

$\|AC\| \approx 21.7 \text{ N}$

$\|BC\|(\cos 20° + \tan 40° + \sin 20°) = 20$

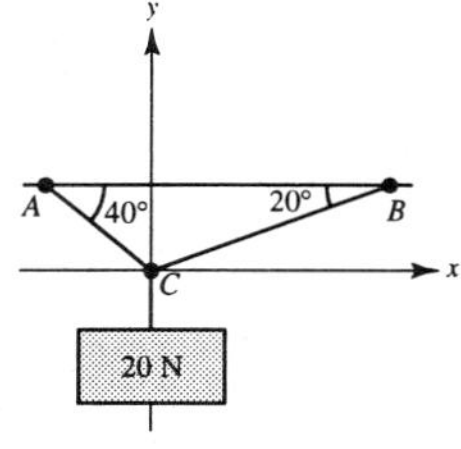

35. $\mathbf{F} = c(\cos 20°\mathbf{j} + \sin 20°\mathbf{k})$

$\overrightarrow{PQ} = 2\mathbf{k}$

$$\overrightarrow{PQ} \times \mathbf{F} = \begin{vmatrix} \mathbf{i} & \mathbf{j} & \mathbf{k} \\ 0 & 0 & 2 \\ 0 & c\cos 20° & c\sin 20° \end{vmatrix} = -2c\cos 20°\mathbf{i}$$

$200 = \|\overrightarrow{PQ} \times \mathbf{F}\| = 2c \cos 20°$

$$c = \frac{100}{\cos 20°}$$

$$\mathbf{F} = \frac{100}{\cos 20°}(\cos 20°\mathbf{j} + \sin 20°\mathbf{k}) = 100(\mathbf{j} + \tan 20°\mathbf{k})$$

$\|\mathbf{F}\| = 100\sqrt{1 + \tan^2 20°} = 100 \sec 20° \approx 106.4 \text{ lb}$

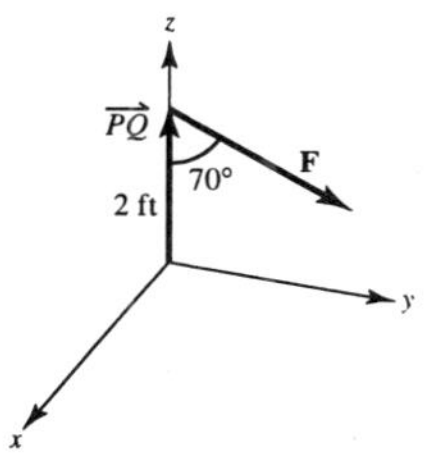

37. $\mathbf{v} = \mathbf{j}$

(a) $x = 1,\ y = 2 + t,\ z = 3$

(b) None

39. $3x - 3y - 7z = -4,\ x - y + 2z = 3$

Solving simultaneously, we have $z = 1$. Substituting $z = 1$ into the second equation we have $y = x - 1$. Substituting for x in this equation we obtain two points on the line of intersection, $(0, -1, 1)$, $(1, 0, 1)$. The direction vector of the line of intersection is $\mathbf{v} = \mathbf{i} + \mathbf{j}$.

(a) $x = t,\ y = -1 + t,\ z = 1$

(b) $x = y + 1,\ z = 1$

41. The two lines are parallel as they have the same direction numbers, $-2, 1, 1$. Therefore, a vector parallel to the plane is $\mathbf{v} = -2\mathbf{i} + \mathbf{j} + \mathbf{k}$. A point on the first line is $(1, 0, -1)$ and a point on the second line is $(-1, 1, 2)$. The vector $\mathbf{u} = 2\mathbf{i} - \mathbf{j} - 3\mathbf{k}$ connecting these two points is also parallel to the plane. Therefore, a normal to the plane is

$$\mathbf{v} \times \mathbf{u} = \begin{vmatrix} \mathbf{i} & \mathbf{j} & \mathbf{k} \\ -2 & 1 & 1 \\ 2 & -1 & -3 \end{vmatrix}$$

$$= -2\mathbf{i} - 4\mathbf{j} = -2(\mathbf{i} + 2\mathbf{j}).$$

Equation of the plane: $(x - 1) + 2y = 0$

$$x + 2y = 1$$

43. $Q = (1, 0, 2)$

$2x - 3y + 6z = 6$

A point P on the plane is $(3, 0, 0)$.

$\overrightarrow{PQ} = \langle -2, 0, 2 \rangle$

$\mathbf{n} = \langle 2, -3, 6 \rangle$

$$D = \frac{|\overrightarrow{PQ} \cdot \mathbf{n}|}{\|\mathbf{n}\|} = \frac{8}{7}$$

45. Direction vectors for the two lines:

$\mathbf{u} = \langle 1, 2, 3 \rangle, \ \mathbf{v} = \langle -1, 3, 2 \rangle$

A vector perpendicular to the two lines:

$$\mathbf{u} \times \mathbf{v} = \begin{vmatrix} \mathbf{i} & \mathbf{j} & \mathbf{k} \\ 1 & 2 & 3 \\ -1 & 3 & 2 \end{vmatrix} = \langle -5, -5, 5 \rangle$$

Point on Line 1: $(1, 2, 3)$

Point on Line 2: $(0, -3, -4)$

The vector connecting the two points: $\mathbf{w} = \langle -1, -5, -7 \rangle$

$$D = \|\text{proj}_{\mathbf{w}}(\mathbf{u} \times \mathbf{v})\|$$
$$= \frac{|\mathbf{w} \cdot (\mathbf{u} \times \mathbf{v})|}{\|\mathbf{u} \times \mathbf{v}\|^2} \|\mathbf{u} \times \mathbf{v}\|$$
$$= \frac{|\mathbf{w} \cdot (\mathbf{u} \times \mathbf{v})|}{\|\mathbf{u} \times \mathbf{v}\|} = \frac{5}{5\sqrt{3}} = \frac{\sqrt{3}}{3}$$

47. $x + 2y + 3z = 6$

Plane

Intercepts: $(6, 0, 0)$, $(0, 3, 0)$, $(0, 0, 2)$

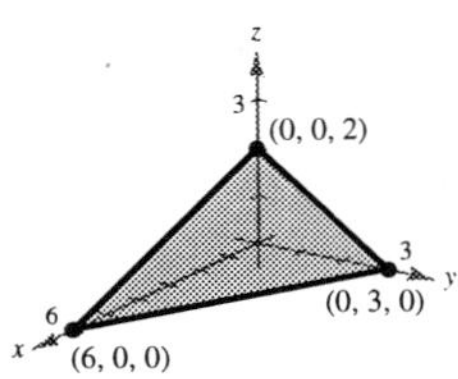

49. $y = \frac{1}{2}z$

Plane with rulings parallel to the x-axis

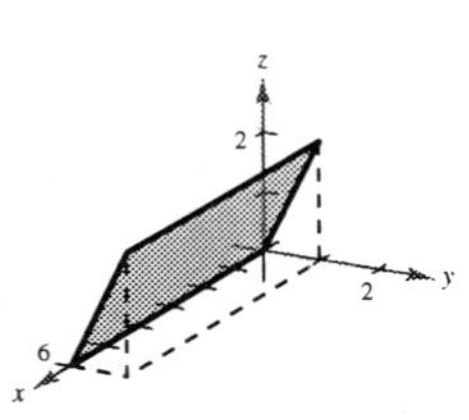

51. $\dfrac{x^2}{16} + \dfrac{y^2}{9} + z^2 = 1$

Ellipsoid

xy-trace: $\dfrac{x^2}{16} + \dfrac{y^2}{9} = 1$

xz-trace: $\dfrac{x^2}{16} + z^2 = 1$

yz-trace: $\dfrac{y^2}{9} + z^2 = 1$

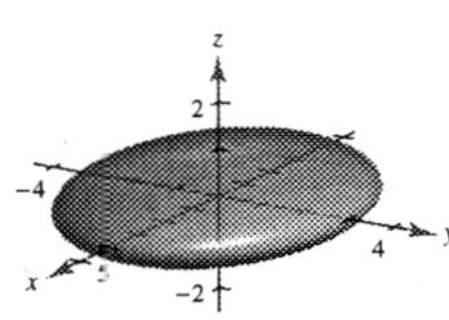

53. $\dfrac{x^2}{16} - \dfrac{y^2}{9} + z^2 = -1$

Hyperboloid of two sheets

xy-trace: $\dfrac{y^2}{4} - \dfrac{x^2}{16} = 1$

xz-trace: None

yz-trace: $\dfrac{y^2}{9} - z^2 = 1$

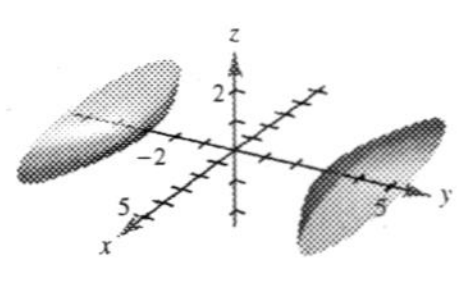

55. (a)
$$x^2 + y^2 = [r(z)]^2 = \left[\sqrt{2(z-1)}\right]^2$$
$$x^2 + y^2 - 2z + 2 = 0$$

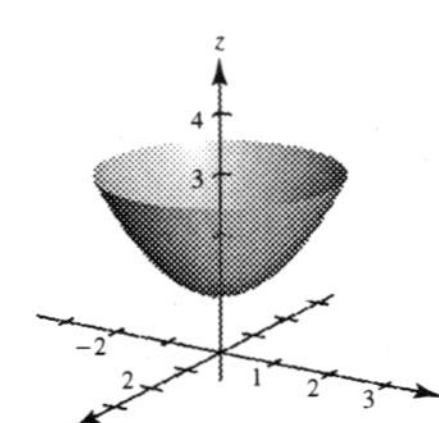

(b)
$$V = 2\pi \int_0^2 x\left[3 - \left(\frac{1}{2}x^2 + 1\right)\right] dx$$
$$= 2\pi \int_0^2 \left(2x - \frac{1}{2}x^3\right) dx$$
$$= 2\pi \left[x^2 - \frac{x^4}{8}\right]_0^2 = 4\pi \approx 12.6 \text{ cm}^3$$
$$= 4\pi - \frac{31\pi}{64} = \frac{225\pi}{64} \approx 11.04$$

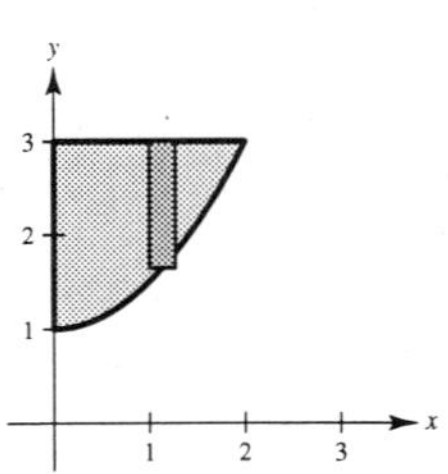

(c)
$$V = 2\pi \int_{1/2}^2 x\left[3 - \left(\frac{1}{2}x^2 + 1\right)\right] dx$$
$$= 2\pi \int_{1/2}^2 \left(2x - \frac{1}{2}x^3\right) dx$$
$$= 2\pi \left[x^2 - \frac{x^4}{8}\right]_{1/2}^2$$
$$= 4\pi - \frac{31\pi}{64} = \frac{225\pi}{64} \approx 11.04$$

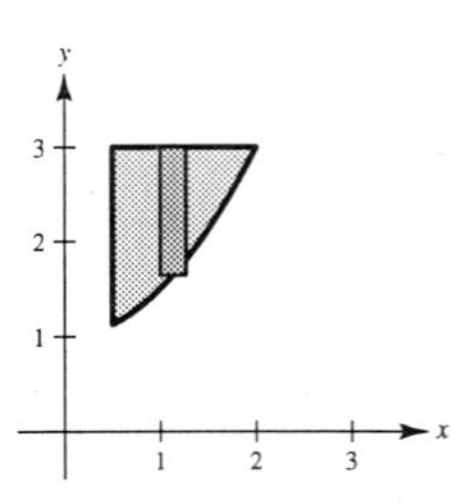

57. $\left(-2\sqrt{2}, 2\sqrt{2}, 2\right)$, rectangular

(a) $r = \sqrt{\left(-2\sqrt{2}\right)^2 + \left(2\sqrt{2}\right)^2} = 4,$

$\theta = \arctan(-1) = \dfrac{3\pi}{4},$

$z = 2,$

$\left(4, \dfrac{3\pi}{4}, 2\right)$, cylindrical

(b) $\rho = \sqrt{\left(-2\sqrt{2}\right)^2 + \left(2\sqrt{2}\right)^2 + (2)^2} = 2\sqrt{5},$

$\theta = \dfrac{3\pi}{4},$

$\phi = \arccos \dfrac{2}{2\sqrt{5}} = \arccos \dfrac{1}{\sqrt{5}},$

$\left(2\sqrt{5}, \dfrac{3\pi}{4}, \arccos \dfrac{\sqrt{5}}{5}\right)$, spherical

59. $x^2 - y^2 = 2z$

(a) Cylindrical: $r^2 \cos^2 \theta - r^2 \sin^2 \theta = 2z,$

$r^2 \cos 2\theta = 2z$

(b) Spherical: $\rho^2 \sin^2 \phi \cos^2 \theta - \rho^2 \sin^2 \phi \sin^2 \theta = 2\rho \cos \phi,$

$\rho \sin^2 \phi \cos 2\theta - 2 \cos \phi = 0,$

$\rho = 2 \sec 2\theta \cos \phi \csc^2 \phi$

CHAPTER 11
Vector-Valued Functions

CHAPTER 11
Vector-Valued Functions

Section 11.1 Vector-Valued Functions

Solutions to Odd-Numbered Exercises

1. $\mathbf{r}(t) = 5t\mathbf{i} - 4t\mathbf{j} - \dfrac{1}{t}\mathbf{k}$

Component functions: $f(t) = 5t$

$g(t) = -4t$

$h(t) = -\dfrac{1}{t}$

Domain: $(-\infty, 0) \cup (0, \infty)$

3. $\mathbf{r}(t) = \ln t\mathbf{i} - e^t\mathbf{j} - t\mathbf{k}$

Component functions: $f(t) = \ln t$

$g(t) = -e^t$

$h(t) = -t$

Domain: $(0, \infty)$

5. $\mathbf{r}(t) = \mathbf{F}(t) + \mathbf{G}(t) = (\cos t\mathbf{i} - \sin t\mathbf{j} + \sqrt{t}\mathbf{k}) + (\cos t\mathbf{i} + \sin t\mathbf{j}) = 2\cos t\mathbf{i} + \sqrt{t}\mathbf{k}$

Component functions: $f(t) = 2\cos t$

$g(t) = 0$

$h(t) = \sqrt{t}$

Domain: $[0, \infty)$

7. $\mathbf{r}(t) = \mathbf{F}(t) \times \mathbf{G}(t) = \begin{vmatrix} \mathbf{i} & \mathbf{j} & \mathbf{k} \\ \sin t & \cos t & 0 \\ 0 & \sin t & \cos t \end{vmatrix} = \cos^2 t\mathbf{i} - \sin t\cos t\mathbf{j} + \sin^2 t\mathbf{k}$

Component functions: $f(t) = \cos^2 t$

$g(t) = -\sin t\cos t$

$h(t) = \sin^2 t$

Domain: $(-\infty, \infty)$

9. $\mathbf{r}(t) = \frac{1}{2}t^2\mathbf{i} - (t - 1)\mathbf{j}$

(a) $\mathbf{r}(1) = \frac{1}{2}\mathbf{i}$

(b) $\mathbf{r}(0) = \mathbf{j}$

(c) $\mathbf{r}(s + 1) = \frac{1}{2}(s + 1)^2\mathbf{i} - (s + 1 - 1)\mathbf{j} = \frac{1}{2}(s + 1)^2\mathbf{i} - s\mathbf{j}$

(d) $$\begin{aligned}\mathbf{r}(2 + \Delta t) - \mathbf{r}(2) &= \tfrac{1}{2}(2 + \Delta t)^2\mathbf{i} - (2 + \Delta t - 1)\mathbf{j} - (2\mathbf{i} - \mathbf{j}) \\ &= \left(2 + 2\Delta t + \tfrac{1}{2}(\Delta t)^2\right)\mathbf{i} - (1 + \Delta t)\mathbf{j} - 2\mathbf{i} + \mathbf{j} \\ &= \left(2\Delta t + \tfrac{1}{2}(\Delta t)^2\right)\mathbf{i} - (\Delta t)\mathbf{j}\end{aligned}$$

11. $\mathbf{r}(t) = \ln t\mathbf{i} + \dfrac{1}{t}\mathbf{j} + 3t\mathbf{k}$

(a) $\mathbf{r}(2) = \ln 2\mathbf{i} + \dfrac{1}{2}\mathbf{j} + 6\mathbf{k}$

(b) $\mathbf{r}(-3)$ is not defined. ($\ln(-3)$ does not exist.)

(c) $\mathbf{r}(t - 4) = \ln(t - 4)\mathbf{i} + \dfrac{1}{t - 4}\mathbf{j} + 3(t - 4)\mathbf{k}$

(d) $$\begin{aligned}\mathbf{r}(1 + \Delta t) - \mathbf{r}(1) &= \ln(1 + \Delta t)\mathbf{i} + \frac{1}{1 + \Delta t}\mathbf{j} + 3(1 + \Delta t)\mathbf{k} - (0\mathbf{i} + \mathbf{j} + 3\mathbf{k}) \\ &= \ln(1 + \Delta t)\mathbf{i} + \left(\frac{1}{1 + \Delta t} - 1\right)\mathbf{j} + (3\Delta t)\mathbf{k}\end{aligned}$$

13. $\mathbf{r}(t) = \sin 3t\mathbf{i} + \cos 3t\mathbf{j} + t\mathbf{k}$

$\|\mathbf{r}(t)\| = \sqrt{(\sin 3t)^2 + (\cos 3t)^2 + t^2} = \sqrt{1 + t^2}$

15. $\mathbf{r}(t) \cdot \mathbf{u}(t) = (3t - 1)(t^2) + \left(\frac{1}{4}t^3\right)(-8) + 4(t^3)$

$= 3t^3 - t^2 - 2t^3 + 4t^3 = 5t^3 - t^2$, a scalar.

The dot product is a scalar-valued function.

17. $\mathbf{r}(t) = t\mathbf{i} + 2t\mathbf{j} + t^2\mathbf{k}, \ -2 \le t \le 2$

$x = t, y = 2t, z = t^2$

Thus, $z = x^2$. Matches (b)

19. $\mathbf{r}(t) = t\mathbf{i} + t^2\mathbf{j} + e^{0.75t}\mathbf{k}, \ -2 \le t \le 2$

$x = t, y = t^2, z = e^{0.75t}$

Thus, $y = x^2$. Matches (d)

21. (a) View from the negative x-axis: $(-20, 0, 0)$

(b) View from above the first octant: $(10, 20, 10)$

(c) View from the z-axis: $(0, 0, 20)$

(d) View from the positive x-axis: $(20, 0, 0)$

23. $x = 3t$

$y = t - 1$

$y = \frac{x}{3} - 1$

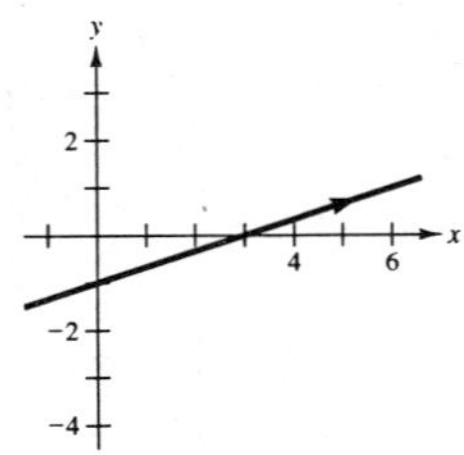

25. $x = -t + 1$

$y = 4t + 2$

$z = 2t + 3$

Line passing through the points:

$(0, 6, 5), \ (1, 2, 3)$

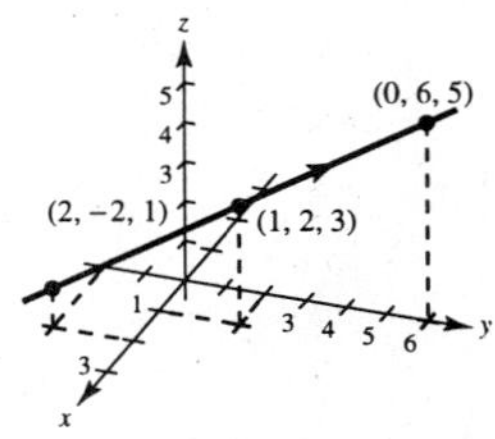

27. $x = 2\cos t, \ y = 2\sin t, \ z = t$

$\frac{x^2}{4} + \frac{y^2}{4} = 1$

$z = t$

Circular helix

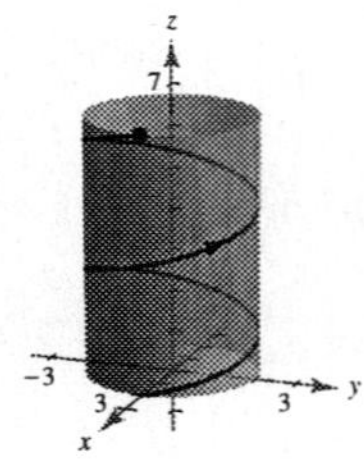

29. $x = 2\sin t, \ y = 2\cos t, \ z = e^{-t}$

$x^2 + y^2 = 4$

$z = e^{-t}$

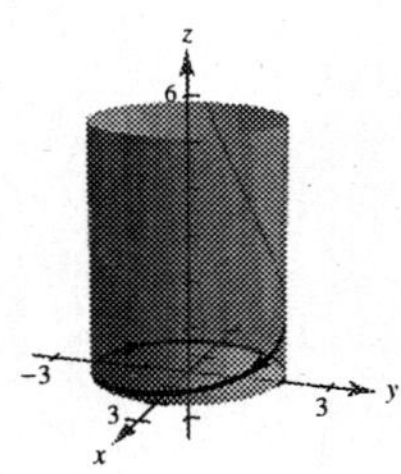

31. $x = t, \ y = t^2, \ z = \frac{2}{3}t^3$

$y = x^2, \ z = \frac{2}{3}x^3$

t	-2	-1	0	1	2
x	-2	-1	0	1	2
y	4	1	0	1	4
z	$-\frac{16}{3}$	$-\frac{2}{3}$	0	$\frac{2}{3}$	$\frac{16}{3}$

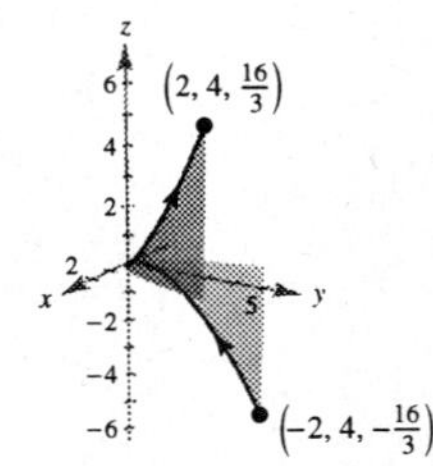

33. $\mathbf{r}(t) = -\frac{1}{2}t^2\mathbf{i} + t\mathbf{j} - \frac{\sqrt{3}}{2}t^2\mathbf{k}$

Parabola

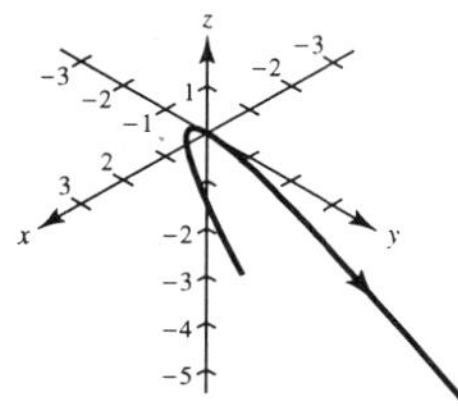

35. $\mathbf{r}(t) = \sin t\mathbf{i} + \left(\frac{\sqrt{3}}{2}\cos t - \frac{1}{2}t\right)\mathbf{j} + \left(\frac{1}{2}\cos t + \frac{\sqrt{3}}{2}\right)\mathbf{k}$

Helix

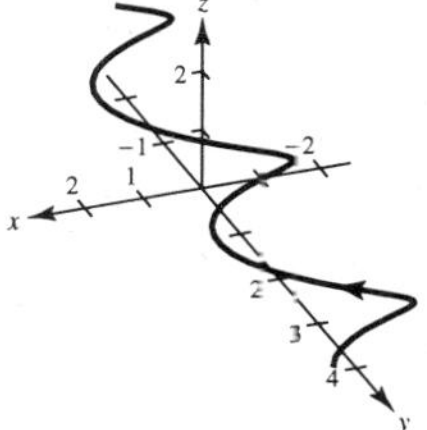

37.

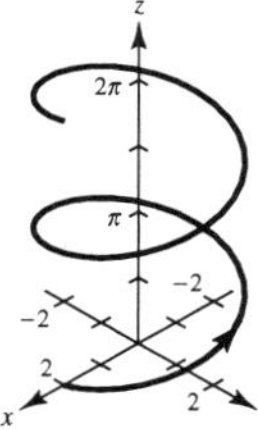

(a)

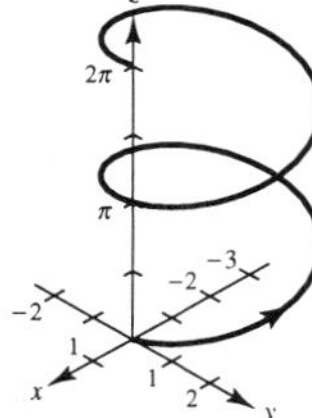

The helix is translated 2 units back on the x-axis.

(b)

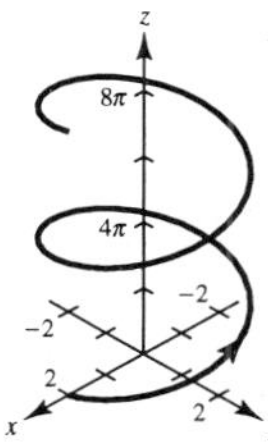

The height of the helix increases at a faster rate.

(c)

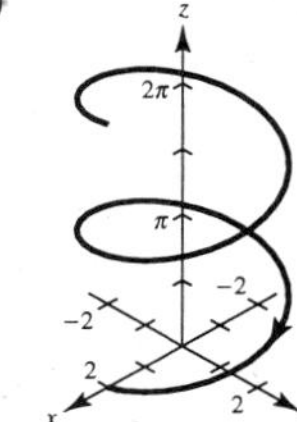

The orientation of the helix is reversed.

(d)

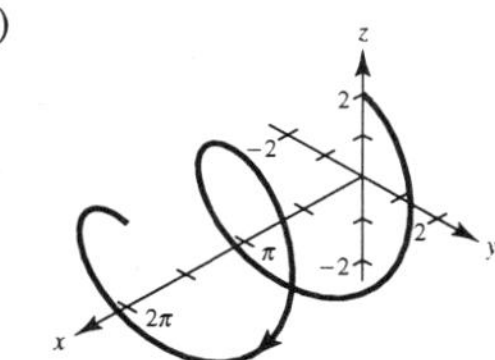

The axis of the helix is the x-axis.

(e)

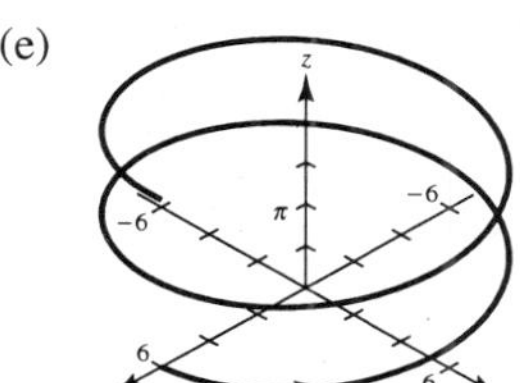

The radius of the helix is increased from 2 to 6.

39. $y = 4 - x$

Let $x = t$, then $y = 4 - t$.

$$\mathbf{r}(t) = t\mathbf{i} + (4 - t)\mathbf{j}$$

41. $x^2 + y^2 = 25$

Let $x = 5\cos t$, then $y = 5\sin t$.

$$\mathbf{r}(t) = 5\cos t\mathbf{i} + 5\sin t\mathbf{j}$$

43. The parametric equations for the line are

$$x = 2 - 2t,\ y = 3 + 5t,\ z = 8t.$$

One possible answer is

$$\mathbf{r}(t) = (2 - 2t)\mathbf{i} + (3 + 5t)\mathbf{j} + 8t\mathbf{k}.$$

45. $\mathbf{r}_1(t) = t\mathbf{i},\quad 0 \le t \le 4\quad (\mathbf{r}_1(0) = \mathbf{0}, \mathbf{r}_1(4) = 4\mathbf{i})$

$\mathbf{r}_2(t) = (4 - 4t)\mathbf{i} + 6t\mathbf{j},\quad 0 \le t \le 1\quad (\mathbf{r}_2(0) = 4\mathbf{i}, \mathbf{r}_2(1) = 6\mathbf{j})$

$\mathbf{r}_3(t) = (6 - t)\mathbf{j},\quad 0 \le t \le 6\quad (\mathbf{r}_3(0) = 6\mathbf{j}, \mathbf{r}_3(6) = \mathbf{0})$

(Other answers possible)

47. $\mathbf{r}_1(t) = t\mathbf{i} + t^2\mathbf{j},\quad 0 \le t \le 2\ (y = x^2)$

$\mathbf{r}_2(t) = (2 - t)\mathbf{i},\quad 0 \le t \le 2$

$\mathbf{r}_3(t) = (4 - t)\mathbf{j},\quad 0 \le t \le 4$

(Other answers possible)

49. $z = x^2 + y^2,\ x + y = 0$

Let $x = t$, then $y = -x = -t$ and $z = x^2 + y^2 = 2t^2$. Therefore,

$$x = t,\ y = -t,\ z = 2t^2.$$

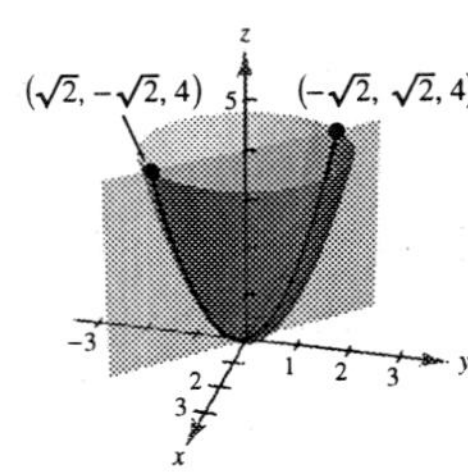

51. $x^2 + y^2 = 4,\ z = x^2$

$x = 2 \sin t,\ y = 2 \cos t$

$z = x^2 = 4 \sin^2 t$

t	0	$\frac{\pi}{6}$	$\frac{\pi}{4}$	$\frac{\pi}{2}$	$\frac{3\pi}{4}$	π
x	0	1	$\sqrt{2}$	2	$\sqrt{2}$	0
y	2	$\sqrt{3}$	$\sqrt{2}$	0	$-\sqrt{2}$	-2
z	0	1	2	4	2	0

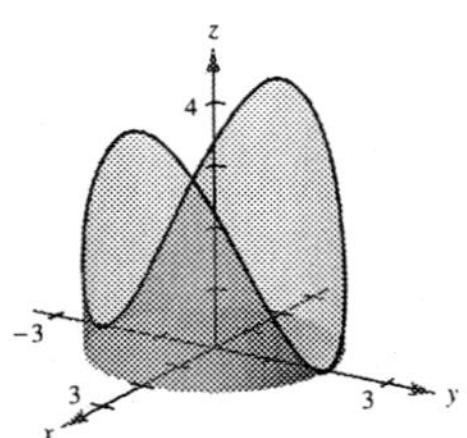

53. $x^2 + y^2 + z^2 = 4,\ x + z = 2$

Let $x = 1 + \sin t$, then $z = 2 - x = 1 - \sin t$ and $x^2 + y^2 + z^2 = 4$.

$$(1 + \sin t)^2 + y^2 + (1 - \sin t)^2 = 2 + 2\sin^2 t + y^2 = 4$$

$y^2 = 2\cos^2 t,\quad y = \pm\sqrt{2} \cos t$

$x = 1 + \sin t,\ y = \pm\sqrt{2} \cos t$

$z = 1 - \sin t$

t	$-\frac{\pi}{2}$	$-\frac{\pi}{6}$	0	$\frac{\pi}{6}$	$\frac{\pi}{2}$
x	0	$\frac{1}{2}$	1	$\frac{3}{2}$	2
y	0	$\pm\frac{\sqrt{6}}{2}$	$\pm\sqrt{2}$	$\pm\frac{\sqrt{6}}{2}$	0
z	2	$\frac{3}{2}$	1	$\frac{1}{2}$	0

55. $x^2 + z^2 = 4,\ y^2 + z^2 = 4$

Subtracting, we have

$$x^2 - y^2 = 0 \quad \text{or} \quad y = \pm x.$$

Therefore, in the first octant, if we let $x = t$, then

$$x = t,\ y = t,\ z = \sqrt{4 - t^2}.$$

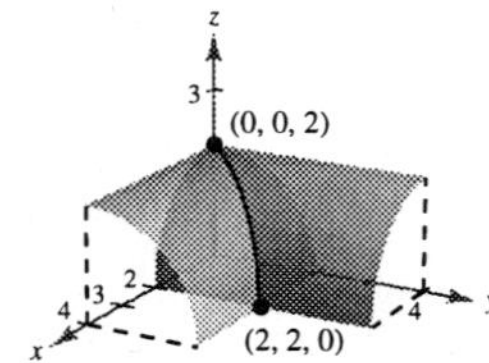

57. $\lim_{t\to 2}\left[t\mathbf{i} + \frac{t^2-4}{t^2-2t}\mathbf{j} + \frac{1}{t}\mathbf{k}\right] = 2\mathbf{i} + 2\mathbf{j} + \frac{1}{2}\mathbf{k}$

since

$$\lim_{t\to 2}\frac{t^2-4}{t^2-2t} = \lim_{t\to 2}\frac{2t}{2t-2} = 2. \quad \text{(L'Hôpital's Rule)}$$

59. $\lim_{t\to 0}\left[t^2\mathbf{i} + 3t\mathbf{j} + \frac{1-\cos t}{t}\mathbf{k}\right] = \mathbf{0}$

since

$$\lim_{t\to 0}\frac{1-\cos t}{t} = \lim_{t\to 0}\frac{\sin t}{1} = 0. \quad \text{(L'Hôpital's Rule)}$$

61. $\lim_{t\to 0}\left[\frac{1}{t}\mathbf{i} + \cos t\mathbf{j} + \sin t\mathbf{k}\right]$

does not exist since $\lim_{t\to 0}\frac{1}{t}$ does not exist.

63. $\mathbf{r}(t) = t\mathbf{i} + \frac{1}{t}\mathbf{j}$

Continuous on $(-\infty, 0),\ (0, \infty)$

65. $\mathbf{r}(t) = t\mathbf{i} + \arcsin t\mathbf{j} + (t-1)\mathbf{k}$

Continuous on $[-1, 1]$

67. $\mathbf{r}(t) = \langle e^{-t}, t^2, \tan t\rangle$

Discontinuous at $t = \frac{\pi}{2} + n\pi$

Continuous on $\left(-\frac{\pi}{2} + n\pi, \frac{\pi}{2} + n\pi\right)$

69. Let $\mathbf{r}(t) = x_1(t) + y_1(t)\mathbf{j} + z_1(t)\mathbf{k}$ and $\mathbf{u}(t) = x_2(t)\mathbf{i} + y_2(t)\mathbf{j} + z_2(t)\mathbf{k}$. Then:

$$\begin{aligned}\lim_{t\to c}[\mathbf{r}(t)\times\mathbf{u}(t)] &= \lim_{t\to c}\{[y_1(t)z_2(t) - y_2(t)z_1(t)]\mathbf{i} - [x_1(t)z_2(t) - x_2(t)z_1(t)]\mathbf{j} + [x_1(t)y_2(t) - x_2(t)y_1(t)]\mathbf{k}\}\\ &= \left[\lim_{t\to c}y_1(t)\lim_{t\to c}z_2(t) - \lim_{t\to c}y_2(t)\lim_{t\to c}z_1(t)\right]\mathbf{i} - \left[\lim_{t\to c}x_1(t)\lim_{t\to c}z_2(t) - \lim_{t\to c}x_2(t)\lim_{t\to c}z_1(t)\right]\mathbf{j}\\ &\quad + \left[\lim_{t\to c}x_1(t)\lim_{t\to c}y_2(t) - \lim_{t\to c}x_2(t)\lim_{t\to c}y_1(t)\right]\mathbf{k}\\ &= \left[\lim_{t\to c}x_1(t)\mathbf{i} + \lim_{t\to c}y_1(t)\mathbf{j} + \lim_{t\to c}z_1(t)\mathbf{k}\right]\times\left[\lim_{t\to c}x_2(t)\mathbf{i} + \lim_{t\to c}y_2(t)\mathbf{j} + \lim_{t\to c}z_2(t)\mathbf{k}\right]\\ &= \lim_{t\to c}\mathbf{r}(t)\times\lim_{t\to c}\mathbf{u}(t)\end{aligned}$$

71. Let $\mathbf{r}(t) = x(t)\mathbf{i} + y(t)\mathbf{j} + z(t)\mathbf{k}$. Since $\mathbf{r}$ is continuous at $t = c$, then $\lim_{t\to c}\mathbf{r}(t) = \mathbf{r}(c)$.

$\mathbf{r}(c) = x(c)\mathbf{i} + y(c)\mathbf{j} + z(c)\mathbf{k} \Rightarrow x(c),\ y(c),\ z(c)$

are defined at c.

$$\|\mathbf{r}\| = \sqrt{(x(t))^2 + (y(t))^2 + (z(t))^2}$$

$$\lim_{t\to c}\|\mathbf{r}\| = \sqrt{(x(c))^2 + (y(c))^2 + (z(c))^2} = \|\mathbf{r}(c)\|$$

Therefore, $\|\mathbf{r}\|$ is continuous at c.

73. True

Section 11.2 Differentiation and Integration of Vector-Valued Functions

1. $\mathbf{r}(t) = t^2\mathbf{i} + t\mathbf{j},\ t_0 = 2$

$x(t) = t^2,\ y(t) = t$

$x = y^2$

$\mathbf{r}(2) = 4\mathbf{i} + 2\mathbf{j}$

$\mathbf{r}'(t) = 2t\mathbf{i} + \mathbf{j}$

$\mathbf{r}'(2) = 4\mathbf{i} + \mathbf{j}$

$\mathbf{r}'(t_0)$ is tangent to the curve.

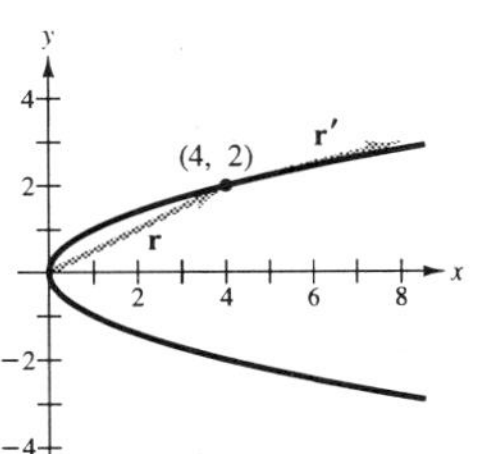

3. $\mathbf{r}(t) = \cos t\mathbf{i} + \sin t\mathbf{j},\ t_0 = \frac{\pi}{2}$

$x(t) = \cos t,\ y(t) = \sin t$

$x^2 + y^2 = 1$

$\mathbf{r}\left(\frac{\pi}{2}\right) = \mathbf{j}$

$\mathbf{r}'(t) = -\sin t\mathbf{i} + \cos t\mathbf{j}$

$\mathbf{r}'\left(\frac{\pi}{2}\right) = -\mathbf{i}$

$\mathbf{r}'(t_0)$ is tangent to the curve.

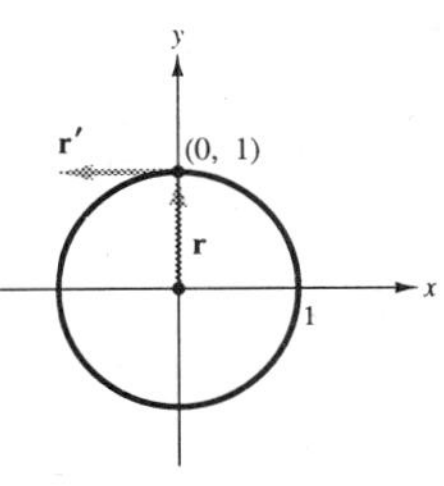

5. $\mathbf{r}(t) = t\mathbf{i} + t^2\mathbf{j}$

(a)

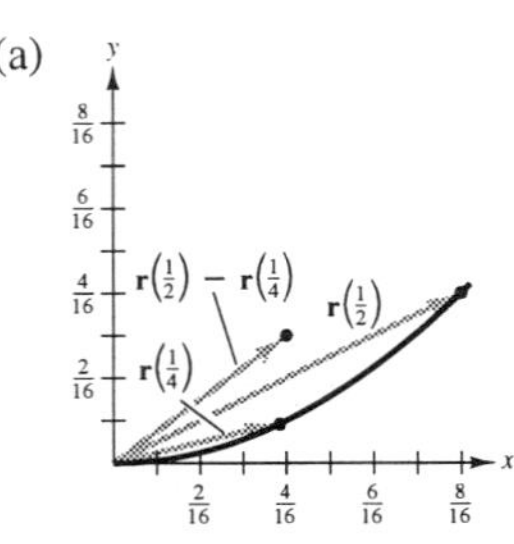

(b) $$\mathbf{r}\left(\frac{1}{4}\right) = \frac{1}{4}\mathbf{i} + \frac{1}{16}\mathbf{j}$$

$$\mathbf{r}\left(\frac{1}{2}\right) = \frac{1}{2}\mathbf{i} + \frac{1}{4}\mathbf{j}$$

$$\mathbf{r}\left(\frac{1}{2}\right) - \mathbf{r}\left(\frac{1}{4}\right) = \frac{1}{4}\mathbf{i} + \frac{3}{16}\mathbf{j}$$

(c) $$\mathbf{r}'(t) = \mathbf{i} + 2t\mathbf{j}$$

$$\mathbf{r}'\left(\frac{1}{4}\right) = \mathbf{i} + \frac{1}{2}\mathbf{j}$$

$$\frac{\mathbf{r}(1/2) - \mathbf{r}(1/4)}{(1/2) - (1/4)} = \frac{(1/4)\mathbf{i} + (3/16)\mathbf{j}}{1/4} = \mathbf{i} + \frac{3}{4}\mathbf{j}$$

This vector approximates $\mathbf{r}'\left(\frac{1}{4}\right)$.

7. $\mathbf{r}(t) = 2\cos t\mathbf{i} + 2\sin t\mathbf{j} + t\mathbf{k},\ t_0 = \dfrac{3\pi}{2}$

$x^2 + y^2 = 4,\ z = t$

$$\mathbf{r}'(t) = -2\sin t\mathbf{i} + 2\cos t\mathbf{j} + \mathbf{k}$$

$$\mathbf{r}\left(\frac{3\pi}{2}\right) = -2\mathbf{j} + \frac{3\pi}{2}\mathbf{k}$$

$$\mathbf{r}'\left(\frac{3\pi}{2}\right) = 2\mathbf{i} + \mathbf{k}$$

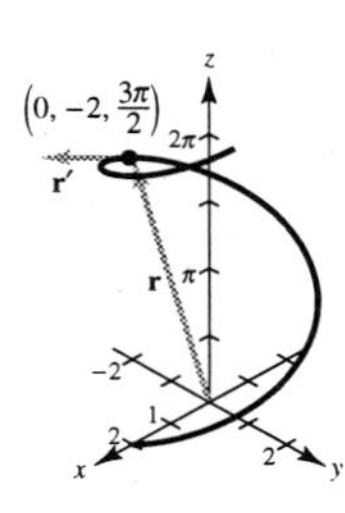

9. $$\mathbf{r}(t) = \cos(\pi t)\mathbf{i} + \sin(\pi t)\mathbf{j} + t^2\mathbf{k},\ t_0 = -\frac{1}{4}$$

$$\mathbf{r}'(t) = -\pi\sin(\pi t)\mathbf{i} + \pi\cos(\pi t)\mathbf{j} + 2t\mathbf{k}$$

$$\mathbf{r}'\left(-\frac{1}{4}\right) = \frac{\sqrt{2}\pi}{2}\mathbf{i} + \frac{\sqrt{2}\pi}{2}\mathbf{j} - \frac{1}{2}\mathbf{k}$$

$$\left\|\mathbf{r}'\left(\frac{1}{4}\right)\right\| = \sqrt{\left(\frac{\sqrt{2}\pi}{2}\right)^2 + \left(\frac{\sqrt{2}\pi}{2}\right)^2 + \left(-\frac{1}{2}\right)^2} = \sqrt{\pi^2 + \frac{1}{4}} = \frac{\sqrt{4\pi^2+1}}{2}$$

$$\frac{\mathbf{r}'(-1/4)}{\|\mathbf{r}'(-1/4)\|} = \frac{1}{\sqrt{4\pi^2+1}}\left(\sqrt{2}\pi\mathbf{i} + \sqrt{2}\pi\mathbf{j} - \mathbf{k}\right)$$

$$\mathbf{r}''(t) = -\pi^2\cos(\pi t)\mathbf{i} - \pi^2\sin(\pi t)\mathbf{j} + 2\mathbf{k}$$

$$\mathbf{r}''\left(-\frac{1}{4}\right) = -\frac{\sqrt{2}\pi^2}{2}\mathbf{i} + \frac{\sqrt{2}\pi^2}{2}\mathbf{j} + 2\mathbf{k}$$

$$\left\|\mathbf{r}''\left(-\frac{1}{4}\right)\right\| = \sqrt{\left(-\frac{\sqrt{2}\pi^2}{2}\right)^2 + \left(\frac{\sqrt{2}\pi^2}{2}\right)^2 + (2)^2} = \sqrt{\pi^4+4}$$

$$\frac{\mathbf{r}''(-1/4)}{\|\mathbf{r}''(-1/4)\|} = \frac{1}{2\sqrt{\pi^4+4}}\left(-\sqrt{2}\pi^2\mathbf{i} + \sqrt{2}\pi^2\mathbf{j} + 4\mathbf{k}\right)$$

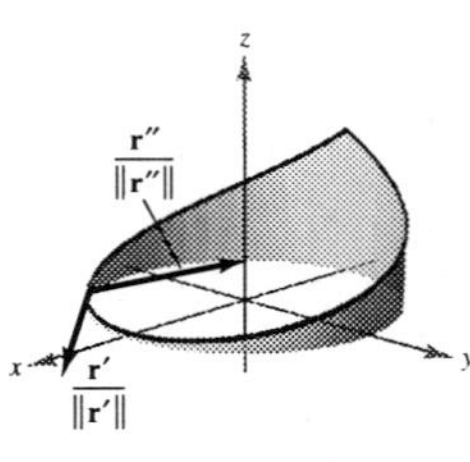

11. $\mathbf{r}(t) = 6t\mathbf{i} - 7t^2\mathbf{j} + t^3\mathbf{k}$

$\mathbf{r}'(t) = 6\mathbf{i} - 14t\mathbf{j} + 3t^2\mathbf{k}$

13. $\mathbf{r}(t) = a\cos^3 t\mathbf{i} + a\sin^3 t\mathbf{j} + \mathbf{k}$

$\mathbf{r}'(t) = -3a\cos^2 t\sin t\mathbf{i} + 3a\sin^2 t\cos t\mathbf{j}$

15. $\mathbf{r}(t) = e^{-t}\mathbf{i} + 4\mathbf{j}$

$\mathbf{r}'(t) = -e^{-t}\mathbf{i}$

17. $\mathbf{r}(t) = \langle t\sin t, t\cos t, t\rangle$

$\mathbf{r}'(t) = \langle \sin t + t\cos t, \cos t - t\sin t, 1\rangle$

19. $\mathbf{r}(t) = t\mathbf{i} + 3t\mathbf{j} + t^2\mathbf{k}$, $\mathbf{u}(t) = 4t\mathbf{i} + t^2\mathbf{j} + t^3\mathbf{k}$

(a) $\mathbf{r}'(t) = \mathbf{i} + 3\mathbf{j} + 2t\mathbf{k}$

(b) $\mathbf{r}''(t) = 2\mathbf{k}$

(c) $\mathbf{r}(t)\cdot\mathbf{u}(t) = 4t^2 + 3t^3 + t^5$

$D_t[\mathbf{r}(t)\cdot\mathbf{u}(t)] = 8t + 9t^2 + 5t^4$

(d) $3\mathbf{r}(t) - \mathbf{u}(t) = -t\mathbf{i} + (9t - t^2)\mathbf{j} + (3t^2 - t^3)\mathbf{k}$

$D_t[3\mathbf{r}(t) - \mathbf{u}(t)] = -\mathbf{i} + (9 - 2t)\mathbf{j} + (6t - 3t^2)\mathbf{k}$

(e) $\mathbf{r}(t)\times\mathbf{u}(t) = 2t^4\mathbf{i} - (t^4 - 4t^3)\mathbf{j} + (t^3 - 12t^2)\mathbf{k}$

$D_t[\mathbf{r}(t)\times\mathbf{u}(t)] = 8t^3\mathbf{i} + (12t^2 - 4t^3)\mathbf{j} + (3t^2 - 24t)\mathbf{k}$

(f) $\|\mathbf{r}(t)\| = \sqrt{10t^2 + t^4} = t\sqrt{10 + t^2}$

$$D_t[\|\mathbf{r}(t)\|] = \frac{10 + 2t^2}{\sqrt{10 + t^2}}$$

21. $\mathbf{r}(t) = 3\sin t\mathbf{i} + 4\cos t\mathbf{j}$

$\mathbf{r}'(t) = 3\cos t\mathbf{i} - 4\sin t\mathbf{j}$

$\mathbf{r}(t)\cdot\mathbf{r}'(t) = 9\sin t\cos t - 16\cos t\sin t = -7\sin t\cos t$

$$\cos\theta = \frac{\mathbf{r}(t)\cdot\mathbf{r}'(t)}{\|\mathbf{r}(t)\|\,\|\mathbf{r}'(t)\|} = \frac{-7\sin t\cos t}{\sqrt{9\sin^2 t + 16\cos^2 t}\sqrt{9\cos^2 t + 16\sin^2 t}}$$

$$\theta = \arccos\left[\frac{-7\sin t\cos t}{\sqrt{(9\sin^2 t + 16\cos^2 t)(9\cos^2 t + 16\sin^2 t)}}\right]$$

$\theta = 1.855$ maximum at $t = 3.927\left(\frac{5\pi}{4}\right)$ and $t = 0.785\left(\frac{\pi}{4}\right)$.

$\theta = 1.287$ minimum at $t = 2.356\left(\frac{3\pi}{4}\right)$ and $t = 5.498\left(\frac{7\pi}{4}\right)$.

$\theta = \frac{\pi}{2}(1.571)$ for $t = n\frac{\pi}{2}$, $n = 0, 1, 2, 3, \ldots$

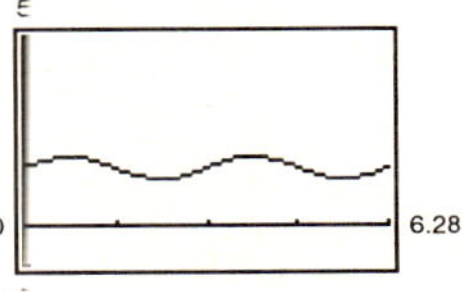

23. $\mathbf{r}(t) = t^2\mathbf{i} + t^3\mathbf{j}$

$\mathbf{r}'(t) = 2t\mathbf{i} + 3t^2\mathbf{j}$

$\mathbf{r}'(0) = \mathbf{0}$

Smooth on $(-\infty, 0)$, $(0, \infty)$

25. $\mathbf{r}(\theta) = 2\cos^3\theta\mathbf{i} + 3\sin^3\theta\mathbf{j}$

$\mathbf{r}'(\theta) = -6\cos^2\theta\sin\theta\mathbf{i} + 9\sin^2\theta\cos\theta\mathbf{j}$

$\mathbf{r}'\left(\frac{n\pi}{2}\right) = \mathbf{0}$

Smooth on $\left(\frac{n\pi}{2}, \frac{(n + 1)\pi}{2}\right)$, n any integer.

27. $\mathbf{r}(\theta) = (\theta - 2\sin\theta)\mathbf{i} + (1 - 2\cos\theta)\mathbf{j}$

$\mathbf{r}'(\theta) = (1 - 2\cos\theta)\mathbf{i} + (1 + 2\sin\theta)\mathbf{j}$

$\mathbf{r}'(\theta) \neq \mathbf{0}$ for any value of θ

Smooth on $(-\infty, \infty)$

29. $\mathbf{r}(t) = (t - 1)\mathbf{i} + \frac{1}{t}\mathbf{j} - t^2\mathbf{k}$

$\mathbf{r}'(t) = \mathbf{i} - \frac{1}{t^2}\mathbf{j} - 2t\mathbf{k} \neq \mathbf{0}$

r is smooth for all $t \neq 0$: $(-\infty, 0), \cup (0, \infty)$

31. $\mathbf{r}(t) = t\mathbf{i} - 3t\mathbf{j} + \tan t\mathbf{k}$

$\mathbf{r}'(t) = \mathbf{i} - 3\mathbf{j} + \sec^2 t\mathbf{k} \neq \mathbf{0}$

r is smooth for all $t \neq \frac{\pi}{2} + n\pi = \frac{2n + 1}{2}\pi$.

Smooth on intervals of form $\left(-\frac{\pi}{2} + n\pi, \frac{\pi}{2} + n\pi\right)$

33. $$\mathbf{r}'(t) = \lim_{\Delta t \to 0} \frac{\mathbf{r}(t + \Delta t) - \mathbf{r}(t)}{\Delta t}$$

$$= \lim_{\Delta t \to 0} \frac{[3(t + \Delta t) + 2]\mathbf{i} + [1 - (t + \Delta t)^2]\mathbf{j} - (3t + 2)\mathbf{i} - (1 - t^2)\mathbf{j}}{\Delta t}$$

$$= \lim_{\Delta t \to 0} \frac{(3\Delta t)\mathbf{i} - (2t(\Delta t) + (\Delta t)^2)\mathbf{j}}{\Delta t}$$

$$= \lim_{\Delta t \to 0} 3\mathbf{i} - (2t + \Delta t)\mathbf{j} = 3\mathbf{i} - 2t\mathbf{j}$$

35. At $t = t_0$, the graph of $\mathbf{u}(t)$ is increasing in the x, y, and z directions simultaneously.

37. $$\int (2t\mathbf{i} + \mathbf{j} + \mathbf{k})\, dt = t^2\mathbf{i} + t\mathbf{j} + t\mathbf{k} + \mathbf{C}$$

39. $$\int \left(\frac{1}{t}\mathbf{i} + \mathbf{j} - t^{3/2}\mathbf{k}\right) dt = \ln t\mathbf{i} + t\mathbf{j} - \frac{2}{5}t^{5/2}\mathbf{k} + \mathbf{C}$$

41. $$\int \left[(2t - 1)\mathbf{i} + 4t^3\mathbf{j} + 3\sqrt{t}\mathbf{k}\right] dt = (t^2 - t)\mathbf{i} + t^4\mathbf{j} + 2t^{3/2}\mathbf{k} + \mathbf{C}$$

43. $$\int \left[\sec^2 t\mathbf{i} + \frac{1}{1 + t^2}\mathbf{j}\right] dt = \tan t\mathbf{i} + \arctan t\mathbf{j} + \mathbf{C}$$

45. $$\mathbf{r}(t) = \int (4e^{2t}\mathbf{i} + 3e^t\mathbf{j})\, dt = 2e^{2t}\mathbf{i} + 3e^t\mathbf{j} + \mathbf{C}$$

$$\mathbf{r}(0) = 2\mathbf{i} + 3\mathbf{j} + \mathbf{C} = 2\mathbf{i} \implies \mathbf{C} = -3\mathbf{j}$$

$$\mathbf{r}(t) = 2e^{2t}\mathbf{i} + 3(e^t - 1)\mathbf{j}$$

47. $$\mathbf{r}'(t) = \int -32\mathbf{j}\, dt = -32t\mathbf{j} + \mathbf{C}_1$$

$$\mathbf{r}'(0) = \mathbf{C}_1 = 600\sqrt{3}\mathbf{i} + 600\mathbf{j}$$

$$\mathbf{r}'(t) = 600\sqrt{3}\mathbf{i} + (600 - 32t)\mathbf{j}$$

$$\mathbf{r}(t) = \int \left[600\sqrt{3}\mathbf{i} + (600 - 32t)\mathbf{j}\right] dt = 600\sqrt{3}t\mathbf{i} + (600t - 16t^2)\mathbf{j} + \mathbf{C}$$

$$\mathbf{r}(0) = \mathbf{C} = \mathbf{0}$$

$$\mathbf{r}(t) = 600\sqrt{3}t\mathbf{i} + (600t - 16t^2)\mathbf{j}$$

49. $$\mathbf{r}(t) = \int (te^{-t^2}\mathbf{i} - e^{-t}\mathbf{j} + \mathbf{k})\, dt = -\frac{1}{2}e^{-t^2}\mathbf{i} + e^{-t}\mathbf{j} + t\mathbf{k} + \mathbf{C}$$

$$\mathbf{r}(0) = -\frac{1}{2}\mathbf{i} + \mathbf{j} + \mathbf{C} = \frac{1}{2}\mathbf{i} - \mathbf{j} + \mathbf{k} \implies \mathbf{C} = \mathbf{i} - 2\mathbf{j} + \mathbf{k}$$

$$\mathbf{r}(t) = \left(1 - \frac{1}{2}e^{-t^2}\right)\mathbf{i} + (e^{-t} - 2)\mathbf{j} + (t + 1)\mathbf{k} = \left(\frac{2 - e^{-t^2}}{2}\right)\mathbf{i} + (e^{-t} - 2)\mathbf{j} + (t + 1)\mathbf{k}$$

51. $$\int_0^1 (8t\mathbf{i} + t\mathbf{j} - \mathbf{k})\, dt = \Big[4t^2\mathbf{i}\Big]_0^1 + \left[\frac{t^2}{2}\mathbf{j}\right]_0^1 - \Big[t\mathbf{k}\Big]_0^1 = 4\mathbf{i} + \frac{1}{2}\mathbf{j} - \mathbf{k}$$

53. $$\int_0^{\pi/2} [(a\cos t)\mathbf{i} + (a\sin t)\mathbf{j} + \mathbf{k}]\, dt = \Big[a\sin t\mathbf{i}\Big]_0^{\pi/2} - \Big[a\cos t\mathbf{j}\Big]_0^{\pi/2} + \Big[t\mathbf{k}\Big]_0^{\pi/2} = a\mathbf{i} + a\mathbf{j} + \frac{\pi}{2}\mathbf{k}$$

55. Let $\mathbf{r}(t) = x(t)\mathbf{i} + y(t)\mathbf{j} + z(t)\mathbf{k}$. Then $c\mathbf{r}(t) = cx(t)\mathbf{i} + cy(t)\mathbf{j} + cz(t)\mathbf{k}$ and

$$\begin{aligned} D_t[c\mathbf{r}(t)] &= cx'(t)\mathbf{i} + cy'(t)\mathbf{j} + cz'(t)\mathbf{k} \\ &= c[x'(t)\mathbf{i} + y'(t)\mathbf{j} + z'(t)\mathbf{k}] = c\mathbf{r}'(t). \end{aligned}$$

57. Let $\mathbf{r}(t) = x(t)\mathbf{i} + y(t)\mathbf{j} + z(t)\mathbf{k}$, then $f(t)\mathbf{r}(t) = f(t)x(t)\mathbf{i} + f(t)y(t)\mathbf{j} + f(t)z(t)\mathbf{k}$.

$$\begin{aligned} D_t[f(t)\mathbf{r}(t)] &= [f(t)x'(t) + f'(t)x(t)]\mathbf{i} + [f(t)y'(t) + f'(t)y(t)]\mathbf{j} + [f(t)z'(t) + f'(t)z(t)]\mathbf{k} \\ &= f(t)[x'(t)\mathbf{i} + y'(t)\mathbf{j} + z'(t)\mathbf{k}] + f'(t)[x(t)\mathbf{i} + y(t)\mathbf{j} + z(t)\mathbf{k}] \\ &= f(t)\mathbf{r}'(t) + f'(t)\mathbf{r}(t) \end{aligned}$$

59. Let $\mathbf{r}(t) = x(t)\mathbf{i} + y(t)\mathbf{j} + z(t)\mathbf{k}$. Then $\mathbf{r}(f(t)) = x(f(t))\mathbf{i} + y(f(t))\mathbf{j} + z(f(t))\mathbf{k}$ and

$$\begin{aligned} D_t[\mathbf{r}(f(t))] &= x'(f(t))f'(t)\mathbf{i} + y'(f(t))f'(t)\mathbf{j} + z'(f(t))f'(t)\mathbf{k} \quad \text{(Chain Rule)} \\ &= f'(t)[x'(f(t))i + y'(f(t))\mathbf{j} + z'(f(t))\mathbf{k}] = f'(t)\mathbf{r}'(f(t)). \end{aligned}$$

61. Let $\mathbf{r}(t) = x_1(t)\mathbf{i} + y_1(t)\mathbf{j} + z_1(t)\mathbf{k}$, $\mathbf{u}(t) = x_2(t)\mathbf{i} + y_2(t)\mathbf{j} + z_2(t)\mathbf{k}$, and $\mathbf{v}(t) = x_3(t)\mathbf{i} + y_3(t)\mathbf{j} + z_3(t)\mathbf{k}$. Then:

$$\mathbf{r}(t) \cdot [\mathbf{u}(t) \times \mathbf{v}(t)] = x_1(t)[y_2(t)z_3(t) - z_2(t)y_3(t)] - y_1(t)[x_2(t)z_3(t) - z_2(t)x_3(t)] + z_1(t)[x_2(t)y_3(t) - y_2(t)x_3(t)]$$

$$\begin{aligned} D_t[\mathbf{r}(t) \cdot (\mathbf{u}(t) \times \mathbf{v}(t))] &= x_1(t)y_2(t)z_3'(t) + x_1(t)y_2'(t)z_3(t) + x_1'(t)y_2(t)z_3(t) - x_1(t)y_3(t)z_2'(t) - \\ &\quad x_1(t)y_3'(t)z_2(t) - x_1'(t)y_3(t)z_2(t) - y_1(t)x_2(t)z_3'(t) - y_1(t)x_2'(t)z_3(t) - y_1'(t)x_2(t)z_3(t) + \\ &\quad y_1(t)z_2(t)x_3'(t) + y_1(t)z_2'(t)x_3(t) + y_1'(t)z_2(t)x_3(t) + z_1(t)x_2(t)y_3'(t) - z_1(t)x_2'(t)y_3(t) + \\ &\quad z_1'(t)x_2(t)y_3(t) - z_1(t)y_2(t)x_3'(t) - z_1(t)y_2'(t)x_3(t) - z_1'(t)y_2(t)x_3(t) \\ &= \{x_1'(t)[y_2(t)z_3(t) - y_3(t)z_2(t)] + y_1'(t)[-x_2(t)z_3(t) + z_2(t)x_3(t)] + z_1'(t)[x_2(t)y_3(t) - y_2(t)x_3(t)]\} + \\ &\quad \{x_1(t)[y_2'(t)z_3(t) - y_3(t)z_2'(t)] + y_1(t)[-x_2'(t)z_3(t) + z_2'(t)x_3(t)] + z_1(t)[x_2'(t)y_3(t) - y_2'(t)x_3(t)]\} + \\ &\quad \{x_1(t)[y_2(t)z_3'(t) - y_3'(t)z_2(t)] + y_1(t)[-x_2(t)z_3'(t) + z_2(t)x_3'(t)] + z_1(t)[x_2(t)y_3'(t) - y_2(t)x_3'(t)]\} \\ &= \mathbf{r}'(t) \cdot [\mathbf{u}(t) \times \mathbf{v}(t)] + \mathbf{r}(t) \cdot [\mathbf{u}'(t) \times \mathbf{v}(t)] + \mathbf{r}(t) \cdot [\mathbf{u}(t) \times \mathbf{v}'(t)] \end{aligned}$$

63. False. Let $\mathbf{r}(t) = \cos t\mathbf{i} + \sin t\mathbf{j} + \mathbf{k}$.

$$\|\mathbf{r}(t)\| = \sqrt{2}$$

$$\frac{d}{dt}[\|\mathbf{r}(t)\|] = 0$$

$$\mathbf{r}'(t) = -\sin t\mathbf{i} + \cos t\mathbf{j}$$

$$\|\mathbf{r}'(t)\| = 1$$

Section 11.3 Velocity and Acceleration

1. $\mathbf{r}(t) = 3t\mathbf{i} + (t - 1)\mathbf{j}$

$\mathbf{v}(t) = \mathbf{r}'(t) = 3\mathbf{i} + \mathbf{j}$

$\mathbf{a}(t) = \mathbf{r}''(t) = \mathbf{0}$

$x = 3t,\ y = t - 1,\ y = \dfrac{x}{3} - 1$

At $(3, 0)$, $t = 1$.

$\mathbf{v}(1) = 3\mathbf{i} + \mathbf{j},\ \mathbf{a}(1) = \mathbf{0}$

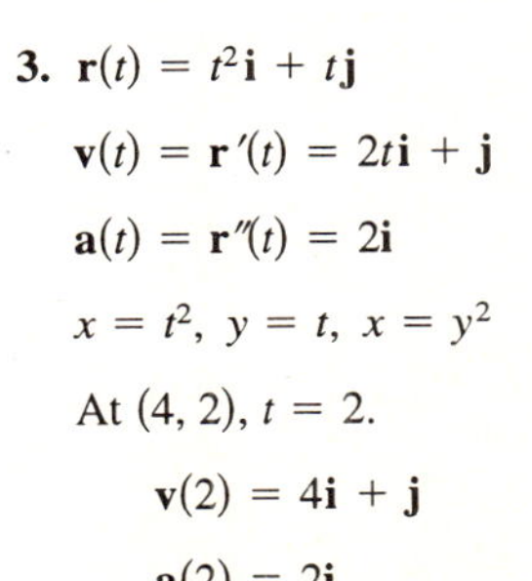

3. $\mathbf{r}(t) = t^2\mathbf{i} + t\mathbf{j}$

$\mathbf{v}(t) = \mathbf{r}'(t) = 2t\mathbf{i} + \mathbf{j}$

$\mathbf{a}(t) = \mathbf{r}''(t) = 2\mathbf{i}$

$x = t^2,\ y = t,\ x = y^2$

At $(4, 2)$, $t = 2$.

$\mathbf{v}(2) = 4\mathbf{i} + \mathbf{j}$

$\mathbf{a}(2) = 2\mathbf{i}$

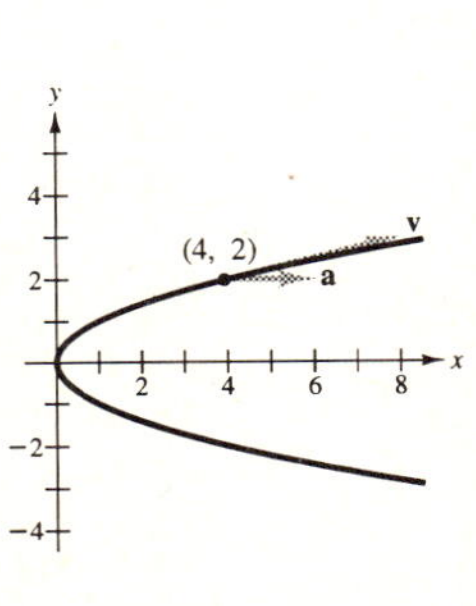

5. $\mathbf{r}(t) = 2\cos t\mathbf{i} + 2\sin t\mathbf{j}$

$\mathbf{v}(t) = \mathbf{r}'(t) = -2\sin t\mathbf{i} + 2\cos t\mathbf{j}$

$\mathbf{a}(t) = \mathbf{r}''(t) = -2\cos t\mathbf{i} - 2\sin t\mathbf{j}$

$x = 2\cos t,\ y = 2\sin t,\ x^2 + y^2 = 4$

At $\left(\sqrt{2}, \sqrt{2}\right)$, $t = \dfrac{\pi}{4}$.

$$\mathbf{v}\left(\frac{\pi}{4}\right) = -\sqrt{2}\mathbf{i} + \sqrt{2}\mathbf{j}$$

$$\mathbf{a}\left(\frac{\pi}{4}\right) = -\sqrt{2}\mathbf{i} - \sqrt{2}\mathbf{j}$$

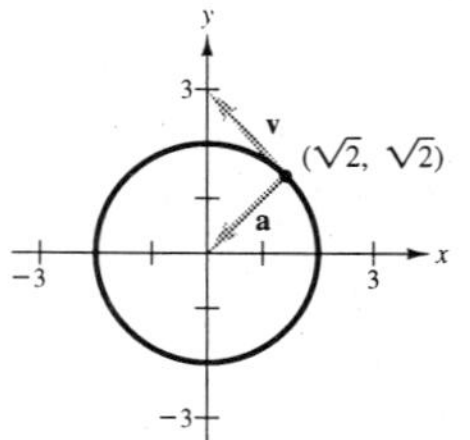

7. $\mathbf{r}(t) = \langle t - \sin t, 1 - \cos t\rangle$

$\mathbf{v}(t) = \mathbf{r}'(t) = \langle 1 - \cos t, \sin t\rangle$

$\mathbf{a}(t) = \mathbf{r}''(t) = \langle \sin t, \cos t\rangle$

$x = t - \sin t,\ y = 1 - \cos t$ (cycloid)

At $(\pi, 2)$, $t = \pi$.

$\mathbf{v}(\pi) = \langle 2, 0\rangle = 2\mathbf{i}$

$\mathbf{a}(\pi) = \langle 0, -1\rangle = -\mathbf{j}$

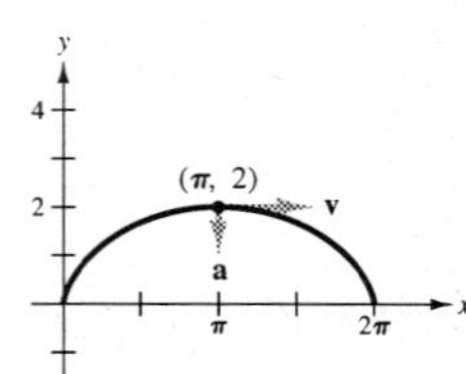

9. $\mathbf{r}(t) = t\mathbf{i} + (2t - 5)\mathbf{j} + 3t\mathbf{k}$

$\mathbf{v}(t) = \mathbf{i} + 2\mathbf{j} + 3\mathbf{k}$

$s(t) = \|\mathbf{v}(t)\| = \sqrt{1 + 4 + 9} = \sqrt{14}$

$\mathbf{a}(t) = \mathbf{0}$

11. $\mathbf{r}(t) = t\mathbf{i} + t^2\mathbf{j} + \dfrac{t^2}{2}\mathbf{k}$

$\mathbf{v}(t) = \mathbf{i} + 2t\mathbf{j} + t\mathbf{k}$

$s(t) = \sqrt{1 + 4t^2 + t^2} = \sqrt{1 + 5t^2}$

$\mathbf{a}(t) = 2\mathbf{j} + \mathbf{k}$

13. $\mathbf{r}(t) = t\mathbf{i} + t\mathbf{j} + \sqrt{9 - t^2}\mathbf{k}$

$$\mathbf{v}(t) = \mathbf{i} + \mathbf{j} - \frac{t}{\sqrt{9 - t^2}}\mathbf{k}$$

$$s(t) = \sqrt{1 + 1 + \frac{t^2}{9 - t^2}} = \sqrt{\frac{18 - t^2}{9 - t^2}}$$

$$\mathbf{a}(t) = -\frac{9}{(9 - t^2)^{3/2}}\mathbf{k}$$

15. $\mathbf{r}(t) = \langle 4t, 3\cos t, 3\sin t\rangle$

$\mathbf{v}(t) = \langle 4, -3\sin t, 3\cos t\rangle = 4\mathbf{i} - 3\sin t\mathbf{j} + 3\cos t\mathbf{k}$

$s(t) = \sqrt{16 + 9\sin^2 t + 9\cos^2 t} = 5$

$\mathbf{a}(t) = \langle 0, -3\cos t, -3\sin t\rangle = -3\cos t\mathbf{j} - 3\sin t\mathbf{k}$

17. (a) $\mathbf{r}(t) = \left\langle t, -t^2, \dfrac{t^3}{4}\right\rangle,\ t_0 = 1$

$$\mathbf{r}'(t) = \left\langle 1, -2t, \frac{3t^2}{4}\right\rangle$$

$$\mathbf{r}'(1) = \left\langle 1, -2, \frac{3}{4}\right\rangle$$

$$x = 1 + t,\ y = -1 - 2t,\ z = \frac{1}{4} + \frac{3}{4}t$$

(b) $\mathbf{r}(1 + 0.1) \approx \left\langle 1 + 0.1, -1 - 2(0.1), \dfrac{1}{4} + \dfrac{3}{4}(0.1)\right\rangle$

$= \langle 1.100, -1.200, 0.325\rangle$

19. $\mathbf{a}(t) = \mathbf{i} + \mathbf{j} + \mathbf{k},\ \mathbf{v}(0) = \mathbf{0},\ \mathbf{r}(0) = \mathbf{0}$

$$\mathbf{v}(t) = \int (\mathbf{i} + \mathbf{j} + \mathbf{k})\,dt = t\mathbf{i} + t\mathbf{j} + t\mathbf{k} + \mathbf{C}$$

$\mathbf{v}(0) = \mathbf{C} = \mathbf{0},\ \mathbf{v}(t) = t\mathbf{i} + t\mathbf{j} + t\mathbf{k},\ \mathbf{v}(t) = t(\mathbf{i} + \mathbf{j} + \mathbf{k})$

$$\mathbf{r}(t) = \int (t\mathbf{i} + t\mathbf{j} + t\mathbf{k})\,dt = \frac{t^2}{2}(\mathbf{i} + \mathbf{j} + \mathbf{k}) + \mathbf{C}$$

$$\mathbf{r}(0) = \mathbf{C} = \mathbf{0},\ \mathbf{r}(t) = \frac{t^2}{2}(\mathbf{i} + \mathbf{j} + \mathbf{k}),$$

$\mathbf{r}(2) = 2(\mathbf{i} + \mathbf{j} + \mathbf{k}) = 2\mathbf{i} + 2\mathbf{j} + 2\mathbf{k}$

21. $\mathbf{a}(t) = t\mathbf{j} + t\mathbf{k},\ \mathbf{v}(1) = 5\mathbf{j},\ \mathbf{r}(1) = \mathbf{0}$

$$\mathbf{v}(t) = \int (t\mathbf{j} + t\mathbf{k})\,dt = \frac{t^2}{2}\mathbf{j} + \frac{t^2}{2}\mathbf{k} + \mathbf{C}$$

$$\mathbf{v}(1) = \frac{1}{2}\mathbf{j} + \frac{1}{2}\mathbf{k} + \mathbf{C} = 5\mathbf{j} \Rightarrow \mathbf{C} = \frac{9}{2}\mathbf{j} - \frac{1}{2}\mathbf{k}$$

$$\mathbf{v}(t) = \left(\frac{t^2}{2} + \frac{9}{2}\right)\mathbf{j} + \left(\frac{t^2}{2} - \frac{1}{2}\right)\mathbf{k}$$

$$\mathbf{r}(t) = \int\left[\left(\frac{t^2}{2} + \frac{9}{2}\right)\mathbf{j} + \left(\frac{t^2}{2} - \frac{1}{2}\right)\mathbf{k}\right]dt$$

$$= \left(\frac{t^3}{6} + \frac{9}{2}t\right)\mathbf{j} + \left(\frac{t^3}{6} - \frac{1}{2}t\right)\mathbf{k} + \mathbf{C}$$

$$\mathbf{r}(1) = \frac{14}{3}\mathbf{j} - \frac{1}{3}\mathbf{k} + \mathbf{C} = \mathbf{0} \Rightarrow \mathbf{C} = -\frac{14}{3}\mathbf{j} + \frac{1}{3}\mathbf{k}$$

$$\mathbf{r}(t) = \left(\frac{t^3}{6} + \frac{9}{2}t - \frac{14}{3}\right)\mathbf{j} + \left(\frac{t^3}{6} - \frac{1}{2}t + \frac{1}{3}\right)\mathbf{k}$$

$$\mathbf{r}(2) = \frac{17}{3}\mathbf{j} + \frac{2}{3}\mathbf{k}$$

23. $\mathbf{r}(t) = (88 \cos 30°)t\mathbf{i} + [10 + (88 \sin 30°)t - 16t^2]\mathbf{j}$

$= 44\sqrt{3}t\mathbf{i} + (10 + 44t - 16t^2)\mathbf{j}$

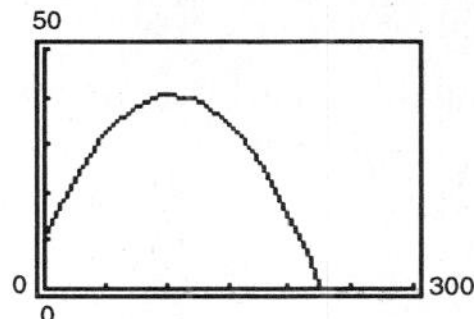

25. $\mathbf{r}(t) = (v_0 \cos \theta)t\mathbf{i} + \left[h + (v_0 \sin \theta)t - \frac{1}{2}gt^2\right]\mathbf{j} = \frac{v_0}{\sqrt{2}}t\mathbf{i} + \left(3 + \frac{v_0}{\sqrt{2}}t - 16t^2\right)\mathbf{j}$

$\frac{v_0}{\sqrt{2}}t = 300$ when $3 + \frac{v_0}{\sqrt{2}}t - 16t^2 = 3.$

$$t = \frac{300\sqrt{2}}{v_0}, \quad \frac{v_0}{\sqrt{2}}\left(\frac{300\sqrt{2}}{v_0}\right) - 16\left(\frac{300\sqrt{2}}{v_0}\right)^2 = 0, \quad 300 - \frac{300^2(32)}{{v_0}^2} = 0$$

$${v_0}^2 = 300(32), \quad v_0 = \sqrt{9600} = 40\sqrt{6}, \quad v_0 = 40\sqrt{6} \approx 97.98 \text{ ft/sec}$$

The maximum height is reached when the derivative of the vertical component is zero.

$$y(t) = 3 + \frac{tv_0}{\sqrt{2}} - 16t^2 = 3 + \frac{40\sqrt{6}}{\sqrt{2}}t - 16t^2 = 3 + 40\sqrt{3}t - 16t^2$$

$$y'(t) = 40\sqrt{3} - 32t = 0$$

$$t = \frac{40\sqrt{3}}{32} = \frac{5\sqrt{3}}{4}$$

Maximum height: $y\left(\frac{5\sqrt{3}}{4}\right) = 3 + 40\sqrt{3}\left(\frac{5\sqrt{3}}{4}\right) - 16\left(\frac{5\sqrt{3}}{4}\right)^2 = 78$ feet

27. $h = 7$ feet, $\theta = 35°$, 30 yards $= 90$ feet

$\mathbf{r}(t) = (v_0 \cos 35°)t\mathbf{i} + [7 + (v_0 \sin 35°)t - 16t^2]\mathbf{j}$

(a) $v_0 \cos 35° t = 90$ when $7 + (v_0 \sin 35°)t - 16t^2 = 4.$

$$t = \frac{90}{v_0 \cos 35°}$$

$$7 + (v_0 \sin 35°)\left(\frac{90}{v_0 \cos 35°}\right) - 16\left(\frac{90}{v_0 \cos 35°}\right)^2 = 4$$

$$90 \tan 35° + 3 = \frac{129{,}600}{{v_0}^2 \cos^2 35°}$$

$${v_0}^2 = \frac{129{,}600}{\cos^2 35°(90 \tan 35° + 3)}$$

$$v_0 \approx 54.088 \text{ feet per second}$$

(b) The maximum height occurs when

$y'(t) = v_0 \sin 35° - 32t = 0.$

$$t = \frac{v_0 \sin 35°}{32} \approx 0.969 \text{ second}$$

At this time, the height is $y(0.969) \approx 22.0$ feet.

(c) $x(t) = 90 \Rightarrow (v_0 \cos 35°)t = 90$

$$t = \frac{90}{54.088 \cos 35°} \approx 2.0 \text{ seconds}$$

29. $\mathbf{r}(t) = (v\cos\theta)t\mathbf{i} + [(v\sin\theta)t - 16t^2]\mathbf{j}$

We want to find the minimum initial speed v as a function of the angle θ. Since the bale must be thrown to the position (16, 8), we have

$$16 = (v\cos\theta)t$$
$$8 = (v\sin\theta)t - 16t^2.$$

$t = 16/(v\cos\theta)$ from the first equation. Substituting into the second equation and solving for v, we obtain:

$$8 = (v\sin\theta)\left(\frac{16}{v\cos\theta}\right) - 16\left(\frac{16}{v\cos\theta}\right)^2$$

$$1 = 2\frac{\sin\theta}{\cos\theta} - 512\left(\frac{1}{v^2\cos^2\theta}\right)$$

$$512\frac{1}{v^2\cos^2\theta} = 2\frac{\sin\theta}{\cos\theta} - 1$$

$$\frac{1}{v^2} = \left(2\frac{\sin\theta}{\cos\theta} - 1\right)\frac{\cos^2\theta}{512} = \frac{2\sin\theta\cos\theta - \cos^2\theta}{512}$$

$$v^2 = \frac{512}{2\sin\theta\cos\theta - \cos^2\theta}$$

We minimize $f(\theta) = \dfrac{512}{2\sin\theta\cos\theta - \cos^2\theta}$.

$$f'(\theta) = -512\frac{2\cos^2\theta - 2\sin^2\theta + 2\sin\theta\cos\theta}{(2\sin\theta\cos\theta - \cos^2\theta)^2}$$

$$f'(\theta) = 0 \implies 2\cos(2\theta) + \sin(2\theta) = 0$$

$$\tan(2\theta) = -2$$

$$\theta \approx 1.01722 \approx 58.28°$$

Substituting into the equation for v, $v \approx 28.78$ feet per second.

31. $\mathbf{r}(t) = (v_0\cos\theta)t\mathbf{i} + [(v_0\sin\theta)t - 16t^2]\mathbf{j}$

$(v_0\sin\theta)t - 16t^2 = 0$ when $t = 0$ and $t = \dfrac{v_0\sin\theta}{16}$.

The range is

$$x = (v_0\cos\theta)t = (v_0\cos\theta)\frac{v_0\sin\theta}{16} = \frac{v_0^2}{32}\sin 2\theta.$$

Hence,

$$x = \frac{1200^2}{32}\sin(2\theta) = 3000 \implies \sin 2\theta = \frac{1}{15} \implies \theta \approx 1.91°.$$

33. (a) $\theta = 10°$, $v_0 = 66$ ft/sec

$\mathbf{r}(t) = (66\cos 10°)t\mathbf{i} + [0 + (66\sin 10°)t - 16t^2]\mathbf{j}$

$\mathbf{r}(t) \approx (65t)\mathbf{i} + (11.46t - 16t^2)\mathbf{j}$

Maximum height: 2.052 feet

Range: 46.557 feet

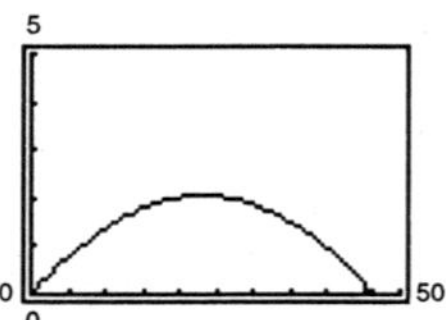

(b) $\theta = 10°$, $v_0 = 146$ ft/sec

$\mathbf{r}(t) = (146\cos 10°)t\mathbf{i} + [0 + (146\sin 10°)t - 16t^2]\mathbf{j}$

$\mathbf{r}(t) \approx (143.78t)\mathbf{i} + (25.35t - 16t^2)\mathbf{j}$

Maximum height: 10.043 feet

Range: 227.828 feet

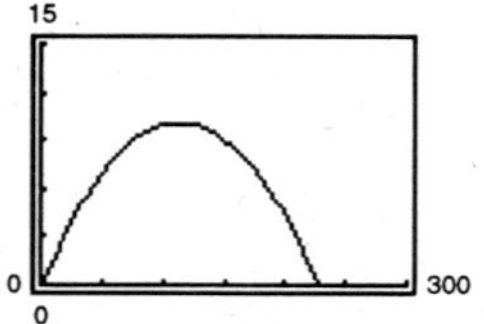

—CONTINUED—

33. —CONTINUED—

(c) $\theta = 45°$, $v_0 = 66$ ft/sec

$\mathbf{r}(t) = (66 \cos 45°)t\mathbf{i} + [0 + (66 \sin 45°)t - 16t^2]\mathbf{j}$

$\mathbf{r}(t) \approx (46.67t)\mathbf{i} + (46.67t - 16t^2)\mathbf{j}$

Maximum height: 34.031 feet

Range: 136.125 feet

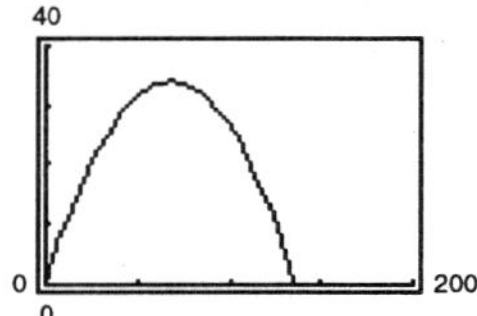

(d) $\theta = 45°$, $v_0 = 146$ ft/sec

$\mathbf{r}(t) = (146 \cos 45°)t\mathbf{i} + [0 + (146 \sin 45°)t - 16t^2]\mathbf{j}$

$\mathbf{r}(t) \approx (103.24t)\mathbf{i} + (103.24t - 16t^2)\mathbf{j}$

Maximum height: 166.531 feet

Range: 666.125 feet

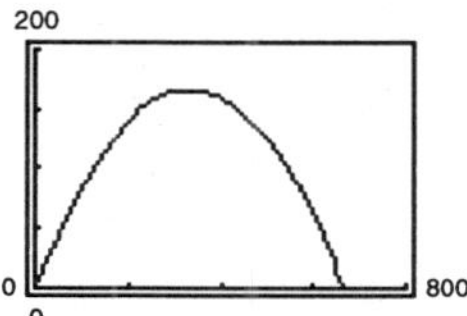

(e) $\theta = 60°$, $v_0 = 66$ ft/sec

$\mathbf{r}(t) = (66 \cos 60°)t\mathbf{i} + [0 + (66 \sin 60°)t - 16t^2]\mathbf{j}$

$\mathbf{r}(t) \approx (33t)\mathbf{i} + (57.16t - 16t^2)\mathbf{j}$

Maximum height: 51.074 feet

Range: 117.888 feet

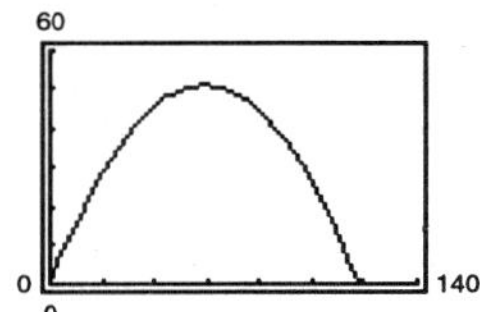

(f) $\theta = 60°$, $v_0 = 146$ ft/sec

$\mathbf{r}(t) = (146 \cos 60°)t\mathbf{i} + [0 + (146 \sin 60°)t - 16t^2]\mathbf{j}$

$\mathbf{r}(t) \approx (73t)\mathbf{i} + (126.44t - 16t^2)\mathbf{j}$

Maximum height: 249.797 feet

Range: 576.881 feet

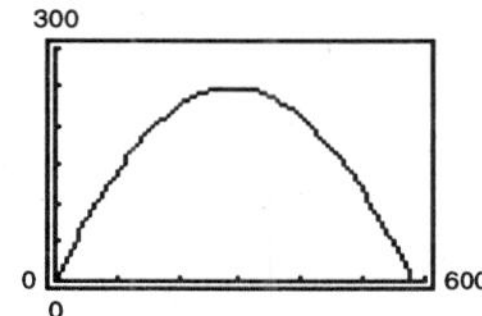

35. $y = x - 0.005x^2$

From Exercise 34 we know that $\tan\theta$ is the coefficient of x. Therefore, $\tan\theta = 1$, $\theta = (\pi/4)$ rad $= 45°$. Also

$$\frac{16}{v_0^2}\sec^2\theta = \text{negative of coefficient of } x^2$$

$$\frac{16}{v_0^2}(2) = 0.005 \text{ or } v_0 = 80 \text{ ft/sec}$$

$\mathbf{r}(t) = \left(40\sqrt{2}t\right)\mathbf{i} + \left(40\sqrt{2}t - 16t^2\right)\mathbf{j}$. Position function.

When $40\sqrt{2}t = 60$,

$$t = \frac{60}{40\sqrt{2}} = \frac{3\sqrt{2}}{4}$$

$$\mathbf{v}(t) = 40\sqrt{2}\mathbf{i} + \left(40\sqrt{2} - 32t\right)\mathbf{j}$$

$$\mathbf{v}\left(\frac{3\sqrt{2}}{4}\right) = 40\sqrt{2}\mathbf{i} + \left(40\sqrt{2} - 24\sqrt{2}\right)\mathbf{j} = 8\sqrt{2}(5\mathbf{i} + 2\mathbf{j}) \text{ direction}$$

$$\text{Speed} = \left\|\mathbf{v}\left(\frac{3\sqrt{2}}{4}\right)\right\| = 8\sqrt{2}\sqrt{25 + 4} = 8\sqrt{58} \text{ ft/sec}$$

37. $\mathbf{r}(t) = (v_0 \cos \theta)t\mathbf{i} + [h + (v_0 \sin \theta)t - 4.9t^2]\mathbf{j}$

$= (100 \cos 30°)t\mathbf{i} + [1.5 + (100 \sin 30°)t - 4.9t^2]\mathbf{j}$

The projectile hits the ground when $-4.9t^2 + 100(\frac{1}{2})t + 1.5 = 0 \Rightarrow t \approx 10.234$ seconds.

The range is therefore $(100 \cos 30°)(10.234) \approx 886.3$ meters.

The maximum height occurs when $dy/dt = 0$.

$100 \sin 30 = 9.8t \Rightarrow t \approx 5.102$ sec

The maximum height is

$y = 1.5 + (100 \sin 30°)(5.102) - 4.9(5.102)^2 \approx 129.1$ meters.

39. $\mathbf{r}(t) = b(\omega t - \sin \omega t)\mathbf{i} + b(1 - \cos \omega t)\mathbf{j}$

$\mathbf{v}(t) = b(\omega - \omega \cos \omega t)\mathbf{i} + b\omega \sin \omega t\,\mathbf{j} = b\omega(1 - \cos \omega t)\mathbf{i} + b\omega \sin \omega t\mathbf{j}$

$\mathbf{a}(t) = (b\omega^2 \sin \omega t)\mathbf{i} + (b\omega^2 \cos \omega t)\mathbf{j} = b\omega^2[\sin(\omega t)\mathbf{i} + \cos(\omega t)\mathbf{j}]$

$\|\mathbf{v}(t)\| = \sqrt{2}\, b\omega\sqrt{1 - \cos(\omega t)}$

$\|\mathbf{a}(t)\| = b\omega^2$

(a) $\|\mathbf{v}(t)\| = 0$ when $\omega t = 0, 2\pi, 4\pi, \ldots$.

(b) $\|\mathbf{v}(t)\|$ is maximum when $\omega t = \pi, 3\pi, \ldots$, then $\|\mathbf{v}(t)\| = 2b\omega$.

41. $\mathbf{v}(t) = -b\omega \sin(\omega t)\mathbf{i} + b\omega \cos(\omega t)\mathbf{j}$

$\mathbf{r}(t) \cdot \mathbf{v}(t) = -b^2\omega \sin(\omega t) \cos(\omega t) + b^2\omega \sin(\omega t) \cos(\omega t) = 0$

Therefore, $\mathbf{r}(t)$ and $\mathbf{v}(t)$ are orthogonal.

43. $\mathbf{a}(t) = -b\omega^2 \cos(\omega t)\mathbf{i} - b\omega^2 \sin(\omega t)\mathbf{j}$

$= -b\omega^2[\cos(\omega t)\mathbf{i} + \sin(\omega t)\mathbf{j}] = -\omega^2\mathbf{r}(t)$

$\mathbf{a}(t)$ is a negative multiple of a unit vector from $(0, 0)$ to $(\cos \omega t, \sin \omega t)$ and thus $\mathbf{a}(t)$ is directed toward the origin.

45. $\|\mathbf{a}(t)\| = \omega^2 b$

$1 = m(32)$

$F = m(\omega^2 b) = \frac{1}{32}(2\omega^2) = 10$

$\omega = 4\sqrt{10}$ rad/sec

$\|\mathbf{v}(t)\| = b\omega = 8\sqrt{10}$ ft/sec

47. To find the range, set $y(t) = h + (v_0 \sin \theta)t - \frac{1}{2}gt^2 = 0$ then $0 = (\frac{1}{2}g)t^2 - (v_0 \sin \theta)t - h$. By the Quadratic Formula, (discount the negative value)

$$t = \frac{v_0 \sin \theta + \sqrt{(-v_0 \sin \theta)^2 - 4[(1/2)g](-h)}}{2[(1/2)g]} = \frac{v_0 \sin \theta + \sqrt{v_0^2 \sin^2 \theta + 2gh}}{g}.$$

At this time,

$$x(t) = v_0 \cos \theta\left(\frac{v_0 \sin \theta + \sqrt{v_0^2 \sin^2 \theta + 2gh}}{g}\right) = \frac{v_0 \cos \theta}{g}\left(v_0 \sin \theta + \sqrt{v_0^2\left(\sin^2 \theta + \frac{2gh}{v_0^2}\right)}\right)$$

$$= \frac{v_0^2 \cos \theta}{g}\left(\sin \theta + \sqrt{\sin^2 \theta + \frac{2gh}{v_0^2}}\right).$$

49. $\mathbf{r}(t) = x(t)\mathbf{i} + y(t)\mathbf{j} + z(t)\mathbf{k}$ Position vector

$\mathbf{v}(t) = x'(t)\mathbf{i} + y'(t)\mathbf{j} + z'(t)\mathbf{k}$ Velocity vector

$\mathbf{a}(t) = x''(t)\mathbf{i} + y''(t)\mathbf{j} + z''(t)\mathbf{k}$ Acceleration vector

$$\text{Speed} = \|\mathbf{v}(t)\| = \sqrt{(x'(t)^2 + y'(t)^2 + z'(t)^2} = C,\ C \text{ is a constant.}$$

$$\frac{d}{dt}[x'(t)^2 + y'(t)^2 + z'(t)^2] = 0$$

$$2x'(t)x''(t) + 2y'(t)y''(t) + 2z'(t)z''(t) = 0$$

$$2[x'(t)x''(t) + y'(t)y''(t) + z'(t)z''(t)] = 0$$

$$\mathbf{v}(t) \cdot \mathbf{a}(t) = 0$$

Orthogonal

51. $\mathbf{r}(t) = 6\cos t\mathbf{i} + 3\sin t\mathbf{j}$

(a) $\mathbf{v}(t) = \mathbf{r}'(t) = -6\sin t\mathbf{i} + 3\cos t\mathbf{j}$

$$\|\mathbf{v}(t)\| = \sqrt{36\sin^2 t + 9\cos^2 t} = 3\sqrt{4\sin^2 t + \cos^2 t} = 3\sqrt{3\sin^2 t + 1}$$

$\mathbf{a}(t) = \mathbf{v}'(t) = -6\cos t\mathbf{i} - 3\sin t\mathbf{j}$

(b)

t	0	$\frac{\pi}{4}$	$\frac{\pi}{2}$	$\frac{2\pi}{3}$	π
Speed	3	$\frac{3}{2}\sqrt{10}$	6	$\frac{3}{2}\sqrt{13}$	3

(c)

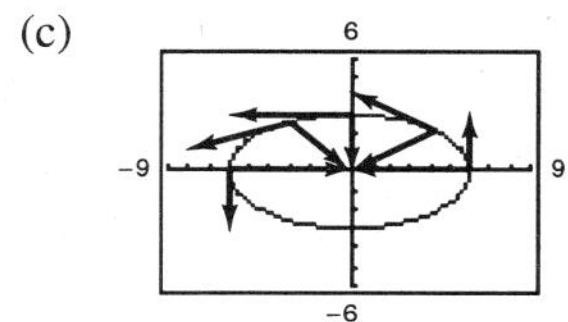

(d) The speed is increasing when the angle between $\mathbf{v}$ and $\mathbf{a}$ is in the interval

$$\left[0, \frac{\pi}{2}\right).$$

The speed is decreasing when the angle is in the interval

$$\left(\frac{\pi}{2}, \pi\right].$$

Section 11.4 Tangent Vectors and Normal Vectors

1. $\mathbf{r}(t) = t\mathbf{i} + t^2\mathbf{j} + t\mathbf{k}$

$\mathbf{r}'(t) = \mathbf{i} + 2t\mathbf{j} + \mathbf{k}$

When $t = 0$, $\mathbf{r}'(0) = \mathbf{i} + \mathbf{k}$, $[t = 0 \text{ at } (0, 0, 0)]$.

$$\mathbf{T}(0) = \frac{\mathbf{r}'(0)}{\|\mathbf{r}'(0)\|} = \frac{\sqrt{2}}{2}(\mathbf{i} + \mathbf{k})$$

Direction numbers: $a = 1,\ b = 0,\ c = 1$

Parametric equations: $x = t,\ y = 0,\ z = t$

3. $\mathbf{r}(t) = 2\cos t\mathbf{i} + 2\sin t\mathbf{j} + t\mathbf{k}$

$\mathbf{r}'(t) = -2\sin t\mathbf{i} + 2\cos t\mathbf{j} + \mathbf{k}$

When $t = 0$, $\mathbf{r}'(0) = 2\mathbf{j} + \mathbf{k}$, $[t = 0 \text{ at } (2, 0, 0)]$.

$$\mathbf{T}(0) = \frac{\mathbf{r}'(0)}{\|\mathbf{r}'(0)\|} = \frac{\sqrt{5}}{5}(2\mathbf{j} + \mathbf{k})$$

Direction numbers: $a = 0,\ b = 2,\ c = 1$

Parametric equations: $x = 2,\ y = 2t,\ z = t$

5. $\mathbf{r}(t) = \langle 2\cos t, 2\sin t, 4\rangle$

$\mathbf{r}'(t) = \langle -2\sin t, 2\cos t, 0\rangle$

When $t = \frac{\pi}{4}$, $\mathbf{r}'\left(\frac{\pi}{4}\right) = \langle -\sqrt{2}, \sqrt{2}, 0\rangle$, $\left[t = \frac{\pi}{4} \text{ at } (\sqrt{2}, \sqrt{2}, 4)\right]$.

$$\mathbf{T}\left(\frac{\pi}{4}\right) = \frac{\mathbf{r}'(\pi/4)}{\|\mathbf{r}'(\pi/4)\|} = \frac{1}{2}\langle -\sqrt{2}, \sqrt{2}, 0\rangle$$

Direction numbers: $a = -\sqrt{2},\ b = \sqrt{2},\ c = 0$

Parametric equations: $x = -\sqrt{2}t + \sqrt{2},\ y = \sqrt{2}t + \sqrt{2},\ z = 4$

7. $\mathbf{r}(t) = \left\langle t, t^2, \frac{2}{3}t^3 \right\rangle$

$\mathbf{r}'(t) = \langle 1, 2t, 2t^2 \rangle$

When $t = 3$, $\mathbf{r}'(3) = \langle 1, 6, 18 \rangle$, $[t = 3 \text{ at } (3, 9, 18)]$.

$$\mathbf{T}(3) = \frac{\mathbf{r}'(3)}{\|\mathbf{r}'(3)\|} = \frac{1}{19}\langle 1, 6, 18 \rangle$$

Direction numbers: $a = 1,\ b = 6,\ c = 18$

Parametric equations: $x = t + 3,\ y = 6t + 9,\ z = 18t + 18$

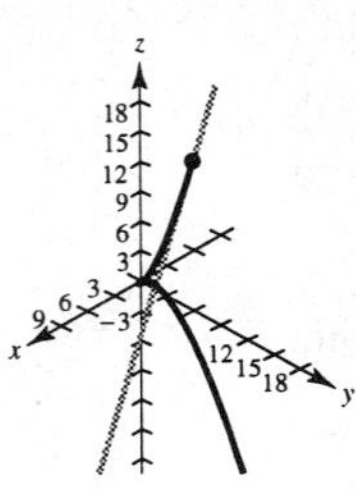

9. $\mathbf{r}(t) = t\mathbf{i} + \ln t\mathbf{j} + \sqrt{t}\mathbf{k}, \quad t_0 = 1$

$\mathbf{r}'(t) = \mathbf{i} + \frac{1}{t}\mathbf{j} + \frac{1}{2\sqrt{t}}\mathbf{k} \cdot \mathbf{r}'(1) = \mathbf{i} + \mathbf{j} + \frac{1}{2}\mathbf{k}$

$$\mathbf{T}(1) = \frac{\mathbf{r}'(t)}{\|\mathbf{r}'(t)\|} = \frac{\mathbf{i} + \mathbf{j} + (1/2)\mathbf{k}}{\sqrt{1 + 1 + (1/4)}} = \frac{2}{3}\mathbf{i} + \frac{2}{3}\mathbf{j} + \frac{1}{3}\mathbf{k}$$

Tangent line: $x = 1 + t,\ y = t,\ z = 1 + \frac{1}{2}t$

$\mathbf{r}(t_0 + 0.1) = \mathbf{r}(1.1) \approx 1.1\mathbf{i} + 0.1\mathbf{j} + 1.05\mathbf{k} = \langle 1.1, 0.1, 1.05 \rangle$

11. $\mathbf{r}(4) = \langle 2, 16, 2 \rangle$

$\mathbf{u}(8) = \langle 2, 16, 2 \rangle$

Hence the curves intersect.

$$\mathbf{r}'(t) = \left\langle 1, 2t, \frac{1}{2} \right\rangle,\ \mathbf{r}'(4) = \left\langle 1, 8, \frac{1}{2} \right\rangle$$

$$\mathbf{u}'(s) = \left\langle \frac{1}{4}, 2, \frac{1}{3}s^{-2/3} \right\rangle,\ \mathbf{u}'(8) = \left\langle \frac{1}{4}, 2, \frac{1}{12} \right\rangle$$

$$\cos\theta = \frac{\mathbf{r}'(4) \cdot \mathbf{u}'(8)}{\|\mathbf{r}'(4)\|\,\|\mathbf{u}'(8)\|} \approx \frac{16.29167}{16.29513} \implies \theta \approx 1.2°$$

13. $\mathbf{r}(t) = 4t\mathbf{i}$

$\mathbf{v}(t) = 4\mathbf{i}$

$\mathbf{a}(t) = \mathbf{O}$

$$\mathbf{T}(t) = \frac{\mathbf{v}(t)}{\|\mathbf{v}(t)\|} = \frac{4\mathbf{i}}{4} = \mathbf{i}$$

$\mathbf{T}'(t) = \mathbf{O}$

$\mathbf{N}(t) = \frac{\mathbf{T}'(t)}{\|\mathbf{T}'(t)\|}$ is undefined.

The path is a line and the speed is constant.

15. $\mathbf{r}(t) = 4t^2\mathbf{i}$

$\mathbf{v}(t) = 8t\mathbf{i}$

$\mathbf{a}(t) = 8\mathbf{i}$

$$\mathbf{T}(t) = \frac{\mathbf{v}(t)}{\|\mathbf{v}(t)\|} = \frac{8t\mathbf{i}}{8t} = \mathbf{i}$$

$\mathbf{T}'(t) = \mathbf{O}$

$\mathbf{N}(t) = \frac{\mathbf{T}'(t)}{\|\mathbf{T}'(t)\|}$ is undefined.

The path is a line and the speed is variable.

17. $\mathbf{r}(t) = t\mathbf{i} + \frac{1}{t}\mathbf{j},\ \mathbf{v}(t) = \mathbf{i} - \frac{1}{t^2}\mathbf{j},\ \mathbf{v}(1) = \mathbf{i} - \mathbf{j},$

$\mathbf{a}(t) = \frac{2}{t^3}\mathbf{j},\ \mathbf{a}(1) = 2\mathbf{j}$

$$\mathbf{T}(t) = \frac{\mathbf{v}(t)}{\|\mathbf{v}(t)\|} = \frac{t^2}{\sqrt{t^4 + 1}}\left(\mathbf{i} - \frac{1}{t^2}\mathbf{j}\right) = \frac{1}{\sqrt{t^4 + 1}}(t^2\mathbf{i} - \mathbf{j})$$

$$\mathbf{T}(1) = \frac{1}{\sqrt{2}}(\mathbf{i} - \mathbf{j}) = \frac{\sqrt{2}}{2}(\mathbf{i} - \mathbf{j})$$

$$\mathbf{N}(t) = \frac{\mathbf{T}'(t)}{\|\mathbf{T}'(t)\|} = \frac{\dfrac{2t}{(t^4 + 1)^{3/2}}\mathbf{i} + \dfrac{2t^3}{(t^4 + 1)^{3/2}}\mathbf{j}}{\dfrac{2t}{(t^4 + 1)}}$$

$$= \frac{1}{\sqrt{t^4 + 1}}(\mathbf{i} + t^2\mathbf{j})$$

$$\mathbf{N}(1) = \frac{1}{\sqrt{2}}(\mathbf{i} + \mathbf{j}) = \frac{\sqrt{2}}{2}(\mathbf{i} + \mathbf{j})$$

$a_{\mathbf{T}} = \mathbf{a} \cdot \mathbf{T} = -\sqrt{2}$

$a_{\mathbf{N}} = \mathbf{a} \cdot \mathbf{N} = \sqrt{2}$

19. $\mathbf{r}(t) = (e^t \cos t)\mathbf{i} + (e^t \sin t)\mathbf{j}$

$\mathbf{v}(t) = e^t(\cos t - \sin t)\mathbf{i} + e^t(\cos t + \sin t)\mathbf{j}$

$\mathbf{a}(t) = e^t(-2 \sin t)\mathbf{i} + e^t(2 \cos t)\mathbf{j}$

At $t = \frac{\pi}{2}$, $\mathbf{T} = \frac{\mathbf{v}}{\|\mathbf{v}\|} = \frac{1}{\sqrt{2}}(-\mathbf{i} + \mathbf{j}) = \frac{\sqrt{2}}{2}(-\mathbf{i} + \mathbf{j})$.

Motion along $\mathbf{r}$ is counterclockwise. Therefore,

$$\mathbf{N} = \frac{1}{\sqrt{2}}(-\mathbf{i} - \mathbf{j}) = -\frac{\sqrt{2}}{2}(\mathbf{i} + \mathbf{j}).$$

$$a_{\mathbf{T}} = \mathbf{a} \cdot \mathbf{T} = \sqrt{2}e^{\pi/2}$$

$$a_{\mathbf{N}} = \mathbf{a} \cdot \mathbf{N} = \sqrt{2}e^{\pi/2}$$

21. $\mathbf{r}(t_0) = (\cos \omega t_0 + \omega t_0 \sin \omega t_0)\mathbf{i} + (\sin \omega t_0 - \omega t_0 \cos \omega t_0)\mathbf{j}$

$\mathbf{v}(t_0) = (\omega^2 t_0 \cos \omega t_0)\mathbf{i} + (\omega^2 t_0 \sin \omega t_0)\mathbf{j}$

$\mathbf{a}(t_0) = \omega^2[(\cos \omega t_0 - \omega t_0 \sin \omega t_0)\mathbf{i} + (\omega t_0 \cos \omega t_0 + \sin \omega t_0)\mathbf{j}]$

$\mathbf{T}(t_0) = \frac{\mathbf{v}}{\|\mathbf{v}\|} = (\cos \omega t_0)\mathbf{i} + (\sin \omega t_0)\mathbf{j}$

Motion along $\mathbf{r}$ is counterclockwise. Therefore

$$\mathbf{N}(t_0) = (-\sin \omega t_0)\mathbf{i} + (\cos \omega t_0)\mathbf{j}.$$

$$a_{\mathbf{T}} = \mathbf{a} \cdot \mathbf{T} = \omega^2$$

$$a_{\mathbf{N}} = \mathbf{a} \cdot \mathbf{N} = \omega^2(\omega t_0) = \omega^3 t_0$$

23. $\mathbf{r}(t) = a \cos(\omega t)\mathbf{i} + a \sin(\omega t)\mathbf{j}$

$\mathbf{v}(t) = -a\omega \sin(\omega t)\mathbf{i} + a\omega \cos(\omega t)\mathbf{j}$

$\mathbf{a}(t) = -a\omega^2 \cos(\omega t)\mathbf{i} - a\omega^2 \sin(\omega t)\mathbf{j}$

$$\mathbf{T}(t) = \frac{\mathbf{v}(t)}{\|\mathbf{v}(t)\|} = -\sin(\omega t)\mathbf{i} + \cos(\omega t)\mathbf{j}$$

$$\mathbf{N}(t) = \frac{\mathbf{T}'(t)}{\|\mathbf{T}'(t)\|} = -\cos(\omega t)\mathbf{i} - \sin(\omega t)\mathbf{j}$$

$$a_{\mathbf{T}} = \mathbf{a} \cdot \mathbf{T} = 0$$

$$a_{\mathbf{N}} = \mathbf{a} \cdot \mathbf{N} = a\omega^2$$

25. Speed: $\|\mathbf{v}(t)\| = a\omega$

The speed is constant since $a_{\mathbf{T}} = 0$.

27. $\mathbf{r}(t) = t\mathbf{i} + \frac{1}{t}\mathbf{j}$, $t_0 = 2$

$$x = t,\ y = \frac{1}{t} \Rightarrow xy = 1$$

$$\mathbf{r}'(t) = \mathbf{i} - \frac{1}{t^2}\mathbf{j}$$

$$\mathbf{T}(t) = \frac{t^2\mathbf{i} - \mathbf{j}}{\sqrt{t^4 + 1}}$$

$$\mathbf{N}(t) = \frac{\mathbf{i} + t^2\mathbf{j}}{\sqrt{t^4 + 1}}$$

$$\mathbf{r}(2) = 2\mathbf{i} + \frac{1}{2}\mathbf{j}$$

$$\mathbf{T}(2) = \frac{\sqrt{17}}{17}(4\mathbf{i} - \mathbf{j})$$

$$\mathbf{N}(2) = \frac{\sqrt{17}}{17}(\mathbf{i} + 4\mathbf{j})$$

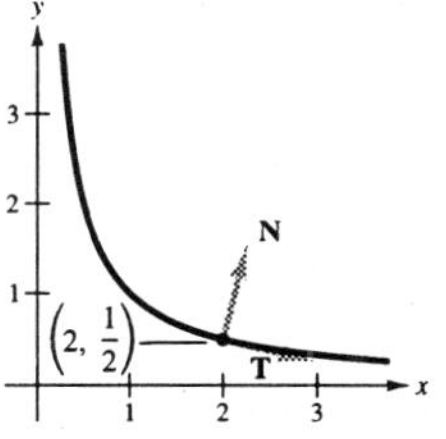

29. $\mathbf{r}(t) = \langle \pi t - \sin \pi t, 1 - \cos \pi t \rangle$

The graph is a cycloid.

(a) $\mathbf{r}(t) = \langle \pi t - \sin \pi t, 1 - \cos \pi t \rangle$

$\mathbf{v}(t) = \langle \pi - \pi \cos \pi t, \pi \sin \pi t \rangle$

$\mathbf{a}(t) = \langle \pi^2 \sin \pi t, \pi^2 \cos \pi t \rangle$

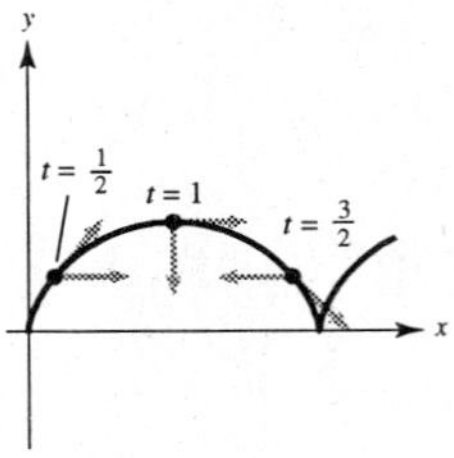

$$\mathbf{T}(t) = \frac{\mathbf{v}(t)}{\|\mathbf{v}(t)\|} = \frac{1}{\sqrt{2(1 - \cos \pi t)}}\langle 1 - \cos \pi t, \sin \pi t \rangle$$

$$\mathbf{N}(t) = \frac{\mathbf{T}'(t)}{\|\mathbf{T}'(t)\|} = \frac{1}{\sqrt{2(1 - \cos \pi t)}}\langle \sin \pi t, -1 + \cos \pi t \rangle$$

$$a_{\mathbf{T}} = \mathbf{a} \cdot \mathbf{T} = \frac{1}{\sqrt{2(1 - \cos \pi t)}}[\pi^2 \sin \pi t(1 - \cos \pi t) + \pi^2 \cos \pi t \sin \pi t] = \frac{\pi^2 \sin \pi t}{\sqrt{2(1 - \cos \pi t)}}$$

$$a_{\mathbf{N}} = \mathbf{a} \cdot \mathbf{N} = \frac{1}{\sqrt{2(1 - \cos \pi t)}}[\pi^2 \sin^2 \pi t + \pi^2 \cos \pi t(-1 + \cos \pi t)] = \frac{\pi^2(1 - \cos \pi t)}{\sqrt{2(1 - \cos \pi t)}} = \frac{\pi^2\sqrt{2(1 - \cos \pi t)}}{2}$$

When $t = \frac{1}{2}$: $a_{\mathbf{T}} = \frac{\pi^2}{\sqrt{2}} = \frac{\sqrt{2}\pi^2}{2}$, $a_{\mathbf{N}} = \frac{\sqrt{2}\pi^2}{2}$

When $t = 1$: $a_{\mathbf{T}} = 0$, $a_{\mathbf{N}} = \pi^2$

When $t = \frac{3}{2}$: $a_{\mathbf{T}} = -\frac{\sqrt{2}\pi^2}{2}$, $a_{\mathbf{N}} = \frac{\sqrt{2}\pi^2}{2}$

(b) Speed: $s = \|\mathbf{v}(t)\| = \pi\sqrt{2(1 - \cos \pi t)}$

$$\frac{ds}{dt} = \frac{\pi^2 \sin \pi t}{\sqrt{2(1 - \cos \pi t)}} = a_{\mathbf{T}}$$

When $t = \frac{1}{2}$: $a_{\mathbf{T}} = \frac{\sqrt{2}\pi^2}{2} > 0 \implies$ the speed in increasing.

When $t = 1$: $a_{\mathbf{T}} = 0 \implies$ the height is maximum.

When $t = \frac{3}{2}$: $a_{\mathbf{T}} = -\frac{\sqrt{2}\pi^2}{2} < 0 \implies$ the speed is decreasing.

31. $\mathbf{r}(t) = t\mathbf{i} + 2t\mathbf{j} - 3t\mathbf{k}$

$\mathbf{v}(t) = \mathbf{i} + 2\mathbf{j} - 3\mathbf{k}$

$\mathbf{a}(t) = \mathbf{0}$

$$\mathbf{T}(t) = \frac{\mathbf{v}}{\|\mathbf{v}\|} = \frac{1}{\sqrt{14}}(\mathbf{i} + 2\mathbf{j} - 3\mathbf{k}) = \frac{\sqrt{14}}{14}(\mathbf{i} + 2\mathbf{j} - 3\mathbf{k})$$

$\mathbf{N}(t) = \frac{\mathbf{T}'}{\|\mathbf{T}'\|}$ is undefined.

$a_{\mathbf{T}}$, $a_{\mathbf{N}}$ are not defined.

33. $\mathbf{r}(t) = t\mathbf{i} + t^2\mathbf{j} + \frac{t^2}{2}\mathbf{k}$

$\mathbf{v}(t) = \mathbf{i} + 2t\mathbf{j} + t\mathbf{k}$

$\mathbf{v}(1) = \mathbf{i} + 2\mathbf{j} + \mathbf{k}$

$\mathbf{a}(t) = 2\mathbf{j} + \mathbf{k}$

$$\mathbf{T}(t) = \frac{\mathbf{v}}{\|\mathbf{v}\|} = \frac{1}{\sqrt{1 + 5t^2}}(\mathbf{i} + 2t\mathbf{j} + t\mathbf{k})$$

$$\mathbf{T}(1) = \frac{\sqrt{6}}{6}(\mathbf{i} + 2\mathbf{j} + \mathbf{k})$$

$$\mathbf{N}(t) = \frac{\mathbf{T}'}{\|\mathbf{T}'\|} = \frac{\dfrac{-5t\mathbf{i} + 2\mathbf{j} + \mathbf{k}}{(1 + 5t^2)^{3/2}}}{\dfrac{\sqrt{5}}{1 + 5t^2}} = \frac{-5t\mathbf{i} + 2\mathbf{j} + \mathbf{k}}{\sqrt{5}\sqrt{1 + 5t^2}}$$

$$\mathbf{N}(1) = \frac{\sqrt{30}}{30}(-5\mathbf{i} + 2\mathbf{j} + \mathbf{k})$$

$$a_{\mathbf{T}} = \mathbf{a} \cdot \mathbf{T} = \frac{5\sqrt{6}}{6}$$

$$a_{\mathbf{N}} = \mathbf{a} \cdot \mathbf{N} = \frac{\sqrt{30}}{6}$$

35. $\mathbf{r}(t) = 4t\mathbf{i} + 3\cos t\mathbf{j} + 3\sin t\mathbf{k}$

$\mathbf{v}(t) = 4\mathbf{i} - 3\sin t\mathbf{j} + 3\cos t\mathbf{k}$

$\mathbf{v}\left(\frac{\pi}{2}\right) = 4\mathbf{i} - 3\mathbf{j}$

$\mathbf{a}(t) = -3\cos t\mathbf{j} - 3\sin t\mathbf{k}$

$\mathbf{a}\left(\frac{\pi}{2}\right) = -3\mathbf{k}$

$\mathbf{T}(t) = \frac{\mathbf{v}}{\|\mathbf{v}\|} = \frac{1}{5}(4\mathbf{i} - 3\sin t\mathbf{j} + 3\cos t\mathbf{k})$

$\mathbf{T}\left(\frac{\pi}{2}\right) = \frac{1}{5}(4\mathbf{i} - 3\mathbf{j})$

$\mathbf{N}(t) = \frac{\mathbf{T}'}{\|\mathbf{T}'\|} = -\cos t\mathbf{j} - \sin t\mathbf{k}$

$\mathbf{N}\left(\frac{\pi}{2}\right) = -\mathbf{k}$

$a_{\mathbf{T}} = \mathbf{a} \cdot \mathbf{T} = 0$

$a_{\mathbf{N}} = \mathbf{a} \cdot \mathbf{N} = 3$

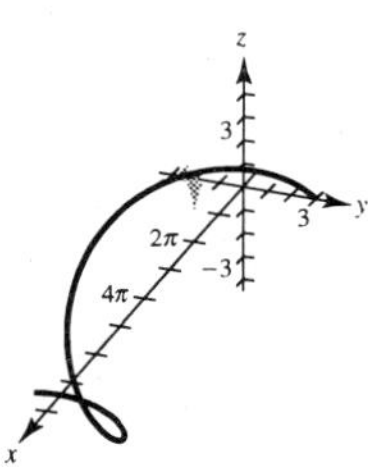

37. $\mathbf{r}(t) = 2\cos t\mathbf{i} + 2\sin t\mathbf{j} + \frac{t}{2}\mathbf{k},\ t_0 = \frac{\pi}{2}$

$\mathbf{r}'(t) = -2\sin t\mathbf{i} + 2\cos t\mathbf{j} + \frac{1}{2}\mathbf{k}$

$\mathbf{T}(t) = \frac{2\sqrt{17}}{17}\left(-2\sin t\mathbf{i} + 2\cos t\mathbf{j} + \frac{1}{2}\mathbf{k}\right)$

$\mathbf{N}(t) = -\cos t\mathbf{i} - \sin t\mathbf{j}$

$\mathbf{r}\left(\frac{\pi}{2}\right) = 2\mathbf{j} + \frac{\pi}{4}\mathbf{k}$

$\mathbf{T}\left(\frac{\pi}{2}\right) = \frac{2\sqrt{17}}{17}\left(-2\mathbf{i} + \frac{1}{2}\mathbf{k}\right) = \frac{\sqrt{17}}{17}(-4\mathbf{i} + \mathbf{k})$

$\mathbf{N}\left(\frac{\pi}{2}\right) = -\mathbf{j}$

$$\mathbf{B}\left(\frac{\pi}{2}\right) = \mathbf{T}\left(\frac{\pi}{2}\right) \times \mathbf{N}\left(\frac{\pi}{2}\right) = \begin{vmatrix} \mathbf{i} & \mathbf{j} & \mathbf{k} \\ -\frac{4\sqrt{17}}{17} & 0 & \frac{\sqrt{17}}{17} \\ 0 & -1 & 0 \end{vmatrix}$$

$$= \frac{\sqrt{17}}{17}\mathbf{i} + \frac{4\sqrt{17}}{17}\mathbf{k} = \frac{\sqrt{17}}{17}(\mathbf{i} + 4\mathbf{k})$$

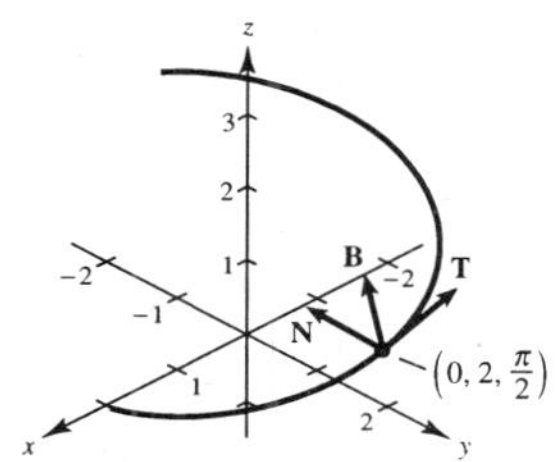

39. $\mathbf{r}(t) = \langle 10\cos 10\pi t, 10\sin 10\pi t, 4 + 4t\rangle,\ 0 \le t \le \frac{1}{20}$

(a) $\mathbf{r}'(t) = \langle -100\pi\sin(10\pi t), 100\pi\cos(10\pi t), 4\rangle$

$\|\mathbf{r}'(t)\| = \sqrt{(100\pi)^2\sin^2(10\pi t) + (100\pi)^2\cos^2(10\pi t) + 16}$

$= \sqrt{(100\pi)^2 + 16} = 4\sqrt{625\pi^2 + 1} \approx 314$ mi/hr

(b) $a_{\mathbf{T}} = 0$ and $a_{\mathbf{N}} = 1000\pi^2$

$a_{\mathbf{T}} = 0$ because the speed is constant.

41. From Theorem 11.3 we have:

$\mathbf{r}(t) = (v_0 t\cos\theta)\mathbf{i} + (h + v_0 t\sin\theta - 16t^2)\mathbf{j}$

$\mathbf{v}(t) = v_0\cos\theta\mathbf{i} + (v_0\sin\theta - 32t)\mathbf{j}$

$\mathbf{a}(t) = -32\mathbf{j}$

$$\mathbf{T}(t) = \frac{(v_0\cos\theta)\mathbf{i} + (v_0\sin\theta - 32t)\mathbf{j}}{\sqrt{v_0^2\cos^2\theta + (v_0\sin\theta - 32t)^2}}$$

$$\mathbf{N}(t) = \frac{(v_0\sin\theta - 32t)\mathbf{i} - v_0\cos\theta\mathbf{j}}{\sqrt{v_0^2\cos^2\theta + (v_0\sin\theta - 32t)^2}}$$ (Motion is clockwise.)

$$a_{\mathbf{T}} = \mathbf{a} \cdot \mathbf{T} = \frac{-32(v_0\sin\theta - 32t)}{\sqrt{v_0^2\cos^2\theta + (v_0\sin\theta - 32t)^2}}$$

$$a_{\mathbf{N}} = \mathbf{a} \cdot \mathbf{N} = \frac{32v_0\cos\theta}{\sqrt{v_0^2\cos^2\theta + (v_0\sin\theta - 32t)^2}}$$

Maximum height when $v_0\sin\theta - 32t = 0$; (vertical component of velocity)

At maximum height, $a_{\mathbf{T}} = 0$ and $a_{\mathbf{N}} = 32$.

43. $\mathbf{r}(t) = (a\cos\omega t)\mathbf{i} + (a\sin\omega t)\mathbf{j}$

From Exercise 23, we know $\mathbf{a}\cdot\mathbf{T} = 0$ and $\mathbf{a}\cdot\mathbf{N} = a\omega^2$.

(a) Let $\omega_0 = 2\omega$. Then

$$\mathbf{a}\cdot\mathbf{N} = a\omega_0^2 = a(2\omega)^2 = 4a\omega^2$$

or the centripetal acceleration is increased by a factor of 4 when the velocity is doubled.

(b) Let $a_0 = a/2$. Then

$$\mathbf{a}\cdot\mathbf{N} = a_0\omega^2 = \left(\frac{a}{2}\right)\omega^2 = \left(\frac{1}{2}\right)a\omega^2$$

or the centripetal acceleration is halved when the radius is halved.

45. $v = \sqrt{\dfrac{9.56\times 10^4}{4100}} \approx 4.83$ mi/sec

47. $v = \sqrt{\dfrac{9.56\times 10^4}{4385}} \approx 4.67$ mi/sec

49. Let $\mathbf{T}(t) = \cos\phi\mathbf{i} + \sin\phi\mathbf{j}$ be the unit tangent vector. Then

$$\mathbf{T}'(t) = \frac{d\mathbf{T}}{dt} = \frac{d\mathbf{T}}{d\phi}\frac{d\phi}{dt} = -(\sin\phi\mathbf{i} + \cos\phi\mathbf{j})\frac{d\phi}{dt} = \mathbf{M}\frac{d\phi}{dt}.$$

$\mathbf{M} = -\sin\phi\mathbf{i} + \cos\phi\mathbf{j} = \cos[\phi + (\pi/2)]\mathbf{i} + \sin[\phi + (\pi/2)]\mathbf{j}$ and is rotated counterclockwise through an angle of $\pi/2$ from $\mathbf{T}$.

If $d\phi/dt > 0$, then the curve bends to the left and $\mathbf{M}$ has the same direction as $\mathbf{T}'$. Thus, $\mathbf{M}$ has the same direction as

$$\mathbf{N} = \frac{\mathbf{T}'}{\|\mathbf{T}'\|},$$

which is toward the concave side of the curve.

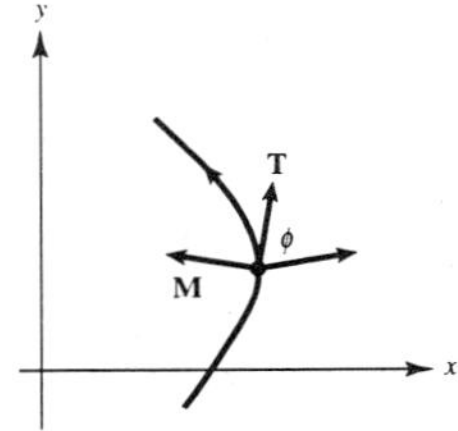

If $d\phi/dt < 0$, then the curve bends to the right and $\mathbf{M}$ has the opposite direction as $\mathbf{T}'$. Thus,

$$\mathbf{N} = \frac{\mathbf{T}'}{\|\mathbf{T}'\|}$$

again points to the concave side of the curve.

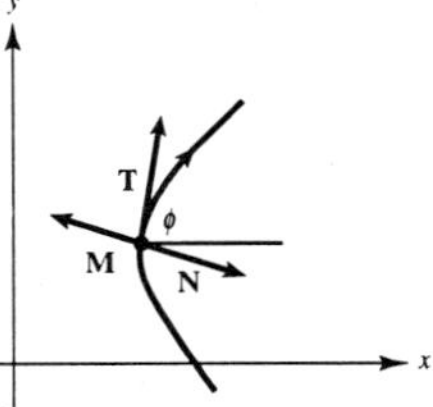

51. Using $\mathbf{a} = a_{\mathbf{T}}\mathbf{T} + a_{\mathbf{N}}\mathbf{N}$, $\mathbf{T}\times\mathbf{T} = \mathbf{O}$, and $\|\mathbf{T}\times\mathbf{N}\| = 1$, we have:

$$\begin{aligned}\mathbf{v}\times\mathbf{a} &= \|\mathbf{v}\|\mathbf{T}\times(a_{\mathbf{T}}\mathbf{T} + a_{\mathbf{N}}\mathbf{N})\\ &= \|\mathbf{v}\|a_{\mathbf{T}}(\mathbf{T}\times\mathbf{T}) + \|\mathbf{v}\|a_{\mathbf{N}}(\mathbf{T}\times\mathbf{N})\\ &= \|\mathbf{v}\|a_{\mathbf{N}}(\mathbf{T}\times\mathbf{N})\end{aligned}$$

$$\|\mathbf{v}\times\mathbf{a}\| = \|\mathbf{v}\|a_{\mathbf{N}}\|\mathbf{T}\times\mathbf{N}\| = \|\mathbf{v}\|a_{\mathbf{N}}$$

Thus, $a_{\mathbf{N}} = \dfrac{\|\mathbf{v}\times\mathbf{a}\|}{\|\mathbf{v}\|}$.

Section 11.5 Arc Length and Curvature

1. $\mathbf{r}(t) = t\mathbf{i} + 3t\mathbf{j}$

$$\frac{dx}{dt} = 1, \frac{dy}{dt} = 3, \frac{dz}{dt} = 0$$

$$s = \int_0^4 \sqrt{1 + 9}\, dt$$

$$= \sqrt{10}\int_0^4 dt$$

$$= \Big[\sqrt{10}t\Big]_0^4 = 4\sqrt{10}$$

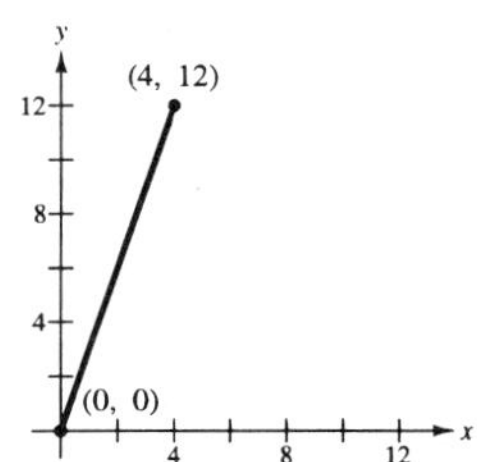

3. $\mathbf{r}(t) = a\cos^3 t\mathbf{i} + a\sin^3 t\mathbf{j}$

$$\frac{dx}{dt} = -3a\cos^2 t\sin t, \frac{dy}{dt} = 3a\sin^2 t\cos t$$

$$s = 4\int_0^{\pi/2} \sqrt{[3a\cos^2 t(-\sin t)]^2 + [3a\sin^2 t\cos t]^2}\, dt$$

$$= 12a\int_0^{\pi/2} \sin t\cos t\, dt$$

$$= 3a\int_0^{\pi/2} 2\sin 2t\, dt = \Big[-3a\cos 2t\Big]_0^{\pi/2} = 6a$$

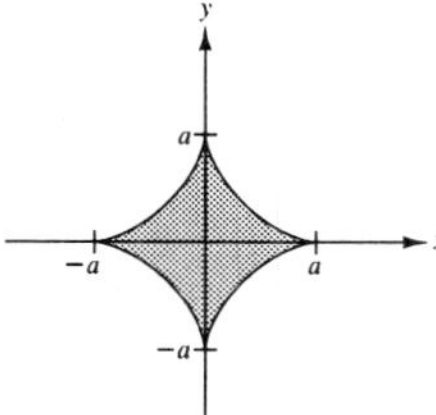

5. $\mathbf{r}(t) = 2t\mathbf{i} - 3t\mathbf{j} + t\mathbf{k}$

$$\frac{dx}{dt} = 2\ \frac{dy}{dt} = -3, \frac{dz}{dt} = 1$$

$$s = \int_0^2 \sqrt{2^2 + (-3)^2 + 1^2}\, dt$$

$$= \int_0^2 \sqrt{14}\, dt = \Big[\sqrt{14}t\Big]_0^2 = 2\sqrt{14}$$

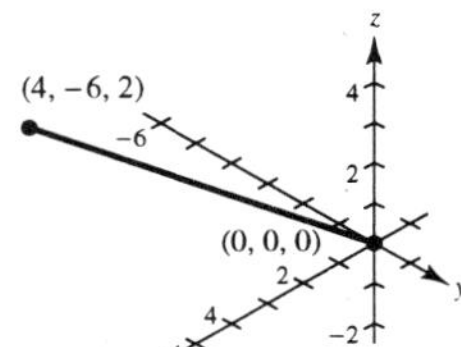

7. $\mathbf{r}(t) = a\cos t\mathbf{i} + a\sin t\mathbf{j} + bt\mathbf{k}$

$$\frac{dx}{dt} = -a\sin t, \frac{dy}{dt} = a\cos t, \frac{dz}{dt} = b$$

$$s = \int_0^{2\pi} \sqrt{a^2\sin^2 t + a^2\cos^2 t + b^2}\, dt$$

$$= \int_0^{2\pi} \sqrt{a^2 + b^2}\, dt = \Big[\sqrt{a^2 - b^2}t\Big]_0^{2\pi} = 2\pi\sqrt{a^2 + b^2}$$

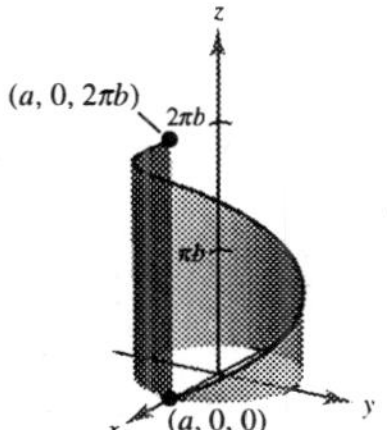

9. $\mathbf{r}(t) = t^2\mathbf{i} + t\mathbf{j} + \ln t\mathbf{k}$

$$\frac{dx}{dt} = 2t, \frac{dy}{dt} = 1, \frac{dz}{dt} = \frac{1}{t}$$

$$s = \int_1^3 \sqrt{(2t)^2 + (1)^2 + \left(\frac{1}{t}\right)^2}\, dt = \int_1^3 \sqrt{\frac{4t^4 + t^2 + 1}{t^2}}\, dt$$

$$= \int_1^3 \frac{\sqrt{4t^4 + t^2 + 1}}{t}\, dt \approx 8.37$$

11. $\mathbf{r}(t) = t\mathbf{i} + (4 - t^2)\mathbf{j} + t^3\mathbf{k}, \quad 0 \le t \le 2$

(a) $\mathbf{r}(0) = \langle 0, 4, 0\rangle, \ \mathbf{r}(2) = \langle 2, 0, 8\rangle$

$$\text{distance} = \sqrt{2^2 + 4^2 + 8^2} = \sqrt{84} = 2\sqrt{21} \approx 9.165$$

(b) $\mathbf{r}(0) = \langle 0, 4, 0\rangle$

$\mathbf{r}(0.5) = \langle 0.5, 3.75, .125\rangle$

$\mathbf{r}(1) = \langle 1, 3, 1\rangle$

$\mathbf{r}(1.5) = \langle 1.5, 1.75, 3.375\rangle$

$\mathbf{r}(2) = \langle 2, 0, 8\rangle$

$$\text{distance} \approx \sqrt{(0.5)^2 + (.25)^2 + (.125)^2} + \sqrt{(.5)^2 + (.75)^2 + (.875)^2} + \sqrt{(0.5)^2 + (1.25)^2 + (2.375)^2} + \sqrt{(0.5)^2 + (1.75)^2 + (4.625)^2}$$

$$\approx 0.5728 + 1.2562 + 2.7300 + 4.9702 \approx 9.529$$

(c) Increase the number of line segments.

(d) Using a graphing utility, you obtain 9.57057.

13. $\mathbf{r}(t) = \langle 2\cos t, 2\sin t, t\rangle$

(a)
$$s = \int_0^t \sqrt{[x'(u)]^2 + [y'(u)]^2 + [z'(u)]^2}\,du = \int_0^t \sqrt{(-2\sin u)^2 + (2\cos u)^2 + (1)^2}\,du = \int_0^t \sqrt{5}\,du = \Big[\sqrt{5}u\Big]_0^t = \sqrt{5}t$$

(b) $\dfrac{s}{\sqrt{5}} = t$

$$x = 2\cos\left(\frac{s}{\sqrt{5}}\right),\ y = 2\sin\left(\frac{s}{\sqrt{5}}\right),\ z = \frac{s}{\sqrt{5}}$$

$$\mathbf{r}(s) = 2\cos\left(\frac{s}{\sqrt{5}}\right)\mathbf{i} + 2\sin\left(\frac{s}{\sqrt{5}}\right)\mathbf{j} + \frac{s}{\sqrt{5}}\mathbf{k}$$

(c) When $s = \sqrt{5}$: $x = 2\cos 1 \approx 1.081$

$y = 2\sin 1 \approx 1.683$

$z = 1$

$(1.081, 1.683, 1.000)$

When $s = 4$: $x = 2\cos\dfrac{4}{\sqrt{5}} \approx -0.433$

$y = 2\sin\dfrac{4}{\sqrt{5}} \approx 1.953$

$z = \dfrac{4}{\sqrt{5}} \approx 1.789$

$(-0.433, 1.953, 1.789)$

(d)
$$\|\mathbf{r}'(s)\| = \sqrt{\left(-\frac{2}{\sqrt{5}}\sin\left(\frac{s}{\sqrt{5}}\right)\right)^2 + \left(\frac{2}{\sqrt{5}}\cos\left(\frac{s}{\sqrt{5}}\right)\right)^2 + \left(\frac{1}{\sqrt{5}}\right)^2} = \sqrt{\frac{4}{5} + \frac{1}{5}} = 1$$

15. $\mathbf{r}(s) = \left(1 + \dfrac{\sqrt{2}}{2}s\right)\mathbf{i} + \left(1 - \dfrac{\sqrt{2}}{2}s\right)\mathbf{j}$

$$\mathbf{r}'(s) = \frac{\sqrt{2}}{2}\mathbf{i} - \frac{\sqrt{2}}{2}\mathbf{j} \quad \text{and} \quad \|\mathbf{r}'(s)\| = \sqrt{\frac{1}{2} + \frac{1}{2}} = 1$$

$$\mathbf{T}(s) = \frac{\mathbf{r}'(s)}{\|\mathbf{r}'(s)\|} = \mathbf{r}'(s)$$

$\mathbf{T}'(s) = \mathbf{0} \Rightarrow K = \|\mathbf{T}'(s)\| = 0$ (The curve is a line.)

17. $\mathbf{r}(s) = 2\cos\left(\dfrac{s}{\sqrt{5}}\right)\mathbf{i} + 2\sin\left(\dfrac{s}{\sqrt{5}}\right)\mathbf{j} + \dfrac{s}{\sqrt{5}}\mathbf{k}$

$$\mathbf{T}(s) = \mathbf{r}'(s) = -\frac{2}{\sqrt{5}}\sin\left(\frac{s}{\sqrt{5}}\right)\mathbf{i} + \frac{2}{\sqrt{5}}\cos\left(\frac{s}{\sqrt{5}}\right)\mathbf{j} + \frac{1}{\sqrt{5}}\mathbf{k}$$

$$\mathbf{T}'(s) = -\frac{2}{5}\cos\left(\frac{s}{\sqrt{5}}\right)\mathbf{i} - \frac{2}{5}\sin\left(\frac{s}{\sqrt{5}}\right)\mathbf{j}$$

$$K = \|\mathbf{T}'(s)\| = \frac{2}{5}$$

19. $\mathbf{r}(t) = 4t\mathbf{i} - 2t\mathbf{j}$

$\mathbf{v}(t) = 4\mathbf{i} - 2\mathbf{j}$

$$\mathbf{T}(t) = \frac{1}{\sqrt{5}}(2\mathbf{i} - \mathbf{j})$$

$\mathbf{T}'(t) = 0$

$$K = \frac{\|\mathbf{T}'(t)\|}{\|\mathbf{r}'(t)\|} = 0 \quad \text{(The curve is a line.)}$$

21. $\mathbf{r}(t) = t\mathbf{i} + \frac{1}{t}\mathbf{j}$

$\mathbf{v}(t) = \mathbf{i} - \frac{1}{t^2}\mathbf{j}$

$\mathbf{v}(1) = \mathbf{i} - \mathbf{j}$

$\mathbf{a}(t) = \frac{2}{t^3}\mathbf{j}$

$\mathbf{a}(1) = 2\mathbf{j}$

$\mathbf{T}(t) = \frac{t^2\mathbf{i} - \mathbf{j}}{\sqrt{t^4 + 1}}$

$\mathbf{N}(t) = \frac{1}{(t^4 + 1)^{1/2}}(\mathbf{i} + t^2\mathbf{j})$

$\mathbf{N}(1) = \frac{1}{\sqrt{2}}(\mathbf{i} + \mathbf{j})$

$K = \frac{\mathbf{a} \cdot \mathbf{N}}{\|\mathbf{v}\|^2} = \frac{\sqrt{2}}{2}$

23. $\mathbf{r}(t) = 4\cos(2\pi t)\mathbf{i} + 4\sin(2\pi t)\mathbf{j}$

$\mathbf{r}'(t) = -8\pi\sin(2\pi t)\mathbf{i} + 8\pi\cos(2\pi t)\mathbf{j}$

$\mathbf{T}(t) = -\sin(2\pi t)\mathbf{i} + \cos(2\pi t)\mathbf{j}$

$\mathbf{T}'(t) = -2\pi\cos(2\pi t)\mathbf{i} - 2\pi\sin(2\pi t)\mathbf{j}$

$K = \frac{\|\mathbf{T}'(t)\|}{\|\mathbf{r}'(t)\|} = \frac{2\pi}{8\pi} = \frac{1}{4}$

25. $\mathbf{r}(t) = a\cos(\omega t)\mathbf{i} + a\sin(\omega t)\mathbf{j}$

$\mathbf{r}'(t) = -a\omega\sin(\omega t)\mathbf{i} + a\omega\cos(\omega t)\mathbf{j}$

$\mathbf{T}(t) = -\sin(\omega t)\mathbf{i} + \cos(\omega t)\mathbf{j}$

$\mathbf{T}'(t) = -\omega\cos(\omega t)\mathbf{i} - \omega\sin(\omega t)\mathbf{j}$

$K = \frac{\|\mathbf{T}'(t)\|}{\|\mathbf{r}'(t)\|} = \frac{\omega}{a\omega} = \frac{1}{a}$

27. $\mathbf{r}(t) = e^t\cos t\mathbf{i} + e^t\sin t\mathbf{j}$

$\mathbf{r}'(t) = (-e^t\sin t + e^t\cos t)\mathbf{i} - (e^t\cos t + e^t\sin t)\mathbf{j}$

$\mathbf{T}(t) = \frac{1}{\sqrt{2}}[(-\sin t + \cos t)\mathbf{i} + (\cos t + \sin t)\mathbf{j}]$

$\mathbf{T}'(t) = \frac{1}{\sqrt{2}}[(-\cos t - \sin t)\mathbf{i} + (-\sin t + \cos t)\mathbf{j}]$

$K = \frac{\|\mathbf{T}'(t)\|}{\|\mathbf{r}'(t)\|} = \frac{1}{\sqrt{2}e^t} = \frac{\sqrt{2}}{2}e^{-t}$

29. $\mathbf{r}(t) = \langle\cos\omega t + \omega t\sin\omega t, \sin\omega t - \omega t\cos\omega t\rangle$

From Exercise 21, Section 11.4, we have:

$\mathbf{a} \cdot \mathbf{N} = \omega^3 t$

$K = \frac{\mathbf{a}(t) \cdot \mathbf{N}(t)}{\|\mathbf{v}\|^2} = \frac{\omega^3 t}{\omega^4 t^2} = \frac{1}{\omega t}$

31. $\mathbf{r}(t) = t\mathbf{i} + t^2\mathbf{j} + \frac{t^2}{2}\mathbf{k}$

$\mathbf{r}'(t) = \mathbf{i} + 2t\mathbf{j} + t\mathbf{k}$

$\mathbf{T}(t) = \frac{\mathbf{i} + 2t\mathbf{j} + t\mathbf{k}}{\sqrt{1 + 5t^2}}$

$\mathbf{T}'(t) = \frac{-5t\mathbf{i} + 2\mathbf{j} + \mathbf{k}}{(1 + 5t^2)^{3/2}}$

$K = \frac{\|\mathbf{T}'(t)\|}{\|\mathbf{r}'(t)\|}$

$= \frac{\frac{\sqrt{5}}{(1 + 5t^2)}}{\sqrt{1 + 5t^2}} = \frac{\sqrt{5}}{(1 + 5t^2)^{3/2}}$

33. $\mathbf{r}(t) = 4t\mathbf{i} + 3\cos t\mathbf{j} + 3\sin t\mathbf{k}$

$\mathbf{r}'(t) = 4\mathbf{i} - 3\sin t\mathbf{j} + 3\cos t\mathbf{k}$

$\mathbf{T}(t) = \frac{1}{5}[4\mathbf{i} - 3\sin t\mathbf{j} + 3\cos t\mathbf{k}]$

$\mathbf{T}'(t) = \frac{1}{5}[-3\cos t\mathbf{j} - 3\sin t\mathbf{k}]$

$K = \frac{\|\mathbf{T}'(t)\|}{\|\mathbf{r}'(t)\|} = \frac{3/5}{5} = \frac{3}{25}$

35. $y = 3x - 2$

Since $y'' = 0$, $K = 0$, and the radius of curvature is undefined.

37. $y = 2x^2 + 3$

$y' = 4x$

$y'' = 4$

$$K = \frac{4}{[1 + (-4)^2]^{3/2}} = \frac{4}{17^{3/2}} \approx 0.057$$

$$\frac{1}{K} = \frac{17^{3/2}}{4} \approx 17.523 \quad \text{(radius of curvature)}$$

39. $y = \sqrt{a^2 - x^2}$

$$y' = \frac{-x}{\sqrt{a^2 - x^2}}$$

$$y'' = \frac{-(2x^2 - a^2)}{(a^2 - x^2)^{3/2}}$$

At $x = 0$: $y' = 0$

$$y'' = \frac{1}{a}$$

$$K = \frac{1/a}{(1 + 0^2)^{3/2}} = \frac{1}{a}$$

$$\frac{1}{K} = a \quad \text{(radius of curvature)}$$

41. (a) Point on circle: $\left(\frac{\pi}{2}, 1\right)$

Center: $\left(\frac{\pi}{2}, 0\right)$

Equation: $\left(x - \frac{\pi}{2}\right)^2 + y^2 = 1$

(b) The circles have different radii since the curvature is different and

$$r = \frac{1}{K}.$$

43. $y = x + \dfrac{1}{x}$

From Exercise 38, $r = \frac{1}{2}$ when $x = 1$. Since the tangent line is horizontal at $(1, 2)$, the normal line is vertical. The center of the circle is $\frac{1}{2}$ unit above the point $(1, 2)$ at $\left(1, \frac{5}{2}\right)$.

Equation of circle: $(x - 1)^2 + \left(y - \frac{5}{2}\right)^2 = \frac{1}{4}$

45. $y = e^x, \quad x = 0$

$y' = e^x, \quad y'' = e^x$

$y'(0) = 1, \quad y''(0) = 1$

$$K = \frac{1}{(1 + 1^2)^{3/2}} = \frac{1}{2^{3/2}} = \frac{1}{2\sqrt{2}}, \quad r = \frac{1}{K} = 2\sqrt{2}$$

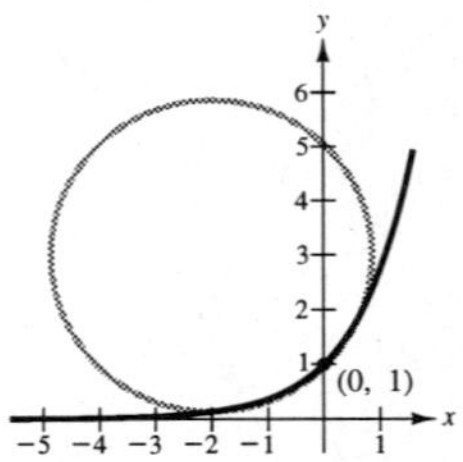

The slope of the tangent line at $(0, 1)$ is $y'(0) = 1$.

The slope of the normal line is -1.

Equation of normal line: $y - 1 = -x$ or $y = -x + 1$

The center of the circle is on the normal line $2\sqrt{2}$ units away from the point $(0, 1)$.

$$\sqrt{(0 - x)^2 + (1 - y)^2} = 2\sqrt{2}$$

$$x^2 + x^2 = 8$$

$$x^2 = 4$$

$$x = \pm 2$$

Since the circle is above the curve, $x = -2$ and $y = 3$.

Center of circle: $(-2, 3)$

Equation of circle: $(x + 2)^2 + (y - 3)^2 = 8$

47. $y = (x - 1)^2 + 3,\ y' = 2(x - 1),\ y'' = 2$

$$K = \frac{2}{(1 + [2(x - 1)]^2)^{3/2}} = \frac{2}{[1 + 4(x - 1)^2]^{3/2}}$$

(a) K is maximum when $x = 1$ or at the vertex $(1, 3)$.

(b) $\lim_{x\to\infty} K = 0$

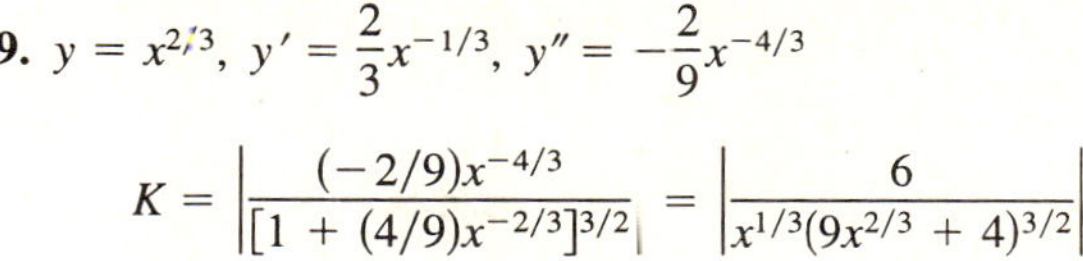

49. $y = x^{2/3},\ y' = \frac{2}{3}x^{-1/3},\ y'' = -\frac{2}{9}x^{-4/3}$

$$K = \left|\frac{(-2/9)x^{-4/3}}{[1 + (4/9)x^{-2/3}]^{3/2}}\right| = \left|\frac{6}{x^{1/3}(9x^{2/3} + 4)^{3/2}}\right|$$

(a) $K \Rightarrow \infty$ as $x \Rightarrow 0$. No maximum

(b) $\lim_{x\to\infty} K = 0$

51. $y = (x - 1)^3 + 3,\ y' = 3(x - 1)^2,\ y'' = 6(x - 1)$

$$K = \left|\frac{6(x - 1)}{[1 + 9(x - 1)^4]^{3/2}}\right|$$

$K = 0$ when $x = 1$.

Point: $(1, 3)$

53. Endpoints of the major axis: $(\pm 2, 0)$

Endpoints of the minor axis: $(0, \pm 1)$

$$x^2 + 4y^2 = 4$$

$$2x + 8yy' = 0$$

$$y' = -\frac{x}{4y}$$

$$y'' = \frac{(4y)(-1) - (-x)(4y')}{16y^2} = \frac{-4y - (x^2/y)}{16y^2} = \frac{-(4y^2 + x^2)}{16y^3} = \frac{-1}{4y^3}$$

$$K = \frac{|-1/4y^3|}{[1 + (-x/4y)^2]^{3/2}} = \frac{|-16|}{(16y^2 + x^2)^{3/2}} = \frac{16}{(12y^2 + 4)^{3/2}} = \frac{16}{(16 - 3x^2)^{3/2}}$$

Therefore, since $-2 \le x \le 2$, K is largest when $x = \pm 2$ and smallest when $x = 0$.

55. $f(x) = x^4 - x^2$

(a) $K = \dfrac{2|6x^2 - 1|}{[16x^6 - 16x^4 + 4x^2 + 1]^{3/2}}$

(b) For $x = 0$, $K = 2$. $f(0) = 0$. At $(0, 0)$, the circle of curvature has radius $\frac{1}{2}$. Using the symmetry of the graph of f, you obtain

$$x^2 + \left(y + \frac{1}{2}\right)^2 = \frac{1}{4}.$$

For $x = 1$, $K = (2\sqrt{5})/5$. $f(1) = 0$. At $(1, 0)$, the circle of curvature has radius

$$\frac{\sqrt{5}}{2} = \frac{1}{K}.$$

Using the graph of f, you see that the center of curvature is $\left(0, \frac{1}{2}\right)$. Thus,

$$x^2 + \left(y - \frac{1}{2}\right)^2 = \frac{5}{4}.$$

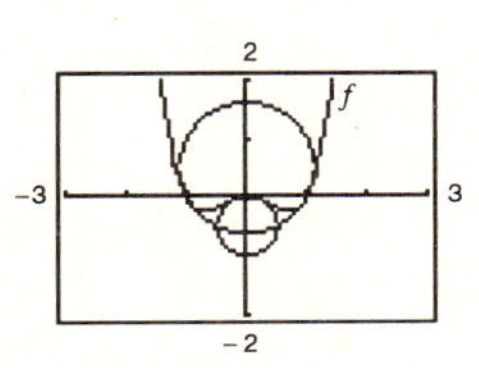

To graph these circles, use

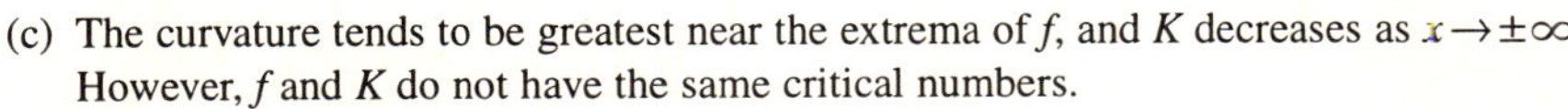

$$y = -\frac{1}{2} \pm \sqrt{\frac{1}{4} - x^2} \quad \text{and} \quad y = \frac{1}{2} \pm \sqrt{\frac{5}{4} - x^2}.$$

(c) The curvature tends to be greatest near the extrema of f, and K decreases as $x \to \pm\infty$. However, f and K do not have the same critical numbers.

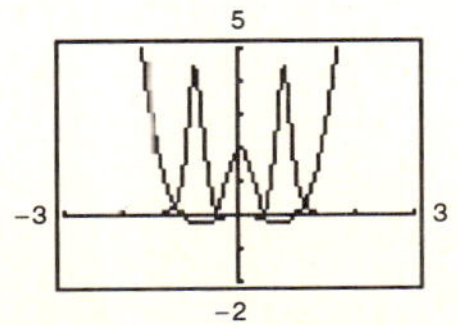

Critical numbers of f: $x = 0, \pm\frac{\sqrt{2}}{2} \approx \pm 0.7071$

Critical numbers of K: $x = 0, \pm.7647, \pm 0.4082$

57. (a) Imagine dropping the circle $x^2 + (y - k)^2 = 16$ into the parabola $y = x^2$. The circle will drop to the point where the tangents to the circle and parabola are equal.

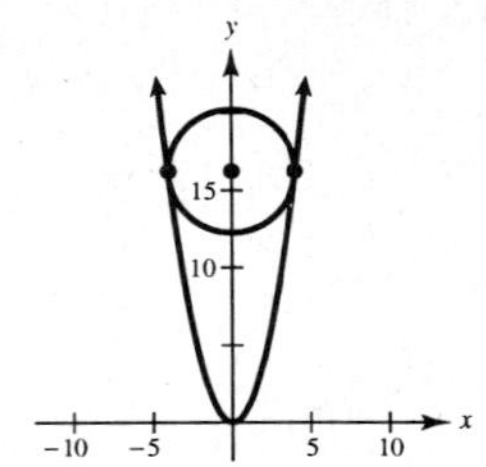

$$y = x^2 \quad \text{and} \quad x^2 + (y - k)^2 = 16 \Rightarrow x^2 + (x^2 - k)^2 = 16$$

Taking derivatives, $2x + 2(y - k)y' = 0$ and $y' = 2x$. Hence,

$$(y - k)y' = -x \Rightarrow y' = \frac{-x}{y - k}.$$

Thus,

$$\frac{-x}{y - k} = 2x \Rightarrow -x = 2x(y - k) \Rightarrow -1 = 2(x^2 - k) \Rightarrow x^2 - k = -\frac{1}{2}.$$

Thus,

$$x^2 + (x^2 - k)^2 = x^2 + \left(-\frac{1}{2}\right)^2 = 16 \Rightarrow x^2 = 15.75.$$

Finally, $k = x^2 + \frac{1}{2} = 16.25$, and the center of the circle is 16.25 units from the vertex of the parabola. Since the radius of the circle is 4, the circle is 12.25 units from the vertex.

(b) In 2-space, the parabola $z = y^2$ (or $z = x^2$) has a curvature of $K = 2$ at $(0, 0)$. The radius of the largest sphere that will touch the vertex has radius $= 1/K = \frac{1}{2}$.

59. $K = \dfrac{|y''|}{[1 + (y')^2]^{3/2}}$

At the smooth relative extremum $y' = 0$, so $K = |y''|$. Yes, for example, $y = x^4$ has a curvature of 0 at its relative minimum $(0, 0)$. The curvature is positive for any other point of the curvature.

61. Given $y = f(x)$: $K = \dfrac{|y''|}{(1 + [y']^2)^{3/2}}$

$$R = \frac{1}{K}$$

The center of the circle is on the normal line at a distance of R from (x, y).

Equation of normal line: $y - y_0 = -\dfrac{1}{y'}(x - x_0)$

$$\sqrt{(x - x_0)^2 + \left[-\frac{1}{y'}(x - x_0)\right]^2} = \frac{(1 + [y']^2)^{3/2}}{|y''|}$$

$$(x - x_0)^2\left[1 + \frac{1}{(y')^2}\right] = \frac{(1 + [y']^2)^3}{(y'')^2}$$

$$(x - x_0)^2 = \frac{(y')^2(1 + [y']^2)^2}{(y'')^2}$$

$$x - x_0 = \frac{y'(1 + [y']^2)}{y''} = y'z$$

$$x_0 = x - y'z$$

$$y - y_0 = -\frac{1}{y'}(x - (x - y'z)) = -z$$

$$y_0 = y + z$$

Thus, $(x_0, y_0) = (x - y'z, y + z)$.

For $y = e^x$, $y' = e^x$, $y'' = e^x$, $z = \dfrac{1 + e^{2x}}{e^x} = e^{-x} + e^x$.

When $x = 0$: $x_0 = x - y'z = 0 - (1)(2) = -2$

$y_0 = y + z = 1 + 2 = 3$

Center of curvature: $(-2, 3)$

(See Exercise 45)

63. $r = 1 + \sin\theta$

$r' = \cos\theta$

$r'' = -\sin\theta$

$$K = \frac{|2(r')^2 - rr'' + r^2|}{[(r')^2 + r^2]^{3/2}}$$

$$= \frac{|2\cos^2\theta - (1 + \sin\theta)(-\sin\theta) + (1 + \sin\theta)^2|}{\sqrt{[\cos^2\theta + (1 + \sin\theta)^2]^3}}$$

$$= \frac{3(1 + \sin\theta)}{\sqrt{8(1 + \sin\theta)^3}} = \frac{3}{2\sqrt{2(1 + \sin\theta)}}$$

65. $r = a\sin\theta$

$r' = a\cos\theta$

$r'' = -a\sin\theta$

$$K = \frac{|2(r\omega)^2 - rr'' + r^2|}{[(r')^2 + r^2]^{3/2}}$$

$$= \frac{|2a^2\cos^2\theta + a^2\sin^2\theta + a^2\sin^2\theta|}{\sqrt{[a^2\cos^2\theta + a^2\sin^2\theta]^3}}$$

$$= \frac{2a^2}{a^3} = \frac{2}{a}, a > 0$$

67. $r = e^{a\theta}, a > 0$

$r' = ae^{a\theta}$

$r'' = a^2e^{a\theta}$

$$K = \frac{|2(r')^2 - rr'' + r^2|}{[(r')^2 + r^2]^{3/2}} = \frac{|2a^2e^{2a\theta} - a^2e^{2a\theta} + e^{2a\theta}|}{[a^2e^{2a\theta} + e^{2a\theta}]^{3/2}} = \frac{1}{e^{a\theta}\sqrt{a^2 + 1}}$$

(a) As $\theta \Rightarrow \infty$, $K \Rightarrow 0$.

(b) As $a \Rightarrow \infty$, $K \Rightarrow 0$.

69. $r = 4\sin 2\theta$

$r' = 8\cos 2\theta$

At the pole: $K = \dfrac{2}{|r'(0)|} = \dfrac{2}{8} = \dfrac{1}{4}$

71. $x = f(t)$

$y = g(t)$

$$y' = \frac{dy}{dx} = \frac{\dfrac{dy}{dt}}{\dfrac{dx}{dt}} = \frac{g'(t)}{f'(t)}$$

$$y'' = \frac{\dfrac{d}{dt}\left[\dfrac{g'(t)}{f'(t)}\right]}{\dfrac{dx}{dt}} = \frac{\dfrac{f'(t)g''(t) - g'(t)f''(t)}{[f'(t)]^2}}{f'(t)} = \frac{f'(t)g''(t) - g'(t)f''(t)}{[f'(t)]^3}$$

$$K = \frac{|y''|}{[1 + (y')^2]^{3/2}} = \frac{\left|\dfrac{f'(t)g''(t) - g'(t)f''(t)}{[f'(t)]^3}\right|}{\left[1 + \left(\dfrac{g'(t)}{f'(t)}\right)^2\right]^{3/2}} = \frac{\left|\dfrac{f'(t)g''(t) - g'(t)f''(t)}{[f'(t)]^3}\right|}{\sqrt{\left\{\dfrac{[f'(t)]^2 + [g'(t)]^2}{[f'(t)]^2}\right\}^3}} = \frac{|f'(t)g''(t) - g'(t)f''(t)|}{([f'(t)]^2 + [g'(t)]^2)^{3/2}}$$

73. $x(\theta) = a(\theta - \sin\theta)$ $\quad y(\theta) = a(1 - \cos\theta)$

$x'(\theta) = a(1 - \cos\theta)$ $\quad y'(\theta) = a\sin\theta$

$x''(\theta) = a\sin\theta$ $\quad y''(\theta) = a\cos\theta$

$$K = \frac{|x'(\theta)y''(\theta) - y'(\theta)x''(\theta)|}{[x'(\theta)^2 + y'(\theta)^2]^{3/2}} = \frac{|a^2(1 - \cos\theta)\cos\theta - a^2\sin^2\theta|}{[a^2(1 - \cos\theta)^2 + a^2\sin^2\theta]^{3/2}} = \frac{1}{a}\frac{|\cos\theta - 1|}{[2 - 2\cos\theta]^{3/2}}$$

$$= \frac{1}{a}\frac{1 - \cos\theta}{2\sqrt{2}[1 - \cos\theta]^{3/2}} \quad (1 - \cos \geq 0)$$

$$= \frac{1}{2a\sqrt{2 - 2\cos\theta}} = \frac{1}{4a}\csc\left(\frac{\theta}{2}\right)$$

Minimum: $\dfrac{1}{4a}$ $\qquad (\theta = \pi)$

Maximum: none $\qquad (K \rightarrow \infty \text{ as } \theta \rightarrow 0)$

75. $a_{\mathbf{N}} = mK\left(\dfrac{ds}{dt}\right)^2 = \left(\dfrac{4000 \text{ lb}}{32 \text{ ft/sec}^2}\right)\left(\dfrac{1}{100 \text{ ft}}\right)\left(\dfrac{30(5280) \text{ ft}}{3600 \text{ sec}}\right)^2 = 2420 \text{ lb}$

77. Let $\mathbf{r} = x(t)\mathbf{i} + y(t)\mathbf{j} + z(t)\mathbf{k}$. Then $r = \|\mathbf{r}\| = \sqrt{[x(t)]^2 + [y(t)]^2 + [z(t)]^2}$ and $r' = x'(t)\mathbf{i} + y'(t)\mathbf{j} + z'(t)\mathbf{k}$. Then,

$$r\left(\frac{dr}{dt}\right) = \sqrt{[x(t)]^2 + [y(t)]^2 + [z(t)]^2}\left[\frac{1}{2}\{[x(t)]^2 + [y(t)]^2 + [z(t)]^2\}^{-1/2} \cdot (2x(t)x'(t) + 2y(t)y'(t) + 2z(t)z'(t))\right]$$

$$= x(t)x'(t) + y(t)y'(t) + z(t)z'(t) = \mathbf{r} \cdot \mathbf{r}'.$$

79. Let $\mathbf{r} = x\mathbf{i} + y\mathbf{j} + z\mathbf{k}$ where x, y, and z are functions of t, and $r = \|\mathbf{r}\|$.

$$\frac{d}{dt}\left[\frac{\mathbf{r}}{r}\right] = \frac{r\mathbf{r}' - \mathbf{r}(dr/dt)}{r^2} = \frac{r\mathbf{r}' - \mathbf{r}[(\mathbf{r} \cdot \mathbf{r}')/r]}{r^2} = \frac{r^2\mathbf{r}' - (\mathbf{r} \cdot \mathbf{r}')\mathbf{r}}{r^3} \quad \text{(using Exercise 77)}$$

$$= \frac{(x^2 + y^2 + z^2)(x'\mathbf{i} + y'\mathbf{j} + z'\mathbf{k}) - (xx' + yy' + zz')(x\mathbf{i} + y\mathbf{j} + z\mathbf{k})}{r^3}$$

$$= \frac{1}{r^3}[(x'y^2 + x'z^2 - xyy' - xzz')\mathbf{i} + (x^2y' + z^2y' - xx'y - zz'y)\mathbf{j} + (x^2z' + y^2z' - xx'z - yy'z)\mathbf{k}]$$

$$= \frac{1}{r^3}\begin{vmatrix} \mathbf{i} & \mathbf{j} & \mathbf{k} \\ yz' - y'z & -(xz' - x'z) & xy' - x'y \\ x & y & z \end{vmatrix} = \frac{1}{r^3}\{[\mathbf{r} \times \mathbf{r}'] \times \mathbf{r}\}$$

81. From Exercise 78, we have concluded that planetary motion is planar. Assume that the planet moves in the xy-plane with the sun at the origin. From Exercise 80, we have

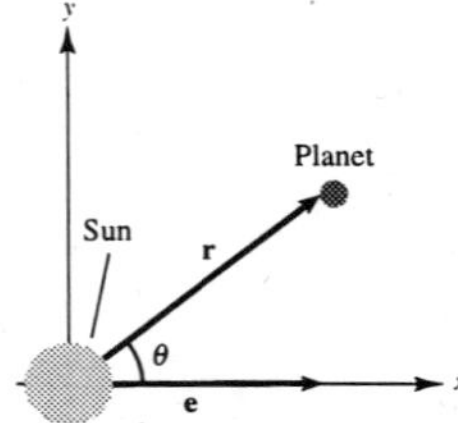

$$\mathbf{r}' \times \mathbf{L} = GM\left(\frac{\mathbf{r}}{r} + \mathbf{e}\right).$$

Since $\mathbf{r}' \times \mathbf{L}$ and $\mathbf{r}$ are both perpendicular to $\mathbf{L}$, so is $\mathbf{e}$. Thus, $\mathbf{e}$ lies in the xy-plane. Situate the coordinate system so that $\mathbf{e}$ lies along the positive x-axis and θ is the angle between $\mathbf{e}$ and $\mathbf{r}$. Let $e = \|\mathbf{e}\|$. Then $\mathbf{r} \cdot \mathbf{e} = \|\mathbf{r}\|\,\|\mathbf{e}\| \cos\theta = re\cos\theta$. Also,

$$\|\mathbf{L}\|^2 = \mathbf{L} \cdot \mathbf{L} = (\mathbf{r} \times \mathbf{r}') \cdot \mathbf{L}$$

$$= \mathbf{r} \cdot (\mathbf{r}' \times \mathbf{L}) = \mathbf{r} \cdot \left[GM\left(\mathbf{e} + \frac{\mathbf{r}}{r}\right)\right] = GM\left[\mathbf{r} \cdot \mathbf{e} + \frac{\mathbf{r} \cdot \mathbf{r}}{r}\right] = GM[re\cos\theta + r]$$

Thus,

$$\frac{\|\mathbf{L}\|^2/GM}{1 + e\cos\theta} = r$$

and the planetary motion is a conic section. Since the planet returns to its initial position periodically, the conic is an ellipse.

83. $A = \dfrac{1}{2}\displaystyle\int_\alpha^\beta r^2\, d\theta$

Thus,

$$\frac{dA}{dt} = \frac{dA}{d\theta}\,\frac{d\theta}{dt} = \frac{1}{2}r^2\frac{d\theta}{dt} = \frac{1}{2}\|\mathbf{L}\|$$

and $\mathbf{r}$ sweeps out area at a constant rate.

85. s is the arc length parameter.

(a) $s = \displaystyle\int_0^a \sqrt{\left(\cos\frac{\pi s^2}{2}\right)^2 + \left(\sin\frac{\pi s^2}{2}\right)^2}\, ds = a$

(b) $\mathbf{r}'(s) = \left(\cos\dfrac{\pi s^2}{2}\right)\mathbf{i} + \left(\sin\dfrac{\pi s^2}{2}\right)\mathbf{j}$

$\mathbf{r}''(s) = -s\pi\left(\sin\dfrac{\pi s^2}{2}\right)\mathbf{i} + s\pi\left(\cos\dfrac{\pi s^2}{2}\right)\mathbf{j}$

$K = \|\mathbf{r}''(s)\| = \pi s$

When $s = a$: $K = \pi a$

(c) $K = \pi a$

Review Exercises for Chapter 11

1. $\mathbf{r}(t) = t\mathbf{i} + \csc t\mathbf{k}$

(a) Domain: $t \neq n\pi$, n an integer

(b) Continuous except at $t = n\pi$, n an integer

3. $\mathbf{r}(t) = \ln t\mathbf{i} + t\mathbf{j} + t\mathbf{k}$

(a) Domain: $(0, \infty)$

(b) Continuous for all $t > 0$

5. (a) $\mathbf{r}(0) = \mathbf{i}$

(b) $\mathbf{r}(-2) = -3\mathbf{i} + 4\mathbf{j} + \frac{8}{3}\mathbf{k}$

(c) $\mathbf{r}(c-1) = (2(c-1)+1)\mathbf{i} + (c-1)^2\mathbf{j} - \frac{1}{3}(c-1)^3\mathbf{k} = (2c-1)\mathbf{i} + (c-1)^2\mathbf{j} - \frac{1}{3}(c-1)^3\mathbf{k}$

(d) $\mathbf{r}(1+\Delta t) - \mathbf{r}(1) = ([2(1+\Delta t)+1]\mathbf{i} + [1+\Delta t]^2\mathbf{j} - \frac{1}{3}[1+\Delta t]^3\mathbf{k}) - (3\mathbf{i} + \mathbf{j} - \frac{1}{3}\mathbf{k})$

$= 2\Delta t\mathbf{i} + \Delta t(\Delta t + 2)\mathbf{j} - \frac{1}{3}(\Delta t^3 + 3\Delta t^2 + 3)\mathbf{k}$

7. $\mathbf{r}(t) \cdot \mathbf{u}(t) = [\tan t\mathbf{i} - \sec t\mathbf{j} + (1-t)\mathbf{k}] \cdot [2\cos t\mathbf{i} + 3\sin t\cos t\mathbf{j} + \mathbf{k}]$

$= 2\tan t\cos t - 3\sec t\sin t\cos t + (1-t) = 2\sin t - 3\sin t + 1 - t = 1 - t - \sin t$

No, the dot product is a scalar, not a vector.

9. $\mathbf{r}(t) = \cos t\mathbf{i} + 2\sin^2 t\mathbf{j}$

$x(t) = \cos t,\ y(t) = 2\sin^2 t$

$x^2 + \frac{y}{2} = 1$

$y = 2(1 - x^2)$

$-1 \le x \le 1$

11. $\mathbf{r}(t) = \mathbf{i} + t\mathbf{j} + t^2\mathbf{k}$

$x = 1$

$y = t$

$z = t^2 \Rightarrow z = y^2$

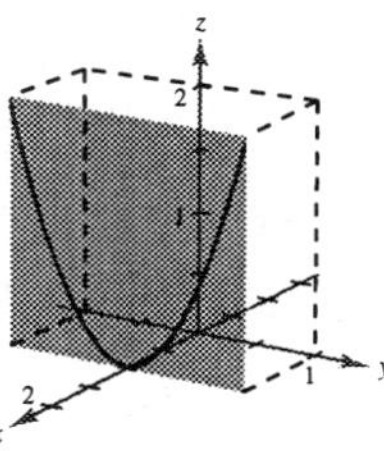

13. $\mathbf{r}(t) = \mathbf{i} + \sin t\mathbf{j} + \mathbf{k}$

$x = 1,\ y = \sin t,\ z = 1$

t	0	$\frac{\pi}{2}$	π	$\frac{3\pi}{2}$
x	1	1	1	1
y	0	1	0	-1
z	1	1	1	1

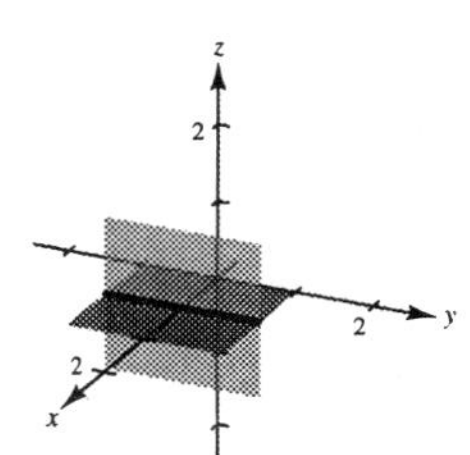

15. $\mathbf{r}(t) = t\mathbf{i} + \ln t\mathbf{j} + \frac{1}{2}t^2\mathbf{k}$

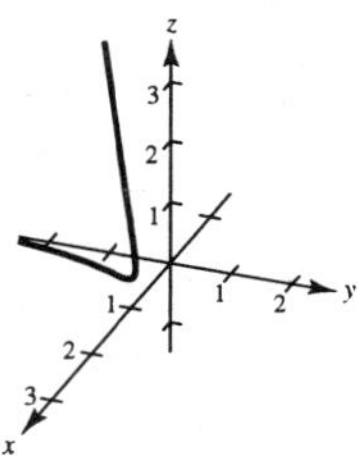

17. One possible answer is:

$\mathbf{r}_1(t) = 4t\mathbf{i} + 3t\mathbf{j}, \quad 0 \le t \le 1$

$\mathbf{r}_2(t) = 4\mathbf{i} + (3-t)\mathbf{j}, \quad 0 \le t \le 3$

$\mathbf{r}_3(t) = (4-t)\mathbf{i}, \quad 0 \le t \le 4$

19. The vector joining the points is $\langle 7, 4, -10\rangle$. One path is

$\mathbf{r}(t) = \langle -2 + 7t, -3 + 4t, 8 - 10t\rangle$.

21. $z = x^2 + y^2,\ x + y = 0,\ t = x$

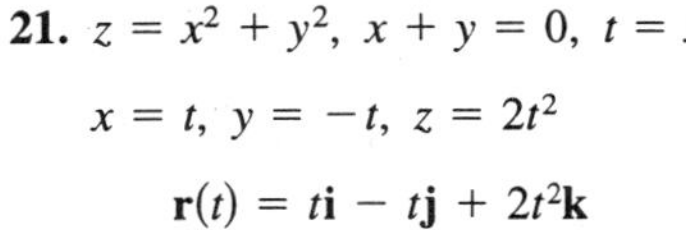

$x = t,\ y = -t,\ z = 2t^2$

$\mathbf{r}(t) = t\mathbf{i} - t\mathbf{j} + 2t^2\mathbf{k}$

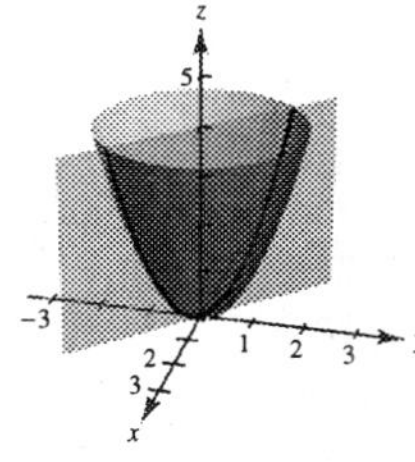

23. $\lim_{t\to 2^-} (t^2\mathbf{i} + \sqrt{4 - t^2}\mathbf{j} + \mathbf{k}) = 4\mathbf{i} + \mathbf{k}$

25. $\mathbf{r}(t) = 3t\mathbf{i} + (t - 1)\mathbf{j},\ \mathbf{u}(t) = t\mathbf{i} + t^2\mathbf{j} + \frac{2}{3}t^3\mathbf{k}$

(a) $\mathbf{r}'(t) = 3\mathbf{i} + \mathbf{j}$

(b) $\mathbf{r}''(t) = \mathbf{0}$

(c) $\mathbf{r}(t) \cdot \mathbf{u}(t) = 3t^2 + t^2(t - 1) = t^3 + 2t^2$

$D_t[\mathbf{r}(t) \cdot \mathbf{u}(t)] = 3t^2 + 4t$

(d) $\mathbf{u}(t) - 2\mathbf{r}(t) = -5t\mathbf{i} + (t^2 - 2t + 2)\mathbf{j} + \frac{2}{3}t^3\mathbf{k}$

$D_t[\mathbf{u}(t) - 2\mathbf{r}(t)] = -5\mathbf{i} + (2t - 2)\mathbf{j} + 2t^2\mathbf{k}$

(e) $\|\mathbf{r}(t)\| = \sqrt{10t^2 - 2t + 1}$

$D_t[\|\mathbf{r}(t)\|] = \dfrac{10t - 1}{\sqrt{10t^2 - 2t + 1}}$

(f) $\mathbf{r}(t) \times \mathbf{u}(t) = \frac{2}{3}(t^4 - t^3)\mathbf{i} - 2t^4\mathbf{j} + (3t^3 - t^2 + t)\mathbf{k}$

$D_t[\mathbf{r}(t) \times \mathbf{u}(t)] = \left(\frac{8}{3}t^3 - 2t^2\right)\mathbf{i} - 8t^3\mathbf{j} + (9t^2 - 2t + 1)\mathbf{k}$

27. $\mathbf{r}(t) = 2\cos t\mathbf{i} + 2\sin t\mathbf{j} + t\mathbf{k},\ x = 2\cos t,\ y = 2\sin t,\ z = t$

When $t = \frac{3\pi}{4},\ x = -\sqrt{2},\ y = \sqrt{2},\ z = \frac{3\pi}{4}.$

$\mathbf{r}'(t) = -2\sin t\mathbf{i} + 2\cos t\mathbf{j} + \mathbf{k}$

Direction numbers when $t = \frac{3\pi}{4},\ a = -\sqrt{2},\ b = -\sqrt{2},\ c = 1$

$x = -\sqrt{2}t - \sqrt{2},\ y = -\sqrt{2}t + \sqrt{2},\ z = t + \frac{3\pi}{4}$

29. $x(t)$ and $y(t)$ are increasing functions at $t = t_0$, and $z(t)$ is a decreasing function at $t = t_0$.

31. $\mathbf{r}(t) = \left\langle \ln(t - 3), t^2, \frac{1}{2}t \right\rangle,\ t_0 = 4$

$\mathbf{r}'(t) = \left\langle \frac{1}{t - 3}, 2t, \frac{1}{2} \right\rangle$

$\mathbf{r}'(4) = \left\langle 1, 8, \frac{1}{2} \right\rangle$ direction numbers

Since $\mathbf{r}(4) = \langle 0, 16, 2 \rangle$, the parametric equations are $x = t,\ y = 16 + 8t,\ z = 2 + \frac{1}{2}t.$

$\mathbf{r}(t_0 + 0.1) = \mathbf{r}(4.1) \approx \langle 0.1, 16.8, 2.05 \rangle$

33. $\int (\cos t\mathbf{i} + t\cos t\mathbf{j})\,dt = \sin t\mathbf{i} + (t\sin t + \cos t)\mathbf{j} + \mathbf{C}$

35. $\int \|\cos t\mathbf{i} + \sin t\mathbf{j} + t\mathbf{k}\|\,dt = \int \sqrt{1 + t^2}\,dt = \frac{1}{2}[t\sqrt{1 + t^2} + \ln|t + \sqrt{1 + t^2}|] + \mathbf{C}$

37. $\mathbf{r}(t) = \int (2t\mathbf{i} + e^t\mathbf{j} + e^{-t}\mathbf{k})\,dt = t^2\mathbf{i} + e^t\mathbf{j} - e^{-t}\mathbf{k} + \mathbf{C}$

$\mathbf{r}(0) = \mathbf{j} - \mathbf{k} + \mathbf{C} = \mathbf{i} + 3\mathbf{j} - 5\mathbf{k} \Rightarrow \mathbf{C} = \mathbf{i} + 2\mathbf{j} - 4\mathbf{k}$

$\mathbf{r}(t) = (t^2 + 1)\mathbf{i} + (e^t + 2)\mathbf{j} - (e^{-t} + 4)\mathbf{k}$

39. $\int_{-2}^{2} (3t\mathbf{i} + 2t^2\mathbf{j} - t^3\mathbf{k})\,dt = \left[\frac{3t^2}{2}\mathbf{i} + \frac{2t^3}{3}\mathbf{j} - \frac{t^4}{4}\mathbf{k}\right]_{-2}^{2} = \frac{32}{3}\mathbf{j}$

41. $\mathbf{r}(t) = \langle \cos^3 t, \sin^3 t, 3t \rangle$

$\mathbf{v}(t) = \mathbf{r}'(t) = \langle -3\cos^2 t \sin t, 3\sin^2 t \cos t, 3 \rangle$

$\|\mathbf{v}(t)\| = \sqrt{9\cos^4 t \sin^2 t + 9\sin^4 t \cos^2 t + 9}$

$= 3\sqrt{\cos^2 t \sin^2 t(\cos^2 t + \sin^2 t) + 1}$

$= 3\sqrt{\cos^2 t \sin^2 t + 1}$

$\mathbf{a}(t) = \mathbf{v}'(t) = \langle -6\cos t(-\sin^2 t) + (-3\cos^2 t)\cos t, 6\sin t\cos^2 t + 3\sin^2 t(-\sin t), 0 \rangle$

$= \langle 3\cos t(2\sin^2 t - \cos^2 t),\ 3\sin t(2\cos^2 t - \sin^2 t), 0 \rangle$

43. Range $= x = \dfrac{v_0^2}{32}\sin 2\theta = \dfrac{(75)^2}{32}\sin 60° \approx 152$ feet

(See Exercise 31, Section 11.3)

45. Range $= x = \dfrac{v_0^2}{9.8}\sin 2\theta = 80$

$\Rightarrow v_0 = \sqrt{\dfrac{(80)(9.8)}{\sin 40°}} \approx 34.9$ m/sec

47. $\mathbf{r}(t) = 5t\mathbf{i}$

$\mathbf{v}(t) = 5\mathbf{i}$

$\|\mathbf{v}(t)\| = 5$

$\mathbf{a}(t) = \mathbf{0}$

$\mathbf{T}(t) = \mathbf{i}$

$\mathbf{N}(t)$ does not exist

$\mathbf{a} \cdot \mathbf{T} = 0$

$\mathbf{a} \cdot \mathbf{N}$ does not exist

$K = 0$

(The curve is a line.)

49. $\mathbf{r}(t) = t\mathbf{i} + \sqrt{t}\mathbf{j}$

$\mathbf{v}(t) = \mathbf{i} + \dfrac{1}{2\sqrt{t}}\mathbf{j}$

$\|\mathbf{v}(t)\| = \dfrac{\sqrt{4t+1}}{2\sqrt{t}}$

$\mathbf{a}(t) = -\dfrac{1}{4t\sqrt{t}}\mathbf{j}$

$\mathbf{T}(t) = \dfrac{\mathbf{i} + \left(1/2\sqrt{t}\right)\mathbf{j}}{\left(\sqrt{4t+1}\right)/2\sqrt{t}} = \dfrac{2\sqrt{t}\mathbf{i} + \mathbf{j}}{\sqrt{4t+1}}$

$\mathbf{N}(t) = \dfrac{\mathbf{i} - 2\sqrt{t}\mathbf{j}}{\sqrt{4t+1}}$

$\mathbf{a} \cdot \mathbf{T} = \dfrac{-1}{4t\sqrt{t}\sqrt{4t+1}}$

$\mathbf{a} \cdot \mathbf{N} = \dfrac{1}{2t\sqrt{4t+1}}$

$K = \dfrac{2}{(4t+1)^{3/2}}$

51. $\mathbf{r}(t) = e^t\mathbf{i} + e^{-t}\mathbf{j}$

$\mathbf{v}(t) = e^t\mathbf{i} - e^{-t}\mathbf{j}$

$\|\mathbf{v}(t)\| = \sqrt{e^{2t} + e^{-2t}}$

$\mathbf{a}(t) = e^t\mathbf{i} + e^{-t}\mathbf{j}$

$\mathbf{T}(t) = \dfrac{e^t\mathbf{i} - e^{-t}\mathbf{j}}{\sqrt{e^{2t} + e^{-2t}}}$

$\mathbf{N}(t) = \dfrac{e^{-t}\mathbf{i} + e^t\mathbf{j}}{\sqrt{e^{2t} + e^{-2t}}}$

$\mathbf{a} \cdot \mathbf{T} = \dfrac{e^{2t} - e^{-2t}}{\sqrt{e^{2t} + e^{-2t}}}$

$\mathbf{a} \cdot \mathbf{N} = \dfrac{2}{\sqrt{e^{2t} + e^{-2t}}}$

$K = \dfrac{\mathbf{a} \cdot \mathbf{N}}{\|\mathbf{v}\|^2} = \dfrac{2}{(e^{2t} + e^{-2t})^{3/2}}$

53. $\mathbf{r}(t) = t\mathbf{i} + t^2\mathbf{j} + \dfrac{1}{2}t^2\mathbf{k}$

$\mathbf{v}(t) = \mathbf{i} + 2t\mathbf{j} + t\mathbf{k}$

$\|\mathbf{v}\| = \sqrt{1 + 5t^2}$

$\mathbf{a}(t) = 2\mathbf{j} + \mathbf{k}$

$\mathbf{T}(t) = \dfrac{\mathbf{i} + 2t\mathbf{j} + t\mathbf{k}}{\sqrt{1 + 5t^2}}$

$\mathbf{N}(t) = \dfrac{-5t\mathbf{i} + 2\mathbf{j} + \mathbf{k}}{\sqrt{5}\sqrt{1 + 5t^2}}$

$\mathbf{a} \cdot \mathbf{T} = \dfrac{5t}{\sqrt{1 + 5t^2}}$

$\mathbf{a} \cdot \mathbf{N} = \dfrac{5}{\sqrt{5}\sqrt{1 + 5t^2}} = \dfrac{\sqrt{5}}{\sqrt{1 + 5t^2}}$

$K = \dfrac{\mathbf{a} \cdot \mathbf{N}}{\|\mathbf{v}\|^2} = \dfrac{\sqrt{5}}{(1 + 5t^2)^{3/2}}$

55. $\mathbf{r}(t) = \frac{1}{2}t\mathbf{i} + \sin t\mathbf{j} + \cos t\mathbf{k}, \quad 0 \le t \le \pi$

$\mathbf{r}'(t) = \frac{1}{2}\mathbf{i} + \cos t\mathbf{j} - \sin t\mathbf{k}$

$$s = \int_0^{\pi} \|\mathbf{r}'(t)\|\, dt = \int_0^{\pi} \sqrt{\frac{1}{4} + \cos^2 t + \sin^2 t}\, dt = \frac{\sqrt{5}}{2}\int_0^{\pi} dt = \left[\frac{\sqrt{5}}{2}t\right]_0^{\pi} = \frac{\sqrt{5}}{2}\pi$$

57. $v = \sqrt{\frac{9.56 \times 10^4}{4600}} \approx 4.56$ mi/sec

(11.4 Exercise 44)

59. The curvature changes abruptly from zero to a nonzero constant at the points B and C.

61. $\mathbf{r}(t) = \langle t \cos \pi t, t \sin \pi t\rangle, \quad 0 \le t \le 2$

(a)

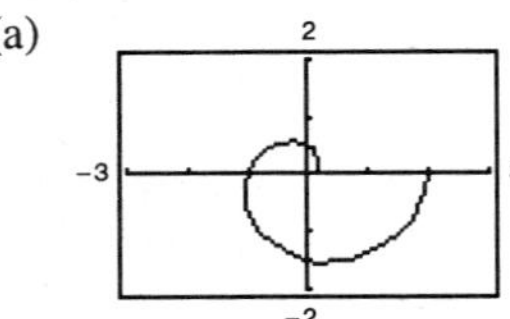

(b) Length $= \int_0^2 \|\mathbf{r}'(t)\|\, dt$

$= \int_0^2 \sqrt{\pi^2 t^2 + 1}\, dt \approx 6.766$ (graphing utility)

(c) $K = \frac{\pi(\pi^2 t^2 + 2)}{[\pi^2 t^2 + 1]^{3/2}}$

$K(0) = 2\pi$

$K(1) = \frac{\pi(\pi^2 + 2)}{(\pi^2 + 1)^{3/2}} \approx 1.04$

$K(2) \approx 0.51$

(d)

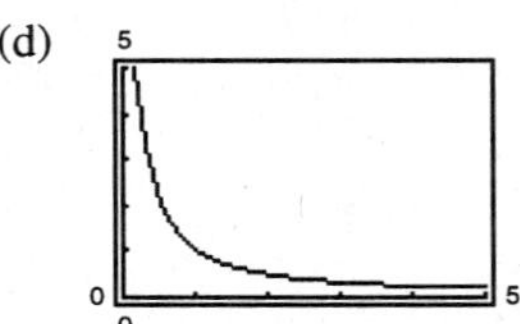

(e) $\lim_{t\to\infty} K = 0$

(f) As $t \to \infty$, the graph spirals outward and the curvature decreases.

CHAPTER 12
Functions of Several Variables

CHAPTER 12
Functions of Several Variables

Section 12.1 Introduction to Functions of Several Variables

Solutions to Odd-Numbered Exercises

1. $x^2z + yz - xy = 10$

$z(x^2 + y) = 10 + xy$

$z = \dfrac{10 + xy}{x^2 + y}$

Yes, z is a function of x and y.

3. $\dfrac{x^2}{4} + \dfrac{y^2}{9} + z^2 = 1$

No, z is not a function of x and y. For example, $(x, y) = (0, 0)$ corresponds to both $z = \pm 1$.

5. $f(x, y) = \dfrac{x}{y}$

(a) $f(3, 2) = \dfrac{3}{2}$ (b) $f(-1, 4) = -\dfrac{1}{4}$ (c) $f(30, 5) = \dfrac{30}{5} = 6$

(d) $f(5, y) = \dfrac{5}{y}$ (e) $f(x, 2) = \dfrac{x}{2}$ (f) $f(5, t) = \dfrac{5}{t}$

7. $f(x, y) = xe^y$

(a) $f(5, 0) = 5e^0 = 5$

(b) $f(3, 2) = 3e^2$

(c) $f(2, -1) = 2e^{-1} = \dfrac{2}{e}$

(d) $f(5, y) = 5e^y$

(e) $f(x, 2) = xe^2$

(f) $f(t, t) = te^t$

9. $h(x, y, z) = \dfrac{xy}{z}$

(a) $h(2, 3, 9) = \dfrac{(2)(3)}{9} = \dfrac{2}{3}$

(b) $h(1, 0, 1) = \dfrac{(1)(0)}{1} = 0$

11. $f(x, y) = x \sin y$

(a) $f\left(2, \dfrac{\pi}{4}\right) = 2 \sin \dfrac{\pi}{4} = \sqrt{2}$

(b) $f(3, 1) = 3 \sin 1$

13. $g(x, y) = \displaystyle\int_x^y (2t - 3)\, dt$

(a) $g(0, 4) = \displaystyle\int_0^4 (2t - 3)\, dt = \Big[t^2 - 3t\Big]_0^4 = 4$

(b) $g(1, 4) = \displaystyle\int_1^4 (2t - 3)\, dt = \Big[t^2 - 3t\Big]_1^4 = 6$

15. $f(x, y) = x^2 - 2y$

(a) $$\frac{f(x + \Delta x, y) - f(x, y)}{\Delta x} = \frac{[(x + \Delta x)^2 - 2y] - (x^2 - 2y)}{\Delta x}$$
$$= \frac{x^2 + 2x(\Delta x) + (\Delta x)^2 - 2y - x^2 + 2y}{\Delta x} = \frac{\Delta x(2x + \Delta x)}{\Delta x} = 2x + \Delta x,\ \Delta x \neq 0$$

(b) $$\frac{f(x, y + \Delta y) - f(x, y)}{\Delta y} = \frac{[x^2 - 2(y + \Delta y)] - (x^2 - 2y)}{\Delta y} = \frac{x^2 - 2y - 2\Delta y - x^2 + 2y}{\Delta y} = \frac{-2\Delta y}{\Delta y} = -2,\ \Delta y \neq 0$$

17. $f(x, y) = \sqrt{4 - x^2 - y^2}$

Domain: $4 - x^2 - y^2 \geq 0$

$x^2 + y^2 \leq 4$

$\{(x, y): x^2 + y^2 \leq 4\}$

Range: $0 \leq z \leq 2$

19. $f(x, y) = \arcsin(x + y)$

Domain:
$\{(x, y): -1 \leq x + y \leq 1\}$

Range: $-\frac{\pi}{2} \leq z \leq \frac{\pi}{2}$

21. $f(x, y) = \ln(4 - x - y)$

Domain: $4 - x - y > 0$

$x + y < 4$

$\{(x, y): y < -x + 4\}$

Range: all real numbers

23. $z = \frac{x + y}{xy}$

Domain: $\{(x, y): x \neq 0 \text{ and } y \neq 0\}$

Range: all real numbers

25. $f(x, y) = e^{x/y}$

Domain: $\{(x, y): y \neq 0\}$

Range: $z > 0$

27. $g(x, y) = \frac{1}{xy}$

Domain: $\{(x, y): x \neq 0 \text{ and } y \neq 0\}$

Range: all real numbers except zero

29. $f(x, y) = \frac{-4x}{x^2 + y^2 + 1}$

(a) View from the positive x-axis: $(20, 0, 0)$

(b) View where x is negative, y and z are positive: $(-15, 10, 20)$

(c) View from the first octant: $(20, 15, 25)$

(d) View from the line $y = x$ in the xy-plane: $(20, 20, 0)$

31. $f(x, y) = 5$

Plane: $z = 5$

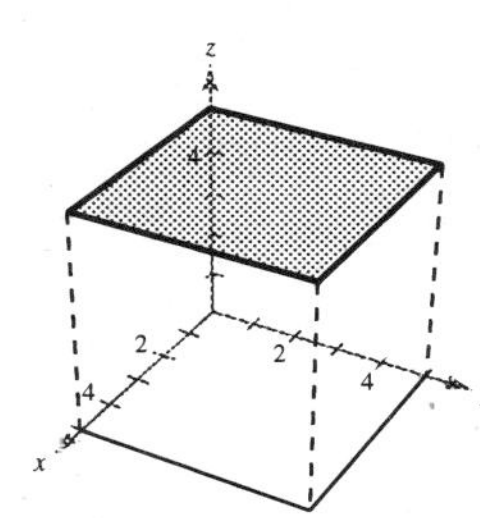

33. $f(x, y) = y^2$

Since the variable x is missing, the surface is a cylinder with rulings parallel to the x-axis. The generating curve is $z = y^2$. The domain is the entire xy-plane and the range is $z \geq 0$.

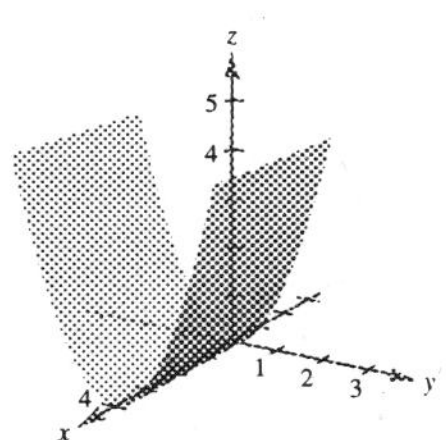

35. $z = 4 - x^2 - y^2$

Paraboloid

Domain: entire xy-plane

Range: $z \leq 4$

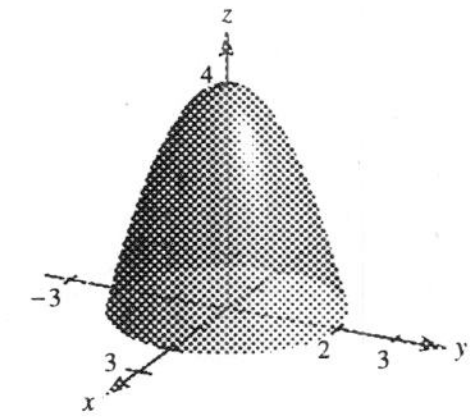

37. $f(x, y) = e^{-x}$

Since the variable y is missing, the surface is a cylinder with rulings parallel to the y-axis. The generating curve is $z = e^{-x}$. The domain is the entire xy-plane and the range is $z > 0$.

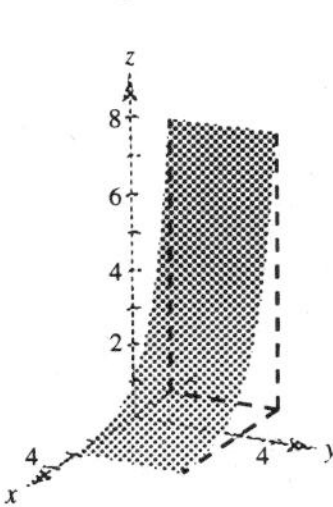

39. $z = y^2 - x^2 + 1$

Hyperbolic paraboloid

Domain: entire xy-plane

Range: $-\infty < z < \infty$

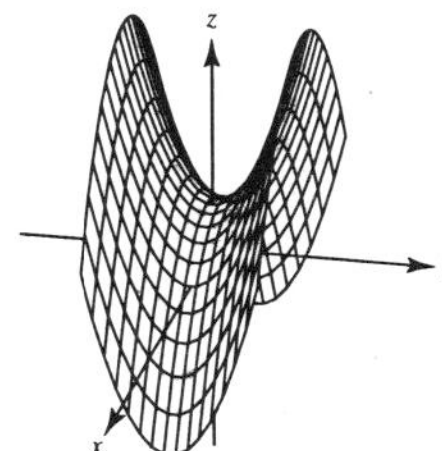

41. $f(x, y) = x^2 e^{(-xy/2)}$

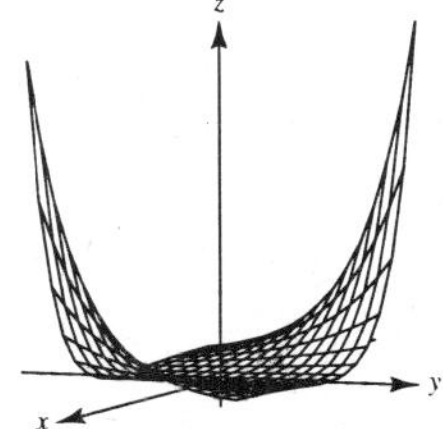

43. $f(x, y) = x^2 + y^2$

(a)

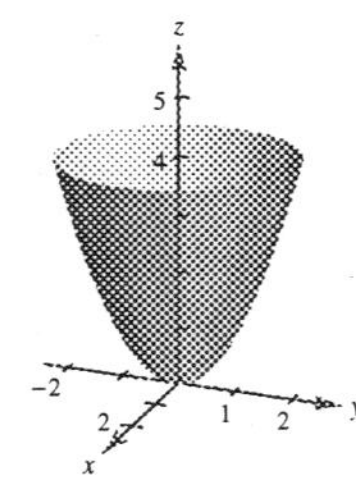

(b) g is a vertical translation of f two units upward

(c) g is a horizontal translation of f two units to the right. The vertex moves from $(0, 0, 0)$ to $(0, 2, 0)$.

(d) g is a reflection of f in the xy-plane followed by a vertical translation 4 units upward.

(e)

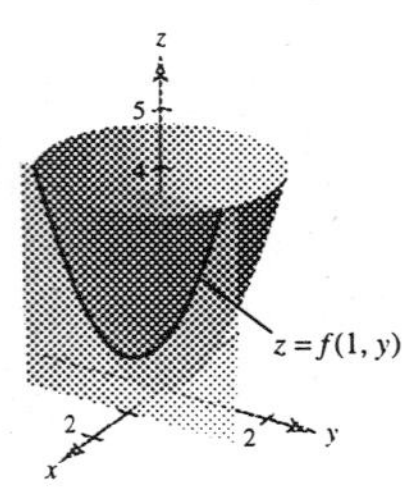

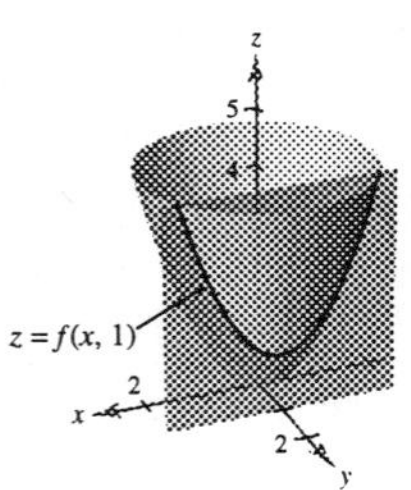

45. $z = e^{1-x^2-y^2}$

Level curves:

$$c = e^{1-x^2-y^2}$$

$$\ln c = 1 - x^2 - y^2$$

$$x^2 + y^2 = 1 - \ln c$$

Circles centered at $(0, 0)$

Matches (c)

47. $z = \ln|y - x^2|$

Level curves:

$$c = \ln|y - x^2|$$

$$\pm e^c = y - x^2$$

$$y = x^2 \pm e^c$$

Parabolas

Matches (b)

49. $z = x + y$

Level curves are parallel lines of the form $x + y = c$.

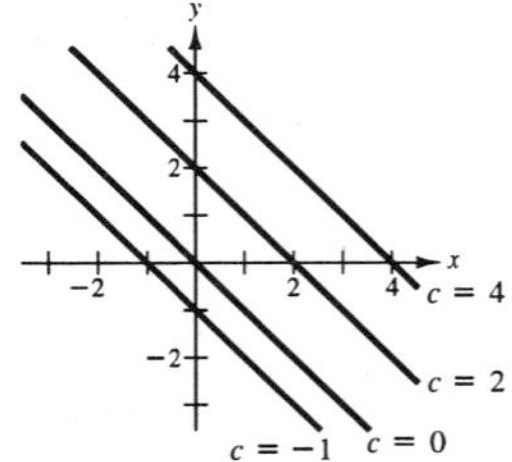

51. $f(x, y) = \sqrt{25 - x^2 - y^2}$

The level curves are of the form

$c = \sqrt{25 - x^2 - y^2}$,

$x^2 + y^2 = 25 - c^2$.

Thus, the level curves are circles of radius 5 or less, centered at the origin.

53. $f(x, y) = xy$

The level curves are hyperbolas of the form $xy = c$.

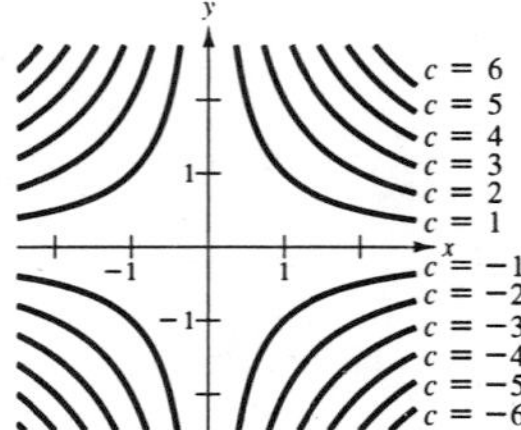

55. $f(x, y) = \dfrac{x}{x^2 + y^2}$

The level curves are of the form

$$c = \frac{x}{x^2 + y^2}$$

$$x^2 - \frac{x}{c} + y^2 = 0$$

$$\left(x - \frac{1}{2c}\right)^2 + y^2 = \left(\frac{1}{2c}\right)^2$$

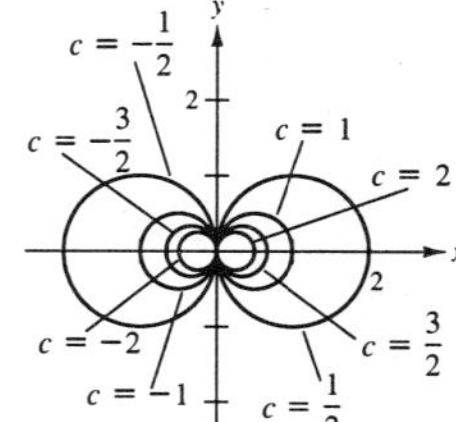

Thus, the level curves are circles passing through the origin and centered at $(1/(2c), 0)$.

57. No The following graphs are not hemispheres.

$z = e^{-(x^2+y^2)}$

$z = x^2 + y^2$

59. The surface is sloped like a saddle. The graph is not unique. Any vertical translation would have the same level curves. One possible function is

$$f(x, y) = x^2 - y^2.$$

61. $V(I, R) = 1000\left[\dfrac{1 + 0.10(1 - R)}{1 + I}\right]^{10}$

	Inflation Rate		
Tax Rate	0	0.03	0.05
0	2593.74	1929.99	1592.33
0.28	2004.23	1491.34	1230.42
0.35	1877.14	1396.77	1152.40

63. $f(x, y, z) = x - 2y + 3z$

$c = 6$

$6 = x - 2y + 3z$

Plane

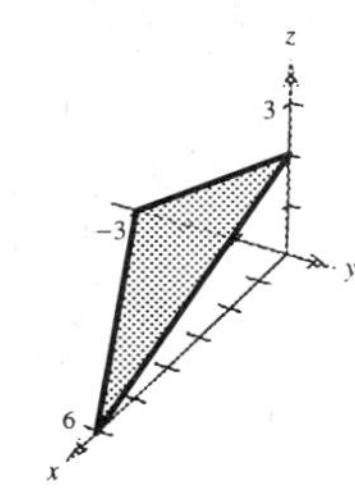

65. $f(x, y, z) = x^2 + y^2 + z^2$

$c = 9$

$9 = x^2 + y^2 + z^2$

Sphere

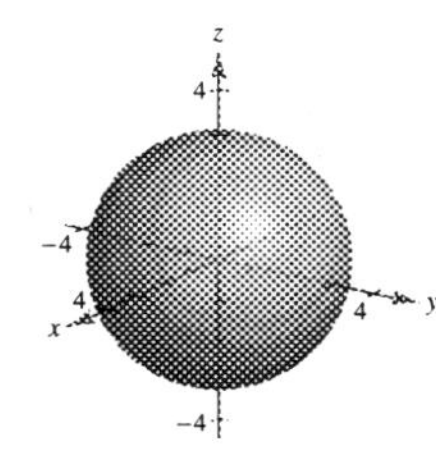

67. $f(x, y, z) = 4x^2 + 4y^2 - z^2$

$c = 0$

$0 = 4x^2 + 4y^2 - z^2$

Elliptic cone

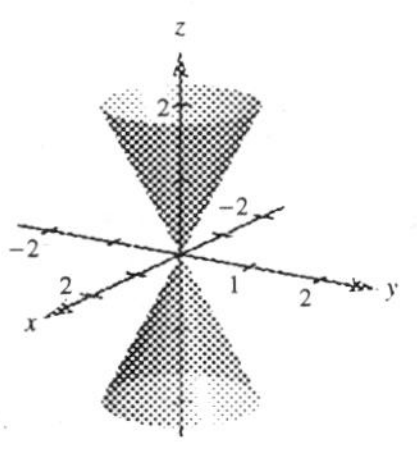

69. $N(d, L) = \left(\dfrac{d - 4}{4}\right)^2 L$

(a) $N(22, 12) = \left(\dfrac{22 - 4}{4}\right)^2(12) = 243$ board-feet

(b) $N(30, 12) = \left(\dfrac{30 - 4}{4}\right)^2(12) = 507$ board-feet

71. $T = 600 - 0.75x^2 - 0.75y^2$

The level curves are of the form

$$c = 600 - 0.75x^2 - 0.75y^2$$

$$x^2 + y^2 = \frac{600 - c}{0.75}.$$

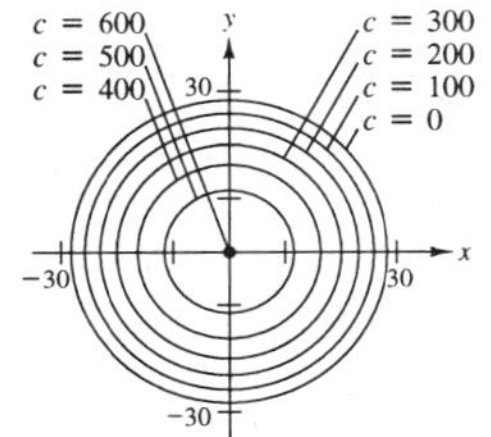

The level curves are circles centered at the origin.

73. $C = 0.75xy + 2(0.40)xz + 2(0.40)yz$

base + front & back + two ends

$= 0.75xy + 0.80(xz + yz)$

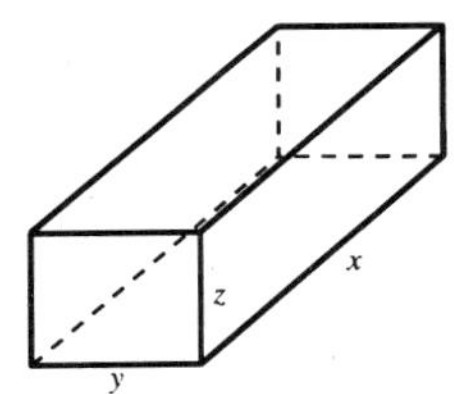

75. $PV = kT$, $20(2600) = k(300)$

(a) $k = \dfrac{20(2600)}{300} = \dfrac{520}{3}$

(b) $P = \dfrac{kT}{V} = \dfrac{520}{3}\left(\dfrac{T}{V}\right)$

The level curves are of the form: $c = \left(\dfrac{520}{3}\right)\left(\dfrac{T}{V}\right)$

$$V = \frac{520}{3c}T$$

Thus, the level curves are lines through the origin with slope $520/(3c)$.

77. (a) Highest pressure at C

(b) Lowest pressure at A

(c) Highest wind velocity at B

79. (a) The boundaries between colors represent level curves

(b) No, the colors represent intervals of different lengths, as indicated in the box

(c) You could use more colors, which means using smaller intervals

81. False. Let

$f(x, y) = 2xy$

$f(1, 2) = f(2, 1)$, but $1 \neq 2$

83. False. Let

$f(x, y) = 5.$

Then, $f(2x, 2y) = 5 \neq 2^2 f(x, y)$

Section 12.2 Limits and Continuity

1. $\lim_{(x, y)\to(a, b)} [f(x, y) - g(x, y)] = \lim_{(x, y)\to(a, b)} f(x, y) - \lim_{(x, y)\to(a, b)} g(x, y) = 5 - 3 = 2$

3. $\lim_{(x, y)\to(a, b)} [f(x, y)g(x, y)] = \left[\lim_{(x, y)\to(a, b)} f(x, y)\right]\left[\lim_{(x, y)\to(a, b)} g(x, y)\right] = 5(3) = 15$

5. $\lim_{(x, y)\to(2, 1)} (x + 3y^2) = 2 + 3(1)^2 = 5$

Continuous everywhere

7. $\lim_{(x, y)\to(2, 4)} \frac{x + y}{x - y} = \frac{2 + 4}{2 - 4} = -3$

Continuous for $x \neq y$

9. $\lim_{(x, y)\to(0, 1)} \frac{\arcsin(x/y)}{1 + xy} = \arcsin 0 = 0$

Continuous for $xy \neq -1, y \neq 0, |x/y| \leq 1$

11. $\lim_{(x, y)\to(0, 0)} e^{xy} = e^0 = 1$

Continuous everywhere

13. $\lim_{(x, y, z)\to(1, 2, 5)} \sqrt{x + y + z} = \sqrt{8} = 2\sqrt{2}$

Continuous for $x + y + z \geq 0$

15. $\lim_{(x, y)\to(0, 0)} e^{xy} = 1$

Continuous everywhere

17. $\lim_{(x, y)\to(0, 0)} \ln(x^2 + y^2) = \ln(0) = -\infty$

The limit does not exist.

Continuous except at $(0, 0)$

19. $f(x, y) = \frac{xy}{x^2 + y^2}$

Continuous except at $(0, 0)$

Path: $y = 0$

(x, y)	$(1, 0)$	$(.5, 0)$	$(.1, 0)$	$(.01, 0)$	$(.001, 0)$
$f(x, y)$	0	0	0	0	0

Path: $y = x$

(x, y)	$(1, 1)$	$(.5, .5)$	$(.1, .1)$	$(.01, .01)$	$(.001, .001)$
$f(x, y)$	$\frac{1}{2}$	$\frac{1}{2}$	$\frac{1}{2}$	$\frac{1}{2}$	$\frac{1}{2}$

The limit does not exist because along the path $y = 0$ the function equals 0, whereas along the path $y = x$ the function equals $\frac{1}{2}$.

21. $f(x, y) = -\frac{xy^2}{x^2 + y^4}$

Continuous except at $(0, 0)$

Path: $x = y^2$

(x, y)	$(1, 1)$	$(.25, .5)$	$(.01, .1)$	$(.0001, 0.01)$	$(.000001, .001)$
$f(x, y)$	$-\frac{1}{2}$	$-\frac{1}{2}$	$-\frac{1}{2}$	$-\frac{1}{2}$	$-\frac{1}{2}$

Path: $x = -y^2$

(x, y)	$(-1, 1)$	$(-.25, .5)$	$(-.01, .1)$	$(-.0001, .01)$	$(-.000001, .001)$
$f(x, y)$	$\frac{1}{2}$	$\frac{1}{2}$	$\frac{1}{2}$	$\frac{1}{2}$	$\frac{1}{2}$

The limit does not exist because along the path $x = y^2$ the function equals $-\frac{1}{2}$, whereas along the path $x = -y^2$ the function equals $\frac{1}{2}$.

23. $\lim_{(x,y)\to(0,0)} (\sin x + \sin y) = 0$

25. $\lim_{(x,y)\to(0,0)} \dfrac{x^2y}{x^4+4y^2}$

Does not exist

27. $f(x, y) = \dfrac{xy^3}{x^2+2y^6}$

The limit does not exist. You can see this analytically by using the paths $y = 0$ and $x = y^3$.

29. $$\lim_{(x,y)\to(0,0)} \frac{\sin(x^2+y^2)}{x^2+y^2} = \lim_{r\to0}\frac{\sin r^2}{r^2} = \lim_{r\to0}\frac{2r\cos r^2}{2r} = \lim_{r\to0}\cos r^2 = 1$$

31. $$\lim_{(x,y)\to(0,0)} \frac{x^3+y^3}{x^2+y^2} = \lim_{r\to0}\frac{r^3(\cos^3\theta+\sin^3\theta)}{r^2} = \lim_{r\to0} r(\cos^3\theta+\sin^3\theta) = 0$$

33. $f(x, y, z) = \dfrac{1}{\sqrt{x^2+y^2+z^2}}$

Continuous except at $(0, 0, 0)$

35. $f(x, y, z) = \dfrac{\sin z}{e^x+e^y}$

Continuous everywhere

37. $$\begin{aligned} f(t) &= t^2 \\ g(x, y) &= 3x - 2y \\ f(g(x, y)) &= f(3x - 2y) \\ &= (3x - 2y)^2 \\ &= 9x^2 - 12xy + 4y^2 \end{aligned}$$

Continuous everywhere

39. $$\begin{aligned} f(t) &= \frac{1}{t} \\ g(x, y) &= 3x - 2y \\ f(g(x, y)) &= f(3x - 2y) = \frac{1}{3x-2y} \end{aligned}$$

Continuous for $y \neq \dfrac{3x}{2}$

41. $f(x, y) = x^2 - 4y$

(a) $$\lim_{\Delta x\to0}\frac{f(x+\Delta x, y) - f(x, y)}{\Delta x} = \lim_{\Delta x\to0}\frac{[(x+\Delta x)^2 - 4y] - (x^2 - 4y)}{\Delta x}$$
$$= \lim_{\Delta x\to0}\frac{2x\Delta x - (\Delta x)^2}{\Delta x} = \lim_{\Delta x\to0}(2x - \Delta x) = 2x$$

(b) $$\lim_{\Delta y\to0}\frac{f(x, y+\Delta y) - f(x, y)}{\Delta y} = \lim_{\Delta y\to0}\frac{[x^2 - 4(y+\Delta y)] - (x^2 - 4y)}{\Delta y}$$
$$= \lim_{\Delta y\to0}\frac{-4\Delta y}{\Delta y} = \lim_{\Delta y\to0}(-4) = -4$$

43. $f(x, y) = 2x + xy - 3y$

(a) $$\lim_{\Delta x\to0}\frac{f(x+\Delta x, y) - f(x, y)}{\Delta x} = \lim_{\Delta x\to0}\frac{[2(x+\Delta x) + (x+\Delta x)y - 3y] - (2x + xy - 3y)}{\Delta x}$$
$$= \lim_{\Delta x\to0}\frac{2\Delta x + \Delta xy}{\Delta x} = \lim_{\Delta x\to0}(2 + y) = 2 + y$$

(b) $$\lim_{\Delta y\to0}\frac{f(x, y+\Delta y) - f(x, y)}{\Delta y} = \lim_{\Delta y\to0}\frac{[2x + x(y+\Delta y) - 3(y+\Delta y)] - (2x + xy - 3y)}{\Delta y}$$
$$= \lim_{\Delta y\to0}\frac{x\Delta y - 3\Delta y}{\Delta y} = \lim_{\Delta y\to0}(x - 3) = x - 3$$

45. Since $\lim_{(x, y)\to(a, b)} f(x, y) = L_1$, then for $\varepsilon/2 > 0$, there corresponds $\delta_1 > 0$ such that $|f(x, y) - L_1| < \varepsilon/2$ whenever

$0 < \sqrt{(x-a)^2 + (y-b)^2} < \delta_1$.

Since $\lim_{(x, y)\to(a, b)} g(x, y) = L_2$, then for $\varepsilon/2 > 0$, there corresponds $\delta_2 > 0$ such that $|g(x, y) - L_2| < \varepsilon/2$ whenever

$0 < \sqrt{(x-a)^2 + (y-b)^2} < \delta_2$.

Let δ be the smaller of δ_1 and δ_2. By the triangle inequality, whenever $\sqrt{(x-a)^2 + (y-b)^2} < \delta$, we have

$$\begin{aligned} |f(x, y) + g(x, y) - (L_1 + L_2)| &= |(f(x, y) - L_1) + (g(x, y) - L_2)| \\ &\le |f(x, y) - L_1| + |g(x, y) - L_2| < \frac{\varepsilon}{2} + \frac{\varepsilon}{2} = \varepsilon. \end{aligned}$$

Therefore, $\lim_{(x, y)\to(a, b)} [f(x, y) + g(x, y)] = L_1 + L_2$.

47. No.

The existence of $f(2, 3)$ has no bearing on the existence of the limit as $(x, y) \to (2, 3)$.

49. True

51. False. Let

$$f(x, y) = \begin{cases} \ln(x^2 + y^2), & (x, y) \neq (0, 0) \\ 0, & x = 0, y = 0 \end{cases}$$

See Exercise 17.

Section 12.3 Partial Derivatives

1. $f_x(4, 1) < 0$

3. $f_y(4, 1) > 0$

5. $f(x, y) = 2x - 3y + 5$

$f_x(x, y) = 2$

$f_y(x, y) = -3$

7. $z = x\sqrt{y}$

$\dfrac{\partial z}{\partial x} = \sqrt{y}$

$\dfrac{\partial z}{\partial y} = \dfrac{x}{2\sqrt{y}}$

9. $z = x^2 e^{2y}$

$\dfrac{\partial z}{\partial x} = 2xe^{2y}$

$\dfrac{\partial z}{\partial y} = 2x^2 e^{2y}$

11. $z = \ln(x^2 + y^2)$

$\dfrac{\partial z}{\partial x} = \dfrac{2x}{x^2 + y^2}$

$\dfrac{\partial z}{\partial y} = \dfrac{2y}{x^2 + y^2}$

13. $z = \ln\left(\dfrac{x+y}{x-y}\right) = \ln(x + y) - \ln(x - y)$

$\dfrac{\partial z}{\partial x} = \dfrac{1}{x+y} - \dfrac{1}{x-y} = -\dfrac{2y}{x^2 - y^2}$

$\dfrac{\partial z}{\partial y} = \dfrac{1}{x+y} + \dfrac{1}{x-y} = \dfrac{2x}{x^2 - y^2}$

15. $h(x, y) = e^{-(x^2+y^2)}$

$h_x(x, y) = -2xe^{-(x^2+y^2)}$

$h_y(x, y) = -2ye^{-(x^2+y^2)}$

17. $f(x, y) = \sqrt{x^2 + y^2}$

$f_x(x, y) = \dfrac{1}{2}(x^2 + y^2)^{-1/2}(2x) = \dfrac{x}{\sqrt{x^2 + y^2}}$

$f_y(x, y) = \dfrac{1}{2}(x^2 + y^2)^{-1/2}(2y) = \dfrac{y}{\sqrt{x^2 + y^2}}$

19. $z = \tan(2x - y)$

$$\frac{\partial z}{\partial x} = 2\sec^2(2x - y)$$

$$\frac{\partial z}{\partial y} = -\sec^2(2x - y)$$

21. $z = e^y \sin xy$

$$\frac{\partial z}{\partial x} = ye^y \cos xy$$

$$\frac{\partial z}{\partial y} = e^y \sin xy + xe^y \cos xy$$

$$= e^y(x \cos xy + \sin xy)$$

23. $f(x, y) = \displaystyle\int_x^y (t^2 - 1)\, dt$

$$= \left[\frac{t^3}{3} - t\right]_x^y = \left(\frac{y^3}{3} - y\right) - \left(\frac{x^3}{3} - x\right)$$

$f_x(x, y) = -x^2 + 1 = 1 - x^2$

$f_y(x, y) = y^2 - 1$

[You could also use the Second Fundamental Theorem of Calculus.]

25. $f(x, y) = 2x + 3y$

$$\frac{\partial f}{\partial x} = \lim_{\Delta x \to 0} \frac{f(x + \Delta x, y) - f(x, y)}{\Delta x} = \lim_{\Delta x \to 0} \frac{2(x + \Delta x) + 3y - 2x - 3y}{\Delta x} = \lim_{\Delta x \to 0} \frac{2\Delta x}{\Delta x} = 2$$

$$\frac{\partial f}{\partial y} = \lim_{\Delta y \to 0} \frac{f(x, y + \Delta y) - f(x, y)}{\Delta y} = \lim_{\Delta y \to 0} \frac{2x + 3(y + \Delta y) - 2x - 3y}{\Delta y} = \lim_{\Delta y \to 0} \frac{3\Delta y}{\Delta y} = 3$$

27. $f(x, y) = \sqrt{x + y}$

$$\frac{\partial f}{\partial x} = \lim_{\Delta x \to 0} \frac{f(x + \Delta x, y) - f(x, y)}{\Delta x} = \lim_{\Delta x \to 0} \frac{\sqrt{x + \Delta x + y} - \sqrt{x + y}}{\Delta x}$$

$$= \lim_{\Delta x \to 0} \frac{\left(\sqrt{x + \Delta x + y} - \sqrt{x + y}\right)\left(\sqrt{x + \Delta x + y} + \sqrt{x + y}\right)}{\Delta x\left(\sqrt{x + \Delta x + y} + \sqrt{x + y}\right)}$$

$$= \lim_{\Delta x \to 0} \frac{1}{\sqrt{x + \Delta x + y} + \sqrt{x + y}} = \frac{1}{2\sqrt{x + y}}$$

$$\frac{\partial f}{\partial y} = \lim_{\Delta y \to 0} \frac{f(x, y + \Delta y) - f(x, y)}{\Delta y} = \lim_{\Delta y \to 0} \frac{\sqrt{x + y + \Delta y} - \sqrt{x + y}}{\Delta y}$$

$$= \lim_{\Delta y \to 0} \frac{\left(\sqrt{x + y + \Delta y} - \sqrt{x + y}\right)\left(\sqrt{x + y + \Delta y} + \sqrt{x + y}\right)}{\Delta y\left(\sqrt{x + y + \Delta y} + \sqrt{x + y}\right)}$$

$$= \lim_{\Delta y \to 0} \frac{1}{\sqrt{x + y + \Delta y} + \sqrt{x + y}} = \frac{1}{2\sqrt{x + y}}$$

29. $g(x, y) = 4 - x^2 - y^2$

$g_x(x, y) = -2x$

At $(1, 1)$: $g_x(1, 1) = -2$

$g_y(x, y) = -2y$

At $(1, 1)$: $g_y(1, 1) = -2$

31. $z = e^{-x} \cos y$

$$\frac{\partial z}{\partial x} = -e^{-x} \cos y$$

At $(0, 0)$: $\dfrac{\partial z}{\partial x} = -1$

$$\frac{\partial z}{\partial y} = -e^{-x} \sin y$$

At $(0, 0)$: $\dfrac{\partial z}{\partial y} = 0$

33. $f(x, y) = \arctan \dfrac{y}{x}$

$$f_x(x, y) = \frac{1}{1 + (y^2/x^2)}\left(-\frac{y}{x^2}\right) = \frac{-y}{x^2 + y^2}$$

At $(2, -2)$: $f_x(2, -2) = \dfrac{1}{4}$

$$f_y(x, y) = \frac{1}{1 + (y^2/x^2)}\left(\frac{1}{x}\right) = \frac{x}{x^2 + y^2}$$

At $(2, -2)$: $f_y(2, -2) = \dfrac{1}{4}$

35. $f(x, y) = \dfrac{xy}{x - y}$

$$f_x(x, y) = \frac{y(x - y) - xy}{(x - y)^2} = \frac{-y^2}{(x - y)^2}$$

At $(2, -2)$: $f_x(2, -2) = -\dfrac{1}{4}$

$$f_y(x, y) = \frac{x(x - y) + xy}{(x - y)^2} = \frac{x^2}{(x - y)^2}$$

At $(2, -2)$: $f_y(2, -2) = \dfrac{1}{4}$

37. The plane $z = x + y = f(x, y)$ satisfies

$$\frac{\partial f}{\partial x} > 0 \text{ and } \frac{\partial f}{\partial y} > 0.$$

39. $z = \sqrt{49 - x^2 - y^2}$, $x = 2$,
$x = 2$, $(2, 3, 6)$

Intersecting curve: $z = \sqrt{45 - y^2}$

$$\frac{\partial z}{\partial y} = \frac{-y}{\sqrt{45 - y^2}}$$

At $(2, 3, 6)$: $\dfrac{\partial z}{\partial y} = \dfrac{-3}{\sqrt{45 - 9}} = -\dfrac{1}{2}$

41. $z = 9x^2 - y^2$, $y = 3$, $(1, 3, 0)$

Intersecting curve: $z = 9x^2 - 9$

$$\frac{\partial z}{\partial x} = 18x$$

At $(1, 3, 0)$: $\dfrac{\partial z}{\partial x} = 18(1) = 18$

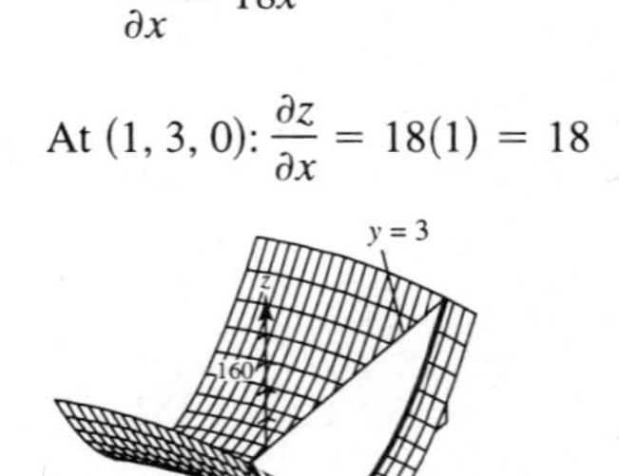

43. $f_x(x, y) = 2x + 4y - 4$, $f_y(x, y) = 4x + 2y + 16$

$f_x = f_y = 0$: $2x + 4y = 4$

$$4x + 2y = -16$$

Solving for x and y,

$$x = -6 \text{ and } y = 4.$$

45. $f_x(x, y) = -\dfrac{1}{x^2} + y$, $f_y(x, y) = -\dfrac{1}{y^2} + x$

$f_x = f_y = 0$: $-\dfrac{1}{x^2} + y = 0$ and $-\dfrac{1}{y^2} + x = 0$

$$y = \frac{1}{x^2} \text{ and } x = \frac{1}{y^2}$$

$y = y^4 \Rightarrow y = 1 = x$

Points: $(1, 1)$

47. $z = x^2 - 2xy + 3y^2$

$$\frac{\partial z}{\partial x} = 2x - 2y$$

$$\frac{\partial^2 z}{\partial x^2} = 2$$

$$\frac{\partial^2 z}{\partial y \partial x} = -2$$

$$\frac{\partial z}{\partial y} = -2x + 6y$$

$$\frac{\partial^2 z}{\partial y^2} = 6$$

$$\frac{\partial^2 z}{\partial x \partial y} = -2$$

49. $z = \sqrt{x^2 + y^2}$

$$\frac{\partial z}{\partial x} = \frac{x}{\sqrt{x^2 + y^2}}$$

$$\frac{\partial^2 z}{\partial x^2} = \frac{y^2}{(x^2 + y^2)^{3/2}}$$

$$\frac{\partial^2 z}{\partial y \partial x} = \frac{-xy}{(x^2 + y^2)^{3/2}}$$

$$\frac{\partial z}{\partial y} = \frac{y}{\sqrt{x^2 + y^2}}$$

$$\frac{\partial^2 z}{\partial y^2} = \frac{x^2}{(x^2 + y^2)^{3/2}}$$

$$\frac{\partial^2 z}{\partial x \partial y} = \frac{-xy}{(x^2 + y^2)^{3/2}}$$

51. $z = e^x \tan y$

$$\frac{\partial z}{\partial x} = e^x \tan y$$

$$\frac{\partial^2 z}{\partial x^2} = e^x \tan y$$

$$\frac{\partial^2 z}{\partial y \partial x} = e^x \sec^2 y$$

$$\frac{\partial z}{\partial y} = e^x \sec^2 y$$

$$\frac{\partial^2 z}{\partial y^2} = 2e^x \sec^2 y \tan y$$

$$\frac{\partial^2 z}{\partial x \partial y} = e^x \sec^2 y$$

53. $z = \arctan \dfrac{y}{x}$

$$\frac{\partial z}{\partial x} = \frac{1}{1 + (y^2/x^2)}\left(-\frac{y}{x^2}\right) = \frac{-y}{x^2 + y^2}$$

$$\frac{\partial^2 z}{\partial x^2} = \frac{2xy}{(x^2 + y^2)^2}$$

$$\frac{\partial^2 z}{\partial y \partial x} = \frac{-(x^2 + y^2) + y(2y)}{(x^2 + y^2)^2} = \frac{y^2 - x^2}{(x^2 + y^2)^2}$$

$$\frac{\partial z}{\partial y} = \frac{1}{1 + (y^2/x^2)}\left(\frac{1}{x}\right) = \frac{x}{x^2 + y^2}$$

$$\frac{\partial^2 z}{\partial y^2} = \frac{-2xy}{(x^2 + y^2)^2}$$

$$\frac{\partial^2 z}{\partial x \partial y} = \frac{(x^2 + y^2) - x(2x)}{(x^2 + y^2)^2} = \frac{y^2 - x^2}{(x^2 + y^2)^2}$$

55. $z = x \sec y$

$$\frac{\partial z}{\partial x} = \sec y$$

$$\frac{\partial^2 z}{\partial x^2} = 0$$

$$\frac{\partial^2 z}{\partial y \partial x} = \sec y \tan y$$

$$\frac{\partial z}{\partial y} = x \sec y \tan y$$

$$\frac{\partial^2 z}{\partial y^2} = x \sec y(\sec^2 y + \tan^2 y)$$

$$\frac{\partial^2 z}{\partial x \partial y} = \sec y \tan y$$

Therefore, $\dfrac{\partial^2 z}{\partial y \partial x} = \dfrac{\partial^2 z}{\partial x \partial y}$.

There are no points for which $z_x = 0 = z_y$, because

$$\frac{\partial z}{\partial x} = \sec y \neq 0.$$

57. $z = \ln\left(\dfrac{x}{x^2 + y^2}\right) = \ln x - \ln(x^2 + y^2)$

$$\frac{\partial z}{\partial x} = \frac{1}{x} - \frac{2x}{x^2 + y^2} = \frac{y^2 - x^2}{x(x^2 + y^2)}$$

$$\frac{\partial^2 z}{\partial x^2} = \frac{x^4 - 4x^2y^2 - y^4}{x^2(x^2 + y^2)^2}$$

$$\frac{\partial^2 z}{\partial y \partial x} = \frac{4xy}{(x^2 + y^2)^2}$$

$$\frac{\partial z}{\partial y} = -\frac{2y}{x^2 + y^2}$$

$$\frac{\partial^2 z}{\partial y^2} = \frac{2(y^2 - x^2)}{(x^2 + y^2)^2}$$

$$\frac{\partial^2 z}{\partial x \partial y} = \frac{4xy}{(x^2 + y^2)^2}$$

There are no points for which $z_x = z_y = 0$.

59. $w = \sqrt{x^2 + y^2 + z^2}$

$$\frac{\partial w}{\partial x} = \frac{x}{\sqrt{x^2 + y^2 + z^2}}$$

$$\frac{\partial w}{\partial y} = \frac{y}{\sqrt{x^2 + y^2 + z^2}}$$

$$\frac{\partial w}{\partial z} = \frac{z}{\sqrt{x^2 + y^2 + z^2}}$$

61. $F(x, y, z) = \ln \sqrt{x^2 + y^2 + z^2}$

$$= \frac{1}{2}\ln(x^2 + y^2 + z^2)$$

$$F_x(x, y, z) = \frac{x}{x^2 + y^2 + z^2}$$

$$F_y(x, y, z) = \frac{y}{x^2 + y^2 + z^2}$$

$$F_z(x, y, z) = \frac{z}{x^2 + y^2 + z^2}$$

63. $H(x, y, z) = \sin(x + 2y + 3z)$

$$H_x(x, y, z) = \cos(x + 2y + 3z)$$

$$H_y(x, y, z) = 2\cos(x + 2y + 3z)$$

$$H_z(x, y, z) = 3\cos(x + 2y + 3z)$$

65. $f(x, y, z) = xyz$

$$f_x(x, y, z) = yz$$

$$f_y(x, y, z) = xz$$

$$f_{yy}(x, y, z) = 0$$

$$f_{xy}(x, y, z) = z$$

$$f_{yx}(x, y, z) = z$$

$$f_{yyx}(x, y, z) = 0$$

$$f_{xyy}(x, y, z) = 0$$

$$f_{yxy}(x, y, z) = 0$$

Therefore, $f_{xyy} = f_{yxy} = f_{yyx} = 0$.

67. $f(x, y, z) = e^{-x} \sin yz$

$f_x(x, y, z) = -e^{-x} \sin yz$

$f_y(x, y, z) = ze^{-x} \cos yz$

$f_{yy}(x, y, z) = -z^2e^{-x} \sin yz$

$f_{xy}(x, y, z) = -ze^{-x} \cos yz$

$f_{yx}(x, y, z) = -ze^{-x} \cos yz$

$f_{yyx}(x, y, z) = z^2e^{-x} \sin yz$

$f_{xyy}(x, y, z) = z^2e^{-x} \sin yz$

$f_{yxy}(x, y, z) = z^2e^{-x} \sin yz$

Therefore, $f_{xyy} = f_{yxy} = f_{yyx}$.

69. $z = 5xy$

$$\frac{\partial z}{\partial x} = 5y$$

$$\frac{\partial^2 z}{\partial x^2} = 0$$

$$\frac{\partial z}{\partial y} = 5x$$

$$\frac{\partial^2 z}{\partial y^2} = 0$$

Therefore, $\dfrac{\partial^2 z}{\partial x^2} + \dfrac{\partial^2 z}{\partial y^2} = 0 + 0 = 0.$

71. $z = e^x \sin y$

$$\frac{\partial z}{\partial x} = e^x \sin y$$

$$\frac{\partial^2 z}{\partial x^2} = e^x \sin y$$

$$\frac{\partial z}{\partial y} = e^x \cos y$$

$$\frac{\partial^2 z}{\partial y^2} = -e^x \sin y$$

Therefore, $\dfrac{\partial^2 z}{\partial x^2} + \dfrac{\partial^2 z}{\partial y^2} = e^x \sin y - e^x \sin y = 0.$

73. $z = \sin(x - ct)$

$$\frac{\partial z}{\partial t} = -c \cos(x - ct)$$

$$\frac{\partial^2 z}{\partial t^2} = -c^2 \sin(x - ct)$$

$$\frac{\partial z}{\partial x} = \cos(x - ct)$$

$$\frac{\partial^2 z}{\partial x^2} = -\sin(x - ct)$$

Therefore, $\dfrac{\partial^2 z}{\partial t^2} = c^2 \dfrac{\partial^2 z}{\partial x^2}.$

75. $z = e^{-t} \cos \dfrac{x}{c}$

$$\frac{\partial z}{\partial t} = -e^{-t} \cos \frac{x}{c}$$

$$\frac{\partial z}{\partial x} = -\frac{1}{c} e^{-t} \sin \frac{x}{c}$$

$$\frac{\partial^2 z}{\partial x^2} = -\frac{1}{c^2} e^{-t} \cos \frac{x}{c}$$

Therefore, $\dfrac{\partial z}{\partial t} = c^2 \dfrac{\partial^2 z}{\partial x^2}.$

77. (a) $C = 32\sqrt{xy} + 175x + 205y + 1050$

$$\frac{\partial C}{\partial x} = 16\sqrt{\frac{y}{x}} + 175$$

$$\left.\frac{\partial C}{\partial x}\right]_{(80, 20)} = 16\sqrt{\frac{1}{4}} + 175 = 183$$

$$\frac{\partial C}{\partial y} = 16\sqrt{\frac{x}{y}} + 205$$

$$\left.\frac{\partial C}{\partial y}\right]_{(80, 20)} = 16\sqrt{4} + 205 = 237$$

(b) The fireplace-insert stove results in the cost increasing at a faster rate because

$$\frac{\partial C}{\partial y} > \frac{\partial C}{\partial x}.$$

79. An increase in either price will cause a decrease in demand.

81. $T = 500 - 0.6x^2 - 1.5y^2$

$$\frac{\partial T}{\partial x} = -1.2x, \frac{\partial T}{\partial x}(2, 3) = -2.4°/\text{m}$$

$$\frac{\partial T}{\partial y} = -3y = \frac{\partial T}{\partial y}(2, 3) = -9°/\text{m}$$

83. $U = -5x^2 + xy - 3y^2$

(a) $U_x = -10x + y$

(b) $U_y = x - 6y$

(c) $U_x(2, 3) = -17$ and $U_y(2, 3) = -16$. The person should consume one more unit of y because the rate of decrease of satisfaction is less for y.

(d)

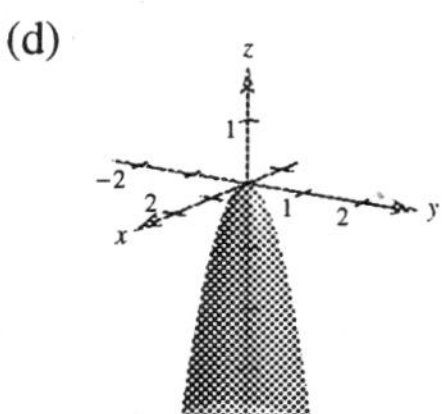

85. $f(x, y) = \begin{cases} \dfrac{xy(x^2 - y^2)}{x^2 + y^2}, & (x, y) \neq (0, 0) \\ 0, & (x, y) = (0, 0) \end{cases}$

(a) $f_x(x, y) = \dfrac{(x^2 + y^2)(3x^2y - y^3) - (x^3y - xy^3)(2x)}{(x^2 + y^2)^2} = \dfrac{y(x^4 + 4x^2y^2 - y^4)}{(x^2 + y^2)^2}$

$f_y(x, y) = \dfrac{(x^2 + y^2)(x^3 - 3xy^2) - (x^3y - xy^3)(2y)}{(x^2 + y^2)^2} = \dfrac{x(x^4 - 4x^2y^2 - y^4)}{(x^2 + y^2)^2}$

(b) $f_x(0, 0) = \lim\limits_{\Delta x \to 0} \dfrac{f(\Delta x, 0) - f(0, 0)}{\Delta x} = \lim\limits_{\Delta x \to 0} \dfrac{0/[(\Delta x)^2] - 0}{\Delta x} = 0$

$f_y(0, 0) = \lim\limits_{\Delta y \to 0} \dfrac{f(0, \Delta y) - f(0, 0)}{\Delta y} = \lim\limits_{\Delta y \to 0} \dfrac{0/[(\Delta y)^2] - 0}{\Delta y} = 0$

(c) $f_{xy}(0, 0) = \dfrac{\partial}{\partial y}\left(\dfrac{\partial f}{\partial x}\right)\Bigg|_{(0,0)} = \lim\limits_{\Delta y \to 0} \dfrac{f_x(0, \Delta y) - f_x(0, 0)}{\Delta y} = \lim\limits_{\Delta y \to 0} \dfrac{\Delta y(-(\Delta y)^4)}{((\Delta y)^2)^2(\Delta y)} = \lim\limits_{\Delta y \to 0} (-1) = -1$

$f_{yx}(0, 0) = \dfrac{\partial}{\partial x}\left(\dfrac{\partial f}{\partial y}\right)\Bigg|_{(0,0)} = \lim\limits_{\Delta x \to 0} \dfrac{f_y(\Delta x, 0) - f_y(0, 0)}{\Delta x} = \lim\limits_{\Delta x \to 0} \dfrac{\Delta x((\Delta x)^4)}{((\Delta x)^2)^2(\Delta x)} = \lim\limits_{\Delta x \to 0} 1 = 1$

(d) f_{yx} or f_{xy} or both are not continuous at (0, 0).

87. True

89. True

Section 12.4 Differentials

1. $z = 3x^2y^3$

$dz = 6xy^3\,dx + 9x^2y^2\,dy$

3. $z = \dfrac{-1}{x^2 + y^2}$

$dz = \dfrac{2x}{(x^2 + y^2)^2}\,dx + \dfrac{2y}{(x^2 + y^2)^2}\,dy$

$= \dfrac{2}{(x^2 + y^2)^2}(x\,dx + y\,dy)$

5. $z = x\cos y - y\cos x$

$dz = (\cos y + y\sin x)\,dx + (-x\sin y - \cos x)\,dy = (\cos y + y\sin x)\,dx - (x\sin y + \cos x)\,dy$

7. $w = 2z^3y \sin x$

$dw = 2z^3y \cos x\, dx + 2z^3 \sin x\, dy + 6z^2y \sin x\, dz$

9. $w = \dfrac{x + y}{z - 2y}$

$$dw = \frac{1}{z - 2y}\, dx + \frac{z + 2x}{(z - 2y)^2}\, dy - \frac{x + y}{(z - 2y)^2}\, dz$$

11. (a) $f(1, 2) = 4$

$f(1.05, 2.1) = 3.4875$

$\Delta z = f(1.05, 2.1) - f(1, 2) = -0.5125$

(b) $dz = -2x\, dx - 2y\, dy$

$= -2(0.05) - 4(0.1) = -0.5$

13. (a) $f(1, 2) = \sin 2$

$f(1.05, 2.1) = 1.05 \sin 2.1$

$\Delta z = f(1.05, 2.1) - f(1, 2) \approx -0.00293$

(b) $dz = \sin y\, dx + x \cos y\, dy$

$= (\sin 2)(0.05) + (\cos 2)(0.1) \approx 0.00385$

15. (a) $f(1, 2) = -5$

$f(1.05, 2.1) = -5.25$

$\Delta z = -0.25$

(b) $dz = 3\, dx - 4\, dy$

$= 3(0.05) - 4(0.1) \approx -0.25$

17. Let $z = \sqrt{x^2 + y^2}$, $x = 5$, $y = 3$, $dx = 0.05$, $dy = 0.1$. Then: $dz = \dfrac{x}{\sqrt{x^2 + y^2}}\, dx + \dfrac{y}{\sqrt{x^2 + y^2}}\, dy$

$$\sqrt{(5.05)^2 + (3.1)^2} - \sqrt{5^2 + 3^2} \approx \frac{5}{\sqrt{5^2 + 3^2}}(0.05) + \frac{3}{\sqrt{5^2 + 3^2}}(0.1) = \frac{0.55}{\sqrt{34}} \approx 0.094$$

19. Let $z = (1 - x^2)/y^2$, $x = 3$, $y = 6$, $dx = 0.05$, $dy = -0.05$. Then: $dz = -\dfrac{2x}{y^2}\, dx + \dfrac{-2(1 - x^2)}{y^3}\, dy$

$$\frac{1 - (3.05)^2}{(5.95)^2} - \frac{1 - 3^2}{6^2} \approx -\frac{2(3)}{6^2}(0.05) - \frac{2(1 - 3^2)}{6^3}(-0.05) \approx -0.012$$

21. $A = lh$

$dA = l\, dh + h\, dl$

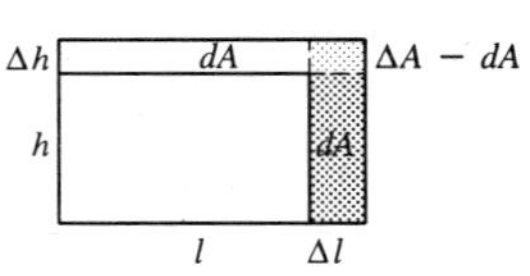

23. $V = \dfrac{\pi r^2 h}{3}$

$r = 3$

$h = 6$

$$dV = \frac{2\pi rh}{3}\, dr + \frac{\pi r^2}{3}\, dh = \frac{\pi r}{3}(2h\, dr + r\, dh)$$

Δr	Δh	dV	ΔV	$\Delta V - dV$
0.1	0.1	4.7124	4.8391	0.1267
0.1	−0.1	2.8274	2.8264	−0.0010
0.001	0.002	0.0565	0.0566	0.0001
−0.0001	0.0002	−0.0019	−0.0019	0.0000

25. $V = \pi r^2 h = dV = (2\pi rh)\, dr + (\pi r^2)\, dh$

$$\frac{dV}{V} = 2\frac{dr}{r} + \frac{dh}{h} = 2(0.04) + (0.02) = 0.10 = 10\%$$

27. $A = \frac{1}{2}ab \sin C$

$dA = \frac{1}{2}[(b \sin C)\, da + (a \sin C)\, db + (ab \cos C)\, dC]$

$= \frac{1}{2}[4(\sin 45°)(\pm\frac{1}{16}) + 3(\sin 45°)(\pm\frac{1}{16}) + 12(\cos 45°)(\pm 0.02)] \approx \pm 0.24 \text{ in.}^2$

29. $P = \dfrac{E^2}{R}$

$$dP = \frac{2E}{R}\,dE - \frac{E^2}{R^2}\,dR$$

$$\frac{dP}{P} = 2\frac{dE}{E} - \frac{dR}{R} = 2(0.02) - (-0.03) = 0.07 = 7\%$$

31. (a) Using the Law of Cosines:

$$a^2 = b^2 + c^2 - 2bc\cos A$$
$$= 330^2 + 420^2 - 2(330)(420)\cos 9^\circ$$
$$a \approx 107.3 \text{ ft.}$$

(b) $a = \sqrt{b^2 + 420^2 - 2b(420)\cos\theta}$

$$da = \frac{1}{2}\left[b^2 + 420^2 - 840b\cos\theta\right]^{-1/2}\left[(2b - 840\cos\theta)\,db + 840b\sin\theta\,d\theta\right]$$

$$= \frac{1}{2}\left[330^2 + 420^2 - 840(330)\left(\cos\frac{\pi}{20}\right)\right]^{-1/2}\left[\left(2(330) - 840\cos\frac{\pi}{20}\right)(6) + 840(330)\left(\sin\frac{\pi}{20}\right)\left(\frac{\pi}{180}\right)\right]$$

$$\approx \frac{1}{2}[11512.79]^{-1/2}[\pm 1774.79] \approx \pm 8.27 \text{ ft}$$

33. $L = 0.00021\left(\ln\dfrac{2h}{r} - 0.75\right)$

$$dL = 0.00021\left[\frac{dh}{h} - \frac{dr}{r}\right] = 0.00021\left[\frac{(\pm 1/100)}{100} - \frac{(\pm 1/16)}{2}\right] \approx (\pm 6.6) \times 10^{-6}$$

$$L = 0.00021(\ln 100 - 0.75) \approx 8.096 \times 10^{-4} \pm dL = 8.096 \times 10^{-4} \pm 6.6 \times 10^{-6} \text{ micro–henrys}$$

35. (a)

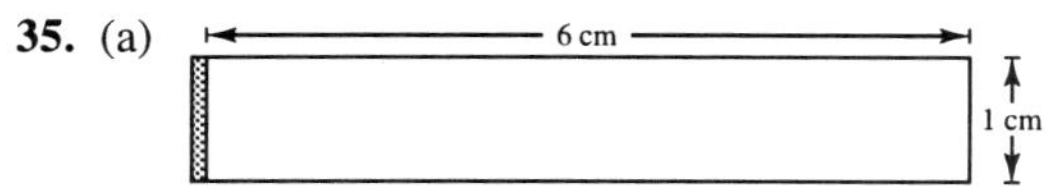

(b)

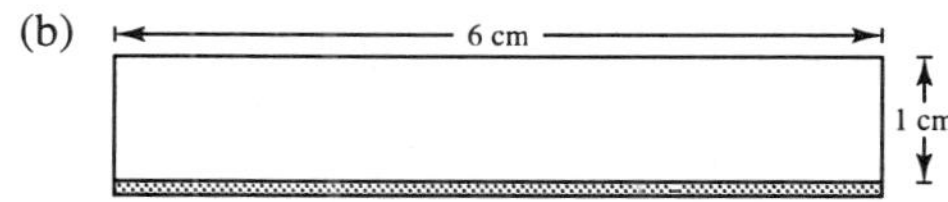

(c) The height has more effect since the shaded region in (b) is larger than the shaded region in (a).

(d) $A = hl \implies dA = l\,dh + h\,dl$

If $dl = 0.01$ and $dh = 0$, then $dA = 1(0.01) = 0.01$.

If $dh = 0.01$ and $dl = 0$, then $dA = 6(0.01) = 0.06$.

37. Essay

39. $z = f(x, y) = x^2 - 2x + y$

$$\Delta z = f(x + \Delta x, y + \Delta y) - f(x, y)$$
$$= (x^2 + 2x(\Delta x) + (\Delta x)^2 - 2x - 2(\Delta x) + y + (\Delta y)) - (x^2 - 2x + y)$$
$$= 2x(\Delta x) + (\Delta x)^2 - 2(\Delta x) + (\Delta y)$$
$$= (2x - 2)\Delta x + \Delta y + \Delta x(\Delta x) + 0(\Delta y)$$
$$= f_x(x, y)\Delta x + f_y(x, y)\Delta y + \epsilon_1\Delta x + \epsilon_2\Delta y \text{ where } \epsilon_1 = \Delta x \text{ and } \epsilon_2 = 0.$$

As $(\Delta x, \Delta y) \to (0, 0)$, $\epsilon_1 \to 0$ and $\epsilon_2 \to 0$.

41. $z = f(x, y) = x^2y$

$$\begin{aligned}\Delta z &= f(x + \Delta x,\ y + \Delta y) - f(x, y)\\ &= (x^2 + 2x(\Delta x) + (\Delta x)^2)(y + \Delta y) - x^2y\\ &= 2xy(\Delta x) + y(\Delta x)^2 + x^2\Delta y + 2x(\Delta x)(\Delta y) + (\Delta x)^2\,\Delta y\\ &= 2xy(\Delta x) + x^2\Delta y + (y\Delta x)\,\Delta x + [2x\Delta x + (\Delta x)^2]\,\Delta y\\ &= f_x(x, y)\,\Delta x + f_y(x, y)\,\Delta y + \epsilon_1\Delta x + \epsilon_2\Delta y \text{ where } \epsilon_1 = y(\Delta x) \text{ and } \epsilon_2 = 2x\Delta x + (\Delta x)^2.\end{aligned}$$

As $(\Delta x, \Delta y) \to (0, 0)$, $\epsilon_1 \to 0$ and $\epsilon_2 \to 0$.

43. $f(x, y) = \begin{cases} \dfrac{3x^2y}{x^4 + y^2}, & (x, y) \neq (0, 0)\\ 0, & (x, y) = (0, 0)\end{cases}$

(a) $$f_x(0, 0) = \lim_{\Delta x \to 0} \frac{f(\Delta x, 0) - f(0, 0)}{\Delta x} = \lim_{\Delta x \to 0} \frac{\dfrac{0}{(\Delta x)^4} - 0}{\Delta x} = 0$$

$$f_y(0, 0) = \lim_{\Delta y \to 0} \frac{f(0, \Delta y) - f(0, 0)}{\Delta y} = \lim_{\Delta y \to 0} \frac{\dfrac{0}{(\Delta y)^2} - 0}{\Delta y} = 0$$

Thus, the partial derivatives exist at $(0, 0)$.

(b) Along the line $y = x$: $\displaystyle\lim_{(x, y) \to (0, 0)} f(x, y) = \lim_{x \to 0} \frac{3x^3}{x^4 + x^2} = \lim_{x \to 0} \frac{3x}{x^2 + 1} = 0$

Along the curve $y = x^2$: $\displaystyle\lim_{(x, y) \to (0, 0)} f(x, y) = \frac{3x^4}{2x^4} = \frac{3}{2}$

f is not continuous at $(0, 0)$. Therefore, f is not differentiable at $(0, 0)$. (See Theroem 12.5)

Section 12.5 Chain Rules for Functions of Several Variables

1. $w = x^2 + y^2$

$x = e^t$

$y = e^{-t}$

$$\frac{dw}{dt} = 2xe^t + 2y(-e^{-t}) = 2(e^{2t} - e^{-2t})$$

3. $w = x \sec y$

$x = e^t$

$y = \pi - t$

$$\begin{aligned}\frac{dw}{dt} &= (\sec y)(e^t) + (x \sec y \tan y)(-1)\\ &= e^t \sec(\pi - t)[1 - \tan(\pi - t)]\\ &= -e^t(\sec t + \sec t \tan t)\end{aligned}$$

5. $w = xy,\ x = 2 \sin t,\ y = \cos t$

(a) $\dfrac{dw}{dt} = 2y \cos t + x(-\sin t) = 2y \cos t - x \sin t = 2(\cos^2 t - \sin^2 t) = 2 \cos 2t$

(b) $w = 2 \sin t \cos t = \sin 2t,\ \dfrac{dw}{dt} = 2 \cos 2t$

7. $w = x^2 + y^2 + z^2$

$x = e^t \cos t$

$y = e^t \sin t$

$z = e^t$

(a) $\dfrac{dw}{dt} = 2x(-e^t \sin t + e^t \cos t) + 2y(e^t \cos t + e^t \sin t) + 2ze^t = 4e^{2t}$

(b) $w = 2e^{2t}, \ \dfrac{dw}{dt} = 4e^{2t}$

9. $w = xy + xz + yz, \ x = t - 1, \ y = t^2 - 1, \ z = t$

(a) $\dfrac{dw}{dt} = (y + z) = (x + z)(2t) + (x + y)$

$= (t^2 - 1 + t) + (t - 1 + 1)(2t) + (t - 1 + t^2 - 1) = 3(2t^2 - 1)$

(b) $w = (t - 1)(t^2 - 1) + (t - 1)t + (t^2 - 1)t$

$\dfrac{dw}{dt} = 2t(t - 1) + (t^2 - 1) + 2t - 1 + 3t^2 - 1 = 3(2t^2 - 1)$

11. $w = \arctan(2xy), \ x = \cos t, \ y = \sin t, \ t = 0$

$$\frac{dw}{dt} = \frac{\partial w}{\partial x}\frac{dx}{dt} + \frac{\partial w}{\partial y}\frac{dy}{dt}$$

$$= \frac{2y}{1 + (4x^2y^2)}(-\sin t) + \frac{2x}{1 + (4x^2y^2)}(\cos t)$$

$$= \frac{2 \sin t}{1 + 4\cos^2 t \sin^2 t}(-\sin t) + \frac{2 \cos t}{1 + 4\cos^2 t \sin^2 t}(\cos t)$$

$$= \frac{2\cos^2 t - 2\sin^2 t}{1 + 4\cos^2 t \sin^2 t}$$

$$\frac{d^2w}{dt^2} = \frac{(1 + 4\cos^2 t \sin^2 t)(-8\cos t \sin t) - (2\cos^2 t - 2\sin^2 t)(8\cos^3 t \sin t - 8\sin^3 t \cos t)}{(1 + 4\cos^2 t \sin^2 t)}$$

At $t = 0, \dfrac{d^2w}{dt^2} = 0.$

13. $w = x^2 + y^2$

$x = s + t$

$y = s - t$

$\dfrac{\partial w}{\partial s} = 2x + 2y = 2(x + y) = 4s$

$\dfrac{\partial w}{\partial t} = 2x + 2y(-1) = 2(x - y) = 4t$

When $s = 2$ and $t = -1$,

$\dfrac{\partial w}{\partial s} = 8$ and $\dfrac{\partial w}{\partial t} = -4.$

15. $w = x^2 - y^2$

$x = s \cos t$

$y = s \sin t$

$\dfrac{\partial w}{\partial s} = 2x \cos t - 2y \sin t$

$= 2s \cos^2 t - 2s \sin^2 t = 2s \cos 2t$

$\dfrac{\partial w}{\partial t} = 2x(-s \sin t) - 2y(s \cos t) = -2s^2 \sin 2t$

When $s = 3$ and $t = \dfrac{\pi}{4}, \dfrac{\partial w}{\partial s} = 0$ and $\dfrac{\partial w}{\partial t} = -18.$

17. $w = x^2 - 2xy + y^2,\ x = r + \theta,\ y = r - \theta$

(a) $\dfrac{\partial w}{\partial r} = (2x - 2y)(1) + (-2x + 2y)(1) = 0$

$$\frac{\partial w}{\partial \theta} = (2x - 2y)(1) + (-2x + 2y)(-1)$$
$$= 4x - 4y = 4(x - y)$$
$$= 4[(r + \theta) - (r - \theta)] = 8\theta$$

(b) $w = (r + \theta)^2 - 2(r + \theta)(r - \theta) + (r - \theta)^2$

$$= (r^2 + 2r\theta + \theta^2) - 2(r^2 - \theta^2) + (r^2 - 2r\theta + \theta^2)$$
$$= 4\theta^2$$
$$\frac{\partial w}{\partial r} = 0$$
$$\frac{\partial w}{\partial \theta} = 8\theta$$

19. $w = \arctan \dfrac{y}{x},\ x = r \cos \theta,\ y = r \sin \theta$

(a) $\dfrac{\partial w}{\partial r} = \dfrac{-y}{x^2 + y^2} \cos \theta + \dfrac{x}{x^2 + y^2} \sin \theta = \dfrac{-r \sin \theta \cos \theta}{r^2} + \dfrac{r \cos \theta \sin \theta}{r^2} = 0$

$$\frac{\partial w}{\partial \theta} = \frac{-y}{x^2 + y^2}(-r \sin \theta) + \frac{x}{x^2 + y^2}(r \cos \theta) = \frac{-(r \sin \theta)(-r \sin \theta)}{r^2} + \frac{(r \cos \theta)(r \cos \theta)}{r^2} = 1$$

(b) $w = \arctan \dfrac{r \sin \theta}{r \cos \theta} = \arctan(\tan \theta) = \theta$

$$\frac{\partial w}{\partial r} = 0$$
$$\frac{\partial w}{\partial \theta} = 1$$

21. $x^2 - 3xy + y^2 - 2x + y - 5 = 0$

$$\frac{dy}{dx} = -\frac{F_x(x, y)}{F_y(x, y)} = -\frac{2x - 3y - 2}{-3x + 2y + 1}$$
$$= \frac{3y - 2x + 2}{2y - 3x + 1}$$

23. $\ln \sqrt{x^2 + y^2} + xy = 4$

$$\frac{1}{2} \ln(x^2 + y^2) + xy - 4 = 0$$

$$\frac{dy}{dx} = -\frac{F_x(x, y)}{F_y(x, y)} = -\frac{\dfrac{x}{x^2 + y^2} + y}{\dfrac{y}{x^2 + y^2} + x} = -\frac{x + x^2y + y^3}{y + xy^2 + x^3}$$

25. $F(x, y, z) = x^2 + y^2 + z^2 - 25$

$$F_x = 2x$$
$$F_y = 2y$$
$$F_z = 2z$$
$$\frac{\partial z}{\partial x} = -\frac{F_x}{F_z} = -\frac{x}{z}$$
$$\frac{\partial z}{\partial y} = -\frac{F_y}{F_z} = -\frac{y}{z}$$

27. $F(x, y, z) = \tan(x + y) + \tan(y + z) - 1$

$$F_x = \sec^2(x + y)$$
$$F_y = \sec^2(x + y) + \sec^2(y + z)$$
$$F_z = \sec^2(y + z)$$
$$\frac{\partial z}{\partial x} = -\frac{F_x}{F_z} = -\frac{\sec^2(x + y)}{\sec^2(y + z)}$$
$$\frac{\partial z}{\partial y} = -\frac{F_y}{F_z} = -\frac{\sec^2(x + y) + \sec^2(y + z)}{\sec^2(y + z)}$$
$$= -\left(\frac{\sec^2(x + y)}{\sec^2(y + z)} + 1\right)$$

29. $x^2 + 2yz + z^2 - 1 = 0$

(i) $2x + 2y\dfrac{\partial z}{\partial x} + 2z\dfrac{\partial z}{\partial x} = 0$ implies $\dfrac{\partial z}{\partial x} = -\dfrac{x}{y + z}$.

(ii) $2y\dfrac{\partial z}{\partial y} + 2z + 2z\dfrac{\partial z}{\partial y} = 0$ implies $\dfrac{\partial z}{\partial y} = -\dfrac{z}{y + z}$.

31. $e^{xz} + xy - 0 = 0$

$$\frac{\partial z}{\partial x} = -\frac{F_x(x, y, z)}{F_z(x, y, z)} = -\frac{ze^{xz} + y}{xe^{xz}}$$
$$\frac{\partial z}{\partial y} = -\frac{F_y(x, y, z)}{F_z(x, y, z)} = \frac{-x}{xe^{xz}} = \frac{-1}{e^{xz}} = -e^{-xz}$$

33. $F(x, y, z, w) = xyz + xzw - yzw + w^2 - 5$

$$F_x = yz + zw$$
$$F_y = xz - zw$$
$$F_z = xy + xw - yw$$
$$F_w = xz - yz + 2w$$
$$\frac{\partial w}{\partial x} = -\frac{F_x}{F_w} = -\frac{z(y + w)}{xz - yz + 2w}$$
$$\frac{\partial w}{\partial y} = -\frac{F_y}{F_w} = -\frac{z(x - w)}{xz - yz + 2w}$$
$$\frac{\partial w}{\partial z} = -\frac{F_z}{F_w} = -\frac{xy + xw - yw}{xz - yz + 2w}$$

35. $F(x, y, z, w) = \cos xy + \sin yz + wz - 20$

$$\frac{\partial w}{\partial x} = \frac{-F_x}{F_w} = \frac{y \sin xy}{z}$$
$$\frac{\partial w}{\partial y} = \frac{-F_y}{F_w} = \frac{x \sin xy - z \cos yz}{z}$$
$$\frac{\partial w}{\partial z} = \frac{-F_z}{F_w} = -\frac{y \cos zy + w}{z}$$

37. $f(x, y) = \dfrac{xy}{\sqrt{x^2 + y^2}}$

$$f(tx, ty) = \frac{(tx)(ty)}{\sqrt{(tx)^2 + (ty)^2}} = t\left(\frac{xy}{\sqrt{x^2 + y^2}}\right) = tf(x, y)$$

Degree: 1

$$xf_x(x, y) + yf_y(x, y) = x\left(\frac{y^3}{(x^2 + y^2)^{3/2}}\right) + y\left(\frac{x^3}{(x^2 + y^2)^{3/2}}\right)$$
$$= \frac{xy}{\sqrt{x^2 + y^2}} = 1f(x, y)$$

39. $f(x, y) = e^{x/y}$

$$f(tx, ty) = e^{tx/ty} = e^{x/y} = f(x, y)$$

Degree: 0

$$xf_x(x, y) + yf_y(x, y) = x\left(\frac{1}{y}e^{x/y}\right) + y\left(-\frac{x}{y^2}e^{x/y}\right) = 0$$

41. $A = \dfrac{1}{2}bh = \left(x \sin\dfrac{\theta}{2}\right)\left(x \cos\dfrac{\theta}{2}\right) = \dfrac{x^2}{2}\sin\theta$

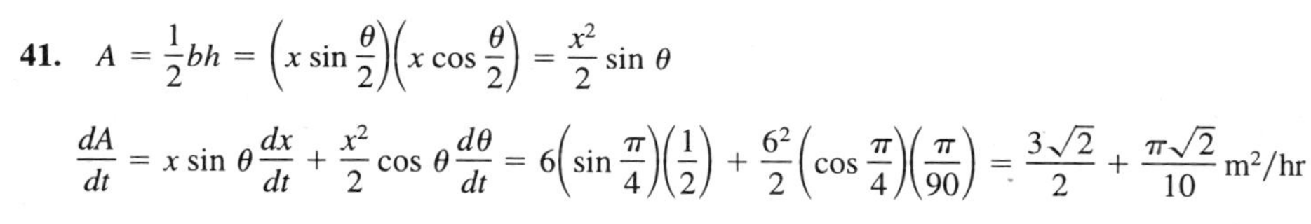

$$\frac{dA}{dt} = x \sin\theta\frac{dx}{dt} + \frac{x^2}{2}\cos\theta\frac{d\theta}{dt} = 6\left(\sin\frac{\pi}{4}\right)\left(\frac{1}{2}\right) + \frac{6^2}{2}\left(\cos\frac{\pi}{4}\right)\left(\frac{\pi}{90}\right) = \frac{3\sqrt{2}}{2} + \frac{\pi\sqrt{2}}{10}\text{ m}^2/\text{hr}$$

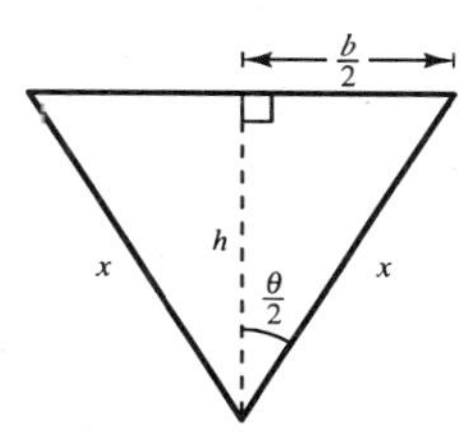

43. (a) $V = \dfrac{1}{3}\pi r^2 h$

$$\frac{dV}{dt} = \frac{1}{3}\pi\left(2rh\frac{dr}{dt} + r^2\frac{dh}{dt}\right) = \frac{1}{3}\pi[2(12)(36)(6) + (12)^2(-4)] = 1536\pi\text{ in.}^3/\text{min}$$

(b) $S = \pi r\sqrt{r^2 + h^2} + \pi r^2$ (Surface area includes base.)

$$\frac{dS}{dt} = \pi\left[\left(\sqrt{r^2 + h^2} + \frac{r^2}{\sqrt{r^2 + h^2}} + 2r\right)\frac{dr}{dt} + \frac{rh}{\sqrt{r^2 + h^2}}\frac{dh}{dt}\right]$$
$$= \pi\left[\left(\sqrt{12^2 + 36^2} + \frac{144}{\sqrt{12^2 + 36^2}} + 2(12)\right)(6) + \frac{36(12)}{\sqrt{12^2 + 36^2}}(-4)\right]$$
$$= \pi\left[\left(12\sqrt{10} + \frac{12}{\sqrt{10}}\right)(6) + 144 + \frac{36}{\sqrt{10}}(-4)\right] = \frac{648\pi}{\sqrt{10}} + 144\pi\text{ in.}^2/\text{min} = \frac{36\pi}{5}(20 + 9\sqrt{10})\text{ in.}^2/\text{min}$$

45. $I = \dfrac{1}{2}m(r_1^2 + r_2^2)$

$$\frac{dI}{dt} = \frac{1}{2}m\left[2r_1\frac{dr_1}{dt} + 2r_2\frac{dr_2}{dt}\right] = m[(6)(2) + (8)(2)] = 28m\text{ cm}^2/\text{sec}$$

47. (a) $x = 64(\cos 45°)t = 32\sqrt{2}t$

$y = 64(\sin 45°)t - 16t^2 = 32\sqrt{2}t - 16t^2$

(b) $\tan \alpha = \dfrac{y}{x + 50}$

$$\alpha = \arctan\left(\frac{y}{x+50}\right) = \arctan\left(\frac{32\sqrt{2}t - 16t^2}{32\sqrt{2}t + 50}\right)$$

(c) $$\frac{d\alpha}{dt} = \frac{1}{1 + \left(\dfrac{32\sqrt{2}t - 16t^2}{32\sqrt{2}t + 50}\right)^2} \cdot \frac{-64(8\sqrt{2}t^2 + 25t - 25\sqrt{2})}{(32\sqrt{2}t + 50)^2} = \frac{-16(8\sqrt{2}t^2 + 25t - 25\sqrt{2})}{64t^4 - 256\sqrt{2}t^3 + 1024t^2 + 800\sqrt{2}t + 625}$$

(d)

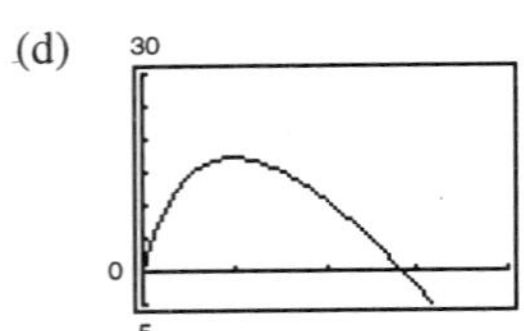

No. The rate of change of α is greatest when the projectile is closest to the camera.

(e) $\dfrac{d\alpha}{dt} = 0$ when

$$8\sqrt{2}t^2 + 25t - 25\sqrt{2} = 0$$

$$t = \frac{-25 + \sqrt{25^2 - 4(8\sqrt{2})(-25\sqrt{2})}}{2(8\sqrt{2})} \approx 0.98 \text{ second.}$$

The projectile is at its maximum height when $dy/dt = 32\sqrt{2} - 32t = 0$ or $t = \sqrt{2} \approx 1.41$ seconds.

49. $g(t) = f(xt, yt) = t^n f(x, y)$

Let $u = xt,\ v = yt,$ then

$$g'(t) = \frac{\partial f}{\partial u}\cdot\frac{du}{dt} + \frac{\partial f}{\partial v}\cdot\frac{dv}{dt} = \frac{\partial f}{\partial u}x + \frac{\partial f}{\partial v}y$$

and

$$g'(t) = nt^{n-1}f(x, y).$$

Now, let $t = 1$ and we have $u = x,\ v = y$. Thus,

$$\frac{\partial f}{\partial x}x + \frac{\partial f}{\partial y}y = nf(x, y).$$

51. $$w = (x - y)\sin(y - x)$$

$$\frac{\partial w}{\partial x} = -(x - y)\cos(y - x) + \sin(y - x)$$

$$\frac{\partial w}{\partial y} = (x - y)\cos(y - x) - \sin(y - x)$$

$$\frac{\partial w}{\partial x} + \frac{\partial w}{\partial y} = 0$$

53. $w = \arctan\dfrac{y}{x},\ x = r\cos\theta,\ y = r\sin\theta$

$$= \arctan\left(\frac{r\sin\theta}{r\cos\theta}\right) = \arctan(\tan\theta) = \theta \text{ for } -\frac{\pi}{2} < \theta < \frac{\pi}{2}$$

$$\frac{\partial w}{\partial x} = \frac{-y}{x^2 + y^2},\quad \frac{\partial w}{\partial y} = \frac{x}{x^2 + y^2},\quad \frac{\partial w}{\partial r} = 0,\quad \frac{\partial w}{\partial \theta} = 1$$

$$\left(\frac{\partial w}{\partial x}\right)^2 + \left(\frac{\partial w}{\partial y}\right)^2 = \frac{y^2}{(x^2 + y^2)^2} + \frac{x^2}{(x^2 + y^2)^2} = \frac{1}{x^2 + y^2} = \frac{1}{r^2}$$

$$\left(\frac{\partial w}{\partial r}\right)^2 + \left(\frac{1}{r^2}\right)\left(\frac{\partial w}{\partial \theta}\right)^2 = 0 + \frac{1}{r^2}(1) = \frac{1}{r^2}$$

Therefore, $\left(\dfrac{\partial w}{\partial x}\right)^2 + \left(\dfrac{\partial w}{\partial y}\right)^2 = \left(\dfrac{\partial w}{\partial r}\right)^2 + \dfrac{1}{r^2}\left(\dfrac{\partial w}{\partial \theta}\right)^2.$

55. Note first that

$$\frac{\partial u}{\partial x} = \frac{\partial v}{\partial y} = \frac{x}{x^2 + y^2}$$

$$\frac{\partial u}{\partial y} = -\frac{\partial v}{\partial x} = \frac{y}{x^2 + y^2}.$$

$$\frac{\partial u}{\partial r} = \frac{x}{x^2 + y^2}\cos\theta + \frac{y}{x^2 + y^2}\sin\theta = \frac{r\cos^2\theta + r\sin^2\theta}{r^2} = \frac{1}{r}$$

$$\frac{\partial v}{\partial \theta} = \frac{-y}{x^2 + y^2}(-r\sin\theta) + \frac{x}{x^2 + y^2}(r\cos\theta) = \frac{r^2\sin^2\theta + r^2\cos^2\theta}{r^2} = 1$$

Thus, $\dfrac{\partial u}{\partial r} = \dfrac{1}{r}\dfrac{\partial v}{\partial \theta}$.

$$\frac{\partial v}{\partial r} = \frac{-y}{x^2 + y^2}\cos\theta + \frac{x}{x^2 + y^2}\sin\theta = \frac{-r\sin\theta\cos\theta + r\sin\theta\cos\theta}{r^2} = 0$$

$$\frac{\partial u}{\partial \theta} = \frac{x}{x^2 + y^2}(-r\sin\theta) + \frac{y}{x^2 + y^2}(r\cos\theta) = \frac{-r^2\sin\theta\cos\theta + r^2\sin\theta\cos\theta}{r^2} = 0$$

Thus, $\dfrac{\partial v}{\partial r} = -\dfrac{1}{r}\dfrac{\partial u}{\partial \theta}$.

Section 12.6 Directional Derivatives and Gradients

1. $f(x, y) = 3x - 4xy + 5y$

$$\mathbf{v} = \frac{1}{2}\left(\mathbf{i} + \sqrt{3}\,\mathbf{j}\right)$$

$$\nabla f(x, y) = (3 - 4y)\mathbf{i} + (-4x + 5)\mathbf{j}$$

$$\nabla f(1, 2) = -5\mathbf{i} + \mathbf{j}$$

$$\mathbf{u} = \frac{\mathbf{v}}{\|\mathbf{v}\|} = \frac{1}{2}\mathbf{i} + \frac{\sqrt{3}}{2}\mathbf{j}$$

$$D_{\mathbf{u}}f(1, 2) = \nabla f(1, 2)\cdot\mathbf{u} = \frac{1}{2}\left(-5 + \sqrt{3}\right)$$

3. $f(x, y) = xy$

$$\mathbf{v} = \mathbf{i} + \mathbf{j}$$

$$\nabla f(x, y) = y\mathbf{i} + x\mathbf{j}$$

$$\nabla f(2, 3) = 3\mathbf{i} + 2\mathbf{j}$$

$$\mathbf{u} = \frac{\mathbf{v}}{\|\mathbf{v}\|} = \frac{\sqrt{2}}{2}\mathbf{i} + \frac{\sqrt{2}}{2}\mathbf{j}$$

$$D_{\mathbf{u}}f(2, 3) = \nabla f(2, 3)\cdot\mathbf{u} = \frac{5\sqrt{2}}{2}$$

5. $g(x, y) = \sqrt{x^2 + y^2}$

$$\mathbf{v} = 3\mathbf{i} - 4\mathbf{j}$$

$$\nabla g = \frac{x}{\sqrt{x^2 + y^2}}\mathbf{i} + \frac{y}{\sqrt{x^2 + y^2}}\mathbf{j}$$

$$\nabla g(3, 4) = \frac{3}{5}\mathbf{i} + \frac{4}{5}\mathbf{j}$$

$$\mathbf{u} = \frac{\mathbf{v}}{\|\mathbf{v}\|} = \frac{3}{5}\mathbf{i} - \frac{4}{5}\mathbf{j}$$

$$D_{\mathbf{u}}g(3, 4) = \nabla g(3, 4)\cdot\mathbf{u} = -\frac{7}{25}$$

7. $h(x, y) = e^x\sin y$

$$\mathbf{v} = -\mathbf{i}$$

$$\nabla h = e^x\sin y\,\mathbf{i} + e^x\cos y\,\mathbf{j}$$

$$h\left(1, \frac{\pi}{2}\right) = e\mathbf{i}$$

$$\mathbf{u} = \frac{\mathbf{v}}{\|\mathbf{v}\|} = -\mathbf{i}$$

$$D_{\mathbf{u}}h\left(1, \frac{\pi}{2}\right) = \nabla h\left(1, \frac{\pi}{2}\right)\cdot\mathbf{u} = -e$$

9. $f(x, y, z) = xy + yz + xz$

$$\mathbf{v} = 2\mathbf{i} + \mathbf{j} - \mathbf{k}$$

$$\nabla f(x, y, z) = (y + z)\mathbf{i} + (x + z)\mathbf{j} + (x + y)\mathbf{k}$$

$$\nabla f(1, 1, 1) = 2\mathbf{i} + 2\mathbf{j} + 2\mathbf{k}$$

$$\mathbf{u} = \frac{\mathbf{v}}{\|\mathbf{v}\|} = \frac{\sqrt{6}}{3}\mathbf{i} + \frac{\sqrt{6}}{6}\mathbf{j} - \frac{\sqrt{6}}{6}\mathbf{k}$$

$$D_{\mathbf{u}} f(1, 1, 1) = \nabla f(1, 1, 1) \cdot \mathbf{u} = \frac{2\sqrt{6}}{3}$$

11. $h(x, y, z) = x \arctan yz$

$$\mathbf{v} = \langle 1, 2, -1 \rangle$$

$$\nabla h(x, y, z) = \arctan yz\mathbf{i} + \frac{xz}{1 + (yz)^2}\mathbf{j} + \frac{xy}{1 + (yz)^2}\mathbf{k}$$

$$\nabla h(4, 1, 1) = \frac{\pi}{4}\mathbf{i} + 2\mathbf{j} + 2\mathbf{k}$$

$$\mathbf{u} = \frac{\mathbf{v}}{\|\mathbf{v}\|} = \left\langle \frac{1}{\sqrt{6}}, \frac{2}{\sqrt{6}}, -\frac{1}{\sqrt{6}} \right\rangle$$

$$D_{\mathbf{u}} h(4, 1, 1) = \nabla h(4, 1, 1) \cdot \mathbf{u} = \frac{\pi + 8}{4\sqrt{6}} = \frac{(\pi + 8)\sqrt{6}}{24}$$

13. $f(x, y) = x^2 + y^2$

$$\mathbf{u} = \frac{1}{\sqrt{2}}\mathbf{i} + \frac{1}{\sqrt{2}}\mathbf{j}$$

$$\nabla f = 2x\mathbf{i} + 2y\mathbf{j}$$

$$D_{\mathbf{u}} f = \nabla f \cdot \mathbf{u} = \frac{2}{\sqrt{2}}x + \frac{2}{\sqrt{2}}y = \sqrt{2}(x + y)$$

15. $f(x, y) = \sin(2x - y)$

$$\mathbf{u} = \frac{1}{2}\mathbf{i} - \frac{\sqrt{3}}{2}\mathbf{j}$$

$$\nabla f = 2\cos(2x - y)\mathbf{i} - \cos(2x - y)\mathbf{j}$$

$$D_{\mathbf{u}} f = \nabla f \cdot \mathbf{u} = \cos(2x - y) + \frac{\sqrt{3}}{2}\cos(2x - y)$$

$$= \left(\frac{2 + \sqrt{3}}{2}\right)\cos(2x - y)$$

17. $f(x, y) = x^2 + 4y^2$

$$\mathbf{v} = -2\mathbf{i} - 2\mathbf{j}$$

$$\nabla f = 2x\mathbf{i} + 8y\mathbf{j}$$

$$\mathbf{u} = \frac{\mathbf{v}}{\|\mathbf{v}\|} = -\frac{1}{\sqrt{2}}\mathbf{i} - \frac{1}{\sqrt{2}}\mathbf{j}$$

$$D_{\mathbf{u}} f = -\frac{2}{\sqrt{2}}x - \frac{8}{\sqrt{2}}y = -\sqrt{2}(x + 4y)$$

At $P = (3, 1)$, $D_{\mathbf{u}} f = -7\sqrt{2}$.

19. $h(x, y, z) = \ln(x + y + z)$

$$\mathbf{v} = 3\mathbf{i} + 3\mathbf{j} + \mathbf{k}$$

$$\nabla h = \frac{1}{x + y + z}(\mathbf{i} + \mathbf{j} + \mathbf{k})$$

At $(1, 0, 0)$, $\nabla h = \mathbf{i} + \mathbf{j} + \mathbf{k}$.

$$\mathbf{u} = \frac{\mathbf{v}}{\|\mathbf{v}\|} = \frac{1}{\sqrt{19}}(3\mathbf{i} + 3\mathbf{j} + \mathbf{k})$$

$$D_{\mathbf{u}} h = \nabla h \cdot \mathbf{u} = \frac{7}{\sqrt{19}} = \frac{7\sqrt{19}}{19}$$

21. (a) In the direction of the vector $-4\mathbf{i} + \mathbf{j}$.

(c) $-\nabla f = \frac{2}{5}\mathbf{i} - \frac{1}{10}\mathbf{j}$, the direction opposite that of the gradient.

(b) $\nabla f = \frac{1}{10}(2x - 3y)\mathbf{i} + \frac{1}{10}(-3x + 2y)\mathbf{j}$

$\nabla f(1, 2) = \frac{1}{10}(-4)\mathbf{i} + \frac{1}{10}(1)\mathbf{j} = -\frac{2}{5}\mathbf{i} + \frac{1}{10}\mathbf{j}$

(Same direction as in part (a).)

23. $f(x, y) = x^2 - y^2$, $(4, -3, 7)$

(a)

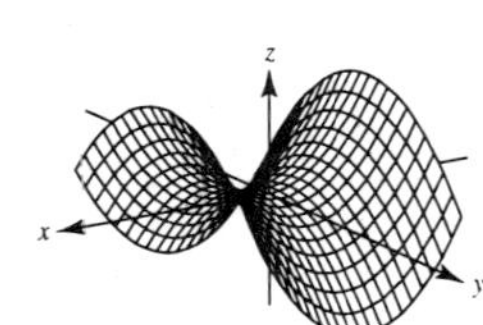

(c) Zeros: $\theta \approx 2.21, 5.36$

These are the angles θ for which $D_{\mathbf{u}} f(4, 3)$ equals zero.

—CONTINUED—

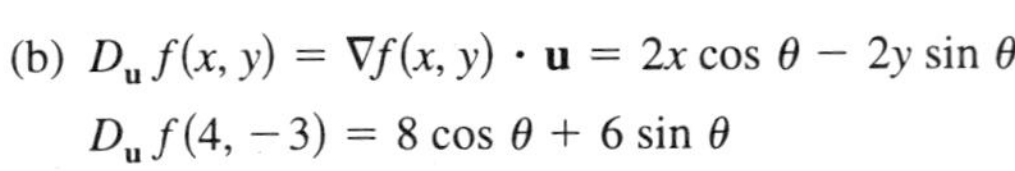

(b) $D_{\mathbf{u}} f(x, y) = \nabla f(x, y) \cdot \mathbf{u} = 2x\cos\theta - 2y\sin\theta$

$D_{\mathbf{u}} f(4, -3) = 8\cos\theta + 6\sin\theta$

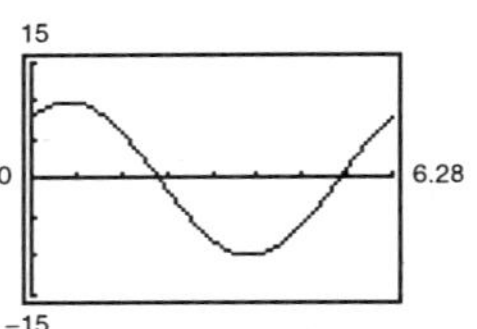

(d) $g(\theta) = D_{\mathbf{u}} f(4, -3) = 8\cos\theta + 6\sin\theta$

$g'(\theta) = -8\sin\theta + 6\cos\theta$

Critical numbers: $\theta \approx 0.64, 3.79$

These are the angles for which $D_{\mathbf{u}} f(4, -3)$ is a maximum (0.64) and minimum (3.79).

23. —CONTINUED—

(e) $\|\nabla f(4, -3)\| = \|2(4)\mathbf{i} - 2(3)\mathbf{j}\| = \sqrt{64 + 36} = 10$, the maximum value of $D_{\mathbf{u}} f(4, -3)$, at $\theta = 0.64$.

(f) $f(x, y) = x^2 - y^2 = 7$

$\nabla f(4, -3) = 8\mathbf{i} + 6\mathbf{j}$ is perpendicular to the level curve at $(4, -3)$.

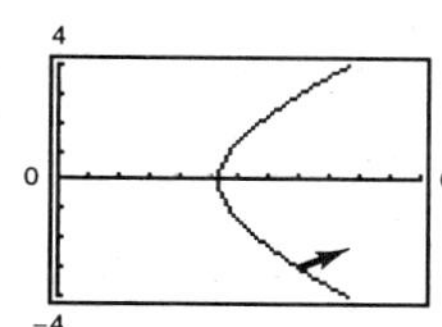

25. $h(x, y) = x \tan y$

$\nabla h(x, y) = \tan y\mathbf{i} + x \sec^2 y\mathbf{j}$

$\nabla h\left(2, \frac{\pi}{4}\right) = \mathbf{i} + 4\mathbf{j}$

$\left\| \nabla h\left(2, \frac{\pi}{4}\right) \right\| = \sqrt{17}$

27. $g(x, y) = \ln \sqrt[3]{x^2 + y^2} = \frac{1}{3} \ln(x^2 + y^2)$

$\nabla g(x, y) = \frac{1}{3}\left[\frac{2x}{x^2 + y^2}\mathbf{i} + \frac{2y}{x^2 - y^2}\mathbf{j}\right]$

$\nabla g(1, 2) = \frac{1}{3}\left(\frac{2}{5}\mathbf{i} + \frac{4}{5}\mathbf{j}\right) = \frac{2}{15}(\mathbf{i} + 2\mathbf{j})$

$\|\nabla g(1, 2)\| = \frac{2\sqrt{5}}{15}$

29. $f(x, y, z) = \sqrt{x^2 + y^2 + z^2}$

$\nabla f(x, y, z) = \frac{1}{\sqrt{x^2 + y^2 + z^2}}(x\mathbf{i} + y\mathbf{j} + z\mathbf{k})$

$\nabla f(1, 4, 2) = \frac{1}{\sqrt{21}}(\mathbf{i} + 4\mathbf{j} + 2\mathbf{k})$

$\|\nabla f(1, 4, 2)\| = 1$

31. $w = \frac{1}{\sqrt{1 - x^2 - y^2 - z^2}}$

$\nabla w = \frac{1}{\left(\sqrt{1 - x^2 - y^2 - z^2}\right)^3}(x\mathbf{i} + y\mathbf{j} + z\mathbf{k})$

$\nabla w(0, 0, 0) = \mathbf{0}$

$\|\nabla w(0, 0, 0)\| = 0$

For Exercises 33–39, $f(x, y) = 3 - \frac{x}{3} - \frac{y}{2}$ and $D_\theta f(x, y) = -\left(\frac{1}{3}\right)\cos\theta - \left(\frac{1}{2}\right)\sin\theta$.

33. $f(x, y) = 3 - \frac{x}{3} - \frac{y}{2}$

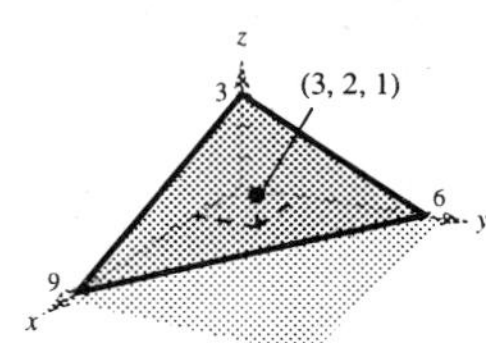

35. (a) $D_{4\pi/3} f(3, 2) = -\left(\frac{1}{3}\right)\left(-\frac{1}{2}\right) - \left(\frac{1}{2}\right)\left(-\frac{\sqrt{3}}{2}\right)$

$= \frac{2 + 3\sqrt{3}}{12}$

(b) $D_{-\pi/6} f(3, 2) = -\left(\frac{1}{3}\right)\left(\frac{\sqrt{3}}{2}\right) - \left(\frac{1}{2}\right)\left(-\frac{1}{2}\right)$

$= \frac{3 - 2\sqrt{3}}{12}$

37. (a) $\mathbf{v} = -3\mathbf{i} + 4\mathbf{j}$

$\|\mathbf{v}\| = \sqrt{9 + 16} = 5$

$\mathbf{u} = -\frac{3}{5}\mathbf{i} + \frac{4}{5}\mathbf{j}$

$D_{\mathbf{u}} f = \nabla f \cdot \mathbf{u} = \frac{1}{5} - \frac{2}{5} = -\frac{1}{5}$

(b) $\mathbf{v} = \mathbf{i} + 3\mathbf{j}$

$\|\mathbf{v}\| = \sqrt{10}$

$\mathbf{u} = \frac{1}{\sqrt{10}}\mathbf{i} + \frac{3}{\sqrt{10}}\mathbf{j}$

$D_{\mathbf{u}} f = \nabla f \cdot \mathbf{u} = \frac{-11}{6\sqrt{10}} = -\frac{11\sqrt{10}}{60}$

39. $\|\nabla f\| = \sqrt{\frac{1}{9} + \frac{1}{4}} = \frac{1}{6}\sqrt{13}$

41. $f(x, y) = 9 - x^2 - y^2$

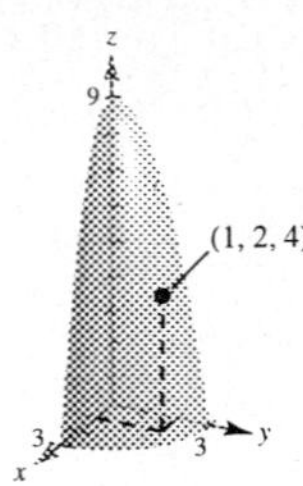

For Exercises 42–44, $f(x, y) = 9 - x^2 - y^2$ and $D_\theta f(x, y) = -2x \cos\theta - 2y \sin\theta = -2(x \cos\theta + y \sin\theta)$.

43. $\nabla f(1, 2) = -2\mathbf{i} - 4\mathbf{j}$

$\|\nabla f(1, 2)\| = \sqrt{4 + 16} = \sqrt{20} = 2\sqrt{5}$

45. $f(x, y) = x^2 + y^2$

$c = 25, \ P = (3, 4)$

$\nabla f(x, y) = 2x\mathbf{i} + 2y\mathbf{j}$

$x^2 + y^2 = 25$

$\nabla f(3, 4) = 6\mathbf{i} + 8\mathbf{j}$

47. $f(x, y) = \dfrac{x}{x^2 + y^2}$

$c = \dfrac{1}{2}, \ P = (1, 1)$

$\nabla f(x, y) = \dfrac{y^2 - x^2}{(x^2 + y^2)^2}\mathbf{i} - \dfrac{2xy}{(x^2 + y^2)^2}\mathbf{j}$

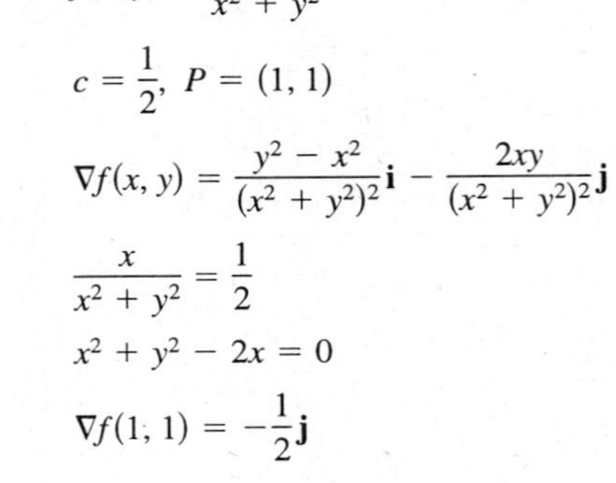

$\dfrac{x}{x^2 + y^2} = \dfrac{1}{2}$

$x^2 + y^2 - 2x = 0$

$\nabla f(1, 1) = -\dfrac{1}{2}\mathbf{j}$

49. $4x^2 - y = 6$

$f(x, y) = 4x^2 - y$

$\nabla f(x, y) = 8x\mathbf{i} - \mathbf{j}$

$\nabla f(2, 10) = 16\mathbf{i} - \mathbf{j}$

$\dfrac{\nabla f(2, 10)}{\|\nabla f(2, 10)\|} = \dfrac{1}{\sqrt{257}}(16\mathbf{i} - \mathbf{j})$

$= \dfrac{\sqrt{257}}{257}(16\mathbf{i} - \mathbf{j})$

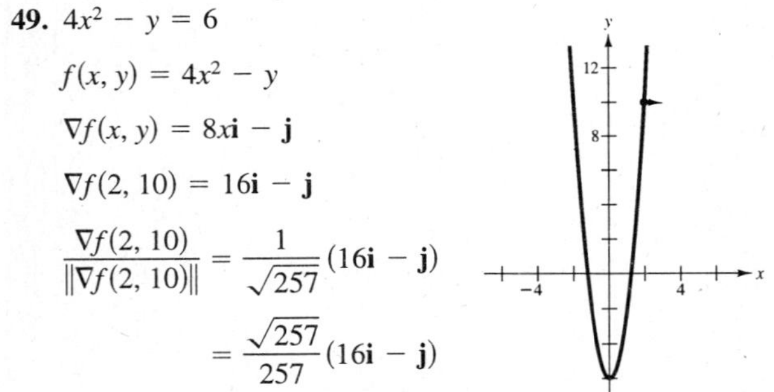

51. $9x^2 + 4y^2 = 40$

$f(x, y) = 9x^2 + 4y^2$

$\nabla f(x, y) = 18x\mathbf{i} + 8y\mathbf{j}$

$\nabla f(2, -1) = 36\mathbf{i} - 8\mathbf{j}$

$\dfrac{\nabla f(2, -1)}{\|\nabla f(2, -1)\|} = \dfrac{1}{\sqrt{85}}(9\mathbf{i} - 2\mathbf{j})$

$= \dfrac{\sqrt{85}}{85}(9\mathbf{i} - 2\mathbf{j})$

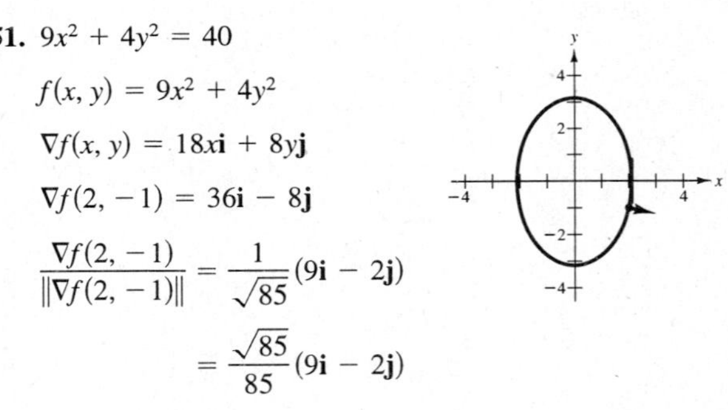

53. $T = \dfrac{x}{x^2 + y^2}$

$\nabla T = \dfrac{y^2 - x^2}{(x^2 + y^2)^2}\mathbf{i} - \dfrac{2xy}{(x^2 + y^2)^2}\mathbf{j}$

$\nabla T(3, 4) = \dfrac{7}{625}\mathbf{i} - \dfrac{24}{625}\mathbf{j} = \dfrac{1}{625}(7\mathbf{i} - 24\mathbf{j})$

55. $T(x, y) = 400 - 2x^2 - y^2, \qquad P = (10, 10)$

$\dfrac{dx}{dt} = -4x \qquad \dfrac{dy}{dt} = -2y$

$x(t) = C_1 e^{-4t} \qquad y(t) = C_2 e^{-2t}$

$10 = x(0) = C_1 \qquad 10 = y(0) = C_2$

$x(t) = 10e^{-4t} \qquad y(t) = 10e^{-2t}$

$x = \dfrac{y^2}{10} \qquad y^2(t) = 100e^{-4t}$

$y^2 = 10x$

57. (a)

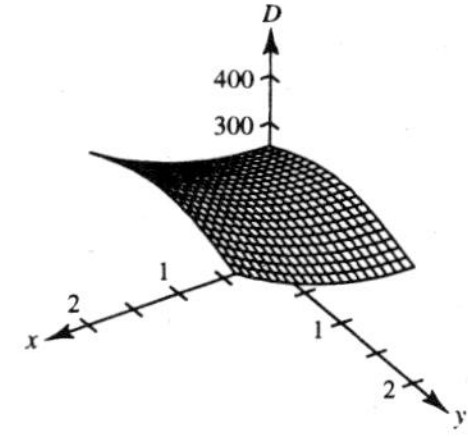

(b) The graph of $-D = -250 - 30x^2 - 50\sin(\pi y/2)$ would model the ocean floor.

(c) $D(1, 0.5) = 250 + 30(1) + 50\sin\frac{\pi}{4} \approx 315.4$ ft

(d) $\frac{\partial D}{\partial x} = 60x$ and $\frac{\partial D}{\partial x}(1, 0.5) = 50$

(e) $\frac{\partial D}{\partial y} = 25\pi\cos\frac{\pi y}{2}$ and $\frac{\partial D}{\partial y}(1, 0.5) = 25\pi\cos\frac{\pi}{4} \approx 55.5$

(f) $\nabla D = 60x\mathbf{i} + 25\pi\cos\left(\frac{\pi y}{2}\right)\mathbf{j}$

$\nabla D(1, 0.5) = 60\mathbf{i} + 55.5\mathbf{j}$

59. The wind speed is greatest at A.

61. True

63. True

65. Let $f(x, y, z) = e^x\cos y + \frac{z^2}{2} + C$. Then $\nabla f(x, y, z) = e^x\cos y\mathbf{i} - e^x\sin y\mathbf{j} + z\mathbf{k}$.

Section 12.7 Tangent Planes and Normal Lines

1. $F(x, y, z) = x + y + z - 4$

$$\nabla F = \mathbf{i} + \mathbf{j} + \mathbf{k}$$

$$\mathbf{n} = \frac{\nabla F}{\|\nabla F\|} = \frac{1}{\sqrt{3}}(\mathbf{i} + \mathbf{j} + \mathbf{k}) = \frac{\sqrt{3}}{3}(\mathbf{i} + \mathbf{j} + \mathbf{k})$$

3. $F(x, y, z) = \sqrt{x^2 + y^2} - z$

$$\nabla F(x, y, z) = \frac{x}{\sqrt{x^2 + y^2}}\mathbf{i} + \frac{y}{\sqrt{x^2 + y^2}}\mathbf{j} - \mathbf{k}$$

$$\nabla F(3, 4, 5) = \frac{3}{5}\mathbf{i} + \frac{4}{5}\mathbf{j} - \mathbf{k}$$

$$\mathbf{n} = \frac{\nabla F}{\|\nabla F\|} = \frac{5}{5\sqrt{2}}\left(\frac{3}{5}\mathbf{i} + \frac{4}{5}\mathbf{j} - \mathbf{k}\right) = \frac{1}{5\sqrt{2}}(3\mathbf{i} + 4\mathbf{j} - 5\mathbf{k}) = \frac{\sqrt{2}}{10}(3\mathbf{i} + 4\mathbf{j} - 5\mathbf{k})$$

5. $F(x, y, z) = x^2y^4 - z$

$$\nabla F(x, y, z) = 2xy^4\mathbf{i} + 4x^2y^3\mathbf{j} - \mathbf{k}$$

$$\nabla F(1, 2, 16) = 32\mathbf{i} + 32\mathbf{j} - \mathbf{k}$$

$$\mathbf{n} = \frac{\nabla F}{\|\nabla F\|} = \frac{1}{\sqrt{2049}}(32\mathbf{i} + 32\mathbf{j} - \mathbf{k}) = \frac{\sqrt{2049}}{2049}(32\mathbf{i} + 32\mathbf{j} - \mathbf{k})$$

7. $F(x, y, z) = -x\sin y + z - 4$

$$\nabla F(x, y, z) = -\sin y\mathbf{i} - x\cos y\mathbf{j} + \mathbf{k}$$

$$\nabla F\left(6, \frac{\pi}{6}, 7\right) = -\frac{1}{2}\mathbf{i} - 3\sqrt{3}\mathbf{j} - \mathbf{k}$$

$$\mathbf{n} = \frac{\nabla F}{\|\nabla F\|} = \frac{2}{\sqrt{113}}\left(-\frac{1}{2}\mathbf{i} - 3\sqrt{3}\mathbf{j} + \mathbf{k}\right) = \frac{1}{\sqrt{113}}\left(-\mathbf{i} - 6\sqrt{3}\mathbf{j} + 2\mathbf{k}\right) = \frac{\sqrt{113}}{113}\left(-\mathbf{i} - 6\sqrt{3}\mathbf{j} + 2\mathbf{k}\right)$$

9. $F(x, y, z) = \ln\left(\frac{x}{y - z}\right) = \ln x - \ln(y - z)$

$\nabla F(x, y, z) = \frac{1}{x}\mathbf{i} - \frac{1}{y - z}\mathbf{j} + \frac{1}{y - z}\mathbf{k}$

$\nabla F(1, 4, 3) = \mathbf{i} - \mathbf{j} + \mathbf{k}$

$\mathbf{n} = \frac{\nabla F}{\|\nabla F\|} = \frac{1}{\sqrt{3}}(\mathbf{i} - \mathbf{j} + \mathbf{k}) = \frac{\sqrt{3}}{3}(\mathbf{i} - \mathbf{j} + \mathbf{k})$

11. $f(x, y) = 25 - x^2 - y^2, \ (3, 1, 15)$

$F(x, y, z) = 25 - x^2 - y^2 - z$

$F_x(x, y, z) = -2x \qquad F_y(x, y, z) = -2y \qquad F_z(x, y, z) = -1$

$F_x(3, 1, 15) = -6 \qquad F_y(3, 1, 15) = -2 \qquad F_z(3, 1, 15) = -1$

$-6(x - 3) - 2(y - 1) - (z - 15) = 0$

$0 = 6x + 2y + z - 35$

$6x + 2y + z = 35$

13. $f(x, y) = \frac{y}{x}, \ (1, 2, 2)$

$F(x, y, z) = \frac{y}{x} - z$

$F_x(x, y, z) = -\frac{y}{x^2} \qquad F_y(x, y, z) = \frac{1}{x} \qquad F_z(x, y, z) = -1$

$F_x(1, 2, 2) = -2 \qquad F_y(1, 2, 2) = 1 \qquad F_z(1, 2, 2) = -1$

$-2(x - 1) + (y - 2) - (z - 2) = 0$

$-2x + y - z + 2 = 0$

$2x - y + z = 2$

15. $g(x, y) = x^2 - y^2, \ (5, 4, 9)$

$G(x, y, z) = x^2 - y^2 - z$

$G_x(x, y, z) = 2x \qquad G_y(x, y, z) = -2y \qquad G_z(x, y, z) = -1$

$G_x(5, 4, 9) = 10 \qquad G_y(5, 4, 9) = -8 \qquad G_z(5, 4, 9) = -1$

$10(x - 5) - 8(y - 4) - (z - 9) = 0$

$10x - 8y - z = 9$

17. $z = e^x(\sin y + 1), \ \left(0, \frac{\pi}{2}, 2\right)$

$F(x, y, z) = e^x(\sin y + 1) - z$

$F_x(x, y, z) = e^x(\sin y + 1) \qquad F_y(x, y, z) = e^x \cos y \qquad F_z(x, y, z) = -1$

$F_x\left(0, \frac{\pi}{2}, 2\right) = 2 \qquad F_y\left(0, \frac{\pi}{2}, 2\right) = 0 \qquad F_z\left(0, \frac{\pi}{2}, 2\right) = -1$

$2x - z = -2$

19. $h(x, y) = \ln \sqrt{x^2 + y^2}, \ (3, 4, \ln 5)$

$H(x, y, z) = \ln \sqrt{x^2 + y^2} - z = \frac{1}{2}\ln(x^2 + y^2) - z$

$H_x(x, y, z) = \frac{x}{x^2 + y^2}$ $\quad H_y(x, y, z) = \frac{y}{x^2 + y^2}$ $\quad H_z(x, y, z) = -1$

$H_x(3, 4, \ln 5) = \frac{3}{25}$ $\quad H_y(3, 4, \ln 5) = \frac{4}{25}$ $\quad H_z(3, 4, \ln 5) = -1$

$\frac{3}{25}(x - 3) + \frac{4}{25}(y - 4) - (z - \ln 5) = 0$

$3(x - 3) + 4(y - 4) - 25(z - \ln 5) = 0$

$3x + 4y - 25z = 25(1 - \ln 5)$

21. $x^2 + 4y^2 + z^2 = 36, \ (2, -2, 4)$

$F(x, y, z) = x^2 + 4y^2 + z^2 - 36$

$F_x(x, y, z) = 2x$ $\quad F_y(x, y, z) = 8y$ $\quad F_z(x, y, z) = 2z$

$F_x(2, -2, 4) = 4$ $\quad F_y(2, -2, 4) = -16$ $\quad F_z(2, -2, 4) = 8$

$4(x - 2) - 16(y + 2) + 8(z - 4) = 0$

$(x - 2) - 4(y + 2) + 2(z - 4) = 0$

$x - 4y + 2z = 18$

23. $xy^2 + 3x - z^2 = 4, \ (2, 1, -2)$

$F(x, y, z) = xy^2 + 3x - z^2 - 4$

$F_x(x, y, z) = y^2 + 3$ $\quad F_y(x, y, z) = 2xy$ $\quad F_z(x, y, z) = -2z$

$F_x(2, 1, -2) = 4$ $\quad F_y(2, 1, -2) = 4$ $\quad F_Z(2, 1, -2) = 4$

$4(x - 2) + 4(y - 1) + 4(z + 2) = 0$

$x + y + z = 1$

25. $x^2 + y^2 + z = 9, \ (1, 2, 4)$

$F(x, y, z) = x^2 + y^2 + z - 9$

$F_x(x, y, z) = 2x$ $\quad F_y(x, y, z) = 2y$ $\quad F_z(x, y, z) = 1$

$F_x(1, 2, 4) = 2$ $\quad F_y(1, 2, 4) = 4$ $\quad F_z(1, 2, 4) = 1$

Direction numbers: 2, 4, 1

Plane: $2(x - 1) + 4(y - 2) + (z - 4) = 0, \ 2x + 4y + z = 14$

Line: $\frac{x - 1}{2} = \frac{y - 2}{4} = \frac{z - 4}{1}$

27. $xy - z = 0, \ (-2, -3, 6)$

$F(x, y, z) = xy - z$

$F_x(x, y, z) = y$ $\quad F_y(x, y, z) = x$ $\quad F_z(x, y, z) = -1$

$F_x(-2, -3, 6) = -3$ $\quad F_y(-2, -3, 6) = -2$ $\quad F_z(-2, -3, 6) = -1$

Direction numbers: 3, 2, 1

Plane: $3(x + 2) + 2(y + 3) + (z - 6) = 0, \ 3x + 2y + z = -6$

Line: $\frac{x + 2}{3} = \frac{y + 3}{2} = \frac{z - 6}{1}$

29. $z = \arctan \dfrac{y}{x}, \left(1, 1, \dfrac{\pi}{4}\right)$

$F(x, y, z) = \arctan \dfrac{y}{x} - z$

$F_x(x, y, z) = \dfrac{-y}{x^2 + y^2}$ $\qquad F_y(x, y, z) = \dfrac{x}{x^2 + y^2}$ $\qquad F_z(x, y, z) = -1$

$F_x\left(1, 1, \dfrac{\pi}{4}\right) = -\dfrac{1}{2}$ $\qquad F_y\left(1, 1, \dfrac{\pi}{4}\right) = \dfrac{1}{2}$ $\qquad F_z\left(1, 1, \dfrac{\pi}{4}\right) = -1$

Direction numbers: $1, -1, 2$

Plane: $(x - 1) - (y - 1) + 2\left(z - \dfrac{\pi}{4}\right) = 0, \; x - y + 2z = \dfrac{\pi}{2}$

Line: $\dfrac{x - 1}{1} = \dfrac{y - 1}{-1} = \dfrac{z - (\pi/4)}{2}$

31. $z = f(x, y) = \dfrac{4xy}{(x^2 + 1)(y^2 + 1)}, \quad -2 \le x \le z, \; 0 \le y \le 3$

(a) Let $F(x, y, z) = \dfrac{4xy}{(x^2 + 1)(y^2 + 1)} - z$

$$\nabla F(x, y, z) = \frac{4y}{y^2 + 1}\left(\frac{x^2 + 1 - 2x^2}{(x^2 + 1)^2}\right)\mathbf{i} + \frac{4x}{x^2 + 1}\left(\frac{y^2 + 1 - 2y^2}{(y^2 + 1)^2}\right)\mathbf{j} - \mathbf{k}$$

$$= \frac{4y(1 - x^2)}{(y^2 + 1)(x^2 + 1)^2}\mathbf{i} + \frac{4x(1 - y^2)}{(x^2 + 1)(y^2 + 1)^2}\mathbf{j} - \mathbf{k}$$

$\nabla F(1, 1, 1) = -\mathbf{k}.$

Direction numbers: $0, 0, -1.$

Line: $x = 1, \; y = 1, \; z = 1 - t$

Tangent plane: $0(x - 1) + 0(y - 1) - 1(z - 1) = 0 \Rightarrow z = 1$

(b) $\nabla F\left(-1, 2, -\dfrac{4}{5}\right) = 0\mathbf{i} + \dfrac{-4(-3)}{(2)(5)^2}\mathbf{j} - \mathbf{k} = \dfrac{6}{25}\mathbf{j} - \mathbf{k}$

Line: $x = -1, \; y = 2 + \dfrac{6}{25}t, \; z = -\dfrac{4}{5} - t$

Plane: $0(x + 1) + \dfrac{6}{25}(y - 2) - 1\left(z + \dfrac{4}{5}\right) = 0$

$$6y - 12 - 25z - 20 = 0$$

$$6y - 25z - 32 = 0$$

(c)

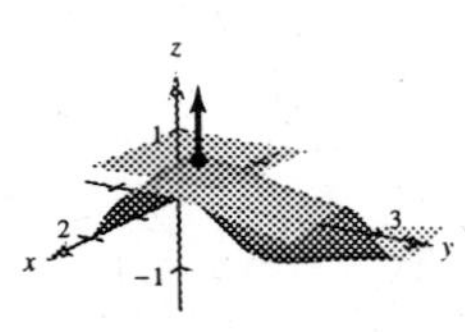

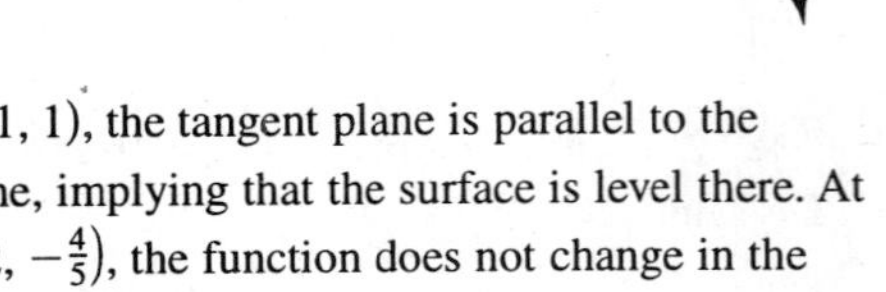

(d) At $(1, 1, 1)$, the tangent plane is parallel to the xy-plane, implying that the surface is level there. At $\left(-1, 2, -\frac{4}{5}\right)$, the function does not change in the x-direction.

33. $F(x, y, z) = x^2 + y^2 - 5$ $\qquad G(x, y, z) = x - z$

$\nabla F(x, y, z) = 2x\mathbf{i} + 2y\mathbf{j}$ $\qquad \nabla G(x, y, z) = \mathbf{i} - \mathbf{k}$

$\nabla F(2, 1, 2) = 4\mathbf{i} + 2\mathbf{j}$ $\qquad \nabla G(2, 1, 2) = \mathbf{i} - \mathbf{k}$

(a) $\nabla F \times \nabla G = \begin{vmatrix} \mathbf{i} & \mathbf{j} & \mathbf{k} \\ 4 & 2 & 0 \\ 1 & 0 & -1 \end{vmatrix} = -2\mathbf{i} + 4\mathbf{j} - 2\mathbf{k} = -2(\mathbf{i} - 2\mathbf{j} + \mathbf{k})$

Direction numbers: $1, -2, 1, \; \dfrac{x - 2}{1} = \dfrac{y - 1}{-2} = \dfrac{z - 2}{1}$

(b) $\cos \theta = \dfrac{|\nabla F \cdot \nabla G|}{\|\nabla F\| \, \|\nabla G\|} = \dfrac{4}{\sqrt{20}\sqrt{2}} = \dfrac{2}{\sqrt{10}} = \dfrac{\sqrt{10}}{5}$; not orthogonal

35. $F(x, y, z) = x^2 + z^2 - 25$ $\qquad G(x, y, z) = y^2 + z^2 - 25$

$\nabla F = 2x\mathbf{i} + 2z\mathbf{k}$ $\qquad \nabla G = 2y\mathbf{j} + 2z\mathbf{k}$

$\nabla F(3, 3, 4) = 6\mathbf{i} + 8\mathbf{k}$ $\qquad \nabla G(3, 3, 4) = 6\mathbf{j} + 8\mathbf{k}$

(a) $\nabla F \times \nabla G = \begin{vmatrix} \mathbf{i} & \mathbf{j} & \mathbf{k} \\ 6 & 0 & 8 \\ 0 & 6 & 8 \end{vmatrix} = -48\mathbf{i} - 48\mathbf{j} + 36\mathbf{k} = -12(4\mathbf{i} + 4\mathbf{j} - 3\mathbf{k})$

Direction numbers: $4, 4, -3$, $\dfrac{x-3}{4} = \dfrac{y-3}{4} = \dfrac{z-4}{-3}$

(b) $\cos\theta = \dfrac{|\nabla F \cdot \nabla G|}{\|\nabla F\|\,\|\nabla G\|} = \dfrac{64}{(10)(10)} = \dfrac{16}{25}$; not orthogonal

37. $F(x, y, z) = x^2 + y^2 + z^2 - 6$ $\qquad G(x, y, z) = x - y - z$

$\nabla F(x, y, z) = 2x\mathbf{i} + 2y\mathbf{j} + 2z\mathbf{k}$ $\qquad \nabla G(x, y, z) = \mathbf{i} - \mathbf{j} - \mathbf{k}$

$\nabla F(2, 1, 1) = 4\mathbf{i} + 2\mathbf{j} + 2\mathbf{k}$ $\qquad \nabla G(2, 1, 1) = \mathbf{i} - \mathbf{j} - \mathbf{k}$

(a) $\nabla F \times \nabla G = \begin{vmatrix} \mathbf{i} & \mathbf{j} & \mathbf{k} \\ 4 & 2 & 2 \\ 1 & -1 & -1 \end{vmatrix} = 6\mathbf{j} - 6\mathbf{k} = 6(\mathbf{j} - \mathbf{k})$

(b) $\cos\theta = \dfrac{|\nabla F \cdot \nabla G|}{\|\nabla F\|\,\|\nabla G\|} = 0$; orthogonal

Direction numbers: $0, 1, -1$, $x = 2$, $\dfrac{y-1}{1} = \dfrac{z-1}{-1}$

39. $f(x, y) = 6 - x^2 - \dfrac{y^2}{4}$, $g(x, y) = 2x + y$

(a) $F(x, y, z) = z + x^2 + \dfrac{y^2}{4} - 6$ $\qquad G(x, y, z) = z - 2x - y$

$\nabla F(x, y, z) = 2x\mathbf{i} + \dfrac{1}{2}y\mathbf{j} + \mathbf{k}$ $\qquad \nabla G(x, y, z) = -2\mathbf{i} - \mathbf{j} + \mathbf{k}$

$\nabla G(1, 2, 4) = -2\mathbf{i} - \mathbf{j} + \mathbf{k}$

$\nabla F(1, 2, 4) = 2\mathbf{i} + \mathbf{j} + \mathbf{k}$

The cross product of these gradients is parallel to the curve of intersection.

$\nabla F(1, 2, 4) \times \nabla G(1, 2, 4) = \begin{vmatrix} \mathbf{i} & \mathbf{j} & \mathbf{k} \\ 2 & 1 & 1 \\ -2 & -1 & 1 \end{vmatrix} = 2\mathbf{i} - 4\mathbf{j}$

Using direction numbers $1, -2, 0$, you get $x = 1 + t$, $y = 2 - 2t$, $z = 4$.

$\cos\theta = \dfrac{\nabla F \cdot \nabla G}{\|\nabla F\|\,\|\nabla G\|} = \dfrac{-4 - 1 + 1}{\sqrt{6}\sqrt{6}} = \dfrac{-4}{6} \Rightarrow \theta \approx 48.2°$

(b)

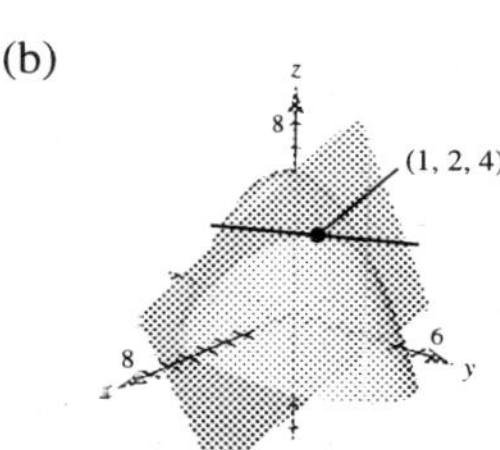

41. $F(x, y, z) = 3x^2 + 2y^2 - z - 15$, $(2, 2, 5)$

$\nabla F(x, y, z) = 6x\mathbf{i} + 4y\mathbf{j} - \mathbf{k}$

$\nabla F(2, 2, 5) = 12\mathbf{i} + 8\mathbf{j} - \mathbf{k}$

$\cos\theta = \dfrac{|\nabla F(2, 2, 5) \cdot \mathbf{k}|}{\|\nabla F(2, 2, 5)\|} = \dfrac{1}{\sqrt{209}}$

$\theta = \arccos\left(\dfrac{1}{\sqrt{209}}\right) \approx 86.03°$

43. $F(x, y, z) = x^2 - y^2 + z$, $(1, 2, 3)$

$\nabla F(x, y, z) = 2x\mathbf{i} - 2y\mathbf{j} + \mathbf{k}$

$\nabla F(1, 2, 3) = 2\mathbf{i} - 4\mathbf{j} + \mathbf{k}$

$\cos\theta = \dfrac{|\nabla F(1, 2, 3) \cdot \mathbf{k}|}{\|\nabla F(1, 2, 3)\|} = \dfrac{1}{\sqrt{21}}$

$\theta = \arccos \dfrac{1}{\sqrt{21}} \approx 77.40°$

45. $F(x, y, z) = 3 - x^2 - y^2 + 6y - z$

$\nabla F(x, y, z) = -2x\mathbf{i} + (-2y + 6)\mathbf{j} - \mathbf{k}$

$-2x = 0,\ x = 0$

$-2y + 6 = 0,\ y = 3$

$z = 3 - 0^2 - 3^2 + 6(3) = 12$

$(0, 3, 12)$ (vertex of paraboloid)

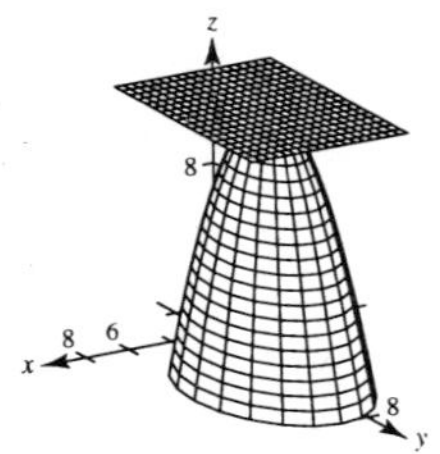

47. $T(x, y, z) = 400 - 2x^2 - y^2 - 4z^2,\ (4, 3, 10)$

$\dfrac{dx}{dt} = -4kx \qquad \dfrac{dy}{dt} = -2ky \qquad \dfrac{dz}{dt} = -8kz$

$x(t) = C_1e^{-4kt} \qquad y(t) = C_2e^{-2kt} \qquad z(t) = C_3e^{-8kt}$

$x(0) = C_1 = 4 \qquad y(0) = C_2 = 3 \qquad z(0) = C_3 = 10$

$x = 4e^{-4kt} \qquad y = 3e^{-2kt} \qquad z = 10e^{-8kt}$

49. $F(x, y, z) = \dfrac{x^2}{a^2} + \dfrac{y^2}{b^2} + \dfrac{z^2}{c^2} - 1$

$F_x(x, y, z) = \dfrac{2x}{a^2}$

$F_y(x, y, z) = \dfrac{2y}{b^2}$

$F_z(x, y, z) = \dfrac{2z}{c^2}$

Plane: $\dfrac{2x_0}{a^2}(x - x_0) + \dfrac{2y_0}{b^2}(y - y_0) + \dfrac{2z_0}{c^2}(z - z_0) = 0$

$$\frac{x_0x}{a^2} + \frac{y_0y}{b^2} + \frac{z_0z}{c^2} = \frac{x_0^2}{a^2} + \frac{y_0^2}{b^2} + \frac{z_0^2}{c^2} = 1$$

51. $F(x, y, z) = a^2x^2 + b^2y^2 - z^2$

$F_x(x, y, z) = 2a^2x$

$F_y(x, y, z) = 2b^2y$

$F_z(x, y, z) = -2z$

Plane:

$$2a^2x_0(x - x_0) + 2b^2y_0(y - y_0) - 2z_0(z - z_0) = 0$$

$$a^2x_0x + b^2y_0y - z_0z = a^2x_0^2 + b^2y_0^2 - z_0^2 = 0$$

Hence, the plane passes through the origin.

53. $f(x, y) = e^{x-y}$

$f_x(x, y) = e^{x-y}, \qquad f_y(x, y) = -e^{x-y}$

$f_{xx}(x, y) = e^{x-y}, \qquad f_{yy}(x, y) = e^{x-y}, \qquad f_{xy}(x, y) = -e^{x-y}$

(a) $P_1(x, y) \approx f(0, 0) + f_x(0, 0)x + f_y(0, 0)y = 1 + x - y$

(b) $P_2(x, y) \approx f(0, 0) + f_x(0, 0)x + f_y(0,0)y + \frac{1}{2}f_{xx}(0, 0)x^2 + f_{xy}(0, 0)xy + \frac{1}{2}f_{yy}(0, 0)y^2$

$= 1 + x - y + \frac{1}{2}x^2 - xy + \frac{1}{2}y^2$

(c) If $x = 0$, $P_2(0, y) = 1 - y + \frac{1}{2}y^2$. This is the second–degree Taylor polynomial for e^{-y}.
If $y = 0$, $P_2(x, 0) = 1 + x + \frac{1}{2}x^2$. This is the second–degree Taylor polynomial for e^x.

(d)

x	y	$f(x, y)$	$P_1(x, y)$	$P_2(x, y)$
0	0	1	1	1
0	0	0.9048	0.9000	0.9050
0.2	0.1	1.1052	1.1000	1.1050
0.2	0.5	0.7408	0.7000	0.7450
1	0.5	1.6487	1.5000	1.6250

(e)

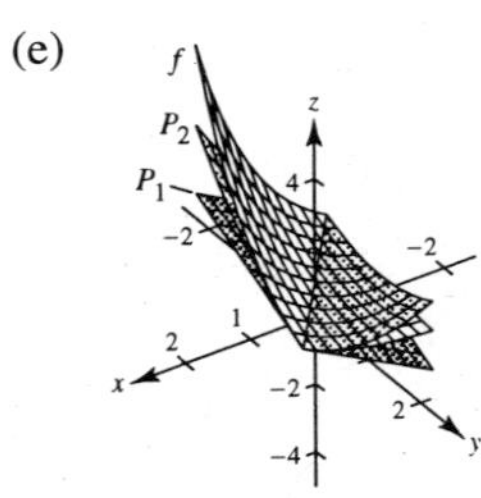

55. Given $w = F(x, y, z)$ where F is differentiable at

$$(x_0, y_0, z_0) \text{ and } \nabla F(x_0, y_0, z_0) \neq \mathbf{0},$$

the level surface of F at (x_0, y_0, z_0) is of the form $F(x, y, z) = C$ for some constant C. Let

$$G(x, y, z) = F(x, y, z) - C = 0.$$

Then $\nabla G(x_0, y_0, z_0) = \nabla F(x_0, y_0, z_0)$ where $\nabla G(x_0, y_0, z_0)$ is normal to $F(x_0, y_0, z_0) - C = 0$. Therefore, $\nabla F(x_0, y_0 z_0)$ is normal to $F(x_0, y_0, z_0) = C$.

Section 12.8 Extrema of Functions of Two Variables

1. $g(x, y) = (x - 1)^2 + (y - 3)^2 \geq 0$

Relative minimum: $(1, 3, 0)$

$g_x = 2(x - 1) = 0 \Rightarrow x = 1$

$g_y = 2(y - 3) = 0 \Rightarrow y = 3$

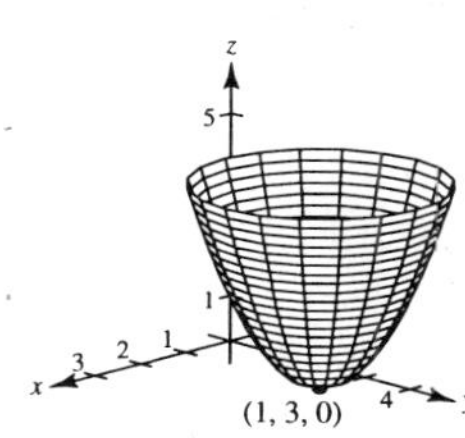

3. $f(x, y) = \sqrt{x^2 + y^2 + 1} \geq 1$

Relative minimum: $(0, 0, 1)$

Check: $f_x = \dfrac{x}{\sqrt{x^2 + y^2 + 1}} = 0 \Rightarrow x = 0$

$f_y = \dfrac{y}{\sqrt{x^2 + y^2 + 1}} = 0 \Rightarrow y = 0$

$$f_{xx} = \frac{y^2 + 1}{(x^2 + y^2 + 1)^{3/2}},\ f_{yy} = \frac{x^2 + 1}{(x^2 + y^2 + 1)^{3/2}},\ f_{xy} = \frac{-xy}{(x^2 + y^2 + 1)^{3/2}}$$

At the critical point $(0, 0)$, $f_{xx} > 0$ and $f_{xx} f_{yy} - (f_{xy})^2 > 0$. Therefore, $(0, 0, 1)$ is a relative minimum.

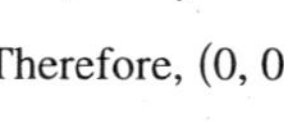

5. $f(x, y) = x^2 + y^2 + 2x - 6y + 6 = (x + 1)^2 + (y - 3)^2 - 4 \geq -4$

Relative minimum: $(-1, 3, -4)$

Check: $f_x = 2x + 2 = 0 \Rightarrow x = -1$

$f_y = 2y - 6 = 0 \Rightarrow y = 3$

$f_{xx} = 2,\ f_{yy} = 2,\ f_{xy} = 0$

At the critical point $(-1, 3)$, $f_{xx} > 0$ and $f_{xx} f_{yy} - (f_{xy})^2 > 0$. Therefore, $(-1, 3, -4)$ is a relative minimum.

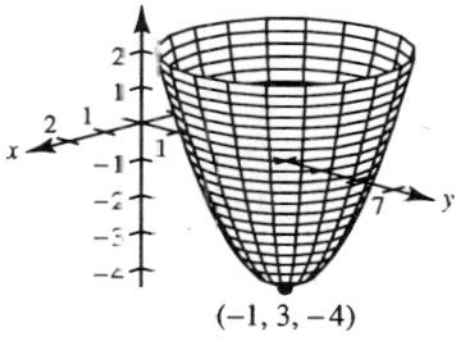

7. $f(x, y) = 2x^2 + 2xy + y^2 + 2x - 3$

$\left.\begin{array}{l} f_x = 4x + 2y + 2 = 0 \\ f_y = 2x + 2y = 0 \end{array}\right\}$ Solving simultaneously yields $x = -1$ and $y = 1$.

$f_{xx} = 4,\ f_{yy} = 2,\ f_{xy} = 2$

At the critical point $(-1, 1)$, $f_{xx} > 0$ and $f_{xx} f_{yy} - (f_{xy})^2 > 0$. Therefore, $(-1, 1, -4)$ is a relative minimum.

9. $f(x, y) = -5x^2 + 4xy - y^2 + 16x + 10$

$\left.\begin{array}{l} f_x = -10x + 4y + 16 = 0 \\ f_y = 4x - 2y = 0 \end{array}\right\}$ Solving simultaneously yields $x = 8$ and $y = 16$.

$f_{xx} = -10,\ f_{yy} = -2,\ f_{xy} = 4$

At the critical point $(8, 16)$, $f_{xx} < 0$ and $f_{xx}f_{yy} - (f_{xy})^2 > 0$. Therefore, $(8, 16, 74)$ is a relative maximum.

11. $f(x, y) = 2x^2 + 3y^2 - 4x - 12y + 13$

$f_x = 4x - 4 = 4(x - 1) = 0$ when $x = 1$.

$f_y = 6y - 12 = 6(y - 2) = 0$ when $y = 2$.

$f_{xx} = 4,\ f_{yy} = 6,\ f_{xy} = 0$

At the critical point $(1, 2)$, $f_{xx} > 0$ and $f_{xx}f_{yy} - (f_{xy})^2 > 0$. Therefore, $(1, 2, -1)$ is a relative minimum.

13. $h(x, y) = x^2 - y^2 - 2x - 4y - 4$

$h_x = 2x - 2 = 2(x - 1) = 0$ when $x = 1$.

$h_y = -2y - 4 = -2(y + 2) = 0$ when $y = -2$.

$h_{xx} = 2,\ h_{yy} = -2,\ h_{xy} = 0$

At the critical point $(1, -2)$, $h_{xx}h_{yy} - (h_{xy})^2 < 0$. Therefore, $(1, -2, -1)$ is a saddle point.

15. $h(x, y) = x^2 - 3xy - y^2$

$\left.\begin{array}{l} h_x = 2x - 3y = 0 \\ h_y = -3x - 2y = 0 \end{array}\right\}$ Solving simultaneously yields $x = 0$ and $y = 0$.

$h_{xx} = 2,\ h_{yy} = -2,\ h_{xy} = -3$

At the critical point $(0, 0)$, $h_{xx}h_{yy} - (h_{xy})^2 < 0$. Therefore, $(0, 0, 0)$ is a saddle point.

17. $f(x, y) = x^3 - 3xy + y^3$

$\left.\begin{array}{l} f_x = 3(x^2 - y) = 0 \\ f_y = 3(-x + y^2) = 0 \end{array}\right\}$ Solving by substitution yields two critical points $(0, 0)$ and $(1, 1)$.

$f_{xx} = 6x,\ f_{yy} = 6y,\ f_{xy} = -3$

At the critical point $(0, 0)$, $f_{xx}f_{yy} - (f_{xy})^2 < 0$. Therefore, $(0, 0, 0)$ is a saddle point. At the critical point $(1, 1)$, $f_{xx} = 6 > 0$ and $f_{xx}f_{yy} - (f_{xy})^2 > 0$. Therefore, $(1, 1, -1)$ is a relative minimum.

19. $f(x, y) = e^{-x}\sin y$

$\left.\begin{array}{l} f_x = -e^{-x}\sin y = 0 \\ f_y = e^{-x}\cos y = 0 \end{array}\right\}$ Since $e^{-x} > 0$ for all x and $\sin y$ and $\cos y$ are never both zero for a given value of y, there are no critical points.

21. $z = \dfrac{-4x}{x^2 + y^2 + 1}$

Relative minimum: $(1, 0, -2)$

Relative maximum: $(-1, 0, 2)$

23. $f(x, y) = y^3 - 3yx^2 - 3y^2 - 3x^2 + 1$

Relative maximum: $(0, 0, 1)$

Saddle points: $(0, 2, -3)$, $(\pm\sqrt{3}, -1, -3)$

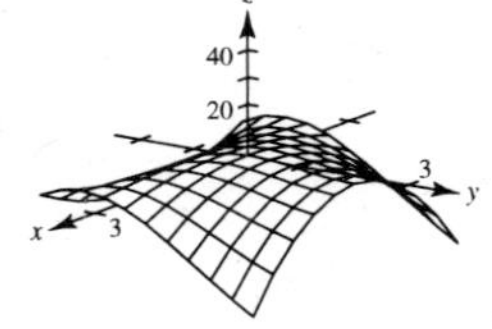

25. $f_{xx}f_{yy} - (f_{xy})^2 = (9)(4) - 6^2 = 0$

Insufficient information.

27. $f_{xx}f_{yy} - (f_{xy})^2 = (-9)(6) - 10^2 < 0$

f has a saddle point at (x_0, y_0).

29.

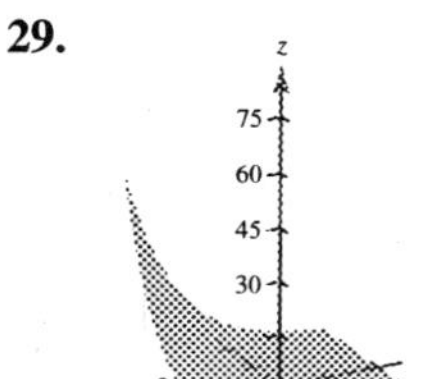

No extrema

31.

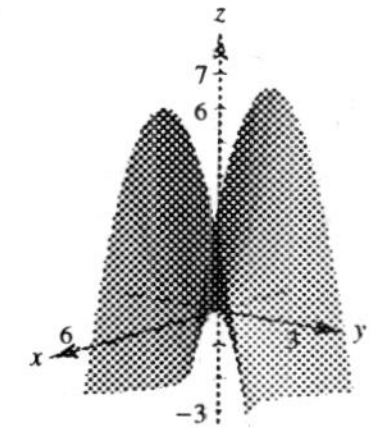

Saddle point

33. $d = f_{xx}f_{yy} - f_{xy}{}^2 = (2)(8) - f_{xy}{}^2 = 16 - f_{xy}{}^2 > 0$

$\Rightarrow f_{xy}{}^2 < 16 \Rightarrow -4 < f_{xy} < 4$

35. $f(x, y) = 12 - 3x - 2y$ has no critical points. On the line $y = x + 1, 0 \le x \le 1$,

$$f(x, y) = f(x) = 12 - 3x - 2(x + 1) = -5x + 10$$

and the maximum is 10, the minimum is 5. On the line $y = -2x + 4, 1 \le x \le 2$,

$$f(x, y) = f(x) = 12 - 3x - 2(-2x + 4) = x + 4$$

and the maximum is 6, the minimum is 5. On the line $y = -\frac{1}{2}x + 1, 0 \le x \le 2$,

$$f(x, y) = f(x) = 12 - 3x - 2\left(-\tfrac{1}{2}x + 1\right) = -2x + 10$$

and the maximum is 10, the minimum is 6.

Absolute maximum: 10 at $(0, 1)$

Absolute minimum: 5 at $(1, 2)$

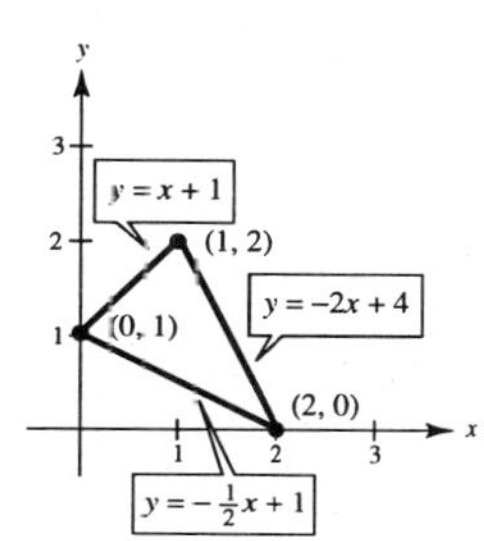

37. $f(x, y) = 3x^2 + 2y^2 - 4y$

$$\left.\begin{array}{l} f_x = 6x = 0 \quad \Rightarrow x = 0 \\ f_y = 4y - 4 = 0 \Rightarrow y = 1 \end{array}\right\} \quad f(0, 1) = -2$$

On the line $y = 4, -2 \le x \le 2$,

$$f(x, y) = f(x) = 3x^2 + 32 - 16 = 3x^2 + 16$$

and the maximum is 28, the minimum is 16. On the curve $y = x^2, -2 \le x \le 2$,

$$f(x, y) = f(x) = 3x^2 + 2(x^2)^2 - 4x^2 = 2x^4 - x^2 = x^2(2x^2 - 1)$$

and the maximum is 28, the minimum is $-\frac{1}{8}$.

Absolute maximum: 28 at $(\pm 2, 4)$

Absolute minimum: -2 at $(0, 1)$

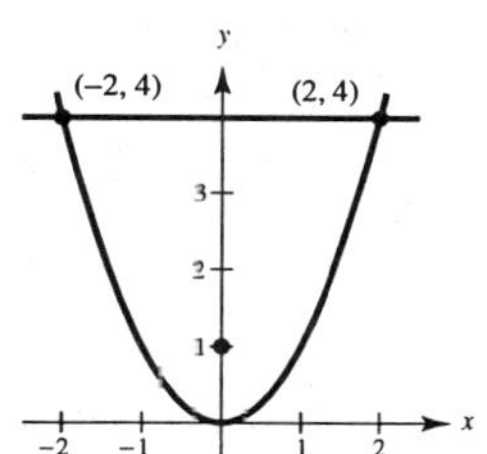

39. $f(x, y) = x^2 + xy, R = \{(x, y): |x| \le 2, |y| \le 1\}$

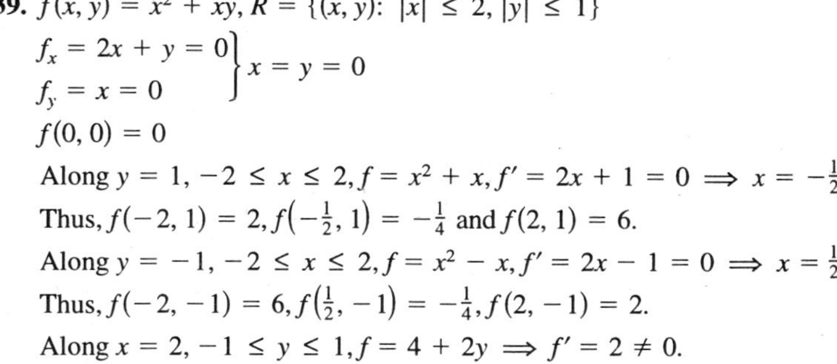

$$\left.\begin{array}{l} f_x = 2x + y = 0 \\ f_y = x = 0 \end{array}\right\} x = y = 0$$

$f(0, 0) = 0$

Along $y = 1, -2 \le x \le 2, f = x^2 + x, f' = 2x + 1 = 0 \Rightarrow x = -\frac{1}{2}$.

Thus, $f(-2, 1) = 2, f\left(-\frac{1}{2}, 1\right) = -\frac{1}{4}$ and $f(2, 1) = 6$.

Along $y = -1, -2 \le x \le 2, f = x^2 - x, f' = 2x - 1 = 0 \Rightarrow x = \frac{1}{2}$.

Thus, $f(-2, -1) = 6, f\left(\frac{1}{2}, -1\right) = -\frac{1}{4}, f(2, -1) = 2$.

Along $x = 2, -1 \le y \le 1, f = 4 + 2y \Rightarrow f' = 2 \ne 0$.

Along $x = -2, -1 \le y \le 1, f = 4 - 2y \Rightarrow f' = -2 \ne 0$.

Thus, the maxima are $f(2, 1) = 6$ and $f(-2, -1) = 6$ and the minima are $f\left(-\frac{1}{2}, 1\right) = -\frac{1}{4}$ and $f\left(\frac{1}{2}, -1\right) = -\frac{1}{4}$.

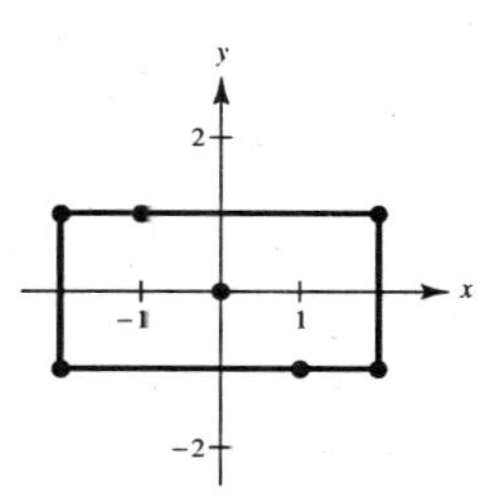

41. $f(x, y) = x^2 + 2xy + y^2, R = \{(x, y): x^2 + y^2 \leq 8\}$

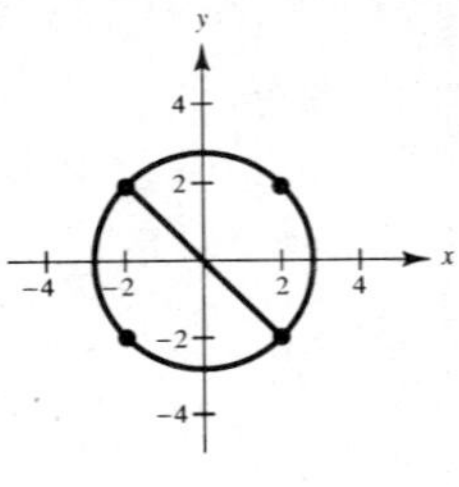

$$\left.\begin{aligned} f_x &= 2x + 2y = 0 \\ f_y &= 2x + 2y = 0 \end{aligned}\right\} y = -x$$

$f(x, -x) = x^2 - 2x^2 + x^2 = 0$

On the boundary $x^2 + y^2 = 8$, we have $y^2 = 8 - x^2$ and $y = \pm\sqrt{8 - x^2}$. Thus,

$$f = x^2 \pm 2x\sqrt{8 - x^2} + (8 - x^2) = 8 \pm 2x\sqrt{8 - x^2}$$

$$f' = \pm(8 - x^2)^{-1/2}(-2x) + 2(8 - x^2)^{1/2}) = \pm\frac{16 - 4x^2}{\sqrt{8 - x^2}}.$$

Then, $f' = 0$ implies $16 = 4x^2$ or $x = \pm 2$.

$$f(2, 2) = f(-2, -2) = 16 \quad \text{and} \quad f(2, -2) = f(-2, 2) = 0$$

Thus, the maxima are $f(2, 2) = 16$ and $f(-2, -2) = 16$, and the minima are $f(x, -x) = 0$, $|x| \leq 2$.

43. $f(x, y) = \dfrac{4xy}{(x^2 + 1)(y^2 + 1)}, R = \{(x, y): 0 \leq x \leq 1, 0 \leq y \leq 1\}$

$$f_x = \frac{4(1 - x^2)y}{(y^2 + 1)(x^2 + 1)} = 0 \Rightarrow x = 1 \text{ or } y = 0$$

$$f_y = \frac{4(1 - y^2)x}{(x^2 + 1)(y^2 + 1)^2} \Rightarrow x = 0 \text{ or } y = 1$$

For $x = 0$, $y = 0$, also, and $f(0, 0) = 0$.

For $x = 1$, $y = 1$, $f(1, 1) = 1$.

The absolute maximum is $1 = f(1, 1)$.

The absolute minimum is $0 = f(0, 0)$. (In fact, $f(0, y) = f(x, 0) = 0$)

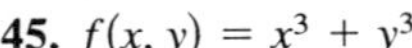

45. $f(x, y) = x^3 + y^3$

$$\left.\begin{aligned} f_x &= 3x^2 = 0 \\ f_y &= 3y^2 = 0 \end{aligned}\right\} \text{Solving yields } x = y = 0.$$

$f_{xx} = 6x$, $f_{yy} = 6y$, $f_{xy} = 0$

At $(0, 0)$, $f_{xx}f_{yy} - (f_{xy})^2 = 0$ and the test fails. $(0, 0, 0)$ is a saddle point.

47. $f(x, y) = (x - 1)^2(y + 4)^2 \geq 0$

$$\left.\begin{aligned} f_x &= 2(x - 1)(y + 4)^2 = 0 \\ f_y &= 2(x - 1)^2(y + 4) = 0 \end{aligned}\right\} \text{Solving yields the critical points } (1, a) \text{ and } (b, -4).$$

$f_{xx} = 2(y + 4)^2$, $f_{yy} = 2(x - 1)^2$, $f_{xy} = 4(x - 1)(y + 4)$

At both $(1, a)$ and $(b, -4)$, $f_{xx}f_{yy} - (f_{xy})^2 = 0$ and the test fails.

Absolute minima: $(1, a, 0)$ and $(b, -4, 0)$

49. $f(x, y) = x^{2/3} + y^{2/3} \geq 0$

$$\left.\begin{aligned} f_x &= \frac{2}{3\sqrt[3]{x}} \\ f_y &= \frac{2}{3\sqrt[3]{y}} \end{aligned}\right\} f_x \text{ and } f_y \text{ are undefined at } x = 0, y = 0. \text{ The critical point is } (0, 0).$$

$$f_{xx} = -\frac{2}{9x\sqrt[3]{3}}, f_{yy} = -\frac{2}{9y\sqrt[3]{y}}, f_{xy} = 0$$

At $(0, 0)$, $f_{xx}f_{yy} - (f_{xy})^2$ is undefined and the test fails.

Absolute minimum: $(0, 0, 0)$

51. $f(x, y, z) = x^2 + (y - 3)^2 + (z + 1)^2 \geq 0$

$$\left.\begin{aligned} f_x &= 2x = 0 \\ f_y &= 2(y - 3) = 0 \\ f_z &= 2(z + 1) = 0 \end{aligned}\right\} \text{ Solving yields the critical point } (0, 3, -1).$$

Absolute minimum: $(0, 3, -1, 0)$

53. The point A will be a saddle point. The function could be

$$f(x, y) = x^2 - y^2.$$

55. (a) There is a minimum at $(0, 0, 0)$, maxima at $(0, \pm 1, 2/e)$ and saddle point at $(\pm 1, 0, 1/e)$:

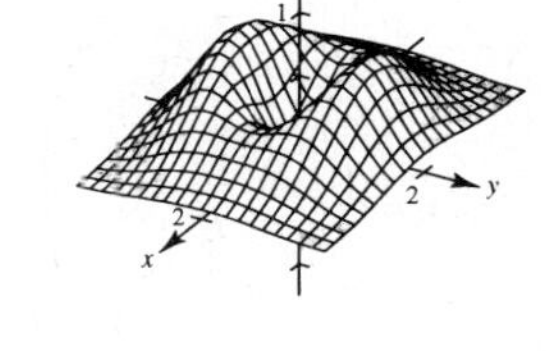

$$\begin{aligned} f_x &= (x^2 + 2y^2)e^{-(x^2+y^2)}(-2x) + (2x)e^{-(x^2+y^2)} \\ &= e^{-(x^2+y^2)}[(x^2 + 2y^2)(-2x) + 2x] \\ &= e^{-(x^2+y^2)}[-2x^3 + 4xy^2 + 2x] = 0 \Rightarrow x^3 + 2xy^2 - x = 0 \end{aligned}$$

$$\begin{aligned} f_y &= (x^2 + 2y^2)e^{-(x^2+y^2)}(-2y) + (4y)e^{-(x^2+y^2)} \\ &= e^{-(x^2+y^2)}[(x^2 + 2y^2)(-2y) + 4y] \\ &= e^{-(x^2+y^2)}[-4y^3 - 2x^2y + 4y] = 0 \Rightarrow 2y^3 + x^2y - 2y = 0 \end{aligned}$$

Solving the two equations $x^3 + 2xy^2 - x = 0$ and $2y^3 + x^2y - 2y = 0$, you obtain the following critical points: $(0, \pm 1)$, $(\pm 1, 0)$, $(0, 0)$. Using the second derivative test, you obtain the results above.

(b) As in part (a), you obtain

$$f_x = e^{-(x^2+y^2)}[2x(x^2 - 1 - 2y^2)]$$
$$f_y = e^{-(x^2+y^2)}[2y(2 + x^2 - 2y^2)]$$

The critical numbers are $(0, 0)$, $(0, \pm 1)$, $(\pm 1, 0)$. These yield

$(\pm 1, 0, -1/e)$ minima
$(0, \pm 1, 2/e)$ maxima
$(0, 0, 0)$ saddle

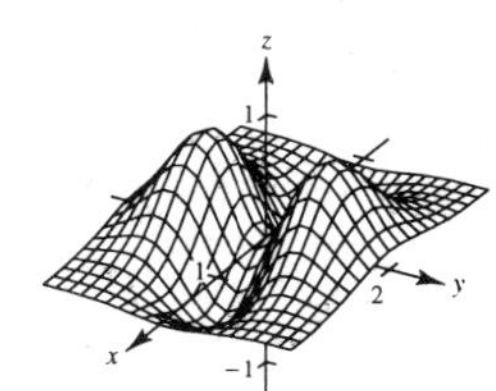

(c) In general, for $\alpha > 0$ you obtain

$(0, 0, 0)$ minimum
$(0, \pm 1, \beta/e)$ maxima
$(\pm 1, 0, \alpha/e)$ saddle

For $\alpha < 0$, you obtain

$(\pm 1, 0, \alpha/e)$ minima
$(0, \pm 1, \beta/e)$ maxima
$(0, 0, 0)$ saddle

57. False

Let $f(x, y) = |1 - x - y|$.

$(0, 0, 1)$ is a relative maximum, but $f_x(0, 0)$ and $f_y(0, 0)$ do not exist.

59. False

Let $f(x, y) = x^4 - 2x^2 + y^2$

Relative minima: $(\pm 1, 0, -1)$

Saddle point: $(0, 0, 0)$

Section 12.9 Applications of Extrema of Functions of Two Variables

1. A point on the plane is given by $(x, y, 12 - 2x - 3y)$. The square of the distance from the origin to this point is

$$S = x^2 + y^2 + (12 - 2x - 3y)^2$$
$$S_x = 2x + 2(12 - 2x - 3y)(-2)$$
$$S_y = 2y + 2(12 - 2x - 3y)(-3).$$

From the equations $S_x = 0$ and $S_y = 0$, we obtain the system

$$5x + 6y = 24$$
$$3x + 5y = 18.$$

Solving simultaneously, we have $x = \frac{12}{7}$, $y = \frac{18}{7}$ $z = 12 - \frac{24}{7} - \frac{54}{7} = \frac{6}{7}$. Therefore, the distance from the origin to $\left(\frac{12}{7}, \frac{18}{7}, \frac{6}{7}\right)$ is

$$\sqrt{\left(\frac{12}{7}\right)^2 + \left(\frac{18}{7}\right)^2 + \left(\frac{6}{7}\right)^2} = \frac{6\sqrt{14}}{7}.$$

3. A point on the paraboloid is given by $(x, y, x^2 + y^2)$. The square of the distance from $(5, 5, 0)$ to a point on the paraboloid is given by

$$S = (x - 5)^2 + (y - 5)^2 + (x^2 + y^2)^2$$

$$S_x = 2(x - 5) + 4x(x^2 + y^2) = 0$$

$$S_y = 2(y - 5) + 4y(x^2 + y^2) = 0.$$

From the equations $S_x = 0$ and $S_y = 0$, we obtain the system

$$2x^3 + 2xy^2 + x - 5 = 0$$

$$2y^3 + 2x^2y + y - 5 = 0$$

Multiply the first equation by y and the second equation by x, and subtract to obtain $x = y$. Then, we have $x = 1, y = 1, z = 2$ and the distance is

$$\sqrt{(1 - 5)^2 + (1 - 5)^2 + (2 - 0)^2} = 6.$$

5. Let x, y and z be the numbers. Since $x + y + z = 30$, $z = 30 - x - y$.

$$P = xyz = 30xy - x^2y - xy^2$$

$$\left.\begin{aligned} P_x &= 30y - 2xy - y^2 = y(30 - 2x - y) = 0 \\ P_y &= 30x - x^2 - 2xy = x(30 - x - 2y) = 0 \end{aligned}\right\} \begin{aligned} 2x + y &= 30 \\ x + 2y &= 30 \end{aligned}$$

Solving simultaneously yields $x = 10$, $y = 10$, and $z = 10$.

7. Let x, y, and z be the numbers and let $S = x^2 + y^2 + z^2$. Since $x + y + z = 30$, we have

$$S = x^2 + y^2 + (30 - x - y)^2$$

$$\left.\begin{aligned} S_x &= 2x + 2(30 - x - y)(-1) = 0 \\ S_y &= 2y + 2(30 - x - y)(-1) = 0 \end{aligned}\right\} \begin{aligned} 2x + y &= 30 \\ x + 2y &= 30. \end{aligned}$$

Solving simultaneously yields $x = 10$, $y = 10$, and $z = 10$.

9. Let x, y, and z be the length, width, and height, respectively. Then the sum of the length and girth is given by $x + (2y + 2z) = 108$ or $x = 108 - 2y - 2z$. The volume is given by

$$V = xyz = 108zy - 2zy^2 - 2yz^2$$

$$V_y = 108z - 4yz - 2z^2 = z(108 - 4y - 2z) = 0$$

$$V_z = 108y - 2y^2 - 4yz = y(108 - 2y - 4z) = 0.$$

Solving the system $4y + 2z = 108$ and $2y + 4z = 108$, we obtain the solution $x = 36$ inches, $y = 18$ inches, and $z = 18$ inches.

11. Let $a + b + c = k$. Then

$$V = \frac{4\pi\, abc}{3} = \frac{4}{3}\pi\, ab(k - a - b) = \frac{4}{3}\pi(kab - a^2b - ab^2)$$

$$\left.\begin{aligned} V_a &= \frac{4\pi}{3}(kb - 2ab - b^2) = 0 \\ V_b &= \frac{4\pi}{3}(ka - a^2 - 2ab) = 0 \end{aligned}\right\} \begin{aligned} kb - 2ab - b^2 &= 0 \\ ka - a^2 - 2ab &= 0. \end{aligned}$$

Solving this system simultaneously yields $a = b$ and substitution yields $b = k/3$. Therefore, the solution is $a = b = c = k/3$.

13. Let x, y, and z be the length, width, and height, respectively and let V_0 be the given volume. Then $V_0 = xyz$ and $z = V_0/xy$. The surface area is

$$S = 2xy + 2yz + 2xz = 2\left(xy + \frac{V_0}{x} + \frac{V_0}{y}\right)$$

$$\left.\begin{aligned} S_x &= 2\left(y - \frac{V_0}{x^2}\right) = 0 \\ S_y &= 2\left(x - \frac{V_0}{y^2}\right) = 0 \end{aligned}\right\} \begin{aligned} x^2y - V_0 &= 0 \\ xy^2 - V_0 &= 0. \end{aligned}$$

Solving simultaneously yields $x = \sqrt[3]{V_0}$, $y = \sqrt[3]{V_0}$, and $z = \sqrt[3]{V_0}$.

15. The distance from P to Q is $\sqrt{x^2 + 4}$. The distance from Q to R is $\sqrt{(y - x)^2 + 1}$. The distance from R to S is $10 - y$.

$$C = 3k\sqrt{x^2 + 4} + 2k\sqrt{(y - x)^2 + 1} + k(10 - y)$$

$$C_x = 3k\left(\frac{x}{\sqrt{x^2 + 4}}\right) + 2k\left(\frac{-(y - x)}{\sqrt{(y - x)^2 + 1}}\right) = 0$$

$$C_y = 2k\left(\frac{y - x}{\sqrt{(y - x)^2 + 1}}\right) - k = 0 \Rightarrow \frac{y - x}{\sqrt{(y - x)^2 + 1}} = \frac{1}{2}$$

$$3k\left(\frac{x}{\sqrt{x^2 + 4}}\right) + 2k\left(-\frac{1}{2}\right) = 0$$

$$\frac{x}{\sqrt{x^2 + 4}} = \frac{1}{3}$$

$$3x = \sqrt{x^2 + 4}$$

$$9x^2 = x^2 + 4$$

$$x^2 = \frac{1}{2}$$

$$x = \frac{\sqrt{2}}{2}$$

$$2(y - x) = \sqrt{(y - x)^2 + 1}$$

$$4(y - x)^2 = (y - x)^2 + 1$$

$$(y - x)^2 = \frac{1}{3}$$

$$y = \frac{1}{\sqrt{3}} + \frac{1}{\sqrt{2}} = \frac{2\sqrt{3} + 3\sqrt{2}}{6}$$

Therefore, $x = \dfrac{\sqrt{2}}{2} \approx 0.707$ km and $y = \dfrac{2\sqrt{3} + 3\sqrt{2}}{6} \approx 1.284$ kms.

17. Let h be the height of the trough and r the length of the slanted sides. We observe that the area of a trapezoidal cross section is given by

$$A = h\left[\frac{(w - 2r) + [(w - 2r) + 2x]}{2}\right] = (w - 2r + x)h$$

where $x = r\cos\theta$ and $h = r\sin\theta$. Substituting these expressions for x and h, we have

$$A(r, \theta) = (w - 2r + r\cos\theta)(r\sin\theta) = wr\sin\theta - 2r^2\sin\theta + r^2\sin\theta\cos\theta$$

Now

$$A_r(r, \theta) = w\sin\theta - 4r\sin\theta + 2r\sin\theta\cos\theta = \sin\theta(w - 4r + 2r\cos\theta) = 0 \Rightarrow w = r(4 - 2\cos\theta)$$

$$A_\theta(r, \theta) = wr\cos\theta - 2r^2\cos\theta + r^2\cos 2\theta = 0.$$

Substituting the expression for w from $A_r(r, \theta) = 0$ into the equation $A_\theta(r, \theta) = 0$, we have

$$r^2(4 - 2\cos\theta)\cos\theta - 2r^2\cos\theta + r^2(2\cos^2\theta - 1) = 0$$

$$r^2(2\cos\theta - 1) = 0 \text{ or } \cos\theta = \frac{1}{2}.$$

Therefore, the first partial derivatives are zero when $\theta = \pi/3$ and $r = w/3$. (Ignore the solution $r = \theta = 0$.) Thus, the trapezoid of maximum area occurs when each edge of width $w/3$ is turned up 60° from the horizontal.

19. (a) $S(x, y) = d_1 + d_2 + d_3$

$$= \sqrt{(x-0)^2 + (y-0)^2} + \sqrt{(x+2)^2 + (y-2)^2} + \sqrt{(x-4)^2 + (y-2)^2}$$

$$= \sqrt{x^2 + y^2} + \sqrt{(x+2)^2 + (y-2)^2} + \sqrt{(x-4)^2 + (y-2)^2}$$

From the graph we see that the surface has a minimum.

(b) $S_x(x, y) = \dfrac{x}{\sqrt{x^2 + y^2}} + \dfrac{x+2}{\sqrt{(x+2)^2 + (y-2)^2}} + \dfrac{x-4}{\sqrt{(x-4)^2 + (y-2)^2}}$

$S_y(x, y) = \dfrac{y}{\sqrt{x^2 + y^2}} + \dfrac{y-2}{\sqrt{(x+2)^2 + (y-2)^2}} + \dfrac{y-2}{\sqrt{(x-4)^2 + (y-2)^2}}$

(c) $-\nabla S(1, 1) = -S_x(1, 1)\mathbf{i} - S_y(1, 1)\mathbf{j} = -\dfrac{1}{\sqrt{2}}\mathbf{i} - \left(\dfrac{1}{\sqrt{2}} - \dfrac{2}{\sqrt{10}}\right)\mathbf{j}$

$\tan\theta = \dfrac{(2/\sqrt{10}) - (1/\sqrt{2})}{-1/\sqrt{2}} = 1 - \dfrac{2}{\sqrt{5}} \Rightarrow \theta \approx 186.027°$

(d) $(x_2, y_2) = (x_1 - S_x(x_1, y_1)t, y_1 - S_y(x_1, y_1)t) = \left(1 - \dfrac{1}{\sqrt{2}}t, 1 + \left(\dfrac{2}{\sqrt{10}} - \dfrac{1}{\sqrt{2}}\right)t\right)$

$$S\left(1 - \frac{1}{\sqrt{2}}t, 1 + \left(\frac{2}{\sqrt{10}} - \frac{1}{\sqrt{2}}\right)t\right) = \sqrt{2 + \left(\frac{2\sqrt{10}}{5} - 2\sqrt{2}\right)t + \left(1 - \frac{2\sqrt{5}}{5} + \frac{2}{5}\right)t^2} + \sqrt{10 - \left(\frac{2\sqrt{10}}{5} + 2\sqrt{2}\right)t + \left(1 - \frac{2\sqrt{5}}{5} + \frac{2}{5}\right)t^2} + \sqrt{10 - \left(\frac{2\sqrt{10}}{5} - 4\sqrt{2}\right)t + \left(1 - \frac{2\sqrt{5}}{5} + \frac{2}{5}\right)t^2}$$

Using a computer algebra system, we find that the minimum occurs when $t \approx 1.344$. Thus, $(x_2, y_2) \approx (0.05, 0.90)$.

(e) $(x_3, y_3) = (x_2 - S_x(x_2, y_2)t, y_2 - S_y(x_2, y_2)t) \approx (0.05 + 0.03t, 0.90 - 0.26t)$

$$S(0.05 + 0.03t, 0.90 - 0.26t) = \sqrt{(0.05 + 0.03t)^2 + (0.90 - 0.26t)^2} + \sqrt{(2.05 + 0.03t)^2 + (-1.10 - 0.26t)^2} + \sqrt{(-3.95 + 0.03t)^2 + (-1.10 - 0.26t)^2}$$

Using a computer algebra system, we find that the minimum occurs when $t \approx 1.78$. Thus $(x_3, y_3) \approx (0.10, 0.44)$.

$(x_4, y_4) = (x_3 - S_x(x_3, y_3)t, y_3 - S_y(x_3, y_3)t) \approx (0.10 - 0.09t, 0.44 - 0.01t)$

$$S(0.10 - 0.09t, 0.45 - 0.01t) = \sqrt{(0.10 - 0.09t)^2 + (0.45 - 0.01t)^2} + \sqrt{(2.10 - 0.09t)^2 + (-1.55 - 0.01t)^2} + \sqrt{(-3.90 - 0.09t)^2 + (-1.55 - 0.01t)^2}$$

Using a computer algebra system, we find that the minimum occurs when $t \approx 0.44$. Thus, $(x_4, y_4) \approx (0.06, 0.44)$.

Note: The minimum occurs at $(x, y) = (0.0555, 0.3992)$

(f) $-\nabla S(x, y)$ points in the direction that S *decreases* most rapidly. You would use $\nabla S(x, y)$ for maximization problems.

21. $P(x_1, x_2) = 15(x_1 + x_2) - C_1 - C_2$

$$= 15x_1 + 15x_2 - (0.02x_1^2 + 4x_1 + 500) - (0.05x_2^2 + 4x_2 + 275)$$

$$= -0.02x_1^2 - 0.05x_2^2 + 11x_1 + 11x_2 - 775$$

$P_{x_1} = -0.04x_1 + 11 = 0, x_1 = 275$

$P_{x_2} = -0.10x_2 + 11 = 0, x_2 = 110$

$P_{x_1x_1} = -0.04$

$P_{x_1x_2} = 0$

$P_{x_2x_2} = -0.10$

$P_{x_1x_1} < 0$ and $P_{x_1x_1}P_{x_2x_2} - (P_{x_1x_2})^2 > 0$

Therefore, profit is maximized when $x_1 = 275$ and $x_2 = 110$.

23. $R(x_1, x_2) = -5x_1^2 - 8x_2^2 - 2x_1x_2 + 42x_1 + 102x_2$

$R_{x_1} = -10x_1 - 2x_2 + 42 = 0,\ 5x_1 + x_2 = 21$

$R_{x_2} = -16x_2 - 2x_1 + 102 = 0,\ x_1 + 8x_2 = 51$

Solving this system yields $x_1 = 3$ and $x_2 = 6$.

$R_{x_1x_1} = -10$

$R_{x_1x_2} = -2$

$R_{x_2x_2} = -16$

$R_{x_1x_1} < 0$ and $R_{x_1x_1}R_{x_2x_2} - (R_{x_1x_2})^2 > 0$

Thus, revenue is maximized when $x_1 = 3$ and $x_2 = 6$.

25. (a)

x	y	xy	x^2
-2	0	0	4
0	1	0	0
2	3	6	4
$\sum x_i = 0$	$\sum y_i = 4$	$\sum x_iy_i = 6$	$\sum x_i^2 = 8$

$$a = \frac{3(6) - 0(4)}{3(8) - 0^2} = \frac{3}{4},\ b = \frac{1}{3}\left[4 - \frac{3}{4}(0)\right] = \frac{4}{3},$$

$$y = \frac{3}{4}x + \frac{4}{3}$$

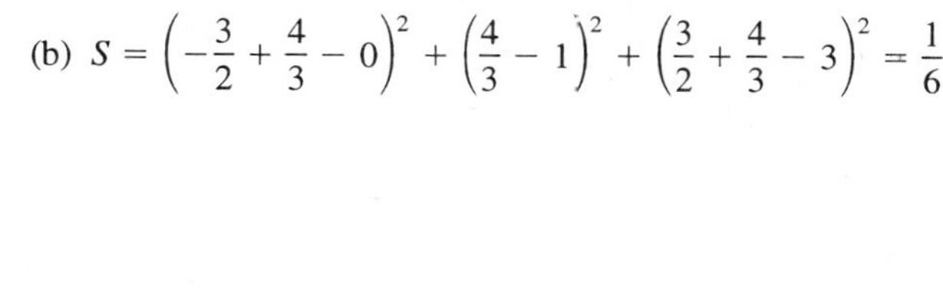

(b) $S = \left(-\frac{3}{2} + \frac{4}{3} - 0\right)^2 + \left(\frac{4}{3} - 1\right)^2 + \left(\frac{3}{2} + \frac{4}{3} - 3\right)^2 = \frac{1}{6}$

27. (a)

x	y	xy	x^2
0	4	0	0
1	3	3	1
1	1	1	1
2	0	0	4
$\sum x_i = 4$	$\sum y_i = 8$	$\sum x_iy_i = 4$	$\sum x_i^2 = 6$

$$a = \frac{4(4) - 4(8)}{4(6) - 4^2} = -2,\ b = \frac{1}{4}[8 + 2(4)] = 4,\ y = -2x + 4$$

(b) $S = (4 - 4)^2 + (2 - 3)^2 + (2 - 1)^2 + (0 - 0)^2 = 2$

29. (0, 0), (1, 1), (3, 4), (4, 2), (5, 5)

$\sum x_i = 13,\qquad \sum y_i = 12,$

$\sum x_iy_i = 46,\qquad \sum x_i^2 = 51$

$$a = \frac{5(46) - 13(12)}{5(51) - (13)^2} = \frac{74}{86} = \frac{37}{43}$$

$$b = \frac{1}{5}\left[12 - \frac{37}{43}(13)\right] = \frac{7}{43}$$

$$y = \frac{37}{43}x + \frac{7}{43}$$

31. (0, 6), (4, 3), (5, 0), (8, −4), (10, −5)Í

$\sum x_i = 27,\qquad \sum y_i = 0,$

$\sum x_iy_i = -70,\qquad \sum x_i^2 = 205$

$$a = \frac{5(-70) - (27)(0)}{5(205) - (27)^2} = \frac{-350}{296} = -\frac{175}{148}$$

$$b = \frac{1}{5}\left[0 - \left(-\frac{175}{148}\right)(27)\right] = \frac{945}{148}$$

$$y = -\frac{175}{148}x + \frac{945}{148}$$

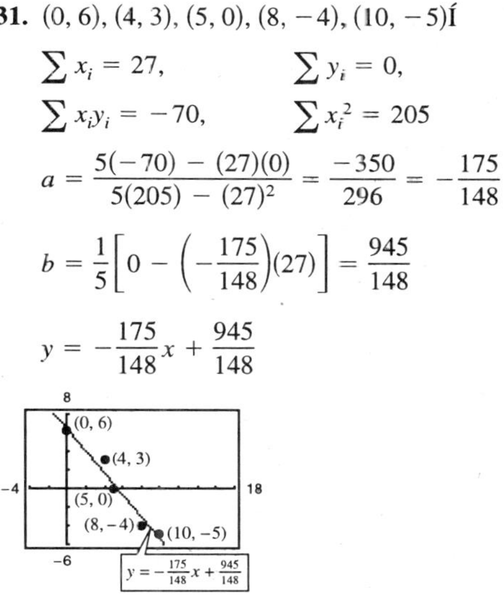

33. (a) $y = 0.03x - 6.29$

(b)

(c) If x increases 1 unit, then y increases

$(0.03)(1) = \$0.03.$

35. (1.0, 32), (1.5, 41), (2.0, 48), (2.5, 53)

$\sum x_i = 7, \sum y_i = 174, \sum x_iy_i = 322, \sum x_i^2 = 13.5$

$a = 14, b = 19, y = 14x + 19$

When $x = 1.6$, $y = 41.4$ bushels per acre.

37. $S(a, b, c) = \sum_{i=1}^{n} (y_i - ax_i^2 - bx_i - c)^2$

$$\frac{\partial S}{\partial a} = \sum_{i=1}^{n} -2x_i^2(y_i - ax_i^2 - bx_i - c) = 0$$

$$\frac{\partial S}{\partial b} = \sum_{i=1}^{n} -2x_i(y_i - ax_i^2 - bx_i - c) = 0$$

$$\frac{\partial S}{\partial c} = -2\sum_{i=1}^{n} (y_i - ax_i^2 - bx_i - c) = 0$$

$$a\sum_{i=1}^{n} x_i^4 + b\sum_{i=1}^{n} x_i^3 + c\sum_{i=1}^{n} x_i^2 = \sum_{i=1}^{n} x_i^2 y_i$$

$$a\sum_{i=1}^{n} x_i^3 + b\sum_{i=1}^{n} x_i^2 + c\sum_{i=1}^{n} x_i = \sum_{i=1}^{n} x_i y_i$$

$$a\sum_{i=1}^{n} x_i^2 + b\sum_{i=1}^{n} x_i + cn = \sum_{i=1}^{n} y_i$$

39. $(-2, 0), (-1, 0), (0, 1), (1, 2), (2, 5)$

$\sum x_i = 0$

$\sum y_i = 8$

$\sum x_i^2 = 10$

$\sum x_i^3 = 0$

$\sum x_i^4 = 34$

$\sum x_i y_i = 12$

$\sum x_i^2 y_i = 22$

$34a + 10c = 22,\ 10b = 12,\ 10a + 5c = 8$

$a = \frac{3}{7},\ b = \frac{6}{5},\ c = \frac{26}{35},\ y = \frac{3}{7}x^2 + \frac{6}{5}x + \frac{26}{35}$

41. $(0, 0), (2, 2), (3, 6), (4, 12)$

$\sum x_i = 9$ $\sum y_i = 20$ $\sum x_i^2 = 29$ $\sum x_i^3 = 99$

$\sum x_i^4 = 353$ $\sum x_i y_i = 70$ $\sum x_i^2 y_i = 254$

$353a + 99b + 29c = 254$

$99a + 29b + 9c = 70$

$29a + 9b + 4c = 20$

$a = 1,\ b = -1,\ c = 0,\ y = x^2 - x$

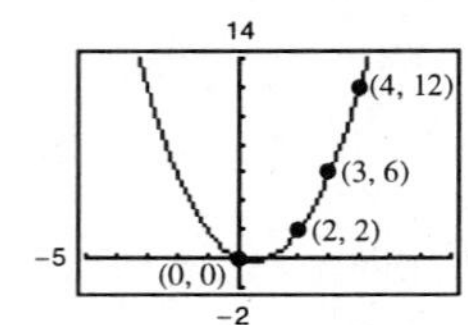

43. $(0, 0), (2, 15), (4, 30), (6, 50), (8, 65), (10, 70)$

$\sum x_i = 30,$ $\sum y_i = 230,$ $\sum x_i^2 = 220,$ $\sum x_i^3 = 1{,}800,$

$\sum x_i^4 = 15{,}664,$ $\sum x_i y_i = 1{,}670,$ $\sum x_i^2 y_i = 13{,}500$

$15{,}664a + 1{,}800b + 220c = 13{,}500$

$1{,}800a + 220b + 30c = 1{,}670$

$220a + 30b + 6c = 230$

$y = -\frac{25}{112}x^2 + \frac{541}{56}x - \frac{25}{14} \approx -0.22x^2 + 9.66x - 1.79$

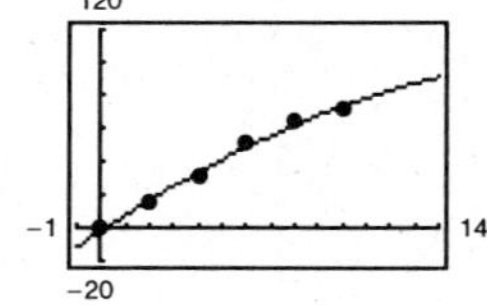

45. (a) $\ln P = -0.1499h + 9.3018$

(b) $\ln P = -0.1499h + 9.3018$

$P = e^{-0.1499h + 9.3018} = 10{,}957.7e^{-0.1499h}$

(c)

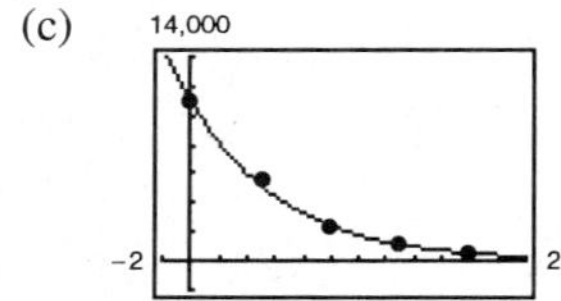

Section 12.10 Lagrange Multipliers

1. Maximize $f(x, y) = xy$.

Constraint: $x + y = 10$

$\nabla f = \lambda \nabla g$

$y\mathbf{i} + x\mathbf{j} = \lambda(\mathbf{i} + \mathbf{j})$

$$\left.\begin{array}{l} y = \lambda \\ x = \lambda \end{array}\right\} x = y$$

$x + y = 10 = \Rightarrow x = y = 5$

$f(5, 5) = 25$

3. Minimize $f(x, y) = x^2 + y^2$.

Constraint: $x + y = 4$

$\nabla f = \lambda \nabla g$

$2x\mathbf{i} + 2y\mathbf{j} = \lambda\mathbf{i} + \lambda\mathbf{j}$

$$\left.\begin{array}{l} 2x = \lambda \\ 2y = \lambda \end{array}\right\} x = y$$

$x + y = 4 \Rightarrow x = y = 2$

$f(2, 2) = 8$

5. Minimize $f(x, y) = x^2 - y^2$.

Constraint: $x - 2y = -6$

$\nabla f = \lambda \nabla g$

$2x\mathbf{i} - 2y\mathbf{j} = \lambda\mathbf{i} - 2\lambda\mathbf{j}$

$2x = \lambda \Rightarrow x = \dfrac{\lambda}{2}$

$-2y = -2\lambda \Rightarrow y = \lambda$

$x - 2y = -6 \Rightarrow -\dfrac{3}{2}\lambda = -6$

$\lambda = 4,\ x = 2,\ y = 4$

$f(2, 4) = -12$

7. Maximize $f(x, y) = 2x + 2xy + y$.

Constraint: $2x + y = 100$

$\nabla f = \lambda \nabla g$

$(2 + 2y)\mathbf{i} + (2x + 1)\mathbf{j} = 2\lambda\mathbf{i} + \lambda\mathbf{j}$

$$\left.\begin{array}{l} 2 + 2y = 2\lambda \Rightarrow y = \lambda - 1 \\ 2x + 1 = \lambda \Rightarrow x = \dfrac{\lambda - 1}{2} \end{array}\right\} y = 2x$$

$2x + y = 100 \Rightarrow 4x = 100$

$x = 25,\ y = 50$

$f(25, 50) = 2600$

9. **Note:** $f(x, y) = \sqrt{6 - x^2 - y^2}$ is maximum when $g(x, y)$ is maximum.

Maximize $g(x, y) = 6 - x^2 - y^2$.

Constraint: $x + y = 2$

$$\left.\begin{array}{l} -2x = \lambda \\ -2y = \lambda \end{array}\right\} x = y$$

$x + y = 2 \Rightarrow x = y = 1$

$f(1, 1) = \sqrt{g(1, 1)} = 2$

11. Maximize $f(x, y) = e^{xy}$.

Constraint: $x^2 + y^2 = 8$

$$\left.\begin{array}{l} ye^{xy} = 2x\lambda \\ xe^{xy} = 2y\lambda \end{array}\right\} x = y$$

$x^2 + y^2 = 8 \Rightarrow 2x^2 = 8$

$x = y = 2$

$f(2, 2) = e^4$

13. Minimize $f(x, y, z) = x^2 + y^2 + z^2$.

Constraint: $x + y + z = 6$

$$\left.\begin{array}{l} 2x = \lambda \\ 2y = \lambda \\ 2z = \lambda \end{array}\right\} x = y = z$$

$x + y + z = 6 \Rightarrow x = y = z = 2$

$f(2, 2, 2) = 12$

15. Minimize $f(x, y, z) = x^2 + y^2 + z^2$.

Constraint: $x + y + z = 1$

$$\left.\begin{array}{l} 2x = \lambda \\ 2y = \lambda \\ 2z = \lambda \end{array}\right\} x = y = z$$

$x + y + z = 1 \Rightarrow x = y = z = \frac{1}{3}$

$f\left(\frac{1}{3}, \frac{1}{3}, \frac{1}{3}\right) = \frac{1}{3}$

17. Maximize $f(x, y, z) = xyz$.

Constraints: $x + y + z = 32$

$$x - y + z = 0$$

$\nabla f = \lambda \nabla g + \mu \nabla h$

$yz\mathbf{i} + xz\mathbf{j} + xy\mathbf{k} = \lambda(\mathbf{i} + \mathbf{j} + \mathbf{k}) + \mu(\mathbf{i} - \mathbf{j} + \mathbf{k})$

$$\left.\begin{aligned} yz &= \lambda + \mu \\ xz &= \lambda - \mu \\ xy &= \lambda + \mu \end{aligned}\right\} yz = xy \Rightarrow x = z$$

$$\left.\begin{aligned} x + y + z &= 32 \\ x - y + z &= 0 \end{aligned}\right\} 2x + 2z = 32 \Rightarrow x = z = 8$$

$$y = 16$$

$f(8, 16, 8) = 1024$

19. Maximize $f(x, y, z) = xy + yz$.

Constraints: $x + 2y = 6$

$$x - 3z = 0$$

$\nabla f = \lambda \nabla g + \mu \nabla h$

$y\mathbf{i} + (x + z)\mathbf{j} + y\mathbf{k} = \lambda(\mathbf{i} + 2\mathbf{j}) + \mu(\mathbf{i} - 3\mathbf{k})$

$$\left.\begin{aligned} y &= \lambda + \mu \\ x + z &= 2\lambda \\ y &= -3\mu \end{aligned}\right\} y = \frac{3}{4}\lambda \Rightarrow x + z = \frac{8}{3}y$$

$$x + 2y = 6 \Rightarrow y = 3 - \frac{x}{2}$$

$$x - 3z = 0 \Rightarrow z = \frac{x}{3}$$

$$x + \frac{x}{3} = \frac{8}{3}\left(3 - \frac{x}{2}\right)$$

$$x = 3, y = \frac{3}{2}, z = 1$$

$f\left(3, \frac{3}{2}, 1\right) = 6$ $\quad f\left(3, \frac{3}{2}, 1\right) = 3 \cdot \frac{3}{2} + \frac{3}{2} \cdot 1 = \frac{9}{4} + \frac{3}{2} = \frac{9+6}{4} = \frac{15}{4}$

21. Maximize or minimize $f(x, y) = x^2 + 3xy + y^2$.

Constraint: $x^2 + y^2 \le 1$

Case 1: On the circle $x^2 + y^2 = 1$

$$\left.\begin{aligned} 2x + 3y &= 2x\lambda \\ 3x + 2y &= 2y\lambda \end{aligned}\right\} x^2 = y^2$$

$$x^2 + y^2 = 1 \Rightarrow x = \pm\frac{\sqrt{2}}{2}, y = \pm\frac{\sqrt{2}}{2}$$

Maxima: $f\left(\pm\frac{\sqrt{2}}{2}, \pm\frac{\sqrt{2}}{2}\right) = \frac{5}{2}$

Minima: $f\left(\pm\frac{\sqrt{2}}{2}, \mp\frac{\sqrt{2}}{2}\right) = -\frac{1}{2}$

Case 2: Inside the circle

$$\left.\begin{aligned} f_x &= 2x + 3y = 0 \\ f_y &= 3x + 2y = 0 \end{aligned}\right\} x = y = 0$$

$f_{xx} = 2,\ f_{yy} = 2,\ f_{xy} = 3,\ f_{xx}f_{yy} - (f_{xy})^2 \le 0$

Saddle point: $f(0, 0) = 0$

By combining these two cases, we have a maximum of $\frac{5}{2}$ at

$$\left(\pm\frac{\sqrt{2}}{2}, \pm\frac{\sqrt{2}}{2}\right)$$

and a minimum of $-\frac{1}{2}$ at

$$\left(\pm\frac{\sqrt{2}}{2}, \mp\frac{\sqrt{2}}{2}\right).$$

23. Minimize the square of the distance $f(x, y) = x^2 + y^2$ subject to the constraint $2x + 3y = -1$.

$$\left.\begin{aligned} 2x &= 2\lambda \\ 2y &= 3\lambda \end{aligned}\right\} y = \frac{3x}{2}$$

$$2x + 3y = -1 \Rightarrow x = -\frac{2}{13}, y = -\frac{3}{13}$$

The point on the line is $\left(-\frac{2}{13}, -\frac{3}{13}\right)$ and the desired distance is

$$d = \sqrt{\left(-\frac{2}{13}\right)^2 + \left(-\frac{3}{13}\right)^2} = \frac{\sqrt{13}}{13}.$$

25. Minimize the square of the distance

$$f(x, y, z) = (x - 2)^2 + (y - 1)^2 + (z - 1)^2$$

subject to the constraint $x + y + z = 1$.

$$\left.\begin{aligned} 2(x - 2) &= \lambda \\ 2(y - 1) &= \lambda \\ 2(z - 1) &= \lambda \end{aligned}\right\} y = z \text{ and } y = x - 1$$

$$x + y + z = 1 \Rightarrow x + 2(x - 1) = 1$$

$$x = 1, y = z = 0$$

The point on the plane is $(1, 0, 0)$ and the desired distance is

$$d = \sqrt{(1 - 2)^2 + (0 - 1)^2 + (0 - 1)^2} = \sqrt{3}.$$

27. Maximize $f(x, y, z) = z$ subject to the constraints $x^2 + y^2 + z^2 = 36$ and $2x + y - z = 2$.

$$\left.\begin{aligned} 0 &= 2x\lambda + 2\mu \\ 0 &= 2y\lambda + \mu \\ 1 &= 2z\lambda - \mu \end{aligned}\right\} x = 2y$$

$$x^2 + y^2 + z^2 = 36$$

$$2x + y - z = 2 \implies z = 2x + y - 2 = 5y - 2$$

$$(2y)^2 + y^2 + (5y - 2)^2 = 36$$

$$30y^2 - 20y - 32 = 0$$

$$15y^2 - 10y - 16 = 0$$

$$y = \frac{5 \pm \sqrt{265}}{15}$$

Choosing the positive value for y we have the point

$$\left(\frac{10 + 2\sqrt{265}}{15}, \frac{5 + \sqrt{265}}{15}, \frac{-1 + \sqrt{265}}{3}\right).$$

29. Maximize $V(x, y, z) = xyz$ subject to the constraint $x + 2y + 2z = 108$.

$$\left.\begin{aligned} yz &= \lambda \\ xz &= 2\lambda \\ xy &= 2\lambda \end{aligned}\right\} y = z \text{ and } x = 2y$$

$$x + 2y + 2z = 108 \implies 6y = 108, y = 18$$

$$x = 36, y = z = 18$$

Volume is maximum when the dimensions are $36 \times 18 \times 18$ inches.

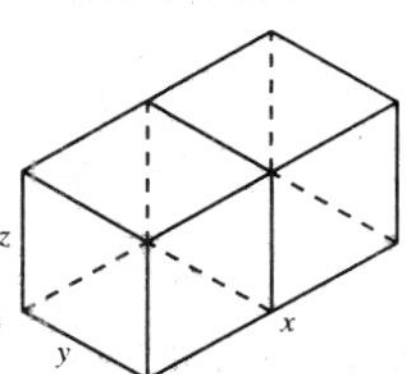

31. Minimize $C(x, y, z) = 5xy + 3(2xz + 2yz + xy)$ subject to the constraint $xyz = 480$.

$$\left.\begin{aligned} 8y + 6z &= yz\lambda \\ 8x + 6z &= xz\lambda \\ 6x + 6y &= xy\lambda \end{aligned}\right\} x = y, 4y = 3z$$

$$xyz = 480 \implies \tfrac{4}{3}y^3 = 480$$

$$x = y = \sqrt[3]{360}, z = \tfrac{4}{3}\sqrt[3]{360}$$

Dimensions: $\sqrt[3]{360} \times \sqrt[3]{360} \times \frac{4}{3}\sqrt[3]{360}$ feet

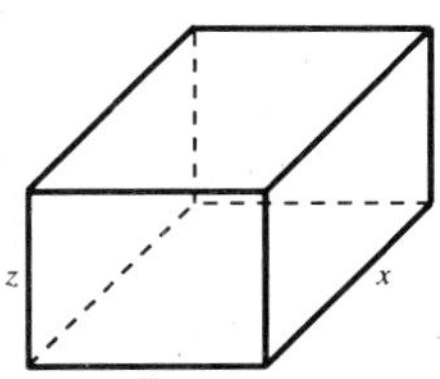

33. Maximize $V(x, y, z) = (2x)(2y)(2z) = 8xyz$ subject to the constraint $\dfrac{x^2}{a^2} + \dfrac{y^2}{b^2} + \dfrac{z^2}{c^2} = 1$.

$$\left.\begin{aligned} 8yz &= \frac{2x}{a^2}\lambda \\ 8xz &= \frac{2y}{b^2}\lambda \\ 8xy &= \frac{2z}{c^2}\lambda \end{aligned}\right\} \frac{x^2}{a^2} = \frac{y^2}{b^2} = \frac{z^2}{c^2}$$

$$\frac{x^2}{a^2} + \frac{y^2}{b^2} + \frac{z^2}{c^2} = 1 \implies \frac{3x^2}{a^2} = 1, \frac{3y^2}{b^2} = 1, \frac{3z^2}{c^2} = 1$$

$$x = \frac{a}{\sqrt{3}}, y = \frac{b}{\sqrt{3}}, z = \frac{c}{\sqrt{3}}$$

Therefore, the dimensions of the box are

$$\frac{2\sqrt{3}a}{3} \times \frac{2\sqrt{3}b}{3} \times \frac{2\sqrt{3}c}{3}.$$

35. Using the formula $\text{Time} = \dfrac{\text{Distance}}{\text{Rate}}$, minimize $T(x, y) = \dfrac{\sqrt{d_1^2 + x^2}}{v_1} + \dfrac{\sqrt{d_2^2 + y^2}}{v_2}$ subject to the constraint $x + y = a$.

$$\left.\begin{aligned} \frac{x}{v_1\sqrt{d_2^2 + x^2}} &= \lambda \\ \frac{y}{v_2\sqrt{d_2^2 + y^2}} &= \lambda \end{aligned}\right\} \frac{x}{v_1\sqrt{d_1^2 + x^2}} = \frac{y}{v_2\sqrt{d_2^2 + y^2}}$$

$$x + y = a$$

Since $\sin\theta_1 = \dfrac{x}{\sqrt{d_1^2 + x^2}}$ and $\sin\theta_2 = \dfrac{y}{\sqrt{d_2^2 + y^2}}$, we have

$$\frac{x/\sqrt{d_1^2 + x^2}}{v_1} = \frac{y/\sqrt{d_2^2 + y^2}}{v_2} \quad\text{or}\quad \frac{\sin\theta_1}{v_1} = \frac{\sin\theta_2}{v_2}.$$

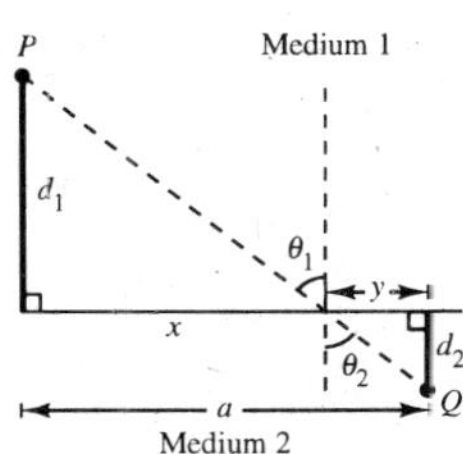

37. Maximize $P(p, q, r) = 2pq + 2pr + 2qr$.

Constraint: $p + q + r = 1$

$\nabla P = \lambda \nabla g$

$$\left.\begin{aligned} 2q + 2r &= \lambda \\ 2p + 2r &= \lambda \\ 2p + 2q &= \lambda \end{aligned}\right\} \Rightarrow 3\lambda = 4(p + q + r) = 4(1) \Rightarrow \lambda = \tfrac{4}{3}$$

$p + q + r = 1$

$$\left.\begin{aligned} q + r &= \tfrac{2}{3} \\ p + q + r &= 1 \end{aligned}\right\} \Rightarrow p = \tfrac{1}{3}, q = \tfrac{1}{3}, r = \tfrac{1}{3}$$

$P\left(\tfrac{1}{3}, \tfrac{1}{3}, \tfrac{1}{3}\right) = 2\left(\tfrac{1}{3}\right)\left(\tfrac{1}{3}\right) + 2\left(\tfrac{1}{3}\right)\left(\tfrac{1}{3}\right) + 2\left(\tfrac{1}{3}\right)\left(\tfrac{1}{3}\right) = \tfrac{2}{3}$.

39. Maximize $P(x, y) = 100x^{0.25}y^{0.75}$ subject to the constraint $48x + 36y = 100{,}000$.

$$25x^{-0.75}y^{0.75} = 48\lambda \Rightarrow \left(\frac{y}{x}\right)^{0.75} = \frac{48\lambda}{25}$$

$$75x^{0.25}y^{-0.25} = 36\lambda \Rightarrow \left(\frac{x}{y}\right)^{0.25} = \frac{36\lambda}{75}$$

$$\left(\frac{y}{x}\right)^{0.75}\left(\frac{y}{x}\right)^{0.25} = \left(\frac{48\lambda}{25}\right)\left(\frac{75}{36\lambda}\right)$$

$$\frac{y}{x} = 4$$

$$y = 4x$$

$$48x + 36y = 100{,}000 \Rightarrow 192x = 100{,}000$$

$$x = \frac{3125}{6}, y = \frac{6250}{3}$$

Therefore, $P\left(\frac{3125}{6}, \frac{6250}{3}\right) \approx 147{,}314$.

41. Minimize $C(x, y) = 48x + 36y$ subject to the constraint $100x^{0.25}y^{0.75} = 20{,}000$.

$$48 = 25x^{-0.75}y^{0.75}\lambda \Rightarrow \left(\frac{y}{x}\right)^{0.75} = \frac{48}{25\lambda}$$

$$36 = 75x^{0.25}y^{-0.25}\lambda \Rightarrow \left(\frac{x}{y}\right)^{0.25} = \frac{36}{75\lambda}$$

$$\left(\frac{y}{x}\right)^{0.75}\left(\frac{y}{x}\right)^{0.25} = \left(\frac{48}{25\lambda}\right)\left(\frac{75\lambda}{36}\right)$$

$$\frac{y}{x} = 4 \Rightarrow y = 4x$$

$$100x^{0.25}y^{0.75} = 20{,}000 \Rightarrow x^{0.25}(4x)^{0.75} = 200$$

$$x = \frac{200}{4^{0.75}} = \frac{200}{2\sqrt{2}} = 50\sqrt{2}$$

$$y = 4x = 200\sqrt{2}$$

Therefore, $C(50\sqrt{2}, 200\sqrt{2}) \approx \$13{,}576.45$.

43. (a) Maximize $g(\alpha, \beta, \gamma) = \cos\alpha\cos\beta\cos\gamma$ subject to the constraint $\alpha + \beta + \gamma = \pi$.

$$\left.\begin{aligned} -\sin\alpha\cos\beta\cos\gamma &= \lambda \\ -\cos\alpha\sin\beta\cos\gamma &= \lambda \\ -\cos\alpha\cos\beta\sin\gamma &= \lambda \end{aligned}\right\} \tan\alpha = \tan\beta = \tan\gamma \Rightarrow \alpha = \beta = \gamma$$

$$\alpha + \beta + \gamma = \pi \Rightarrow \alpha = \beta = \gamma = \frac{\pi}{3}$$

$$g\left(\frac{\pi}{3}, \frac{\pi}{3}, \frac{\pi}{3}\right) = \frac{1}{8}$$

(b) $\alpha + \beta + \gamma = \pi \Rightarrow \gamma = \pi - (\alpha + \beta)$

$$\begin{aligned} g(\alpha + \beta) &= \cos\alpha\cos\beta\cos(\pi - (\alpha + \beta)) \\ &= \cos\alpha\cos\beta[\cos\pi\cos(\alpha + \beta) + \sin\pi\sin(\alpha + \beta)] \\ &= -\cos\alpha\cos\beta\cos(\alpha + \beta) \end{aligned}$$

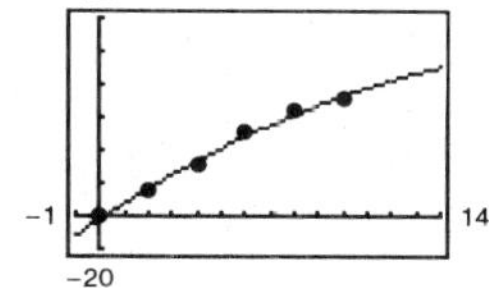

Review Exercises for Chapter 12

1. No, it is not the graph of a function.

3. $f(x, y) = e^{x^2+y^2}$

The level curves are of the form

$$c = e^{x^2+y^2}$$

$$\ln c = x^2 + y^2.$$

The level curves are circles centered at the origin.

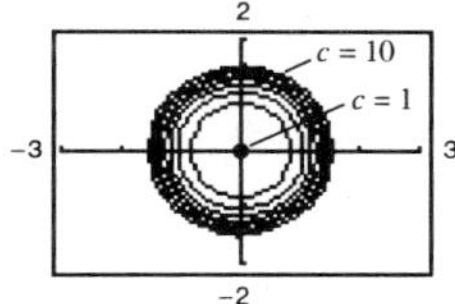

5. $f(x, y) = x^2 - y^2$

The level curves are of the form

$$c = x^2 - y^2$$

$$1 = \frac{x^2}{c} - \frac{y^2}{c}.$$

The level curves are hyperbolas.

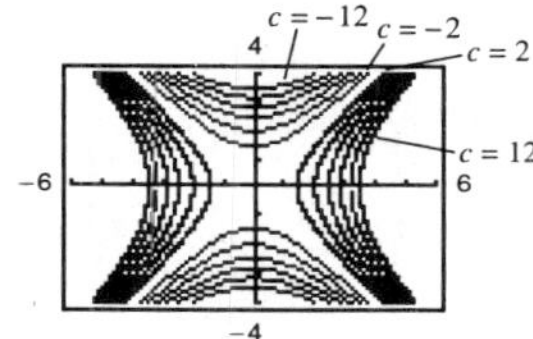

7. $f(x, y) = e^{-(x^2+y^2)}$

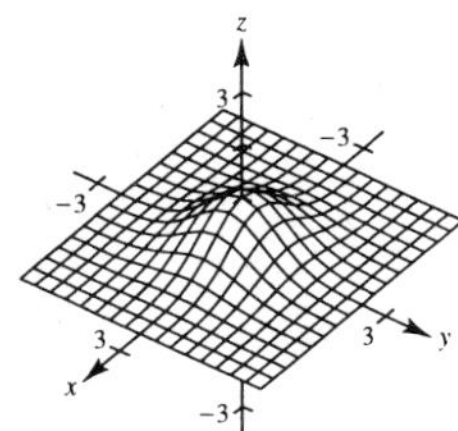

9. $\displaystyle\lim_{(x, y)\to(1, 1)} \frac{xy}{x^2 + y^2} = \frac{1}{2}$

Continuous except at $(0, 0)$.

11. $\displaystyle\lim_{(x, y)\to(0, 0)} \frac{-4x^2y}{x^4 + y^2}$

For $y = x^2$, $\dfrac{-4x^2y}{x^4 + y^2} = \dfrac{-4x^4}{x^4 + x^4} = -2$, for $x \neq 0$

For $y = 0$, $\dfrac{-4x^2y}{x^4 + y^2} = 0$, for $x \neq 0$

Thus, the limit does not exist. Continuous except at $(0, 0)$.

13. $f(x, y) = e^x \cos y$

$$f_x = e^x \cos y$$

$$f_y = -e^x \sin y$$

15. $z = xe^y + ye^x$

$$\frac{\partial z}{\partial x} = e^y + ye^x$$

$$\frac{\partial z}{\partial y} = xe^y + e^x$$

17. $g(x, y) = \dfrac{xy}{x^2 + y^2}$

$$g_x = \frac{y(x^2 + y^2) - xy(2x)}{(x^2 + y^2)^2} = \frac{y(y^2 - x^2)}{(x^2 + y^2)^2}$$

$$g_y = \frac{x(x^2 - y^2)}{(x^2 + y^2)^2}$$

19. $f(x, y, z) = z \arctan \dfrac{y}{x}$

$$f_x = \frac{z}{1 + (y^2/x^2)}\left(-\frac{y}{x^2}\right) = \frac{-yz}{x^2 + y^2}$$

$$f_y = \frac{z}{1 + (y^2/x^2)}\left(\frac{1}{x}\right) = \frac{xz}{x^2 + y^2}$$

$$f_z = \arctan \frac{y}{x}$$

21. $u(x, t) = ce^{-n^2t} \sin(nx)$

$$\frac{\partial u}{\partial x} = cne^{-n^2t} \cos(nx)$$

$$\frac{\partial u}{\partial t} = -cn^2e^{-n^2t} \sin(nx)$$

23. $x^2y - 2yz - xz - z^2 = 0$

$$2xy - 2y\frac{\partial z}{\partial x} - x\frac{\partial z}{\partial x} - z - 2z\frac{\partial z}{\partial x} = 0$$

$$\frac{\partial z}{\partial x} = \frac{-2xy + z}{-2y - x - 2z} = \frac{2xy - z}{x + 2y + 2z}$$

$$x^2 - 2y\frac{\partial z}{\partial y} - 2z - x\frac{\partial z}{\partial y} - 2z\frac{\partial z}{\partial y} = 0$$

$$\frac{\partial z}{\partial y} = \frac{-x^2 + 2z}{-2y - x - 2z} = \frac{x^2 - 2z}{x + 2y + 2z}$$

25. $f(x, y) = 3x^2 - xy + 2y^3$

$$f_x = 6x - y$$
$$f_y = -x + 6y^2$$
$$f_{xx} = 6$$
$$f_{yy} = 12y$$
$$f_{xy} = -1$$
$$f_{yx} = -1$$

27. $h(x, y) = x \sin y + y \cos x$

$$h_x = \sin y - y \sin x$$
$$h_y = x \cos y + \cos x$$
$$h_{xx} = -y \cos x$$
$$h_{yy} = -x \sin y$$
$$h_{xy} = \cos y - \sin x$$
$$h_{yx} = \cos y - \sin x$$

29. $z = x^2 - y^2$

$$\frac{\partial z}{\partial x} = 2x$$

$$\frac{\partial^2 z}{\partial x^2} = 2$$

$$\frac{\partial z}{\partial y} = -2y$$

$$\frac{\partial^2 z}{\partial y^2} = -2$$

Therefore, $\dfrac{\partial^2 z}{\partial x^2} + \dfrac{\partial^2 z}{\partial y^2} = 0.$

31. $z = \dfrac{y}{x^2 + y^2}$

$$\frac{\partial z}{\partial x} = \frac{-2xy}{(x^2 + y^2)^2}$$

$$\frac{\partial^2 z}{\partial x^2} = -2y\left[\frac{-4x^2}{(x^2 + y^2)^3} + \frac{1}{(x^2 + y^2)^2}\right] = 2y\frac{3x^2 - y^2}{(x^2 + y^2)^3}$$

$$\frac{\partial z}{\partial y} = \frac{(x^2 + y^2) - 2y}{(x^2 + y^2)^2} = \frac{x^2 - y^2}{(x^2 + y^2)^2}$$

$$\frac{\partial^2 z}{\partial y^2} = \frac{(x^2 + y^2)^2(-2y) - 2(x^2 - y^2)(x^2 + y^2)(2y)}{(x^2 + y^2)^4}$$

$$= -2y\frac{3x^2 - y^2}{(x^2 + y^2)^3}$$

Therefore, $\dfrac{\partial^2 z}{\partial x^2} + \dfrac{\partial^2 z}{\partial y^2} = 0.$

33. $u = x^2 + y^2 + z^2,\ x = r\cos t,\ y = r\sin t,\ z = t$

Chain Rule: $\dfrac{\partial u}{\partial r} = \dfrac{\partial u}{\partial x}\dfrac{\partial x}{\partial r} + \dfrac{\partial u}{\partial y}\dfrac{\partial y}{\partial r} + \dfrac{\partial u}{\partial z}\dfrac{\partial z}{\partial r}$

$$= 2x\cos t + 2y\sin t + 2z(0)$$

$$= 2(r\cos^2 t + r\sin^2 t) = 2r$$

$$\frac{\partial u}{\partial t} = \frac{\partial u}{\partial x}\frac{\partial x}{\partial t} + \frac{\partial u}{\partial y}\frac{\partial y}{\partial t} + \frac{\partial u}{\partial z}\frac{\partial z}{\partial t}$$

$$= 2x(-r\sin t) + 2y(r\cos t) + 2z$$

$$= 2(-r^2\sin t\cos t + r^2\sin t\cos t) + 2t$$

$$= 2t$$

Substitution:

$u(r, t) = r^2\cos^2 t + r^2\sin^2 t + t^2 = r^2 + t^2$

$$\frac{\partial u}{\partial r} = 2r$$

$$\frac{\partial u}{\partial t} = 2t$$

35. $f(x, y) = x^2 y$

$$\nabla f = 2xy\mathbf{i} + x^2\mathbf{j}$$

$$\nabla f(2, 1) = 4\mathbf{i} + 4\mathbf{j}$$

$$\mathbf{u} = \frac{1}{\sqrt{2}}\mathbf{v} = \frac{\sqrt{2}}{2}\mathbf{i} - \frac{\sqrt{2}}{2}\mathbf{j}$$

$$D_{\mathbf{u}}f(2, 1) = \nabla f(2, 1)\cdot\mathbf{u} = 2\sqrt{2} - 2\sqrt{2} = 0$$

37. $w = y^2 + xz$

$$\nabla w = z\mathbf{i} + 2y\mathbf{j} + x\mathbf{k}$$

$$\nabla w(1, 2, 2) = 2\mathbf{i} + 4\mathbf{j} + \mathbf{k}$$

$$\mathbf{u} = \frac{1}{3}\mathbf{v} = \frac{2}{3}\mathbf{i} - \frac{1}{3}\mathbf{j} + \frac{2}{3}\mathbf{k}$$

$$D_{\mathbf{u}}w(1, 2, 2) = \nabla w(1, 2, 2)\cdot\mathbf{u} = \frac{4}{3} - \frac{4}{3} + \frac{2}{3} = \frac{2}{3}$$

39. $z = \dfrac{y}{x^2 + y^2}$

$$\nabla z = -\frac{2xy}{(x^2 + y^2)^2}\mathbf{i} + \frac{x^2 - y^2}{(x^2 + y^2)^2}\mathbf{j}$$

$$\nabla z(1, 1) = -\frac{1}{2}\mathbf{i} = \left\langle -\frac{1}{2}, 0\right\rangle$$

$$\|\nabla z(1, 1)\| = \frac{1}{2}$$

41. $z = e^{-x}\cos y$

$$\nabla z = -e^{-x}\cos y\mathbf{i} - e^{-x}\sin y\mathbf{j}$$

$$\nabla z\left(0, \frac{\pi}{4}\right) = -\frac{\sqrt{2}}{2}\mathbf{i} - \frac{\sqrt{2}}{2}\mathbf{j} = \left\langle -\frac{\sqrt{2}}{2}, -\frac{\sqrt{2}}{2}\right\rangle$$

$$\left\|\nabla z\left(0, \frac{\pi}{4}\right)\right\| = 1$$

43. $F(x, y, z) = x^2 y - z = 0$

$$\nabla F = 2xy\mathbf{i} + x^2\mathbf{j} - \mathbf{k}$$

$$\nabla F(2, 1, 4) = 4\mathbf{i} + 4\mathbf{j} - \mathbf{k}$$

Therefore, the equation of the tangent plane is

$$4(x - 2) + 4(y - 1) - (z - 4) = 0 \quad \text{or}$$

$$4x + 4y - z = 8,$$

and the equation of the normal line is

$$\frac{x - 2}{4} = \frac{y - 1}{4} = \frac{z - 4}{-1}.$$

45. $F(x, y, z) = x^2 + y^2 - 4x + 6y + z + 9 = 0$

$$\nabla F = (2x - 4)\mathbf{i} + (2y + 6)\mathbf{j} + \mathbf{k}$$

$$\nabla F(2, -3, 4) = \mathbf{k}$$

Therefore, the equation of the tangent plane is

$$z - 4 = 0 \quad \text{or} \quad z = 4,$$

and the equation of the normal line is

$$x = 2,\ y = -3,\ z = 4 + t.$$

47. $F(x, y, z) = x^2 - y^2 - z = 0$

$$G(x, y, z) = 3 - z = 0$$

$$\nabla F = 2x\mathbf{i} - 2y\mathbf{j} - \mathbf{k}$$

$$\nabla G = -\mathbf{k}$$

$$\nabla F(2, 1, 3) = 4\mathbf{i} - 2\mathbf{j} - \mathbf{k}$$

$$\nabla F \times \nabla G = \begin{vmatrix} \mathbf{i} & \mathbf{j} & \mathbf{k} \\ 4 & -2 & -1 \\ 0 & 0 & -1 \end{vmatrix} = 2(\mathbf{i} + 2\mathbf{j})$$

Therefore, the equation of the tangent line is

$$\frac{x - 2}{1} = \frac{y - 1}{2},\ z = 3.$$

49. $f(x, y) = x^3 - 3xy + y^2$

$$f_x = 3x^2 - 3y = 3(x^2 - y) = 0$$
$$f_y = -3x + 2y = 0$$
$$f_{xx} = 6x$$
$$f_{yy} = 2$$
$$f_{xy} = -3$$

From $f_x = 0$, we have $y = x^2$. Substituting this into $f_y = 0$, we have $-3x + 2x^2 = x(2x - 3) = 0$. Thus, $x = 0$ or $\frac{3}{2}$.
At the critical point $(0, 0)$, $f_{xx}f_{yy} - (f_{xy})^2 < 0$. Therefore, $(0, 0, 0)$ is a saddle point.
At the critical point $\left(\frac{3}{2}, \frac{9}{4}\right)$, $f_{xx}f_{yy} - (f_{xy})^2 > 0$ and $f_{xx} > 0$. Therefore, $\left(\frac{3}{2}, \frac{9}{4}, -\frac{27}{16}\right)$ is a relative minimum.

51. $f(x, y) = xy + \dfrac{1}{x} + \dfrac{1}{y}$

$$f_x = y - \frac{1}{x^2} = 0, \; x^2y = 1$$
$$f_y = x - \frac{1}{y^2} = 0, \; xy^2 = 1$$

Thus, $x^2y = xy^2$ or $x = y$ and substitution yields the critical point $(1, 1)$.

$$f_{xx} = \frac{2}{x^3}$$
$$f_{xy} = 1$$
$$f_{yy} = \frac{2}{y^3}$$

At the critical point $(1, 1)$, $f_{xx} = 2 > 0$ and $f_{xx}f_{yy} - (f_{xy})^2 = 3 > 0$. Thus, $(1, 1, 3)$ is a relative minimum.

53. The level curves are hyperbolas. There is a critical point at $(0, 0)$, but there are no relative extrema. The gradient is normal to the level curve at any given point at (x_0, y_0).

55.

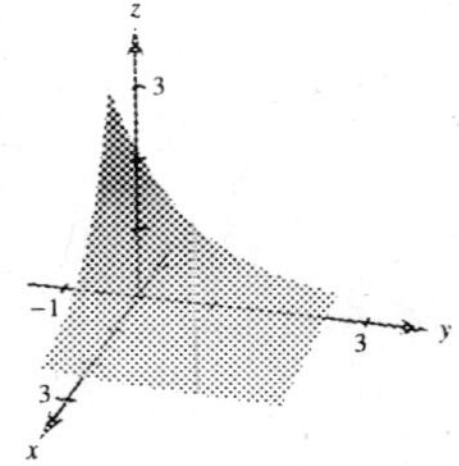

57. $z = x \sin \dfrac{y}{x}$

$$dz = \frac{\partial z}{\partial x}\,dx + \frac{\partial z}{\partial y}\,dy = \left(\sin \frac{y}{x} - \frac{y}{x}\cos \frac{y}{x}\right)dx + \left(\cos \frac{y}{x}\right)dy$$

59. $z^2 = x^2 + y^2$

$$2z\,dz = 2x\,dx + 2y\,dy$$
$$dz = \frac{x}{z}\,dx + \frac{y}{z}\,dy = \frac{5}{13}\left(\frac{1}{2}\right) + \frac{12}{13}\left(\frac{1}{2}\right) = \frac{17}{26} \approx 0.654 \text{ cm}$$

Percentage error: $\dfrac{dz}{z} = \dfrac{17/26}{13} \approx 0.0503 \approx 5\%$

61. $V = \frac{1}{3}\pi r^2 h$

$$dV = \tfrac{2}{3}\pi rh\,dr + \tfrac{1}{3}\pi r^2\,dh = \tfrac{2}{3}\pi(2)(5)\left(\pm\tfrac{1}{8}\right) + \tfrac{1}{3}\pi(2)^2\left(\pm\tfrac{1}{8}\right) = \pm\tfrac{5}{6}\pi \pm \tfrac{1}{6}\pi = \pm\pi \text{ in}^3$$

63. $P(x_1, x_2) = R - C_1 - C_2$

$$= [225 - 0.4(x_1 + x_2)](x_1 + x_2) - (0.05x_1^2 + 15x_1 + 5400) - (0.03x_2^2 + 15x_2 + 6100)$$

$$= -0.45x_1^2 - 0.43x_2^2 - 0.8x_1x_2 + 210x_1 + 210x_2 - 11{,}500$$

$$P_{x_1} = -0.9x_1 - 0.8x_2 + 210 = 0$$

$$0.9x_1 + 0.8x_2 = 210$$

$$P_{x_2} = -0.86x_2 - 0.8x_1 + 210 = 0$$

$$0.8x_1 + 0.86x_2 = 210$$

Solving this system yields $x_1 \approx 94$ and $x_2 \approx 157$.

$$P_{x_1x_1} = -0.9$$

$$P_{x_1x_2} = -0.8$$

$$P_{x_2x_2} = -0.86$$

$$P_{x_1x_1} < 0$$

$$P_{x_1x_1}P_{x_2x_2} - (P_{x_1x_2})^2 > 0$$

Therefore, profit is maximum when $x_1 \approx 94$ and $x_2 \approx 157$.

65. Maximize $f(x, y) = 4x + xy + 2y$ subject to the constraint $20x + 4y = 2000$.

$$\left.\begin{aligned} 4 + y &= 20\lambda \\ x + 2 &= 4\lambda \end{aligned}\right\} 5x - y = -6$$

$$20x + 4y = 2000 \implies \quad 5x + y = 500$$

$$5x - y = -6$$

$$10x = 494$$

$$x = 49.4$$

$$y = 253$$

$f(49.4, 253) = 13{,}201.8$

67. (a) $y = 2.29t + 2.34$

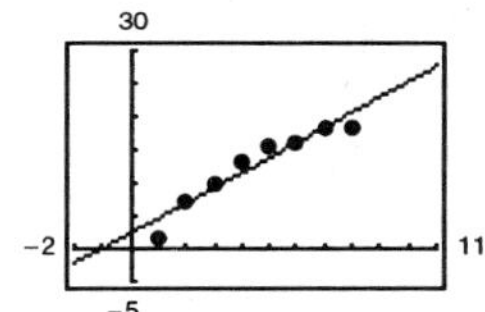

(b)

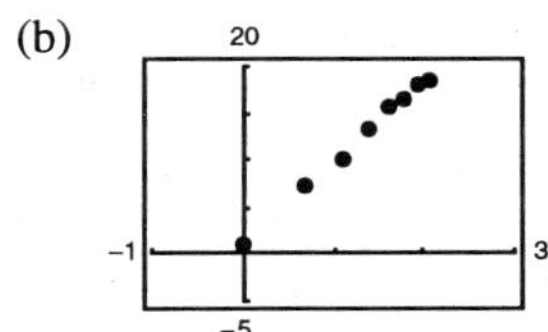

Yes, the data appears more linear.

(c) $y = 8.37 \ln t + 1.54$

(d)

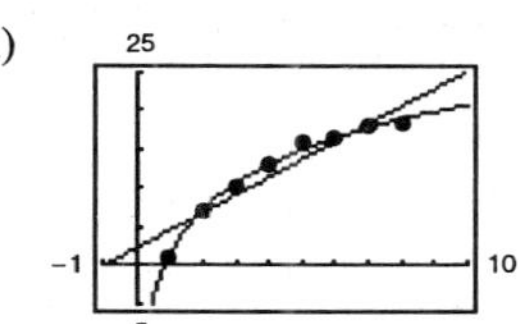

The logarithmic model is a better fit.

69. Optimize $f(x, y, z) = xy + yz + xz$ subject to the constraint $x + y + z = 1$.

$$\left.\begin{aligned} y + z &= \lambda \\ x + z &= \lambda \\ x + y &= \lambda \end{aligned}\right\} x = y = z$$

$$x + y + z = 1 \implies x = y = z = \tfrac{1}{3}$$

Maximum: $f\left(\tfrac{1}{3}, \tfrac{1}{3}, \tfrac{1}{3}\right) = \tfrac{1}{3}$

71. False, $\nabla F(x_0, y_0, z_0)$ is normal to the surface.

CHAPTER 13
Multiple Integration

CHAPTER 13
Multiple Integration

Section 13.1 Iterated Integrals and Area in the Plane

Solutions to Odd-Numbered Exercises

1. $\displaystyle\int_0^x (2x - y)\, dy = \left[2xy - \frac{1}{2}y^2\right]_0^x = \frac{3}{2}x^2$

3. $\displaystyle\int_1^{2y} \frac{y}{x}\, dx = \left[y \ln x\right]_1^{2y} = y \ln 2y - 0 = y \ln 2y$

5. $\displaystyle\int_0^{\sqrt{4-x^2}} x^2 y\, dy = \left[\frac{1}{2}x^2y^2\right]_0^{\sqrt{4-x^2}} = \frac{4x^2 - x^4}{2}$

7. $\displaystyle\int_{e^y}^{y} \frac{y \ln x}{x}\, dx = \left[\frac{1}{2}y \ln^2 x\right]_{e^y}^{y}$

$\displaystyle= \frac{1}{2}y[\ln^2 y - \ln^2 e^y] = \frac{y}{2}[(\ln y)^2 - y^2]$

9. $\displaystyle\int_0^{x^3} ye^{-y/x}\, dy = \left[-xye^{-y/x}\right]_0^{x^3} + x\int_0^{x^3} e^{-y/x}\, dy = -x^4 e^{-x^2} - \left[x^2e^{-y/x}\right]_0^{x^3} = x^2(1 - e^{-x^2} - x^2e^{-x^2})$

$u = y,\ du = dy,\ dv = e^{-y/x}\, dy,\ v = -xe^{-y/x}$

11. $\displaystyle\int_0^1\int_0^2 (x + y)\, dy\, dx = \int_0^1 \left[xy + \frac{1}{2}y^2\right]_0^2 dx = \int_0^1 (2x + 2)\, dx = \left[x^2 + 2x\right]_0^1 = 3$

13. $\displaystyle\int_1^2\int_0^4 (x^2 - 2y^2 + 1)\, dx\, dy = \int_1^2 \left[\frac{1}{3}x^3 - 2xy^2 + x\right]_0^4 dy$

$\displaystyle= \int_1^2 \left(\frac{64}{3} - 8y^2 + 4\right) dy = \frac{4}{3}\int_1^2 (19 - 6y^2)\, dy = \left[\frac{4}{3}(19y - 2y^3)\right]_1^2 = \frac{20}{3}$

15. $\displaystyle\int_0^1\int_0^{\sqrt{1-y^2}} (x + y)\, dx\, dy = \int_0^1 \left[\frac{1}{2}x^2 + xy\right]_0^{\sqrt{1-y^2}} dy$

$\displaystyle= \int_0^1 \left[\frac{1}{2}(1 - y^2) + y\sqrt{1 - y^2}\right] dy = \left[\frac{1}{2}y - \frac{1}{6}y^3 - \frac{1}{2}\left(\frac{2}{3}\right)(1 - y^2)^{3/2}\right]_0^1 = \frac{2}{3}$

17. $\displaystyle\int_0^2\int_0^{\sqrt{4-y^2}} \frac{2}{\sqrt{4 - y^2}}\, dx\, dy = \int_0^2 \left[\frac{2x}{\sqrt{4 - y^2}}\right]_0^{\sqrt{4-y^2}} dy = \int_0^2 2\, dy = \left[2y\right]_0^2 = 4$

19. $\displaystyle\int_0^{\pi/2}\int_0^{\sin\theta} \theta r\, dr\, d\theta = \int_0^{\pi/2} \left[\theta\frac{r^2}{2}\right]_0^{\sin\theta} d\theta = \int_0^{\pi/2} \frac{1}{2}\theta \sin^2\theta\, d\theta$

$\displaystyle= \frac{1}{4}\int_0^{\pi/2} (\theta - \theta\cos 2\theta)\, d\theta = \frac{1}{4}\left[\frac{\theta^2}{2} - \left(\frac{1}{4}\cos 2\theta + \frac{\theta}{2}\sin 2\theta\right)\right]_0^{\pi/2} = \frac{\pi^2}{32} + \frac{1}{8}$

21. $\displaystyle\int_1^\infty\int_0^{1/x} y\, dy\, dx = \int_1^\infty \left[\frac{y^2}{2}\right]_0^{1/x} dx = \frac{1}{2}\int_1^\infty \frac{1}{x^2}\, dx = \left[-\frac{1}{2x}\right]_1^\infty = 0 + \frac{1}{2} = \frac{1}{2}$

23. $\displaystyle\int_1^{\infty}\int_1^{\infty} \frac{1}{xy}\,dx\,dy = \int_1^{\infty}\left[\frac{1}{y}\ln x\right]_1^{\infty} dy = \int_1^{\infty}\left[\frac{1}{y}(\infty) - \frac{1}{y}(0)\right] dy$

Diverges

25. $\displaystyle\int_0^4\int_0^y f(x, y)\,dx\,dy,\ 0 \le x \le y,\ 0 \le y \le 4$

$\displaystyle = \int_0^4\int_x^4 f(x, y)\,dy\,dx$

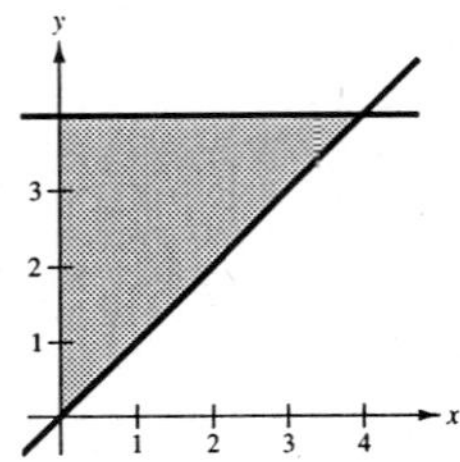

27. $\displaystyle\int_{-1}^1\int_{x^2}^1 f(x, y)\,dy\,dx, x^2 \le y \le 1, 1 \le x \le 1$

$\displaystyle = \int_0^1\int_{-\sqrt{y}}^{\sqrt{y}} f(x, y)\,dx\,dy$

29. $\displaystyle\int_0^1\int_0^2 dy\,dx = \int_0^2\int_0^1 dx\,dy = 2$

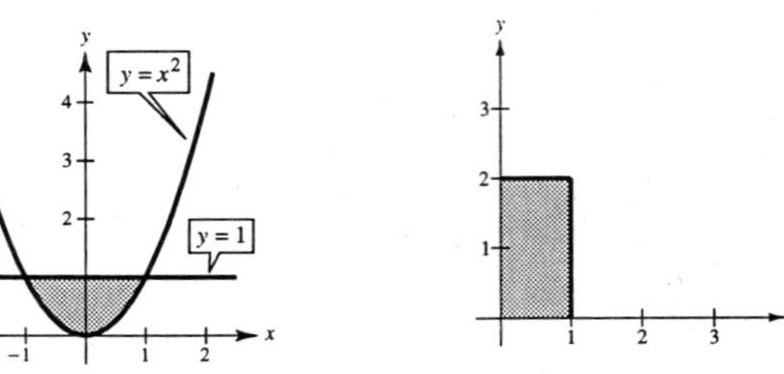

31. $\displaystyle\int_0^1\int_{-\sqrt{1-y^2}}^{\sqrt{1-y^2}} dx\,dy = \int_{-1}^1\int_0^{\sqrt{1-x^2}} dy\,dx = \frac{\pi}{2}$

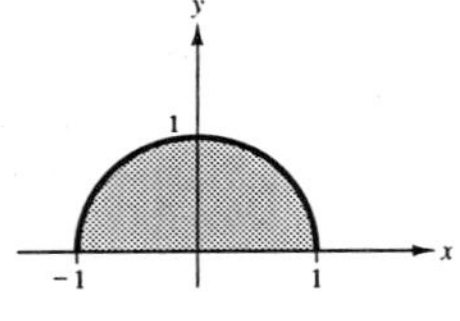

33. $\displaystyle\int_0^2\int_{x/2}^1 dy\,dx = \int_0^1\int_0^{2y} dx\,dy = 1$

35. $\displaystyle\int_0^1\int_{y^2}^{\sqrt[3]{y}} dx\,dy = \int_0^1\int_{x^3}^{\sqrt{x}} dy\,dx = \frac{5}{12}$

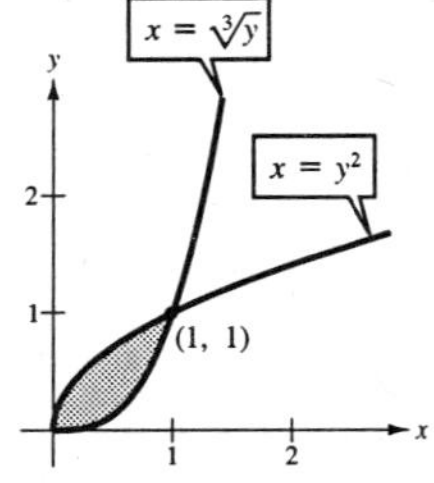

37. $\displaystyle A = \int_0^8\int_0^3 dy\,dx = \int_0^8\left[y\right]_0^3 dx = \int_0^8 3\,dx = \left[3x\right]_0^8 = 24$

$\displaystyle A = \int_0^3\int_0^8 dx\,dy = \int_0^3\left[x\right]_0^8 dy = \int_0^3 8\,dy = \left[8y\right]_0^3 = 24$

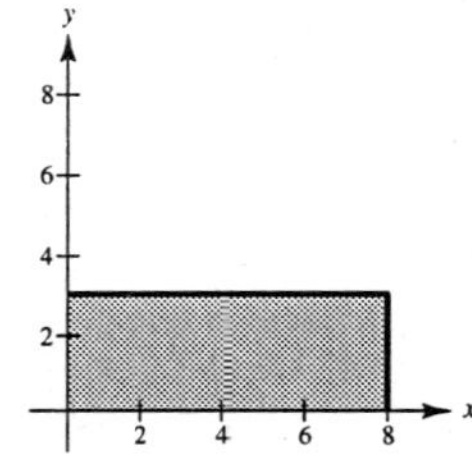

39. $A = \int_0^2 \int_0^{4-x^2} dy\,dx = \int_0^2 \left[y\right]_0^{4-x^2} dx$

$= \int_0^2 (4 - x^2)\,dx$

$= \left[4x - \frac{x^3}{3}\right]_0^2 = \frac{16}{3}$

$A = \int_0^4 \int_0^{\sqrt{4-y}} dx\,dy$

$= \int_0^4 \left[x\right]_0^{\sqrt{4-y}} dy$

$= \int_0^4 \sqrt{4-y}\,dy$

$= -\int_0^4 (4-y)^{1/2}(-1)\,dy$

$= \left[-\frac{2}{3}(4 - y^{3/2})\right]_0^4 = \frac{2}{3}(8) = \frac{16}{3}$

41. $A = \int_{-2}^1 \int_{x+2}^{4-x^2} dy\,dx$

$= \int_{-2}^1 \left[y\right]_{x+2}^{4-x^2} dx$

$= \int_{-2}^1 (4 - x^2 - x - 2)\,dx$

$= \int_{-2}^1 (2 - x - x^2)\,dx$

$= \left[2x - \frac{1}{2}x^2 - \frac{1}{3}x^3\right]_{-2}^1 = \frac{9}{2}$

$A = \int_0^3 \int_{-\sqrt{4-y}}^{y-2} dx\,dy + 2\int_3^4 \int_0^{\sqrt{4-y}} dx\,dy$

$= \int_0^3 \left[x\right]_{-\sqrt{4-y}}^{y-2} dy + 2\int_3^4 \left[x\right]_0^{\sqrt{4-y}} dy$

$= \int_0^3 \left(y - 2 + \sqrt{4-y}\right) dy + 2\int_3^4 \sqrt{4-y}\,dy$

$= \left[\frac{1}{2}y^2 - 2y - \frac{2}{3}(4-y)^{3/2}\right]_0^3 - \left[\frac{4}{3}(4-y)^{3/2}\right]_3^4 = \frac{9}{2}$

43. $\int_0^4 \int_0^{(2-\sqrt{x})^2} dy\,dx = \int_0^4 \left[y\right]_0^{(2-\sqrt{x})^2} dx$

$= \int_0^4 \left(4 - 4\sqrt{x} + x\right) dx$

$= \left[4x - \frac{8}{3}x\sqrt{x} + \frac{x^2}{2}\right]_0^4 = \frac{8}{3}$

$\int_0^4 \int_0^{(2-\sqrt{y})^2} dx\,dy = \frac{8}{3}$

Integration steps are similar to those above.

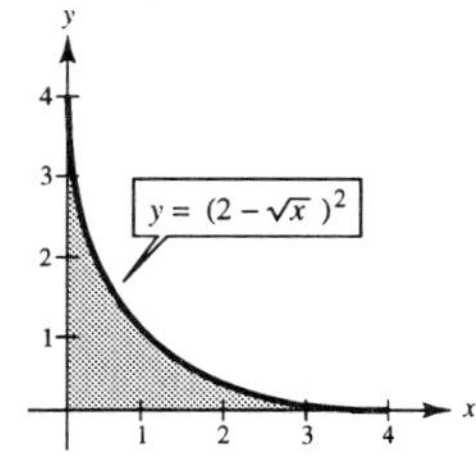

45. $A = \int_0^3 \int_0^{2x/3} dy\,dx + \int_3^5 \int_0^{5-x} dy\,dx$

$= \int_0^3 \left[y\right]_0^{2x/3} dx + \int_3^5 \left[y\right]_0^{5-x} dx$

$= \int_0^3 \frac{2x}{3}\,dx + \int_3^5 (5 - x)\,dx$

$= \left[\frac{1}{3}x^2\right]_0^3 + \left[5x - \frac{1}{2}x^2\right]_3^5 = 5$

$A = \int_0^2 \int_{3y/2}^{5-y} dx\,dy$

$= \int_0^2 \left[x\right]_{3y/2}^{5-y} dy$

$= \int_0^2 \left(5 - y - \frac{3y}{2}\right) dy$

$= \int_2^2 \left(5 - \frac{5y}{2}\right) dy = \left[5y - \frac{5}{4}y^2\right]_0^2 = 5$

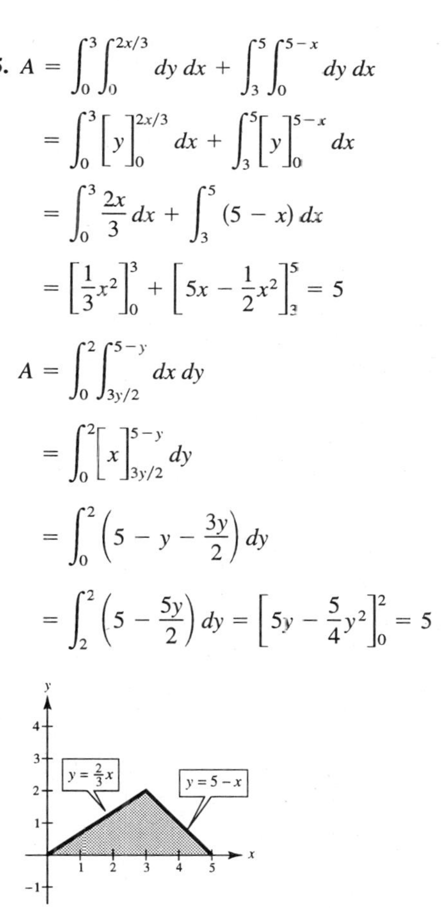

47. $\displaystyle \frac{A}{4} = \int_0^a \int_0^{(b/a)\sqrt{a^2-x^2}} dy\, dx = \int_0^a \Big[y \Big]_0^{(b/a)\sqrt{a^2-x^2}} dx$

$\displaystyle = \frac{b}{a}\int_0^a \sqrt{a^2 - x^2}\, dx = ab\int_0^{\pi/2} \cos^2\theta\, d\theta (x = a\sin\theta,\ dx = a\cos\theta\, d\theta)$

$\displaystyle = \frac{ab}{2}\int_0^{\pi/2} (1 + \cos 2\theta)\, d\theta = \left[\frac{ab}{2}\left(\theta + \frac{1}{2}\sin 2\theta\right)\right]_0^{\pi/2} = \frac{\pi ab}{4}$

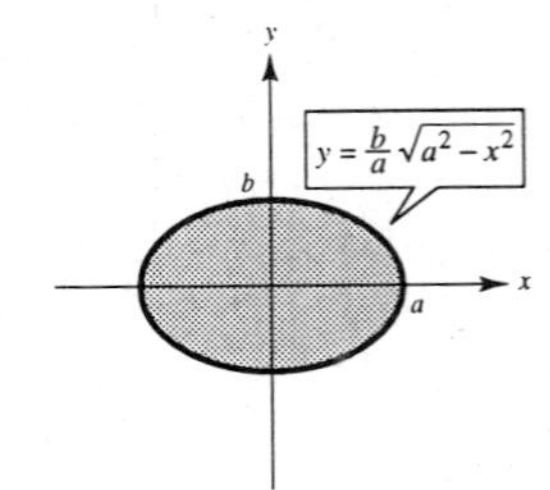

Therefore, $A = \pi ab$.

$$\frac{A}{4} = \int_0^b \int_0^{(a/b)\sqrt{b^2-y^2}} dx\, dy = \frac{\pi ab}{4}$$

Therefore, $A = \pi ab$. Integration steps are similar to those above.

49. The first integral arises using vertical representative rectangles. The second two integrals arise using horizontal representative rectangles.

$$\int_0^5 \int_x^{\sqrt{50-x^2}} x^2y^2\, dy\, dx = \int_0^5 \left[\frac{1}{3}x^2(50 - x^2)^{3/2} - \frac{1}{3}x^5\right] dx = \frac{15625}{24}\pi$$

$$\int_0^5 \int_0^y x^2y^2\, dx\, dy + \int_5^{5\sqrt{2}} \int_0^{\sqrt{50-y^2}} x^2y^2\, dx\, dy = \int_0^5 \frac{1}{3}y^5\, dy + \int_5^{5\sqrt{2}} \frac{1}{3}(50 - y^2)^{3/2}\, y^2\, dy)$$

$$= \frac{1526}{18} + \left(\frac{1526}{18}\pi - \frac{1526}{18}\right) = \frac{15625}{24}\pi$$

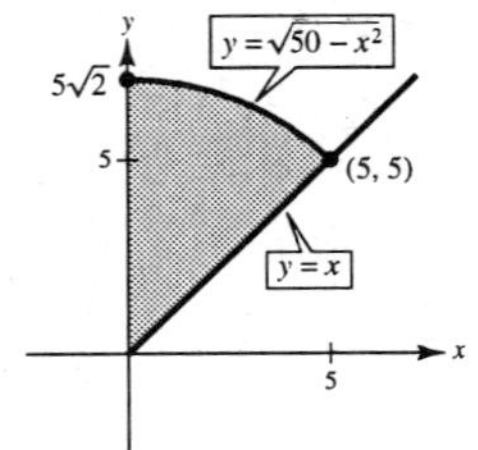

51. $\displaystyle \int_0^2 \int_x^2 x\sqrt{1 + y^3}\, dy\, dx = \int_0^2 \int_0^y x\sqrt{1 + y^3}\, dx\, dy = \int_0^2 \left[\sqrt{1 + y^3} \cdot \frac{x^2}{2}\right]_0^y dy$

$\displaystyle = \frac{1}{2}\int_0^2 \sqrt{1 + y^3}\, y^2\, dy = \left[\frac{1}{2} \cdot \frac{1}{3} \cdot \frac{2}{3}(1 + y^3)^{3/2}\right]_0^2 = \frac{1}{9}(27) - \frac{1}{9}(1) = \frac{26}{9}$

53. $\displaystyle \int_0^1 \int_y^1 \sin(x^2)\, dx\, dy = \int_0^1 \int_0^x \sin(x^2)\, dy\, dx = \int_0^1 \Big[y\sin(x^2) \Big]_0^x dx$

$\displaystyle = \int_0^1 x\sin(x^2)\, dx = \left[-\frac{1}{2}\cos(x^2)\right]_0^1 = -\frac{1}{2}\cos 1 + \frac{1}{2}(1) = \frac{1}{2}(1 - \cos 1) \approx 0.2298$

55. $\displaystyle \int_0^2 \int_{x^2}^{2x} (x^3 + 3y^2)\, dy\, dx = \frac{1664}{105} \approx 15.848$

57. $\displaystyle \int_0^4 \int_0^y \frac{2}{(x + 1)(y + 1)}\, dx\, dy = (\ln 5)^2 \approx 2.590$

59. (a) $x = y^3 \Leftrightarrow y = x^{1/3}$

$x = 4\sqrt{2y} \Leftrightarrow x^2 = 32y$

(b) $\displaystyle \int_0^8 \int_{x^2/32}^{x^{1/3}} (x^2y - xy^2)\, dy\, dx$

(c) Both integrals equal $67520/693 \approx 97.43$

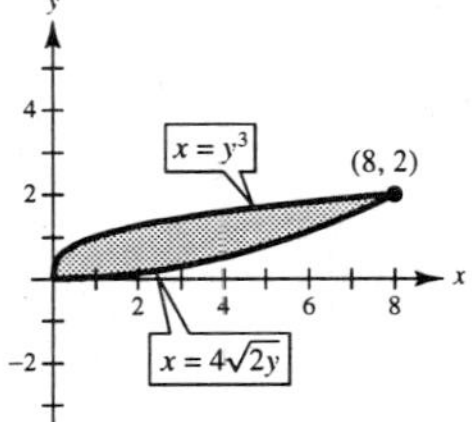

61. $\displaystyle \int_0^2 \int_0^{4-x^2} e^{xy}\, dy\, dx \approx 20.5648$

63. An iterated integral is a double integral of a function of two variables. First integrate with respect to one variable while holding the other variable constant. Then integrate with respect to the second variable.

65. True

Section 13.2 Double Integrals and Volume

For Exercise 1 and 3, $\Delta x_i = \Delta y_i = 1$ and the midpoints of the squares are

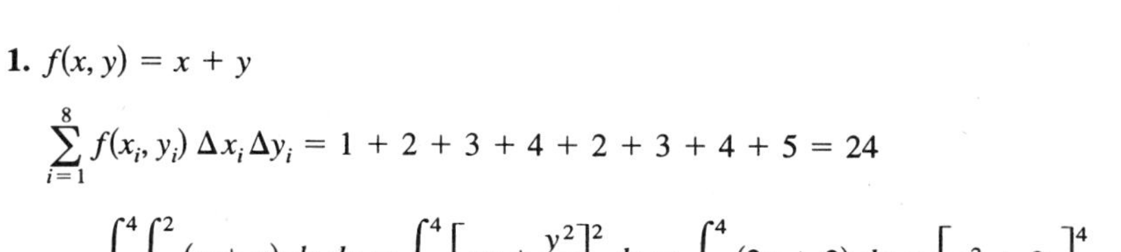

$$\left(\frac{1}{2}, \frac{1}{2}\right), \left(\frac{3}{2}, \frac{1}{2}\right), \left(\frac{5}{2}, \frac{1}{2}\right), \left(\frac{7}{2}, \frac{1}{2}\right), \left(\frac{1}{2}, \frac{3}{2}\right), \left(\frac{3}{2}, \frac{3}{2}\right), \left(\frac{5}{2}, \frac{3}{2}\right), \left(\frac{7}{2}, \frac{3}{2}\right).$$

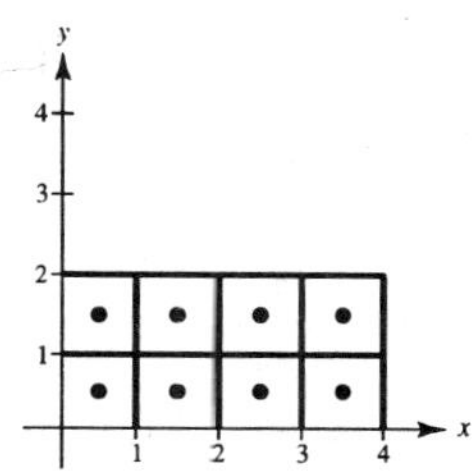

1. $f(x, y) = x + y$

$$\sum_{i=1}^{8} f(x_i, y_i)\,\Delta x_i \Delta y_i = 1 + 2 + 3 + 4 + 2 + 3 + 4 + 5 = 24$$

$$\int_0^4 \int_0^2 (x + y)\,dy\,dx = \int_0^4 \left[xy + \frac{y^2}{2}\right]_0^2 dx = \int_0^4 (2x + 2)\,dx = \left[x^2 + 2x\right]_0^4 = 24$$

3. $f(x, y) = x^2 + y^2$

$$\sum_{i=1}^{8} f(x_i, y_i)\,\Delta x_i \Delta y_i = \frac{2}{4} + \frac{10}{4} + \frac{26}{4} + \frac{50}{4} + \frac{10}{4} + \frac{18}{4} + \frac{34}{4} + \frac{58}{4} = 52$$

$$\int_0^4 \int_0^2 (x^2 + y^2)\,dy\,dx = \int_0^4 \left[x^2 y + \frac{y^3}{3}\right]_0^2 dx = \int_0^4 \left(2x^2 + \frac{8}{3}\right) dx = \left[\frac{2x^3}{3} + \frac{8x}{3}\right]_0^4 = \frac{160}{3}$$

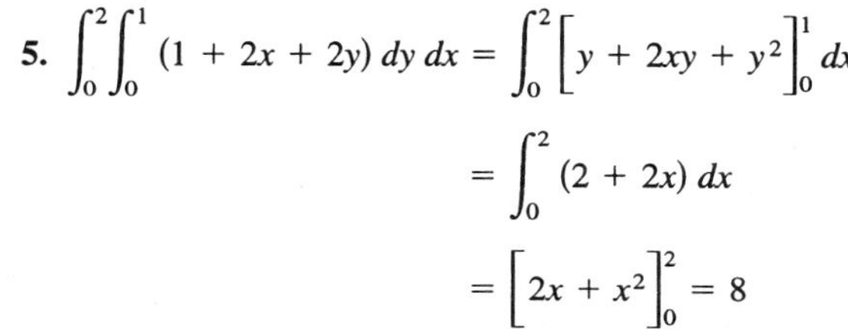

5.
$$\int_0^2 \int_0^1 (1 + 2x + 2y)\,dy\,dx = \int_0^2 \left[y + 2xy + y^2\right]_0^1 dx = \int_0^2 (2 + 2x)\,dx = \left[2x + x^2\right]_0^2 = 8$$

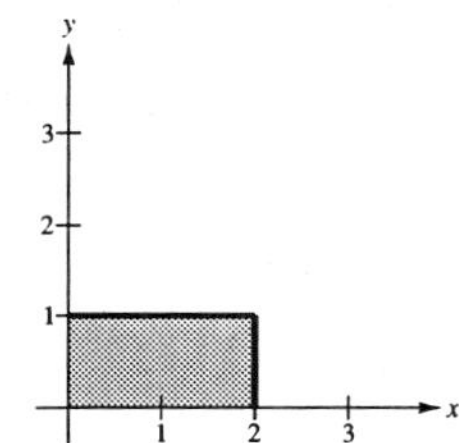

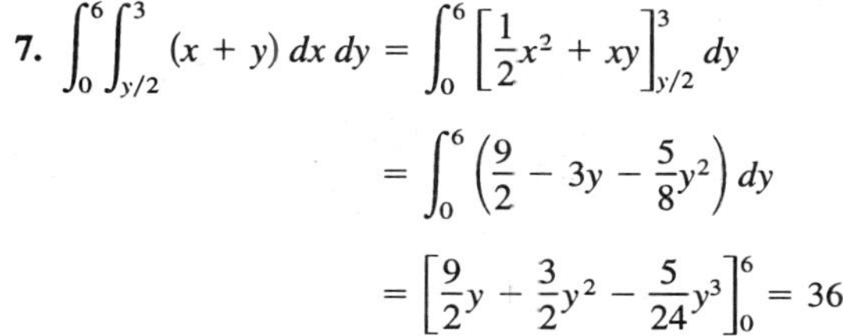

7.
$$\int_0^6 \int_{y/2}^3 (x + y)\,dx\,dy = \int_0^6 \left[\frac{1}{2}x^2 + xy\right]_{y/2}^3 dy = \int_0^6 \left(\frac{9}{2} - 3y - \frac{5}{8}y^2\right) dy = \left[\frac{9}{2}y - \frac{3}{2}y^2 - \frac{5}{24}y^3\right]_0^6 = 36$$

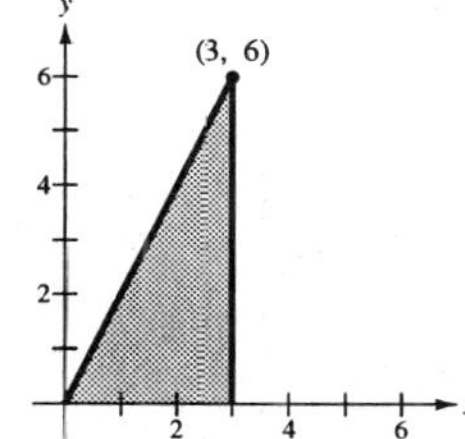

9.
$$\int_{-a}^{a} \int_{-\sqrt{a^2-x^2}}^{\sqrt{a^2-x^2}} (x + y)\,dy\,dx = \int_{-a}^{a} \left[xy + \frac{1}{2}y^2\right]_{-\sqrt{a^2-x^2}}^{\sqrt{a^2-x^2}} dx = \int_{-a}^{a} 2x\sqrt{a^2 - x^2}\,dx = \left[-\frac{2}{3}(a^2 - x^2)^{3/2}\right]_{-a}^{a} = 0$$

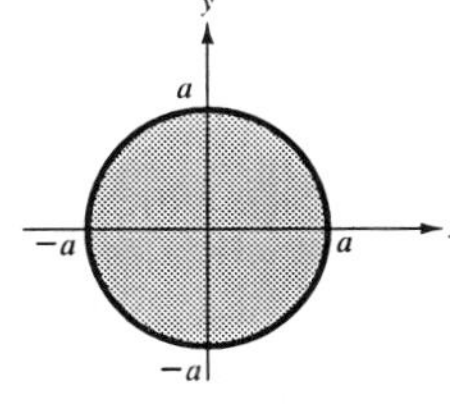

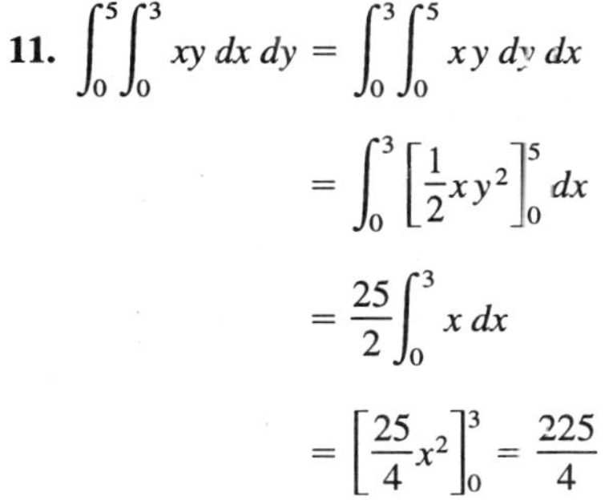

11.
$$\int_0^5 \int_0^3 xy\,dx\,dy = \int_0^3 \int_0^5 xy\,dy\,dx = \int_0^3 \left[\frac{1}{2}xy^2\right]_0^5 dx = \frac{25}{2}\int_0^3 x\,dx = \left[\frac{25}{4}x^2\right]_0^3 = \frac{225}{4}$$

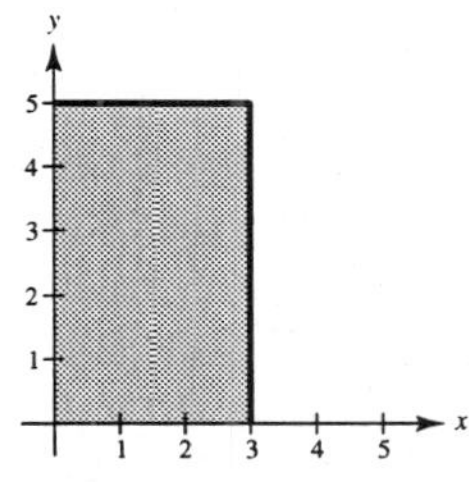

13. $$\int_0^2\int_{y/2}^{y} \frac{y}{x^2+y^2}\,dx\,dy + \int_2^4\int_{y/2}^{2} \frac{y}{x^2+y^2}\,dx\,dy = \int_0^2\int_x^{2x} \frac{y}{x^2+y^2}\,dy\,dx$$

$$= \frac{1}{2}\int_0^2 \Big[\ln(x^2+y^2)\Big]_x^{2x}\,dx$$

$$= \frac{1}{2}\int_0^2 (\ln 5x^2 - \ln 2x^2)\,dx$$

$$= \frac{1}{2}\ln\frac{5}{2}\int_0^2 dx$$

$$= \left[\frac{1}{2}\left(\ln\frac{5}{2}\right)x\right]_0^2 = \ln\frac{5}{2}$$

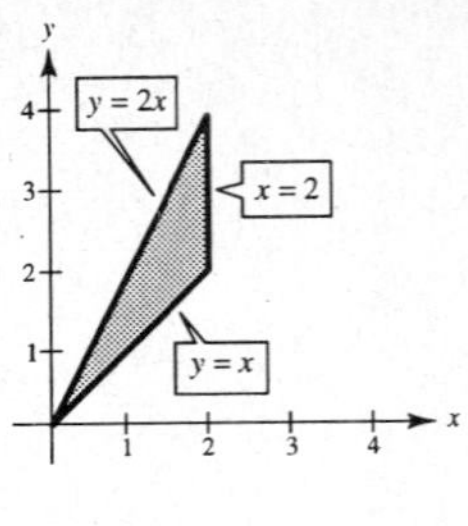

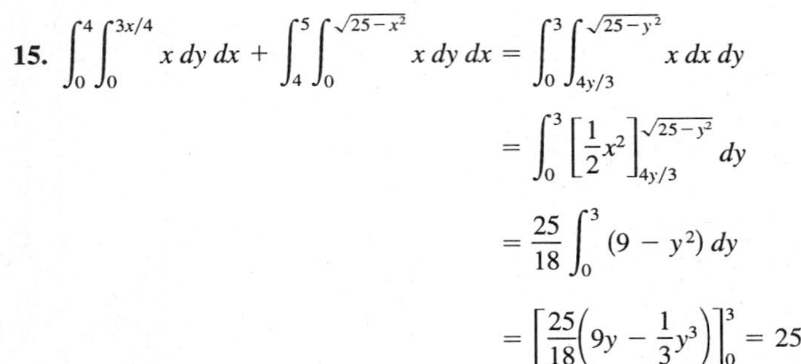

15. $$\int_0^4\int_0^{3x/4} x\,dy\,dx + \int_4^5\int_0^{\sqrt{25-x^2}} x\,dy\,dx = \int_0^3\int_{4y/3}^{\sqrt{25-y^2}} x\,dx\,dy$$

$$= \int_0^3 \left[\frac{1}{2}x^2\right]_{4y/3}^{\sqrt{25-y^2}}\,dy$$

$$= \frac{25}{18}\int_0^3 (9-y^2)\,dy$$

$$= \left[\frac{25}{18}\left(9y - \frac{1}{3}y^3\right)\right]_0^3 = 25$$

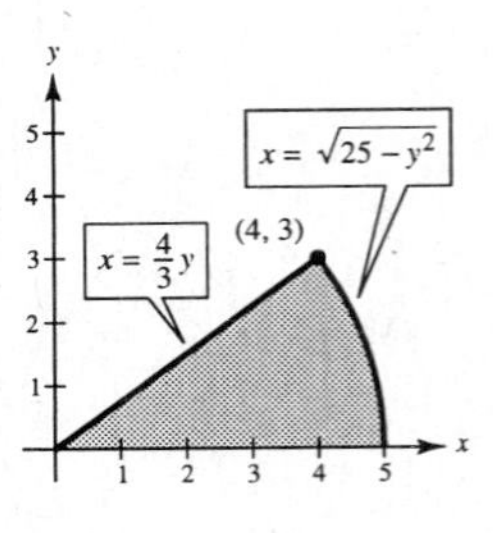

17. $$\int_0^4\int_0^2 \frac{y}{2}\,dy\,dx = \int_0^4 \left[\frac{y^4}{4}\right]_0^2 dx$$

$$= \int_0^4 dx = 4$$

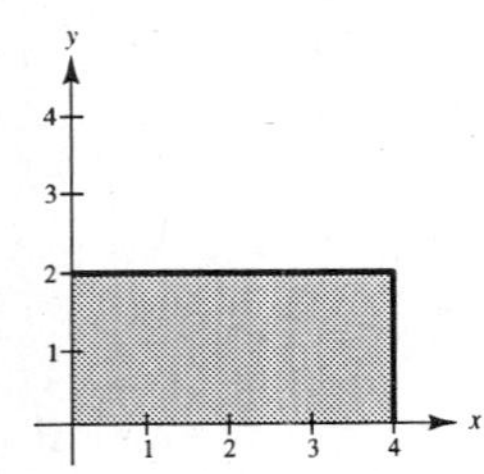

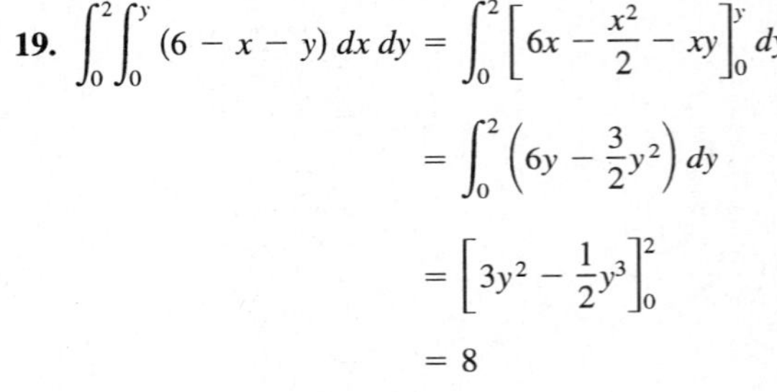

19. $$\int_0^2\int_0^y (6-x-y)\,dx\,dy = \int_0^2 \left[6x - \frac{x^2}{2} - xy\right]_0^y dy$$

$$= \int_0^2 \left(6y - \frac{3}{2}y^2\right) dy$$

$$= \left[3y^2 - \frac{1}{2}y^3\right]_0^2$$

$$= 8$$

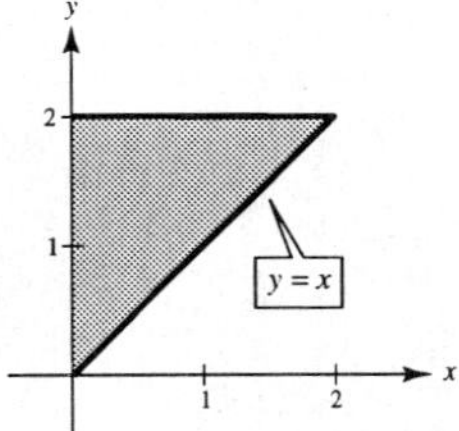

21. $$\int_0^6\int_0^{(-2/3)x+4} \left(\frac{12-2x-3y}{4}\right) dy\,dx = \int_0^6 \left[\frac{1}{4}\left(12y - 2xy - \frac{3}{2}y^2\right)\right]_0^{(-2/3)x+4} dx$$

$$= \int_0^6 \left(\frac{1}{6}x^2 - 2x + 6\right) dx$$

$$= \left[\frac{1}{18}x^3 - x^2 + 6x\right]_0^6$$

$$= 12$$

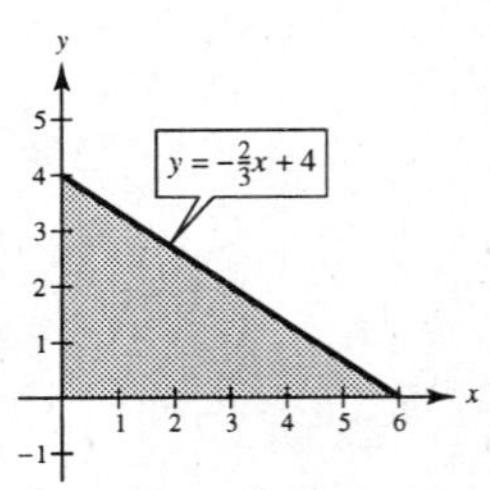

23. $\displaystyle\int_0^1\int_0^y (1 - xy)\,dx\,dy = \int_0^1 \left[x - \frac{x^2y}{2}\right]_0^y dy$

$\displaystyle= \int_0^1 \left(y - \frac{y^3}{2}\right) dy$

$\displaystyle= \left[\frac{y^2}{2} - \frac{y^4}{8}\right]_0^1 = \frac{3}{8}$

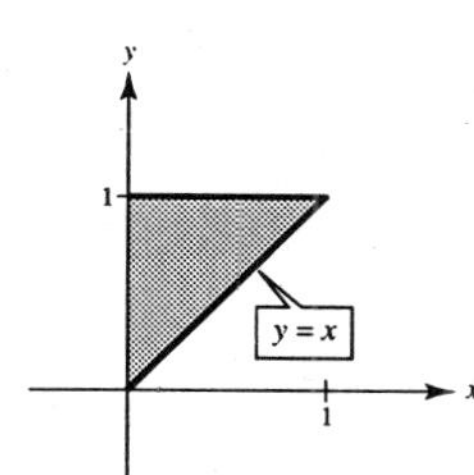

25. $\displaystyle\int_0^\infty\int_0^\infty \frac{1}{(x+1)^2(y+1)^2}\,dy\,dx = \int_0^\infty \left[-\frac{1}{(x+1)^2(y+1)}\right]_0^\infty dx$

$\displaystyle= \int_0^\infty \frac{1}{(x+1)^2}\,dx = \left[-\frac{1}{(x+1)}\right]_0^\infty = 1$

27. $\displaystyle 4\int_0^2\int_0^{\sqrt{4-x^2}} (4 - x^2 - y^2)\,dy\,dx = 8\pi$

29. $\displaystyle V = \int_0^1\int_0^x xy\,dy\,dx$

$\displaystyle= \int_0^1 \left[\frac{1}{2}xy^2\right]_0^x dx = \frac{1}{2}\int_0^1 x^3\,dx$

$\displaystyle= \left[\frac{1}{8}x^4\right]_0^1 = \frac{1}{8}$

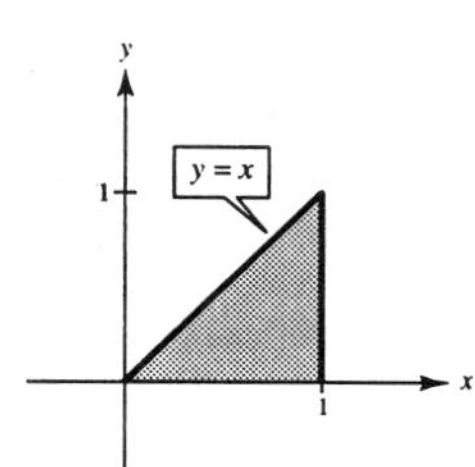

31. $\displaystyle V = \int_0^2\int_0^4 x^2\,dy\,dx$

$\displaystyle= \int_0^2 \left[x^2y\right]_0^4 dx = \int_0^2 4x^2\,dx$

$\displaystyle= \left[\frac{4x^3}{3}\right]_0^2 = \frac{32}{3}$

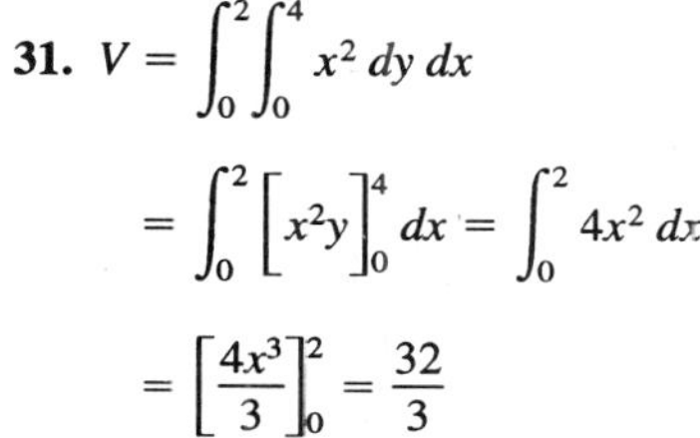

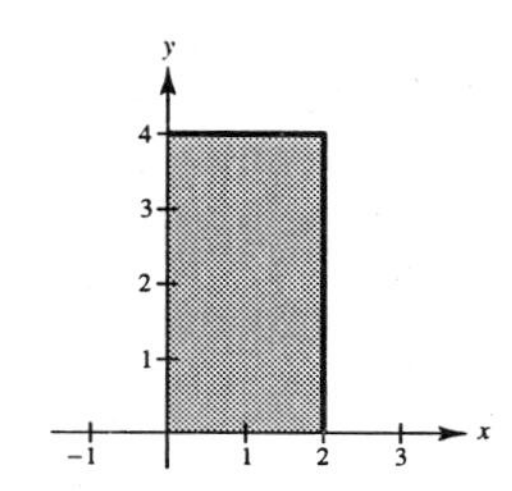

33. Divide the solid into two equal parts.

$\displaystyle V = 2\int_0^1\int_0^x \sqrt{1 - x^2}\,dy\,dx$

$\displaystyle= 2\int_0^1 \left[y\sqrt{1 - x^2}\right]_0^x dx$

$\displaystyle= 2\int_0^1 x\sqrt{1 - x^2}\,dx = \left[-\frac{2}{3}(1 - x^2)^{3/2}\right]_0^1 = \frac{2}{3}$

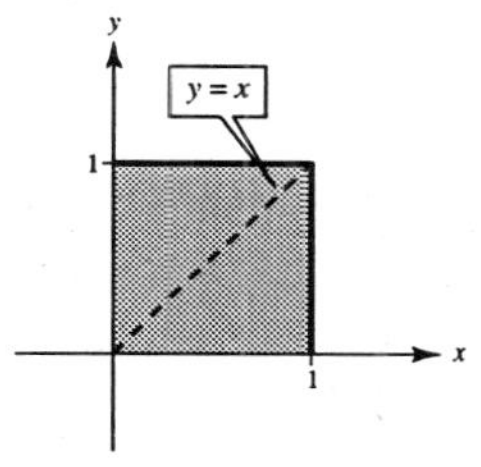

35. $\displaystyle V = \int_0^2\int_0^{\sqrt{4-x^2}} (x + y)\,dy\,dx$

$\displaystyle= \int_0^2 \left[xy + \frac{1}{2}y^2\right]_0^{\sqrt{4-x^2}} dx$

$\displaystyle= \int_0^2 \left(x\sqrt{4 - x^2} + 2 - \frac{1}{2}x^2\right) dx$

$\displaystyle= \left[-\frac{1}{3}(4 - x^2)^{3/2} + 2x - \frac{1}{6}x^3\right]_0^2 = \frac{16}{3}$

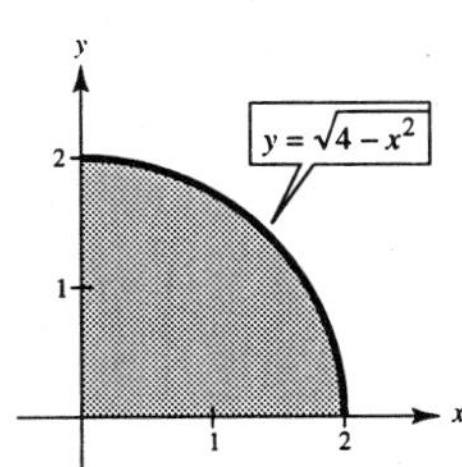

37. $\displaystyle V = 4\int_0^2\int_0^{\sqrt{4-x^2}} (x^2 + y^2)\,dy\,dx$

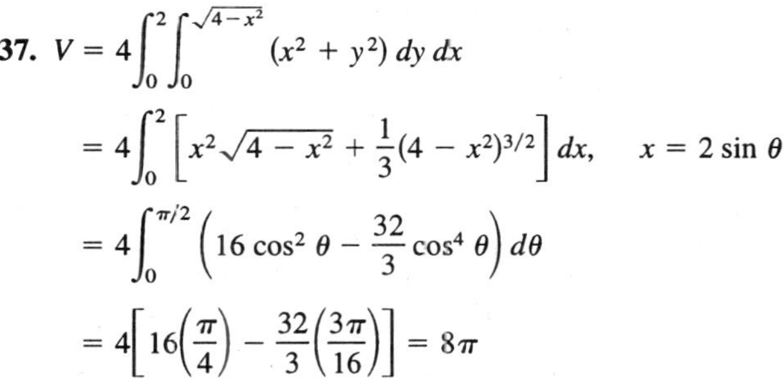

$\displaystyle= 4\int_0^2 \left[x^2\sqrt{4 - x^2} + \frac{1}{3}(4 - x^2)^{3/2}\right] dx, \qquad x = 2\sin\theta$

$\displaystyle= 4\int_0^{\pi/2} \left(16\cos^2\theta - \frac{32}{3}\cos^4\theta\right) d\theta$

$\displaystyle= 4\left[16\left(\frac{\pi}{4}\right) - \frac{32}{3}\left(\frac{3\pi}{16}\right)\right] = 8\pi$

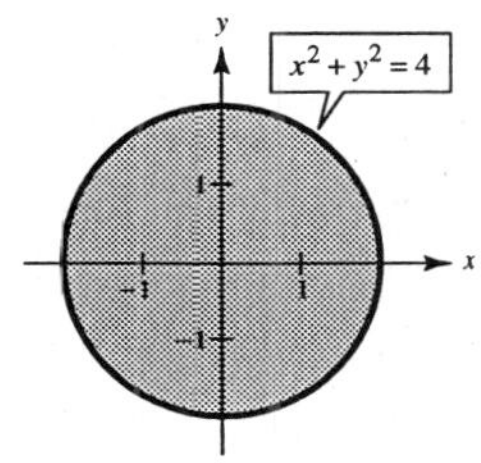

39. $V = 4\int_0^2\int_0^{\sqrt{4-x^2}} (4 - x^2 - y^2)\, dy\, dx = 8\pi$

41. $V = \int_0^2\int_0^{-0.5x+1} \frac{2}{1 + x^2 + y^2}\, dy\, dx \approx 1.2315$

43. f is a continuous function such that $0 \le f(x, y) \le 1$ over a region R of area 1. Let $f(m, n)$ = the minimum value of f over R and $f(M, N)$ = the maximum value of f over R. Then

$$f(m, n)\int_R\int dA \le \int_R\int f(x, y)\, dA \le f(M, N)\int_R\int dA.$$

Since $\int_R\int dA = 1$ and $0 \le f(m, n) \le f(M, N) \le 1$, we have $0 \le f(m, n)(1) \le \int_R\int f(x, y)\, dA \le f(M, N)(1) \le 1$.

Therefore, $0 \le \int_R\int f(x, y)\, dA \le 1$.

45.
$$\int_0^1\int_{y/2}^{1/2} e^{-x^2}\, dx\, dy = \int_0^{1/2}\int_0^{2x} e^{-x^2}\, dy\, dx$$
$$= \int_0^{1/2} 2xe^{-x^2}\, dx$$
$$= \Big[-e^{-x^2}\Big]_0^{1/2}$$
$$= -e^{-1/4} + 1$$
$$= 1 - e^{-1/4} \approx 0.221$$

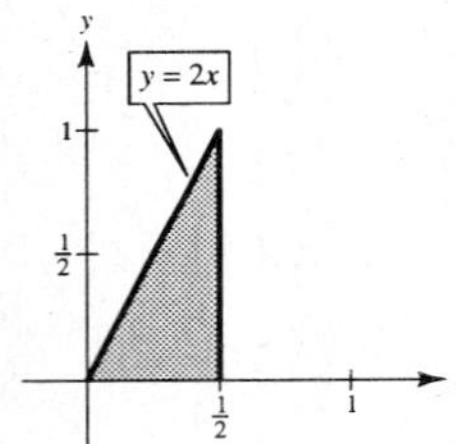

47.
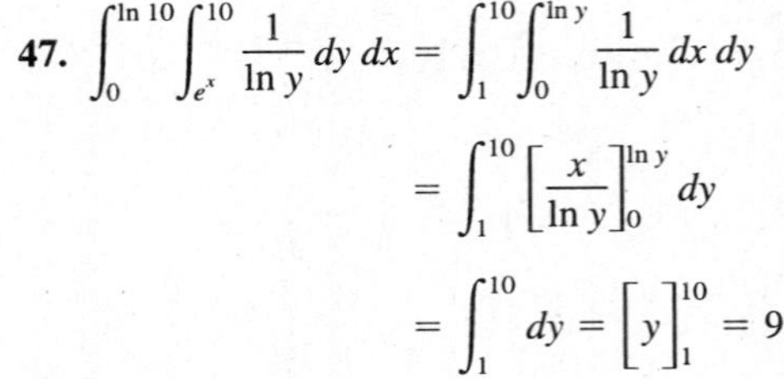
$$\int_0^{\ln 10}\int_{e^x}^{10} \frac{1}{\ln y}\, dy\, dx = \int_1^{10}\int_0^{\ln y} \frac{1}{\ln y}\, dx\, dy$$
$$= \int_1^{10}\left[\frac{x}{\ln y}\right]_0^{\ln y} dy$$
$$= \int_1^{10} dy = \Big[y\Big]_1^{10} = 9$$

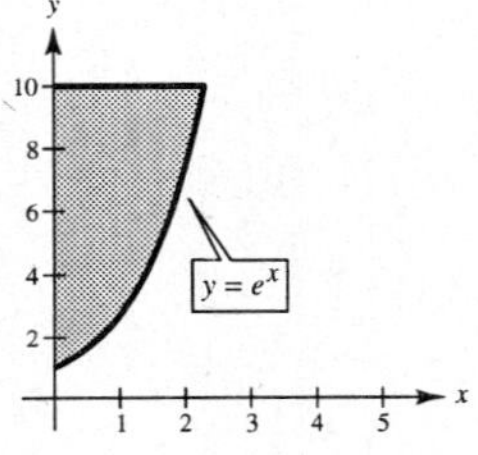

49. Average $= \frac{1}{8}\int_0^4\int_0^2 x\, dy\, dx = \frac{1}{8}\int_0^4 2x\, dx = \left[\frac{x^2}{8}\right]_0^4 = 2$

51.
$$\text{Average} = \frac{1}{4}\int_0^2\int_0^2 (x^2 + y^2)\, dx\, dy$$
$$= \frac{1}{4}\int_0^2\left[\frac{x^3}{3} + xy^2\right]_0^2 dy = \frac{1}{4}\int_0^2\left(\frac{8}{3} + 2y^2\right) dy$$
$$= \left[\frac{1}{4}\left(\frac{8}{3}y + \frac{2}{3}y^3\right)\right]_0^2 = \frac{8}{3}$$

53.
$$\text{Average} = \frac{1}{1250}\int_{300}^{325}\int_{200}^{250} 100x^{0.6}y^{0.4}\, dx\, dy$$
$$= \frac{1}{1250}\int_{300}^{325}\left[(100y^{0.4})\frac{x^{1.6}}{1.6}\right]_{200}^{250} dy = \frac{128{,}844.1}{1250}\int_{300}^{325} y^{0.4}\, dy = 103.0753\left[\frac{y^{1.4}}{1.4}\right]_{300}^{325} \approx 25{,}645.24$$

55. The value of $\int_R\int f(x, y)\, dA$ would be kB.

57. $f(x, y) \geq 0$ for all (x, y) and

$$\int_{-\infty}^{\infty}\int_{-\infty}^{\infty} f(x, y)\, dA = \int_0^5\int_0^2 \frac{1}{10}\, dy\, dx = \int_0^5 \frac{1}{5}\, dx = 1$$

$$P(0 \leq x \leq 2, 1 \leq y \leq 2) = \int_0^2\int_1^2 \frac{1}{10}\, dy\, dx = \int_0^2 \frac{1}{10}\, dx = \frac{1}{5}.$$

59. $f(x, y) \geq 0$ for all (x, y) and

$$\int_{-\infty}^{\infty}\int_{-\infty}^{\infty} f(x, y)\, dA = \int_0^3\int_3^6 \frac{1}{27}(9 - x - y)\, dy\, dx$$

$$= \int_0^3 \frac{1}{27}\left[9y - xy - \frac{y^2}{2}\right]_3^6 dx = \int_0^3 \left(\frac{1}{2} - \frac{1}{9}x\right) dx = \left[\frac{x}{2} - \frac{x^2}{18}\right]_0^3 = 1$$

$$P(0 \leq x \leq 1, 4 \leq y \leq 6) = \int_0^1\int_4^6 \frac{1}{27}(9 - x - y)\, dy\, dx = \int_0^1 \frac{2}{27}(4 - x)\, dx = \frac{7}{27}.$$

61. Divide the base into six squares, and assume the height at the center of each square is the height of the entire square. Thus,

$$V \approx (4 + 3 + 6 + 7 + 3 + 2)(100) = 2500 m^3.$$

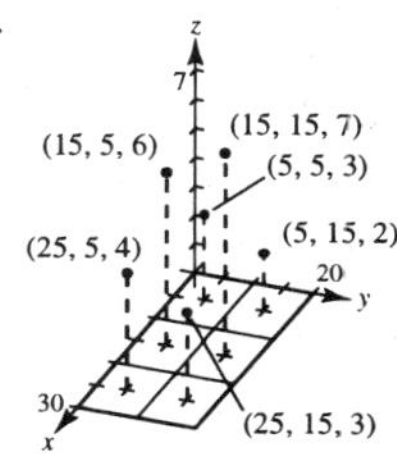

63. $\displaystyle\int_0^1\int_0^2 \sin\sqrt{x + y}\, dy\, dx \qquad m = 4, n = 8$

(a) 1.78435

(b) 1.7879

65. $\displaystyle\int_4^6\int_0^2 y\cos\sqrt{x}\, dx\, dy \qquad m = 4, n = 8$

(a) 11.0571

(b) 11.0414

67. Essay

69. $V \approx 50$

Matches a.

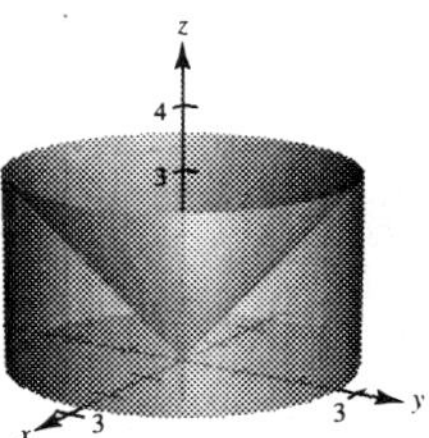

71. True

73. $\displaystyle\int_1^2 e^{-xy}\, dy = \left[-\frac{1}{x}e^{-xy}\right]_1^2 = \frac{e^{-x} - e^{-2x}}{x}$

Thus,

$$\int_0^{\infty} \frac{e^{-x} - e^{-2x}}{x}\, dx = \int_0^{\infty}\int_1^2 e^{-xy}\, dy\, dx = \int_1^2\int_0^{\infty} e^{-xy}\, dx\, dy$$

$$= \int_1^2 \left[-\frac{e^{-xy}}{y}\right]_0^{\infty} dy = \int_1^2 \frac{1}{y}\, dy = \Big[\ln y\Big]_1^2 = \ln 2.$$

Section 13.3 Change of Variables: Polar Coordinates

1. $$\int_0^{2\pi}\int_0^6 3r^2 \sin\theta \, dr \, d\theta = \int_0^{2\pi}\left[r^3 \sin\theta\right]_0^6 d\theta = \int_0^{2\pi} 216 \sin\theta \, d\theta = \left[-216\cos\theta\right]_0^{2\pi} = 0$$

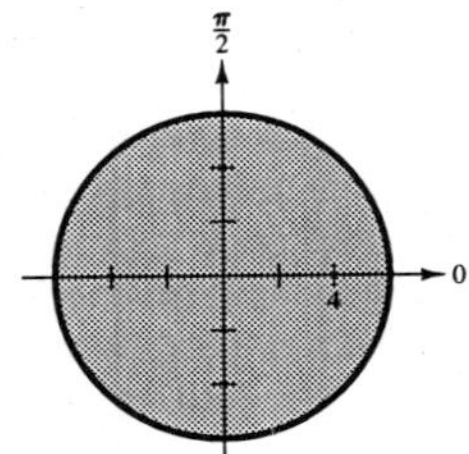

3. $$\int_0^{\pi/2}\int_2^3 \sqrt{9-r^2}\, r \, dr \, d\theta = \int_0^{\pi/2}\left[-\frac{1}{3}(9-r^2)^{3/2}\right]_2^3 d\theta = \left[\frac{5\sqrt{5}}{3}\theta\right]_0^{\pi/2} = \frac{5\sqrt{5}\pi}{6}$$

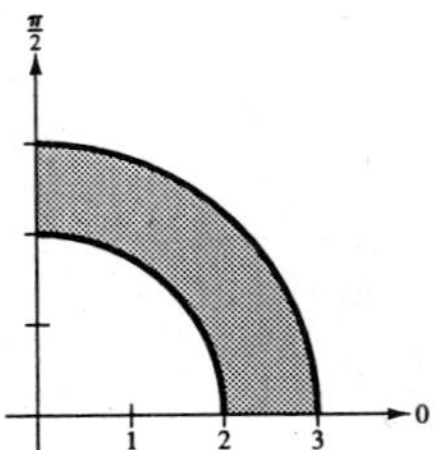

5. $$\int_0^{\pi/2}\int_0^{1+\sin\theta} \theta r \, dr \, d\theta = \int_0^{\pi/2}\left[\frac{\theta r^2}{2}\right]_0^{1+\sin\theta} d\theta = \int_0^{\pi/2}\frac{1}{2}\theta(1+\sin\theta)^2 \, d\theta$$
$$= \left[\frac{1}{8}t^2 + \sin t - t\cos t + \frac{1}{2}t\left(-\frac{1}{2}\cos t \cdot \sin t + \frac{1}{2}t\right) - \frac{1}{8}\cos^2 t\right]_0^{\pi/2} = \frac{3}{32}\pi^2 + \frac{9}{8}$$

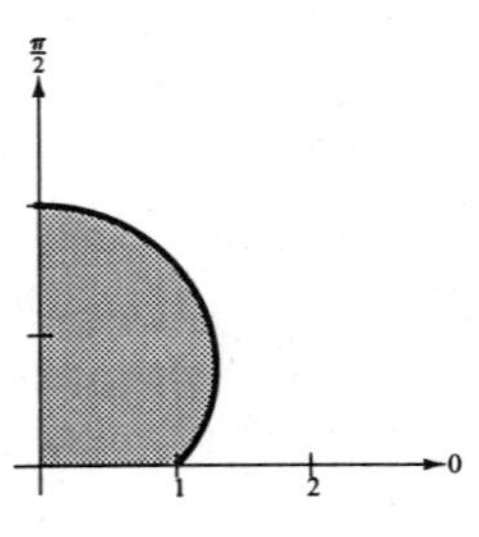

7. $$A = \int_0^{\pi}\int_0^{6\cos\theta} r \, dr \, d\theta = \int_0^{\pi} 18\cos^2\theta \, d\theta = 9\int_0^{\pi}(1+\cos 2\theta)\, d\theta = \left[9\left(\theta + \frac{1}{2}\sin 2\theta\right)\right]_0^{\pi} = 9\pi$$

9. $$\int_0^{2\pi}\int_0^{1+\cos\theta} r \, dr \, d\theta = \frac{1}{2}\int_0^{2\pi}(1 + 2\cos\theta + \cos^2\theta)\, d\theta = \frac{1}{2}\int_0^{2\pi}\left(1 + 2\cos\theta + \frac{1+\cos 2\theta}{2}\right) d\theta$$
$$= \frac{1}{2}\left[\theta + 2\sin\theta + \frac{1}{2}\left(\theta + \frac{1}{2}\sin 2\theta\right)\right]_0^{2\pi} = \frac{3\pi}{2}$$

11. $$3\int_0^{\pi/3}\int_0^{2\sin 3\theta} r \, dr \, d\theta = \frac{3}{2}\int_0^{\pi/3} 4\sin^2 3\theta \, d\theta = 3\int_0^{\pi/3}(1-\cos 6\theta)\, d\theta = 3\left[\theta - \frac{1}{6}\sin 6\theta\right]_0^{\pi/3} = \pi$$

13. $$\int_0^a\int_0^{\sqrt{a^2-y^2}} y \, dx \, dy = \int_0^{\pi/2}\int_0^a r^2\sin\theta \, dr \, d\theta = \frac{a^3}{3}\int_0^{\pi/2}\sin\theta \, d\theta = \left[\frac{a^3}{3}(-\cos\theta)\right]_0^{\pi/2} = \frac{a^3}{3}$$

15. $$\int_0^3\int_0^{\sqrt{9-x^2}} (x^2+y^2)^{3/2}\, dy \, dx = \int_0^{\pi/2}\int_0^3 r^4 \, dr \, d\theta = \frac{243}{5}\int_0^{\pi/2} d\theta = \frac{243\pi}{10}$$

17. $$\int_0^2\int_0^{\sqrt{2x-x^2}} xy \, dy \, dx = \int_0^{\pi/2}\int_0^{2\cos\theta} r^3\cos\theta\sin\theta \, dr \, d\theta = 4\int_0^{\pi/2}\cos^5\theta\sin d\theta = \left[-\frac{4\cos^6\theta}{6}\right]_0^{\pi/2} = \frac{2}{3}$$

19. $\displaystyle\int_0^2\int_0^x \sqrt{x^2+y^2}\,dy\,dx + \int_2^{2\sqrt{2}}\int_0^{\sqrt{8-x^2}} \sqrt{x^2+y^2}\,dy\,dx = \int_0^{\pi/4}\int_0^{2\sqrt{2}} r^2\,dr\,d\theta$

$\displaystyle = \int_0^{\pi/4} \frac{16\sqrt{2}}{3}\,d\theta$

$\displaystyle = \frac{4\sqrt{2}\pi}{3}$

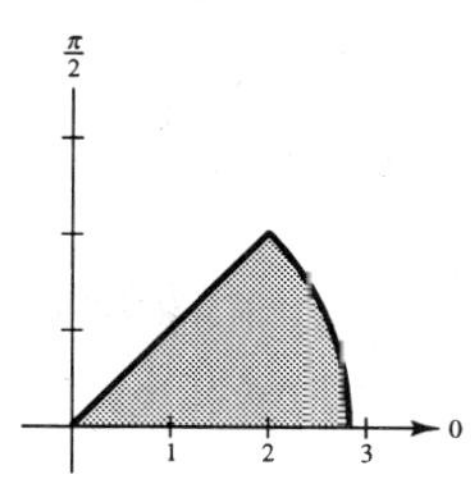

21. $\displaystyle\int_0^2\int_0^{\sqrt{4-x^2}} (x+y)\,dy\,dx = \int_0^{\pi/2}\int_0^2 (r\cos\theta + r\sin\theta)r\,dr\,d\theta = \int_0^{\pi/2}\int_0^2 (\cos\theta+\sin\theta)r^2\,dr\,d\theta$

$\displaystyle = \frac{8}{3}\int_0^{\pi/2} (\cos\theta + \sin\theta)\,d\theta = \left[\frac{8}{3}(\sin\theta - \cos\theta)\right]_0^{\pi/2} = \frac{16}{3}$

23. $\displaystyle\int_0^{1/\sqrt{2}}\int_{\sqrt{1-y^2}}^{\sqrt{4-y^2}} \arctan\frac{y}{x}\,dx\,dy + \int_{1/\sqrt{2}}^{\sqrt{2}}\int_y^{\sqrt{4-y^2}} \arctan\frac{y}{x}\,dx\,dy$

$\displaystyle = \int_0^{\pi/4}\int_1^2 \theta r\,dr\,d\theta$

$\displaystyle = \int_0^{\pi/4} \frac{3}{2}\theta\,d\theta = \left[\frac{3\theta^2}{4}\right]_0^{\pi/4} = \frac{3\pi^2}{64}$

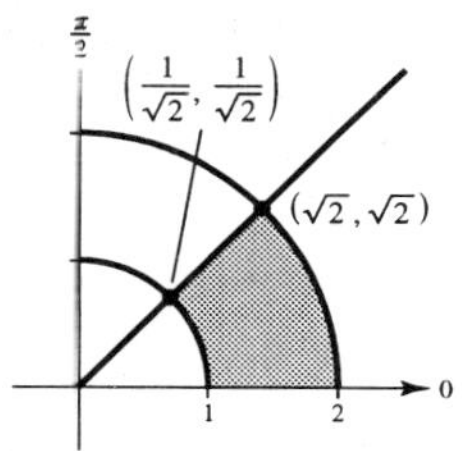

25. $\displaystyle V = \int_0^{\pi/2}\int_0^1 (r\cos\theta)(r\sin\theta)r\,dr\,d\theta$

$\displaystyle = \frac{1}{2}\int_0^{\pi/2}\int_0^1 r^3 \sin 2\theta\,dr\,d\theta$

$\displaystyle = \frac{1}{8}\int_0^{\pi/2} \sin 2\theta\,d\theta = \left[-\frac{1}{16}\cos 2\theta\right]_0^{\pi/2} = \frac{1}{8}$

27. $\displaystyle V = \int_0^{2\pi}\int_0^5 r^2\,dr\,d\theta = \frac{250\pi}{3}$

29. $\displaystyle V = 2\int_0^{\pi/2}\int_0^{4\cos\theta} \sqrt{16-r^2}\,r\,dr\,d\theta = 2\int_0^{\pi/2}\left[-\frac{1}{3}\left(\sqrt{16-r^2}\right)^3\right]_0^{4\cos\theta} d\theta = -\frac{2}{3}\int_0^{\pi/2} (64\sin^3\theta - 64)\,d\theta$

$\displaystyle = \frac{128}{3}\int_0^{\pi/2} [1 - \sin\theta(1-\cos^2\theta)]\,d\theta = \frac{128}{3}\left[\theta + \cos\theta - \frac{\cos^3\theta}{3}\right]_0^{\pi/2} = \frac{64}{9}(3\pi - 4)$

31. $\displaystyle V = \int_0^{2\pi}\int_a^4 \sqrt{16-r^2}\,r\,dr\,d\theta = \int_0^{2\pi}\left[-\frac{1}{3}\left(\sqrt{16-r^2}\right)^3\right]_a^4 d\theta = \frac{1}{3}\left(\sqrt{16-a^2}\right)^3(2\pi)$

One-half the volume of the hemisphere is $(64\pi)/3$.

$$\frac{2\pi}{3}(16-a^2)^{3/2} = \frac{64\pi}{3}$$

$$(16-a^2)^{3/2} = 32$$

$$16 - a^2 = 32^{2/3}$$

$$a^2 = 16 - 32^{2/3} = 16 - 8\sqrt[3]{2}$$

$$a = \sqrt{4\left(4 - 2\sqrt[3]{2}\right)} = 2\sqrt{4 - 2\sqrt[3]{2}} \approx 2.4332$$

33. Total Volume $= V = \int_0^{2\pi}\int_0^4 25e^{-r^2/4}\,r\,dr\,d\theta$

$$= \int_0^{2\pi}\left[-50e^{-r^2/4}\right]_0^4 d\theta$$

$$= \int_0^{2\pi} -50(e^{-4} - 1)\,d\theta$$

$$= (1 - e^{-4})\,100\pi \approx 308.40524$$

Let c be the radius of the hole that is removed.

$$\frac{1}{10}V = \int_0^{2\pi}\int_0^c 25e^{-r^2/4}\,r\,dr\,d\theta = \int_0^{2\pi}\left[-50e^{-r^2/4}\right]_0^c d\theta$$

$$= \int_0^{2\pi} -50(e^{-c^2/4} - 1)\,d\theta \Rightarrow 30.84052 = 100\pi(1 - e^{-c^2/4})$$

$$\Rightarrow e^{-c^2/4} = 0.90183$$

$$-\frac{c}{4} = -0.10333$$

$$c^2 = 0.41331$$

$$c = 0.6429$$

$$\Rightarrow \text{diameter} = 2c = 1.2858$$

35. You would need to insert a factor of r because of the $r\,dr\,d\theta$ nature of polar coordinate integrals. The plane regions would be sectors of circles.

37. $\int_{\pi/4}^{\pi/2}\int_0^5 r\sqrt{1 + r^3}\sin\sqrt{\theta}\,dr\,d\theta \approx 56.051$

$\left[\textbf{Note:} \text{ This integral equals } \left(\int_{\pi/4}^{\pi/2} \sin\sqrt{\theta}\,d\theta\right)\left(\int_0^5 r\sqrt{1 + r^3}\,dr\right)\right]$

39. Volume = base × height

$$\approx 8\pi \times 12 \approx 300$$

Answer (c)

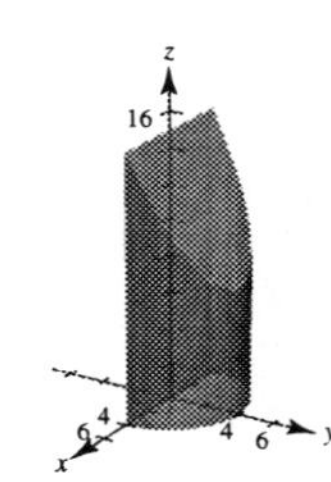

41. False

Let $f(r, \theta) = r - 1$ where R is the circular sector $0 \le r \le 6$ and $0 \le \theta \le \pi$. Then,

$$\int_R\int (r - 1)\,dA > 0 \quad \text{but} \quad r - 1 \not> 0 \text{ for all } r.$$

43. (a) $I^2 = \int_{-\infty}^{\infty}\int_{-\infty}^{\infty} e^{-(x^2+y^2)/2}\,dA = 4\int_0^{\pi/2}\int_0^{\infty} e^{-r^2/2}\,r\,dr\,d\theta = 4\int_0^{\pi/2}\left[-e^{-r^2/2}\right]_0^{\infty} d\theta = 4\int_0^{\pi/2} d\theta = 2\pi$

(b) Therefore, $I = \sqrt{2\pi}$.

45. $\int_{-7}^{7}\int_{-\sqrt{49-x^2}}^{\sqrt{49-x^2}} 4000e^{-0.01(x^2+y^2)}\,dy\,dx = \int_0^{2\pi}\int_0^7 4000e^{-0.01r^2}\,r\,dr\,d\theta = \int_0^{2\pi}\left[-200{,}000e^{-0.01r^2}\right]_0^7 d\theta \approx 486{,}788$

$$= 2\pi(-200{,}000)(e^{-0.49} - 1) = 400{,}000\pi(1 - e^{-0.49}) \approx 486{,}788$$

47. (a) $\int_2^4\int_{y/\sqrt{3}}^{y} f\,dx\,dy$

(b) $\int_{2/\sqrt{3}}^{2}\int_2^{\sqrt{3}x} f\,dy\,dx + \int_2^{4/\sqrt{3}}\int_x^{\sqrt{3}x} f\,dy\,dx + \int_{4/\sqrt{3}}^{4}\int_x^4 f\,dy\,dx$

(c) $\int_{\pi/4}^{\pi/3}\int_{2\csc\theta}^{4\csc\theta} fr\,dr\,d\theta$

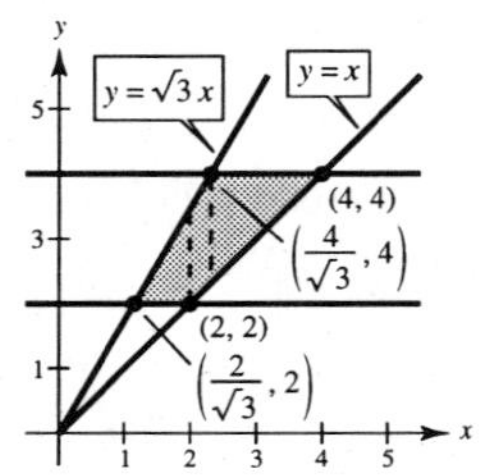

49. $A = \frac{\Delta\theta r_2^2}{2} - \frac{\Delta\theta r_1^2}{2} = \Delta\theta\left(\frac{r_1 + r_2}{2}\right)(r_2 - r_1) = r\Delta r\Delta\theta$

Section 13.4 Center of Mass and Moments of Inertia

1. (a) $m = \int_0^a \int_0^b k\,dy\,dx = kab$

$M_x = \int_0^a \int_0^b ky\,dy\,dx = \frac{kab^2}{2}$

$M_y = \int_0^a \int_0^b kx\,dy\,dx = \frac{ka^2b}{2}$

$\bar{x} = \frac{M_y}{m} = \frac{ka^2b/2}{kab} = \frac{a}{2}$

$\bar{y} = \frac{M_x}{m} = \frac{kab^2/2}{kab} = \frac{b}{2}$

(b) $m = \int_0^a \int_0^b ky\,dy\,dx = \frac{kab^2}{2}$

$M_x = \int_0^a \int_0^b ky^2\,dy\,dx = \frac{kab^3}{3}$

$M_y = \int_0^a \int_0^b kxy\,dy\,dx = \frac{ka^2b^2}{4}$

$\bar{x} = \frac{M_y}{m} = \frac{ka^2b^2/4}{kab^2/2} = \frac{a}{2}$

$\bar{y} = \frac{M_x}{m} = \frac{kab^3/3}{kab^2/2} = \frac{2}{3}b$

(c) $m = \int_0^a \int_0^b kx\,dy\,dx = k\int_0^a xb\,dx = \frac{1}{2}ka^2b$

$M_x = \int_0^a \int_0^b kxy\,dy\,dx = \frac{ka^2b^2}{4}$

$M_y = \int_0^a \int_0^b kx^2\,dy\,dx = \frac{ka^3b}{3}$

$\bar{x} = \frac{M_y}{m} = \frac{ka^3b/3}{ka^2b/2} = \frac{2}{3}a$

$\bar{y} = \frac{M_x}{m} = \frac{ka^2b^2/4}{ka^2b/2} = \frac{b}{2}$

$(\bar{x}, \bar{y}) = \left(\frac{2}{3}a, \frac{b}{2}\right)$

3. (a) $m = \frac{k}{2}bh$

$\bar{x} = \frac{b}{2}$ by symmetry

$M_x = \int_0^{b/2} \int_0^{2hx/b} ky\,dy\,dx + \int_{b/2}^{b} \int_0^{-2h(x-b)/b} ky\,dy\,dx$

$= \frac{kbh^2}{12} + \frac{kbh^2}{12} = \frac{kbh^2}{6}$

$\bar{y} = \frac{M_x}{m} = \frac{kbh^2/6}{kbh/2} = \frac{h}{3}$

(b) $m = \int_0^{b/2} \int_0^{2hx/b} ky\,dy\,dx + \int_{b/2}^{b} \int_0^{-2h(x-b)/b} ky\,dy\,dx = \frac{kbh^2}{6}$

$M_x = \int_0^{b/2} \int_0^{2hx/b} ky^2\,dy\,dx + \int_{b/2}^{b} \int_0^{-2h(x-b)/b} ky^2\,dy\,dx = \frac{kbh^2}{12}$

$M_y = \int_0^{b/2} \int_0^{2hx/b} kxy\,dy\,dx + \int_{b/2}^{b} \int_0^{-2h(x-b)/b} kxy\,dy\,dx = \frac{kb^2h^2}{12}$

$\bar{x} = \frac{M_y}{m} = \frac{kb^2h^2/12}{kbh^2/6} = \frac{b}{2}$

$\bar{y} = \frac{M_x}{m} = \frac{kbh^3/12}{kbh^2/6} = \frac{h}{2}$

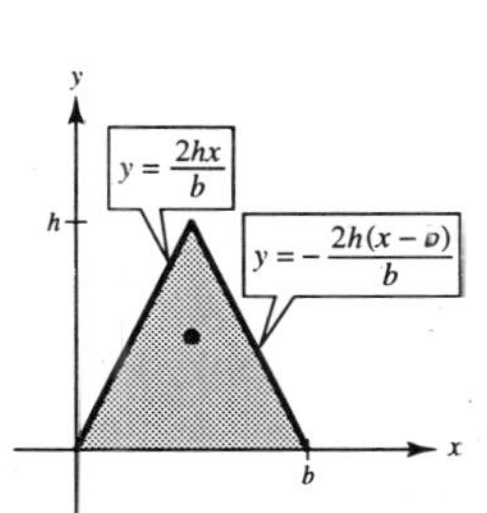

—CONTINUED—

3. —CONTINUED—

(c) $m = \int_0^{b/2}\int_0^{2hx/b} kx\,dy\,dx + \int_{b/2}^{b}\int_0^{-2h(x-b)/b} kx\,dy\,dx$

$= \frac{1}{12}kb^2h + \frac{1}{6}kb^2h = \frac{1}{4}kb^2h$

$M_x = \int_0^{b/2}\int_0^{2hx/b} kxy\,dy\,dx + \int_{b/2}^{b}\int_0^{-2h(x-b)/b} kxy\,dy\,dx$

$= \frac{1}{32}kh^2b^2 + \frac{5}{96}kh^2b^2 = \frac{1}{12}kh^2b^2$

$M_y = \int_0^{b/2}\int_0^{2hx/b} kx^2\,dy\,dx + \int_{b/2}^{b}\int_0^{-2h(x-b)/b} kx^2\,dy\,dx$

$= \frac{1}{32}kb^3h + \frac{11}{96}kb^3h = \frac{7}{48}kb^3h$

$\bar{x} = \frac{M_y}{m} = \frac{7kb^3h/48}{kb^2h/4} = \frac{7}{12}b$

$\bar{y} = \frac{M_x}{m} = \frac{kh^2b^2/12}{kb^2h/4} = \frac{h}{3}$

5. (a) The x-coordinate changes by 5: $(\bar{x}, \bar{y}) = \left(\frac{a}{2} + 5, \frac{b}{2}\right)$

(b) The x-coordinate changes by 5: $(\bar{x}, \bar{y}) = \left(\frac{a}{2} + 5, \frac{2b}{3}\right)$

(c) $m = \int_5^{a+5}\int_0^{b} kx\,dy\,dx = \frac{1}{2}k(a+5)^2b^2 - \frac{25}{4}kb^2$

$M_x = \int_5^{a+5}\int_0^{b} kxy\,dy\,dx = \frac{1}{4}k(a+5)^2b^2 - \frac{25}{4}kb^2$

$M_y = \int_5^{a+5}\int_0^{b} kx^2\,dy\,dx = \frac{1}{3}k(a+5)^3\,b - \frac{125}{3}kb$

$\bar{x} = \frac{M_y}{m} = \frac{2(a^2 + 15a + 75)}{3(a+10)}$

$\bar{y} = \frac{M_x}{m} = \frac{b}{2}$

7. (a) $\bar{x} = 0$ by symmetry

$m = \frac{\pi a^2 k}{2}$

$M_x = \int_{-a}^{a}\int_0^{\sqrt{a^2-x^2}} yk\,dy\,dx = \frac{2a^3k}{3}$

$\bar{y} = \frac{M_x}{m} = \frac{2a^3k}{3}\cdot\frac{2}{\pi a^2 k} = \frac{4a}{3\pi}$

(b) $m = \int_{-a}^{a}\int_0^{\sqrt{a^2-x^2}} k(a-y)\,y\,dy\,dx = \frac{a^4k}{24}(16 - 3\pi)$

$M_x = \int_{-a}^{a}\int_0^{\sqrt{a^2-x^2}} k(a-y)\,y^2\,dy\,dx = \frac{a^5k}{120}(15\pi - 32)$

$M_y = \int_{-a}^{a}\int_0^{\sqrt{a^2-x^2}} kx(a-y)\,y\,dy\,dx = 0$

$\bar{x} = \frac{M_y}{m} = 0$

$\bar{y} = \frac{M_x}{m} = \frac{a}{5}\left[\frac{15\pi - 32}{16 - 3\pi}\right]$

9. $m = \int_0^4\int_0^{\sqrt{x}} kxy\,dy\,dx = \frac{32k}{3}$

$M_x = \int_0^4\int_0^{\sqrt{x}} kxy^2\,dy\,dx = \frac{256k}{21}$

$M_y = \int_0^4\int_0^{\sqrt{x}} kx^2y\,dy\,dx = 32k$

$\bar{x} = \frac{M_y}{m} = \frac{32k}{1}\cdot\frac{3}{32k} = 3$

$\bar{y} = \frac{M_x}{m} = \frac{256k}{21}\cdot\frac{3}{32k} = \frac{8}{7}$

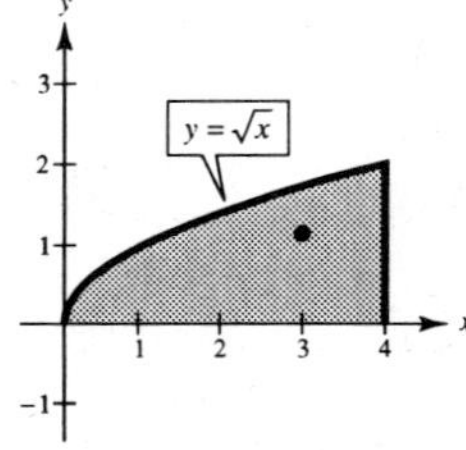

11. $\bar{x} = 0$ by symmetry

$$m = \int_{-1}^{1}\int_{0}^{1/(1+x^2)} k\,dy\,dx = \frac{k\pi}{2}$$

$$M_x = \int_{-1}^{1}\int_{0}^{1/(1+x^2)} ky\,dy\,dx = \frac{k}{8}(2 + \pi)$$

$$\bar{y} = \frac{M_x}{m} = \frac{k}{8}(2 + \pi) \cdot \frac{2}{k\pi} = \frac{2 + \pi}{4\pi}$$

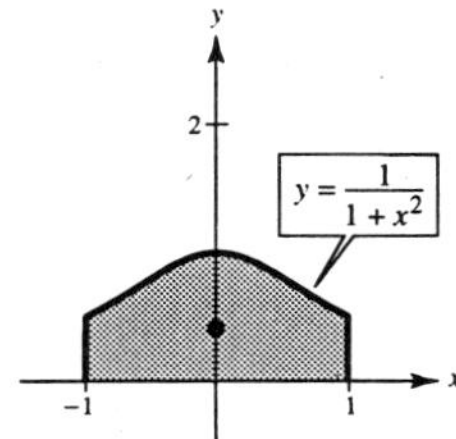

13. $\bar{y} = 0$ by symmetry

$$m = \int_{-4}^{4}\int_{0}^{16-y^2} kx\,dx\,dy = \frac{8192k}{15}$$

$$M_y = \int_{-4}^{4}\int_{0}^{16-y^2} kx^2\,dx\,dy = \frac{524{,}288k}{105}$$

$$\bar{x} = \frac{M_y}{m} = \frac{524{,}288k}{105} \cdot \frac{15}{8192k} = \frac{64}{7}$$

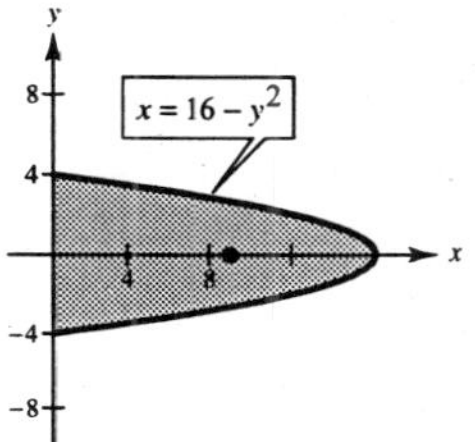

15. $\bar{x} = \dfrac{L}{2}$ by symmetry

$$m = \int_{0}^{L}\int_{0}^{\sin \pi x/L} ky\,dy\,dx = \frac{kL}{4}$$

$$M_x = \int_{0}^{L}\int_{0}^{\sin \pi x/L} ky^2\,dy\,dx = \frac{4kL}{9\pi}$$

$$\bar{y} = \frac{M_x}{m} = \frac{4kL}{9\pi} \cdot \frac{4}{kL} = \frac{16}{9\pi}$$

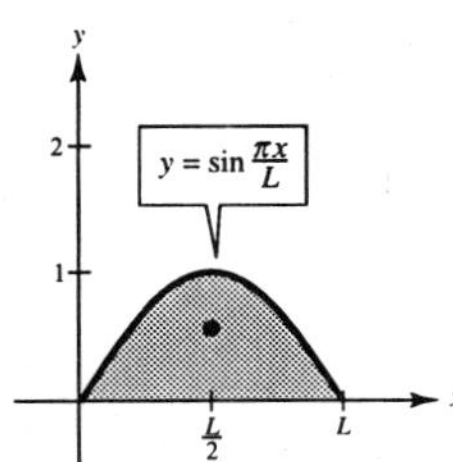

17. $m = \dfrac{\pi a^2 k}{8}$

$$M_x = \iint_R ky\,dA = \int_{0}^{\pi/4}\int_{0}^{a} kr^2 \sin\theta\,dr\,d\theta = \frac{ka^3(2 - \sqrt{2})}{6}$$

$$M_y = \iint_R kx\,dA = \int_{0}^{\pi/4}\int_{0}^{a} kr^2 \cos\theta\,dr\,d\theta = \frac{ka^3\sqrt{2}}{6}$$

$$\bar{x} = \frac{M_y}{m} = \frac{ka^3\sqrt{2}}{6} \cdot \frac{8}{\pi a^2 k} = \frac{4a\sqrt{2}}{3\pi}$$

$$\bar{y} = \frac{M_x}{m} = \frac{ka^3(2 - \sqrt{2})}{6} \cdot \frac{8}{\pi a^2 k} = \frac{4a(2 - \sqrt{2})}{3\pi}$$

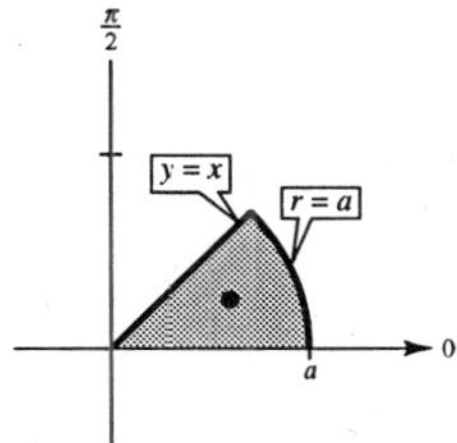

19. $$m = \int_{0}^{2}\int_{0}^{e^{-x}} ky\,dy\,dx = \frac{k}{4}(1 - e^{-4})$$

$$M_x = \int_{0}^{2}\int_{0}^{e^{-x}} ky^2\,dy\,dx = \frac{k}{9}(1 - e^{-6})$$

$$M_y = \int_{0}^{2}\int_{0}^{e^{-x}} kxy\,dy\,dx = \frac{k(1 - 5e^{-4})}{8}$$

$$\bar{x} = \frac{M_y}{m} = \frac{k(e^4 - 5)}{8e^4} \cdot \frac{4e^4}{k(e^4 - 1)} = \frac{e^4 - 5}{2(e^4 - 1)} \approx 0.46$$

$$\bar{y} = \frac{M_x}{m} = \frac{k(e^6 - 1)}{9e^6} \cdot \frac{4e^4}{k(e^4 - 1)} = \frac{4}{9}\left[\frac{e^6 - 1}{e^6 - e^2}\right] \approx 0.45$$

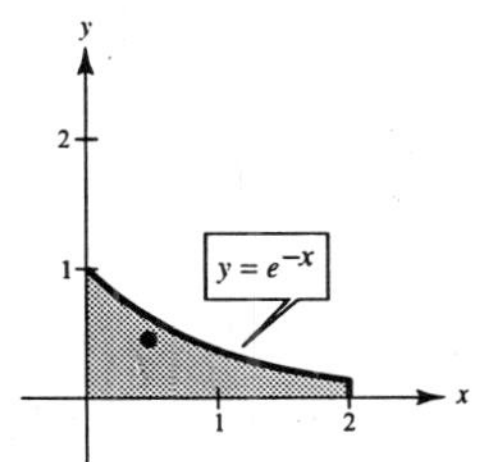

21. $\bar{y} = 0$ by symmetry

$$m = \iint_R k\,dA = \int_{-\pi/6}^{\pi/6}\int_0^{2\cos 3\theta} kr\,dr\,d\theta = \frac{k\pi}{3}$$

$$M_y = \iint_R kx\,dA = \int_{-\pi/6}^{\pi/6}\int_0^{2\cos 3\theta} kr^2\cos\theta\,dr\,d\theta \approx 1.17k$$

$$\bar{x} = \frac{M_y}{m} \approx 1.17k\left(\frac{3}{\pi k}\right) \approx 1.12$$

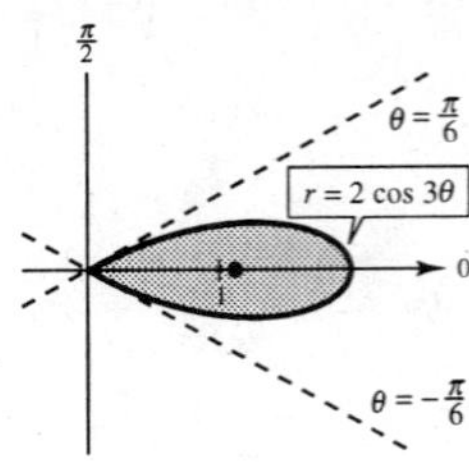

23. $e(x, y) = ky$. $\bar{y}$ will increase.

25. $\rho(x, y) = kxy$.

Both $\bar{x}$ and $\bar{y}$ will increase

27. $m = bh$

$$I_x = \int_0^b\int_0^h y^2\,dy\,dx = \frac{bh^3}{3}$$

$$I_y = \int_0^b\int_0^h x^2 dy\,dx = \frac{b^3h}{3}$$

$$\bar{\bar{x}} = \sqrt{\frac{I_y}{m}} = \sqrt{\frac{b^3h}{3}\cdot\frac{1}{bh}} = \sqrt{\frac{b^2}{3}} = \frac{b}{\sqrt{3}}$$

$$\bar{\bar{y}} = \sqrt{\frac{I_x}{m}} = \sqrt{\frac{bh^3}{3}\cdot\frac{1}{bh}} = \sqrt{\frac{h^2}{3}} = \frac{h}{\sqrt{3}}$$

29. $m = \pi a^2$

$$I_x = \iint_R y^2\,dA = \int_0^{2\pi}\int_0^a r^3\sin^2\theta\,dr\,d\theta = \frac{a^4\pi}{4}$$

$$I_y = \iint_R x^2\,dA = \int_0^{2\pi}\int_0^a r^3\cos^2\theta\,dr\,d\theta = \frac{a^4\pi}{4}$$

$$I_0 = I_x + I_y = \frac{a^4\pi}{4} + \frac{a^4\pi}{4} = \frac{a^4\pi}{2}$$

$$\bar{\bar{x}} = \bar{\bar{y}} = \sqrt{\frac{I_x}{m}} = \sqrt{\frac{a^4\pi}{4}\cdot\frac{1}{\pi a^2}} = \frac{a}{2}$$

31. $m = \dfrac{\pi a^2}{4}$

$$I_x = \iint_R y^2\,dA = \int_0^{\pi/2}\int_0^a r^3\sin^2\theta\,dr\,d\theta = \frac{\pi a^4}{16}$$

$$I_y = \iint_R x^2\,dA = \int_0^{\pi/2}\int_0^a r^3\cos^2\theta\,dr\,d\theta = \frac{\pi a^4}{16}$$

$$I_0 = I_x + I_y = \frac{\pi a^4}{16} + \frac{\pi a^4}{16} = \frac{\pi a^4}{8}$$

$$\bar{\bar{x}} = \bar{\bar{y}} = \sqrt{\frac{I_x}{m}} = \sqrt{\frac{\pi a^4}{16}\cdot\frac{4}{\pi a^4}} = \frac{a}{2}$$

33. $\rho = ky$

$$m = k\int_0^a\int_0^b y\,dy\,dx = \frac{kab^2}{2}$$

$$I_x = k\int_0^a\int_0^b y^3\,dy\,dx = \frac{kab^4}{4}$$

$$I_y = k\int_0^a\int_0^b x^2y\,ydy\,dx = \frac{ka^3b^2}{6}$$

$$I_0 = I_x + I_y = \frac{3kab^4 + 2kb^2a^3}{12}$$

$$\bar{\bar{x}} = \sqrt{\frac{I_y}{m}} = \sqrt{\frac{ka^3b^2/6}{kab^2/2}} = \sqrt{\frac{a^2}{3}} = \frac{a}{\sqrt{3}}$$

$$\bar{\bar{y}} = \sqrt{\frac{I_x}{m}} = \sqrt{\frac{kab^4/4}{kab^2/2}} = \sqrt{\frac{b^2}{2}} = \frac{b}{\sqrt{2}}$$

35. $\rho = kx$

$$m = k\int_0^2\int_0^{4-x^2} x\,dy\,dx = 4k$$

$$I_x = k\int_0^2\int_0^{4-x^2} xy^2\,dy\,dx = \frac{32k}{3}$$

$$I_y = k\int_0^2\int_0^{4-x^2} x^3\,dy\,dx = \frac{16k}{3}$$

$$I_0 = I_x + I_y = 16k$$

$$\bar{\bar{x}} = \sqrt{\frac{I_y}{m}} = \sqrt{\frac{16k/3}{4k}} = \sqrt{\frac{4}{3}} = \frac{2}{\sqrt{3}}$$

$$\bar{\bar{y}} = \sqrt{\frac{I_x}{m}} = \sqrt{\frac{32k/3}{4k}} = \sqrt{\frac{8}{3}} = \frac{4}{\sqrt{6}}$$

37. $\rho = kxy$

$$m = \int_0^4\int_0^{\sqrt{x}} kxy\,dy\,dx = \frac{32k}{3}$$

$$I_x = \int_0^4\int_0^{\sqrt{x}} kxy^3\,dy\,dx = 16k$$

$$I_y = \int_0^4\int_0^{\sqrt{x}} kx^3y\,dy\,dx = \frac{512k}{5}$$

$$I_0 = I_x + I_y = \frac{592k}{5}$$

$$\bar{\bar{x}} = \sqrt{\frac{I_y}{m}} = \sqrt{\frac{512k}{5}\cdot\frac{3}{32k}} = \sqrt{\frac{48}{5}} = \frac{4\sqrt{15}}{5}$$

$$\bar{\bar{y}} = \sqrt{\frac{I_x}{m}} = \sqrt{\frac{16k}{1}\cdot\frac{3}{32k}} = \sqrt{\frac{3}{2}} = \frac{\sqrt{6}}{2}$$

39. $\rho = kx$

$$m = \int_0^1\int_{x^2}^{\sqrt{x}} kx\,dy\,dx = \frac{3k}{20}$$

$$I_x = \int_0^1\int_{x^2}^{\sqrt{x}} kxy^2\,dy\,dx = \frac{3k}{56}$$

$$I_y = \int_0^1\int_{x^2}^{\sqrt{x}} kx^3\,dy\,dx = \frac{k}{18}$$

$$I_0 = I_x + I_y = \frac{55k}{504}$$

$$\bar{\bar{x}} = \sqrt{\frac{I_y}{m}} = \sqrt{\frac{k}{18}\cdot\frac{20}{3k}} = \frac{\sqrt{30}}{9}$$

$$\bar{\bar{y}} = \sqrt{\frac{I_x}{m}} = \sqrt{\frac{3k}{56}\cdot\frac{20}{3k}} = \frac{\sqrt{70}}{14}$$

41. $$I = 2k\int_{-b}^{b}\int_0^{\sqrt{b^2-x^2}} (x-a)^2\,dy\,dx = 2k\int_{-b}^{b}(x-a)^2\sqrt{b^2-x^2}\,dx$$

$$= 2k\left[\int_{-b}^{b} x^2\sqrt{b^2-x^2}\,dx - 2a\int_{-b}^{b} x\sqrt{b^2-x^2}\,dx + a^2\int_{-b}^{b}\sqrt{b^2-x^2}\,dx\right]$$

$$= 2k\left[\frac{\pi b^4}{8} + 0 + \frac{\pi a^2b^2}{2}\right] = \frac{k\pi b^2}{4}(b^2 + 4a^2)$$

43. $$I = \int_0^4\int_0^{\sqrt{x}} kx(x-6)^2\,dy\,dx = \int_0^4 kx\sqrt{x}(x^2 - 12x + 36)\,dx$$

$$= k\left[\frac{2}{9}x^{9/2} - \frac{24}{7}x^{7/2} + \frac{72}{5}x^{5/2}\right]_0^4$$

$$= \frac{42{,}752k}{315}$$

45. $I = \int_0^a \int_0^{\sqrt{a^2-x^2}} k(a-y)(y-a)^2\,dy\,dx = \int_0^a \int_0^{\sqrt{a^2-x^2}} k(a-y)^3\,dy\,dx = \int_0^a \left[-\frac{k}{a}(a-y)^4\right]_0^{\sqrt{a^2-x^2}} dx$

$$= -\frac{k}{a}\int_0^a \left[a^4 - 4a^3y + 4ay^3 + y^4\right]_0^{\sqrt{a^2-x^2}} dx$$

$$= -\frac{k}{4}\int_0^a \left[a^4 - 4a^3\sqrt{a^2-x^2} + 6a^2(a^2-x^2) - 4a(a^2-x^2)\sqrt{a^2-x^2} + (a^4 - 2a^2x^2 + x^4) - a^4\right] dx$$

$$= -\frac{k}{a}\int_0^a \left[a^4 - 8a^2x^2 + x^4 - 8a^3\sqrt{a^2-x^2} + 4ax^2 = \sqrt{a^2-x^2}\right] dx$$

$$= -\frac{k}{a}\left[7a^4x - \frac{8a^2}{3}x^3 + \frac{x^5}{5} - 4a^3\left(x\sqrt{a^2-x^2} + a^2 \arcsin\frac{x}{a}\right) + \frac{a}{2}\left(x(2x^2-a^2)\sqrt{a^2-x^2} + a^4 \arcsin\frac{x}{a}\right)\right]_0^a$$

$$= -\frac{k}{a}\left(7a^5 - \frac{8}{3}a^5 + \frac{1}{5}a^5 - 2a^5\pi + \frac{1}{4}a^5\pi\right) = a^5k\left(\frac{7\pi}{16} - \frac{17}{15}\right)$$

47. Orient the xy-coordinate system so that L is along the y-axis and R is in the first quadrant. Then the volume of the solid is

$$V = \int_R\int 2\pi x\,dA$$

$$= 2\pi \int_R\int x\,dA$$

$$= 2\pi\left(\frac{\int_R\int x\,dA}{\int_R\int dA}\right)\int_R\int dA$$

$$= 2\pi\bar{x}A.$$

By our positioning, $\bar{x} = r$. Therefore, $V = 2\pi rA$.

49. $\bar{y} = \frac{a}{2}, A = ab, h = L - \frac{a}{2}$

$$I_{\bar{y}} = \int_0^b \int_0^a \left(y - \frac{a}{2}\right)^2 dy\,dx = \frac{a^3b}{12}$$

$$y_a = \frac{a}{2} - \frac{a^3b/12}{[L-(a/2)]ab} = \frac{a(3L-2a)}{3(2L-a)}$$

51. $\bar{y} = 0, A = \pi a^2, h = L$

$$I_{\bar{y}} = \int_{-a}^a \int_{-\sqrt{a^2-x^2}}^{\sqrt{a^2-x^2}} y^2\,dy\,dx$$

$$= \int_0^{2\pi}\int_0^a r^3\sin^2\theta\,dr\,d\theta$$

$$= \int_0^{2\pi} \frac{a^4}{4}\sin^2\theta\,d\theta$$

$$= \frac{a^4\pi}{4}$$

$$y_a = -\frac{(a^4\pi/4)}{L\pi a^2} = -\frac{a^2}{4L}$$

Section 13.5 Surface Area

1. $f(x, y) = 2x + 2y$

R = triangle with vertices $(0, 0)$, $(2, 0)$, $(0, 2)$

$f_x = 2, f_y = 2$

$\sqrt{1 + (f_x)^2 + (f_y)^2} = 3$

$$S = \int_0^2 \int_0^{2-x} 3\, dy\, dx = 3 \int_0^2 (2 - x)\, dx$$

$$= \left[3\left(2x - \frac{x^2}{2}\right)\right]_0^2 = 6$$

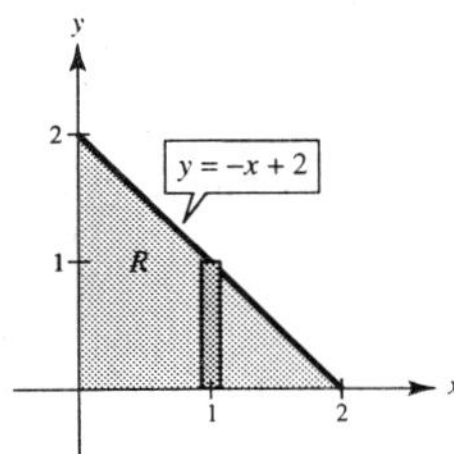

3. $f(x, y) = 8 + 2x + 2y$

$R = \{(x, y): x^2 + y^2 \le 4\}$

$f_x = 2, f_y = 2$

$\sqrt{1 + (f_x)^2 + (f_y)^2} = 3$

$$S = \int_{-2}^{2} \int_{-\sqrt{4-x^2}}^{\sqrt{4-x^2}} 3\, dy\, dx = \int_0^{2\pi} \int_0^2 3r\, dr\, d\theta = 12\pi$$

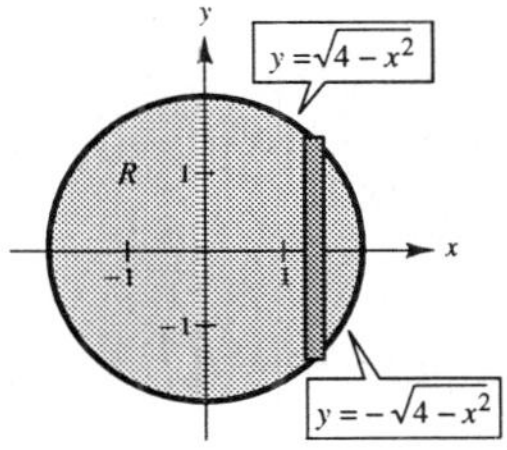

5. $f(x, y) = 9 - x^2$

R = square with vertices, $(0, 0)$, $(3, 0)$, $(0, 3)$, $(3, 3)$

$f_x = -2x,\ f_y = 0$

$\sqrt{1 + (f_x)^2 + (f_y)^2} = \sqrt{1 + 4x^2}$

$$S = \int_0^3 \int_0^3 \sqrt{1 + 4x^2}\, dy\, dx = \int_0^3 3\sqrt{1 + 4x^2}\, dx$$

$$= \left[\frac{3}{4}\left(2x\sqrt{1 + 4x^2} + \ln\left|2x + \sqrt{1 + 4x^2}\right|\right)\right]_0^3 = \frac{3}{4}\left(6\sqrt{37} + \ln\left|6 + \sqrt{37}\right|\right)$$

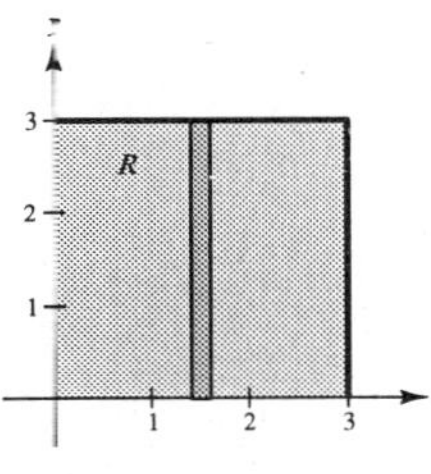

7. $f(x, y) = 2 + x^{3/2}$

R = rectangle with vertices $(0, 0)$, $(0, 4)$, $(3, 4)$, $(3, 0)$

$f_x = \frac{3}{2}x^{1/2}, f_y = 0$

$$\sqrt{1 + (f_x)^2 + (f_y)^2} = \sqrt{1 + \left(\frac{9}{4}\right)x} = \frac{\sqrt{4 + 9x}}{2}$$

$$S = \int_0^3 \int_0^4 \frac{\sqrt{4 + 9x}}{2}\, dy\, dx = \int_0^3 4\left(\frac{\sqrt{4 + 9x}}{2}\right) dx$$

$$= \left[\frac{4}{27}(4 + 9)^{3/2}\right]_0^3 = \frac{4}{27}\left(3\sqrt{31} - 8\right)$$

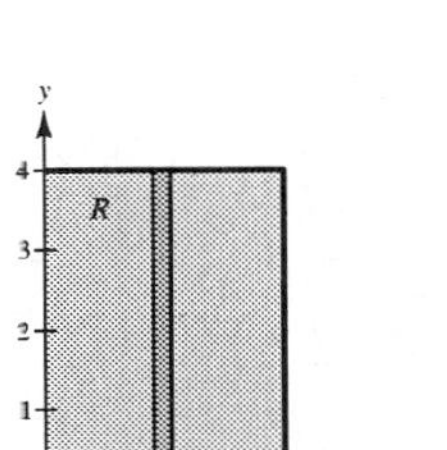

9. $f(x, y) = \ln|\sec x|$

$$R = \left\{(x, y): 0 \le x \le \frac{\pi}{4},\ 0 \le y \le \tan x\right\}$$

$f_x = \tan x,\ f_y = 0$

$\sqrt{1 + (f_x)^2 + (f_y)^2} = \sqrt{1 + \tan^2 x} = \sec x$

$$S = \int_0^{\pi/4} \int_0^{\tan x} \sec x\, dy\, dx = \int_0^{\pi/4} \sec x \tan x\, dx = \Big[\sec x\Big]_0^{\pi/4} = \sqrt{2} - 1$$

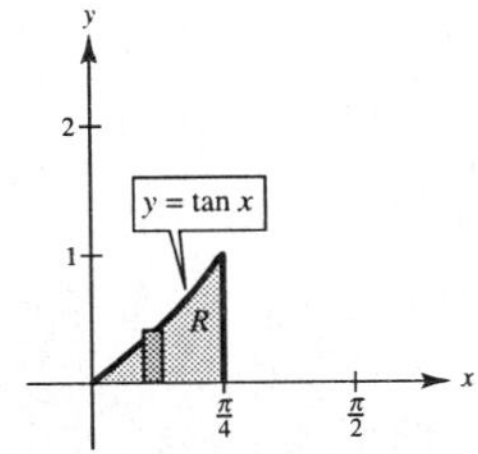

11. $f(x, y) = \sqrt{x^2 + y^2}$

$R = \{(x, y): 0 \le f(x, y) \le 1\}$

$0 \le \sqrt{x^2 + y^2} \le 1,\ x^2 + y^2 \le 1$

$$f_x = \frac{x}{\sqrt{x^2 + y^2}},\ f_y = \frac{y}{\sqrt{x^2 + y^2}}$$

$$\sqrt{1 + (f_x)^2 + (f_y)^2} = \sqrt{1 + \frac{x^2}{x^2 + y^2} + \frac{y^2}{x^2 + y^2}} = \sqrt{2}$$

$$S = \int_{-1}^{1} \int_{-\sqrt{1-x^2}}^{\sqrt{1-x^2}} \sqrt{2}\, dy\, dx = \int_0^{2\pi} \int_0^1 \sqrt{2}\, r\, dr\, d\theta = \sqrt{2}\pi$$

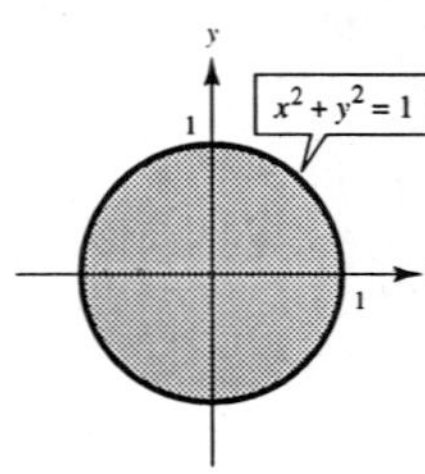

13. $f(x, y) = \sqrt{a^2 - x^2 - y^2}$

$R = \{(x, y): x^2 + y^2 \le b^2, b < a\}$

$$f_x = \frac{-x}{\sqrt{a^2 - x^2 - y^2}},\ f_y = \frac{-y}{\sqrt{a^2 - x^2 - y^2}}$$

$$\sqrt{1 + (f_x)^2 + (f_y)^2} = \sqrt{1 + \frac{x^2}{a^2 - x^2 - y^2} + \frac{y^2}{a^2 - x^2 - y^2}} = \frac{a}{\sqrt{a^2 - x^2 - y^2}}$$

$$S = \int_{-b}^{b} \int_{-\sqrt{b^2-x^2}}^{\sqrt{b^2-x^2}} \frac{a}{\sqrt{a^2 - x^2 - y^2}}\, dy\, dx = \int_0^{2\pi} \int_0^b \frac{a}{\sqrt{a^2 - r^2}}\, r\, dr\, d\theta = 2\pi a\left(a - \sqrt{a^2 - b^2}\right)$$

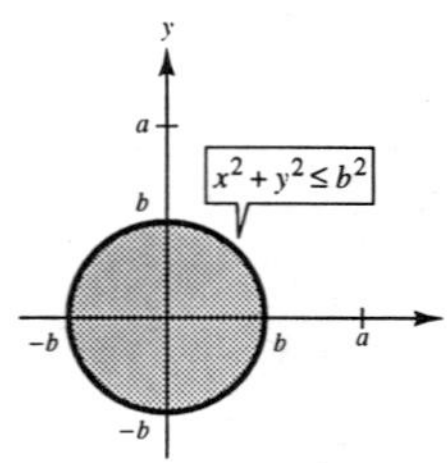

15. $z = 24 - 3x - 2y$

$$\sqrt{1 + (f_x)^2 + (f_y)^2} = \sqrt{14}$$

$$S = \int_0^8 \int_0^{-(3/2)x + 12} \sqrt{14}\, dy\, dx = 48\sqrt{14}$$

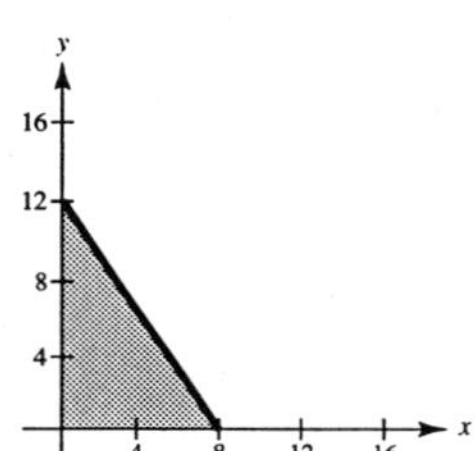

17. $z = \sqrt{25 - x^2 - y^2}$

$$\sqrt{1 + (f_x)^2 + (f_y)^2} = \sqrt{1 + \frac{x^2}{25 - x^2 - y^2} + \frac{y^2}{25 - x^2 - y^2}} = \frac{5}{\sqrt{25 - x^2 - y^2}}$$

$$S = 2\int_{-3}^{3} \int_{-\sqrt{9-x^2}}^{\sqrt{9-x^2}} \frac{5}{\sqrt{25 - (x^2 + y^2)}}\, dy\, dx$$

$$= 2\int_0^{2\pi} \int_0^3 \frac{5}{\sqrt{25 - r^2}}\, r\, dr\, d\theta = 20\pi$$

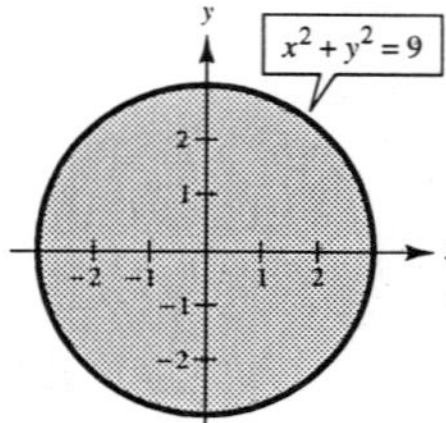

19. $f(x, y) = 2y + x^2$

R = triangle with vertices $(0, 0), (1, 0), (1, 1)$

$$\sqrt{1 + (f_x)^2 + (f_y)^2} = \sqrt{5 + 4x^2}$$

$$S = \int_0^1 \int_0^x \sqrt{5 + 4x^2}\, dy\, dx = \frac{1}{12}(27 - 5\sqrt{5})$$

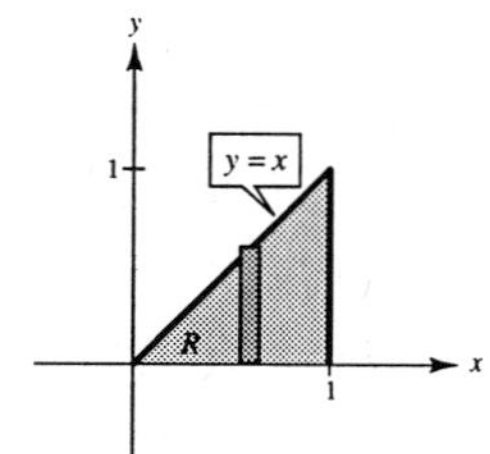

21. $f(x, y) = 4 - x^2 - y^2$

$R = \{(x, y): 0 \le f(x, y)\}$

$0 \le 4 - x^2 - y^2,\ x^2 + y^2 \le 4$

$f_x = -2x,\ f_y = -2y$

$\sqrt{1 + (f_x)^2 + (f_y)^2} = \sqrt{1 + 4x^2 + 4y^2}$

$$S = \int_{-2}^{2}\int_{-\sqrt{4-x^2}}^{\sqrt{4-x^2}} \sqrt{1 + 4x^2 + 4y^2}\,dy\,dx = \int_0^{2\pi}\int_0^2 \sqrt{1 + 4r^2}\,r\,dr\,d\theta = \frac{(17\sqrt{17} - 1)\pi}{6}$$

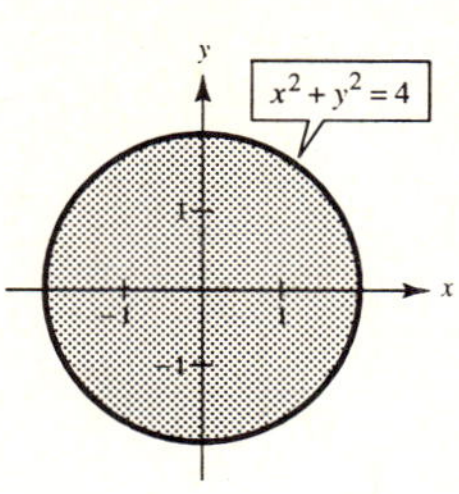

23. $f(x, y) = 4 - x^2 - y^2$

$R = \{(x, y): 0 \le x \le 1,\ 0 \le y \le 1\}$

$f_x = -2x,\ f_y = -2y$

$\sqrt{1 + (f_x)^2 + (f_y)^2} = \sqrt{1 + 4x^2 + 4y^2}$

$$S = \int_0^1\int_0^1 \sqrt{(1 + 4x^2) + 4y^2}\,dy\,dx \approx 1.81616$$

25. Surface area > (4) · (6) = 24.

Matches (e)

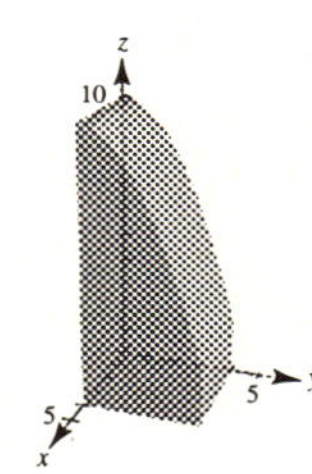

27. $f(x, y) = e^x$

$R = \{(x, y): 0 \le x \le 1,\ 0 \le y \le 1\}$

$f_x = e^x,\ f_y = 0$

$\sqrt{1 + (f_x)^2 + (f_y)^2} = \sqrt{1 + e^{2x}}$

$$S = \int_0^1\int_0^1 \sqrt{1 + e^{2x}}\,dy\,dx = \int_0^1 \sqrt{1 + e^{2x}} \approx 2.0035$$

29. $f(x, y) = x^3 - 3xy^3 + y^3$

$R =$ square with vertices $(1, 1), (-1, 1), (-1, -1), (1, -1)$

$f_x = 3x^2 - 3y = 3(x^2 - y), f_y = -3x + 3y^2 = 3(y^2 - x)$

$$S = \int_{-1}^1\int_{-1}^1 \sqrt{1 + 9(x^2 - y)^2 + 9(y^2 - x)^2}\,dy\,dx$$

31. $f(x, y) = e^{-x}\sin y$

$R = \{(x, y): x^2 + y^2 \le 4\}$

See Exercise 30.

$$S = \int_{-2}^{2}\int_{-\sqrt{4-x^2}}^{\sqrt{4-x^2}} \sqrt{1 + e^{-2x}}\,dy\,dx$$

33. $f(x, y) = e^{xy}$

$R = \{(x, y): 0 \le x \le 4,\ 0 \le y \le 10\}$

$f_x = ye^{xy},\ f_y = xe^{xy}$

$\sqrt{1 + (f_x)^2 + (f_y)^2} = \sqrt{1 + y^2e^{2xy} + x^2e^{2xy}} = \sqrt{1 + e^{2xy}(x^2 + y^2)}$

$$S = \int_0^4\int_0^{10} \sqrt{1 + e^{2x}(x^2 + y^2)}\,dy\,dx$$

35. $f(x, y) = \sqrt{1 - x^2};\ f_x = \dfrac{-x}{\sqrt{1^2 - x^2}}, f_y = 0$

$$S = \iint_R \sqrt{1 + f_x^2 + f_y^2}\,dA$$

$$= 16\int_0^1\int_0^x \frac{1}{\sqrt{1 - x^2}}\,dy\,dx = 16\int_0^1 \frac{x}{\sqrt{1 - x^2}}\,dx = \Big[-16(1 - x^2)^{1/2}\Big]_0^1 = 16$$

37. (a) $V = \int_0^{50}\int_0^{\sqrt{50^2-x^2}} \left(20 + \frac{xy}{100} - \frac{x+y}{5}\right) dy\, dx$

$$= \int_0^{50} \left[20\sqrt{50^2 - x^2} + \frac{x}{200}(50^2 - x^2) - \frac{x}{5}\sqrt{50^2 - x^2} - \frac{50^2 - x^2}{10}\right] dy$$

$$= \left[10\left(x\sqrt{50 - x^2} + 50^2 \arcsin\frac{x}{50}\right) + \frac{25}{4}x^2 - \frac{x^4}{800} + \frac{1}{15}(50^2 - x^2)^{3/2} - 250x + \frac{x^3}{30}\right]_0^{50}$$

$$\approx 30{,}415.74 \text{ ft}^3$$

(b) $z = 20 + \frac{xy}{100}$

$$\sqrt{1 + (f_x)^2 + (f_y)^2} = \sqrt{1 + \frac{y^2}{100^2} + \frac{x^2}{100^2}} = \frac{\sqrt{100^2 + x^2 + y^2}}{100}$$

$$S = \frac{1}{100}\int_0^{50}\int_0^{\sqrt{50^2-x^2}} \sqrt{100^2 + x^2 + y^2}\, dy\, dx$$

$$= \frac{1}{100}\int_0^{\pi/2}\int_0^{50} \sqrt{100^2 + r^2}\, r\, dr\, d\theta \approx 2081.53 \text{ ft}^2$$

39. The greater the angle between the given plane and the xy-plane, the greater the surface area. Hence:

$z_2 < z_1 < z_4 < z_3$

41. False. The surface area will remain the same for any vertical translation.

Section 13.6 Triple Integrals and Applications

1. $\int_0^3\int_0^2\int_0^1 (x + y + z)\, dx\, dy\, dx = \int_0^3\int_0^2 \left[\frac{1}{2}x^2 + xy + xz\right]_0^1 dy\, dx$

$$= \int_0^3\int_0^2 \left(\frac{1}{2} + y + z\right) dy\, dz = \int_0^3 \left[\frac{1}{2}y + \frac{1}{2}y^2 + yz\right]_0^2 dz = \left[3z + z^2\right]_0^3 = 18$$

3. $\int_0^1\int_0^x\int_0^{xy} x\, dz\, dy\, dx = \int_0^1\int_0^x \left[xy\right]_0^{xy} dy\, dx$

$$= \int_0^1\int_0^x x^2y\, dy\, dx = \int_0^1 \left[\frac{x^2y^2}{2}\right]_0^x dx = \int_0^1 \frac{x^4}{2}\, dx = \left[\frac{x^5}{10}\right]_0^1 = \frac{1}{10}$$

5. $\int_1^4\int_0^1\int_0^x 2ze^{-x^2}\, dy\, dx\, dz = \int_1^4\int_0^1 \left[(2ze^{-x^2})y\right]_0^x dx\, dz = \int_1^4\int_0^1 2zxe^{-x^2}\, dx\, dz$

$$= \int_1^4 \left[-ze^{-x^2}\right]_0^1 dz = \int_1^4 z(1 - e^{-1})\, dz = \left[(1 - e^{-1})\frac{z^2}{2}\right]_1^4 = \frac{15}{2}\left(1 - \frac{1}{e}\right)$$

7. $\int_0^9\int_0^{y/3}\int_0^{\sqrt{y^2-9x^2}} z\, dz\, dx\, dy = \frac{1}{2}\int_0^9\int_0^{y/3} (y^2 - 9x^2)\, dx\, dy$

$$= \frac{1}{2}\int_0^9 \left[xy^2 - 3x^3\right]_0^{y/3} dy = \frac{2}{18}\int_0^9 y^3\, dy = \left[\frac{1}{36}y^4\right]_0^9 = \frac{729}{4}$$

9. $$\int_0^2 \int_{-\sqrt{4-x^2}}^{\sqrt{4-x^2}} \int_0^{x^2} x \, dz \, dy \, dx = \int_0^2 \int_{-\sqrt{4-x^2}}^{\sqrt{4-x^2}} x^3 \, dy \, dx = \frac{128}{15}$$

11. $$\int_0^2 \int_0^{\sqrt{4-x^2}} \int_1^4 \frac{x^2 \sin y}{z} \, dz \, dy \, dx = \int_0^2 \int_0^{\sqrt{4-x^2}} \left[x^2 \sin y \ln |z| \right]_1^4 dy \, dx$$

$$= \int_0^2 \left[x^2 \ln 4(-\cos y) \right]_0^{\sqrt{4-x^2}} dx = \int_0^2 x^2 \ln 4\left[1 - \cos\sqrt{4 - x^2}\right] dx \approx 2.44167$$

13. Plane: $3x + 6y + 4z = 12$

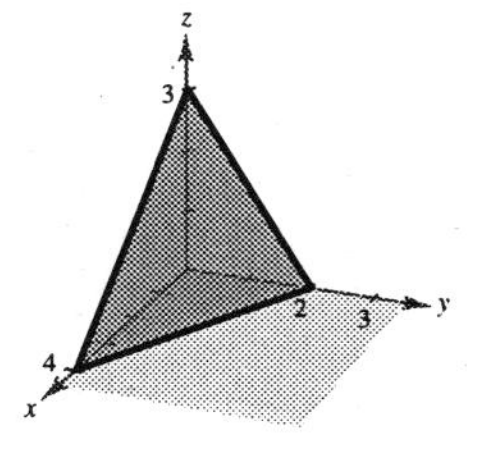

$$\int_0^3 \int_0^{(12-4z)/3} \int_0^{(12-4z-3x)/6} dy \, dx \, dz$$

15. Top cylinder: $y^2 + z^2 = 1$

Side plane: $x = y$

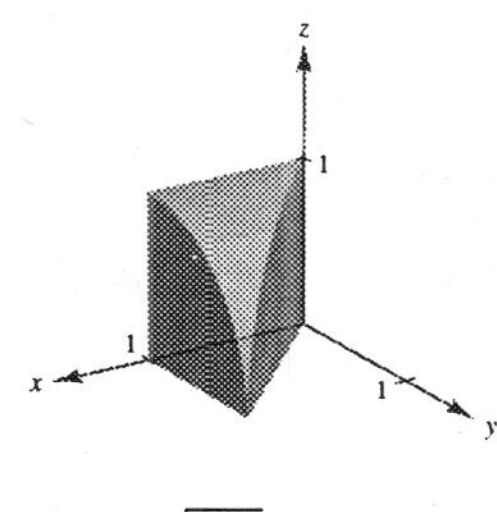

$$\int_0^1 \int_0^x \int_0^{\sqrt{1-y^2}} dz \, dy \, dx$$

17. $Q = \{(x, y, z): 0 \le x \le 1, 0 \le y \le x, 0 \le z \le 3\}$

$$\iiint_Q xyz \, dV = \int_0^3 \int_0^1 \int_y^1 xyz \, dx \, dy \, dz = \int_0^3 \int_0^1 \int_0^x xyz \, dy \, dx \, dz$$

$$= \int_0^1 \int_0^3 \int_y^1 xyz \, dx \, dz \, dy$$

$$= \int_0^1 \int_0^3 \int_0^x xyz \, dy \, dz \, dx$$

$$= \int_0^1 \int_y^1 \int_0^3 xyz \, dz \, dx \, dy$$

$$= \int_0^1 \int_0^x \int_0^3 xyz \, dz \, dy \, dx \left(= \frac{9}{16} \right)$$

19. $$\int_{-2}^2 \int_0^{4-y^2} \int_0^x dz \, dx \, dy = \int_{-2}^2 \int_0^{4-y^2} x \, dx \, dy$$

$$= \frac{1}{2}\int_{-2}^2 (4 - y^2)^2 \, dy = \int_0^2 (16 - 8y^2 + y^4) \, dy = \left[16y - \frac{8}{3}y^3 + \frac{1}{5}y^5 \right]_0^2 = \frac{256}{15}$$

21. $$8\int_0^a \int_0^{\sqrt{a^2-x^2}} \int_0^{\sqrt{a^2-x^2-y^2}} dz \, dy \, dx = 8\int_0^a \int_0^{\sqrt{a^2-x^2}} \sqrt{a^2 - x^2 - y^2} \, dy \, dx$$

$$= 4\int_0^a \left[y\sqrt{a^2 - x^2 - y^2} + (a^2 - x^2) \arcsin\left(\frac{y}{\sqrt{a^2 - x^2}} \right) \right]_0^{\sqrt{a^2-x^2}} dx$$

$$= 4\left(\frac{\pi}{2}\right) \int_0^a (a^2 - x^2) \, dx = \left[2\pi\left(a^2x - \frac{1}{3}x^3 \right) \right]_0^a = \frac{4}{3}\pi a^3$$

23. $\displaystyle\int_0^2\int_0^{4-x^2}\int_0^{4-x^2} dz\,dy\,dx = \int_0^2 (4-x^2)^2\,dx$

$\displaystyle = \int_0^2 (16 - 8x^2 + x^4)\,dx$

$\displaystyle = \left[16 - \frac{8}{3}x^3 + \frac{1}{5}x^5\right]_0^2 = \frac{265}{15}$

25. $\displaystyle m = k\int_0^6\int_0^{4-(2x/3)}\int_0^{2-(y/2)-(x/3)} dz\,dy\,dx$

$= 8k$

$\displaystyle M_{yz} = k\int_0^6\int_0^{4-(2x/3)}\int_0^{2-(y/2)-(x/3)} x\,dz\,dy\,dx$

$= 12k$

$\displaystyle \bar{x} = \frac{M_{yz}}{m} = \frac{12k}{8k} = \frac{3}{2}$

27. $\displaystyle m = k\int_0^4\int_0^4\int_0^{4-x} x\,dz\,dy\,dx = k\int_0^4\int_0^4 x(4-x)\,dy\,dx$

$\displaystyle = 4k\int_0^4 (4x - x^2)\,dx = \frac{128k}{3}$

$\displaystyle M_{xy} = k\int_0^4\int_0^4\int_0^{4-x} xz\,dz\,dy\,dx = k\int_0^4\int_0^4 x\frac{(4-x)^2}{2}\,dy\,dx$

$\displaystyle = 2k\int_0^4 (16x - 8x^2 + x^3)\,dx = \frac{128k}{3}$

$\displaystyle \bar{z} = \frac{M_{xy}}{m} = 1$

29. $\displaystyle m = k\int_0^b\int_0^b\int_0^b xy\,dz\,dy\,dx = \frac{kb^5}{4}$

$\displaystyle M_{yz} = k\int_0^b\int_0^b\int_0^b x^2y\,dz\,dy\,dx = \frac{kb^6}{6}$

$\displaystyle M_{xz} = k\int_0^b\int_0^b\int_0^b xy^2\,dz\,dy\,dx = \frac{kb^6}{6}$

$\displaystyle M_{xy} = k\int_0^b\int_0^b\int_0^b xyz\,dz\,dy\,dx = \frac{kb^6}{8}$

$\displaystyle \bar{x} = \frac{M_{yz}}{m} = \frac{kb^6/6}{kb^5/4} = \frac{2b}{3}$

$\displaystyle \bar{y} = \frac{M_{xz}}{m} = \frac{kb^6/6}{kb^5/4} = \frac{2b}{3}$

$\displaystyle \bar{z} = \frac{M_{xy}}{m} = \frac{kb^6/8}{kb^5/4} = \frac{b}{2}$

31. $\bar{x}$ will be greater than 2, whereas $\bar{y}$ and $\bar{z}$ will be unchanged.

33. $\bar{y}$ will be greater than 0, whereas $\bar{x}$ and $\bar{z}$ will be unchanged.

35. $\displaystyle m = \frac{1}{3}k\pi r^2 h$

$\bar{x} = \bar{y} = 0$ by symmetry

$\displaystyle M_{xy} = 4k\int_0^r\int_0^{\sqrt{r^2-x^2}}\int_{h\sqrt{x^2+y^2}/r}^{h} z\,dz\,dy\,dx$

$\displaystyle = \frac{3kh^2}{r^2}\int_0^r\int_0^{\sqrt{r^2-x^2}} (r^2 - x^2 - y^2)\,dy\,dx$

$\displaystyle = \frac{4kh^2}{3r^2}\int_0^r (r^2 - x^2)^{3/2}\,dx$

$\displaystyle = \frac{k\pi r^2h^2}{4}$

$\displaystyle \bar{z} = \frac{M_{xy}}{m} = \frac{k\pi r^2h^2/4}{k\pi r^2h/3} = \frac{3h}{4}$

37. $\displaystyle m = \frac{128k\pi}{3}$

$\bar{x} = \bar{y} = 0$ by symmetry

$z = \sqrt{4^2 - x^2 - y^2}$

$\displaystyle M_{xy} = 4k\int_0^4\int_0^{\sqrt{4^2-x^2}}\int_0^{\sqrt{4^2-x^2-y^2}} z\,dz\,dy\,dx$

$\displaystyle = 2k\int_0^4\int_0^{\sqrt{4^2-x^2}} (4^2 - x^2 - y^2)\,dy\,dx$

$\displaystyle = 2k\int_0^4 \left[16y - x^2y - \frac{1}{3}y^3\right]_0^{\sqrt{4^2-x^2}} dx$

$\displaystyle = \frac{4k}{3}\int_0^4 (4^2 - x^2)^{3/2}\,dx$

$\displaystyle = \frac{1024k}{3}\int_0^{\pi/2} \cos^4\theta\,d\theta \qquad (\text{let } x = 4\sin\theta)$

$= 64\pi k$ by Wallis's Formula

$\displaystyle \bar{z} = \frac{M_{xy}}{m} = \frac{64k\pi}{1}\cdot\frac{3}{128k\pi} = \frac{3}{2}$

39. $f(x, y) = \frac{5}{12}y$

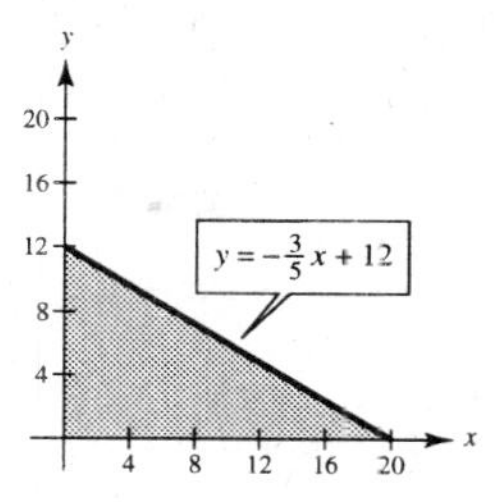

$$m = k\int_0^{20}\int_0^{-(3/5)x+12}\int_0^{(5/12)y} dz\,dy\,dx = 200k$$

$$M_{yz} = k\int_0^{20}\int_0^{-(3/5)x+12}\int_0^{(5/12)y} x\,dz\,dy\,dx = 1000k$$

$$M_{xz} = k\int_0^{20}\int_0^{-(3/5)+x+12}\int_0^{(5/12)y} y\,dz\,dy\,dx = 1200k$$

$$M_{xy} = k\int_0^{20}\int_0^{-(3/5)x+12}\int_0^{(5/12)y} z\,dz\,dy\,dx = 250k$$

$$\bar{x} = \frac{M_{yz}}{m} = \frac{1000k}{200k} = 5$$

$$\bar{y} = \frac{M_{xz}}{m} = \frac{1200k}{200k} = 6$$

$$\bar{z} = \frac{M_{xy}}{m} = \frac{250k}{200k} = \frac{5}{4}$$

41. (a) $I_x = k\int_0^a\int_0^a\int_0^a (y^2 + z^2)\,dx\,dy\,dz = ka\int_0^a\int_0^a (y^2 + z^2)\,dy\,dz$

$$= ka\int_0^a \left[\frac{1}{3}y^3 + z^2y\right]_0^a dz = ka\int_0^a \left(\frac{1}{3}a^3 + az^2\right) dz = \left[ka\left(\frac{1}{3}a^3k + \frac{1}{3}az^3\right)\right]_0^a = \frac{2ka^5}{3}$$

$I_x = I_y = I_z = \frac{2ka^5}{3}$ by symmetry

(b) $I_x = k\int_0^a\int_0^a\int_0^a (y^2 + z^2)xyz\,dx\,dy\,dz = \frac{ka^2}{a}\int_0^a\int_0^a (y^3z + yz^3)\,dy\,dz$

$$= \frac{ka^2}{2}\int_0^a \left[\frac{y^4a}{4} + \frac{y^2z^3}{2}\right]_0^a dz = \frac{ka^4}{8}\int_0^a (a^2z + 2z^3)\,dz = \left[\frac{ka^4}{8}\left(\frac{a^2z^2}{2} + \frac{2z^4}{4}\right)\right]_0^a = \frac{ka^8}{8}$$

$I_x = I_y = I_z = \frac{ka^8}{8}$ by symmetry

43. (a) $I_x = k\int_0^4\int_0^4\int_0^{4-x} (y^2 + z^2)\,dz\,dy\,dx = k\int_0^4\int_0^4 \left[y^2(4 - x) + \frac{1}{3}(4 - x)^3\right] dy\,dx$

$$= k\int_0^4 \left[\frac{y^3}{3}(4 - x) + \frac{y}{3}(4 - x)^3\right]_0^4 dx = k\int_0^4 \left[\frac{64}{3}(4 - x) + \frac{4}{3}(4 - x)^3\right] dx$$

$$= k\left[-\frac{32}{3}(4 - x)^2 - \frac{1}{3}(4 - x)^4\right]_0^4 = 256k$$

$$I_y = k\int_0^4\int_0^4\int_0^{4-x} (x^2 + z^2)\,dz\,dy\,dx = k\int_0^4\int_0^4 \left[x^2(4 - x) + \frac{1}{3}(4 - x)^3\right] dy\,dx$$

$$= 4k\int_0^4 \left[4x^2 - x^3 + \frac{1}{3}(4 - x)^3\right] dx = 4k\left[\frac{4}{3}x^3 - \frac{1}{4}x^4 - \frac{1}{12}(4 - x)^4\right]_0^4 = \frac{512k}{3}$$

$$I_z = k\int_0^4\int_0^4\int_0^{4-x} (x^2 + y^2)\,dz\,dy\,dx = k\int_0^4\int_0^4 (x^2 + y^2)(4 - x)\,dy\,dx$$

$$= k\int_0^4 \left[\left(x^2y + \frac{y^3}{3}\right)(4 - x)\right]_0^4 dx = k\int_0^4 \left(4x^2 + \frac{64}{3}\right)(4 - x)\,dx = 256k$$

—CONTINUED—

43. —CONTINUED—

(b) $I_x = k\int_0^4\int_0^4\int_0^{4-x} y(y^2+z^2)\,dz\,dy\,dx = k\int_0^4\int_0^4\left[y^3(4-x)+\frac{1}{3}y(4-x)^3\right]dy\,dx$

$$= k\int_0^4\left[\frac{y^4}{4}(4-x)+\frac{y^2}{6}(4-x)^3\right]_0^4 dx = k\int_0^4\left[64(4-x)+\frac{8}{3}(4-x)^3\right]dx$$

$$= k\left[-32(4-x)^2-\frac{2}{3}(4-x)^4 = \frac{2048k}{3}\right.$$

$$I_y = k\int_0^4\int_0^4\int_0^{4-x} y(x^2+z^2)\,dz\,dy\,dx = k\int_0^4\int_0^4\left[x^2y(4-x)+\frac{1}{3}y(4-x)^3\right]dy\,dx$$

$$= 8k\int_0^4\left[4x^2-x^3+\frac{1}{3}(4-x)^3\right]dx = 8k\left[\frac{4}{3}x^3-\frac{1}{4}x^4-\frac{1}{12}(4-x)^4\right]_0^4 = \frac{1024k}{3}$$

$$I_z = k\int_0^4\int_0^4\int_0^{4-x} y(x^2+y^2)\,dz\,dy\,dx = k\int_0^4\int_0^4 (x^2y+y^3)(4-x)\,dx$$

$$= k\int_0^4\left[\left(\frac{x^2y^2}{2}+\frac{y^4}{4}\right)(4-x)\right]_0^4 dx = k\int_0^4 (8x^2+64)(4-x)\,dx$$

$$= 8k\int_0^4 (32-8x+4x^2-x^3)\,dx = \left[8k\left(32x-4x^2+\frac{4}{3}x^3-\frac{1}{4}x^4\right)\right]_0^4 = \frac{2048k}{3}$$

45. $I_{xy} = k\int_{-L/2}^{L/2}\int_{-a}^{a}\int_{-\sqrt{a^2-x^2}}^{\sqrt{a^2-x^2}} z^2\,dz\,dx\,dy = k\int_{-L/2}^{L/2}\int_{-a}^{a}\frac{2}{3}(a^2-x^2)\sqrt{a^2-x^2}\,dx\,dy$

$$= \frac{2}{3}\int_{-L/2}^{L/2}\left[\frac{a^2}{2}\left(x\sqrt{a^2-x^2}+a^2\arcsin\frac{x}{a}\right)-\frac{1}{8}\left(x(2x^2-a^2)\sqrt{x^2-a^2}+a^4\arcsin\frac{x}{a}\right)\right]_{-a}^{a} dy$$

$$= \frac{2k}{3}\int_{-L/2}^{L/2} 2\left(\frac{a^4\pi}{4}-\frac{a^4\pi}{16}\right)dy = \frac{a^4\pi Lk}{4}$$

Since $m = \pi a^2 Lk$, $I_{xy} = ma^2/4$.

$$I_{xz} = k\int_{-L/2}^{L/2}\int_{-a}^{a}\int_{-\sqrt{a^2-x^2}}^{\sqrt{a^2-x^2}} y^2\,dz\,dx\,dy = 2k\int_{-L/2}^{L/2}\int_{-a}^{a} y^2\sqrt{a^2-x^2}\,dx\,dy$$

$$= 2k\int_{-L/2}^{L/2}\left[\frac{y^2}{2}\left(x\sqrt{a^2-x^2}+a^2\arcsin\frac{x}{a}\right)\right]_{-a}^{a} dy = k\pi a^2\int_{-L/2}^{L/2} y^2\,dy = \frac{2k\pi a^2}{3}\left(\frac{L^3}{8}\right) = \frac{1}{12}mL^2$$

$$I_{yz} = k\int_{-L/2}^{L/2}\int_{-a}^{a}\int_{-\sqrt{a^2-x^2}}^{\sqrt{a^2-x^2}} x^2\,dz\,dx\,dy = 2k\int_{-L/2}^{L/2}\int_{-a}^{a} x^2\sqrt{a^2-x^2}\,dx\,dy$$

$$= 2k\int_{-L/2}^{L/2}\frac{1}{8}\left[x(2x^2-a^2)\sqrt{a^2-x^2}+a^4\arcsin\frac{x}{a}\right]_{-a}^{a} dy = \frac{ka^4\pi}{4}\int_{-L/2}^{L/2} dy = \frac{ka^4\pi L}{4} = \frac{ma^2}{4}$$

$$I_x = I_{xy}+I_{xz} = \frac{ma^2}{4}+\frac{mL^2}{12} = \frac{m}{12}(3a^2+L^2)$$

$$I_y = I_{xy}+I_{yz} = \frac{ma^2}{4}+\frac{ma^2}{4} = \frac{ma^2}{2}$$

$$I_z = I_{xz}+I_{yz} = \frac{mL^2}{12}+\frac{ma^2}{4} = \frac{m}{12}(3a^2+L^2)$$

47. $\int_{-1}^{1}\int_{-1}^{1}\int_{0}^{1-x} (x^2+y^2)\sqrt{x^2+y^2+z^2}\,dz\,dy\,dx$

49. Because the density increases as you move away from the axis of symmetry, the moment of inertia will increase.

Section 13.7   Triple Integrals in Cylindrical and Spherical Coordinates

1. $\displaystyle\int_0^4\int_0^{\pi/2}\int_0^2 r\cos\theta\,dr\,d\theta\,dz = \int_0^4\int_0^{\pi/2}\left[\frac{r^2}{2}\cos\theta\right]_0^2 d\theta\,dz$

$\displaystyle= \int_0^4\int_0^{\pi/2} 2\cos\theta\,d\theta\,dz = \int_0^4\left[2\sin\theta\right]_0^{\pi/2} dz = \int_0^4 2\,dz = 8$

3. $\displaystyle\int_0^{\pi/2}\int_0^{2\cos^2\theta}\int_0^{4-r^2} r\sin\theta\,dz\,dr\,d\theta = \int_0^{\pi/2}\int_0^{2\cos^2\theta} r(4-r^2)\sin\theta\,dr\,d\theta = \int_0^{\pi/2}\left[\left(2r^2-\frac{r^4}{4}\right)\sin\theta\right]_0^{-\cos^2\theta} d\theta$

$\displaystyle= \int_0^{\pi/2}[8\cos^4\theta - 4\cos^8\theta]\sin\theta\,d\theta = \left[-\frac{8\cos^5\theta}{5}+\frac{4\cos^9\theta}{9}\right]_0^{\pi/2} = \frac{52}{45}$

5. $\displaystyle\int_0^{2\pi}\int_0^{\pi/4}\int_0^{\cos\phi} \rho^2\sin\phi\,d\rho\,d\phi\,d\theta = \frac{1}{3}\int_0^{2\pi}\int_0^{\pi/4}\cos^3\phi\sin\phi\,d\phi\,d\theta = -\frac{1}{12}\int_0^{2\pi}\left[\cos^4\phi\right]_0^{\pi/4} d\theta = \frac{\pi}{8}$

7. $\displaystyle\int_0^4\int_0^z\int_0^{\pi/2} re^r\,d\theta\,dr\,dz = \pi(e^4+3)$

9. $\displaystyle\int_0^{\pi/2}\int_0^3\int_0^{e^{-r^2}} r\,dz\,dr\,d\theta = \int_0^{\pi/2}\int_0^3 re^{-r^2}\,dr\,d\theta$

$\displaystyle= \int_0^{\pi/2}\left[-\frac{1}{2}e^{-r^2}\right]_0^3 d\theta$

$\displaystyle= \int_0^{\pi/2}\frac{1}{2}(1-e^{-9})\,d\theta$

$\displaystyle= \frac{\pi}{4}(1-e^{-9})$

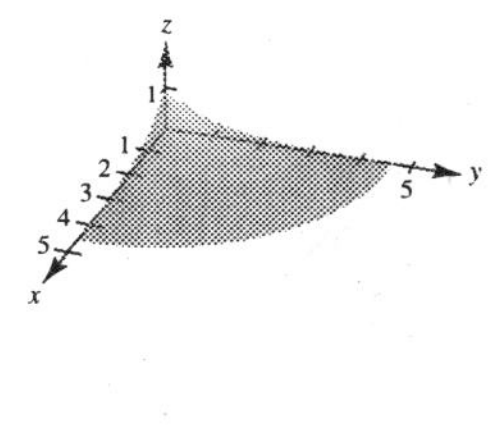

11. $\displaystyle\int_0^{2\pi}\int_{\pi/6}^{\pi/2}\int_0^4 \rho^2\sin\phi\,d\rho\,d\phi\,d\theta = \frac{64}{3}\int_0^{2\pi}\int_{\pi/6}^{\pi/2}\sin\phi\,d\phi\,d\theta$

$\displaystyle= \frac{64}{3}\int_0^{2\pi}\left[-\cos\phi\right]_{\pi/6}^{\pi/2} d\theta$

$\displaystyle= \frac{32\sqrt{3}}{3}\int_0^{2\pi} d\theta$

$\displaystyle= \frac{64\sqrt{3}\pi}{3}$

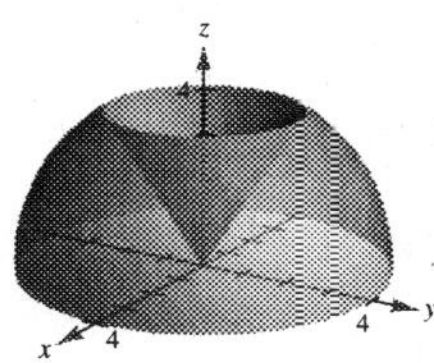

13. (a) $\displaystyle\int_0^{2\pi}\int_0^2\int_{r^2}^4 r^2\cos\theta\,dz\,dr\,d\theta = 0$

(b) $\displaystyle\int_0^{2\pi}\int_0^{\arctan(1/2)}\int_0^{4\sec\phi} \rho^3\sin^2\phi\cos\theta\,d\rho\,d\phi\,d\theta + \int_0^{2\pi}\int_{\arctan(1/2)}^{\pi/2}\int_0^{\cot\phi\csc\phi} \rho^3\sin^2\phi\cos\theta\,d\rho\,d\phi\,d\theta = 0$

15. (a) $\displaystyle\int_0^{2\pi}\int_0^a\int_a^{a+\sqrt{a^2-r^2}} r^2\cos\theta\,dz\,dr\,d\theta = 0$

(b) $\displaystyle\int_0^{\pi/4}\int_0^{2\pi}\int_{a\sec\phi}^{2a\cos\phi} \rho^3\sin^2\phi\cos\theta\,d\rho\,d\theta\,d\phi = 0$

17. $z = h - \dfrac{h}{r_0}\sqrt{x^2+y^2} = \dfrac{h}{r_0}(r_0 - r)$

$$V = 4\int_0^{\pi/2}\int_0^{r_0}\int_0^{h(r_0-r)/r_0} r\,dz\,dr\,d\theta$$
$$= \frac{4h}{r_0}\int_0^{\pi/2}\int_0^{r_0}(r_0r - r^2)\,dr\,d\theta$$
$$= \frac{4h}{r_0}\int_0^{\pi/2}\frac{r_0^3}{6}\,d\theta$$
$$= \frac{4h}{r_0}\left(\frac{r_0^3}{6}\right)\left(\frac{\pi}{2}\right) = \frac{1}{3}\pi r_0^2 h$$

19. $\rho = k\sqrt{x^2+y^2} = kr$

$\bar{x} = \bar{y} = 0$ by symmetry

$$m = 4k\int_0^{\pi/2}\int_0^{r_0}\int_0^{h(r_0-r)/r_0} r^2\,dz\,dr\,d\theta$$
$$= \frac{1}{6}k\pi r_0^3 h$$
$$M_{xy} = 4k\int_0^{\pi/2}\int_0^{r_0}\int_0^{h(r_0-r)/r_0} r^2 z\,dz\,dr\,d\theta$$
$$= \frac{1}{30}k\pi r_0^3 h^2$$
$$\bar{z} = \frac{M_{xy}}{m} = \frac{k\pi r_0^3h^2/30}{k\pi r_0^3h/6} = \frac{h}{5}$$

21. $\displaystyle I_z = 4k\int_0^{\pi/2}\int_0^{r_0}\int_0^{h(r_0-r)/r_0} r^3\,dz\,dr\,d\theta$

$$= \frac{4kh}{r_0}\int_0^{\pi/2}\int_0^{r_0}(r_0r^3 - r^4)\,dr\,d\theta$$
$$= \frac{4kh}{r_0}\left(\frac{r_0^5}{20}\right)\left(\frac{\pi}{2}\right)$$
$$= \frac{1}{10}k\pi r_0^4 h$$
$$= \left(\frac{1}{3}k\pi r_0^2 h\right)\left(\frac{3}{10}r_0^2\right)$$
$$= \frac{3}{10}mr_0^2$$

23. $m = k(\pi b^2h - \pi a^2h) = k\pi h(b^2 - a^2)$

$$I_z = 4k\int_0^{\pi/2}\int_a^b\int_0^h r^3\,dz\,dr\,d\theta$$
$$= 4kh\int_0^{\pi/2}\int_a^b r^3\,dr\,d\theta$$
$$= kh\int_0^{\pi/2}(b^4 - a^4)\,d\theta$$
$$= \frac{k\pi(b^4 - a^4)h}{2}$$
$$= \frac{k\pi(b^2-a^2)(b^2+a^2)}{2}$$
$$= \frac{1}{2}m(a^2 + b^2)$$

25. $\displaystyle V = 4\int_0^{\pi/2}\int_0^{a\cos\theta}\int_0^{\sqrt{a^2-r^2}} r\,dz\,dr\,d\theta$

$$= 4\int_0^{\pi/2}\int_0^{a\cos\theta} r\sqrt{a^2-r^2}\,dr\,d\theta$$
$$= \frac{4}{3}a^3\int_0^{\pi/2}(1 - \sin^3\theta)\,d\theta$$
$$= \frac{4}{3}a^3\left[\theta + \frac{1}{3}\cos\theta(\sin^2\theta + 2)\right]_0^{\pi/2}$$
$$= \frac{4}{3}a^3\left(\frac{\pi}{2} - \frac{2}{3}\right)$$

27. $V = 2\int_0^{\pi}\int_0^{a\cos\theta}\int_0^{\sqrt{a^2-r^2}} r\,dz\,dr\,d\theta$

$= 2\int_0^{\pi}\int_0^{a\cos\theta} r\sqrt{a^2-r^2}\,dr\,d\theta$

$= 2\int_0^{\pi}\left[-\frac{1}{3}(a^2-r^2)^{3/2}\right]_0^{a\cos\theta} d\theta$

$= \frac{2a^3}{3}\int_0^{\pi}(1-\sin^3\theta)\,d\theta$

$= \frac{2a^3}{3}\left[\theta + \cos\theta - \frac{\cos^3\theta}{3}\right]_0^{\pi} = \frac{2a^3}{9}(3\pi - 4)$

29. $V = \int_0^{2\pi}\int_0^{\pi}\int_0^{4\sin\phi} \rho^2\sin\phi\,d\rho\,d\phi\,d\theta = 16\pi^2$

31. $m = 8k\int_0^{\pi/2}\int_0^{\pi/2}\int_0^{a} \rho^3\sin\phi\,d\rho\,d\theta\,d\phi$

$= 2ka^4\int_0^{\pi/2}\int_0^{\pi/2}\sin\phi\,d\theta\,d\phi$

$= k\pi a^4\int_0^{\pi/2}\sin\phi\,d\phi$

$= \left[k\pi a^4(-\cos\phi)\right]_0^{\pi/2}$

$= k\pi a^4$

33. $m = \frac{2}{3}k\pi r^3$

$\bar{x} = \bar{y} = 0$ by symmetry

$M_{xy} = 4k\int_0^{\pi/2}\int_0^{\pi/2}\int_0^{r} \rho^3\cos\phi\sin\phi\,d\rho\,d\theta\,d\phi$

$= \frac{1}{2}kr^4\int_0^{\pi/2}\int_0^{\pi/2}\sin 2\phi\,d\theta\,d\phi$

$= \frac{kr^4\pi}{4}\int_0^{\pi/2}\sin 2\phi\,d\phi$

$= \left[-\frac{1}{8}k\pi r^4\cos 2\phi\right]_0^{\pi/2} = \frac{1}{4}k\pi r^4$

$\bar{z} = \frac{M_{xy}}{m} = \frac{k\pi r^4/4}{2k\pi r^3/3} = \frac{3r}{8}$

35. $I_z = 4k\int_{\pi/4}^{\pi/2}\int_0^{\pi/2}\int_0^{\cos\phi} \rho^4\sin^3\phi\,d\rho\,d\theta\,d\phi$

$= \frac{4}{5}k\int_{\pi/4}^{\pi/2}\int_0^{\pi/2}\cos^5\phi\sin^3\phi\,d\theta\,d\phi$

$= \frac{2}{5}k\pi\int_{\pi/4}^{\pi/2}\cos^5\phi(1-\cos^2\phi)\sin\phi\,d\phi$

$= \left[\frac{2}{5}k\pi\left(-\frac{1}{6}\cos^6\phi + \frac{1}{8}\cos^8\phi\right)\right]_{\pi/4}^{\pi/2} = \frac{k\pi}{192}$

37. (a) $r = r_0$: right circular cylinder about z-axis

$\theta = \theta_0$: plane parallel to z-axis

$z = z_0$: plane parallel to xy-plane

(b) $\rho = \rho_0$: sphere of radius ρ_0

$\theta = \theta_0$: plane parallel to z-axis

$\phi = \phi_0$: cone

39. $16\int_0^{a}\int_0^{\sqrt{a^2-x^2}}\int_0^{\sqrt{a^2-x^2-y^2}}\int_0^{\sqrt{a^2-x^2-y^2-z^2}} dw\,dz\,dy\,dx$

$= 16\int_0^{a}\int_0^{\sqrt{a^2-x^2}}\int_0^{\sqrt{a^2-x^2-y^2}} \sqrt{a^2-x^2-y^2-z^2}\,dz\,dy\,dx$

$= 16\int_0^{\pi/2}\int_0^{a}\int_0^{\sqrt{a^2-r^2}} \sqrt{(a^2-r^2)-z^2}\,dz(r\,dr\,d\theta)$

$= 16\int_0^{\pi/2}\int_0^{a}\frac{1}{2}\left[z\sqrt{(a^2-r^2)-z^2} + (a^2-r^2)\arcsin\frac{z}{\sqrt{a^2-r^2}}\right]_0^{\sqrt{a^2-r^2}} r\,dr\,d\theta$

$= 8\int_0^{\pi/2}\int_0^{a}\frac{\pi}{2}(a^2-r^2)r\,dr\,d\theta$

$= 4\pi\int_0^{\pi/2}\left[\frac{a^2r^2}{2} - \frac{r^4}{4}\right]_0^{a} d\theta = a^4\pi\int_0^{\pi/2} d\theta = \frac{a^4\pi^2}{2}$

Section 13.8 Change of Variables: Jacobians

1. $x = -\frac{1}{2}(u - v)$

$y = \frac{1}{2}(u + v)$

$$\frac{\partial x}{\partial u}\frac{\partial y}{\partial v} - \frac{\partial y}{\partial u}\frac{\partial x}{\partial v} = \left(-\frac{1}{2}\right)\left(\frac{1}{2}\right) - \left(\frac{1}{2}\right)\left(\frac{1}{2}\right)$$

$$= -\frac{1}{2}$$

3. $x = u - v^2$

$y = u + v$

$$\frac{\partial x}{\partial u}\frac{\partial y}{\partial v} - \frac{\partial y}{\partial u}\frac{\partial x}{\partial v} = (1)(1) - (1)(-2v) = 1 + 2v$$

5. $x = u \cos\theta - v \sin\theta$

$y = u \sin\theta + v \cos\theta$

$$\frac{\partial x}{\partial u}\frac{\partial y}{\partial v} - \frac{\partial y}{\partial u}\frac{\partial x}{\partial v} = \cos^2\theta + \sin^2\theta = 1$$

7. $x = e^u \sin v$

$y = e^u \cos v$

$$\frac{\partial x}{\partial u}\frac{\partial y}{\partial v} - \frac{\partial y}{\partial u}\frac{\partial x}{\partial v} = (e^u \sin v)(-e^u \sin v) - (e^u \cos v)(e^u \cos v) = -e^{2u}$$

9. $x = 3u + 2v$

$y = 3v$

$v = \frac{y}{3}$

$$u = \frac{x - 2v}{3} = \frac{x - 2(y/3)}{3} = \frac{x}{3} - \frac{2y}{9}$$

(x, y)	(u, v)
$(0, 0)$	$(0, 0)$
$(3, 0)$	$(1, 0)$
$(2, 3)$	$(0, 1)$

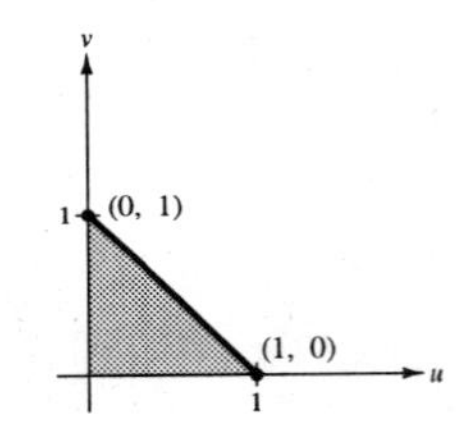

11. $x = \frac{1}{2}(u + v)$

$y = \frac{1}{2}(u - v)$

$$\frac{\partial x}{\partial u}\frac{\partial y}{\partial v} - \frac{\partial y}{\partial u}\frac{\partial x}{\partial v} = \left(\frac{1}{2}\right)\left(-\frac{1}{2}\right) - \left(\frac{1}{2}\right)\left(\frac{1}{2}\right) = -\frac{1}{2}$$

$$\int_R\int 4(x^2 + y^2)\,dx\,dy = \int_{-1}^{1}\int_{-1}^{1} 4\left[\frac{1}{4}(u + v)^2 + \frac{1}{4}(u - v)^2\right]\left(\frac{1}{2}\right) dv\,du$$

$$= \int_{-1}^{1}\int_{-1}^{1} (u^2 + v^2)\,dv\,du = \int_{-1}^{1} 2\left(u^2 + \frac{1}{3}\right) du = \left[2\left(\frac{u^3}{3} + \frac{u}{3}\right)\right]_{-1}^{1} = \frac{8}{3}$$

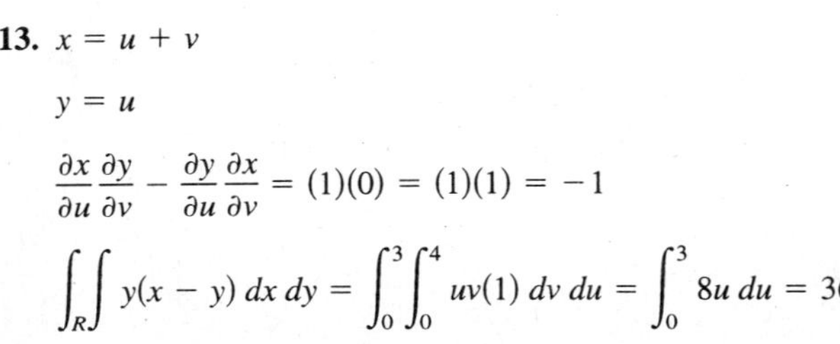

13. $x = u + v$

$y = u$

$$\frac{\partial x}{\partial u}\frac{\partial y}{\partial v} - \frac{\partial y}{\partial u}\frac{\partial x}{\partial v} = (1)(0) = (1)(1) = -1$$

$$\int_R\int y(x - y)\,dx\,dy = \int_0^3\int_0^4 uv(1)\,dv\,du = \int_0^3 8u\,du = 36$$

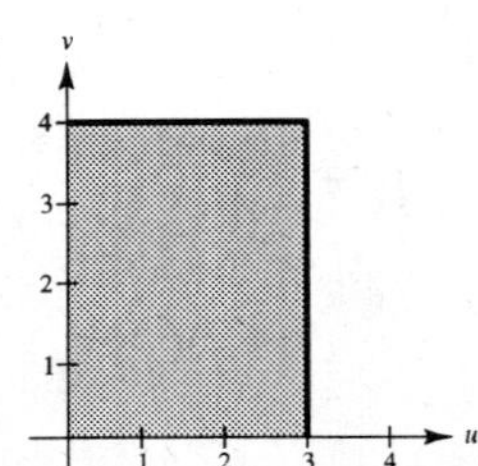

15. $\displaystyle\iint_R e^{-xy/2}\,dA$

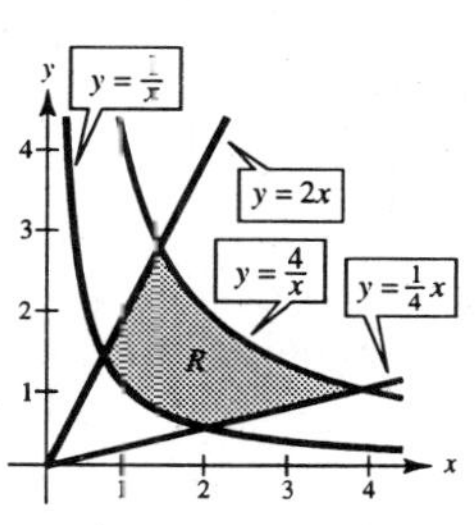

R: $y = \dfrac{x}{4}, y = 2x, y = \dfrac{1}{x}, y = \dfrac{4}{x}$

$x = \sqrt{v/u}, y = \sqrt{uv} \Rightarrow u = \dfrac{y}{x}, \; v = xy$

$$\frac{\partial(x, y)}{\partial(u, v)} = \begin{vmatrix} \dfrac{\partial x}{\partial u} & \dfrac{\partial x}{\partial v} \\ \dfrac{\partial y}{\partial u} & \dfrac{\partial y}{\partial v} \end{vmatrix} = \begin{vmatrix} -\dfrac{1}{2}\dfrac{v^{1/2}}{u^{3/2}} & \dfrac{1}{2}\dfrac{1}{u^{1/2}v^{1/2}} \\ \dfrac{1}{2}\dfrac{v^{1/2}}{u^{1/2}} & \dfrac{1}{2}\dfrac{u^{1/2}}{v^{1/2}} \end{vmatrix} = -\frac{1}{4}\left(\frac{1}{u} + \frac{1}{u}\right) = -\frac{1}{2u}$$

Transformed Region:

$y = \dfrac{1}{x} \Rightarrow yx = 1 \Rightarrow v = 1$

$y = \dfrac{4}{x} \Rightarrow yx = 4 \Rightarrow v = 4$

$y = 2x \Rightarrow \dfrac{y}{x} = 2 \Rightarrow u = 2$

$y = \dfrac{x}{4} \Rightarrow \dfrac{y}{x} = \dfrac{1}{4} \Rightarrow u = \dfrac{1}{4}$

$$\iint_R e^{-xy/2}\,dA = \int_{1/2}^{2}\int_{1}^{4} e^{-v/2}\left(\frac{1}{2u}\right)dv\,du = \int_{1/2}^{2}\left[\frac{-1}{u}e^{-v/2}\right]_1^4 du$$

$$= \int_{1/2}^{2} -\frac{1}{u}(e^{-2} - e^{-1/2})\,du = -(e^{-2} - e^{-1/2})\left(\ln 2 - \ln\frac{1}{2}\right) = (e^{-1/2} - e^{-2})\ln 4 \approx 0.6532$$

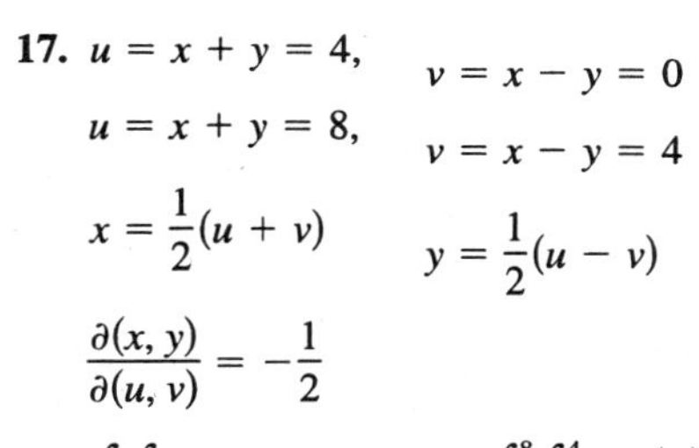

17. $u = x + y = 4, \quad v = x - y = 0$

$u = x + y = 8, \quad v = x - y = 4$

$x = \dfrac{1}{2}(u + v) \quad y = \dfrac{1}{2}(u - v)$

$\dfrac{\partial(x, y)}{\partial(u, v)} = -\dfrac{1}{2}$

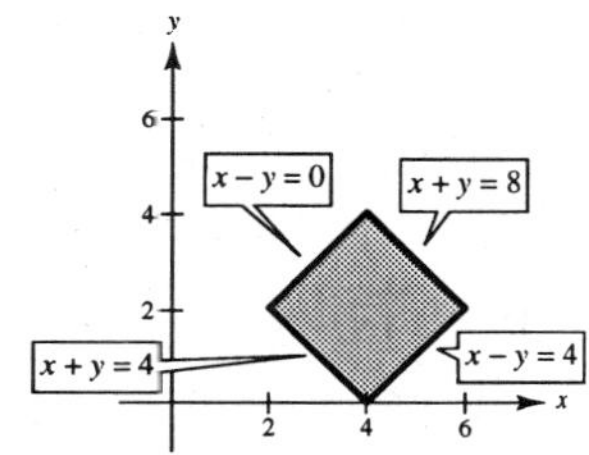

$$\iint_R (x + y)e^{x-y}\,dA = \int_4^8\int_0^4 ue^v\left(\frac{1}{2}\right)dv\,du$$

$$= \frac{1}{2}\int_4^8 u(e^4 - 1)\,du = \left[\frac{1}{4}u^2(e^4 - 1)\right]_4^8 = 12(e^4 - 1)$$

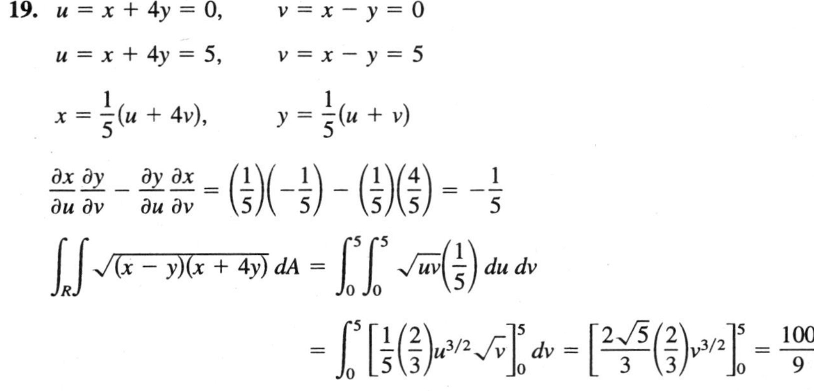

19. $u = x + 4y = 0, \quad v = x - y = 0$

$u = x + 4y = 5, \quad v = x - y = 5$

$x = \dfrac{1}{5}(u + 4v), \quad y = \dfrac{1}{5}(u + v)$

$\dfrac{\partial x}{\partial u}\dfrac{\partial y}{\partial v} - \dfrac{\partial y}{\partial u}\dfrac{\partial x}{\partial v} = \left(\dfrac{1}{5}\right)\left(-\dfrac{1}{5}\right) - \left(\dfrac{1}{5}\right)\left(\dfrac{4}{5}\right) = -\dfrac{1}{5}$

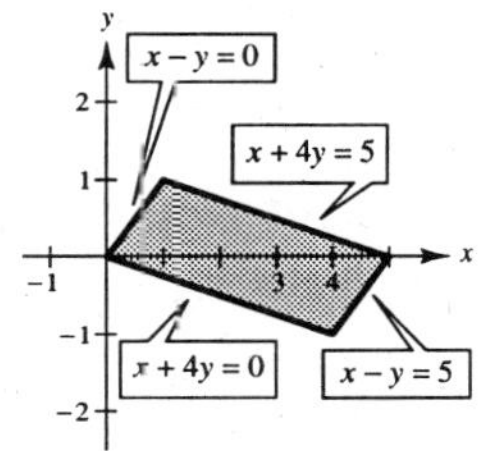

$$\iint_R \sqrt{(x - y)(x + 4y)}\,dA = \int_0^5\int_0^5 \sqrt{uv}\left(\frac{1}{5}\right)du\,dv$$

$$= \int_0^5\left[\frac{1}{5}\left(\frac{2}{3}\right)u^{3/2}\sqrt{v}\right]_0^5 dv = \left[\frac{2\sqrt{5}}{3}\left(\frac{2}{3}\right)v^{3/2}\right]_0^5 = \frac{100}{9}$$

21. $u = x + y, v = x - y, x = \frac{1}{2}(u + v), y = \frac{1}{2}(u - v)$

$$\frac{\partial x}{\partial u}\frac{\partial y}{\partial v} - \frac{\partial y}{\partial u}\frac{\partial x}{\partial v} = -\frac{1}{2}$$

$$\iint_R \sqrt{x + y}\, dA = \int_0^a \int_{-u}^u \sqrt{u}\left(\frac{1}{2}\right) dv\, du$$

$$= \int_0^a u\sqrt{u}\, du = \left[\frac{2}{5}u^{5/2}\right]_0^a = \frac{2}{5}a^{5/2}$$

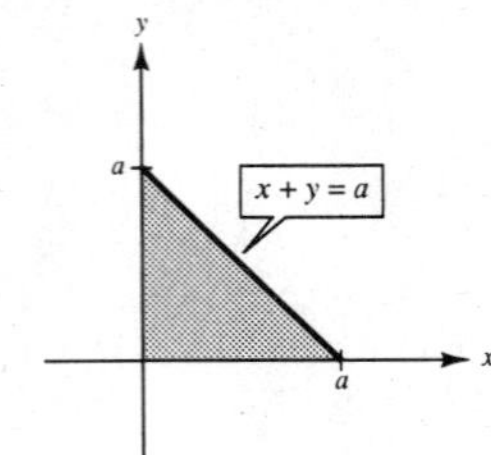

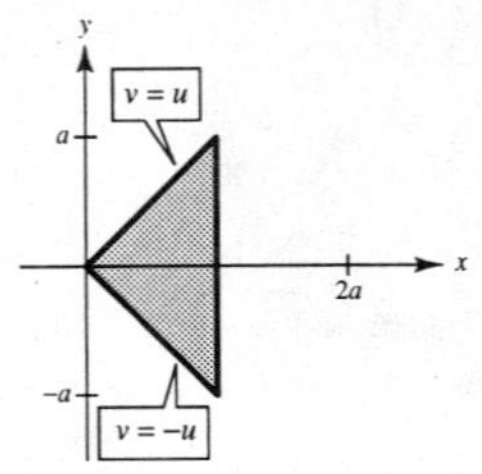

23. $\frac{x^2}{a^2} + \frac{y^2}{b^2} = 1, x = au, y = bv$

$$\frac{(au)^2}{a^2} + \frac{(bv)^2}{b^2} = 1$$

$$u^2 + v^2 = 1$$

(a) $\frac{x^2}{a^2} + \frac{y^2}{b^2} = 1$ $\qquad u^2 + v^2 = 1$

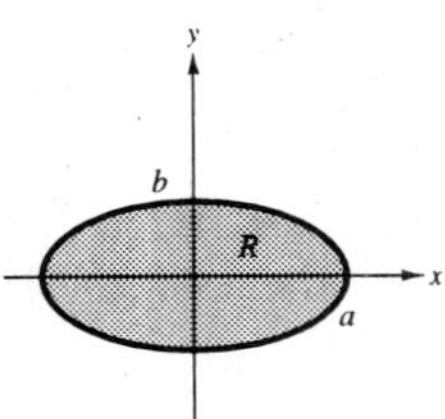

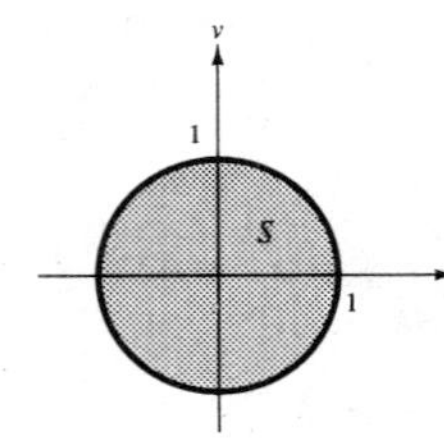

(b) $\frac{\partial(x, y)}{\partial(u, v)} = \frac{\partial x}{\partial u}\frac{\partial y}{\partial v} - \frac{\partial y}{\partial u}\frac{\partial x}{\partial v}$

$$= (a)(b) - (0)(0) = ab$$

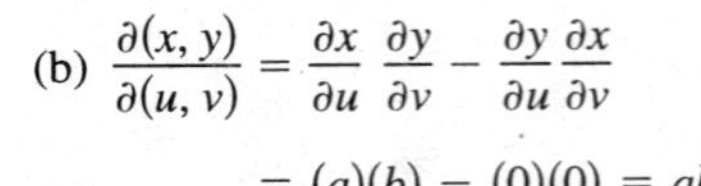

(c) $A = \iint_S ab\, du\, dv$

$$= ab(\pi(1)^2) = \pi ab$$

25. $x = u(1 - v),\ y = uv(1 - w),\ z = uvw$

$$\frac{\partial(x, y, z)}{\partial(u, v, w)} = \begin{vmatrix} 1 - v & -u & 0 \\ v(1 - w) & u(1 - w) & -uv \\ vw & uw & uv \end{vmatrix} = 1(1 - v)[u^2v(1 - w) + u^2vw] + u[uv^2(1 - w) + uv^2w]$$

$$= (1 - v)(u^2v) + u(uv^2) = u^2v$$

27. $x = \rho \sin \phi \cos \theta,\ y = \rho \sin \sin \theta,\ z = \rho \cos \phi$

$$\frac{\partial(x, y, z)}{\partial(\rho, \theta, \phi)} = \begin{vmatrix} \sin \phi \cos \theta & -\rho \sin \phi \sin \theta & \rho \cos \phi\, \cos \theta \\ \sin \phi \sin \theta & \rho \sin \phi \cos \theta & \rho \cos \phi \sin \theta \\ \cos \phi & 0 & -\rho \sin \phi \end{vmatrix}$$

$$= \cos \phi[-\rho^2 \sin \phi \cos \phi \sin^2 \theta - \rho^2 \sin \phi \cos \phi \cos^2 \theta] - \rho \sin \phi[\rho \sin^2 \phi \cos^2\theta + \rho \sin^2 \phi \sin^2 \theta]$$

$$= \cos \phi[-\rho^2 \sin \phi \cos \phi(\sin^2 \theta + \cos^2 \theta)] - \rho \sin \phi[\rho \sin^2 \phi(\cos^2 \theta + \sin^2 \theta)]$$

$$= -\rho^2 \sin \phi \cos^2 \phi - \rho^2 \sin^3 \phi$$

$$= -\rho^2 \sin \phi(\cos^2 \phi + \sin^2 \theta) = -\rho^2 \sin \phi$$

Review Exercises for Chapter 13

1. $\int_1^{x^2} x \ln y\, dy = \left[xy(-1 + \ln y)\right]_0^{x^2}$

$$= x^3(-1 + \ln x^2) + x$$

$$= x - x^3 + x^3 \ln x^2$$

3. $\int_0^1 \int_0^{1+x} (3x + 2y)\, dy\, dx = \int_0^1 \left[3xy + y^2\right]_0^{1+x} dx$

$$= \int_0^1 (4x^2 + 5x + 1)\, dx$$

$$= \left[\frac{4}{3}x^3 + \frac{5}{2}x^2 + x\right]_0^1 = \frac{29}{6}$$

5. $\displaystyle\int_0^3\int_0^{\sqrt{9-x^2}} 4x\,dy\,dx = \int_0^3 4x\sqrt{9-x^2}\,dx = \left[-\frac{4}{3}(9-x^2)^{3/2}\right]_0^3 = 36$

7. $\displaystyle\int_0^h\int_0^x \sqrt{x^2+y^2}\,dy\,dx = \int_0^{\pi/4}\int_0^{h\sec\theta} r^2\,dr\,d\theta = \frac{h^3}{3}\int_0^{\pi/4}\sec^3\theta\,d\theta$

$$= \frac{h^3}{6}\Big[\sec\theta\tan\theta + \ln|\sec\theta+\tan\theta|\Big]_0^{\pi/4} = \frac{h^3}{6}\left[\sqrt{2} + \ln\left(\sqrt{2}+1\right)\right]$$

9. $\displaystyle\int_{-3}^3\int_{-\sqrt{9-x^2}}^{\sqrt{9-x^2}}\int_{x^2+y^2}^9 \sqrt{x^2+y^2}\,dz\,dy\,dx = \int_0^{2\pi}\int_0^3\int_{r^2}^9 r^2\,dz\,dr\,d\theta$

$$= \int_0^{2\pi}\int_0^3 (9r^2 - r^4)\,dr\,d\theta = \int_0^{2\pi}\left[3r^3 - \frac{r^5}{5}\right]_0^3 d\theta = \frac{162}{5}\int_0^{2\pi} d\theta = \frac{324\pi}{5}$$

11. $\displaystyle\int_0^a\int_0^b\int_0^c (x^2+y^2+z^2)\,dx\,dy\,dz = \int_0^a\int_0^b\left(\frac{1}{3}c^3 + cy^2 + cz^2\right)dy\,dz$

$$= \int_0^a\left(\frac{1}{3}bc^3 + \frac{1}{3}b^3c + bcz^2\right)dz = \frac{1}{3}abc^3 + \frac{1}{3}ab^3c + \frac{1}{3}a^3bc = \frac{1}{3}abc(a^2+b^2+c^2)$$

13. $\displaystyle\int_{-2}^4\int_{y^2/4}^{(4+y)/2} (x-y)\,dx\,dy = \frac{27}{5}$

15. $\displaystyle\int_{-\sqrt{1-x^2}}^{\sqrt{1-x^2}}\int_{-\sqrt{1-x^2-y^2}}^{\sqrt{1-x^2-y^2}} (x^2+y^2)\,dz\,dy\,dx = \int_0^{2\pi}\int_0^1\int_{-\sqrt{1-r^2}}^{\sqrt{1-r^2}} r^3\,dz\,dr\,d\theta = \frac{8\pi}{15}$

17. $\displaystyle\int_0^3\int_0^{(3-x)/3} dy\,dx = \int_0^1\int_0^{3-3y} dx\,dy$

$$A = \int_0^1\int_0^{3-3y} dx\,dy = \int_0^1 (3-3y)\,dy = \left[3y - \frac{3}{2}y^2\right]_0^1 = \frac{3}{2}$$

19. $\displaystyle\int_{-5}^3\int_{-\sqrt{25-x^2}}^{\sqrt{25-x^2}} dy\,dx = \int_{-5}^{-4}\int_{-\sqrt{25-y^2}}^{\sqrt{25-y^2}} dx\,dy + \int_{-4}^4\int_{-\sqrt{25-y^2}}^3 dx\,dy + \int_4^5\int_{-\sqrt{25-y^2}}^{\sqrt{25-y^2}} dx\,dy$

$$A = 2\int_{-5}^3\int_0^{\sqrt{25-x^2}} dy\,dx = 2\int_{-5}^3 \sqrt{25-x^2}\,dx = \left[x\sqrt{25-x^2} + 25\arcsin\frac{x}{5}\right]_{-5}^3 = \frac{25\pi}{2} + 12 + 25\arcsin\frac{3}{5} \approx 67.36$$

21. $\displaystyle A = 4\int_0^1\int_0^{x\sqrt{1-x^2}} dy\,dx = 4\int_0^1 x\sqrt{1-x^2}\,dx = \left[-\frac{4}{3}(1-x^2)^{3/2}\right]_0^1 = \frac{4}{3}$

$$A = 4\int_0^{1/2}\int_{\sqrt{(1-\sqrt{1-4y^2})/2}}^{\sqrt{(1+\sqrt{1-4y^2})/2}} dx\,dy$$

23. $\displaystyle A = \int_2^5\int_{x-3}^{\sqrt{x-1}} dy\,dx + 2\int_1^2\int_0^{\sqrt{x-1}} dy\,dx = \int_{-1}^2\int_{y^2+1}^{y+3} dx\,dy = \frac{9}{2}$

25. Both integrations are over the common region R shown in the figure. Analytically,

$$\int_0^1 \int_{2y}^{2\sqrt{2-y^2}} (x + y)\, dx\, dy = \frac{4}{3} + \frac{4}{3}\sqrt{2}$$

$$\int_0^2 \int_0^{x/2} (x + y)\, dy\, dx + \int_2^{2\sqrt{2}} \int_0^{\sqrt{8-x^2}/2} (x + y)\, dy\, dx = \frac{5}{3} + \left(\frac{4}{3}\sqrt{2} - \frac{1}{3}\right) = \frac{4}{3} + \frac{4}{3}\sqrt{2}$$

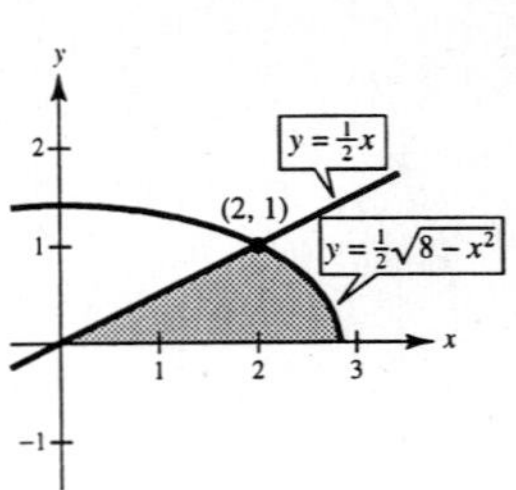

27. $V = \int_0^4 \int_0^{x^2+4} (x^2 - y + 4)\, dy\, dx$

$$= \int_0^4 \left[x^2y - \frac{1}{2}y^2 + 4y\right]_0^{x^2+4} dx$$

$$= \int_0^4 \left(\frac{1}{2}x^4 + 4x^2 + 8\right) dx$$

$$= \left[\frac{1}{10}x^5 + \frac{4}{3}x^3 + 8x\right]_0^4 = \frac{3296}{15}$$

29. $V = 4\int_0^h \int_0^{\pi/2} \int_1^{\sqrt{1+z^2}} r\, dr\, d\theta\, dz$

$$= 2\int_0^h \int_0^{\pi/2} (1 + z^2 - 1)\, d\theta\, dz$$

$$= \pi \int_0^h z^2\, dz$$

$$= \left[\pi\left(\frac{1}{3}z^3\right)\right]_0^h = \frac{\pi h^3}{3}$$

31. $V = 4\int_0^{\pi/2} \int_0^{2\cos\theta} \int_0^{\sqrt{4-r^2}} r\, dz\, dr\, d\theta$

$$= 4\int_0^{\pi/2} \int_0^{2\cos\theta} r\sqrt{4 - r^2}\, dr\, d\theta$$

$$= -\int_0^{\pi/2} \left[\frac{4}{3}(4 - r^2)^{3/2}\right]_0^{2\cos\theta} d\theta$$

$$= \frac{32}{3}\int_0^{\pi/2} (1 - \sin^3\theta)\, d\theta$$

$$= \frac{32}{3}\left[\theta + \cos\theta - \frac{1}{3}\cos^3\theta\right]_0^{\pi/2} = \frac{32}{3}\left(\frac{\pi}{2} - \frac{2}{3}\right)$$

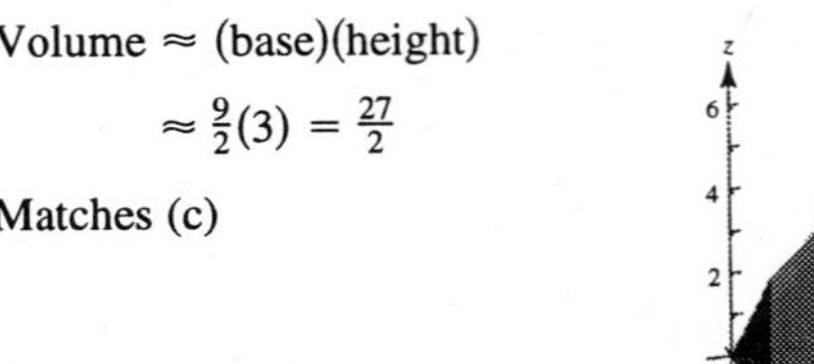

33. Volume $\approx$ (base)(height)

$$\approx \tfrac{9}{2}(3) = \tfrac{27}{2}$$

Matches (c)

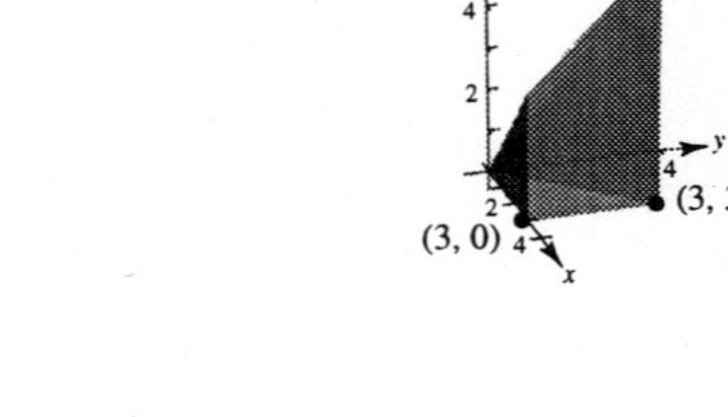

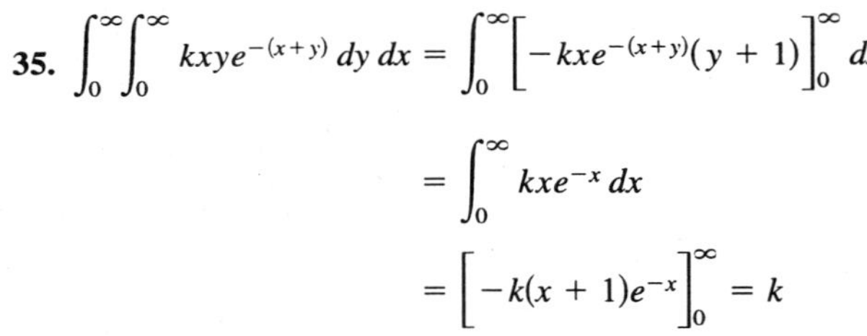

35. $\int_0^\infty \int_0^\infty kxye^{-(x+y)}\, dy\, dx = \int_0^\infty \left[-kxe^{-(x+y)}(y + 1)\right]_0^\infty dx$

$$= \int_0^\infty kxe^{-x}\, dx$$

$$= \left[-k(x + 1)e^{-x}\right]_0^\infty = k$$

Therefore, $k = 1$.

$$P = \int_0^1 \int_0^1 xye^{-(x+y)}\, dy\, dx \approx 0.070$$

37. (a) $(x^2 + y^2) = 9(x^2 - y^2)$

$$(r^2)^2 = 9(r^2\cos^2\theta - r^2\sin^2\theta)$$

$$r^2 = 9(\cos^2\theta - \sin^2\theta) = 9\cos 2\theta$$

$$r = 3\sqrt{\cos 2\theta}$$

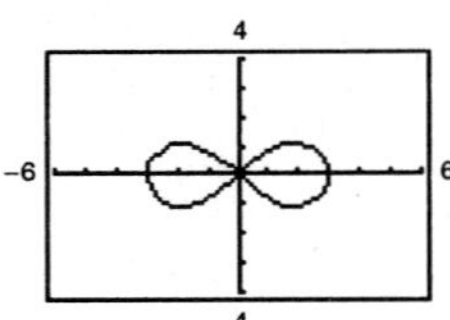

(b) $A = 4\int_0^{\pi/4} \int_0^{3\sqrt{\cos 2\theta}} r\, dr\, d\theta = 9$

(c) $V = 4\int_0^{\pi/4} \int_0^{3\sqrt{\cos 2\theta}} \sqrt{9 - r^2}\, r\, dr\, d\theta \approx 20.392$

39. (a) $m = k\int_0^1\int_{2x^3}^{2x} xy\,dy\,dx = \frac{k}{4}$

$M_x = k\int_0^1\int_{2x^3}^{2x} xy^2\,dy\,dx = \frac{16k}{55}$

$M_y = k\int_0^1\int_{2x^3}^{2x} x^2y\,dy\,dx = \frac{8k}{45}$

$\bar{x} = \frac{M_y}{m} = \frac{32}{45}$

$\bar{y} = \frac{M_x}{m} = \frac{64}{55}$

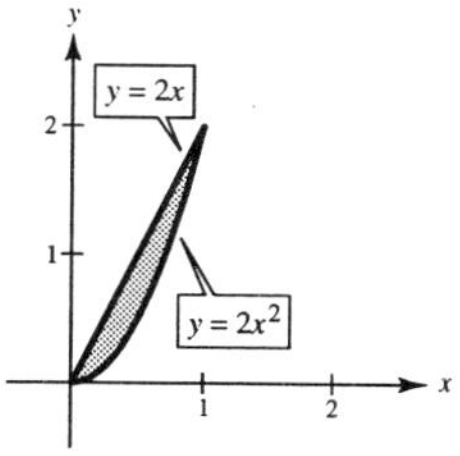

(b) $m = k\int_0^1\int_{2x^3}^{2x} (x^2 + y^2)dy\,dx = \frac{17k}{30}$

$M_x = k\int_0^1\int_{2x^3}^{2x} y(x^2 + y^2)dy\,dx = \frac{392k}{585}$

$M_y = k\int_0^1\int_{2x^3}^{2x} x(x^2 + y^2)dy\,dx = \frac{156k}{385}$

$\bar{x} = \frac{M_y}{m} = \frac{936}{1309}$

$\bar{y} = \frac{M_x}{m} = \frac{784}{663}$

41. $S = \int_R\int \sqrt{1 + (f_x)^2 + (f_y)^2}\,dA$

$= 4\int_0^4\int_0^{\sqrt{16-x^2}} \sqrt{1 + 4x^2 + 4y^2}\,dy\,dx$

$= 4\int_0^{\pi/2}\int_0^4 \sqrt{1 + 4r^2}\,r\,dr\,d\theta$

$= \left[\frac{1}{3}(65^{3/2} - 1)\theta\right]_0^{\pi/2} = \frac{\pi}{6}\left(65\sqrt{65} - 1\right)$

43. $f(x, y) = 9 - y^2$

$f_x = 0,\ f_y = -2y$

$S = \int_R\int \sqrt{1 + f_x^2 + f_y^2}\,dA$

$= \int_0^3\int_{-y}^{y} \sqrt{1 + 4y^2}\,dx\,dy$

$= \int_0^3\left[\sqrt{1 + 4y^2}\,x\right]_{-y}^{y} dy$

$= \int_0^3 2\sqrt{1 + 4y^2}\,dy = \frac{1}{4}\frac{2}{3}(1 + 4y^2)\Big]_0^3 = \frac{1}{6}[(37)^{3/2} - 1]$

45. $m = 4k\int_{\pi/4}^{\pi/2}\int_0^{\pi/2}\int_0^{\cos\phi} \rho^2\sin\phi\,d\rho\,d\theta\,d\phi$

$= \frac{4}{3}k\int_{\pi/4}^{\pi/2}\int_0^{\pi/2} \cos^3\phi\sin\phi\,d\theta\,d\phi = \frac{2}{3}k\pi\int_{\pi/4}^{\pi/2}\cos^3\phi\sin\phi\,d\phi = \left[-\frac{2}{3}k\pi\left(\frac{1}{4}\cos^4\phi\right)\right]_{\pi/4}^{\pi/2} = \frac{k\pi}{24}$

$M_{xy} = 4k\int_{\pi/4}^{\pi/2}\int_0^{\pi/2}\int_0^{\cos\phi} \rho^3\cos\phi\,d\rho\sin d\rho\,d\theta\,d\phi$

$= k\int_{\pi/4}^{\pi/2}\int_0^{\pi/2} \cos^5\phi\sin\phi\,d\rho\,d\theta\,d\phi = \frac{1}{2}k\pi\int_{\pi/4}^{\pi/2}\cos^5\phi\sin\phi\,d\phi = \left[-\frac{1}{12}k\pi\cos^6\phi\right]_{\pi/4}^{\pi/2} = \frac{k\pi}{96}$

$\bar{z} = \frac{M_{xy}}{m} = \frac{k\pi/96}{k\pi/24} = \frac{1}{4}$

$\bar{x} = \bar{y} = 0$ by symmetry

47. $m = k\int_0^{\pi/2}\int_0^{\pi/2}\int_0^a \rho^2\sin\phi\,d\rho\,d\theta\,d\phi = \frac{k\pi a^3}{6}$

$M_{xy} = k\int_0^{\pi/2}\int_0^{\pi/2}\int_0^a (\rho\cos\phi)\rho^2\sin\phi\,d\rho\,d\theta\,d\phi = \frac{k\pi a^4}{16}$

$\bar{x} = \bar{y} = \bar{z} = \frac{M_{xy}}{m} = \frac{k\pi a^4}{16}\left(\frac{6}{k\pi a^3}\right) = \frac{3a}{8}$

49. $I_z = 4k\int_0^{\pi/2}\int_3^4\int_0^{16-r^2} r^3\,dz\,dr\,d\theta$

$= 4k\int_0^{\pi/2}\int_3^4 (16r^3 - r^5)\,dr\,d\theta = \frac{833\pi k}{3}$

51. $z = f(x, y) = \sqrt{a^2 - x^2 - y^2}$

$= \sqrt{a^2 - r^2}$

$0 \le r \le \sqrt{2ah - h^2}$

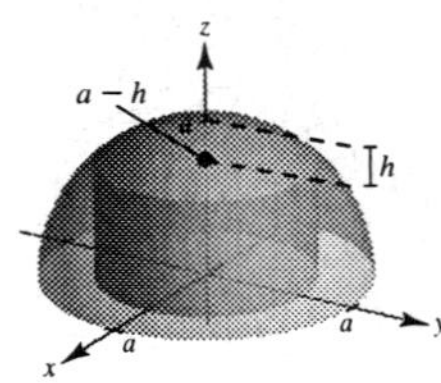

(a) Disc Method

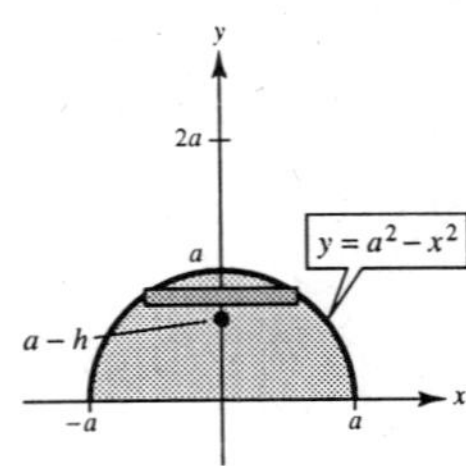

$$V = \pi\int_{a-h}^{a} (a^2 - y^2)dy$$

$$= \pi\left[a^2y - \frac{y^3}{3}\right]_{a-h}^{a} = \pi\left[\left(a^3 - \frac{a^3}{3}\right) - \left(a^2(a-h) - \frac{(a-h)^3}{3}\right)\right]$$

$$= \pi\left[a^3 - \frac{a^3}{3} - a^3 + a^2h + \frac{a^3}{3} - a^2h + ah^2 - \frac{h^3}{3}\right] = \pi\left[ah^2 - \frac{h^3}{3}\right] = \frac{1}{3}\pi h^2[3a - h]$$

Equivalently, use spherical coordinates

$$V = \int_0^{2\pi}\int_0^{\cos^{-1}(a-h/a)}\int_{(a-h)\sec\phi}^{a} \rho^2 \sin\phi \, d\rho \, d\phi \, d\theta$$

(b) $$M_{xy} = \int_0^{2\pi}\int_0^{\cos^{-1}(a-h/a)}\int_{(a-h)\sec\phi}^{a} (\rho\cos\phi)\rho^2 \sin\phi \, d\rho \, d\phi \, d\theta = \frac{1}{4}h^2\pi(2a-h)^2$$

$$\bar{z} = \frac{M_{xy}}{V} = \frac{\frac{1}{4}h^2\pi(2a-h)^2}{\frac{1}{3}h^2\pi(3a-h)} = \frac{3}{4}\frac{(2a-h)^2}{3a-h}$$

centroid: $\left(0, 0, \dfrac{3(2a-h)^2}{4(3a-h)}\right)$

(c) If $h = a$, $\bar{z} = \dfrac{3(a)^2}{4(2a)} = \dfrac{3}{8}a$

centroid of hemisphere: $\left(0, 0, \dfrac{3}{8}a\right)$

(d) $$\lim_{h\to 0} \bar{z} = \lim_{h\to 0}\frac{3(2a-h)^2}{4(3a-h)} = \frac{3(4a^2)}{12a} = a$$

(e) $x^2 + y^2 = \rho^2 \sin^2\phi$

$$I_z = \int_0^{2\pi}\int_0^{\cos^{-1}(a-h/a)}\int_{(a-h)\sec\phi}^{a} (\rho^2\sin^2\phi)\rho^2 \sin\phi \, d\rho \, d\phi \, d\theta = \frac{h^3}{30}(20a^2 - 15ah + 3h^2)\pi$$

(f) If $h = a$, $I_z = \dfrac{a^3\pi}{30}(20a^2 - 15a^2 + 3a^2) = \dfrac{4}{15}a^5\pi$

53. $$\int_0^{2\pi}\int_0^{\pi}\int_0^{6\sin\phi} \rho^2 \sin\phi \, d\rho \, d\phi \, d\theta$$

Since $\rho = 6\sin\phi$ represents (in the yz-plane) a circle of radius 3 centered at $(0, 3, 0)$, the integral represents the volume of the torus formed by revolving $(0 < \theta < 2\pi)$ this circle about the z-axis.

55. True

57. True

CHAPTER 14
Vector Analysis

CHAPTER 14
Vector Analysis

Section 14.1 Vector Fields

Solutions to Odd-Numbered Exercises

1. All vectors are parallel to y-axis.

Matches (c)

3. All vectors point outwards.

Matches (b)

5. Vectors are parallel to x-axis for $y = n\pi$.

Matches (a)

7. $\mathbf{F}(x, y) = \mathbf{i} + \mathbf{j}$

$\|\mathbf{F}\| = \sqrt{2}$

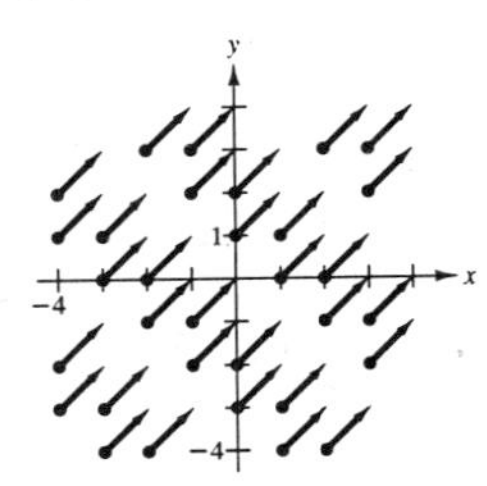

9. $\mathbf{F}(x, y) = x\mathbf{i} + y\mathbf{j}$

$\|\mathbf{F}\| = \sqrt{x^2 + y^2} = c$

$x^2 + y^2 = c^2$

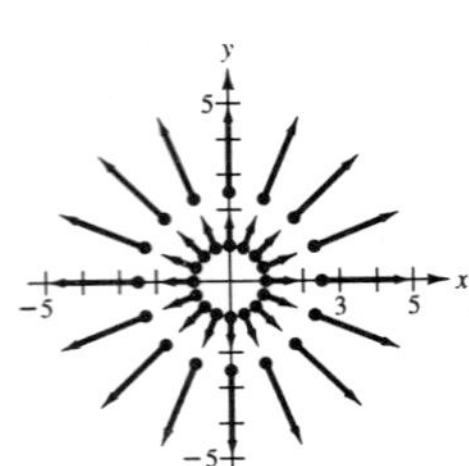

11. $\mathbf{F}(x, y, z) = 3y\mathbf{j}$

$\|\mathbf{F}\| = 3|y| = c$

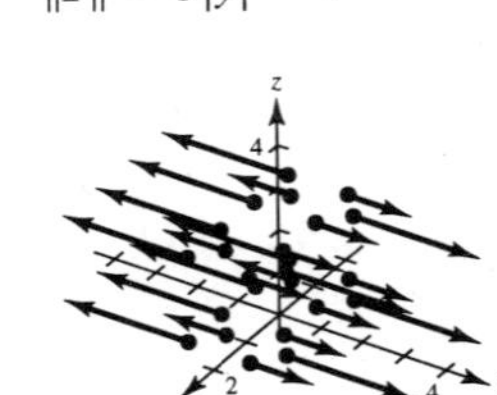

13. $\mathbf{F}(x, y) = 4x\mathbf{i} + y\mathbf{j}$

$\|\mathbf{F}\| = \sqrt{16x^2 + y^2} = c$

$\dfrac{x^2}{c^2/16} + \dfrac{y^2}{c^2} = 1$

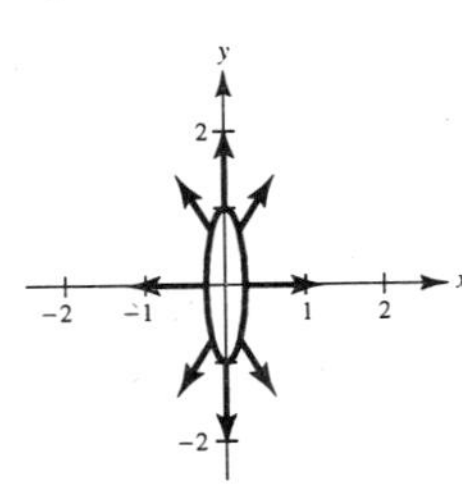

15. $\mathbf{F}(x, y, z) = \mathbf{i} + \mathbf{j} + \mathbf{k}$

$\|\mathbf{F}\| = \sqrt{3}$

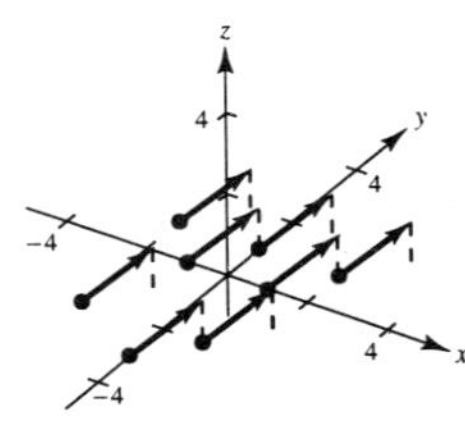

17.

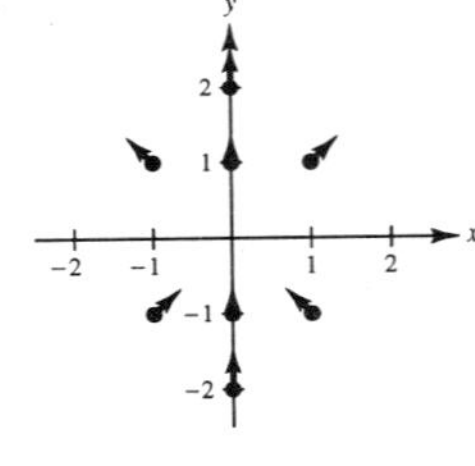

19.

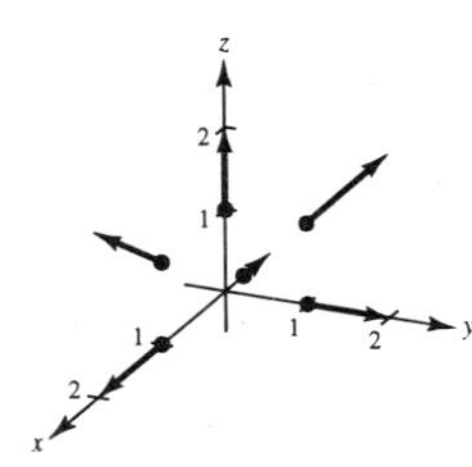

21. $f(x, y) = 5x^2 + 3xy + 10y^2$

$f_x(x, y) = 10x + 3y$

$f_y(x, y) = 3x + 20y$

$\mathbf{F}(x, y) = (10x + 3y)\mathbf{i} + (3x + 20y)\mathbf{j}$

23. $f(x, y, z) = z - ye^{x^2}$

$f_x(x, y, z) = -2xye^{x^2}$

$f_y(x, y, z) = -e^{x^2}$

$f_z = 1$

$\mathbf{F}(x, y, z) = -2xye^{x^2}\mathbf{i} - e^{x^2}\mathbf{j} + \mathbf{k}$

25. $g(x, y, z) = xy \ln(x + y)$

$$g_x(x, y, z) = y \ln(x + y) + \frac{xy}{x + y}$$

$$g_y(x, y, z) = x \ln(x + y) + \frac{xy}{x + y}$$

$$g_z(x, y, z) = 0$$

$$\mathbf{G}(x, y, z) = \left[\frac{xy}{x + y} + y \ln(x + y)\right]\mathbf{i} + \left[\frac{xy}{x + y} + x \ln(x + y)\right]\mathbf{j}$$

27. $\mathbf{F}(x, y) = 2xy\mathbf{i} + x^2\mathbf{j}$

$$\frac{\partial}{\partial y}[2xy] = 2x$$

$$\frac{\partial}{\partial x}[x^2] = 2x$$

Conservative

$$f_x(x, y) = 2xy$$

$$f_y(x, y) = 2x$$

$$f(x, y) = x^2y + K$$

29. $\mathbf{F}(x, y) = xe^{x^2y}(2y\mathbf{i} + x\mathbf{j})$

$$\frac{\partial}{\partial y}[2xye^{x^2y}] = 2xe^{x^2y} + 2x^3ye^{x^2y}$$

$$\frac{\partial}{\partial x}[x^2e^{x^2y}] = 2xe^{x^2y} + 2x^3ye^{x^2y}$$

Conservative

$$f_x(x, y) = 2xye^{x^2y}$$

$$f_y(x, y) = x^2e^{x^2y}$$

$$f(x, y) = e^{x^2y} + K$$

31. $\mathbf{F}(x, y) = \frac{x}{x^2 + y^2}\mathbf{i} + \frac{y}{x^2 + y^2}\mathbf{j}$

$$\frac{\partial}{\partial y}\left[\frac{x}{x^2 + y^2}\right] = -\frac{2xy}{(x^2 + y^2)^2}$$

$$\frac{\partial}{\partial x}\left[\frac{y}{x^2 + y^2}\right] = -\frac{2xy}{(x^2 + y^2)^2}$$

Conservative

$$f_x(x, y) = \frac{x}{x^2 + y^2}$$

$$f_y(x, y) = \frac{y}{x^2 + y^2}$$

$$f(x, y) = \frac{1}{2}\ln(x^2 + y^2) + K$$

33. $\mathbf{F}(x, y) = e^x(\cos y\mathbf{i} + \sin y\mathbf{j})$

$$\frac{\partial}{\partial y}[e^x \cos y] = -e^x \sin y$$

$$\frac{\partial}{\partial x}[e^x \sin y] = e^x \sin y$$

Not conservative

35. $\mathbf{F}(x, y, z) = xyz\mathbf{i} + y\mathbf{j} + z\mathbf{k}$, $(1, 2, 1)$

$$\textbf{curl } \mathbf{F} = \begin{vmatrix} \mathbf{i} & \mathbf{j} & \mathbf{k} \\ \frac{\partial}{\partial x} & \frac{\partial}{\partial y} & \frac{\partial}{\partial z} \\ xyz & y & z \end{vmatrix} = xy\mathbf{j} - xz\mathbf{k}$$

$$\textbf{curl } \mathbf{F}\,(1, 2, 1) = 2\mathbf{j} - \mathbf{k}$$

37. $\mathbf{F}(x, y, z) = e^x \sin y\mathbf{i} - e^x \cos y\mathbf{j}$, $(0, 0, 3)$

$$\textbf{curl } \mathbf{F} = \begin{vmatrix} \mathbf{i} & \mathbf{j} & \mathbf{k} \\ \frac{\partial}{\partial x} & \frac{\partial}{\partial y} & \frac{\partial}{\partial z} \\ e^x \sin y & -e^x \cos y & 0 \end{vmatrix} = -2e^x \cos y\mathbf{k}$$

$$\textbf{curl } \mathbf{F}\,(0, 0, 3) = -2\mathbf{k}$$

39. $\mathbf{F}(x, y, z) = \arctan\left(\frac{x}{y}\right)\mathbf{i} + \ln\sqrt{x^2 + y^2}\,\mathbf{j} + \mathbf{k}$

$$\textbf{curl } \mathbf{F} = \begin{vmatrix} \mathbf{i} & \mathbf{j} & \mathbf{k} \\ \frac{\partial}{\partial x} & \frac{\partial}{\partial y} & \frac{\partial}{\partial z} \\ \arctan\left(\frac{x}{y}\right) & \frac{1}{2}\ln(x^2 + y^2) & 1 \end{vmatrix}$$

$$= \left[\frac{x}{x^2 + y^2} - \frac{(-x/y^2)}{1 + (x/y)^2}\right]\mathbf{k} = \frac{2x}{x^2 + y^2}\mathbf{k}$$

41. $\mathbf{F}(x, y, z) = \sin(x - y)\mathbf{i} + \sin(y - z)\mathbf{j} + \sin(z - x)\mathbf{k}$

$$\textbf{curl } \mathbf{F} = \begin{vmatrix} \mathbf{i} & \mathbf{j} & \mathbf{k} \\ \frac{\partial}{\partial x} & \frac{\partial}{\partial y} & \frac{\partial}{\partial z} \\ \sin(x - y) & \sin(y - z) & \sin(z - x) \end{vmatrix} = \cos(y - z)\mathbf{i} + \cos(z - x)\mathbf{j} + \cos(x - y)\mathbf{k}$$

43. $\mathbf{F}(x, y, z) = \sin y\mathbf{i} - x\cos y\mathbf{j} + \mathbf{k}$

$$\textbf{curl F} = \begin{vmatrix} \mathbf{i} & \mathbf{j} & \mathbf{k} \\ \dfrac{\partial}{\partial x} & \dfrac{\partial}{\partial y} & \dfrac{\partial}{\partial z} \\ \sin y & -x\cos y & 1 \end{vmatrix} = -2\cos y\mathbf{k} \neq \mathbf{0}$$

Not conservative

45. $\mathbf{F}(x, y, z) = e^z(y\mathbf{i} + x\mathbf{j} + xy\mathbf{k})$

$$\textbf{curl F} = \begin{vmatrix} \mathbf{i} & \mathbf{j} & \mathbf{k} \\ \dfrac{\partial}{\partial x} & \dfrac{\partial}{\partial y} & \dfrac{\partial}{\partial z} \\ ye^z & xe^z & xye^z \end{vmatrix} = \mathbf{0}$$

Conservative

$$f_x(x, y, z) = ye^z$$

$$f_y(x, y, z) = xe^z$$

$$f_z(x, y, z) = xye^z$$

$$f(x, y, z) = xye^z + K$$

47. $\mathbf{F}(x, y, z) = \dfrac{1}{y}\mathbf{i} - \dfrac{x}{y^2}\mathbf{j} + (2z - 1)\mathbf{k}$

$$\textbf{curl F} = \begin{vmatrix} \mathbf{i} & \mathbf{j} & \mathbf{k} \\ \dfrac{\partial}{\partial x} & \dfrac{\partial}{\partial y} & \dfrac{\partial}{\partial z} \\ \dfrac{1}{y} & -\dfrac{x}{y^2} & 2z - 1 \end{vmatrix} = \mathbf{0}$$

Conservative

$$f_x(x, y, z) = \frac{1}{y}$$

$$f_y(x, y, z) = -\frac{x}{y^2}$$

$$f_z(x, y, z) = 2z - 1$$

$$f(x, y, z) = \int \frac{1}{y}\,dx = \frac{x}{y} + g(y, z) + K_1$$

$$f(x, y, z) = \int -\frac{x}{y^2}\,dy = \frac{x}{y} + h(x, z) + K_2$$

$$f(x, y, z) = \int (2z - 1)\,dz$$

$$= z^2 - z + p(x, y) + K_3$$

$$f(x, y, z) = \frac{x}{y} + z^2 - z + K$$

49. $\mathbf{F}(x, y, z) = \mathbf{i} + 2x\mathbf{j} + 3y\mathbf{k}$

$\mathbf{G}(x, y, z) = x\mathbf{i} - y\mathbf{j} + z\mathbf{k}$

$$\mathbf{F} \times \mathbf{G} \begin{vmatrix} \mathbf{i} & \mathbf{j} & \mathbf{k} \\ 1 & 2x & 3y \\ x & -y & z \end{vmatrix} = (2xz + 3y^2)\mathbf{i} - (z - 3xy)\mathbf{j} + (-y - 2x^2)\mathbf{k}$$

$$\textbf{curl}\,(\mathbf{F} \times \mathbf{G}) = \begin{vmatrix} \mathbf{i} & \mathbf{j} & \mathbf{k} \\ \dfrac{\partial}{\partial x} & \dfrac{\partial}{\partial y} & \dfrac{\partial}{\partial z} \\ 2xz + 3y^2 & 3xy - z & -y - 2x^2 \end{vmatrix} = (-1 + 1)\mathbf{i} - (-4x - 2x)\mathbf{j} + (3y - 6y)\mathbf{k} = 6x\mathbf{j} - 3y\mathbf{k}$$

51. $\mathbf{F}(x, y, z) = xyz\mathbf{i} + y\mathbf{j} + z\mathbf{k}$

$$\textbf{curl F} = \begin{vmatrix} \mathbf{i} & \mathbf{j} & \mathbf{k} \\ \dfrac{\partial}{\partial x} & \dfrac{\partial}{\partial y} & \dfrac{\partial}{\partial z} \\ xyz & y & z \end{vmatrix} = xy\mathbf{j} - xz\mathbf{k}$$

$$\textbf{curl}(\textbf{curl F}) = \begin{vmatrix} \mathbf{i} & \mathbf{j} & \mathbf{k} \\ \dfrac{\partial}{\partial x} & \dfrac{\partial}{\partial y} & \dfrac{\partial}{\partial z} \\ 0 & xy & -xz \end{vmatrix} = z\mathbf{j} + y\mathbf{k}$$

53. $\mathbf{F}(x, y) = 6x^2\mathbf{i} - xy^2\mathbf{j}$

$$\text{div}\,\mathbf{F}(x, y) = \frac{\partial}{\partial x}[6x^2] + \frac{\partial}{\partial y}[-xy^2]$$

$$= 12x - 2xy$$

55. $\mathbf{F}(x, y, z) = \sin x\mathbf{i} + \cos y\mathbf{j} + z^2\mathbf{k}$

$$\text{div } \mathbf{F}(x, y, z) = \frac{\partial}{\partial x}[\sin x] + \frac{\partial}{\partial y}[\cos y] + \frac{\partial}{\partial z}[z^2] = \cos x - \sin y + 2z$$

57. $\mathbf{F}(x, y, z) = xyz\mathbf{i} + y\mathbf{j} + z\mathbf{k}$

$\text{div } \mathbf{F}(x, y, z) = yz + 1 + 1 = yz + 2$

$\text{div } \mathbf{F}(1, 2, 1) = 4$

59. $\mathbf{F}(x, y, z) = e^x \sin y\mathbf{i} - e^x \cos y\mathbf{j}$

$\text{div } \mathbf{F}(x, y, z) = e^x \sin y + e^x \sin y$

$\text{div } \mathbf{F}(0, 0, 3) = 0$

61. $\mathbf{F}(x, y, z) = \mathbf{i} + 2x\mathbf{j} + 3y\mathbf{k}$

$\mathbf{G}(x, y, z) = x\mathbf{i} - y\mathbf{j} + z\mathbf{k}$

$$\mathbf{F} \times \mathbf{G} = \begin{vmatrix} \mathbf{i} & \mathbf{j} & \mathbf{k} \\ 1 & 2x & 3y \\ x & -y & z \end{vmatrix}$$

$$= (2xz + 3y^2)\mathbf{i} - (z - 3xy)\mathbf{j} + (-y - 2x^2)\mathbf{k}$$

$\text{div}(\mathbf{F} \times \mathbf{G}) = 2z + 3x$

63. $\mathbf{F}(x, y, z) = xyz\mathbf{i} + y\mathbf{j} + z\mathbf{k}$

$$\mathbf{curl\ F} = \begin{vmatrix} \mathbf{i} & \mathbf{j} & \mathbf{k} \\ \dfrac{\partial}{\partial x} & \dfrac{\partial}{\partial y} & \dfrac{\partial}{\partial z} \\ xyz & y & z \end{vmatrix} = xy\mathbf{j} - xz\mathbf{k}$$

$\text{div}(\mathbf{curl\ F}) = x - x = 0$

65. Let $\mathbf{F} = M\mathbf{i} + N\mathbf{j} + P\mathbf{k}$ and $\mathbf{G} = Q\mathbf{i} + R\mathbf{j} + S\mathbf{k}$ where $M, N, P, Q, R,$ and S have continuous partial derivatives.

$\mathbf{F} + \mathbf{G} = (M + Q)\mathbf{i} + (N + R)\mathbf{j} + (P + S)\mathbf{k}$

$$\mathbf{curl}(\mathbf{F} + \mathbf{G}) = \begin{vmatrix} \mathbf{i} & \mathbf{j} & \mathbf{k} \\ \dfrac{\partial}{\partial x} & \dfrac{\partial}{\partial y} & \dfrac{\partial}{\partial z} \\ M + Q & N + R & P + S \end{vmatrix}$$

$$= \left[\frac{\partial}{\partial y}(P + S) - \frac{\partial}{\partial z}(N + R)\right]\mathbf{i} - \left[\frac{\partial}{\partial x}(P + S) - \frac{\partial}{\partial z}(M + Q)\right]\mathbf{j} + \left[\frac{\partial}{\partial x}(N + R) - \frac{\partial}{\partial y}(M + Q)\right]\mathbf{k}$$

$$= \left(\frac{\partial P}{\partial y} - \frac{\partial N}{\partial z}\right)\mathbf{i} - \left(\frac{\partial P}{\partial x} - \frac{\partial M}{\partial z}\right)\mathbf{j} + \left(\frac{\partial N}{\partial x} - \frac{\partial M}{\partial y}\right)\mathbf{k} + \left(\frac{\partial S}{\partial y} - \frac{\partial R}{\partial z}\right)\mathbf{i} - \left(\frac{\partial S}{\partial x} - \frac{\partial Q}{\partial z}\right)\mathbf{j} + \left(\frac{\partial R}{\partial x} - \frac{\partial Q}{\partial y}\right)\mathbf{k}$$

$$= \mathbf{curl\ F} + \mathbf{curl\ G}$$

67. Let $\mathbf{F} = M\mathbf{i} + N\mathbf{j} + P\mathbf{k}$ and $\mathbf{G} = R\mathbf{i} + S\mathbf{j} + T\mathbf{k}$.

$$\text{div}(\mathbf{F} + \mathbf{G}) = \frac{\partial}{\partial x}(M + R) + \frac{\partial}{\partial y}(N + S) + \frac{\partial}{\partial z}(P + T) = \frac{\partial M}{\partial x} + \frac{\partial R}{\partial x} + \frac{\partial N}{\partial y} + \frac{\partial S}{\partial y} + \frac{\partial P}{\partial z} + \frac{\partial T}{\partial z}$$

$$= \left[\frac{\partial M}{\partial x} + \frac{\partial N}{\partial y} + \frac{\partial P}{\partial z}\right] + \left[\frac{\partial R}{\partial x} + \frac{\partial S}{\partial y} + \frac{\partial T}{\partial z}\right]$$

$$= \text{div } \mathbf{F} + \text{div } \mathbf{G}$$

69. $\mathbf{F} = M\mathbf{i} + N\mathbf{j} + P\mathbf{k}$

$\nabla \times [\nabla f + (\nabla \times \mathbf{F})] = \mathbf{curl}(\nabla f + (\nabla \times \mathbf{F}))$

$= \mathbf{curl}(\nabla f) + \mathbf{curl}(\nabla \times \mathbf{F})$ (Exercise 65)

$= \mathbf{curl}(\nabla \times \mathbf{F})$ (Exercise 66)

$= \nabla \times (\nabla \times \mathbf{F})$

71. Let $\mathbf{F} = M\mathbf{i} + N\mathbf{j} + P\mathbf{k}$, then $f\mathbf{F} = fM\mathbf{i} + fN\mathbf{j} + fP\mathbf{k}$.

$$\operatorname{div}(f\mathbf{F}) = \frac{\partial}{\partial x}(fM) + \frac{\partial}{\partial y}(fN) + \frac{\partial}{\partial z}(fP) = f\frac{\partial M}{\partial x} + M\frac{\partial f}{\partial x} + f\frac{\partial N}{\partial y} + N\frac{\partial f}{\partial y} + f\frac{\partial P}{\partial z} + P\frac{\partial f}{\partial z}$$

$$= f\left(\frac{\partial M}{\partial x} + \frac{\partial N}{\partial y} + \frac{\partial N}{\partial z}\right) + \left(\frac{\partial f}{\partial x}M + \frac{\partial f}{\partial y}N + \frac{\partial f}{\partial z}P\right)$$

$$= f \operatorname{div} \mathbf{F} + \nabla f \cdot \mathbf{F}$$

In Exercises 73–77, $\mathbf{F}(x, y, z) = x\mathbf{i} + y\mathbf{j} + z\mathbf{k}$ and $f(x, y, z) = \|\mathbf{F}(x, y, z)\| = \sqrt{x^2 + y^2 + z^2}$.

73. $\ln f = \frac{1}{2}\ln(x^2 + y^2 + z^2)$

$$\nabla(\ln f) = \frac{x}{x^2 + y^2 + z^2}\mathbf{i} + \frac{y}{x^2 + y^2 + z^2}\mathbf{j} + \frac{z}{x^2 + y^2 + z^2}\mathbf{k} = \frac{x\mathbf{i} + y\mathbf{j} + z\mathbf{k}}{x^2 + y^2 + z^2} = \frac{\mathbf{F}}{f^2}$$

75. $f^n = \left(\sqrt{x^2 + y^2 + z^2}\right)^n$

$$\nabla f^n = n\left(\sqrt{x^2 + y^2 + z^2}\right)^{n-1}\frac{x}{\sqrt{x^2 + y^2 + z^2}}\mathbf{i} + n\left(\sqrt{x^2 + y^2 + z^2}\right)^{n-1}\frac{y}{\sqrt{x^2 + y^2 + z^2}}\mathbf{j}$$

$$+ n\left(\sqrt{x^2 + y^2 + z^2}\right)^{n-1}\frac{z}{\sqrt{x^2 + y^2 + z^2}}\mathbf{k}$$

$$= n\left(\sqrt{x^2 + y^2 + z^2}\right)^{n-2}(x\mathbf{i} + y\mathbf{j} + z\mathbf{k}) = nf^{n-2}\mathbf{F}$$

77. $\mathbf{F}(x, y) = M(x, y)\mathbf{i} + N(x, y)\mathbf{j} = \dfrac{m}{(x^2 + y^2)^{5/2}}[3xy\mathbf{i} + (2y^2 - x^2)\mathbf{j}]$

$$M = \frac{3mxy}{(x^2 + y^2)^{5/2}} = 3mxy(x^2 + y^2)^{-5/2}$$

$$\frac{\partial M}{\partial y} = 3mxy\left[-\frac{5}{2}(x^2 + y^2)^{-7/2}(2y)\right] + (x^2 + y^2)^{-5/2}(3mx)$$

$$= 3mx(x^2 + y^2)^{-7/2}[-5y^2 + (x^2 + y^2)] = \frac{3mx(x^2 - 4y^2)}{(x^2 + y^2)^{7/2}}$$

$$N = \frac{m(2y^2 - x^2)}{(x^2 + y^2)^{5/2}} = m(2y^2 - x^2)(x^2 + y^2)^{-5/2}$$

$$\frac{\partial N}{\partial x} = m(2y^2 - x^2)\left[-\frac{5}{2}(x^2 + y^2)^{-7/2}(2x)\right] + (x^2 + y^2)^{-5/2}(-2mx)$$

$$= mx(x^2 + y^2)^{-7/2}[(2y^2 - x^2)(-5) + (x^2 + y^2(-2)]$$

$$= mx(x^2 + y^2)^{-7/2}(3x^2 - 12y^2) = \frac{3mx(x^2 - 4y^2)}{(x^2 + y^2)^{7/2}}$$

Therefore, $\dfrac{\partial N}{\partial x} = \dfrac{\partial M}{\partial y}$ and $\mathbf{F}$ is conservative.

Section 14.2 Line Integrals

1. $x^2 + y^2 = 9$

$$\frac{x^2}{9} + \frac{y^2}{9} = 1$$

$$\cos^2 t + \sin^2 t = 1$$

$$\cos^2 t = \frac{x^2}{9}$$

$$\sin^2 t = \frac{y^2}{9}$$

$$x = 3\cos t$$

$$y = 3\sin t$$

$$\mathbf{r}(t) = 3\cos t\mathbf{i} + 3\sin t\mathbf{j}$$

$$0 \le t \le 2\pi$$

3. $\mathbf{r}(t) = \begin{cases} t\mathbf{i}, & 0 \le t \le 3 \\ 3\mathbf{i} + (t-3)\mathbf{j}, & 3 \le t \le 6 \\ (9-t)\mathbf{i} + 3\mathbf{j}, & 6 \le t \le 9 \\ (12-t)\mathbf{j}, & 9 \le t \le 12 \end{cases}$

5. $\mathbf{r}(t) = \begin{cases} t\mathbf{i} + \sqrt{t}\mathbf{j}, & 0 \le t \le 1 \\ (2-t)\mathbf{i} + (2-t)\mathbf{j}, & 1 \le t \le 2 \end{cases}$

7. $\mathbf{r}(t) = 4t\mathbf{i} + 3t\mathbf{j},\ 0 \le t \le 2;\ \mathbf{r}'(t) = 4\mathbf{i} + 3\mathbf{j}$

$$\int_C (x - y)\,ds = \int_0^2 (4t - 3t)\sqrt{(4)^2 + (3)^2}\,dt = \int_0^2 5t\,dt = \left[\frac{5t^2}{2}\right]_0^2 = 10$$

9. $\mathbf{r}(t) = \sin t\mathbf{i} + \cos t\mathbf{j} + 8t\mathbf{k},\ 0 \le t \le \frac{\pi}{2};\ \mathbf{r}'(t) = \cos t\mathbf{i} - \sin t\mathbf{j} + 8\mathbf{k}$

$$\int_C (x^2 + y^2 + z^2)\,ds = \int_0^{\pi/2} (\sin^2 t + \cos^2 t + 64t^2)\sqrt{(\cos t)^2 + (-\sin t)^2 + 64}\,dt$$

$$= \int_0^{\pi/2} \sqrt{65}(1 + 64t^2)\,dt = \left[\sqrt{65}\left(t + \frac{64t^3}{3}\right)\right]_0^{\pi/2} = \sqrt{65}\left(\frac{\pi}{2} + \frac{8\pi^3}{3}\right) = \frac{\sqrt{65}\pi}{6}(3 + 16\pi^2)$$

11. $\mathbf{r}(t) = t\mathbf{i},\ 0 \le t \le 3$

$$\int_C (x^2 + y^2)\,ds = \int_0^3 [t^2 + 0^2]\sqrt{1 + 0}\,dt$$

$$= \int_0^3 t^2\,dt$$

$$= \left[\frac{1}{3}t^3\right]_0^3 = 9$$

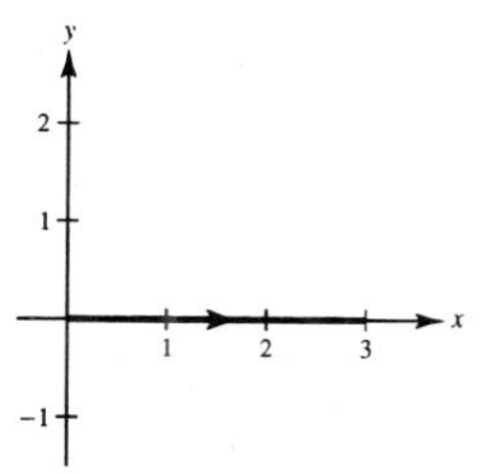

13. $\mathbf{r}(t) = \cos t\mathbf{i} + \sin t\mathbf{j},\ 0 \le t \le \frac{\pi}{2}$

$$\int_C (x^2 + y^2)\,ds = \int_0^{\pi/2} [\cos^2 t + \sin^2 t]\sqrt{(-\sin t)^2 + (\cos t)^2}\,dt$$

$$= \int_0^{\pi/2} dt = \frac{\pi}{2}$$

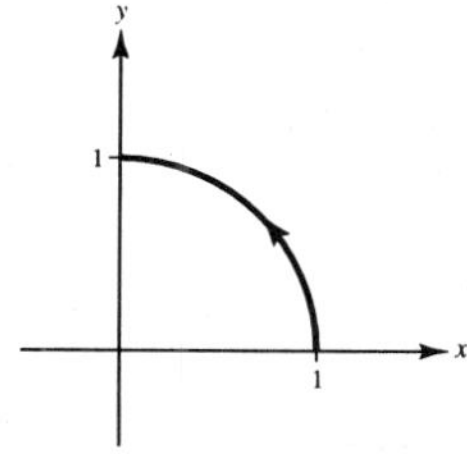

15. $\mathbf{r}(t) = t\mathbf{i} + t\mathbf{j},\ 0 \le t \le 1$

$$\int_C (x + 4\sqrt{y})\,ds = \int_0^1 (t + 4\sqrt{t})\sqrt{1+1}\,dt$$

$$= \left[\sqrt{2}\left(\frac{t^2}{2} + \frac{8}{3}t^{3/2}\right)\right]_0^1 = \frac{19\sqrt{2}}{6}$$

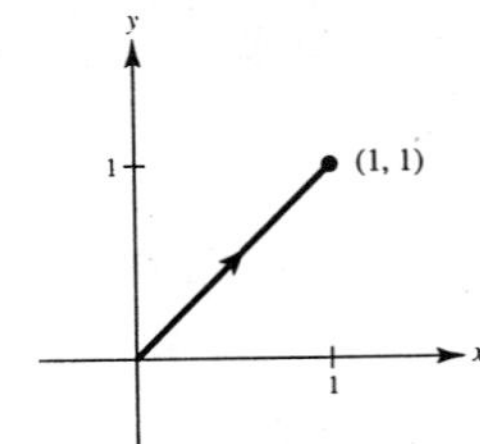

17. $\mathbf{r}(t) = \begin{cases} t\mathbf{i}, & 0 \le t \le 1 \\ (2-t)\mathbf{i} + (t-1)\mathbf{j}, & 1 \le t \le 2 \\ (3-t)\mathbf{j}, & 2 \le t \le 3 \end{cases}$

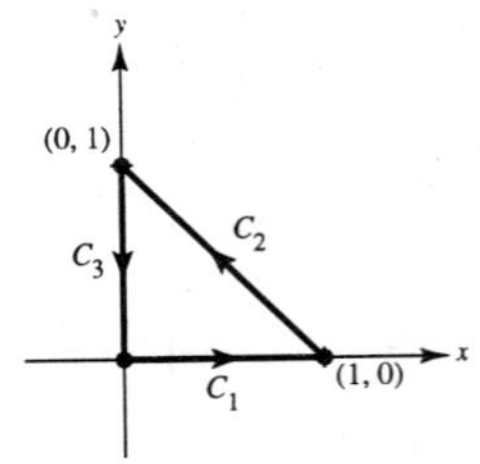

$$\int_{C_1} (x + 4\sqrt{y})\,ds = \int_0^1 t\,dt = \frac{1}{2}$$

$$\int_{C_2} (x + 4\sqrt{y})\,ds = \int_1^2 \left[(2-t) + 4\sqrt{t-1}\right]\sqrt{1+1}\,dt$$

$$= \sqrt{2}\left[2t - \frac{t^2}{2} + \frac{8}{3}(t-1)^{3/2}\right]_1^2 = \frac{19\sqrt{2}}{6}$$

$$\int_{C_3} (x + 4\sqrt{y})\,ds = \int_2^3 4\sqrt{3-t}\,dt = \left[-\frac{8}{3}(3-t)^{3/2}\right]_2^3 = \frac{8}{3}$$

$$\int_C (x + 4\sqrt{y})\,ds = \frac{1}{2} + \frac{19\sqrt{2}}{6} + \frac{8}{3} = \frac{19 + 19\sqrt{2}}{6} = \frac{19(1+\sqrt{2})}{6}$$

19. $\rho(x, y, z) = \frac{1}{2}(x^2 + y^2 + z^2)$

$$\mathbf{r}(t) = 3\cos t\mathbf{i} + 3\sin t\mathbf{j} + 2t\mathbf{k},\ 0 \le t \le 4\pi$$

$$\mathbf{r}'(t) = -3\sin t\mathbf{i} + 3\cos t\mathbf{j} + 2\mathbf{k}$$

$$\|\mathbf{r}'(t)\| = \sqrt{(-3\sin t)^2 + (3\cos t)^2 + (2)^2} = \sqrt{13}$$

$$\text{Mass} = \int_C \rho(x, y, z)\,ds = \int_0^{4\pi} \frac{1}{2}[(3\cos t)^2 + (3\sin t)^2 + (2t)^2]\sqrt{13}\,dt$$

$$= \frac{\sqrt{13}}{2}\int_0^{4\pi} (9 + 4t^2)\,dt = \left[\frac{\sqrt{13}}{2}\left(9t + \frac{4t^3}{3}\right)\right]_0^{4\pi}$$

$$= \frac{2\sqrt{13}\pi}{3}(27 + 64\pi^2) \approx 4973.8$$

21. $\mathbf{F}(x, y) = xy\mathbf{i} + y\mathbf{j}$

C: $\mathbf{r}(t) = 4t\mathbf{i} + t\mathbf{j},\ 0 \le t \le 1$

$\mathbf{F}(t) = 4t^2\mathbf{i} + t\mathbf{j}$

$\mathbf{r}'(t) = 4\mathbf{i} + \mathbf{j}$

$$\int_C \mathbf{F} \cdot d\mathbf{r} = \int_0^1 (16t^2 + t)\,dt$$

$$= \left[\frac{16}{3}t^3 + \frac{1}{2}t^2\right]_0^1 = \frac{35}{6}$$

23. $\mathbf{F}(x, y) = 3x\mathbf{i} + 4y\mathbf{i}$

C: $\mathbf{r}(t) = 2\cos t\mathbf{i} + 2\sin t\mathbf{j},\ 0 \le t \le \frac{\pi}{2}$

$\mathbf{F}(t) = 6\cos t\mathbf{i} + 8\sin t\mathbf{j}$

$\mathbf{r}'(t) = -2\sin t\mathbf{i} + 2\cos t\mathbf{j}$

$$\int_C \mathbf{F} \cdot d\mathbf{r} = \int_0^{\pi/2} (-12\sin t\cos t + 16\sin t\cos t)\,dt$$

$$= \left[2\sin^2 t\right]_0^{\pi/2} = 2$$

25. $\mathbf{F}(x, y, z) = x^2y\mathbf{i} + (x - z)\mathbf{j} + xyz\mathbf{k}$

C: $\mathbf{r}(t) = t\mathbf{i} + t^2\mathbf{j} + 2\mathbf{k},\ 0 \le t \le 1$

$\mathbf{F}(t) = t^4\mathbf{i} + (t - 2)\mathbf{j} + 2t^3\mathbf{k}$

$\mathbf{r}'(t) = \mathbf{i} + 2t\mathbf{j}$

$$\int_C \mathbf{F} \cdot d\mathbf{r} = \int_0^1 [t^4 + 2t(t - 2)]\,dt = \left[\frac{t^5}{5} + \frac{2t^3}{3} - 2t^2\right]_0^1 = -\frac{17}{15}$$

27. $\mathbf{F}(x, y, z) = x^2z\mathbf{i} + 6y\mathbf{j} + yz^2\mathbf{k}$

$\mathbf{r}(t) = t\mathbf{i} + t^2\mathbf{j} + \ln t\mathbf{k},\ 1 \le t \le 3$

$\mathbf{F}(t) = t^2 \ln t\mathbf{i} + 6t^2\mathbf{j} + t^2 \ln^2 t\mathbf{k}$

$$d\mathbf{r} = \left(\mathbf{i} + 2t\mathbf{j} + \frac{1}{t}\mathbf{k}\right)dt$$

$$\int_C \mathbf{F} \cdot d\mathbf{r} = \int_1^3 [t^2 \ln t + 12t^3 + t(\ln t)^2]\,dt \approx 249.49$$

29. $\mathbf{F}(x, y) = -x\mathbf{i} - 2y\mathbf{j}$

C: $y = x^3$ from $(0, 0)$ to $(2, 8)$

$\mathbf{r}(t) = t\mathbf{i} + t^3\mathbf{j},\ 0 \le t \le 2$

$\mathbf{r}'(t) = \mathbf{i} + 3t^2\mathbf{j}$

$\mathbf{F}(t) = -t\mathbf{i} - 2t^3\mathbf{j}$

$\mathbf{F} \cdot \mathbf{r}' = -t - 6t^5$

$$\text{Work} = \int_C \mathbf{F} \cdot d\mathbf{r} = \int_0^2 (-t - 6t^5)\,dt = \left[-\frac{1}{2}t^2 - t^6\right]_0^2 = -66$$

31. $\mathbf{F}(x, y) = 2x\mathbf{i} + y\mathbf{j}$

C: counterclockwise around the triangle whose vertices are $(0, 0)$, $(1, 0)$, $(1, 1)$

$$\mathbf{r}(t) = \begin{cases} t\mathbf{i}, & 0 \le t \le 1 \\ \mathbf{i} + (t - 1)\mathbf{j}, & 1 \le t \le 2 \\ (3 - t)\mathbf{i} + (3 - t)\mathbf{j}, & 2 \le t \le 3 \end{cases}$$

On C_1: $\mathbf{F}(t) = 2t\mathbf{i},\ \mathbf{r}'(t) = \mathbf{i}$

$$\text{Work} = \int_{C_1} \mathbf{F} \cdot d\mathbf{r} = \int_0^1 2t\,dt = 1$$

On C_2: $\mathbf{F}(t) = 2\mathbf{i} + (t - 1)\mathbf{j},\ \mathbf{r}'(t) = \mathbf{j}$

$$\text{Work} = \int_{C_2} \mathbf{F} \cdot d\mathbf{r} = \int_1^2 (t - 1)\,dt = \frac{1}{2}$$

On C_3: $\mathbf{F}(t) = 2(3 - t)\mathbf{i} + (3 - t)\mathbf{j},\ \mathbf{r}'(t) = -\mathbf{i} - \mathbf{j}$

$$\text{Work} = \int_{C_3} \mathbf{F} \cdot d\mathbf{r} = \int_2^3 [-2(3 - t) - (3 - t)]\,dt = -\frac{3}{2}$$

$$\text{Total work} = \int_C \mathbf{F} \cdot d\mathbf{r} = 1 + \frac{1}{2} - \frac{3}{2} = 0$$

33. $\mathbf{F}(x, y, z) = x\mathbf{i} + y\mathbf{j} - 5z\mathbf{k}$

C: $\mathbf{r}(t) = 2\cos t\mathbf{i} + 2\sin t\mathbf{j} + t\mathbf{k},\ 0 \le t \le 2\pi$

$\mathbf{r}'(t) = -2\sin t\mathbf{i} + 2\cos t\mathbf{j} + \mathbf{k}$

$\mathbf{F}(t) = 2\cos t\mathbf{i} + 2\sin t\mathbf{j} - 5t\mathbf{k}$

$\mathbf{F} \cdot \mathbf{r}' = -5t$

$$\text{Work} = \int_C \mathbf{F} \cdot d\mathbf{r} = \int_0^{2\pi} -5t\,dt = -10\pi^2$$

35. $\mathbf{r}(t) = 3\sin t\mathbf{i} + 3\cos t\mathbf{j} + \dfrac{10}{2\pi}t\mathbf{k},\ 0 \le t \le 2\pi$

$\mathbf{F} = 150\mathbf{k}$

$$d\mathbf{r} = \left(3\cos t\mathbf{i} - 3\sin t\mathbf{j} + \frac{10}{2\pi}\mathbf{k}\right)dt$$

$$\int_C \mathbf{F} \cdot d\mathbf{r} = \int_0^{2\pi} \frac{1500}{2\pi}\,dt = \left[\frac{1500}{2\pi}t\right]_0^{2\pi} = 1500 \text{ ft} \cdot \text{lb}$$

37. $\mathbf{F}(x, y) = y\mathbf{i} - x\mathbf{j}$

C: $\mathbf{r}(t) = t\mathbf{i} - 2t\mathbf{j}$

$\mathbf{r}'(t) = \mathbf{i} - 2\mathbf{j}$

$\mathbf{F}(t) = -2t\mathbf{i} - t\mathbf{j}$

$\mathbf{F} \cdot \mathbf{r}' = -2t + 2t = 0$

Thus, $\displaystyle\int_C \mathbf{F} \cdot d\mathbf{r} = 0.$

39. $\mathbf{F}(x, y) = (x^3 - 2x^2)\mathbf{i} + \left(x - \dfrac{y}{2}\right)\mathbf{j}$

C: $\mathbf{r}(t) = t\mathbf{i} + t^2\mathbf{j}$

$\mathbf{r}'(t) = \mathbf{i} + 2t\mathbf{j}$

$\mathbf{F}(t) = (t^3 - 2t^2)\mathbf{i} + \left(t - \dfrac{t^2}{2}\right)\mathbf{j}$

$\mathbf{F} \cdot \mathbf{r}' = (t^3 - 2t^2) + 2t\left(t - \dfrac{t^2}{2}\right) = 0$

Thus, $\displaystyle\int_C \mathbf{F} \cdot d\mathbf{r} = 0.$

41. $W = \displaystyle\int_C \mathbf{F} \cdot d\mathbf{r} = \int_C M\,dx + N\,dy$

$M = 15(4 - x^2y) = 60 - 15x^2(c - cx^2)$

$N = -15xy = -15x(c - cx^2)$

$dx = dx,\ dy = -2cx\,dx$

$W = \displaystyle\int_{-1}^{1} [60 - 15x^2(c - cx^2) + (-15x(c - cx^2))(-2\,cx)]\,dx$

$= 120 - 4c + 8c^2$ (parabola)

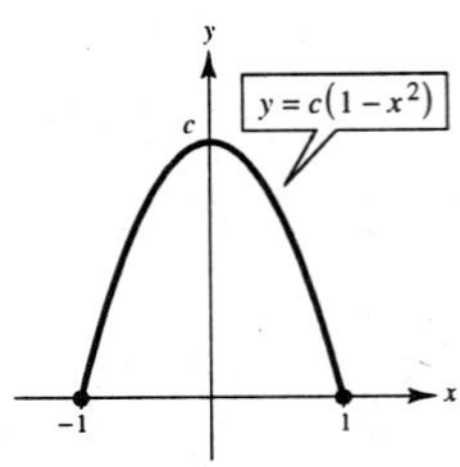

$w' = 16c - 4 = 0 \Rightarrow c = \frac{1}{4}$ yields the minimum work, 119.5. Along the straight line path, $y = 0$, the work is 120.

43. $x = 2t,\ y = 10t,\ 0 \le t \le 1 \Rightarrow y = 5x$ or $x = \dfrac{y}{5},\ 0 \le y \le 10$

$\displaystyle\int_C (x + 3y^2)\,dy = \int_0^{10}\left(\frac{y}{5} + 3y^2\right)dy = \left[\frac{y^2}{10} + y^3\right]_0^{10} = 1010$

45. $x = 2t,\ y = 10t,\ 0 \le t \le 1 \Rightarrow x = \dfrac{y}{5},\ 0 \le y \le 10,\ dx = \dfrac{1}{5}\,dy$

$\displaystyle\int_C xy\,dx + y\,dy = \int_0^{10}\left(\frac{y^2}{25} + y\right)dy = \left[\frac{y^3}{75} + \frac{y^2}{2}\right]_0^{10} = \frac{190}{3}$ OR

$y = 5x,\ dy = 5\,dx,\ 0 \le x \le 2$

$\displaystyle\int_C xy\,dx + y\,dy = \int_0^{2}(5x^2 + 25x)\,dx = \left[\frac{5x^3}{3} + \frac{25x^2}{2}\right]_0^{2} = \frac{190}{3}$

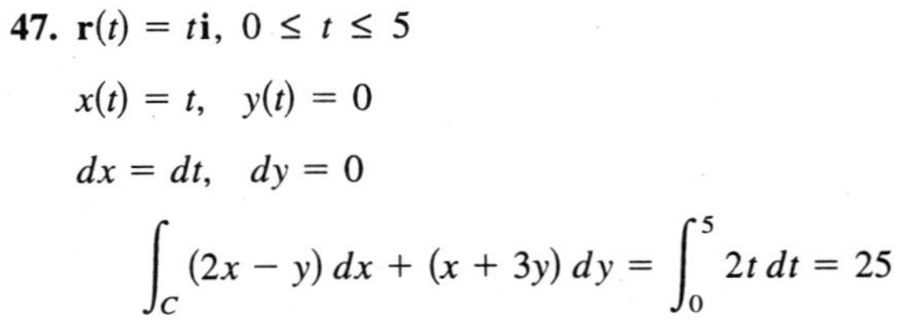

47. $\mathbf{r}(t) = t\mathbf{i},\ 0 \le t \le 5$

$x(t) = t,\quad y(t) = 0$

$dx = dt,\quad dy = 0$

$\displaystyle\int_C (2x - y)\,dx + (x + 3y)\,dy = \int_0^{5} 2t\,dt = 25$

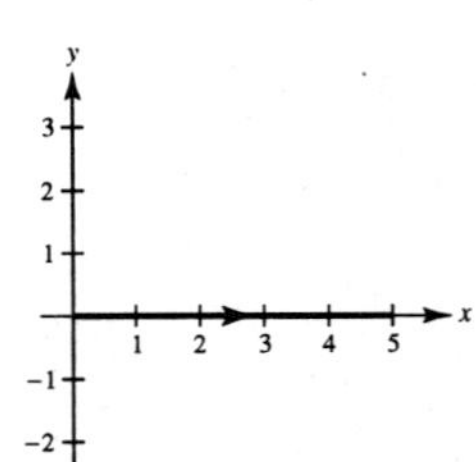

49. $\mathbf{r}(t) = \begin{cases} t\mathbf{i}, & 0 \le t \le 3 \\ 3\mathbf{i} + (t-3)\mathbf{j}, & 3 \le t \le 6 \end{cases}$

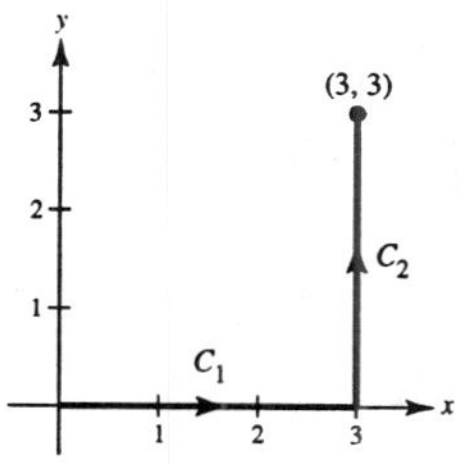

C_1: $x(t) = t,\ y(t) = 0,$

$dx = dt,\ dy = 0$

$$\int_{C_1} (2x - y)\,dx + (x + 3y)\,dy = \int_0^3 2t\,dt = 9$$

C_2: $x(t) = 3,\ y = t - 3$

$dx = 0,\ dy = dt$

$$\int_{C_2} (2x - y)\,dx + (x + 3y)\,dy = \int_3^6 [3 + 3(t-3)]\,dt = \left[\frac{3t^2}{2} - 6t\right]_3^6 = \frac{45}{2}$$

$$\int_C (2x - y)\,dx + (x + 3y)\,dy = 9 + \frac{45}{2} = \frac{63}{2}$$

51. $x(t) = t,\ y(t) = 2t^2,\ 0 \le t \le 2$

$dx = dt,\ dy = 4t\,dt$

$$\int_C (2x - y)\,dx + (x + 3y)\,dy = \int_0^2 (2t - 2t^2)\,dt + (t + 6t^2)4t\,dt$$

$$= \int_0^2 (24t^3 + 2t^2 + 2t)\,dt = \left[6t^4 + \frac{2}{3}t^3 + t^2\right]_0^2 = \frac{316}{3}$$

53. $f(x, y) = h$

C: line from $(0, 0)$ to $(3, 4)$

$\mathbf{r} = 3t\mathbf{i} + 4t\mathbf{j},\ 0 \le t \le 1$

$\mathbf{r}'(t) = 3\mathbf{i} + 4\mathbf{j}$

$\|\mathbf{r}'(t)\| = 5$

Lateral surface area:

$$\int_C f(x, y)\,ds = \int_0^1 5h\,dt = 5h$$

55. $f(x, y) = xy$

C: $x^2 + y^2 = 1$ from $(1, 0)$ to $(0, 1)$

$\mathbf{r}(t) = \cos t\mathbf{i} + \sin t\mathbf{j},\quad 0 \le t \le \frac{\pi}{2}$

$\mathbf{r}'(t) = -\sin t\mathbf{i} + \cos t\mathbf{j}$

$\|\mathbf{r}'(t)\| = 1$

Lateral surface area:

$$\int_C f(x, y)\,ds = \int_0^{\pi/2} \cos t \sin t\,dt$$

$$= \left[\frac{\sin^2 t}{2}\right]_0^{\pi/2} = \frac{1}{2}$$

57. $f(x, y) = h$

C: $y = 1 - x^2$ from $(1, 0)$ to $(0, 1)$

$\mathbf{r}(t) = (1 - t)\mathbf{i} + [1 - (1 - t)^2]\mathbf{j},\ 0 \le t \le 1$

$\mathbf{r}'(t) = -\mathbf{i} + 2(1 - t)\mathbf{j}$

$\|\mathbf{r}'(t)\| = \sqrt{1 + 4(1 - t)^2}$

Lateral surface area:

$$\int_C f(x, y)\,ds = \int_0^1 h\sqrt{1 + 4(1 - t)^2}\,dt$$

$$= -\frac{h}{4}\left[2(1 - t)\sqrt{1 + 4(1 - t)^2} + \ln\left|2(1 - t) + \sqrt{1 + 4(1 - t)^2}\right|\right]_0^1$$

$$= \frac{h}{4}\left[2\sqrt{5} + \ln\left(2 + \sqrt{5}\right)\right] \approx 1.4789h$$

59. $f(x, y) = xy$

C: $y = 1 - x^2$ from $(1, 0)$ to $(0, 1)$

You could parameterize the curve C as in Exercises 57 and 58. Alternatively, let $x = \cos t$, then:

$$y = 1 - \cos^2 t = \sin^2 t$$

$$\mathbf{r}(t) = \cos t\mathbf{i} + \sin^2 t\mathbf{j},\ 0 \le t \le \frac{\pi}{2}$$

$$\mathbf{r}'(t) = -\sin t\mathbf{i} + 2\sin t\cos t\mathbf{j}$$

$$\|\mathbf{r}'(t)\| = \sqrt{\sin^2 t + 4\sin^2 t\cos^2 t} = \sin t\sqrt{1 + 4\cos^2 t}$$

Lateral surface area:

$$\int_C f(x, y)\,dx \int_0^{\pi/2} \cos t\sin^2 t\left(\sin t\sqrt{1 + 4\cos^2 t}\right) dt = \int_0^{\pi/2} \sin^2 t[(1 + 4\cos^2 t)^{1/2}\sin t\cos t]\,dt$$

Let $u = \sin^2 t$ and $dv = (1 + 4\cos^2 t)^{1/2}\sin t\cos t$, then $du = 2\sin t\cos t\,dt$ and $v = -\frac{1}{12}(1 + 4\cos^2 t)^{3/2}$.

$$\int_C f(x, y)\,ds = \left[-\frac{1}{2}\sin^2 t(1 + 4\cos^2 t)^{3/2}\right]_0^{\pi/2} + \frac{1}{6}\int_0^{\pi/2} (1 + 4\cos^2 t)^{3/2}\sin t\cos t\,dt$$

$$= \left[-\frac{1}{12}\sin^2 t(1 + 4\cos^2 t)^{3/2} - \frac{1}{120}(1 + 4\cos^2 t)^{5/2}\right]_0^{\pi/2}$$

$$= \left(-\frac{1}{12} - \frac{1}{120}\right) + \frac{1}{120}(5)^{5/2} = \frac{1}{120}\left(25\sqrt{5} - 11\right) \approx 0.3742$$

61. (a) $f(x, y) = 1 + y^2$

$$\mathbf{r}(t) = 2\cos t\mathbf{i} + 2\sin t\mathbf{j},\ 0 \le t \le 2\pi$$

$$\mathbf{r}'(t) = -2\sin t\mathbf{i} + 2\cos t\mathbf{j}$$

$$\|\mathbf{r}'(t)\| = 2$$

$$S = \int_C f(x, y)\,ds = \int_0^{2\pi} (1 + 4\sin^2 t)(2)\,dt$$

$$= \left[2t + 4(t - \sin t\cos t)\right]_0^{2\pi} = 12\pi \approx 37.70 \text{ cm}^2$$

(b) $0.2(12\pi) = \dfrac{12\pi}{5} \approx 7.54 \text{ cm}^3$

(c)

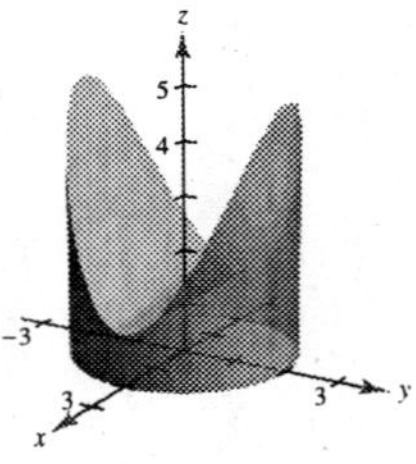

63. The greater the height of the surface over the curve, the greater the lateral surface area. Hence,

$z_3 < z_1 < z_2 < z_4$.

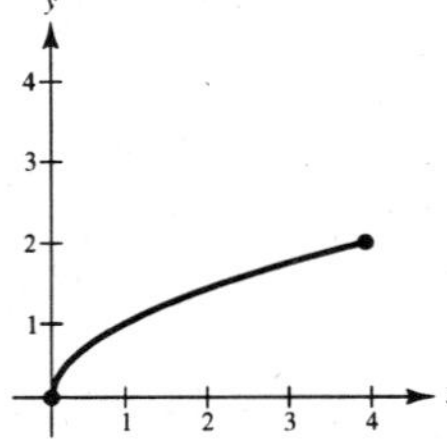

65. $S \approx 25$

Matches b

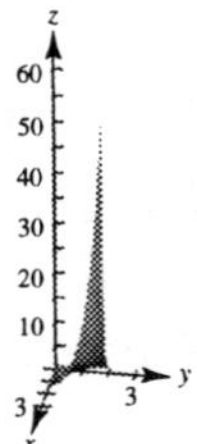

67. False

$$\int_C xy\,ds = \sqrt{2}\int_0^1 t^2\,dt$$

69. False, the orientations are different.

Section 14.3 Conservative Vector Fields and Independence of Path

1. $\mathbf{F}(x, y) = x^2\mathbf{i} + xy\mathbf{j}$

(a) $\mathbf{r}_1(t) = t\mathbf{i} + t^2\mathbf{j},\ 0 \le t \le 1$

$\mathbf{r}_1'(t) = \mathbf{i} + 2t\,\mathbf{j}$

$\mathbf{F}(t) = t^2\mathbf{i} + t^3\mathbf{j}$

$$\int_C \mathbf{F} \cdot d\mathbf{r} = \int_0^1 (t^2 + 2t^4)\,dt = \frac{11}{15}$$

(b) $\mathbf{r}_2(\theta) = \sin\theta\mathbf{i} + \sin^2\theta\mathbf{j},\ 0 \le \theta \le \frac{\pi}{2}$

$\mathbf{r}_2'(\theta) = \cos\theta\mathbf{i} + 2\sin\theta\cos 6\mathbf{j}$

$\mathbf{F}(t) = \sin^2\theta\mathbf{i} + \sin^3\theta\mathbf{j}$

$$\int_C \mathbf{F} \cdot d\mathbf{r} = \int_0^{\pi/2} (\sin^2\theta\cos\theta + 2\sin^4\theta\cos\theta)\,d\theta$$

$$= \left[\frac{\sin^3\theta}{3} + \frac{2\sin^5\theta}{5}\right]_0^{\pi/2} = \frac{11}{15}$$

3. $\mathbf{F}(x, y) = y\mathbf{i} - x\mathbf{j}$

(a) $\mathbf{r}_1(\theta) = \sec\theta\mathbf{i} + \tan\theta\mathbf{j},\ 0 \le \theta \le \frac{\pi}{3}$

$\mathbf{r}_1'(\theta) = \sec\theta\tan\theta\mathbf{i} + \sec^2\theta\mathbf{j}$

$\mathbf{F}(\theta) = \tan\theta\mathbf{i} - \sec\theta\mathbf{j}$

$$\int_C \mathbf{F} \cdot d\mathbf{r} = \int_0^{\pi/3} (\sec\theta\tan^2\theta - \sec^3\theta)\,d\theta = \int_0^{\pi/3} [\sec\theta(\sec^2\theta - 1) - \sec^3\theta]\,d\theta$$

$$= -\int_0^{\pi/3} \sec\theta\,d\theta = \Big[-\ln|\sec\theta + \tan\theta|\Big]_0^{\pi/3} = -\ln(2 + \sqrt{3}) \approx -1.317$$

(b) $\mathbf{r}_2(t) = \sqrt{t+1}\mathbf{i} + \sqrt{t}\mathbf{j},\quad 0 \le t \le 3$

$\mathbf{r}_2'(t) = \dfrac{1}{2\sqrt{t+1}}\mathbf{i} + \dfrac{1}{2\sqrt{t}}\mathbf{j}$

$\mathbf{F}(t) = \sqrt{t}\mathbf{i} - \sqrt{t+1}\mathbf{j}$

$$\int_C \mathbf{F} \cdot d\mathbf{r} = \int_0^3 \left[\frac{\sqrt{t}}{2\sqrt{t+1}} - \frac{\sqrt{t+1}}{2\sqrt{t}}\right] dt = -\frac{1}{2}\int_0^3 \frac{1}{\sqrt{t}\sqrt{t+1}}\,dt = -\frac{1}{2}\int_0^3 \frac{1}{\sqrt{t^2 + t + (1/4) - (1/4)}}\,dt$$

$$= -\frac{1}{2}\int_0^3 \frac{1}{\sqrt{[t + (1/2)]^2 - (1/4)}}\,dt = \left[-\frac{1}{2}\ln\left|\left(t + \frac{1}{2}\right) + \sqrt{t^2 + t}\right|\right]_0^3$$

$$= -\frac{1}{2}\left[\ln\left(\frac{7}{2} + 2\sqrt{3}\right) - \ln\left(\frac{1}{2}\right)\right] = -\frac{1}{2}\ln(7 + 4\sqrt{3}) \approx -1.317$$

5. $\mathbf{F}(x, y) = e^x\sin y\mathbf{i} + e^x\cos y\mathbf{j}$

$\dfrac{\partial N}{\partial x} = e^x\cos y \qquad \dfrac{\partial M}{\partial y} = e^x\cos y$

Since $\dfrac{\partial N}{\partial x} = \dfrac{\partial M}{\partial y}$, $\mathbf{F}$ is conservative.

7. $\mathbf{F}(x, y) = \dfrac{1}{y}\mathbf{i} + \dfrac{x}{y^2}\mathbf{j}$

$\dfrac{\partial N}{\partial x} = \dfrac{1}{y^2} \qquad \dfrac{\partial M}{\partial y} = -\dfrac{1}{y^2}$

Since $\dfrac{\partial N}{\partial x} \ne \dfrac{\partial M}{\partial y}$, $\mathbf{F}$ is not conservative.

9. $\mathbf{F}(x, y, z) = y^2z\mathbf{i} + 2xyz\mathbf{j} + xy^2\mathbf{k}$

curl F $= \mathbf{0} \implies \mathbf{F}$ is conservative.

11. $\mathbf{F}(x, y) = 2xy\mathbf{i} + x^2\mathbf{j}$

(a) $\mathbf{r}_1(t) = t\mathbf{i} + t^2\mathbf{j},\ 0 \le t \le 1$

$\mathbf{r}_1'(t) = \mathbf{i} + 2t\mathbf{j}$

$\mathbf{F}(t) = 2t^3\mathbf{i} + t^2\mathbf{j}$

$$\int_C \mathbf{F} \cdot d\mathbf{r} = \int_0^1 4t^3\,dt = 1$$

(b) $\mathbf{r}_2(t) = t\mathbf{i} + t^3\mathbf{j},\ 0 \le t \le 1$

$\mathbf{r}_2'(t) = \mathbf{i} + 3t^2\mathbf{j}$

$\mathbf{F}(t) = 2t^4\mathbf{i} + t^2\mathbf{j}$

$$\int_C \mathbf{F} \cdot d\mathbf{r} = \int_0^1 5t^4\,dt = 1$$

13. $\mathbf{F}(x, y) = y\mathbf{i} - x\mathbf{j}$

(a) $\mathbf{r}_1(t) = t\mathbf{i} + t\mathbf{j},\quad 0 \le t \le 1$

$\mathbf{r}_1'(t) = \mathbf{i} + \mathbf{j}$

$\mathbf{F}(t) = t\mathbf{i} - t\mathbf{j}$

$$\int_C \mathbf{F} \cdot d\mathbf{r} = 0$$

(b) $\mathbf{r}_2(t) = t\mathbf{i} + t^2\mathbf{j},\quad 0 \le t \le 1$

$\mathbf{r}_2'(t) = \mathbf{i} + 2t\mathbf{j}$

$\mathbf{F}(t) = t^2\mathbf{i} - t\mathbf{j}$

$$\int_C \mathbf{F} \cdot d\mathbf{r} = \int_0^1 -t^2\,dt = -\frac{1}{3}$$

(c) $\mathbf{r}_3(t) = t\mathbf{i} + t^3\mathbf{j},\quad 0 \le t \le 1$

$\mathbf{r}_3'(t) = \mathbf{i} + 3t^2\mathbf{j}$

$\mathbf{F}(t) = t^3\mathbf{i} - t\mathbf{j}$

$$\int_C \mathbf{F} \cdot d\mathbf{r} = \int_0^1 -2t^3\,dt = -\frac{1}{2}$$

15. $\displaystyle\int_C y^2\,dx + 2xy\,dy$

Since $\partial M/\partial y = \partial N/\partial x = 2y$, $\mathbf{F}(x, y) = y^2\mathbf{i} + 2xy\mathbf{j}$ is conservative. The potential function is $f(x, y) = xy^2 + k$. Therefore, we can use the Fundamental Theorem of Line Integrals.

(a) $\displaystyle\int_C y^2\,dx + 2xy\,dy = \Big[x^2y\Big]_{(0,0)}^{(4,4)} = 64$

(b) $\displaystyle\int_C y^2\,dx + 2xy\,dy = \Big[x^2y\Big]_{(-1,0)}^{(1,0)} = 0$

(c) and (d) Since C is a closed curve, $\displaystyle\int_C y^2\,dx + 2xy\,dy = 0$.

17. $\displaystyle\int_C 2xy\,dx + (x^2 + y^2)\,dy$

Since $\partial M/\partial y = \partial N/\partial x = 2x$,

$\mathbf{F}(x, y) = 2xy\mathbf{i} + (x^2 + y^2)\mathbf{j}$

is conservative. The potential function is 0

(a) $\displaystyle\int_C 2xy\,dx + (x^2 + y^2)\,dy = \left[x^2y + \frac{y^3}{3}\right]_{(5,0)}^{(0,4)} = \frac{64}{3}$

(b) $\displaystyle\int_C 2xy\,dx + (x^2 + y^2)\,dy = \left[x^2y + \frac{y^3}{3}\right]_{(2,0)}^{(0,4)} = \frac{64}{3}$

19. $\mathbf{F}(x, y, z) = yz\mathbf{i} + xz\mathbf{j} + xy\mathbf{k}$

Since **curl F** $= \mathbf{0}$, $\mathbf{F}(x, y, z)$ is conservative. The potential function is $f(x, y, z) = xyz + k$.

(a) $\mathbf{r}_1(t) = t\mathbf{i} + 2\mathbf{j} + t\mathbf{k},\ 0 \le t \le 4$

$$\int_C \mathbf{F} \cdot d\mathbf{r} = \Big[xyz\Big]_{(0,2,0)}^{(4,2,4)} = 32$$

(b) $\mathbf{r}_2(t) = t^2\mathbf{i} + t\mathbf{j} + t^2\mathbf{k},\ 0 \le t \le 2$

$$\int_C \mathbf{F} \cdot d\mathbf{r} = \Big[xyz\Big]_{(0,0,0)}^{(4,2,4)} = 32$$

21. $\mathbf{F}(x, y, z) = (2y + x)\mathbf{i} + (x^2 - z)\mathbf{j} + (2y - 4z)\mathbf{k}$

$\mathbf{F}(x, y, z)$ is not conservative.

(a) $\mathbf{r}_1(t) = t\mathbf{i} + t^2\mathbf{j} + \mathbf{k},\ 0 \le t \le 1$

$\mathbf{r}_1'(t) = \mathbf{i} + 2t\mathbf{j}$

$\mathbf{F}(t) = (2t^2 + t)\mathbf{i} + (t^2 - 1)\mathbf{j} + (2t^2 - 4)\mathbf{k}$

$$\int_C \mathbf{F} \cdot d\mathbf{r} = \int_0^1 (2t^3 + 2t^2 - t)\,dt = \frac{2}{3}$$

—CONTINUED—

21. —CONTINUED—

(b) $\mathbf{r}_2(t) = t\mathbf{i} + t\mathbf{j} + (2t-1)^2\mathbf{k},\ 0 \le t \le 1$

$\mathbf{r}_2'(t) = \mathbf{i} + \mathbf{j} + 4(2t-1)\mathbf{k}$

$\mathbf{F}(t) = 3t\mathbf{i} + [t^2 - (2t-1)^2]\mathbf{j} + [2t - 4(2t-1)^2\mathbf{k}$

$$\int_C \mathbf{F}\cdot d\mathbf{r} = \int_0^1 [3t + t^2 - (2t-1)^2 + 8t(2t-1) - 16(2t-1)^3]\,dt$$

$$= \int_0^1 [17t^2 - 5t - (2t-1)^2 - 16(2t-1)^3]\,dt = \left[\frac{17t^3}{3} - \frac{5t^2}{2} - \frac{(2t-1)^3}{6} - 2(2t-1)^4\right]_0^1 = \frac{17}{6}$$

23. $\mathbf{F}(x, y, z) = e^z(y\mathbf{i} + x\mathbf{j} + xy\mathbf{k})$

$\mathbf{F}(x, y, z)$ is conservative. The potential function is $f(x, y, z) = xye^z + k$.

(a) $\mathbf{r}_1(t) = 4\cos t\mathbf{i} + 4\sin t\mathbf{j} + 3\mathbf{k},\ 0 \le t \le \pi$

$$\int_C \mathbf{F}\cdot d\mathbf{r} = \Big[xye^z\Big]_{(4,0,3)}^{(-4,0,3)} = 0$$

(b) $\mathbf{r}_2(t) = (4 - 8t)\mathbf{i} + 3\mathbf{k},\ 0 \le t \le 1$

$$\int_C \mathbf{F}\cdot d\mathbf{r} = \Big[xye^z\Big]_{(4,0,3)}^{(-4,0,3)} = 0$$

25. $$\int_C (y\mathbf{i} + x\mathbf{j})\cdot d\mathbf{r} = \Big[xy\Big]_{(0,0)}^{(3,8)} = 24$$

27. $$\int_C \cos x\sin y\,dx + \sin x\cos y\,dy = \Big[\sin x\sin y\Big]_{(0,-\pi)}^{(3\pi/2,\,\pi/2)} = -1$$

29. $$\int_C e^x\sin y\,dx + e^x\cos y\,dy = \Big[e^x\sin y\Big]_{(0,0)}^{(2\pi,0)} = 0$$

31. $$\int_C (z + 2y)\,dx + (2x - z)\,dy + (x - y)\,dz$$

Note: Since $\mathbf{F}(x, y, z) = (z + 2y)\mathbf{i} + (2x - z)\mathbf{j} + (x - y)\mathbf{k}$ is conservative and the potential function is $f(x, y, z) = xz + 2xy - yz + k$, the integral is independent of path as illustrated below.

(a) $\Big[xz + 2xy - yz\Big]_{(0,0,0)}^{(1,1,1)} = 2$

(b) $\Big[xz + 2xy - yz\Big]_{(0,0,0)}^{(0,0,1)} + \Big[xz + 2xy - yz\Big]_{(0,0,1)}^{(1,1,1)} = 0 + 2 = 2$

(c) $\Big[xz + 2xy - yz\Big]_{(0,0,0)}^{(1,0,0)} + \Big[xz + 2xy - yz\Big]_{(1,0,0)}^{(1,1,0)} + \Big[xz + 2xy - yz\Big]_{(1,1,0)}^{(1,1,1)} = 0 + 2 + 0 = 2$

33. $$\int_C -\sin x\,dx + z\,dy + y\,dz = \Big[\cos x + yz\Big]_{(0,0,0)}^{(\pi/2,\,3,\,4)} = 12 - 1 = 11$$

35. $\mathbf{F}(x, y) = 9x^2y^2\mathbf{i} + (6x^3y - 1)\mathbf{j}$ is conservative.

$$\text{Work} = \Big[3x^3y^2 - y\Big]_{(0,0)}^{(5,9)} = 30{,}366$$

37. $\mathbf{r}(t) = 2\cos 2\pi t\mathbf{i} + 2\sin 2\pi t\mathbf{j}$

$\mathbf{r}'(t) = -4\pi\sin 2\pi t\mathbf{i} + 4\pi\cos 2\pi t\mathbf{j}$

$\mathbf{a}(t) = -8\pi^2\cos 2\pi t\mathbf{i} - 8\pi^2\sin 2\pi t\mathbf{j}$

$\mathbf{F}(t) = m \cdot \mathbf{a}(t) = \dfrac{1}{32}\mathbf{a}(t) = -\dfrac{\pi^2}{4}(\cos 2\pi t\mathbf{i} + \sin 2\pi t\mathbf{j})$

$$W = \int_C \mathbf{F} \cdot d\mathbf{r} = \int_C -\frac{\pi^2}{4}(\cos 2\pi t\mathbf{i} + \sin 2\pi t\mathbf{j}) \cdot 4\pi(-\sin 2\pi t\mathbf{i} + \cos 2\pi t\mathbf{j})\,dt = -\pi^3\int_C 0\,dt = 0$$

39. Since the sum of the potential and kinetic energies remains constant from point to point, if the kinetic energy is decreasing at a rate of 10 units per minute, then the potential energy is increasing at a rate of 10 units per minute.

41. No. The force field is conservative.

43. False, it would be true if $\mathbf{F}$ were conservative.

45. True

47. Let

$$\mathbf{F} = M\mathbf{i} + N\mathbf{j} = \frac{\partial f}{\partial y}\mathbf{i} - \frac{\partial f}{\partial x}\mathbf{j}.$$

Then $\dfrac{\partial M}{\partial y} = \dfrac{\partial}{\partial y}\left(\dfrac{\partial f}{\partial y}\right) = \dfrac{\partial^2 f}{\partial y^2}$ and $\dfrac{\partial N}{\partial x} = \dfrac{\partial}{\partial x}\left(-\dfrac{\partial f}{\partial x}\right) = -\dfrac{\partial^2 f}{\partial x^2}$. Since

$\dfrac{\partial^2 f}{\partial x^2} + \dfrac{\partial^2 f}{\partial y^2} = 0$ we have $\dfrac{\partial M}{\partial y} = \dfrac{\partial N}{\partial x}$.

Thus, $\mathbf{F}$ is conservative. Therefore, by Theorem 14.7, we have

$$\int_C \left(\frac{\partial f}{\partial y}dx - \frac{\partial f}{\partial x}dy\right) = \int_C (M\,dx + N\,dy) = \int_C \mathbf{F} \cdot d\mathbf{r} = 0$$

for every closed curve in the plane.

49. No, the amount of fuel required depends on the flight path. Fuel consumption is dependent on wind speed and direction. The vector field is not conservative.

Section 14.4 Green's Theorem

1. $\mathbf{r}(t) = \begin{cases} t\mathbf{i}, & 0 \le t \le 4 \\ 4\mathbf{i} + (t-4)\mathbf{j}, & 4 \le t \le 8 \\ (12-t)\mathbf{i} + 4\mathbf{j}, & 8 \le t \le 12 \\ (16-t)\mathbf{j}, & 12 \le t \le 16 \end{cases}$

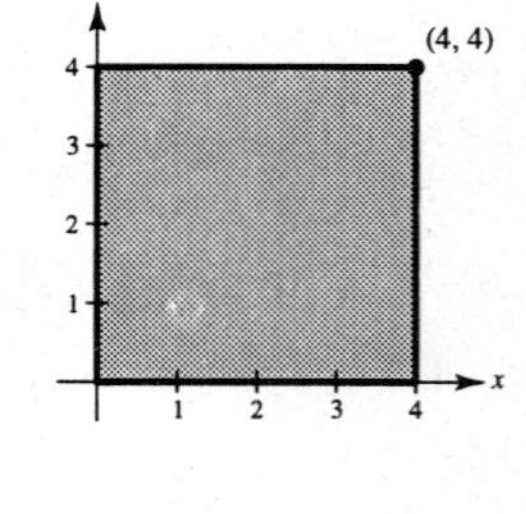

$$\int_C y^2\,dx + x^2\,dy = \int_0^4 [0\,dt + t^2(0)] + \int_4^8 [(t-4)^2(0) + 16\,dt]$$

$$+ \int_8^{12} [16(-dt) + (12-t)^2(0)] + \int_{12}^{16} [(16-t)^2(0) + 0(-dt)]$$

$$= 0 + 64 - 64 + 0 = 0$$

By Green's Theorem, $\displaystyle\int_R\!\!\int \left(\frac{\partial N}{\partial x} - \frac{\partial M}{\partial y}\right)dA = \int_0^4\!\!\int_0^4 (2x - 2y)\,dy\,dx = \int_0^4 (8x - 16)\,dx = 0.$

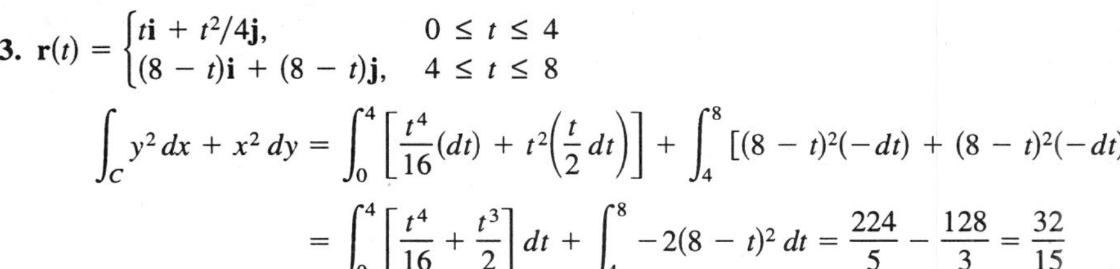

3. $\mathbf{r}(t) = \begin{cases} t\mathbf{i} + t^2/4\mathbf{j}, & 0 \le t \le 4 \\ (8-t)\mathbf{i} + (8-t)\mathbf{j}, & 4 \le t \le 8 \end{cases}$

$$\int_C y^2\,dx + x^2\,dy = \int_0^4 \left[\frac{t^4}{16}(dt) + t^2\left(\frac{t}{2}\,dt\right)\right] + \int_4^8 [(8-t)^2(-dt) + (8-t)^2(-dt)]$$

$$= \int_0^4 \left[\frac{t^4}{16} + \frac{t^3}{2}\right] dt + \int_4^8 -2(8-t)^2\,dt = \frac{224}{5} - \frac{128}{3} = \frac{32}{15}$$

By Green's Theorem,

$$\int_R\!\!\int \left(\frac{\partial N}{\partial x} - \frac{\partial M}{\partial y}\right) dA = \int_0^4\!\int_{x^2/4}^{x} (2x - 2y)\,dy\,dx = \int_0^4 \left(x^2 - \frac{x^3}{2} + \frac{x^4}{16}\right) dx = \frac{32}{15}.$$

5. C: $x^2 + y^2 = 4$

Let $x = 2\cos t$ and $y = 2\sin t$, $0 \le t \le 2\pi$.

$$\int_C xe^y\,dx + e^x\,dy = \int_0^{2\pi} [2\cos t e^{2\sin t}(-2\sin t) + e^{2\cos t}(2\cos t)]\,dt \approx 19.99$$

$$\int_R\!\!\int \left(\frac{\partial N}{\partial x} - \frac{\partial M}{\partial y}\right) dA = \int_{-2}^{2}\int_{-\sqrt{4-x^2}}^{\sqrt{4-x^2}} (e^x - xe^y)\,dy\,dx = \int_{-2}^{2} \left[2\sqrt{4-x^2}\,e^x - xe^{\sqrt{4-x^2}} + xe^{-\sqrt{4-x^2}}\right] dx \approx 19.99$$

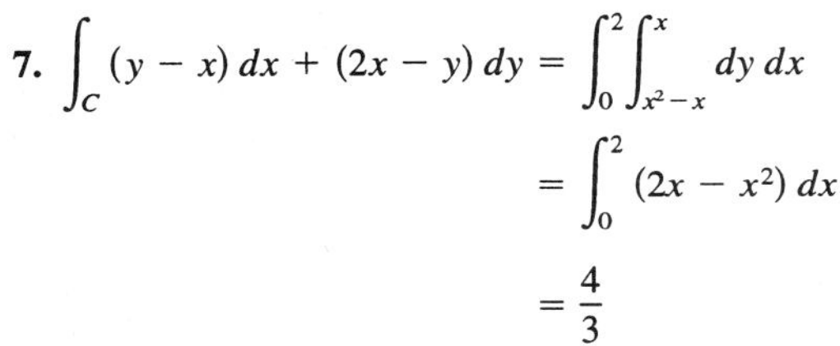

7. $$\int_C (y-x)\,dx + (2x-y)\,dy = \int_0^2\!\int_{x^2-x}^{x} dy\,dx$$

$$= \int_0^2 (2x - x^2)\,dx$$

$$= \frac{4}{3}$$

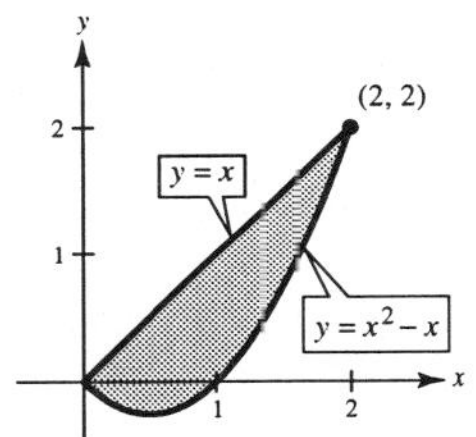

9. From the accompanying figure, we see that R is the shaded region. Thus, Green's Theorem yields

$$\int_C (y-x)\,dx + (2x-y)\,dy = \int_R\!\!\int 1\,dA$$

$$= \text{Area of } R$$

$$= 6(10) - 2(2)$$

$$= 56.$$

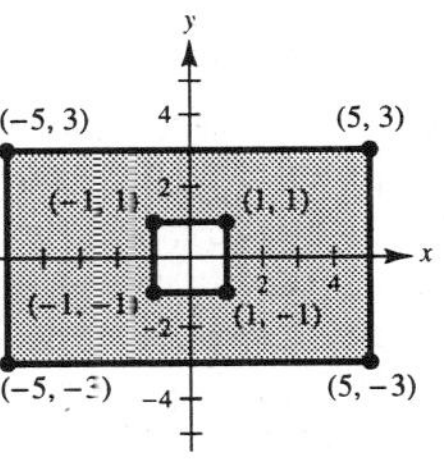

11. Since the curves $y = 0$ and $y = 4 - x^2$ intersect at $(-2, 0)$ and $(2, 0)$, Green's Theorem yields

$$\int_C 2xy\,dx + (x+y)\,dy = \int_R\!\!\int (1-2x)\,dA = \int_{-2}^{2}\int_0^{4-x^2} (1-2x)\,dy\,dx$$

$$= \int_{-2}^{2} \left[y - 2xy\right]_0^{4-x^2} dx$$

$$= \int_{-2}^{2} (4 - 8x - x^2 + 2x^3)\,dx$$

$$= \left[4x - 4x^2 - \frac{x^3}{3} + \frac{x^4}{2}\right]_{-2}^{2}$$

$$= -\frac{8}{3} - \frac{8}{3} + 16 = \frac{32}{3}.$$

13. Since R is the interior of the circle $x^2 + y^2 = a^2$, Green's Theorem yields

$$\int_C (x^2 - y^2)\,dx + 2xy\,dy = \iint_R (2y + 2y)\,dA$$

$$= \int_{-a}^{a}\int_{-\sqrt{a^2-x^2}}^{\sqrt{a^2-x^2}} 4y\,dy\,dx$$

$$= 4\int_{-a}^{a} 0\,dx = 0.$$

15. Since $\dfrac{\partial M}{\partial y} = \dfrac{2x}{x^2 + y^2} = \dfrac{\partial N}{\partial x}$,

we have path independence and

$$\iint_R \left(\frac{\partial N}{\partial x} - \frac{\partial M}{\partial y}\right) dA = 0.$$

17. By Green's Theorem,

$$\int_C \sin x \cos y\,dx + (xy + \cos x \sin y)\,dy = \iint_R [(y - \sin x \sin y) - (-\sin x \sin y)]\,dA$$

$$= \int_0^1\int_x^{\sqrt{x}} y\,dy\,dx = \frac{1}{2}\int_0^1 (x - x^2)\,dx = \frac{1}{2}\left[\frac{x^2}{2} - \frac{x^3}{3}\right]_0^1 = \frac{1}{12}.$$

19. By Green's Theorem,

$$\int_C xy\,dx + (x + y)\,dy = \iint_R (1 - x)\,dA$$

$$= \int_0^{2\pi}\int_1^3 (1 - r\cos\theta)r\,dr\,d\theta = \int_0^{2\pi}\left(4 - \frac{26}{3}\cos\theta\right)d\theta = 8\pi.$$

21. $\mathbf{F}(x, y) = xy\mathbf{i} + (x + y)\mathbf{j}$

C: $x^2 + y^2 = 4$

$$\text{Work} = \int_C xy\,dx + (x + y)\,dy = \iint_R (1 - x)\,dA = \int_0^{2\pi}\int_0^2 (1 - r\cos\theta)r\,dr\,d\theta = \int_0^{2\pi}\left(2 - \frac{8}{3}\cos\theta\right)d\theta = 4\pi$$

23. $\mathbf{F}(x, y) = (x^{3/2} - 3y)\mathbf{i} + (6x + 5\sqrt{y})\mathbf{j}$

C: boundary of the triangle with vertices $(0, 0)$, $(5, 0)$, $(0, 5)$

$$\text{Work} = \int_C (x^{3/2} - 3y)\,dx + \left(6x + 5\sqrt{y}\right)dy = \iint_R 9\,dA = 9\left(\frac{1}{2}\right)(5)(5) = \frac{225}{2}$$

25. C: let $x = a\cos t$, $y = a\sin t$, $0 \le t \le 2\pi$. By Theorem 14.9, we have

$$A = \frac{1}{2}\int_C x\,dy - y\,dx = \frac{1}{2}\int_0^{2\pi} [a\cos t(a\cos t) - a\sin t(-a\sin t)]\,dt = \frac{1}{2}\int_0^{2\pi} a^2\,dt = \left[\frac{a^2}{2}t\right]_0^{2\pi} = \pi a^2.$$

27. From the accompanying figure we see that

C_1: $y = 2x + 1$, $dy = 2\,dx$

C_2: $y = 4 - x^2$, $dy = -2x\,dx$.

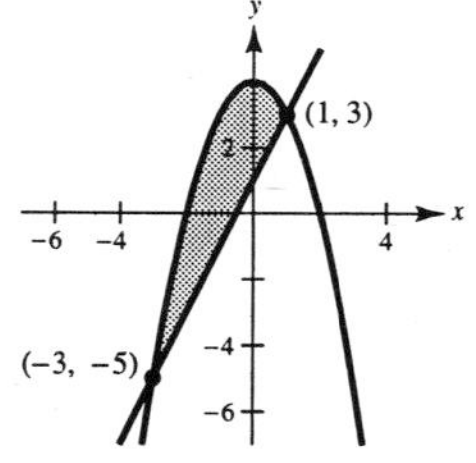

Thus, by Theorem 14.9, we have

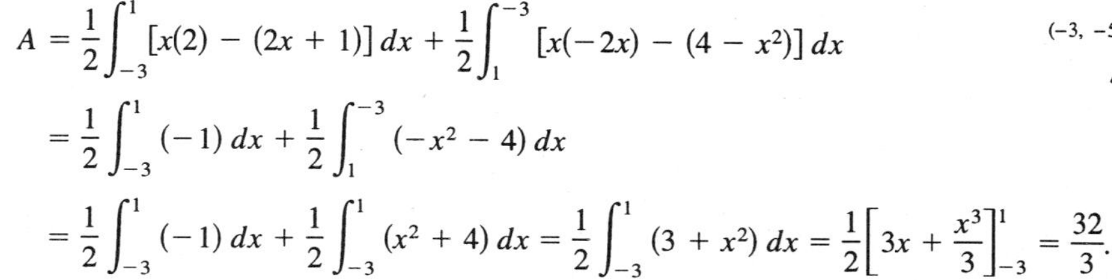

$$A = \frac{1}{2}\int_{-3}^{1} [x(2) - (2x + 1)]\,dx + \frac{1}{2}\int_1^{-3} [x(-2x) - (4 - x^2)]\,dx$$

$$= \frac{1}{2}\int_{-3}^{1} (-1)\,dx + \frac{1}{2}\int_1^{-3} (-x^2 - 4)\,dx$$

$$= \frac{1}{2}\int_{-3}^{1} (-1)\,dx + \frac{1}{2}\int_{-3}^{1} (x^2 + 4)\,dx = \frac{1}{2}\int_{-3}^{1} (3 + x^2)\,dx = \frac{1}{2}\left[3x + \frac{x^3}{3}\right]_{-3}^{1} = \frac{32}{3}.$$

29. For the moment about the x-axis, $M_x = \iint_R y\, dA$. Let $N = 0$ and $M = -y^2/2$. By Green's Theorem,

$$M_x = \int_C -\frac{y^2}{2}\, dx = -\frac{1}{2}\int_C y^2\, dx \text{ and } \bar{y} = \frac{M_x}{2A} = -\frac{1}{2A}\int_C y^2\, dx.$$

For the moment about the y-axis, $M_y = \iint_R x\, dA$. Let $N = x^2/2$ and $M = 0$. By Green's Theorem,

$$M_y = \int_C \frac{x^2}{2}\, dy = \frac{1}{2}\int_C x^2\, dy \text{ and } \bar{x} = \frac{M_y}{2A} = \frac{1}{2A}\int_C x^2\, dy.$$

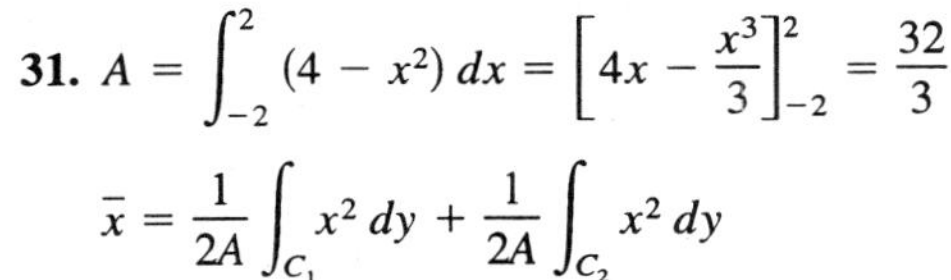

31. $A = \int_{-2}^{2} (4 - x^2)\, dx = \left[4x - \frac{x^3}{3}\right]_{-2}^{2} = \frac{32}{3}$

$$\bar{x} = \frac{1}{2A}\int_{C_1} x^2\, dy + \frac{1}{2A}\int_{C_2} x^2\, dy$$

For C_1, $dy = -2x\, dx$ and for C_2, $dy = 0$. Thus,

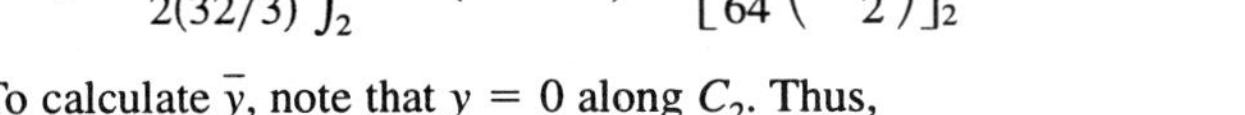

$$\bar{x} = \frac{1}{2(32/3)}\int_{2}^{-2} x^2(-2x\, dx) = \left[\frac{3}{64}\left(-\frac{x^4}{2}\right)\right]_{2}^{-2} = 0.$$

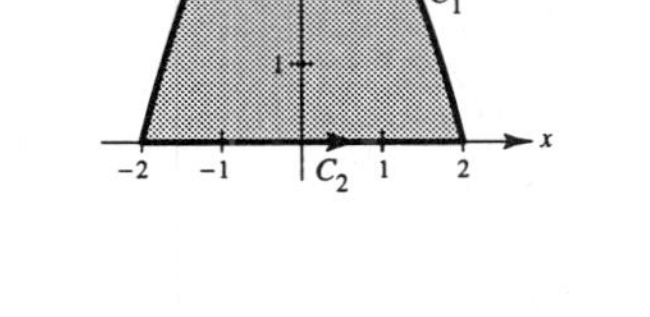

To calculate $\bar{y}$, note that $y = 0$ along C_2. Thus,

$$\bar{y} = \frac{-1}{2(32/3)}\int_{2}^{-2} (4 - x^2)^2\, dx = \frac{3}{64}\int_{-2}^{2} (16 - 8x^2 + x^4)\, dx = \frac{3}{64}\left[16x - \frac{8x^3}{3} + \frac{x^5}{5}\right]_{-2}^{2} = \frac{8}{5}.$$

33. Since $A = \int_0^1 (x - x^3)\, dx = \left[\frac{x^2}{2} - \frac{x^4}{4}\right]_0^1 = \frac{1}{4}$, we have $\frac{1}{2A} = 2$. On C_1 we have $y = x^3$, $dy = 3x^2\, dx$ and on C_2 we have $y = x$, $dy = dx$. Thus,

$$\bar{x} = 2\int_C x^2\, dy = 2\int_{C_1} x^2(3x^2\, dx) + 2\int_{C_2} x^2\, dx$$

$$= 6\int_0^1 x^4\, dx + 2\int_1^0 x^2\, dx = \frac{6}{5} - \frac{2}{3} = \frac{8}{15}$$

$$\bar{y} = -2\int_C y^2\, dx$$

$$= -2\int_0^1 x^6\, dx - 2\int_1^0 x^2\, dx = -\frac{2}{7} + \frac{2}{3} = \frac{8}{21}.$$

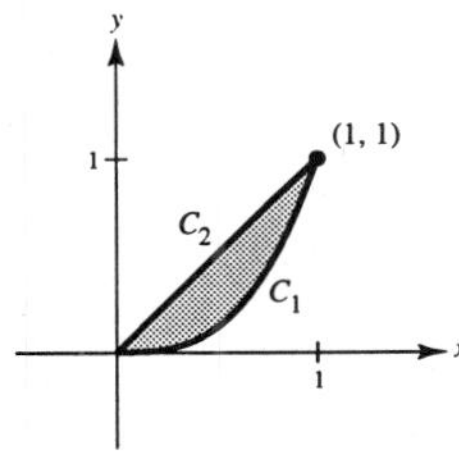

35. $A = \frac{1}{2}\int_0^{2\pi} a^2(1 - \cos\theta)^2\, d\theta$

$$= \frac{a^2}{2}\int_0^{2\pi}\left(1 - 2\cos\theta + \frac{1}{2} + \frac{\cos 2\theta}{2}\right) d\theta = \frac{a^2}{2}\left[\frac{3\theta}{2} - 2\sin\theta + \frac{1}{2}\sin 2\theta\right]_0^{2\pi} = \frac{a^2}{2}(3\pi) = \frac{3\pi a^2}{2}$$

37. In this case the inner loop has domain $\frac{2\pi}{3} \le \theta \le \frac{4\pi}{3}$. Thus,

$$A = \frac{1}{2}\int_{2\pi/3}^{4\pi/3} (1 + 4\cos\theta + 4\cos^2\theta)\, d\theta$$

$$= \frac{1}{2}\int_{2\pi/3}^{4\pi/3} (3 + 4\cos\theta + 2\cos 2\theta)\, d\theta = \frac{1}{2}\Big[3\theta + 4\sin\theta + \sin 2\theta\Big]_{2\pi/3}^{4\pi/3} = \pi - \frac{3\sqrt{3}}{2}.$$

39. $I = \int_C \dfrac{y\,dx - x\,dy}{x^2 + y^2}$

(a) Let $\mathbf{F} = \dfrac{y}{x^2+y^2}\mathbf{i} - \dfrac{x}{x^2+y^2}\mathbf{j}$.

$\mathbf{F}$ is conservative since $\dfrac{\partial N}{\partial x} = \dfrac{\partial M}{\partial y} = \dfrac{x^2 - y^2}{(x^2+y^2)^2}$.

$\mathbf{F}$ is defined and has continuous first partials everywhere except at the origin. If C is a circle (a closed path) that does not contain the origin, then

$$\int_C \mathbf{F}\cdot d\mathbf{r} = \int_C M\,dx + N\,dy = \int\!\!\int_R \left(\frac{\partial N}{\partial x} - \frac{\partial M}{\partial y}\right) dA = 0.$$

(b) Let $\mathbf{r} = a\cos t\mathbf{i} - a\sin t\mathbf{j}$, $0 \le t \le 2\pi$ be a circle C_1 oriented clockwise inside C (see figure). Introduce line segments C_2 and C_3 as illustrated in Example 6 of this section in the text. For the region inside C and outside C_1, Green's Theorem applies. Note that since C_2 and C_3 have opposite orientations, the line integrals over them cancel. Thus, $C_4 = C_1 + C_2 + C + C_3$ and

$$\int_{C_4} \mathbf{F}\cdot d\mathbf{r} = \int_{C_1} \mathbf{F}\cdot d\mathbf{r} + \int_C \mathbf{F}\cdot d\mathbf{r} = 0.$$

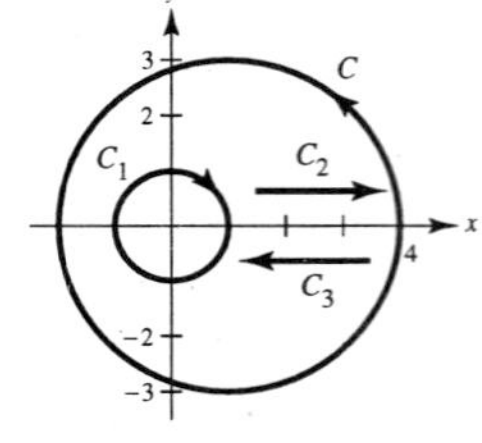

But,

$$\int_{C_1} \mathbf{F}\cdot d\mathbf{r} = \int_0^{2\pi} \left[\frac{(-a\sin t)(-a\sin t)}{a^2\cos^2 t + a^2\sin^2 t} + \frac{(-a\cos t)(-a\cos t)}{a^2\cos^2 t + a^2\sin^2 t}\right] dt$$

$$= \int_0^{2\pi} (\sin^2 t + \cos^2 t)\,dt = \Big[t\Big]_0^{2\pi} = 2\pi.$$

Finally, $\displaystyle\int_C \mathbf{F}\cdot d\mathbf{r} = -\int_{C_1} \mathbf{F}\cdot d\mathbf{r} = -2\pi.$

Note: If C were orientated clockwise, then the answer would have been 2π.

41. Pentagon: $(0, 0), (2, 0), (3, 2), (1, 4), (-1, 1)$

$A = \frac{1}{2}[(0-0) + (4-0) + (12-2) + (1+4) + (0-0)] = \frac{19}{2}$

43. $\displaystyle\int_C y^n\,dx + x^n\,dy = \int\!\!\int_R \left(\frac{\partial N}{\partial x} - \frac{\partial M}{\partial y}\right) dA$

For the line integral, use the two paths

C_1: $\mathbf{r}_1(x) = x\mathbf{i}$, $-a \le x \le a$

C_2: $\mathbf{r}_2(x) = x\mathbf{i} + \sqrt{a^2 - x^2}\mathbf{j}$, $x = a$ to $x = -a$

$$\int_{C_1} y^n\,dx + x^n\,dy = 0$$

$$\int_{C_2} y^n\,dx + x^n\,dy = \int_a^{-a} \left[(a^2 - x^2)^{n/2} + x^n \frac{-x}{\sqrt{a^2 - x^2}}\right] dx$$

$$\int\!\!\int_R \left(\frac{\partial N}{\partial x} - \frac{\partial M}{\partial y}\right) dA = \int_{-a}^{a}\int_0^{\sqrt{a^2-x^2}} [nx^{n-1} - ny^{n-1}]\,dy\,dx$$

(a) For $n = 1, 3, 5, 7$, both integrals give 0.

(b) For n even, you obtain

$n = 2: -\frac{4}{3}a^3$ $\quad n = 4: -\frac{16}{15}a^5$ $\quad n = 6: -\frac{32}{35}a^7$ $\quad n = 8: -\frac{256}{315}a^9$

(c) If n is odd and $0 < a < 1$, then the integral equals 0.

45. $\int_C (fD_N g - gD_N f)\, ds = \int_C fD_N g\, ds - \int_C gD_N f\, ds$

$$= \iint_R (f\nabla^2 g + \nabla f \cdot \nabla g)\, dA - \iint_R (g\nabla^2 f + \nabla g \cdot \nabla f)\, dA = \iint_R (f\nabla^2 g - g\nabla^2 f)\, dA$$

47. $\mathbf{F} = M\mathbf{i} + N\mathbf{j}$

$$\textbf{curl F} = \left(\frac{\partial N}{\partial x} - \frac{\partial M}{\partial y}\right)\mathbf{k} = 0 \Rightarrow \frac{\partial N}{\partial x} = \frac{\partial M}{\partial y}$$

$$\int_C \mathbf{F} \cdot d\mathbf{r} = \int_C M\, dx + N\, dy = \iint_R \left(\frac{\partial N}{\partial x} - \frac{\partial M}{\partial y}\right) dA = \iint_R (0)\, dA = 0$$

Section 14.5 Parametric Surfaces

1. $\mathbf{r}(u, v) = u\mathbf{i} + v\mathbf{j} + uv\mathbf{k}$

$z = xy$

Matches c.

3. $\mathbf{r}(u, v) = 2\cos v \cos u\mathbf{i} + 2\cos v \sin u\mathbf{j} + 2\sin v\mathbf{k}$

$x^2 + y^2 + z^2 = 4$

Matches b.

5. $\mathbf{r}(u, v) = u\mathbf{i} + v\mathbf{j} + \frac{v}{2}\mathbf{k}$

$y - 2z = 0$

Plane

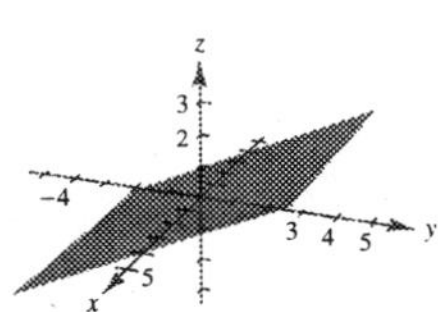

7. $\mathbf{r}(u, v) = 2\cos u\mathbf{i} + v\mathbf{j} + 2\sin u\mathbf{k}$

$x^2 + z^2 = 4$

Cylinder

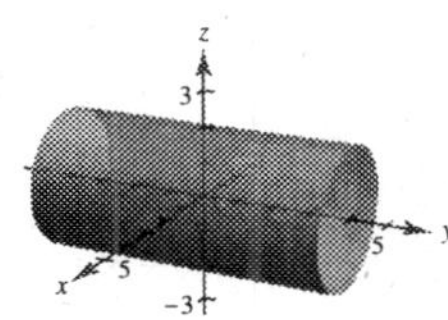

For Exercises 9 and 11,

$$\mathbf{r}(u, v) = u\cos v\mathbf{i} + u\sin v\mathbf{j} + u^2\mathbf{k},\ 0 \le u \le 2,\ 0 \le v \le 2\pi.$$

Eliminating the parameter yields

$$z = x^2 + y^2,\ 0 \le z \le 4.$$

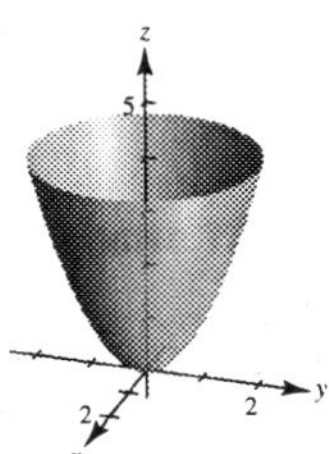

9. $\mathbf{s}(u, v) = u\cos v\mathbf{i} + u\sin v\mathbf{j} - u^2\mathbf{k},\ 0 \le u \le 2,\ 0 \le v \le 2\pi$

$z = -(x^2 + y^2)$

The paraboloid is reflected (inverted) through the xy-plane.

11. $\mathbf{s}(u, v) = u\cos v\mathbf{i} - u\sin v\mathbf{j} - u^2\mathbf{k},\ 0 \le u \le 3,\ 0 \le v \le 2\pi$

The height of the paraboloid is increased from 4 to 9.

13. $\mathbf{r}(u, v) = 2u \cos v\mathbf{i} + 2u \sin v\mathbf{j} + u^4\mathbf{k}$,

$0 \le u \le 1,\ 0 \le v \le 2\pi$

$z = \dfrac{(x^2 + y^2)^2}{16}$

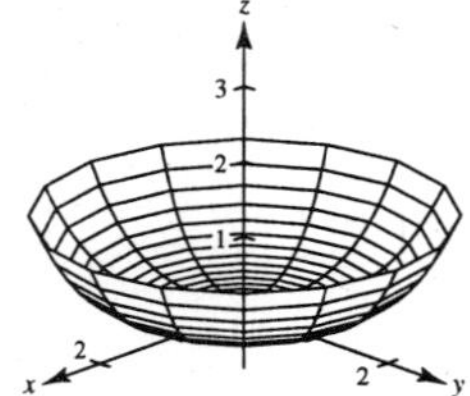

15. $\mathbf{r}(u, v) = 2 \sinh u \cos v\mathbf{i} + \sinh u \sin v\mathbf{j} + \cosh u\mathbf{k}$,

$0 \le u \le 2,\ 0 \le v \le 2\pi$

$\dfrac{z^2}{1} - \dfrac{x^2}{4} - \dfrac{y^2}{1} = 1$

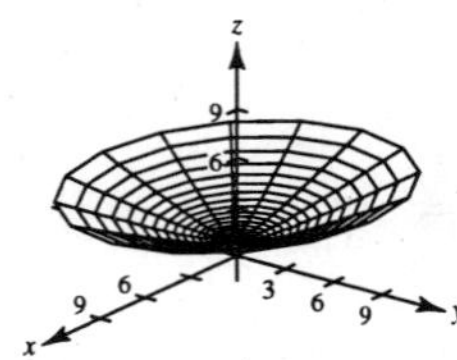

17. $\mathbf{r}(u, v) = (u - \sin u) \cos v\mathbf{i} + (1 - \cos u) \sin v\mathbf{j} + u\mathbf{k}$,

$0 \le u \le \pi,\ 0 \le v \le 2\pi$

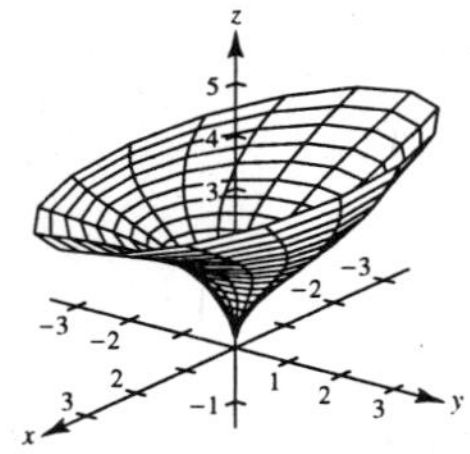

19. (a) From $(-10, 10, 0)$

(b) From $(10, 10, 10)$

(c) From $(0, 10, 0)$

(d) From $(10, 0, 0)$

21. (a) $\mathbf{r}(u, v) = (4 + \cos v) \cos u\mathbf{i} + (4 + \cos v) \sin u\mathbf{j} + \sin v\mathbf{k}$,

$0 \le u \le 2\pi,\ 0 \le v \le 2\pi$

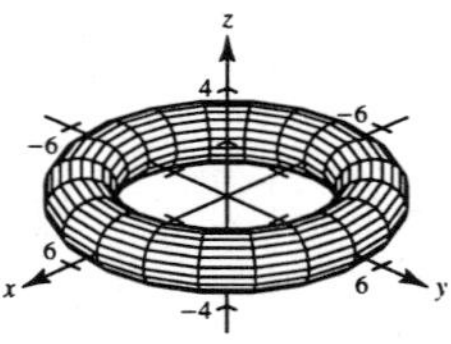

(b) $\mathbf{r}(u, v) = (4 + 2 \cos v) \cos u\mathbf{i} + (4 + 2 \cos v) \sin u\mathbf{j} + 2 \sin v\mathbf{k}$,

$0 \le u \le 2\pi,\ 0 \le v \le 2\pi$

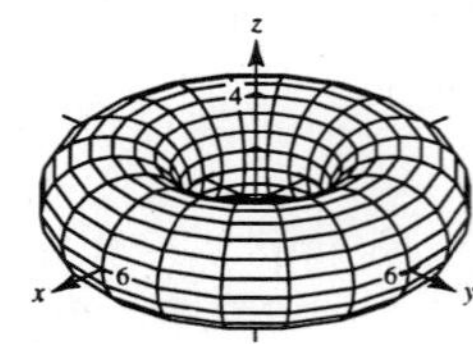

(c) $\mathbf{r}(u, v) = (8 + \cos v) \cos u\mathbf{i} + (8 + \cos v) \sin u\mathbf{j} + \sin v\mathbf{k}$,

$0 \le u \le 2\pi,\ 0 \le v \le 2\pi$

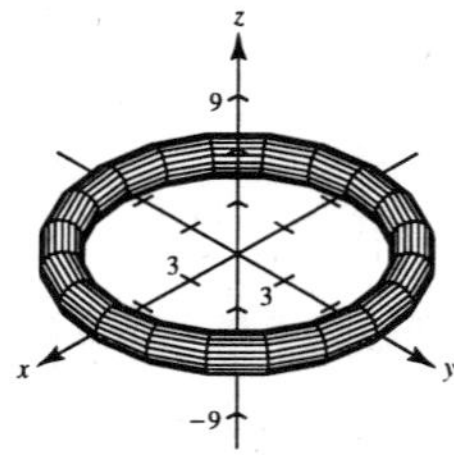

(d) $\mathbf{r}(u, v) = (8 + 3 \cos v) \cos u\mathbf{i} + (8 + 3 \cos v) \sin u\mathbf{j} + 3 \sin v\mathbf{k}$,

$0 \le u \le 2\pi,\ 0 \le v \le 2\pi$

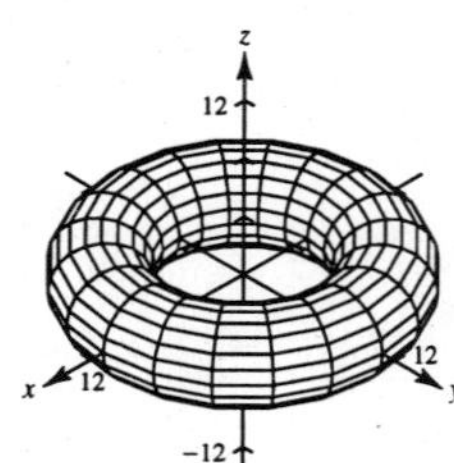

The radius of the generating circle that is revolved about the z-axis is b, and its center is a units from the axis of revolution.

23. $z = y$

$\mathbf{r}(u, v) = u\mathbf{i} + v\mathbf{j} + v\mathbf{k}$

25. $x^2 + y^2 = 16$

$\mathbf{r}(u, v) = 4\cos u\mathbf{i} + 4\sin u\mathbf{j} + v\mathbf{k}$

27. $z = x^2$

$\mathbf{r}(u, v) = u\mathbf{i} + v\mathbf{j} + u^2\mathbf{k}$

29. $z = 4$ inside $x^2 + y^2 = 9$.

$\mathbf{r}(u, v) = v\cos u\mathbf{i} + v\sin u\mathbf{j} + 4\mathbf{k},\ 0 \le v \le 3$

31. Function: $y = \dfrac{x}{2},\ 0 \le x \le 6$

Axis of revolution: x-axis

$x = u,\ y = \dfrac{u}{2}\cos v,\ z = \dfrac{u}{2}\sin v$

$0 \le u \le 6,\ 0 \le v \le 2\pi$

33. Function: $x = \sin z,\ 0 \le z \le \pi$

Axis of revolution: z-axis

$x = \sin u \cos v,\ y = \sin u \sin v,\ z = u$

$0 \le u \le \pi,\ 0 \le v \le 2\pi$

35. $\mathbf{r}(u, v) = (u + v)\mathbf{i} + (u - v)\mathbf{j} + v\mathbf{k},\ (1, -1, 1)$

$\mathbf{r}_u(u, v) = \mathbf{i} + \mathbf{j},\ \mathbf{r}_v(u, v) = \mathbf{i} - \mathbf{j} + \mathbf{k}$

At $(1, -1, 1)$, $u = 0$ and $v = 1$.

$\mathbf{r}_u(0, 1) = \mathbf{i} + \mathbf{j},\ \mathbf{r}_v(0, 1) = \mathbf{i} - \mathbf{j} + \mathbf{k}$

$$\mathbf{N} = \mathbf{r}_u(0, 1) \times \mathbf{r}_v(0, 1) = \begin{vmatrix} \mathbf{i} & \mathbf{j} & \mathbf{k} \\ 1 & 1 & 0 \\ 1 & -1 & 1 \end{vmatrix} = \mathbf{i} - \mathbf{j} - 2\mathbf{k}$$

Tangent plane: $(x - 1) - (y + 1) - 2(z - 1) = 0$

$$x - y - 2z = 0$$

(The original plane!)

37. $\mathbf{r}(u, v) = 2u\cos v\mathbf{i} + 3u\sin v\mathbf{j} + u^2\mathbf{k},\ (0, 6, 4)$

$\mathbf{r}_u(u, v) = 2\cos v\mathbf{i} + 3\sin v\mathbf{j} + 2u\mathbf{k}$

$\mathbf{r}_v(u, v) = -2u\sin v\mathbf{i} + 3u\cos v\mathbf{j}$

At $(0, 6, 4)$, $u = 2$ and $v = \pi/2$.

$\mathbf{r}_u\left(2, \dfrac{\pi}{2}\right) = 3\mathbf{j} + 4\mathbf{k},\ \mathbf{r}_v\left(2, \dfrac{\pi}{2}\right) = -4\mathbf{i}$

$$\mathbf{N} = \mathbf{r}_u\left(2, \frac{\pi}{2}\right) \times \mathbf{r}_v\left(2, \frac{\pi}{2}\right)$$

$$= \begin{vmatrix} \mathbf{i} & \mathbf{j} & \mathbf{k} \\ 0 & 3 & 4 \\ -4 & 0 & 0 \end{vmatrix} = -16\mathbf{j} + 12\mathbf{k}$$

Direction numbers: $0, 4, -3$

Tangent plane: $4(y - 6) - 3(z - 4) = 0$

$$4y - 3z = 12$$

39. $\mathbf{r}(u, v) = 2u\mathbf{i} - \dfrac{v}{2}\mathbf{j} + \dfrac{v}{2}\mathbf{k},\ 0 \le u \le 2,\ 0 \le v \le 1$

$\mathbf{r}_u(u, v) = 2\mathbf{i},\ \mathbf{r}_v(u, v) = -\dfrac{1}{2}\mathbf{j} + \dfrac{1}{2}\mathbf{k}$

$$\mathbf{r}_u \times \mathbf{r}_v = \begin{vmatrix} \mathbf{i} & \mathbf{j} & \mathbf{k} \\ 2 & 0 & 0 \\ 0 & -\frac{1}{2} & \frac{1}{2} \end{vmatrix} = -\mathbf{j} - \mathbf{k}$$

$\|\mathbf{r}_u \times \mathbf{r}_v\| = \sqrt{2}$

$$A = \int_0^1 \int_0^2 \sqrt{2}\, du\, dv = 2\sqrt{2}$$

41. $\mathbf{r}(u, v) = a\cos u\mathbf{i} + a\sin u\mathbf{j} + v\mathbf{k},\ 0 \le u \le 2\pi,\ 0 \le v \le b$

$\mathbf{r}_u(u, v) = -a\sin u\mathbf{i} + a\cos u\mathbf{j}$

$\mathbf{r}_v(u, v) = \mathbf{k}$

$$\mathbf{r}_u \times \mathbf{r}_v = \begin{vmatrix} \mathbf{i} & \mathbf{j} & \mathbf{k} \\ -a\sin u & a\cos u & 0 \\ 0 & 0 & 1 \end{vmatrix} = a\cos u\mathbf{i} + a\sin u\mathbf{j}$$

$\|\mathbf{r}_u \times \mathbf{r}_v\| = a$

$$A = \int_0^b \int_0^{2\pi} a\, du\, dv = 2\pi ab$$

43. $\mathbf{r}(u, v) = au\cos v\mathbf{i} + au\sin v\mathbf{j} + u\mathbf{k},\ 0 \le u \le b,\ 0 \le v \le 2\pi$

$\mathbf{r}_u(u, v) = a\cos v\mathbf{i} + a\sin v\mathbf{j} + \mathbf{k}$

$\mathbf{r}_v(u, v) = -au\sin v\mathbf{i} + au\cos v\mathbf{j}$

$$\mathbf{r}_u \times \mathbf{r}_v = \begin{vmatrix} \mathbf{i} & \mathbf{j} & \mathbf{k} \\ a\cos v & a\sin v & 1 \\ -au\sin v & au\cos v & 0 \end{vmatrix} = -au\cos v\mathbf{i} - au\sin v\mathbf{j} + a^2u\mathbf{k}$$

$\|\mathbf{r}_u \times \mathbf{r}_v\| = au\sqrt{1 + a^2}$

$$A = \int_0^{2\pi}\int_0^b a\sqrt{1 + a^2}u\,du\,dv = \pi ab^2\sqrt{1 + a^2}$$

45. $\mathbf{r}(u, v) = \sqrt{u}\cos v\mathbf{i} + \sqrt{u}\sin v\mathbf{j} + u\mathbf{k},\ 0 \le u \le 4,\ 0 \le v \le 2\pi$

$\mathbf{r}_u(u, v) = \dfrac{\cos v}{2\sqrt{u}}\mathbf{i} + \dfrac{\sin v}{2\sqrt{u}}\mathbf{j} + \mathbf{k}$

$\mathbf{r}_v(u, v) = -\sqrt{u}\sin v\mathbf{i} + \sqrt{u}\cos v\mathbf{j}$

$$\mathbf{r}_u \times \mathbf{r}_v = \begin{vmatrix} \mathbf{i} & \mathbf{j} & \mathbf{k} \\ \dfrac{\cos v}{2\sqrt{u}} & \dfrac{\sin v}{2\sqrt{u}} & 1 \\ -\sqrt{u}\sin v & \sqrt{u}\cos v & 0 \end{vmatrix} = -\sqrt{u}\cos v\mathbf{i} - \sqrt{u}\sin v\mathbf{j} + \frac{1}{2}\mathbf{k}$$

$\|\mathbf{r}_u \times \mathbf{r}_v\| = \sqrt{u + \dfrac{1}{4}}$

$$A = \int_0^{2\pi}\int_0^4 \sqrt{u + \frac{1}{4}}\,du\,dv = \frac{\pi}{6}\left(17\sqrt{17} - 1\right) \approx 36.177$$

47. $\mathbf{r}(u, v) = 20\sin(u)\cos(v)\mathbf{i} + 20\sin(u)\sin(v)\mathbf{j} + 20\cos(u)\mathbf{k}\ \ 0 \le u \le \pi/3,\ 0 \le v \le 2\pi$

$\mathbf{r}_u = 20\cos u\cos v\mathbf{i} + 20\cos u\sin v\mathbf{j} - 20\sin u\mathbf{k}$

$\mathbf{r}_v = -20\sin u\sin v\mathbf{i} + 20\sin u\cos v\mathbf{j}$

$$\begin{aligned}\mathbf{r}_u \times \mathbf{r}_v &= \begin{vmatrix} \mathbf{i} & \mathbf{j} & \mathbf{k} \\ 20\cos u\cos v & 20\cos u\sin v & -20\sin u \\ -20\sin u\sin v & 20\sin u\cos v & 0 \end{vmatrix} \\ &= 400\sin^2 u\cos v\mathbf{i} + 400\sin^2 u\sin v\mathbf{j} + 400(\cos u\sin u\cos^2 v + \cos u\sin u\sin^2 v)\mathbf{k} \\ &= 400[\sin^2 u\cos v\mathbf{i} + \sin^2 u\sin v\mathbf{j} + \cos u\sin u\mathbf{k}]\end{aligned}$$

$$\begin{aligned}\|\mathbf{r}_u \times \mathbf{r}_v\| &= 400\sqrt{\sin^4 u\cos^2 v + \sin^4 u\sin^2 v + \cos^2 u\sin^2 u} \\ &= 400\sqrt{\sin^4 u + \cos^2 u\sin^2 u} \\ &= 400\sqrt{\sin^2 u} = 400\sin u\end{aligned}$$

$$\begin{aligned}S = \iint_S dS &= \int_0^{2\pi}\int_0^{\pi/3} 400\sin u\,du\,dv = \int_0^{2\pi}\Big[-400\cos u\Big]_0^{\pi/3}dv \\ &= \int_0^{2\pi} 200\,dv = 400\pi\text{ m}^2\end{aligned}$$

49. $\mathbf{r}(u, v) = u \cos v\mathbf{i} + u \sin v\mathbf{j} + 2v\mathbf{k},\ 0 \le u \le 3,\ 0 \le v \le 2\pi$

$\mathbf{r}_u(u, v) = \cos v\mathbf{i} + \sin v\mathbf{j}$

$\mathbf{r}_v(u, v) = -u \sin v\mathbf{i} + u \cos v\mathbf{j} + 2\mathbf{k}$

$$\mathbf{r}_u \times \mathbf{r}_v = \begin{vmatrix} \mathbf{i} & \mathbf{j} & \mathbf{k} \\ \cos v & \sin v & 0 \\ -u \sin v & u \cos v & 2 \end{vmatrix} = 2 \sin v\mathbf{i} - 2 \cos v\mathbf{j} + u\mathbf{k}$$

$\|\mathbf{r}_u \times \mathbf{r}_v\| = \sqrt{4 + u^2}$

$$A = \int_0^{2\pi}\int_0^3 \sqrt{4 + u^2}\, du\, dv = \pi\left[3\sqrt{13} + 4 \ln\left(\frac{3 + \sqrt{13}}{2}\right)\right]$$

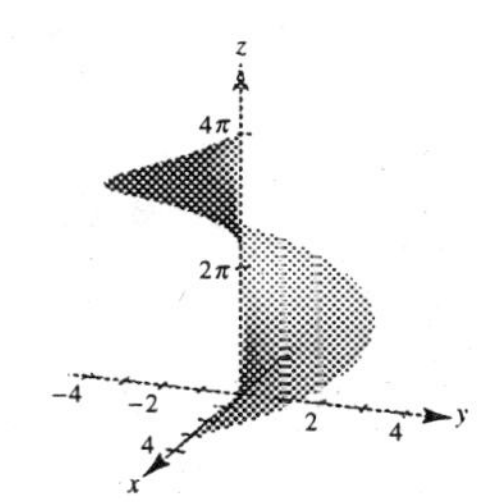

51. Essay

Section 14.6 Surface Integrals

1. S: $z = 4 - x,\ 0 \le x \le 4,\ 0 \le y \le 4,\ \dfrac{\partial z}{\partial x} = -1,\ \dfrac{\partial z}{\partial y} = 0$

$$\iint_S (x - 2y + z)\, dS = \int_0^4\int_0^4 (x - 2y + 4 - x)\sqrt{1 + (-1)^2 + (0)^2}\, dy\, dx$$

$$= \sqrt{2}\int_0^4\int_0^4 (4 - 2y)\, dy\, dx = 0$$

3. S: $z = 10,\ x^2 + y^2 \le 1,\ \dfrac{\partial z}{\partial x} = \dfrac{\partial z}{\partial y} = 0$

$$\iint_S (x - 2y + z)\, dS = \int_{-1}^1\int_{-\sqrt{1-x^2}}^{\sqrt{1-x^2}} (x - 2y + 10)\sqrt{1 + (0)^2 + (0)^2}\, dy\, dx$$

$$= \int_0^{2\pi}\int_0^1 (r \cos\theta - 2r \sin\theta + 10)r\, dr\, d\theta$$

$$= \int_0^{2\pi}\left(\frac{1}{3}\cos\theta - \frac{2}{3}\sin\theta + 5\right) d\theta = \left[\frac{1}{3}\sin\theta + \frac{2}{3}\cos\theta + 5\theta\right]_0^{2\pi} = 10\pi$$

5. S: $z = 6 - x - 2y$, (first octant) $\dfrac{\partial z}{\partial x} = -1,\ \dfrac{\partial z}{\partial y} = -2$

$$\iint_S xy\, dS = \int_0^6\int_0^{3-(x/2)} xy\sqrt{1 + (-1)^2 + (-2)^2}\, dy\, dx$$

$$= \sqrt{6}\int_0^6\left[\frac{xy^2}{2}\right]_0^{3-(x/2)} dx$$

$$= \frac{\sqrt{6}}{2}\int_0^6 x\left(9 - 3x + \frac{1}{4}x^2\right) dx$$

$$= \frac{\sqrt{6}}{2}\left[\frac{9x^2}{2} - x^3 + \frac{x^4}{16}\right]_0^6 = \frac{27\sqrt{6}}{2}$$

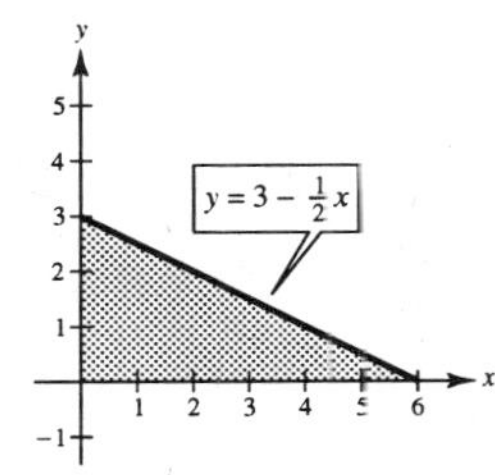

7. $S: z = 9 - x^2,\ 0 \le x \le 2,\ 0 \le y \le x,$

$$\frac{\partial z}{\partial x} = -2x,\ \frac{\partial z}{\partial y} = 0$$

$$\iint_S xy\, dS = \int_0^2 \int_y^2 xy\sqrt{1 + 4x^2}\, dx\, dy$$

$$= \frac{391\sqrt{17} + 1}{240}$$

9. $S:\ z = 10 - x^2 - y^2,\ 0 \le x \le 2,\ 0 \le y \le 2$

$$\iint_S (x^2 - 2xy)\, dS = \int_0^2 \int_0^2 (x^2 - 2xy)\sqrt{1 + 4x^2 + 4y^2}\, dy\, dx \approx -11.47$$

11. $S:\ \mathbf{r}(u, v) = u\mathbf{i} + v\mathbf{j} + \frac{v}{2}\mathbf{k},\ 0 \le u \le 1,\ 0 \le v \le 2$

$$\|\mathbf{r}_u \times \mathbf{r}_v\| = \left\| -\frac{1}{2}\mathbf{j} + \mathbf{k} \right\| = \frac{\sqrt{5}}{2}$$

$$\iint_S (y + 5)\, dS = \int_0^2 \int_0^1 (v + 5)\frac{\sqrt{5}}{2}\, du\, dv = 6\sqrt{5}$$

13. $S:\ \mathbf{r}(u, v) = 2\cos u\mathbf{i} + 2\sin u\mathbf{j} + v\mathbf{k},\ 0 \le u \le \frac{\pi}{2},$
$0 \le v \le 2$

$$\|\mathbf{r}_u \times \mathbf{r}_v\| = \|2\cos u\mathbf{i} + 2\sin u\mathbf{j}\| = 2$$

$$\iint_S xy\, dS = \int_0^2 \int_0^{\pi/2} 8\cos u \sin u\, du\, dv = 8$$

15. $f(x, y, z) = x^2 + y^2 + z^2$

$S:\ z = x + 2,\ x^2 + y^2 \le 1$

$$\iint_S f(x, y, z)\, dS = \int_{-1}^1 \int_{-\sqrt{1-x^2}}^{\sqrt{1-x^2}} [x^2 + y^2 + (x + 2)^2]\sqrt{1 + (1)^2 + (0)^2}\, dy\, dx$$

$$= \sqrt{2}\int_0^{2\pi}\int_0^1 [r^2 + (r\cos\theta + 2)^2]r\, dr\, d\theta$$

$$= \sqrt{2}\int_0^{2\pi}\int_0^1 [r^2 + r^2\cos^2\theta + 4r\cos\theta + 4]r\, dr\, d\theta$$

$$= \sqrt{2}\int_0^{2\pi} \left[\frac{r^4}{4} + \frac{r^4}{4}\cos^2\theta + \frac{4r^3}{3}\cos\theta + 2r^2\right]_0^1 d\theta$$

$$= \sqrt{2}\int_0^{2\pi} \left[\frac{9}{4} + \left(\frac{1}{4}\right)\frac{1 + \cos 2\theta}{2} + \frac{4}{3}\cos\theta\right] d\theta$$

$$= \sqrt{2}\left[\frac{9}{4}\theta + \frac{1}{8}\left(\theta + \frac{1}{2}\sin 2\theta\right) + \frac{4}{3}\sin\theta\right]_0^{2\pi} = \sqrt{2}\left[\frac{18\pi}{4} + \frac{\pi}{4}\right] = \frac{19\sqrt{2}\pi}{4}$$

17. $f(x, y, z) = \sqrt{x^2 + y^2 + z^2}$

$S:\ z = \sqrt{x^2 + y^2},\ x^2 + y^2 \le 4$

$$\iint_S f(x, y, z)\, dS = \int_{-2}^2 \int_{-\sqrt{4-x^2}}^{\sqrt{4-x^2}} \sqrt{x^2 + y^2 + \left(\sqrt{x^2 + y^2}\right)^2}\sqrt{1 + \left(\frac{x}{\sqrt{x^2 + y^2}}\right)^2 + \left(\frac{y}{\sqrt{x^2 + y^2}}\right)^2}\, dy\, dx$$

$$= \sqrt{2}\int_{-2}^2 \int_{-\sqrt{4-x^2}}^{\sqrt{4-x^2}} \sqrt{x^2 + y^2}\sqrt{\frac{x^2 + y^2 + x^2 + y^2}{x^2 + y^2}}\, dy\, dx$$

$$= 2\int_{-2}^2 \int_{-\sqrt{4-x^2}}^{\sqrt{4-x^2}} \sqrt{x^2 + y^2}\, dy\, dx$$

$$= 2\int_0^{2\pi}\int_0^2 r^2\, dr\, d\theta = 2\int_0^{2\pi}\left[\frac{r^3}{3}\right]_0^2 d\theta = \left[\frac{16}{3}\theta\right]_0^{2\pi} = \frac{32\pi}{3}$$

19. $f(x, y, z) = x^2 + y^2 + z^2$

S: $x^2 + y^2 = 9,\ 0 \le x \le 3,\ 0 \le y \le 3, 0 \le z \le 9$

Project the solid onto the yz-plane; $x = \sqrt{9 - y^2},\ 0 \le y \le 3,\ 0 \le z \le 9.$

$$\int_S\int f(x, y, z)\,dS = \int_0^3\int_0^9 [(9 - y^2) + y^2 + z^2]\sqrt{1 + \left(\frac{y}{\sqrt{9 - y^2}}\right)^2 + (0)^2}\,dz\,dy$$

$$= \int_0^3\int_0^9 (9 + z^2)\frac{3}{\sqrt{9 - y^2}}\,dz\,dy = \int_0^3\left[\frac{3}{\sqrt{9 - y^2}}\left(9z + \frac{z^3}{3}\right)\right]_0^9 dy$$

$$= 324\int_0^3 \frac{3}{\sqrt{9 - y^2}}\,dy = \left[972 \arcsin\left(\frac{y}{3}\right)\right]_0^3 = 972\left(\frac{\pi}{2} - 0\right) = 486\pi$$

21. $\mathbf{F}(x, y, z) = 3z\mathbf{i} - 4\mathbf{j} + y\mathbf{k}$

S: $x + y + z = 1$ (first octant)

$G(x, y, z) = x + y + z - 1$

$\nabla G(x, y, z) = \mathbf{i} + \mathbf{j} + \mathbf{k}$

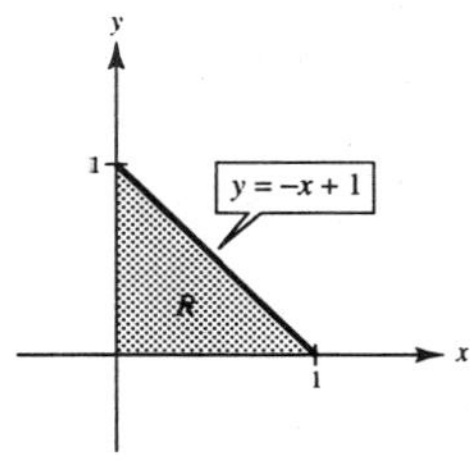

$$\int_S\int \mathbf{F}\cdot\mathbf{N}\,dS = \int_R\int \mathbf{F}\cdot\nabla G\,dA = \int_0^1\int_0^{1-x} (3z - 4 + y)\,dy\,dx$$

$$= \int_0^1\int_0^{1-x} [3(1 - x - y) - 4 + y]\,dy\,dx$$

$$= \int_0^1\int_0^{1-x} (-1 - 3x - 2y)\,dy\,dx$$

$$= \int_0^1\left[-y - 3xy - y^2\right]_0^{1-x} dx$$

$$= -\int_0^1 [(1 - x) + 3x(1 - x) + (1 - x)^2]\,dx$$

$$= -\int_0^1 (2 - 2x^2)\,dx = -\frac{4}{3}$$

23. $\mathbf{F}(x, y, z) = x\mathbf{i} + y\mathbf{j} + z\mathbf{k}$

S: $z = 9 - x^2 - y^2,\ 0 \le z$

$G(x, y, z) = x^2 + y^2 + z - 9$

$\nabla G(x, y, z) = 2x\mathbf{i} + 2y\mathbf{j} + \mathbf{k}$

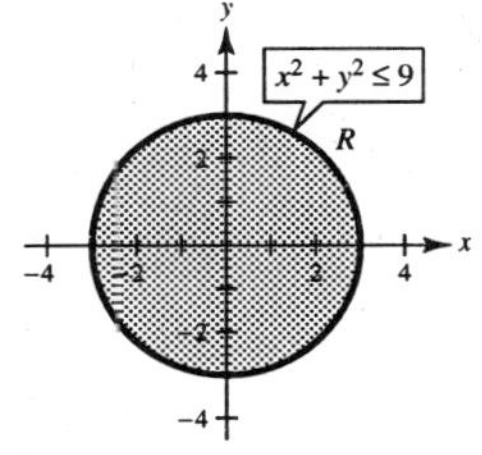

$$\int_S\int \mathbf{F}\cdot\mathbf{N}\,dS = \int_R\int \mathbf{F}\cdot\nabla G\,dA = \int_R\int (2x^2 + 2y^2 + z)\,dA$$

$$= \int_R\int [2x^2 + 2y^2 + (9 - x^2 - y^2)]\,dA$$

$$= \int_R\int (x^2 + y^2 + 9)\,dA$$

$$= \int_0^{2\pi}\int_0^3 (r^2 + 9)r\,dr\,d\theta$$

$$= \int_0^{2\pi}\left[\frac{r^4}{4} + \frac{9r^2}{2}\right]_0^3 d\theta = \frac{243\pi}{2}$$

25. $\mathbf{F}(x, y, z) = 4\mathbf{i} - 3\mathbf{j} + 5\mathbf{k}$

S: $z = x^2 + y^2,\ x^2 + y^2 \le 4$

$G(x, y, z) = -x^2 - y^2 + z$

$\nabla G(x, y, z) = -2x\mathbf{i} - 2y\mathbf{j} + \mathbf{k}$

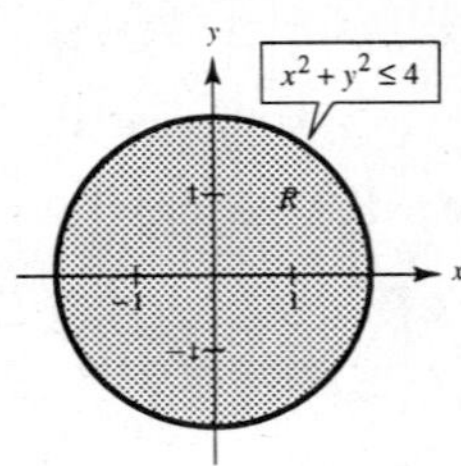

$$\int_S\int \mathbf{F}\cdot\mathbf{N}\,dS = \int_R\int \mathbf{F}\cdot\nabla G\,dA = \int_R\int(-8x + 6y + 5)\,dA$$

$$= \int_0^{2\pi}\int_0^2 [-8r\cos\theta + 6r\sin\theta + 5]r\,dr\,d\theta$$

$$= \int_0^{2\pi}\left[-\frac{8}{3}r^3\cos\theta + 2r^3\sin\theta + \frac{5}{2}r^2\right]_0^2 d\theta$$

$$= \int_0^{2\pi}\left[-\frac{64}{3}\cos\theta + 16\sin\theta + 10\right] d\theta$$

$$= \left[-\frac{64}{3}\sin\theta - 16\cos\theta + 10\theta\right]_0^{2\pi} = 20\pi$$

27. $\mathbf{F}(x, y, z) = 4xy\mathbf{i} + z^2\mathbf{j} + yz\mathbf{k}$

S: unit cube bounded by $x = 0,\ x = 1,\ y = 0,\ y = 1,\ z = 0,\ z = 1$

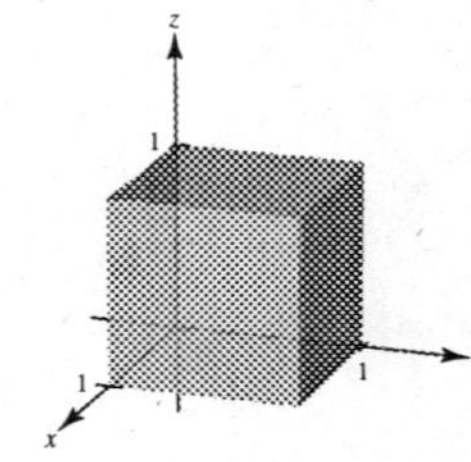

S_1: The top of the cube

$\mathbf{N} = \mathbf{k},\ z = 1$

$$\int_{S_1}\int \mathbf{F}\cdot\mathbf{N}\,dS = \int_0^1\int_0^1 y(1)\,dy\,dx = \frac{1}{2}$$

S_2: The bottom of the cube

$\mathbf{N} = -\mathbf{k},\ z = 0$

$$\int_{S_2}\int \mathbf{F}\cdot\mathbf{N}\,dS = \int_0^1\int_0^1 -y(0)\,dy\,dx = 0$$

S_4: The back of the cube

$\mathbf{N} = -\mathbf{i},\ x = 0$

$$\int_{S_4}\int \mathbf{F}\cdot\mathbf{N}\,dS = \int_0^1\int_0^1 -4(0)y\,dy\,dx = 0$$

S_6: The left side of the cube

$\mathbf{N} = -\mathbf{j},\ y = 0$

$$\int_{S_6}\int \mathbf{F}\cdot\mathbf{N}\,dS = \int_0^1\int_0^1 -z^2\,dz\,dx = -\frac{1}{3}$$

S_3: The front of the cube

$\mathbf{N} = \mathbf{i},\ x = 1$

$$\int_{S_3}\int \mathbf{F}\cdot\mathbf{N}\,dS = \int_0^1\int_0^1 4(1)y\,dy\,dz = 2$$

S_5: The right side of the cube

$\mathbf{N} = \mathbf{j},\ y = 1$

$$\int_{S_5}\int \mathbf{F}\cdot\mathbf{N}\,dS = \int_0^1\int_0^1 z^2\,dz\,dx = \frac{1}{3}$$

$$\int_S\int \mathbf{F}\cdot\mathbf{N}\,dS = \frac{1}{2} + 0 + 2 + 0 + \frac{1}{3} - \frac{1}{3} = \frac{5}{2}$$

29. S: $2x + 3y + 6z = 12$ (first octant) $\Rightarrow z = 2 - \frac{1}{3}x - \frac{1}{2}y$

$\rho(x, y, z) = x^2 + y^2$

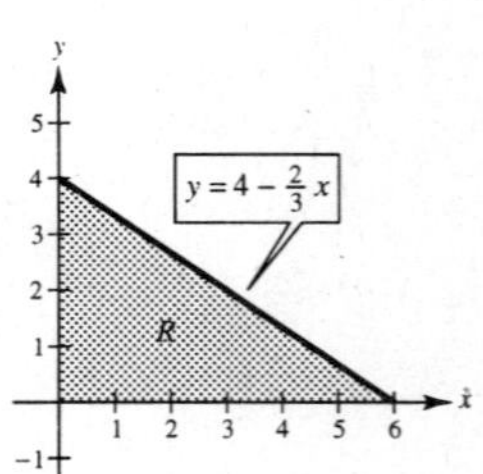

$$m = \int_R\int(x^2 + y^2)\sqrt{1 + \left(-\frac{1}{3}\right)^3 + \left(-\frac{1}{2}\right)^2}\,dA$$

$$= \frac{7}{6}\int_0^6\int_0^{4-(2x/3)}(x^2 + y^2)\,dy\,dx$$

$$= \frac{7}{6}\int_0^6\left[x^2\left(4 - \frac{2}{3}x\right) + \frac{1}{3}\left(4 - \frac{2}{3}x\right)^3\right]dx = \frac{7}{6}\left[\frac{4}{3}x^3 - \frac{1}{6}x^4 - \frac{1}{8}\left(4 - \frac{2}{3}x\right)^4\right]_0^6 = \frac{364}{3}$$

31. $z = \sqrt{x^2 + y^2},\ 0 \le z \le a$

$$m = \iint_S k\,dS = k\iint_R \sqrt{1 + \left(\frac{x}{\sqrt{x^2+y^2}}\right)^2 + \left(\frac{y}{\sqrt{x^2+y^2}}\right)^2}\,dA = k\iint_R \sqrt{2}\,dA = \sqrt{2}\,k\pi a^2$$

$$I_z = \iint_S k(x^2+y^2)\,dS = \iint_R k(x^2+y^2)\sqrt{2}\,dA$$

$$= \sqrt{2}k\int_0^{2\pi}\int_0^a r^3\,dr\,d\theta = \frac{\sqrt{2}ka^4}{4}(2\pi)$$

$$= \frac{\sqrt{2}k\pi a^4}{2} = \frac{a^2}{2}\left(\sqrt{2}k\pi a^2\right) = \frac{a^2 m}{2}$$

33. $x^2 + y^2 = a^2,\ 0 \le z \le h$

$\rho(x, y, z) = 1$

$y = \pm\sqrt{a^2 - x^2}$

Project the solid onto the xz-plane.

$$I_z = 4\iint_S (x^2 + y^2)(1)\,dS$$

$$= 4\int_0^h\int_0^a [x^2 + (a^2 - x^2)]\sqrt{1 + \left(\frac{-x}{\sqrt{a^2-x^2}}\right)^2 + (0)^2}\,dx\,dz$$

$$= 4a^3\int_0^h\int_0^a \frac{1}{\sqrt{a^2-x^2}}\,dx\,dz$$

$$= 4a^3\int_0^h \left[\arcsin\frac{x}{a}\right]_0^a dz = 4a^3\left(\frac{\pi}{2}\right)(h) = 2\pi a^3 h$$

35. $S\colon z = 16 - x^2 - y^2,\ z \ge 0$

$\mathbf{F}(x, y, z) = 0.5z\mathbf{k}$

$$\iint_S \rho\mathbf{F}\cdot\mathbf{N}\,dS = \iint_R \rho\mathbf{F}\cdot(-g_x(x,y)\mathbf{i} - g_y(x,y)\mathbf{j} + \mathbf{k})\,dA = \iint_R 0.5\rho z\mathbf{k}\cdot(2x\mathbf{i} + 2y\mathbf{j} + \mathbf{k})\,dA$$

$$= \iint_R 0.5\rho z\,dA = \iint_R 0.5\rho(16 - x^2 - y^2)\,dA = 0.5\rho\int_0^{2\pi}\int_0^4 (16 - r^2)r\,dr\,d\theta = 0.5\rho\int_0^{2\pi} 64\,d\theta = 64\pi\rho$$

37. (a)

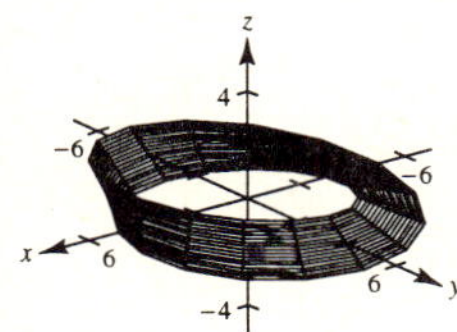

(b) If a normal vector at a point P on the surface is moved around the Möbius strip once, it will point in the opposite direction.

(c) $\mathbf{r}(u, 0) = 4\cos(2u)\mathbf{i} + 4\sin(2u)\mathbf{j}$

This is a circle.

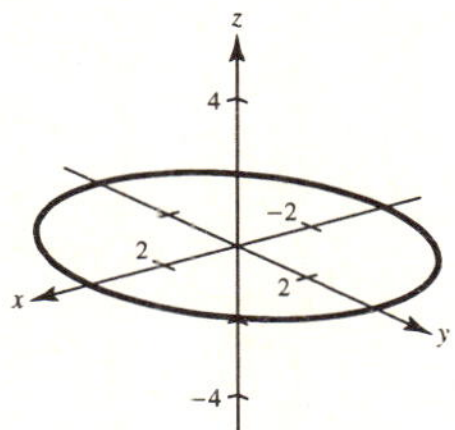

(d) (construction)

(e) You obtain a strip with a double twist and twice as long as the original Möbius strip.

39. $\mathbf{E} = yz\mathbf{i} + xz\mathbf{j} + xy\mathbf{k}$

$S\colon z = \sqrt{1 - x^2 - y^2}$

$$\iint_S \mathbf{E} \cdot \mathbf{N}\, dS = \iint_R \mathbf{E} \cdot (-g_x(x, y)\mathbf{i} - g_y(x, y)\mathbf{j} + \mathbf{k})\, dA$$

$$= \iint_R (yz\mathbf{i} + xz\mathbf{j} + xy\mathbf{k}) \cdot \left(\frac{x}{\sqrt{1 - x^2 - y^2}}\mathbf{i} + \frac{y}{\sqrt{1 - x^2 - y^2}}\mathbf{j} + \mathbf{k}\right) dA$$

$$= \iint_R \left(\frac{2xyz}{\sqrt{1 - x^2 - y^2}} + xy\right) dA = \iint_R 3xy\, dA = \int_{-1}^{1}\int_{-\sqrt{1-x^2}}^{\sqrt{1-x^2}} 3xy\, dy\, dx = 0$$

Section 14.7 Divergence Theorem

1. Surface Integral: There are six surfaces to the cube, each with $dS = \sqrt{1}\, dA$.

$z = 0, \quad \mathbf{N} = -\mathbf{k}, \quad \mathbf{F} \cdot \mathbf{N} = -z^2, \quad \displaystyle\iint_{S_1} 0\, dA = 0$

$z = a, \quad \mathbf{N} = \mathbf{k}, \quad \mathbf{F} \cdot \mathbf{N} = z^2, \quad \displaystyle\iint_{S_2} a^2\, dA = \int_0^a\int_0^a a^2\, dx\, dy = a^4$

$x = 0, \quad \mathbf{N} = -\mathbf{i}, \quad \mathbf{F} \cdot \mathbf{N} = -2x, \quad \displaystyle\iint_{S_3} 0\, dA = 0$

$x = a, \quad \mathbf{N} = \mathbf{i}, \quad \mathbf{F} \cdot \mathbf{N} = 2x, \quad \displaystyle\iint_{S_4} 2a\, dy\, dz = \int_0^a\int_0^a 2a\, dy\, dz = 2a^3$

$y = 0, \quad \mathbf{N} = -\mathbf{j}, \quad \mathbf{F} \cdot \mathbf{N} = 2y, \quad \displaystyle\iint_{S_5} 0\, dA = 0$

$y = a, \quad \mathbf{N} = \mathbf{j}, \quad \mathbf{F} \cdot \mathbf{N} = -2y, \quad \displaystyle\iint_{S_6} -2a\, dA = \int_0^a\int_0^a -2a\, dz\, dx = -2a^3$

Therefore, $\displaystyle\iint_S \mathbf{F} \cdot \mathbf{N}\, dS = a^4 + 2a^3 - 2a^3 = a^4$.

Divergence Theorem: Since div $\mathbf{F} = 2z$, the Divergence Theorem yields

$$\iiint_Q \text{div}\,\mathbf{F}\, dV = \int_0^a\int_0^a\int_0^a 2z\, dz\, dy\, dx = \int_0^a\int_0^a a^2\, dy\, dx = a^4.$$

3. Surface Integral: There are four surfaces to this solid.

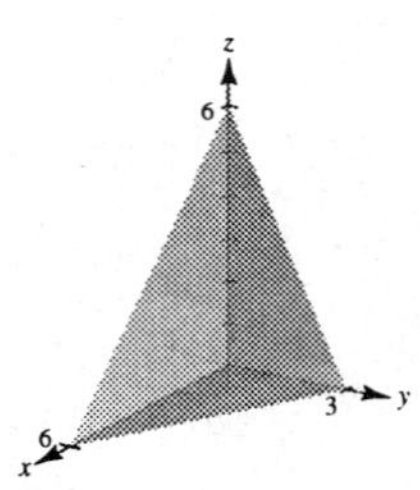

$z = 0, \ \mathbf{N} = -\mathbf{k}, \ \mathbf{F} \cdot \mathbf{N} = -z$

$$\iint_{S_1} 0\, dS = 0$$

$y = 0, \ \mathbf{N} = -\mathbf{j}, \ \mathbf{F} \cdot \mathbf{N} = 2y - z, \ dS = dA = dx\, dz$

$$\iint_{S_2} -z\, dS = \int_0^6\int_0^{6-z} -z\, dx\, dz = \int_0^6 (z^2 - 6z)\, dz = -36$$

$x = 0, \ \mathbf{N} = -\mathbf{i}, \ \mathbf{F} \cdot \mathbf{N} = y - 2x, \ dS = dA = dz\, dy$

$$\iint_{S_3} y\, dS = \int_0^3\int_0^{6-2y} y\, dz\, dy = \int_0^3 (6y - 2y^2)\, dy = 9$$

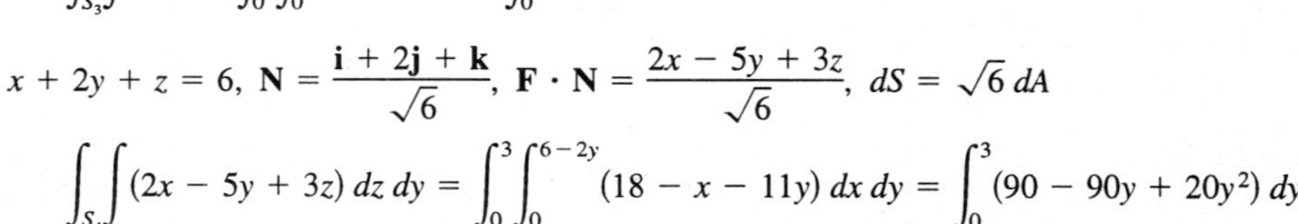

$x + 2y + z = 6, \ \mathbf{N} = \dfrac{\mathbf{i} + 2\mathbf{j} + \mathbf{k}}{\sqrt{6}}, \ \mathbf{F} \cdot \mathbf{N} = \dfrac{2x - 5y + 3z}{\sqrt{6}}, \ dS = \sqrt{6}\, dA$

$$\iint_{S_4} (2x - 5y + 3z)\, dz\, dy = \int_0^3\int_0^{6-2y} (18 - x - 11y)\, dx\, dy = \int_0^3 (90 - 90y + 20y^2)\, dy = 45$$

Therefore, $\displaystyle\iint_S \mathbf{F} \cdot \mathbf{N}\, dS = 0 - 36 + 9 + 45 = 18$.

Divergence Theorem: Since div $\mathbf{F} = 1$, we have

$$\iiint_Q dV = (\text{Volume of solid}) = \frac{1}{3}(\text{Area of base}) \times (\text{Height}) = \frac{1}{3}(9)(6) = 18.$$

5. Since div **F** = $2x + 2y + 2z$, we have

$$\iiint_Q \text{div}\,\mathbf{F}\,dV = \int_0^a\int_0^a\int_0^a (2x + 2y + 2z)\,dz\,dy\,dx$$

$$= \int_0^a\int_0^a (2ax + 2ay + a^2)\,dy\,dx = \int_0^a (2a^2x + 2a^3)\,dx = \Big[a^2x^2 + 2a^3x\Big]_0^a = 3a^4.$$

7. Since div **F** = $2x - 2x + 2xyz = 2xyz$

$$\iiint_Q \text{div}\,\mathbf{F}\,dV = \iiint_Q 2xyz\,dV = \int_0^a\int_0^{2\pi}\int_0^{\pi/2} 2(\rho\sin\phi\cos\theta)(\rho\sin\phi\sin\theta)(\rho\cos\phi)\rho^2\sin\phi\,d\phi\,d\theta\,d\rho$$

$$= \int_0^a\int_0^{2\pi}\int_0^{\pi/2} 2\rho^5(\sin\theta\cos\theta)(\sin^3\phi\cos\phi)\,d\phi\,d\theta\,d\rho$$

$$= \int_0^a\int_0^{2\pi} \frac{1}{2}\rho^5\sin\theta\cos\theta\,d\theta\,d\rho = \int_0^a\left[\left(\frac{\rho^5}{2}\right)\frac{\sin^2\theta}{2}\right]_0^{2\pi} d\rho = 0.$$

9. Since div **F** = 3, we have

$$\iiint_Q 3\,dV = 3(\text{Volume of sphere}) = 3\left[\frac{4}{3}\pi(2)^3\right] = 32\pi.$$

11. Since div **F** = $1 + 2y - 1 = 2y$, we have

$$\iiint_Q 2y\,dV = \int_0^4\int_{-3}^3\int_{-\sqrt{9-y^2}}^{\sqrt{9-y^2}} 2y\,dx\,dy\,dz = \int_0^4\int_{-3}^3 4y\sqrt{9-y^2}\,dy\,dz = \int_0^4\left[-\frac{4}{3}(9-y^2)^{3/2}\right]_{-3}^3 dz = 0.$$

13. Since div **F** = $3x^2 + x^2 + 0 = 4x^2$, we have

$$\iiint_Q 4x^2\,dV = \int_0^6\int_0^4\int_0^{4-y} 4x^2\,dz\,dy\,dx = \int_0^6\int_0^4 4x^2(4-y)\,dy\,dx = \int_0^6 32x^2\,dx = 2304.$$

15. $\mathbf{F}(x, y, z) = xy\mathbf{i} + 4y\mathbf{j} + xz\mathbf{k}$

div **F** = $y + 4 + x$

$$\iint_S \mathbf{F}\cdot\mathbf{N}\,dS = \iiint_Q \text{div}\,\mathbf{F}\,dV = \iiint_Q (y + x + 4)\,dV$$

$$= \int_0^2\int_0^\pi\int_0^{2\pi} (\rho\sin\phi\sin\theta + \rho\sin\phi\cos\theta + 4)\rho^2\sin\phi\,d\theta\,d\phi\,d\rho$$

$$= \int_0^2\int_0^\pi\int_0^{2\pi} [\rho^3\sin^2\phi\sin\theta + \rho^3\sin^2\phi\cos\theta + 4\rho^2\sin\phi]\,d\theta\,d\phi\,d\rho$$

$$= \int_0^2\int_0^\pi\Big[-\rho^3\sin^2\phi\cos\theta + \rho^3\sin^2\phi\sin\theta + 4\rho^2\sin\phi\cdot\theta\Big]_0^{2\pi} d\phi\,d\rho$$

$$= \int_0^2\int_0^\pi 8\pi\rho^2\sin\phi\,d\phi\,d\rho$$

$$= \int_0^2\Big[-8\pi\rho^2\cos\phi\Big]_0^\pi d\rho$$

$$= \int_0^2 16\pi\rho^2\,d\rho = \left[\frac{16\pi\rho^3}{3}\right]_0^2 = \frac{128\pi}{3}.$$

17. Using the triple integral to find volume, we need **F** so that

$$\text{div}\,\mathbf{F} = \frac{\partial M}{\partial x} + \frac{\partial N}{\partial y} + \frac{\partial P}{\partial z} = 1.$$

Hence, we could have $\mathbf{F} = x\mathbf{i}$, $\mathbf{F} = y\mathbf{j}$, or $\mathbf{F} = z\mathbf{k}$.

For $dA = dy\,dz$ consider $\mathbf{F} = x\mathbf{i}$, $x = f(y, z)$, then $\mathbf{N} = \dfrac{\mathbf{i} + f_y\mathbf{j} + f_z\mathbf{k}}{\sqrt{1 + f_y^2 + f_z^2}}$ and $dS = \sqrt{1 + f_y^2 + f_z^2}\,dy\,dz$.

For $dA = dz\,dx$ consider $\mathbf{F} = y\mathbf{j}$, $y = f(x, z)$, then $\mathbf{N} = \dfrac{f_x\mathbf{i} + \mathbf{j} + f_z\mathbf{k}}{\sqrt{1 + f_x^2 + f_z^2}}$ and $dS = \sqrt{1 + f_x^2 + f_z^2}\,dz\,dx$.

For $dA = dx\,dy$ consider $\mathbf{F} = z\mathbf{k}$, $z = f(x, y)$, then $\mathbf{N} = \dfrac{f_x\mathbf{i} + f_y\mathbf{j} + \mathbf{k}}{\sqrt{1 + f_x^2 + f_y^2}}$ and $dS = \sqrt{1 + f_x^2 + f_y^2}\,dx\,dy$.

Correspondingly, we then have $V = \displaystyle\iint_S \mathbf{F} \cdot \mathbf{N}\,dS = \iint_S x\,dy\,dz = \iint_S y\,dz\,dx = \iint_S z\,dx\,dy.$

19. Using the Divergence Theorem, we have

$$\iint_S \textbf{curl F} \cdot \mathbf{N}\,dS = \iiint_Q \text{div}\,(\textbf{curlF})\,dV$$

$$\textbf{curl F}(x, y, z) = \begin{vmatrix} \mathbf{i} & \mathbf{j} & \mathbf{k} \\ \dfrac{\partial}{\partial x} & \dfrac{\partial}{\partial y} & \dfrac{\partial}{\partial z} \\ 4xy + z^2 & 2x^2 + 6yz & 2xz \end{vmatrix} = -6y\mathbf{i} - (2z - 2z)\mathbf{j} + (4x - 4x)\mathbf{k} = -6y\mathbf{i}$$

$$\text{div}\,(\textbf{curl F}) = 0.$$

Therefore, $\displaystyle\iiint_Q \text{div}\,(\textbf{curl F})\,dV = 0.$

21. Using the Divergence Theorem, we have $\displaystyle\iint_S \textbf{curl F} \cdot \mathbf{N}\,dS = \iiint_Q \text{div}\,(\textbf{curl F})\,dV$. Let

$$\mathbf{F}(x, y, z) = M\mathbf{i} + N\mathbf{j} + P\mathbf{k}$$

$$\textbf{curl F} = \left(\frac{\partial P}{\partial y} - \frac{\partial N}{\partial z}\right)\mathbf{i} - \left(\frac{\partial P}{\partial x} - \frac{\partial M}{\partial z}\right)\mathbf{j} + \left(\frac{\partial N}{\partial x} - \frac{\partial M}{\partial y}\right)\mathbf{k}$$

$$\text{div}\,(\textbf{curl F}) = \frac{\partial^2 P}{\partial x \partial y} - \frac{\partial^2 N}{\partial x \partial z} - \frac{\partial^2 P}{\partial y \partial x} + \frac{\partial^2 M}{\partial y \partial z} + \frac{\partial^2 N}{\partial z \partial x} - \frac{\partial^2 M}{\partial z \partial y} = 0.$$

Therefore, $\displaystyle\iint_S \textbf{curl F} \cdot \mathbf{N}\,dS = \iiint_Q 0\,dV = 0.$

23. If $\mathbf{F}(x, y, z) = x\mathbf{i} + y\mathbf{j} + z\mathbf{k}$, then $\text{div}\,\mathbf{F} = 3$.

$$\iint_S \mathbf{F} \cdot \mathbf{N}\,dS = \iiint_Q \text{div}\,\mathbf{F}\,dV = \iiint_Q 3\,dV = 3V.$$

25. $\displaystyle\iint_S f D_{\mathbf{N}} g\,dS = \iint_S f\nabla g \cdot \mathbf{N}\,dS$

$$= \iiint_Q \text{div}\,(f\nabla g)\,dV = \iiint_Q (f\,\text{div}\,\nabla g + \nabla f \cdot \nabla g)\,dV = \iiint_Q (f\nabla^2 g + \nabla f \cdot \nabla g)\,dV$$

Section 14.8 Stokes's Theorem

1. $\mathbf{F}(x, y, z) = (2y - z)\mathbf{i} + xyz\mathbf{j} + e^z\mathbf{k}$

$$\textbf{curl F} = \begin{vmatrix} \mathbf{i} & \mathbf{j} & \mathbf{k} \\ \frac{\partial}{\partial x} & \frac{\partial}{\partial y} & \frac{\partial}{\partial z} \\ 2y - z & xyz & e^z \end{vmatrix} = -xy\mathbf{i} - \mathbf{j} + (yz - 2)\mathbf{k}$$

3. $\mathbf{F}(x, y, z) = 2z\mathbf{i} - 4x^2\mathbf{j} + \arctan x\mathbf{k}$

$$\textbf{curl F} = \begin{vmatrix} \mathbf{i} & \mathbf{j} & \mathbf{k} \\ \frac{\partial}{\partial x} & \frac{\partial}{\partial y} & \frac{\partial}{\partial z} \\ 2z & -4x^2 & \arctan x \end{vmatrix} = \left(2 - \frac{1}{1 + x^2}\right)\mathbf{j} - 8x\mathbf{k}$$

5. $\mathbf{F}(x, y, z) = e^{x^2+y^2}\mathbf{i} + e^{y^2+z^2}\mathbf{j} + xyz\mathbf{k}$

$$\textbf{curl F} = \begin{vmatrix} \mathbf{i} & \mathbf{j} & \mathbf{k} \\ \frac{\partial}{\partial x} & \frac{\partial}{\partial y} & \frac{\partial}{\partial z} \\ e^{x^2+y^2} & e^{y^2+z^2} & xyz \end{vmatrix} = (xz - 2ze^{y^2+z^2})\mathbf{i} - yz\mathbf{j} - 2ye^{x^2+y^2}\mathbf{k} = z(x - 2e^{y^2+z^2})\mathbf{i} - yz\mathbf{j} - 2ye^{x^2+y^2}\mathbf{k}$$

7. In this case, $M = -y + z,\ N = x - z,\ P = x - y$ and C is the circle $x^2 + y^2 = 1,\ z = 0,\ dz = 0$.

Line Integral: $\displaystyle\int_C \mathbf{F} \cdot d\mathbf{r} = \int_C -y\,dx + x\,dy$

Letting $x = \cos t,\ y = \sin t,$ we have $dx = -\sin t\,dt,\ dy = \cos t\,dt$ and

$$\int_C -y\,dx + x\,dy = \int_0^{2\pi} (\sin^2 t + \cos^2 t)dt = 2\pi.$$

Double Integral: Consider $F(x, y, z) = x^2 + y^2 + z^2 - 1$. Then

$$\mathbf{N} = \frac{\nabla F}{\|\nabla F\|} = \frac{2x\mathbf{i} + 2y\mathbf{j} + 2z\mathbf{k}}{2\sqrt{x^2 + y^2 + z^2}} = x\mathbf{i} + y\mathbf{j} + z\mathbf{k}.$$

Since

$$z^2 = 1 - x^2 - y^2,\ z_x = \frac{-2x}{2z} = \frac{-x}{z},\ \text{and } z_y = \frac{-y}{z},\ dS = \sqrt{1 + \frac{x^2}{z^2} + \frac{y^2}{z^2}}\,dA = \frac{1}{z}\,dA.$$

Now, since **curl F** $= 2\mathbf{k}$, we have

$$\int_S\int (\textbf{curl F}) \cdot \mathbf{N}\,dS = \int_R\int 2z\left(\frac{1}{z}\right) dA = \int_R\int 2\,dA = 2(\text{Area of circle of radius 1}) = 2\pi.$$

9. Line Integral: From the accompanying figure we see that for

$C_1: z = 0,\ dz = 0$

$C_2: x = 0,\ dx = 0$

$C_3: y = 0,\ dy = 0.$

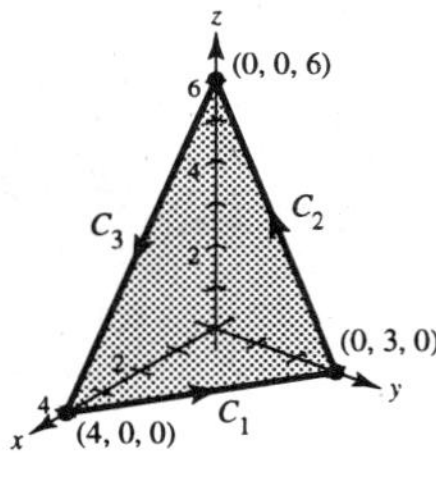

Hence, $\displaystyle\int_C \mathbf{F} \cdot d\mathbf{r} = \int_C xyz\,dx + y\,dy + z\,dz = \int_{C_1} y\,dy + \int_{C_2} y\,dy + z\,dz + \int_{C_3} z\,dz$

$$= \int_0^3 y\,dy + \int_3^0 y\,dy + \int_0^6 z\,dz + \int_6^0 z\,dz = 0.$$

Double Integral: **curl F** $= xy\mathbf{j} - xz\mathbf{k}$

Considering $F(x, y, z) = 3x + 4y + 2z - 12$, then $\mathbf{N} = \dfrac{\nabla F}{\|\nabla F\|} = \dfrac{3\mathbf{i} + 4\mathbf{j} + 2\mathbf{k}}{\sqrt{29}}$ and $dS = \sqrt{29}\,dA$.

Thus,

$$\int_S\int (\textbf{curl F}) \cdot \mathbf{N}\,dS = \int_R\int (4xy - 2xz)\,dy\,dx = \int_0^4\int_0^{(-3x+12)/4} \left[4xy - 2x\left(6 - 2y - \frac{3}{2}x\right)\right] dy\,dx$$

$$= \int_0^4\int_0^{(12-3x)/4} (8xy + 3x^2 - 12x)\,dy\,dx = \int_0^4 0\,dx = 0.$$

11. Let $A = (0, 0, 0)$, $B = (1, 1, 1)$ and $C = (0, 2, 0)$. Then $\mathbf{U} = \overrightarrow{AB} = \mathbf{i} + \mathbf{j} + \mathbf{k}$ and $\mathbf{V} = \overrightarrow{AC} = 2\mathbf{j}$. Thus,

$$\mathbf{N} = \frac{\mathbf{U} \times \mathbf{V}}{\|\mathbf{U} \times \mathbf{V}\|} = \frac{-2\mathbf{i} + 2\mathbf{k}}{2\sqrt{2}} = \frac{-\mathbf{i} + \mathbf{k}}{\sqrt{2}}.$$

Surface S has direction numbers $-1, 0, 1$, with equation $z - x = 0$ and $dS = \sqrt{2}\,dA$. Since $\mathbf{curl\ F} = -3\mathbf{i} + \mathbf{j} - 2\mathbf{k}$, we have

$$\int_S\!\!\int (\mathbf{curl\ F}) \cdot \mathbf{N}\,dS = \int_R\!\!\int \frac{1}{\sqrt{2}}(\sqrt{2})\,dA = \int_R\!\!\int dA = (\text{Area of triangle with } a = 1, b = 2) = 1.$$

13. $\mathbf{F}(x, y, z) = z^2\mathbf{i} + x^2\mathbf{j} + y^2\mathbf{k}$, $S\colon z = 4 - x^2 - y^2,\ 0 \le z$

$$\mathbf{curl\ F} = \begin{vmatrix} \mathbf{i} & \mathbf{j} & \mathbf{k} \\ \dfrac{\partial}{\partial x} & \dfrac{\partial}{\partial y} & \dfrac{\partial}{\partial z} \\ z^2 & x^2 & y^2 \end{vmatrix} = 2y\mathbf{i} + 2z\mathbf{j} + 2x\mathbf{k}$$

$G(x, y, z) = x^2 + y^2 + z - 4$

$\nabla G(x, y, z) = 2x\mathbf{i} + 2y\mathbf{j} + \mathbf{k}$

$$\begin{aligned}\int_S\!\!\int (\mathbf{curl\ F}) \cdot \mathbf{N}\,dS = \int_R\!\!\int (4xy + 4yz + 2x)\,dA &= \int_{-2}^{2}\int_{-\sqrt{4-x^2}}^{\sqrt{4-x^2}} [4xy + 4y(4 - x^2 - y^2) + 2x]\,dy\,dx \\ &= \int_{-2}^{2}\int_{-\sqrt{4-x^2}}^{\sqrt{4-x^2}} [4xy + 16y - 4x^2y - 4y^3 + 2x]\,dy\,dx \\ &= \int_{-2}^{2} 4x\sqrt{4 - x^2}\,dx = 0\end{aligned}$$

15. $\mathbf{F}(x, y, z) = z^2\mathbf{i} + y\mathbf{j} + xz\mathbf{k}$, $S\colon z = \sqrt{4 - x^2 - y^2}$

$$\mathbf{curl\ F} = \begin{vmatrix} \mathbf{i} & \mathbf{j} & \mathbf{k} \\ \dfrac{\partial}{\partial x} & \dfrac{\partial}{\partial y} & \dfrac{\partial}{\partial z} \\ z^2 & y & xz \end{vmatrix} = z\mathbf{j}$$

$G(x, y, z) = z - \sqrt{4 - x^2 - y^2}$

$$\nabla G(x, y, z) = \frac{x}{\sqrt{4 - x^2 - y^2}}\mathbf{i} + \frac{y}{\sqrt{4 - x^2 - y^2}}\mathbf{j} + \mathbf{k}$$

$$\int_S\!\!\int (\mathbf{curl\ F}) \cdot \mathbf{F}\,dS = \int_R\!\!\int \frac{yz}{\sqrt{4 - x^2 - y^2}}\,dA = \int_R\!\!\int \frac{y\sqrt{4 - x^2 - y^2}}{\sqrt{4 - x^2 - y^2}}\,dA = \int_{-2}^{2}\int_{-\sqrt{4-x^2}}^{\sqrt{4-x^2}} y\,dy\,dx = 0$$

17. $\mathbf{F}(x, y, z) = -\ln\sqrt{x^2 + y^2}\,\mathbf{i} + \arctan\dfrac{x}{y}\mathbf{j} + \mathbf{k}$

$$\mathbf{curl\ F} = \begin{vmatrix} \mathbf{i} & \mathbf{j} & \mathbf{k} \\ \dfrac{\partial}{\partial x} & \dfrac{\partial}{\partial y} & \dfrac{\partial}{\partial z} \\ -\dfrac{1}{2}\ln(x^2 + y^2) & \arctan\dfrac{x}{y} & 1 \end{vmatrix} = \left[\frac{(1/y)}{1 + (x^2/y^2)} + \frac{y}{x^2 + y^2}\right]\mathbf{k} = \left[\frac{2y}{x^2 + y^2}\right]\mathbf{k}$$

$S\colon z = 9 - 2x - 3y$ over one petal of $r = 2\sin 2\theta$ in the first octant.

$G(x, y, z) = 2x + 3y + z - 9$

$\nabla G(x, y, z) = 2\mathbf{i} + 3\mathbf{j} + \mathbf{k}$

$$\begin{aligned}\int_S\!\!\int (\mathbf{curl\ F}) \cdot \mathbf{N}\,dS &= \int_R\!\!\int \frac{2y}{x^2 + y^2}\,dA = \int_0^{\pi/2}\int_0^{2\sin 2\theta} \frac{2r\sin\theta}{r^2}\,r\,dr\,d\theta \\ &= \int_0^{\pi/2}\int_0^{4\sin\theta\cos\theta} 2\sin\theta\,dr\,d\theta = \int_0^{\pi/2} 8\sin^2\theta\cos\theta\,d\theta = \left[\frac{8\sin^3\theta}{3}\right]_0^{\pi/2} = \frac{8}{3}\end{aligned}$$

19. From Exercise 10, we have $\mathbf{N} = \dfrac{2x\mathbf{i} - \mathbf{k}}{\sqrt{1 + 4x^2}}$ and

$dS = \sqrt{1 + 4x^2}\, dA$. Since $\mathbf{curl\, F} = xy\mathbf{j} - xz\mathbf{k}$, we have

$$\int_S\int (\mathbf{curl\, F}) \cdot \mathbf{N}\, dS = \int_R\int xz\, dA$$

$$= \int_0^a \int_0^a x^3\, dy\, dx$$

$$= \int_0^a ax^3\, dx = \left[\frac{ax^4}{4}\right]_0^a = \frac{a^5}{4}.$$

21. $\mathbf{F}(x, y, z) = \mathbf{i} + \mathbf{j} - 2\mathbf{k}$

$$\mathbf{curl\, F} = \begin{vmatrix} \mathbf{i} & \mathbf{j} & \mathbf{k} \\ \dfrac{\partial}{\partial x} & \dfrac{\partial}{\partial y} & \dfrac{\partial}{\partial z} \\ 1 & 1 & -2 \end{vmatrix} = \mathbf{0}$$

Letting $\mathbf{N} = \mathbf{k}$, we have $\displaystyle\int_S\int (\mathbf{curl\, F}) \cdot \mathbf{N}\, dS = 0.$

23. (a) $\displaystyle\int_C f\nabla g \cdot d\mathbf{r} = \int_S\int \mathbf{curl}[f\nabla g] \cdot \mathbf{N}\, dS$ (Stoke's Theorem)

$$f\nabla g = f\frac{\partial g}{\partial x}\mathbf{i} + f\frac{\partial g}{\partial y}\mathbf{j} + f\frac{\partial g}{\partial z}\mathbf{k}$$

$$\mathbf{curl}\,(f\nabla g) = \begin{vmatrix} \mathbf{i} & \mathbf{j} & \mathbf{k} \\ \dfrac{\partial}{\partial x} & \dfrac{\partial}{\partial y} & \dfrac{\partial}{\partial z} \\ f\dfrac{\partial g}{\partial x} & f\dfrac{\partial y}{\partial z} & f\dfrac{\partial g}{\partial z} \end{vmatrix}$$

$$= \left[\left[f\left(\frac{\partial^2 g}{\partial y \partial z}\right) + \left(\frac{\partial f}{\partial y}\right)\left(\frac{\partial g}{\partial z}\right)\right] - \left[f\left(\frac{\partial^2 g}{\partial z \partial y}\right) + \left(\frac{\partial f}{\partial z}\right)\left(\frac{\partial g}{\partial y}\right)\right]\right]\mathbf{i}$$

$$- \left[\left[f\left(\frac{\partial^2 g}{\partial x \partial z}\right) + \left(\frac{\partial f}{\partial x}\right)\left(\frac{\partial g}{\partial z}\right)\right] - \left[f\left(\frac{\partial^2 g}{\partial z \partial x}\right) + \left(\frac{\partial f}{\partial z}\right)\left(\frac{\partial g}{\partial x}\right)\right]\right]\mathbf{j}$$

$$+ \left[\left[f\left(\frac{\partial^2 g}{\partial x \partial y}\right) + \left(\frac{\partial f}{\partial x}\right)\left(\frac{\partial g}{\partial y}\right)\right] - \left[f\left(\frac{\partial^2 g}{\partial y \partial x}\right) + \left(\frac{\partial f}{\partial y}\right)\left(\frac{\partial g}{\partial x}\right)\right]\right]\mathbf{k}$$

$$= \left[\left(\frac{\partial f}{\partial y}\right)\left(\frac{\partial g}{\partial z}\right) - \left(\frac{\partial f}{\partial z}\right)\left(\frac{\partial g}{\partial y}\right)\right]\mathbf{i} - \left[\left(\frac{\partial f}{\partial x}\right)\left(\frac{\partial g}{\partial z}\right) - \left(\frac{\partial f}{\partial z}\right)\left(\frac{\partial g}{\partial x}\right)\right]\mathbf{j} + \left[\left(\frac{\partial f}{\partial x}\right)\left(\frac{\partial g}{\partial y}\right) - \left(\frac{\partial f}{\partial y}\right)\left(\frac{\partial g}{\partial x}\right)\right]\mathbf{k}$$

$$= \begin{vmatrix} \mathbf{i} & \mathbf{j} & \mathbf{k} \\ \dfrac{\partial f}{\partial x} & \dfrac{\partial f}{\partial y} & \dfrac{\partial f}{\partial z} \\ \dfrac{\partial g}{\partial x} & \dfrac{\partial g}{\partial y} & \dfrac{\partial g}{\partial z} \end{vmatrix} = \nabla f \times \nabla g$$

Therefore, $\displaystyle\int_C f\nabla g \cdot d\mathbf{r} = \int_S\int \mathbf{curl}[f\nabla g] \cdot \mathbf{N}\, dS = \int_S\int [\nabla f \times \nabla g] \cdot \mathbf{N}\, dS.$

25. Let $\mathbf{C} = a\mathbf{i} + b\mathbf{j} + c\mathbf{k}$, then

$$\frac{1}{2}\int_C (\mathbf{C} \times \mathbf{r}) \cdot d\mathbf{r} = \frac{1}{2}\int_S\int \mathbf{curl}\,(\mathbf{C} \times \mathbf{r}) \cdot \mathbf{N}\, dS = \frac{1}{2}\int_S\int 2\mathbf{C} \cdot \mathbf{N}\, dS = \int_S\int \mathbf{C} \cdot \mathbf{N}\, dS$$

since

$$\mathbf{C} \times \mathbf{r} = \begin{vmatrix} \mathbf{i} & \mathbf{j} & \mathbf{k} \\ a & b & c \\ x & y & z \end{vmatrix} = (bz - cy)\mathbf{i} - (az - cx)\mathbf{j} + (ay - bx)\mathbf{k}$$

and

$$\mathbf{curl}(\mathbf{C} \times \mathbf{r}) = \begin{vmatrix} \mathbf{i} & \mathbf{j} & \mathbf{k} \\ \dfrac{\partial}{\partial x} & \dfrac{\partial}{\partial y} & \dfrac{\partial}{\partial z} \\ bz - cy & cx - az & ay - bx \end{vmatrix} = 2(a\mathbf{i} + b\mathbf{j} + c\mathbf{k}) = 2\mathbf{C}.$$

Review Exercises for Chapter 14

1. $\mathbf{F}(x, y, z) = x\mathbf{i} + \mathbf{j} + 2\mathbf{k}$

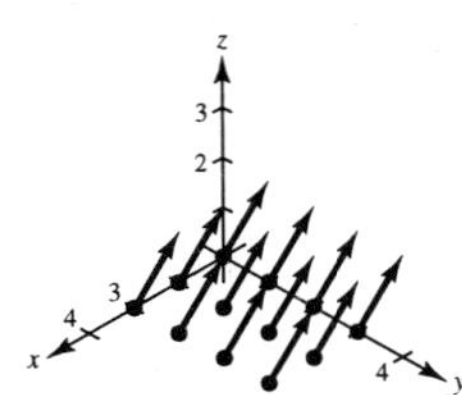

3. $f(x, y, z) = 8x^2 + xy + z^2$

$\mathbf{F}(x, y, z) = (16x + y)\mathbf{i} + x\mathbf{j} + 2z\mathbf{k}$

5. Since $\partial M/\partial y = -1/y^2 \neq \partial N/\partial x$, $\mathbf{F}$ is not conservative.

7. Since $\partial M/\partial y = 12xy = \partial N/\partial x$, $\mathbf{F}$ is conservative. From $M = \partial U/\partial x = 6xy^2 - 3x^2$ and $N = \partial U/\partial y = 6x^2y + 3y^2 - 7$, partial integration yields $U = 3x^2y^2 - x^3 + h(y)$ and $U = 3x^2y^2 + y^3 - 7y + g(x)$ which suggests $h(y) = y^3 - 7y$, $g(x) = -x^3$, and $U(x, y) = 3x^2y^2 - x^3 + y^3 - 7y + C$.

9. Since

$$\frac{\partial M}{\partial y} = 4x = \frac{\partial N}{\partial x},$$

$$\frac{\partial M}{\partial z} = 1 \neq \frac{\partial P}{\partial x}.$$

$\mathbf{F}$ is not conservative.

11. Since

$$\frac{\partial M}{\partial y} = \frac{-1}{y^2z} = \frac{\partial N}{\partial x}, \quad \frac{\partial M}{\partial z} = \frac{-1}{yz^2} = \frac{\partial P}{\partial x}, \quad \frac{\partial N}{\partial z} = \frac{x}{y^2z^2} = \frac{\partial P}{\partial y},$$

$\mathbf{F}$ is conservative. From

$$M = \frac{\partial U}{\partial x} = \frac{1}{yz}, \quad N = \frac{\partial U}{\partial y} = \frac{-x}{y^2z}, \quad P = \frac{\partial U}{\partial z} = \frac{-x}{yz^2}$$

we obtain

$$U = \frac{x}{yz} + f(y, z), \quad U = \frac{x}{yz} + g(x, z), \quad U = \frac{x}{yz} + h(x, y)$$

which suggests that $f(y, z) = C_1$, $g(x, z) = C_2$, $h(x, y) = C_3$. Thus, $U(x, y, z) = (x/yz) + C$.

13. Since $\mathbf{F} = x^2\mathbf{i} + y^2\mathbf{j} + z^2\mathbf{k}$:

(a) $\text{div}\,\mathbf{F} = 2x + 2y + 2z$

(b) $\textbf{curl F} = \left(\frac{\partial P}{\partial y} - \frac{\partial N}{\partial z}\right)\mathbf{i} - \left(\frac{\partial P}{\partial x} - \frac{\partial M}{\partial z}\right)\mathbf{j} + \left(\frac{\partial N}{\partial x} - \frac{\partial M}{\partial y}\right)\mathbf{k} = 0\mathbf{i} - 0\mathbf{j} + 0\mathbf{k} = \mathbf{0}$

15. Since $\mathbf{F} = (\cos y + y\cos x)\mathbf{i} + (\sin x - x\sin y)\mathbf{j} + xyz\mathbf{k}$:

(a) $\text{div}\,\mathbf{F} = -y\sin x - x\cos y + xy$

(b) $\textbf{curl F} = xz\mathbf{i} - yz\mathbf{j} + (\cos x - \sin y + \sin y - \cos x)\mathbf{k} = xz\mathbf{i} - yz\mathbf{j}$

17. Since $\mathbf{F} = \arcsin x\mathbf{i} + xy^2\mathbf{j} + yz^2\mathbf{k}$:

(a) $\text{div}\,\mathbf{F} = \dfrac{1}{\sqrt{1-x^2}} + 2xy + 2yz$

(b) $\textbf{curl}\,\mathbf{F} = z^2\mathbf{i} + y^2\mathbf{k}$

19. Since $\mathbf{F} = \ln(x^2 + y^2)\mathbf{i} + \ln(x^2 + y^2)\mathbf{j} + z\mathbf{k}$:

(a) $$\text{div}\,\mathbf{F} = \frac{2x}{x^2+y^2} + \frac{2y}{x^2+y^2} + 1$$
$$= \frac{2x+2y}{x^2+y^2} + 1$$

(b) $$\textbf{curl}\,\mathbf{F} = \frac{2x-2y}{x^2+y^2}\mathbf{k}$$

21. (a) Let $x = t$, $y = t$, $-1 \le t \le 2$, then $ds = \sqrt{2}\,dt$.

$$\int_C (x^2 + y^2)\,ds = \int_{-1}^{2} 2t^2\sqrt{2}\,dt = \left[2\sqrt{2}\left(\frac{t^3}{3}\right)\right]_{-1}^{2} = 6\sqrt{2}$$

(b) Let $x = 4\cos t$, $y = 4\sin t$, $0 \le t \le 2\pi$, then $ds = 4\,dt$.

$$\int_C (x^2 + y^2)\,ds = \int_0^{2\pi} 16(4\,dt) = 128\pi$$

23. $x = \cos t + t\sin t$, $y = \sin t - t\cos t$, $0 \le t \le 2\pi$, $\dfrac{dx}{dt} = t\cos t$, $\dfrac{dy}{dt} = t\sin t$

$$\int_C (x^2 + y^2)\,ds = \int_0^{2\pi} [(\cos t + t\sin t)^2 + (\sin t - t\cos t)^2]\sqrt{t^2\cos^2 t + t^2\sin^2 t}\,dt$$
$$= \int_0^{2\pi} (1 + t^2)t\,dt = \left[\frac{t^2}{2} + \frac{t^4}{4}\right]_0^{2\pi} = 2\pi^2 + 4\pi^4 = 2\pi^2(1 + 2\pi^2)$$

25. (a) Let $x = 2t$, $y = -3t$, $0 \le t \le 1$

$$\int_C (2x - y)\,dx + (x + 3y)\,dy = \int_0^1 [7t(2) + (-7t)(-3)]\,dt = \int_0^1 35t\,dt = \frac{35}{2}$$

(b) $x = 3\cos t$, $y = 3\sin t$, $dx = -3\sin t\,dt$, $dy = 3\cos t\,dt$, $0 \le t \le 2\pi$

$$\int_C (2x - y)dx + (x + 3y)\,dy = \int_0^{2\pi} (9 + 9\sin t\cos t)\,dt = 18\pi$$

27. $\displaystyle\int_C (2x + y)\,ds$, $\mathbf{r}(t) = a\cos^3 t\mathbf{i} + \mathbf{a}\sin^3 t\,\mathbf{j}$, $0 \le t \le \dfrac{\pi}{2}$

$x'(t) = -3a \cdot \cos^2 t\sin t$

$y'(t) = 3a \cdot \sin^2 t\cos t$

$$\int_C (2x + y)\,ds = \int_0^{\pi/2} (2(a \cdot \cos^3 t) + a \cdot \sin^3 t)\sqrt{x'(t)^2 + y'(t)^2}\,dt = \frac{9a^2}{5}$$

29. $f(x, y) = 5 + \sin(x + y)$

C: $y = 3x$ from $(0, 0)$ to $(2, 6)$

$\mathbf{r}(t) = t\mathbf{i} + 3t\mathbf{j}$, $0 \le t \le 2$

$\mathbf{r}'(t) = \mathbf{i} + 3\mathbf{j}$

$\|\mathbf{r}'(t)\| = \sqrt{10}$

Lateral surface area:

$$\int_{C_2} f(x, y)\,ds = \int_0^2 [5 + \sin(t + 3t)]\sqrt{10}\,dt = \sqrt{10}\int_0^2 (5 + \sin 4t)\,dt = \frac{\sqrt{10}}{4}(41 - \cos 8) \approx 32.528$$

31. $d\mathbf{r} = (2t\mathbf{i} + 3t^2\mathbf{j})\,dt$

$\mathbf{F} = t^5\mathbf{i} + t^4\mathbf{j}, 0 \le t \le 1$

$$\int_C \mathbf{F} \cdot d\mathbf{r} = \int_0^1 5t^6\,dt = \frac{5}{7}$$

33. $d\mathbf{r} = [(-2\sin t)\mathbf{i} + (2\cos t)\mathbf{j} + \mathbf{k}]\,dt$

$\mathbf{F} = (2\cos t)\mathbf{i} + (2\sin t)\mathbf{j} + t\mathbf{k}, 0 \le t \le 2\pi$

$$\int_C \mathbf{F} \cdot d\mathbf{r} = \int_0^{2\pi} t\,dt = 2\pi^2$$

35. Let $x = t, y = -t, z = 2t^2, -2 \le t \le 2, d\mathbf{r} = [\mathbf{i} - \mathbf{j} + 4t\mathbf{k}]\,dt.$

$\mathbf{F} = (-t - 2t^2)\mathbf{i} + (2t^2 - t)\mathbf{j} + (2t)\mathbf{k}$

$$\int_C \mathbf{F} \cdot d\mathbf{r} = \int_{-2}^2 4t^2\,dt = \left[\frac{4t^3}{3}\right]_{-2}^2 = \frac{64}{3}$$

37. For $y = x^2$, $\mathbf{r}_1(t) = t\mathbf{i} + t^2\mathbf{j}, 0 \le t \le 2$

For $y = 2x$, $\mathbf{r}_2(t) = (2 - t)\mathbf{i} + (4 - 2t)\mathbf{j}, 0 \le t \le 2$

$$\int_C xy\,dx + (x^2 + y^2)\,dy = \int_{C_1} xy\,dx + (x^2 + y^2)\,dy + \int_{C_2} xy\,dx + (x^2 + y^2)\,dy$$

$$= \frac{100}{3} + (-32) = \frac{4}{3}$$

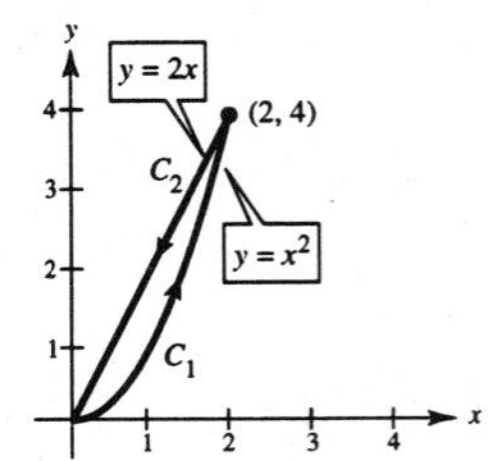

39. $\mathbf{F} = x\mathbf{i} - \sqrt{y}\mathbf{j}$ is conservative.

$$\text{Work} = \left[\frac{1}{2}x^2 - \frac{2}{3}y^{3/2}\right]_{(0,\,0)}^{(4,\,8)} = \frac{1}{2}(16) - \left(\frac{2}{3}\right)8^{3/2} = \frac{8}{3}(3 - 4\sqrt{2})$$

41. $$\int_C 2xyz\,dx + x^2z\,dy + x^2y\,dz = \Big[x^2yz\Big]_{(0,\,0,\,0)}^{(1,\,4,\,3)} = 12$$

43. (a) $$\int_C y^2\,dx + 2xy\,dy = \int_0^1 [(1 + t)^2(3) + 2(1 + 3t)(1 + t)]\,dt$$

$$= \int_0^1 3(t^2 + 2t + 1) + 2(3t^2 + 4t + 1)]\,dt$$

$$= \int_0^1 (9t^2 + 14t + 5)\,dt$$

$$= \Big[3t^3 + 7t^2 + 5t\Big]_0^1 = 15$$

(b) $$\int_C y^2\,dx + 2xy\,dy = \int_1^4 \left[t(1) + 2(t)(\sqrt{t})\frac{1}{2\sqrt{t}}\right]dt$$

$$= \int_1^4 (t + t)\,dt$$

$$= \Big[t^2\Big]_1^4 = 15$$

(c) $\mathbf{F}(x, y) = y^2\mathbf{i} + 2xy\,\mathbf{j} = \nabla f$ where $f(x, y) = xy^2$.

Hence,

$$\int_C \mathbf{F} \cdot d\mathbf{r} = 4(2)^2 - 1(1)^2 = 15$$

45. $\displaystyle\int_C y\,dx + 2x\,dy = \int_0^2\int_0^2 (2-1)\,dy\,dx = \int_0^2 2\,dx = 4$

47. $\displaystyle\int_C xy^2\,dx + x^2y\,dy = \iint_R (2xy - 2xy)\,dA = 0$

49. $\displaystyle\int_C xy\,dx + x^2dy = \int_0^1\int_{x^2}^x x\,dy\,dx = \int_0^1 (x^2 - x^3)\,dx = \frac{1}{12}$

51. $\mathbf{r}(u, v) = \sec u \cos v\mathbf{i} + (1 + 2\tan u)\sin v\,\mathbf{j} + 2u\mathbf{k}$

$0 \le u \le \dfrac{\pi}{3},\quad 0 \le v \le 2\pi$

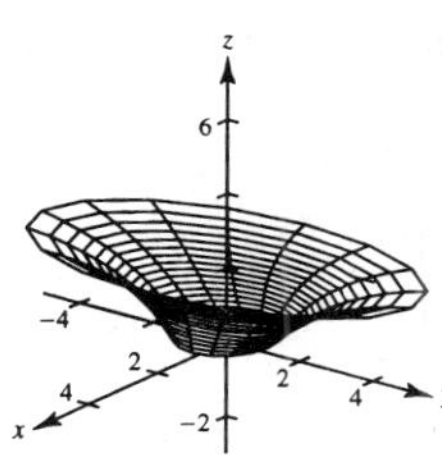

53. (a)

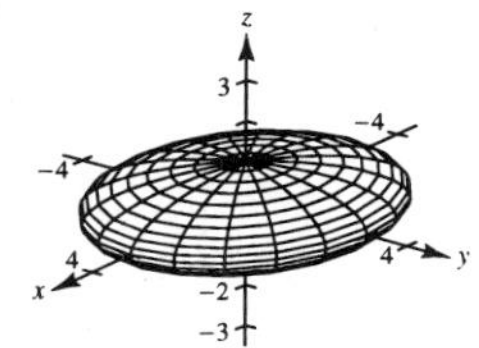

(b)

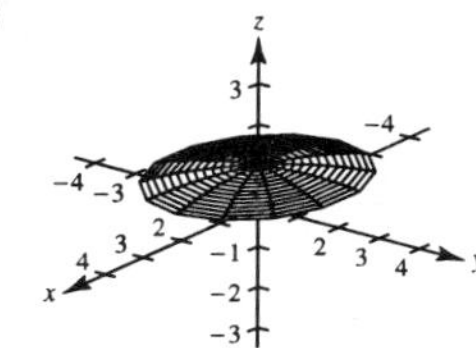

(c)

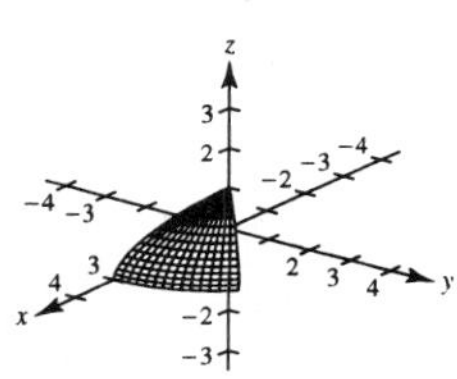

(d)

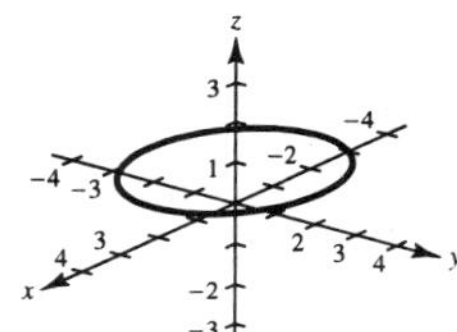

The space curve is a circle:

$$\mathbf{r}\left(u, \frac{\pi}{4}\right) = \frac{3\sqrt{2}}{2}\cos u\mathbf{i} + \frac{3\sqrt{2}}{2}\sin u\mathbf{j} + \frac{\sqrt{2}}{2}\mathbf{k}$$

(e) $\mathbf{r}_u = -3\cos v \sin u\,\mathbf{i} + 3\cos v\cos u\,\mathbf{j}$

$\mathbf{r}_v = -3\sin v\cos u\,\mathbf{i} - 3\sin v\sin u\sin u\,\mathbf{j} + \cos v\,\mathbf{k}$

$$\mathbf{r}_u \times \mathbf{r}_v = \begin{vmatrix} \mathbf{i} & \mathbf{j} & \mathbf{k} \\ -3\cos v\sin u & 3\cos v\cos u & 0 \\ -3\sin v\cos u & -3\sin v\sin u & \cos v \end{vmatrix}$$

$$= (3\cos^2 v\cos u)\mathbf{i} + (3\cos^2 v\sin u)\mathbf{j} + 9\cos v\sin v\sin^2 u + 9\cos v\sin v\cos^2 u)\mathbf{k}$$

$$= (3\cos^2 v\cos u)\mathbf{i} + (3\cos^2 v\sin u)\mathbf{j} + (9\cos v\sin v)\mathbf{k}$$

$$\|\mathbf{r}_u \times \mathbf{r}_v\| = \sqrt{9\cos^4 v\cos^2 u + 9\cos^4 v\sin^2 u + 81\cos^2 v\sin\sin^2 v}$$

$$= \sqrt{9\cos^4 v + 81\cos^2 v\sin^2 v}$$

Using a Symbolic integration utility,

$$\int_{\pi/4}^{\pi/2}\int_0^{2\pi} \|\mathbf{r}_u \times \mathbf{r}_v\|\,dv\,du \approx 14.44$$

(f) Similarly,

$$\int_0^{\pi/4}\int_0^{\pi/2} \|\mathbf{r}_u \times \mathbf{r}_v\|\,dv\,du \approx 4.27$$

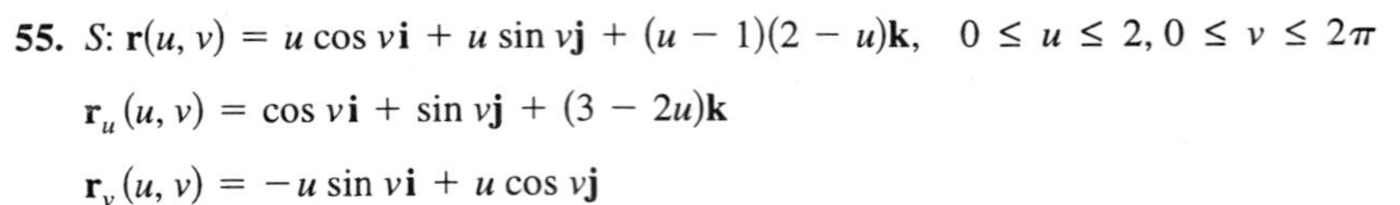

55. S: $\mathbf{r}(u, v) = u\cos v\mathbf{i} + u\sin v\mathbf{j} + (u - 1)(2 - u)\mathbf{k}, \quad 0 \le u \le 2, 0 \le v \le 2\pi$

$\mathbf{r}_u(u, v) = \cos v\mathbf{i} + \sin v\mathbf{j} + (3 - 2u)\mathbf{k}$

$\mathbf{r}_v(u, v) = -u\sin v\mathbf{i} + u\cos v\mathbf{j}$

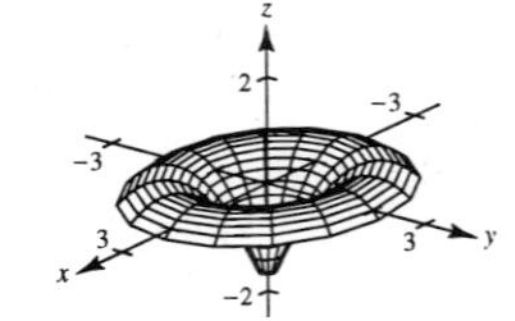

$$\mathbf{r}_u \times \mathbf{r}_u = \begin{vmatrix} \mathbf{i} & \mathbf{j} & \mathbf{k} \\ \cos v & \sin v & 3 - 2u \\ -u\sin v & u\cos v & 0 \end{vmatrix} = (2v - 3)u\cos v\mathbf{i} + (2u - 3)u\sin v\mathbf{j} + u\mathbf{k}$$

$$\|\mathbf{r}_u \times \mathbf{r}_v\| = u\sqrt{(2u - 3)^2 + 1}$$

$$\int_S\int (x + y)\,dS = \int_0^{2\pi}\int_0^2 (u\cos v + u\sin v)\,u\sqrt{(2u - 3)^2 + 1}\,du\,dv$$

$$= \int_0^2\int_0^{2\pi} (\cos v + \sin v)u^2\sqrt{(2u - 3)^2 + 1}\,dv\,du = 0$$

57. $\mathbf{F}(x, y, z) = x^2\mathbf{i} + xy\mathbf{j} + z\mathbf{k}$

Q: solid region bounded by the coordinates planes and the plane $2x + 3y + 4z = 12$

Surface Integral: There are four surfaces for this solid.

$z = 0 \quad \mathbf{N} = -\mathbf{k}, \quad \mathbf{F}\cdot\mathbf{N} = -z, \quad \int_{S_1}\int 0\,dS = 0$

$y = 0, \quad \mathbf{N} = -\mathbf{j}, \quad \mathbf{F}\cdot\mathbf{N} = -xy, \quad \int_{S_2}\int 0\,dS = 0$

$x = 0, \quad \mathbf{N} = -\mathbf{i}, \quad \mathbf{F}\cdot\mathbf{N} = -x^2, \quad \int_{S_3}\int 0\,dS = 0$

$$2x + 3y + 4z = 12, \mathbf{N} = \frac{2\mathbf{i} + 3\mathbf{j} + 4\mathbf{k}}{\sqrt{29}}, \; dS = \sqrt{1 + \left(\frac{1}{4}\right) + \left(\frac{9}{16}\right)}dA = \frac{\sqrt{29}}{4}\,dA$$

$$\int_{S_4}\int \mathbf{F}\cdot\mathbf{N}\,dS = \frac{1}{4}\int_R\int (2x^2 + 3xy + 4z)\,dA$$

$$= \frac{1}{4}\int_0^6\int_0^{4-(2x/3)} (2x^2 + 3xy + 12 - 2x - 3y)\,dy\,dx$$

$$= \frac{1}{4}\int_0^6\left[2x^2\left(\frac{12 - 2x}{3}\right) + \frac{3x}{2}\left(\frac{12 - 2x}{3}\right)^2 + 12\left(\frac{12 - 2x}{3}\right) - 2x\left(\frac{12 - 2x}{3}\right) - \frac{3}{2}\left(\frac{12 - 2x}{3}\right)^2\right]dx$$

$$= \frac{1}{6}\int_0^6 (-x^3 + x^2 + 24x + 36)\,dx$$

$$= \frac{1}{6}\left[-\frac{x^4}{4} + \frac{x^3}{3} + 12x^2 + 36x\right]_0^6 = 66$$

Divergence Theorem: Since div $\mathbf{F} = 2x + x + 1 = 3x + 1$, Divergence Theorem yields

$$\iiint_Q \text{div}\,\mathbf{F}\,dV = \int_0^6\int_0^{(12-2x)/3}\int_0^{(12-2x-3y)/4} (3x + 1)\,dz\,dy\,dx$$

$$= \int_0^6\int_0^{(12-2x)/3} (3x + 1)\left(\frac{12 - 2x - 3y}{4}\right)dy\,dx$$

$$= \frac{1}{4}\int_0^6 (3x + 1)\left[12y - 2xy - \frac{3}{2}y^2\right]_0^{(12-2x)/3} dx$$

$$= \frac{1}{4}\int_0^6 (3x + 1)\left[4(12 - 2x) - 2x\left(\frac{12 - 2x}{3}\right) - \frac{3}{2}\left(\frac{12 - 2x}{3}\right)^2\right]dx$$

$$= \frac{1}{4}\int_0^6 \frac{2}{3}(3x^3 - 35x^2 + 96x + 36)\,dx = \frac{1}{6}\left[\frac{3x^4}{4} - \frac{35x^3}{3} + 48x^2 + 36x\right]_0^6 = 66.$$

59. $\mathbf{F}(x, y, z) = (\cos y + y\cos x)\mathbf{i} + (\sin x - x\sin y)\mathbf{j} + xyz\mathbf{k}$

S: portion of $z = y^2$ over the square in the xy-plane with vertices $(0, 0)$, $(a, 0)$, (a, a), $(0, a)$

Line Integral: Using the line integral we have:

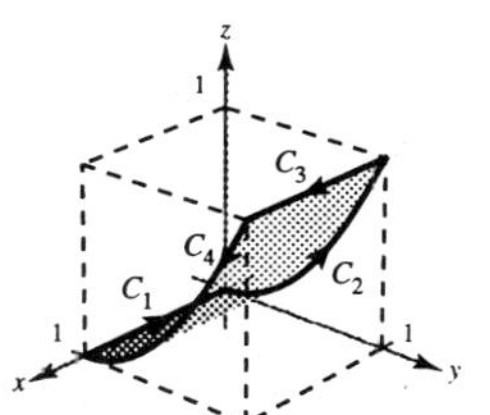

C_1: $y = 0$, $dy = 0$

C_2: $x = 0$, $dx = 0$, $z = y^2$, $dz = 2y\,dy$

C_3: $y = a$, $dy = 0$, $z = a^2$, $dz = 0$

C_4: $x = a$, $dx = 0$, $z = y^2$, $dz = 2y\,dy$

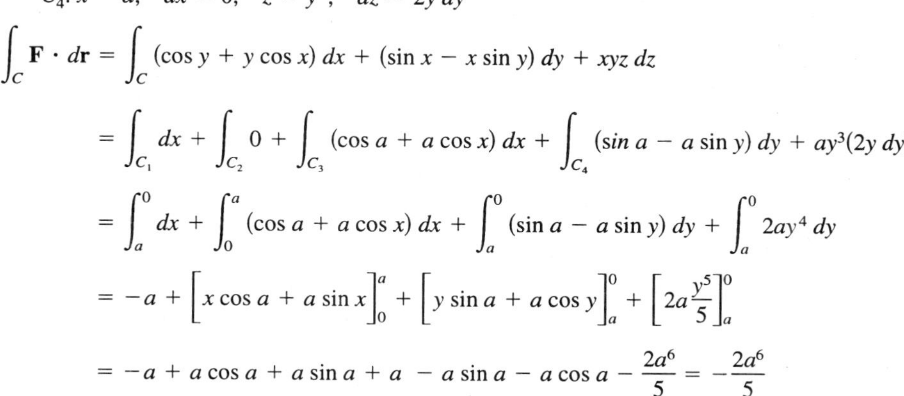

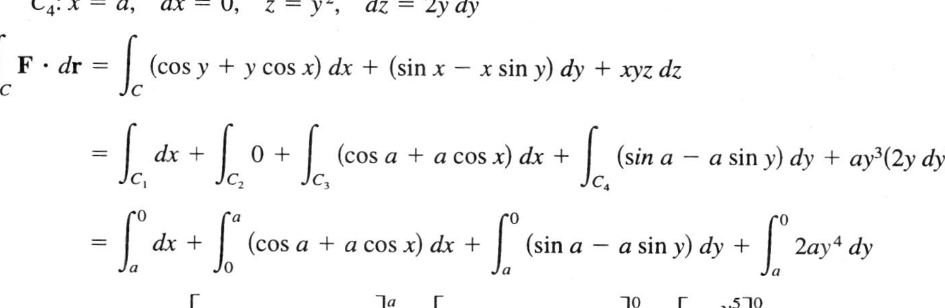

$$\int_C \mathbf{F}\cdot d\mathbf{r} = \int_C (\cos y + y\cos x)\,dx + (\sin x - x\sin y)\,dy + xyz\,dz$$

$$= \int_{C_1} dx + \int_{C_2} 0 + \int_{C_3} (\cos a + a\cos x)\,dx + \int_{C_4} (\sin a - a\sin y)\,dy + ay^3(2y\,dy)$$

$$= \int_a^0 dx + \int_0^a (\cos a + a\cos x)\,dx + \int_a^0 (\sin a - a\sin y)\,dy + \int_a^0 2ay^4\,dy$$

$$= -a + \Big[x\cos a + a\sin x\Big]_0^a + \Big[y\sin a + a\cos y\Big]_a^0 + \left[2a\frac{y^5}{5}\right]_a^0$$

$$= -a + a\cos a + a\sin a + a - a\sin a - a\cos a - \frac{2a^6}{5} = -\frac{2a^6}{5}$$

Double Integral: Considering $f(x, y, z) = y^2 - z$, we have:

$$\mathbf{N} = \frac{\nabla f}{\|\nabla f\|} = \frac{2y\mathbf{j} + \mathbf{k}}{\sqrt{1 + 4y^2}},\quad dS = \sqrt{1 + 4y^2}\,dA, \text{ and } \mathbf{curl\ F} = xz\mathbf{i} - yz\mathbf{j}.$$

Hence,

$$\int\!\!\int_S (\mathbf{curl\ F})\cdot\mathbf{N}\,dS = \int_0^a\!\!\int_0^a -2y^2z\,dy\,dx = \int_0^a\!\!\int_0^a -2y^4\,dy\,dx = \int_0^a -\frac{2a^5}{5}\,dx = -\frac{2a^6}{5}.$$

CHAPTER 15
Differential Equations

CHAPTER 15
Differential Equations

Section 15.1 Exact First-Order Equations

Solutions to Odd-Numbered Exercises

1. $(2x - 3y)dx + (2y - 3x)dy = 0$

$\frac{\partial M}{\partial y} = -3 = \frac{\partial N}{\partial x}$ exact.

$U(x, y) = x^2 - 3xy + f(y)$

$f'(y) = 2y$

$f(y) = y^2 + C_1$

$x^2 - 3xy + y^2 = C$

3. $(3y^2 + 10xy^2)dx + (6xy - 2 + 10x^2y)dy = 0$

$\frac{\partial M}{\partial y} = 6y + 20xy = \frac{\partial N}{\partial x}$ exact.

$U(x, y) = 3y^2x + 5x^2y^2 + f(y)$

$f'(y) = -2$

$f(y) = -2y + C_1$

$3y^2x + 5x^2y^2 - 2y = C$

5. $(4x^3 - 6xy^2)dx + (4y^3 - 6xy)dy = 0$

$\frac{\partial M}{\partial y} = -12xy$

$\frac{\partial N}{\partial x} = -6y$

Not exact

7. $\frac{-y}{x^2 + y^2}dx + \frac{x}{x^2 + y^2}dy = 0$

$\frac{\partial M}{\partial y} = \frac{y^2 - x^2}{(x^2 + y^2)^2} = \frac{\partial N}{\partial x}$ exact.

$U(x, y) = -\arctan\frac{x}{y} + f(y)$

$f'(y) = 0$

$f(y) = C_1$

$\arctan\frac{x}{y} = C$

9. $\left(\frac{y}{x - y}\right)^2 dx + \left(\frac{x}{x - y}\right)^2 dy = 0$

$\frac{\partial M}{\partial y} = \frac{2xy}{(x - y)^3}$

$\frac{\partial N}{\partial x} = \frac{-2xy}{(x - y)^3}$

Not exact

11. (a, c)

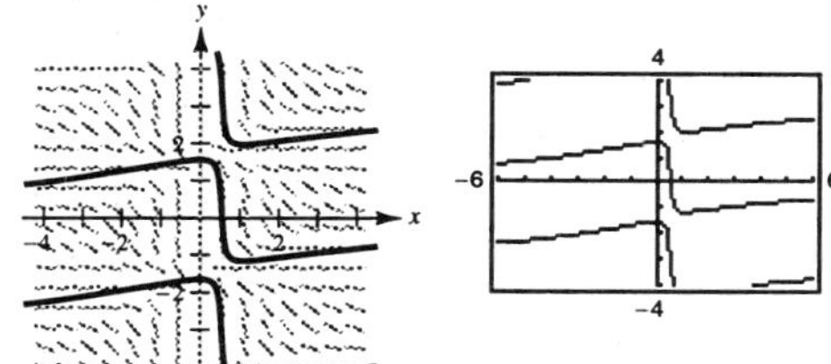

(b) $(2x \tan y + 5)\,dx + (x^2 \sec^2 y)\,dy = 0,\ y\left(\frac{1}{2}\right) = \frac{\pi}{4}$

$\frac{\partial M}{\partial y} = 2x \sec^2 y = \frac{\partial N}{\partial x} \Rightarrow$ Exact.

$f(x, y) = \int M(x, y)\,dx = \int (2x \tan y + 5)\,dx$

$= x^2 \tan y + 5x + g(y)$

$f_y(x, y) = x^2 \sec^2 y + g'(y) \Rightarrow g(y) = C$

$f(x, y) = x^2 \tan y + 5x = C$

$f\left(\frac{1}{2}, \frac{\pi}{4}\right) = \frac{1}{4} + \frac{5}{2} = \frac{11}{4} = C$

Answer: $x^2 \tan y + 5x = \frac{11}{4}$

13. $\frac{y}{x-1}\,dx + [\ln(x-1) + 2y]\,dy = 0$

$$\frac{\partial M}{\partial y} = \frac{1}{x-1} = \frac{\partial N}{\partial x}$$

$U(x, y) = y\ln(x-1) + f(y)$

$f'(y) = 2y$

$f(y) = y^2 + C_1$

$y\ln(x-1) + y^2 = C$

Initial condition: $y(2) = 4$, $4\ln(1) + 16 = C$, $C = 16$

Particular solution: $y\ln(x-1) + y^2 = 16$

15. $(e^{3x}\sin 3y)\,dx + (e^{3x}\cos 3y)\,dy = 0$

$$\frac{\partial M}{\partial y} = 3e^{3x}\cos 3y = \frac{\partial N}{\partial x}$$

$U(x, y) = \frac{1}{3}e^{3x}\sin 3y + f(y)$

$f'(y) = 0$

$f(y) = C_1$

$e^{3x}\sin 3y = C$

Initial condition: $y(0) = \pi$, $C = 0$

Particular solution: $e^{3x}\sin 3y = 0$

17. $y\,dx - (x + 6y^2)\,dy = 0$

$$\frac{(\partial N/\partial x) - (\partial M/\partial y)}{M} = -\frac{2}{y} = k(y)$$

Integrating factor: $e^{\int k(y)\,dy} = e^{\ln y^{-2}} = \frac{1}{y^2}$

Exact equation: $\frac{1}{y}\,dx - \left(\frac{x}{y^2} + 6\right)dy = 0$

$U(x, y) = \frac{x}{y} + f(y)$

$f'(y) = -6$

$f(y) = -6y + C_1$

$\frac{x}{y} - 6y = C$

19. $(5x^2 - y)\,dx + x\,dy = 0$

$$\frac{(\partial M/\partial y) - (\partial N/\partial x)}{N} = \frac{-2}{x} = h(x)$$

Integrating factor: $e^{\int h(x)\,dx} = e^{\ln x^{-2}} = \frac{1}{x^2}$

Exact equation: $\left(5 - \frac{y}{x^2}\right)dx + \frac{1}{x}\,dy = 0$

$U(x, y) = 5x + \frac{y}{x} + f(y)$

$f'(y) = 0$

$f(y) = C_1$

$5x + \frac{y}{x} = C$

21. $(x + y)\,dx + (\tan x)\,dy = 0$

$$\frac{(\partial M/\partial y) - (\partial N/\partial x)}{N} = -\tan x = h(x)$$

Integrating factor: $e^{\int h(x)\,dx} = e^{\ln\cos x} = \cos x$

Exact equation: $(x + y)\cos x\,dx + \sin x\,dy = 0$

$U(x, y) = x\sin x + \cos x + y\sin x + f(y)$

$f'(y) = 0$

$f(y) = C_1$

$x\sin x + \cos x + y\sin x = C$

23. $y^2\,dx + (xy - 1)\,dy = 0$

$$\frac{(\partial N/\partial x) - (\partial M/\partial y)}{M} = -\frac{1}{y} = k(y)$$

Integrating factor: $e^{\int k(y)\,dy} = e^{\ln(1/y)} = \frac{1}{y}$

Exact equation: $y\,dx + \left(x - \frac{1}{y}\right)dy = 0$

$U(x, y) = xy + f(y)$

$f'(y) = -\frac{1}{y}$

$f(y) = -\ln|y| + C_1$

$xy - \ln|y| = C$

25. $2y\,dx + (x - \sin\sqrt{y})\,dy = 0$

$$\frac{(\partial N/\partial x) - (\partial M/\partial y)}{M} = \frac{-1}{2y} = k(y)$$

Integrating factor: $e^{\int k(y)\,dy} = e^{\ln(1/\sqrt{y})} = \dfrac{1}{\sqrt{y}}$

Exact equation: $2\sqrt{y}\,dx + \left(\dfrac{x}{\sqrt{y}} - \dfrac{\sin\sqrt{y}}{\sqrt{y}}\right)dy = 0$

$U(x, y) = 2\sqrt{y}\,x + f(y)$

$f'(y) = -\dfrac{\sin\sqrt{y}}{\sqrt{y}}$

$f(y) = 2\cos\sqrt{y} + C_1$

$\sqrt{y}\,x + \cos\sqrt{y} = C$

27. $(4x^2y + 2y^2)\,dx + (3x^3 + 4xy)\,dy = 0$

Integrating factor: xy^2

Exact equation:

$$(4x^3y^3 + 2xy^4)\,dx + (3x^4y^2 + 4x^2y^3)\,dy = 0$$

$U(x, y) = x^4y^3 + x^2y^4 + f(y)$

$f'(y) = 0$

$f(y) = C_1$

$x^4y^3 + x^2y^4 = C$

29. $(-y^5 + x^2y)\,dx + (2xy^4 - 2x^3)\,dy = 0$

Integrating factor: $x^{-2}y^{-3}$

Exact equation: $\left(-\dfrac{y^2}{x^2} + \dfrac{1}{y^2}\right)dx + \left(2\dfrac{y}{x} - 2\dfrac{x}{y^3}\right)dy = 0$

$U(x, y) = \dfrac{y^2}{x} + \dfrac{x}{y^2} + f(y)$

$f'(y) = 0$

$f(y) = C_1$

$\dfrac{y^2}{x} + \dfrac{x}{y^2} = C$

31. $y\,dx - x\,dy = 0$

(a) $\dfrac{1}{x^2}, \dfrac{y}{x^2}\,dx - \dfrac{1}{x}\,dy = 0,\ \dfrac{\partial M}{\partial y} = \dfrac{1}{x^2} = \dfrac{\partial N}{\partial x}$

(b) $\dfrac{1}{y^2}, \dfrac{1}{y}\,dx - \dfrac{x}{y^2}\,dy = 0,\ \dfrac{\partial M}{\partial y} = \dfrac{-1}{y^2} = \dfrac{\partial N}{\partial x}$

(c) $\dfrac{1}{xy}, \dfrac{1}{x}\,dx - \dfrac{1}{y}\,dy = 0,\ \dfrac{\partial M}{\partial y} = 0 = \dfrac{\partial N}{\partial x}$

(d) $\dfrac{1}{x^2 + y^2}, \dfrac{y}{x^2 + y^2}\,dx - \dfrac{x}{x^2 + y^2}\,dy = 0,$

$$\frac{\partial M}{\partial y} = \frac{x^2 - y^2}{(x^2 + y^2)^2} = \frac{\partial N}{\partial x}$$

33. $\mathbf{F}(x, y) = \dfrac{y}{\sqrt{x^2 + y^2}}\mathbf{i} - \dfrac{x}{\sqrt{x^2 + y^2}}\mathbf{j}$

$\dfrac{dy}{dx} = -\dfrac{x}{y}$

$y\,dy + x\,dx = 0$

$y^2 + x^2 = C$

Family of circles

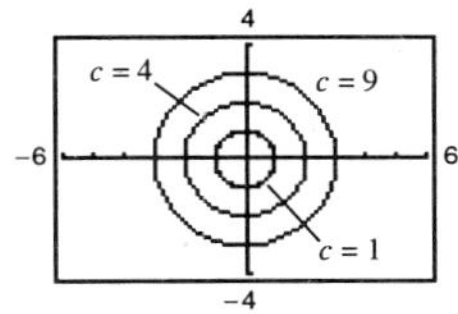

35. $\mathbf{F}(x, y) = 4x^2y\mathbf{i} - \left(2xy^2 + \dfrac{x}{y^2}\right)\mathbf{j}$

$\dfrac{dy}{dx} = \dfrac{-y}{2x} - \dfrac{1}{4xy^3}$

$\dfrac{8y^3}{2y^4 + 1}\,dy = -\dfrac{2}{x}\,dx$

$\ln(2y^4 + 1) = \ln\left(\dfrac{1}{x^2}\right) + \ln C$

$2y^4 + 1 = \dfrac{C}{x^2}$

$2x^2y^4 + x^2 = C$

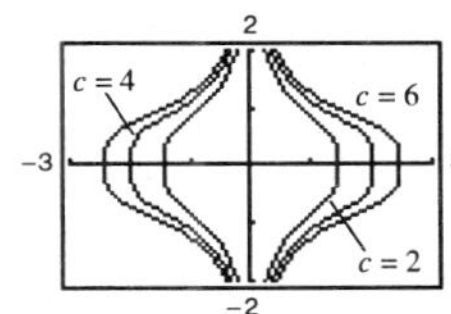

37. $\dfrac{dy}{dx} = \dfrac{y - x}{3y - x}$

$(x - y)\,dx + (3y - x)\,dy = 0$

$\dfrac{\partial M}{\partial y} = -1 = \dfrac{\partial N}{\partial x}$

$U(x, y) = \dfrac{x^2}{2} - xy + f(y)$

$f'(y) = 3y$

$f(y) = \dfrac{3y^2}{2} + C_1$

$x^2 - 2xy + 3y^2 = C$

Initial condition: $y(2) = 1,\ 4 - 4 + 3 = C,\ C = 3$

Particular solution: $x^2 - 2xy + 3y^2 = 3$

39. $E(x) = \dfrac{20x - y}{2y - 10x} = \dfrac{x}{y}\dfrac{dy}{dx}$

$(20xy - y^2)\,dx + (10x^2 - 2xy)\,dy = 0$

$\dfrac{\partial M}{\partial y} = 20x - 2y = \dfrac{\partial N}{\partial x}$

$U(x, y) = 10x^2y - xy^2 + f(y)$

$f'(y) = 0$

$f(y) = C_1$

$10x^2y - xy^2 = K$

Initial condition: $C(100) = 500,\ 100 \le x,\ K = 25{,}000{,}000$

$10x^2y - xy^2 = 25{,}000{,}000$

$xy^2 - 10x^2y + 25{,}000{,}000 = 0$ Quadratic Formula

$$y = \frac{10x^2 + \sqrt{100x^4 - 4x(25{,}000{,}000)}}{2x}$$

$$= \frac{5\left(x^2 + \sqrt{x^4 - 1{,}000{,}000x}\right)}{x}$$

41. $y' = x + \sqrt{y},\ y(0) = 2$ and $\Delta x = 0.50$

$y_1 = 2 + \left(0 + \sqrt{2}\right)(0.50) \approx 2.7071$

$y_2 = 2.7071 + \left(0.5 + \sqrt{2.7071}\right)(0.50) \approx 3.7798$

$y(1) \approx 3.7798$

$y' = x + \sqrt{y},\ y(0) = 2$ and $\Delta x = 0.25$

$y_1 = 2 + \left(0 + \sqrt{2}\right)(0.25) \approx 2.3536$

$y_2 = 2.3536 + \left(0.25 + \sqrt{2.3536}\right)(0.25) \approx 2.7996$

$y_3 = 2.7996 + \left(0.50 + \sqrt{2.7996}\right)(0.25) \approx 3.3429$

$y_4 = 3.3429 + \left(0.75 + \sqrt{3.3429}\right)(0.25) \approx 3.9875$

$y(1) \approx 3.9875$

y
5
4
3
2
1
-4 -3 -2 -1 1 2
x

$y' = x + \sqrt{y},\ y(0) = 2$ and $\Delta x = 0.10$

$y_1 = 2 + \left(0 + \sqrt{2}\right)(0.10) \approx 2.1414$

$y_2 = 2.1414 + \left(0.10 + \sqrt{2.1414}\right)(0.10) \approx 2.2977$

$y_3 = 2.2977 + \left(0.20 + \sqrt{2.2977}\right)(0.10) \approx 2.4693$

$y_4 = 2.4693 + \left(0.30 + \sqrt{2.4693}\right)(0.10) \approx 2.6565$

$y_5 = 2.6564 + \left(0.40 + \sqrt{2.6564}\right)(0.10) \approx 2.8595$

$y_6 = 2.8594 + \left(0.50 + \sqrt{2.8594}\right)(0.10) \approx 3.0786$

$y_7 = 3.0785 + \left(0.60 + \sqrt{3.0785}\right)(0.10) \approx 3.3140$

$y_8 = 3.3140 + \left(0.70 + \sqrt{3.3140}\right)(0.10) \approx 3.5661$

$y_9 = 3.5660 + \left(0.80 + \sqrt{3.5660}\right)(0.10) \approx 3.8349$

$y_{10} = 3.8348 + \left(0.90 + \sqrt{3.8348}\right)(0.10) \approx 4.1207$

$y(1) \approx 4.1207$

Δx	0.50	0.25	0.10
Estimate	3.7798	3.9875	4.1207

43. (a) (a) $y(2) \approx 2.8177$

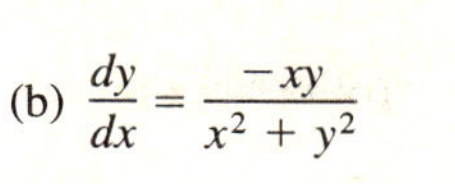

(b)
$$\frac{dy}{dx} = x\sqrt[3]{y}$$
$$\int y^{-1/3}\,dy = \int x\,dx$$
$$\frac{3}{2}y^{2/3} = \frac{x^2}{2} + C_1$$
$$y^{2/3} = \frac{1}{3}x^2 + C$$
$$y = \left[\frac{1}{3}x^2 + C\right]^{3/2}$$

Initial condition: $y(1) = 1, 1 = \left[\frac{1}{3} + C\right]^{3/2} \Rightarrow C = \frac{2}{3}$

Particular solution: $y = \left[\frac{1}{3}x^2 + \frac{2}{3}\right]^{3/2}$ or

$$3y^{2/3} - x^2 = 2$$

Thus, $y(2) = 2^{3/2} \approx 2.8284$.

(c)

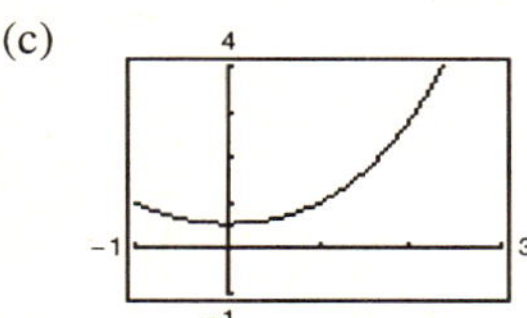

45. (a) $y(4) \approx 0.5231$

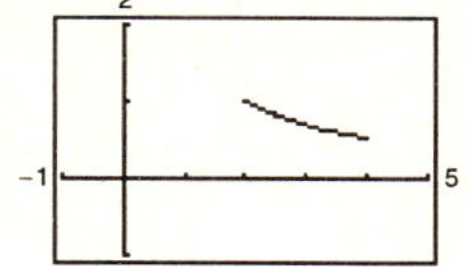

(b) $\dfrac{dy}{dx} = \dfrac{-xy}{x^2 + y^2}$

$xy\,dx + (x^2 + y^2)\,dy = 0$

$\dfrac{1}{M}[N_x - M_y] = \dfrac{1}{xy}[2x - x] = \dfrac{1}{y}$ function of y alone.

Integrating factor $e^{\int 1/y\,dy} = e^{\ln y} = y$

$xy^2\,dx + (x^2y + y^3)\,dy = 0$

$f(x, y) = \displaystyle\int xy^2\,dx = \frac{x^2y^2}{2} + g(y)$

$f_y(x, y) = x^2y + g'(y) \Rightarrow g(y) = \dfrac{y^4}{4}$

$f(x, y) = \dfrac{x^2y^2}{2} + \dfrac{y^4}{4} = C$

Initial condition: $y(2) = 1, \dfrac{4}{2} + \dfrac{1}{4} = \dfrac{9}{4} = C$

Particular solution: $\dfrac{x^2y^2}{2} + \dfrac{y^4}{4} = \dfrac{9}{4}$ or $2x^2y^2 + y^4 = 9.$

For $x = 4$, $32y^2 + y^4 = 9 \Rightarrow y(4) = 0.528$

(c)

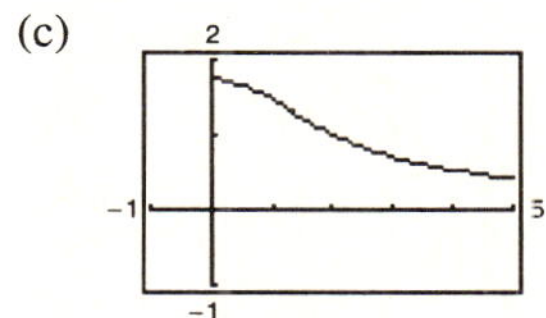

47. (a)

(b) $\dfrac{dy}{dx} = \dfrac{-xy}{x^2 + y^2}$

$xy\,dx + (x^2 + y^2)\,dy = 0$

$\dfrac{1}{M}[N_x - M_y] = \dfrac{1}{xy}[2x - x] = \dfrac{1}{y}$ function of y alone.

Integrating factor $e^{\int 1/y\,dy} = e^{\ln y} = y$

$xy^2\,dx + (x^2y + y^3)\,dy = 0$

$f(x, y) = \displaystyle\int xy^2\,dx = \frac{x^2y^2}{2} + g(y)$

$f_y(x, y) = x^2y + g^1(y) \Rightarrow g(y) = \dfrac{y^4}{4}$

$f(x, y) = \dfrac{x^2y^2}{2} + \dfrac{y^4}{4} = C$

Initial condition: $y(2) = 1, \dfrac{4}{2} + \dfrac{1}{4} = \dfrac{9}{4} = C$

Particular solution: $\dfrac{x^2y^2}{2} + \dfrac{y^4}{4} = \dfrac{9}{4}$ or

$$2x^2y^2 + y^4 = 9.$$

For $x = 4$, $32y^2 + y^4 = 9 \Rightarrow y(4) = 0.528$

(c)

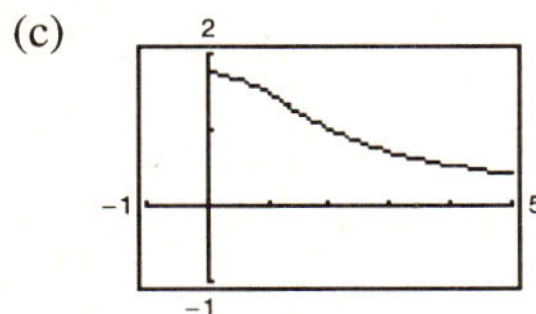

The solution is less accurate. For #45, Euler's Method gives $y(4) \approx 0.523$, whereas in #47 you obtain $y(4) \approx 0.408$. The errors are $0.528 - 0.523 = 0.005$ and $0.528 - 0.408 = 0.120$.

49. False

$\dfrac{\partial M}{\partial y} = 2x$ and $\dfrac{\partial N}{\partial x} = -2x$

51. True

$\dfrac{\partial}{\partial y}[f(x) + M] = \dfrac{\partial M}{\partial y}$ and $\dfrac{\partial}{\partial x}[g(y) + N] = \dfrac{\partial N}{\partial x}$

Section 15.2 First-Order Linear Differential Equations

1. False

$y' + xy = x^2$ is first-order linear.

3. (a),(c)

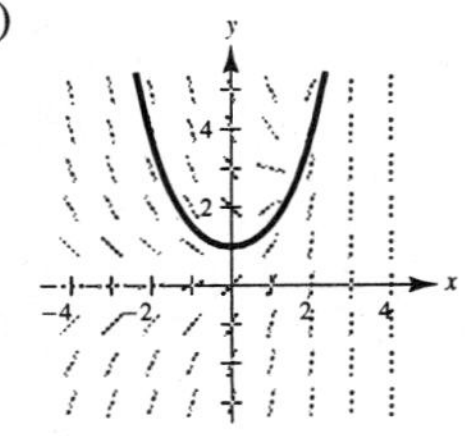

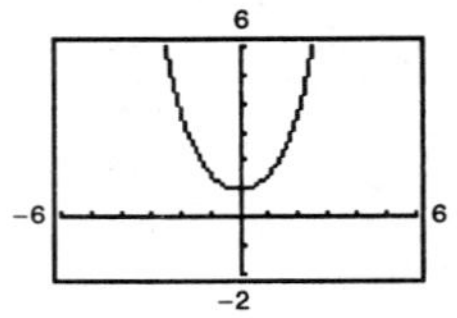

(b) $\dfrac{dy}{dx} = e^x - y$

$\dfrac{dy}{dx} + y = e^x$ Integrating factor: $e^{\int dx} = e^x$

$e^x y' + e^x y = e^{2x}$

$$(ye^x) = \int e^{2x}\, dx$$

$$ye^x = \frac{1}{2}e^{2x} + C$$

$$y(0) = 1 \Rightarrow 1 = \frac{1}{2} + C \Rightarrow C = \frac{1}{2}$$

$$ye^x = \frac{1}{2}e^{2x} + \frac{1}{2}$$

$$y = \frac{1}{2}e^x + \frac{1}{2}e^{-x} = \frac{1}{2}(e^x + e^{-x})$$

5. $\dfrac{dy}{dx} + \left(\dfrac{1}{x}\right)y = 3x + 4$

Integrating factor: $e^{\int (1/x)\, dx} = e^{\ln x} = x$

$$xy = \int x(3x + 4)\, dx = x^3 + 2x^2 + C$$

$$y = x^2 + 2x + \frac{C}{x}$$

7. $y' - y = \cos x$

Integrating factor: $e^{\int -1\, dx} = e^{-x}$

$$ye^{-x} = \int e^{-x} \cos x\, dx$$

$$= \frac{1}{2}e^{-x}(-\cos x + \sin x) + C$$

$$y = \frac{1}{2}(\sin x - \cos x) + Ce^x$$

9. $(3y + \sin 2x)\, dx - dy = 0$

$y' - 3y = \sin 2x$

Integrating factor: $e^{\int -3\, dx} = e^{-3x}$

$$ye^{-3x} = \int e^{-3x} \sin 2x\, dx$$

$$= \frac{1}{13}e^{-3x}(-3 \sin 2x - 2 \cos 2x) + C$$

$$y = -\frac{1}{13}(3 \sin 2x + 2 \cos 2x) + Ce^{3x}$$

11. $(x - 1)y' + y = x^2 - 1$

$$y' + \left(\frac{1}{x - 1}\right)y = x + 1$$

Integrating factor: $e^{\int [1/(x-1)]\, dx} = e^{\ln|x-1|} = x - 1$

$$y(x - 1) = \int (x^2 - 1)\, dx = \frac{1}{3}x^3 - x + C_1$$

$$y = \frac{x^3 - 3x + C}{3(x - 1)}$$

13. $y'\cos^2 x + y - 1 = 0$

$y' + (\sec^2 x)y = \sec^2 x$

Integrating factor: $e^{\int \sec^2 x\, dx} = e^{\tan x}$

$$ye^{\tan x} = \int \sec^2 x e^{\tan x}\, dx = e^{\tan x} + C$$

$$y = 1 + Ce^{-\tan x}$$

Initial condition: $y(0) = 5, C = 4$

Particular solution: $y = 1 + 4e^{-\tan x}$

15. $y' + y\tan x = \sec x + \cos x$

Integrating factor: $e^{\int \tan x\, dx} = e^{\ln|\sec x|} = \sec x$

$$y\sec x = \int \sec x(\sec x + \cos x)\, dx = \tan x + x + C$$

$$y = \sin x + x\cos x + C\cos x$$

Initial condition: $y(0) = 1, 1 = C$

Particular solution: $y = \sin x + (x + 1)\cos x$

17. $y' + \left(\frac{1}{x}\right)y = 0$

Integrating factor: $e^{\int (1/x)\, dx} = e^{\ln|x|} = x$

Separation of variables:

$$\frac{dy}{dx} = -\frac{y}{x}$$

$$\int \frac{1}{y}\, dy = \int -\frac{1}{x}\, dx$$

$$\ln y = -\ln x + \ln C$$

$$\ln xy = \ln C$$

$$xy = C$$

Initial condition: $y(2) = 2, C = 4$

Particular solution: $xy = 4$

19. $y' + 3x^2y = x^2y^3$

$n = 3, Q = x^2, P = 3x^2$

$$y^{-2}e^{\int (-2)3x^2\, dx} = \int (-2)x^2e^{\int (-2)3x^2\, dx}\, dx$$

$$y^{-2}e^{-2x^3} = -\int 2x^2e^{-2x^3}\, dx$$

$$y^{-2}e^{-2x^3} = \frac{1}{3}e^{-2x^3} + C$$

$$y^{-2} = \frac{1}{3} + Ce^{2x^3}$$

$$\frac{1}{y^2} = Ce^{2x^3} + \frac{1}{3}$$

21. $y' + \left(\frac{1}{x}\right)y = xy^2$

$n = 2, Q = x, P = x^{-1}$

$e^{\int -(1/x)\, dx} = e^{-\ln|x|} = x^{-1}$

$$y^{-1}x^{-1} = \int -x(x^{-1})\, dx = -x + C$$

$$\frac{1}{y} = -x^2 + Cx$$

$$y = \frac{1}{Cx - x^2}$$

23. $y' - y = x^3\sqrt[3]{y}$

$n = \frac{1}{3}, Q = x^3, P = -1$

$e^{-(2/3)\, dx} = e^{-(2/3)x}$

$$y^{2/3}e^{-(2/3)x} = \int \frac{2}{3}x^3e^{-(2/3)x}\, dx$$

$$y^{2/3} = -\frac{1}{4}(4x^3 + 18x^2 + 54x + 81) + Ce^{2x/3}$$

25. (a)

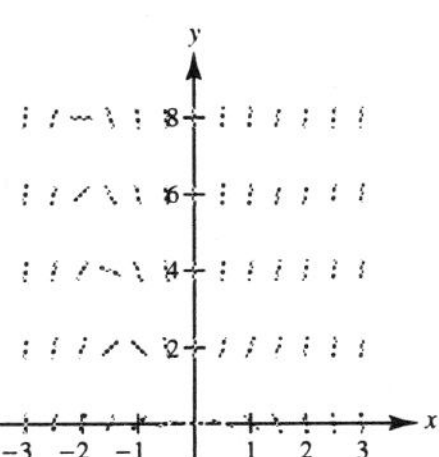

(c)

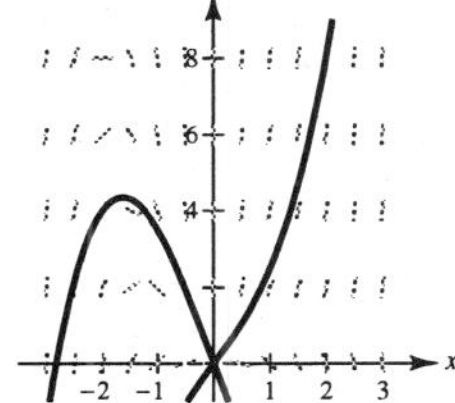

—CONTINUED—

25. —CONTINUED—

(b) $\dfrac{dy}{dx} - \dfrac{1}{x}y = x^2$

Integrating factor $e^{\int -1/x\,dx} = e^{-\ln x} = \dfrac{1}{x}$

$$\frac{1}{x}y' - \frac{1}{x^2}y = x$$

$$\left(\frac{1}{x}y\right) = \int x\,dx = \frac{x^2}{2} + C$$

$$y = \frac{x^3}{2} + Cx$$

$$(-2, 4): 4 = \frac{-8}{2} - 2C \Rightarrow C = -4 \Rightarrow y = \frac{x^3}{2} - 4x = \frac{1}{2}x(x^2 - 8)$$

$$(2, 8): 8 = \frac{8}{2} + 2C \Rightarrow C = 2 \Rightarrow y = \frac{x^3}{2} + 2x = \frac{1}{2}x(x^2 + 4)$$

27. (a)

(c)

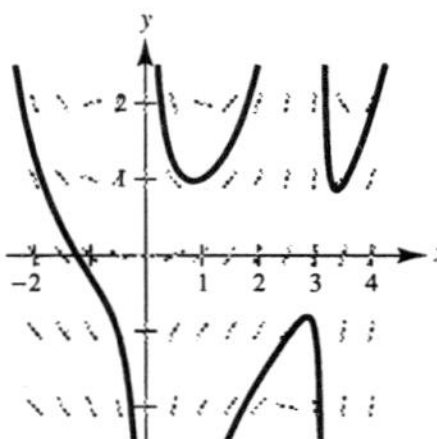

(b) $y^1 + (\cot x)y = x$

Integrating factor $e^{\int \cot x\,dx} = e^{\ln \sin x} = \sin x$

$$y' \sin x + (\cos x)y = x \sin x$$

$$y \sin x = \int x \sin x\,dx = \sin x - x \cos x + C$$

$$y = 1 - x \cot x + C - \csc x$$

$$(1, 1): 1 = 1 - \cot(1) + C\csc(\perp) \Rightarrow C = \frac{\cot(1)}{\csc(1)} = \cos(1)$$

$$y = 1 - x \cot x + \cos(1) \csc x$$

$$(3, -1): -1 = 1 - 3 \cot 3 + C \cdot \csc 3 \Rightarrow C \cdot \csc 3 = 3 \cot 3 - 2$$

$$C = \frac{3 \cot 3 - 2}{\csc 3} = 3 \cos 3 - 2 \sin 3$$

$$y = 1 - x \cot x + (3 \cos 3 - 2 \sin 3) \csc x$$

29. $L\dfrac{dI}{dt} + RI = E_0,\ I' + \dfrac{R}{L}I = \dfrac{E_0}{L}$

Integrating factor: $e^{\int (R/L)\,dt} = e^{Rt/L}$

$$I e^{Rt/L} = \int \frac{E_0}{L} e^{Rt/L}\,dt = \frac{E_0}{R} e^{Rt/L} + C$$

$$I = \frac{E_0}{R} + Ce^{-Rt/L}$$

31. $L\dfrac{dI}{dt} + RI = E_0 \sin \omega t$

$$\frac{dI}{dt} + \frac{R}{L}I = \frac{E_0}{L} \sin \omega t$$

Integrating factor: $e^{\int (R/L)\,dt} = e^{Rt/L}$

$$Ie^{Rt/L} = \int \frac{E_0}{L} e^{Rt/L} \sin \omega t\,dt$$

$$= \frac{E_0}{L}\left[\frac{L^2 e^{Rt/L}}{R^2 + L^2\omega^2}\left(\frac{R}{L} \sin \omega t - \omega \cos \omega t\right)\right] + C$$

$$= \frac{E_0 e^{Rt/L}}{R^2 + \omega^2 L^2}(R \sin \omega t - \omega L \cos \omega t) + C$$

$$I = \frac{E_0}{R^2 + \omega^2 L^2}(R \sin \omega t - \omega L \cos \omega t) + Ce^{-Rt/L}$$

33. $$\frac{dP}{dt} = kP + N, N \text{ constant}$$

$$\frac{dP}{kP + N} = dt$$

$$\int \frac{1}{kP + N}\,dP = \int dt$$

$$\frac{1}{k}\ln(kP + N) = t + C_1$$

$$\ln(kP + N) = kt + C_2$$

$$kP + N = e^{kt + C_2}$$

$$P = \frac{C_3e^{kt} - N}{k}$$

$$P = Ce^{kt} - \frac{N}{k}$$

When $t = 0$: $P = P_0$

$$P_0 = C - \frac{N}{k} \Rightarrow C = P_0 + \frac{N}{k}$$

$$P = \left(P_0 + \frac{N}{k}\right)e^{kt} - \frac{N}{k}$$

35. (a) $A = \dfrac{P}{r}(e^{rt} - 1)$

$$A = \frac{100{,}000}{0.06}(e^{0.06(5)} - 1) \approx 583{,}098.01$$

(b) $A = \dfrac{250{,}000}{0.05}(e^{0.05(10)} - 1) \approx 3{,}243{,}606.35$

37. (a) $\dfrac{dQ}{dt} = q - kQ$, q constant

(b) $Q' + kQ = q$

Let $P(t) = k$, $Q(t) = q$, then the integrating factor is $u(t) = e^{kt}$.

$$Q = e^{-kt}\int qe^{kt}\,dt = e^{-kt}\left(\frac{q}{k}e^{kt} + C\right) = \frac{q}{k} + Ce^{-kt}$$

When $t = 0$: $Q = Q_0$

$$Q_0 = \frac{q}{k} + C \Rightarrow C = Q_0 - \frac{q}{k}$$

$$Q = \frac{q}{k} + \left(Q_0 - \frac{q}{k}\right)e^{-kt}$$

(c) $\displaystyle\lim_{t\to\infty} Q = \frac{q}{k}$

39. Let Q be the number of pounds of concentrate in the solution at any time t. Since the number of gallons of solution in the tank at any time t is $v_0 + (r_1 - r_2)t$ and since the tank loses r_2 gallons of solution per minute, it must lose concentrate at the rate

$$\left[\frac{Q}{v_0 + (r_1 - r_2)t}\right]r_2.$$

The solution gains concentrate at the rate r_1q_1. Therefore, the net rate of change is

$$\frac{dQ}{dt} = q_1r_1 - \left[\frac{Q}{v_0 + (r_1 - r_2)t}\right]r_2 \quad \text{or} \quad \frac{dQ}{dt} + \frac{r_2Q}{v_0 + (r_1 - r_2)t} = q_1r_1.$$

41. (a) $Q' + \dfrac{r^2 Q}{v_0 + (r_1 - r_2)t} = q_1 r_1$

$Q(0) = q_0, q_0 = 25, q_1 = 0, v_0 = 200, r_1 = 10, r_2 = 10, Q' + \dfrac{1}{20}Q = 0$

$$\int \frac{1}{Q}\,dQ = \int -\frac{1}{20}\,dt$$

$$\ln Q = -\frac{1}{20}t + \ln C_1$$

$$Q = Ce^{-(1/20)t}$$

Initial condition: $Q(0) = 25, C = 25$

Particular solution: $Q = 25e^{-(1/20)t}$

(b) $15 = 25e^{-(1/20)t}$

$$\ln\left(\frac{3}{5}\right) = -\frac{1}{20}t$$

$$t = -20\ln\left(\frac{3}{5}\right) \approx 10.2 \text{ min}$$

(c) $\lim_{t\to\infty} 25e^{-(1/20)t} = 0$

43. (a) The volume of the solution in the tank is given by $v_0 + (r_1 - r_2)t$. Therefore, $100 + (5 - 3)t = 200$ or $t = 50$ minutes.

(b) $Q' + \dfrac{r_2 Q}{v_0 + (r_1 - r_2)t} = q_1 r_1$

$Q(0) = q_0, q_0 = 0, q_1 = 0.5, v_0 = 100, r_1 = 5, r_2 = 3, Q' + \dfrac{3}{100 + 2t}Q = 2.5$

Integrating factor: $e^{\int [3/(100+2t)]\,dt} = (50 + t)^{3/2}$

$$Q(50 + t)^{3/2} = \int 2.5(50 + t)^{3/2}\,dt = (50 + t)^{5/2} + C$$

$$Q = (50 + t) + C(50 + t)^{-3/2}$$

Initial condition: $Q(0) = 0, 0 = 50 + C(50^{-3/2}), C = -50^{5/2}$

Particular solution: $Q = (50 + t) - 50^{-5/2}(50 + t)^{-3/2}$

$$Q(50) = 100 - 50^{5/2}(100)^{-3/2} = 100 - \frac{25}{\sqrt{2}} \approx 82.32 \text{ lbs}$$

45. $y' - 2x = 0$

$$\int dy = \int 2x\,dx$$

$$y = x^2 + C$$

Matches c.

47. $y' - 2xy = 0$

$$\int \frac{dy}{y} = \int 2x\,dx$$

$$\ln y = x^2 + C_1$$

$$y = Ce^{x^2}$$

Matches a.

49. $e^{2x+y}\,dx - e^{x-y}\,dy = 0$

Separation of variables:

$$e^{2x}e^{y}\,dx = e^{x}e^{-y}\,dy$$

$$\int e^{x}\,dx = \int e^{-2y}\,dy$$

$$e^{x} = -\frac{1}{2}e^{-2y} + C_1$$

$$2e^{x} + e^{-2y} = C$$

51. $(1 + y^2)\,dx + (2xy + y + 2)\,dy = 0$

Exact: $\dfrac{\partial M}{\partial y} = 2y = \dfrac{\partial N}{\partial x}$

$$U(x, y) = \int (1 + y^2)\,dx = x + xy^2 + f(y)$$

$$U_y(x, y) = 2xy + f'(y)$$

$$= 2xy + y + 2 \text{ or } f'(y) = y + 2$$

$$f(y) = \frac{1}{2}y^2 + 2y + C_1$$

$$U(x, y) = x + xy^2 + \frac{1}{2}y^2 + 2y + C_1$$

$$x + xy^2 + \frac{1}{2}y^2 + 2y = C$$

53. $(y\cos x - \cos x)\,dx + dy = 0$

Separation of variables:

$$\int \cos x\,dx = \int \frac{-1}{y-1}\,dy$$

$$\sin x = -\ln(y-1) + \ln C$$

$$\ln(y-1) = -\sin x + \ln C$$

$$y = Ce^{-\sin x} + 1$$

55. $2xy\,dx + (x^2 + \cos y)\,dy = 0$

Exact: $\dfrac{\partial M}{\partial y} = 2x = \dfrac{\partial N}{\partial x}$

$$U(x, y) = \int 2xy\,dx = x^2y + f(y)$$

$$U_y(x, y) = x^2 + f'(y) = x^2 + \cos y$$

$$\text{or } f'(y) = \cos y$$

$$f(y) = \sin y + C_1$$

$$U(x, y) = x^2y + \sin y + C_1$$

$$x^2y + \sin y = C$$

57. $(3y^2 + 4xy)\,dx + (2xy + x^2)\,dy = 0$

Homogeneous: $y = vx$, $dy = v\,dx + x\,dv$

$$(3v^2x^2 + 4vx^2)\,dx + (2vx^2 + x^2)(v\,dx + x\,dv) = 0$$

$$\int \frac{5}{x}\,dx + \int \left(\frac{2v+1}{v^2+v}\right)dv = 0$$

$$\ln x^5 + \ln|v^2 + v| = \ln C$$

$$x^5(v^2 + v) = C$$

$$x^3y^2 + x^4y = C$$

59. $(2y - e^{x})\,dx + x\,dy = 0$

Linear: $y' + \left(\dfrac{2}{x}\right)y = \dfrac{1}{x}e^{x}$

Integrating factor: $e^{\int (2/x)\,dx} = e^{\ln x^2} = x^2$

$$yx^2 = \int x^2\frac{1}{x}e^{x}\,dx = e^{x}(x-1) + C$$

$$y = \frac{e^{x}}{x^2}(x-1) + \frac{C}{x^2}$$

61. $(x^2y^4 - 1)\,dx + x^3y^3\,dy = 0$

$$y' + \left(\frac{1}{x}\right)y = x^{-3}y^{-3}$$

Bernoulli: $n = -3, Q = x^{-3}, P = x^{-1}, e^{\int (4/x)\,dx} = e^{\ln x^4} = x^4$

$$y^4x^4 = \int 4(x^{-3})(x^4)\,dx = 2x^2 + C$$

$$x^4y^4 - 2x^2 = C$$

63. $3y\,dx - (x^2 + 3x + y^2)\,dy = 0$

Multiplying by the integrating factor, $1/(x^2 + y^2)$, and regrouping, we have

$$3\left[\frac{y\,dx - x\,dy}{x^2 + y^2}\right] - dy = 0$$

$$\int 3d\left[\arctan\frac{x}{y}\right] - \int dy = 0$$

$$3\arctan\frac{x}{y} - y = C.$$

Section 15.3 Second-Order Homogeneous Linear Equations

1. $$y = C_1e^{-3x} + C_2xe^{-3x}$$
$$y' = -3C_1e^{-3x} + C_2e^{-3x} - 3C_2xe^{-3x}$$
$$y'' = 9C_1e^{-3x} - 6C_2e^{-3x} + 9C_2xe^{-3x}$$
$$y'' + 6y' + 9y = (9C_1e^{-3x} - 6C_2e^{-3x} + 9C_2xe^{-3x}) + (-18C_1e^{-3x} + 6C_2e^{-3x} - 18C_2xe^{-3x}) + (9C_1e^{-3x} + 9C_2xe^{-3x}) = 0$$

3. $$y = C_1 \cos 2x + C_2 \sin 2x$$
$$y' = -2C_1 \sin 2x + 2C_2 \cos 2x$$
$$y'' = -4C_1 \cos 2x - 4C_2 \sin 2x = -4y$$
$$y'' + 4y = -4y + 4y = 0$$

5. $y'' - y' = 0$

Characteristic equation: $m^2 - m = 0$

Roots: $m = 0, 1$

$y = C_1 + C_2e^x$

7. $y'' - y' - 6y = 0$

Characteristic equation: $m^2 - m - 6 = 0$

Roots: $m = 3, -2$

$y = C_1e^{3x} + C_2e^{-2x}$

9. $2y'' + 3y' - 2y = 0$

Characteristic equation: $2m^2 + 3m - 2 = 0$

Roots: $m = \frac{1}{2}, -2$

$y = C_1e^{(1/2)x} + C_2e^{-2x}$

11. $y'' + 6y' + 9y = 0$

Characteristic equation: $m^2 + 6m + 9 = 0$

Roots: $m = -3, -3$

$y = C_1e^{-3x} + C_2xe^{-3x}$

13. $16y'' - 8y' + y = 0$

Characteristic equation: $16m^2 - 8m + 1 = 0$

Roots: $m = \frac{1}{4}, \frac{1}{4}$

$y = C_1e^{(1/4)x} + C_2xe^{(1/4)x}$

15. $y'' + y = 0$

Characteristic equation: $m^2 + 1 = 0$

Roots: $m = -i, i$

$y = C_1 \cos x + C_2 \sin x$

17. $y'' - 9y = 0$

Characteristic equation: $m^2 - 9 = 0$

Roots: $m = -3, 3$

$y = C_1e^{3x} + C_2e^{-3x}$

19. $y'' - 2y' + 4y = 0$

Characteristic equation: $m^2 - 2m + 4 = 0$

Roots: $m = 1 - \sqrt{3}i, 1 + \sqrt{3}i$

$y = e^x(C_1 \cos \sqrt{3}x + C_2 \sin \sqrt{3}x)$

21. $y'' - 3y' + y = 0$

Characteristic equation: $m^2 - 3m + 1 = 0$

Roots: $m = \dfrac{3 - \sqrt{5}}{2}, \dfrac{3 + \sqrt{5}}{2}$

$y = C_1e^{[(3+\sqrt{5})/2]x} + C_2e^{[(3-\sqrt{5})/2]x}$

23. $9y'' - 12y' + 11y = 0$

Characteristic equation: $9m^2 - 12m + 11 = 0$

Roots: $m = \dfrac{2 + \sqrt{7}i}{3}, \dfrac{2 - \sqrt{7}i}{3}$

$y = e^{(2/3)x}\left[C_1 \cos\left(\dfrac{\sqrt{7}}{3}x\right) + C_2 \sin\left(\dfrac{\sqrt{7}}{3}x\right)\right]$

25. $y^{(4)} - y = 0$

Characteristic equation: $m^4 - 1 = 0$

Roots: $m = -1, 1, -i, i$

$y = C_1e^x + C_2e^{-x} + C_3 \cos x + C_4 \sin x$

27. $y''' - 6y'' + 11y' - 6y = 0$

Characteristic equation: $m^3 - 6m^2 + 11m - 6 = 0$

Roots: $m = 1, 2, 3$

$y = C_1e^x + C_2e^{2x} + C_3e^{3x}$

29. $y''' - 3y'' + 7y' - 5y = 0$

Characteristic equation: $m^3 - 3m^2 + 7m - 5 = 0$

Roots: $m = 1, 1 - 2i, 1 + 2i$

$y = C_1e^x + e^x(C_2 \cos 2x + C_3 \sin 2x)$

31. $y'' + 100y = 0$

$y = C_1 \cos 10x + C_2 \sin 10x$

$y' = -10C_1 \sin 10x + 10C_2 \cos 10x$

(a) $y(0) = 2: 2 = C_1$

$y'(0) = 0: 0 = 10C_2 \Rightarrow C_2 = 0$

Particular solution: $y = 2 \cos 10x$

(b) $y(0) = 0: 0 = C_1$

$y'(0) = 2: 2 = 10C_2 \Rightarrow C_2 = \frac{1}{5}$

Particular solution: $y = \frac{1}{5} \sin 10x$

(c) $y(0) = -1: -1 = C_1$

$y'(0) = 3: 3 = 10C_2 \Rightarrow C_2 = \frac{3}{10}$

Particular solution: $y = -\cos 10x + \frac{3}{10} \sin 10x$

33. $y'' - y' - 30y = 0, y(0) = 1, y'(0) = -4$

Characteristic equation: $m^2 - m - 30 = 0$

Roots: $m = 6, -5$

$y = C_1e^{6x} + C_2e^{-5x}, y' = 6C_1e^{6x} - 5C_2e^{-5x}$

Initial conditions: $y(0) = 1, y'(0) = -4, 1 = C_1 + C_2, -4 = 6C_1 - 5C_2$

Solving simultaneously: $C_1 = \frac{1}{11}, C_2 = \frac{10}{11}$

Particular solution: $y = \frac{1}{11}(e^{6x} + 10e^{-5x})$

35. $y'' + 16y = 0, y(0) = 0, y'(0) = 2$

Characteristic equation: $m^2 + 16 = 0$

Roots: $m = \pm 4i$

$y = C_1 \cos 4x + C_2 \sin 4x$

$y' = -4C_1 \sin 4x + 4C_2 \cos 4x$

Initial conditions: $y(0) = 0 = C_1$

$y'(0) = 2 = 4C_2 \Rightarrow C_2 = \frac{1}{2}$

Particular solution: $y = \frac{1}{2} \sin 4x$

37. y'' and y' are not equal for $x < 0$. $y'' > 0$ for all x, since graph is concave upwards, but $y' < 0$ for $x < 0$ since y is decreasing there.

39. By Hooke's Law, $F = kx$

$$k = \frac{F}{x} = \frac{32}{2/3} = 48.$$

Also, $F = ma$, and

$$m = \frac{F}{a} = \frac{32}{32} = 1.$$

Therefore, $y = \frac{1}{2} \cos\left(4\sqrt{3}t\right)$

41. $y = C_1 \cos\left(\sqrt{\frac{k}{m}}\,t\right) + C_2 \sin\left(\sqrt{\frac{k}{m}}\,t\right),\ \sqrt{\frac{k}{m}} = \sqrt{48} = 4\sqrt{3}$

Initial conditions: $y(0) = \frac{2}{3},\ y'(0) = -\frac{1}{2}$

$$y = C_1 \cos(4\sqrt{3}\,t) + C_2 \sin(4\sqrt{3}\,t)$$

$$y(0) = C_1 = \frac{2}{3}$$

$$y'(t) = -4\sqrt{3}\,C_1 \sin(4\sqrt{3}\,t) + 4\sqrt{3}\,C_2 \cos(4\sqrt{3}\,t)$$

$$y'(0) = 4\sqrt{3}\,C_2 = -\frac{1}{2} \Rightarrow C_2 = -\frac{1}{8\sqrt{3}} = -\frac{\sqrt{3}}{24}$$

$$y(t) = \frac{2}{3}\cos(4\sqrt{3}\,t) - \frac{\sqrt{3}}{24}\sin(4\sqrt{3}\,t)$$

43. By Hooke's Law, $32 = k(2/3)$ so that $k = 48$. Moreover, since the weight w is given by mg, it follows that $m = w/g = 32/32 = 1$. Also, the damping force is given by $(-1/8)(dy/dt)$. Thus, the differential equation for the oscillations of the weight is

$$m\left(\frac{d^2y}{dt^2}\right) = -\frac{1}{8}\left(\frac{dy}{dt}\right) - 48y$$

$$m\left(\frac{d^2y}{dt^2}\right) + \frac{1}{8}\left(\frac{dy}{dt}\right) + 48y = 0.$$

In this case the characteristic equation is $8m^2 + m + 384 = 0$ with complex roots $m = (-1/16) \pm (\sqrt{12{,}287}/16)i$. Therefore, the general solution is

$$y(t) = e^{-t/16}\left(C_1 \cos\frac{\sqrt{12{,}287}\,t}{16} + C_2 \sin\frac{\sqrt{12{,}287}\,t}{16}\right).$$

Using the initial conditions, we have

$$y(0) = C_1 = \frac{1}{2}$$

$$y'(t) = e^{-t/16}\left[\left(-\frac{\sqrt{12{,}287}}{16}C_1 - \frac{C_2}{16}\right)\sin\frac{\sqrt{12{,}287}\,t}{16} + \left(\frac{\sqrt{12{,}287}}{16}C_2 - \frac{C_1}{16}\right)\cos\frac{\sqrt{12{,}287}\,t}{16}\right]$$

$$y'(0) = \frac{\sqrt{12{,}287}}{16}C_2 - \frac{C_1}{16} = 0 \Rightarrow C_2 = \frac{\sqrt{12{,}287}}{24{,}574}$$

and the particular solution is

$$y(t) = \frac{e^{-t/16}}{2}\left(\cos\frac{\sqrt{12{,}287}\,t}{16} + \frac{\sqrt{12{,}287}}{12{,}287}\sin\frac{\sqrt{12{,}287}\,t}{16}\right).$$

45. $y'' + 9y = 0$

Undamped vibration

Period: $\frac{2\pi}{3}$

Matches (b)

47. $y'' + 2y' + 10y = 0$

Damped vibration

Matches (c)

49. Since $m = -a/2$ is a double root of the characteristic equation, we have

$$\left(m + \frac{a}{2}\right)^2 = m^2 + am + \frac{a^2}{4} = 0$$

and the differential equation is $y'' + ay' + (a^2/4)y = 0$. The solution is

$$y = (C_1 + C_2x)e^{-(a/2)x}$$

$$y' = \left(-\frac{C_1a}{2} + C_2 - \frac{C_2a}{2}x\right)e^{-(a/2)x}$$

$$y'' = \left(\frac{C_1a^2}{4} - aC_2 + \frac{C_2a^2}{4}x\right)e^{-(a/2)x}$$

$$y'' + ay' + \frac{a^2}{4}y = \left(\frac{C_1a^2}{4} - C_2a + \frac{C_2a^2}{4}x\right) + \left(-\frac{C_1a^2}{2} + C_2a - \frac{C_2a^2}{2}x\right)e^{-(a/2)x} + \left(\frac{C_1a^2}{4} + \frac{C_2a^2}{4}x\right)e^{-(a/2)x} = 0.$$

51. False. The general solution is $y = C_1e^{3x} + C_2xe^{3x}$.

53. True

55. $y_1 = e^{ax}, \quad y_2 = e^{bx}, \quad a \neq b$

$$W(y_1, y_2) = \begin{vmatrix} e^{ax} & e^{bx} \\ ae^{ax} & be^{bx} \end{vmatrix}$$

$$= (b - a)e^{ax+bx} \neq 0 \text{ for any value of } x.$$

57. $y_1 = e^{ax}\sin bx, \quad y_2 = e^{ax}\cos bx, \quad b \neq 0$

$$W(y_1, y_2) = \begin{vmatrix} e^{ax}\sin bx & e^{ax}\cos bx \\ ae^{ax}\sin bx + be^{ax}\cos bx & ae^{ax}\cos bx - be^{ax}\sin bx \end{vmatrix}$$

$$= -be^{2ax}\sin^2 bx - be^{2ax}\cos^2 bx$$

$$= -be^{2ax} \neq 0 \text{ for any value of } x.$$

59. $x^2y'' + axy' + by = 0, \quad x > 0$

Let $x = e^t$.

(a) $\dfrac{dy}{dx} = \dfrac{dy/dt}{dx/dt} = e^{-t}\dfrac{dy}{dt}$

$$\frac{d^2y}{dx^2} = \frac{d/dt[e^{-t}(dy/dt)]}{e^t} = e^{-t}\left[e^{-t}\frac{d^2y}{dt^2} - e^{-t}\frac{dy}{dt}\right] = e^{-2t}\left[\frac{d^2y}{dt^2} - \frac{dy}{dt}\right]$$

$$x^2y'' + axy' + by = 0$$

$$e^{2t}\left[e^{-2t}\left(\frac{d^2y}{dt^2} - \frac{dy}{dt}\right)\right] + ae^t\left(e^{-t}\frac{dy}{dt}\right) + by = 0$$

$$\frac{d^2y}{dt^2} + (a - 1)\frac{dy}{dt} + by = 0$$

(b) $x^2y'' + 6xy' + 6y = 0$

Let $x = e^t$. From part (a), we have:

$$\frac{d^2y}{dt^2} + 5\frac{dy}{dt} + 6y = 0$$

$$m^2 + 5m + 6 = 0$$

$$(m + 3)(m + 2) = 0$$

$$m_1 = -3, \quad m_2 = -2$$

$$y = C_1e^{-3t} + C_2e^{-2t} = C_1e^{-3\ln x} + C_2e^{-2\ln x} = C_1e^{\ln(1/x^3)} + C_2e^{\ln(1/x^2)} = \frac{C_1}{x^3} + \frac{C_2}{x^2}$$

Section 15.4 Second-Order Nonhomogeneous Linear Equations

1.
$$y = 2e^{2x} - 2\cos x$$
$$y' = 4e^{2x} + 2\sin x$$
$$y'' = 8e^{2x} + 2\cos x$$
$$y'' + y = (8e^{2x} + 2\cos x) + (2e^{2x} - \cos x) = 10e^{2x}$$

3.
$$y = 3\sin x - \cos x \ln|\sec x + \tan x|$$
$$y' = 3\cos x - 1 + \sin x \ln|\sec x + \tan x|$$
$$y'' = -3\sin x + \tan x + \cos x \ln|\sec x + \tan x|$$
$$y'' + y = (-3\sin x + \tan x + \cos x \ln|\sec x \tan x|) + (3\sin x - \cos x \ln|\sec x + \tan x|) = \tan x$$

5. $y'' - 3y' + 2y = 2x$

$y'' - 3y' + 2y = 0$

$m^2 - 3m + 2 = 0$ when $m = 1, 2$.
$$y_h = C_1e^x + C_2e^{2x}$$
$$y_p = A_0 + A_1x$$
$$y_p' = A_1$$
$$y_p'' = 0$$
$$y_p'' - 3y_p' + 2y_p = (2A_0 - 3A_1) + 2A_1x = 2x$$
$$\left.\begin{array}{l} 2A_0 - 3A_1 = 0 \\ 2A_1 = 2 \end{array}\right\} A_1 = 1, \quad A_0 = \tfrac{3}{2}$$
$$y = C_1e^x + C_2e^{2x} + x + \tfrac{3}{2}$$

7. $y'' + y = x^3, \; y(0) = 1, \; y'(0) = 0$

$y'' + y = 0$

$m^2 + 1 = 0$ when $m = i, \; -i$.
$$y_h = C_1\cos x + C_2\sin x$$
$$y_p = A_0 + A_1x + A_2x^2 + A_3x^3$$
$$y_p' = A_1 + 2A_2x + 3A_3x^2$$
$$y_p'' = 2A_2 + 6A_3x$$
$$y_p'' + y_p = A_3x^3 + A_2x^2 + (A_1 + 6A_3)x + (A_0 + 2A_2) = x^3 \text{ or } A_3 = 1, \; A_2 = 0, \; A_1 = -6, \; A_0 = 0$$
$$y = C_1\cos x + C_2\sin x + x^3 - 6x$$
$$y' = -C_1\sin x + C_2\cos x + 3x^2 - 6$$

Initial conditions: $y(0) = 1, \; y'(0) = 0, \; 1 = C_1, \; 0 = C_2 - 6, \; C_2 = 6$

Particular solution: $y = \cos x + 6\sin x + x^3 - 6x$

9. $y'' + 2y' = 2e^x$

$y'' + 2y' = 0$

$m^2 + 2m = 0$ when $m = 0, -2.$

$$y_h = C_1 + C_2e^{-2x}$$

$$y_p = Ae^x = y_p' = y_p''$$

$$y_p'' + 2y_p' = 3Ae^x = 2e^x \text{ or } A = \tfrac{2}{3}$$

$$y = C_1 + C_2e^{-2x} + \tfrac{2}{3}e^x$$

11. $y'' - 10y' + 25y = 5 + 6e^x$

$y'' - 10y' + 25y = 0$

$m^2 - 10m + 25 = 0$ when $m = 5, 5.$

$$y_h = C_1e^{5x} + C_2xe^{5x}$$

$$y_p = A_0 + A_1e^x$$

$$y_p' = y_p'' = A_1e^x$$

$$y_p'' - 10y_p' + 25y_p = 25A_0 + 16A_1e^x = 5 + 6e^x$$

$$\text{or } A_0 = \tfrac{1}{5},\ A_1 = \tfrac{3}{8}$$

$$y = (C_1 + C_2x)e^{5x} + \tfrac{3}{8}e^x + \tfrac{1}{5}$$

13. $y'' + y' = 2\sin x,\ y(0) = 0,\ y'(0) = -3$

$y'' + y' = 0$

$m^2 + m = 0$ when $m = 0, -1.$

$$y_h = C_1 + C_2e^{-x}$$

$$y_p = A\cos x + B\sin x$$

$$y_p' = -A\sin x + B\cos x$$

$$y_p'' = -A\cos x - B\sin x$$

$$y_p'' + y_p' = (-A + B)\cos x + (-A - B)\sin x$$

$$= 2\sin x$$

$$\left.\begin{aligned} -A + B &= 0 \\ -A - B &= 2 \end{aligned}\right\} A = -1,\ B = -1$$

$y = C_1 + C_2e^x - (\cos x + \sin x)$

$y' = -C_2e^{-x} - (-\sin x + \cos x)$

Initial conditions: $y(0) = 0,\ y'(0) = -3,$

$0 = C_1 + C_2 - 1,\ -3 = -C_2 - 1,$

$C_2 = 2,\ C_1 = -1$

Particular solution: $y = -1 + 2e^{-x} - (\cos x + \sin x)$

15. $y'' + 9y = \sin 3x$

$y'' + 9y = 0$

$m^2 + 9 = 0$ when $m = -3i, 3i.$

$$y_h = C_1\cos 3x + C_2\sin 3x$$

$$y_p = A_0\sin 3x + A_1x\sin 3x + A_2\cos 3x + A_3x\cos 3x$$

$$y_p'' = (-9A_0 - 6A_3)\sin 3x - 9A_1x\sin 3x + (6A_1 - 9A_2)\cos 3x - 9A_3x\cos 3x$$

$$y_p'' + 9y_p = -6A_3\sin 3x + 6A_1\cos 3x = \sin 3x,\ A_1 = 0,\ A_3 = -\tfrac{1}{6}$$

$$y = \left(C_1 - \tfrac{1}{6}x\right)\cos 3x + C_2\sin 3x$$

17. $y''' - 3y' + 2y = 2e^{-2x}$

$y''' - 3y' + 2y = 0$

$m^3 - 3m + 2 = 0$ when $m = 1, 1, -2.$

$$y_h = C_1e^x + C_2xe^x + C_3e^{-2x}$$

$$y_p = A_0e^{-2x} + A_1xe^{-2x}$$

$$y_p' = (-2A_0 + A_1)e^{-2x} - 2A_1xe^{-2x}$$

$$y_p'' = (4A_0 - 4A_1)e^{-2x} + 4A_1xe^{-2x}$$

$$y_p''' = (-8A_0 + 12A_1)e^{-2x} - 8A_1xe^{-2x}$$

$$y_p''' - 3y_p' + 2y_p = 9A_1e^{-2x} = 2e^{-2x} \text{ or } A_1 = \tfrac{2}{9}$$

$$y = C_1e^x + C_2xe^x + \left(C_3 + \tfrac{2}{9}x\right)e^{-2x}$$

19. $y' - 4y = xe^x - xe^{4x}, y(0) = \frac{1}{3}$

$y' - 4y = 0$

$m - 4 = 0$ when $m = 4.$

$$y_h = Ce^{4x}$$

$$y_p = (A_0 + A_1x)e^x + (A_2x + A_3x^2)e^{4x}$$

$$y_p' = (A_0 = A_1x)e^x + A_1e^x + 4(A_2x + A_3x^2)e^{4x} + (A_2 + 2A_3x)e^{4x}$$

$$y_p' - 4y_p = (-3A_0 - 3A_1x)e^x + A_1e^x + A_2e^{4x} + 2A_3xe^{4x} = xe^x - xe^{4x}$$

$$A_0 = -\tfrac{1}{9}, A_1 = -\tfrac{1}{3}, A_2 = 0, A_3 = -\tfrac{1}{2}$$

$$y = \left(C - \tfrac{1}{2}x^2\right)e^{4x} - \tfrac{1}{9}(1 + 3x)e^x$$

Initial conditions: $y(0) = \frac{1}{3}, \frac{1}{3} = C - \frac{1}{9}, C = \frac{4}{9}$

Particular solution: $y = \left(\frac{4}{9} - \frac{1}{2}x^2\right)e^{4x} - \frac{1}{9}(1 + 3x)e^x$

21. (a) Because $y_p'' = 0$ and $3(y_p) = 3(4) = 12$ (b) $y_p = 2$ (c) $y_p = 4$

23. $y'' + y = \sec x$

$y'' + y = 0$

$m^2 + 1 = 0$ when $m = -i, i.$

$y_h = C_1 \cos x + C_2 \sin x$

$y_p = v_1 \cos x + v_2 \sin x$

$$v_1' \cos x + v_2' \sin x = 0$$

$$v_1'(-\sin x) + v_2'(\cos x) = \sec x$$

$$v_1' = \frac{\begin{vmatrix} 0 & \sin x \\ \sec x & \cos x \end{vmatrix}}{\begin{vmatrix} \cos x & \sin x \\ -\sin x & \cos x \end{vmatrix}} = -\tan x$$

$$v_1 = \int -\tan x\, dx = \ln|\cos x|$$

$$v_2' = \frac{\begin{vmatrix} \cos x & 0 \\ -\sin x & \sec x \end{vmatrix}}{\begin{vmatrix} \cos x & \sin x \\ -\sin x & \cos x \end{vmatrix}} = 1$$

$$v_2 = \int dx = x$$

$$y = (C_1 + \ln|\cos x|)\cos x + (C_2 + x)\sin x$$

25. $y'' + 4y = \csc 2x$

$y'' + 4y = 0$

$m^2 + 4 = 0$ when $m = -2i, 2i.$

$y_h = C_1 \cos 2x + C_2 \sin 2x$

$y_p = v_1 \cos 2x + v_2 \sin 2x = 0$

$$v_1' \cos 2x + v_2' \sin 2x = 0$$

$$v_1'(-2 \sin 2x) + v_2'(2 \cos 2x) = \csc 2x$$

$$v_1' = \frac{\begin{vmatrix} 0 & \sin 2x \\ \csc 2x & 2\cos 2x \end{vmatrix}}{\begin{vmatrix} \cos 2x & \sin 2x \\ -2\sin 2x & 2\cos 2x \end{vmatrix}} = -\frac{1}{2}$$

$$v_1 = \int -\frac{1}{2}dx = -\frac{1}{2}x$$

$$v_2' = \frac{\begin{vmatrix} \cos 2x & 0 \\ -2\sin 2x & \csc 2x \end{vmatrix}}{\begin{vmatrix} \cos 2x & \sin 2x \\ -2\sin 2x & 2\cos 2x \end{vmatrix}} = \frac{1}{2}\cot 2x$$

$$v_2 = \int \frac{1}{2}\cot 2x\, dx = \frac{1}{4}\ln|\sin 2x|$$

$$y = \left(C_1 - \frac{1}{2}x\right)\cos 2x + \left(C_2 + \frac{1}{4}\ln|\sin 2x|\right)\sin 2x$$

27. $y'' - 2y' + y = e^x \ln x$

$y'' - 2y' + y = 0$

$m^2 - 2m + 1 = 0$ when $m = 1, 1$.

$y_h = (C_1 + C_2 x)e^x$

$y_p = (v_1 + v_2 x)e^x$

$v_1' e^x + v_2' x e^x = 0$

$v_1' e^x + v_2'(x + 1)e^x = e^x \ln x$

$v_1' = -x \ln x$

$$v_1 = \int -x \ln x \, dx = -\frac{x^2}{2}\ln x + \frac{x^2}{4}$$

$v_2' = \ln x$

$$v_2 = \int \ln x \, dx = x \ln x - x$$

$$y = (C_1 + C_2 x)e^x + \frac{x^2 e^x}{4}(\ln x^2 - 3)$$

29. $q'' + 10q' + 25q = 6 \sin 5t,\ q(0) = 0,\ q'(0) = 0$

$m^2 + 10m + 25 = 0$ when $m = -5, -5$.

$q_h = (C_1 + C_2 t)e^{-5t}$

$q_p = A \cos 5t + B \sin 5t$

$q_p' = -5A \sin 5t + 5B \cos 5t$

$q_p'' = -25A \cos 5t - 25B \sin 5t$

$q_p'' + 10q_p' + 25q_p = 50B \cos 5t - 50A \sin 5t = 6 \sin 5t,\ A = -\frac{3}{25},\ B = 0$

$q = (C_1 + C_2 t)e^{-5t} - \frac{3}{25}\cos 5t$

Initial conditions: $q(0) = 0,\ q'(0) = 0,\ C_1 - \frac{3}{25} = 0,\ -5C_1 + C_2 = 0,\ C_1 = \frac{3}{25},\ C_2 = \frac{3}{5}$

Particular solution: $q = \frac{3}{25}(e^{-5t} + 5te^{-5t} - \cos 5t)$

31. $\frac{24}{32}y'' + 48y = \frac{24}{32}(48 \sin 4t),\ y(0) = \frac{1}{4},\ y'(0) = 0$

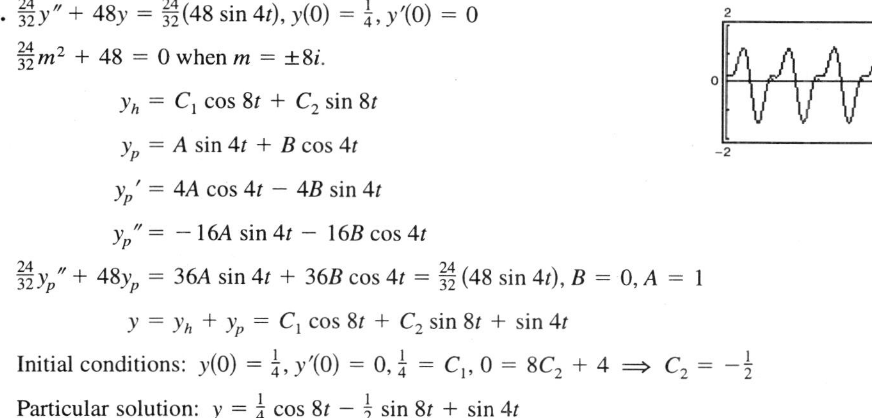

$\frac{24}{32}m^2 + 48 = 0$ when $m = \pm 8i$.

$y_h = C_1 \cos 8t + C_2 \sin 8t$

$y_p = A \sin 4t + B \cos 4t$

$y_p' = 4A \cos 4t - 4B \sin 4t$

$y_p'' = -16A \sin 4t - 16B \cos 4t$

$\frac{24}{32}y_p'' + 48y_p = 36A \sin 4t + 36B \cos 4t = \frac{24}{32}(48 \sin 4t),\ B = 0,\ A = 1$

$y = y_h + y_p = C_1 \cos 8t + C_2 \sin 8t + \sin 4t$

Initial conditions: $y(0) = \frac{1}{4},\ y'(0) = 0,\ \frac{1}{4} = C_1,\ 0 = 8C_2 + 4 \Rightarrow C_2 = -\frac{1}{2}$

Particular solution: $y = \frac{1}{4}\cos 8t - \frac{1}{2}\sin 8t + \sin 4t$

33. $\frac{2}{32}y'' + y' + 4y = \frac{2}{32}(4 \sin 8t),\ y(0) = \frac{1}{4},\ y'(0) = -3$

$\frac{1}{16}m^2 + m + 4 = 0$ when $m = -8, -8$.

$$y_h = (C_1 + C_2 t)e^{-8t}$$
$$y_p = A \sin 8t + B \cos 8t$$
$$y_p' = 8A \cos 8t - 8B \sin 8t$$
$$y_p'' = -64A \sin 8t - 64B \cos 8t$$

$$\frac{2}{32}y_p'' + y_p' + 4y_p = -8B \sin 8t + 8A \cos 8t = \frac{2}{32}(4 \sin 8t) - 8B = \frac{1}{4} \Rightarrow B = -\frac{1}{32},\ 8A = 0 \Rightarrow A = 0$$

$$y = y_h + y_p = (C_1 + C_2 t)e^{-8t} - \frac{1}{32}\cos 8t$$

Initial conditions: $y(0) = \frac{1}{4},\ y'(0) = -3,\ \frac{1}{4} = C_1 - \frac{1}{32} \Rightarrow C_1 = \frac{9}{32},\ -3 = -8C_1 + C_2 \Rightarrow C_2 = -\frac{3}{4}$

Particular solution: $y = \left(\frac{9}{32} - \frac{3}{4}t\right)e^{-8t} - \frac{1}{32}\cos 8t$

35. In Exercise 31,

$$y_h = \frac{1}{4}\cos 8t - \frac{1}{2}\sin 8t = \frac{\sqrt{5}}{4}\sin\left[8t + \arctan\left(-\frac{1}{2}\right)\right] = \frac{\sqrt{5}}{4}\sin\left(8t - \arctan\frac{1}{2}\right) \approx \frac{\sqrt{5}}{4}\sin(8t - 0.4636).$$

37. $-5y'' - 8y' = 160$

$-5m^2 - 8m = 0$ when $m = 0, -\frac{8}{5}$.

$$y_h = C_1 + C_2 e^{-1.6t}$$
$$y_p = At + B$$
$$y_p' = A$$
$$y_p'' = 0$$

$-5y'' - 8y' = -8A = 160 \Rightarrow A = -20$

$$y = C_1 + C_2 e^{-1.6t} - 20t$$

Initial conditions: $y(0) = 2000,\ y'(0) = -100,$

$2000 = C_1 + C_2,\ -100 = -1.6C_2 - 20,$

$C_2 = 50 \Rightarrow C_1 = 1950$

Particular solution: $y = 1950 + 50e^{-1.6t} - 20t$

39. $x^2y'' - xy' + y = 4x \ln x$

$y_1 = x$ and $y_2 = x \ln x$

$u_1'x + u_2'x \ln x = 0 \Rightarrow u_1' = -u_2' \ln x$

$u_1' + u_2'(1 + \ln x) = \frac{4}{x}\ln x \Rightarrow u_2' = \frac{4}{x}\ln x$

and $u_1' = -\frac{4}{x}(\ln x)^2$

$u_1 = -\frac{4}{3}(\ln x)^3$ and $u_2 = 2(\ln x)^2$

$$y_p = -\frac{4}{3}x(\ln x)^3 + 2x(\ln x)^3 = \frac{2}{3}x(\ln x)^3$$

$$y = y_n + y_p = C_1 x + C_2 x \ln x + \frac{2}{3}x(\ln x)^3$$

Section 15.5 Series Solutions of Differential Equations

1. $y' - y = 0$. Letting $y = \sum_{n=0}^{\infty} a_n x^n$:

$$y' - y = \sum_{n=1}^{\infty} n a_n x^{n-1} - \sum_{n=0}^{\infty} a_n x^n$$
$$= \sum_{n=0}^{\infty} (n+1)a_{n+1}x^n - \sum_{n=0}^{\infty} a_n x^n = 0$$

$$(n+1)a_{n+1} = a_n$$

$$a_{n+1} = \frac{a_n}{n+1}$$

$$a_1 = a_0,\ a_2 = \frac{a_1}{2} = \frac{a_0}{2},\ a_3 = \frac{a_2}{3} = \frac{a_0}{1 \cdot 2 \cdot 3}, \ldots, a_n = \frac{a_0}{n!}$$

$$y = \sum_{n=0}^{\infty} \frac{a_0}{n!}x^n = a_0 e^x$$

Check: By separation of variables, we have:

$$\int \frac{dy}{y} = \int dx$$
$$\ln y = x + C_1$$
$$y = Ce^x$$

3. $y'' - 9y = 0$. Letting $y = \sum_{n=0}^{\infty} a_n x^n$:

$$y'' - 9y = \sum_{n=2}^{\infty} n(n-1)a_n x^{n-2} - 9\sum_{n=0}^{\infty} a_n x^n = \sum_{n=0}^{\infty} (n+2)(n+1)a_{n+2}x^n - \sum_{n=0}^{\infty} 9a_n x^n = 0$$

$$(n+2)(n+1)a_{n+2} = 9a_n$$

$$a_{n+2} = \frac{9a_n}{(n+2)(n+1)}$$

$$a_0 = a_0 \qquad a_1 = a_1$$

$$a_2 = \frac{9a_0}{2} \qquad a_3 = \frac{9a_1}{3 \cdot 2}$$

$$a_4 = \frac{9a_2}{4 \cdot 3} = \frac{9^2 a_0}{4 \cdot 3 \cdot 2 \cdot 1} \qquad a_5 = \frac{9a_3}{5 \cdot 4} = \frac{9^2 a_1}{5 \cdot 4 \cdot 3 \cdot 2 \cdot 1}$$

$$\vdots \qquad \vdots$$

$$a_{2n} = \frac{9^n a_0}{(2n)!} \qquad a_{2n+1} = \frac{9^n a_1}{(2n+1)!}$$

$$y = \sum_{n=0}^{\infty} \frac{9^n a_0}{(2n)!}x^{2n} + \sum_{n=0}^{\infty} \frac{9^n a_1}{(2n+1)!}x^{2n+1} = a_0 \sum_{n=0}^{\infty} \frac{(3x)^{2n}}{(2n)!} + \frac{a_1}{3}\sum_{n=0}^{\infty} \frac{(3x)^{2n+1}}{(2n+1)!} = C_0 \sum_{n=0}^{\infty} \frac{(3x)^n}{n!} + C_1 \sum_{n=0}^{\infty} \frac{(-3x)^n}{n!}$$

$$= C_0 e^{3x} + C_1 e^{-3x} \text{ where } C_0 + C_1 = a_0 \text{ and } C_0 - C_1 = \frac{a_1}{3}.$$

Check: $y'' - 9y = 0$ is a second-order homogeneous linear equation.

$$m^2 - 9 = 0 \implies m_1 = 3 \text{ and } m_2 = -3$$

$$y = C_1 e^{3x} + C_2 e^{-3x}$$

5. $y'' + 4y = 0$. Letting $y = \sum_{n=0}^{\infty} a_n x^n$:

$$y'' + 4y = \sum_{n=2}^{\infty} n(n-1)a_n x^{n-2} + 4\sum_{n=0}^{\infty} a_n x^n = \sum_{n=0}^{\infty} (n+2)(n+1)a_{n+2}x^n + \sum_{n=0}^{\infty} 4a_n x^n = 0$$

$$(n+2)(n+1)a_{n+2} = -4a_n$$

$$a_{n+2} = \frac{-4a_n}{(n+2)(n+1)}$$

$$a_0 = a_0 \qquad a_1 = a_1$$

$$a_2 = \frac{-4a_0}{2} \qquad a_3 = \frac{-4a_1}{3 \cdot 2}$$

$$a_4 = \frac{-4a_2}{4 \cdot 3} = \frac{(-4)^2}{4!}a_0 \qquad a_5 = \frac{-4a_3}{5 \cdot 4} = \frac{(-4)^2 a_1}{5!}$$

$$\vdots \qquad \vdots$$

$$a_{2n} = \frac{(-1)^n 4^n}{(2n)!}a_0 \qquad a_{2n+1} = \frac{(-1)^n 4^n}{(2n+1)!}a_1$$

$$y = \sum_{n=0}^{\infty} \frac{(-1)^n 4^n a_0}{(2n)!}x^{2n} + \sum_{n=0}^{\infty} \frac{(-1)^n 4^n a_1}{(2n+1)!}x^{2n+1} = a_0 \sum_{n=0}^{\infty} \frac{(-1)^n (2x)^{2n}}{(2n)!} + \frac{a_1}{2}\sum_{n=0}^{\infty} \frac{(-1)^n (2x)^{2n+1}}{(2n+1)!}$$

$$= C_0 \cos 2x + C_1 \sin 2x$$

Check: $y'' + 4y = 0$ is a second-order homogeneous linear equation.

$$m^2 + 4 = 0 \implies m = \pm 2i$$

$$y = C_1 \cos 2x + C_2 \sin 2x$$

7. $y' + 3xy = 0$. Letting $y = \sum_{n=0}^{\infty} a_n x^n$:

$$y' + 3xy = \sum_{n=1}^{\infty} na_n x^{n-1} + \sum_{n=0}^{\infty} 3a_n x^{n+1} = 0$$

$$\sum_{n=-1}^{\infty} (n+2)a_{n+2}x^{n+1} = \sum_{n=0}^{\infty} -3a_n x^{n+1} \Rightarrow a_1 = 0 \text{ and}$$

$$a_{n+2} = \frac{-3a_n}{n+2}$$

$$a_0 = a_0 \qquad a_1 = 0$$

$$a_2 = -\frac{3a_0}{2} \qquad a_3 = -\frac{3a_1}{3} = 0$$

$$a_4 = -\frac{3}{4}\left(-\frac{3a_0}{2}\right) = \frac{3^2}{2^3}a_0 \qquad a_5 = -\frac{3}{5}\left(-\frac{3a_1}{3}\right) = 0$$

$$a_6 = -\frac{3}{6}\left(\frac{3^2}{2^3}a_0\right) = -\frac{3^3 a_0}{2^3(3 \cdot 2)} \qquad a_7 = -\frac{3}{7}\left(\frac{3^2 a_1}{3 \cdot 5}\right) = 0$$

$$a_8 = -\frac{3}{8}\left(-\frac{3^3 a_0}{2^3(3 \cdot 2)}\right) = \frac{3^4 a_0}{2^4(4 \cdot 3 \cdot 2)} \qquad a_9 = -\frac{3}{9}\left(-\frac{3^3 a_1}{3 \cdot 5 \cdot 7}\right) = 0$$

$$y = a_0 \sum_{n=0}^{\infty} \frac{(-3)^n x^{2n}}{2^n n!}$$

$$\lim_{n\to\infty}\left|\frac{u_{n+1}}{u_n}\right| = \lim_{n\to\infty}\left|\frac{(-3)^{n+1}x^{2n+2}}{2^{n+1}(n+1)!} \cdot \frac{2^n n!}{(-3)^n x^{2n}}\right| = \lim_{n\to\infty}\frac{3x^2}{2(n+1)} = 0$$

The interval of covergence for the solution is $(-\infty, \infty)$.

9. $y'' - xy' = 0$. Letting $y = \sum_{n=0}^{\infty} a_n x^n$:

$$y'' - xy' = \sum_{n=2}^{\infty} n(n-1)a_n x^{n-2} - x\sum_{n=1}^{\infty} na_n x^{n-1} = 0$$

$$\sum_{n=2}^{\infty} n(n-1)a_n x^{n-2} = \sum_{n=0}^{\infty} na_n x^n$$

$$\sum_{n=0}^{\infty} (n+2)(n+1)a_{n+2}x^n = \sum_{n=0}^{\infty} na_n x^n$$

$$a_{n+2} = \frac{na_n}{(n+2)(n+1)}$$

$$a_0 = a_0 \qquad a_1 = a_1$$

$$a_2 = 0 \qquad a_3 = \frac{a_1}{3 \cdot 2}$$

There are no even-powered terms.

$$a_5 = \frac{3a_3}{5 \cdot 4} = \frac{3a_1}{5!}$$

$$a_7 = \frac{5a_5}{7 \cdot 6} = \frac{5 \cdot 3a_1}{7!}$$

$$y = a_0 + a_1 \sum_{n=0}^{\infty} \frac{1 \cdot 3 \cdot 5 \cdot 7 \cdots (2n-1)x^{2n+1}}{(2n+1)!} = a_0 + a_1 \sum_{n=0}^{\infty} \frac{(2n)!x^{2n+1}}{2^n n!(2n+1)!} = a_0 + a_1 \sum_{n=0}^{\infty} \frac{x^{2n+1}}{2^n n!(2n+1)}$$

$$\lim_{n\to\infty}\left|\frac{u_{n+1}}{u_n}\right| = \lim_{n\to\infty}\left|\frac{x^{2n+3}}{2^{n+1}(n+1)!(2n+3)} \cdot \frac{2^n n!(2n+1)}{x^{2n+1}}\right| = \lim_{n\to\infty}\frac{(2n+1)x^2}{2(n+1)(2n+3)} = 0$$

Interval of covergence: $(-\infty, \infty)$

11. $(x^2 + 4)y'' + y = 0$. Letting $y = \sum_{n=0}^{\infty} a_n x^n$:

$$(x^2 + 4)y'' + y = \sum_{n=2}^{\infty} n(n-1)a_n x^n + 4\sum_{n=2}^{\infty} n(n-1)a_n x^{n-2} + \sum_{n=0}^{\infty} a_n x^n$$

$$= \sum_{n=0}^{\infty} (n^2 - n + 1)a_n x^n + \sum_{n=0}^{\infty} 4(n+2)(n+1)a_{n+2}x^n = 0$$

$$a_{n+2} = \frac{-(n^2 - n + 1)a_n}{4(n+2)(n+1)}$$

$a_0 = a_0$ $\qquad$ $a_1 = a_1$

$a_2 = \dfrac{-a_0}{4 \cdot 2 \cdot 1}$ $\qquad$ $a_3 = \dfrac{-a_1}{4(3)(2)} = \dfrac{-a_1}{24}$

$a_4 = \dfrac{-3a_2}{4(4)(3)} = \dfrac{a_0}{128}$ $\qquad$ $a_5 = \dfrac{-7a_3}{4(5)(4)} = \dfrac{7a_1}{1920}$

$$y = a_0\left(1 - \frac{x^2}{8} + \frac{x^4}{128} - \cdots\right) + a_1\left(x - \frac{x^3}{24} + \frac{7x^5}{1920} - \cdots\right)$$

13. $y' + (2x - 1)y = 0,\ y(0) = 2$

$y' = (1 - 2x)y$ $\qquad$ $y'(0) = 2$

$y'' = (1 - 2x)y' - 2y$ $\qquad$ $y''(0) = -2$

$y''' = (1 - 2x)y'' - 4y'$ $\qquad$ $y'''(0) = -10$

$y^{(4)} = (1 - 2x)y''' - 6y''$ $\qquad$ $y^{(4)}(0) = 2$

$\vdots$ $\qquad$ $\vdots$

$$y(x) = 2 + \frac{2}{1!}x - \frac{2}{2!}x^2 - \frac{10}{3!}x^3 + \frac{2}{4!}x^4 + \cdots$$

Using the first five terms of the series, $y\left(\frac{1}{2}\right) = \frac{163}{64} \approx 2.547$.

Using Euler's Method with $\Delta x = 0.1$ we have $y' = (1 - 2x)y$.

i	x_i	y_i
0	0	2
1	0.1	2.2
2	0.2	2.376
3	0.3	2.51856
4	0.4	2.61930
5	0.5	2.67169

Therefore, $y\left(\frac{1}{2}\right) \approx 2.672$.

15. (a) $m^2 + 9 = 0 \Rightarrow m = \pm 3i$

$y = C_1 \cos 3x + C_2 \sin 3x$

$y(0) = 2 = C_1$

$y' = -3C_1 \sin 3x + 3C_2 \cos 3x$

$y'(0) = 6 = 3C_2 \Rightarrow C_2 = 2$

$y_p = 2 \cos 3x + 2 \sin 3x$

(b) Let $y = \sum_{n=0}^{\infty} a_n x^n$, $y' = \sum_{n=1}^{\infty} na_n x^{n-1}$, $y'' = \sum_{n=2}^{\infty} n(n-1)a_n x^{n-2}$

$$y'' + 9y = \sum_{n=2}^{\infty} n(n-1)a_n x^{n-2} + 9\sum_{n=0}^{\infty} a_n x^n = 0$$

$$\sum_{n=0}^{\infty} (n+2)(n+1)a_{n+2}x^n + 9\sum_{n=0}^{\infty} a_n x^n = 0$$

$$(n+2)(n+1)a_{n+2} = -9a_n$$

$$a_{n+2} = \frac{-9}{(n+2)(n+1)}a_n$$

For n even,

$$a_2 = \frac{-9}{2}a_0$$

$$a_4 = \frac{-9}{4 \cdot 3}a_2 = \frac{(-9)^2}{4!}a_0$$

$$a_6 = \frac{(-9)^3}{6!}a_0$$

and in general, $a_{2n} = \dfrac{(-9)^n}{(2n)!}a_0$

For n odd,

$$a_3 = \frac{-9}{3 \cdot 2}a_1$$

$$a_5 = \frac{-9}{5 \cdot 4}a_3 = \frac{(-9)^2}{5!}a_1$$

$$a_7 = \frac{(-9)^3}{7!}a_1$$

and in general, $a_{2n+1} = \dfrac{(-9)^n}{(2n+1)!}a_1$

Hence,

$$\begin{aligned} y &= \sum_{n=0}^{\infty} a_n x^n \\ &= \sum_{n=0}^{\infty} a_{2n}x^{2n} + \sum_{n=0}^{\infty} a_{2n+1}x^{2n+1} \\ &= \sum_{n=0}^{\infty} \frac{(-9)^n}{(2n)!}a_0 x^{2n} + \sum_{n=0}^{\infty} \frac{(-9)^n}{(2n+1)!}a_1 x^{2n+1} \\ &= a_0 \sum_{n=0}^{\infty} \frac{(-1)^n(3x)^{2n}}{(2n)!} + a_1 \sum_{n=0}^{\infty} \frac{(-1)^n(3x)^{2n+1}}{(2n+1)!} \end{aligned}$$

Applying the initial conditions, $a_0 = a_1 = 2$, and $y = 2\left[\sum_{n=0}^{\infty} \frac{(-1)^n(3x)^{2n}}{(2n)!} + \sum_{n=0}^{\infty} \frac{(-1)^n(3x)^{2n+1}}{(2n+1)!}\right]$.

(c)

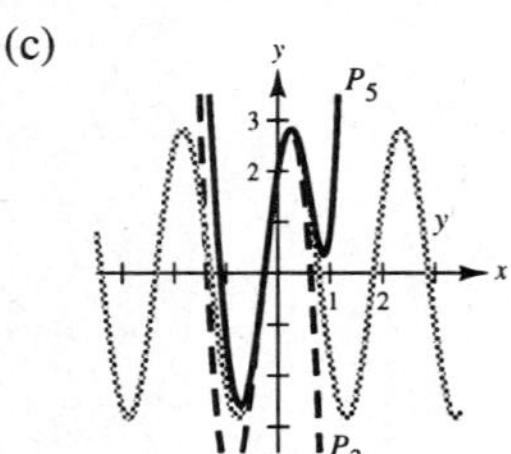

17. $y'' - 2xy = 0,\ y(0) = 1,\ y'(0) = -3$

$$
\begin{aligned}
y'' &= 2xy & y''(0) &= 0\\
y''' &= 2(xy' + y) & y'''(0) &= 2\\
y^{(4)} &= 2(xy'' + 2y') & y^{(4)}(0) &= -12\\
y^{(5)} &= 2(xy''' + 3y'') & y^{(5)}(0) &= 0\\
y^{(6)} &= 2(xy^{(4)} + 4y''') & y^{(6)}(0) &= 16\\
y^{(7)} &= 2(xy^{(5)} + 5y^{(4)}) & y^{(7)}(0) &= -120\\
&\vdots & &\vdots
\end{aligned}
$$

$$y(x) = 1 - \frac{3}{1!}x + \frac{2}{3!}x^3 - \frac{12}{4!}x^4 + \frac{16}{6!}x^6 - \frac{120}{7!}x^7 + \cdots$$

Using the first six terms of the series, $y\left(\frac{1}{4}\right) \approx 0.253$.

19. $f(x) = e^x,\ f'(x) = e^x,\ y' - y = 0$. Assume $y = \sum_{n=0}^{\infty} a_n x^n$, then:

$$y' = \sum_{n=0}^{\infty} na_n x^{n-1}$$

$$\sum_{n=1}^{\infty} na_n x^{n-1} = \sum_{n=0}^{\infty} a_n x^n$$

$$\sum_{n=0}^{\infty} (n+1)a_{n+1}x^n = \sum_{n=0}^{\infty} a_n x^n$$

$$a_{n+1} = \frac{a_n}{n+1},\quad n \ge 0$$

$n = 0,\quad a_1 = a_0$

$n = 1,\quad a_2 = \frac{a_1}{2} = \frac{a_0}{2}$

$n = 2,\quad a_3 = \frac{a_2}{3} = \frac{a_0}{2(3)}$

$n = 3,\quad a_4 = \frac{a_3}{4} = \frac{a_0}{2(3)(4)}$

$n = 4,\quad a_5 = \frac{a_4}{5} = \frac{a_0}{2(3)(4)(5)}$

$\vdots$

$$a_{n+1} = \frac{a_0}{(n+1)!} \Rightarrow a_n = \frac{a_0}{n!}$$

$y = a_0 \sum_{n=0}^{\infty} \frac{x^n}{n!}$ which converges on $(-\infty, \infty)$. When $a_0 = 1$, we have the Maclaurin Series for $f(x) = e^x$.

21.

$$f(x) = \arctan x$$

$$f'(x) = \frac{1}{1 + x^2}$$

$$f''(x) = \frac{-2x}{(1 + x^2)^2}$$

$$y'' = \frac{-2x}{1 + x^2}y'$$

$(1 + x^2)y'' + 2xy' = 0$

Assume $y = \sum_{n=0}^{\infty} a_n x^n$, then:

$$y' = \sum_{n=0}^{\infty} na_n x^{n-1}$$

$$y'' = \sum_{n=2}^{\infty} n(n-1)a_n x^{n-2}$$

$$(1 + x^2)y'' + 2xy' = \sum_{n=2}^{\infty} n(n-1)a_n x^{n-2} + \sum_{n=0}^{\infty} n(n-1)a_n x^n + \sum_{n=0}^{\infty} 2na_n x^n = 0$$

$$\sum_{n=2}^{\infty} n(n-1)a_n x^{n-2} = -\sum_{n=0}^{\infty} n(n-1)a_n x^n - \sum_{n=0}^{\infty} 2na_n x^n$$

$$\sum_{n=0}^{\infty} (n+2)(n+1)a_{n+2} x^n = -\sum_{n=0}^{\infty} n(n+1)a_n x^n$$

$$(n+2)(n+1)a_{n+2} = -n(n+1)a_n$$

$$a_{n+2} = -\frac{n}{n+2}a_n, \quad n \geq 0$$

$n = 0 \implies a_2 = 0 \implies$ all the even-powered terms have a coefficient of 0.

$n = 1, \qquad a_3 = -\frac{1}{3}a_1$

$n = 3, \qquad a_5 = -\frac{3}{5}a_3 = \frac{1}{5}a_1$

$n = 5, \qquad a_7 = -\frac{5}{7}a_5 = -\frac{1}{7}a_1$

$n = 7, \qquad a_9 = -\frac{7}{9}a_7 = \frac{1}{9}a_1$

$$\vdots$$

$$a_{2n+1} = \frac{(-1)^n a_1}{2n + 1}$$

$y = a_1 \sum_{n=0}^{\infty} \frac{(-1)^n x^{2n+1}}{2n + 1}$ which converges on $(-1, 1)$. When $a_1 = 1$, we have the Maclaurin Series for $f(x) = \arctan x$.

23. $y'' - xy = 0$. Let $y = \sum_{n=0}^{\infty} a_n x^n$.

$$y'' - xy = \sum_{n=2}^{\infty} n(n-1)a_n x^{n-2} - x\sum_{n=0}^{\infty} a_n x^n = \sum_{n=-1}^{\infty} (n+3)(n+2)a_{n+3}x^{n+1} - \sum_{n=0}^{\infty} a_n x^{n+1} = 0$$

$$2a_2 + \sum_{n=0}^{\infty} [(n+3)(n+2)a_{n+3} - a_n]x^{n+1} = 0$$

Hence, $a_2 = 0$ and

$$a_{n+3} = \frac{a_n}{(n+3)(n+2)} \text{ for } n = 0, 1, 2, \ldots.$$

The constants a_0 and a_1 are arbitrary.

$$a_0 = a_0 \qquad a_1 = a_1$$

$$a_3 = \frac{a_0}{3 \cdot 2} \qquad a_4 = \frac{a_1}{4 \cdot 3}$$

$$a_6 = \frac{a_3}{6 \cdot 5} = \frac{a_0}{6 \cdot 5 \cdot 3 \cdot 2} \qquad a_7 = \frac{a_4}{7 \cdot 6} = \frac{a_1}{7 \cdot 6 \cdot 4 \cdot 3}$$

Therefore, $y = a_0 + a_1 x + \frac{a_0}{6}x^3 + \frac{a_1}{12}x^4 + \frac{a_0}{180}x^6 + \frac{a_1}{504}x^7$.

Chapter 15 Review Exercises

	Differential Equation	*Type*	*Order*
1.	$\frac{\partial^2 u}{\partial t^2} = c^2 \frac{\partial^2 u}{\partial x^2}$	Partial	2
3.	$y'' + 3y' - 10 = 0$	Ordinary	2

5. (a)

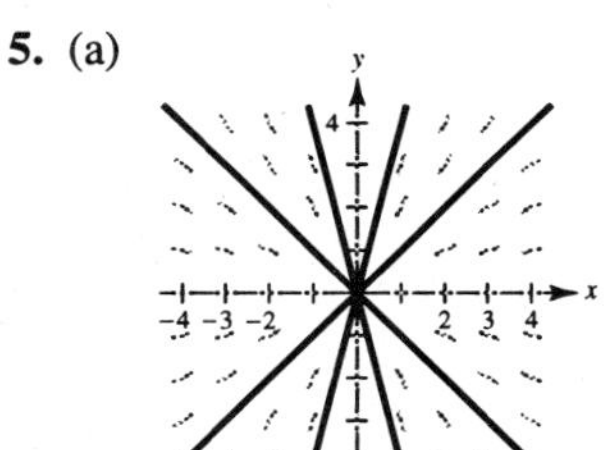

(b) $\frac{dy}{dx} = \frac{y}{x}$

$$\int \frac{dy}{y} = \int \frac{dx}{x}$$

$$\ln y = \ln x + \ln C$$

$$y = Cx$$

7. $y' - 4 = 0$

$$\int dy = \int 4\, dx$$

$$y = 4x + C$$

Matches b.

9. $y'' - 4y = 0$

$$m^2 - 4 = 0$$

$$m_1 = 2,\ m_2 = -2$$

$$y = C_1 e^{2x} + C_2 e^{-2x}$$

Matches a.

11. $\frac{dy}{dx} - \frac{y}{x} = 2 + \sqrt{x}$

Integrating factor: $e^{-\int(1/x)\,dx} = e^{-\ln|x|} = \frac{1}{x}$

$$y\left(\frac{1}{x}\right) = \int \frac{1}{x}(2 + \sqrt{x})dx = \ln x^2 + 2\sqrt{x} + C$$

$$y = x \ln x^2 + 2x^{3/2} + Cx$$

13. $y' - \frac{2y}{x} = \frac{1}{x}y'$

$$(x-1)\frac{dy}{dx} = 2y$$

$$\int \frac{1}{y}\, dy = \int \frac{2}{x-1}\, dx$$

$$\ln|y| = \ln(x-1)^2 + \ln C$$

$$y = C(x-1)^2$$

15. $\dfrac{dy}{dx} - \dfrac{y}{x} = \dfrac{x}{y}$, Bernoulli

$n = -1,\ P = -\dfrac{1}{x},\ Q = x,$

$$e^{\int(-2/x)\,dx} = e^{\ln x^{-2}} = x^{-2}$$

$$y^2x^{-2} = \int 2(x)x^{-2}\,dx = \ln x^2 + C$$

$$y^2 = x^2\ln x^2 + Cx^2$$

17. $(10x + 8y + 2)\,dx + (8x + 5y + 2)\,dy = 0$

Exact: $\dfrac{\partial M}{\partial y} = 8 = \dfrac{\partial N}{\partial x}$

$$U(x, y) = \int (10x + 8y + 2)\,dx = 5x^2 + 8xy + 2x + f(y)$$

$$U_y(x, y) = 8x + f'(y) = 8x + 5y + 2$$

$$f'(y) = 5y + 2$$

$$f(y) = \frac{5}{2}y^2 + 2y + C_1$$

$$U(x, y) = 5x^2 + 8xy + 2x + \frac{5}{2}y^2 + 2y + C_1$$

$$5x^2 + 8xy + 2x + \frac{5}{2}y^2 + 2y = C$$

19. $(2x - 2y^3 + y)\,dx + (x - 6xy^2)\,dy = 0$

Exact: $\dfrac{\partial M}{\partial y} = -6y^2 + 1 = \dfrac{\partial N}{\partial x}$

$$U(x, y) = \int (2x - 2y^3 + y)dx$$

$$= x^2 - 2xy^3 + xy + f(y)$$

$$U_y(x, y) = -6xy^2 + x + f'(y) = x - 6xy^2$$

$$f'(y) = 0$$

$$f(y) = C_1$$

$$U(x, y) = x^2 - 2xy^3 + xy + C_1$$

$$x^2 - 2xy^3 + xy = C$$

21. $dy = (y\tan x + 2e^x)\,dx$

$$\frac{dy}{dx} - (\tan x)y = 2e^x$$

Integrating factor: $e^{-\int \tan x\,dx} = e^{\ln|\cos x|} = \cos x$

$$y\cos x = \int 2e^x\cos x\,dx = e^x(\cos x + \sin x) + C$$

$$y = e^x(1 + \tan x) + C\sec x$$

23. $(x - y - 5)dx - (x + 3y - 2)dy = 0$

Exact: $\dfrac{\partial M}{\partial y} = -1 = \dfrac{\partial N}{\partial x}$

$$U(x, y) = \int (x - y - 5)\,dx = \frac{1}{2}x^2 - xy - 5x + f(y)$$

$$U_y(x, y) = -x + f'(y) = -x - 3y + 2$$

$$f'(y) = -3y + 2$$

$$f(y) = -\frac{3}{2}y^2 + 2y + C_1$$

$$U(x, y) = \frac{1}{2}x^2 - xy - 5x - \frac{3}{2}y^2 + 2y + C_1$$

$$x^2 - 2xy - 10x - 3y^2 + 4y = C$$

25. $x + yy' = \sqrt{x^2 + y^2}$

$$\int \frac{x\,dx + y\,dy}{\sqrt{x^2 + y^2}} = \int dx$$

$$\sqrt{x^2 + y^2} = x + C$$

$$x^2 + y^2 = x^2 + 2Cx + C^2$$

$$y^2 = 2Cx + C^2$$

27. $yy' + y^2 = 1 + x^2$

$$y' + y = \frac{1}{y}(1 + x^2), \text{ Bernoulli}$$

$$n = -1,\ P = 1,\ Q = 1 + x^2,\ e^{\int 2\,dx} = e^{2x}$$

$$y^2e^{2x} = \int 2(1 + x^2)e^{2x}\,dx$$

$$= \left(x^2 - x + \frac{3}{2}\right)e^{2x} + C$$

$$y^2 = x^2 - x + \frac{3}{2} + Ce^{-2x}$$

29. $(1 + x^2)\,dy = (1 + y^2)\,dx$

$$\int \frac{1}{1 + y^2}\,dy = \int \frac{1}{1 + x^2}\,dx$$

$$\arctan y - \arctan x = C_1$$

$$\tan(\arctan y - \arctan x) = \tan C_1$$

$$\frac{y - x}{1 + xy} = C$$

31. $y' - \left(\frac{a}{x}\right)y = bx^3$

Integrating factor: $e^{-\int (a/x)dx} = e^{-a\ln x} = x^{-a}$

$$yx^{-a} = \int bx^3(x^{-a})\,dx = \frac{b}{4 - a}x^{4-a} + C$$

$$y = \frac{bx^4}{4 - a} + Cx^a$$

33. $y' - 2y = e^x$

Integrating factor: $e^{\int -2\,dx} = e^{-2x}$

$$ye^{-2x} = \int e^{-2x}e^x dx + C = -e^{-x} + C$$

$$y = Ce^{2x} - e^x$$

Initial condition: $y(0) = 4$

$$4 = C - 1 \implies C = 5$$

Particular solution: $y = 5e^{2x} - e^x$

35. $x\,dy = (x + y + 2)\,dx$

$$\frac{dy}{dx} - \left(\frac{1}{x}\right)y = \frac{x + 2}{x}$$

Integrating factor: $e^{\int -(1/x)\,dx} = e^{-\ln|x|} = \frac{1}{x}$

$$y\left(\frac{1}{x}\right) = \int \frac{x + 2}{x^2}\,dx = \ln|x| - \frac{2}{x} + C$$

$$y = x\ln|x| - 2 + Cx$$

Initial condition: $y(1) = 10$

$$10 = -2 + C \implies C = 12$$

Particular solution: $y = x\ln|x| - 2 + 12x$

37. $\ln(1 + y)dx + \left(\frac{1}{1 + y}\right)dy = 0$

$$\int dx + \int \frac{1}{(1 + y)\ln(1 + y)}\,dy = C_1$$

$$x + \ln|\ln(1 + y)| = C_1$$

$$\ln|\ln(1 + y)| = C_1 - x$$

$$\ln|1 + y| = e^{C_1 - x} = Ce^{-x}$$

Initial condition: $y(0) = 2$

$$\ln 3 = C$$

Particular solution: $\ln|1 + y| = (\ln 3)e^{-x}$

39.

$$y' = x^2y^2 - 9x^2$$

$$\int \frac{1}{y^2 - 9}\,dy = \int x^2\,dx$$

$$\frac{1}{6}\ln\left|\frac{y-3}{y+3}\right| = \frac{1}{3}x^3 + C_1$$

$$\ln\left|\frac{y-3}{y+3}\right| = 2x^3 + C$$

Initial condition: $y(0) = \dfrac{3(1+e)}{1-e}$

$$C = \ln\left|\frac{\dfrac{3(1+e)}{1-e} - 3}{\dfrac{3(1+e)}{1-e} + 3}\right| = 1$$

Particular solution: $\ln\left|\dfrac{y-3}{y+3}\right| = 2x^3 + 1$

$$\frac{y-3}{y+3} = e^{2x^3+1}$$

$$y = \frac{3(e^{2x^3+1} + 1)}{1 - e^{2x^3+1}}$$

41.

$$(x - C)^2 + y^2 = C^2$$

$$x^2 - 2Cx + C^2 + y^2 = C^2$$

$$\frac{x^2 + y^2}{x} = 2C$$

$$\frac{x(2x + 2yy') - (x^2 + y^2)}{x^2} = 0$$

$$2x^2 + 2xyy' - x^2 - y^2 = 0$$

$$y' = \frac{y^2 - x^2}{2xy}$$

The negative reciprocal of y' is the slope of the orthogonal trajectories.

$$\frac{dy}{dx} = \frac{2xy}{x^2 - y^2}$$

$$2xy\,dx + (y^2 - x^2)dy = 0$$

Homogeneous

$x = vy,\ dx = v\,dy + y\,dv$

$$2vy^2(v\,dy + y\,dv) + (y^2 - v^2y^2)dy = 0$$

$$\int \frac{2v}{1 + v^2}\,dv + \int \frac{1}{y}\,dy = 0$$

$$\ln(1 + v^2) + \ln|y| = \ln K_1$$

$$y^2 + x^2 = K_1 y$$

Circles: $x^2 + (y - K)^2 = K^2$

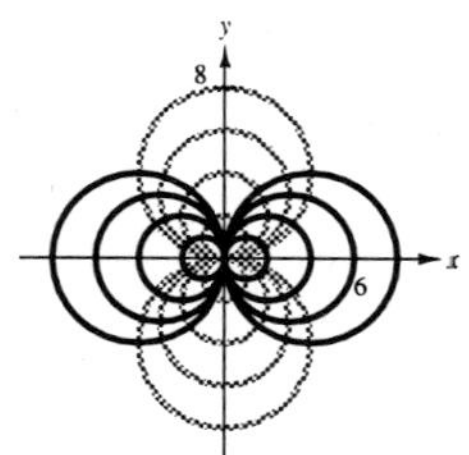

43. (a) $\dfrac{ds}{dh} = \dfrac{k}{h}$

$$\int ds = \int \frac{k}{h}\,dh$$

$$s = k\ln h + C_1 = k\ln Ch = k\ln h + C_1$$

(b) Since $s = 25$ when $h = 2$ and $s = 12$ when $h = 10$, it follows that $25 = k\ln 2C$ and $12 = k\ln 10C$, which implies

$$C = \frac{1}{2}e^{-(25/13)\ln 5} \approx 0.0605 \text{ and } k = \frac{25}{\ln 2C} = \frac{-13}{\ln 5} \approx -8.0774.$$

Therefore, s is given by the following.

$$s = -\frac{13}{\ln 5}\ln\left[\frac{h}{2}e^{-(25/13)\ln 5}\right]$$

$$= -\frac{13}{\ln 5}\left[\ln\frac{h}{2} - \frac{25}{13}\ln 5\right] = -\frac{1}{\ln 5}\left[13\ln\frac{h}{2} - 25\ln 5\right] = 25 - \frac{13\ln(h/2)}{\ln 5},\ 2 \le h \le 15.$$

45. $\dfrac{dN}{dt} = kN(L - N)$ (Logistic Differential Equation)

$$\int \frac{dN}{N(L-N)} = \int k\,dt$$

$$\frac{1}{L}\int\left[\frac{1}{N} + \frac{1}{L-N}\right]dN = \int k\,dt$$

$$\ln|N| - \ln|L - N| = L(kt + C_1)$$

$$\frac{N}{L-N} = e^{Lkt + C_2} = Ce^{Lkt}$$

$$N = \frac{LCe^{Lkt}}{1 + Ce^{Lkt}}$$

When $t = 0$, $N = 100$. Thus, $100 = \dfrac{LC}{1 + C} \Rightarrow C = \dfrac{100}{L - 100}$, and thus, $N = \dfrac{100Le^{Lkt}}{(L - 100) + 100e^{Lkt}}$.

When $t = 4$, $N = 200$. Thus, $200 = \dfrac{100Le^{4Lk}}{(L - 100) + 100e^{4Lk}} \Rightarrow k = \dfrac{1}{4L}\ln\left[\dfrac{2(L - 100)}{L - 200}\right]$.

Therefore, $N = \dfrac{100Le^{(t/4)\ln[(2(L-100))/(L-200)]}}{(L - 100) + 100e^{(t/4)\ln[(2(L-100))/(L-200)]}} = \dfrac{L}{\left(\dfrac{L}{100} - 1\right)e^{-(t/4)\ln[(2(L-100))/(L-200)]} + 1} = \dfrac{500}{1 + 4e^{-0.2452t}}$.

47. $y' = \sin x - 0.5y \quad y(0) = 1$

$y' + 0.5y = \sin x$ linear differential equation

Integrating factor: $e^{\int 0.5dx} = e^{.5x}$

$$y'e^{.5x} + 0.5e^{.5x}y = e^{.5x}\sin x$$

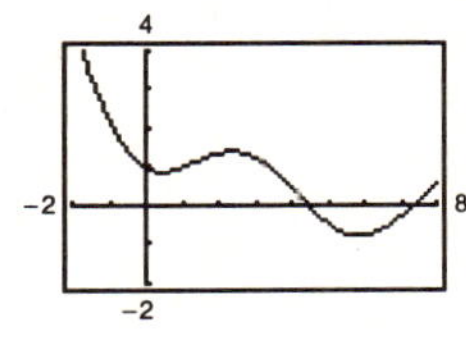

$$ye^{.5x} = \int e^{.5x}\sin x \cdot dx = -\frac{4}{5}e^{.5x}\cos x + \frac{2}{5}e^{.5x}\sin x + C$$

$$y = -\frac{4}{5}\cos x + \frac{2}{5}\sin x + Ce^{-.5x}$$

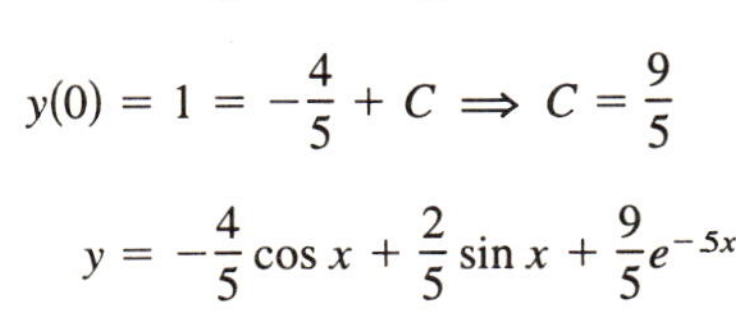

$$y(0) = 1 = -\frac{4}{5} + C \Rightarrow C = \frac{9}{5}$$

$$y = -\frac{4}{5}\cos x + \frac{2}{5}\sin x + \frac{9}{5}e^{-.5x}$$

49. $A_0 = 500{,}000,\ r = 0.10$

(a) $P = 40{,}000$

$$A = \frac{40{,}000}{0.10} + \left(500{,}000 - \frac{40{,}000}{0.10}\right)e^{0.10t} = 100{,}000(4 + e^{0.10t})$$

The balance continues to increase.

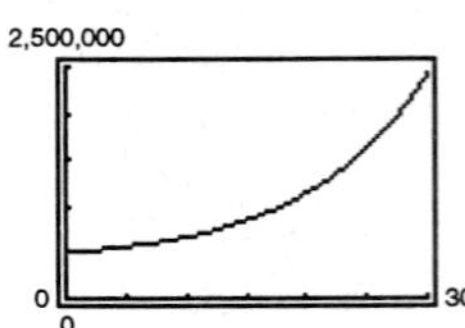

(b) $P = 50{,}000$

$$A = \frac{50{,}000}{0.10} + \left(500{,}000 - \frac{50{,}000}{0.10}\right)e^{0.10t} = 500{,}000$$

The balance remains at $500,000.

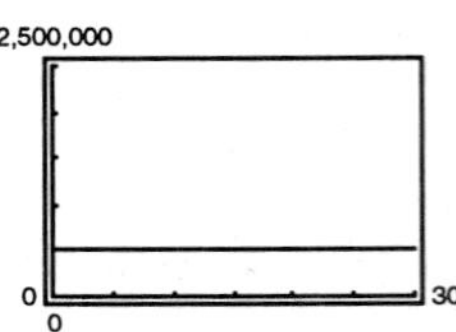

(c) $P = 60{,}000$

$$A = \frac{60{,}000}{0.10} + \left(500{,}000 - \frac{60{,}000}{0.10}\right)e^{0.10t} = 100{,}000(6 - e^{0.10t})$$

The balance decreases and is depleted in $t = (\ln 6)/0.10 \approx 17.9$ years.

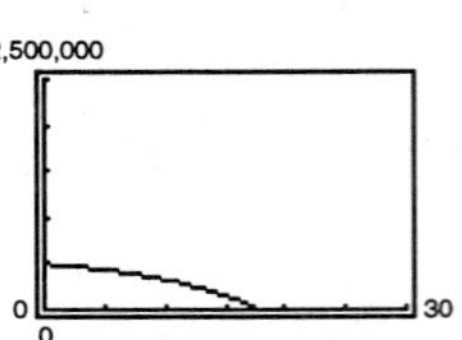

51. $y'' - y' - 2y = 0$

$m^2 - m - 2 = (m - 2)(m + 1) = 0 \qquad m = 2, -1$

$$y = C_1e^{2x} + C_2e^{-x}$$
$$y' = 2C_1e^{2x} - C_2e^{-x}$$
$$y(0) = 0 = C_1 + C_2$$
$$y'(0) = 3 = 2C_1 - C_2$$

Adding these equations $3 = 3C_1 \implies C_1 = 1$ and $C_2 = -1$

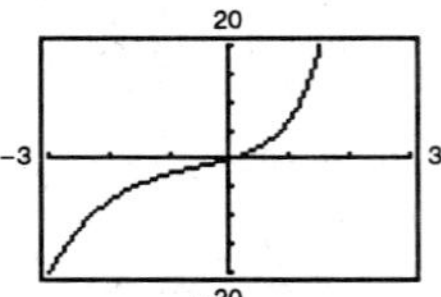

$y = e^{2x} - e^{-x}$

53. $y'' + 2y' - 3y = 0$

$m^2 + 2m - 3 = (m + 3)(m - 1) = 0 \implies m = -3, 1$

$$y = C_1e^{-3x} + C_2e^{x}$$
$$y' = -3C_1e^{-3x} + C_2e^{x}$$
$$y(0) = 2 = C_1 + C_2$$
$$y'(0) = 0 = -3C_1 + C_2$$

Subtracting these equations, $2 = 4C_1 \implies C_1 = \frac{1}{2}$ and $C_2 = \frac{3}{2}$

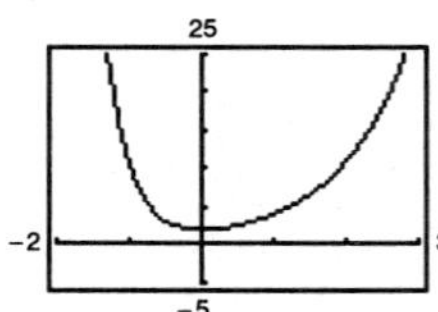

$y = \frac{3}{2}e^{x} + \frac{1}{2}e^{-3x}$

55. $y'' + y = x^3 + x$

$m^2 + 1 = 0$ when $m = -i, i$.

$$y_h = C_1 \cos x + C_2 \sin x$$
$$y_p = A_0 + A_1x + A_2x^2 + A_3x^3$$
$$y_p' = A_1 + 2A_2x + 3A_3x^2$$
$$y_p'' = 2A_2 + 6A_3x$$
$$y_p'' + y_p = (A_0 + 2A_2) + (A_1 + 6A_3)x + A_2x^2 + A_3x^3 = x^3 + x$$
$$A_0 = 0,\ A_1 = -5,\ A_2 = 0,\ A_3 = 1$$
$$y = C_1 \cos x + C_2 \sin x - 5x + x^3$$

57. $y'' + y = 2 \cos x$

$m^2 + 1 = 0$ when $m = -i, i$.

$$y_h = C_1 \cos x + C_2 \sin x$$
$$y_p = Ax \cos x + Bx \sin x$$
$$y_p' = (Bx + A) \cos x + (B - Ax) \sin x$$
$$y_p'' = (2B - Ax) \cos x + (-Bx - 2A) \sin x$$
$$y_p'' + y_p = 2B \cos x - 2A \sin x = 2 \cos x$$
$$A = 0,\ B = 1$$
$$y = C_1 \cos x + (C_2 + x) \sin x$$

59. $y'' - 2y' + y = 2xe^x$

$m^2 - 2m + 1 = 0$ when $m = 1, 1$.

$y_h = (C_1 + C_2x)e^x$

$y_p = (v_1 + v_2x)e^x$

$v_1'e^x + v_2'xe^x = 0$

$v_1'e^x + v_2'(x + 1)e^x = 2xe^x$

$v_1' = -2x^2$

$v_1 = \int -2x^2\,dx = -\frac{2}{3}x^3$

$v_2' = 2x$

$v_2 = \int 2x\,dx = x^2$

$y = (C_1 + C_2x + \frac{1}{3}x^3)e^x$

61. $y'' + y' - 6y = 54, \quad y(0) = 2, y'(0) = 0$

$m^2 - m - 6 = 0$

$(m - 3)(m + 2) = 0$

$m_1 = 3, \ m_2 = -2$

$y_h = C_1e^{3x} + C_2e^{-2x}$

$y_p = -9$ by inspection

$y = y_h + y_p = C_1e^{3x} + C_2e^{-2x} - 9$

Initial conditions:

$y(0) = 2\colon \ 2 = C_1 + C_2 - 9 \Rightarrow C_1 + C_2 = 11$

$y'(0) = 0\colon \ 0 = 3C_1 - 2C_2 \quad \Rightarrow C_1 = \frac{22}{5}, C_2 = \frac{33}{5}$

$y = \frac{11}{5}(2e^{3x} + 3e^{-2x}) - 9$

63. $y'' + 4y = \cos x$

$m^2 + 4 = 0 \Rightarrow m = \pm 2i$

$y_h = C_1 \cos 2x + C_2 \sin 2x$

$y_p = A \cos x + B \sin x$

$y_p' = -A \sin x + B \cos x$

$y_p'' = -A \cos x - B \sin x$

$y_p'' + 4y_p = (-A \cos x - B \sin x) + 4(A \cos x + B \sin x) = \cos x$

$3A \cos x + 3B \sin x = \cos x \Rightarrow A = \frac{1}{3}$ and $B = 0$

$y_p = \frac{1}{3} \cos x$

$y = y_h + y_p = C_1 \cos 2x + C_2 \sin 2x + \frac{1}{3} \cos x$

Initial conditions: $y(0) = 6\colon \quad 6 = C_1 + \frac{1}{3} \Rightarrow C_1 = \frac{17}{3}$

$y'(0) = -6\colon \ -6 = 2C_2 \quad \Rightarrow C_2 = -3$

Particular solution: $y = \frac{17}{3} \cos 2x - 3 \sin 2x + \frac{1}{3} \cos x$

65. By Hooke's Law, $F = kx$, $k = F/x = 64/(4/3) = 48$. Also, $F = ma$ and $m = F/a = 64/32 = 2$. Therefore,

$$\frac{d^2y}{dt^2} + \left(\frac{48}{2}\right)y = 0$$

$y = C_1 \cos(2\sqrt{6}t) + C_2 \sin(2\sqrt{6}t)$.

67. (a) (i) $y = \frac{1}{2}\cos 2t + \frac{12\pi}{\pi^2 - 4}\sin 2t + \frac{24}{4 - \pi^2}\sin \pi t$ (ii) $y = \frac{1}{2}\left[\left(1 - 6\sqrt{2}t\right)\cos\left(2\sqrt{2}t\right) + 3\sin\left(2\sqrt{2}t\right)\right]$

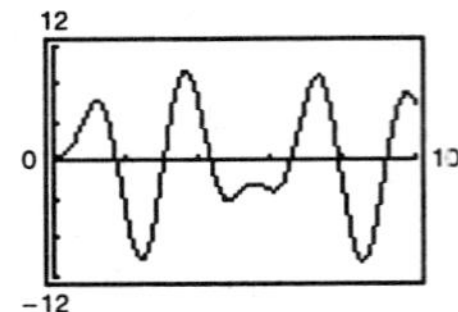

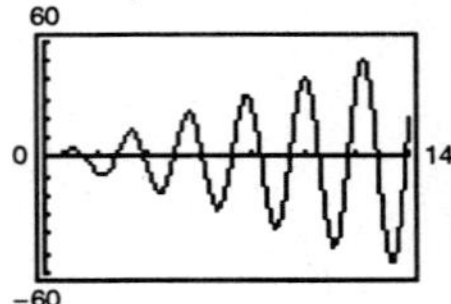

(iii) $y = \frac{e^{-t/5}}{398}\left[199\cos\frac{\sqrt{199}t}{5} + \sqrt{199}\sin\frac{\sqrt{199}t}{5}\right]$ (iv) $y = \frac{1}{2}e^{-2t}(\cos 2t + \sin 2t)$

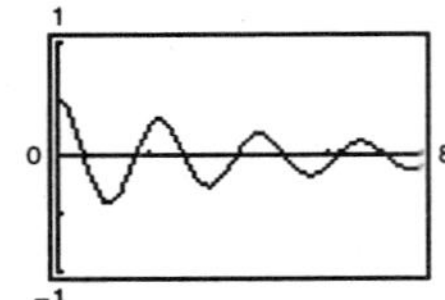

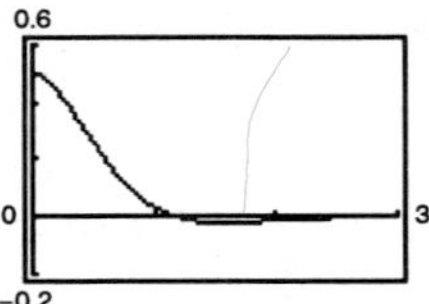

(b) The object comes to rest more quickly. It may not even oscillate, as in part (iv).

(c) It would oscillate more rapidly.

(d) Part (iii). The amplitude becomes increasingly large.

69. $(x - 4)y' + y = 0$. Letting $y = \sum_{n=0}^{\infty} a_n x^n$:

$$xy' - 4y' + y = \sum_{n=0}^{\infty} na_n x^n - 4\sum_{n=1}^{\infty} na_n x^{n-1} + \sum_{n=0}^{\infty} a_n x^n$$

$$= \sum_{n=0}^{\infty}(n + 1)a_n x^n - \sum_{n=1}^{\infty} 4na_n x^{n-1} = \sum_{n=0}^{\infty}(n + 1)a_n x^n - \sum_{n=-1}^{\infty} 4(n + 1)a_{n+1}x^n = 0$$

$$(n + 1)a_n = 4(n + 1)a_{n+1}$$

$$a_{n+1} = \frac{1}{4}a_n$$

$$a_0 = a_0,\ a_1 = \frac{1}{4}a_0,\ a_2 = \frac{1}{4}a_1 = \frac{1}{4^2}a_0, \ldots, a_n = \frac{1}{4^n}a_0$$

$$y = a_0\sum_{n=0}^{\infty}\frac{x^n}{4^n}$$

Appendix A.1

1. $0.7 = \frac{7}{10}$

Rational

3. $\frac{3\pi}{2}$

Irrational (since π is irrational)

5. $4.3451\overline{451}$

Rational

7. $\sqrt[3]{64} = 4$

Rational

9. $4\frac{5}{8} = \frac{45}{8}$

Rational

11. Let $x = 0.36\overline{36}$.

$$100x = 36.36\overline{36}$$
$$-x = -0.36\overline{36}$$
$$99x = 36$$
$$x = \frac{36}{99} = \frac{4}{11}$$

13. Let $x = 0.297\overline{297}$.

$$1{,}000x = 297.297\overline{297}$$
$$-x = -0.297\overline{297}$$
$$999x = 297$$
$$x = \frac{297}{999} = \frac{11}{37}$$

15. Given $a < b$:

(a) $a + 2 < b + 2$; True

(b) $5b < 5a$; False

(c) $5 - a > 5 - b$; True

(d) $\frac{1}{a} < \frac{1}{b}$; False

(e) $(a - b)(b - a) > 0$; False

(f) $a^2 < b^2$; False

17. x is greater than -3 and less than 3.

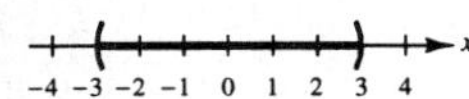

The interval is bounded.

19. x is less than, or equal to, 5.

The interval is unbounded.

21. $y \geq 4$, $[4, \infty)$

23. $0.03 < r \leq 0.07$, $(0.03, 0.07]$

25.
$$2x - 1 \geq 0$$
$$2x \geq 1$$
$$x \geq \tfrac{1}{2}$$

27.
$$-4 < 2x - 3 < 4$$
$$-1 < 2x < 7$$
$$-\tfrac{1}{2} < x < \tfrac{7}{2}$$

29.
$$\frac{x}{2} + \frac{x}{3} > 5$$
$$3x + 2x > 30$$
$$5x > 30$$
$$x > 6$$

31. $|x| < 1 \Rightarrow -1 < x < 1$

33. $\left|\frac{x-3}{2}\right| \geq 5$

$x - 3 \geq 10$ or $x - 3 \leq -10$

$x \geq 13$ $\quad x \leq -7$

35.
$$|x - a| < b$$
$$-b < x - a < b$$
$$a - b < x < a + b$$

37.
$$|2x + 1| < 5$$
$$-5 < 2x + 1 < 5$$
$$-6 < 2x < 4$$
$$-3 < x < 2$$

39.
$$\left|1 - \frac{2x}{3}\right| < 1$$
$$-1 < 1 - \frac{2x}{3} < 1$$
$$-2 < -\frac{2x}{3} < 0$$
$$3 > x > 0$$

41. $x^2 \le 3 - 2x$

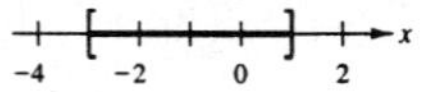

$x^2 + 2x - 3 \le 0$

$(x + 3)(x - 1) \le 0$

Test intervals:

$(-\infty, -3),\ (-3, 1),\ (1, \infty)$

Solution: $-3 \le x \le 1$

43. $x^2 + x - 1 \le 5$

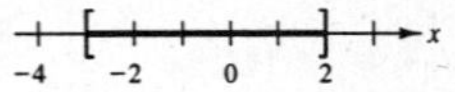

$x^2 + x - 6 \le 0$

$(x + 3)(x - 2) \le 0$

$x = -3$

$x = 2$

Test intervals:

$(-\infty, -3),\ (-3, 2),\ (2, \infty)$

Solution: $-3 \le x \le 2$

45. $a = -1,\ b = 3$

Directed distance from a to b: 4

Directed distance from b to a: -4

Distance between a and b: 4

47. (a) $a = 126,\ b = 75$

Directed distance from a to b: -51

Directed distance from b to a: 51

Distance between a and b: 51

(b) $a = -126,\ b = -75$

Directed distance from a to b: 51

Directed distance from b to a: -51

Distance between a and b: 51

49. $a = -1,\ b = 3$

Midpoint: $\dfrac{-1 + 3}{2} = 1$

51. (a) $[7, 21]$

Midpoint: 14

(b) $[8.6, 11.4]$

Midpoint: 10

53. $a = -2, b = 2$

Midpoint: 0

Distance between midpoint and each endpoint: 2

$|x - 0| \le 2$

$|x| \le 2$

55. $a = 0,\ b = 4$

Midpoint: 2

Distance between midpoint and each endpoint: 2

$|x - 2| > 2$

57. (a) All numbers that are at most 10 units from 12

$|x - 12| \le 10$

(b) All numbers that are at least 10 units from 12

$|x - 12| \ge 10$

59. $R = 115.95x,\ C = 95x + 750,\ R > C$

$115.95x > 95x + 750$

$20.95x > 750$

$x > 35.7995$

$x \ge 36$ units

61. $\left|\dfrac{x - 50}{5}\right| \ge 1.645$

$\dfrac{x - 50}{5} \le -1.645$ or $\dfrac{x - 50}{5} \ge 1.645$

$x - 50 \le -8.225$ $x - 50 \ge 8.225$

$x \le 41.775$ $x \ge 58.225$

$x \le 41$ $x \ge 59$

63. (a) $\pi \approx 3.1415926535$

$\frac{355}{113} = 3.141592920$

$\frac{355}{113} > \pi$

(b) $\pi \approx 3.1415926535$

$\frac{22}{7} \approx 3.142857143$

$\frac{22}{7} > \pi$

65. Speed of light: 2.998×10^8 meters per second

Distance traveled in one year = rate × time

$$d = (2.998 \times 10^8) \times (\underset{\text{days}}{365} \times \underset{\text{hours}}{24} \times \underset{\text{minutes}}{60} \times \underset{\text{seconds}}{60})$$

$$= (2.998 \times 10^8) \times (3.1536 \times 10^7) \approx 9.45 \times 10^{15}$$

This is best estimated by (b).

67. False; 2 is a nonzero integer and the reciprocal of 2 is $\frac{1}{2}$.

69. True

71. True; if $x < 0$, then $|x| = -x = \sqrt{x^2}$.

73. If $a \geq 0$ and $b \geq 0$, then $|ab| = ab = |a|\,|b|$.
If $a < 0$ and $b < 0$, then $|ab| = ab = (-a)(-b) = |a|\,|b|$.
If $a \geq 0$ and $b < 0$, then $|ab| = -ab = a(-b) = |a|\,|b|$.
If $a < 0$ and $b \geq 0$, then $|ab| = -ab = (-a)b = |a|\,|b|$.

75. $$\left|\frac{a}{b}\right| = \left|a\left(\frac{1}{b}\right)\right|$$

$$= |a|\left|\frac{1}{b}\right| = |a| \cdot \frac{1}{|b|} = \frac{|a|}{|b|}, \quad b \neq 0$$

77. $n = 1, \quad |a| = |a|$

$n = 2, \quad |a^2| = |a \cdot a| = |a||a| = |a|^2$

$n = 3, \quad |a^3| = |a^2 \cdot a| = |a^2||a| = |a|^2|a| = |a|^3$

$\vdots$

$|a^n| = |a^{n-1}a| = |a^{n-1}||a| = |a|^{n-1}|a| = |a|^n$

79. $|a| \leq k \Leftrightarrow \sqrt{a^2} \leq k \Leftrightarrow a^2 \leq k^2 \Leftrightarrow a^2 - k^2 \leq 0 \Leftrightarrow (a + k)(a - k) \leq 0 \Leftrightarrow -k \leq a \leq k, \quad k > 0$

81. $$\left.\begin{array}{l}|7 - 12| = |-5| = 5 \\ |7| - |12| = 7 - 12 = -5\end{array}\right\} |7 - 12| > |7| - |12|$$

$$\left.\begin{array}{l}|12 - 7| = |5| = 5 \\ |12| - |7| = 12 - 7 = 5\end{array}\right\} |12 - 7| = |12| - |7|$$

We know that $|a||b| \geq ab$. Thus, $-2|a||b| \leq -2ab$. Since $a^2 = |a|^2$ and $b^2 = |b|^2$, we have

$$|a|^2 + |b|^2 - 2|a||b| \leq a^2 + b^2 - 2ab$$

$$0 \leq (|a| - |b|)^2 \leq (a - b)^2$$

$$\sqrt{(|a| - |b|)^2} \leq \sqrt{(a - b)^2}$$

$$\big||a| - |b|\big| \leq |a - b|.$$

Since $|a| - |b| \leq \big||a| - |b|\big|$, we have $|a| - |b| \leq |a - b|$. Thus, $|a - b| \geq |a| - |b|$.

Appendix A.2

1. $d = \sqrt{(4-2)^2 + (5-1)^2}$

$= \sqrt{4+16} = \sqrt{20} = 2\sqrt{5}$

Midpoint: $\left(\frac{4+2}{2}, \frac{5+1}{2}\right) = (3, 3)$

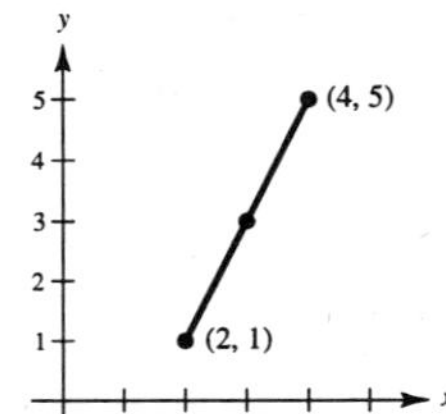

3. $d = \sqrt{\left(\frac{1}{2} + \frac{3}{2}\right)^2 + (1+5)^2}$

$= \sqrt{4+36} = \sqrt{40} = 2\sqrt{10}$

Midpoint: $\left(\frac{(-3/2)+(1/2)}{2}, \frac{-5+1}{2}\right) = \left(-\frac{1}{2}, -2\right)$

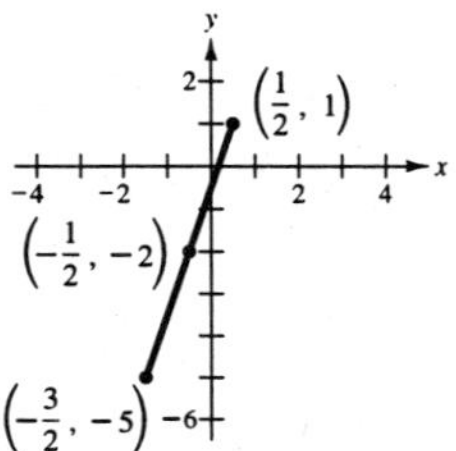

5. $d = \sqrt{(-1-1)^2 + (1-\sqrt{3})^2}$

$= \sqrt{4 + 1 - 2\sqrt{3} + 3} = \sqrt{8 - 2\sqrt{3}}$

Midpoint: $\left(\frac{-1+1}{2}, \frac{1+\sqrt{3}}{2}\right) = \left(0, \frac{1+\sqrt{3}}{2}\right)$

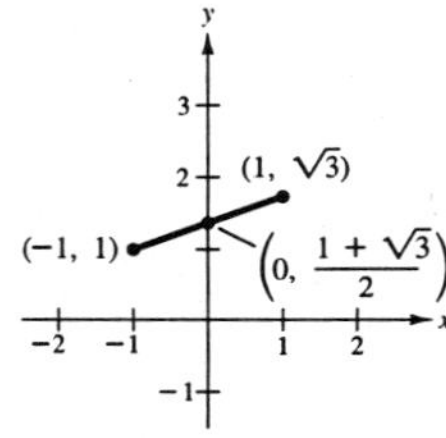

7. $d_1 = \sqrt{9+36} = \sqrt{45}$

$d_2 = \sqrt{4+1} = \sqrt{5}$

$d_3 = \sqrt{25+25} = \sqrt{50}$

$(d_1)^2 + (d_2)^2 = (d_3)^2$

Right triangle

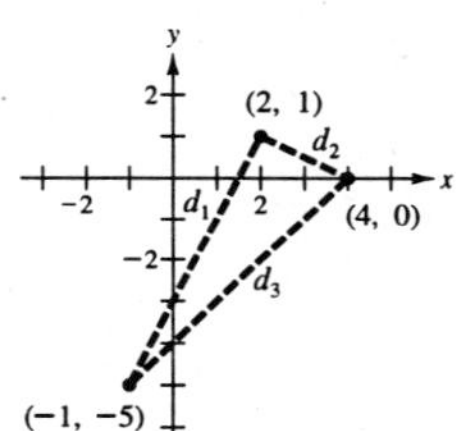

9. $d_1 = d_2 = d_3 = d_4 = \sqrt{5}$

Rhombus

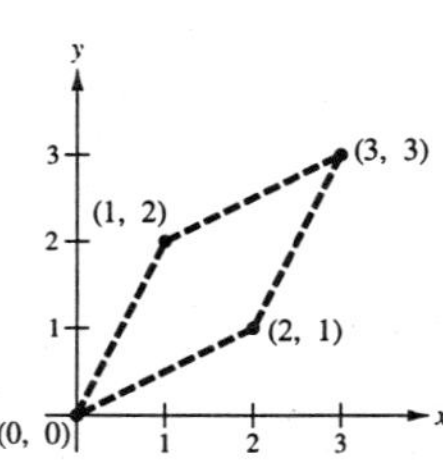

11. $x = -2 \Rightarrow$ quadrants II, III

$y > 0 \Rightarrow$ quadrants I, II

Therefore, quadrant II

13. $xy > 0 \Rightarrow$ quadrants I or III

15.

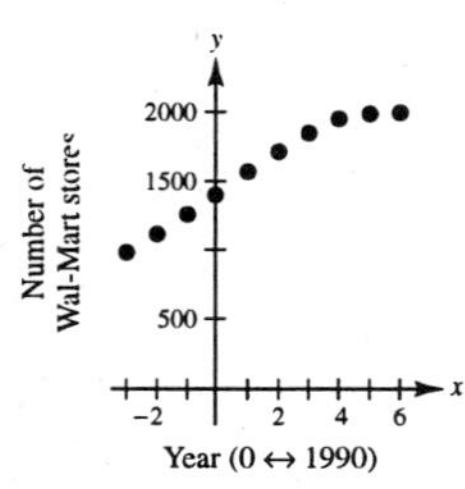

17. $d_1 = \sqrt{4+16} = \sqrt{20} = 2\sqrt{5}$

$d_2 = \sqrt{1+4} = \sqrt{5}$

$d_3 = \sqrt{9+36} = 3\sqrt{5}$

$d_1 + d_2 = d_3$

Collinear

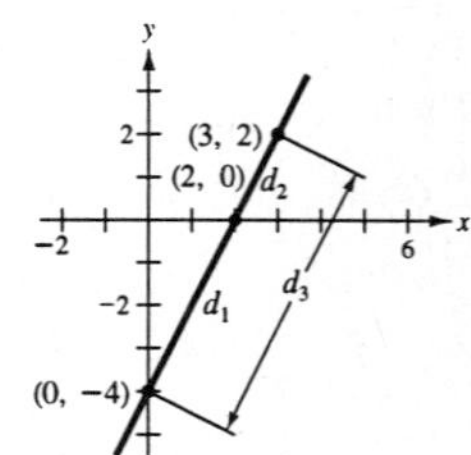

19. $d_1 = \sqrt{1 + 1} = \sqrt{2}$

$d_2 = \sqrt{9 + 4} = \sqrt{13}$

$d_3 = \sqrt{16 + 9} = 5$

$d_1 + d_2 \neq d_3$

Not collinear

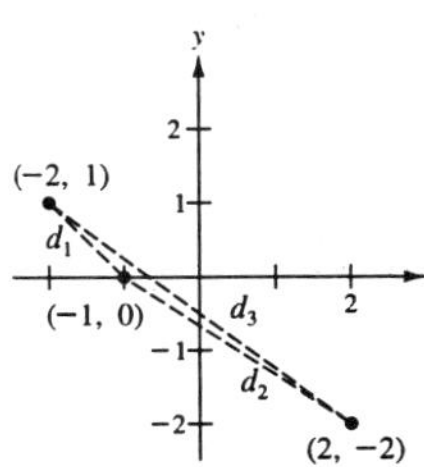

21.
$$\begin{aligned} 5 &= \sqrt{(x - 0)^2 + (-4 - 0)^2} \\ 5 &= \sqrt{x^2 + 16} \\ 25 &= x^2 + 16 \\ 9 &= x^2 \\ x &= \pm 3 \end{aligned}$$

23.
$$\begin{aligned} 8 &= \sqrt{(3 - 0)^2 + (y - 0)^2} \\ 8 &= \sqrt{9 + y^2} \\ 64 &= 9 + y^2 \\ 55 &= y^2 \\ y &= \pm\sqrt{55} \end{aligned}$$

25. The midpoint of the given line segment is $\left(\frac{x_1 + x_2}{2}, \frac{y_1 + y_2}{2}\right)$.

The midpoint between (x_1, y_1) and $\left(\frac{x_1 + x_2}{2}, \frac{y_1 + y_2}{2}\right)$ is $\left(\frac{x_1 + (x_1 + x_2)/2}{2}, \frac{y_1 + (y_1 + y_2)/2}{2}\right) = \left(\frac{3x_1 + x_2}{4}, \frac{3y_1 + y_2}{4}\right)$.

The midpoint between $\left(\frac{x_1 + x_2}{2}, \frac{y_1 + y_2}{2}\right)$ and (x_2, y_2) is $\left(\frac{(x_1 + x_2)/2 + x_2}{2}, \frac{(y_1 + y_2)/2 + y_2}{2}\right) = \left(\frac{x_1 + 3x_2}{4}, \frac{y_1 + 3y_2}{4}\right)$.

Thus, the three points are

$$\left(\frac{3x_1 + x_2}{4}, \frac{3y_1 + y_2}{4}\right), \left(\frac{x_1 + x_2}{2}, \frac{y_1 + y_2}{2}\right), \left(\frac{x_1 + 3x_2}{4}, \frac{y_1 + 3y_2}{4}\right).$$

27. Center: $(0, 0)$

Radius: 1

Matches graph (c)

29. Center: $(1, 0)$

Radius: 0

Matches graph (a)

31. $(x - 0)^2 + (y - 0)^2 = (3)^2$

$x^2 + y^2 - 9 = 0$

33.
$$\begin{aligned} (x - 2)^2 + (y + 1)^2 &= (4)^2 \\ x^2 + y^2 - 4x + 2y - 11 &= 0 \end{aligned}$$

35. Radius $= \sqrt{(-1 - 0)^2 + (2 - 0)^2} = \sqrt{5}$
$$\begin{aligned} (x + 1)^2 + (y - 2)^2 &= 5 \\ x^2 + 2x + 1 + y^2 - 4y + 4 &= 5 \\ x^2 + y^2 + 2x - 4y &= 0 \end{aligned}$$

37. Center = Midpoint = $(3, 2)$

Radius = $\sqrt{10}$
$$\begin{aligned} (x - 3)^2 + (y - 2)^2 &= (\sqrt{10})^2 \\ x^2 - 6x + 9 + y^2 - 4y + 4 &= 10 \\ x^2 + y^2 - 6x - 4y + 3 &= 0 \end{aligned}$$

39. Place the center of the earth at the origin. Then we have
$$\begin{aligned} x^2 + y^2 &= (22,000 + 4,000)^2 \\ x^2 + y^2 &= 26,000^2. \end{aligned}$$

41.
$$\begin{aligned} x^2 + y^2 - 2x + 6y + 6 &= 0 \\ (x^2 - 2x + 1) + (y^2 + 6y + 9) &= -6 + 1 + 9 \\ (x - 1)^2 + (y + 3)^2 &= 4 \end{aligned}$$

Center: $(1, -3)$

Radius: 2

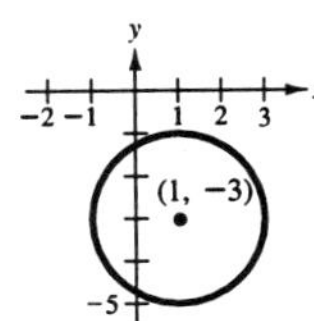

43. $$x^2 + y^2 - 2x + 6y + 10 = 0$$
$$(x^2 - 2x + 1) + (y^2 + 6y + 9) = -10 + 1 + 9$$
$$(x - 1)^2 + (y + 3)^2 = 0$$

Only a point $(1, -3)$

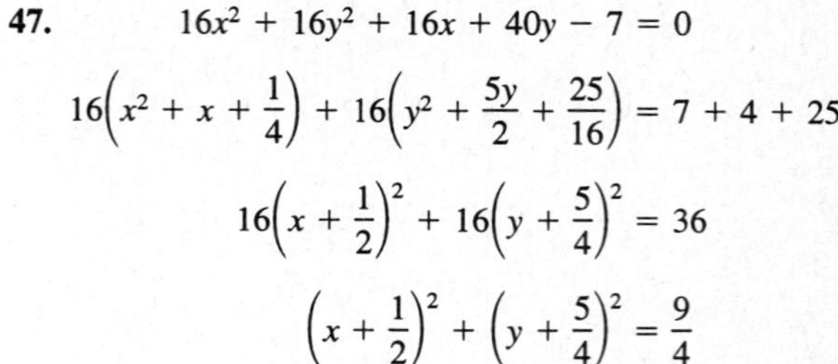

45. $$2x^2 + 2y^2 - 2x - 2y - 3 = 0$$
$$2\left(x^2 - x + \frac{1}{4}\right) + 2\left(y^2 - y + \frac{1}{4}\right) = 3 + \frac{1}{2} + \frac{1}{2}$$
$$\left(x - \frac{1}{2}\right)^2 - \left(y - \frac{1}{2}\right)^2 = 2$$

Center: $\left(\frac{1}{2}, \frac{1}{2}\right)$

Radius: $\sqrt{2}$

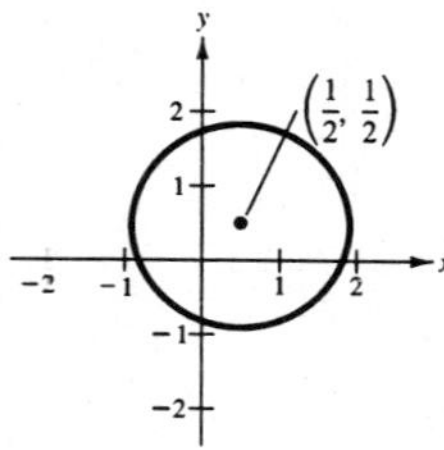

47. $$16x^2 + 16y^2 + 16x + 40y - 7 = 0$$
$$16\left(x^2 + x + \frac{1}{4}\right) + 16\left(y^2 + \frac{5y}{2} + \frac{25}{16}\right) = 7 + 4 + 25$$
$$16\left(x + \frac{1}{2}\right)^2 + 16\left(y + \frac{5}{4}\right)^2 = 36$$
$$\left(x + \frac{1}{2}\right)^2 + \left(y + \frac{5}{4}\right)^2 = \frac{9}{4}$$

Center: $\left(-\frac{1}{2}, -\frac{5}{4}\right)$

Radius: $\frac{3}{2}$

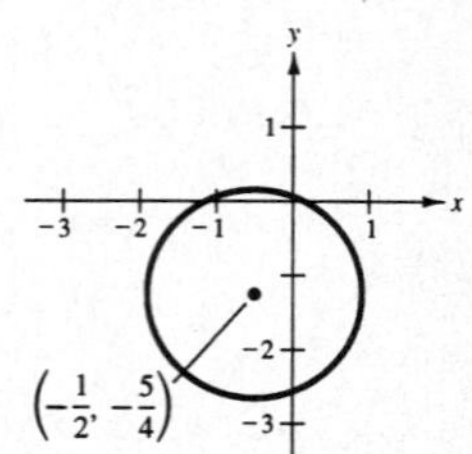

49. (a) $$4x^2 + 4y^2 - 4x + 24y - 63 = 0$$
$$x^2 + y^2 - x + 6y = \frac{63}{4}$$
$$\left(x^2 - x + \frac{1}{4}\right) + (y^2 + 6y + 9) = \frac{63}{4} + \frac{1}{4} + 9$$
$$\left(x - \frac{1}{2}\right)^2 + (y + 3)^2 = 25$$
$$(y + 3)^2 = 25 - \left(x - \frac{1}{2}\right)^2$$
$$y + 3 = \pm\sqrt{25 - \left(x - \frac{1}{2}\right)^2}$$
$$y = -3 \pm \sqrt{25 - \left(x - \frac{1}{2}\right)^2}$$
$$= \frac{-6 \pm \sqrt{99 + 4x - 4x^2}}{2}$$

(b)

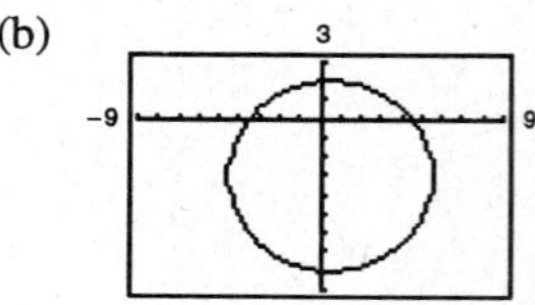

51. $$x^2 + y^2 - 4x + 2y + 1 \leq 0$$
$$(x^2 - 4x + 4) + (y^2 + 2y + 1) \leq -1 + 4 + 1$$
$$(x - 2)^2 + (y + 1)^2 \leq 4$$

Center: $(2, -1)$

Radius: 2

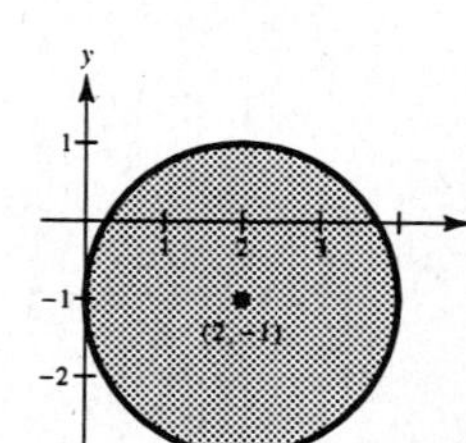

53. The distance between (x_1, y_1) and $\left(\frac{2x_1 + x_2}{3}, \frac{2y_1 + y_2}{3}\right)$ is

$$\begin{aligned} d &= \sqrt{\left(x_1 - \frac{2x_1 + x_2}{3}\right)^2 + \left(y_1 - \frac{2y_1 + y_2}{3}\right)^2} \\ &= \sqrt{\left(\frac{x_1 - x_2}{3}\right)^2 + \left(\frac{y_1 - y_2}{3}\right)^2} \\ &= \sqrt{\frac{1}{9}[(x_1 - x_2)^2 + (y_1 - y_2)^2]} = \frac{1}{3}\sqrt{(x_1 - x_2)^2 + (y_1 - y_2)^2} \end{aligned}$$

which is $\frac{1}{3}$ of the distance between (x_1, y_1) and (x_2, y_2).

$$\left(\frac{\left(\frac{2x_1 + x_2}{3}\right) + x_2}{2}, \frac{\left(\frac{2y_1 + y_2}{3}\right) + y_2}{2}\right) = \left(\frac{x_1 + 2x_2}{3}, \frac{y_1 + 2y_2}{3}\right)$$

is the second point of trisection.

55. True; if $ab < 0$ then either a is positive and b is negative (Quadrant IV) or a is negative and b is positive (Quadrant II).

57. True

59. Let one vertex be at $(0, 0)$ and another at $(a, 0)$.

Midpoint of $(0, 0)$ and (d, e) is $\left(\frac{d}{2}, \frac{e}{2}\right)$.

Midpoint of (b, c) and $(a, 0)$ is $\left(\frac{a + b}{2}, \frac{c}{2}\right)$.

Midpoint of $(0, 0)$ and $(a, 0)$ is $\left(\frac{a}{2}, 0\right)$.

Midpoint of (b, c) and (d, e) is $\left(\frac{b + d}{2}, \frac{c + e}{2}\right)$.

Midpoint of line segment joining $\left(\frac{d}{2}, \frac{e}{2}\right)$ and $\left(\frac{a + b}{2}, \frac{c}{2}\right)$ is $\left(\frac{a + b + d}{4}, \frac{c + e}{4}\right)$.

Midpoint of line segment joining $\left(\frac{a}{2}, 0\right)$ and $\left(\frac{b + d}{2}, \frac{c + e}{2}\right)$ is $\left(\frac{a + b + d}{4}, \frac{c + e}{4}\right)$.

Therefore the line segments intersect at their midpoints.

61. Let (a, b) be a point on the semicircle of radius r, centered at the origin. We will show that the angle at (a, b) is a right angle by verifying that $d_1^2 + d_2^2 = d_3^2$.

$$\begin{aligned} d_1^2 &= (a + r)^2 + (b - 0)^2 \\ d_2^2 &= (a - r)^2 + (b - 0)^2 \\ d_1^2 + d_2^2 &= (a^2 + 2ar + r^2 + b^2) + (a^2 - 2ar + r^2 + b^2) \\ &= 2a^2 + 2b^2 + 2r^2 \\ &= 2(a^2 + b^2) + 2r^2 \\ &= 2r^2 + 2r^2 \\ &= 4r^2 = (2r)^2 = d_3^2 \end{aligned}$$

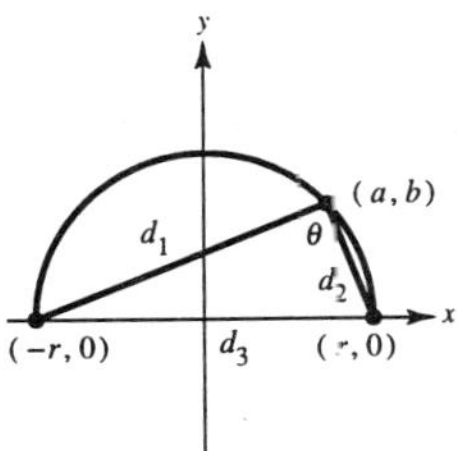

Appendix A.3

1. (a) 396°, −324° (b) 240°, −480°

3. (a) $\frac{19\pi}{9}, -\frac{17\pi}{9}$ (b) $\frac{10\pi}{3}, -\frac{2\pi}{3}$

5. (a) $30\left(\frac{\pi}{180}\right) = \frac{\pi}{6} \approx 0.524$

(b) $150\left(\frac{\pi}{180}\right) = \frac{5\pi}{6} \approx 2.618$

(c) $315\left(\frac{\pi}{180}\right) = \frac{7\pi}{4} \approx 5.498$

(d) $120\left(\frac{\pi}{180}\right) = \frac{2\pi}{3} \approx 2.094$

7. (a) $\frac{3\pi}{2}\left(\frac{180}{\pi}\right) = 270°$

(b) $\frac{7\pi}{6}\left(\frac{180}{\pi}\right) = 210°$

(c) $-\frac{7\pi}{12}\left(\frac{180}{\pi}\right) = -105°$

(d) $-2.637\left(\frac{180}{\pi}\right) \approx -135.6°$

9.

r	8 ft	15 in.	85 cm	24 in.	$\frac{12963}{\pi}$ mi.
s	12 ft.	24 in.	63.72π	96 in.	8642 mi.
θ	1.5	1.6	$\frac{3\pi}{4}$	4	$\frac{2\pi}{3}$

11. (a) $x = 3, y = 4, r = 5$

$\sin\theta = \frac{4}{5}$ $\csc\theta = \frac{5}{4}$

$\cos\theta = \frac{3}{5}$ $\sec\theta = \frac{5}{3}$

$\tan\theta = \frac{4}{3}$ $\cot\theta = \frac{3}{4}$

(b) $x = -12, y = -5, r = 13$

$\sin\theta = -\frac{5}{13}$ $\csc\theta = -\frac{13}{5}$

$\cos\theta = -\frac{12}{13}$ $\sec\theta = -\frac{13}{12}$

$\tan\theta = \frac{5}{12}$ $\cot\theta = \frac{12}{5}$

13. (a) $\sin\theta < 0 \Rightarrow \theta$ is in Quadrant III or IV.

$\cos\theta < 0 \Rightarrow \theta$ is in Quadrant II or III.

$\sin\theta < 0$ **and** $\cos\theta < 0 \Rightarrow \theta$ is in Quadrant III.

(b) $\sec\theta > 0 \Rightarrow \theta$ is in Quadrant I or IV.

$\cot\theta < 0 \Rightarrow \theta$ is in Quadrant II or IV.

$\sec\theta > 0$ **and** $\cot\theta < 0 \Rightarrow \theta$ is in Quadrant IV.

15. $x^2 + 1^2 = 2^2 \Rightarrow x = \sqrt{3}$

$\cos\theta = \frac{x}{2} = \frac{\sqrt{3}}{2}$

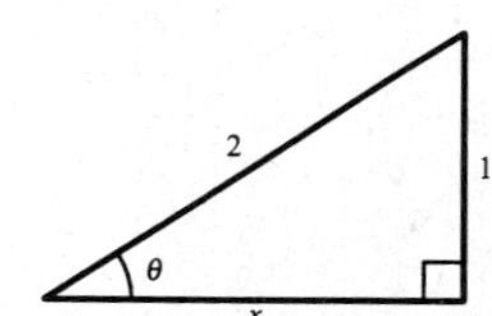

17. $4^2 + y^2 = 5^2 \Rightarrow y = 3$

$\cot\theta = \frac{4}{y} = \frac{4}{3}$

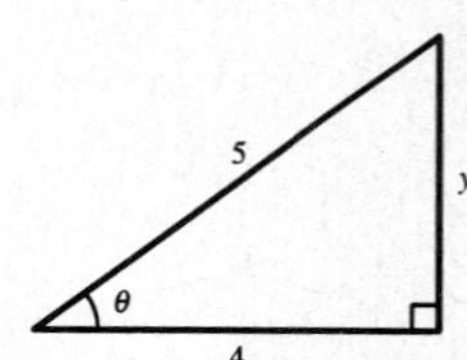

19. (a) $\sin 60° = \frac{\sqrt{3}}{2}$

$\cos 60° = \frac{1}{2}$

$\tan 60° = \sqrt{3}$

(b) $\sin 120° = \sin 60° = \frac{\sqrt{3}}{2}$

$\cos 120° = -\cos 60° = -\frac{1}{2}$

$\tan 120° = -\tan 60° = -\sqrt{3}$

(c) $\sin\frac{\pi}{4} = \frac{\sqrt{2}}{2}$

$\cos\frac{\pi}{4} = \frac{\sqrt{2}}{2}$

$\tan\frac{\pi}{4} = 1$

(d) $\sin\frac{5\pi}{4} - \sin\frac{\pi}{4} = -\frac{\sqrt{2}}{2}$

$\cos\frac{5\pi}{4} = \cos\frac{\pi}{4} = -\frac{\sqrt{2}}{2}$

$\tan\frac{5\pi}{4} = \tan\frac{\pi}{4} = 1$

21. (a) $\sin 225° = -\sin 45° = -\frac{\sqrt{2}}{2}$

$\cos 225° = -\cos 45° = -\frac{\sqrt{2}}{2}$

$\tan 225° = \tan 45° = 1$

(b) $\sin(-225°) = \sin 45° = \frac{\sqrt{2}}{2}$

$\cos(-225°) = -\cos 45° = \frac{\sqrt{2}}{2}$

$\tan(-225°) = -\tan 45 = -1$

(c) $\sin \frac{5\pi}{3} = -\sin \frac{\pi}{3} = -\frac{\sqrt{3}}{2}$

$\cos \frac{5\pi}{3} = \cos \frac{\pi}{3} = \frac{1}{2}$

$\tan \frac{5\pi}{3} = -\tan \frac{\pi}{3} = -\sqrt{3}$

(d) $\sin \frac{11\pi}{6} = -\sin \frac{\pi}{6} = -\frac{1}{2}$

$\cos \frac{11\pi}{6} = \cos \frac{\pi}{6} = \frac{\sqrt{3}}{2}$

$\tan \frac{11\pi}{6} = -\tan \frac{\pi}{6} = -\frac{\sqrt{3}}{3}$

23. (a) $\sin 10° \approx 0.1736$ (b) $\csc 10° \approx 5.759$

25. (a) $\tan \frac{\pi}{9} \approx 0.3640$ (b) $\tan \frac{10\pi}{9} \approx 0.3640$

27. (a) $\cos \theta = \frac{\sqrt{2}}{2}$

$\theta = \frac{\pi}{4}, \frac{7\pi}{4}$

(b) $\cos \theta = -\frac{\sqrt{2}}{2}$

$\theta = \frac{3\pi}{4}, \frac{5\pi}{4}$

29. (a) $\tan \theta = 1$

$\theta = \frac{\pi}{4}, \frac{5\pi}{4}$

(b) $\cot \theta = -\sqrt{3}$

$\theta = \frac{5\pi}{6}, \frac{11\pi}{6}$

31. $2 \sin^2 \theta = 1$

$\sin \theta = \pm\frac{\sqrt{2}}{2}$

$\theta = \frac{\pi}{4}, \frac{3\pi}{4}, \frac{5\pi}{4}, \frac{7\pi}{4}$

33. $\tan^2 \theta = \tan \theta = 0$

$\tan \theta(\tan \theta - 1) = 0$

$\tan \theta = 0$ $\tan \theta = 1$

$\theta = 0, \pi$ $\theta = \frac{\pi}{4}, \frac{5\pi}{4}$

35. $\sec \theta \csc \theta - 2 \csc \theta = 0$

$\csc \theta(\sec \theta - 2) = 0$

($\csc \theta \neq 0$ for any value of θ)

$\sec \theta = 2$

$\theta = \frac{\pi}{3}, \frac{5\pi}{3}$

37. $\cos^2 \theta + \sin \theta = 1$

$1 - \sin^2 \theta + \sin \theta = 1$

$\sin^2 \theta - \sin \theta = 0$

$\sin \theta(\sin \theta - 1) = 0$

$\sin \theta = 0$ $\sin \theta = 1$

$\theta = 0, \pi$ $\theta = \frac{\pi}{2}$

39. (275 ft/sec)(60 sec) = 16,500 feet

$\sin 18° = \frac{a}{16{,}500}$

$a = 16{,}500 \sin 18° \approx 5099$ feet

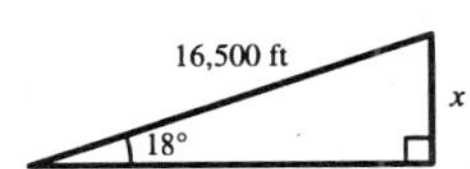

41. (a) Period: π

Amplitude: 2

(b) Period: 2

Amplitude: $\frac{1}{2}$

43. Period: $\frac{1}{2}$

Amplitude: 3

45. Period: $\frac{\pi}{2}$

47. Period: $\frac{2\pi}{5}$

49. (a) $f(x) = c \sin x$; changing c changes the amplitude.

When $c = -2$: $f(x) = -2 \sin x$.

When $c = -1$: $f(x) = -\sin x$.

When $c = 1$: $f(x) = \sin x$.

When $c = 2$: $f(x) = 2 \sin x$.

(b) $f(x) = \cos(cx)$; changing c changes the period.

When $c = -2$: $f(x) = \cos(-2x) = \cos 2x$.

When $c = -1$: $f(x) = \cos(-x) = \cos x$.

When $c = 1$: $f(x) = \cos x$.

When $c = 2$: $f(x) = \cos 2x$.

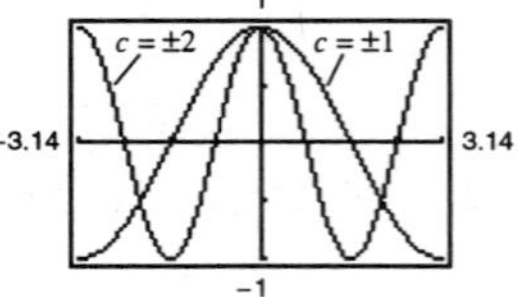

(c) $f(x) = \cos(\pi x - c)$; changing c causes a horizontal shift.

When $c = -2$: $f(x) = \cos(\pi x + 2)$.

When $c = -1$: $f(x) = \cos(\pi x + 1)$.

When $c = 1$: $f(x) = \cos(\pi x - 1)$.

When $c = 2$: $f(x) = \cos(\pi x - 2)$.

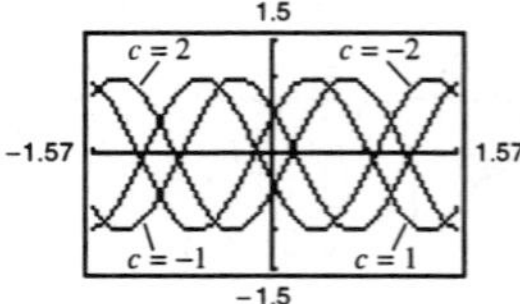

51. $y = \sin \dfrac{x}{2}$

Period: 4π

Amplitude: 1

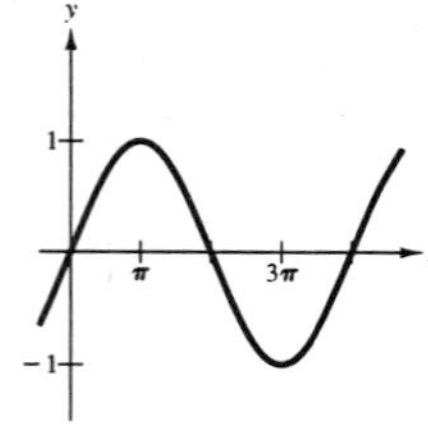

53. $y = -\sin \dfrac{2\pi x}{3}$

Period: 3

Amplitude: 1

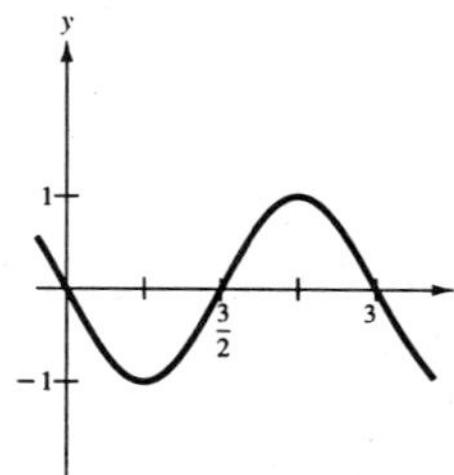

55. $y = \csc \dfrac{x}{2}$

Period: 4π

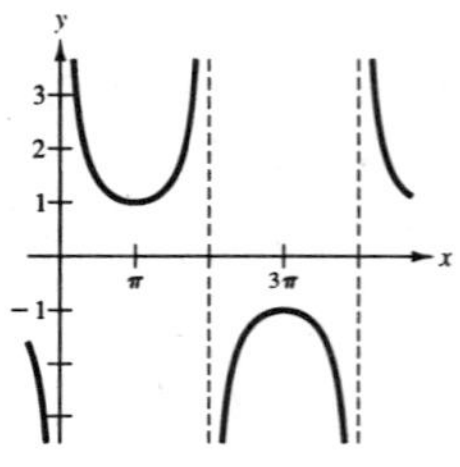

57. $y = 2 \sec 2x$

Period: π

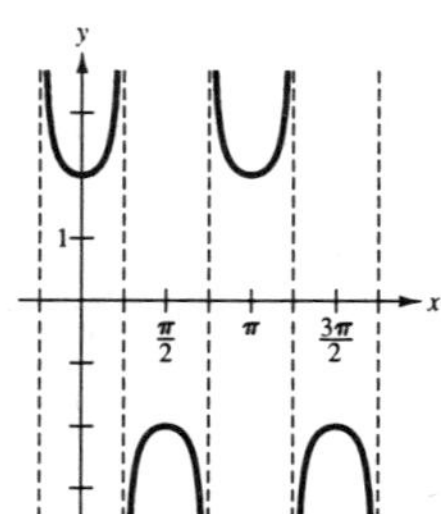

59. $y = \sin(x + \pi)$

Period: 2π

Amplitude: 1

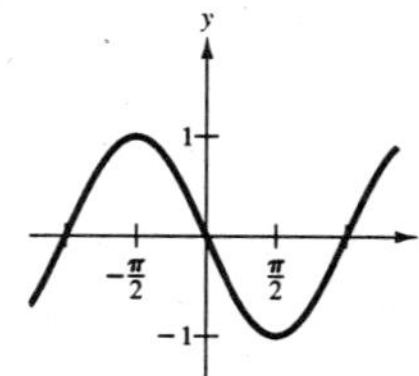

61. $y = 1 + \cos\left(x - \dfrac{\pi}{2}\right)$

Period: 2π

Amplitude: 1

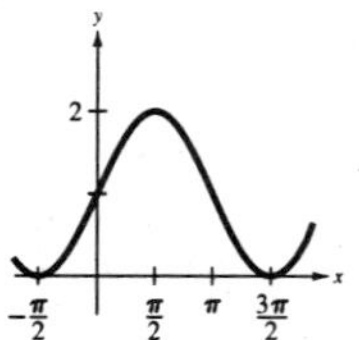

63. $y = a \cos(bx - c)$

From the graph, we see that the amplitude is 3, the period is 4π and the horizontal shift is, π. Thus,

$$a = 3$$

$$\frac{2\pi}{b} = 4\pi \Rightarrow b = \frac{1}{2}$$

$$\frac{c}{d} = \pi \Rightarrow c = \frac{\pi}{2}.$$

Therefore, $y = 3 \cos[(1/2)x - (\pi/2)]$.

65. $f(x) = \sin x$

$g(x) = |\sin x|$

$h(x) = \sin|x|$

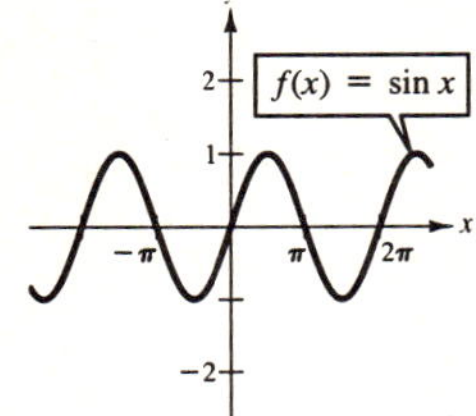

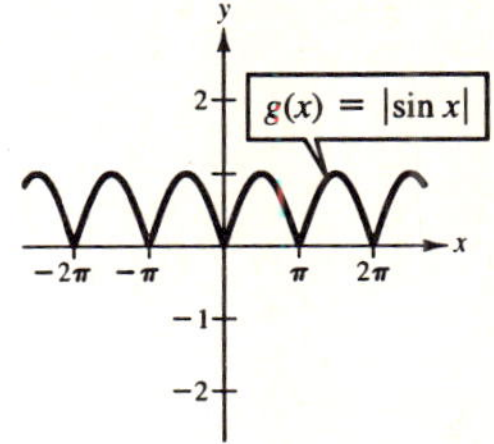

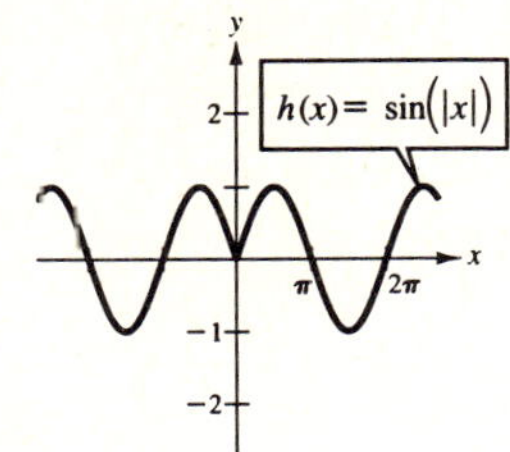

The graph of $|f(x)|$ will reflect any parts of the graph below the x-axis about the y-axis.
The graph of $f(|x|)$ will reflect the part of the graph to the left of the y-axis about the x-axis.

67. $S = 58.3 + 32.5 \cos \dfrac{\pi t}{6}$

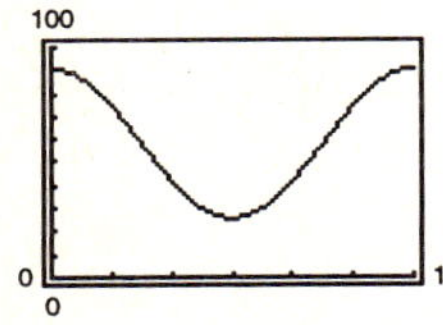

Sales exceed 75,000 during the months of January, November, and December.

69. $f(x) = \dfrac{4}{\pi}\left(\sin \pi x + \dfrac{1}{3} \sin 3\pi x\right)$

$g(x) = \dfrac{4}{\pi}\left(\sin \pi x + \dfrac{1}{3} \sin 3\pi x + \dfrac{1}{5} \sin 5\pi x\right)$

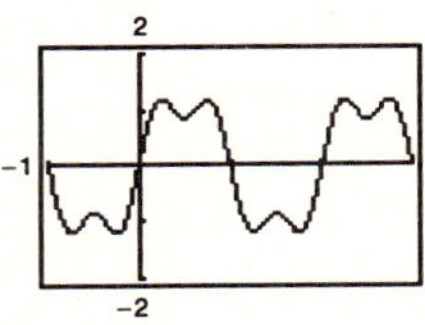

Pattern:

$$f(x) = \frac{4}{\pi}\left(\sin \pi x + \frac{1}{3} \sin 3\pi x + \frac{1}{5} \sin 5\pi x + \cdots + \frac{1}{2n-1} \sin(2n-1)\pi x\right), n = 1, 2, 3 \ldots$$

Appendix E

1. $xy + 1 = 0$

$\cot 2\theta = \dfrac{A - C}{B} = 0,\ 2\theta = \dfrac{\pi}{2} \Rightarrow \theta = \dfrac{\pi}{4}$

$x = x' \cos \dfrac{\pi}{4} - y' \sin \dfrac{\pi}{4} = \dfrac{x' - y'}{\sqrt{2}}$

$y = x' \sin \dfrac{\pi}{4} + y' \cos \dfrac{\pi}{4} = \dfrac{x' + y'}{\sqrt{2}}$

$\left(\dfrac{x' - y'}{\sqrt{2}}\right)\left(\dfrac{x' + y'}{\sqrt{2}}\right) + 1 = 0$

$\dfrac{(y')^2}{2} - \dfrac{(x')^2}{2} = 1$

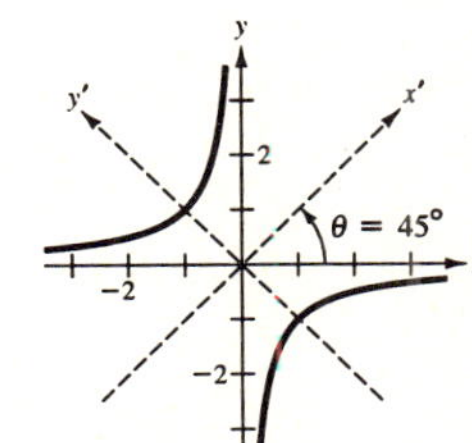

3. $x^2 - 10xy + y^2 + 1 = 0$

$$\cot 2\theta = \frac{A - C}{B} = 0,\ 2\pi = \frac{\pi}{2} \Rightarrow \theta = \frac{\pi}{4}$$

$$x = x'\cos\frac{\pi}{4} - y'\sin\frac{\pi}{4} = \frac{x' - y'}{\sqrt{2}}$$

$$y = x'\sin\frac{\pi}{4} + y'\cos\frac{\pi}{4} = \frac{x' + y'}{\sqrt{2}}$$

$$\left(\frac{x' - y'}{\sqrt{2}}\right)^2 - 10\left(\frac{x' - y'}{\sqrt{2}}\right)\left(\frac{x' + y'}{\sqrt{2}}\right) + \left(\frac{x' + y'}{\sqrt{2}}\right)^2 + 1 = 0$$

$$-4(x')^2 + 6(y')^2 + 1 = 0$$

$$\frac{(x')^2}{1/4} - \frac{(y')^2}{1/6} = 1$$

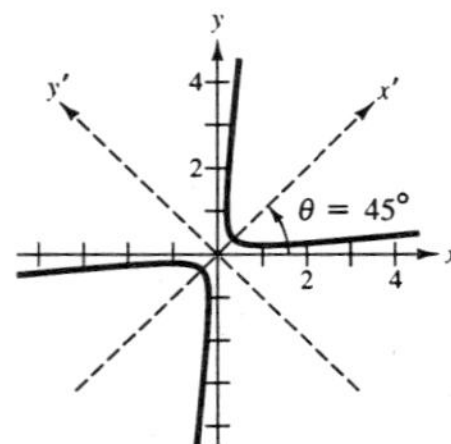

5. $xy - 2y - 4x = 0$

$$\cot 2\theta = \frac{A - C}{B} = 0,\ 2\theta = \frac{\pi}{2} \Rightarrow \theta = \frac{\pi}{4}$$

$$x = x'\cos\frac{\pi}{4} - y'\sin\frac{\pi}{4} = \frac{x' - y'}{\sqrt{2}}$$

$$y = x'\sin\frac{\pi}{4} + y'\cos\frac{\pi}{4} = \frac{x' + y'}{\sqrt{2}}$$

$$\left(\frac{x' - y'}{\sqrt{2}}\right)\left(\frac{x' + y'}{\sqrt{2}}\right) - 2\left(\frac{x' + y'}{\sqrt{2}}\right) - 4\left(\frac{x' - y'}{\sqrt{2}}\right) = 0$$

$$\frac{1}{2}(x')^2 - \frac{1}{2}(y')^2 - 3\sqrt{2}x' + \sqrt{2}y' = 0$$

$$\frac{(x' - 3\sqrt{2})^2}{16} - \frac{(y' - \sqrt{2})^2}{16} = 1$$

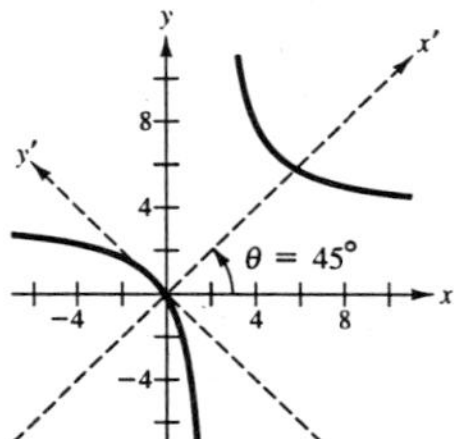

7. $5x^2 - 2xy + 5y^2 - 12 = 0$

$$\cot 2\theta = \frac{A - C}{B} = 0,\ 2\theta = \frac{\pi}{2} \Rightarrow \theta = \frac{\pi}{4}$$

$$x = x'\cos\frac{\pi}{4} - y'\sin\frac{\pi}{4} = \frac{x' - y'}{\sqrt{2}}$$

$$y = x'\sin\frac{\pi}{4} + y'\cos\frac{\pi}{4} = \frac{x' + y'}{\sqrt{2}}$$

$$5\left(\frac{x' - y'}{\sqrt{2}}\right)^2 - 2\left(\frac{x' - y'}{\sqrt{2}}\right)\left(\frac{x' + y'}{\sqrt{2}}\right) + 5\left(\frac{x' + y'}{\sqrt{2}}\right)^2 - 12 = 0$$

$$4(x')^2 + 6(y')^2 - 12 = 0$$

$$\frac{(x')^2}{3} + \frac{(y')^2}{2} = 1$$

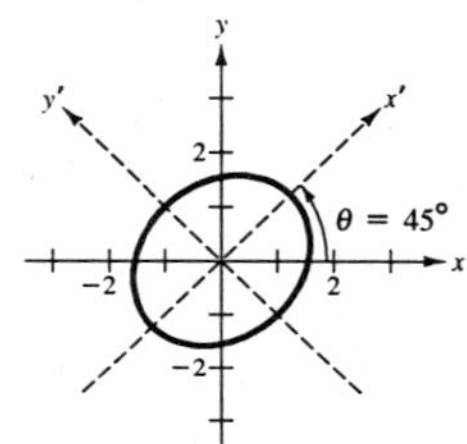

9. $3x^2 - 2\sqrt{3}xy + y^2 + 2x + 2\sqrt{3}y = 0$

$$\cot 2\theta = -\frac{1}{\sqrt{3}},\ 2\theta = \frac{2\pi}{3} \Rightarrow \theta = \frac{\pi}{3}$$

$$x = x'\cos\frac{\pi}{3} - y'\sin\frac{\pi}{3} = \frac{x' - 3y'}{2}$$

$$y = x'\sin\frac{\pi}{3} + y'\cos\frac{\pi}{3} = \frac{\sqrt{3}x' + y'}{2}$$

$$3\left(\frac{x' - \sqrt{3}y'}{2}\right)^2 - 2\sqrt{3}\left(\frac{x' - 3\sqrt{y'}}{2}\right)\left(\frac{\sqrt{3}x' + y'}{2}\right) + \left(\frac{\sqrt{3}x' + y'}{2}\right)^2 + 2\left(\frac{x' - \sqrt{3}y'}{2}\right) + 2\sqrt{3}\left(\frac{\sqrt{3}x' + y'}{2}\right) = 0$$

$$4(y')^2 + 4x' = 0$$

$$x' = -(y')^2$$

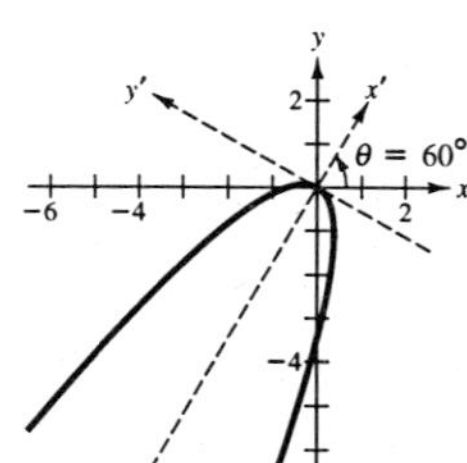

11. $9x^2 + 24xy + 16y^2 + 90x - 130y = 0$

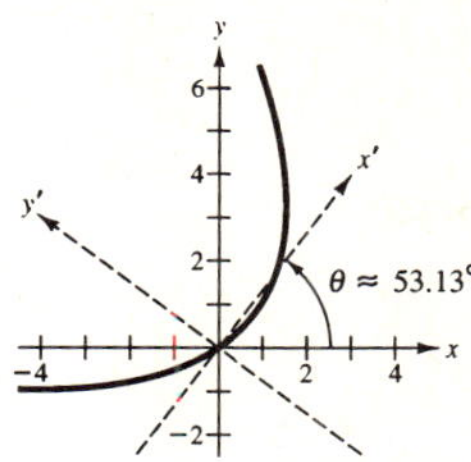

$\cot 2\theta = -\dfrac{7}{24} \Rightarrow \cos \theta = -\dfrac{7}{25}$

$\sin \theta = \sqrt{\dfrac{1 + (7/25)}{2}} = \dfrac{4}{5}, \ \cos \theta = \sqrt{\dfrac{1 - (7/25)}{2}} = \dfrac{3}{5}$

$\theta = \sin^{-1}\dfrac{4}{5} \approx 53.13°$

$x = x' \cos \theta - y' \sin \theta = \dfrac{3x' - 4y'}{5}$

$y = x' \sin \theta + y' \cos \theta = \dfrac{4x' + 3y'}{5}$

$9\left(\dfrac{3x' - 4y'}{5}\right)^2 + 24\left(\dfrac{3x' - 4y'}{5}\right)\left(\dfrac{4x' + 3y'}{5}\right) + 16\left(\dfrac{4x' + 3y'}{5}\right)^2 + 90\left(\dfrac{3x' - 4y'}{5}\right) - 130\left(\dfrac{4x' + 3y'}{5}\right) = 0$

$25(x')^2 - 50x' - 150y' = 0$

$y' = \dfrac{(x)^2}{6} - \dfrac{x'}{3}$

13. $x^2 + xy + y^2 = 10$

$\cot 2\theta = \dfrac{A - C}{B} = 0 \Rightarrow \theta = \dfrac{\pi}{4} = 45°$

Solve for y in terms of x.

$y^2 + xy + \dfrac{x^2}{4} = 10 - x^2 + \dfrac{x^2}{4}$

$\left(y + \dfrac{x}{2}\right)^2 = \dfrac{40 - 3x^2}{4}$

$y = -\dfrac{x}{2} \pm \dfrac{\sqrt{40 - 3x^2}}{2} = \dfrac{-x \pm \sqrt{40 - 3x^2}}{2}$

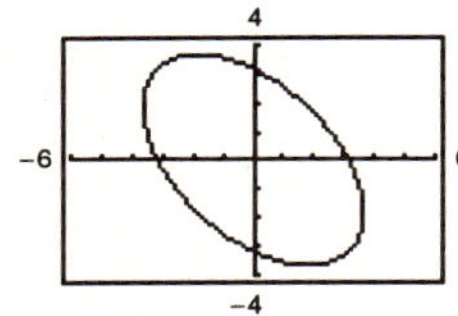

15. $17x^2 + 32xy - 7y^2 = 75$

$\cot 2\theta = \dfrac{3}{4} \Rightarrow \theta = \dfrac{1}{2}\text{arccot}\dfrac{3}{4} \approx 26.57°$

Solve for y in terms of x.

$7\left(y^2 - \dfrac{32}{7}xy + \dfrac{256}{49}x^2\right) = 17x^2 - 75 + \dfrac{256}{7}x^2$

$\left(y - \dfrac{16}{7}x\right)^2 = \dfrac{375x^2 - 525}{49}$

$y = \dfrac{16x \pm \sqrt{375x^2 - 525}}{7}$

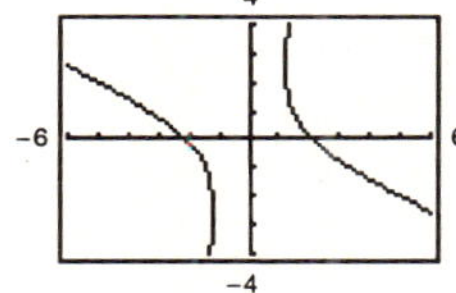

17. $32x^2 + 50xy + 7y^2 = 52$

$\cot 2\theta = \dfrac{A - C}{B} = \dfrac{1}{2} \Rightarrow \theta = \dfrac{1}{2}\text{arccot}\dfrac{1}{2} \approx 31.72°$

Solve for y in terms of x.

$7\left(y^2 + \dfrac{50}{7}xy + \dfrac{625}{49}x^2\right) = 52 - 32x^2 + \dfrac{625}{7}x^2$

$\left(y + \dfrac{25}{7}x\right)^2 = \dfrac{364 + 401x^2}{49}$

$y = \dfrac{-25x \pm \sqrt{364 + 401x^2}}{7}$

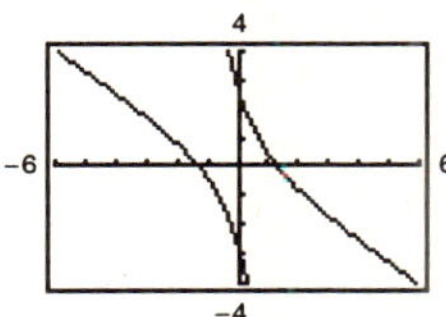

19. $B^2 - 4AC = (-24)^2 - 4(16)(9) = 0$

Parabola

21. $B^2 - 4AC = (-8)^2 - 4(13)(7) = -300$

Ellipse

23. $B^2 - 4AC = (-6)^2 - 4(1)(-5) = 56$

Hyperbola

25. $B^2 - 4AC = (4)^2 - 4(1)(4) = 0$

Parabola

27. $y^2 - 4x^2 = 0$

$y = \pm 2x$

Two intersecting lines

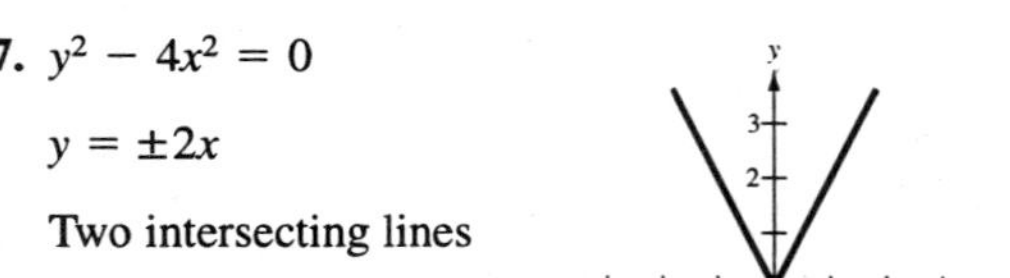

29. $x^2 + 2xy + y^2 - 1 = 0$

$(x + y)^2 = 1$

$x + y = \pm 1$

Two parallel lines

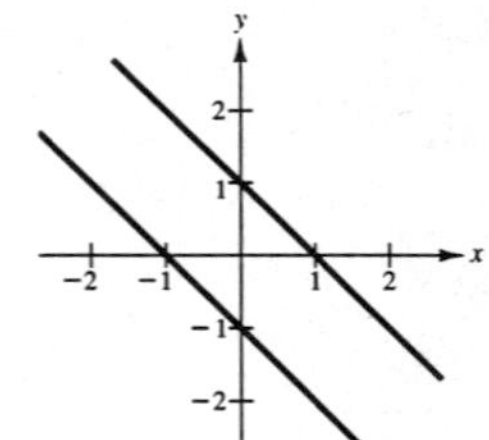

31. $(x - 2y + 1)(x + 2y - 3) = 0$

$x - 2y + 1 = 0$ or $x + 2y - 3 = 0$

$x - 2y = -1$ $\quad x + 2y = 3$

Two intersecting lines

33. $(x')^2 + (y')^2 = (x \cos \theta + y \sin \theta)^2 + (y \cos \theta - x \sin \theta)^2$

$= x^2 \cos^2 \theta + 2xy \cos \theta \sin \theta + y^2 \sin^2 \theta + y^2 \cos^2 \theta - 2xy \cos \theta \sin \theta + x^2 \sin^2 \theta$

$= x^2(\cos^2 \theta + \sin^2 \theta) + y^2(\sin^2 \theta + \cos^2 \theta) = x^2 + y^2 = r^2$

Appendix F

1. $(5 + i) + (6 - 2i) = 11 - i$

3. $(8 - i) - (4 - i) = 8 - i - 4 + i = 4$

5. $\left(-2 + \sqrt{-8}\right) + \left(5 - \sqrt{-50}\right) = -2 + 2\sqrt{2}i + 5 - 5\sqrt{2}i$

$= 3 - 3\sqrt{2}i$

7. $13i - (14 - 7i) = 13i - 14 + 7i = -14 + 20i$

9. $-\left(\frac{3}{2} + \frac{5}{2}i\right) + \left(\frac{5}{3} + \frac{11}{3}i\right) = -\frac{3}{2} - \frac{5}{2}i + \frac{5}{3} + \frac{11}{3}i$

$= -\frac{9}{6} - \frac{15}{6}i + \frac{10}{6} + \frac{22}{6}i$

$= \frac{1}{6} + \frac{7}{6}i$

11. $\sqrt{-6} \cdot \sqrt{-2} = \left(\sqrt{6}i\right)\left(\sqrt{2}i\right)$

$= \sqrt{12}i^2 = \left(2\sqrt{3}\right)(-1) = -2\sqrt{3}$

13. $\left(-\sqrt{10}\right)^2 = \left(\sqrt{10}i\right)^2 = 10i^2 = -10$

15. $(1 + i)(3 - 2i) = 3 - 2i + 3i - 2i^2$

$= 3 + i + 2 = 5 + i$

17. $6i(5 - 2i) = 30i - 12i^2 = 30i + 12 = 12 + 30i$

19. $\left(\sqrt{14} + \sqrt{10}i\right)\left(\sqrt{14} - \sqrt{10}i\right) = 14 - 10i^2 = 14 + 10 = 24$

21. $(4 + 5i)^2 = 16 + 40i + 25i^2 = 16 + 40i - 25$

$= -9 + 40i$

23. $(2 + 3i)^2 + (2 - 3i)^2 = 4 + 12i + 9i^2 + 4 - 12i + 9i^2$

$= 4 + 12i - 9 + 4 - 12i - 9$

$= -10$

25. The complex of $5 + 3i$ is $5 - 3i$.

$(5 + 3i)(5 - 3i) = 25 - 9i^2 = 25 + 9 = 34$

27. The complex conjugate of $-2 - \sqrt{5}i$ is $-2 + \sqrt{5}i$.

$(-2 - \sqrt{5}i)(-2 + \sqrt{5}i) = 4 - 5i^2 = 4 + 5 = 9$

29. The complex conjugate of $20i$ is $-20i$.

$(20i)(-20i) = -400i^2 = 400$

31. The complex conjugate of $\sqrt{8}$ is $\sqrt{8}$.

$(\sqrt{8})(\sqrt{8}) = 8$

33. $\dfrac{6}{i} = \dfrac{6}{i} \cdot \dfrac{-i}{-i} = \dfrac{-6i}{-i^2} = \dfrac{-6i}{1} = -6i$

35. $\dfrac{4}{4 - 5i} = \dfrac{4}{4 - 5i} \cdot \dfrac{4 + 5i}{4 + 5i}$

$$= \frac{4(4 + 5i)}{16 + 25} = \frac{16 - 20i}{41} = \frac{16}{41} + \frac{20}{41}i$$

37. $\dfrac{2 + i}{2 - i} = \dfrac{2 + i}{2 - i} \cdot \dfrac{2 + i}{2 + i} = \dfrac{4 + 4i + i^2}{4 + 1} = \dfrac{3 + 4i}{5} = \dfrac{3}{5} + \dfrac{4}{5}i$

39. $\dfrac{6 - 7i}{i} = \dfrac{6 - 7i}{i} \cdot \dfrac{-6\text{i} - 7}{1} = -7 - 6\text{i}$

41. $\dfrac{1}{(4 - 5i)^2} = \dfrac{1}{16 - 40i + 25i^2} = \dfrac{1}{-9 - 40i} \cdot \dfrac{-9 + 40i}{-9 + 40i}$

$$= \frac{-9 + 40i}{81 + 1600} = \frac{-9 + 40i}{1681} = -\frac{9}{1681} + \frac{40}{1681}i$$

43. $\dfrac{2}{1 - i} - \dfrac{3}{1 - i} = \dfrac{2(1 - i) - 3(1 + i)}{(1 - i)(1 - i)}$

$$= \frac{2 - 2i - 3 - 3i}{1 - 1}$$

$$= \frac{-1 - 5i}{2} = -\frac{1}{2} - \frac{5}{2}i$$

45. $\dfrac{i}{3 - 2i} + \dfrac{2i}{3 + 8i} = \dfrac{i(3 + 8i) + 2i(3 - 2i)}{(3 - 2i)(3 + 8i)}$

$$= \frac{3i + 8i^2 + 6i - 4i^2}{9 + 24i - 6i - 16i^2}$$

$$= \frac{4i^2 + 9i}{9 + 18i + 16}$$

$$= \frac{-4 + 9i}{25 + 18i} \cdot \frac{25 - 18i}{25 - 18i}$$

$$= \frac{-100 + 72i + 225i - 162i^2}{625 + 324}$$

$$= \frac{-100 + 297i + 162}{949}$$

$$= \frac{62 + 297i}{949} = \frac{62}{949} + \frac{297}{949}i$$

47. $x^2 - 2x + 2 = 0;\ a = 1,\ b = -2,\ c = 2$

$$x = \frac{-(-2) \pm \sqrt{(-2)^2 - 4(1)(2)}}{2(1)}$$

$$= \frac{2 \pm \sqrt{-4}}{2}$$

$$= \frac{2 \pm 2i}{2} = 1 \pm i$$

49. $4x^2 + 16x + 17 = 0;\ a = 4,\ b = 16,\ c = 17$

$$x = \frac{-16 \pm \sqrt{(16)^2 - 4(4)(17)}}{2(4)}$$

$$= \frac{-16 \pm \sqrt{-16}}{8}$$

$$= \frac{-16 \pm 4i}{8} = -2 \pm \frac{1}{2}i$$

51. $4x^2 + 16x + 15 = 0;\ a = 4,\ b = 16,\ c = 15$

$$x = \frac{-16 \pm \sqrt{(16)^2 - 4(4)(15)}}{2(4)}$$

$$= \frac{-16 \pm \sqrt{16}}{8} = \frac{-16 \pm 4}{8}$$

$$x = -\frac{12}{8} = -\frac{3}{2} \quad \text{or} \quad x = \frac{-20}{8} = -\frac{5}{2}$$

53. $16t^2 - 4t + 3 = 0;\ a = 16,\ b = -4,\ c = 3$

$$t = \frac{-(-4) \pm \sqrt{(-4)^2 - 4(16)(3)}}{2(16)}$$

$$= \frac{4 \pm \sqrt{-176}}{32} = \frac{4 \pm 4\sqrt{11}i}{32} = \frac{1}{8} \pm \frac{\sqrt{11}}{8}i$$

55. $-6i^3 + i^2 = -6i^2i + i^2$

$$= -6(-1)i + (-1)$$

$$= 6i - 1$$

$$= -1 + 6i$$

57. $-5i^5 = -5i^2i^2i$

$$= -5(-1)(-1)i = -5i$$

59. $\left(\sqrt{-75}\right)^3 = \left(5\sqrt{3}i\right)^3 = 5^3\left(\sqrt{3}\right)^3i^3$

$$= 125\left(3\sqrt{3}\right)(-i)$$

$$= -375\sqrt{3}i$$

61. $\dfrac{1}{i^3} = \dfrac{1}{-i} = \dfrac{1}{-i} \cdot \dfrac{i}{i} = \dfrac{1}{-i^2} = \dfrac{i}{1} = i$

63. $|-5i| = \sqrt{0^2 + (-5)^2}$

$$= \sqrt{25} = 5$$

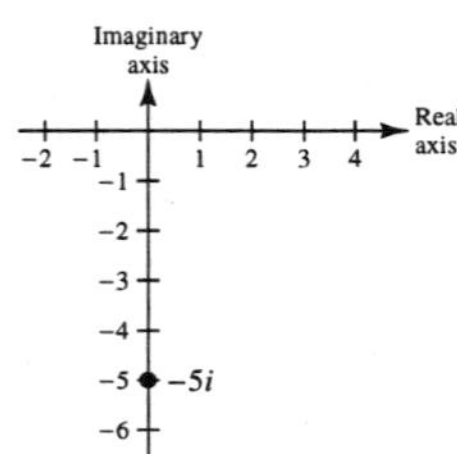

65. $|-4 + 4i| = \sqrt{(-4)^2 + (4)^2}$

$$= \sqrt{32} = 4\sqrt{2}$$

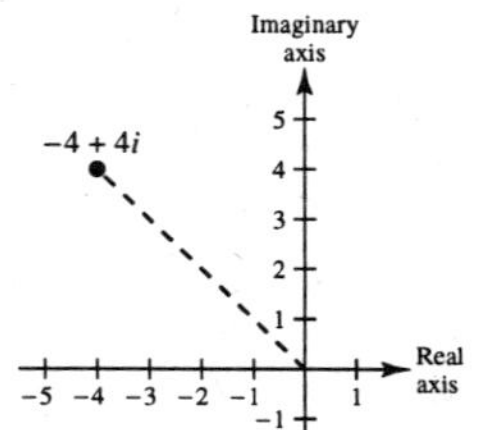

67. $|6 - 7i| = \sqrt{6^2 + (-7)^2}$

$$= \sqrt{85}$$

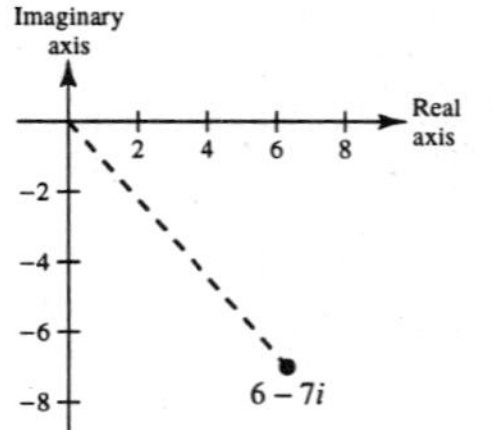

69. $z = 3 - 3i$

$$r = \sqrt{3^2 + (-3)^2} = \sqrt{18} = 3\sqrt{2}$$

$$\tan\theta = \frac{-3}{3} = -1,\ \theta \text{ is in Quadrant IV} \Rightarrow \theta = \frac{7\pi}{4}.$$

$$z = 3\sqrt{2}\left(\cos\frac{7\pi}{4} + i\sin\frac{7\pi}{4}\right)$$

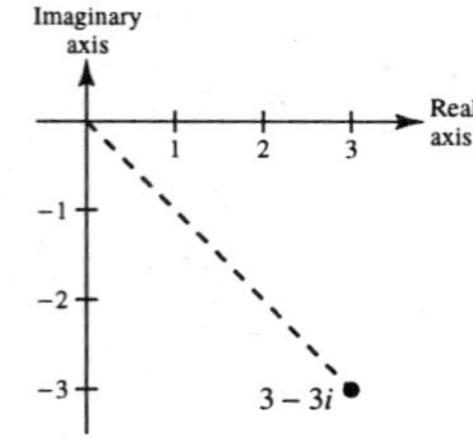

71. $z = \sqrt{3} + i$

$$r = \sqrt{\left(\sqrt{3}\right)^2 + 1^2} = \sqrt{4} = 2$$

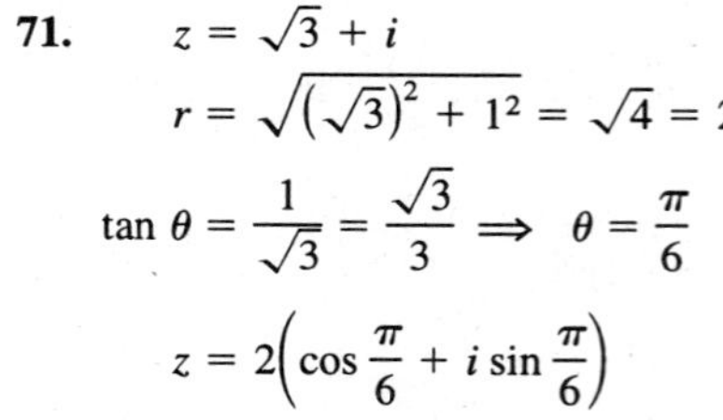

$$\tan\theta = \frac{1}{\sqrt{3}} = \frac{\sqrt{3}}{3} \Rightarrow \theta = \frac{\pi}{6}$$

$$z = 2\left(\cos\frac{\pi}{6} + i\sin\frac{\pi}{6}\right)$$

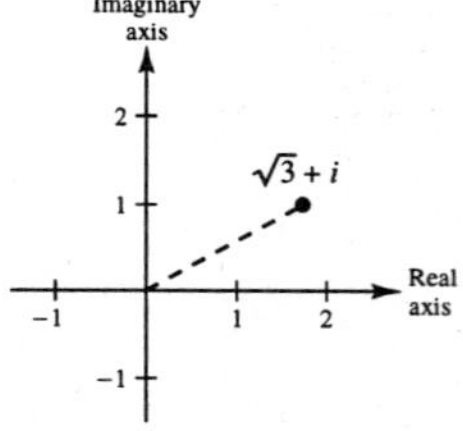

73. $z = -2(1 + \sqrt{3}i)$

$$r = \sqrt{(-2)^2 + (-2\sqrt{3})^2} = \sqrt{16} = 4$$

$$\tan\theta = \frac{\sqrt{3}}{1} = \sqrt{3},\ \theta \text{ is in Quadrant III} \Rightarrow \theta = \frac{4\pi}{3}.$$

$$z = 4\left(\cos\frac{4\pi}{3} + i\sin\frac{4\pi}{3}\right)$$

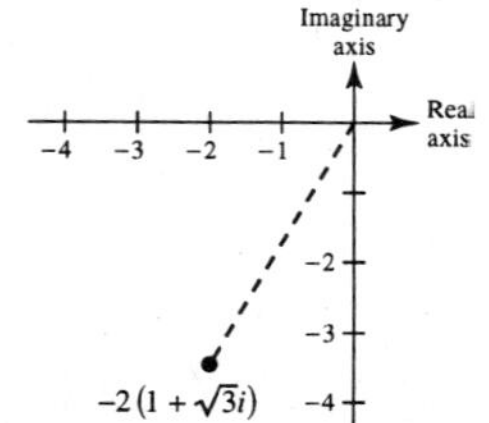

75. $z = 0 + 6i$

$$r = \sqrt{0^2 + (6)^2} = \sqrt{36} = 6$$

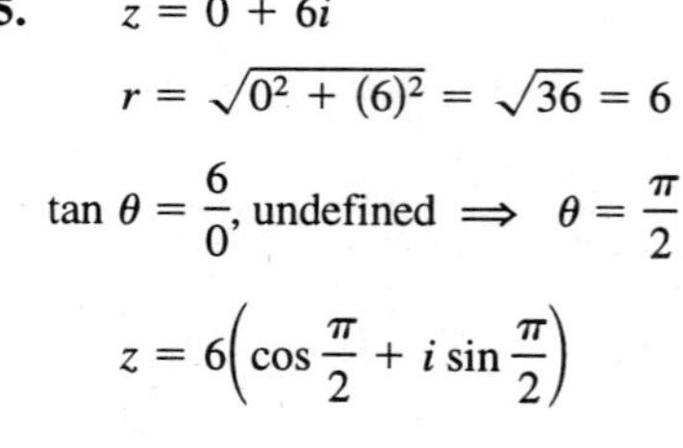

$$\tan\theta = \frac{6}{0},\ \text{undefined} \Rightarrow \theta = \frac{\pi}{2}$$

$$z = 6\left(\cos\frac{\pi}{2} + i\sin\frac{\pi}{2}\right)$$

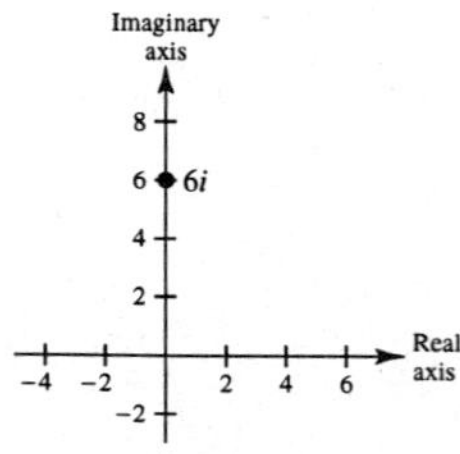

77. $2(\cos 150° + i \sin 150°) = 2\left[-\frac{\sqrt{3}}{2} + i\left(\frac{1}{2}\right)\right]$

$= -\sqrt{3} + i$

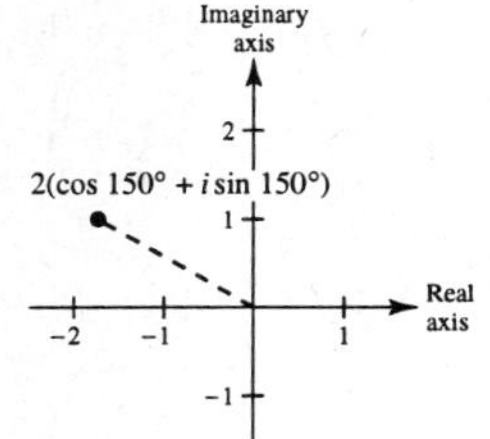

79. $\frac{3}{2}(\cos 300° + i \sin 300°) = \frac{3}{2}\left[\frac{1}{2} + i\left(-\frac{\sqrt{3}}{2}\right)\right]$

$= \frac{3}{4} - \frac{3\sqrt{3}}{4}i$

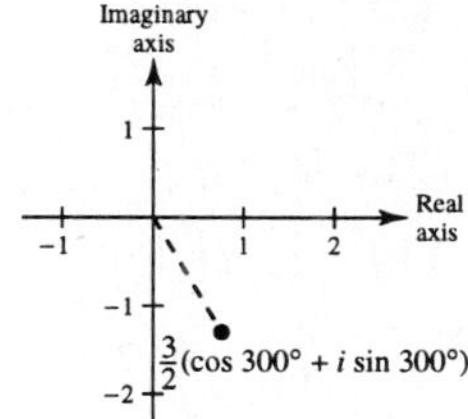

81. $3.75\left(\cos\frac{3\pi}{4} + i \sin\frac{3\pi}{4}\right) = -\frac{15\sqrt{2}}{8} + \frac{15\sqrt{2}}{8}i$

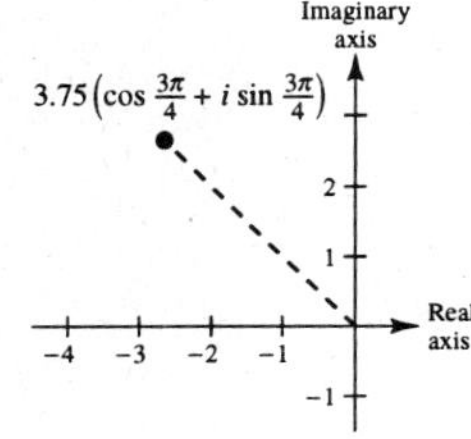

83. $\left[3\left(\cos\frac{\pi}{3} + i \sin\frac{\pi}{3}\right)\right]\left[4\left(\cos\frac{\pi}{6} + i \sin\frac{\pi}{6}\right)\right] = (3)(4)\left[\cos\left(\frac{\pi}{3} + \frac{\pi}{6}\right) + i \sin\left(\frac{\pi}{6} + \frac{\pi}{3}\right)\right] = 12\left(\cos\frac{\pi}{2} + i \sin\frac{\pi}{2}\right)$

85. $\left[\frac{5}{3}(\cos 140° + i \sin 140°)\right]\left[\frac{2}{3}(\cos 60° + i \sin 60°)\right] = \left(\frac{5}{3}\right)\left(\frac{2}{3}\right)[\cos(140° + 60°) + i \sin(140° + 60°)]$

87. $(1 + i)^5 = \left[\sqrt{2}\left(\cos\frac{\pi}{4} + i \sin\frac{\pi}{4}\right)\right]^5$

$= (\sqrt{2})^5\left(\cos\frac{5\pi}{4} + i \sin\frac{5\pi}{4}\right)$

$= 4\sqrt{2}\left(-\frac{\sqrt{2}}{2} - \frac{\sqrt{2}}{2}i\right)$

$= -4 - 4i$

89. $(-1 + i)^{10} = \left[\sqrt{2}\left(\cos\frac{3\pi}{4} + i \sin\frac{3\pi}{4}\right)\right]^{10}$

$= (\sqrt{2})^{10}\left(\cos\frac{30\pi}{4} + i \sin\frac{30\pi}{4}\right)$

$= 32\left[\cos\left(\frac{3\pi}{2} + 6\pi\right) + i \sin\left(\frac{3\pi}{2} + 6\pi\right)\right]$

$= 32\left(\cos\frac{3\pi}{2} + i \sin\frac{3\pi}{2}\right)$

$= 32[0 + i(-1)]$

$= -32i$

91. $2(\sqrt{3} + i)^7 = 2\left[2\left(\cos\frac{\pi}{6} + i \sin\frac{\pi}{6}\right)\right]^7 = 2\left[2^7\left(\cos\frac{7\pi}{6} + i \sin\frac{7\pi}{6}\right)\right] = 256\left(-\frac{\sqrt{3}}{2} - \frac{1}{2}i\right) = -128\sqrt{3} - 128i$

93. $\left(\cos\frac{5\pi}{4} + i \sin\frac{5\pi}{4}\right)^{10} = \cos\frac{25\pi}{2} + i \sin\frac{25\pi}{2} = \cos\left(12\pi + \frac{\pi}{2}\right) + i \sin\left(12\pi + \frac{\pi}{2}\right) = \cos\frac{\pi}{2} + i \sin\frac{\pi}{2} = i$

95. (a) Square roots of $5(\cos 120° + i \sin 120°)$:

$$\sqrt{5}\left[\cos\left(\frac{120° + 360°k}{2}\right) + i \sin\left(\frac{120° + 360°k}{2}\right)\right],\ k = 0, 1$$

$k = 0$: $\sqrt{5}(\cos 60° + i \sin 60°)$

$k = 1$: $\sqrt{5}(\cos 240° + i \sin 240°)$

(b)

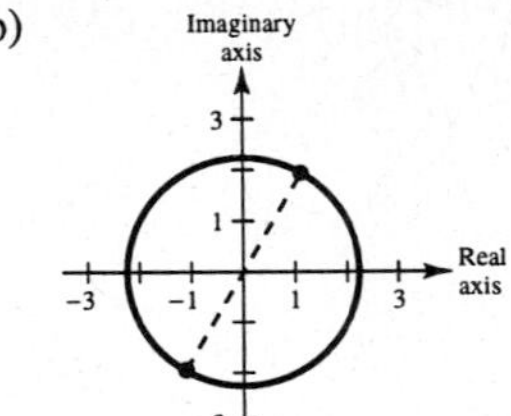

(c) $\dfrac{\sqrt{5}}{2} + \dfrac{\sqrt{15}}{2}i,\ -\dfrac{\sqrt{5}}{2} - \dfrac{\sqrt{15}}{2}i$

97. (a) Fourth roots of $16\left(\cos\dfrac{4\pi}{3} + i \sin\dfrac{4\pi}{3}\right)$:

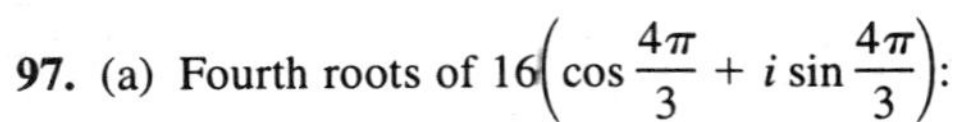

$$\sqrt[4]{16}\left[\cos\left(\frac{(4\pi/3) + 2k\pi}{4}\right) + i \sin\left(\frac{(4\pi/3) + 2k\pi}{4}\right)\right],\ k = 0, 1, 2, 3$$

$k = 0$: $2\left(\cos\dfrac{\pi}{3} + i \sin\dfrac{\pi}{3}\right)$

$k = 1$: $2\left(\cos\dfrac{5\pi}{6} + i \sin\dfrac{5\pi}{6}\right)$

$k = 2$: $2\left(\cos\dfrac{4\pi}{3} + i \sin\dfrac{4\pi}{3}\right)$

$k = 3$: $2\left(\cos\dfrac{11\pi}{6} + i \sin\dfrac{11\pi}{6}\right)$

(b)

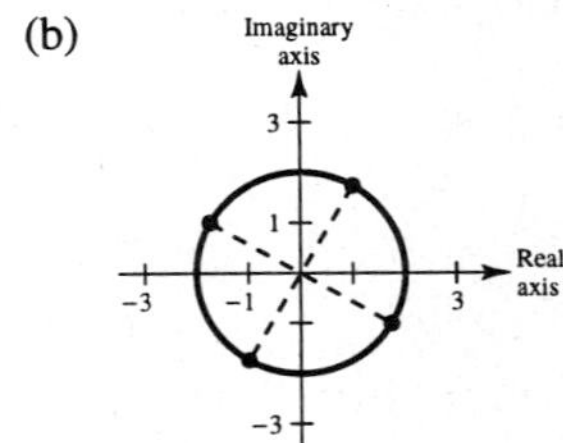

(c) $1 + \sqrt{3}i,\ -\sqrt{3} + i,\ -1 - \sqrt{3}i,\ \sqrt{3} - i$

99. (a) Cube roots of $-\dfrac{125}{3}(1 + \sqrt{3}i) = 125\left(\cos\dfrac{4\pi}{3} + i \sin\dfrac{4\pi}{3}\right)$:

$$\sqrt[3]{125}\left[\cos\left(\frac{(4\pi/3) + 2k\pi}{3}\right) + i \sin\left(\frac{(4\pi/3) + 2k\pi}{3}\right)\right],\ k = 0, 1, 2$$

$k = 0$: $5\left(\cos\dfrac{4\pi}{9} + i \sin\dfrac{4\pi}{9}\right)$

$k = 1$: $5\left(\cos\dfrac{10\pi}{9} + i \sin\dfrac{10\pi}{9}\right)$

$k = 2$: $5\left(\cos\dfrac{16\pi}{9} + i \sin\dfrac{16\pi}{9}\right)$

(b)

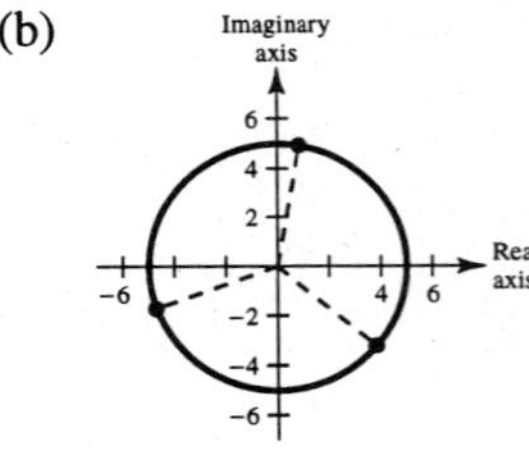

(c) $0.8682 + 4.924i,\ -4.698 - 1.710i,\ 3.830 - 3.214i$

101. $x^4 - i = 0$

$x^4 = i$

The solutions are the fourth roots of $i = \cos\dfrac{\pi}{2} + i \sin\dfrac{\pi}{2}$:

$$\sqrt[4]{1}\left[\cos\left(\frac{(\pi/2) + 2k\pi}{4}\right) + i \sin\left(\frac{(\pi/2) + 2k\pi}{4}\right)\right],\ k = 0, 1, 2, 3$$

$k = 0$: $\cos\dfrac{\pi}{8} + i \sin\dfrac{\pi}{8}$

$k = 1$: $\cos\dfrac{5\pi}{8} + i \sin\dfrac{5\pi}{8}$

$k = 2$: $\cos\dfrac{9\pi}{8} + i \sin\dfrac{9\pi}{8}$

$k = 3$: $\cos\dfrac{13\pi}{8} + i \sin\dfrac{13\pi}{8}$

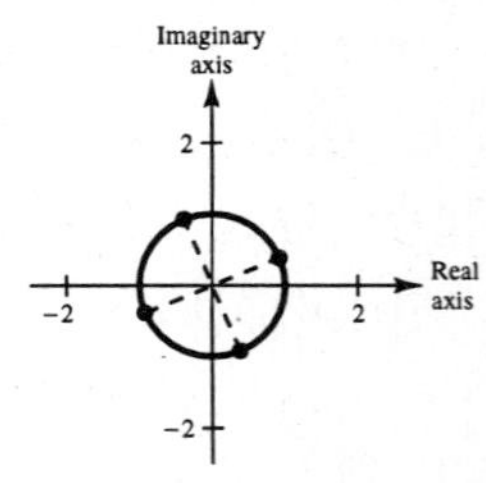

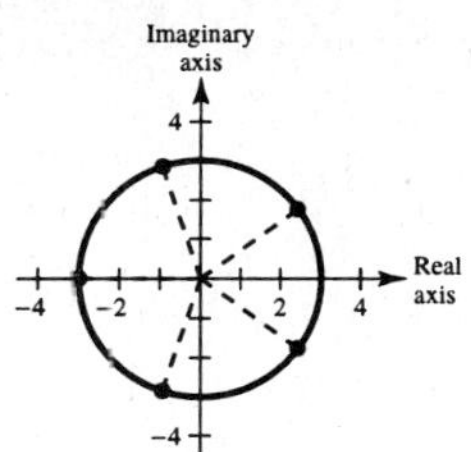

103. $x^5 + 243 = 0$

$$x^5 = -243$$

The solutions are the fifth roots of $-243 = 243(\cos \pi + i \sin \pi)$:

$$\sqrt[5]{243}\left[\cos\left(\frac{\pi + 2k\pi}{5}\right) + i \sin\left(\frac{\pi + 2k\pi}{5}\right)\right], \quad k = 0, 1, 2, 3, 4$$

$k = 0$: $3\left(\cos \frac{\pi}{5} + i \sin \frac{\pi}{5}\right)$

$k = 1$: $3\left(\cos \frac{3\pi}{5} + i \sin \frac{3\pi}{5}\right)$

$k = 2$: $3(\cos \pi + i \sin \pi) = -3$

$k = 3$: $3\left(\cos \frac{7\pi}{5} + i \sin \frac{7\pi}{5}\right)$

$k = 4$: $3\left(\cos \frac{9\pi}{5} + i \sin \frac{9\pi}{5}\right)$

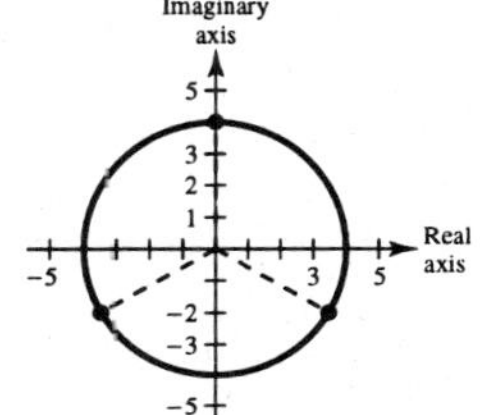

105. $x^3 + 64i = 0$

$$x^3 = -64i$$

The solutions are the cube roots of $-64i = 64\left(\cos\frac{3\pi}{2} + i \sin\frac{3\pi}{2}\right)$:

$$\sqrt[3]{64}\left[\cos\left(\frac{(3\pi/2) + 2k\pi}{3}\right) + i \sin\left(\frac{(3\pi/2) + 2k\pi}{3}\right)\right], \quad k = 0, 1, 2$$

$k = 0$: $4\left(\cos \frac{\pi}{2} + i \sin \frac{\pi}{2}\right) = 4i$

$k = 1$: $4\left(\cos \frac{7\pi}{6} + i \sin \frac{7\pi}{6}\right) = -2\sqrt{3} - 2i$

$k = 2$: $4\left(\cos \frac{11\pi}{6} + i \sin \frac{11\pi}{6}\right) = 2\sqrt{3} - 2i$

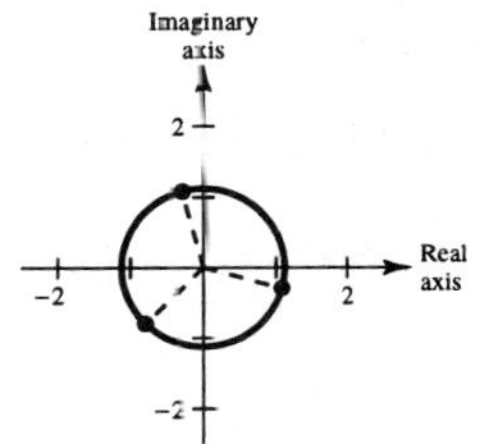

107. $x^3 - (1 - i) = 0$

$$x^3 = 1 - i = \sqrt{2}(\cos 315° + i \sin 315°)$$

The solutions are the cube roots of $1 - i$:

$$\sqrt[3]{\sqrt{2}}\left[\cos\left(\frac{315° + 360°k}{3}\right) + i \sin\left(\frac{315° + 360°k}{3}\right)\right], \quad k = 0, 1, 2$$

$k = 0$: $\sqrt[6]{2}(\cos 105° + i \sin 105°)$

$k = 1$: $\sqrt[6]{2}(\cos 225° + i \sin 225°)$

$k = 2$: $\sqrt[6]{2}(\cos 345° + i \sin 345°)$